Library of Congress Cataloging-in-Publication Data

Monier, Charles J.
Electric circuit analysis / Charles J. Monier.
p. cm.
Includes bibliographical references and index.
ISBN 0-13-014410-X
1. Electric circuit analysis. I. Title.

TK454 .M645 2001
621.319'2–dc21

Vice President and Publisher: Dave Garza
Editor in Chief: Stephen Helba
Acquisitions Editor: Scott Sambucci
Production Editor: Rex Davidson
Production Coordinator: Lisa Garboski, bookworks
Design Coordinator: Karrie Converse-Jones
Cover Designer: Jason Moore
Production Manager: Pat Tonneman
Marketing Manager: Ben Leonard

This book was set by York Graphic Services. It was printed and bound by Courier/Kendallville, Inc. The cover was printed by Phoenix Color Corp.

9 8 7 6 5 4 3 2 1
ISBN 0-13-014410-X

ELECTRIC CIRCUIT ANALYSIS

Charles J. Monier
Nicholls State University

Upper Saddle River, New Jersey
Columbus, Ohio

TO

Charles, Jr.
Michael
Daniel
Paul

Preface

The material presented in this textbook is intended to establish a clear relationship between the basic principles of electric circuit analysis and the problem-solving procedures for analyzing electric circuits. Although the material is presented with a reasonable amount of mathematical rigor, the reader needs no more than a working knowledge of algebra and trigonometry to benefit from the textbook. Clarity, conciseness, completeness, and, above all, readability are intended as the major characteristics of this textbook.

Each chapter of the textbook contains an abundance of examples with varying degrees of difficulty to illustrate the problem-solving procedures. Exercises are provided at the end of each chapter to allow the student to complete the learning process by applying the newly learned material. The exercises and examples are designed to promote analytical reasoning and improve the student's problem-solving ability. The examples are very detailed and should be treated as an integral part of the text material.

Electric circuit analysis is one of the first courses of study required in electrical engineering technology programs, and a complete mastery of the material is a prerequisite for the electrical courses that follow. This text provides material for a traditional two-semester course of study in dc and ac circuit analysis along with many advanced topics. The text is designed to allow the engineering technology student to complete the required work in electric circuit analysis as early as possible in his or her academic career with a minimum of prerequisites.

In addition to the topics that are traditionally included in electric circuit–analysis textbooks at the technology level, matrix methods for solving systems of algebraic equations for simultaneous solutions, derivatives and integrals, differential equations, and Laplace transforms are also included. Matrix methods are included to enhance the multiloop circuit-analysis methods. Derivatives and integrals are included to increase the student's understanding of and ability to analyze circuits containing inductors and capacitors. Differential equations are introduced to facilitate the study of transients in electric circuits. Laplace transforms are introduced to enable the student to solve electric circuit problems by using simple algebraic techniques.

The matrix methods, derivatives and integrals, the methods for solving differential equations, and Laplace transforms are presented from a conceptual viewpoint with emphasis on methodology and provide the student with the mathematical concepts necessary for proceeding with the study of electric circuits in a meaningful manner. Although mathematics courses scheduled later in the student's academic program will extend and enhance the mathematical concepts presented in this text, the student needs to know many of these concepts, at least from a procedural viewpoint, while enrolled in electric circuit–analysis courses.

Because this textbook is primarily concerned with the student developing an understanding of the basic principles of electric circuits and the related problem-solving procedures, computer programs for the analysis and design of electric circuits are not included. This is in no way meant to diminish the vast potential of the digital computer for analyzing and designing electric circuits. Computer software packages such as PSPICE are presently in widespread use, and software packages of this type should be an integral part of the student's training. However, the author has found that teaching the principles of electric circuits and related problem-solving procedures separate from computer methods leads to a better understanding of the principles of electric circuit analysis. Ideally, the principles and problem-solving procedures are presented in lecture courses, which are accompanied by laboratory courses where the software packages are presented along with the usual laboratory experiments. A short demonstration on the use of PSPICE to analyze electric circuits is presented in Appendix A.

The textbook is intended for two three-semester-hour lecture courses. The recommended coverage for the first course is Chapters 1 through 15, and for the second course, Chapters 16 through 23. Chapter 15 is placed immediately after Chapters 12, 13, and 14, because it relates so well to those chapters. However, Chapter 15 can be delayed until the second course and covered prior to Chapter 21 without any loss of continuity.

ACKNOWLEDGMENTS

I wish to thank my colleagues at Nicholls State University, particularly, Professors Badiollah Asrabodi, Donald M. Bardwell, R. J. Yakupzack, and Professor Emeritus Glenn R. Swetman for their valuable assistance. Senior student David P. Sanchez also deserves my thanks for assisting me with the exercise problems. I would like to thank the following reviewers: George Fredericks, Northeast State TCC; Kenneth Reid, Indiana University/Purdue University Indianapolis (IUPUI); and Sohail Anwar, Penn State University, Altoona College. I also wish to express my deepest appreciation to my wife, Lora Jane, for typing and editing this manuscript.

Charles J. Monier

Contents

CHAPTER

1

Preliminary Concepts

After completing this chapter the student should know

* the basic terms associated with the analysis of electric circuits
* what constitutes an electric circuit

Electrical energy plays a dominant role in our lives. We encounter it, directly or indirectly, in computers, control systems, communications, heating and cooling our homes, industrial processes, lasers, electronic devices, and in a myriad of other instances on a daily basis. Understanding the principles of electrical energy and the methods of harnessing it for the betterment of humankind are the primary concerns of the electrical engineering technologist. Fundamental to this understanding is the course of study known as electric circuit analysis. The discussion of electric circuit analysis in this chapter is for introductory purposes only and is enhanced and enlarged upon in later chapters.

An electric circuit consists of electrical elements and/or devices connected in such a manner that the elements and devices are electrically energized as evidenced by currents and voltages being associated with them. Electric current involves the flow of charged particles (electrons) through the elements and devices, and voltage involves the work or energy that causes the current. A simple example of an electric circuit is the basic flashlight. A 6-V battery (source of voltage) causes current to flow through an incandescent lamp (bulb), heating it to incandescence, which causes the bulb to glow and provide light.

1.1 ELECTRIC CHARGE

All matter is made up of atoms, each of which consists of a dense nucleus at the center surrounded by orbiting electrons. The nucleus of an atom is composed of protons and neutrons. Protons are positively charged; electrons are negatively charged; and neutrons are neutral. The electrons of an atom are held in orbit by the attractive force between the nucleus of the atom and the electrons.

The property of charge on a body is distinguishable by the force on the body caused by the presence of another charged body. Bodies with like charges repel each other with a repulsive force and bodies with unlike charges attract each other with an attractive force. A body becomes charged by losing or gaining electrons. If a body loses electrons, the absence of negative charge makes the body positively charged. If a body gains electrons, the excess of negative charge makes the body negatively charged. Electric charge, the excess or absence of electrons, is significant and plays an important role in the principles of electricity. Electric charge is measured in coulombs named in honor of the French physicist Charles Coulomb (1736–1806). The charge of one electron is negative and equal to 1.6021×10^{-18} coulombs (C).

1.2 CONDUCTORS

Materials with atomic structures that readily release electrons from their orbits for free movement within the material are called *conductors,* whereas materials that greatly restrict the release of electrons from their orbits are called *insulators,* or dielectrics. Copper and aluminum, along with most metals, are good conductors, whereas glass, plastic, rubber, and similar materials are good insulators. Materials that fall between conductors and insulators with a moderate amount of free electrons available for movement within the material are called *semiconductors.* Silicon and germanium are examples of good semiconductors. Conductors are used where a flow of electrons is desired; insulators are used where it is desired to suppress the flow of electrons; and semiconductors are useful in the manufacture of electronic devices such as transistors. The flow of electrons through a conductor has the desirable effect of allowing us to efficiently transfer energy from one place to another. An example is an electric power station generating electrical energy and delivering it to consumers located miles away by the flow of electrons through wires (conductors) connected between the power station and the consumers.

1.3 CURRENT

The flow of electrons through a conductor is called electric current. The unit of measurement for electric current is the ampere, named in honor of French physicist André-Marie Ampère (1775–1836). One ampere equals the flow of one coulomb of charge per second. The free electrons in a conductor move about in random directions but do not constitute an electric current until there is a net flow of electrons in the same direction. The free electrons in a conductor can be forced to flow in the same direction by the application of a voltage source across the conductor.

1.4 VOLTAGE

The voltage source is an external energy-giving device that causes forces to act on the free electrons in the conductor, driving them through the conductor from one terminal of the voltage source to the other and thereby causing an electric current through the conductor. The voltage source placed across a conductor has an imbalance of accumulated charge within itself, negative charge at one terminal and positive charge at the other.

The negatively charged free electrons in the conductor are repelled by the accumulation of negative charge and attracted by the accumulation of positive charge created by the voltage source. The resulting flow of electrons through the conductor constitutes electric current. Typical voltage sources are batteries and electric generators. A battery provides the energy that causes the flow of electrons through a conductor by chemical action and an electric generator uses mechanical action.

When a voltage source is placed across two points, a voltage is said to exist across the two points. The voltage across two points is an indication of the energy available to move electrons through a conductor placed between the two points. Voltage is measured in volts named in honor of Italian physicist Alessandro Volta (1745–1827). The voltage across two points is equal to one volt if one joule of energy is expended in moving one coulomb of charge from one point to the other. One joule is the amount of energy used to move a body a distance of one meter with a constant force of one newton. A voltage source providing voltage across a conductor is illustrated in Figure 1.1.

The terminals of the voltage source are marked with + and − signs to indicate the positively and negatively charged sides of the voltage source. Electrons flowing through the conductor are repelled by the negatively charged side of the voltage source and attracted by the positively charged side of the voltage source. The resulting current (flow of electrons) is indicated in Figure 1.1.

Figure 1.1 Electron current.

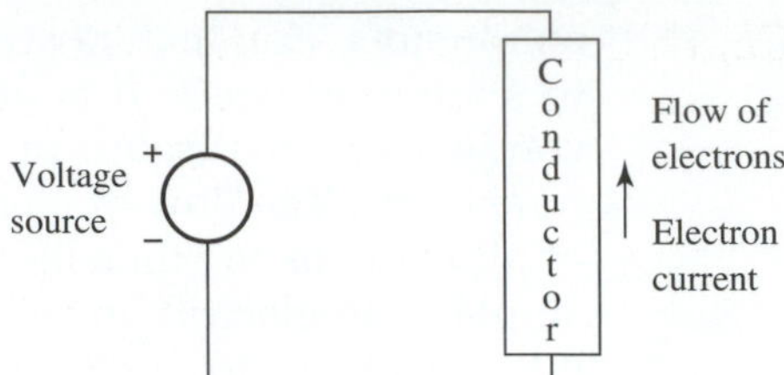

The American experimenter Benjamin Franklin (1706–1790) and other early scientists investigating electricity were not aware that electric current consisted of electron flow and erroneously supposed that current flowed from + to − instead of − to + through a conductor, as illustrated in Figure 1.1. Electric current flow through a conductor as envisioned by the early investigators is illustrated in Figure 1.2.

Figure 1.2 Conventional current.

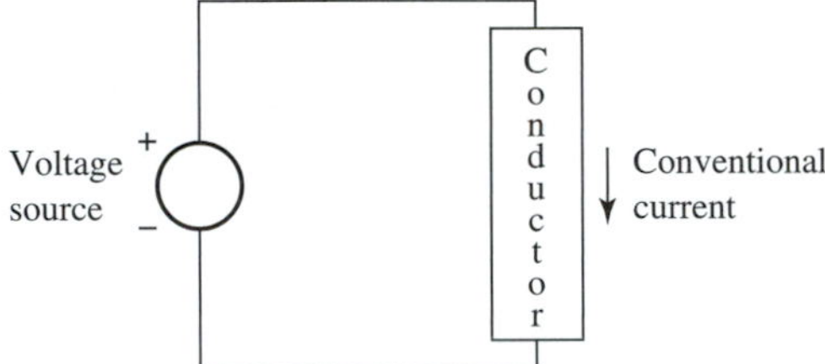

Current as illustrated in Figure 1.1 is called electron current, and current as illustrated in Figure 1.2 is called conventional current. Most engineering and engineering technology textbooks use conventional current, and this textbook is no exception. The predominance of the use of conventional current in electric circuits and related sciences stems from the large accumulation of knowledge and written material developed prior to the presentation of the electron flow theory and also from the fact that the use of electron current provides little or no advantage in electric circuit analysis.

1.5 BASIC ELECTRICAL ELEMENTS

There are five electrical elements considered basic to the study of electric circuits. These are the *resistor, inductor, capacitor, voltage source,* and *current source.* The voltage and current sources can be *independent* or *dependent;* that is, the performance of the voltage and current sources may or may not depend on the circuit in which they are connected.

A resistor is a circuit element that limits the flow of current and converts electrical energy into heat energy.

An inductor is a circuit element that consists of a conducting wire wound around a core material, usually made of air or iron. The most significant properties of the inductor are that it opposes changes in current and is capable of storing energy in a magnetic field.

A capacitor is a circuit element that consists of plates made of a conducting material such as copper with dielectric material between (separating) the plates. The most significant properties of the capacitor are that it opposes changes in voltage and is capable of storing energy in an electric field.

A voltage source is a device that maintains a specific voltage across its terminals regardless of the current flowing through it.

A current source is a device that maintains a specific current output regardless of the voltage across its terminals. Current sources are usually created by combining voltage sources with other electrical elements.

1.6 RESISTANCE, INDUCTANCE, AND CAPACITANCE

The resistor, inductor, and capacitor are constructed to display particular current-voltage relationships when energized and are measured in terms of *resistance, inductance,* and *capacitance,* respectively.

The voltage across a resistor is proportional to the current flowing through the resistor. The constant of proportionality relating the current and voltage is called the resistance of the resistor; that is,

$$\begin{pmatrix}\text{voltage across}\\ \text{a resistor}\end{pmatrix} \alpha \begin{pmatrix}\text{current flowing}\\ \text{through the resistor}\end{pmatrix}$$

and because resistance is the proportionality constant,

$$\begin{pmatrix}\text{voltage across}\\ \text{a resistor}\end{pmatrix} = (\text{resistance}) \times \begin{pmatrix}\text{current flowing}\\ \text{through the resistor}\end{pmatrix}$$

The voltage across an inductor is proportional to the rate at which the current through the inductor varies. The constant of proportionality relating the voltage to the rate of current variation is called the inductance of the inductor. That is,

$$\begin{pmatrix}\text{voltage across}\\ \text{an inductor}\end{pmatrix} \alpha \begin{pmatrix}\text{change in current flowing through}\\ \text{the inductor with respect to time}\end{pmatrix}$$

and because inductance is the proportionality constant

$$\begin{pmatrix}\text{voltage}\\ \text{across an}\\ \text{inductor}\end{pmatrix} = (\text{inductance}) \times \begin{pmatrix}\text{change in current flowing}\\ \text{through the inductor with}\\ \text{respect to time}\end{pmatrix}$$

The current flowing through a capacitor is proportional to the rate at which the voltage across the capacitor varies. The constant or proportionality relating the current to the rate of voltage variation is called the capacitance of the capacitor:

$$\begin{pmatrix}\text{current flowing}\\ \text{through a capacitor}\end{pmatrix} \alpha \begin{pmatrix}\text{change in voltage across the}\\ \text{capacitor with respect to time}\end{pmatrix}$$

and because capacitance is the proportionality constant,

$$\begin{pmatrix}\text{current flowing}\\ \text{through a}\\ \text{capacitor}\end{pmatrix} = (\text{capacitance}) \times \begin{pmatrix}\text{change in voltage}\\ \text{across the}\\ \text{capacitor with}\\ \text{respect to time}\end{pmatrix}$$

Resistance, inductance, and capacitance can be present without being associated with a deliberately constructed resistor, inductor, or capacitor. As an example, the conducting wires used to connect the basic elements and devices of a circuit have resistance but are not considered resistors.

1.7 ELECTRICAL DEVICES

Electrical devices (transformers, electric motors, operational amplifiers, electric generators, transistors, etc.) can usually be represented for analysis purposes in terms of equivalents consisting of one or more of the five basic electrical elements. A transistor, for example, can be represented by a current source and two resistors for some analysis purposes. A circuit containing electrical devices can generally be reduced for analysis purposes to a circuit consisting of only basic electrical elements. For the preceding reason and because the basic electrical elements are primal to the study of electric circuits, electric circuit–analysis courses usually emphasize the analysis of circuits that are restricted to the five basic electrical elements.

1.8 ACTIVE AND PASSIVE ELEMENTS

The five basic elements are classified as active or passive. An active electrical element is one that is capable of adding net energy to a circuit. A passive electrical element is one that does not add but instead receives energy from a circuit. Resistors, inductors, and capacitors are passive elements, whereas voltage and current sources are active elements.

1.9 LINEARITY

Each of the basic passive elements (resistor, inductor, and capacitor) is considered linear for the following reasons. The resistor is considered a linear electrical element because a straight line results when the voltage across the resistor is plotted versus the current flowing through the resistor. The capacitor is considered a linear electrical element because a straight line results when the change in voltage across the capacitor is plotted versus the current flowing into the capacitor. The inductor is considered a linear electrical element because a straight line results when the change in current flowing through the inductor is plotted versus the voltage across the inductor. The linearity of the passive elements greatly decreases the level of mathematical difficulty that would otherwise be associated with the analysis of electric circuits.

1.10 BILATERAL ELEMENTS

A bilateral electrical element exhibits the same voltage-current relationship regardless of the direction of current flow through the element. The resistor, inductor, and capacitor each exhibit the same linear characteristic previously described regardless of the direction of current flowing through them and are bilateral elements.

1.11 TWO-TERMINAL ELEMENTS

Each of the basic electrical elements has two terminals. That is, each has two connecting leads (as shown in Figure 1.3) used to connect the element to other elements or devices to form an electric circuit, as illustrated in Figure 1.4.

1.12 SYMBOLS FOR BASIC ELECTRIC ELEMENTS

The symbols for the five basic elements are shown in Figure 1.3. Two symbols are required for both current and voltage sources to indicate if the source is independent or dependent. Recall that the operation of voltage and current

Figure 1.3 Basic electrical elements.

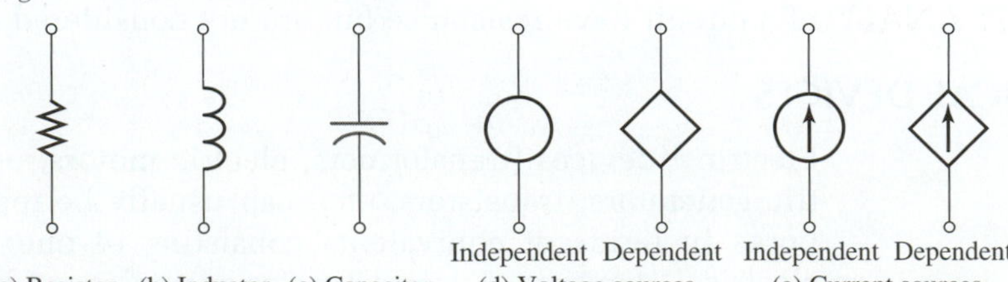

sources may or may not depend on the circuit in which they are connected. The voltage across a dependent voltage source can depend on a voltage or current in another part of the circuit; and in a similar manner, the current flowing through a dependent current source can depend on a voltage or current in another part of the circuit. In contrast, the voltage across an independent voltage source and the current furnished by an independent current source are independent of the circuit in which they are connected.

1.13 CIRCUIT EXAMPLE

The connection of electrical elements, as shown in Figure 1.4 with the use of the symbols in Figure 1.3, is an example of an electric circuit.

Figure 1.4 Example of an electric circuit.

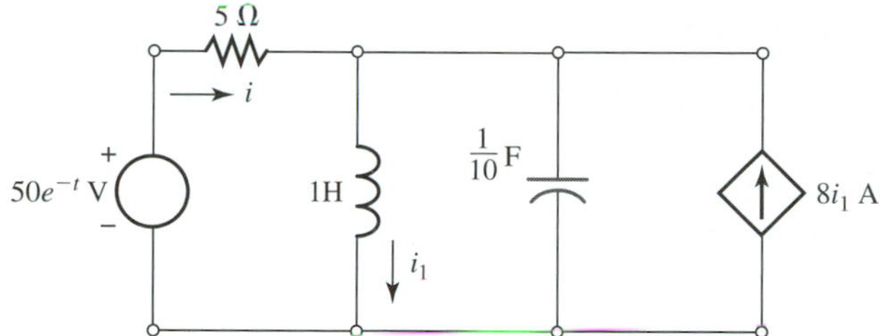

The values of resistance, inductance, and capacitance are indicated next to the element symbols in terms of ohms (Greek letter omega Ω), henrys (H), and farads (F). Ohms, henrys, and farads are the units of measurement for resistance, inductance, and capacitance. Each is discussed in a later chapter.

The value of voltage across the independent voltage source is indicated next to the symbol as $50e^{-t}$ volts (V), where volt is the unit of measurement for voltage and the + and − signs indicate the polarity associated with the voltage. Since the voltage source is independent, the voltage across it is independent of the circuit and varies exponentially with time as indicated.

The value of the current through the dependent current source is given as $8i_1$ amperes (A), where ampere is the unit of measurement for current and the arrow in the symbol indicates the positive direction of current through the current source. Notice that unlike the voltage source in the circuit, the current source is a dependent source, and the current flowing through it depends on the current i_1 through the inductor.

1.14 LUMPED-PARAMETER CIRCUITS

It takes connectors (conductors) to connect the elements that form a circuit, as shown in Figure 1.4. These connectors have resistance, which can affect the currents and voltages in the circuit. If the effect of the connectors on the currents and voltages of the circuit is considered negligible, the circuit is said to be a *lumped-parameter form.*

1.15 CIRCUIT ANALYSIS

If it is desired to determine the currents, voltages, or other quantities associated with a circuit, the principles and laws of electric circuit analysis are applied to the circuit, which will result in a mathematical description of the circuit called a *math model.* The math model will consist of one or more mathematical equations that can be solved for the desired quantities. This procedure is generally referred to as *electric circuit analysis.*

1.16 DC AND AC

Most voltages and currents appearing in electric circuits can be categorized as *dc* or *ac*. The symbol dc stands for direct current and is used to describe currents that do not change direction and voltages that do not change polarity with the passage of time. A constant current or voltage (one that does not vary with time) is a primary example of dc, and many reserve the term dc to describe only constant voltages and currents.

The symbol ac stands for alternating current and is used to describe currents that alternately change direction and voltages that alternately change polarity with the passage of time. A current or voltage that changes sinusoidally is a primary example of ac, and many reserve the term ac to describe only voltages and currents that change sinusoidally with time.

1.17 DESCRIPTION OF SOURCES

In this text, we shall use the symbols for current and voltage sources given in Figures 1.3(d) and (e) and let the information placed next to the source symbol describe the variation of current or voltage associated with the source. For example, in Figure 1.5(a) the voltage across the independent voltage source remains constant at 10 V as time elapses, and in Figure 1.5(b) the voltage across the independent voltage source varies as a sinusoid as time elapses. In Figure 1.5(c) the voltage across the dependent voltage source is equal to $50v_1$ volts, where v_1 is the voltage across a particular element or elements in the circuit.

Figure 1.5 Example of voltage sources.

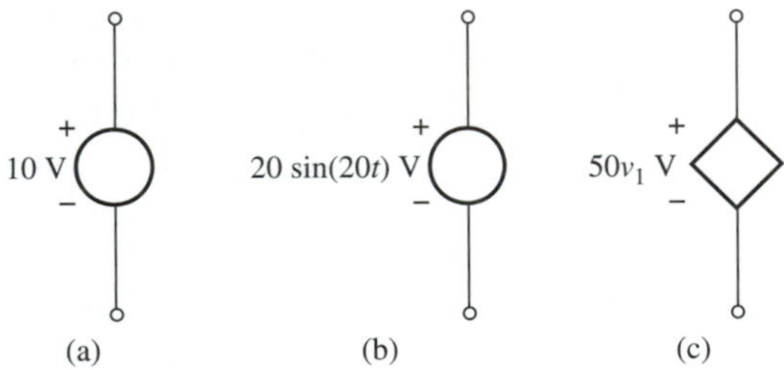

1.18 UNITS

In analyzing an electric circuit, we assign units of measurement to the measurable physical quantities describing the circuit. The system of units currently in use throughout the world is the International System of Units (commonly called the SI system). The SI system of units was adopted in 1960 by the International Conference on Weights and Measures and was officially endorsed by the United States in 1975. We generally use the SI system of units throughout this text.

The SI system consists of six basic units, the meter (measurement of distance), the second (measurement of time), the kilogram (measurement of mass), the coulomb (measurement of electric charge), the kelvin (measurement of temperature), and the candela (measurement of luminous intensity). All other units are derived from the six basic units. For example, an important quantity in electric circuit analysis is electric current. The unit for electric current is the ampere, which can be expressed as coulombs per second. That is, one ampere equals one coulomb per second, where coulombs and seconds are basic SI units. The units for the electric quantities used in this text are discussed as they are encountered.

1.19 SYMBOLS

The representation of physical quantities by symbols and names is necessary for conciseness and the accountability of variables in problem solving. The use of Greek letters as symbols to represent electrical quantities is traditional and in widespread use. The units, symbols, and Greek letters used in this text are listed in Tables 1.1 and 1.2.

TABLE 1.1 Units and Symbols for Basic Electrical Quantities

Quantity (symbol)	Unit (symbol)
Admittance (Y)	Siemens (S)
Angular frequency (ω)	Radians/second (r/s)
Angular measurement (arbitrary symbol)	Radian (r) and degree (°)
Apparent power (S)	Volt-amperes (VA)
Average power (P)	Watts (W)
Capacitance (C)	Farads (F)
Conductance (G)	Siemens (S)
Current (i)	Amperes (A)
Electric charge (q)	Coulombs (c)
Electric field strength (ξ)	Volts/meter (V/m)
Electric flux density (D)	Coulombs/square meter (C/m^2)
Energy (w)	Joules (J)
Frequency (f)	Hertz (Hz)
Impedance (Z)	Ohms (Ω)
Inductance (L)	Henrys (H)
Instantaneous power (p)	Watts (W)
Mutual inductance (M)	Henrys (H)
Reactance (X)	Ohms (Ω)
Reactive power (Q)	Volt-amperes reactive (VARs)
Resistance (R)	Ohms (Ω)
Susceptance (B)	Siemens (S)
Temperature (T)	Degree celsius (°C)
Time (t)	Seconds (s)
Voltage (v)	Volts (V)

TABLE 1.2 Greek Letters Used in this Text

Alpha	α
Beta	β
Gamma	γ
Delta	δ, Δ
Epsilon	ε
Eta	η
Theta	θ
Lambda	λ
Mu	μ
Xi	φ
Pi	π
Rho	ρ
Sigma	Σ, σ
Tau	τ
Phi	Φ, ϕ
Psi	χ
Omega	Ω, ω

1.20 CONVERSION OF UNITS

Although the SI system of units is used predominately throughout the Western World, other systems—such as the U.S. customary system of units—are often encountered. Conversion from one system of units to another is necessary for comparison purposes and to ensure the dimensional integrity of mathematical terms and equations. A simple procedure for converting units involves the introduction of terms that cancel out the units to be converted from and result in the units to be converted to. For example, if we want to convert 10 m/s (meters per second) to meters per minute, we introduce a term expressing seconds per minute to cancel out seconds and introduce minutes. Because there are 60 s per minute, the term 60 s/min is introduced as follows.

$$(10 \text{ m/s})\,(60 \text{ s/min}) = 600 \text{ m/min}$$

As another example, convert 88 ft/s to miles per hour (mi/h). We know that 1 mi = 5280 ft and 3600 s = 1 hr.

$$(88 \text{ ft/s})\left(\frac{1}{5280} \text{ mi/ft}\right)(3600 \text{ s/h}) = 60 \text{ mi/h}$$

Notice that the term $\frac{1}{5280}$ mi/ft is used to introduce miles and eliminate feet and the term 3600 s/h is used to introduce hours and eliminate seconds.

1.21 CALCULATIONS AND NOTATION

Using *scientific notation* eases the difficulty of performing mathematical operations, especially with extremely large or extremely small numbers. Numbers in scientific notation are represented in terms of integer powers of ten. As an example, 1,420,000 and 0.0000891 can be expressed concisely in scientific notation as 1.42×10^6 and 8.91×10^{-5}, respectively. Numbers in scientific notation are always expressed with one digit to the left of the decimal point and the remaining digits to the right, as indicated in the previous example. The use of scientific notation expedites mathematical calculations because the laws of exponents readily apply to the integer powers of term. As an example,

$$(6.00 \times 10^5) \times (5.20 \times 10^3) = 31.2 \times 10^{(5+3)} = 31.2 \times 10^8$$

which is written as

$$3.12 \times 10^9$$

in scientific notation.

The powers of ten in scientific notation are given names (prefixes) in the SI system, and the names are abbreviated as indicated in Table 1.3. For ex-

TABLE 1.3 Prefix Abbreviations for the SI System

Power of Ten	Prefix	Abbreviation
10^{-12}	Pico	p
10^{-9}	Nano	n
10^{-6}	Micro	μ
10^{-3}	Milli	m
10^{3}	Kilo	k
10^{6}	Mega	M
10^{9}	Giga	G
10^{12}	Tera	T

ample, the quantity 10^{-3} is called milli and abbreviated by the lowercase m. Thus, 1.6×10^{-3} amperes can be written as 1.6 milliamperes, or 1.6 mA.

Engineering notation is similar to scientific notation except that the exponents of 10 are displayed in multiples of three in order to facilitate the use of the SI prefixes. As an example, the number 0.000281 is expressed as 2.81×10^{-4} in scientific notation and as 281×10^{-6} in engineering notation. Notice from Table 1.3 that the powers of ten for all the prefixes in the SI system are multiples of three.

Significant figures determine the accuracy of a number and include all the digits of the number from 0 to 9. However, zero is not a significant figure if it is used only to indicate the position of the decimal point for the number. As an example, the numbers 0.002130 and 2.130 each have four significant figures. The two zeros immediately after the decimal in 0.002130 are used only to indicate the position of the decimal point, but the zero after the three is a significant figure for both numbers. It is customary to include the significant figures with the decimal part of the number in both scientific and engineering notation. As an example, if the number 60,200,000 has four significant figures, it is expressed as 6.020×10^7 in scientific notation and 60.20×10^6 in engineering notation. In each case there are four significant figures associated with the decimal part of the number. Using scientific or engineering notation clarifies the number of significant figures contained in a number.

1.22 ROUNDING

In performing numerical calculations, we often arrive at answers with more significant figures than the original data. In such cases, the number of significant figures in the resulting number must be reduced by a procedure known as *rounding*. As an example, consider the product of 5.21×10^4 and 4.0×10^3. The result is 2.084×10^8 in scientific notation. However, because the number 4.0×10^3 has only two significant figures, the result must be rounded to two significant figures, which results in 2.1×10^8 in scientific notation and 0.21×10^9 in engineering notation.

The rules generally used for rounding numbers to n significant figures are as follows:

> If the $(n + 1)$st digit in a number is less than 5, drop the $(n + 1)$st digit and the other digits following the $(n + 1)$st digit. As an example, the number 1.0082 rounded to four significant figures is 1.008.
>
> If the $(n + 1)$st digit in a number is equal to 5 with only zeros following the 5, drop the 5 and change the nth number to the next highest even number if the nth number is odd and leave the nth number unchanged if it is even. As an example, the numbers 0.8015 and 0.006545 rounded to three significant figures are 0.802 and 0.00654, respectively.
>
> If the $(n + 1)$st digit is equal to 5 with any nonzero digits following the 5 or if the $(n + 1)$st digit is greater than 5, drop the $(n + 1)$st digit and increase the nth digit by 1. As an example, the numbers 92.568 and 6.248 rounded to three significant figures are 92.6 and 6.25, respectively.

1.23 EXERCISES

1. Name the five basic elements in electric circuits.
2. What is the difference between a passive element and an active element?
3. Explain the meaning of linearity as it pertains to the basic electric elements.
4. Explain the meaning of bilateral as it pertains to the basic electric elements.

5. Explain the difference between dependent and independent voltage sources.
6. Explain the difference between dependent and independent current sources.
7. Explain the difference between resistor and resistance.
8. Explain the difference between inductor and inductance.
9. Explain the difference between capacitor and capacitance.
10. Draw the symbols for a resistor, inductor, and capacitor.
11. Draw the symbols for dependent current and voltage sources.
12. Draw the symbols for independent current and voltage sources.
13. What is meant by a dc current or voltage?
14. What is meant by an ac current or voltage?
15. Explain how the five basic electric elements can be used to simplify the analysis of complex electrical devices.
16. Explain what is meant by an electric current.
17. Name the six basic units in the SI system of units.
18. What is the difference between basic and derived units in the SI system?
19. Why are symbols used in most analytical problem-solving procedures?
20. Express the following numbers in scientific notation. Each number has four significant figures.

(a) 0.0009102 (b) 2,010,000.
(c) 32.08 (d) 9196×10^{-1}
(e) 2.006 (f) 0.6428
(g) $64{,}900. \times 10^{2}$ (h) 0.02140×10^{-1}
(i) $800{,}000. \times 10^{-8}$ (j) 89.11×10^{2}
(k) 3011. (l) 661,000.

21. Express the numbers in Problem 20 in engineering notation.
22. Round the numbers in Problem 20 to three significant figures and express the resulting numbers in scientific notation.
23. Express the acceleration 20 ft/s^2 ((feet per second) per second) in (miles per minute) per minute.
24. Express the following measurements in microfarads (μF).

(a) 0.000192 F (b) 0.058 F (c) $\frac{1}{400}$ F

25. Express the following measurements in kΩ.

(a) 482 Ω (b) 916,000 Ω (c) 0.27 Ω

26. Express the following measurements in millihenrys (mH).

(a) 0.552 H (b) 0.076 H (c) 4.2 H

27. Round each number to three significant figures.

(a) 0.9035 (b) 0.1026 (c) 7.774 (d) 662.5

CHAPTER

2

Ohm's Law and Resistance

After completing this chapter the student should be able to

* use Ohm's law to calculate voltages and currents in resistive circuits
* determine the power associated with basic circuit elements

A good conductor provides a small resistance (opposition) to the flow of electric current, whereas a poor conductor provides a large resistance (opposition) to the flow of electric current. When current flows through a conductor, the free electrons that constitute the current collide with the lattice atoms of the conductor and the current flow is impeded. The amount of impedance to the flow of free electrons and the amount of free electrons available for flow are determined by the physical characteristics of the conductor. A quantity called *resistance* is a measurement of the opposition to the flow of current through a conductor.

The resistance of a conductor depends on the physical dimensions (length, cross-sectional area) of the conductor, the material of which the conductor is made, and the temperature of the conductor. Resistance can be unintended and can occur inadvertently because of the physical structure of a circuit. However, if a device is deliberately constructed as a circuit element with the resistance intended, it is called a *resistor*. The resistance of a resistor is a measurement of the resistor's opposition to the flow of current. The relationship between current and voltage for a resistor is given by Ohm's law.

2.1 OHM'S LAW

The German physicist Georg Ohm (1787–1854) developed the relationship between current and voltage for a resistor and the relationship is called *Ohm's law.*

> **OHM'S LAW** The voltage across a resistor is equal to the resistance of the resistor multiplied by the current flowing through the resistor. The voltage is expressed in volts; the current, in amperes; and the resistance, in ohms.

Consider the circuit consisting of a voltage source and a resistor, as shown in Figure 2.1. Because the voltage across the source is also the voltage across the resistor, Ohm's law for the resistor can be stated as

$$v = Ri \tag{2.1}$$

where v represents voltage, R represents resistance, and i represents current.

Figure 2.1 Ohm's law.

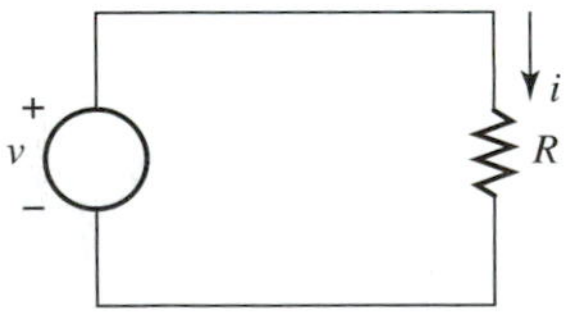

Resistance is measured in ohms, where one ohm equals one volt per ampere. The symbol for resistance is R and the symbol for ohms is the Greek letter omega (Ω).

We further investigate Ohm's law with regard to the notation for current direction and voltage polarity. Consider the resistance R in Figure 2.2.

Figure 2.2 Ohm's law using subscripted variables.

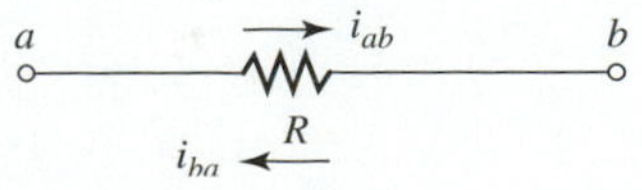

The conventional current flowing from point a to point b is designated i_{ab}, and the conventional current flowing from point b to a is designated i_{ba}, where

$$i_{ab} = -i_{ba}$$

The voltage of point a with respect to point b is designated v_{ab} and the voltage of point b with respect to point a is designated v_{ba}, where

$$v_{ab} = -v_{ba}$$

The proper direction for current and the proper polarity for voltage must be observed when applying Ohm's law.

If the conventional current through a resistance is expressed as flowing from point a to point b, then the voltage across the resistance must be expressed as the voltage of point a with respect to point b. Applying Ohm's law to the resistance in Figure 2.2 results in

$$\left.\begin{aligned} v_{ab} &= Ri_{ab} \\ &\text{or} \\ v_{ba} &= Ri_{ba} \end{aligned}\right\} \tag{2.2}$$

If v_{ab} is positive, then the voltage of point a is positive with respect to point b; and if v_{ab} is negative, then the voltage of point a is negative with respect to point b. For example, if $v_{ab} = 10$ V, then point a is 10 V positive with respect to point b; and if $v_{ab} = -10$ V, then point a is 10 V negative with respect to point b.

Consider the following examples.

EXAMPLE 2.1 Determine i_{ab} and i_{ba} for the resistance shown in Figure 2.3 if $v_{ab} = 60$ V. ■

Figure 2.3 a — 10 Ω — b

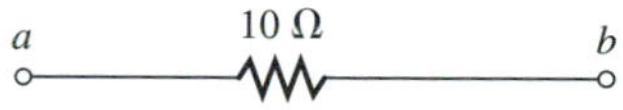

SOLUTION From Ohm's law (Equation 2.2),

$$v_{ab} = Ri_{ab}$$
$$60 = 10i_{ab}$$
$$i_{ab} = 6 \text{ A}$$

Because $v_{ba} = -v_{ab}$, Ohm's law results in

$$v_{ba} = Ri_{ba}$$
$$-60 = 10i_{ba}$$
$$i_{ba} = -6 \text{ A}$$

Alternatively, because $i_{ba} = -i_{ab}$,

$$i_{ba} = -6 \text{ A}$$

The minus sign for i_{ba} indicates that the actual current flows from point a to point b.

EXAMPLE 2.2 Determine v_{ab} and v_{ba} for the resistance shown in Figure 2.4, where $i_{ab} = 2$ A. ■

Figure 2.4

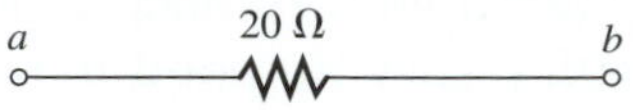

SOLUTION From Ohm's law (Equation 2.2),

$$\begin{aligned} v_{ab} &= Ri_{ab} \\ &= 20(2) \\ &= 40 \text{ V} \end{aligned}$$

Because $i_{ba} = -i_{ab}$, Ohm's law results in

$$\begin{aligned} v_{ba} &= Ri_{ba} \\ &= 20(-2) \\ &= -40 \text{ V} \end{aligned}$$

Alternatively, because $v_{ba} = -v_{ab}$,

$$v_{ba} = -40 \text{ V}$$

Terminal a is 40 V positive with respect to terminal b; equivalently, terminal b is 40 V negative with respect to terminal a.

EXAMPLE 2.3 Determine v_{ab} and v_{ba} for the resistance shown in Figure 2.5 if $i_{ab} = -4$ A. ■

Figure 2.5

SOLUTION From Ohm's law (Equation 2.2),

$$\begin{aligned} v_{ab} &= Ri_{ab} \\ &= 15(-4) \\ &= -60 \text{ V} \end{aligned}$$

Because $i_{ba} = -i_{ab}$, Ohm's law results in

$$\begin{aligned} v_{ba} &= Ri_{ba} \\ &= 15(4) \\ &= 60 \text{ V} \end{aligned}$$

Alternatively, because $v_{ba} = -v_{ab}$,

$$v_{ba} = 60 \text{ V}$$

Terminal a is 60 V negative with respect to terminal b; equivalently, terminal b is 60 V positive with respect to terminal a.

Although the procedure of labeling terminals with letters to designate voltages and currents is convenient, we shall present another commonly used notation that is more concise and lends itself more to circuit analysis procedures. Consider the resistance in Figure 2.6, where the voltage v is the voltage of the positive-marked terminal with respect to the negative-marked terminal. If v is positive, then the voltage of terminal a is positive with respect to terminal b; if v is negative, the voltage of terminal a is negative with respect to terminal b. Also, if i is positive, the actual current flows in the direction of the arrow, and if i is negative, the actual current flows opposite to the arrow direction.

Figure 2.6 Ohm's law using unsubscripted variables.

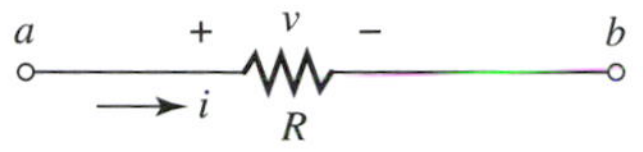

When applying Ohm's law, if the current i is expressed as flowing from the positive-marked terminal to the negative-marked terminal, then v must be expressed as the voltage of the positive-marked terminal with respect to the negative-marked terminal. Ohm's law results in

$$v = Ri$$

for the resistor in Figure 2.6.

As stated previously, the positive and negative markings on the terminals do not necessarily correspond to the actual polarity of the terminals. For instance, if $v = -6$ V in Figure 2.6, the positive-marked terminal is 6 V negative with respect to the negative-marked terminals. Also as stated before, the arrow for current does not necessarily correspond to the actual current direction. For instance, if $i = -2$ A, the current flows in a direction opposite to the arrow.

Consider the following examples.

EXAMPLE 2.4 Determine the current i if $v = 20$ V, as shown in Figure 2.7. ■

Figure 2.7

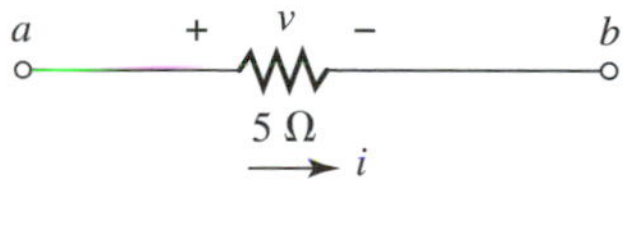

SOLUTION From Ohm's law

$$v = Ri$$
$$20 = 5i$$
$$i = 4\text{ A}$$

EXAMPLE 2.5 Determine the current i if $v = 30$ V, as shown in Figure 2.8. ■

Figure 2.8

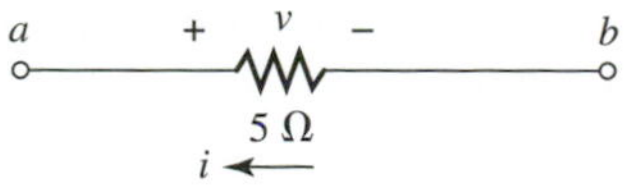

SOLUTION Using Ohm's law and recognizing that the current $-i$ flows from the positive-marked terminal to the negative-marked terminal (opposite direction of i),

$$\begin{aligned} v &= R(-i) \\ 30 &= 5(-i) \\ i &= -6 \text{ A} \end{aligned}$$

Alternatively, because the voltage of the negative-marked terminal with respect to the positive-marked terminal equals $-v$, Ohm's law can be written as

$$\begin{aligned} -v &= R(i) \\ -30 &= 5(i) \\ i &= -6 \text{ A} \end{aligned}$$

The minus sign indicates that the actual current flows in the opposite direction of the arrow.

EXAMPLE 2.6 Determine the voltage v if $i = 2$ A, as shown in Figure 2.9. ■

Figure 2.9

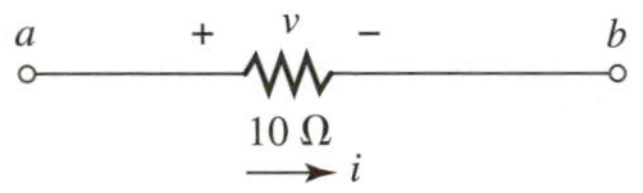

SOLUTION From Ohm's law

$$\begin{aligned} v &= Ri \\ &= 10(2) \\ &= 20 \text{ V} \end{aligned}$$

EXAMPLE 2.7 Determine the voltage v if $i = -3$ A, as shown in Figure 2.10. ■

Figure 2.10

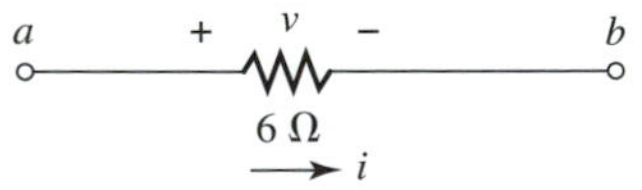

SOLUTION From Ohm's law

$$\begin{aligned} v &= Ri \\ &= 6(-3) \\ &= -18 \text{ V} \end{aligned}$$

The negative sign indicates that the voltage of the positive-marked terminal is 18 V negative with respect to the negative-marked terminal; alternatively, the negative-marked terminal is 18 V positive with respect to the positive-marked terminal.

2.2 CONDUCTANCE

A useful quantity associated with resistance is called *conductance.* Conductance is equal to the reciprocal of resistance and is designated by the symbol G, where

$$G = \frac{1}{R} \tag{2.3}$$

Because the unit for resistance is volts/amperes, the unit for conductance is amperes/volts. Conductance is measured in siemens, which is symbolized by the capital letter S and named in honor of the British engineer Sir William Siemens. Using conductance instead of resistance in Ohm's law—as given in Equation (2.1)—results in

$$i = Gv \tag{2.4}$$

where i is in amperes, v is in volts, and G is in siemens. The advantage of using conductance to relate voltage and current for resistors is illustrated in some of the circuit analysis methods presented in later chapters.

2.3 TYPES OF RESISTORS

Resistors are manufactured as fixed or variable. Most fixed resistors are made of a carbon composition, but other materials such as metal film and metal wire are also used. Variable resistors (also called rheostats and potentiometers, depending on the application) are usually made of metal wire. Resistors are characterized with regard to the precision of the specified resistance, the power-dissipation capability, the probability of resistor failure, and the sensitivity to variations in temperature and other environmental factors. Commercially available fixed resistors range from approximately 0.10 Ω to 22.0 MΩ, with about 200 different standard values available. The resistance of a resistor is usually stamped on the resistor unless a color code indicating the resistance is employed.

The standard color code most widely used for resistors is given in Table 2.1. The first and second bands are the first and second digits of the numerical value of the resistance, and the third band places the decimal point. If the third band is one of the colors in the first column of Table 2.1, the color indicates the number of zeros that follow the second digit. If the third band is gold or silver, the color indicates the factor to be multiplied by the first two digits. The fourth band is the precision tolerance, and the fifth band gives the percentage of resistor failures per 1000 h of use. For example, consider a resistor with the following band markings:

TABLE 2.1 Resistor Color Code

Bands 1, 2, 3	Band 3	Band 4	Band 5
0 Black	0.1 Gold	5% Gold	1% Brown
1 Brown	0.01 Silver	10% Silver	0.1% Red
2 Red		20% Absence of band	0.01% Orange
3 Orange			0.001% Yellow
4 Yellow			
5 Green			
6 Blue			
7 Violet			
8 Gray			
9 White			

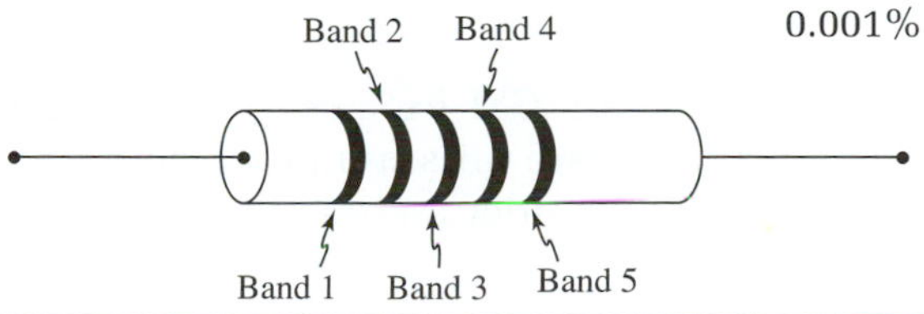

Band 1	Band 2	Band 3	Band 4	Band 5
Green	Violet	Red	Gold	Brown

It has a nominal resistance of 5700 Ω with a precision tolerance of 5%. That is, the resistance should range between 5415 Ω and 5985 Ω. The failure rate percentage is 1% per 1000 h of use.

Because circular conducting wire is prevalent in most electrical circuits, the resistance of circular wire made of copper, aluminum, or other conducting materials is of interest to us. The cross-sectional area of circular conducting wire is usually measured in *circular mils (CM).*

A circular wire with a diameter of 1 mil has an area of 1 CM by definition, where 1 mil = 0.001 in. Consider a circular area with a diameter of 1 mil. The area is

$$\begin{aligned} A &= \frac{\pi d^2}{4} \\ &= \frac{\pi(1)^2}{4} \\ &= \frac{\pi}{4} \text{ square mils} \end{aligned}$$

Because a circular area with a diameter of 1 mil has an area of 1 CM by definition, this area is

$$A = 1 \text{ CM}$$

It follows that

$$1 \text{ CM} = \frac{\pi}{4} \text{ square mils}$$

and

$$1 \text{ square mil} = \frac{4}{\pi} \text{ CM}$$

To obtain CM from square mils, multiply the square mils by $4/\pi$, and to obtain square mils from CM, multiply circular mils by $\pi/4$.

Consider a circle with a diameter of d mils. The area of the circle is

$$A = \frac{\pi}{4} d^2 \text{ square mils}$$

Converting to circular mils results in

$$A = \left\{\frac{\pi}{4} d^2\right\} \frac{4}{\pi} = d^2 \text{ CM}$$

It follows that an area in circular mils is equal to the diameter of the area in mils squared. Consider a circle with a diameter of 0.080 in. Because 1 mil equals 0.001 in., the diameter in mils is

$$d = 80 \text{ mils}$$

and the area in circular mils is

$$\begin{aligned} A &= d^2 = (80)^2 \\ &= 6400 \text{ CM} \end{aligned}$$

2.4 RESISTIVITY

The resistance of a circular wire is inversely proportional to the cross-sectional area and directly proportional to the length of the wire and can be written as

$$R = \rho \frac{l}{A} \tag{2.5}$$

where ρ (Greek letter rho) is called the *resistivity* of the material. The resistivity varies with temperature and is usually specified for room temperature, which is 20¡Celsius (20¡C). If the cross-sectional area of the wire is expressed in circular mils (CM) and the length is expressed in feet, the units for resistivity are CM-Ω/ft. Table 2.2 lists resistivity values at room temperature for some widely used materials.

TABLE 2.2

Material	Resistivity at 20°C $\frac{\text{CM-}\Omega}{\text{ft}}$
Silver	9.9
Copper	10.37
Gold	14.7
Aluminum	17.0
Tungsten	33.0
Nickel	47.0
Iron	74.0
Nichrome	600.0
Carbon	21,000.0

Using Equation (2.5) and Table 2.2, the resistance of circular wire can be expressed in ohms per 1000 feet for given cross-sectional areas. Standard commercially available wires are manufactured with a specified number of ohms per 1000 feet of wire. Based on the number of ohms per 1000 feet of wire, the wire is assigned an identifying number called the AWG (American Wire Gage) number. For example, the AWG numbers range from approximately AWG 0000 (0.0490 Ω/1000 ft at 20°C and a cross-sectional area of 211,600 CM) to AWG 40 (1049.0 Ω/1000 ft at 20°C and a cross-sectional area of 9.89 CM) for solid, round copper wire. Wire tables are available for all commercially available wire.

2.5 TEMPERATURE EFFECTS ON RESISTORS

The resistance of most conductors increases with temperature because the molecular activity within the conductor increases and retards the flow of charge. The relationship between resistance and temperature for conductors is arrived at for a given material by plotting the resistance versus temperature from laboratory measurements and approximating the plot as a straight line, as shown in Figure 2.11(a).

Figure 2.11 Variation of resistance with temperature.

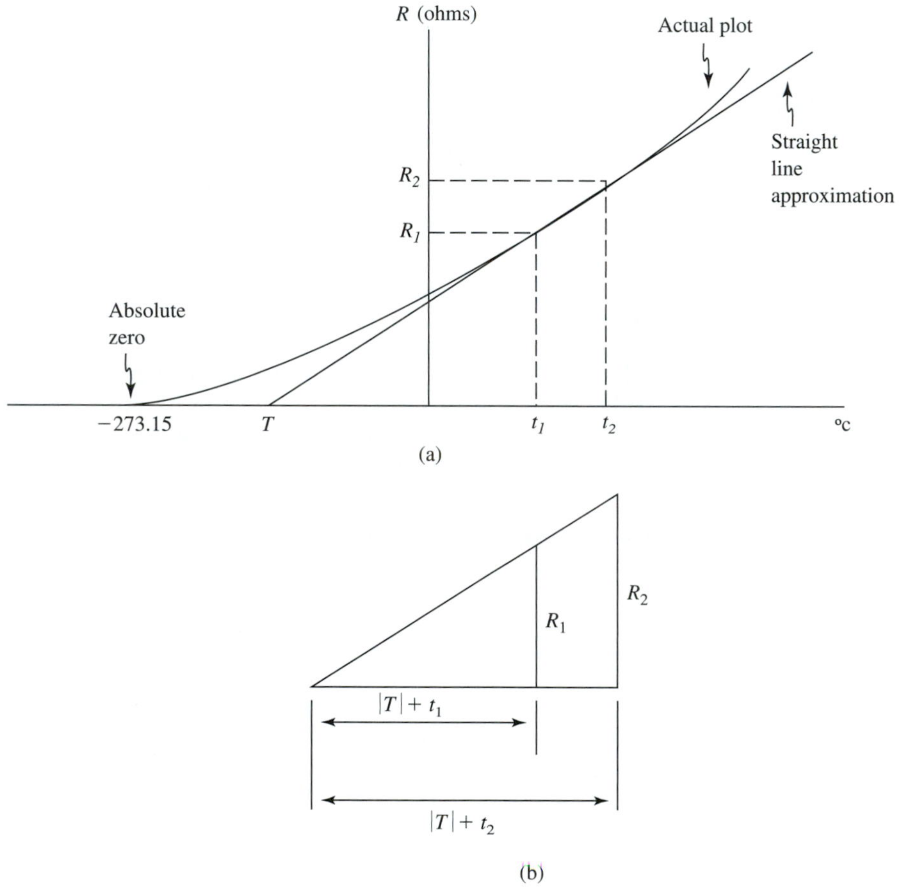

The straight-line approximation allows us to use the similar triangles in Figure 2.11(b) to arrive at the relationship

$$\frac{|T| + t_1}{R_1} = \frac{|T| + t_2}{R_2}$$

where $|T|$ is the absolute value (magnitude) of T. The temperature T is called the inferred absolute temperature and varies with the conducting material. Table 2.3 lists values of T for some widely used conducting materials.

The equation

$$\frac{|T| + t_1}{R_1} = \frac{|T| + t_2}{R_2}$$

TABLE 2.3

Material	Inferred Temperature T (°C)
Silver	−243
Copper	−234.5
Gold	−274
Aluminum	−236
Tungsten	−204
Nickel	−147
Iron	−162
Nichrome	−2,250

and Table 2.3 allow us to determine the resistance of a conductor at any temperature if the resistance of the conductor at a particular temperature is known. For example, if the resistance of a copper wire at 20°C is known to be 100 Ω, the resistance of the wire can be determined at 80°C:

$$\frac{|T| + t_1}{R_1} = \frac{|T| + t_2}{R_2}$$

$$\frac{|-234.5| + 20}{100} = \frac{|-234.5| + 80}{R_2}$$

$$\frac{234.5 + 20}{100} = \frac{234.5 + 80}{R_2}$$

$$R_2 = 123.6\ \Omega$$

Notice that resistance increases with temperature, as previously stated.

2.6 VOLTAGE SOURCES

All physically realizable voltage sources have an internal resistance that drains energy from the voltage source, leaving less energy available to supply an external load (circuit connected to the voltage source). The internal resistance is minimized in voltage source design in order to maximize the energy available for supplying an external load. Although it is impossible to design a voltage source with an internal resistance equal to zero, we often use the theoretical concept of a perfect (ideal) voltage source with zero internal resistance to arrive at a representation for an imperfect voltage source.

A *perfect (ideal) voltage source* is illustrated in Figure 2.12. The terminal

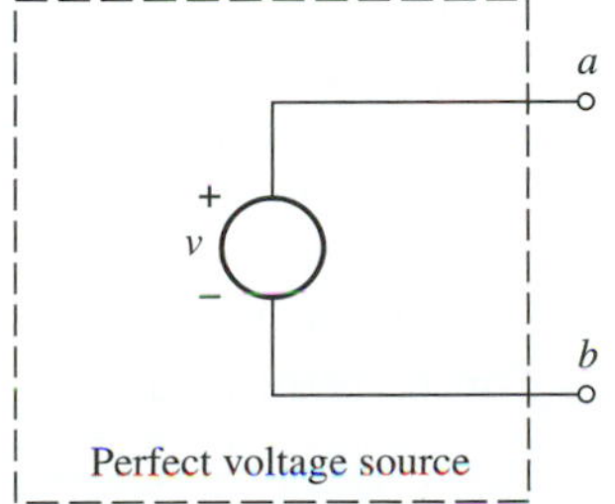

Figure 2.12 Perfect voltage source.

voltage v_{ab} is equal to v regardless of the current flowing through the voltage source. That is, a load (circuit) connected across terminals *a-b* has no effect on the terminal voltage of the perfect voltage source.

The perfect voltage source with a load connected across its terminals is shown in Figure 2.13. The terminal voltage v_{ab} equals v regardless of the value of the current i. The specified voltage v for the voltage source specifies the voltage of the positive-marked terminal with respect to the negative-marked terminal.

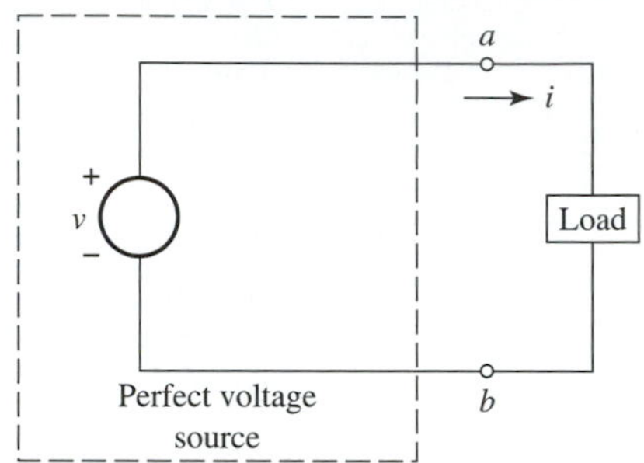

Figure 2.13 Perfect voltage source with load.

An *imperfect (practical) voltage source* consists of a perfect (ideal) voltage source and a resistance connected as shown in Figure 2.14. The resistance R represents the internal resistance of the imperfect voltage source. When no current flows through the imperfect voltage source, the terminal voltage v_{ab} equals v because no voltage appears across the resistance R. However, when a load is connected across terminals a-b, a current i flows through R; from Ohm's law the voltage across R is equal to Ri, with polarity as shown in Figure 2.15. The terminal voltage of the imperfect voltage source v_{ab} is

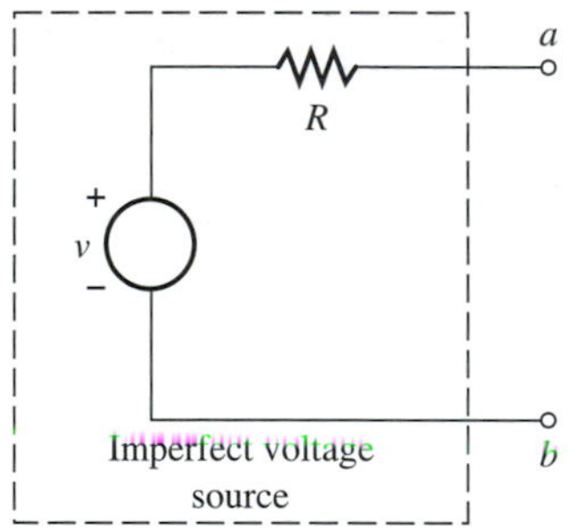

Figure 2.14 Imperfect voltage source.

$$v_{ab} = v - Ri$$

The voltage across the terminals of the imperfect voltage source decreases as the current through the imperfect voltage source increases.

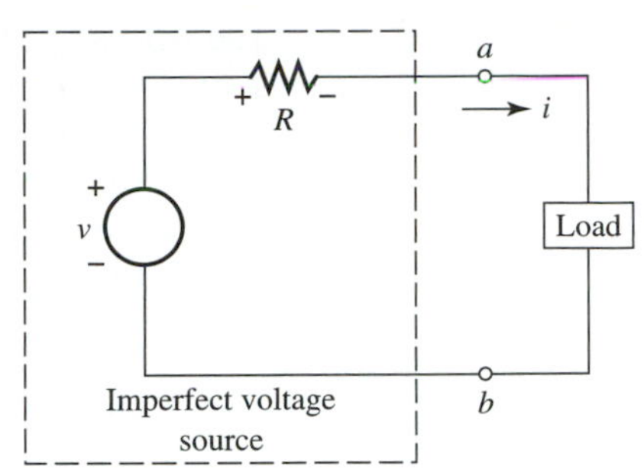

Figure 2.15 Imperfect voltage source with load.

2.7 CURRENT SOURCES

Although voltage sources are more common in everyday life (batteries, dc generators), there are also devices that are known as current sources. A transistor is an example of an electronic device that can be used in conjunction with other elements to provide a current source. Just as a perfect voltage source furnishes a voltage that is independent of the current flowing through the voltage source, a perfect current source furnishes a current that is independent of the voltage across the current source. A perfect current source cannot be physically realized. However, we use the theoretical concept of a perfect (ideal) current source to arrive at a representation for an imperfect (practical) current source.

A *perfect (ideal) current source* is illustrated in Figure 2.16. The current

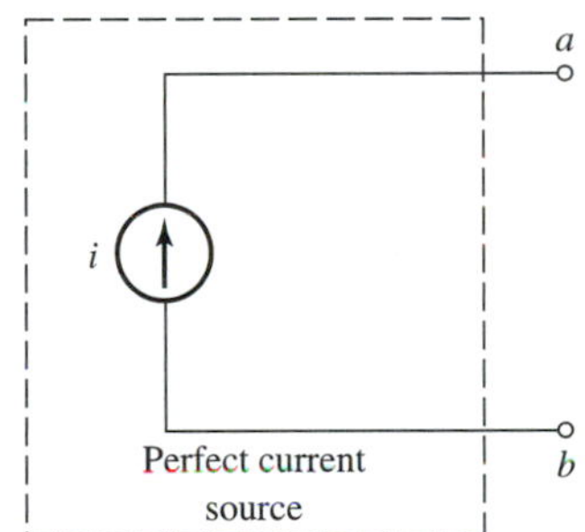

Figure 2.16 Perfect current source.

flowing in the direction of the arrow is always equal to i regardless of the voltage across the current source. That is, the load connected across terminals a-b has no effect on the current flowing through the current source. Although the internal resistance of a perfect voltage source is zero, the internal resistance of a perfect current source is infinitely large (open).

The perfect current source with a load connected across its terminals is shown in Figure 2.17. The current flowing through the source is equal to i

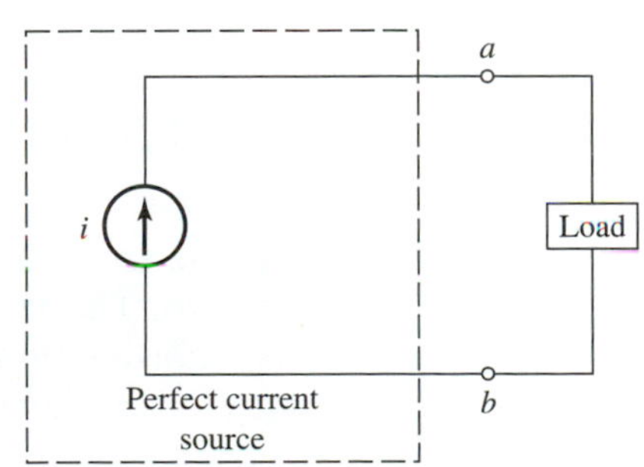

Figure 2.17 Perfect current source with load.

regardless of the voltage v_{ab} across the terminals of the current source. The arrow associated with a current source indicates the actual current direction when i is positive.

An *imperfect (practical) current source* consists of a perfect (ideal) current source and a resistance connected as shown in Figure 2.18. The resistance

Figure 2.18 Imperfect current source.

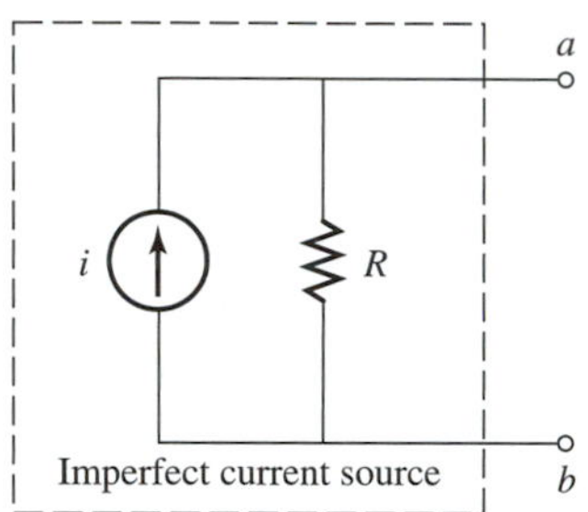

R represents the internal resistance of the imperfect current source. When no load is connected across terminals a-b, all the current i flows through the resistance R. However, when a load is placed across terminals a-b, some of the current from the current source flows through the load, as shown in Figure 2.19. The current delivered to the load by the imperfect current source is

$$i_L = i - i_R$$

Figure 2.19 Imperfect current source with load.

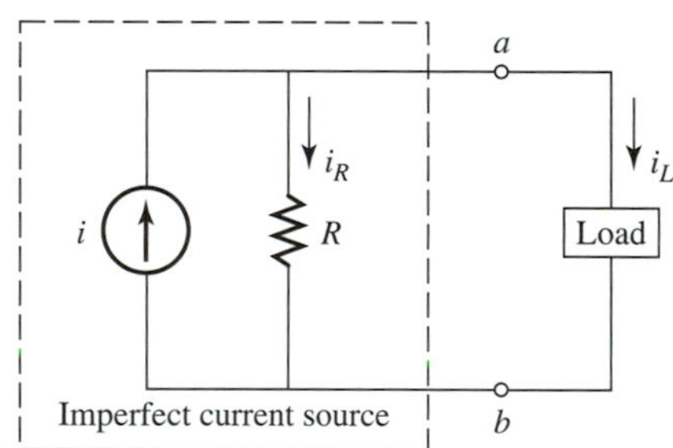

In designing current sources, the internal resistance R is maximized in order to minimize i_R and maximize the current available for delivery to the load.

2.8 SHORT AND OPEN CIRCUITS WITH REGARD TO SOURCES

A *short* is a conductor with zero resistance, and an *open* is an infinitely large resistance. The voltage across a short is zero, and the current through an open is zero. The terminals of imperfect (practical) current and voltage sources can be shorted (connected by a short) or opened (unconnected) without violating any of the laws of circuit analysis. However, the same is not true for perfect sources. A perfect voltage source cannot be shorted, and a perfect current source cannot be left open without violating the laws of circuit analysis and the associated mathematics.

Consider the perfect sources in Figure 2.20, where the terminals of the perfect voltage source in Figure 2.20(a) are shorted and the terminals of the perfect current source in Figure 2.20(b) are left open.

Figure 2.20 Shorted and open sources.

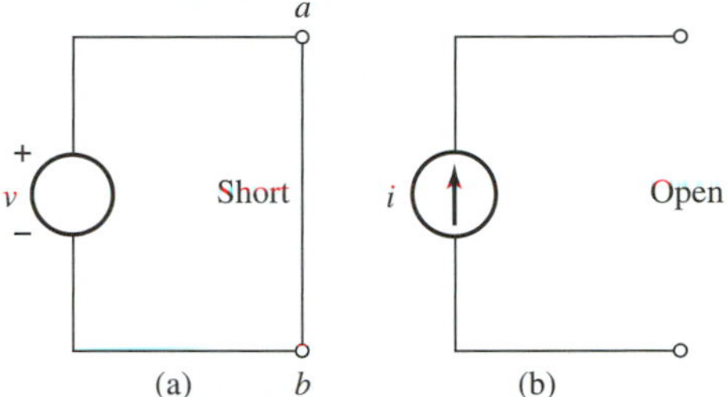

In the case of the perfect voltage source, the short indicates that $v_{ab} = 0$, but the perfect voltage source indicates that $v_{ab} = v$. v_{ab} cannot equal v and zero at the same time except for the trivial condition $v = 0$. Therefore, we cannot short a perfect voltage source without violating the principles of circuit analysis and the associated mathematics.

In the case of the perfect current source, current cannot flow unless there is a closed path. The perfect current source in Figure 2.20(b) indicates that the current through the current source is equal to i, but the open indicates that the current through the current source is equal to zero. The current through the perfect current source cannot equal i and zero at the same time except for the trivial condition $i = 0$. Therefore, we cannot leave a perfect current source open without violating the principles of circuit analysis and the associated mathematics.

2.9 ENERGY

Energy is the capacity for doing work and power is the rate of delivering or receiving energy. Voltage and current sources are sources of energy because they have the capacity for doing work. When a resistor is connected across a voltage source, the source does work by causing current to flow through the resistance. The energy delivered by the source to the resistance is lost from the circuit to the atmosphere in the form of heat. Energy is measured in *joules* (J) in honor of English physicist James Joule (1818–1889). One joule is the amount of energy expended in moving a body a distance of one meter with a constant force of one newton.

2.10 POWER

The rate of receiving or delivering energy is called power. Along with energy, voltage, and current, power is a primal quantity in electric circuit analysis. Power is measured in watts (W) in honor of Scottish engineer James Watt (1736–1819). One watt is equal to one joule per second. In electrical terms, one watt is equal to one volt-ampere. That is, a perfect voltage source of 1 V with a current of 1 A flowing through it is delivering energy at the rate of 1 J/s, or 1 W.

As stated previously, there are five basic electrical elements, and they are classified as active or passive with regard to their ability to deliver net energy to a circuit. Voltage and current sources are active elements and resistors, inductors, and capacitors are passive elements. An active element is capable of supplying net energy to a circuit, whereas a passive element can only receive energy from a circuit.

2.11 INSTANTANEOUS POWER

Consider a voltage v moving a small increment of charge Δq from one point to another in a circuit. From the definition of voltage, the increment of work (Δw) done on Δq can be written as

$$\Delta w = v\,\Delta q$$

We divide both sides of this equation by the increment of time, Δt, that expires during the accomplishment of the increment of work, Δw. This results in

$$\frac{\Delta w}{\Delta t} = v\frac{\Delta q}{\Delta t}$$

If we consider the work instantaneously accomplished (Δt approaching zero), then $\Delta w/\Delta t$ is the instantaneous work rate (power) and $\Delta q/\Delta t$ is the instantaneous rate of charge flow. Because the rate at which charge flows is equal to current,

$$p = vi \tag{2.6}$$

where p is the instantaneous power in watts, v is in volts, and i is in amperes. Instantaneous power refers to the power at a particular instant in time and is based on the current and voltage measured at that instant. If v and i are constant, then p is constant. However, if v or i varies with time, then p varies with time.

The direction of current and the voltage polarity for a pair of terminals must be considered when determining the power associated with the circuit connected across the terminals. Consider the circuits in Figure 2.21, where the circuits may consist of passive and/or active elements.

Figure 2.21 Power and electric circuits.

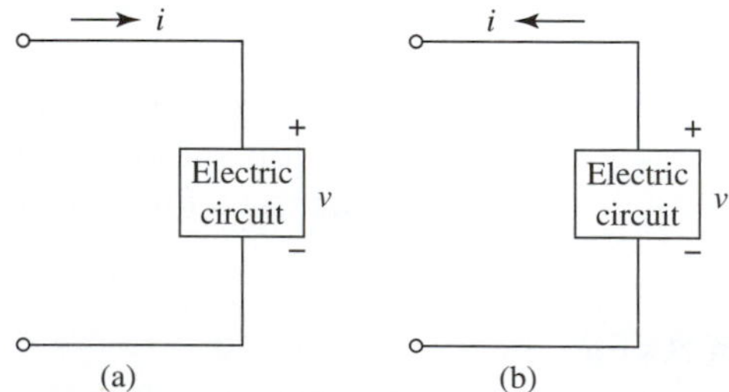

If the current i and the voltage v are positive in Figure 2.21(a), then energy is delivered to the circuit. If the current i and the voltage v are positive in Figure 2.21(b), then energy is delivered by the circuit.

Power is considered positive if energy is delivered and negative if energy is received by a circuit. In Figure 2.21(a) the power is negative power, because it is delivered to the electric circuit; in Figure 2.21(b) the power is positive power, because it is delivered by the electric circuit.

Consider the voltage sources in Figure 2.22. A voltage source is an active element and can receive or deliver energy. For example, a battery that is being charged is receiving energy and a battery that is discharging is delivering energy. If the current i and the voltage v are positive in Figure 2.22(a), the voltage source is delivering energy. If the current i and the voltage v are positive in Figure 2.22(b), the voltage source is receiving energy.

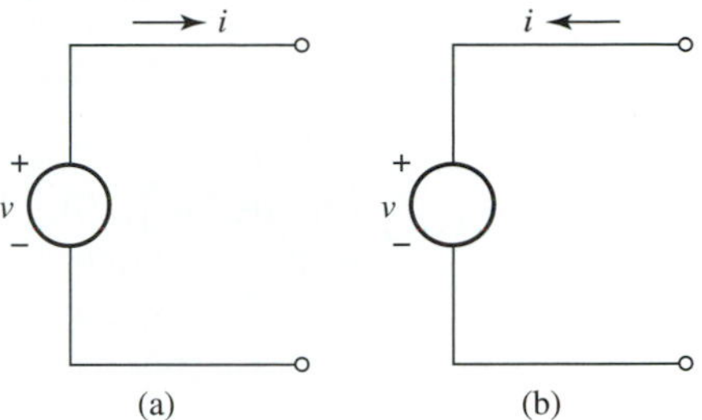

Figure 2.22 Power and voltage sources.

Consider the current sources in Figure 2.23. A current source is an active element and can receive or deliver energy. If the voltage v and the current i are positive as shown in Figure 2.23(a), the current source is delivering energy. If the voltage v and the current i are positive as shown in Figure 2.23(b), the current source is receiving energy.

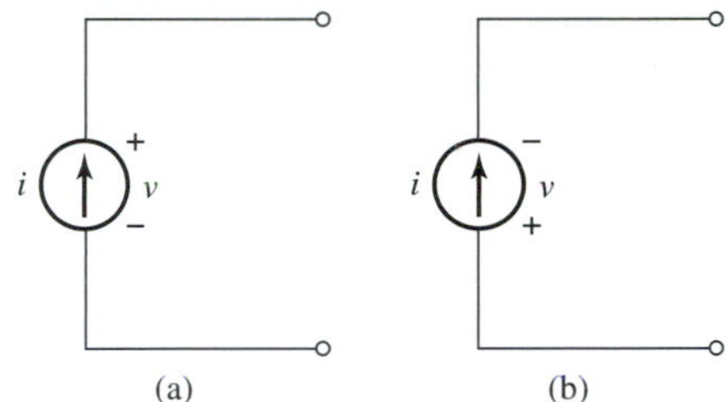

Figure 2.23 Power and current sources.

2.12 POWER AND RESISTANCE

Consider the resistance in Figure 2.24. Resistance is a passive element and can only receive energy. The current must flow from + to − through the resistance, as shown in Figure 2.24. From Ohm's law,

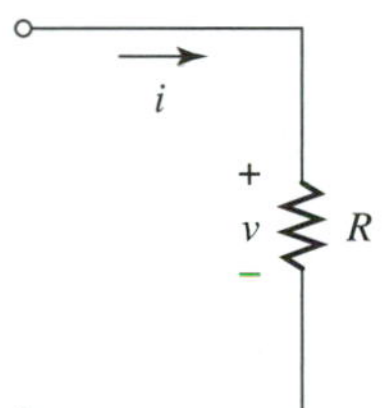

Figure 2.24 Power and resistance.

$$v = Ri$$

and from Equation (2.6), the instantaneous power equation,

$$p = vi$$

Substituting Ri for v in the power equation results in

$$p = i^2R \tag{2.7}$$

Substituting $i = v/R$ for i in the power equation results in

$$p = \frac{v^2}{R} \tag{2.8}$$

Equations (2.6), (2.7), and (2.8) can be used to determine the power delivered to a resistance.

If constant power is delivered or received for a time t, the energy delivered or received is given as

$$w = pt \tag{2.9}$$

where w is in joules, p is in watts, and t is in seconds. If p is not constant, the average value of p for the time interval t must be determined for use in Equation (2.9). Varying power and energy are investigated in a later chapter.

EXAMPLE 2.8 Determine the power delivered by the voltage source v and the energy delivered by the voltage source in 50 s for the circuit in Figure 2.25. ■

Figure 2.25

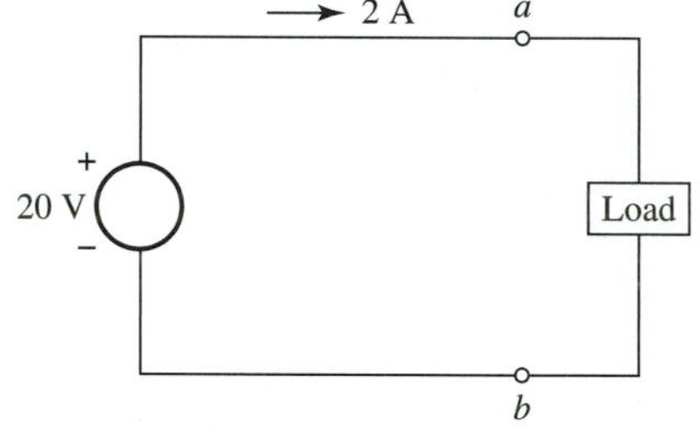

SOLUTION From Equation (2.6)

$$\begin{aligned} p &= vi \\ &= (20)(2) \\ &= 40 \text{ W} \end{aligned}$$

From Equation (2.9) the energy delivered in 50 s is

$$\begin{aligned} w &= pt \\ &= (40)(50) \\ &= 2000 \text{ J} \end{aligned}$$

EXAMPLE 2.9 Determine the power delivered by the voltage source for the circuit in Figure 2.26. ■

Figure 2.26

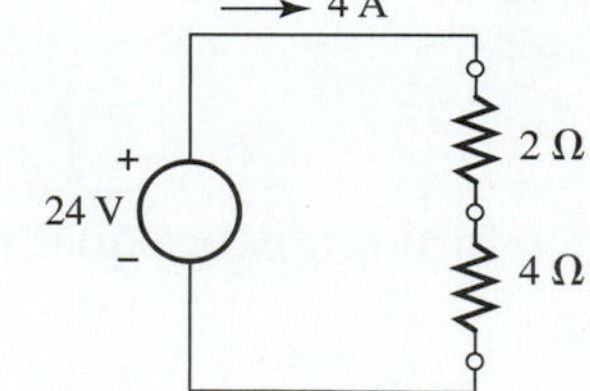

SOLUTION From Equation (2.6)

$$\begin{aligned} p &= vi \\ &= (24)(4) \\ &= 96 \text{ W} \end{aligned}$$

Alternatively, from Equation (2.7) the power delivered to the 2-Ω resistance is

$$\begin{aligned} p &= i^2R \\ &= (4)^2(2) \\ &= 32 \text{ W} \end{aligned}$$

and the power delivered to the 4-Ω resistance is

$$\begin{aligned} p &= i^2R \\ &= (4)^2(4) \\ &= 64 \text{ W} \end{aligned}$$

The total power delivered by the voltage source is the sum of the power delivered to the two resistances.

$$\begin{aligned} p &= 32 + 64 \\ &= 96 \text{ W} \end{aligned}$$

EXAMPLE 2.10 Determine the power delivered by the voltage source for the circuit in Figure 2.27. ■

Figure 2.27

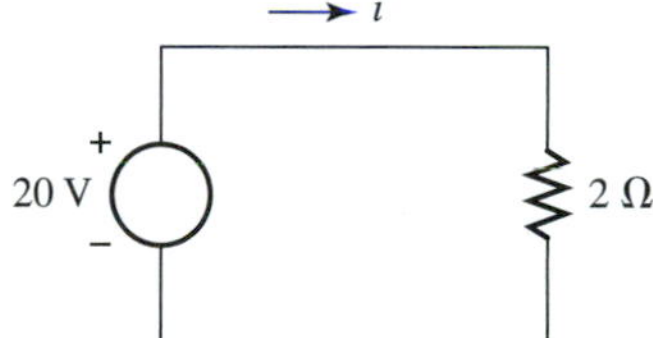

SOLUTION From Ohm's law

$$\begin{aligned} v &= Ri \\ i &= \frac{20}{2} \\ &= 10 \text{ A} \end{aligned}$$

From Equation (2.6)

$$\begin{aligned} p &= vi \\ &= (20)(10) \\ &= 200 \text{ W} \end{aligned}$$

Alternatively, from Equation (2.7)

$$\begin{aligned} p &= i^2R \\ &= (10)^2(20) \\ &= 200 \text{ W} \end{aligned}$$

Alternatively, from Equation (2.8)

$$\begin{aligned} p &= \frac{v^2}{R} \\ &= \frac{(20)^2}{2} \\ &= 200 \text{ W} \end{aligned}$$

EXAMPLE 2.11 Determine the power delivered by the current source for the circuit in Figure 2.28. ■

Figure 2.28

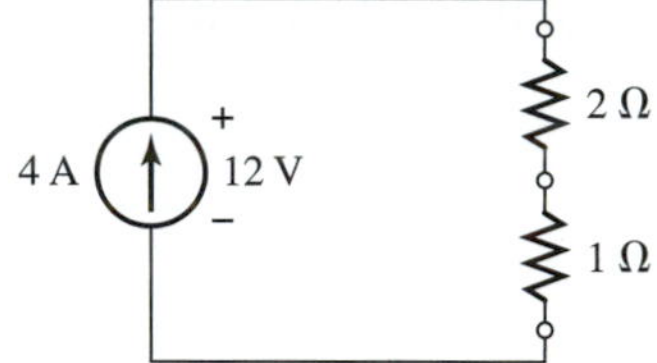

SOLUTION From the Equation (2.6)

$$\begin{aligned} p &= vi \\ &= (12)(4) \\ &= 48 \text{ W} \end{aligned}$$

Alternatively, from Equation (2.7) the power delivered to the 2-Ω resistance is

$$\begin{aligned} p &= i^2R \\ &= (4)^2(2) \\ &= 32 \text{ W} \end{aligned}$$

and the power delivered to the 4-Ω resistance is

$$\begin{aligned} p &= i^2R \\ &= (4)^2(1) \\ &= 16 \text{ W} \end{aligned}$$

The total power delivered by the current source is the sum of the power delivered to the two resistances.

$$p = 48 \text{ W}$$

EXAMPLE 2.12 Determine the power associated with the resistor and each voltage source in Figure 2.29. ■

Figure 2.29

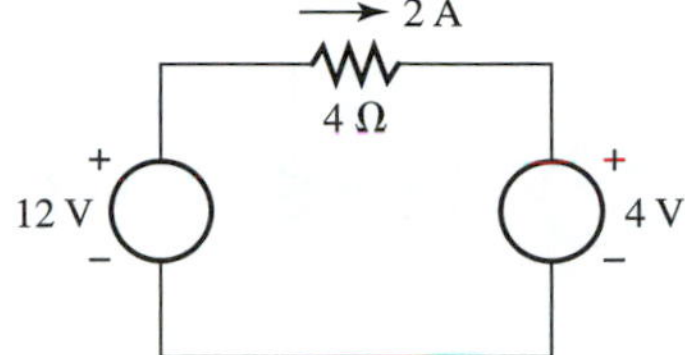

SOLUTION From Equation (2.6), the power delivered by the 12-V voltage source is

$$\begin{aligned} p &= vi \\ &= 12(2) \\ &= 24 \text{ W} \end{aligned}$$

The 4-V voltage source is receiving power, because current is flowing into the positive-marked terminal. From Equation (2.6) the power received by the 4-V voltage source is

$$\begin{aligned} p &= vi \\ &= (4)(2) \\ &= 8 \text{ W} \end{aligned}$$

From Equation (2.7) the power delivered to the 4-Ω resistance is

$$\begin{aligned} p &= i^2R \\ &= (2)^2(4) \\ &= 16 \text{ W} \end{aligned}$$

The 12-V voltage source is supplying power to the 4-Ω resistance and the 4-V voltage source. The following statement verifies the previous result.

$$\begin{pmatrix}\text{power delivered}\\ \text{by 12-V source}\end{pmatrix} = \begin{pmatrix}\text{power received}\\ \text{by 4-V source}\end{pmatrix} + \begin{pmatrix}\text{power received}\\ \text{by 4-}\Omega\text{ resistance}\end{pmatrix}$$

$$24 \text{ W} = 8 \text{ W} + 16 \text{ W}$$

2.13 EXERCISES

1. Explain how the resistance of a resistor varies with temperature.
2. What are the characteristics of a perfect voltage source.
3. What are the characteristics of a perfect current source.
4. What are the characteristics of an imperfect (ideal) voltage source.
5. What are the characteristics of an imperfect (ideal) current source.

6. Determine the magnitude and direction of the current associated with the resistor for each circuit in Figure 2.30.

Figure 2.30

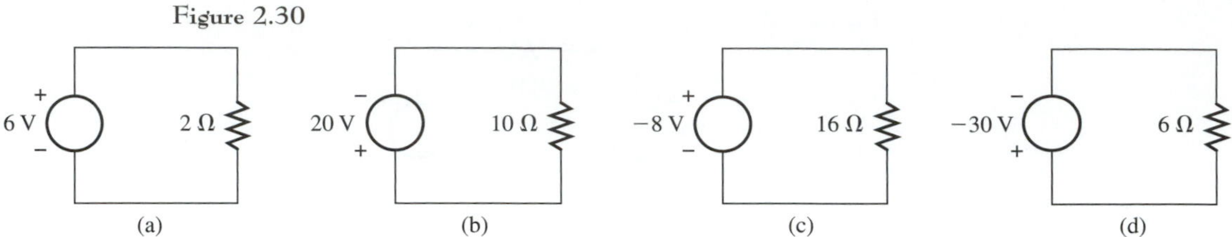

7. Determine the value of the indicated current in mA for each circuit shown in Figure 2.31.

Figure 2.31

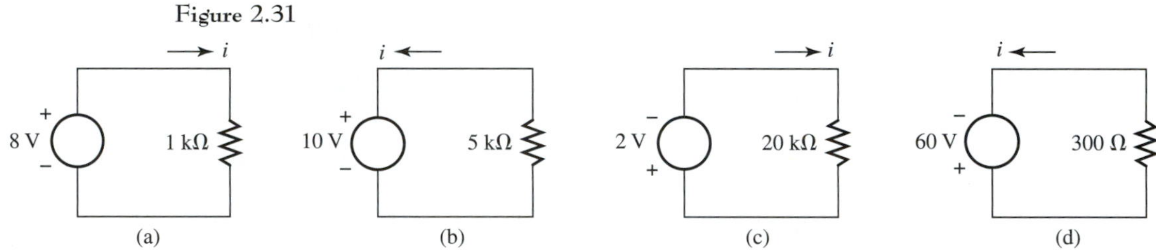

8. Determine the magnitude and polarity of the voltage source for each circuit shown in Figure 2.32.

Figure 2.32

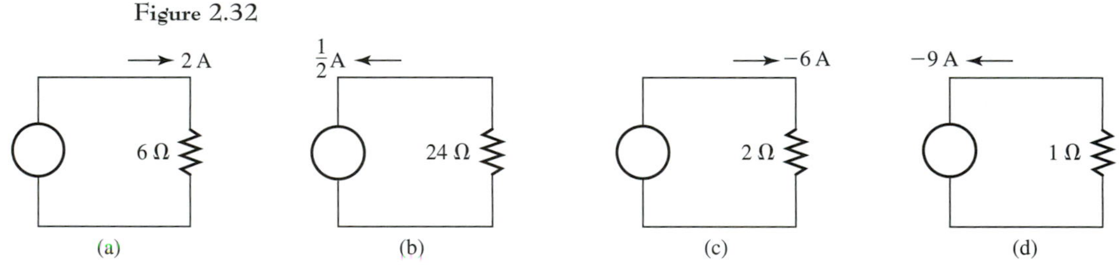

9. Determine the magnitude and polarity of the voltage across the resistor for each circuit in Figure 2.33.

Figure 2.33

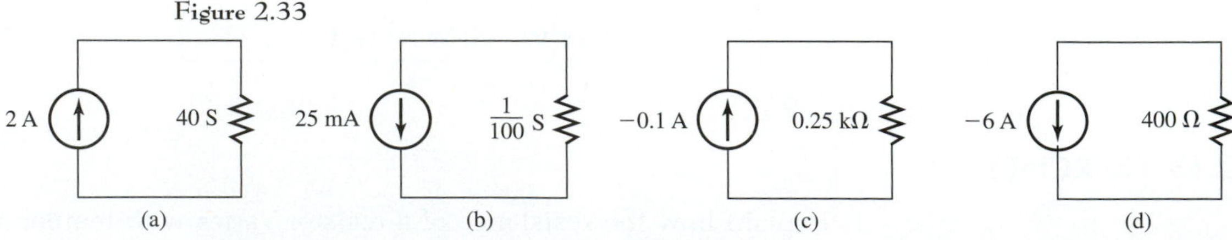

10. Determine the conductance in siemens of each resistance.
(a) 1000 kΩ (b) 660 Ω (c) 72 Ω (d) 0.0912 kΩ

11. Explain the difference between a perfect (ideal) voltage source and an imperfect (practical) voltage source.

12. Explain the difference between a perfect (ideal) current source and an imperfect (practical) current source.

13. Determine the voltage across each resistor in Figure 2.34 and indicate the polarity.

Figure 2.34

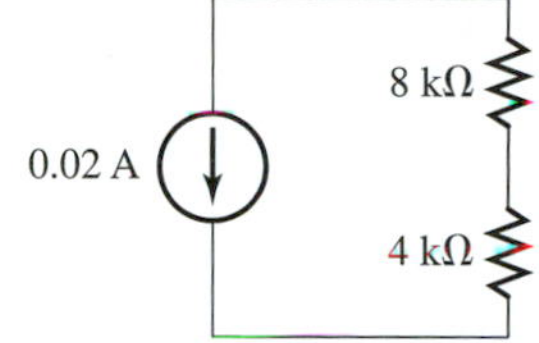

14. Determine i_1, i_2, and i_3 for the circuit in Figure 2.35.

Figure 2.35

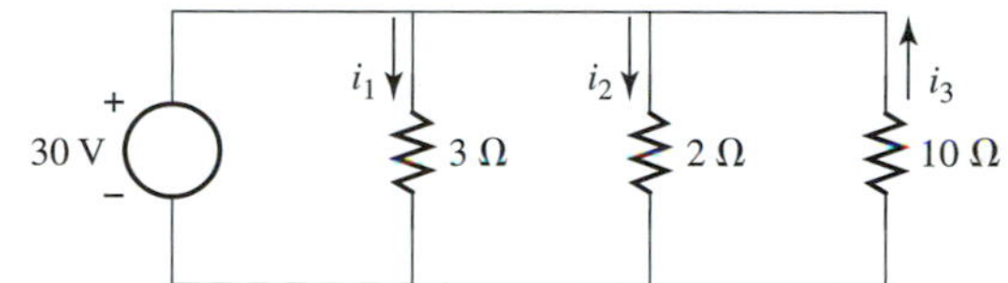

15. Can a perfect voltage source be shorted? Why?

16. Can a perfect current source be left open? Why?

17. Find the power in watts delivered to each resistor by the current source for the circuit in Figure 2.36.

Figure 2.36

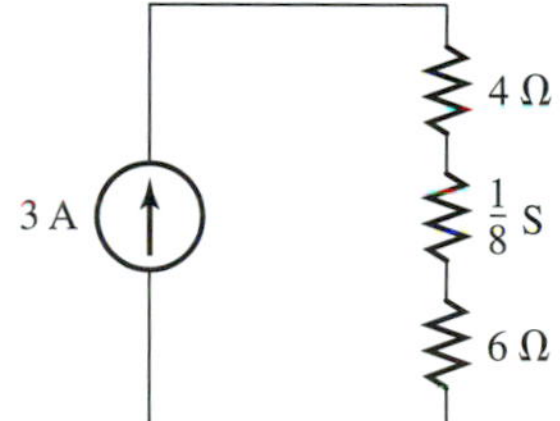

18. Find the power delivered to each resistor in Figure 2.35.

19. Find the power delivered to each resistor by the voltage source in Figure 2.37.

Figure 2.37

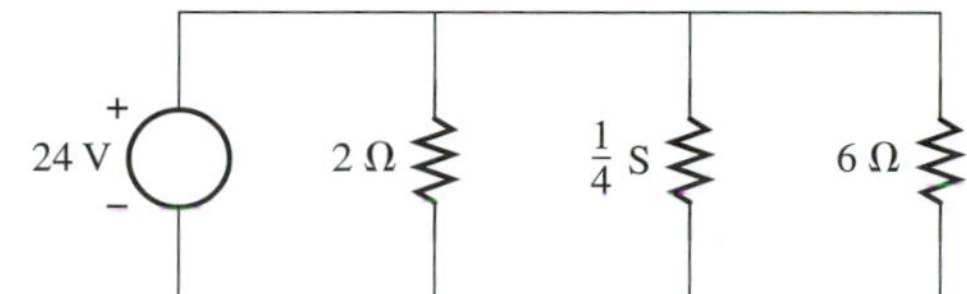

20. Find the power delivered to the resistor if 360 joules of energy are delivered to a 20-Ω resistor in 2 min at a constant rate.

21. What is the voltage across the resistor in Problem 20?

22. What is the current through the resistor in Problem 20?

23. Determine the power associated with each resistor and each source in Figure 2.38.

Figure 2.38

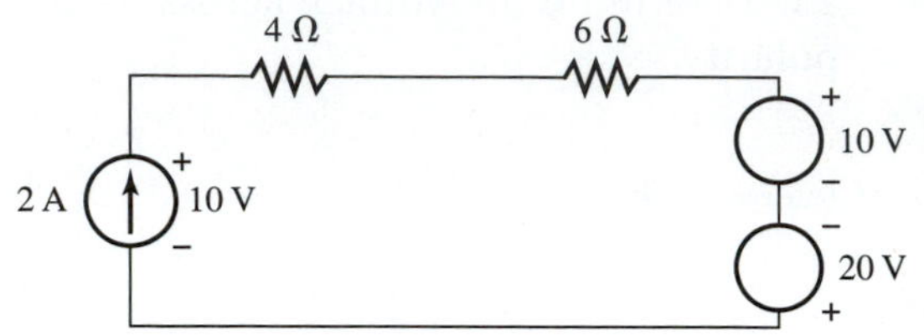

24. Indicate which sources are receiving energy and which are delivering energy in Problem 23.

CHAPTER

3

Kirchhoff's Laws and Resistor Combinations

After completing this chapter the student should be able to

* employ Kirchhoff's voltage and current laws to determine currents and voltages in resistive circuits
* determine a single equivalent resistance for resistors connected in series and/or parallel

Ohm's law, Kirchhoff's current and voltage laws, and the relationship between current and voltage for the basic electrical elements form the foundation for many of the problem-solving procedures used in electric circuit analysis. In this chapter, Kirchhoff's current and voltage laws are presented and used in conjunction with Ohm's law to analyze circuits consisting of voltage sources, current sources, and resistors.

3.1 BRANCHES, NODES, AND LOOPS

When electrical elements are connected to each other in such a manner that currents and voltages appear, they form electric circuits. *Nodes*, *branches*, and *loops* are integral components of electric circuits and are defined as follows.

NODE A node is a point in an electric circuit where two or more electrical elements are connected.

BRANCH A branch is a path between two nodes consisting of one electrical element.

LOOP A loop is a continuous closed path in an electric circuit that does not pass through any node more than once.

Consider the following example.

EXAMPLE 3.1 Determine the nodes, loops, and branches for the circuit in Figure 3.1. ■

Figure 3.1

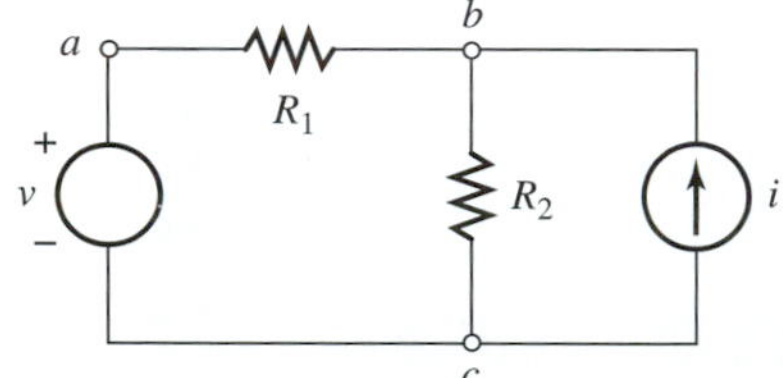

SOLUTION There are three nodes, four branches, and three loops. The three nodes are labled a, b, and c and all four of the branches are between these three nodes. One branch consists of voltage source v; one branch consists of resistance R_1; one branch consists of resistance R_2; and one branch consists of current source i. There are three loops, as illustrated in Figure 3.2. One loop

Figure 3.2

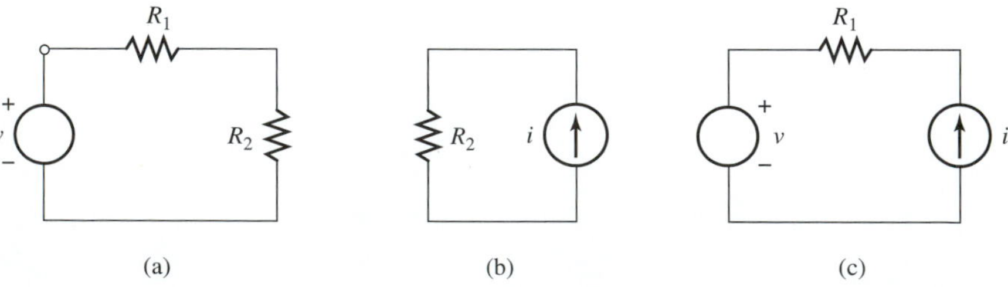

consists of voltage source v and resistances R_1 and R_2; one loop consists of current source i and resistance R_2; and one loop consists of voltage source v, current source i, and resistance R_1.

3.2 KIRCHHOFF'S VOLTAGE LAW

Gustav Kirchhoff (1824–1887), a German physicist, presented his voltage and current laws in 1847. Prior to discussing Kirchhoff's voltage law, consider the resistance in Figure 3.3.

Figure 3.3 Notation for voltage and current.

It is important to recall that if current flows through R as shown in Figure 3.3(a) and we use Ohm's law to write $v = Ri$, then the voltage v must be expressed as the voltage of the positive-marked terminal with respect to the negative-marked terminal, as shown in Figure 3.3(b). We sometimes state this current-voltage relationship by stating that conventional current flows from + to − through a resistance.

In applying Kirchhoff's voltage law it is not important that we know the actual current direction through a resistance. We assume a direction for current and place the current arrow in the assumed direction, along with the associated voltage polarity based on the assumed current. The direction of the current arrow is then the positive direction for the current. If the current is calculated as negative, the actual current is opposite to the assumed current direction.

Consider the resistance in Figure 3.4 to illustrate the procedure for assuming current direction. The current i is unknown and assumed to flow in

Figure 3.4 Current notation.

the direction shown in Figure 3.4(a). Based on the assumed direction for current, the voltage polarity is indicated in Figure 3.4(b), and Ohm's law results in

$$v_{ab} = Ri$$

Kirchhoff's voltage law can be stated as follows.

KIRCHHOFF'S VOLTAGE LAW The algebraic sum of the voltages around a closed path is equal to zero.

Consider the circuit in Figure 3.5. We shall use Kirchhoff's voltage law to determine the current flow in the circuit. We first assume a direction for current,

Figure 3.5 Kirchhoff's voltage law.

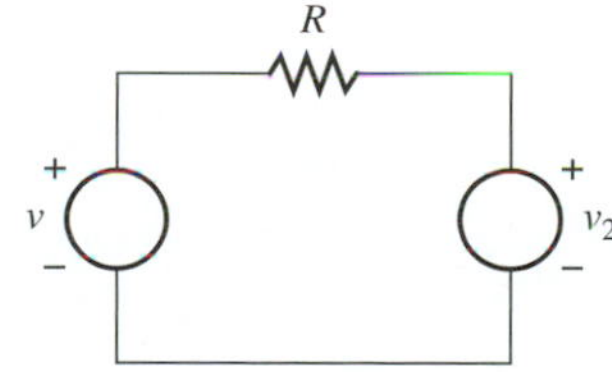

as indicated in Figure 3.6(a). We then assign the voltage polarity to the resistance

Figure 3.6 Assumed current for Kirchhoff's voltage law.

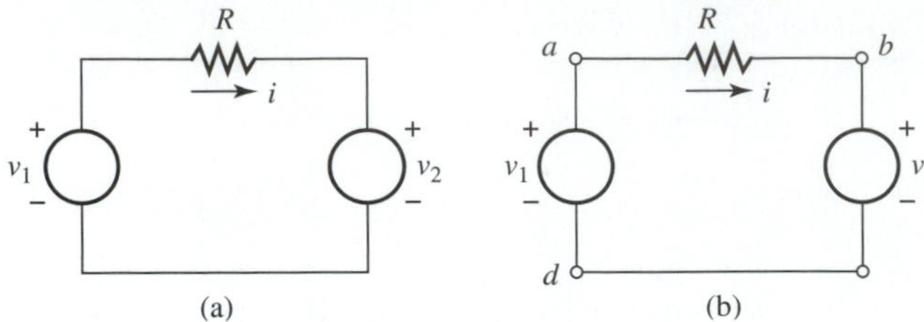

based on the assumed current, as indicated in Figure 3.6(b). Finally, we apply Kirchhoff's voltage law to the circuit in Figure 3.6(b). Using subscripted voltage notation, Kirchhoff's voltage law results in

$$v_{ad} + v_{ba} + v_{db} = 0$$

Because $v_{ad} = v_1$, $v_{db} = -v_2$, and $v_{ba} = R(-i) = -Ri$ from Ohm's law,

$$v_1 - Ri - v_2 = 0$$

and

$$i = \frac{v_1 - v_2}{R}$$

After inserting the values of v_1, v_2, and R in the preceding equation, if the resulting current i is positive, the actual current is in the assumed direction, and if the resulting current i is negative, the actual current is opposite to the assumed direction.

Another commonly used method of applying Kirchhoff's voltage law involves the use of voltage rises and drops. In terms of voltage rises and drops, Kirchhoff's voltage law can be stated as follows.

KIRCHHOFF'S VOLTAGE LAW The algebraic sum of the voltage rises and drops around a closed path is equal to zero.

A voltage rise occurs when we travel from − to + through an electrical element.

A voltage drop occurs when we travel from + to − through an electrical element.

A voltage rise is assigned a positive sign and a voltage drop is assigned a negative sign.

Consider the circuit in Figure 3.5. We assume a direction for current as indicated in Figure 3.6(a) and assign voltage polarity to the resistance based on the assumed current as indicated in Figure 3.6(b). Finally, we apply Kirchhoff's voltage law to the circuit in Figure 3.6(b) by algebraically summing rises and drops as we travel around the closed path in a clockwise direction. The result is

$$v_1 - Ri - v_2 = 0$$

and

$$i = \frac{v_1 - v_2}{R}$$

which is the same result arrived at previously using the subscripted voltage notation. Notice we can travel clockwise or counterclockwise around a loop when applying Kirchhoff's voltage law and the result is the same.

Consider the following examples.

EXAMPLE 3.2 Determine the current flow in the circuit shown in Figure 3.7. ■

Figure 3.7

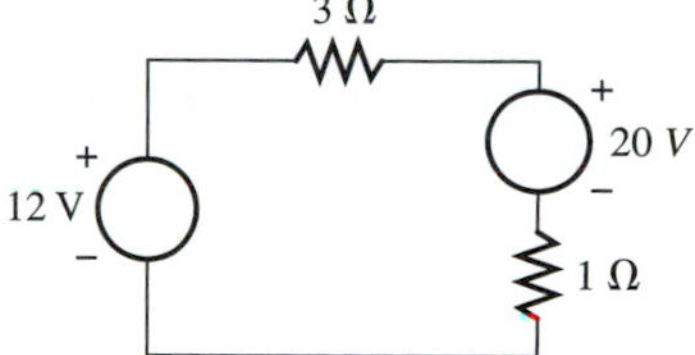

SOLUTION The first step is to assume a current direction and assign voltage polarities to the resistors based on the assumed current as shown in Figure 3.8. From Ohm's law the magnitudes of the voltage across the 3-Ω and 1-Ω resistors, respectively, are $3i$ and $1i$.

Figure 3.8

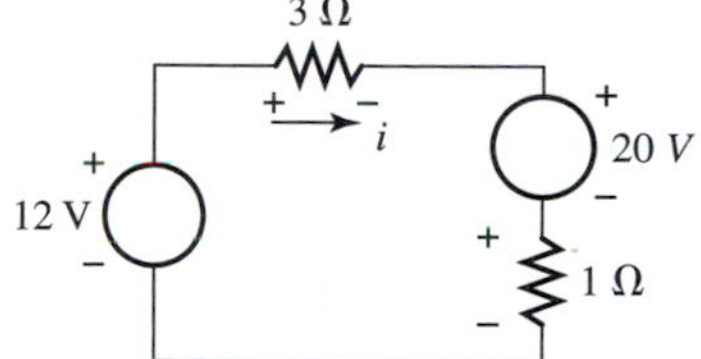

Applying Kirchhoff's voltage law results in

$$12 - 3i - 20 - 1i = 0$$

and

$$i = -2 \text{ A}$$

The minus sign indicates that the current flows in a direction opposite to the assumed direction for current.

EXAMPLE 3.3 Determine the voltages v_1 and v_2 for the circuit shown in Figure 3.9. ■

Figure 3.9

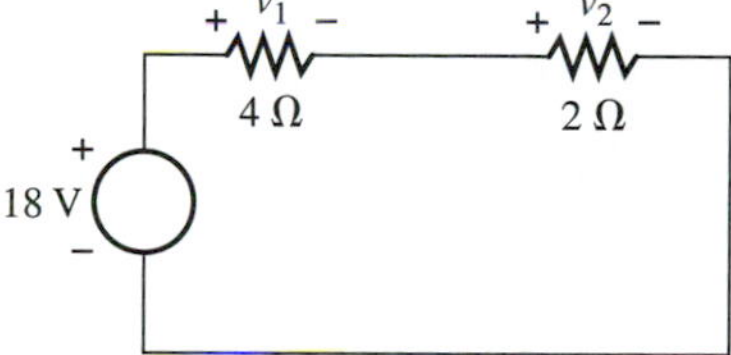

SOLUTION The first step is to assign voltage polarities to the resistors based on the assigned current direction shown in Figure 3.10.

Figure 3.10

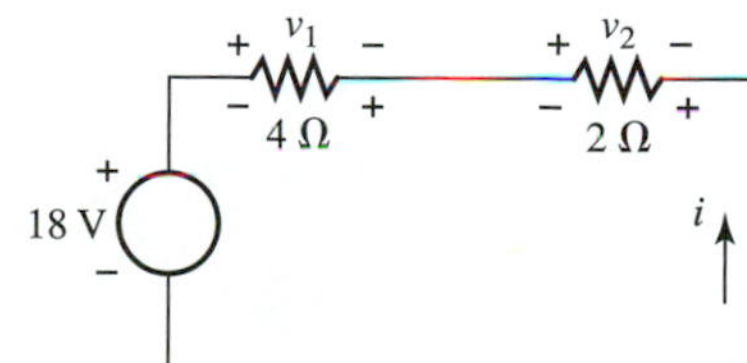

When applying Kirchhoff's voltage law we can use the polarities associated with the assumed current flow and not the polarity markings associated with v_1 and v_2. This results in

$$18 + 4i + 2i = 0$$
$$i = -3\text{ A}$$

From Ohm's law and using the polarity markings associated with v_1 and v_2,

$$\begin{aligned} v_1 &= 4(-i) = -4i \\ &= (-4)(-3) \\ &= 12\text{ V} \end{aligned}$$

and

$$\begin{aligned} v_2 &= 2(-i) = -2i \\ &= (-2)(-3) \\ &= 6\text{ V} \end{aligned}$$

Alternatively, we can use v_1 and v_2 and their associated polarity markings when applying Kirchhoff's voltage law. The result is

$$18 - v_1 - v_2 = 0$$

From Ohm's law, $v_1 = -4i$ and $v_2 = -2i$; it follows that

$$18 - (-4i) - (-2i) = 0$$
$$i = -3\text{ A}$$

EXAMPLE 3.4 Determine the voltages v_1 and v_2 in the circuit shown in Figure 3.11. ■

Figure 3.11

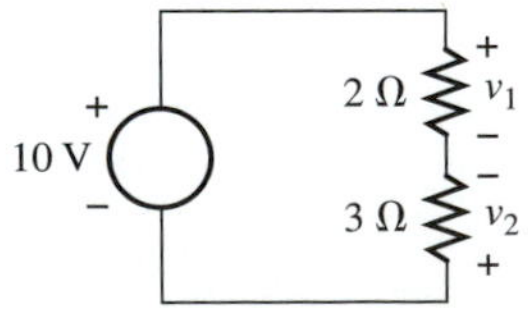

SOLUTION The first step is to assume a current direction and assign voltage polarities based on the assumed current direction, as shown in Figure 3.12.

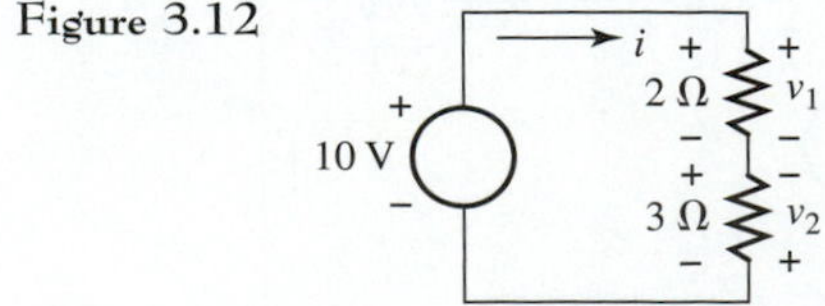

Figure 3.12

Applying Kirchhoff's voltage law results in

$$10 - 2i - 3i = 0$$
$$i = 2\text{ A}$$

From Ohm's law

$$v_1 = 2i = (2)(2) = 4\text{ V}$$

and

$$v_2 = 3(-i) = -3i = -3(2) = -6\text{ V}$$

Alternatively, we can use v_1 and v_2 and their associated polarity markings when applying Kirchhoff's voltage law. The result is

$$10 - v_1 + v_2 = 0$$

From Ohm's law $v_1 = 2i$ and $v_2 = -3i$; it follows that

$$10 - 2i + (-3i) = 0$$
$$i = 2\text{ A}$$

3.3 KIRCHHOFF'S CURRENT LAW

Kirchhoff's current law can be stated as follows.

> **KIRCHHOFF'S CURRENT LAW** The sum of the currents entering a node equals the sum of the currents leaving the node.

If the currents entering a node are assigned a positive sign and the currents leaving a node are assigned a negative sign, Kirchhoff's current law can alternatively be stated as follows.

> **KIRCHHOFF'S CURRENT LAW** The algebraic sum of the currents at a node equals zero.

Consider the partial circuit in Figure 3.13. The sum of the currents entering node a equals

$$i_1 + i_3$$

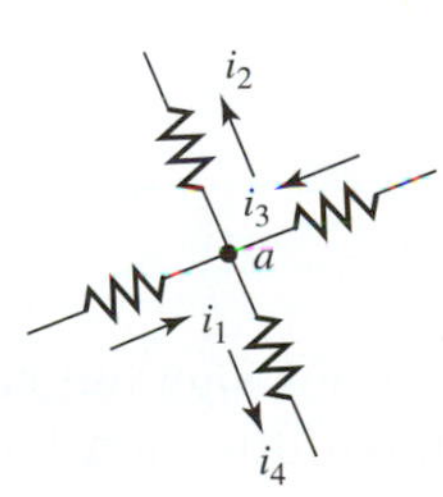

Figure 3.13 Kirchhoff's current law.

and the sum of the currents leaving node a equals

$$i_2 + i_4$$

From Kirchhoff's current law

$$i_1 + i_3 = i_2 + i_4$$

Alternatively, we can assign a positive sign to currents entering node a and a negative sign to currents leaving node a and algebraically sum the currents at node a. This action results in

$$i_1 + i_3 - i_2 - i_4 = 0$$

and

$$i_1 + i_3 = i_2 + i_4$$

which is the same as the previous result, which was obtained by equating the currents entering to the currents leaving the node.

When applying Kirchhoff's current law, the currents assumed are not always known, and we use assumed current directions just as we do when applying Kirchhoff's voltage law. If an assumed current is calculated as negative, then the actual current flows in a direction opposite to the assumed current direction.

EXAMPLE 3.5 Determine the current i in the partial circuit in Figure 3.14. ■

Figure 3.14

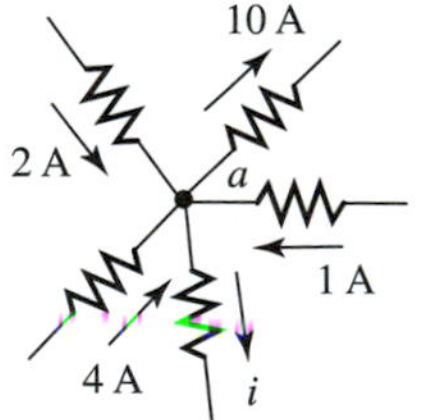

SOLUTION Assigning positive signs to the currents entering node a and negative signs to the currents leaving node a and applying Kirchhoff's current law results in

$$2 + 4 + 1 - 10 - i = 0$$

and

$$i = -3 \text{ A}$$

The minus sign indicates that the current i flows in a direction opposite to the assumed direction for current.

EXAMPLE 3.6 Determine the currents i_1 and i_2 for the partial circuit in Figure 3.15. ■

Figure 3.15

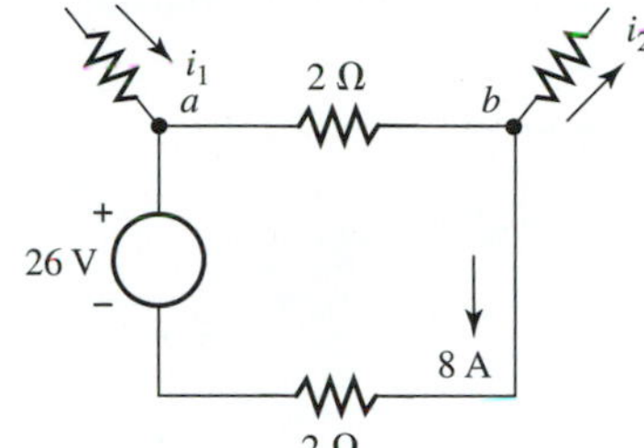

SOLUTION We first assume a current i for the upper 2-Ω resistor and assign voltage polarity, as shown in Figure 3.16. The voltage polarity for the lower 2-Ω resistor is dictated by the 8-A current, as indicated.

Figure 3.16

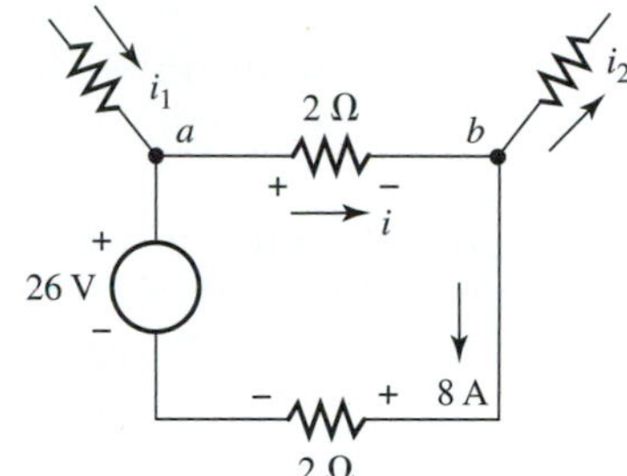

Applying Kirchhoff's voltage law and Ohm's law to the loop containing the 26-V voltage source and the two 2-Ω resistors results in

$$26 - 2i - (2)(8) = 0$$

and

$$i = 5 \text{ A}$$

Applying Kirchhoff's current law to node b results in

$$i - 8 - i_2 = 0$$

and because $i = 5$ A

$$i_2 = -3 \text{ A}$$

Applying Kirchhoff's current law to node a results in

$$i_1 + 8 - i = 0$$

and because $i = 5$ A

$$i_1 = -3 \text{ A}$$

The negative signs for currents i_1 and i_2 indicate that they are in directions opposite to the assigned current directions.

3.4 EQUIVALENT SERIES RESISTANCE

Two or more resistors are said to be connected in series if the same current flows through each resistor. Consider the circuits in Figure 3.17. Resistors R_1 and R_2 in Figure 3.17(a) are in series, because the same current flows through each resistor; for the same reason the n resistors in Figure 3.17(b) are all in series.

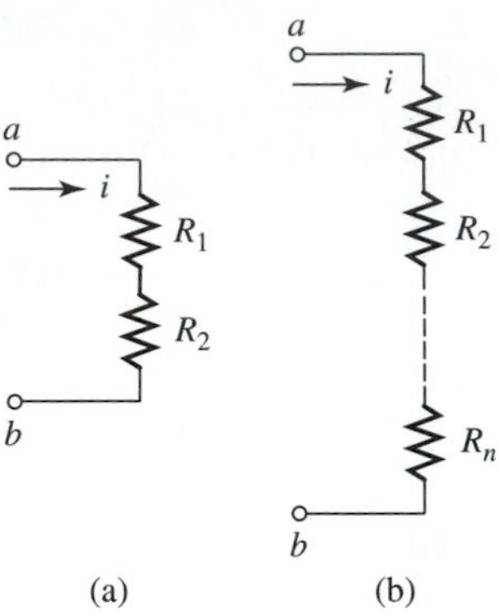

Figure 3.17 Series resistors.

It is often desired to replace two or more resistors in series with a single equivalent resistance. This equivalency is always with respect to a pair of terminals. Consider the circuit in Figure 3.18.

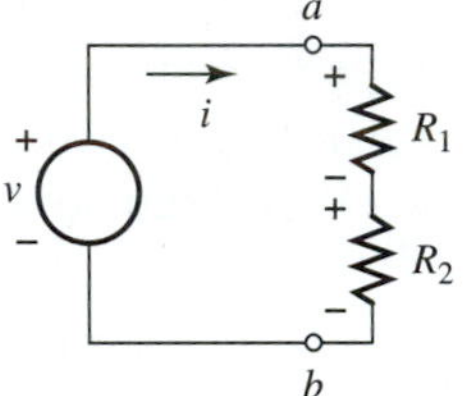

Figure 3.18 Two resistors in series.

Kirchhoff's voltage law and Ohm's law result in

$$v - R_1 i - R_2 i = 0$$

and

$$v = (R_1 + R_2)i$$

Let the two resistors in series (R_1 and R_2) in Figure 3.18 be replaced by one resistor, R_{EQ}, where

$$R_{EQ} = R_1 + R_2$$

as shown in Figure 3.19.

Applying Ohm's law to the circuit in Figure 3.19 results in

$$\begin{aligned} v &= R_{EQ} i \\ &= (R_1 + R_2)i \end{aligned}$$

which is the same voltage-current relationship for terminals a-b arrived at for the circuit in Figure 3.18.

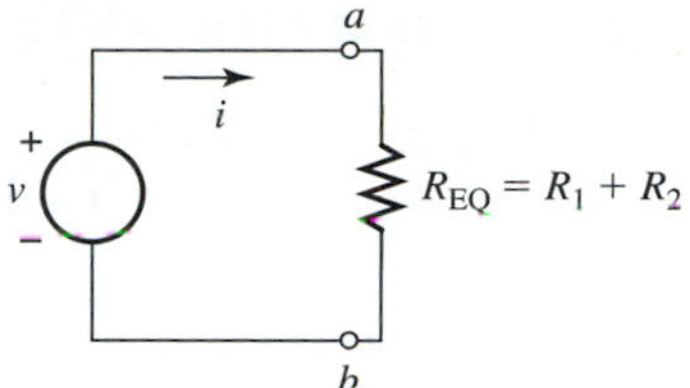

Figure 3.19 Series equivalent.

Because the current-voltage relationship for terminals *a-b* is the same for the circuits in Figure 3.18 and 3.19, R_{EQ} is called the equivalent resistance with respect to terminals *a-b* for the series combination of R_1 and R_2.

These results can be extended for *n* resistors in series, as indicated in Figure 3.20. The equation for combining *n* resistors in series is

$$R_{EQ} = R_1 + R_2 + \cdots + R_n \tag{3.1}$$

Figure 3.20 Combining *n* resistors in series.

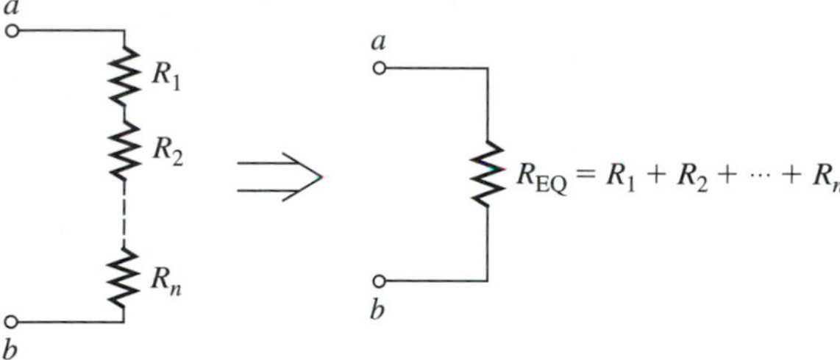

3.5 EQUIVALENT SERIES CONDUCTANCE

We now reconsider the circuit in Figure 3.18 with the resistances expressed in terms of conductance (siemens), as shown in Figure 3.21. Applying Kirchoff's

Figure 3.21 Two conductances in series.

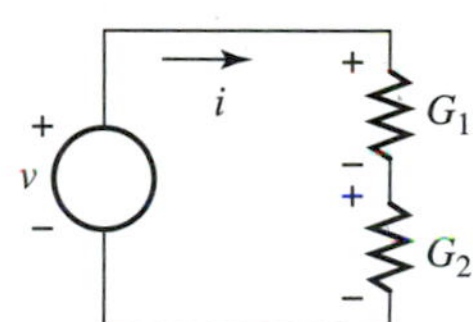

voltage law and recalling that $i = Gv$ and $v = i/G$ results in

$$v - \frac{i}{G_1} - \frac{i}{G_2} = 0$$

and

$$i = \left(\frac{1}{1/G_1 + 1/G_2}\right)(v)$$

Let the two conductances in series (G_1 and G_2) in Figure 3.21 be replaced by one conductance, G_{EQ}, where

$$G_{EQ} = \frac{1}{1/G_1 + 1/G_2}$$

as shown in Figure 3.22.

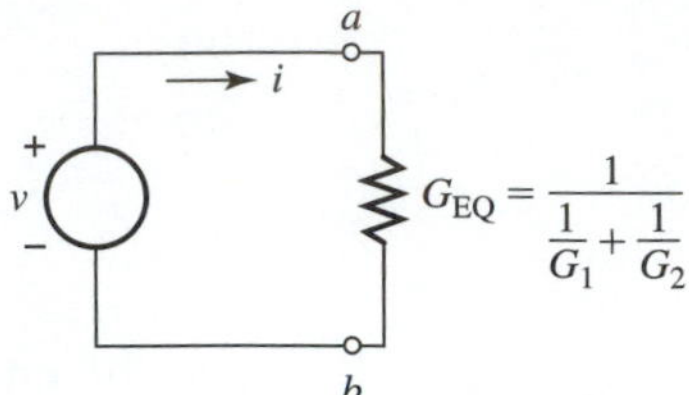

Figure 3.22 Equivalent conductance.

Applying Ohm's law to the circuit in Figure 3.22 results in

$$i = G_{EQ}v = \left(\frac{1}{1/G_1 + 1/G_2}\right)v$$

which is the same voltage-current relationship for terminals *a-b* determined for the circuit in Figure 3.21.

Because the current-voltage relationship for terminals *a-b* is the same for the circuits in Figures 3.21 and 3.22, G_{EQ} is called the equivalent conductance for the series combination of G_1 and G_2.

These results can be extended for *n* conductances in series, as indicated in Figure 3.23. The equation for combining *n* conductances in series is

$$G_{EQ} = \frac{1}{1/G_1 + 1/G_2 + \cdots + 1/G_n} \tag{3.2}$$

Figure 3.23 Combining *n* conductances in series.

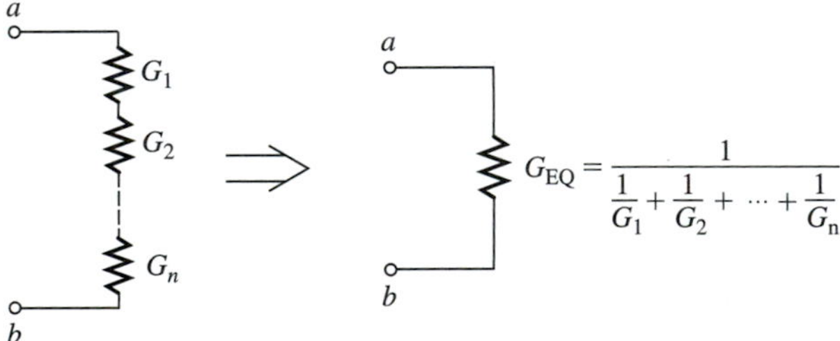

It is often convenient to combine conductances in pairs instead of all together, as indicated in Equation (3.2). For two conductances in series, Equation (3.2) becomes

$$G_{EQ} = \frac{1}{1/G_1 + 1/G_2}$$

$$= \frac{1}{(G_2 + G_1)/(G_1G_2)}$$

or

$$G_{EQ} = \frac{G_1G_2}{G_1 + G_2} \tag{3.3}$$

Consider the following examples.

EXAMPLE 3.7 Replace the three series resistances with a single resistance for the circuit in Figure 3.24. ■

Figure 3.24

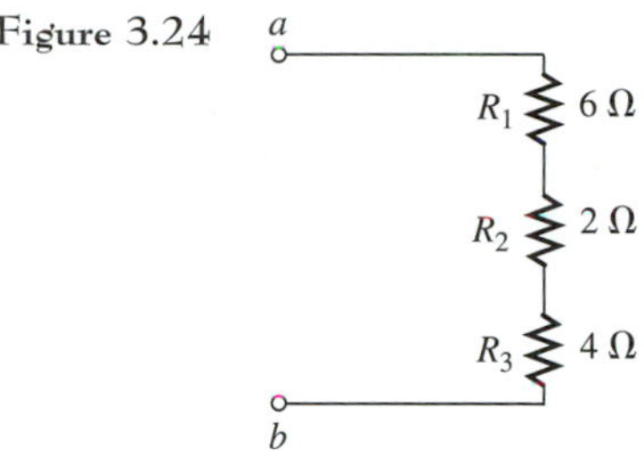

SOLUTION From Equation (3.1)

$$
\begin{aligned}
R_{EQ} &= R_1 + R_2 + R_3 \\
&= 6 + 2 + 4 \\
&= 12\ \Omega
\end{aligned}
$$

The equivalent circuit with respect to terminals *a-b* is shown in Figure 3.25.

Figure 3.25

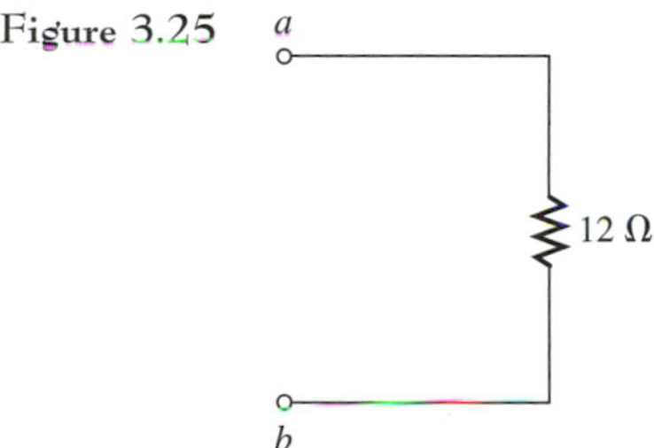

EXAMPLE 3.8 Replace the three series conductances with a single conductance for the circuit in Figure 3.26. ■

Figure 3.26

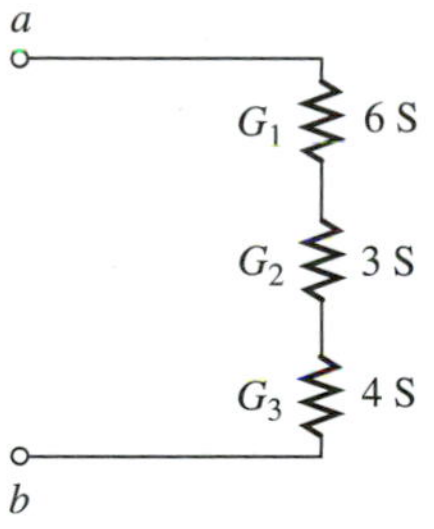

a
G_1 6 S
G_2 3 S
G_3 4 S
b

SOLUTION From Equation (3.2)

$$
\begin{aligned}
G_{EQ} &= \frac{1}{\frac{1}{6} + \frac{1}{3} + \frac{1}{4}} \\
&= \frac{1}{(2 + 4 + 3)/12} \\
&= \frac{4}{3}\ \text{S}
\end{aligned}
$$

Alternatively, from Equation (3.3) we can combine the conductances in pairs. Combining G_1 and G_2 and calling the result G'_{EQ} results in

$$G'_{EQ} = \frac{G_1 G_2}{G_1 + G_2}$$
$$= \frac{(6)(3)}{6 + 3}$$
$$= 2\text{ S}$$

Next combining G'_{EQ} with G_3 results in

$$G_{EQ} = \frac{G'_{EQ} G_3}{G'_{EQ} + G_3}$$
$$= \frac{(2)(4)}{2 + 4}$$
$$= \frac{4}{3}\text{ S}$$

Using another method, we can convert each conductance (siemens) to resistance (ohms), as shown in Figure 3.27, and use Equation (3.1) to combine resistances (ohms) and convert back to siemens.

Figure 3.27

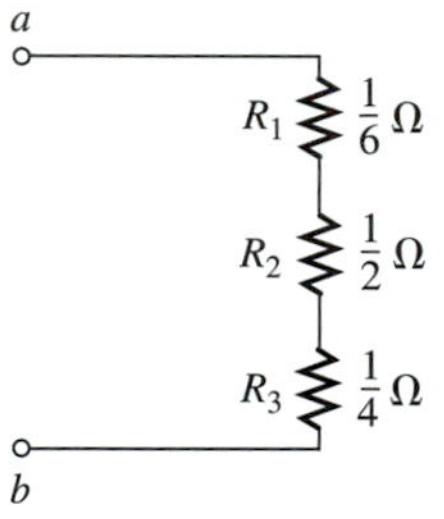

From Equation (3.1)

$$R_{EQ} = \frac{1}{6} + \frac{1}{3} + \frac{1}{4}$$
$$= \frac{3}{4}\,\Omega$$

From Equation (2.3)

$$G_{EQ} = \frac{1}{R_{EQ}}$$
$$= \frac{1}{3/4}$$
$$= \frac{4}{3}\text{ S}$$

An inspection of Equations (3.1) and (3.2) and the preceding examples reveal that when adding resistance in series, it is easier to work with resistance (ohms) than conductance (siemens).

3.6 EQUIVALENT PARALLEL RESISTANCE

Two or more resistances are said to be connected in parallel if the same voltage appears across each resistance. Consider the circuits in Figure 3.28. Resistors

Figure 3.28 Parallel resistors.

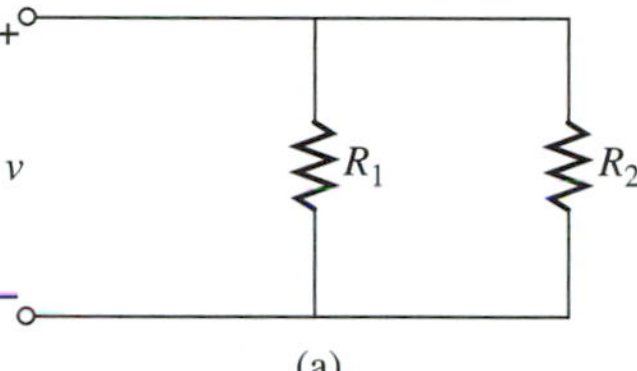

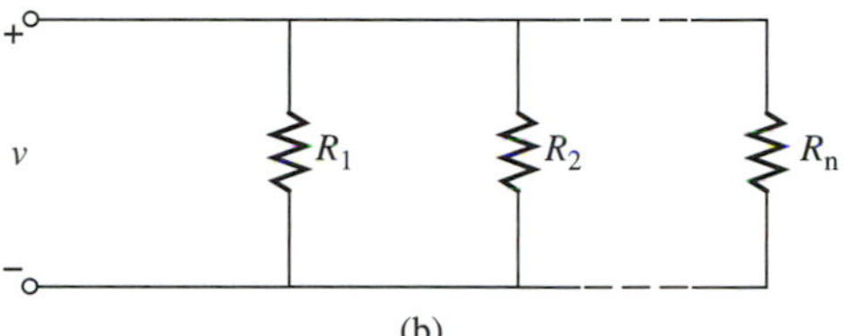

R_1 and R_2 in Figure 3.28(a) are in parallel, because the same voltage appears across each resistor; for the same reason the n resistors in Figure 3.28(b) are in parallel.

It is often desired to replace two or more resistors in parallel with a single equivalent resistance. This equivalency is always with respect to a pair of terminals. Consider the circuit in Figure 3.29.

Figure 3.29 Two resistors in parallel.

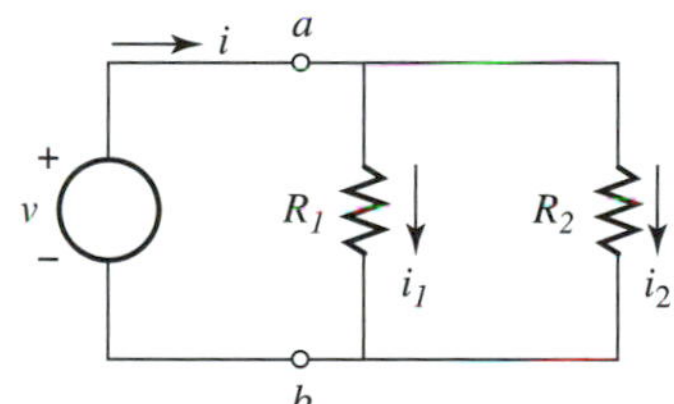

From Kirchhoff's current law

$$i = i_1 + i_2$$

and from Ohm's law

$$i_1 = \frac{v}{R_1}$$

$$i_2 = \frac{v}{R_2}$$

Substituting v/R_1 and v/R_2 for i_1 and i_2, respectively, in the preceding equation results in

$$i = \frac{v}{R_1} + \frac{v}{R_2}$$

$$= \left(\frac{1}{R_1} + \frac{1}{R_2}\right)v$$

or

$$v = \left(\frac{1}{1/R_1 + 1/R_2}\right) i$$

Let the two parallel resistors in Figure 3.29 be replaced by one equivalent resistance, R_{EQ}, where

$$R_{EQ} = \frac{1}{1/R_1 + 1/R_2}$$

as shown in Figure 3.30. Applying Ohm's law to the circuit in Figure 3.30 results in

$$\begin{aligned} v &= R_{EQ} i \\ &= \left(\frac{1}{1/R_1 + 1/R_2}\right) i \end{aligned}$$

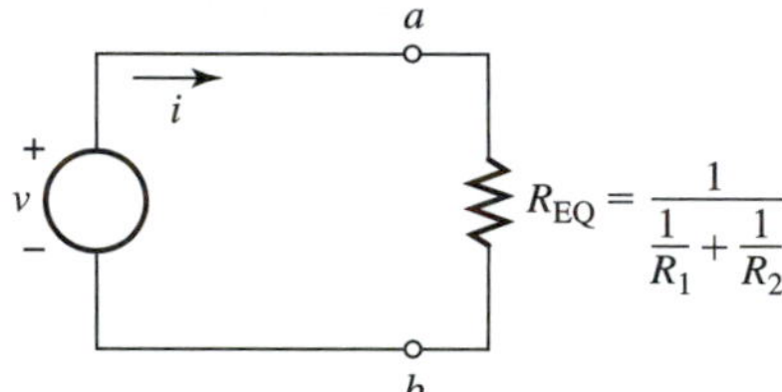

Figure 3.30 Parallel equivalent.

which is the same voltage-current relationship for terminals *a-b* determined for the circuit in Figure 3.29.

Because the current-voltage relationship for terminals *a-b* is the same for the circuits in Figure 3.29 and 3.30, R_{EQ} is called the equivalent resistance with respect to terminals *a-b* for the parallel combination of R_1 and R_2.

These results can be extended for *n* resistors in parallel, as indicated in Figure 3.31. The equation for combining *n* resistors in parallel is

$$R_{EQ} = \frac{1}{1/R_1 + 1/R_2 + \cdots + 1/R_n} \tag{3.4}$$

Figure 3.31 Combining *n* resistors in parallel.

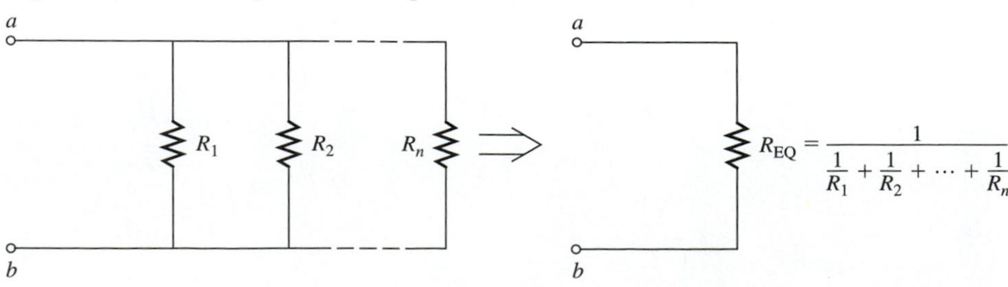

It is often convenient to combine resistors in pairs instead of all together, as in Equation (3.4). For two resistors in parallel, Equation (3.4) becomes

$$R_{EQ} = \frac{1}{1/R_1 + 1/R_2}$$

$$= \frac{1}{(R_1 + R_2)/(R_1R_2)}$$

or

$$R_{EQ} = \frac{R_1R_2}{R_1 + R_2} \tag{3.5}$$

3.7 EQUIVALENT PARALLEL CONDUCTANCE

We now consider the circuit in Figure 3.29 with the resistances expressed in terms of conductance, as shown in Figure 3.32.

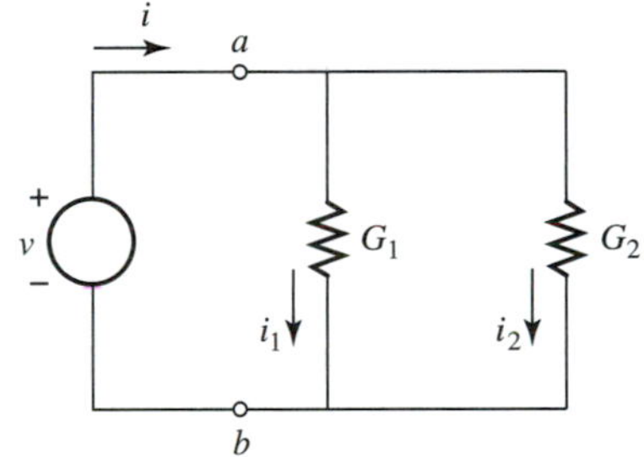

Figure 3.32 Two conductances in parallel.

From Kirchhoff's current law,

$$i = i_1 + i_2$$

Recalling that $i = Gv$,

$$i_1 = G_1v$$

$$i_2 = G_2v$$

Substituting G_1v and G_2v for i_1 and i_2, respectively, in the preceding equation results in

$$i = G_1v + G_2v$$

$$= (G_1 + G_2)v$$

Now, let the two parallel conductances in Figure 3.32 be replaced by one equivalent conductance, G_{EQ}, where

$$G_{EQ} = G_1 + G_2$$

as shown in Figure 3.33. Applying Ohm's law to the circuit in Figure 3.33 results in

$$\begin{aligned} i &= G_{EQ}v \\ &= (G_1 + G_2)v \end{aligned}$$

Figure 3.33 Equivalent conductance.

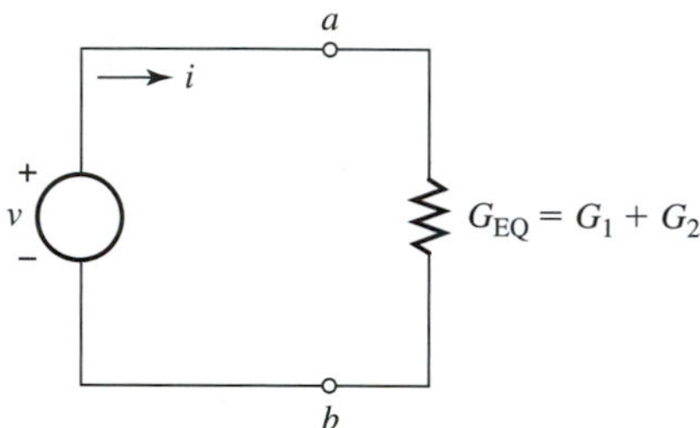

which is the same voltage-current relationship for terminals a-b determined for the circuit in Figure 3.32.

Because the current-voltage relationship for terminals a-b is the same for the circuits in Figures 3.32 and 3.33, G_{EQ} is called the equivalent conductance with respect to terminals a-b for the parallel combination of G_1 and G_2.

These results can be extended for n conductances in parallel, as indicated in Figure 3.34. The equation for combining n conductances in parallel is

$$G_{EQ} = G_1 + G_2 + \cdots + G_n \tag{3.6}$$

Figure 3.34 Combining n conductances in parallel.

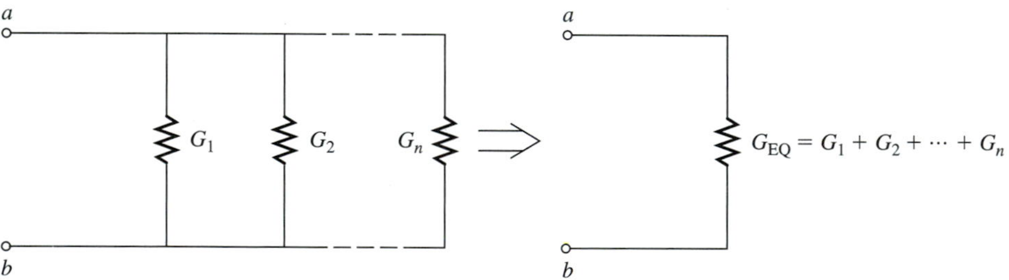

Consider the following examples.

EXAMPLE 3.9 Replace the four parallel conductances with a single conductance for the circuit in Figure 3.35. ■

Figure 3.35

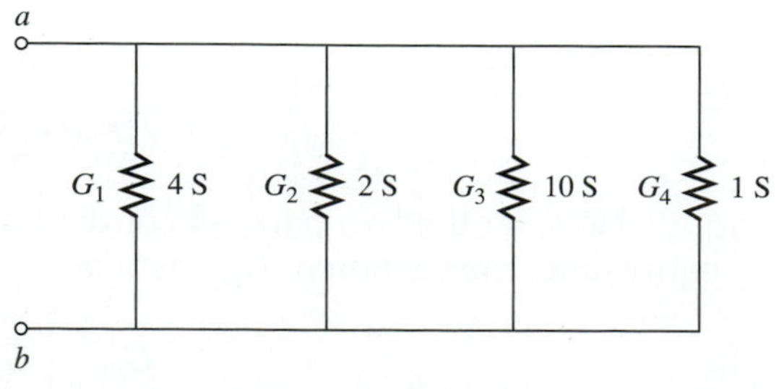

a

G_1 4 S G_2 2 S G_3 10 S G_4 1 S

b

SOLUTION From Equation (3.6)

$$G_{EQ} = G_1 + G_2 + G_3 + G_4$$
$$= 4 + 2 + 10 + 1$$
$$= 17 \text{ S}$$

The equivalent circuit with respect to terminals *a-b* is shown in Figure 3.36.

Figure 3.36

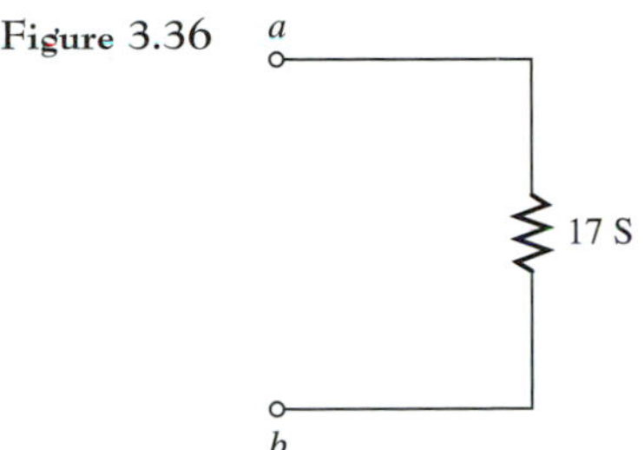

EXAMPLE 3.10 Replace the three parallel resistances with a single resistance for the circuit in Figure 3.37. ■

Figure 3.37

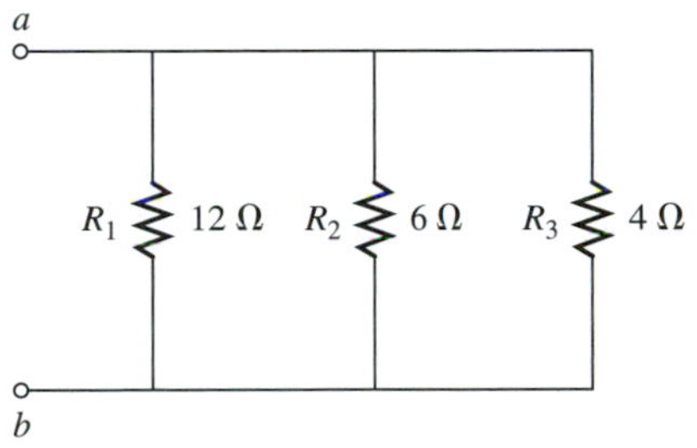

SOLUTION From Equation (3.4)

$$R_{EQ} = \frac{1}{\frac{1}{12} + \frac{1}{6} + \frac{1}{4}}$$
$$= \frac{1}{(1 + 2 + 3)/12}$$
$$= \frac{1}{\frac{6}{12}}$$
$$= 2\ \Omega$$

Alternatively, from Equation (3.5) we can combine the resistances in pairs. Combining R_1 and R_2 and calling the result R'_{EQ} results in

$$R'_{EQ} = \frac{R_1 R_2}{R_1 + R_2}$$
$$= \frac{(12)(6)}{12 + 6}$$
$$= 4\ \Omega$$

Next, combining R'_{EQ} with R_3 results in

$$R_{EQ} = \frac{R'_{EQ} R_3}{R'_{EQ} + R_3}$$

$$= \frac{(4)(4)}{4 + 4}$$

$$= 2\ \Omega$$

Using another method, we can convert each resistance (ohms) to conductance (siemens), as shown in Figure 3.38, use Equation (3.6) to determine G_{EQ}, and use Equation (2.5) to convert back to ohms.

Figure 3.38

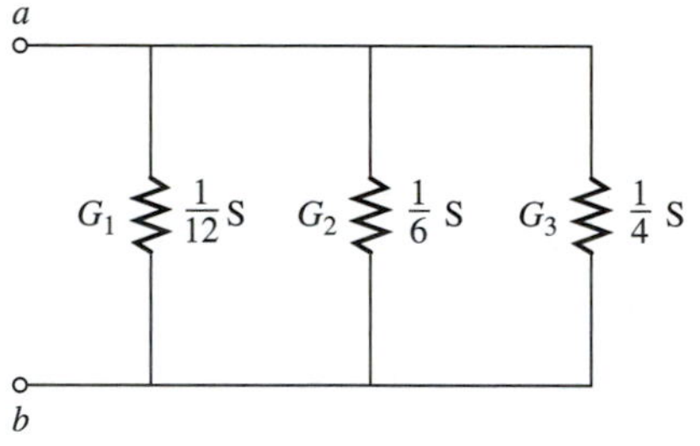

From Equation (3.6)

$$G_{EQ} = \frac{1}{12} + \frac{1}{6} + \frac{1}{4}$$

$$= \frac{1 + 2 + 3}{12}$$

$$= \frac{1}{2}\ \mathrm{S}$$

From Equation (2.5)

$$R_{EQ} = \frac{1}{G_{EQ}}$$

$$= \frac{1}{\frac{1}{12}}$$

$$= 2\ \Omega$$

An inspection of Equations (3.4) and (3.6) and Examples 3.8 and 3.9 reveals that when adding resistances in parallel, it is easier to work with conductance (siemens) than resistance (ohms). As stated previously, it is easier to work with resistance (ohms) than to work with conductance (siemens) when adding resistances in series.

3.8 SERIES-PARALLEL CIRCUITS

Many electric circuits consist of resistors connected in series and in parallel instead of in only one or the other, as previously presented. The following examples illustrate the use of the procedures previously presented to reduce series-parallel combinations to single equivalent resistances and conductances.

It is advantageous to use the following concise notation for representing resistances in parallel and series.

$(R_1 + R_2)$ is read as "R_1 and R_2 are in series."

$(R_1 \| R_2)$ is read as "R_1 and R_2 are in parallel."

As an example,

$$(R_1 + R_2 + R_3) \| (R_4 + R_5)$$

is read as R_1, R_2, and R_3 are in series; R_4 and R_5 are in series; and the series combination of R_4 and R_5 is in parallel with the series combination R_1, R_2, and R_3.

This notation is illustrated in the following examples.

EXAMPLE 3.11 Determine a single equivalent resistance for terminals *a*-*b* in Figure 3.39. ■

Figure 3.39

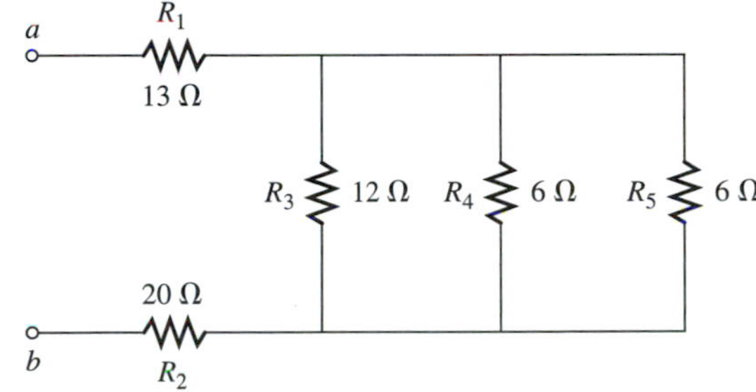

SOLUTION The equivalent resistance R_{EQ} for terminals *a*-*b* is

$$R_{EQ} = R_1 + R_2 + (R_3 \| R_4 \| R_5)$$

which is read as

R_3, R_4, and R_5 are in parallel and the parallel

combination is in series with R_1 and R_2.

From Equations (3.1) and (3.4),

$$\begin{aligned} R_{EQ} &= R_1 + R_2 + \frac{1}{1/R_3 + 1/R_4 + 1/R_5} \\ &= 13 + 20 + \frac{1}{\frac{1}{12} + \frac{1}{6} + \frac{1}{6}} \\ &= 13 + 20 + \frac{1}{\frac{5}{12}} \\ &= 35.4\ \Omega \end{aligned}$$

EXAMPLE 3.12 Determine a single equivalent resistance for terminals *a-b* in Figure 3.40. ■

SOLUTION The equivalent resistance R_{EQ} for terminals *a-b* is

$$R_{EQ} = [\{(R_4 \| R_5) + (R_2 \| R_3) + R_6\} \| \{R_8\}] + R_1 + R_7$$

Figure 3.40

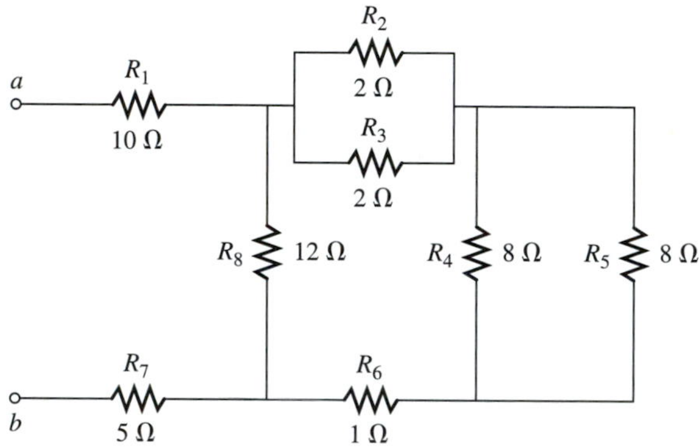

Using Equations (3.1) and (3.4) results in

$$R_4 \| R_5 = \frac{1}{\frac{1}{8} + \frac{1}{8}}$$

$$= 4\ \Omega$$

$$R_2 \| R_3 = \frac{1}{\frac{1}{2} + \frac{1}{2}}$$

$$= 1\ \Omega$$

$$(R_4 \| R_5) + (R_2 \| R_3) + R_6 = 4 + 1 + 1$$

$$= 6\ \Omega$$

$$\{(R_4 \| R_5) + (R_2 \| R_3) + R_6\} \| (R_8) = \frac{1}{\frac{1}{6} + \frac{1}{12}}$$

$$= 4\ \Omega$$

$$[\{(R_4 \| R_5)(R_2 \| R_3) + R_6\} \| (R_8)] + R_1 + R_7 = 4 + 10 + 5$$

$$= 19\ \Omega$$

$$R_{EQ} = 19\ \Omega$$

3.9 EXERCISES

1. Determine the nodes, loops, and branches for the circuits in Figure 3.41.

Figure 3.41

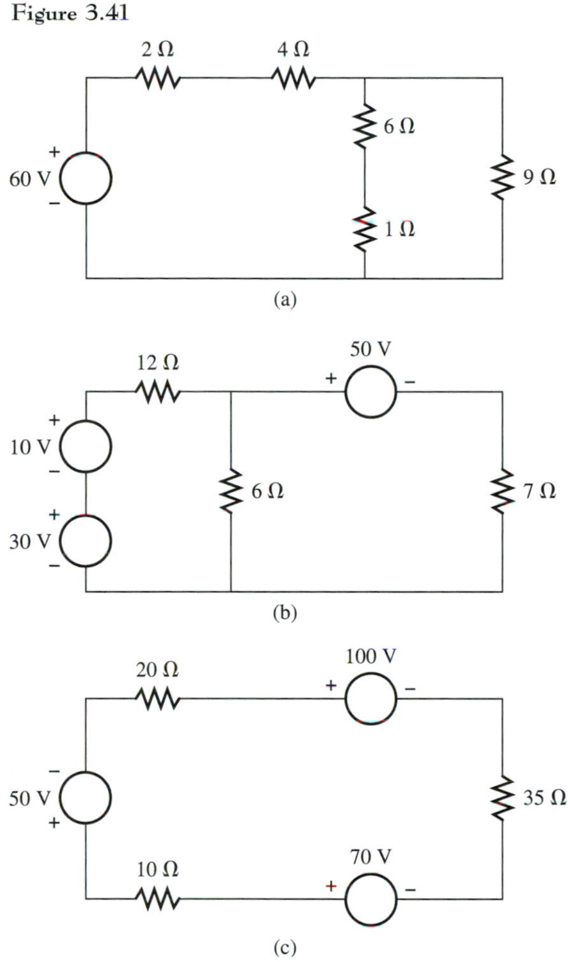

2. Determine the nodes, loops, and branches for the circuits in Figure 3.42.

Figure 3.42

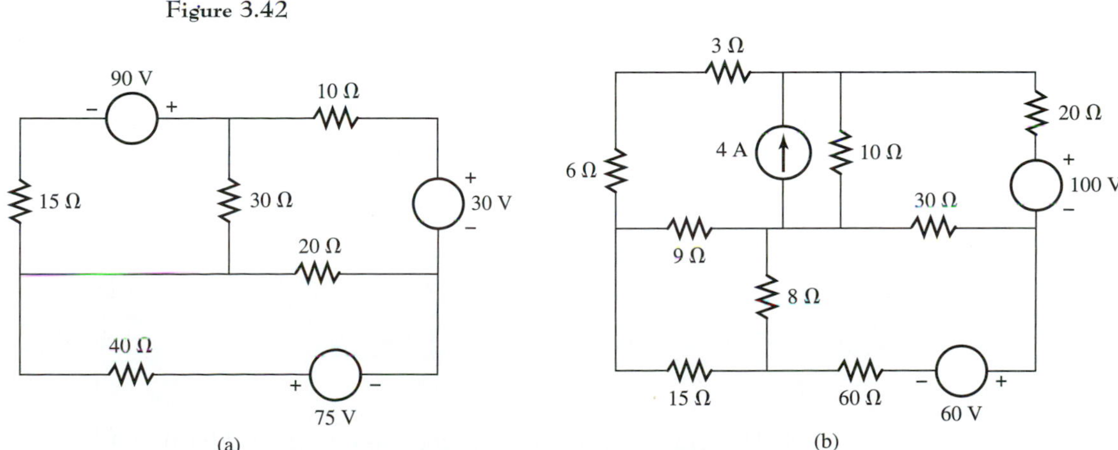

3. Use Kirchhoff's voltage law to determine the current i for the circuit in Figure 3.43.

Figure 3.43

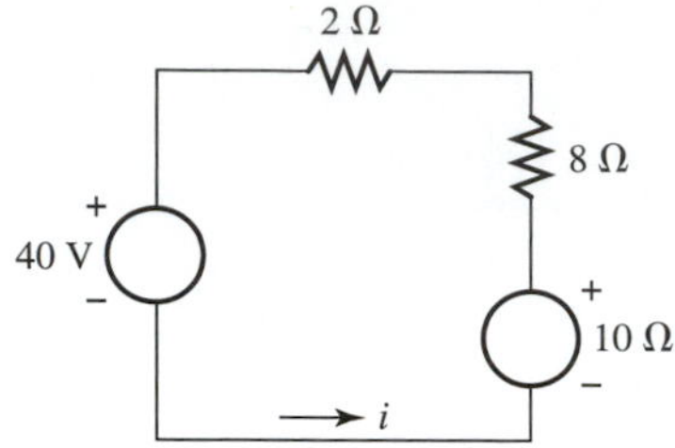

4. Use Kirchhoff's voltage law to determine the current i for the circuit in Figure 3.44.

Figure 3.44

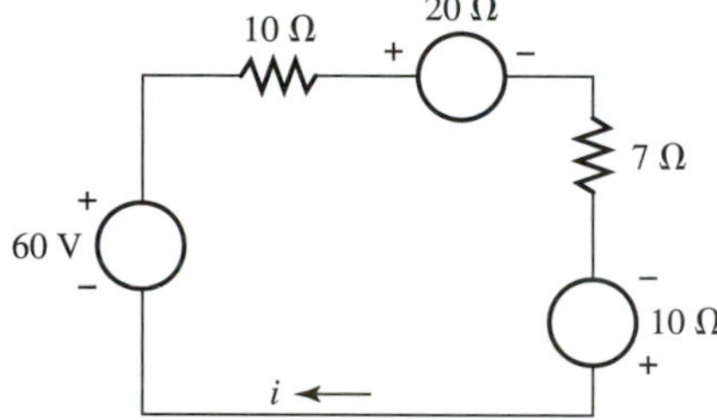

5. Use Kirchhoff's voltage law to determine the voltages v_1 and v_2 for the circuit in Figure 3.45.

Figure 3.45

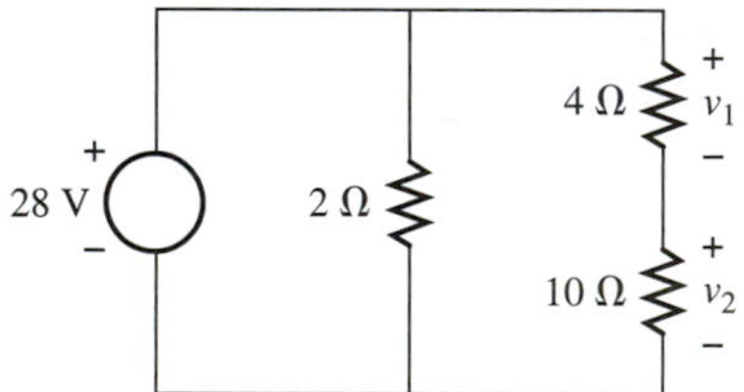

6. Use Kirchhoff's voltage law to determine the current i for the circuit in Figure 3.46.

Figure 3.46

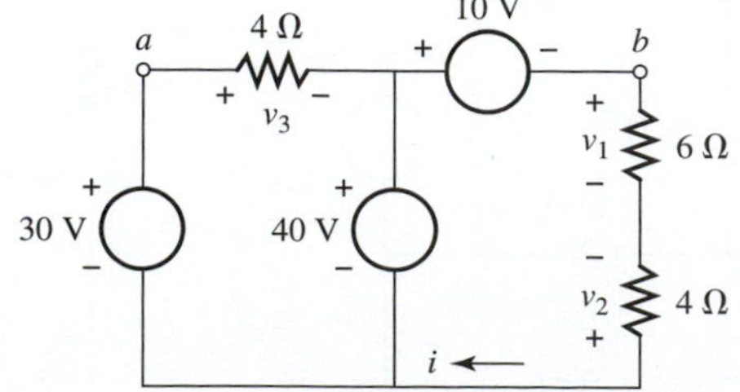

7. Determine v_1, v_2, and v_{ab} for the circuit in Figure 3.46.

8. Verify Kirchhoff's voltage law for each loop for the circuit in Figure 3.47.

Figure 3.47

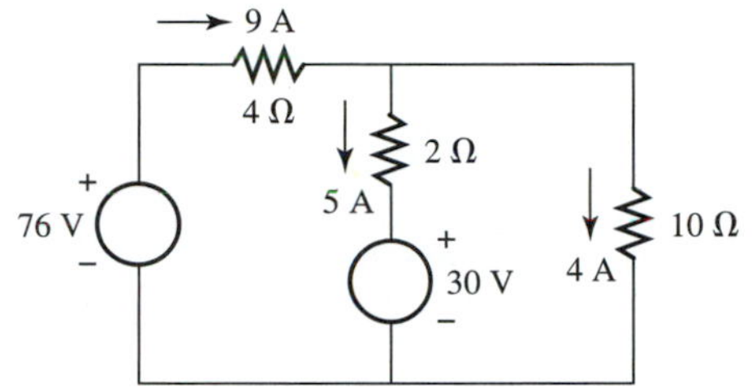

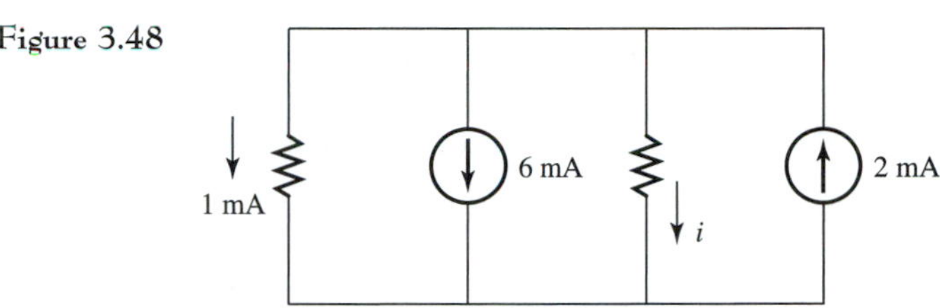

9. Use Kirchhoff's current law to determine i for the circuit in Figure 3.48.

Figure 3.48

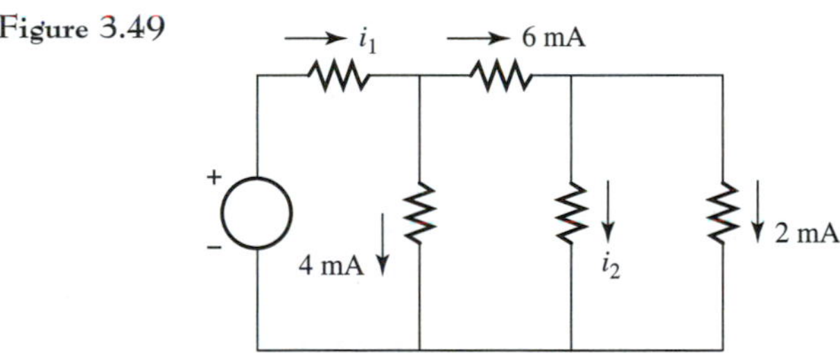

10. Use Kirchhoff's current law to determine i_1 and i_2 for the circuit in Figure 3.49.

Figure 3.49

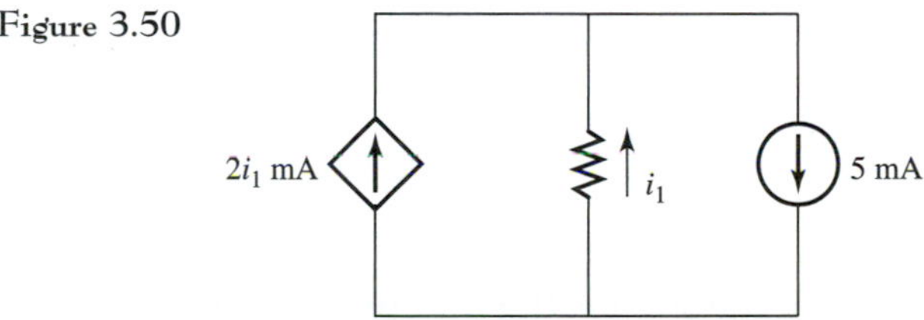

11. Determine the current i_1 for the circuit in Figure 3.50.

Figure 3.50

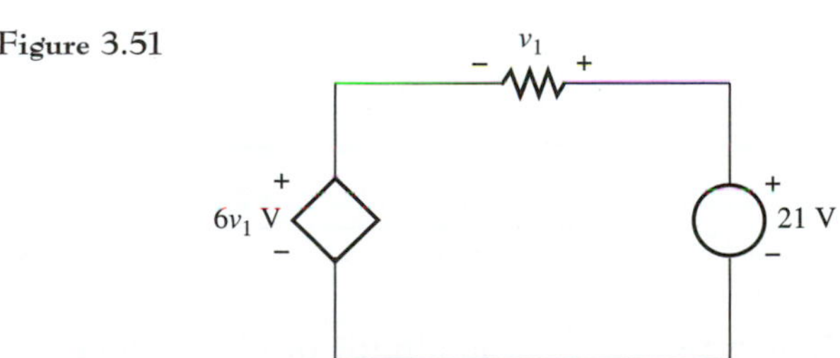

12. Determine the voltage v_1 for the circuit in Figure 3.51.

Figure 3.51

13. Determine v_{ab}, v_{bc} and v_{ac} for the circuit in Figure 3.52.

Figure 3.52

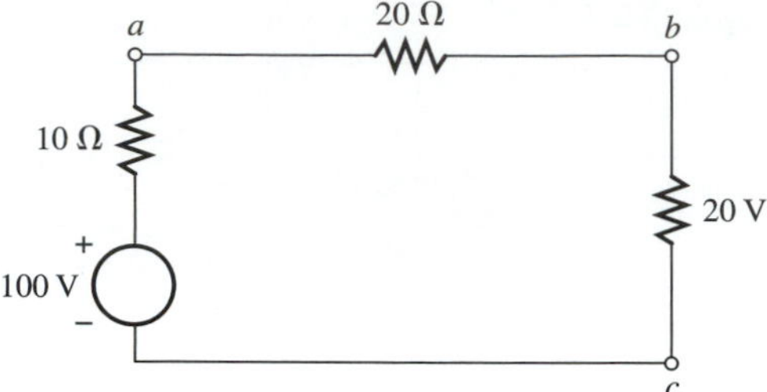

14. Determine i and v for the circuit in Figure 3.53.

Figure 3.53

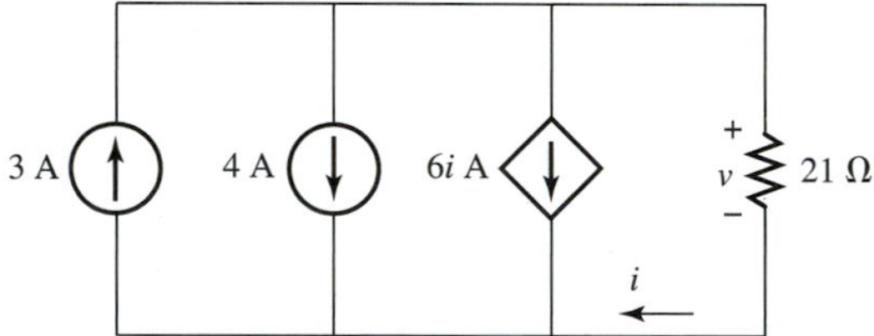

15. Determine the voltage v_{ab} and the current i for the circuit in Figure 3.54.

Figure 3.54

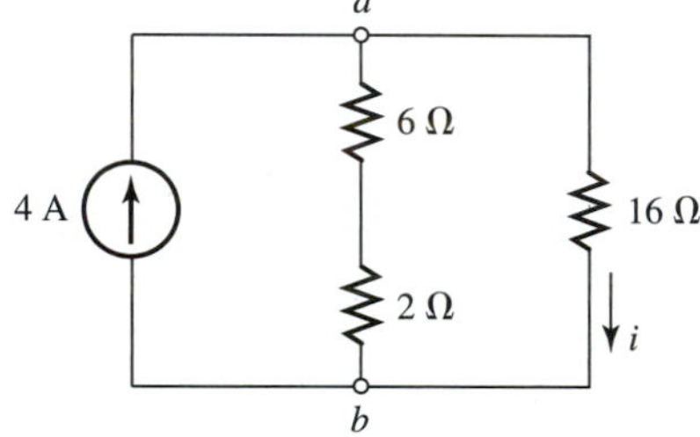

16. Determine the current i for the circuit in Figure 3.55.

Figure 3.55

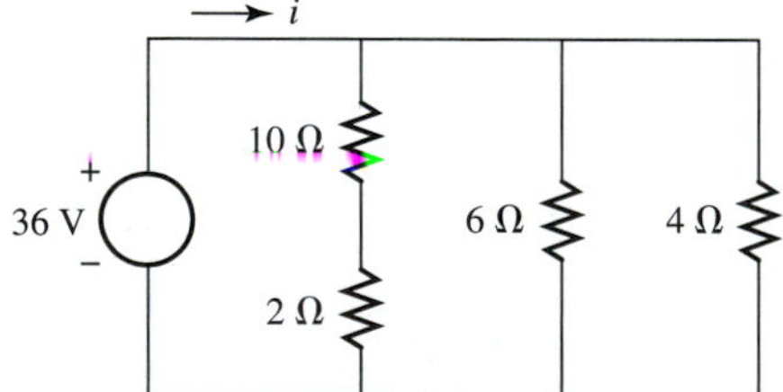

17. Determine R_{ab} for the circuit in Figure 3.56.

Figure 3.56

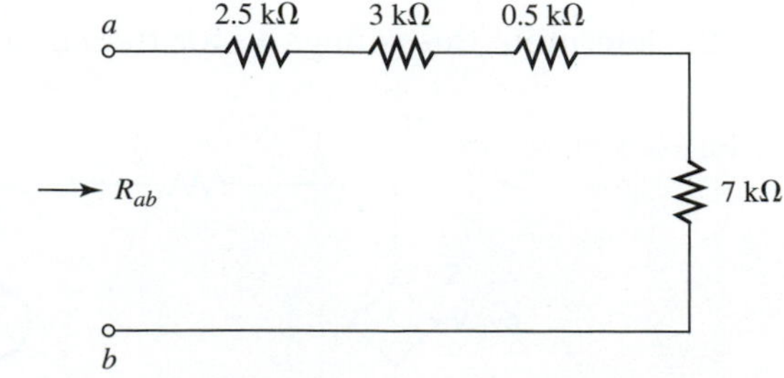

18. Determine G_{ab} for the circuit in Figure 3.56.

19. Determine R_{ab} for the circuit in Figure 3.57.

Figure 3.57

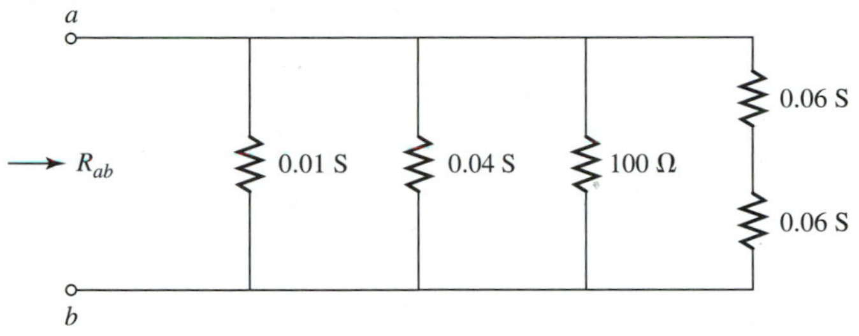

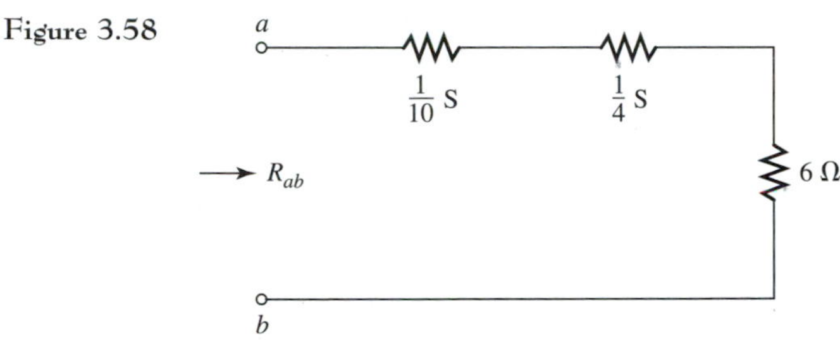

20. Determine G_{ab} for the circuit in Figure 3.57.

21. Determine R_{ab} for the circuit in Figure 3.58.

Figure 3.58

22. Determine G_{ab} for the circuit in Figure 3.58.

23. Determine R_{ab} for the circuit in Figure 3.59.

Figure 3.59

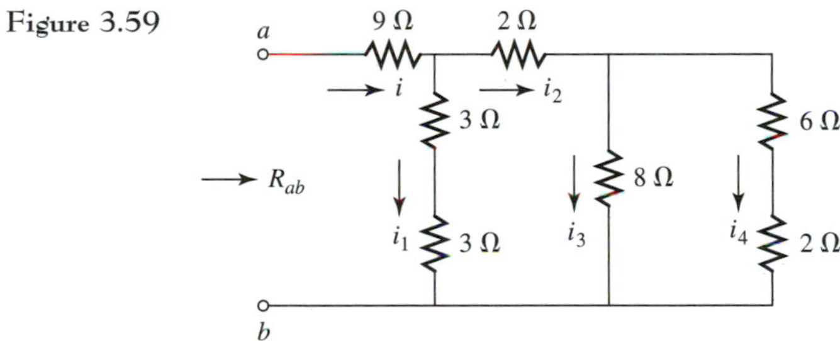

24. Determine i, i_1, i_2, i_3, and i_4 for the circuit in Figure 3.59 if v_{ab} is 24 V.

25. Determine G_{ab} for the circuit in Figure 3.60.

Figure 3.60

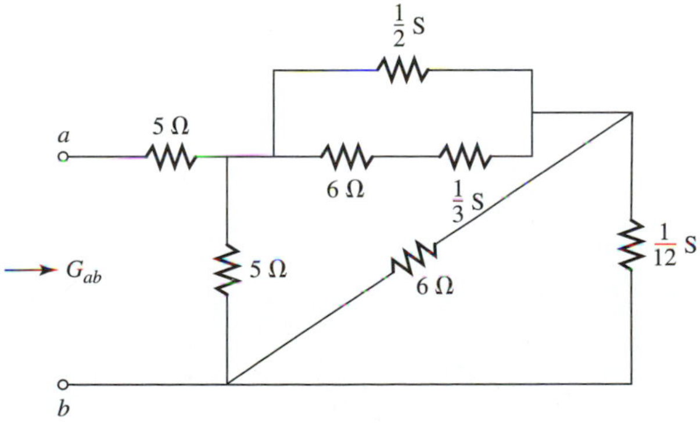

26. Determine v for the circuit in Figure 3.61.

Figure 3.61

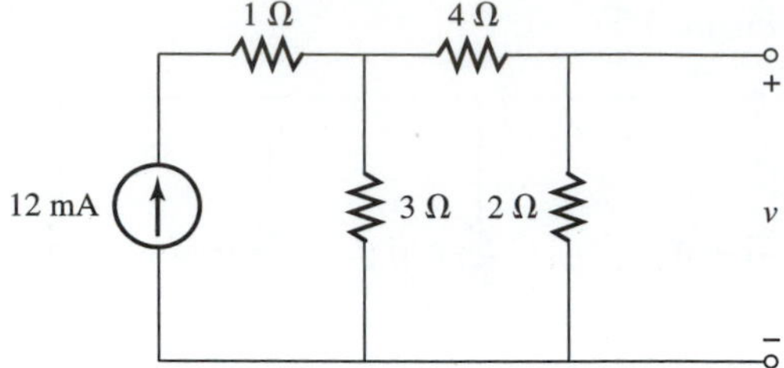

27. Determine v for the circuit in Figure 3.62.

Figure 3.62

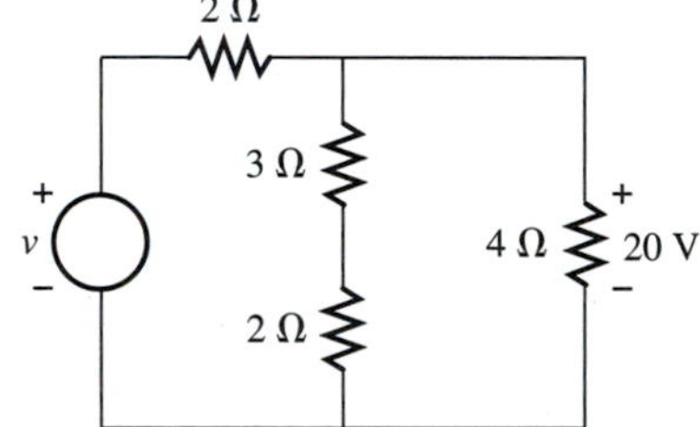

28. Determine R_{ab} for the circuit in Figure 3.63.

Figure 3.63

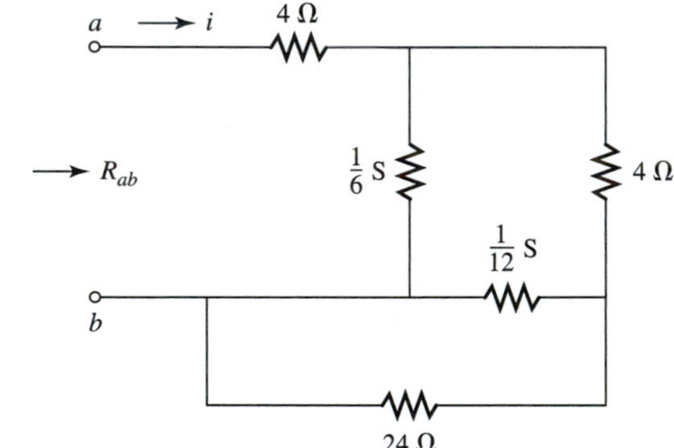

29. Determine the current i if the voltage is v_{ab} = 24 V in Figure 3.63.

CHAPTER

4

Basic Analysis Tools

After completing this chapter the student should be able to

* use the current and voltage division formulas to determine currents and voltages
* combine current and voltage sources in series and parallel
* determine currents and voltages in resistive circuits containing dependent sources
* use the branch current method to analyze resistive circuits

The basic analysis tools presented in this chapter are useful for analyzing particular circuit configurations. Voltage and current division enable us to determine the division of a current at a node and the division of a voltage across series elements. The branch current method serves as an introduction to analyzing circuits with multiple sources. The proportionality property relates the changes in the independent sources of a circuit to the corresponding changes in the currents and voltages of the circuit. Finally, the delta-wye transformation allows us to determine an equivalent resistance when the rules for combining series and parallel resistors are not applicable.

4.1 CURRENT-DIVISION FORMULA USING RESISTANCE

Consider the circuit in Figure 4.1. From Equation (3.4) an equivalent resistance

Figure 4.1 Current division.

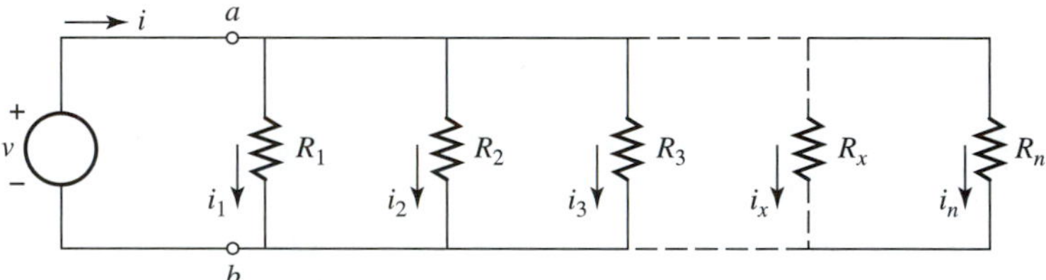

for the circuit to the right of terminals *a*-*b* is

$$R_{EQ} = \frac{1}{1/R_1 + 1/R_2 + \cdots + 1/R_n}$$

and the circuit in Figure 4.1 can be redrawn as shown in Figure 4.2.

Figure 4.2 Equivalent circuit.

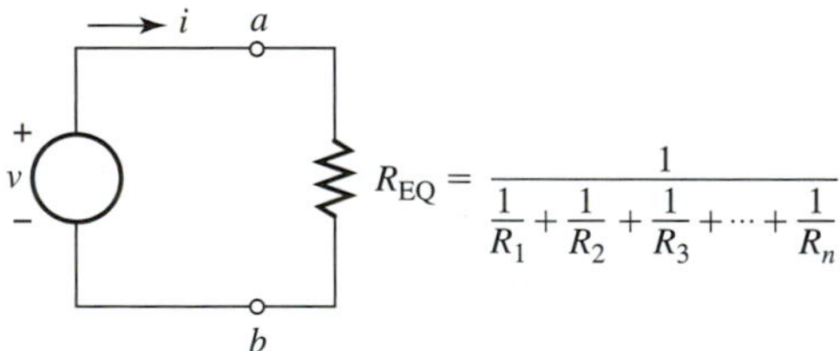

Applying Ohm's law to the circuit in Figure 4.2 results in

$$v = R_{EQ}i$$

Applying Ohm's law to the circuit in Figure 4.1 results in

$$i_1 = \frac{v}{R_1}$$

$$i_2 = \frac{v}{R_2}$$

$$\vdots$$

$$i_x = \frac{v}{R_x}$$

$$\vdots$$

$$i_n = \frac{v}{R_n}$$

Substituting $v = R_{EQ}i$ into these equations results in

$$i_1 = \left(\frac{R_{EQ}}{R_1}\right)i$$

$$i_2 = \left(\frac{R_{EQ}}{R_2}\right)i$$

$$\vdots$$

$$i_x = \left(\frac{R_{EQ}}{R_x}\right)i$$

$$\vdots$$

$$i_n = \left(\frac{R_{EQ}}{R_x}\right)i$$

In general, the current i_x through a resistance R_x in Figure 4.1 is given as

$$i_x = \left(\frac{R_{EQ}}{R_X}\right)i \tag{4.1}$$

where R_{EQ} is the single equivalent resistance for terminals a-b and R_x is the resistance that i_x flows through.

Consider the case where only two resistors are present, as shown in

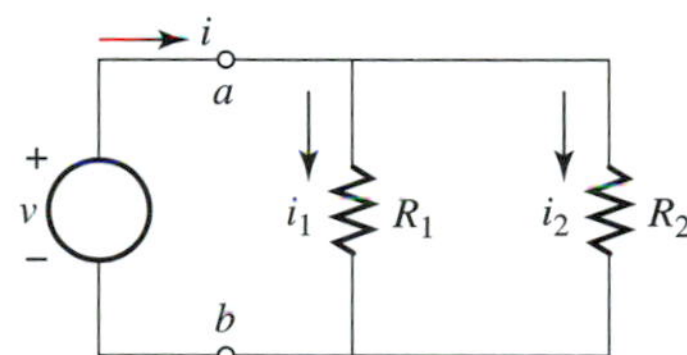

Figure 4.3 Current division for two resistors.

Figure 4.3. The equivalent resistance with respect to terminals a-b is

$$R_{EQ} = \frac{1}{1/R_1 + 1/R_2} = \frac{R_1R_2}{R_1 + R_2}$$

Applying Equation (4.1) to determine i_1 and i_2 results in

$$i_1 = \left(\frac{R_2}{R_1 + R_2}\right)i$$

$$i_2 = \left(\frac{R_1}{R_1 + R_2}\right)i$$

EXAMPLE 4.1 Determine currents i_1 and i_2 for the circuit shown in Figure 4.4. ■

Figure 4.4

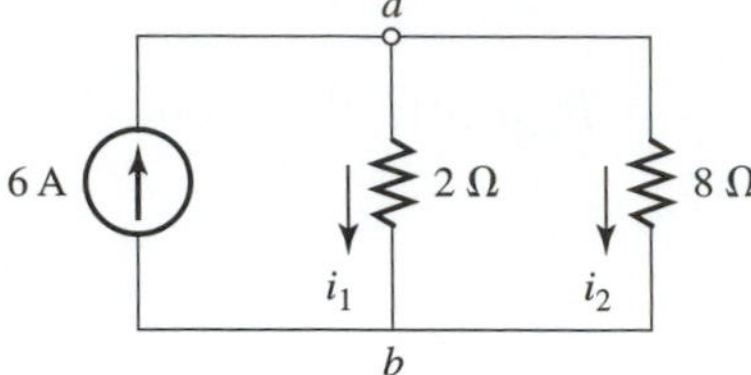

SOLUTION From Equation (4.1)

$$i_1 = \left(\frac{8}{8+2}\right)(6)$$
$$= 4.8 \text{ A}$$

and

$$i_2 = \left(\frac{2}{8+2}\right)(6)$$
$$= 1.2 \text{ A}$$

Notice that the results provided by the current division formula verify Kirchhoff's current law at each node in the circuit. At node a

$$6 = i_1 + i_2$$
$$= 4.8 + 1.2$$
$$= 6$$

4.2 CURRENT-DIVISION FORMULA USING CONDUCTANCE

Using Equation (2.3), R_{EQ} and R_x for the circuit in Figure 4.1 can be expressed in terms of conductance (siemens) as

$$R_{EQ} = \frac{1}{G_{EQ}}$$

$$R_x = \frac{1}{G_x}$$

Substituting $(1/G_{EQ})$ and $(1/G_x)$ for R_{EQ} and R_x, respectively, in Equation (4.1) results in

$$i_x = \left(\frac{G_x}{G_{EQ}}\right)i \tag{4.2}$$

where G_{EQ} is the single equivalent conductance for terminals a-b and G_x is the conductance through which i_x flows. From Equation (3.6)

$$G_{EQ} = G_1 + G_2 + \cdots + G_n$$

Equation (4.2) is the general current-division formula in terms of conductance.

Consider the case where only two conductances are present, as shown in

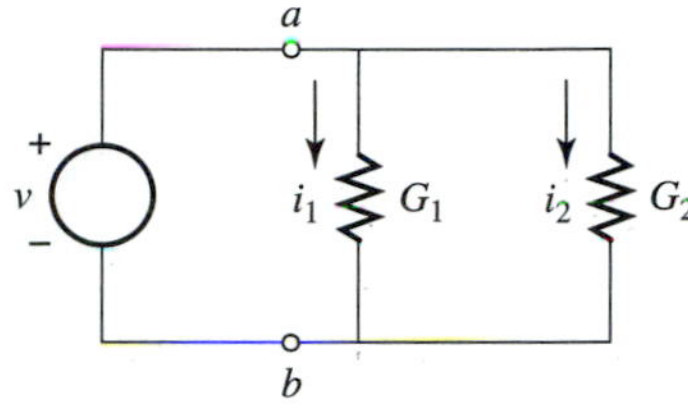

Figure 4.5 Current division for two conductances.

Figure 4.5. The equivalent conductance with respect to terminals *a-b* is

$$G_{EQ} = G_1 + G_2$$

Applying Equation (4.2) to determine i_1 and i_2 results in

$$i_1 = \left(\frac{G_1}{G_1 + G_2}\right)i$$

$$i_2 = \left(\frac{G_2}{G_1 + G_2}\right)i$$

Because G_{EQ} is easier to determine than R_{EQ} for resistors in parallel, it is usually advantageous to employ Equation (4.2) rather than (4.1) when determining current division. Consider the following examples.

EXAMPLE 4.2 Determine the currents i_1, i_2, i_3, and i_4 for the circuit shown in Figure 4.6. ■

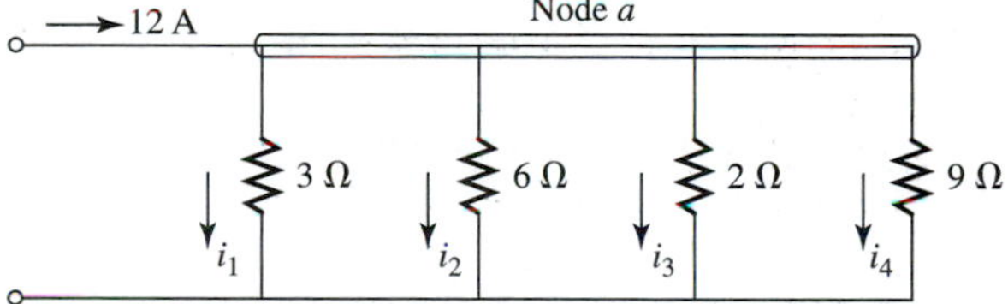

Figure 4.6

SOLUTION From Equation (3.4)

$$R_{EQ} = (3) \parallel (6) \parallel (2) \parallel (9)$$

$$= \frac{1}{\frac{1}{3} + \frac{1}{6} + \frac{1}{2} + \frac{1}{9}}$$

$$= \frac{9}{10} = 0.9\ \Omega$$

From Equation (4.1)

$$i_1 = \left(\frac{R_{EQ}}{3}\right)(12)$$

$$= \left(\frac{0.9}{3}\right)(12)$$

$$= 3.6 \text{ A}$$

$$i_2 = \left(\frac{R_{EQ}}{6}\right)(12)$$

$$= \left(\frac{0.9}{6}\right)(12)$$

$$= 1.8 \text{ A}$$

$$i_3 = \left(\frac{R_{EQ}}{2}\right)(12)$$

$$= \left(\frac{0.9}{2}\right)(12)$$

$$= 5.4 \text{ A}$$

$$i_4 = \left(\frac{R_{EQ}}{9}\right)(12)$$

$$= \left(\frac{0.9}{9}\right)(12)$$

$$= 1.2 \text{ A}$$

Notice that Kirchhoff's current law is verified at node a.

$$12 = i_1 + i_2 + i_3 + i_4$$

$$= 3.6 + 1.8 + 5.4 + 1.2$$

$$= 12$$

EXAMPLE 4.3 Determine the currents i_1, i_2, i_3, and i_4 for the circuit shown in Figure 4.7. ■

Figure 4.7

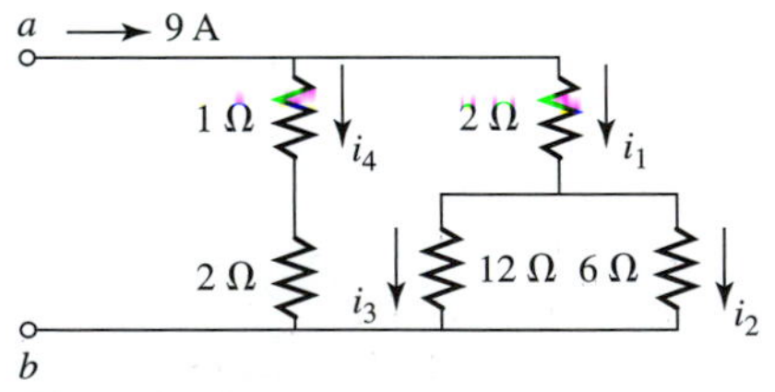

SOLUTION For terminals a-b

$$R_{EQ} = (1 + 2) \parallel \{2 + (12 \parallel 6)\}$$

From Equations (3.1) and (3.5)

$$R_{EQ} = 3 \parallel \left(2 + \frac{6(12)}{6 + 12}\right)$$

$$= 3 \parallel (2 + 4)$$

$$= 3 \parallel 6$$

$$= \frac{6(3)}{6 + 3}$$

$$= 2\,\Omega$$

The circuit can be redrawn as shown in Figure 4.8.

Figure 4.8

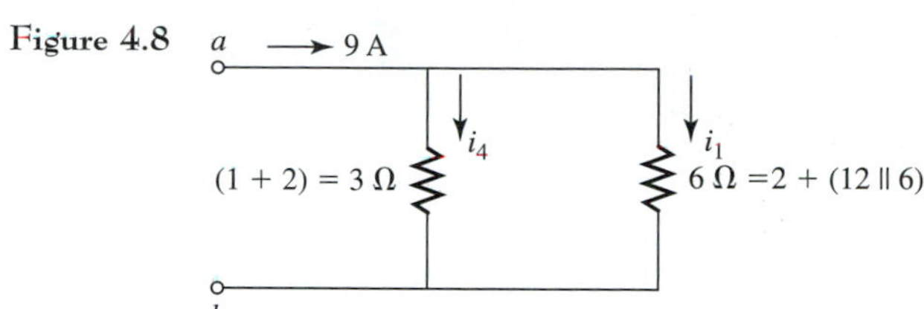

Applying Equation (4.1) to the equivalent circuit results in

$$i_1 = \left(\frac{2}{6}\right)(9)$$

$$= 3\text{ A}$$

$$i_4 = \left(\frac{2}{3}\right)(9)$$

$$= 6\text{ A}$$

The current i_1 divides into i_2 and i_3 in the original circuit, so Equation (4.1) can be used to determine i_2 and i_3. Consider the partial circuit shown in Figure 4.9.

Figure 4.9

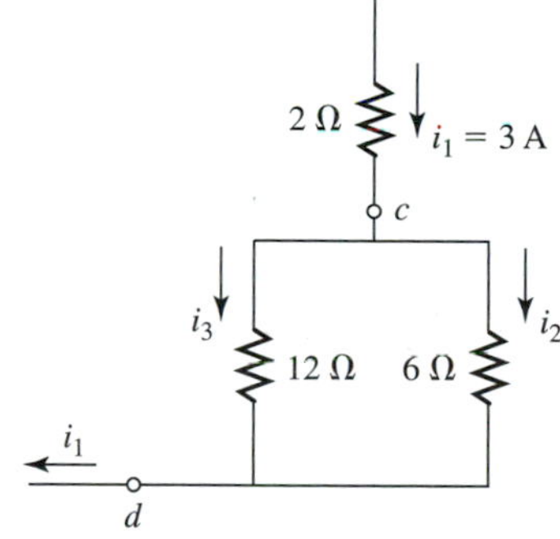

R_{EQ} for terminals *c-d* is

$$R_{EQ} = 6 \| 12 = 4\,\Omega$$

Applying Equation (4.1) to the division of current i_1 results in

$$i_2 = \left(\frac{4}{6}\right)(3)$$

$$= 2\text{ A}$$

$$i_3 = \left(\frac{4}{12}\right)(3)$$

$$= 1\text{ A}$$

Notice that Kirchhoff's current law is verified at nodes a and c, as shown in Figure 4.10.

Figure 4.10

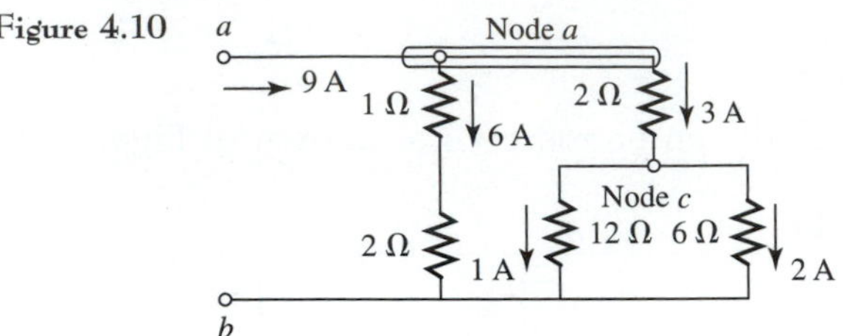

At node (a), 9 A entering equals 9 A leaving.
At node (c), 3 A entering equals 3 A leaving.

4.3 VOLTAGE-DIVISION FORMULA USING RESISTANCE

Consider the circuit in Figure 4.11. From Equation (3.1) an equivalent resistance for terminals a-b is

$$R_{EQ} = R_1 + R_2 + \cdots + R_n$$

Figure 4.11 Voltage division.

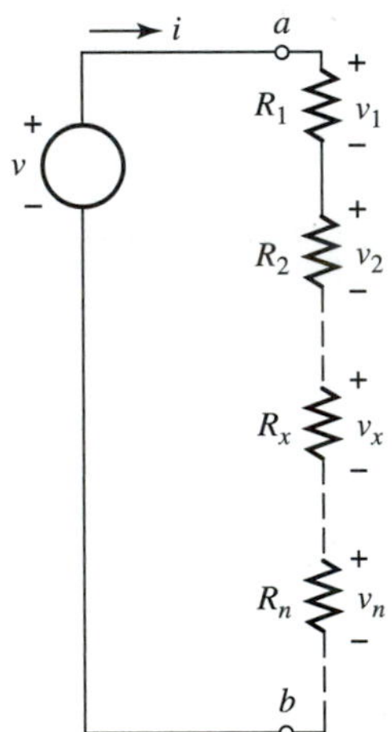

and the circuit in Figure 4.11 can be redrawn as shown in Figure 4.12.

Figure 4.12 Equivalent circuit.

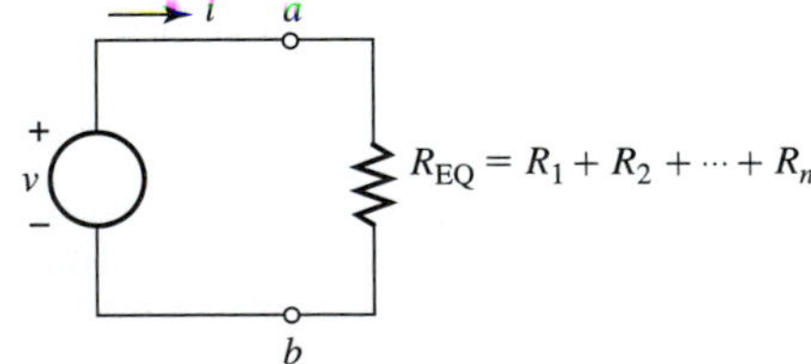

Applying Ohm's law to each resistance in Figure 4.11 results in

$$v_1 = R_1 i$$
$$v_2 = R_2 i$$
$$v_3 = R_3 i$$
$$\vdots$$
$$v_x = R_x i$$
$$\vdots$$
$$v_n = R_n i$$

Applying Ohm's law to the equivalent circuit in Figure 4.12 results in

$$i = \frac{v}{R_{EQ}}$$

Substituting (v/R_{EQ}) for i in the preceding equations results in

$$v_1 = \left(\frac{R_1}{R_{EQ}}\right)v$$

$$v_2 = \left(\frac{R_2}{R_{EQ}}\right)v$$

$$v_3 = \left(\frac{R_3}{R_{EQ}}\right)v$$

$$\vdots$$

$$v_x = \left(\frac{R_x}{R_{EQ}}\right)v$$

$$\vdots$$

$$v_n = \left(\frac{R_n}{R_{EQ}}\right)v$$

In general, the voltage v_x across a resistance R_x in Figure 4.11 is given as

$$v_x = \left(\frac{R_x}{R_{EQ}}\right)v \tag{4.3}$$

where R_{EQ} is the single resistance equivalent for terminals a-b and R_x is the resistance across which v_x appears.

Consider the case where only two resistors are present, as shown in Figure 4.13. The equivalent resistance with respect to terminals a-b is

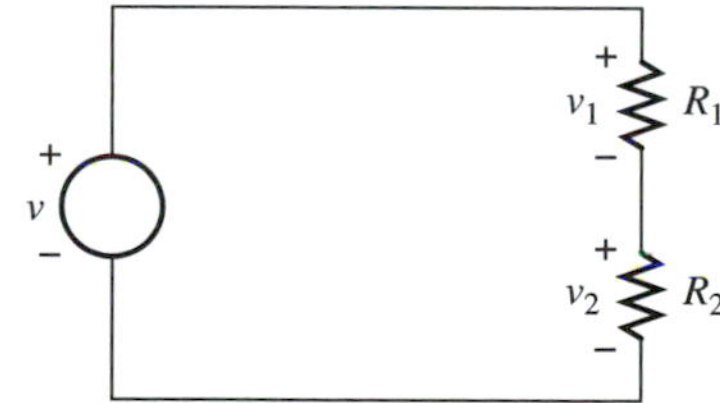

Figure 4.13 Voltage division for two resistors.

$$R_{EQ} = R_1 + R_2$$

Applying Equation (4.3) to determine v_1 and v_2 results in

$$v_1 = \left(\frac{R_1}{R_1 + R_2}\right)v$$

$$v_2 = \left(\frac{R_2}{R_1 + R_2}\right)v$$

4.4 VOLTAGE-DIVISION FORMULA USING CONDUCTANCE

Using Equation (2.3), R_{EQ} and R_x for the circuit in Figure 4.11 can be expressed in terms of conductance (siemens) as

$$R_{EQ} = \frac{1}{G_{EQ}}$$

and

$$R_x = \frac{1}{G_x}$$

Substituting $(1/G_{EQ})$ and $(1/G_x)$ for R_{EQ} and R_x, respectively, in Equation (4.3) results in

$$v_x = \left(\frac{G_{EQ}}{G_x}\right)v \tag{4.4}$$

where G_{EQ} is the single equivalent conductance for terminals a-b and G_x is the conductance across which v_x appears. From Equation (3.2)

$$G_{EQ} = \frac{1}{1/G_1 + 1/G_2 + \cdots + 1/G_n}$$

Equation (4.4) is the general voltage-division formula in terms of conductance.

Consider the case where only two conductors are present, as shown in Figure 4.14. The equivalent resistance with respect to terminals a-b is

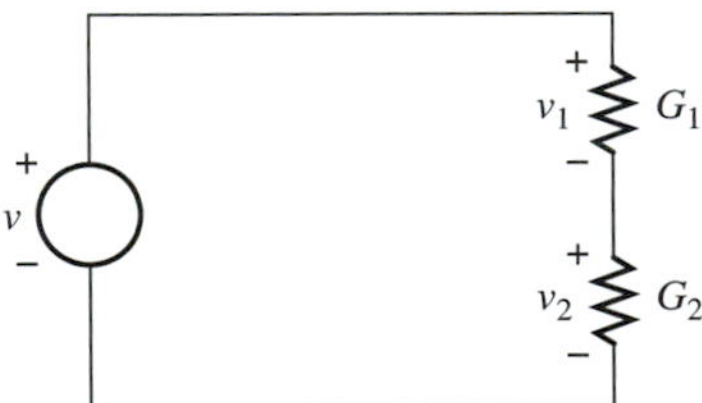

Figure 4.14 Voltage division for two conductances.

$$G_{EQ} = \frac{1}{1/G_1 + 1/G_2} = \frac{G_1G_2}{G_1 + G_2}$$

Applying Equation (4.4) to determine v_1 and v_2 results in

$$v_1 = \left(\frac{G_2}{G_1 + G_2}\right)v$$

$$v_2 = \left(\frac{G_1}{G_1 + G_2}\right)v$$

Because R_{EQ} is easier to determine than G_{EQ} for resistances in series, it is advantageous to employ Equation (4.3) rather than Equation (4.4) when using the voltage-division formula.

Consider the following examples.

EXAMPLE 4.4 Determine voltages v_1 and v_2 for the circuit shown in Figure 4.15. ■

Figure 4.15

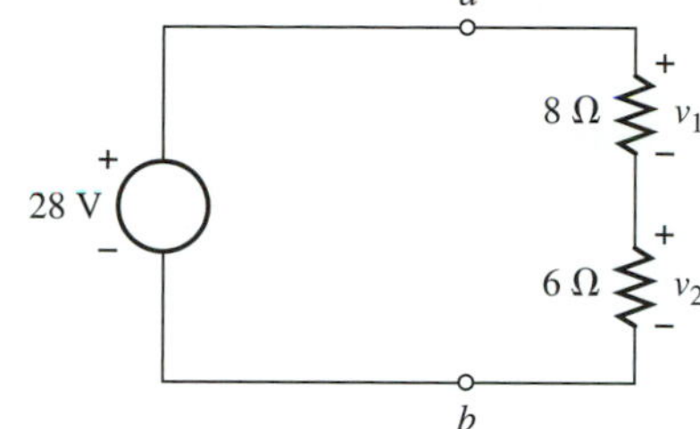

SOLUTION Applying Equation (4.3) to the circuit results in

$$\begin{aligned} v_1 &= \left(\frac{R_1}{R_1 + R_2}\right)(v) \\ &= \left(\frac{8}{14}\right)(28) \\ &= 16 \text{ V} \\ v_2 &= \left(\frac{R_2}{R_1 + R_2}\right)(v) \\ &= \left(\frac{6}{14}\right)(28) \\ &= 12 \text{ V} \end{aligned}$$

Notice that Kirchhoff's voltage law around the loop is verified:

$$\begin{aligned} 28 - v_1 - v_2 &= 0 \\ 28 - 16 - 12 &= 0 \\ 0 &= 0 \end{aligned}$$

EXAMPLE 4.5. Determine v_1, v_2, v_3, and v_4 for the circuit shown in Figure 4.16. ■

Figure 4.16

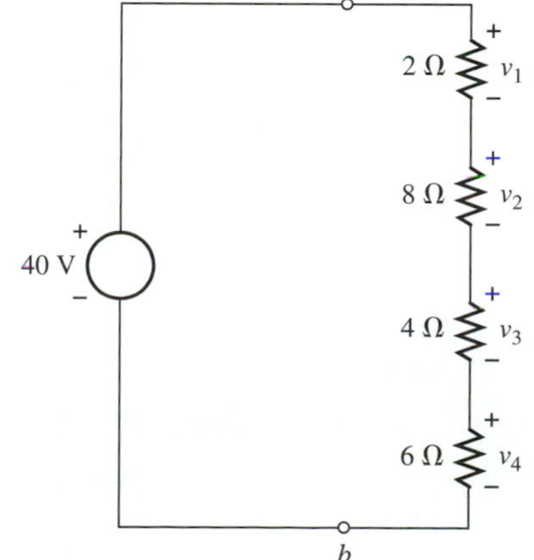

SOLUTION From Equation (3.1), R_{EQ} for terminals a-b is

$$R_{EQ} = 2 + 8 + 4 + 6$$
$$= 20\ \Omega$$

From Equation (4.3)

$$v_1 = \left(\frac{2}{20}\right)(40)$$
$$= 4\ \text{V}$$
$$v_2 = \left(\frac{8}{20}\right)(40)$$
$$= 16\ \text{V}$$
$$v_3 = \left(\frac{4}{20}\right)(40)$$
$$= 8\ \text{V}$$
$$v_4 = \left(\frac{6}{20}\right)(40)$$
$$= 12\ \text{V}$$

Notice that Kirchhoff's voltage law is verified:

$$40 - v_1 - v_2 - v_3 - v_4 = 0$$
$$40 - 4 - 16 - 8 - 12 = 0$$
$$0 = 0$$

EXAMPLE 4.6 Determine v_1, v_2, v_3, and v_4 for the circuit shown in Figure 4.17. ■

Figure 4.17

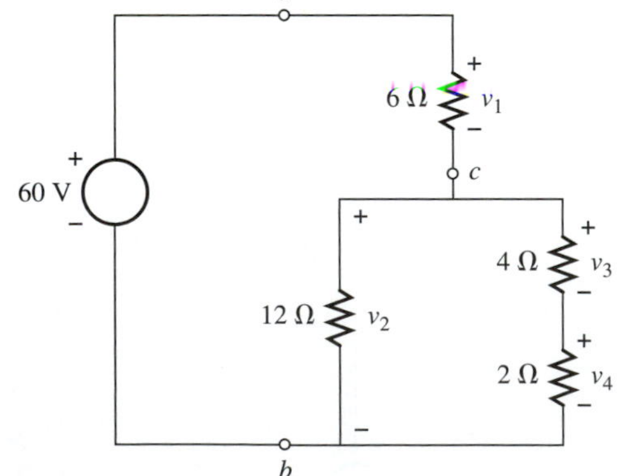

SOLUTION From Equations (3.1) and (3.5), R_{EQ} with respect to terminals a-b is

$$R_{EQ} = 6 + \{(4 + 2) \parallel (12)\}$$
$$= 6 + \{(6) \parallel (12)\}$$
$$= 6 + \left(\frac{(6)(12)}{6 + 12}\right)$$

$$= 6 + 4$$
$$= 10\ \Omega$$

Combining the 12-Ω, 4-Ω, and 2-Ω resistances with respect to terminals c-b and calling the resulting resistance R gives

$$R = (12) \parallel (4 + 2)$$
$$= (12) \parallel (6)$$
$$= \frac{(12)(6)}{12 + 6}$$
$$= 4\ \Omega$$

The circuit can be redrawn as shown in Figure 4.18.

Figure 4.18

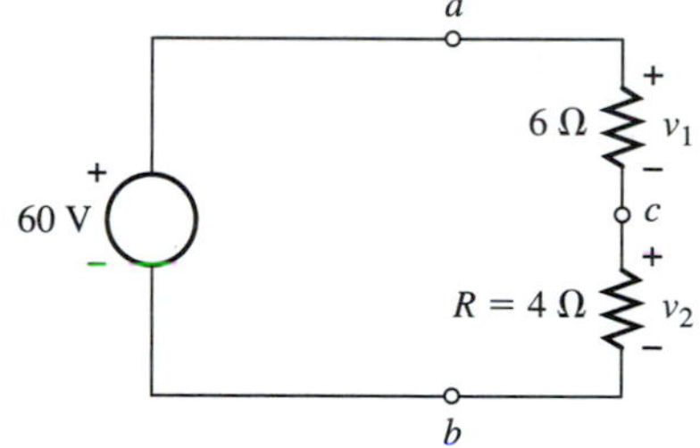

From Equation (4.3)

$$v_1 = \left(\frac{6}{10}\right)(60)$$
$$= 36\ \text{V}$$
$$v_2 = \left(\frac{4}{10}\right)(60)$$
$$= 24\ \text{V}$$

Because v_2 divides across the 4-Ω and 2-Ω resistances as voltages v_3 and v_4, we employ Equation (4.3) a second time to determine v_3 and v_4. The equivalent resistance for the 4-Ω and 2-Ω resistances in series is

$$R_{EQ} = 4 + 2$$
$$= 6\ \Omega$$

From Equation (4.3)

$$v_3 = \left(\frac{4}{6}\right)(24)$$
$$= 16\ \text{V}$$
$$v_4 = \left(\frac{2}{6}\right)(24)$$
$$= 8\ \text{V}$$

4.5 EXAMPLES OF CIRCUIT ANALYSIS

The following examples are given to illustrate the use of the methods previously presented to analyze electric circuits.

EXAMPLE 4.7 Determine currents i_1 and i_2 for the circuit in Figure 4.19. ■

Figure 4.19

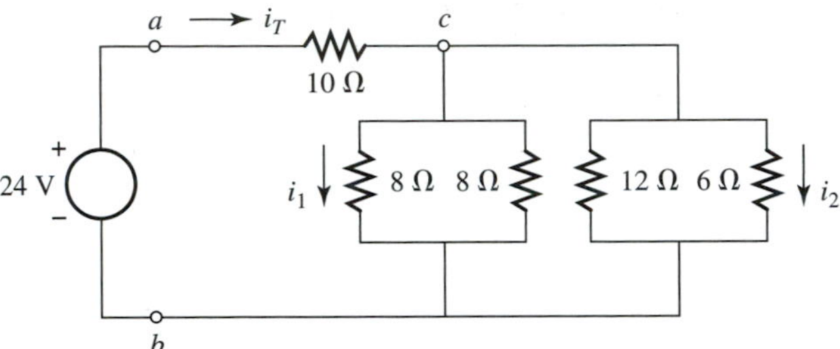

SOLUTION We shall first determine an equivalent single resistance for the circuit to the right of terminals *a-b* (the resistance seen from the 24-V source).

$$R_{EQ} = \{(8 \parallel 8) \parallel (12 \parallel 6)\} + 10$$

From Equations (3.1) and (3.5),

$$\begin{aligned} R_{EQ} &= \left[\left\{\frac{(8)(8)}{8+8}\right\} \parallel \left\{\frac{(12)(6)}{12+6}\right\}\right] + 10 \\ &= \{(4) \parallel (4)\} + 10 \\ &= \left\{\frac{(4)(4)}{4+4}\right\} + 10 \\ &= 12\ \Omega \end{aligned}$$

The circuit can be redrawn as shown in Figure 4.20.

Figure 4.20

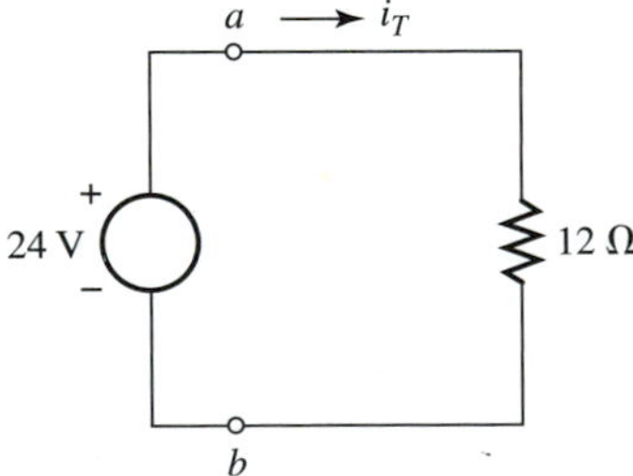

From Ohm's law

$$\begin{aligned} i_T &= \frac{24}{12} \\ &= 2\ \text{A} \end{aligned}$$

The original circuit can also be redrawn as shown in Figure 4.21, where R_3 and R_4 are

$$\begin{aligned} R_3 &= 8 \| 8 \\ &= 4\ \Omega \end{aligned}$$

$$R_4 = 12 \| 6$$
$$= 4\ \Omega$$

Figure 4.21

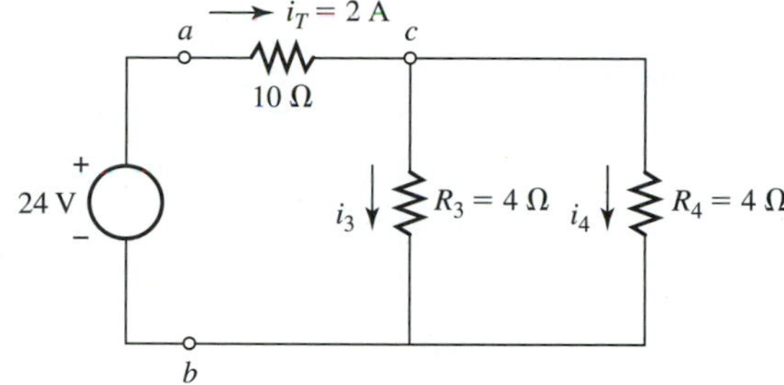

Using Equation (4.1) for the division of i_T at node c results in

$$i_3 = \left(\frac{4}{4+4}\right)(2)$$
$$= 1\text{ A}$$
$$i_4 = \left(\frac{4}{4+4}\right)(2)$$
$$= 1\text{ A}$$

Redrawing the original circuit as shown in Figure 4.22, with i_3 and i_4 indicated, and applying Equation (4.1) a second time results in

$$i_1 = \left(\frac{8}{8+8}\right)i_3$$
$$= \left(\frac{1}{2}\right)(1)$$
$$= \frac{1}{2}\text{ A}$$
$$i_2 = \left(\frac{12}{12+6}\right)i_4$$
$$= \left(\frac{2}{3}\right)(1)$$
$$= \frac{2}{3}\text{ A}$$

Notice that currents divide equally between two equal resistances. This is easily verified from the current-division formula.

Figure 4.22

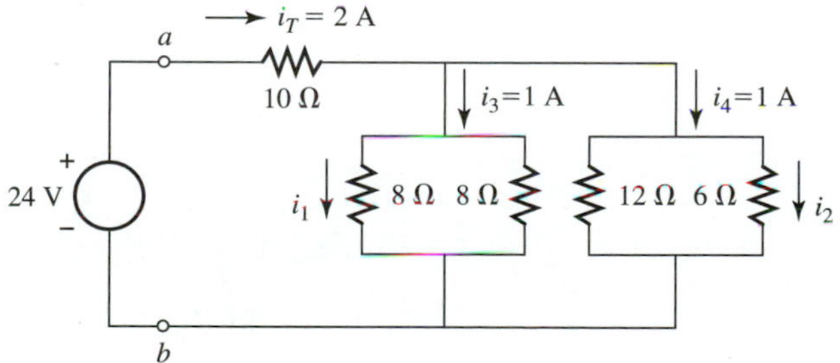

EXAMPLE 4.8 Determine v_1, v_2, v_3, i_1, i_2, and i_T for the circuit shown in Figure 4.23. ■

Figure 4.23

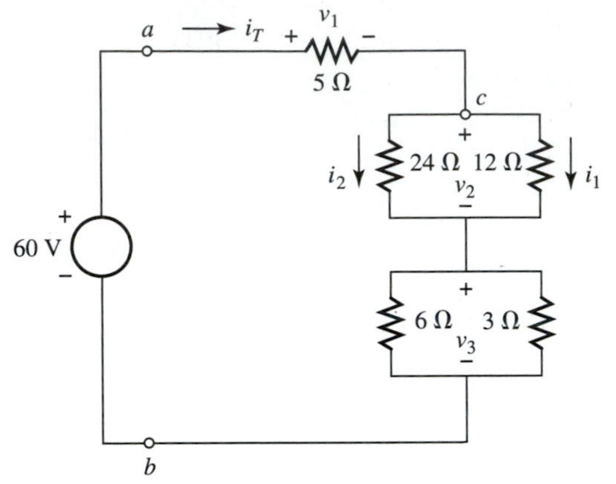

SOLUTION The circuit can be redrawn as shown in Figure 4.24.

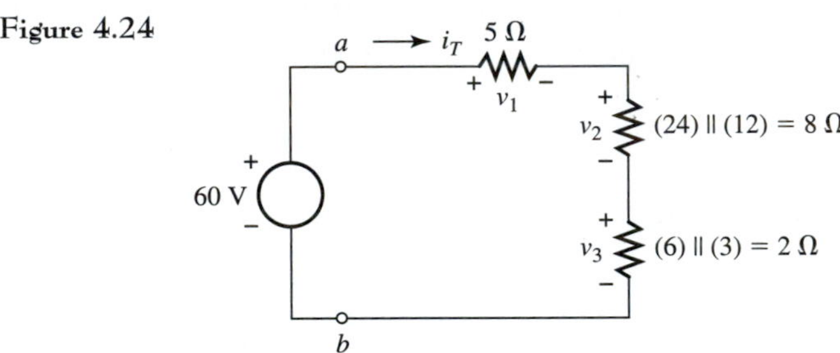

Figure 4.24

The single equivalent resistance, R_{EQ}, for the circuit to the right of terminals a-b is

$$\begin{aligned} R_{EQ} &= \{24 \| 12\} + \{6 \| 3\} + 5 \\ &= 8 + 2 + 5 \\ &= 15\ \Omega \end{aligned}$$

From Equation (4.3) for voltage division,

$$\begin{aligned} v_1 &= \left(\frac{5}{15}\right)(60) \\ &= 20\ \text{V} \\ v_2 &= \left(\frac{8}{15}\right)(60) \\ &= 32\ \text{V} \end{aligned}$$

$$v_3 = \left(\frac{2}{15}\right)(60)$$
$$= 8 \text{ V}$$

Applying Ohm's law to the original circuit results in

$$i_1 = \frac{v_2}{12}$$
$$= \frac{32}{12}$$
$$= \frac{8}{3} \text{ A}$$
$$i_2 = \frac{v_2}{24}$$
$$= \frac{32}{24}$$
$$= \frac{4}{3} \text{ A}$$

From Kirchhoff's current law at node c of the original circuit,

$$i_T = i_1 + i_2$$
$$= \frac{8}{3} + \frac{4}{3}$$
$$= 4 \text{ A}$$

Using R_{EQ} for terminals a-b and Ohm's law, i_T can also be determined as

$$i_T = \frac{60}{15}$$
$$= 4 \text{ A}$$

EXAMPLE 4.9 Determine v_{cd} for the circuit shown in Figure 4.25. ■

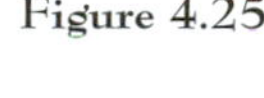

Figure 4.25

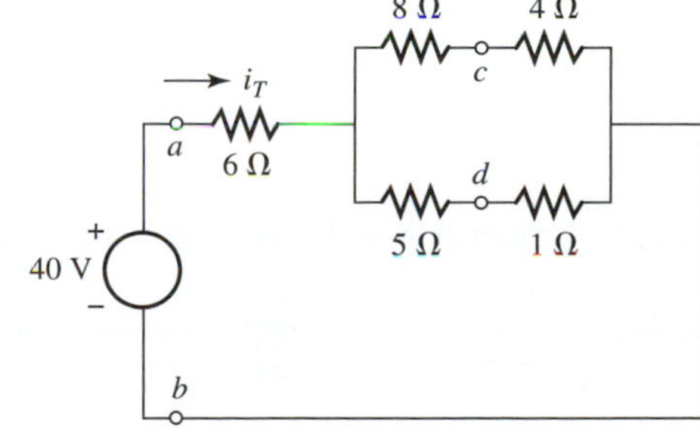

SOLUTION The single equivalent resistance for the circuit to the right of terminals a-b is

$$
\begin{aligned}
R_{\text{EQ}} &= \{(8+4)\|(5+1)\} + 6 \\
&= (12\|6) + 6 \\
&= \left\{\frac{(12)(6)}{12+6}\right\} + 6 \\
&= 10\,\Omega
\end{aligned}
$$

The circuit can be redrawn as shown in Figure 4.26 to determine i_T.

Figure 4.26

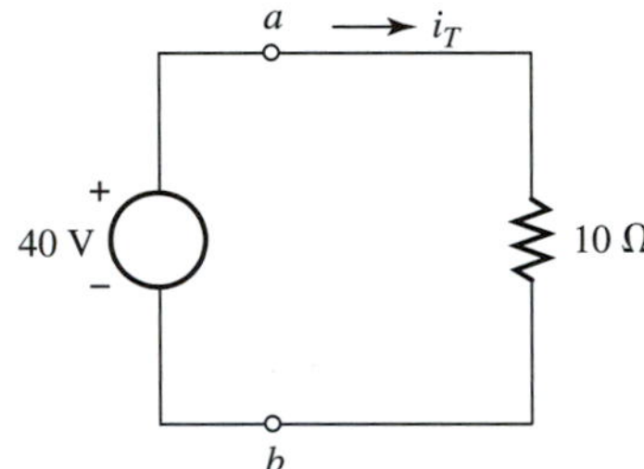

From Ohm's law

$$
\begin{aligned}
i_T &= \frac{40}{10} \\
&= 4\text{ A}
\end{aligned}
$$

Consider the portion of the original circuit shown in Figure 4.27.

Figure 4.27

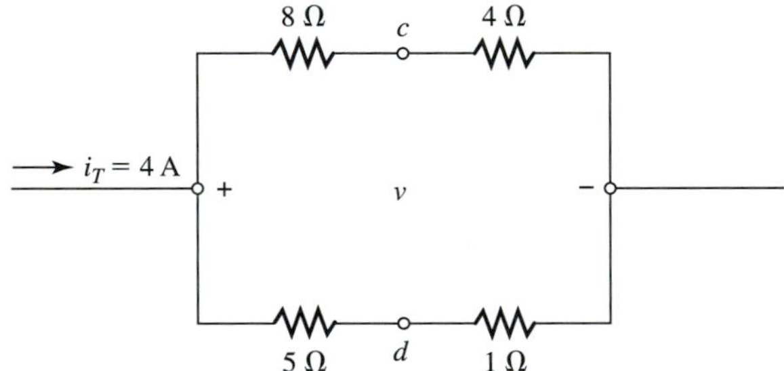

Combining the resistors shown in series and in parallel results in

$$
\begin{aligned}
8 + 4 &= 12\,\Omega \\
5 + 1 &= 6\,\Omega \\
(12)\|(6) &= \frac{6(12)}{6+12} = 4\,\Omega
\end{aligned}
$$

The partial circuit can be redrawn as shown in Figure 4.28.

Figure 4.28

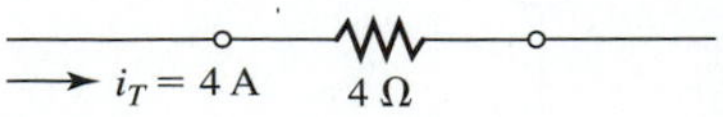

$i_T = 4$ A   4 Ω

From Ohm's law

$$v = 4i_T = 4(4)$$
$$= 16 \text{ V}$$

The original circuit is redrawn as shown in Figure 4.29, with voltage v indicated as 16 V.

Figure 4.29

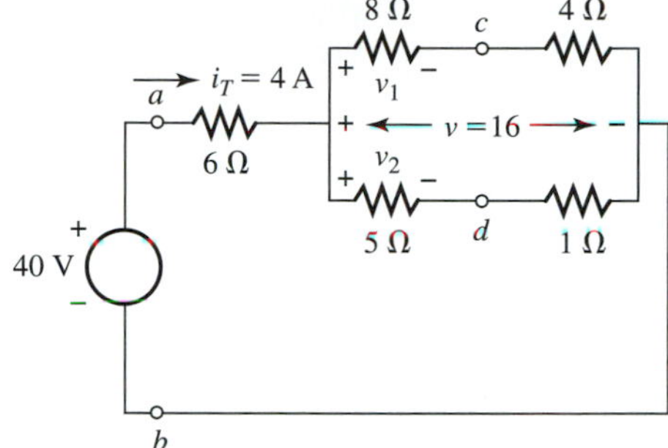

Applying voltage division Equation (4.3) to the partial circuit shown in Figure 4.30 results in

$$v_1 = \left(\frac{8}{8+4}\right)v$$
$$= \left(\frac{8}{12}\right)(16)$$
$$= \frac{32}{3} \text{ V}$$

Figure 4.30

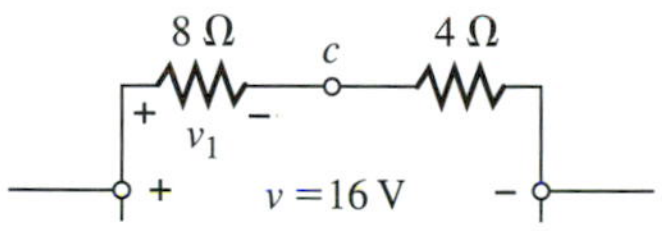

Applying voltage division Equation (4.3) to the partial circuit shown in Figure 4.31 results in

Figure 4.31

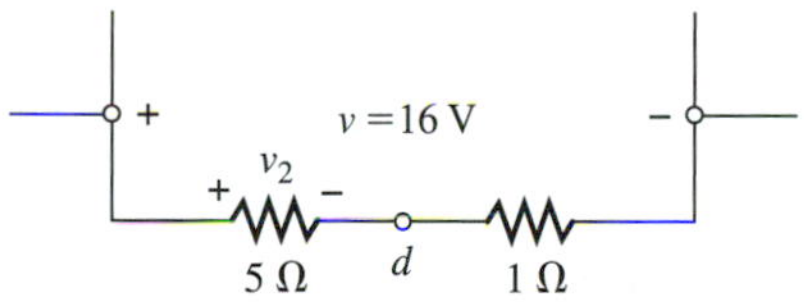

$$v_2 = \left(\frac{5}{5+1}\right)v$$
$$= \left(\frac{5}{6}\right)(16)$$
$$= \frac{40}{3} \text{ V}$$

Applying Kirchhoff's voltage law to the closed loop consisting of v_2, v_1, and v_{cd} in the partial circuit shown in Figure 4.32 results in

$$v_2 - v_1 - v_{cd} = 0$$

Figure 4.32

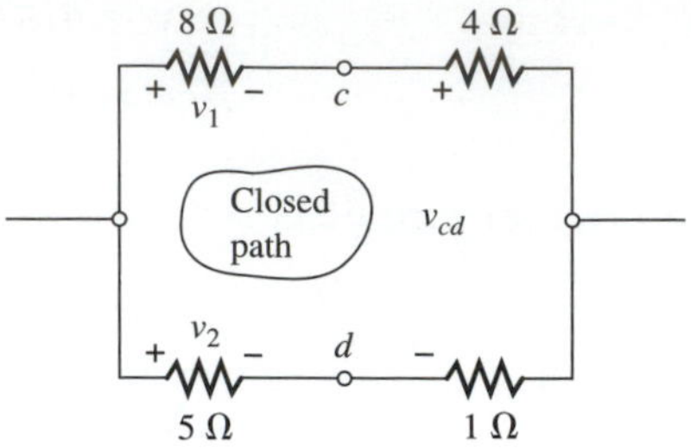

and

$$v_{cd} = v_2 - v_1 = \frac{40}{3} - \frac{32}{3} = \frac{8}{3}\text{ V}$$

EXAMPLE 4.10 Determine the current i for the circuit shown in Figure 4.33. ■

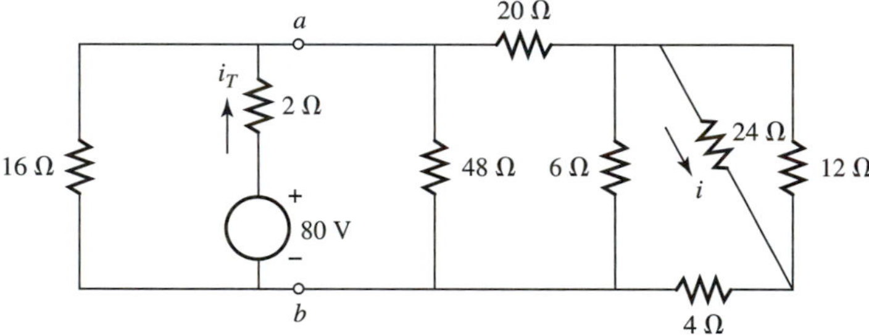

Figure 4.33

SOLUTION An equivalent resistance R_1 for the circuit to the right of terminals a-b is determined by applying the rules for combining resistances in series and parallel:

$$R_1 = \{[\{(12 \| 24) + 4\} \| (6)] + 20\} \| \{48\} = 16\ \Omega$$

The circuit can be redrawn as shown in Figure 4.34.

Figure 4.34

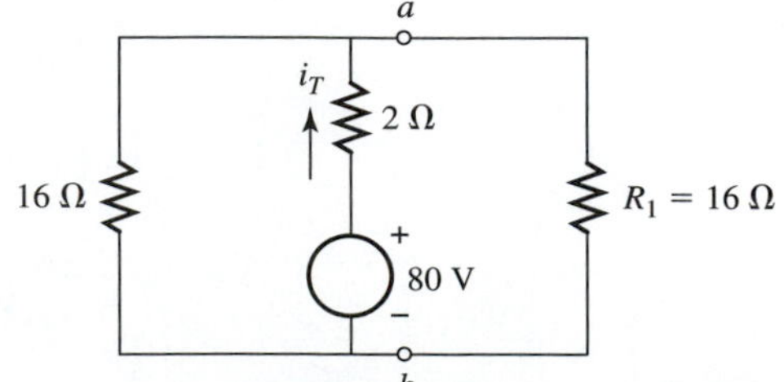

Because the total resistance seen by the source is

$$\{16 \| 16\} + 2 = 10\ \Omega$$

the circuit can be further reduced, as shown in Figure 4.35.

Figure 4.35

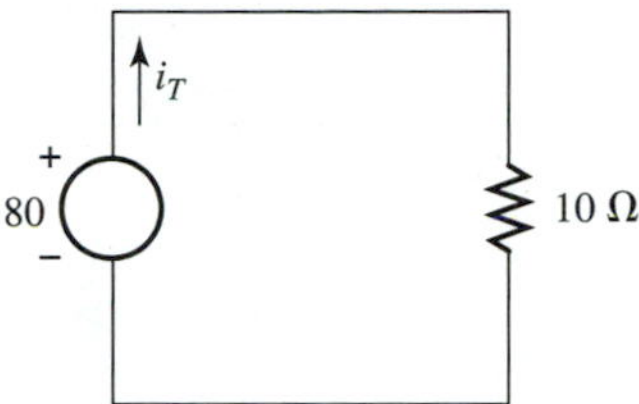

From Ohm's law

$$i_T = \frac{80}{10}$$
$$= 8\text{ A}$$

Applying the current-division Equation (4.1) to the circuit, as shown in Figure 4.36, results in

Figure 4.36

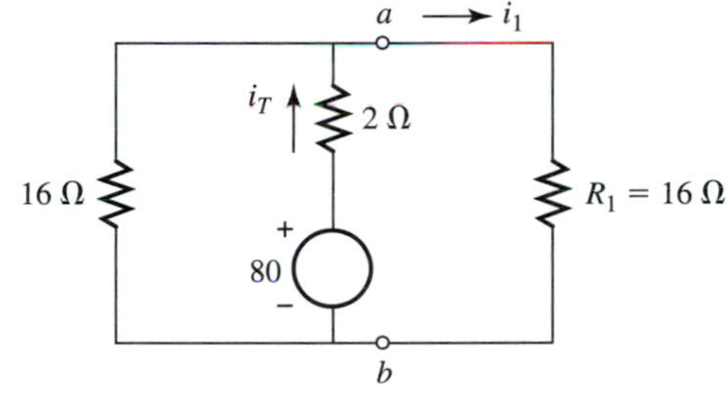

$$R_{EQ} = (16)\|(16) = 8\ \Omega$$
$$i_1 = \left(\frac{R_{EQ}}{R_1}\right)(i_T)$$
$$= \left(\frac{8}{16}\right)(8)$$
$$= 4\text{ A}$$

We now apply the current-division Equation (4.1) for the division of current i_1 in the partial circuit shown in Figure 4.37.

Figure 4.37

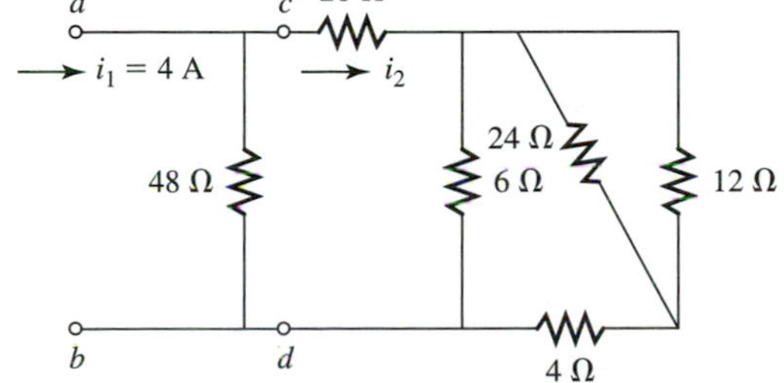

R_2 is the equivalent resistance to the right of terminals *c-d* and is

$$R_2 = [\{\{(12)\|(24)\} + 4\}\|(6)] + 20$$
$$= 24\ \Omega$$

The preceding circuit can be redrawn as shown in Figure 4.38.

Figure 4.38

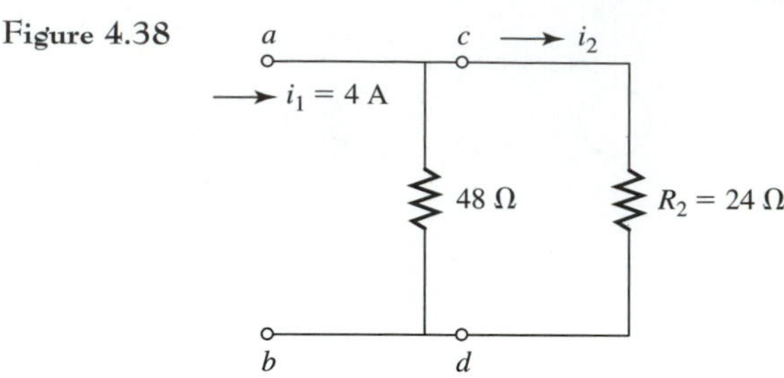

Applying Equation (4.1) to find i_2 results in

$$R_{EQ} = (24)\|(48) = 16\,\Omega$$
$$i_2 = \left(\frac{R_{EQ}}{R_2}\right)(i_2)$$
$$= \left(\frac{16}{24}\right)4$$
$$= \frac{8}{3}\text{ A}$$

The partial circuit in Figure 4.39 can be used to determine the division of current i_2.

Figure 4.39

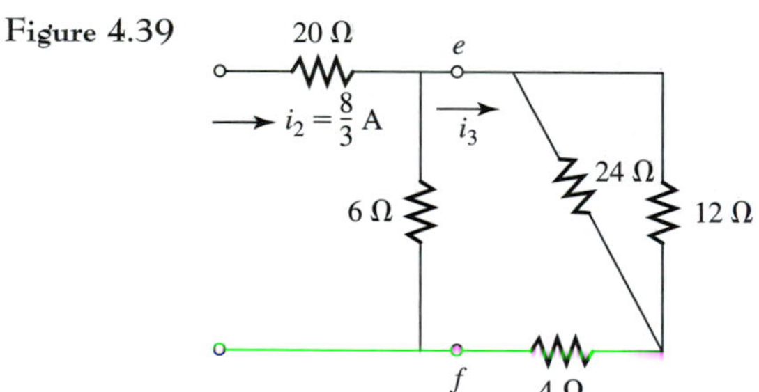

R_3 is the equivalent resistance for the circuit to the right of terminals *e-f*.

$$R_3 = \{24\|12\} + 4$$
$$= 12\,\Omega$$
$$R_{EQ} = (6)\|(12) = 4\,\Omega$$
$$i_3 = \left(\frac{R_{EQ}}{R_3}\right)(i_2)$$
$$= \left(\frac{4}{12}\right)\left(\frac{8}{3}\right)$$
$$= \frac{8}{9}\text{ A}$$

Applying Equation (4.1) for the division of current i_3, as indicated in the partial circuit shown in Figure 4.40, results in

Figure 4.40

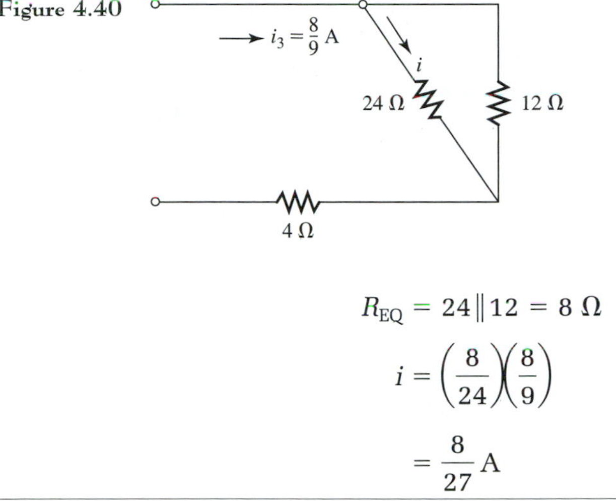

$$R_{EQ} = 24\|12 = 8\ \Omega$$

$$i = \left(\frac{8}{24}\right)\left(\frac{8}{9}\right)$$

$$= \frac{8}{27}\text{ A}$$

4.6 CONVERSION OF IMPERFECT SOURCES

Perfect (ideal) and imperfect (physically realizable) voltage and current sources, as previously presented, are redrawn in Figure 4.41.

Figure 4.41 Current and voltage sources.

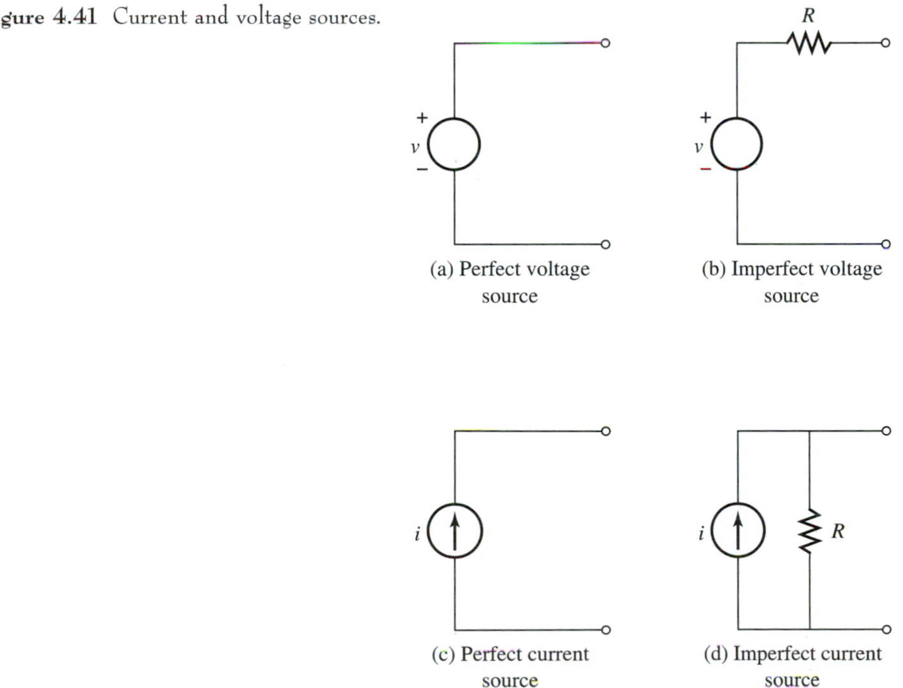

The conversion of a current source to a voltage source or vice versa is useful in circuit analysis. Source conversion can greatly simplify an otherwise complicated circuit for analysis purposes. It can also alter the circuit in a manner that allows us to employ circuit analysis procedures that may not apply to the circuit prior to the source conversion.

Perfect voltage and current sources are primal electrical elements and cannot be converted. Imperfect voltage and current sources can be converted, as illustrated in Figure 4.42. The conversion of sources is always with respect

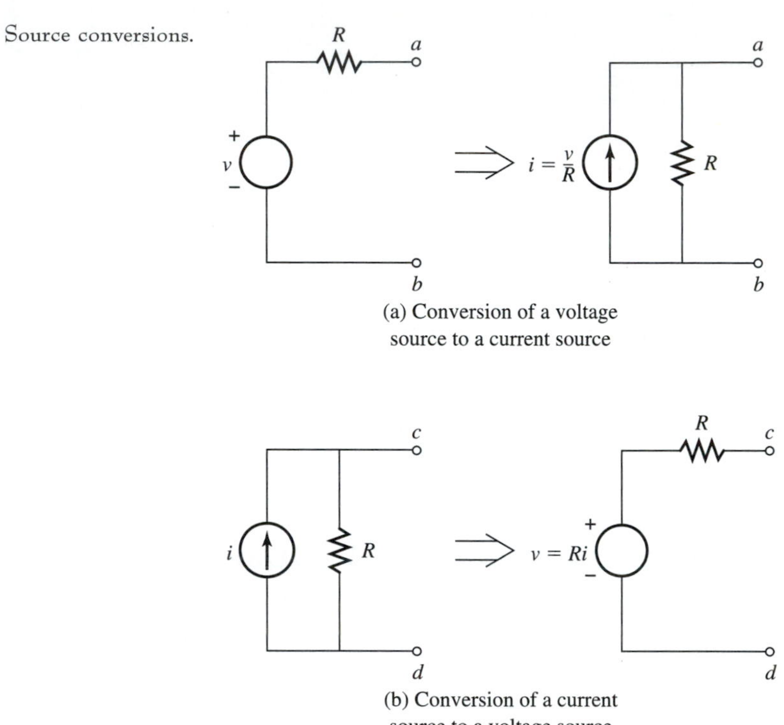

Figure 4.42 Source conversions.

to a pair of terminals. In Figure 4.42(a), the current source equivalent is with respect to terminals *a-b*. In Figure 4.42(b), the voltage source equivalent is with respect to terminals *c-d*. Justification for the source conversion rules indicated in Figure 4.42 is delayed until a later chapter, when the theorems of Norton and Thevenin are presented. These theorems allow us easily to justify the source-conversion rules.

Consider the following examples.

EXAMPLE 4.11 Convert the imperfect voltage source in Figure 4.43 to a current source. ■

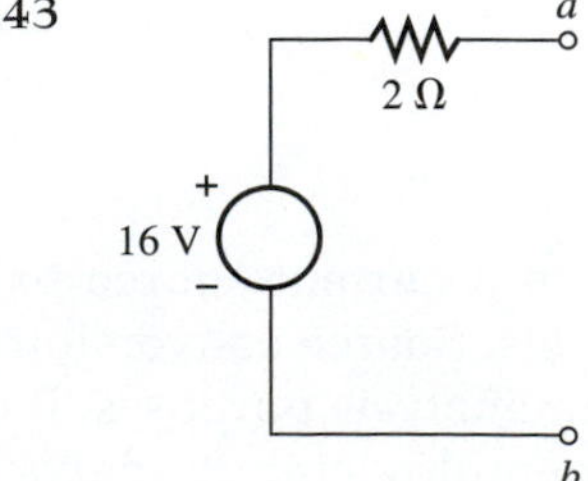

Figure 4.43

SOLUTION From the conversion rule given in Figure 4.42(a), the equivalent current source is as shown in Figure 4.44.

Figure 4.44

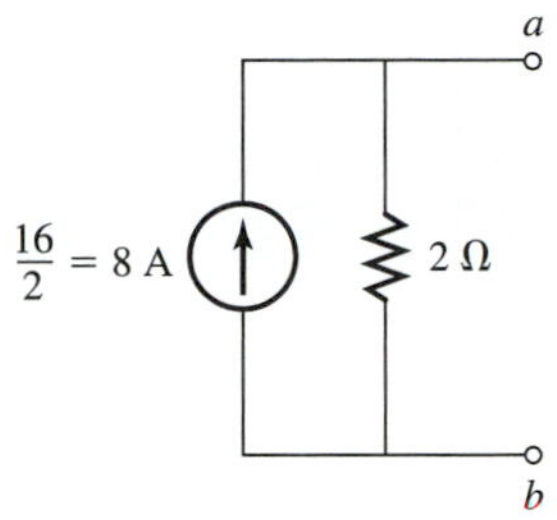

EXAMPLE 4.12 Convert the imperfect current source in Figure 4.45 to a voltage source. ■

Figure 4.45

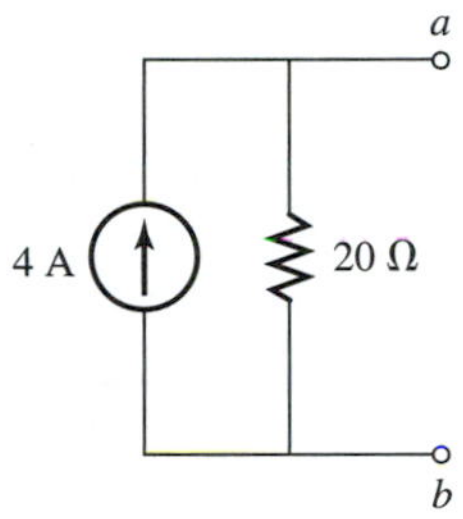

SOLUTION From the conversion rule given in Figure 4.42(b), the equivalent voltage source is as shown in Figure 4.46.

Figure 4.46

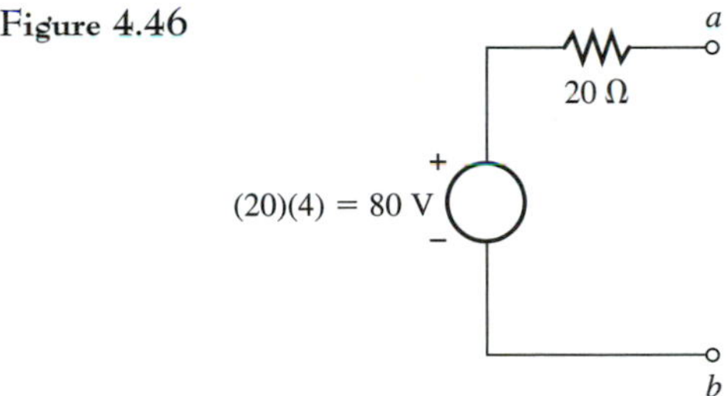

EXAMPLE 4.13 Convert the imperfect voltage source in Figure 4.47 to a current source. ■

Figure 4.47

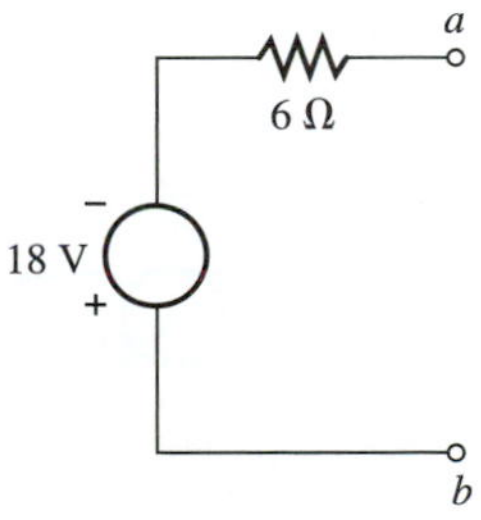

SOLUTION From the conversion rule given in Figure 4.42(a) the equivalent current source is as shown in Figure 4.48.

Figure 4.48

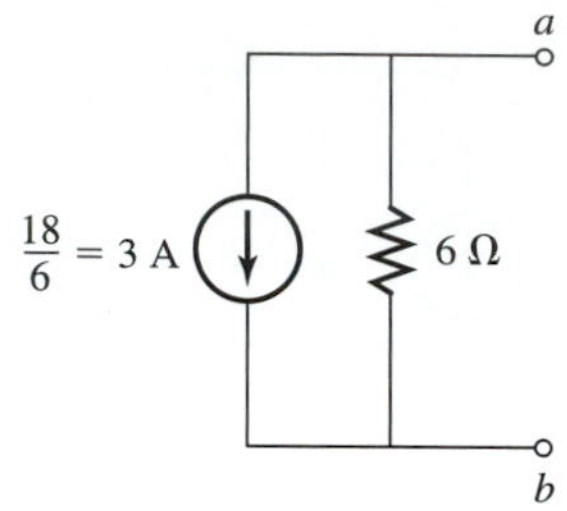

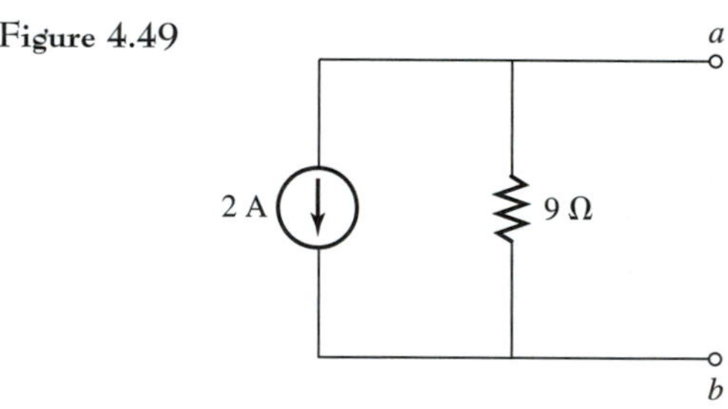

EXAMPLE 4.14 Convert the imperfect current source in Figure 4.49 to a voltage source. ■

Figure 4.49

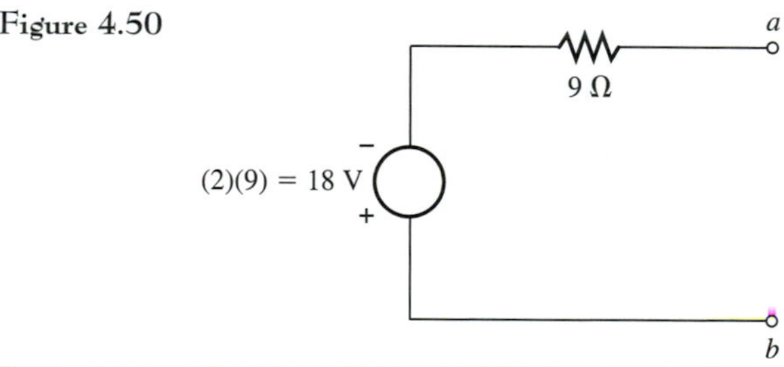

SOLUTION From the conversion rule given in Figure 4.42(b), the equivalent voltage source is as shown in Figure 4.50.

Figure 4.50

4.7 PERFECT SOURCES IN SERIES

Perfect voltage sources in series can be replaced by a single, equivalent voltage source. Consider the circuits in Figure 4.51. Applying Kirchhoff's voltage

Figure 4.51 Series voltage sources.

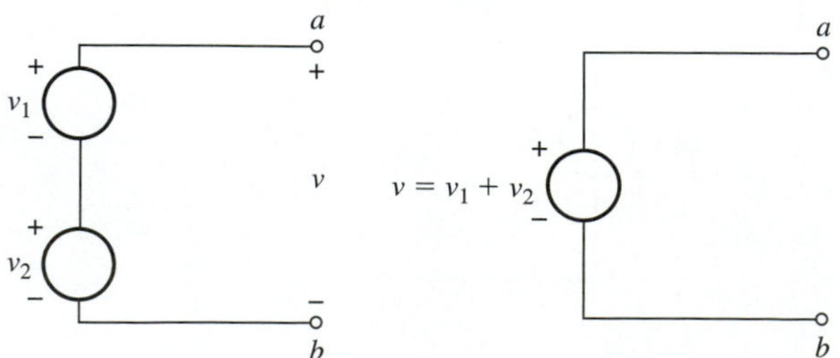

law to the circuit in Figure 4.51(a) while traveling around the loop in a clockwise direction results in

$$v_2 + v_1 - v = 0$$
$$v = v_1 + v_2$$

With respect to terminals *a*-*b*, voltage sources v_1 and v_2 can be replaced by a single voltage source equal to *v*, as shown in Figure 4.51(b).

Consider the following example.

EXAMPLE 4.15 Replace the four voltage sources with a single equivalent voltage source for the circuit shown in Figure 4.52. ■

Figure 4.52

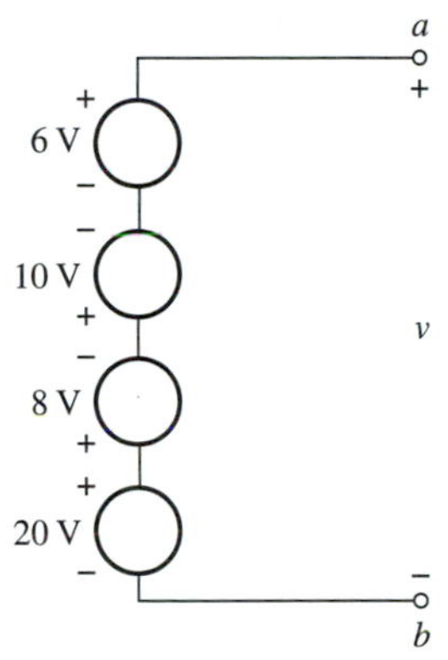

SOLUTION From Kirchhoff's voltage law

$$20 - 8 - 10 + 6 - v = 0$$
$$v = 20 - 8 - 10 + 6$$
$$= 8 \text{ V}$$

The equivalent voltage source with respect to terminals *a*-*b* is shown in Figure 4.53.

Figure 4.53

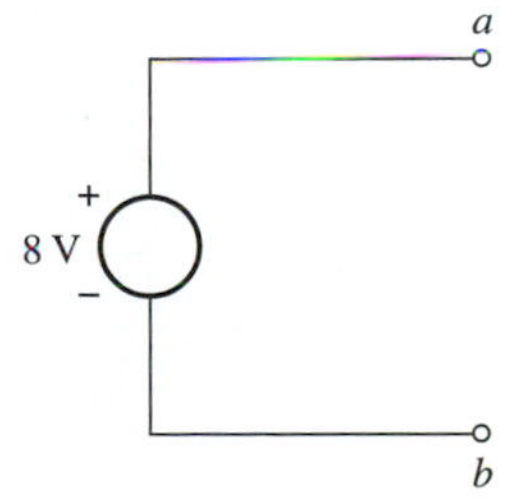

Perfect current sources cannot be placed in series unless they are equal in magnitude and direction. Consider the circuit in Figure 4.54. If $i_1 \neq i_2$, then

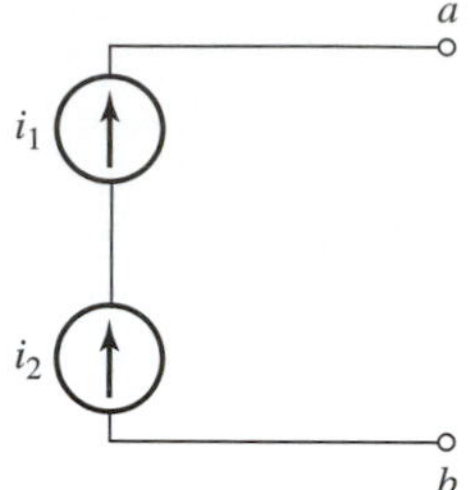

Figure 4.54 Series current sources.

contradictory statements arise. Current source i_1 implies that current i_1 flows in the branch, whereas current source i_2 implies that a different current, i_2, flows in the same branch. The situation is mathematically and physically untenable, and so current sources that are unequal cannot be placed in series.

4.8 PERFECT SOURCES IN PARALLEL

Perfect voltage sources cannot be placed in parallel unless they are equal. Consider the circuit in Figure 4.55. If $v_1 \neq v_2$, then contradictory statements arise. Voltage source v_1 implies that terminal voltage v equals v_1, whereas voltage

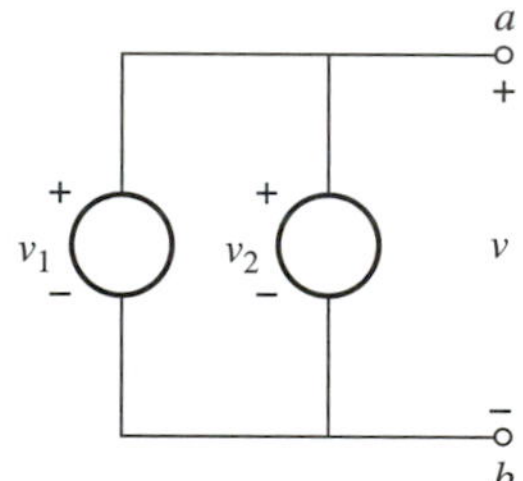

Figure 4.55 Parallel voltage sources.

source v_2 implies that terminal voltage v equals v_2. The situation is mathematically and physically untenable, so perfect voltage sources that are unequal cannot be placed in parallel.

Perfect current sources in parallel can be replaced by a single equivalent current source. Consider the circuits in Figure 4.56. Applying Kirchhoff's cur-

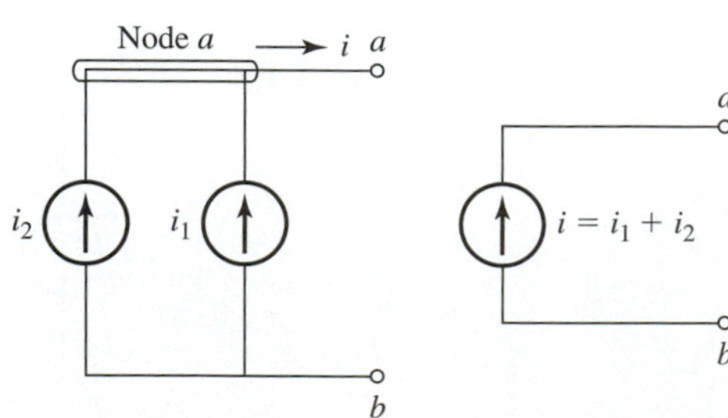

Figure 4.56 Parallel current sources.

rent law to node a of the circuit in Figure 4.56(a) results in

$$i = i_1 + i_2$$

With respect to terminals a-b, current sources i_1 and i_2 can be replaced by a single current source equal to i, as shown in Figure 4.56(b).

Consider the following example.

EXAMPLE 4.16 Replace the three current sources with a single equivalent current source for the circuit shown in Figure 4.57. ■

Figure 4.57

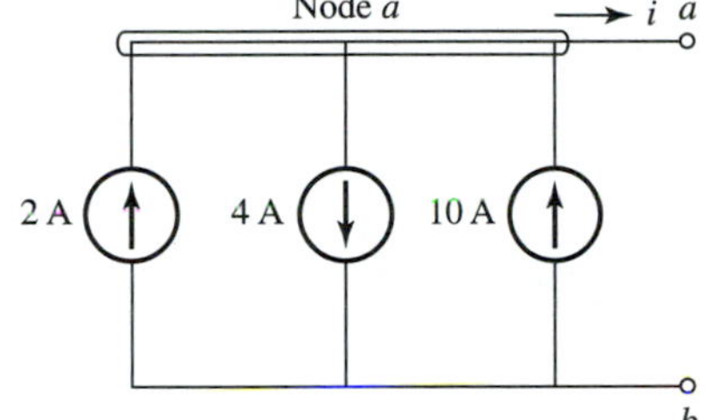

SOLUTION Applying Kirchhoff's current law to node a results in

$$2 - 4 + 10 - i = 0$$

$$i = 2 - 4 + 10$$

$$= 8 \text{ A}$$

The equivalent current source is shown in Figure 4.58.

Figure 4.58

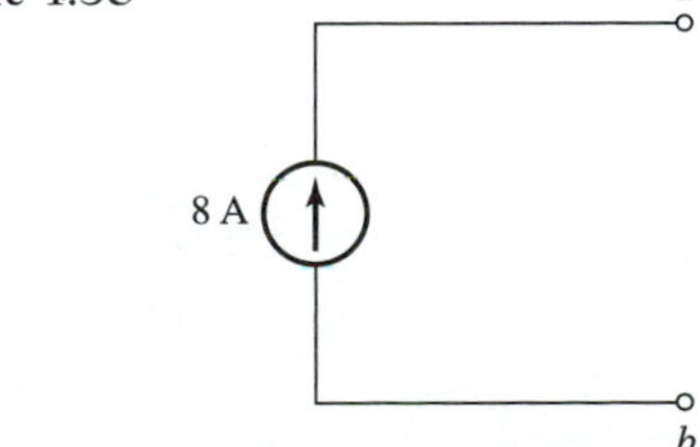

4.9 DEPENDENT SOURCES

The voltage across a dependent voltage source and the current flowing through a dependent current source are controlled by currents or voltages in other parts of the circuit. Consider the examples of dependent sources in Figure 4.59.

Figure 4.59 Dependent sources.

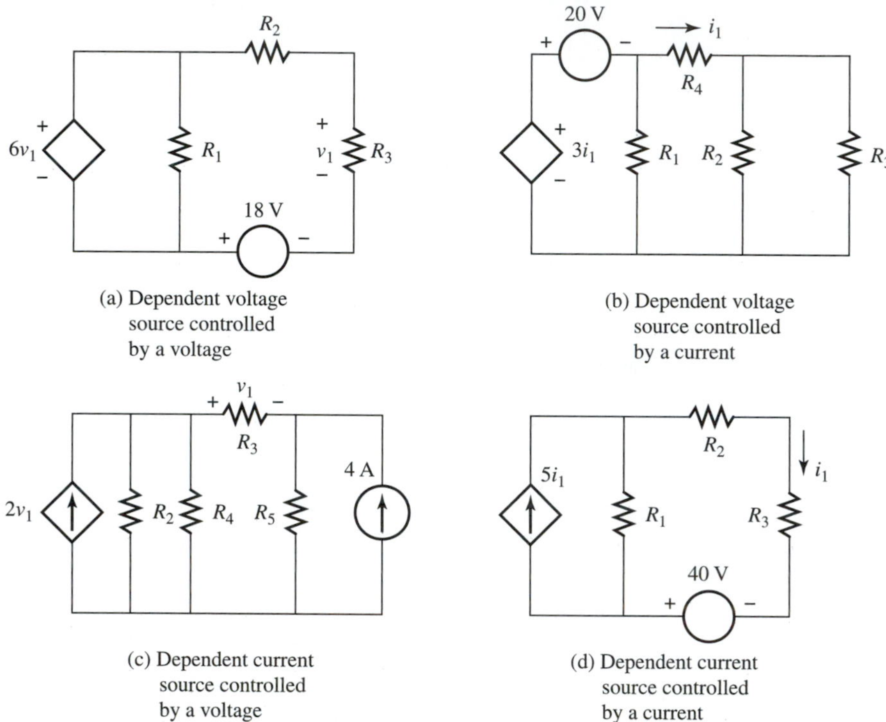

The dependent voltage source in Figure 4.59(a) has a magnitude of $(6v_1)$ volts and is controlled by the voltage across resistance R_3. The dependent voltage source in Figure 4.59(b) has a magnitude of $(3i_1)$ volts and is controlled by the current through resistance R_4.

The dependent current source in Figure 4.59(c) has a magnitude of $(2v_1)$ amperes and is controlled by the voltage across resistance R_3. The dependent current source in Figure 4.59(d) has a magnitude of $(5i_1)$ amperes and is controlled by the current through resistance R_3.

Consider the following examples.

EXAMPLE 4.17 Determine the current i for the circuit shown in Figure 4.60. ■

Figure 4.60

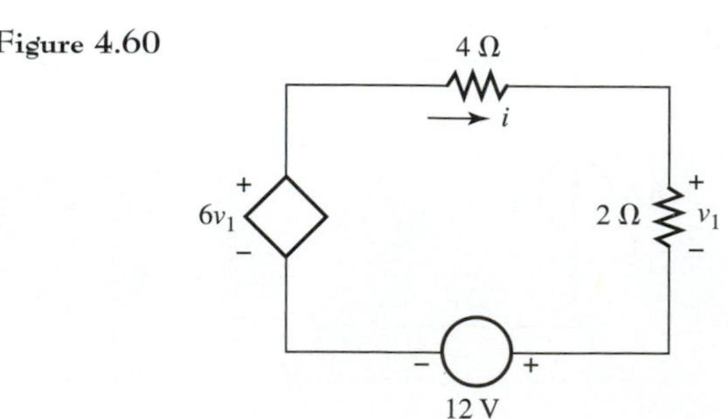

SOLUTION Notice that the dependent voltage source is controlled by the voltage across the 2-Ω resistance.

Assign polarities to the 4-Ω and 2-Ω resistances based on the direction of i, as shown in Figure 4.61. Applying Kirchhoff's voltage law to the closed loop results in

$$6v_1 - 4i - 2i - 12 = 0$$

$$6v_1 - 6i - 12 = 0$$

Figure 4.61

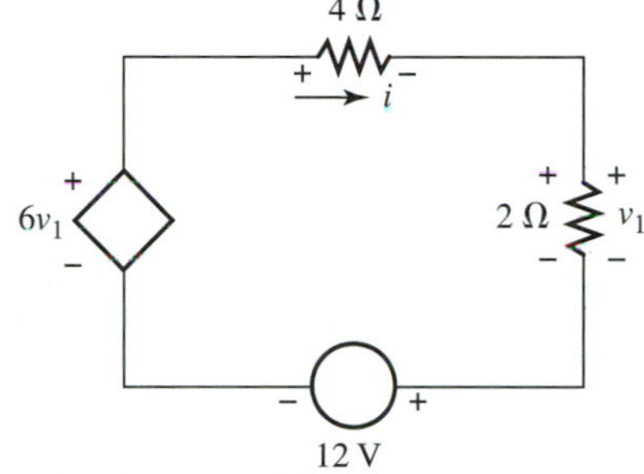

From Ohm's law the controlling voltage v_1 can be expressed as

$$v_1 = 2i$$

Substituting $(2i)$ for v_1 in the preceding equation results in

$$6(2i) - 6i - 12 = 0$$

$$6i = 12$$

$$i = 2\text{ A}$$

EXAMPLE 4.18 Determine the voltage v across the dependent current source for the circuit shown in Figure 4.62. ■

Figure 4.62

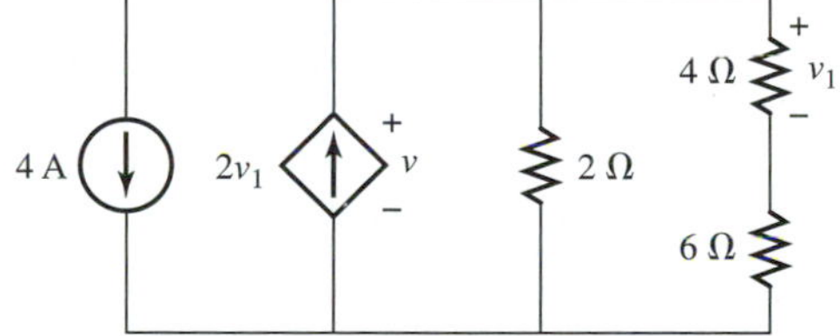

SOLUTION Notice that the dependent current source is controlled by the voltage across the 4-Ω resistance. Assign currents i_1 and i_2 to the branches containing the 2-Ω resistance and the series combination of the 4-Ω and 6-Ω resistances, as shown in Figure 4.63.

Figure 4.63

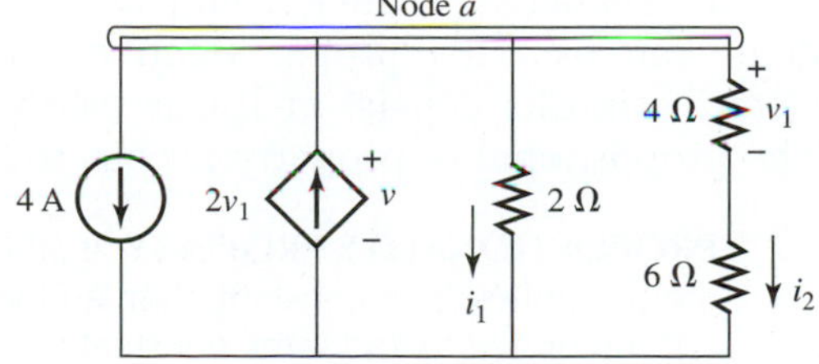

Applying Kirchhoff's current law at node *a* results in

$$-4 + 2v_1 - i_1 - i_2 = 0$$

Recognizing that voltage v appears across the 2-Ω resistance and across the series combination of the 6-Ω and 4-Ω resistances and applying Ohm's law to each branch results in

$$i_1 = \frac{v}{2}$$

and

$$i_2 = \frac{v}{6 + 4}$$
$$= \frac{v}{10}$$

From Ohm's law the controlling voltage, v_1, is

$$v_1 = 4i_2$$

and because $i_2 = v/10$,

$$v_1 = \frac{4v}{10}$$

Substituting $v/2$, $v/10$, and $4v/10$ for i_1, i_2, and v_1, respectively, in the preceding equation, $-4 + 2v_1 - i_1 - i_2 = 0$, results in

$$-4 + 2\left(\frac{4v}{10}\right) - \left(\frac{v}{2}\right) - \left(\frac{v}{10}\right) = 0$$
$$-4 + \left(\frac{8v}{10}\right) - \left(\frac{v}{2}\right) - \left(\frac{v}{10}\right) = 0$$
$$\left(\frac{8v}{10}\right) - \left(\frac{v}{2}\right) - \left(\frac{v}{10}\right) = 0$$
$$\frac{2v}{10} = 4$$
$$2v = 40$$
$$v = 20 \text{ V}$$

4.10 PROPORTIONALITY PROPERTY

In an electric circuit restricted to independent sources, a certain proportionality exists between the currents and voltages in the circuit and the independent sources. This proportionality relationship derives from the fact that electric circuits consist of linear passive elements, as previously discussed. The proportionality property for electric circuits is stated next.

PROPORTIONALITY PROPERTY If all the independent sources in a circuit are multiplied by a constant, then all the currents and voltages in the circuit are multiplied by the same constant.

The following examples illustrate a method for analyzing circuits using the proportionality property.

EXAMPLE 4.19 If the 6-V voltage source is changed to 30 V, determine the resulting current i_1 for the circuit shown in Figure 4.64. ■

Figure 4.64

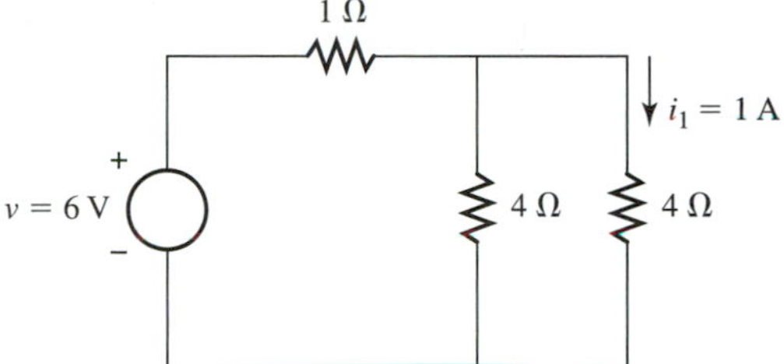

SOLUTION From the proportionality property, if the 6-V voltage source is multiplied by a constant K, then the new voltage source, v', is equal to $6K$. Every voltage and current in the circuit is then multiplied by K, and current i_1 becomes i_1' and is equal to Ki_1, as shown in Figure 4.65.

Figure 4.65

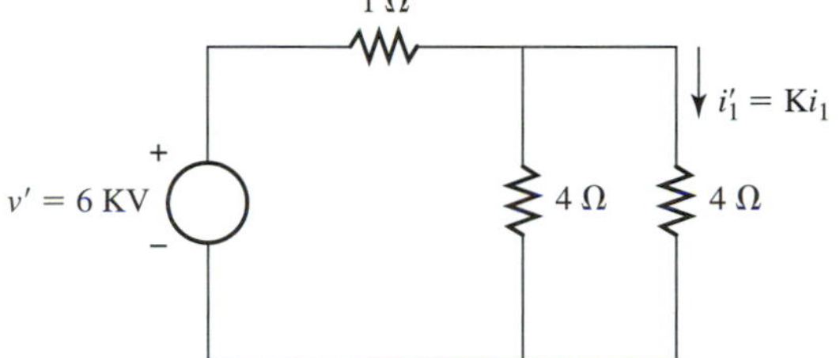

Because the new voltage source v' is equal to

$$v' = 6K = 30$$
$$K = 5$$

Because the new current i_1' equals Ki_1 and i_1 equals 1 A,

$$i_1' = (5)(1)$$
$$= 5\text{ A}$$

EXAMPLE 4.20 It is desired to decrease the current i_1 from 4 A to 2 A for the circuit shown in Figure 4.66. Determine the new values of the voltage and current sources for i_1 to equal 2 A. ■

Figure 4.66

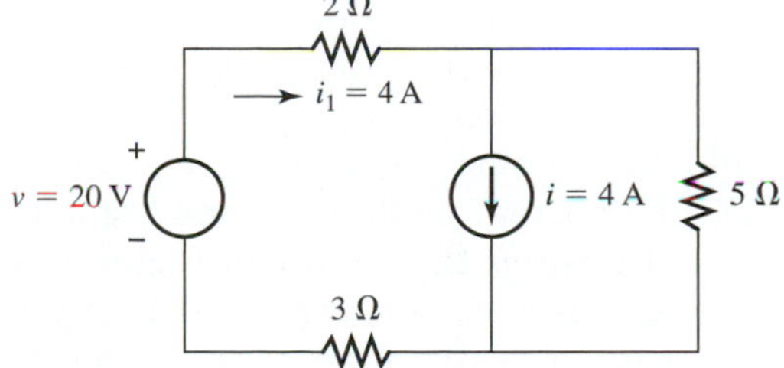

SOLUTION From the proportionality property, if the 20-V voltage source and the 4-A current source are multiplied by a constant K, then the new current i_1' is equal to Ki_1, as shown in Figure 4.67.

Figure 4.67

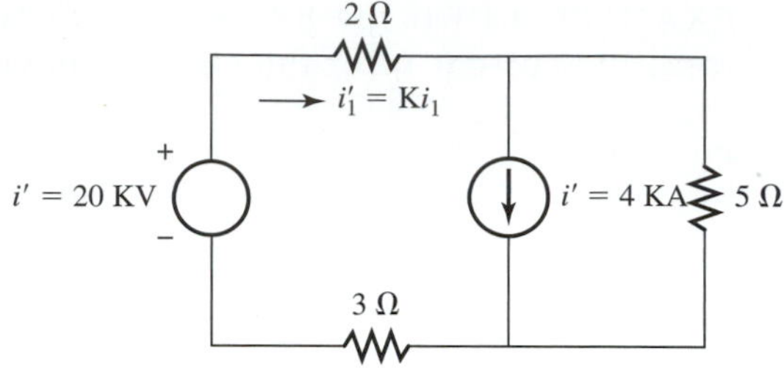

Because $i_1 = 4$

$$i_1' = Ki_1$$
$$= 4K$$

It is desired that $i_1' = 2$ A, so

$$2 = 4K$$
$$K = \frac{1}{2}$$

Because the new voltage is $v' = 20K$,

$$v' = 10 \text{ V}$$

The new current is $i' = 4K$, so

$$i' = 2 \text{ A}$$

The new circuit with $i_1 = 2$ A is shown in Figure 4.68.

Figure 4.68

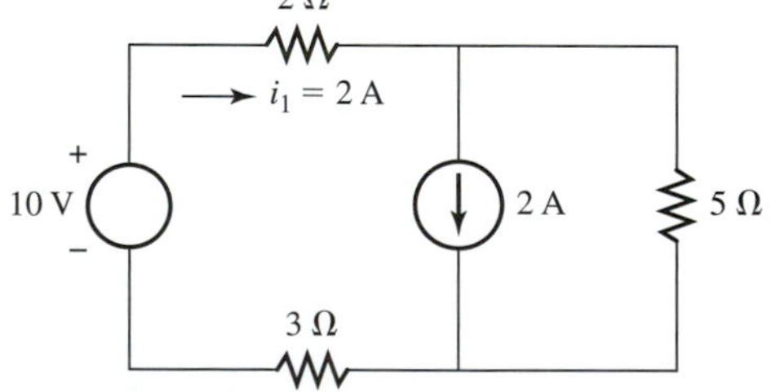

4.11 BRANCH CURRENT METHOD

The branch current method allows us to analyze circuits with multiple sources. The method involves the application of Kirchhoff's current and voltage laws to the junctions and loops in the circuit. The equations resulting from the application of Kirchhoff's laws must result in a solvable set of equations for a unique solution. That is, the equations must be independent (not redundant) and there must be as many equations as unknowns. The unknowns in the branch current method are the branch currents.

Consider the circuit in Figure 4.69. A loop is defined in Chapter 3 as a closed path not passing through any node more than once. If the loop does not geometrically encircle any circuit elements, it is said to encircle a window. As illustrated in Figure 4.69, loop 1, consisting of elements v_1, R_1, and

R_3, encircles window 1; loop 2, consisting of elements R_2, R_3, and v_2, encircles window 2; and loop 3, consisting of elements v_1, R_1, R_2, and v_2, encircles R_3 and therefore does not encircle a window.

Kirchhoff's voltage law is applied around each window in the circuit.

A junction is defined as a point in a circuit where three or more branches are connected. The circuit in Figure 4.69 has two junctions—*a* and *b*.

Kirchhoff's current law is applied to (J – 1) junctions, where J is the number of junctions in the circuit.

Because there are two windows and two junctions in the circuit in Figure 4.69, Kirchhoff's voltage law is applied twice and Kirchhoff's current law is applied once.

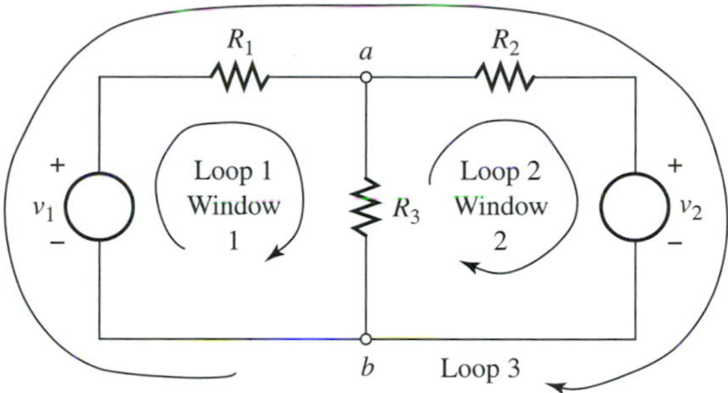

Figure 4.69 Windows and loops.

Consider the circuit in Figure 4.69 as redrawn in Figure 4.70. The *first step* in the branch current method is to assign a current to each branch, as indicated in Figure 4.70.

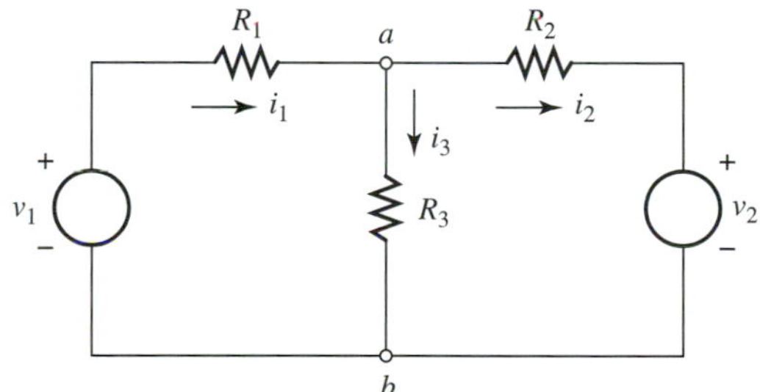

Figure 4.70 Branch current method.

The *second step* in the branch current method is to assign voltage polarity signs to the resistances based on the assumed branch currents, as shown in Figure 4.71. The *third step* in the branch current method is to ap-

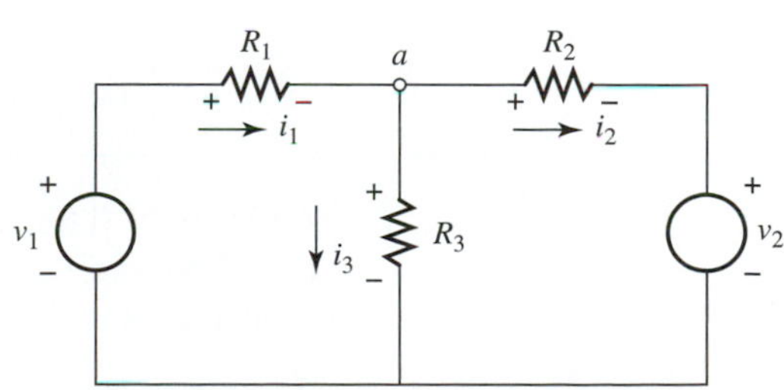

Figure 4.71 Voltage polarities indicated.

ply Kirchhoff's voltage law to all the windows of the circuit and Kirchhoff's current law to $(J - 1)$ junctions.

Applying Kirchhoff's voltage law to window 1 results in

$$v_1 - R_1 i_1 - R_3 i_3 = 0$$

Applying Kirchhoff's voltage law to window 2 results in

$$R_3 i_3 - R_2 i_2 - v_2 = 0$$

Applying Kirchhoff's current law to junction a results in the third equation, which is

$$i_1 - i_2 - i_3 = 0$$

These three equations form the mathematical model for the circuit in Figure 4.70.

The *fourth step* in the branch current method is to solve the mathematical model for the branch currents. Once the branch currents are determined, Ohm's law can be employed to determine the voltage across every element in the circuit.

The four-step procedure for the branch current method is summarized next:

1. Assign a current to each branch in the circuit.
2. Assign voltage polarities to the resistances based on the assigned currents.
3. Apply Kirchhoff's voltage law to all the windows of the circuit and Kirchhoff's current law to $(J - 1)$ junctions.
4. Solve the mathematical model for the branch currents.

Consider the following examples.

EXAMPLE 4.23 Determine the current flowing through each branch of the circuit shown in Figure 4.72. ■

Figure 4.72

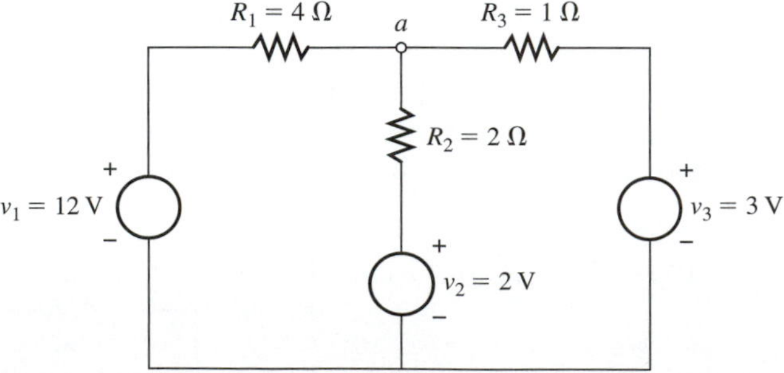

SOLUTION The circuit is redrawn in Figure 4.73 with a branch current assigned to each of the three branches and an associated voltage polarity indicated for each resistance based on the direction of the assigned current.

Figure 4.73

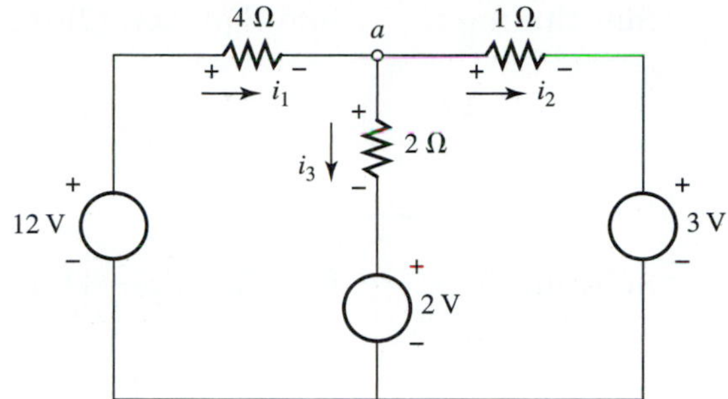

Applying Kirchhoff's voltage law to the loop containing v_1, v_2, R_1, and R_2 results in

$$12 - 4i_1 - 2i_3 - 2 = 0$$

(1)
$$4i_1 + 2i_3 = 10$$

Applying Kirchhoff's voltage law to the loop containing v_2, R_2, R_3, and v_3 results in

$$2 + 2i_3 - 1i_2 - 3 = 0$$

(2)
$$i_2 - 2i_3 = -1$$

Applying Kirchhoff's current law to node a results in

$$i_1 = i_2 + i_3$$

(3)
$$i_1 - i_2 - i_3 = 0$$

We use the method of addition and subtraction to solve Equations (1), (2), and (3). Multiplying Equation (3) by (−4) and adding the result to Equation (1),

$$\begin{aligned} \text{EQ(1)} \qquad & 4i_1 + 2i_3 = 10 \\ (-4) \times \text{EQ(3)} \qquad & -4i_1 + 4i_2 + 4i_3 = 0 \end{aligned}$$

results in

$$4i_2 + 6i_3 = 10$$

$$i_2 + \left(\frac{3}{2}\right)i_3 = \frac{5}{2}$$

Multiplying Equation (2) by (−1) and adding the result to the preceding equation results in

$$\begin{aligned} & i_2 + \left(\frac{3}{2}\right)i_3 = \frac{5}{2} \\ (-1) \times \text{EQ(2)} \qquad & -i_2 + 2i_3 = +1 \\ & \frac{7}{2}i_3 = \frac{7}{2} \\ & i_3 = 1\text{ A} \end{aligned}$$

Substituting $i_3 = 1$ into Equation (2) results in

$$i_2 = 2(1) - 1$$
$$= 1\text{ A}$$

Substituting $i_3 = 1$ into Equation (1) results in

$$4i_1 = 10 - 2(1)$$
$$= 8$$
$$i_1 = 2\text{ A}$$

EXAMPLE 4.24 Determine the current flowing through each branch for the circuit shown in Figure 4.74. ■

Figure 4.74

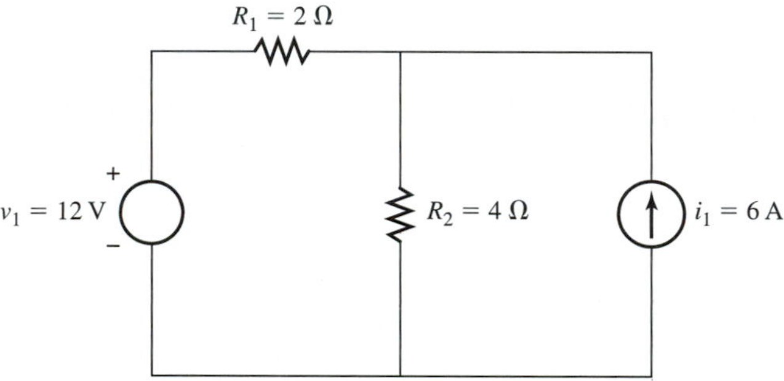

SOLUTION The circuit is redrawn in Figure 4.75 with branch currents assigned to the three branches and the associated voltage polarity indicated for each resistance.

Figure 4.75

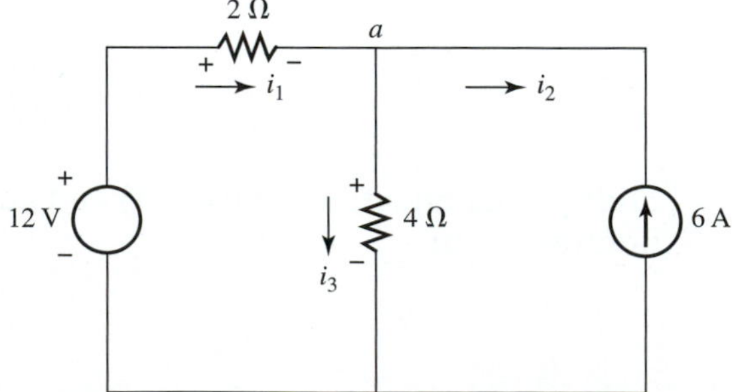

From an inspection of the circuit,

$$i_2 = -6$$

Applying Kirchhoff's voltage law to the loop containing v_1, R_1, and R_2 results in

$$12 - 2i_1 - 4i_3 = 0$$

(1) $$i_1 + 2i_3 = 6$$

Applying Kirchhoff's current law to node a results in

$$i_1 = i_2 + i_3$$

(2) $$i_1 - i_3 = i_2$$

Substituting $i_2 = -6$ into Equation (2) results in

$$i_1 - i_3 = -6$$

Combining this equation ($i_1 - i_3 = -6$) with Equation (1) results in

$$3i_3 = 12$$

$$i_3 = 4 \text{ A}$$

Substituting $i_3 = 4$ into Equation (1) results in

$$i_1 = 6 - 2(4)$$

$$= -2 \text{ A}$$

The minus sign indicates that the current through the 2-Ω resistance flows opposite to the direction of the arrow assigned.

The branch current method serves as an introduction to the solution of multiloop circuits. The methods presented in Chapter 6, along with the methods for solving simultaneous equations presented in Chapter 5, greatly improve and enhance the procedures for solving multiloop circuits.

4.12 DELTA AND WYE CIRCUITS

Resistance circuits are often connected in such a manner that the resistances are neither in series nor in parallel and cannot be combined using the methods previously presented. Such circuits can usually be simplified by employing the delta and wye conversion formulas presented here.

The delta and wye circuits, also denoted by Δ (Greek letter delta) and Y, are indicated in Figure 4.76. The Δ and Y circuits can be redrawn to resem-

Figure 4.76 Delta and wye circuits.

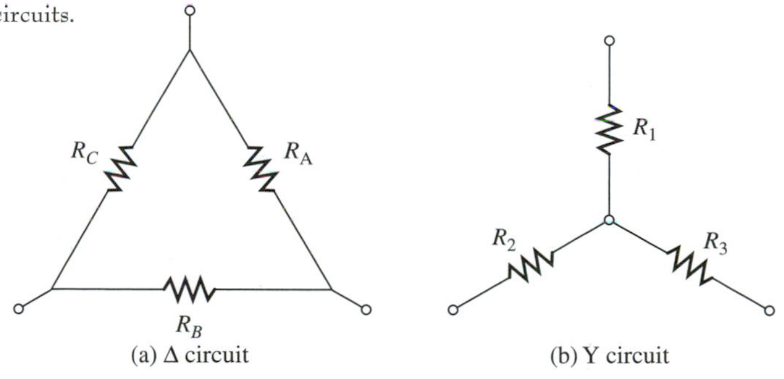

R_C R_A R_B

(a) Δ circuit

R_1 R_2 R_3

(b) Y circuit

ble the Greek letter pi (π) and T, respectively, as shown in Figure 4.77. The π and T circuits are simply redrawn versions of the Δ and Y circuits and are unchanged electrically. The Δ and Y notation is used in electric machinery studies, and the π and T notation is usually used in electronics and communications work.

Figure 4.77 π and T circuits.

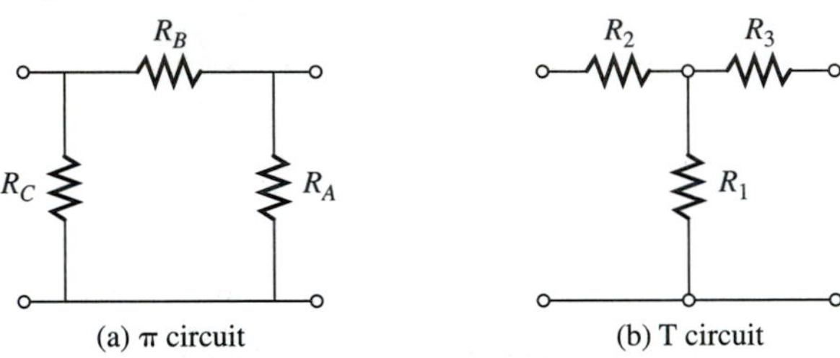

A Δ (delta) circuit can be converted to a Y circuit and a Y circuit can be converted to a Δ circuit using the conversion formulas given in Figure 4.78.

Figure 4.78 Delta-wye transformation rules.

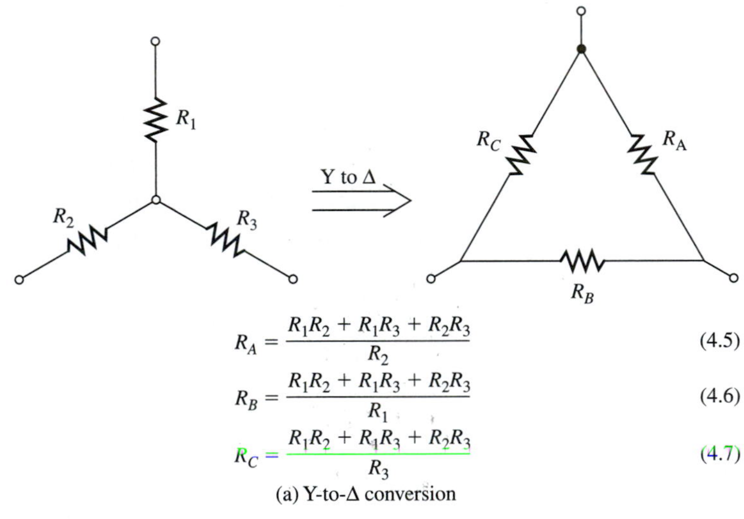

$$R_A = \frac{R_1R_2 + R_1R_3 + R_2R_3}{R_2} \quad (4.5)$$

$$R_B = \frac{R_1R_2 + R_1R_3 + R_2R_3}{R_1} \quad (4.6)$$

$$R_C = \frac{R_1R_2 + R_1R_3 + R_2R_3}{R_3} \quad (4.7)$$

(a) Y-to-Δ conversion

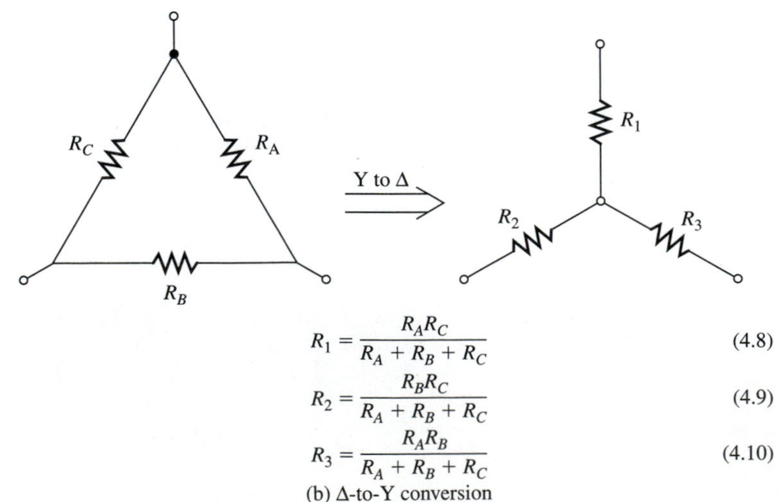

$$R_1 = \frac{R_AR_C}{R_A + R_B + R_C} \quad (4.8)$$

$$R_2 = \frac{R_BR_C}{R_A + R_B + R_C} \quad (4.9)$$

$$R_3 = \frac{R_AR_B}{R_A + R_B + R_C} \quad (4.10)$$

(b) Δ-to-Y conversion

Notice in Figure 4.78(a) that the equivalent Δ resistances are each equal to the sum of the three possible product combinations of the three Y resistances divided by the Y resistance farthest from the Δ resistance to be determined.

Notice also in Figure 4.78(b) that the equivalent Y resistances are each equal to the product of the two adjacent Δ resistances divided by the sum of the Δ resistances.

The words *farthest* and *adjacent* in the preceding rules are easily understood if the Y circuit is placed inside the Δ circuit, as shown in Figure 4.79. For example, it is obvious that R_A and R_C are adjacent to R_1, and R_2 is farthest from R_A.

Figure 4.79 Delta-wye circuits.

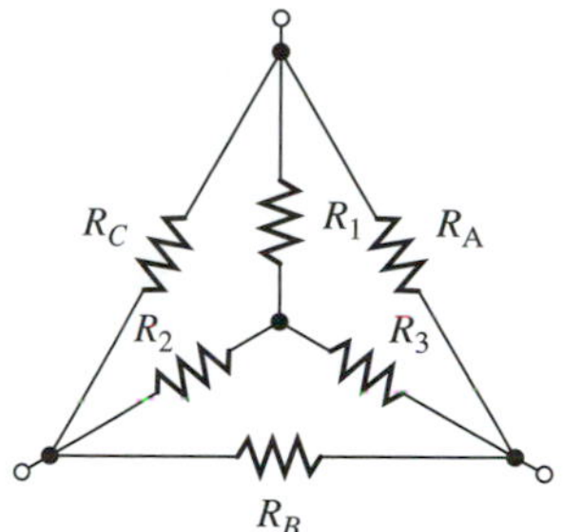

Consider the circuit in Figure 4.80(a). None of the resistances are in series or parallel and the circuit cannot be reduced to a single resistance by the methods previously presented. However, the Δ-Y conversion formulas presented in Figure 4.78 can be employed to reduce the circuit to a single equivalent resistance.

The upper Δ is removed and replaced with an equivalent Y, as indicated in Figure 4.80(b) and (c). The values for the equivalent Y resistances are obtained from Equations (4.8), (4.9), and (4.10) as follows:

$$R_3 = \frac{R_x R_y}{R_x + R_y + R_z}$$

$$R_4 = \frac{R_y R_z}{R_x + R_y + R_z}$$

$$R_5 = \frac{R_x R_z}{R_x + R_y + R_z}$$

The single equivalent resistance for the circuit in Figure 4.80(a) can now be determined by combining series and parallel resistances for the circuit in Figure 4.80(c). The single equivalent resistance is

$$R_{EQ} = \{(R_5 + R_v) \| (R_4 + R_u)\} + R_3$$

as shown in Figure 4.80(d).

Figure 4.80 Reducing a circuit containing a Δ connection.

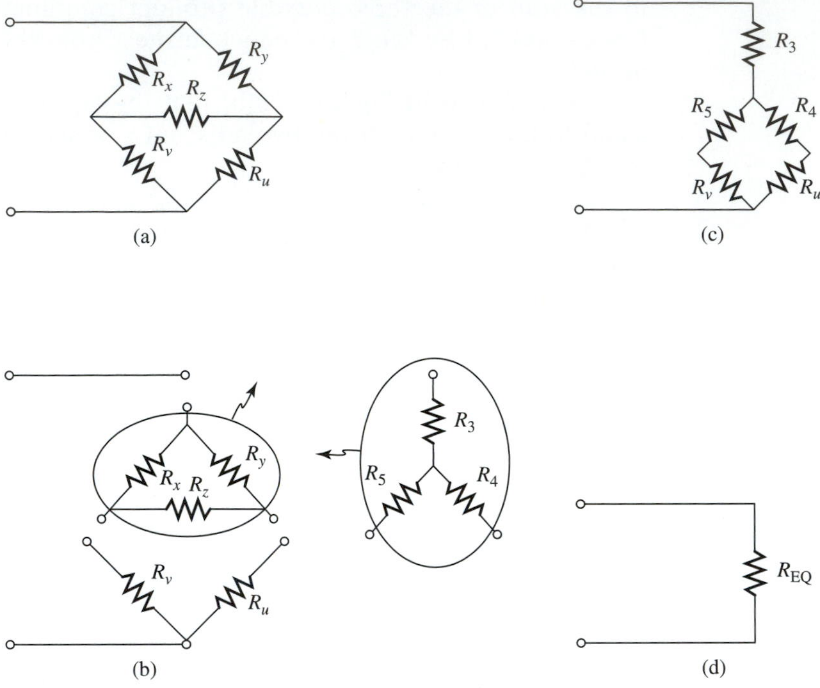

Consider the following examples.

EXAMPLE 4.25 Determine a single equivalent resistance for terminals *a*-*b* in Figure 4.81. ■

Figure 4.81

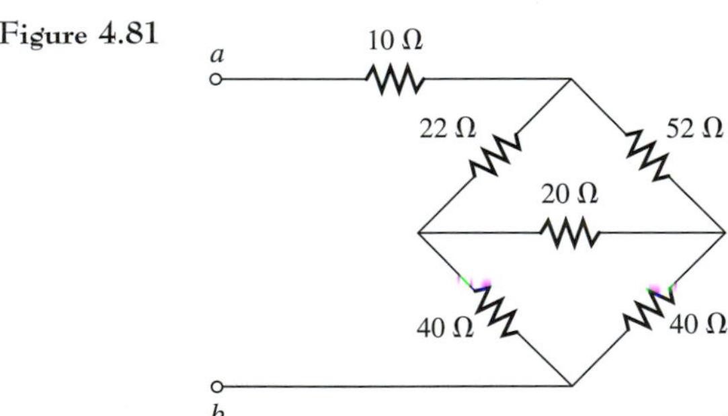

SOLUTION The Δ circuit consisting of the 20-Ω and the two 40-Ω resistances can be converted to an equivalent Y using Equations (4.8), (4.9), and (4.10), as shown in Figure 4.82.

Figure 4.82

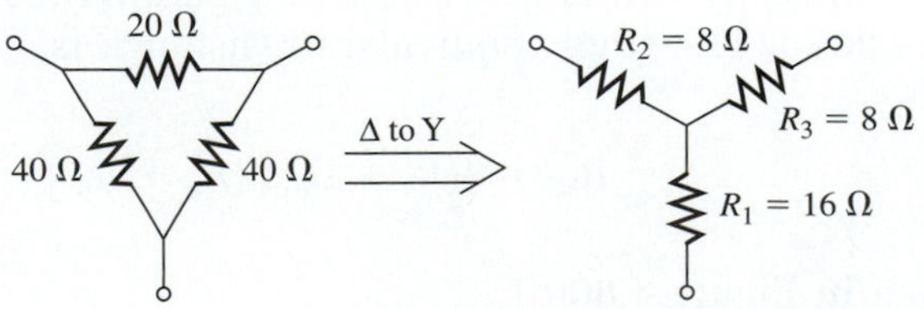

20 Ω, 40 Ω, 40 Ω — Δ to Y — $R_2 = 8\ \Omega$, $R_3 = 8\ \Omega$, $R_1 = 16\ \Omega$

$$R_1 = \frac{(40)(40)}{20 + 40 + 40}$$
$$= \frac{1600}{100}$$
$$= 16\ \Omega$$
$$R_2 = \frac{(20)(40)}{20 + 40 + 40}$$
$$= \frac{800}{100}$$
$$= 8\ \Omega$$
$$R_3 = \frac{(20)(40)}{20 + 40 + 40}$$
$$= \frac{800}{100}$$
$$= 8\ \Omega$$

The original circuit with the Δ replaced by an equivalent Y is shown in

Figure 4.83

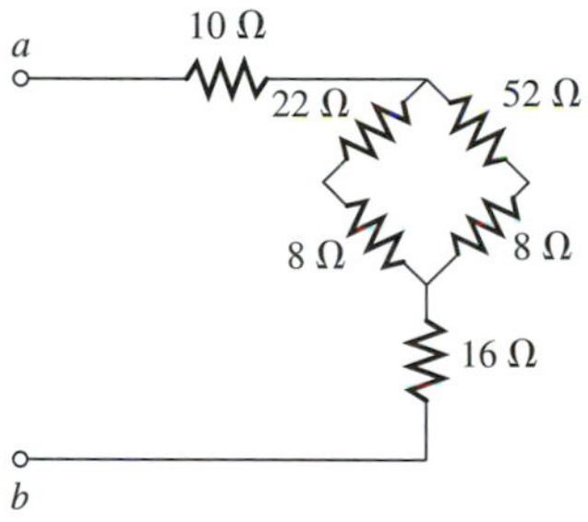

Figure 4.83. All the resistances can now be combined to form a single equivalent resistance as follows.

$$R_{EQ} = \{(22 + 8) \parallel (52 + 8)\} + 16 + 10$$
$$= \{(30 \parallel 60)\} + 16 + 10$$
$$= \left(\frac{(30)(60)}{30 + 60}\right) + 16 + 10$$
$$= 20 + 16 + 10$$
$$= 46\ \Omega$$

The equivalent single resistance for terminals a-b is indicated in Figure 4.84.

Figure 4.84

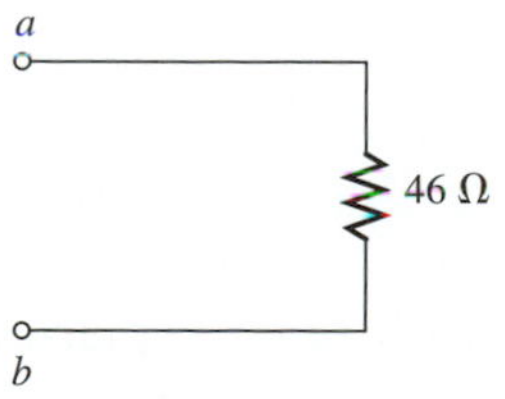

EXAMPLE 4.26 Determine a single equivalent resistance for terminals *a-b* in Figure 4.85. ■

Figure 4.85

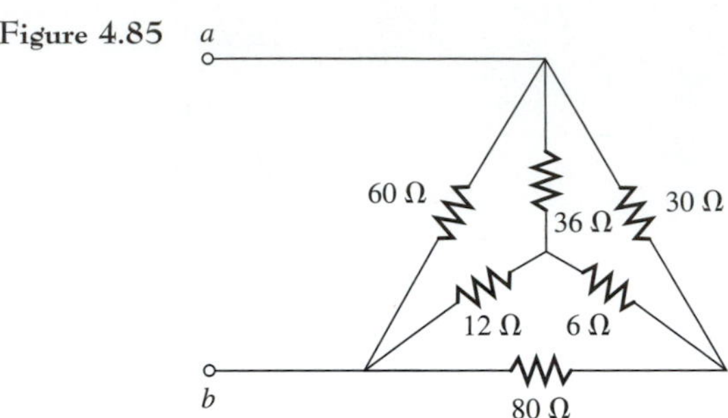

SOLUTION The Y circuit consisting of the 36-Ω, 12-Ω, and 6-Ω resistances can be converted to an equivalent Δ using Equations (4.5), (4.6), and (4.7), as shown in Figure 4.86.

Figure 4.86

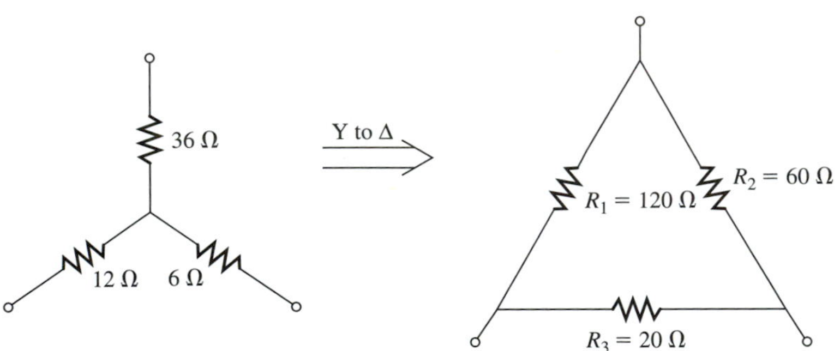

$$R_1 = \frac{(36)(6) + (36)(12) + (6)(12)}{6}$$

$$= 120\ \Omega$$

$$R_2 = \frac{(36)(6) + (36)(12) + (6)(12)}{12}$$

$$= 60\ \Omega$$

$$R_3 = \frac{(36)(6) + (36)(12) + (6)(12)}{36}$$

$$= 20\ \Omega$$

The original circuit with the Y replaced by an equivalent Δ is shown in Figure 4.87.

Figure 4.87

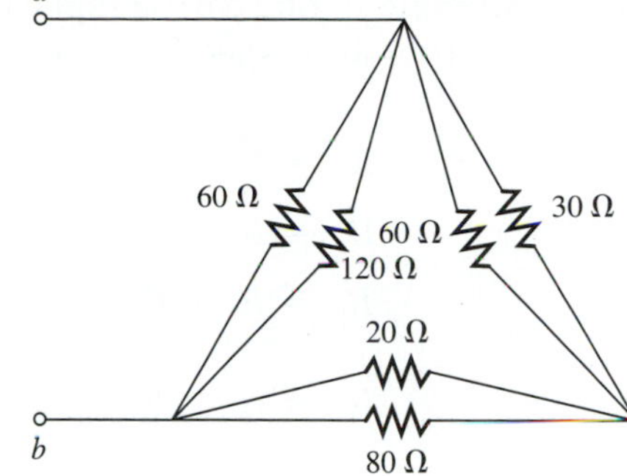

All the resistances can now be combined to form a single equivalent resistance as follows:

$$R_{\text{EQ}} = \{(60 \parallel 30) + (20 \parallel 80)\} \parallel \{60 \parallel 120\}$$

Because

$$60 \parallel 30 = \frac{(60)(30)}{60 + 30} = 20\ \Omega$$

$$20 \parallel 80 = \frac{(20)(80)}{20 + 80} = 16\ \Omega$$

$$60 \parallel 120 = \frac{(60)(120)}{60 + 120} = 40\ \Omega$$

it follows that

$$\begin{aligned} R_{\text{EQ}} &= \{20 + 16\} \parallel \{40\} \\ &= 36 \parallel 40 \\ &= \frac{(36)(40)}{36 + 40} \\ &= 18.9\ \Omega \end{aligned}$$

The Δ-to-Y and Y-to-Δ conversion formulas in Figure 4.78 can be derived as follows. Consider the circuits in Figure 4.88. The Δ and Y circuits in Fig-

Figure 4.88 Delta-wye equivalents.

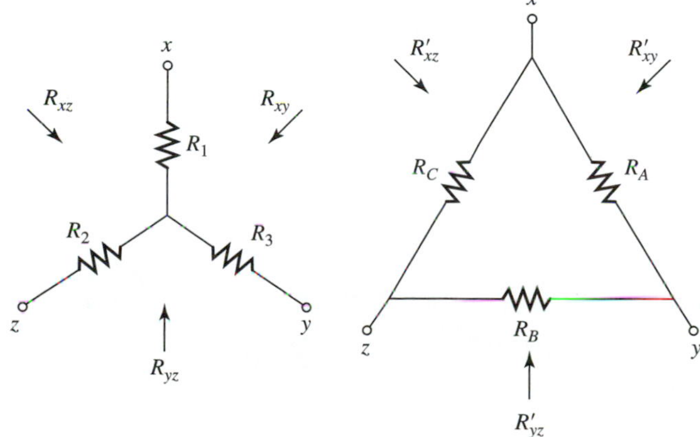

ure 4.88 are equivalent (one can be replaced by the other) if $R_{xz} = R'_{xz}$, $R_{xy} = R'_{xy}$ and $R_{yz} = R'_{yz}$. From an inspection of the circuits in Figure 4.88,

$$R_{xz} = R_1 + R_2$$

$$R_{xy} = R_1 + R_3$$

$$R_{yz} = R_2 + R_3$$

$$R'_{xz} = R_C \parallel (R_A + R_B) = \frac{R_C(R_A + R_B)}{R_A + R_B + R_C}$$

$$R'_{xy} = R_A \parallel (R_B + R_C) = \frac{R_A(R_B + R_C)}{R_A + R_B + R_C}$$

$$R'_{yz} = R_B \parallel (R_A + R_C) = \frac{R_B(R_A + R_C)}{R_A + R_B + R_C}$$

The conditions for the Δ and Y circuits to be equivalent are then

$$R_1 + R_2 = \frac{R_C(R_A + R_B)}{R_A + R_B + R_C}$$

$$R_1 + R_3 = \frac{R_A(R_B + R_C)}{R_A + R_B + R_C}$$

$$R_2 + R_3 = \frac{R_B(R_A + R_C)}{R_A + R_B + R_C}$$

Solving these equations for R_A, R_B, and R_C results in Equations (4.5), (4.6), and (4.7) for converting a Y circuit to a Δ circuit. Solving these equations for R_1, R_2, and R_3 results in Equations (4.8), (4.9), and (4.10) for converting a Δ circuit to a Y circuit. The algebra is left as an exercise for the student.

4.13 EXERCISES

1. Determine i_1 and i_2 for the circuits in Figure 4.89.

Figure 4.89

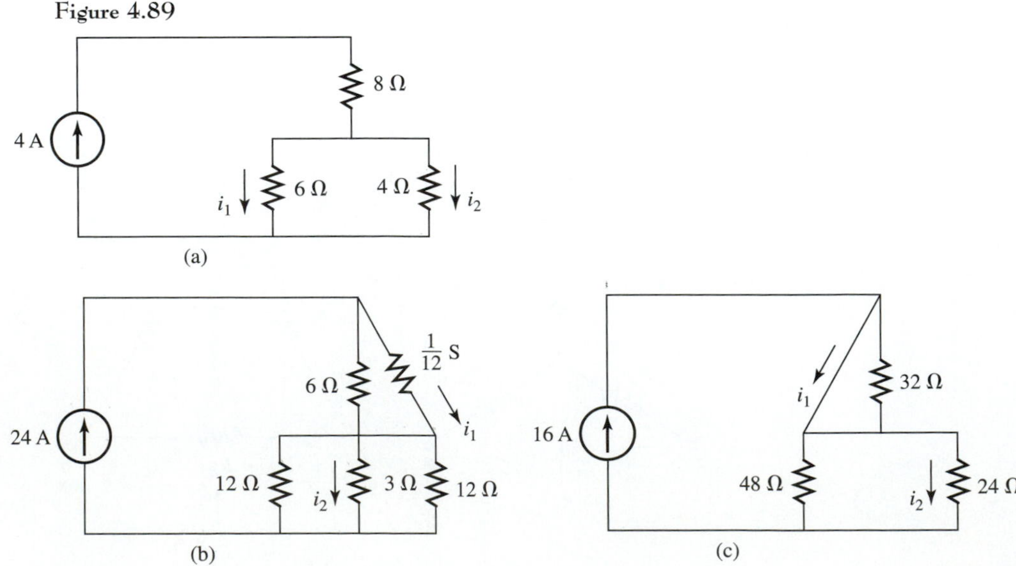

2. Determine i_1, i_2, and i_3 for the circuits in Figure 4.90.

Figure 4.90

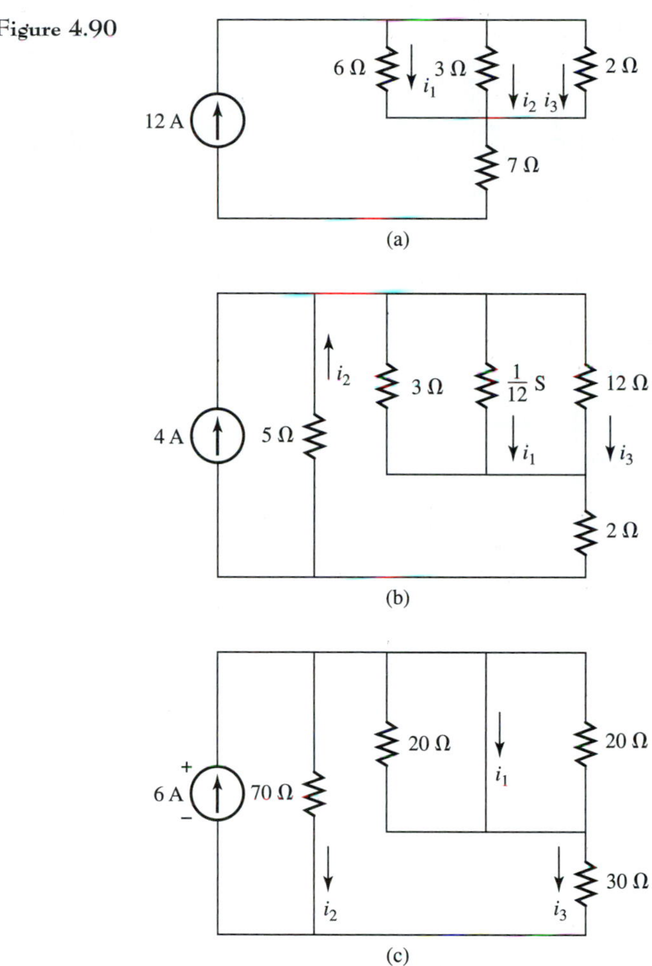

3. Determine i_1, i_2, i_3, and i_4 for the circuits in Figure 4.91.

Figure 4.91

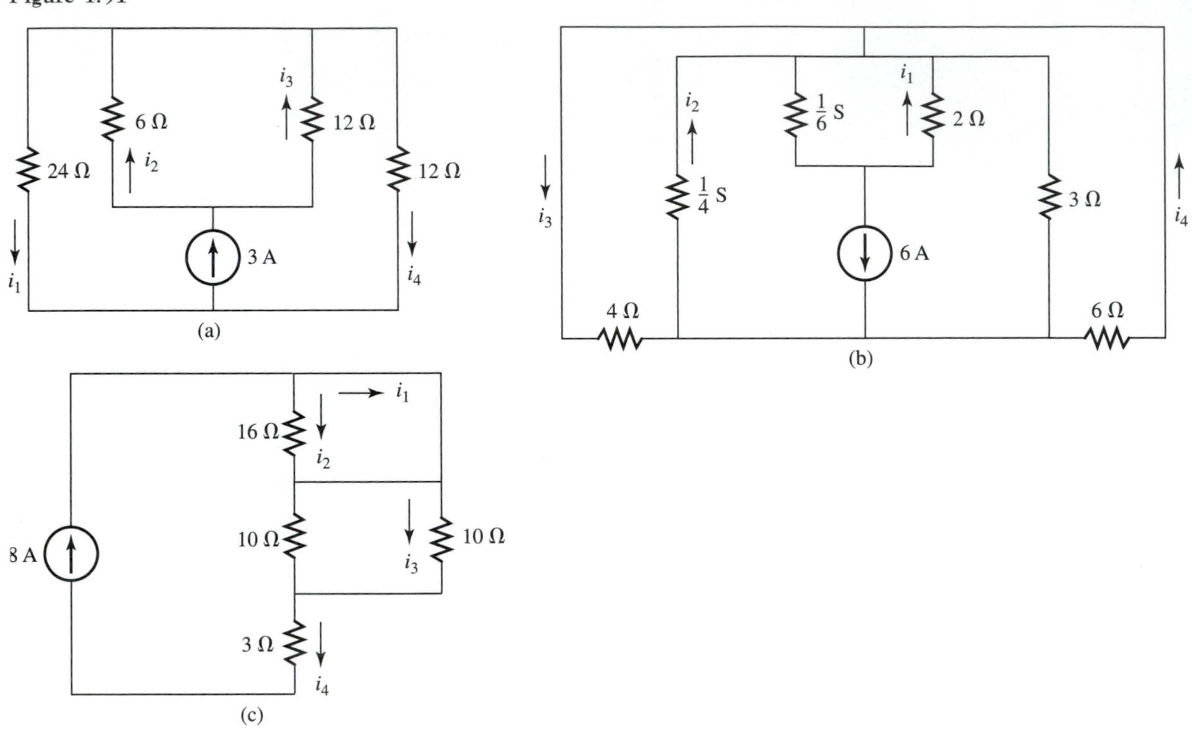

4. Determine i for the circuits in Figure 4.92.

Figure 4.92

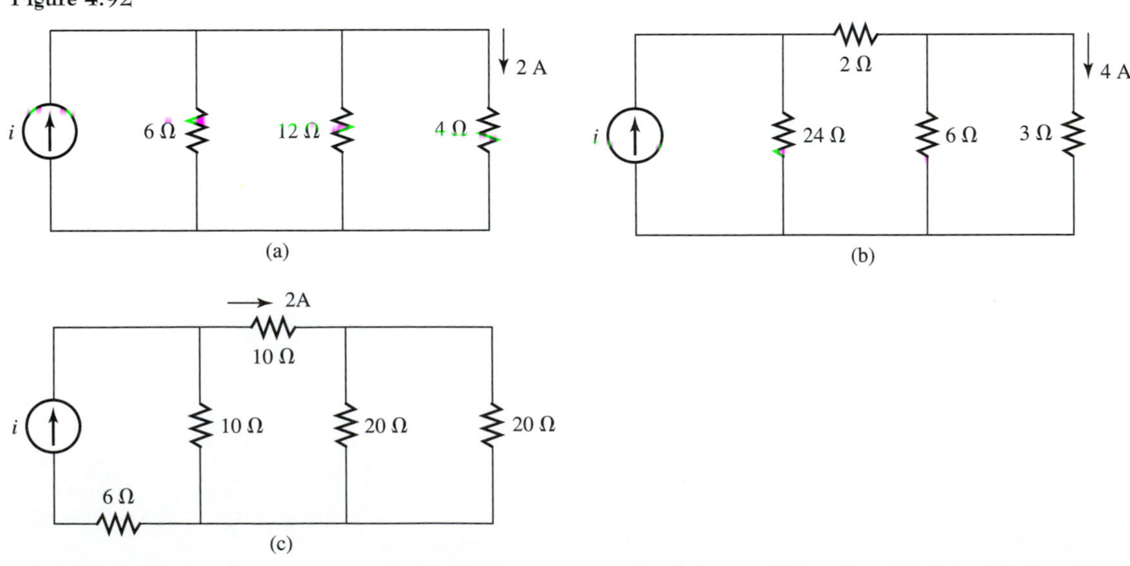

5. Determine i for the circuits in Figure 4.93.

Figure 4.93

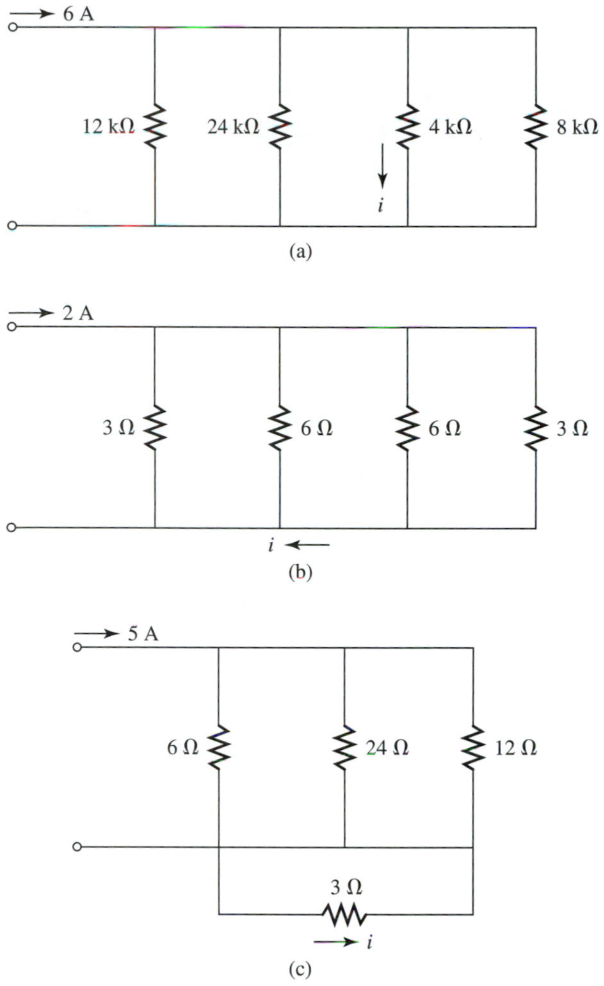

6 A
12 kΩ
24 kΩ
4 kΩ
8 kΩ
i
(a)
2 A
3 Ω
6 Ω
6 Ω
3 Ω
i
(b)
5 A
6 Ω
24 Ω
12 Ω
3 Ω
i
(c)

6. Determine v_1 and v_2 for the circuits in Figure 4.94.

Figure 4.94

40 V
6 Ω
v1
6 Ω
10 Ω
v2
(a)
8 Ω
v2
20 V
8 Ω
v1
24 Ω
48 Ω
(b)
4 Ω
16 V
v2
2 Ω
6 Ω
v1
3 Ω
(c)

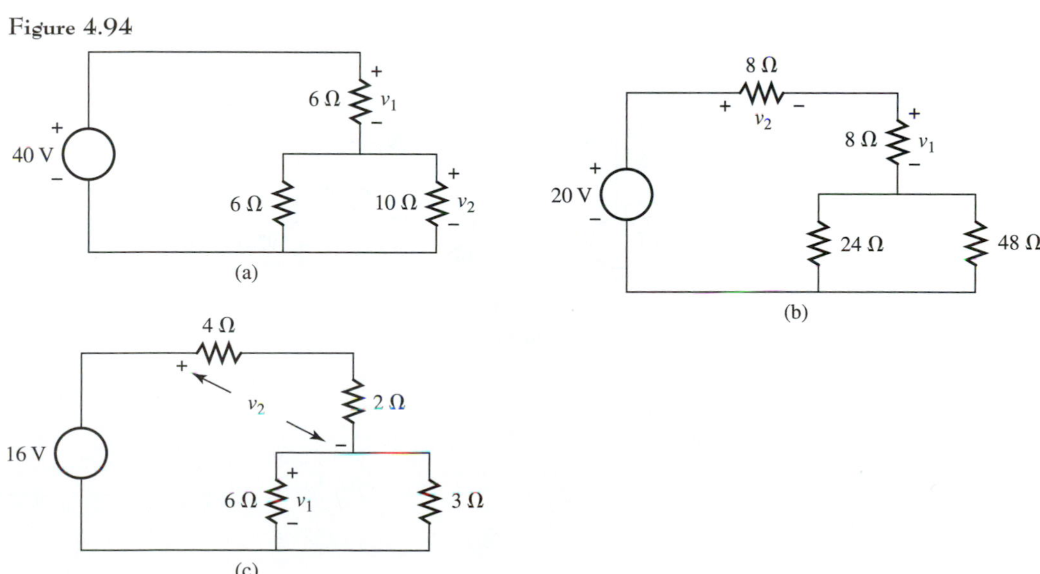

7. Determine v_1 and v_2 for the circuits in Figure 4.95.

Figure 4.95

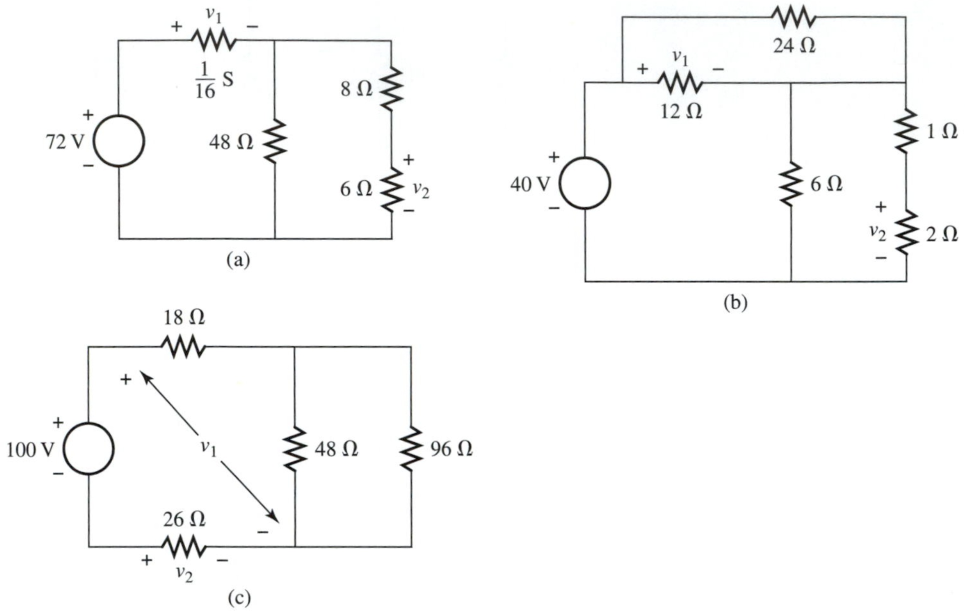

8. Determine v for the circuits in Figure 4.96.

Figure 4.96

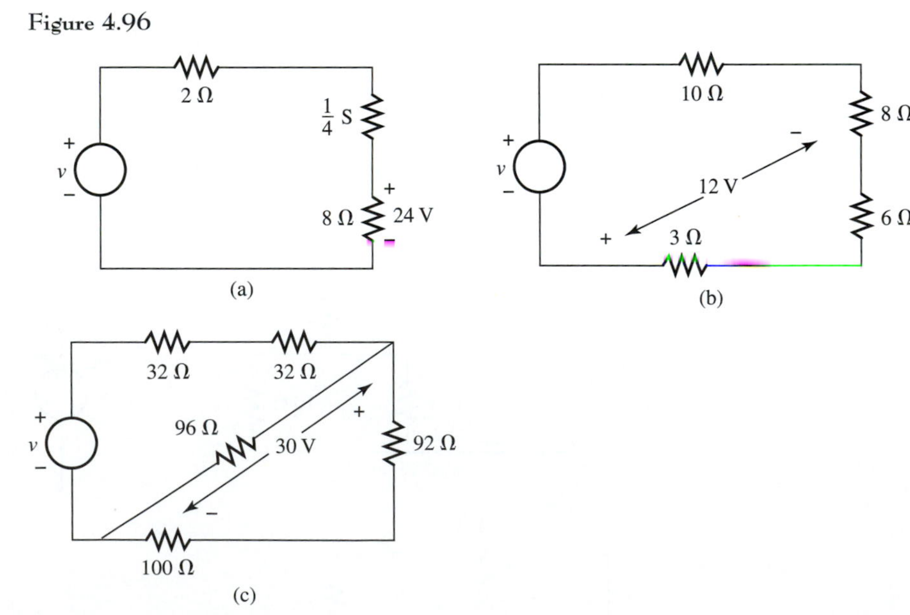

9. Determine v for the circuits in Figure 4.97.

Figure 4.97

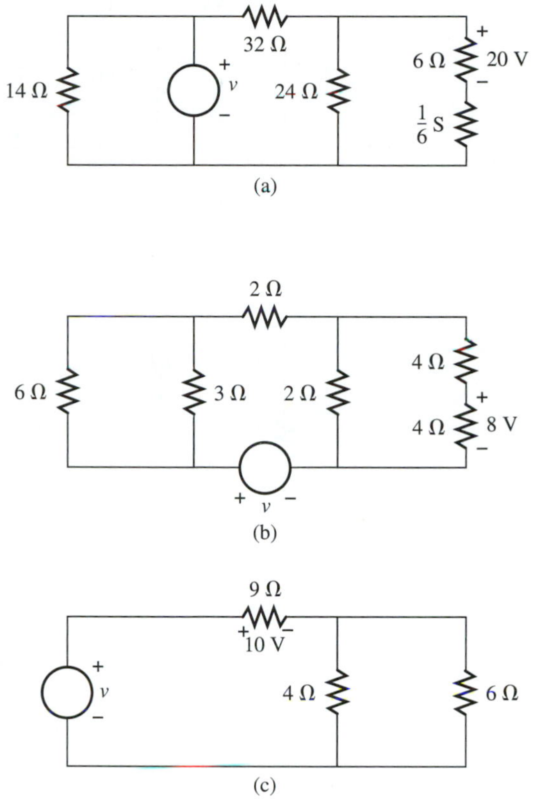

(a)

(b)

(c)

10. Determine v for the circuits in Figure 4.98.

Figure 4.98

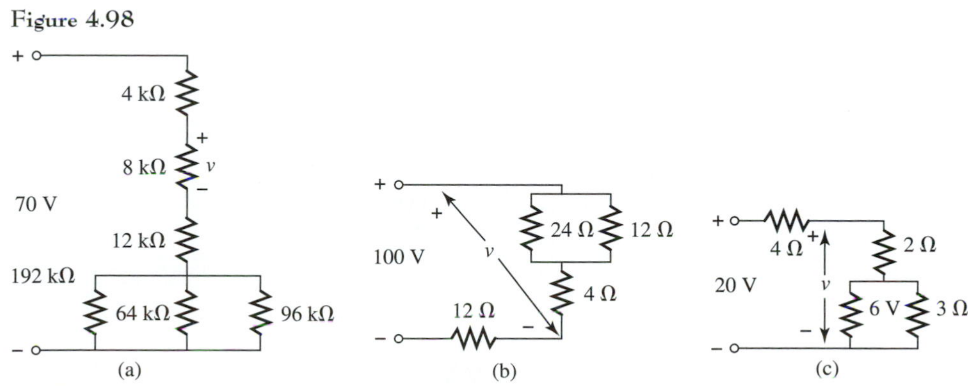

(a) (b) (c)

11. Determine v for the circuits in Figure 4.99.

Figure 4.99

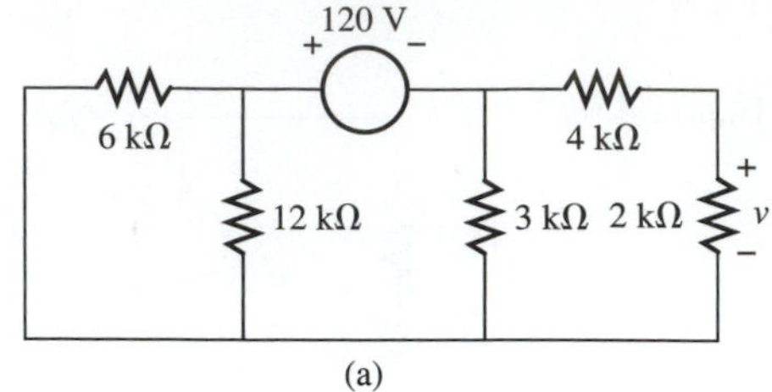

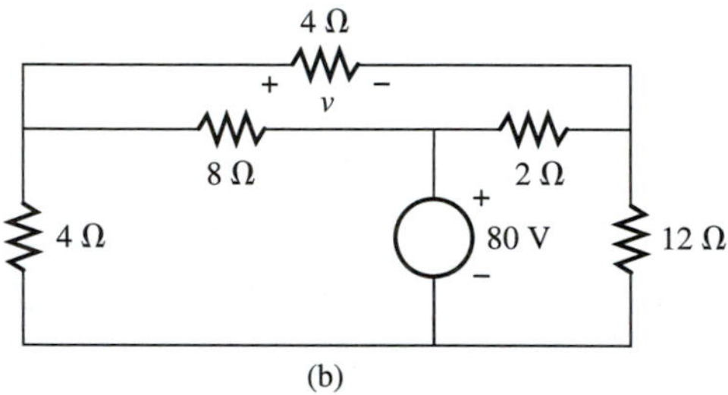

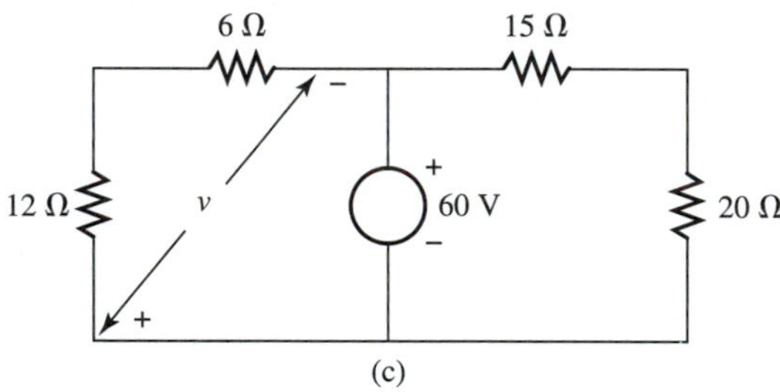

12. Determine i_1, i_2, and i_3 for the circuits in Figure 4.100.

Figure 4.100

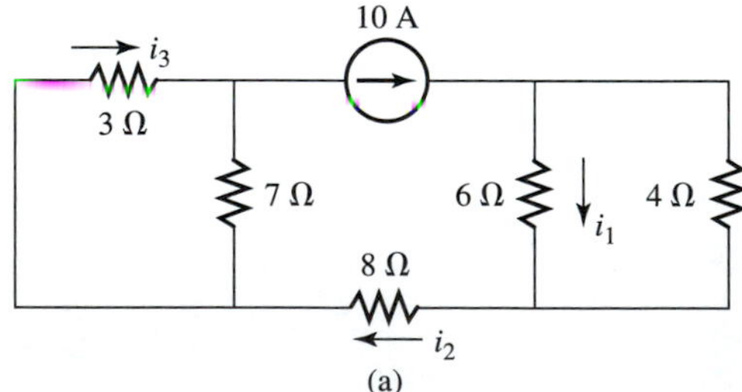

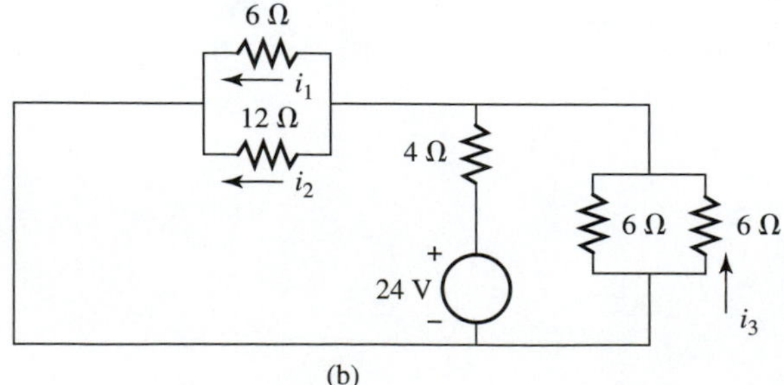

13. Determine v and i for the circuits in Figure 4.101.

Figure 4.101

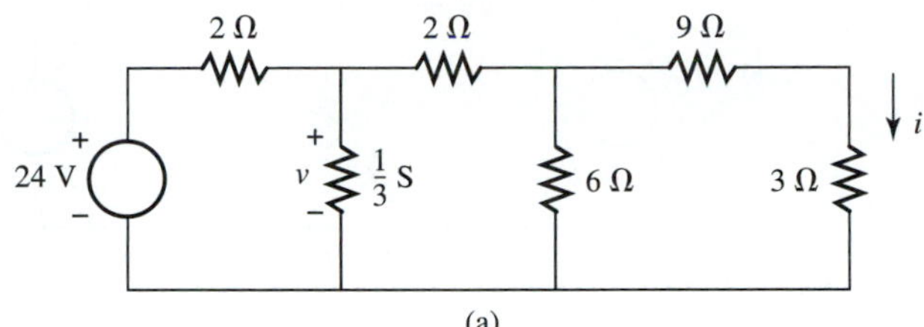

(a)

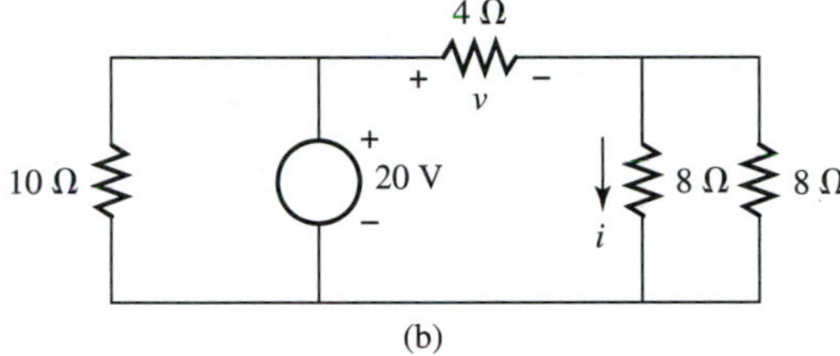

(b)

14. Determine v and i for the circuits in Figure 4.102.

Figure 4.102

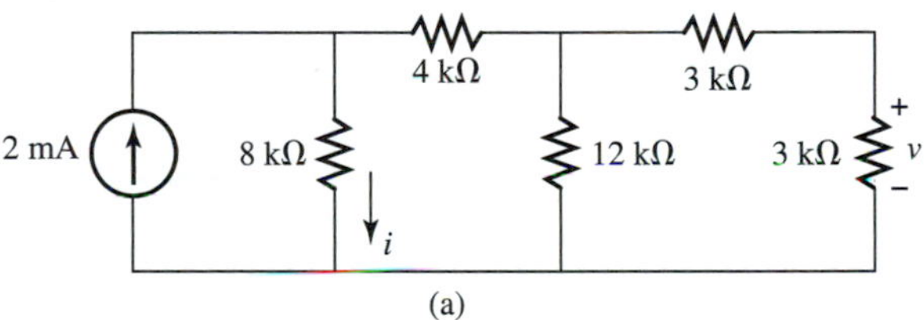

(a)

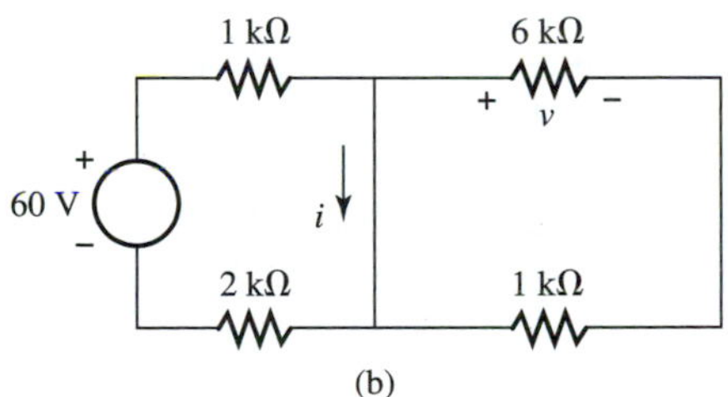

(b)

15. Determine the branch current i for the circuits in Figure 4.103.

Figure 4.103

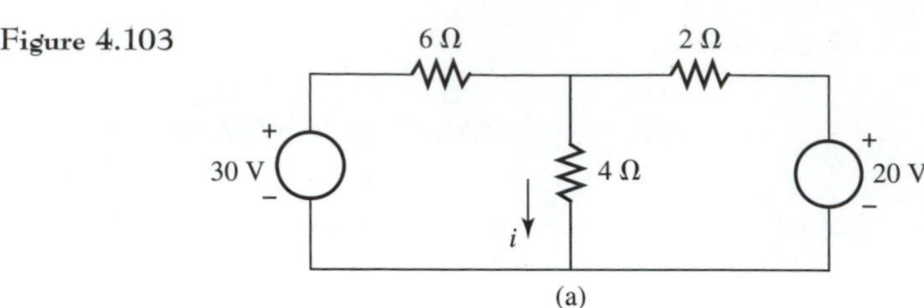

(a)

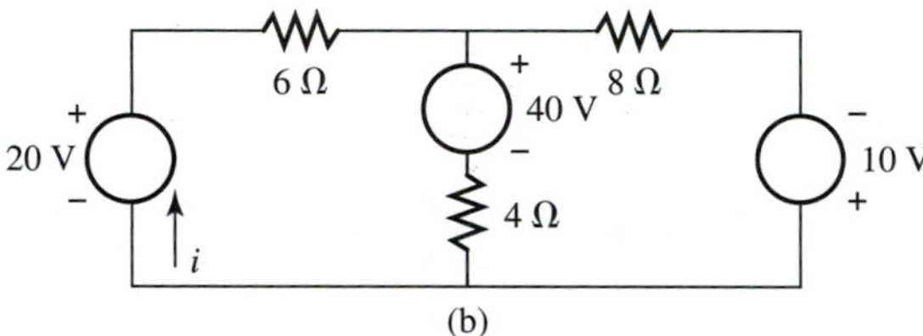

(b)

16. Determine the branch currents i_1, i_2, and i_3 for the circuits in Figure 4.104.

Figure 4.104

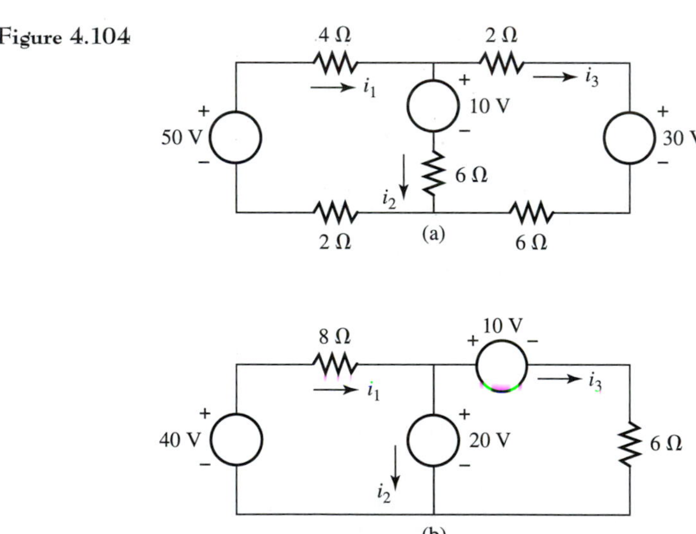

17. Write the equations necessary to solve for the branch currents i_1, i_2, i_3, i_4, i_5, and i_6 for the circuits in Figure 4.105.

Figure 4.105

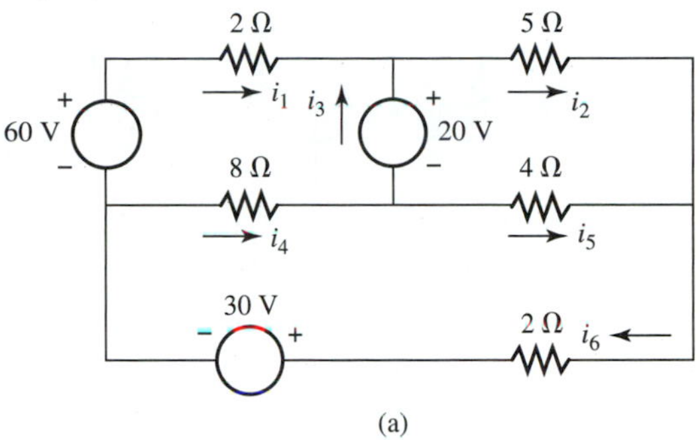

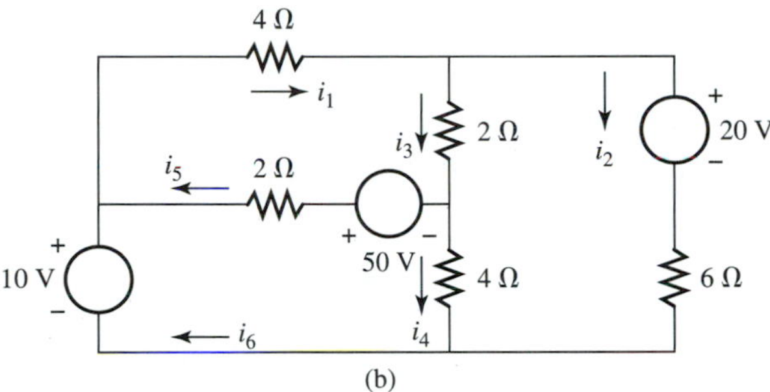

18. Determine the number of equations necessary to solve for all the branch currents in the circuit in Figure 4.106.

Figure 4.106

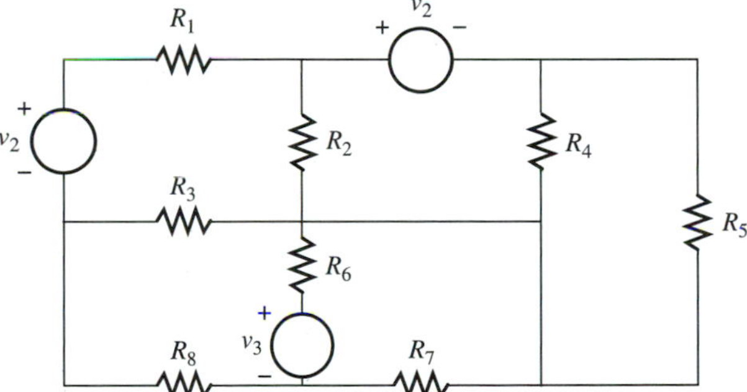

19. Using the branch current method to determine the branch currents for the circuit in Figure 4.106, how many times would Kirchhoff's voltage law and Kirchhoff's current law each be applied?

20. Determine the branch current i for the circuit in Figure 4.107.

Figure 4.107

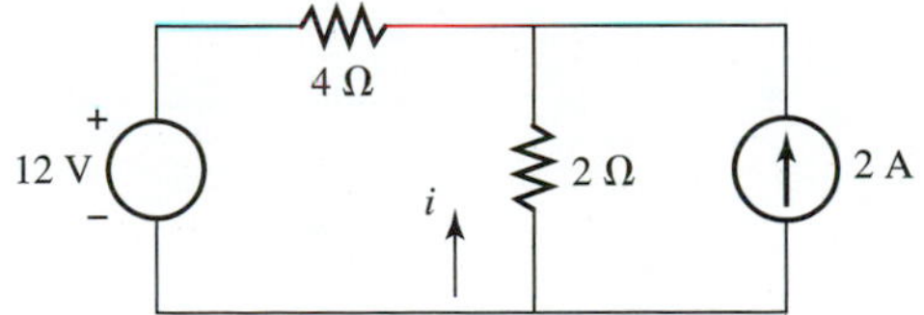

21. The value of i for the circuit in Figure 4.108 is 4 A. Determine the value of i if the voltage and current sources in the circuit are each doubled in value.

Figure 4.108

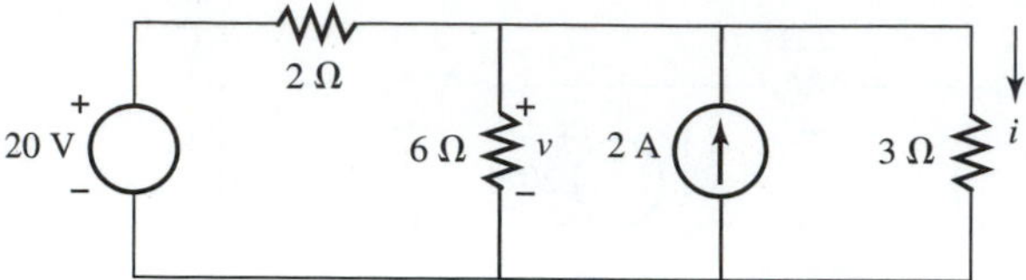

22. Determine R_{ab} for the circuit in Figure 4.109.

Figure 4.109

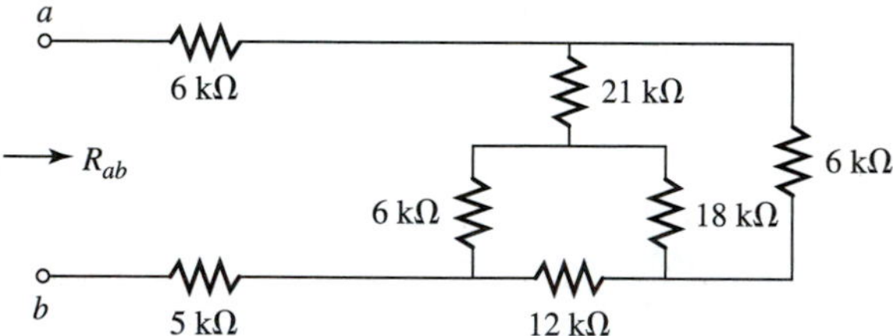

23. Determine R_{ab} for the circuit in Figure 4.110.

Figure 4.110

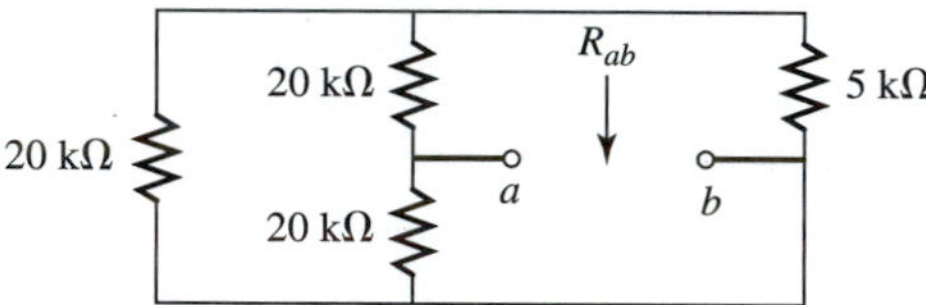

24. Determine i for the circuit in Figure 4.111.

Figure 4.111

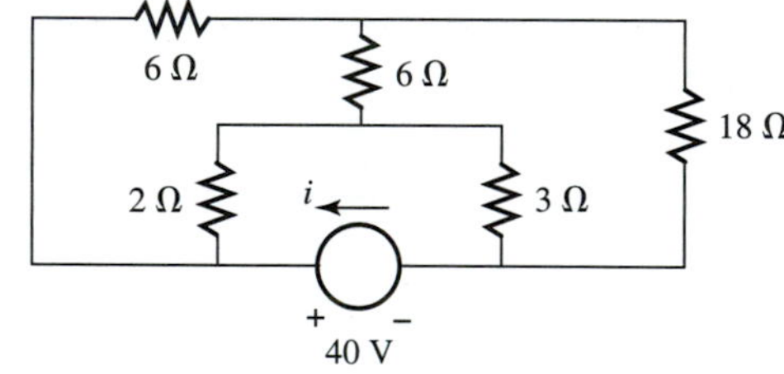

CHAPTER

5

Numerical Methods

After completing this chapter the student should be able to

* evaluate determinants
* solve algebraic equations for simultaneous solutions

The multiple-loop circuits considered in the next chapter are generally described by systems of algebraic equations. In this chapter, we present systematic methods for finding simultaneous solutions to systems of algebraic equations. An important mathematical device used in solving and concisely representing systems of algebraic equations is the matrix.

5.1 MATRICES

A *matrix* is an array of elements arranged in rows and columns, as shown in Figure 5.1. In general, the elements can be numbers, expressions, variables, or anything that we choose. In the circuit analysis methods presented in this text, the elements of the matrices are always numbers (real or complex) or symbols that represent numbers.

Figure 5.1 Matrix A.

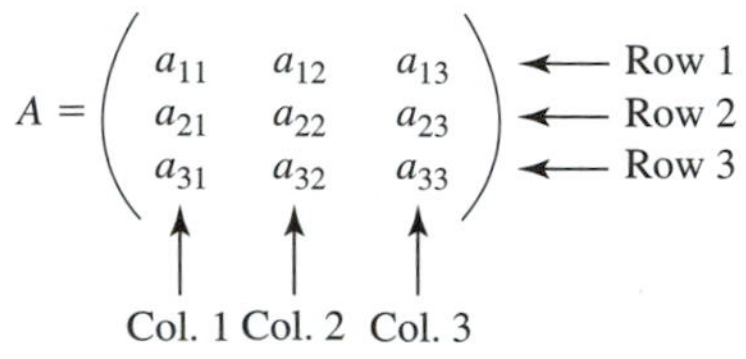

$$A = \begin{pmatrix} a_{11} & a_{12} & a_{13} \\ a_{21} & a_{22} & a_{23} \\ a_{31} & a_{32} & a_{33} \end{pmatrix} \begin{matrix} \leftarrow \text{Row 1} \\ \leftarrow \text{Row 2} \\ \leftarrow \text{Row 3} \end{matrix}$$

The *name* of the matrix in Figure 5.1 is A. The *rows* of the matrix are numbered from the top and the *columns* are numbered from the left, as shown in Figure 5.1. Row 1 consists of elements a_{11}, a_{12}, and a_{13}, and column 1 consists of elements a_{11}, a_{21} and a_{31}. The first subscript of an element indicates the row and the second subscript the column of the element. Element a_{32}, for example, is in the third row and second column. The *dimension* of a matrix is a description of its size. A matrix with the dimension $p \times q$ has p rows and q columns. The matrix named A in Figure 5.1 is a 3×3 matrix.

5.2 DETERMINANTS

A matrix with the same number of rows and columns is called a *square matrix,* and the number of rows or columns is the *order* of the matrix. A square matrix has a *determinant*, which is the numerical value obtained by performing a particular mathematical operation on a square matrix. The *major diagonal* of a square matrix consists of the elements along the line drawn from the upper leftmost element to the lower rightmost element. The major diagonal for matrix A in Figure 5.1 consists of elements a_{11}, a_{22}, and a_{33}.

The mathematical operation for determining the determinant of a 2×2 square matrix is given in Figure 5.2. Notice that the determinant of a square matrix can be represented by the symbol *det* (read as "the determinant of") followed by the square matrix or by two parallel vertical lines containing the square array.

Figure 5.2 Evaluating a 2×2 determinant.

$$A = \begin{pmatrix} a_{11} & a_{12} \\ a_{21} & a_{22} \end{pmatrix}$$

(a) Square matrix named A

$$\det A = \det \begin{pmatrix} a_{11} & a_{12} \\ a_{21} & a_{22} \end{pmatrix} \quad \text{or} \quad \det A = \begin{vmatrix} a_{11} & a_{12} \\ a_{21} & a_{22} \end{vmatrix}$$

(b) Determinant notation

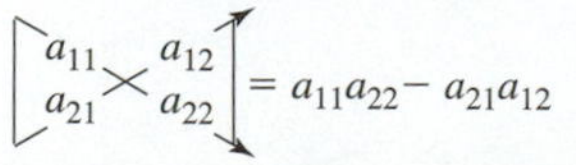

$$\begin{bmatrix} a_{11} & a_{12} \\ a_{21} & a_{22} \end{bmatrix} = a_{11}a_{22} - a_{21}a_{12}$$

(c) Rule for evaluating a 2×2 determinant

The value of det A is the product of the elements along the diagonal pointing downward minus the product of the elements along the diagonal pointing upward, as illustrated in Figure 5.2(c).

A mathematical operation for determining the determinant of a 3×3 square matrix is given in Figure 5.3. The first step is to copy the first and second columns of the array to the immediate right of the array and form the upward and downward diagonals, as shown. The value of det A is the sum of the three products of the elements along the three diagonals pointing downward minus the sum of the three products of the elements along the three diagonals pointing upward, as illustrated in Figure 5.3(c).

Figure 5.3 Evaluating a 3×3 determinant.

$$A = \begin{pmatrix} a_{11} & a_{12} & a_{13} \\ a_{21} & a_{22} & a_{23} \\ a_{31} & a_{32} & a_{33} \end{pmatrix}$$

(a) Square matrix named A

$$\det A = \det \begin{pmatrix} a_{11} & a_{12} & a_{13} \\ a_{21} & a_{22} & a_{23} \\ a_{31} & a_{32} & a_{33} \end{pmatrix} \quad \text{or} \quad \det A = \begin{vmatrix} a_{11} & a_{12} & a_{13} \\ a_{21} & a_{22} & a_{23} \\ a_{31} & a_{32} & a_{33} \end{vmatrix}$$

(b) Determinant notation

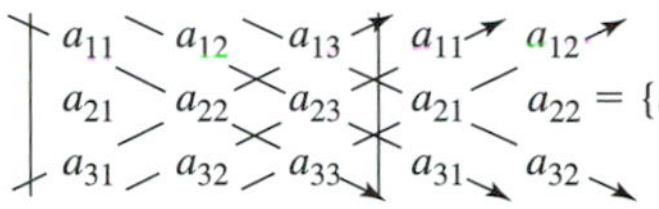

$$\begin{matrix} a_{11} & a_{12} & a_{13} & a_{11} & a_{12} \\ a_{21} & a_{22} & a_{23} & a_{21} & a_{22} \\ a_{31} & a_{32} & a_{33} & a_{31} & a_{32} \end{matrix} = \{a_{11}a_{22}a_{33} + a_{12}a_{23}a_{31} + a_{13}a_{21}a_{32} - a_{31}a_{22}a_{13} - a_{32}a_{23}a_{11} - a_{33}a_{21}a_{12}\}$$

(c) Rule for evaluating a 3×3 determinant

Consider the following examples.

EXAMPLE 5.1 Evaluate the determinant of the 2×2 square matrix B:

$$B = \begin{pmatrix} 2 & -1 \\ 4 & -3 \end{pmatrix}$$ ■

SOLUTION

$$\det B = \begin{vmatrix} 2 & -1 \\ 4 & -3 \end{vmatrix}$$

From Figure 5.2

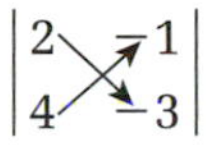

$$\begin{vmatrix} 2 & -1 \\ 4 & -3 \end{vmatrix}$$

$$\det B = 2(-3) - (4)(-1)$$
$$= -2$$

EXAMPLE 5.2 Evaluate the determinant of the 3×3 square matrix C:

$$C = \begin{pmatrix} 1 & -2 & 0 \\ 2 & 4 & -1 \\ -6 & 1 & 0 \end{pmatrix}$$ ■

SOLUTION

$$\det C = \begin{vmatrix} 1 & -2 & 0 \\ 2 & 4 & -1 \\ -6 & 1 & 0 \end{vmatrix}$$

From Figure 5.3

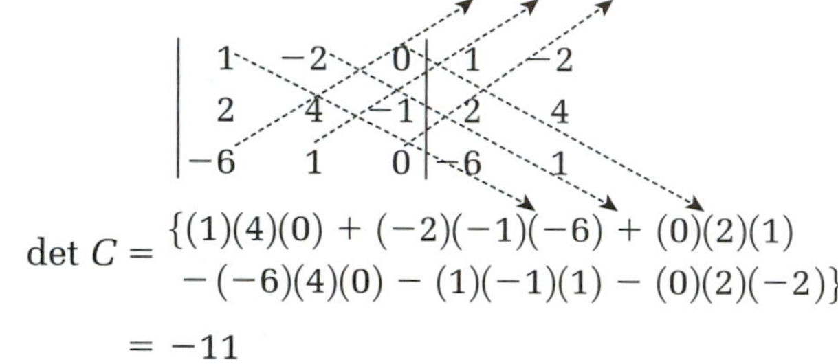

$$\det C = \begin{matrix}\{(1)(4)(0) + (-2)(-1)(-6) + (0)(2)(1) \\ -(-6)(4)(0) - (1)(-1)(1) - (0)(2)(-2)\}\end{matrix}$$
$$= -11$$

Because 2 × 2 determinants are relatively simple to evaluate, we shall present a method for evaluating determinants with higher dimensions by reducing them to 2 × 2 determinants. The resulting 2 × 2 determinants can then be evaluated by the rule in Figure 5.2(c).

5.3 CHIO'S METHOD

In the nineteenth century, the Italian mathematician Felice Chio presented a method to systematically reduce the dimension of a determinant larger than 2 × 2. The method is illustrated for 3 × 3 and 4 × 4 determinants in Figures 5.4 and 5.5.

Figure 5.4 Reduction of a 3 × 3 determinant to a 2 × 2 determinant using Chio's method.

$$A = \begin{pmatrix} a_{11} & a_{12} & a_{13} \\ a_{21} & a_{22} & a_{23} \\ a_{31} & a_{32} & a_{33} \end{pmatrix}$$

$$\det A = \det \begin{vmatrix} a_{11} & a_{12} & a_{13} \\ a_{21} & a_{22} & a_{23} \\ a_{31} & a_{32} & a_{33} \end{vmatrix} = \frac{1}{(a_{11})^{n-2}} \begin{Vmatrix} \begin{vmatrix} a_{11} & a_{12} \\ a_{21} & a_{22} \end{vmatrix} & \begin{vmatrix} a_{11} & a_{13} \\ a_{21} & a_{23} \end{vmatrix} \\ \begin{vmatrix} a_{11} & a_{12} \\ a_{31} & a_{32} \end{vmatrix} & \begin{vmatrix} a_{11} & a_{13} \\ a_{31} & a_{33} \end{vmatrix} \end{Vmatrix}$$

n = number of rows and columns in matrix A
= 3

$$\det A = \frac{1}{a_{11}} \begin{vmatrix} (a_{11}a_{22} - a_{21}a_{12}) & (a_{11}a_{23} - a_{21}a_{13}) \\ (a_{11}a_{32} - a_{31}a_{12}) & (a_{11}a_{33} - a_{31}a_{13}) \end{vmatrix}$$

Figure 5.5 Reduction of a 4 × 4 determinant to a 3 × 3 determinant using Chio's method.

$$A = \begin{pmatrix} a_{11} & a_{12} & a_{13} & a_{14} \\ a_{21} & a_{22} & a_{23} & a_{24} \\ a_{31} & a_{32} & a_{33} & a_{34} \\ a_{41} & a_{42} & a_{43} & a_{44} \end{pmatrix}$$

$$\det A = \det \begin{vmatrix} a_{11} & a_{12} & a_{13} & a_{14} \\ a_{21} & a_{22} & a_{23} & a_{24} \\ a_{31} & a_{32} & a_{33} & a_{34} \\ a_{41} & a_{42} & a_{43} & a_{44} \end{vmatrix} = \frac{1}{(a_{11})^{n-2}} \begin{Vmatrix} \begin{vmatrix} a_{11} & a_{12} \\ a_{21} & a_{22} \end{vmatrix} & \begin{vmatrix} a_{11} & a_{13} \\ a_{21} & a_{23} \end{vmatrix} & \begin{vmatrix} a_{11} & a_{14} \\ a_{21} & a_{24} \end{vmatrix} \\ \begin{vmatrix} a_{11} & a_{12} \\ a_{31} & a_{32} \end{vmatrix} & \begin{vmatrix} a_{11} & a_{13} \\ a_{31} & a_{33} \end{vmatrix} & \begin{vmatrix} a_{11} & a_{14} \\ a_{31} & a_{34} \end{vmatrix} \\ \begin{vmatrix} a_{11} & a_{12} \\ a_{41} & a_{42} \end{vmatrix} & \begin{vmatrix} a_{11} & a_{13} \\ a_{41} & a_{43} \end{vmatrix} & \begin{vmatrix} a_{11} & a_{14} \\ a_{41} & a_{43} \end{vmatrix} \end{Vmatrix}$$

n = number of rows and columns in matrix A
$= 4$

$$\det A = \frac{1}{(a_{11})^2} \begin{vmatrix} (a_{11}a_{22} - a_{21}a_{12}) & (a_{11}a_{23} - a_{21}a_{13}) & (a_{11}a_{24} - a_{21}a_{14}) \\ (a_{11}a_{32} - a_{31}a_{12}) & (a_{11}a_{33} - a_{31}a_{13}) & (a_{11}a_{34} - a_{31}a_{14}) \\ (a_{11}a_{42} - a_{41}a_{13}) & (a_{11}a_{43} - a_{41}a_{13}) & (a_{11}a_{44} - a_{41}a_{14}) \end{vmatrix}$$

Each time Chio's method is applied to a determinant the dimension is reduced by 1. For example, we apply Chio's method to a 6 × 6 determinant three times to reduce it to a 3 × 3 determinant and four times to reduce it to a 2 × 2 determinant.

Notice that in Figures 5.4 and 5.5, Chio's method involves division by a_{11} (the element in the first row, first column of the determinant). Because division by zero is impossible, a_{11} cannot equal zero. If we wish to apply Chio's method to a determinant with $a_{11} = 0$, we must interchange two rows or two columns to make a_{11} a nonzero element using the following rule.

> If two rows or two columns of a determinant are interchanged, the resulting determinant is the negative of the original determinant.

The procedure for applying Chio's method to determinants is illustrated in the following examples.

EXAMPLE 5.3 Use Chio's method to evaluate the following determinant.

$$\begin{vmatrix} 2 & 0 & 1 \\ 1 & -1 & 6 \\ 0 & -1 & -4 \end{vmatrix}$$

■

SOLUTION From Figure 5.4 and the rule in Figure 5.2 for evaluating 2 × 2 determinants

$$\begin{vmatrix} 2 & 0 & 1 \\ 1 & -1 & 6 \\ 0 & -1 & -4 \end{vmatrix} = \frac{1}{(2)^{3-2}} \begin{Vmatrix} \begin{vmatrix} 2 & 0 \\ 1 & -1 \end{vmatrix} & \begin{vmatrix} 2 & 1 \\ 1 & 6 \end{vmatrix} \\ \begin{vmatrix} 2 & 0 \\ 0 & -1 \end{vmatrix} & \begin{vmatrix} 2 & 1 \\ 0 & -4 \end{vmatrix} \end{Vmatrix}$$

$$= \frac{1}{2} \begin{vmatrix} (2)(-1) - (1)(0) & (2)(6) - (1)(1) \\ (2)(-1) - (0)(0) & (2)(-4) - (0)(1) \end{vmatrix}$$

$$= \frac{1}{2}\begin{vmatrix} -2 & 11 \\ -2 & -8 \end{vmatrix}$$

$$= \left(\frac{1}{2}\right)\{(-2)(-8) - (-2)(11)\}$$

$$= 19$$

EXAMPLE 5.4 Use Chio's method to evaluate the following determinant.

$$\begin{vmatrix} 0 & 2 & -10 \\ 8 & -3 & 1 \\ 2 & 0 & 2 \end{vmatrix}$$

■

SOLUTION Because the element in the first column, first row is zero, we must interchange two rows or two columns before proceeding to apply Chio's method.

Interchanging columns 1 and 2 results in

$$\begin{vmatrix} 0 & 2 & -10 \\ 8 & -3 & 1 \\ 2 & 0 & 2 \end{vmatrix} = -\begin{vmatrix} 2 & 0 & -10 \\ -3 & 8 & 1 \\ 0 & 2 & 2 \end{vmatrix}$$

From Figure 5.4 and the rule in Figure 5.2 for evaluating 2×2 determinants,

$$-\begin{vmatrix} 2 & 0 & -10 \\ -3 & 8 & 1 \\ 0 & 2 & 2 \end{vmatrix} = -\frac{(1)}{(2)^{3-2}}\begin{vmatrix} \begin{vmatrix} 2 & 0 \\ -3 & 8 \end{vmatrix} & \begin{vmatrix} 2 & -10 \\ -3 & 1 \end{vmatrix} \\ \begin{vmatrix} 2 & 0 \\ 0 & 2 \end{vmatrix} & \begin{vmatrix} 2 & -10 \\ 0 & 2 \end{vmatrix} \end{vmatrix}$$

$$= -\left(\frac{1}{2}\right)\begin{vmatrix} 16 & -28 \\ 4 & 4 \end{vmatrix}$$

$$= -\left(\frac{1}{2}\right)\{(16)(4) - (4)(-28)\}$$

$$= -88$$

EXAMPLE 5.5 Use Chio's method to evaluate the following determinant.

$$\begin{vmatrix} -4 & 2 & 0 & -1 \\ -1 & 0 & 4 & 0 \\ 0 & 1 & -2 & 4 \\ -2 & 0 & 0 & 5 \end{vmatrix}$$

■

SOLUTION From Figure 5.5 and the rule in Figure 5.2 for evaluating 2×2 determinants,

$$\begin{vmatrix} -4 & 2 & 0 & -1 \\ -1 & 0 & 4 & 0 \\ 0 & 1 & -2 & 4 \\ -2 & 0 & 0 & 5 \end{vmatrix} = \frac{1}{(-4)^{4-2}} \begin{vmatrix} \begin{vmatrix} -4 & 2 \\ -1 & 0 \end{vmatrix} & \begin{vmatrix} -4 & 0 \\ -1 & 4 \end{vmatrix} & \begin{vmatrix} -4 & -1 \\ -1 & 0 \end{vmatrix} \\ \begin{vmatrix} -4 & 2 \\ 0 & 1 \end{vmatrix} & \begin{vmatrix} -4 & 0 \\ 0 & -2 \end{vmatrix} & \begin{vmatrix} -4 & -1 \\ 0 & 4 \end{vmatrix} \\ \begin{vmatrix} -4 & 2 \\ -2 & 0 \end{vmatrix} & \begin{vmatrix} -4 & 0 \\ -2 & 0 \end{vmatrix} & \begin{vmatrix} -4 & -1 \\ -2 & 5 \end{vmatrix} \end{vmatrix}$$

$$= \frac{1}{16} \begin{vmatrix} 2 & -16 & -1 \\ -4 & 8 & -16 \\ 4 & 0 & -22 \end{vmatrix}$$

Applying Chio's method a second time and using Figure 5.4 and the rule in Figure 5.2 for evaluating 2×2 determinants results in

$$= \left(\frac{1}{16}\right) \frac{1}{(2)^{3-2}} \begin{vmatrix} \begin{vmatrix} 2 & -16 \\ -4 & 8 \end{vmatrix} & \begin{vmatrix} 2 & -1 \\ -4 & -16 \end{vmatrix} \\ \begin{vmatrix} 2 & -16 \\ 4 & 0 \end{vmatrix} & \begin{vmatrix} 2 & -1 \\ 4 & -22 \end{vmatrix} \end{vmatrix}$$

$$= \frac{1}{32} \begin{vmatrix} -48 & -36 \\ 64 & -40 \end{vmatrix}$$

$$= \left(\frac{1}{32}\right) \{(-48)(-40) - (64)(-36)\}$$

$$= 132$$

The general rule for reducing an $n \times n$ determinant by Chio's method is given in Figure 5.6.

Figure 5.6 Chio's method for reducing an $n \times n$ determinant.

$$A = \begin{vmatrix} a_{11} & a_{12} & a_{13} \cdots a_{1n} \\ a_{21} & a_{22} & a_{23} \cdots a_{2n} \\ \vdots & \vdots & \vdots \quad\ \vdots \\ a_{n1} & a_{n2} & a_{n3} \cdots a_{nn} \end{vmatrix}$$

$$\det A = \frac{1}{(a_{11})^{n-2}} \begin{vmatrix} \begin{vmatrix} a_{11} & a_{12} \\ a_{21} & a_{22} \end{vmatrix} & \begin{vmatrix} a_{11} & a_{13} \\ a_{21} & a_{23} \end{vmatrix} & \cdots & \begin{vmatrix} a_{11} & a_{1n} \\ a_{21} & a_{2n} \end{vmatrix} \\ \begin{vmatrix} a_{11} & a_{12} \\ a_{31} & a_{32} \end{vmatrix} & \begin{vmatrix} a_{11} & a_{13} \\ a_{31} & a_{33} \end{vmatrix} & \cdots & \begin{vmatrix} a_{11} & a_{1n} \\ a_{31} & a_{3n} \end{vmatrix} \\ \vdots & \vdots & & \vdots \\ \begin{vmatrix} a_{11} & a_{12} \\ a_{n1} & a_{n2} \end{vmatrix} & \begin{vmatrix} a_{11} & a_{13} \\ a_{n1} & a_{n3} \end{vmatrix} & \cdots & \begin{vmatrix} a_{11} & a_{1n} \\ a_{n1} & a_{nn} \end{vmatrix} \end{vmatrix}$$

Recall that a_{11} cannot equal zero and that Chio's method must be applied $(n - 2)$ times to reduce a determinant with dimension $n \times n$ to a determinant with dimension 2×2.

5.4 CRAMER'S RULE

In the eighteenth century Gabriel Cramer (1704–1752), a Swiss mathematician, presented a rule for solving systems of equations. Cramer's rule allows us to systematically solve systems of algebraic equations for simultaneous solutions by using determinants. Consider the following system of equations, where x and y are unknown quantities and the a's and b's are known constants.

$$a_{11}x + a_{12}y = b_1$$
$$a_{21}x + a_{22}y = b_2$$

Notice that the x's, y's, and b's are arranged in columns and all the constant terms (b's) are on the right of the equal signs. It is necessary to arrange the equations in the exact form shown here prior to applying Cramer's rule.

The *first step* in applying Cramer's rule is to form the *coefficient matrix*:

$$A = \begin{pmatrix} a_{11} & a_{12} \\ a_{21} & a_{22} \end{pmatrix}$$

The *second step* is to form the single column matrix consisting of the constants on the right of the equal signs:

$$B = \begin{pmatrix} b_1 \\ b_2 \end{pmatrix}$$

Cramer's rule allows us to write x and y as the ratio of two determinants:

$$x = \frac{\begin{vmatrix} b_1 & a_{12} \\ b_2 & a_{22} \end{vmatrix}}{\begin{vmatrix} a_{11} & a_{12} \\ a_{21} & a_{22} \end{vmatrix}}$$

$$y = \frac{\begin{vmatrix} a_{11} & b_1 \\ a_{21} & b_2 \end{vmatrix}}{\begin{vmatrix} a_{11} & a_{12} \\ a_{21} & a_{22} \end{vmatrix}}$$

Notice that the denominator determinants for x and y are identical, and each is the determinant of the coefficient matrix. Notice also that the numerator determinants for x and y are formed by replacing a single column in the denominator determinant with the column matrix B. The first column of the denominator determinant is replaced by the column matrix B to form the numerator determinant for x because x is in the first column in the original equations. The second column of the denominator determinant is replaced by the column matrix B to form the numerator determinant for y, because y is in the second column in the original equations.

Consider the following examples.

EXAMPLE 5.6 Solve the following set of equations for x and y.

$$2x - y = 4$$
$$x + y = 0$$
■

SOLUTION The coefficient matrix is

$$\begin{pmatrix} 2 & -1 \\ 1 & 1 \end{pmatrix}$$

The column matrix for the elements to the right of the equal signs is

$$\begin{pmatrix} 4 \\ 0 \end{pmatrix}$$

From Cramer's rule and the rule in Figure 5.2 for evaluating 2×2 determinants,

$$x = \frac{\begin{vmatrix} 4 & -1 \\ 0 & 1 \end{vmatrix}}{\begin{vmatrix} 2 & -1 \\ 1 & 1 \end{vmatrix}} = \frac{4(1) - (0)(-1)}{2(1) - (1)(-1)}$$

$$x = \frac{4}{3}$$

$$y = \frac{\begin{vmatrix} 2 & 4 \\ 1 & 0 \end{vmatrix}}{\begin{vmatrix} 2 & -1 \\ 1 & 1 \end{vmatrix}} = \frac{(2)(0) - (1)(4)}{(2)(1) - (1)(-1)}$$

$$y = -\frac{4}{3}$$

The procedure for solving two equations with two unknowns using Cramer's rule is extended to three equations with three unknowns in Example 5.7.

EXAMPLE 5.7 Solve the following set of equations for x, y, and z.

$$\begin{aligned} x - y + z &= 2 \\ -2y - z &= 1 \\ x + 4z &= 0 \end{aligned}$$
■

SOLUTION The coefficient matrix is

$$\begin{pmatrix} 1 & -1 & 1 \\ 0 & -2 & -1 \\ 1 & 0 & 4 \end{pmatrix}$$

The column matrix for the elements to the right of the equal signs is

$$\begin{pmatrix} 2 \\ 1 \\ 0 \end{pmatrix}$$

From Cramer's rule and the rule in Figure 5.3 for evaluating 3×3 determinants,

$$x = \frac{\begin{vmatrix} 2 & -1 & 1 \\ 1 & -2 & -1 \\ 0 & 0 & 4 \end{vmatrix}}{\begin{vmatrix} 1 & -1 & 1 \\ 0 & -2 & -1 \\ 1 & 0 & 4 \end{vmatrix}} = \frac{-12}{-5} = \frac{12}{5}$$

$$y = \frac{\begin{vmatrix} 1 & 2 & 1 \\ 0 & 1 & -1 \\ 1 & 0 & 4 \end{vmatrix}}{\begin{vmatrix} 1 & -1 & 1 \\ 0 & -2 & -1 \\ 1 & 0 & 4 \end{vmatrix}} = \frac{1}{-5} = -\frac{1}{5}$$

$$z = \frac{\begin{vmatrix} 1 & -1 & 2 \\ 0 & -2 & 1 \\ 1 & 0 & 0 \end{vmatrix}}{\begin{vmatrix} 1 & -1 & 1 \\ 0 & -2 & -1 \\ 1 & 0 & 4 \end{vmatrix}} = \frac{3}{-5} = -\frac{3}{5}$$

The application of Cramer's rule to n equations with n unknowns is given next. Consider the following system of n equations with n unknowns.

$$\begin{aligned} a_{11}x_1 + a_{12}x_2 + \cdots + a_{1n}x_n &= b_1 \\ a_{21}x_1 + a_{22}x_2 + \cdots + a_{2n}x_n &= b_2 \\ \vdots \quad & \quad \vdots \\ a_{n1}x_1 + a_{n2}x_2 + \cdots + a_{nn}x_n &= b_n \end{aligned}$$

From Cramer's rule the unknown quantities $x_1, x_2, \cdots, x_n$, are as follows.

$$x_1 = \frac{\begin{vmatrix} b_1 & a_{12} & a_{13} & \cdots & a_{1n} \\ b_2 & a_{22} & a_{23} & \cdots & a_{2n} \\ \vdots & \vdots & \vdots & & \vdots \\ b_n & a_{n2} & a_{n3} & \cdots & a_{nn} \end{vmatrix}}{\begin{vmatrix} a_{11} & a_{12} & a_{13} & \cdots & a_{1n} \\ a_{21} & a_{22} & a_{23} & \cdots & a_{2n} \\ \vdots & \vdots & \vdots & & \vdots \\ a_{n1} & a_{n2} & a_{n3} & \cdots & a_{nn} \end{vmatrix}}$$

$$x_2 = \frac{\begin{vmatrix} a_{11} & b_1 & a_{13} & \cdots & a_{1n} \\ a_{21} & b_2 & a_{23} & \cdots & a_{2n} \\ \vdots & \vdots & \vdots & & \vdots \\ a_{n1} & b_{n2} & a_{n3} & \cdots & a_{nn} \end{vmatrix}}{\begin{vmatrix} a_{11} & a_{12} & a_{13} & \cdots & a_{1n} \\ a_{21} & a_{22} & a_{23} & \cdots & a_{2n} \\ \vdots & \vdots & \vdots & & \vdots \\ a_{n1} & a_{n2} & a_{n3} & \cdots & a_{nn} \end{vmatrix}}$$

$$\vdots$$

$$x_n = \frac{\begin{vmatrix} a_{11} & a_{12} & a_{13} & \cdots & b_1 \\ a_{21} & a_{22} & a_{23} & \cdots & b_2 \\ \vdots & \vdots & \vdots & & \vdots \\ a_{n1} & a_{n2} & a_{n3} & \cdots & b_n \end{vmatrix}}{\begin{vmatrix} a_{11} & a_{12} & a_{13} & \cdots & a_{1n} \\ a_{21} & a_{22} & a_{23} & \cdots & a_{2n} \\ \vdots & \vdots & \vdots & & \vdots \\ a_{n1} & a_{n2} & a_{n3} & \cdots & a_{nn} \end{vmatrix}}$$

5.5 GAUSSIAN ELIMINATION METHOD

Although we make extensive use of Cramer's rule in this text to solve systems of equations, the Gaussian elimination method is more suitable for larger systems of equations and is an excellent alternative for smaller systems.

In the nineteenth century Carl Friedrich Gauss (1777–1855), a German mathematician and physicist, presented a method for solving systems of equations. The method is called the Gaussian elimination method. Prior to discussing the Gaussian elimination method, consider the procedure for solving the system of equations (Equations (1), (2), and (3)) given here.

$$\begin{aligned} &(1) \quad & x_1 + x_2 - x_3 &= 4 \\ &(2) \quad & 2x_1 - x_2 + x_3 &= 2 \\ &(3) \quad & -x_1 + x_2 + 2x_3 &= 8 \end{aligned}$$

We eliminate x_1 from Equation (2) by multiplying Equation (1) by -2 and adding the resulting equation to Equation (2). The result is

$$\begin{aligned} &(1) \quad & x_1 + x_2 - x_3 &= 4 \\ &(2) \quad & -3x_2 + 3x_3 &= -6 \\ &(3) \quad & -x_1 + x_2 - 2x_3 &= 8 \end{aligned}$$

We next eliminate x_1 from Equation (3) by adding Equation (1) to Equation (3). The result is

$$\begin{aligned} &(1) \quad & x_1 + x_2 - x_3 &= 4 \\ &(2) \quad & -3x_2 + 3x_3 &= -6 \\ &(3) \quad & 2x_2 + x_3 &= 12 \end{aligned}$$

We next divide Equation (2) by -3 and Equation (3) by 2. The result is

$$
\begin{aligned}
&(1) \quad x_1 + x_2 - x_3 = 4\\
&(2) \quad x_2 - x_3 = 2\\
&(3) \quad x_2 + \frac{1}{2}x_3 = 6
\end{aligned}
$$

We next multiply Equation (2) by -1 and add the resulting equation to Equation (3). The result is

$$
\begin{aligned}
&(1) \quad x_1 + x_2 - x_3 = 4\\
&(2) \quad x_2 - x_3 = 2\\
&(3) \quad \frac{3}{2}x_3 = 4
\end{aligned}
$$

Finally, we divide Equation (3) by $\frac{3}{2}$. The result is

$$
\begin{aligned}
&(1) \quad x_1 + x_2 - x_3 = 4\\
&(2) \quad x_2 - x_3 = 2\\
&(3) \quad x_3 = \frac{8}{3}
\end{aligned}
$$

Equation (3) states that

$$x_3 = \frac{8}{3}$$

Substituting $x_3 = \frac{8}{3}$ into Equation (2) results in

$$x_2 = \frac{14}{3}$$

and substituting $x_2 = \frac{-14}{3}$ and $x_3 = \frac{8}{3}$ into Equation (1) results in

$$x_1 = 2$$

In this example, x_1 and x_2 are eliminated from Equation (3) and x_1 is eliminated from Equation (2). Equation (3) then yields x_3 directly. x_3 is substituted into Equation (2), yielding x_2. x_2 and x_3 are then substituted into Equation (1), yielding x_1. The elimination of x_1 from Equation (2) and x_1 and x_2 from Equation (3) is accomplished by appropriately adding or subtracting equations and multiplying equations by constants, as illustrated.

The Gaussian elimination method is identical to the procedure just presented, except that we work with arrays that represent the equations instead of the equations themselves. We shall demonstrate the Gaussian elimination method by using the same equations previously considered.

$$\begin{aligned} x_1 + x_2 - x_3 &= 4 \\ 2x_1 - x_2 + x_3 &= 2 \\ -x_1 + x_2 + 2x_3 &= 8 \end{aligned}$$

Although it is not necessary to accompany each array with the equations represented by the array, we shall do so in this illustration for clarity.

The original equations and the corresponding Guassian array are as follows.

$$\begin{pmatrix} 1 & 1 & -1 & 4 \\ 2 & -1 & 1 & 2 \\ -1 & 1 & 2 & 8 \end{pmatrix} \qquad \begin{aligned} x_1 + x_2 - x_3 &= 4 \\ 2x_1 - x_2 + x_3 &= 2 \\ -x_1 + x_2 + 2x_3 &= 8 \end{aligned}$$

Multiplying row 1 by -2 and adding the resulting row to row 2 results in

$$\begin{pmatrix} 1 & 1 & -1 & 4 \\ 0 & -3 & 3 & -6 \\ -1 & 1 & 2 & 8 \end{pmatrix} \qquad \begin{aligned} x_1 + x_2 - x_3 &= 4 \\ -3x_2 + 3x_3 &= -6 \\ -x_1 + x_2 + 2x_3 &= 8 \end{aligned}$$

Adding row 1 to row 3 results in

$$\begin{pmatrix} 1 & 1 & -1 & 4 \\ 0 & -3 & 3 & -6 \\ 0 & 2 & 1 & 12 \end{pmatrix} \qquad \begin{aligned} x_1 + x_2 - x_3 &= 4 \\ -3x_2 + 3x_3 &= -6 \\ 2x_2 + x_3 &= 12 \end{aligned}$$

Dividing row 2 by -3 and dividing row 3 by 2 results in

$$\begin{pmatrix} 1 & 1 & -1 & 4 \\ 0 & 1 & -1 & 2 \\ 0 & 1 & \frac{1}{2} & 6 \end{pmatrix} \qquad \begin{aligned} x_1 + x_2 - x_3 &= 4 \\ x_2 - x_3 &= 2 \\ x_2 + \frac{1}{2}x_3 &= 6 \end{aligned}$$

Multiplying row 2 by -1 and adding the resulting row to row 3 gives

$$\begin{pmatrix} 1 & 1 & -1 & 4 \\ 0 & 1 & -1 & 2 \\ 0 & 0 & \frac{3}{2} & 4 \end{pmatrix} \qquad \begin{aligned} x_1 + x_2 - x_3 &= 4 \\ x_2 - x_3 &= 2 \\ \frac{3}{2}x_3 &= 4 \end{aligned}$$

Dividing row 3 by $\frac{3}{2}$ results in

$$\begin{pmatrix} 1 & 1 & -1 & 4 \\ 0 & 1 & -1 & 2 \\ 0 & 0 & 1 & \frac{8}{3} \end{pmatrix} \qquad \begin{aligned} x_1 + x_2 - x_3 &= 4 \\ x_2 - x_3 &= 2 \\ x_3 &= \frac{8}{3} \end{aligned}$$

Finally,

$$x_3 = \frac{8}{3}$$

$$x_2 - \frac{8}{3} = 2$$

$$x_2 = \frac{14}{3}$$

$$x_1 + \frac{14}{3} - \frac{8}{3} = 4$$

$$x_1 = 2$$

Notice that the procedure consists of operating on the original Guassian array until it is of the form

$$\begin{pmatrix} 1 & a_{12} & a_{13} & a_{14} \\ 0 & 1 & a_{23} & a_{24} \\ 0 & 0 & 1 & a_{34} \end{pmatrix}$$

where the a's are constants determined from the mathematical operations performed on the rows.

Consider the following examples.

EXAMPLE 5.8 Solve the following set of simultaneous equations using the Gaussian elimination method.

$$\begin{aligned} x_1 + 2x_2 &= -6 \\ -x_1 + 4x_2 &= 8 \end{aligned}$$ ■

SOLUTION The Gaussian array is

$$\begin{pmatrix} 1 & 2 & -6 \\ -1 & 4 & 8 \end{pmatrix}$$

Adding row 1 to row 2 results in

$$\begin{pmatrix} 1 & 2 & -6 \\ 0 & 6 & 2 \end{pmatrix}$$

Dividing row 2 by 6 results in

$$\begin{pmatrix} 1 & 2 & -6 \\ 0 & 1 & \frac{1}{3} \end{pmatrix}$$

$$x_2 = \frac{1}{3}$$

$$x_1 + 2\left(\frac{1}{3}\right) = -6$$

$$x_1 = \frac{-20}{3}$$

EXAMPLE 5.9 Solve the following set of simultaneous equations using the Gaussian elimination method.

$$\begin{aligned} x_1 + x_2 - x_3 &= -2 \\ 2x_1 - x_2 + x_3 &= -5 \\ x_1 - 2x_2 + 3x_3 &= 4 \end{aligned}$$

■

SOLUTION The Gaussian array is

$$\begin{pmatrix} 1 & 1 & -1 & -2 \\ 2 & -1 & 1 & -5 \\ 1 & -2 & 3 & 4 \end{pmatrix}$$

Multiplying row 1 by -2 and adding the resulting row to row 2 results in

$$\begin{pmatrix} 1 & 1 & -1 & -2 \\ 0 & -3 & 3 & -1 \\ 1 & -2 & 3 & 4 \end{pmatrix}$$

Multiplying row 1 by -1 and adding the resulting row to row 3 results in

$$\begin{pmatrix} 1 & 1 & -1 & -2 \\ 0 & -3 & 3 & -1 \\ 0 & -3 & 4 & 6 \end{pmatrix}$$

Multiplying row 2 by -1 and adding the resulting row to row 3 results in

$$\begin{pmatrix} 1 & 1 & -1 & -2 \\ 0 & -3 & 3 & -1 \\ 0 & 0 & 1 & 7 \end{pmatrix}$$

Dividing row 2 by -3 results in

$$\begin{pmatrix} 1 & 1 & -1 & -2 \\ 0 & 1 & -1 & \frac{1}{3} \\ 0 & 0 & 1 & 7 \end{pmatrix}$$

$$x_3 = 7$$

$$x_2 - 7 = \frac{1}{3}$$

$$x_2 = \frac{22}{3}$$

$$x_1 + \left(\frac{22}{3}\right) - 7 = -2$$

$$x_1 = \frac{-7}{3}$$

5.6 GAUSS-JORDAN ELIMINATION METHOD

The Gaussian elimination method can be continued until the Gaussian array is in the following form.

$$\begin{pmatrix} 1 & 0 & 0 & a_{14} \\ 0 & 1 & 0 & a_{24} \\ 0 & 0 & 1 & a_{34} \end{pmatrix}$$

The method is then called the Gauss-Jordan elimination method in honor of Camille Jordan (1838–1921), a French mathematician. The method is demonstrated in Example 5.10.

EXAMPLE 5.10 Solve the simultaneous equations in Example 5.9 using the Gauss-Jordan elimination method.

$$\begin{aligned} x_1 + x_2 - x_3 &= -2 \\ 2x_1 - x_2 + x_3 &= -5 \\ x_1 - 2x_2 + 3x_3 &= 4 \end{aligned}$$

■

SOLUTION From Example 5.9, the Gaussian array can be reduced to the following array.

$$\begin{pmatrix} 1 & 1 & -1 & -2 \\ 0 & 1 & -1 & \frac{1}{3} \\ 0 & 0 & 1 & 7 \end{pmatrix}$$

Adding row 3 to row 2 results in

$$\begin{pmatrix} 1 & 1 & -1 & -2 \\ 0 & 1 & 0 & \frac{22}{3} \\ 0 & 0 & 1 & 7 \end{pmatrix}$$

Adding row 3 to row 1 results in

$$\begin{pmatrix} 1 & 1 & 0 & 5 \\ 0 & 1 & 0 & \frac{22}{3} \\ 0 & 0 & 1 & 7 \end{pmatrix}$$

Multiplying row 2 by −1 and adding the resulting row to row 1 results in

$$\begin{pmatrix} 1 & 0 & 0 & \frac{-7}{3} \\ 0 & 1 & 0 & \frac{22}{3} \\ 0 & 0 & 1 & 7 \end{pmatrix}$$

$$x_3 = 7$$

$$x_2 = \frac{22}{3}$$

$$x_1 = \frac{-7}{3}$$

The solution of simultaneous algebraic equations is an integral part of electric circuit analysis procedures. This will become apparent in the next chapter, where methods for solving multiple-loop circuits are presented.

5.7 EXERCISES

1. Evaluate each of the 2 × 2 determinants in Figure 5.7 using the rule in Figure 5.2.

Figure 5.7

$$\begin{vmatrix} 3 & -8 \\ 1 & 2 \end{vmatrix} \quad \begin{vmatrix} 16 & 7 \\ 0 & 2 \end{vmatrix} \quad \begin{vmatrix} -1 & 12 \\ 8 & -6 \end{vmatrix}$$

(a) (b) (c)

2. Evaluate each of the 2 × 2 determinants in Figure 5.8 using the rule in Figure 5.2.

Figure 5.8

$$\begin{vmatrix} -4 & -2 \\ -8 & -1 \end{vmatrix} \quad \begin{vmatrix} 30 & -40 \\ 20 & -10 \end{vmatrix} \quad \begin{vmatrix} 14 & 7 \\ -2 & -4 \end{vmatrix}$$

(a) (b) (c)

3. Evaluate each of the 2 × 2 determinants in Figure 5.9 using the rule in Figure 5.2.

Figure 5.9

$$\begin{vmatrix} 0 & -8 \\ 6 & 0 \end{vmatrix} \quad \begin{vmatrix} 0 & 0 \\ -4 & 20 \end{vmatrix} \quad \begin{vmatrix} 0 & -3 \\ 0 & 42 \end{vmatrix}$$

(a) (b) (c)

4. Evaluate each of the 2 × 2 determinants in Figure 5.10 using the rule in Figure 5.2.

Figure 5.10

$$\begin{vmatrix} -4 & 0 \\ -5 & 8 \end{vmatrix} \quad \begin{vmatrix} 0 & 1 \\ 1 & 1 \end{vmatrix} \quad \begin{vmatrix} 12 & -2 \\ 4 & -1 \end{vmatrix}$$

(a) (b) (c)

5. Evaluate each of the 3 × 3 determinants in Figure 5.11 using the rule in Figure 5.3.

Figure 5.11

$$\begin{vmatrix} 4 & -8 & 10 \\ 0 & 1 & -2 \\ 10 & 0 & 3 \end{vmatrix} \qquad \begin{vmatrix} 6 & -1 & 2 \\ 4 & 0 & 0 \\ 8 & 1 & 6 \end{vmatrix} \qquad \begin{vmatrix} 14 & 16 & 20 \\ 0 & 0 & 1 \\ -2 & 0 & -1 \end{vmatrix}$$

(a) (b) (c)

6. Evaluate each of the 3×3 determinants in Figure 5.12 using the rule in Figure 5.3.

Figure 5.12

$$\begin{vmatrix} 8 & 2 & -1 \\ -2 & -1 & 2 \\ 4 & 6 & -3 \end{vmatrix} \qquad \begin{vmatrix} -1 & -4 & -1 \\ -6 & -6 & -3 \\ -8 & -1 & -5 \end{vmatrix} \qquad \begin{vmatrix} 2 & 16 & -14 \\ 0 & 0 & 0 \\ -1 & 2 & -4 \end{vmatrix}$$

(a) (b) (c)

7. Evaluate each of the 3×3 determinants in Figure 5.13 using the rule in Figure 5.3.

Figure 5.13

$$\begin{vmatrix} -3 & 8 & 9 \\ 2 & 0 & 1 \\ 0 & 0 & -8 \end{vmatrix} \qquad \begin{vmatrix} 14 & 0 & 30 \\ -40 & 0 & 1 \\ -2 & 0 & 8 \end{vmatrix} \qquad \begin{vmatrix} 16 & -7 & -1 \\ 4 & 8 & 2 \\ 16 & -7 & -1 \end{vmatrix}$$

(a) (b) (c)

8. Evaluate each of the 3×3 determinants in Figure 5.14 using the rule in Figure 5.3.

Figure 5.14

$$\begin{vmatrix} 6 & -6 & 6 \\ 1 & -4 & 1 \\ 15 & 5 & -10 \end{vmatrix} \qquad \begin{vmatrix} 4 & 14 & 4 \\ -6 & 18 & -6 \\ 10 & 30 & 10 \end{vmatrix} \qquad \begin{vmatrix} 31 & 6 & -1 \\ 0 & 0 & -2 \\ -2 & 8 & 0 \end{vmatrix}$$

(a) (b) (c)

9. Use Chio's rule to evaluate each of the determinants in Figure 5.15.

Figure 5.15

$$\begin{vmatrix} 1 & 4 & 6 \\ 0 & -1 & 2 \\ 8 & 0 & 1 \end{vmatrix} \qquad \begin{vmatrix} 0 & 6 & 7 \\ 4 & -8 & 2 \\ 1 & -2 & 5 \end{vmatrix}$$

(a) (b)

10. Use Chio's rule to evaluate each of the determinants in Figure 5.16.

Figure 5.16

$$\begin{vmatrix} 0 & 2 & -1 & 5 \\ 9 & 0 & 0 & -2 \\ 4 & -4 & 1 & 1 \\ 10 & 6 & -8 & 0 \end{vmatrix} \qquad \begin{vmatrix} 14 & 1 & 0 & 0 \\ 1 & 10 & -3 & 2 \\ 6 & 1 & 4 & 0 \\ 0 & 0 & 1 & 8 \end{vmatrix}$$

(a) (b)

11. Use Chio's rule to evaluate each of the determinants in Figure 5.17.

Figure 5.17

$$\begin{vmatrix} 6 & 1 & 2 & 4 & 8 \\ 0 & -2 & 7 & 6 & 0 \\ 0 & 0 & 0 & -8 & 1 \\ 1 & 5 & -5 & 8 & -1 \\ 10 & -4 & 0 & 0 & -4 \end{vmatrix} \qquad \begin{vmatrix} 0 & -2 & 3 & 8 & -1 \\ 4 & 8 & 2 & 6 & 4 \\ 5 & 0 & 0 & 1 & 1 \\ 10 & -10 & 8 & 2 & 0 \\ -2 & -4 & 1 & 8 & -1 \end{vmatrix}$$

(a) (b)

12. Use Cramer's rule to determine solutions to the sets of equations in Figure 5.18.

Figure 5.18

$$\begin{array}{rcl} 2x_1 - 10x_2 = 6 & \quad 10x_1 - x_2 = 0 & \quad -3x_1 + x_2 = 0 \\ -x_1 + 6x_2 = 0 & \quad -8x_1 + 9x_2 = 6 & \quad -x_1 + 6x_2 = -4 \\ \text{(a)} & \text{(b)} & \text{(c)} \end{array}$$

13. Use Cramer's rule to determine solutions to the sets of equations in Figure 5.19.

Figure 5.19

$$\begin{array}{ccc} -x_1 + x_2 + 2x_3 = 8 & \quad x_1 + 3x_2 - 6x_3 = 6 & \quad x_1 - x_3 = 4 \\ 2x_1 - x_2 + x_3 = 3 & \quad 2x_1 - x_2 + x_3 = 1 & \quad 2_{x1} + x_2 - 2x_3 = 0 \\ x_1 + x_2 + x_3 = 4 & \quad x_1 - 2x_2 + 2x_3 = -1 & \quad x_1 + x_2 - 3x_3 = -1 \\ \text{(a)} & \text{(b)} & \text{(c)} \end{array}$$

14. Use Cramer's rule to determine solutions to the sets of equations in Figure 5.20.

Figure 5.20

$$\begin{array}{ccc} 4x_1 + x_2 - x_3 = -1 & \quad 4x_1 - 2x_2 + 3x_3 = 4 & \quad x_1 - x_2 + x_3 = -1 \\ x_1 - x_2 = 4 & \quad 3x_1 + 5x_2 + x_3 = 5 & \quad 2x_1 + x_2 - 2x_3 = 0 \\ 6x_1 - 2x_2 + x_3 = 0 & \quad 5x_1 - x_2 + 4x_3 = 5 & \quad 4x_1 - x_2 + 3x_3 = 1 \\ \text{(a)} & \text{(b)} & \text{(c)} \end{array}$$

15. Use Cramer's rule to determine solutions to the sets of equations in Figure 5.21.

Figure 5.21

$$\begin{array}{ccc} 3x_1 + 2x_2 - x_4 = 0 & \quad x_1 - x_2 + 2x_3 + x_4 = 2 & \quad 3x_1 - x_2 + x_3 - 4x_4 = 0 \\ 2x_1 + x_3 + 2x_4 = 4 & \quad x_2 + 2x_3 - x_4 = 4 & \quad x_1 + x_3 = -2 \\ x_1 + 2x_2 - x_3 = -2 & \quad x_1 + x_2 + 2x_4 = 0 & \quad x_2 + 6x_4 = 4 \\ 2x_1 - x_2 + x_3 + x_4 = 2 & \quad 4x_2 - 3x_3 = -1 & \quad x_1 + 2x_2 - 6x_3 + x_4 = 0 \\ \text{(a)} & \text{(b)} & \text{(c)} \end{array}$$

16. Use the Gaussian elimination method to determine solutions to the sets of equations in Figure 5.18.

17. Use the Gaussian elimination method to determine solutions to the sets of equations in Figure 5.19.

18. Use the Gaussian elimination method to determine solutions to the sets of equations in Figure 5.20.

19. Use the Gaussian elimination method to determine solutions to the sets of equations in Figure 5.21.

CHAPTER

6

Multiloop Circuits

After completing this chapter the student should be able to

* use the mesh method of analysis to solve multiple-loop circuits
* use the nodal method of analysis to solve multiple-loop circuits
* use the loop method of analysis to solve multiple-loop circuits

The loop, mesh, and nodal methods of analysis give us systematic procedures for analyzing multiple-loop circuits. These methods allow us to analyze circuits with more than one source, where the sources are neither in series nor parallel. The material presented in Chapter 5 for solving algebraic equations for simultaneous solutions plays an important role in the analysis methods presented in this chapter.

6.1 MESH-ANALYSIS METHOD

Recall that a loop is a continuous path that closes on itself without passing through any nodes more than once. Loops enclosing an area (window) in a circuit, as illustrated in Figure 6.1, are called meshes. A *mesh* is a particular type of loop that does not contain any loops or portions of loops within itself. That is, as we traverse around a mesh we do not enclose any circuit elements.

Figure 6.1 Meshes and windows.

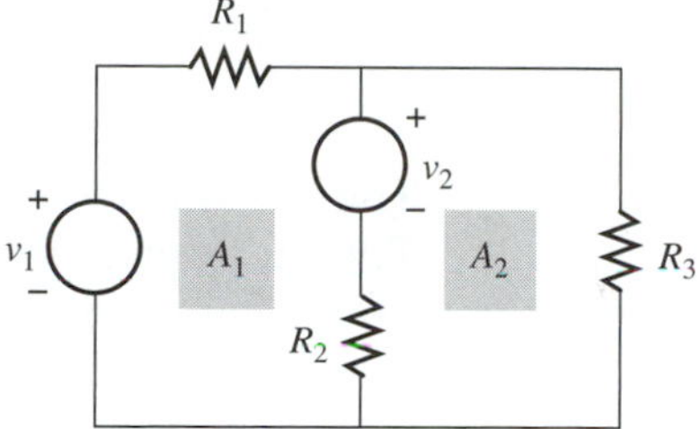

Consider the circuits in Figure 6.1, where the areas enclosed by meshes are crosshatched and labeled A_1 and A_2. The mesh enclosing area A_1 consists of elements v_1, R_1, v_2, and R_2, which form a closed path around area A_1. The mesh enclosing area A_2 consists of elements v_2, R_3, and R_2, which form a closed path around area A_2. Notice that the outermost loop, consisting of elements v_1, R_1, and R_3, is not a mesh, because it encloses elements R_2 and v_2.

Consider the following example.

EXAMPLE 6.1 Determine the meshes for the circuit in Figure 6.2. ■

Figure 6.2

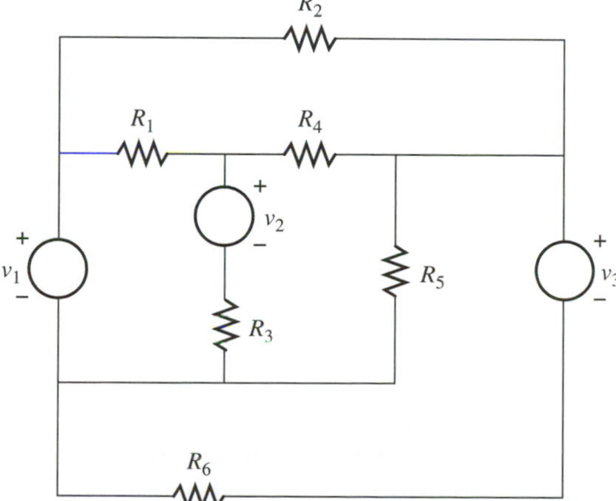

SOLUTION The circuit is redrawn in Figure 6.3 with the areas (or windows) indicated by crosshatching and labeled A_1, A_2, A_3, and A_4. There are four areas and, therefore, four meshes.

Figure 6.3

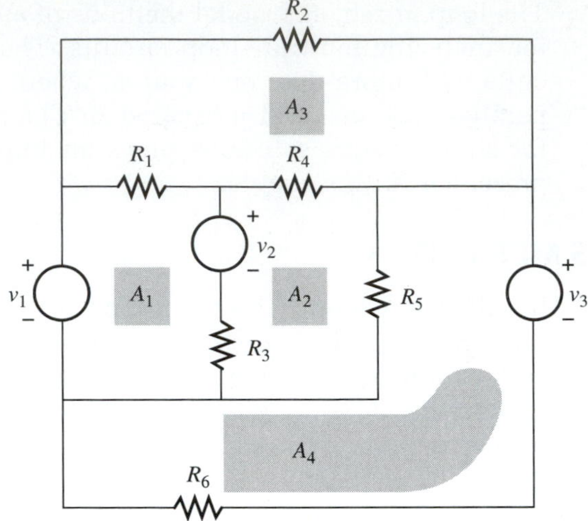

The mesh enclosing area A_1 consists of elements v_1, R_1, v_2, and R_3.
The mesh enclosing area A_2 consists of elements R_3, v_2, R_4, and R_5.
The mesh enclosing area A_3 consists of electrical elements R_1, R_2, and R_4.
The mesh enclosing area A_4 consists of electrical elements R_6, R_5, and v_3.

The *first step* in the mesh analysis method is to identify the meshes in the circuit and the *second step* is to assign a mesh current to each mesh.

A *mesh current* is a current that is common to all of the electrical elements in a mesh.

In Figure 6.4, mesh current i_1 is common to all the elements in mesh 1, which consists of elements R_1, v_1, and R_2. Mesh current i_2 is common to all the elements in mesh 2, which consists of elements R_2, v_1, R_3, and v_2. Although

Figure 6.4 Mesh currents.

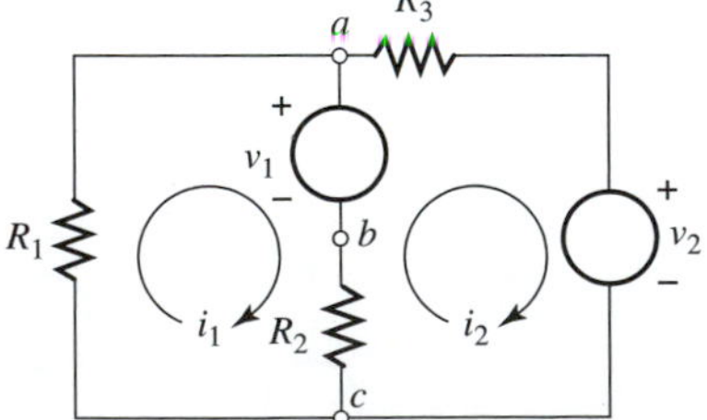

mesh currents can be assigned in a clockwise or counterclockwise direction without affecting the analysis, we shall assign all mesh currents in a clockwise direction for consistency.

Notice that a mesh current may or may not be the actual current (current read by an ammeter) flowing through a particular element. In Figure 6.4 mesh current i_1 is the actual current flowing through resistance R_1 because there are no other mesh currents flowing through R_1; and for the same reason mesh current i_2 is the actual current flowing through elements R_3 and v_2.

Elements v_1 and R_2 in Figure 6.4 each have two mesh currents flowing through them. The actual current flowing through voltage source v_1 from point a to point b is mesh current i_1 minus mesh current i_2; and the actual current flowing through voltage source v_1 from point b to point a is mesh current i_2 minus mesh current i_1. In a similar manner the actual current flowing through resistance R_2 from point b to point c is mesh current i_1 minus mesh current i_2; and the actual current flowing through resistance R_2 from point c to point b is mesh current i_2 minus mesh current i_1.

Mesh currents are treated as algebraic quantities when used to determine the actual current through an element. As an example, consider the actual current flowing from point b to point c for the circuit in Figure 6.4, where mesh currents i_1 and i_2 are arbitrarily given as

$$i_1 = -4 \text{ A}$$
$$i_2 = 2 \text{ A}$$

The actual current flowing from point b to point c is

$$\begin{aligned} i_{bc} &= i_1 - i_2 \\ &= -4 - (2) = -6 \text{ A} \end{aligned}$$

The negative sign indicates that the 6-A current actually flows from point c to point b.

The *third step* in the mesh analysis method is to assign polarities to the resistance elements based on the assigned (assumed) mesh currents. Consider the circuit in Figure 6.5, where the mesh currents are i_1 and i_2. Based on the assigned mesh current, polarities are assigned to the resistive elements, as shown in Figure 6.6. Notice that resistance R_3 is assigned two sets of polarities, one set for mesh current i_1 and one set for mesh current i_2.

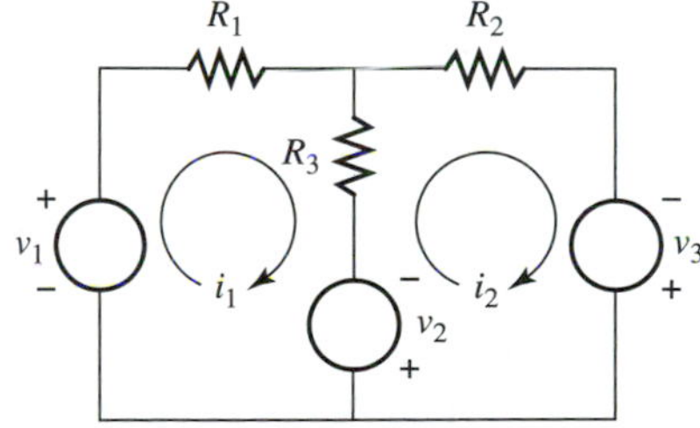

Figure 6.5 Two-mesh circuit.

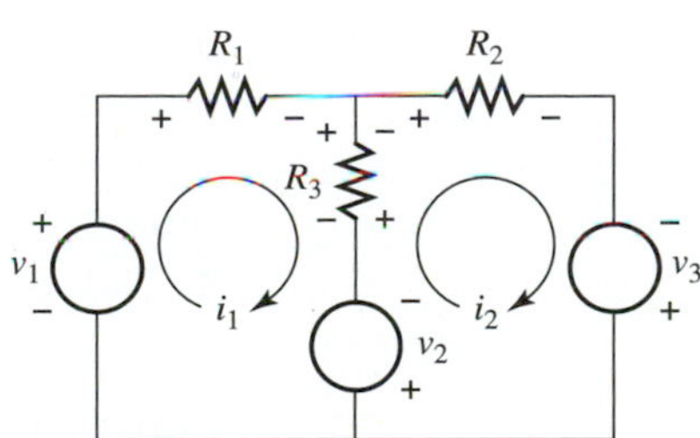

Figure 6.6 Meshes with polarities indicated.

The *fourth step* in the mesh-analysis method is to apply Kirchhoff's voltage law to each mesh in the circuit to arrive at the mathematical equations that describe the circuit. These equations form the mathematical model for the circuit, and the unknown quantities in the equations are the mesh currents.

The number of unknown mesh currents and the number of equations in the mathematical model will equal the number of meshes in the circuit.

Applying Kirchhoff's voltage law around each mesh of the circuit in Figure 6.6 in the direction of the assigned mesh currents results in

$$\text{Mesh 1}^\circ \quad v_1 - R_1 i_1 - R_3 i_1 + R_3 i_2 + v_2 = 0$$
$$\text{Mesh 2}^\circ \; -v_2 - R_3 i_2 + R_3 i_1 - R_2 i_2 + v_3 = 0$$

Notice that there are two voltage terms for resistance R_3 in each equation, because there are two mesh currents flowing through resistance R_3.

The *fifth and final step* in the mesh-analysis method is to solve the equations obtained in step 4 for the mesh currents. The equations can be written in the following form:

$$(R_1 + R_3)i_1 - (R_3)i_2 = v_1 + v_2$$
$$-(R_3)i_1 + (R_3 + R_2)i_2 = v_3 - v_2$$

Using Cramer's rule to solve the equations, the mesh currents in determinant form for the circuit in Figure 6.6 are

$$i_1 = \frac{\begin{vmatrix} (v_1 + v_2) & -(R_3) \\ (v_3 - v_2) & (R_2 + R_3) \end{vmatrix}}{\begin{vmatrix} (R_1 + R_3) & -(R_3) \\ -(R_3) & (R_2 + R_3) \end{vmatrix}}$$

$$i_2 = \frac{\begin{vmatrix} (R_1 + R_3) & (v_1 + v_2) \\ -(R_3) & (v_3 - v_2) \end{vmatrix}}{\begin{vmatrix} (R_1 + R_3) & -(R_3) \\ -(R_3) & (R_2 + R_3) \end{vmatrix}}$$

The five-step procedure for the mesh-analysis method is summarized next.

1. Determine the meshes for the circuit.
2. Assign a mesh current to each mesh.
3. Assign polarities to the resistance elements based on the assigned mesh currents.
4. Apply Kirchhoff's voltage law to each mesh to determine the math model.
5. Solve the math model for the mesh currents.

Consider the following examples.

EXAMPLE 6.2 Determine the current flowing from point a to point b through the 10-Ω resistance in Figure 6.7. ■

Figure 6.7

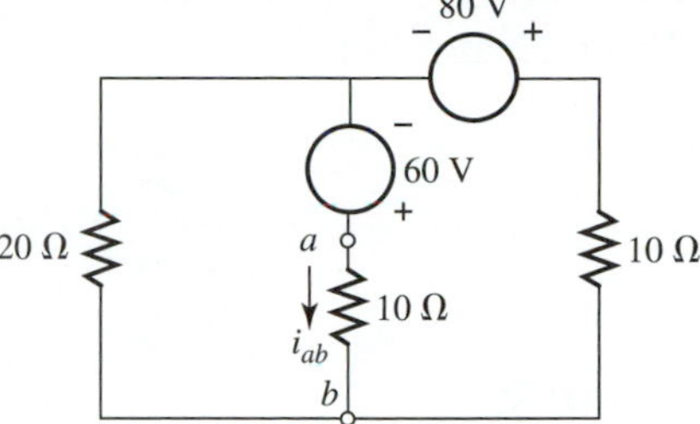

SOLUTION There are two meshes in the circuit and, therefore, two mesh currents. The circuit is shown in Figure 6.8 with the assigned mesh currents and the associated voltage polarities.

Figure 6.8

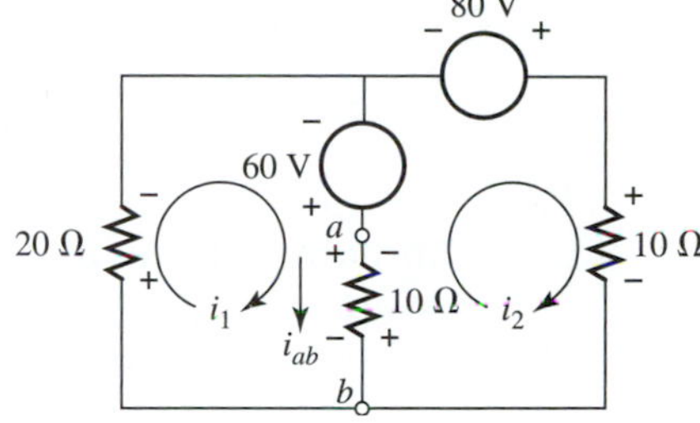

The current required is i_{ab}, where

$$i_{ab} = i_1 - i_2$$

Employing Kirchhoff's voltage law around each mesh in the direction of assigned mesh currents results in

$$\text{Mesh } 1^\circ \qquad -20i_1 + 60 - 10i_1 + 10i_2 = 0$$

$$\text{Mesh } 2^\circ \ -10i_2 + 10i_1 - 60 + 80 - 10i_2 = 0$$

and these equations can be written in the following form:

$$30i_1 - 10i_2 = 60$$

$$-10i_1 + 20i_2 = 20$$

From Cramer's rule, the mesh currents are

$$i_1 = \frac{\begin{vmatrix} 60 & -10 \\ 20 & 20 \end{vmatrix}}{\begin{vmatrix} 30 & -10 \\ -10 & 20 \end{vmatrix}} = \frac{14}{5} \text{ A}$$

$$i_2 = \frac{\begin{vmatrix} 30 & 60 \\ -10 & 20 \end{vmatrix}}{\begin{vmatrix} 30 & -10 \\ -10 & 20 \end{vmatrix}} = \frac{12}{5} \text{ A}$$

and $i_{ab} = i_1 - i_2 = \dfrac{2}{5}$ A.

EXAMPLE 6.3 For the circuit in Example 6.2, determine i_{ba}, i_{cb}, i_{bc}, i_{ad}, v_{bd}, and v_{db}. ■

SOLUTION The circuit is shown in Figure 6.9, with the appropriate points labeled. From Example 6.2,

Figure 6.9

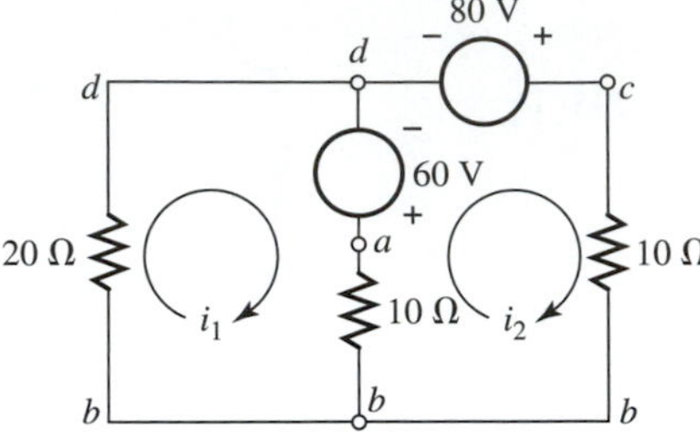

$$\text{mesh current } i_1 = \frac{14}{5} \text{ A}$$

$$\text{mesh current } i_2 = \frac{12}{5} \text{ A}$$

Because i_1 flows from point a to point b and i_2 flows from point b to point a through the 10-Ω resistor,

$$i_{ba} = i_2 - i_1 = \frac{12}{5} - \frac{14}{5} = -\frac{2}{5} \text{ A}$$

The negative sign indicates the actual current flows from point a to point b with a magnitude of $\frac{2}{5}$ A. Because only mesh current i_2 flows through the 10-Ω resistance from point c to point b,

$$i_{cb} = i_2 = \frac{12}{5} \text{ A}$$

and

$$i_{bc} = -i_2 = -\frac{12}{5} \text{ A}$$

Because i_1 flows from point d to point a and i_2 flows from point a to point d through the 60-V voltage source,

$$i_{da} = i_1 - i_2 = \frac{14}{5} - \frac{12}{5} = \frac{2}{5} \text{ A}$$

and

$$i_{ad} = i_2 - i_1 = \frac{12}{5} - \frac{14}{5} = -\frac{2}{5} \text{ A}$$

Because i_1 flows from point b to point d through the 20-Ω resistance, Ohm's law results in

$$v_{bd} = 20i_1 = (20)\left(\frac{14}{5}\right) = 56 \text{ V}$$

and

$$v_{db} = -20i_1 = -(20)\left(\frac{14}{5}\right) = -56 \text{ V}$$

EXAMPLE 6.4 Determine the mesh currents for the circuit in Figure 6.10. ■

Figure 6.10

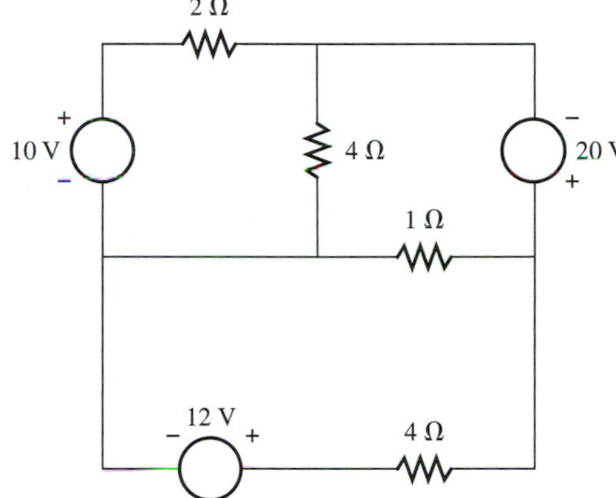

SOLUTION There are three meshes in the circuit. The circuit is shown in Figure 6.11 with the assigned mesh currents and the associated voltage polarities.

Figure 6.11

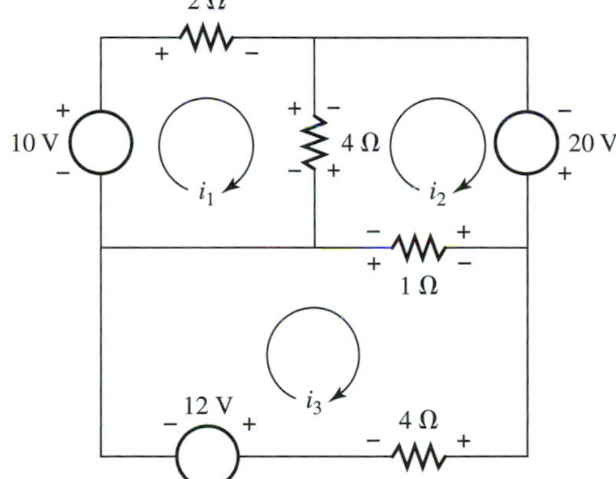

Employing Kirchhoff's voltage law around each mesh in the direction of assigned mesh currents results in

$$\text{mesh } 1^{\circ} \qquad 10 - 2i_1 - 4i_1 + 4i_2 = 0$$

$$\text{mesh } 2^{\circ} \quad -4i_2 + 4i_1 + 20 - 1i_2 + 1i_3 = 0$$

$$\text{mesh } 3^{\circ} \qquad -12 - 1i_3 + 1i_2 - 4i_3 = 0$$

These equations can be written in the following form:

$$\begin{aligned} 6i_1 - 4i_2 \quad\quad &= 10 \\ -4i_1 + 5i_2 - 1i_3 &= 20 \\ -1i_2 + 5i_3 &= -12 \end{aligned}$$

From Cramer's rule the mesh currents are

$$i_1 = \frac{\begin{vmatrix} 10 & -4 & 0 \\ 20 & 5 & -1 \\ -12 & -1 & 5 \end{vmatrix}}{\begin{vmatrix} 6 & -4 & 0 \\ -4 & 5 & -1 \\ 0 & -1 & 5 \end{vmatrix}} = \frac{592}{64} = 9.25 \text{ A}$$

$$i_2 = \frac{\begin{vmatrix} 6 & 10 & 0 \\ -4 & 20 & -1 \\ 0 & -12 & 5 \end{vmatrix}}{\begin{vmatrix} 6 & -4 & 0 \\ -4 & 5 & -1 \\ 0 & -1 & 5 \end{vmatrix}} = \frac{720}{64} = 11.3 \text{ A}$$

$$i_3 = \frac{\begin{vmatrix} 6 & 4 & 10 \\ -4 & 5 & 20 \\ 0 & -1 & -12 \end{vmatrix}}{\begin{vmatrix} 6 & -4 & 0 \\ -4 & 5 & -1 \\ 0 & -1 & 5 \end{vmatrix}} = -\frac{392}{64} = -6.13 \text{ A}$$

Although we did not consider current sources when presenting the mesh-analysis method, the method is applicable to circuits containing current sources. In fact, the presence of current sources in the circuit usually reduces the work required to obtain a solution, as illustrated in Example 6.5.

EXAMPLE 6.5 Determine the mesh currents for the circuit shown in Figure 6.12 using the mesh-analysis method. ■

Figure 6.12

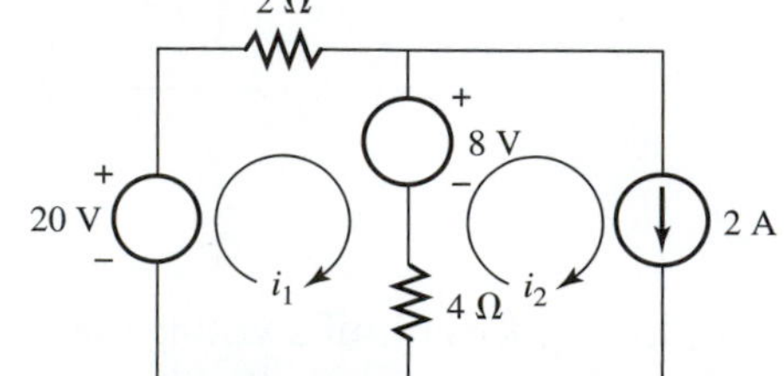

SOLUTION Notice that we do not apply Kirchhoff's voltage law to mesh 2, because we do not know the voltage across the current source. However, a simple observation of the circuit reveals that mesh current $i_2 = 2$ A, and we have only mesh current i_1 to determine.

The circuit with the resistance polarities indicated for mesh 1 is shown in Figure 6.13. Applying Kirchhoff's voltage law to mesh 1 in the direction of the assigned mesh current results in

$$20 - 2i_1 - 8 - 4i_1 + 4i_2 = 0$$

Figure 6.13

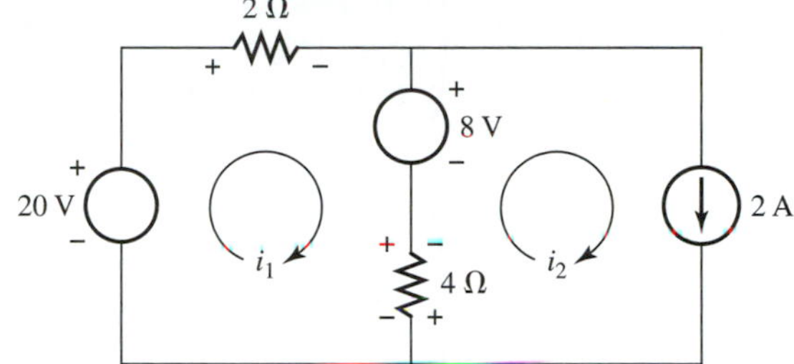

Substituting $i_2 = 2$ into the preceding equation results in

$$i_1 = \frac{10}{3} = 3.3 \text{ A}$$

6.2 UNIFORM APPROACH TO MESH ANALYSIS

The five steps for the mesh-analysis method can be simplified by comparing the mesh currents in determinant form to the circuit described by the mesh currents. The uniform approach to mesh analysis (also called the format method) allows us to write the mesh currents in determinant form by a simple inspection of the circuit after the mesh currents are assigned.

Consider the circuit in Figure 6.5, as redrawn in Figure 6.14. As determined in Section 6.1, the mesh currents in determinant form for the circuit in Figure 6.14 are

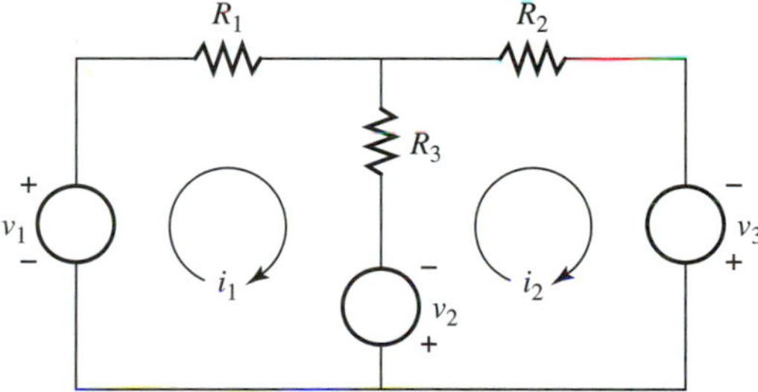

Figure 6.14 Two-mesh circuit.

$$i_1 = \frac{\begin{vmatrix} (v_1+v_2) & -(R_3) \\ (v_3-v_2) & (R_2+R_3) \end{vmatrix}}{\begin{vmatrix} (R_1+R_3) & -(R_3) \\ -(R_3) & (R_2+R_3) \end{vmatrix}}$$

$$i_2 = \frac{\begin{vmatrix} (R_1+R_3) & (v_1+v_2) \\ -(R_3) & (v_3-v_2) \end{vmatrix}}{\begin{vmatrix} (R_1+R_3) & -(R_3) \\ -(R_3) & (R_2+R_3) \end{vmatrix}}$$

Comparing the determinant entries for the mesh currents to the circuit elements for the circuit in Figure 6.14 results in

$$i_1 = \frac{\begin{vmatrix} \begin{pmatrix}\text{algebraic sum of}\\ \text{voltage sources}\\ \text{in mesh 1, pro-}\\ \text{ceeding around}\\ \text{mesh 1 in a cw}\\ \text{direction}\end{pmatrix} & -\begin{pmatrix}\text{resistance common}\\ \text{to meshes 1 and 2}\end{pmatrix} \\ \begin{pmatrix}\text{algebraic sum of}\\ \text{voltage sources}\\ \text{in mesh 2, pro-}\\ \text{ceeding around}\\ \text{mesh 2 in a cw}\\ \text{direction}\end{pmatrix} & \begin{pmatrix}\text{total resistance}\\ \text{in mesh 2}\end{pmatrix} \end{vmatrix}}{\begin{vmatrix} \begin{pmatrix}\text{total resistance}\\ \text{in mesh 1}\end{pmatrix} & -\begin{pmatrix}\text{resistance common}\\ \text{to meshes 1 and 2}\end{pmatrix} \\ -\begin{pmatrix}\text{resistance common}\\ \text{to meshes 1 and 2}\end{pmatrix} & \begin{pmatrix}\text{total resistance in}\\ \text{mesh 2}\end{pmatrix} \end{vmatrix}} \tag{6.1}$$

$$i_2 = \frac{\begin{vmatrix} \begin{pmatrix}\text{total resistance}\\ \text{mesh 1}\end{pmatrix} & \begin{pmatrix}\text{algebraic sum of}\\ \text{voltage sources}\\ \text{in mesh 1, pro-}\\ \text{ceeding around}\\ \text{mesh 1 in a cw}\\ \text{direction}\end{pmatrix} \\ -\begin{pmatrix}\text{resistance common}\\ \text{to meshes 1 and 2}\end{pmatrix} & \begin{pmatrix}\text{algebraic sum of}\\ \text{voltage sources}\\ \text{in mesh 2, pro-}\\ \text{ceeding around}\\ \text{mesh 2 in a cw}\\ \text{direction}\end{pmatrix} \end{vmatrix}}{\begin{vmatrix} \begin{pmatrix}\text{total resistance}\\ \text{in mesh 1}\end{pmatrix} & -\begin{pmatrix}\text{resistance common}\\ \text{to meshes 1 and 2}\end{pmatrix} \\ -\begin{pmatrix}\text{resistance common}\\ \text{to meshes 1 and 2}\end{pmatrix} & \begin{pmatrix}\text{total resistance in}\\ \text{mesh 2}\end{pmatrix} \end{vmatrix}} \tag{6.2}$$

Using Equations (6.1) and (6.2), we can now write the mesh currents in determinant form for two-mesh circuits by a simple inspection of the circuit. Because clockwise mesh currents are used in arriving at Equations (6.1) and (6.2), all mesh currents are assigned in the clockwise direction when using the uniform approach to the mesh-analysis method. This restriction poses no problem, because the choice of mesh current direction is arbitrary.

EXAMPLE 6.6 Determine mesh currents i_1 and i_2 for the circuit shown in Figure 6.15. ■

Figure 6.15

1 Ω, 2 Ω, 4 Ω, 10 Ω; 6 V, 8 V, 20 V, 16 V; i_1, i_2

SOLUTION Using Equations (6.1) and (6.2), the mesh currents are

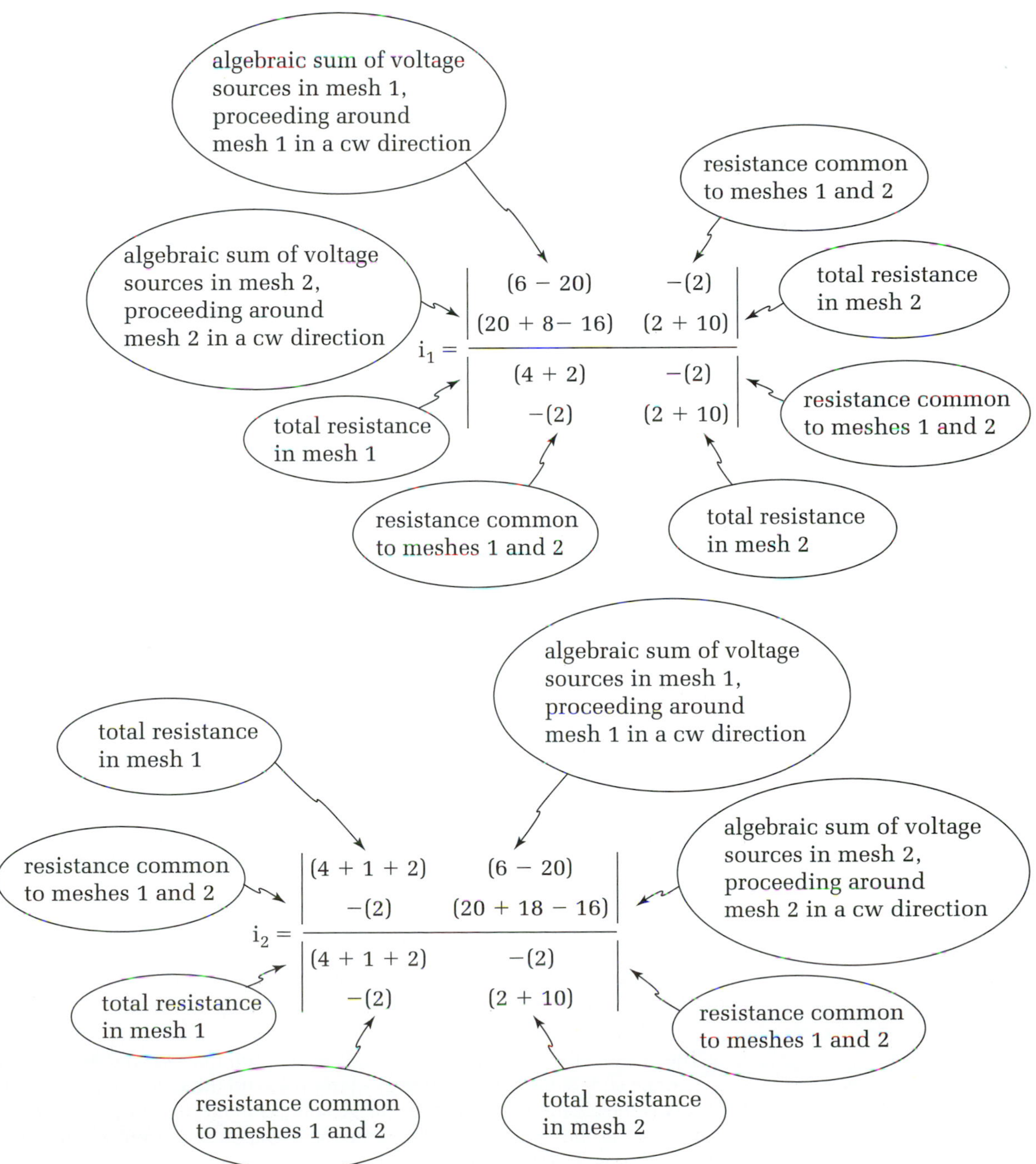

$$i_1 = \frac{\begin{vmatrix} (6-20) & -(2) \\ (20+8-16) & (2+10) \end{vmatrix}}{\begin{vmatrix} (4+2) & -(2) \\ -(2) & (2+10) \end{vmatrix}}$$

$$i_2 = \frac{\begin{vmatrix} (4+1+2) & (6-20) \\ -(2) & (20+18-16) \end{vmatrix}}{\begin{vmatrix} (4+1+2) & -(2) \\ -(2) & (2+10) \end{vmatrix}}$$

$$i_1 = \frac{\begin{vmatrix} -14 & -2 \\ 12 & 12 \end{vmatrix}}{\begin{vmatrix} 7 & -14 \\ -2 & 12 \end{vmatrix}} = \frac{9}{5}\text{ A}$$

$$i_2 = \frac{\begin{vmatrix} 7 & -14 \\ -2 & 12 \end{vmatrix}}{\begin{vmatrix} 7 & -2 \\ -2 & 12 \end{vmatrix}} = \frac{7}{10}\text{ A}$$

Let us reconsider the circuit in Figure 6.14 and the associated mesh currents, which are

$$i_1 = \frac{\begin{vmatrix} (v_1 + v_2) & -(R_3) \\ (v_3 - v_2) & (R_2 + R_3) \end{vmatrix}}{\begin{vmatrix} (R_1 + R_3) & -(R_3) \\ -(R_3) & (R_2 + R_3) \end{vmatrix}}$$

$$i_2 = \frac{\begin{vmatrix} (R_1 + R_3) & (v_1 + v_2) \\ -(R_3) & (v_3 - v_2) \end{vmatrix}}{\begin{vmatrix} (R_1 + R_3) & -(R_3) \\ -(R_3) & (R_2 + R_3) \end{vmatrix}}$$

Notice that the denominator determinants for mesh currents i_1 and i_2 are identical.

> In the uniform approach to mesh analysis, the denominator determinants are the same for all mesh currents.

Notice also the symmetry of the denominator determinants.

> In the uniform approach to mesh analysis, the denominator determinant is symmetrical about the major diagonal.

Notice the algebraic signs of the determinant entries for the denominator determinant.

> In the uniform approach to the mesh-analysis method, all off-diagonal entries in the denominator are negative, and all entries on the major diagonal are positive.

Notice that the numerator determinant for i_1 is the same as the denominator determinant for i_1 except for column 1. And the numerator determinant for i_2 is the same as the denominator determinant for i_2 except for column 2. Notice also that column 1 in the numerator determinant of i_1 and column 2 in the numerator determinant of i_2 are the same, and the entries are determined by algebraically summing the voltage sources around each mesh. The column formed in this manner is called the voltage source summation column.

> The entry in the first row of the voltage source summation column is the algebraic sum of the voltage sources in mesh 1, and the entry in the second row is the algebraic sum of the voltage sources in mesh 2. In each case we proceed around the mesh in the direction of the mesh current when summing the voltage sources.

Using the observations listed here and comparing the circuit to the mesh currents, we can formulate mesh currents i_1 and i_2 for the circuit in Figure 6.5 in the following manner.

First, construct the denominator determinant for mesh currents i_1 and i_2:

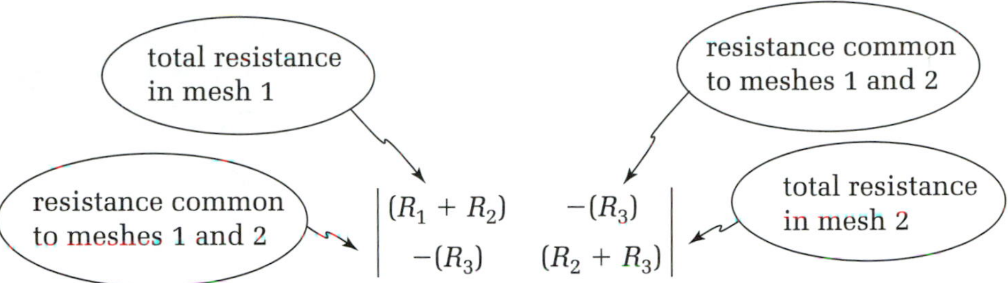

Next construct the voltage source summation column:

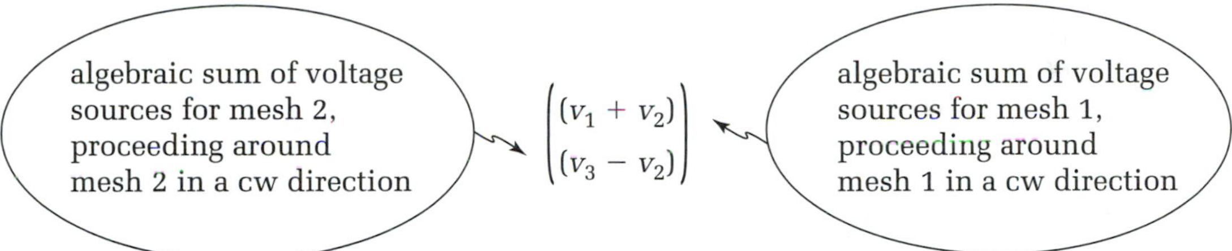

Finally, substitute the voltage source summation column for the first column of the denominator determinant to form the numerator determinant for mesh current i_1 and substitute the voltage source summation column for the second column of the denominator determinant to form the numerator determinant for mesh current i_2. This results in the following numerator determinants for mesh currents i_1 and i_2, respectively.

$$\begin{vmatrix} (v_1 + v_2) & -(R_3) \\ (v_3 - v_2) & (R_2 + R_3) \end{vmatrix}$$

$$\begin{vmatrix} (R_1 + R_2) & (v_1 + v_2) \\ -(R_3) & (v_3 - v_2) \end{vmatrix}$$

Mesh currents i_1 and i_2 for the circuit in Figure 6.5 can then be written as follows:

$$i_1 = \frac{\begin{vmatrix} (v_1 + v_2) & -(R_3) \\ (v_3 - v_2) & (R_2 + R_3) \end{vmatrix}}{\begin{vmatrix} (R_1 + R_2) & -(R_3) \\ -(R_3) & (R_2 + R_3) \end{vmatrix}}$$

$$i_2 = \frac{\begin{vmatrix} (R_1 + R_2) & (v_1 + v_2) \\ -(R_3) & (v_3 - v_2) \end{vmatrix}}{\begin{vmatrix} (R_1 + R_2) & -(R_3) \\ -(R_3) & (R_2 + R_3) \end{vmatrix}}$$

Consider the following example.

EXAMPLE 6.7 Determine the mesh currents for the circuit shown in Figure 6.16. ■

Figure 6.16

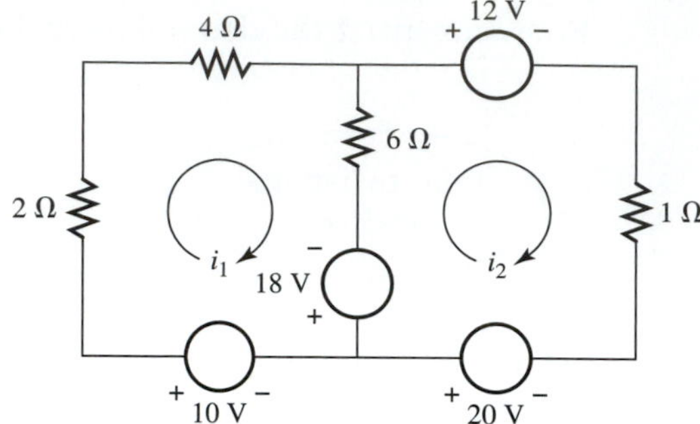

SOLUTION The denominator determinant for mesh currents i_1 and i_2 is

$$\begin{vmatrix} (2 + 4 + 6) & -(6) \\ -(6) & (6 + 1) \end{vmatrix}$$

which results in

$$\begin{vmatrix} 12 & -6 \\ -6 & 7 \end{vmatrix}$$

The voltage source summation column is

$$\begin{pmatrix} (10 + 18) \\ (20 - 18 - 12) \end{pmatrix}$$

which results in

$$\begin{pmatrix} 28 \\ -10 \end{pmatrix}$$

The numerator determinant for mesh currents i_1 and i_2, respectively, are

$$\begin{vmatrix} 28 & -6 \\ -10 & 7 \end{vmatrix}$$

$$\begin{vmatrix} 12 & 28 \\ -6 & -10 \end{vmatrix}$$

and

$$i_1 = \frac{\begin{vmatrix} 28 & -6 \\ -10 & 7 \end{vmatrix}}{\begin{vmatrix} 12 & -6 \\ -6 & 7 \end{vmatrix}} = \frac{136}{8} = 17 \text{ A}$$

$$i_2 = \frac{\begin{vmatrix} 12 & 28 \\ -6 & -10 \end{vmatrix}}{\begin{vmatrix} 12 & -6 \\ -6 & 7 \end{vmatrix}} = \frac{48}{8} = 6 \text{ A}$$

Although we considered only two-mesh circuits when presenting the uniform approach to the mesh-analysis method, the method is valid for circuits with any number of meshes. We shall introduce some symbols for conciseness while extending the method to circuits with more than two meshes.

Consider a circuit with n meshes. The major diagonal entries for the denominator determinant of the mesh currents can be written as

$$\begin{aligned} R_{11} &= \text{total resistance in mesh 1} \\ R_{22} &= \text{total resistance in mesh 2} \\ &\vdots \\ R_{nn} &= \text{total resistance in mesh } n \end{aligned}$$

The off-diagonal entries for the denominator determinant of the mesh currents can be written as

$$\begin{aligned} R_{12} &= \text{resistance common to meshes 1 and 2} \\ R_{21} &= \text{resistance common to meshes 2 and 1} \\ R_{13} &= \text{resistance common to meshes 1 and 3} \\ R_{31} &= \text{resistance common to meshes 3 and 1} \\ &\vdots \\ R_{n-1,\,n} &= \text{resistance common to meshes } n-1 \text{ and } n \\ R_{n,\,n-1} &= \text{resistance common to meshes } n \text{ and } n-1 \end{aligned}$$

Using the notation given previously, the denominator determinant for the mesh currents of an n-mesh circuit can be written as

$$\begin{vmatrix} R_{11} & -R_{12} & -R_{13} & \cdots & -R_{1n} \\ -R_{21} & R_{22} & -R_{23} & \cdots & -R_{2n} \\ \vdots & \vdots & \vdots & & \vdots \\ -R_{n1} & -R_{n2} & -R_{n3} & \cdots & R_{nn} \end{vmatrix}$$

The entries for the voltage source summation column can be written as

$$\begin{aligned} v_1 &= \text{algebraic sum of voltage sources in mesh 1 obtained by proceeding around mesh 1 in the direction of mesh current 1} \\ v_2 &= \text{algebraic sum of voltage sources in mesh 2 obtained by proceeding around mesh 2 in the direction of mesh current 2} \\ &\vdots \\ v_n &= \text{algebraic sum of voltage sources in mesh } n \text{ obtained by proceeding around mesh } n \text{ in the direction of mesh current } n \end{aligned}$$

and the voltage source summation column can then be written as

$$\begin{pmatrix} v_1 \\ v_2 \\ \vdots \\ v_n \end{pmatrix}$$

We can now write the n mesh currents for an n mesh circuit as follows:

$$i_1 = \frac{\begin{vmatrix} v_1 & -R_{12} & -R_{13} & \cdots & -R_{1n} \\ v_2 & R_{22} & -R_{23} & \cdots & -R_{2n} \\ \vdots & \vdots & \vdots & & \vdots \\ v_n & -R_{n2} & -R_{n3} & \cdots & R_{nn} \end{vmatrix}}{\begin{vmatrix} R_{11} & -R_{12} & -R_{13} & \cdots & -R_{1n} \\ -R_{21} & R_{22} & -R_{23} & \cdots & -R_{2n} \\ \vdots & \vdots & \vdots & & \vdots \\ -R_{n1} & -R_{n2} & -R_{n3} & \cdots & R_{nn} \end{vmatrix}} \tag{6.3}$$

$$i_2 = \frac{\begin{vmatrix} R_{11} & v_1 & -R_{13} & \cdots & -R_{1n} \\ -R_{21} & v_2 & -R_{23} & \cdots & -R_{2n} \\ \vdots & \vdots & \vdots & & \vdots \\ -R_{n1} & v_n & -R_{n3} & \cdots & R_{nn} \end{vmatrix}}{\begin{vmatrix} R_{11} & -R_{12} & -R_{13} & \cdots & -R_{1n} \\ -R_{21} & R_{22} & -R_{23} & \cdots & -R_{2n} \\ \vdots & \vdots & \vdots & & \vdots \\ -R_{n1} & -R_{n2} & -R_{n3} & \cdots & R_{nn} \end{vmatrix}} \tag{6.4}$$

$$\vdots$$

$$i_n = \frac{\begin{vmatrix} R_{11} & -R_{12} & -R_{13} & \cdots & v_1 \\ -R_{21} & R_{22} & -R_{23} & \cdots & v_2 \\ \vdots & \vdots & \vdots & & \vdots \\ -R_{n1} & -R_{n2} & -R_{n3} & \cdots & v_n \end{vmatrix}}{\begin{vmatrix} R_{11} & -R_{12} & -R_{13} & \cdots & -R_{1n} \\ -R_{21} & R_{22} & -R_{23} & \cdots & -R_{2n} \\ \vdots & \vdots & \vdots & & \vdots \\ -R_{n1} & -R_{n2} & -R_{n3} & \cdots & R_{nn} \end{vmatrix}} \tag{6.5}$$

Notice that all the denominators for the mesh currents in Equations (6.3), (6.4), and (6.5) are the same and that all the numerator determinants are formed from the denominator determinant by replacing one column of the denominator determinant with the voltage source summation column. The voltage source summation column replaces column 1 of the denominator determinant to form the numerator determinant for mesh current i_1; the voltage source summation column replaces column 2 of the denominator determinant to form the numerator determinant for mesh current i_2; and all the other mesh currents through i_n are formed in a similar manner.

Notice also that the denominator determinant is symmetrical about the major diagonal, because $R_{ij} = R_{ji}$; the offdiagonal elements in the denominator determinant are all negative; the major diagonal elements in the denominator determinant are all positive; and the number of rows and columns in the denominator determinant equals the number of meshes in the circuit and the number of unknown mesh currents.

Consider the following examples.

EXAMPLE 6.8 Determine the mesh currents for the circuit shown in Figure 6.17. ■

Figure 6.17

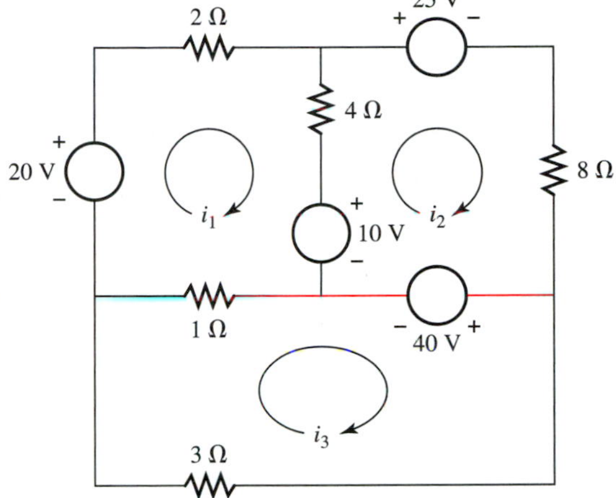

SOLUTION Because there are three meshes, the denominator determinant contains three rows and three columns. The entries for the denominator determinant of the mesh currents are

$$R_{11} = 2 + 4 + 1 = 7$$

$$R_{22} = 4 + 8 = 12$$

$$R_{33} = 1 + 3 = 4$$

$$R_{12} = R_{21} = 4$$

$$R_{13} = R_{31} = 1$$

$$R_{23} = R_{32} = 0$$

and the denominator determinant is

$$\begin{vmatrix} 7 & -4 & -1 \\ -4 & 12 & -0 \\ -1 & -0 & 4 \end{vmatrix}$$

The entries for the voltage summation column are

$$v_1 = 20 - 10 = 10$$

$$v_2 = 10 - 25 - 40 = -55$$

$$v_3 = 40$$

and the voltage summation column is

$$\begin{pmatrix} 10 \\ -55 \\ 40 \end{pmatrix}$$

The numerator determinants for mesh currents i_1, i_2, and i_3, respectively, are

$$\begin{vmatrix} 10 & -4 & -1 \\ -55 & 12 & 0 \\ 40 & 0 & 4 \end{vmatrix}$$

$$\begin{vmatrix} 7 & 10 & -1 \\ -4 & -55 & 0 \\ -1 & 40 & 4 \end{vmatrix}$$

$$\begin{vmatrix} 7 & -4 & 10 \\ -4 & 12 & -55 \\ -1 & 0 & 40 \end{vmatrix}$$

so,

$$i_1 = \frac{\begin{vmatrix} 10 & -4 & -1 \\ -55 & 12 & 0 \\ 40 & 0 & 4 \end{vmatrix}}{\begin{vmatrix} 7 & -4 & -1 \\ -4 & 12 & 0 \\ -1 & 0 & 4 \end{vmatrix}} = \frac{80}{260} = 308 \text{ mA}$$

$$i_2 = \frac{\begin{vmatrix} 7 & 10 & -1 \\ -4 & -55 & 0 \\ -1 & 40 & 4 \end{vmatrix}}{\begin{vmatrix} 7 & -4 & -1 \\ -4 & 12 & 0 \\ -1 & 0 & 4 \end{vmatrix}} = -\frac{1165}{260} = -4.48 \text{ A}$$

$$i_3 = \frac{\begin{vmatrix} 7 & -4 & 10 \\ -4 & 12 & -55 \\ -1 & 0 & 40 \end{vmatrix}}{\begin{vmatrix} 7 & -4 & -1 \\ -4 & 12 & 0 \\ -1 & 0 & 4 \end{vmatrix}} = \frac{2620}{260} = 10.1 \text{ A}$$

EXAMPLE 6.9 Determine the mesh currents for the circuit shown in Figure 6.18. ■

SOLUTION Because there are four meshes, the denominator determinant contains four rows and four columns. The entries of the denominator determinant for the four mesh currents are

Figure 6.18

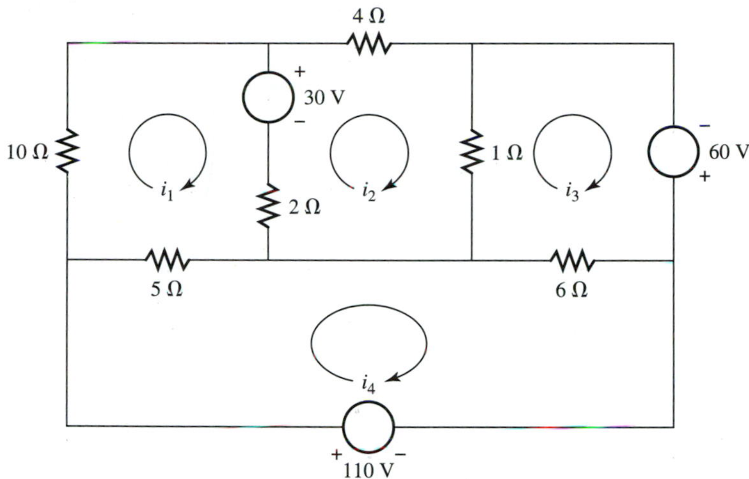

$$R_{11} = 10 + 5 + 2 = 17$$
$$R_{22} = 2 + 4 + 1 = 7$$
$$R_{33} = 1 + 6 = 7$$
$$R_{44} = 5 + 6 = 11$$
$$R_{12} = R_{21} = 2$$
$$R_{13} = R_{31} = 0$$
$$R_{14} = R_{41} = 5$$
$$R_{23} = R_{32} = 1$$
$$R_{24} = R_{42} = 0$$
$$R_{34} = R_{43} = 6$$

and the denominator determinant is

$$\begin{vmatrix} 17 & -2 & -0 & -5 \\ -2 & 7 & -1 & -0 \\ -0 & -1 & 7 & -6 \\ -5 & -0 & -6 & 11 \end{vmatrix}$$

The entries for the voltage summation column are

$$v_1 = -30$$
$$v_2 = 30$$
$$v_3 = 60$$
$$v_4 = 110$$

and the voltage summation column is

$$\begin{pmatrix} -30 \\ 30 \\ 60 \\ 110 \end{pmatrix}$$

The mesh currents are

$$i_1 = \frac{\begin{vmatrix} -30 & -2 & -0 & -5 \\ 30 & 7 & -1 & -0 \\ 60 & -1 & 7 & -6 \\ 110 & -0 & -6 & 11 \end{vmatrix}}{\begin{vmatrix} 17 & -2 & -0 & -5 \\ -2 & 7 & -1 & -0 \\ -0 & -1 & 7 & -6 \\ -5 & -0 & -6 & 11 \end{vmatrix}} = \frac{39{,}720}{3208} = 12.38 \text{ A}$$

$$i_2 = \frac{\begin{vmatrix} 17 & -30 & -0 & -5 \\ -2 & 30 & -1 & -0 \\ -0 & 60 & 7 & -6 \\ -5 & 110 & -6 & 11 \end{vmatrix}}{\begin{vmatrix} 17 & -2 & -0 & -5 \\ -2 & 7 & -1 & -0 \\ -0 & -1 & 7 & -6 \\ -5 & -0 & -6 & 11 \end{vmatrix}} = \frac{44{,}540}{3208} = 13.88 \text{ A}$$

$$i_3 = \frac{\begin{vmatrix} 17 & -2 & -30 & -5 \\ -2 & 7 & 30 & -0 \\ -0 & -1 & 60 & -6 \\ -5 & -0 & 110 & 11 \end{vmatrix}}{\begin{vmatrix} 17 & -2 & -0 & -5 \\ -2 & 7 & -1 & -0 \\ -0 & -1 & 7 & -6 \\ -5 & -0 & -6 & 11 \end{vmatrix}} = \frac{142{,}100}{3208} = 44.3 \text{ A}$$

$$i_4 = \frac{\begin{vmatrix} 17 & -2 & -0 & -30 \\ -2 & 7 & -1 & 30 \\ -0 & -1 & 7 & 60 \\ -5 & -0 & -6 & 110 \end{vmatrix}}{\begin{vmatrix} 17 & -2 & -0 & -5 \\ -2 & 7 & -1 & -0 \\ -0 & -1 & 7 & -6 \\ -5 & -0 & -6 & 11 \end{vmatrix}} = \frac{12{,}680}{3208} = 3.95 \text{ A}$$

6.3 LOOP-ANALYSIS METHOD

The mesh-analysis method is a particular form of the loop-analysis method. Every mesh is a loop, but the reverse statement is not true. The loop-analysis method is more general but less systematic than the mesh-analysis method and offers an alternative approach to the solution of multiple-loop electric circuits. In the mesh-analysis method, the number of meshes is obvious from a simple inspection of the circuit. In the loop-analysis method, the total number of loops for large circuits is sometimes difficult to determine. However, it is not important to determine the total number of loops in a circuit but to determine the number of independent loops necessary to analyze the circuit. Independent loops are loops that result in a set of independent equations that can be solved for a unique solution of the circuit.

The number of independent loops necessary to analyze a circuit using the loop-analysis method is

$$L = B - N + 1 \tag{6.6}$$

where

L = number of independent loops necessary to analyze the circuit

B = number of branches in the circuit

N = number of nodes in the circuit

Recall the definitions of a branch, a node, and a loop.

A *node* is a point that connects two or more electrical elements.

A *branch* is a path between two nodes containing one electrical element.

A *loop* is a continuous path that closes on itself without passing through any node more than once.

The *first step* in the loop-analysis method is to determine the number of independent loops necessary to analyze the circuit. Consider the circuit in Figure 6.19. There are six nodes in the circuit. The six points a, b, c, d, e, and f are nodes, because each point connects two or more electrical elements.

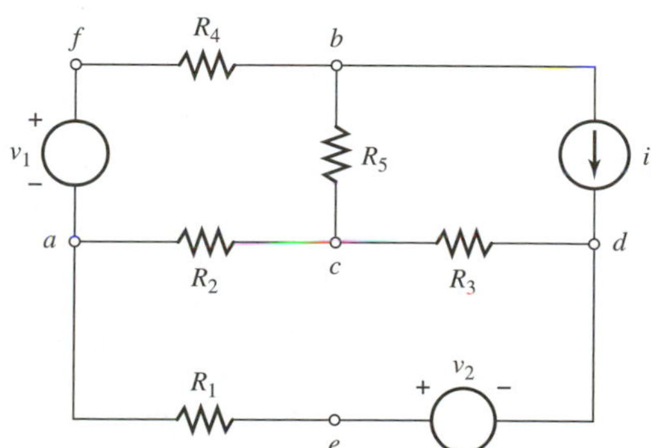

Figure 6.19 Circuit for loop analysis.

There are eight branches and six nodes, so the number of independent loops necessary to analyze the circuit in Figure 6.19 is given by Equation (6.6).

$$L = B - N + 1$$
$$= 8 - 6 + 1 = 3 \text{ loops}$$

The *second step* in the loop-analysis method is to select the loops for use in the loop analysis. There are seven loops in the circuit in Figure 6.19, as illustrated in Figure 6.20, and we must choose three independent loops to use in the loop analysis of the circuit.

We must choose the loops for use in the loop analysis in such a manner that every branch is included in at least one of the loops chosen. This will ensure that the loops chosen result in an independent set of equations.

Figure 6.20 Loops for the circuit in Figure 6.19.

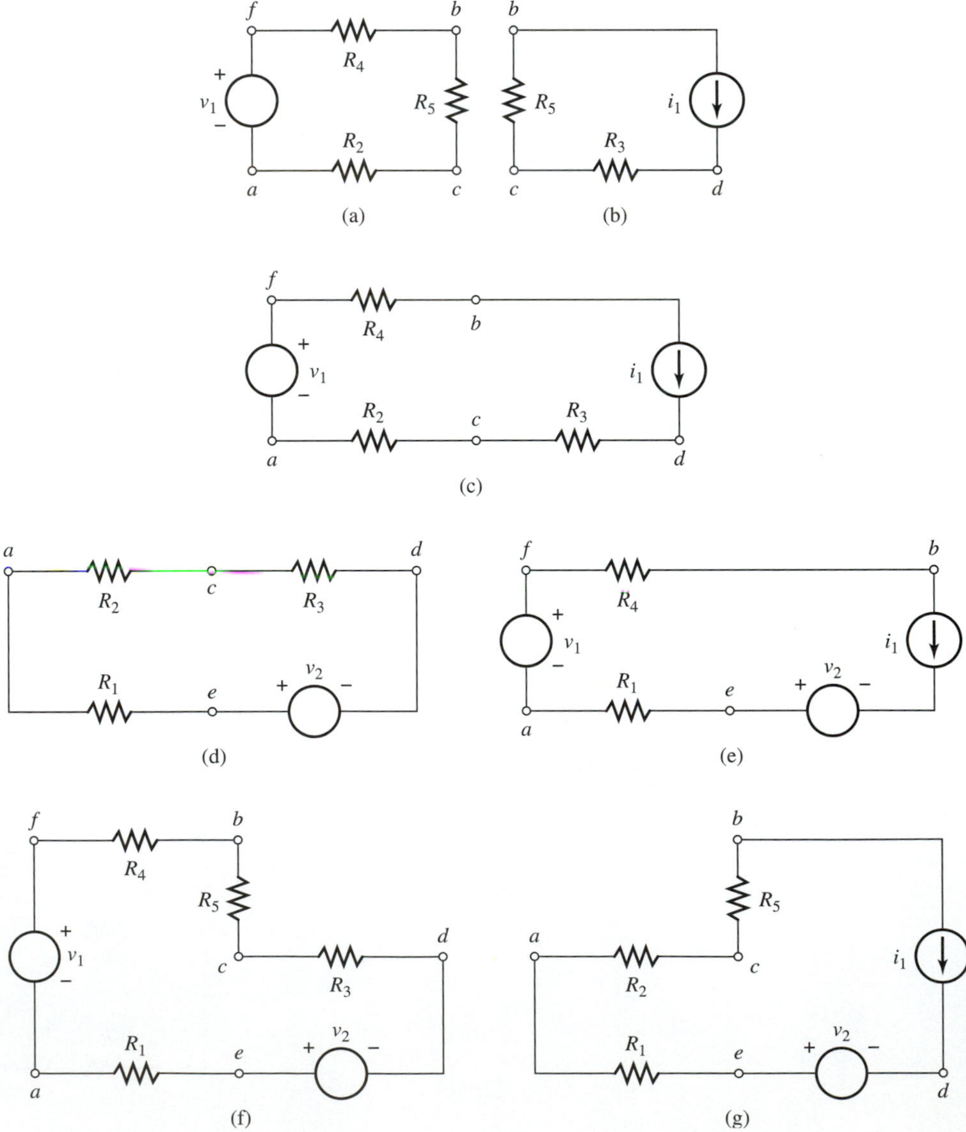

For example, if we choose the three loops from Figure 6.20(a), (b), and (c), the branches from a to e through R_1 and e to d through v_2 are not included in any of the loops, and these three loops will not result in an independent set of equations. However, if we choose the loops in Figure 6.20(a), (b) and (d), each of the branches of the circuit is included in at least one of the loops, and these three loops will result in an independent set of equations and can be used to analyze the circuit.

The *third step* in the loop-analysis method is to assign loop currents to the loops chosen for the loop analysis.

A *loop current* is a current that is common to all the electrical elements in a loop.

A loop current is similar to a mesh current that is common to all the electrical elements of a mesh.

We shall choose the loops in Figure 6.20(a), (d), and (e) for the loop analysis of the circuit in Figure 6.19 and assign loop currents as shown in Figure 6.21. Notice that each of the branches of the circuit in Figure 6.21 is included in at least one of the loops as required.

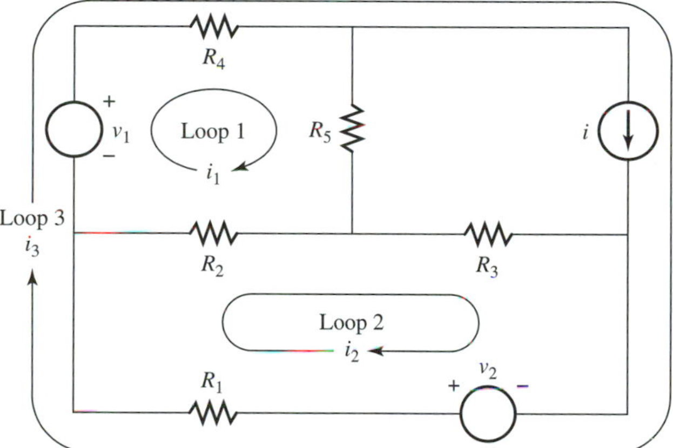

Figure 6.21 Independent loops for the circuits in Figure 6.19.

Notice also that loop current i_3 is the only loop current flowing through the current source labeled i and is, therefore, equal to i.

Although it is not necessary, it is advantageous to choose the loops for loop analysis in such a manner that each current source in the circuit has only one loop current flowing through it. This reduces the complexity of the mathematical model.

In the mesh-analysis method, we assigned all mesh currents in a clockwise direction for consistency. However, in the loop-analysis method, we shall assign loop currents in clockwise and counterclockwise directions for greater flexibility.

The *fourth step* in the loop-analysis method is to assign voltage polarities to the resistances in the circuit based on the assigned loop currents. The circuit in Figure 6.21 is redrawn in Figure 6.22 with the voltage polarities indicated.

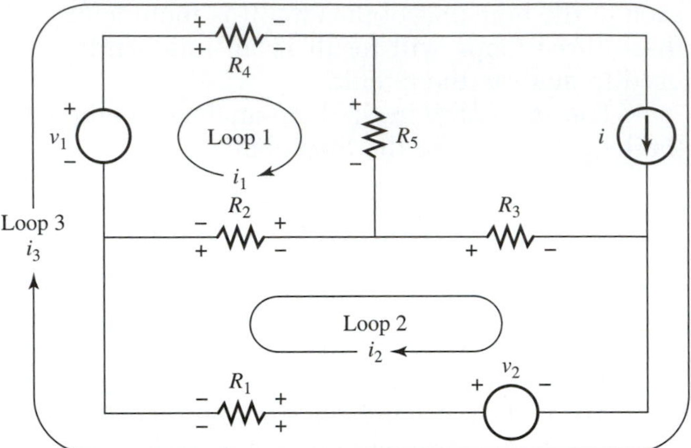

Figure 6.22 Voltage polarities for circuit in Figure 6.21.

The *fifth step* in the loop-analysis method is to apply Kirchhoff's voltage law to loops that are without current sources. We do not apply Kirchhoff's voltage law to the loops with current sources, because we can express the loop current through a current source in terms of the current furnished by the source. Applying Kirchhoff's voltage law to loops 1 and 2 in Figure 6.22 results in

$$\text{loop 1°} \quad v_1 - R_4 i_1 - R_4 i_3 - R_5 i_1 - R_2 i_1 + R_2 i_2 = 0$$
$$\text{loop 2°} \; -R_2 i_2 + R_2 i_1 - R_3 i_2 + v_2 - R_1 i_3 - R_1 i_2 = 0$$

which can be written as

$$(R_4 + R_5 + R_2)i_1 - (R_2)i_2 + (R_4)i_3 = v_1$$
$$-(R_2)i_1 + (R_1 + R_2 + R_3)i_2 + (R_1)i_3 = v_2$$

The *sixth step* in the loop-analysis method is to equate the algebraic sum of the loop currents flowing through each current source to the value of the current source. For the circuit in Figure 6.22, this results in

$$i_3 = i$$

The total number of equations obtained in steps 5 and 6 will equal the number of loop currents assigned to the circuit.

The *seventh and final step* in the loop-analysis method is to solve the mathematical model, which consists of the equations obtained in steps 5 and 6. All the voltage sources, current sources, and resistances in the mathematical model are known, and the loop currents are the unknown quantities. For the circuit in Figure 6.22, the mathematical model is

$$(R_4 + R_5 + R_2)i_1 - (R_2)i_2 + (R_4)i_3 = v_1$$
$$-(R_2)i_1 + (R_1 + R_2 + R_3)i_2 + (R_1)i_3 = v_2$$
$$i_3 = i$$

Cramer's rule can be employed to determine loop currents i_1, i_2, and i_3.

The seven-step procedure for the loop analysis method is summarized here.

1. Determine the number of loops necessary for loop analysis from Equation (6.6).
2. Select the loops for loop analysis in such a manner that each branch is included in at least one of the loops.
3. Assign loop currents to the loops selected.
4. Assign voltage polarities to the resistive elements based on the assigned loop currents.
5. Apply Kirchhoff's voltage law to the loops that are without current sources.
6. For each current source, equate the algebraic sum of the loop currents flowing through the current source to the value of the current source.
7. Solve the mathematical model for the loop currents. The mathematical model consists of the equations obtained in steps 5 and 6.

Consider the following examples of the loop-analysis method.

EXAMPLE 6.10 Determine i_{ab} for the circuit shown in Figure 6.23. ■

Figure 6.23

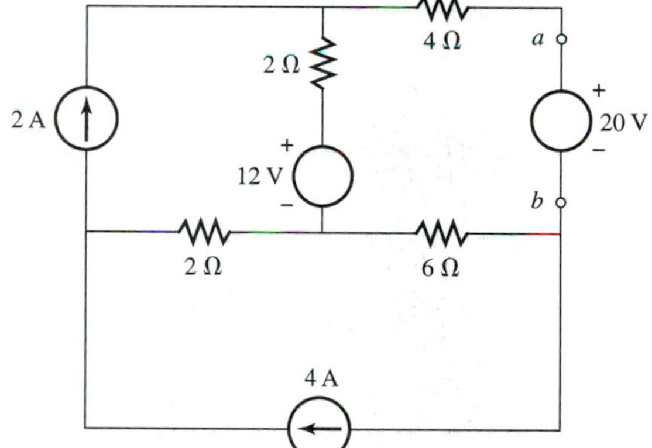

SOLUTION From an inspection of the circuit, there are eight branches and six nodes. Using Equation (6.6), the number of loops necessary to analyze the circuit is

$$\begin{aligned} L &= B - N + 1 \\ &= 8 - 6 + 1 \\ &= 3 \text{ loops} \end{aligned}$$

The loops selected for the circuit in Figure 6.24 meet the requirement of including each branch in at least one loop. The voltage polarities for the resistances in loop 2 are indicated, and it is noted that loop 2 is the only loop without current sources.

Figure 6.24

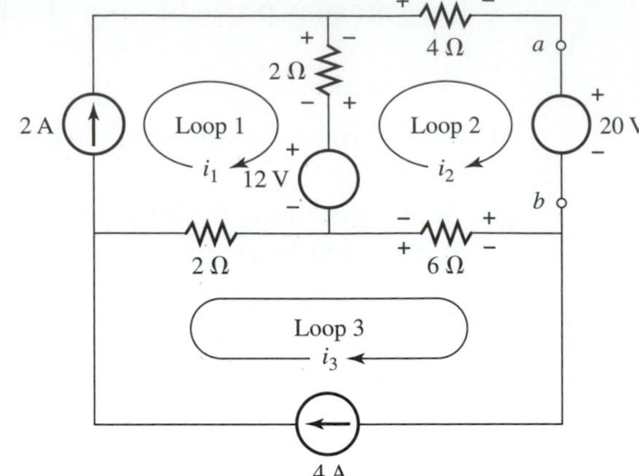

Because i_1 is the only loop current flowing through the 2-A current source (and in the same direction as the current source),

$$i_1 = 2 \text{ A}$$

and because i_3 is the only loop current flowing through the 4-A current source (and in the same direction as the current source),

$$i_3 = 4 \text{ A}$$

Applying Kirchhoff's voltage law to loop 2 results in the following equation:

$$12 - 2i_2 + 2i_1 - 4i_2 - 20 - 6i_2 + 6i_3 = 0$$

Substituting $i_1 = 2$ A and $i_3 = 4$ A into the preceding equation results in

$$i_2 = \frac{5}{3} \text{ A}$$

Because i_{ab} equals i_2,

$$i_{ab} = \frac{5}{3} \text{ A}$$

Notice that the choice of loop currents in Example 6.10 results in only one unknown loop current to determine by applying Kirchhoff's voltage law. This occurs because we assigned only one loop current to each current source in the circuit. If we assign more than one loop current to a current source, it has the undesirable effect of increasing the complexity of the mathematical model, as illustrated in the following example.

EXAMPLE 6.11 Determine i_{ab} for the circuit in Example 6.10 using the loop currents shown in Figure 6.25 for the loop analysis. ■

Figure 6.25

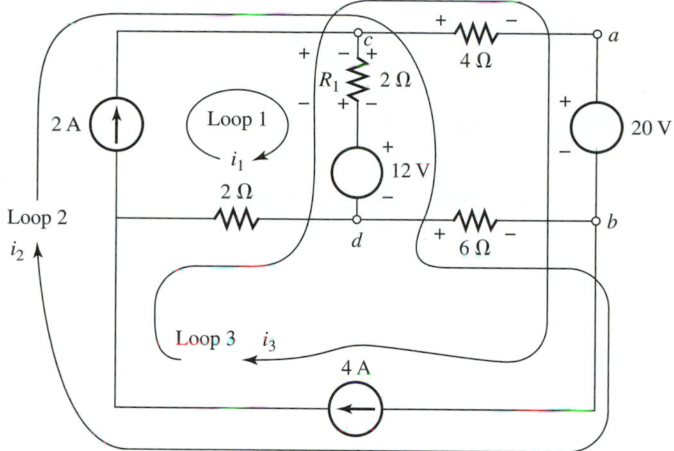

SOLUTION The loops selected are valid for loop-analysis purposes but result in a more complex mathematical model than the one obtained in Example 6.10. Two of the equations for the mathematical model are

$$i_1 + i_2 = 2$$

$$i_2 + i_3 = 4$$

which are obtained by equating each current source to the loop currents flowing through it. The third equation results from applying Kirchhoff's voltage law to the only loop selected that is without current sources, loop *a-b-d-c*. The third equation is

$$12 + 2i_1 + 2i_2 - 2i_3 - 4i_3 - 20 + 6i_2 = 0$$

Notice that there are three voltages across R_1 in this equation, because three loop currents flow through it. The three equations describing the circuit can be written as

$$\begin{aligned} i_1 + i_2 \qquad &= 2 \\ i_2 + i_3 &= 4 \\ -2i_1 - 8i_2 + 6i_3 &= -8 \end{aligned}$$

and from Cramer's rule, the loop currents are

$$i_1 = -\frac{1}{3}\text{ A}$$

$$i_2 = \frac{7}{3}\text{ A}$$

$$i_3 = \frac{5}{3}\text{ A}$$

The application of Cramer's rule is left as an exercise for the student. Because the only loop current flowing from *a* to *b* is i_3,

$$i_{ab} = \frac{5}{3}\text{ A}$$

6.4 NODAL-ANALYSIS METHOD

The loop-current and mesh-analysis methods depend primarily on Kirchhoff's voltage law to obtain the mathematical model for a circuit. The nodal-analysis method depends primarily on Kirchhoff's current law to obtain the mathematical model for a circuit. Recall Kirchhoff's current law.

KIRCHHOFF'S CURRENT LAW The sum of the currents entering a node equals the sum of the currents leaving the node.

In the loop-current and mesh-analysis methods we utilize loops and meshes to analyze circuits, and in the nodal-analysis method we utilize nodes to analyze circuits. Recall the definition of a node.

A *node* is a point that connects two or more electrical elements in a circuit.

The *first step* in the nodal-analysis method is to identify and label the nodes in the circuit. Consider the circuit in Figure 6.26. The nodes are labeled a, b, and c and are easily identified from the definition of a node.

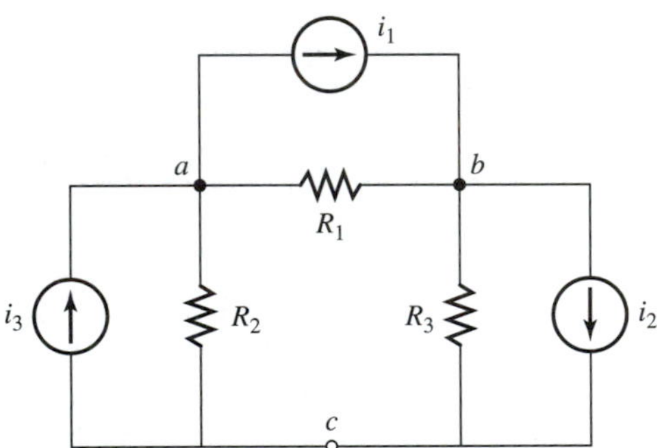

Figure 6.26 Circuit for nodal analysis.

The *second step* in the nodal-analysis method is to select a reference node. The reference node serves as a voltage reference for every nodal voltage in the circuit.

A *nodal voltage* is the voltage of a node with respect to the reference node.

The circuit in Figure 6.26 is redrawn in Figure 6.27, with reference nodes as shown. The reference nodes are marked with the reference symbol.

In Figure 6.27(a), node c is the reference node, and nodal voltages v_a and v_b are the voltages of nodes a and b, respectively, with respect to the reference node.

In Figure 6.27(b), node a is the reference node, and nodal voltages v_b and v_c are the voltages of nodes b and c, respectively, with respect to the reference node.

In Figure 6.27(c), node b is the reference node, and nodal voltages v_a and v_c are the voltages of nodes a and c, respectively, with respect to the reference node.

Figure 6.27 Reference nodes for circuit in Figure 6.26.

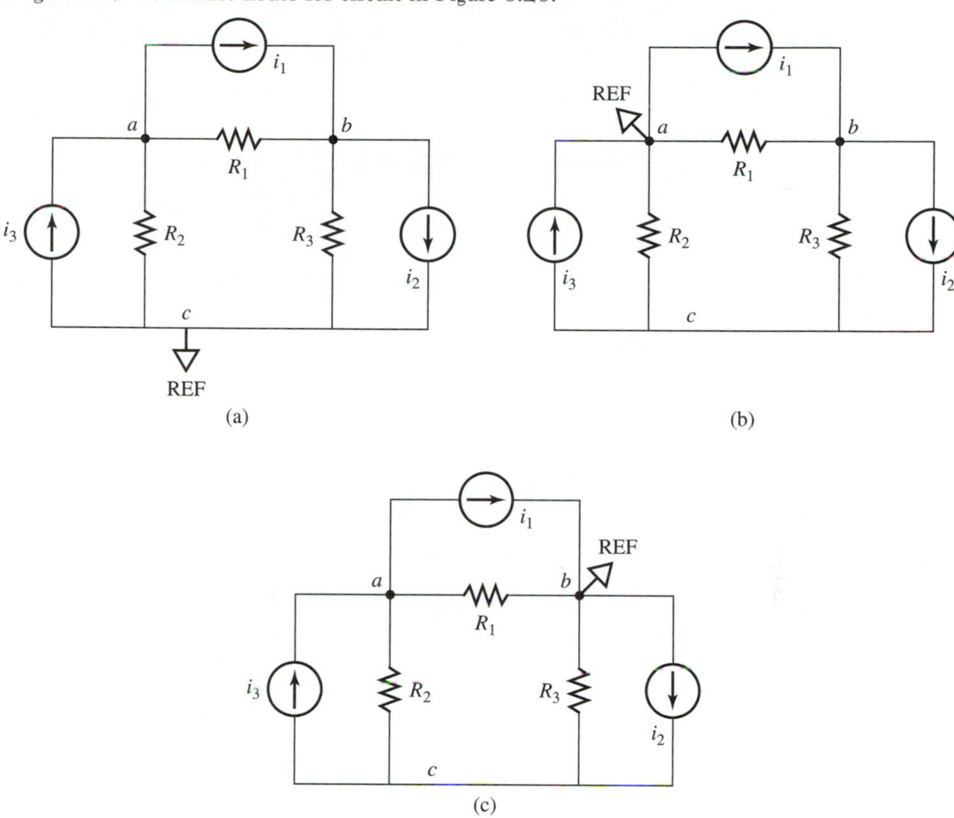

The *third step* in the nodal analysis method is to apply Kirchhoff's current law to each node in the circuit except the reference node. For a circuit with n nodes, we apply Kirchhoff's current law to the $(n - 1)$ nodes that are not reference nodes to determine the mathematical model for the circuit.

In nodal analysis the mathematical model will consist of $(n - 1)$ equations and $(n - 1)$ unknown nodal voltages, where n is the number of nodes in the circuit.

Before proceeding we shall relate the current flowing between two nodes to the nodal voltages. Consider the partial circuit in Figure 6.28, where points

Figure 6.28 Current flow between nodes.

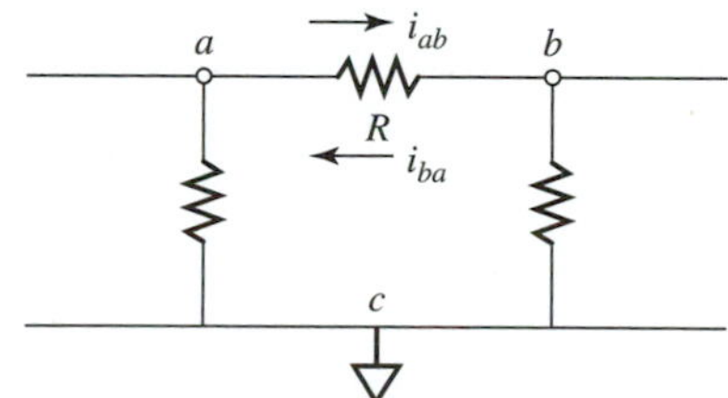

a, b, and c are nodes and point c is taken as the reference node. The nodal voltages v_a, v_b, and v_c can be written as

$$v_a = v_{ac} = \text{ voltage of node } a \text{ with respect to reference node } c$$

$$v_b = v_{bc} = \text{ voltage of node } b \text{ with respect to reference node } c$$

$$v_c = v_{cc} = 0 = \text{voltage of node } c \text{ with respect to itself}$$

Applying Kirchhoff's voltage law to the loop c-a-b-c in Figure 6.28 results in

$$v_{ac} + v_{ba} + v_{cb} = 0$$

Because $v_{ab} = -v_{ba}$ and $v_{bc} = -v_{cb}$,

$$\left.\begin{aligned} v_{ab} &= v_{ac} - v_{bc} = v_a - v_b \\ v_{ba} &= v_{bc} - v_{ac} = v_b - v_a \end{aligned}\right\} \tag{6.7}$$

and from Ohm's law,

$$\left.\begin{aligned} i_{ab} &= \frac{v_{ab}}{R} = \frac{v_a - v_b}{R} \\ i_{ba} &= \frac{v_{ba}}{R} = \frac{v_b - v_a}{R} \end{aligned}\right\} \tag{6.8}$$

If we use conductance instead of resistance in Equation (6.8),

$$\left.\begin{aligned} i_{ab} &= (v_a - v_b)G \\ i_{ba} &= (v_b - v_a)G \end{aligned}\right\} \tag{6.9}$$

where $G = 1/R$.

The current flowing from node a to node b through a conductance G is equal to $(v_a - v_b)G$ and the current flowing from node b to node a through a conductance G is $(v_b - v_a)G$, where v_a and v_b are the nodal voltages of nodes a and b, respectively.

We shall now continue with the third step in the nodal-analysis method—that is, the application of Kirchhoff's current law to all the nodes in the circuit except the reference node. Consider the circuit in Figure 6.26 with node c assigned as the reference node and with the resistances converted to conductances, as shown in Figure 6.29.

Figure 6.29 Circuit in Figure 6.26 with conductances.

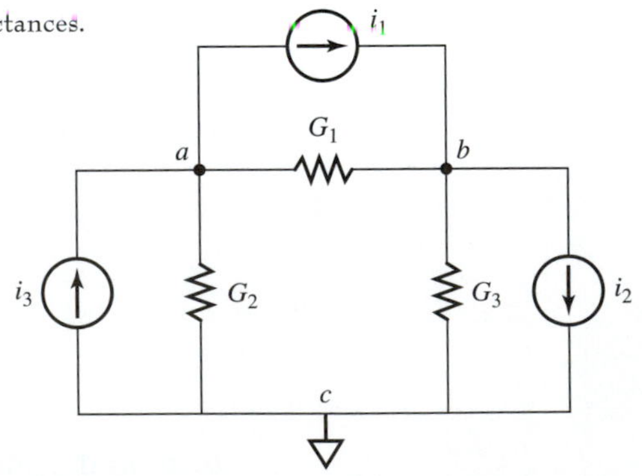

In applying Kirchhoff's current law to a node, we must assume a direction for the currents flowing through the resistances connected to the node. Although these currents can be assumed to enter or leave the node, we shall assume for consistency that all the currents flowing through resistances leave the node under consideration. We shall also use the convention of assigning a positive sign to currents entering a node and a negative sign to currents leaving a node when applying Kirchhoff's current law. Consider nodes a and b for the circuit in Figure 6.29, as shown in the partial circuits of Figure 6.30.

Figure 6.30 Partial circuits for the circuit in Figure 6.29.

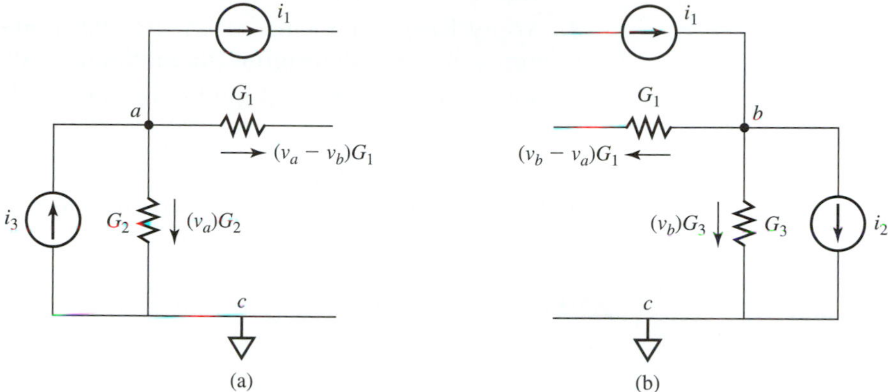

Notice that the assigned currents flowing through the resistances for the partial circuits in Figure 6.30 leave the nodes and are determined from Equation (6.9). The currents flowing through G_2 and G_3 flow out of nodes a and b directly to the reference node, and their expressions are simply v_aG_2 and v_bG_3, respectively, because $v_c = 0$. That is, the voltage of the reference node with respect to the reference node equals zero.

Finally, we complete the third step of the nodal-analysis method by applying Kirchhoff's current law to nodes a and b for the partial circuits in Figure 6.14, which results in

$$\text{node } a\text{:}^\circ\ i_3 - i_1 - (v_a - v_b)G_1 - v_a(G_2) = 0$$

$$\text{node } b\text{:}^\circ\ i_1 - i_2 - (v_b - v_a)G_1 - v_3(G_3) = 0$$

The *fourth and final step* in the nodal-analysis method is to solve the equations obtained in step 3 for the nodal voltages. Notice that the only unknowns in the equations are the nodal voltages. The equations can be written in the following form:

$$\begin{aligned} (G_1 + G_2)v_a - \quad (G_1)v_b &= i_3 - i_1 \\ -(G_1)v_a + (G_1 + G_3)v_b &= i_1 - i_2 \end{aligned}$$

Using Cramer's rule to solve the equations, the nodal voltages in determinant form for the circuit in Figure 6.29 are

$$v_a = \frac{\begin{vmatrix} i_3 - i_1 & -G_1 \\ i_1 - i_2 & G_1 + G_3 \end{vmatrix}}{\begin{vmatrix} G_1 + G_2 & -G_1 \\ -G_1 & G_1 + G_3 \end{vmatrix}}$$

$$v_b = \frac{\begin{vmatrix} -G_1 & i_3 - i_1 \\ G_1 + G_3 & i_1 - i_2 \end{vmatrix}}{\begin{vmatrix} G_1 + G_2 & -G_1 \\ -G_1 & G_1 + G_3 \end{vmatrix}}$$

The four-step procedure for the nodal-analysis method is summarized as follows:

1. Identify the nodes in the circuit.
2. Select a reference node and label the other nodes.
3. Apply Kirchhoff's current law to each node in the circuit except the reference node to determine the mathematical model.
4. Solve the mathematical model for the nodal voltages.

Notice that once the nodal voltages are available, the voltage across each resistive element in the circuit can be determined from Equation (6.7).

Consider the following examples of the nodal-analysis method.

EXAMPLE 6.12 Determine the voltage v_{ab} for the circuit in Figure 6.31. ■

Figure 6.31

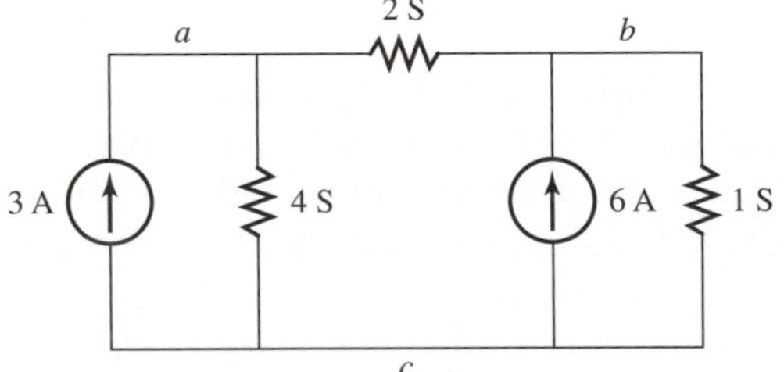

SOLUTION There are three nodes in the circuit, so the mathematical model consists of two equations with two unknown nodal voltages. We shall select node c as the reference node, leaving nodal voltages v_a and v_b as the unknown quantities in the math model. The partial circuits for nodes a and b are drawn in Figure 6.32 with the reference node and the assumed currents flowing through the resistance elements indicated. Equation (6.9) is used to express the resistive current, as shown.

Figure 6.32

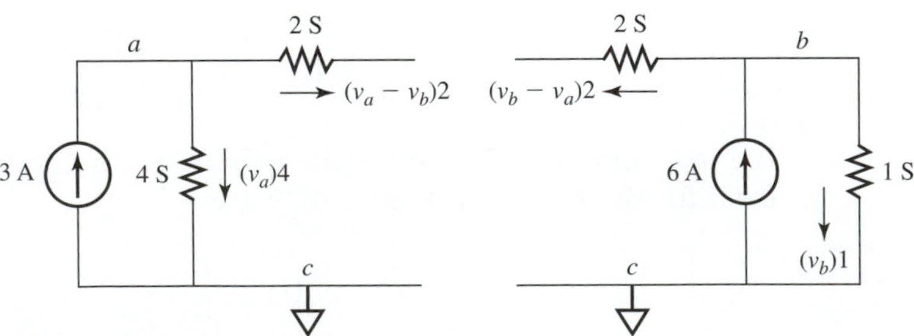

Applying Kirchhoff's current law to nodes a and b results in

$$\text{node } a\text{:} \quad 3 - 4v_a - 2(v_a - v_b) = 0$$

$$\text{node } b\text{:} \quad 6 - 2(v_b - v_a) - 1v_b = 0$$

These equations can be written in the following form:

$$\text{node } a{:}^{\circ} \quad 6v_a - 2v_b = 3$$

$$\text{node } b{:}^{\circ} \; -2v_a + 3v_b = 6$$

From Cramer's rule the nodal voltages are

$$v_a = \frac{\begin{vmatrix} 3 & -2 \\ 6 & 3 \end{vmatrix}}{\begin{vmatrix} 6 & -2 \\ -2 & 3 \end{vmatrix}} = \frac{3}{2}\text{ V}$$

$$v_b = \frac{\begin{vmatrix} 6 & 3 \\ -2 & 6 \end{vmatrix}}{\begin{vmatrix} 6 & -2 \\ -2 & 3 \end{vmatrix}} = \frac{42}{14}\text{ V}$$

and from Equation (6.7),

$$v_{ab} = v_a - v_b = -\frac{3}{2}\text{ V}$$

The selection of node c as the reference node in Example 6.12 results in fewer terms in the mathematical model. Because node c has more branches than the other nodes, there are more currents to consider when applying Kirchhoff's current law to node c. Although it is not necessary, it is advantageous to select the node with the greatest number of branches as the reference node. This selection results in fewer terms in the mathematical model for the circuit.

EXAMPLE 6.13 Determine the nodal voltages v_b, v_c, and v_d for the circuit in Figure 6.33. Use node a as the reference node, as indicated. ■

Figure 6.33

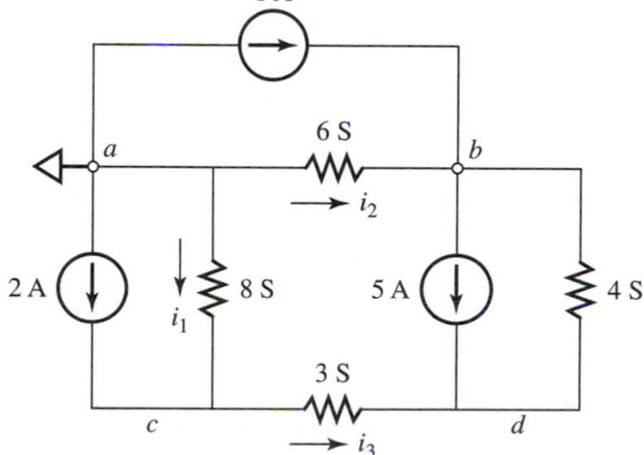

SOLUTION There are four nodes in the circuit and, therefore, three equations and three unknown nodal voltages in the mathematical model. Because node a has been selected as the reference node, nodal voltages v_b, v_c, and v_d are the unknown nodal voltages. Partial circuits for nodes b, c, and d are shown in

Figures 6.34, 6.35, and 6.36, with the assumed resistive currents indicated. Equation (6.9) is used to express the resistive currents in terms of the nodal voltages, as shown.

Figure 6.34

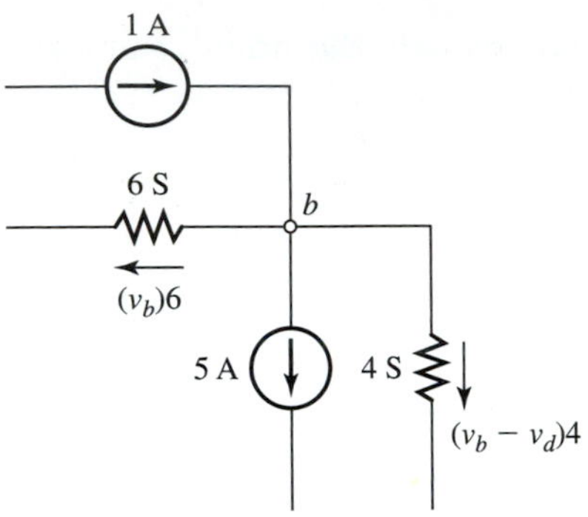

Figure 6.35

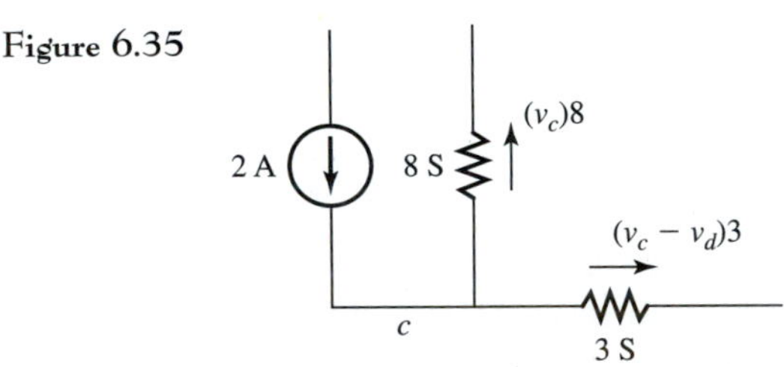

Figure 6.36

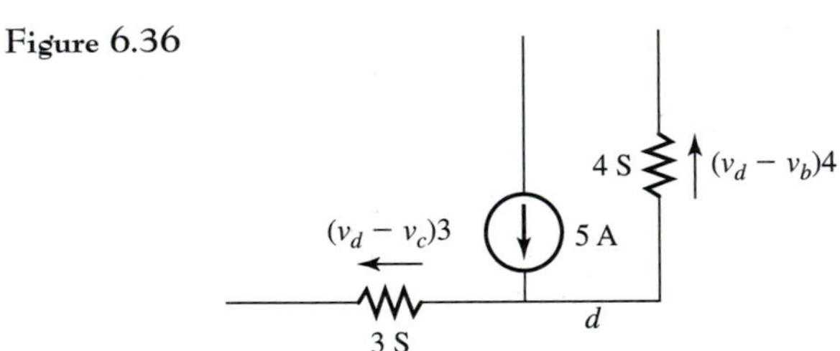

Applying Kirchhoff's current law to nodes b, c, and d results in

$$\begin{aligned} \text{node } b: &\quad 1 - 6(v_b) - 5 - 4(v_b - v_d) = 0 \\ \text{node } c: &\quad 2 - 8(v_c) - 3(v_c - v_d) = 0 \\ \text{node } d: &\quad 5 - 3(v_d - v_c) - 4(v_d - v_b) = 0 \end{aligned}$$

These equations can be written in the following form:

$$\begin{aligned} 10v_b \qquad\quad - 4v_d &= -4 \\ 11v_c - 3v_d &= 2 \\ -4v_b - 3v_c + 7v_d &= 5 \end{aligned}$$

From Cramer's rule, the nodal voltages are

$$v_b = \frac{\begin{vmatrix} -4 & 0 & -4 \\ 2 & 11 & -3 \\ 5 & -3 & 7 \end{vmatrix}}{\begin{vmatrix} 10 & 0 & -4 \\ 0 & 11 & -3 \\ -4 & -3 & 7 \end{vmatrix}} = -\frac{28}{504} = -55.5\text{ mV}$$

$$v_c = \frac{\begin{vmatrix} 10 & -4 & -4 \\ 0 & 2 & -3 \\ -4 & 5 & 7 \end{vmatrix}}{\begin{vmatrix} 10 & 0 & -4 \\ 0 & 11 & -3 \\ -4 & -3 & 7 \end{vmatrix}} = \frac{210}{504} = 416\text{ mV}$$

$$v_d = \frac{\begin{vmatrix} 10 & 0 & -4 \\ 0 & 11 & 2 \\ -4 & -3 & 5 \end{vmatrix}}{\begin{vmatrix} 10 & 0 & -4 \\ 0 & 11 & -3 \\ -4 & -3 & 7 \end{vmatrix}} = \frac{434}{504} = 861\text{ mV}$$

Although we did not consider voltage sources in presenting the nodal-analysis method, the method is applicable to circuits with voltage sources. In fact, the presence of voltage sources in the circuit usually reduces the work required to obtain a solution. Consider the following examples.

EXAMPLE 6.14 Determine the nodal voltages v_a and v_b and branch currents i_1, i_2, and i_3 for the circuit in Figure 6.37. Use node c as a reference node as indicated. ■

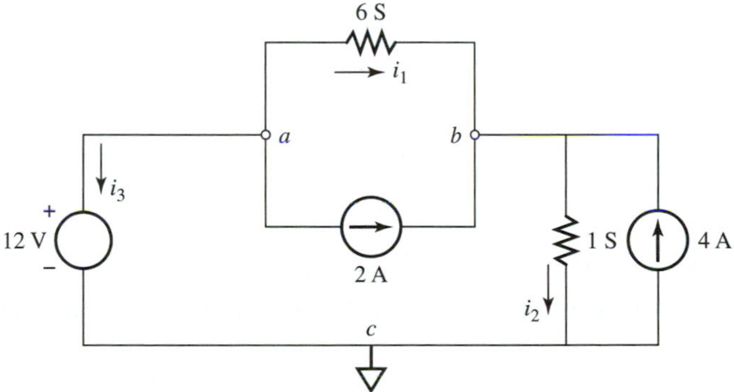

Figure 6.37

SOLUTION There are three nodes and, therefore, two equations and two unknown voltages in the mathematical model. However, in the circuit under consideration, the nodal voltage v_a is obtained from an inspection of the circuit, and the only unknown nodal voltage is v_b. From an inspection of the circuit,

$$v_a = 12\text{ V}$$

Therefore, we need to apply Kirchhoff's current law only to node *b*. A partial circuit for node *b* is shown in Figure 6.38, with the resistive currents in terms of the nodal voltages, as indicated.

Figure 6.38

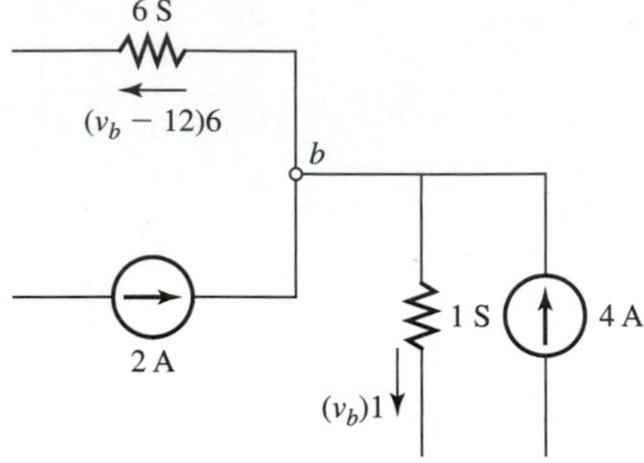

Applying Kirchhoff's current law to node *b* results in

$$-(v_b - 12)6 - (v_b)1 + 2 + 4 = 0$$

and

$$v_b = \frac{78}{7}\,\text{V}$$

Because v_a and v_b are known, i_1 and i_2 can be determined from Equation (6.9).

$$\begin{aligned} i_1 &= 6(v_a - v_b) \\ &= 6\left(12 - \frac{78}{7}\right) \\ &= \frac{36}{7} = 5.1\,\text{A} \\ i_2 &= 1(v_b) \\ &= \frac{78}{7} = 11.1\,\text{A} \end{aligned}$$

Because i_1 is known, i_3 can be determined by applying Kirchhoff's current law to node *a* as follows.

$$-i_3 - i_1 - 2 = 0$$

$$i_3 = -\frac{50}{7} = -7.1\,\text{A}$$

EXAMPLE 6.15 Determine the nodal voltages v_a and v_b for the circuit in Figure 6.39. Use node *c* as the reference node, as indicated. ■

Figure 6.39

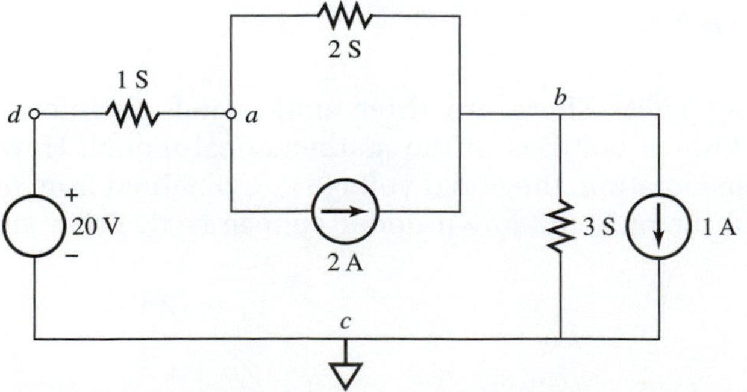

SOLUTION Because there are four nodes in the circuit, the math model consists of three equations with three unknown nodal voltages, v_a, v_b, and v_d. Nodal voltage v_d is equal to 20 V by inspection.

Partial circuits for nodes a and b are shown in Figures 6.40 and 6.41, with the resistive currents in terms of the nodal voltages according to Equation (6.9).

Figure 6.40

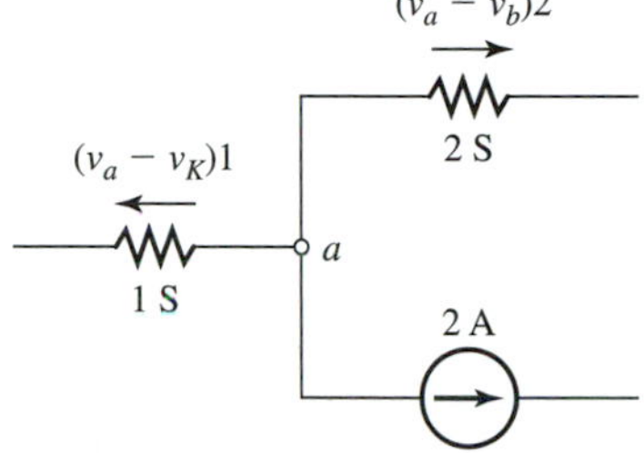

Figure 6.41

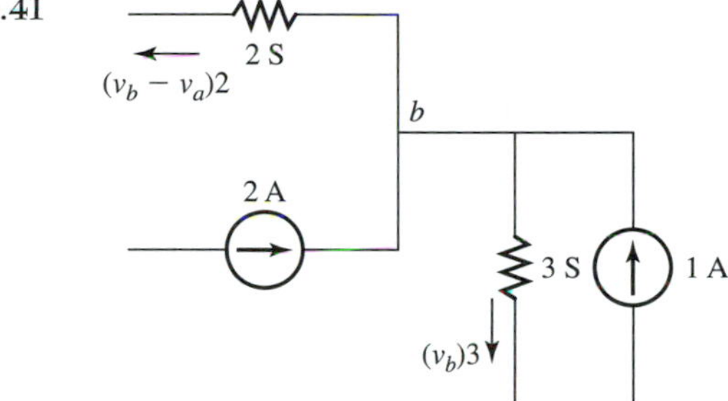

Applying Kirchhoff's current law to nodes a and b results in

$$\text{node } a{:}\ -1(v_a - 20) - 2(v_a - v_b) - 2 = 0$$

$$\text{node } b{:}\ -2(v_b - v_a) + 2 - 1 - 2(v_b) = 0$$

These equations can be written in the following form:

$$\begin{aligned} 3v_a - 2v_b &= +18 \\ -2v_a + 5v_b &= 1 \end{aligned}$$

From Cramer's rule the nodal voltages are

$$v_a = \frac{\begin{vmatrix} 18 & -2 \\ 1 & 5 \end{vmatrix}}{\begin{vmatrix} 3 & -2 \\ -2 & 5 \end{vmatrix}} = \frac{92}{11} = 8.4 \text{ V}$$

$$v_b = \frac{\begin{vmatrix} 3 & 18 \\ -2 & 1 \end{vmatrix}}{\begin{vmatrix} 3 & -2 \\ -2 & 5 \end{vmatrix}} = \frac{39}{11} = 3.6 \text{ V}$$

6.5 UNIFORM APPROACH TO NODAL ANALYSIS

The four steps for the nodal-analysis method presented in Section 6.4 can be reduced significantly by comparing the nodal voltages in determinant form to the circuit described by the nodal voltages. The uniform approach to nodal analysis (also called the format method) is similar to the uniform approach to mesh analysis in that it allows us to write the nodal voltages in determinant form by a simple inspection of the circuit after the reference node is chosen and the other nodes are labeled.

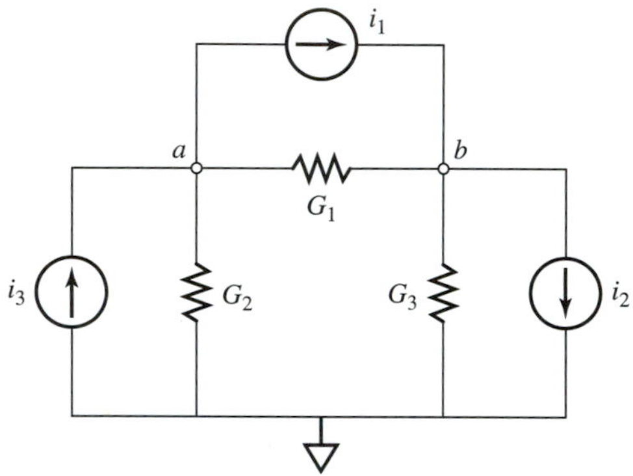

Figure 6.42 Circuit in Figure 6.29.

Consider the circuit in Figure 6.29 as redrawn in Figure 6.42. As determined in Section 6.4, the nodal voltages for the circuit in Figure 6.42 are

$$v_a = \frac{\begin{vmatrix} (i_3 - i_1) & -(G_1) \\ (i_1 - i_2) & (G_1 + G_3) \end{vmatrix}}{\begin{vmatrix} (G_1 + G_2) & -(G_1) \\ -(G_1) & (G_1 + G_3) \end{vmatrix}}$$

$$v_b = \frac{\begin{vmatrix} (G_1 + G_2) & (i_3 - i_1) \\ -(G_1) & (i_1 - i_2) \end{vmatrix}}{\begin{vmatrix} (G_1 + G_2) & -(G_1) \\ -(G_1) & (G_1 + G_3) \end{vmatrix}}$$

Comparing the determinant entries for the nodal voltages to the circuit elements for the circuit in Figure 6.42 results in

$$v_a = \frac{\begin{vmatrix} \begin{pmatrix} \text{net current from} \\ \text{current sources} \\ \text{entering node } a \end{pmatrix} & -\begin{pmatrix} \text{conductance common} \\ \text{to nodes } a \text{ and } b \end{pmatrix} \\ \begin{pmatrix} \text{net current from} \\ \text{current sources} \\ \text{entering node } b \end{pmatrix} & \begin{pmatrix} \text{total conductance} \\ \text{connected to node } b \end{pmatrix} \end{vmatrix}}{\begin{vmatrix} \begin{pmatrix} \text{total conductance} \\ \text{connected to} \\ \text{node } a \end{pmatrix} & -\begin{pmatrix} \text{conductance common} \\ \text{to nodes } a \text{ and } b \end{pmatrix} \\ -\begin{pmatrix} \text{conductance} \\ \text{common to} \\ \text{nodes } a \text{ and } b \end{pmatrix} & \begin{pmatrix} \text{total conductance} \\ \text{connected to node } b \end{pmatrix} \end{vmatrix}} \tag{6.10}$$

$$v_b = \frac{\begin{vmatrix} \begin{pmatrix}\text{total conductance}\\ \text{connected to}\\ \text{node } a\end{pmatrix} & \begin{pmatrix}\text{net current from}\\ \text{current sources}\\ \text{entering node } a\end{pmatrix} \\ -\begin{pmatrix}\text{conductance}\\ \text{common to nodes}\\ a \text{ and } b\end{pmatrix} & \begin{pmatrix}\text{net current from}\\ \text{current sources}\\ \text{entering node } b\end{pmatrix} \end{vmatrix}}{\begin{vmatrix} \begin{pmatrix}\text{total conductance}\\ \text{connected to}\\ \text{node } a\end{pmatrix} & -\begin{pmatrix}\text{conductance common}\\ \text{to nodes } a \text{ and } b\end{pmatrix} \\ -\begin{pmatrix}\text{conductance}\\ \text{common to}\\ \text{nodes } a \text{ and } b\end{pmatrix} & \begin{pmatrix}\text{total conductance}\\ \text{connected to node } b\end{pmatrix} \end{vmatrix}} \tag{6.11}$$

The net current from current sources entering a node is determined by algebraically summing the currents from the current sources connected to the node. We shall use the convention adopted in Section 6.4; that is, a positive sign is assigned to currents entering the node and a negative sign is assigned to currents leaving the node.

We can now write the nodal voltages for a three-node circuit in determinant form by an inspection of the circuit. Consider the following example.

EXAMPLE 6.16 Determine the nodal voltages v_1 and v_2 for the circuit in Figure 6.43. ■

Figure 6.43

SOLUTION Using Equations (6.10) and (6.11), the nodal voltages are

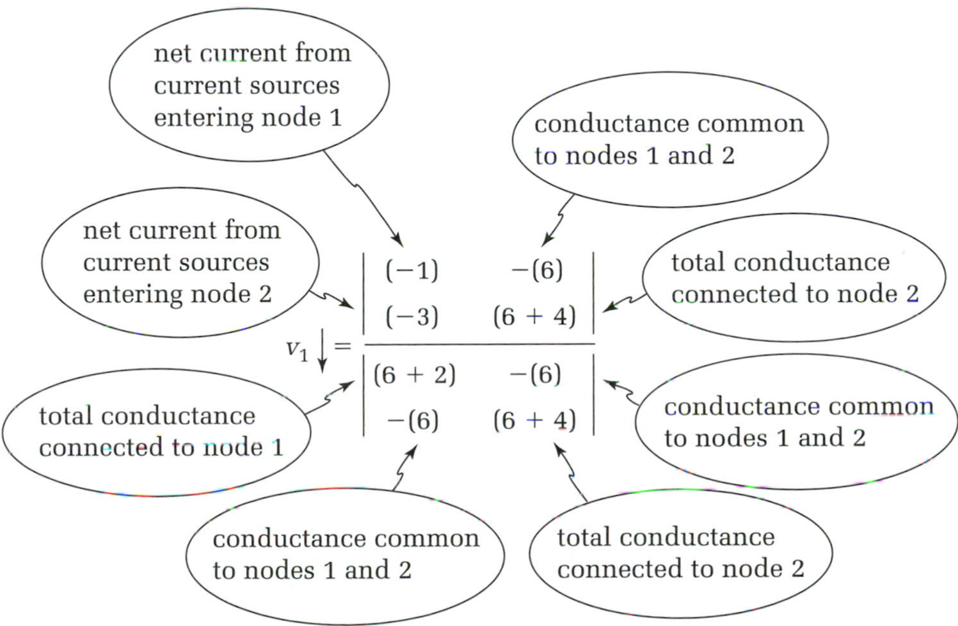

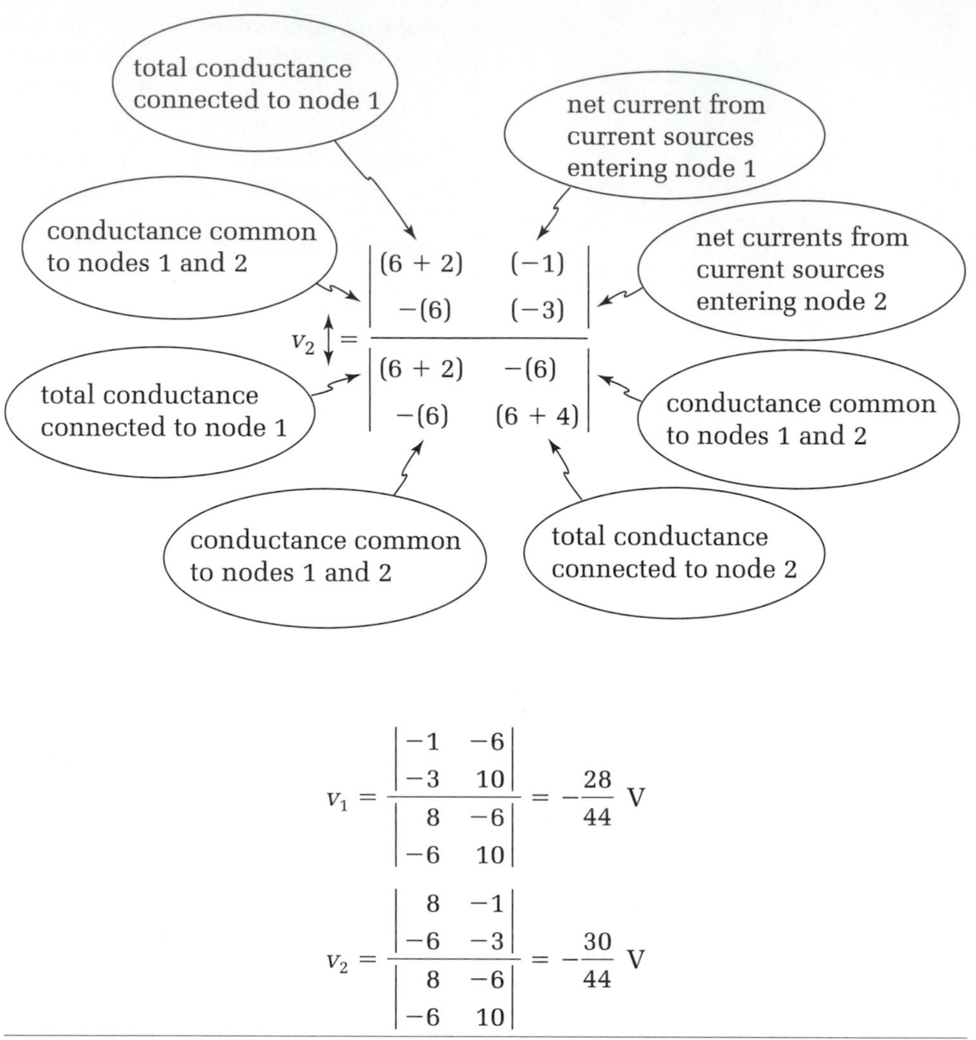

$$v_1 = \frac{\begin{vmatrix} -1 & -6 \\ -3 & 10 \end{vmatrix}}{\begin{vmatrix} 8 & -6 \\ -6 & 10 \end{vmatrix}} = -\frac{28}{44}\ \text{V}$$

$$v_2 = \frac{\begin{vmatrix} 8 & -1 \\ -6 & -3 \end{vmatrix}}{\begin{vmatrix} 8 & -6 \\ -6 & 10 \end{vmatrix}} = -\frac{30}{44}\ \text{V}$$

Referring to the circuit in Figure 6.42 and the determinant form of the solution given in Equations (6.10) and (6.11), the denominator determinants for v_a and v_b are identical.

> In the uniform approach to nodal analysis, the denominator determinants are the same for all nodal voltages.

Notice the symmetry of the denominator determinants.

> In the uniform approach to nodal analysis, the denominator determinant is symmetrical about the major diagonal.

This results from the fact that the conductance common to nodes a and b is also common to nodes b and a.

Notice the algebraic signs of the determinant entries for the denominator determinant.

> In the uniform approach to the nodal-analysis method, all off-diagonal entries in the denominator are negative, and all entries on the major diagonal are positive.

Notice that the numerator determinant for v_a is the same as the denominator determinant for v_a except for column 1. And the numerator determinant for v_b is the same as the denominator determinant for v_b except for column 2. Notice also that column 1 in the numerator of v_a and column 2 in the numerator of v_b are the same and consist of the current source summation column formed by algebraically summing the current sources at each node.

> The entry in the first row of the current source summation column is the algebraic sum of the current sources entering node *a* and the entry in the second row is the algebraic sum of the current sources entering node *b*. In the algebraic summations, we sign current sources entering the node positive and current sources leaving the node negative.

Using these observations, we can formulate nodal voltages v_a and v_b for the circuit in Figure 6.42 in the following manner.

First construct the denominator determinant for nodes *a* and *b*, as follows.

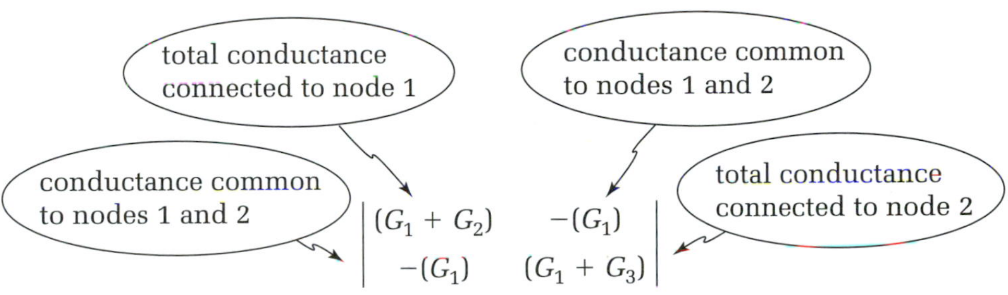

Next construct the current-source summation column:

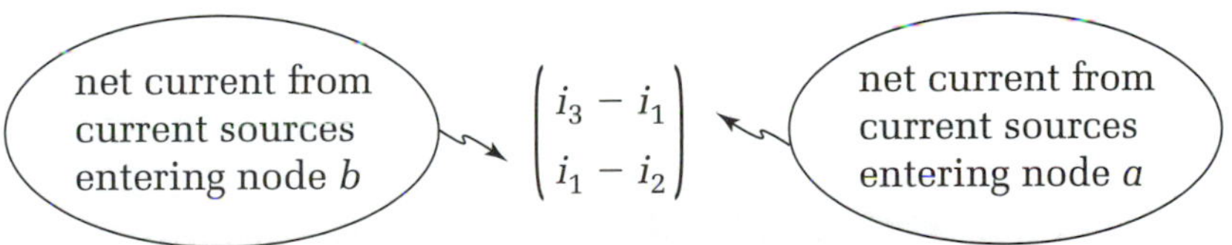

Finally, substitute the current source summation column for the first column of the denominator determinant to form the numerator determinant for nodal voltage v_a and substitute the current source summation column for the second column of the denominator determinant to form the numerator determinant for nodal voltage v_b. This results in the following numerator determinants for nodal voltages v_a and v_b, respectively.

$$\begin{vmatrix} (i_3 - i_1) & -(G_1) \\ (i_1 - i_2) & (G_1 + G_3) \end{vmatrix}$$

$$\begin{vmatrix} (G_1 + G_2) & (i_3 - i_1) \\ -(G_1) & (i_1 - i_2) \end{vmatrix}$$

Nodal voltages v_a and v_b in determinant form for the circuit in Figure 6.42 can then be written as follows:

$$v_a = \frac{\begin{vmatrix} (i_3 - i_1) & -(G_1) \\ (i_1 - i_2) & (G_1 + G_3) \end{vmatrix}}{\begin{vmatrix} (G_1 + G_2) & -(G_1) \\ -(G_1) & (G_1 + G_3) \end{vmatrix}}$$

$$v_b = \frac{\begin{vmatrix} (G_1 + G_2) & (i_3 - i_1) \\ -(G_1) & (i_1 - i_2) \end{vmatrix}}{\begin{vmatrix} (G_1 + G_2) & -(G_1) \\ -(G_1) & (G_1 + G_3) \end{vmatrix}}$$

Consider the following example.

EXAMPLE 6.17 Determine the nodal voltages v_1 and v_2 for the circuit shown in Figure 6.44. ■

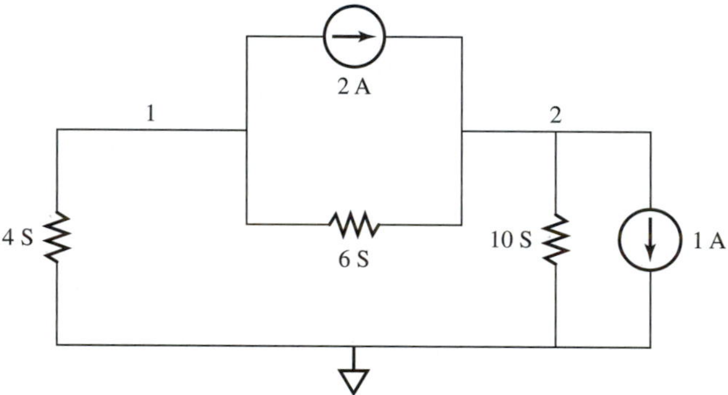

Figure 6.44

SOLUTION The denominator determinant for nodal voltages v_1 and v_2 is

$$\begin{vmatrix} (4 + 6) & -(6) \\ -(6) & (10 + 6) \end{vmatrix}$$

which results in

$$\begin{vmatrix} 10 & -6 \\ -6 & 16 \end{vmatrix}$$

The current-source summation column is

$$\begin{pmatrix} -2 \\ 2 - 1 \end{pmatrix}$$

which results in

$$\begin{pmatrix} -2 \\ 1 \end{pmatrix}$$

The numerator determinants for nodal voltages v_1 and v_2, respectively, are

$$\begin{vmatrix} -2 & -6 \\ 1 & 16 \end{vmatrix}$$

$$\begin{vmatrix} 10 & -2 \\ -6 & 1 \end{vmatrix}$$

so

$$v_1 = \frac{\begin{vmatrix} -2 & -6 \\ 1 & 16 \end{vmatrix}}{\begin{vmatrix} 10 & -6 \\ -6 & 16 \end{vmatrix}} = -\frac{26}{124} = -209 \text{ mV}$$

$$v_2 = \frac{\begin{vmatrix} 10 & -2 \\ -6 & 1 \end{vmatrix}}{\begin{vmatrix} 10 & -6 \\ -6 & 16 \end{vmatrix}} = -\frac{2}{124} = -16.1 \text{ mV}$$

Although we have considered only three-node circuits when presenting the uniform approach to the nodal-analysis method, the method is valid for circuits with any number of nodes. Let us introduce some symbols for conciseness while extending the method to circuits with more than two nodes.

Considering a circuit with $(n + 1)$ nodes, where one node serves as a reference node, the major diagonal elements of the denominator determinants for the n nodal voltages can be written as

$$\begin{aligned}
G_{11} &= \text{total conductance connected to node 1} \\
G_{22} &= \text{total conductance connected to node 2} \\
&\vdots \\
G_{nn} &= \text{total conductance connected to node } n
\end{aligned}$$

The off-diagonal entries for the denominator determinants for the n nodal voltages can be written as

$$\begin{aligned}
G_{12} &= \text{conductance common to nodes 1 and 2} \\
G_{21} &= \text{conductance common to nodes 2 and 1} \\
G_{13} &= \text{conductance common to nodes 1 and 3} \\
G_{31} &= \text{conductance common to nodes 3 and 1} \\
&\vdots \\
G_{n-1,n} &= \text{conductance common to nodes } n-1 \text{ and } n \\
G_{n,n-1} &= \text{conductance common to nodes } n \text{ and } n-1
\end{aligned}$$

The denominator determinant for the nodal voltages of an $(n + 1)$ node circuit can then be written as

$$\begin{vmatrix} G_{11} & -G_{12} & -G_{13} & \cdots & -G_{1n} \\ -G_{21} & G_{22} & -G_{23} & \cdots & -G_{2n} \\ \vdots & \vdots & \vdots & & \vdots \\ -G_{n1} & -G_{n2} & -G_{n3} & \cdots & G_{nn} \end{vmatrix}$$

The entries for the current-source summation column can be written as

$$\begin{aligned} i_1 &= \text{net sum of current sources entering node 1} \\ i_2 &= \text{net sum of current sources entering node 2} \\ &\vdots \\ i_n &= \text{net sum of current sources entering node } n \end{aligned}$$

so the current-source summation column can then be written as

$$\begin{pmatrix} i_1 \\ i_2 \\ \vdots \\ i_n \end{pmatrix}$$

We can now write the nodal voltages for the n nodes in determinant form as follows:

$$v_1 = \frac{\begin{vmatrix} i_1 & -G_{12} & -G_{13} & \cdots & -G_{1n} \\ i_2 & G_{22} & -G_{23} & \cdots & -G_{2n} \\ \vdots & \vdots & \vdots & & \vdots \\ i_n & -G_{n2} & -G_{n3} & \cdots & G_{nn} \end{vmatrix}}{\begin{vmatrix} G_{11} & -G_{12} & -G_{13} & \cdots & -G_{1n} \\ -G_{21} & G_{22} & -G_{23} & \cdots & -G_{2n} \\ \vdots & \vdots & \vdots & & \vdots \\ -G_{n1} & -G_{n2} & -G_{n3} & \cdots & G_{nn} \end{vmatrix}} \tag{6.12}$$

$$v_2 = \frac{\begin{vmatrix} G_{11} & i_1 & -G_{13} & \cdots & -G_{1n} \\ -G_{21} & i_2 & -G_{23} & \cdots & -G_{2n} \\ \vdots & \vdots & \vdots & & \vdots \\ -G_{n1} & i_{n2} & -G_{n3} & \cdots & G_{nn} \end{vmatrix}}{\begin{vmatrix} G_{11} & -G_{12} & -G_{13} & \cdots & -G_{1n} \\ -G_{21} & G_{22} & -G_{23} & \cdots & -G_{2n} \\ \vdots & \vdots & \vdots & & \vdots \\ -G_{n1} & -G_{n2} & -G_{n3} & \cdots & G_{nn} \end{vmatrix}} \tag{6.13}$$

$$\vdots$$

$$v_n = \frac{\begin{vmatrix} G_{22} & -G_{12} & -G_{13} & \cdots & i_1 \\ -G_{21} & G_{22} & -G_{23} & \cdots & i_2 \\ \vdots & \vdots & \vdots & & \vdots \\ -G_{n1} & -G_{n2} & -G_{n3} & \cdots & i_{nn} \end{vmatrix}}{\begin{vmatrix} G_{11} & -G_{12} & -G_{13} & \cdots & -G_{1n} \\ -G_{21} & G_{22} & -G_{23} & \cdots & -G_{2n} \\ \vdots & \vdots & \vdots & & \vdots \\ -G_{n1} & -G_{n2} & -G_{n3} & \cdots & G_{nn} \end{vmatrix}} \tag{6.14}$$

Notice that each denominator for the nodal voltages in Equations (6.12), (6.13), and (6.14) are identical and that all the numerator determinants are formed from the denominator determinant by replacing one column of the denominator determinant with the current-source summation column. The current-source summation column replaces column 1 of the denominator determinant to form the numerator determinant for nodal voltage v_1; the current-source summation column replaces column 2 of the denominator determinant to form the numerator determinant for nodal voltage v_2; and all the other numerator determinants for the nodal voltages through v_n are formed in a similar manner.

Notice also that the denominator determinant is symmetrical about the major diagonal, because $G_{ij} = G_{ji}$; the off-diagonal elements in the denominator determinant are all negative; the major diagonal elements of the denominator determinant are all positive; and the number of rows and columns in the numerator and denominator determinants equal the number of nodes in the circuit minus 1.

Consider the following examples.

EXAMPLE 6.18 Determine the nodal voltages v_1, v_2, and v_3 for the circuit in Figure 6.45. Use the reference node indicated. ■

Figure 6.45

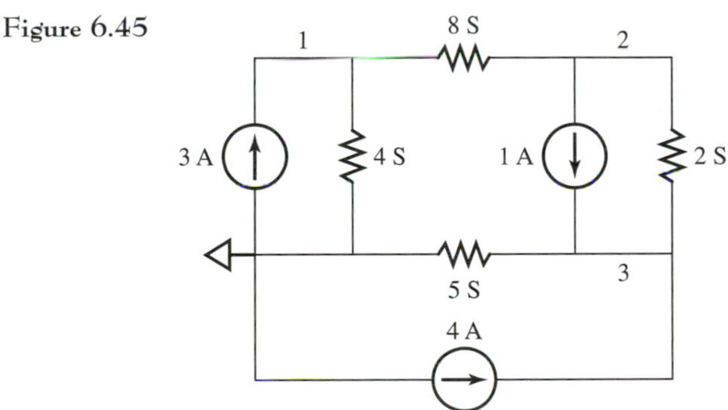

SOLUTION Because there are four nodes, the denominator determinant contains three rows and three columns. The entries for the denominator determinant of the nodal voltages are

$$G_{11} = 8 + 4 = 12$$
$$G_{22} = 8 + 2 = 10$$
$$G_{33} = 5 + 2 = 7$$
$$G_{12} = G_{21} = 8$$
$$G_{13} = G_{31} = 0$$
$$G_{23} = G_{32} = 2$$

and the denominator determinant is

$$\begin{vmatrix} 12 & -8 & 0 \\ -8 & 10 & -2 \\ 0 & -2 & 7 \end{vmatrix}$$

The entries for the current-source summation column are

$$i_1 = 3$$
$$i_2 = -1$$
$$i_3 = 1+4 = 5$$

and the current-source summation column is

$$\begin{pmatrix} 3 \\ -1 \\ 5 \end{pmatrix}$$

The numerator determinants for nodal voltages v_1, v_2, and v_3, respectively, are

$$\begin{vmatrix} 3 & -8 & 0 \\ -1 & 10 & -2 \\ 5 & -2 & 7 \end{vmatrix}$$
$$\begin{vmatrix} 12 & 3 & 0 \\ -8 & -1 & -2 \\ 0 & 5 & 7 \end{vmatrix}$$
$$\begin{vmatrix} 12 & -8 & 3 \\ -8 & 10 & -1 \\ 0 & -2 & 5 \end{vmatrix}$$

and

$$v_1 = \frac{\begin{vmatrix} 3 & -8 & 0 \\ -1 & 10 & -2 \\ 5 & -2 & 7 \end{vmatrix}}{\begin{vmatrix} 12 & -8 & 0 \\ -8 & 10 & -2 \\ 0 & -2 & 7 \end{vmatrix}} = \frac{222}{344} = 650\,\text{mV}$$

$$v_2 = \frac{\begin{vmatrix} 12 & 3 & 0 \\ -8 & -1 & -2 \\ 0 & 5 & 7 \end{vmatrix}}{\begin{vmatrix} 12 & -8 & 0 \\ -8 & 10 & -2 \\ 0 & -2 & 7 \end{vmatrix}} = \frac{204}{344} = 590\,\text{mV}$$

$$v_3 = \frac{\begin{vmatrix} 12 & -8 & 3 \\ -8 & 10 & -1 \\ 0 & -2 & 5 \end{vmatrix}}{\begin{vmatrix} 12 & -8 & 0 \\ -8 & 10 & -2 \\ 0 & -2 & 7 \end{vmatrix}} = \frac{304}{344} = 880\,\text{mV}$$

EXAMPLE 6.19 Determine the nodal voltages v_1, v_2, v_3, and v_4 for the circuit in Figure 6.46. ■

Figure 6.46

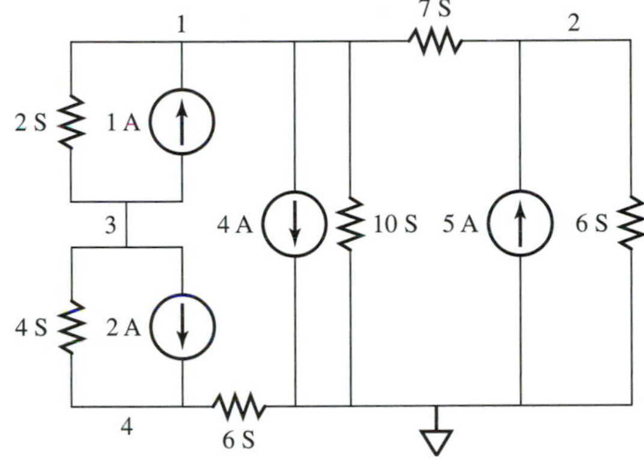

SOLUTION Because there are five nodes, the denominator determinant contains four rows and four columns. The entries for the denominator determinant of the nodal voltages are

$$G_{11} = 2 + 7 + 10 = 19$$

$$G_{22} = 7 + 6 = 13$$

$$G_{33} = 2 + 4 = 6$$

$$G_{44} = 4 + 6 = 10$$

$$G_{12} = G_{21} = 7$$

$$G_{13} = G_{31} = 2$$

$$G_{14} = G_{41} = 0$$

$$G_{23} = G_{32} = 0$$

$$G_{34} = G_{43} = 4$$

$$G_{24} = G_{42} = 0$$

and the denominator determinant is

$$\begin{vmatrix} 19 & -7 & -2 & 0 \\ -7 & 13 & 0 & 0 \\ -2 & 0 & 6 & -4 \\ 0 & 0 & -4 & 10 \end{vmatrix}$$

The entries for the current-source summation column are

$$
\begin{aligned}
i_1 &= 1 - 4 = -3 \\
i_2 &= 5 \\
i_3 &= -1 - 2 = -3 \\
i_4 &= 2
\end{aligned}
$$

and the current-source summation column is

$$\begin{pmatrix} -3 \\ 5 \\ -3 \\ 2 \end{pmatrix}$$

The numerator determinants for the nodal voltages v_1, v_2, v_3, and v_4, respectively, are

$$\begin{vmatrix} -3 & -7 & -2 & 0 \\ 5 & 13 & 0 & 0 \\ -3 & 0 & 6 & -4 \\ 2 & 0 & -4 & 10 \end{vmatrix}$$

$$\begin{vmatrix} 19 & -3 & -2 & 0 \\ -7 & 5 & 0 & 0 \\ -2 & -3 & 6 & -4 \\ 0 & 2 & -4 & 10 \end{vmatrix}$$

$$\begin{vmatrix} 19 & -7 & -3 & 0 \\ -7 & 13 & 5 & 0 \\ -2 & 0 & -3 & -4 \\ 0 & 0 & 2 & 10 \end{vmatrix}$$

$$\begin{vmatrix} 19 & -7 & -2 & -3 \\ -7 & 13 & 0 & 5 \\ -2 & 0 & 6 & -3 \\ 0 & 0 & -4 & 2 \end{vmatrix}$$

so

$$v_1 = \frac{\begin{vmatrix} -3 & -7 & -2 & 0 \\ 5 & 13 & 0 & 0 \\ -3 & 0 & 6 & -4 \\ 2 & 0 & -4 & 10 \end{vmatrix}}{\begin{vmatrix} 19 & -7 & -2 & 0 \\ -7 & 13 & 0 & 0 \\ -2 & 0 & 6 & -4 \\ 0 & 0 & -4 & 10 \end{vmatrix}} = -\frac{748}{8192} = -91.3 \text{ mV}$$

$$V_2 = \frac{\begin{vmatrix} 19 & -3 & -2 & 0 \\ -7 & 5 & 0 & 0 \\ -2 & -3 & 6 & -4 \\ 0 & 2 & -4 & 10 \end{vmatrix}}{\begin{vmatrix} 19 & -7 & -2 & 0 \\ -7 & 13 & 0 & 0 \\ -2 & 0 & 6 & -4 \\ 0 & 0 & -4 & 10 \end{vmatrix}} = \frac{2748}{8192} = 355 \text{ mV}$$

$$V_3 = \frac{\begin{vmatrix} 19 & -7 & -3 & 0 \\ -7 & 13 & 5 & 0 \\ -2 & 0 & -3 & -4 \\ 0 & 0 & 2 & 10 \end{vmatrix}}{\begin{vmatrix} 19 & -7 & -2 & 0 \\ -7 & 13 & 0 & 0 \\ -2 & 0 & 6 & -4 \\ 0 & 0 & -4 & 10 \end{vmatrix}} = -\frac{4436}{8192} = -542 \text{ mV}$$

$$V_4 = \frac{\begin{vmatrix} 19 & -7 & -2 & -3 \\ -7 & 13 & 0 & 5 \\ -2 & 0 & 6 & -3 \\ 0 & 0 & -4 & 2 \end{vmatrix}}{\begin{vmatrix} 19 & -7 & -2 & 0 \\ -7 & 13 & 0 & 0 \\ -2 & 0 & 6 & -4 \\ 0 & 0 & -4 & 10 \end{vmatrix}} = -\frac{136}{8192} = -16.6 \text{ mV}$$

6.6 DEPENDENT SOURCES

In general, the mesh, loop, and nodal methods of analysis can be applied to circuits with dependent sources. However, the more systematic uniform-approach methods are usually restricted to circuits with only independent sources.

In applying multiple-loop methods to circuits with dependent sources, the basic analysis procedures are unchanged; only the algebra associated with the solution of the mathematical model differs. Consider the following examples of circuits with dependent sources.

EXAMPLE 6.20 Use the mesh-analysis method for the circuit in Figure 6.47 to solve for the current i. The circuit contains a dependent voltage source ($v = 5i$). ■

Figure 6.47

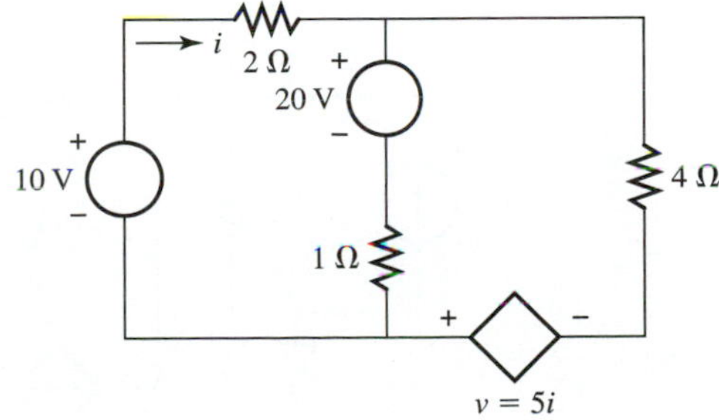

SOLUTION The circuit is shown in Figure 6.48, with assigned mesh currents i_1 and i_2 and the associated voltage polarities.

Figure 6.48

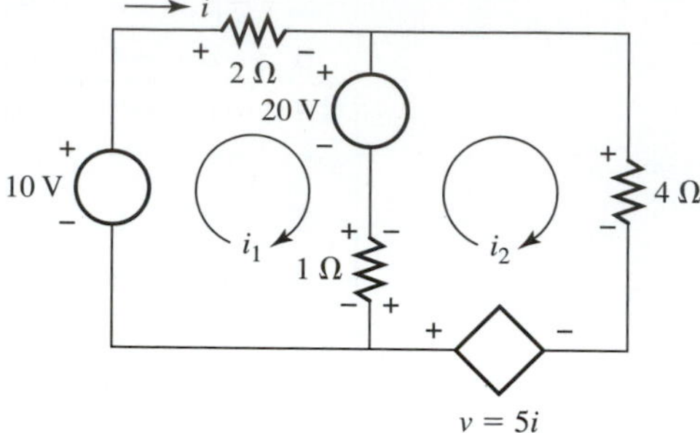

Applying Kirchhoff's voltage law to each mesh results in

$$\text{mesh 1: } 10 - 2i_1 - 20 - 1i_1 + 1i_2 = 0$$
$$\text{mesh 2: } 5i - 1i_2 + 1i_1 + 20 - 4i_2 = 0$$

Notice that there are three unknowns (i_1, i_2, and i) in the mathematical model. However, from an inspection of the circuit, mesh current i_1 equals i.

Substituting i_1 for i in the two mesh equations and combining terms results in

$$3i_1 - i_2 = -10$$
$$-6i_1 + 5i_2 = 20$$

From Cramer's rule,

$$i_1 = \frac{\begin{vmatrix} -10 & -1 \\ 20 & 5 \end{vmatrix}}{\begin{vmatrix} 3 & -1 \\ -6 & 5 \end{vmatrix}} = -\frac{30}{9}\text{ A}$$

Because $i = i_1$,

$$i = -\frac{10}{3}\text{ A}$$

EXAMPLE 6.21 Use the nodal-analysis method for the circuit in Figure 6.49 to solve for the voltage v. The circuit contains a dependent-current source ($i = 2v$). ■

Figure 6.49

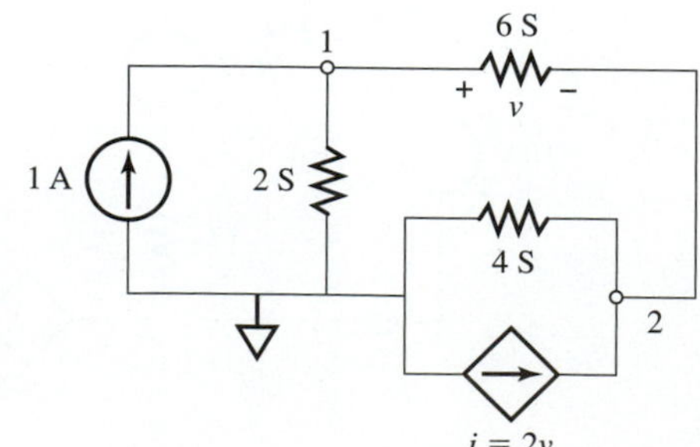

SOLUTION Nodes 1 and 2 are isolated and the resistive currents are indicated for the partial circuits in Figure 6.50.

Figure 6.50

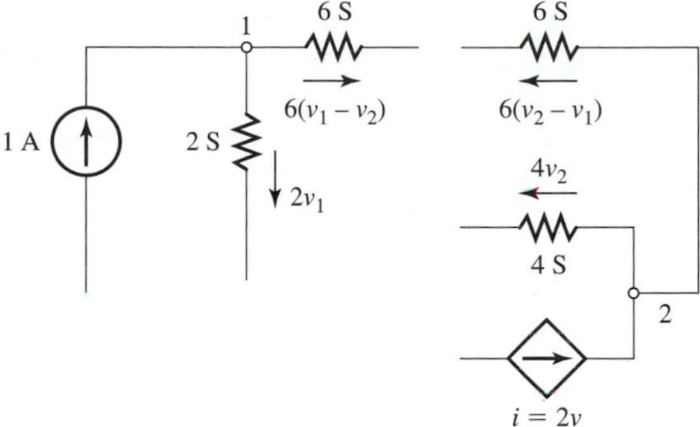

Applying Kirchhoff's current law to nodes 1 and 2 results in

$$\text{node 1: } 1 - 2v_1 - 6(v_1 - v_2) = 0$$
$$\text{node 2: } i - 4v_2 - 6(v_2 - v_1) = 0$$

Notice that there are three unknowns (v_1, v_2, and i) in the mathematical model. Because $i = 2v$, from the dependent-source relationship, and $v = v_1 - v_2$, from Equation (6.7),

$$i = 2(v_1 - v_2)$$

Substituting $2(v_1 - v_2)$ for i in the two nodal equations results in

$$8v_1 - 6v_2 = 1$$
$$-8v_1 + 12v_2 = 0$$

and from Cramer's rule,

$$v_1 = \frac{\begin{vmatrix} 1 & -6 \\ 0 & 12 \end{vmatrix}}{\begin{vmatrix} 8 & -6 \\ -8 & 12 \end{vmatrix}} = \frac{12}{48} = \frac{1}{4}\text{ V}$$

$$v_2 = \frac{\begin{vmatrix} 8 & 1 \\ -8 & 0 \end{vmatrix}}{\begin{vmatrix} 8 & -6 \\ -8 & 12 \end{vmatrix}} = \frac{8}{48} = \frac{1}{6}\text{ V}$$

From Equation (6.7),

$$v = v_1 - v_2$$
$$= \frac{1}{4} - \frac{1}{6} = \frac{1}{12}\text{ V}$$

6.7 COMPARISON OF METHODS

Most multiple-loop circuits can be analyzed by the loop, mesh, or nodal methods of analysis presented in this chapter. The selection of an analysis method is usually based on the complexity of the mathematical model associated with the method. The mathematical model with the least number of equations and unknown quantities involves the least amount of work in arriving at a solution.

In the mesh-analysis method, the number of equations and unknown quantities in the mathematical model is equal to the number of meshes in the circuit. In the loop-analysis method, the number of equations and unknown quantities in the mathematical model is a function of the number of branches and nodes in the circuit. In the nodal-analysis method, the number of equations and unknown quantities in the mathematical model is equal to the number of nodes in the circuit minus 1.

The types of sources (current and voltage) in the circuit should also be considered when selecting an analysis method. The sources in the circuit affect the complexity of the mathematical model for the mesh-, nodal-, and loop-analysis methods and determine if the uniform-approach methods can be employed. If the circuit contains all independent current sources or can be converted to all independent current sources, the uniform approach to nodal analysis can be selected. If the circuit contains all independent voltage sources or can be converted to all independent voltage sources, the uniform approach to mesh analysis can be selected. The uniform-approach methods are systematic and yield solutions with a minimum of work.

6.8 EXERCISES

1. Use the mesh-analysis method to solve for i in Figure 6.51.

Figure 6.51

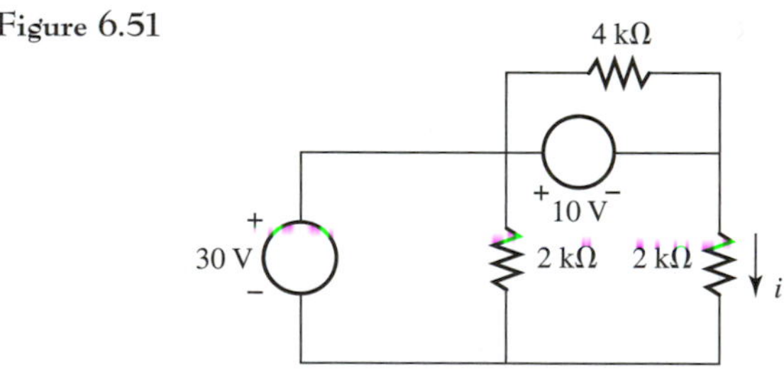

2. Use the mesh-analysis method to solve for the mesh currents (i_1 and i_2) and the branch current i in Figure 6.52.

Figure 6.52

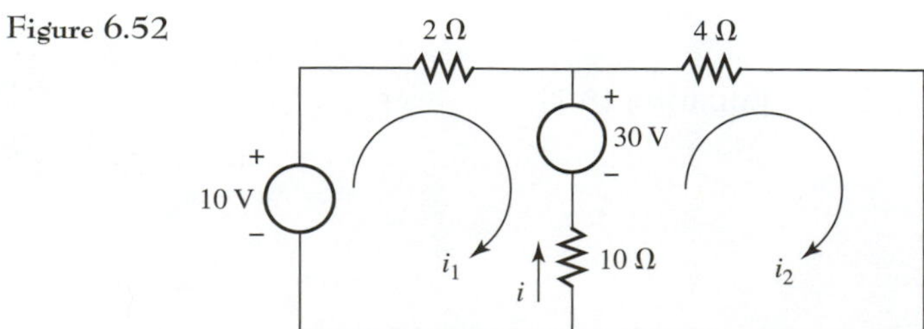

3. Use the mesh-analysis method to solve for i in Figure 6.53.

Figure 6.53

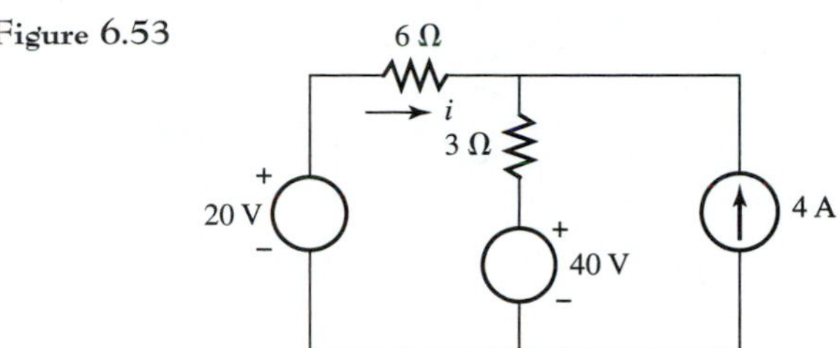

4. Use the mesh-analysis method to solve for i in Figure 6.54.

Figure 6.54

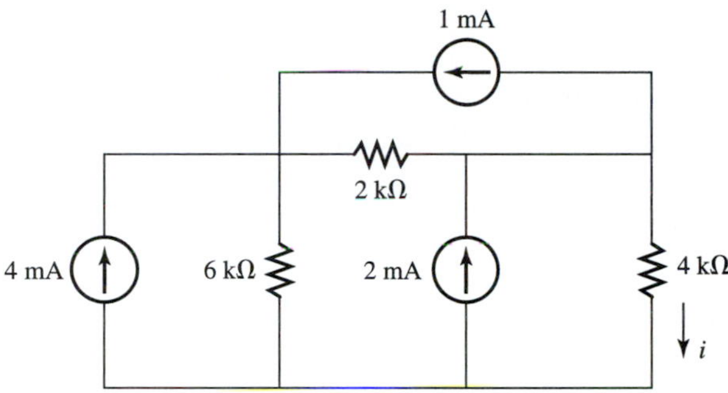

5. Use the mesh-analysis method to solve for i in Figure 6.55.

Figure 6.55

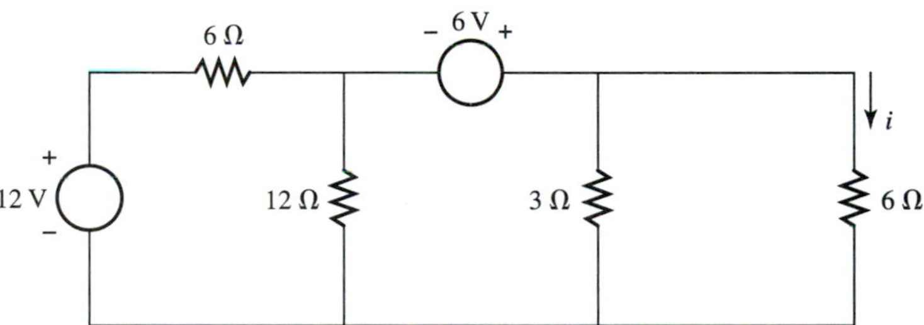

6. Use the uniform method for mesh-analysis to solve for i in Figure 6.56.

Figure 6.56

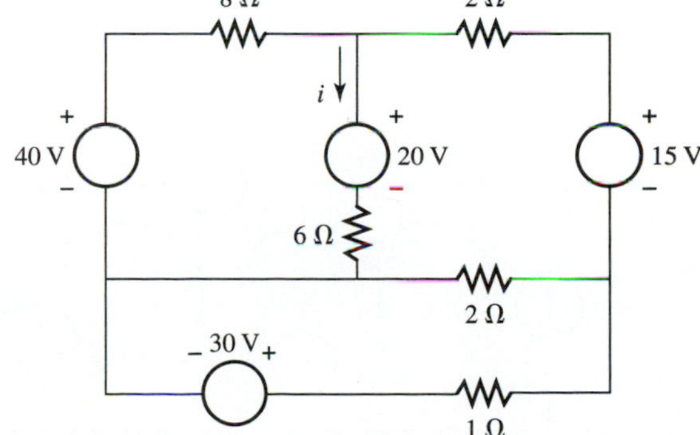

7. Use the uniform approach method for mesh analysis to solve for i in Figure 6.57.

Figure 6.57

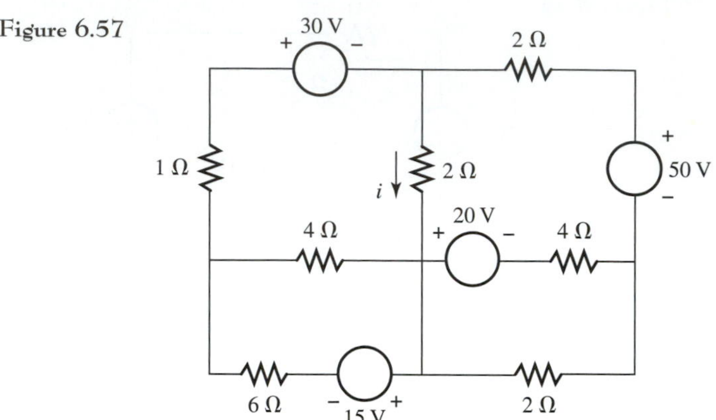

8. Use the uniform method for mesh analysis to solve for the mesh currents (i_1, i_2, i_3, i_4, and i_5) in determinant form for the circuit in Figure 6.58.

Figure 6.58

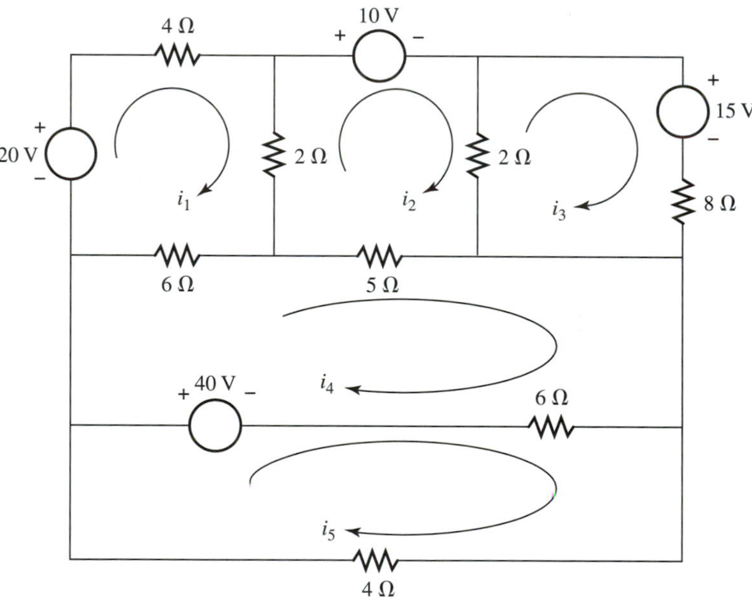

9. Use the loops indicated in Figure 6.59 to solve for i by the loop analysis method.

Figure 6.59

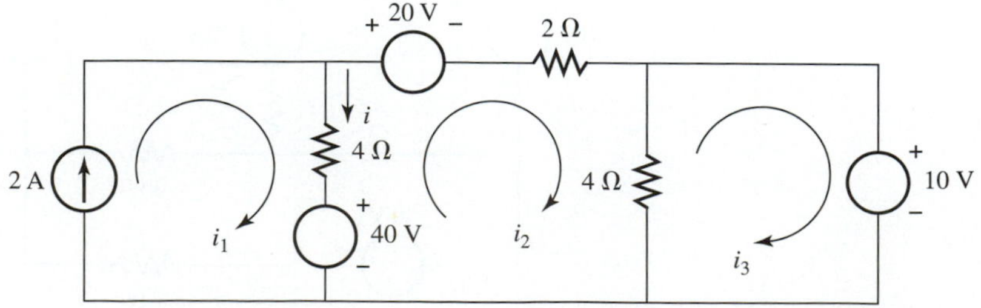

10. Use the loops indicated in Figure 6.60 to solve Problem 9.

Figure 6.60

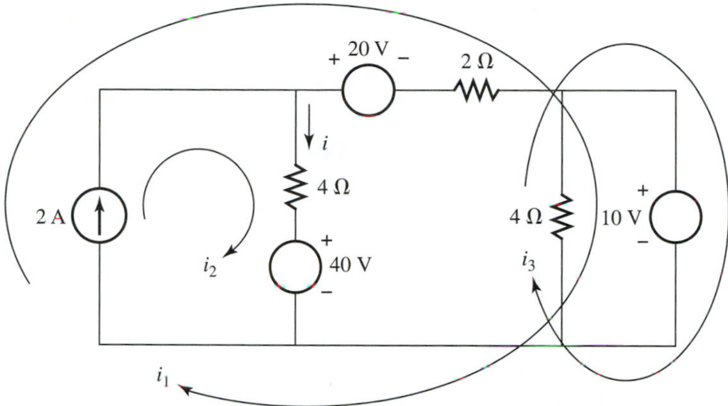

11. Use the loops indicated in Figure 6.61 to solve Problem 9.

Figure 6.61

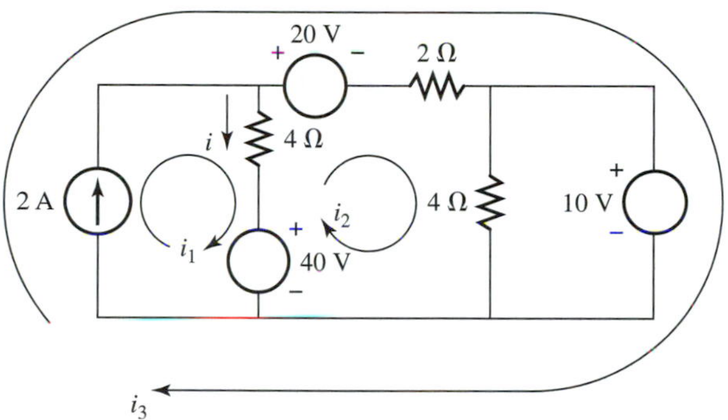

12. Use the nodal-analysis method to solve for the nodal voltages (v_1 and v_2) and the voltage v for the circuit in Figure 6.62.

Figure 6.62

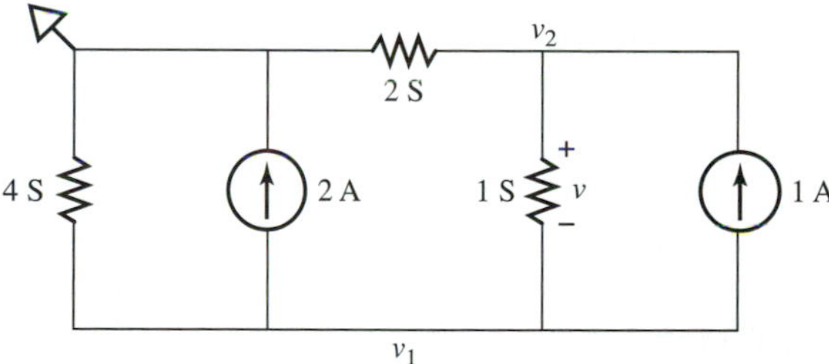

13. Use the nodes indicated in Figure 6.63 to solve Problem 12.

Figure 6.63

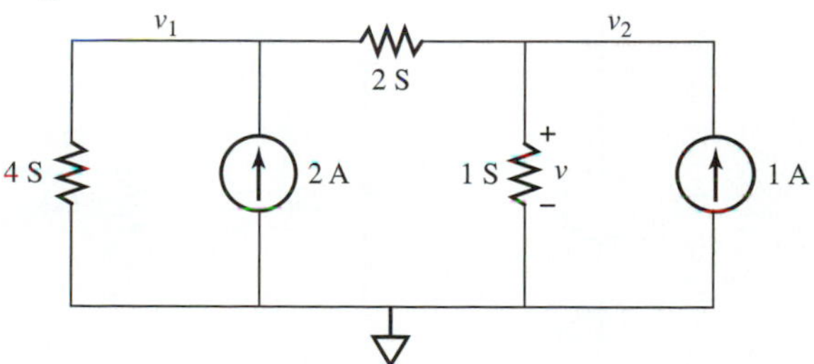

14. Use the nodal-analysis method to solve for the nodal voltages (v_1 and v_2) in Figure 6.64.

Figure 6.64

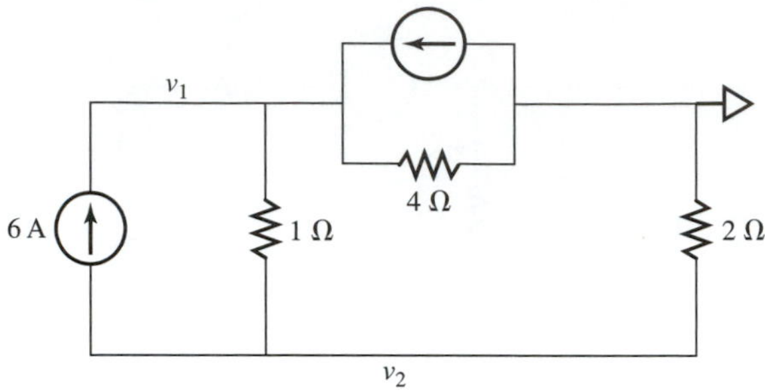

15. Use the nodal-analysis method to solve for the nodal voltages (v_1, v_2, and v_3) for the circuit in Figure 6.65.

Figure 6.65

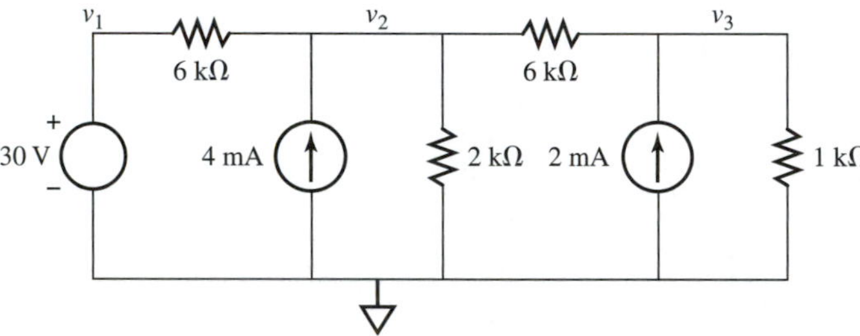

16. Use the nodal-analysis method to solve for all the nodal voltages for the circuit in Figure 6.66.

Figure 6.66

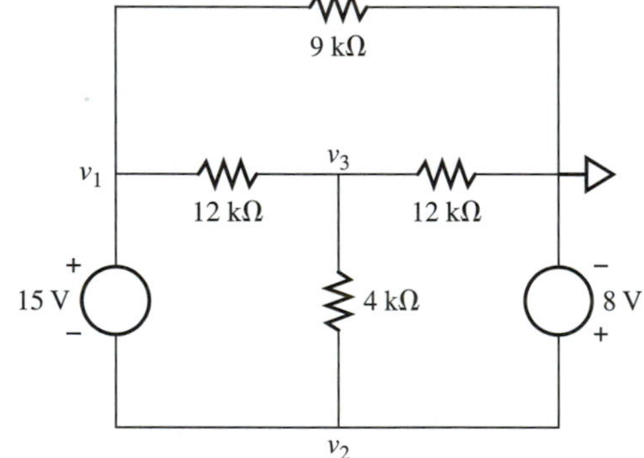

17. Use the nodal-analysis method to solve for all the nodal voltages for the circuit in Figure 6.67.

Figure 6.67

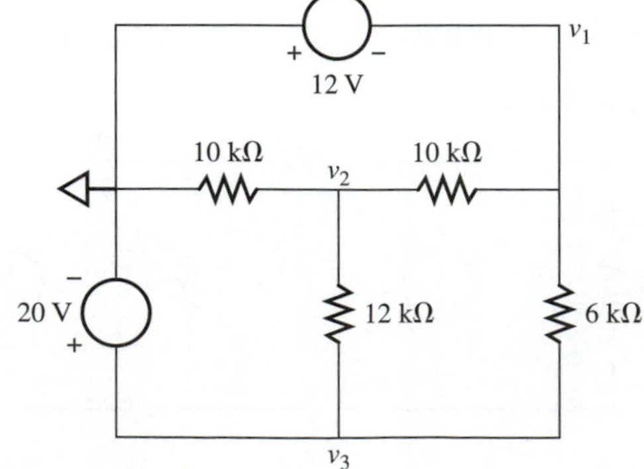

18. Use the uniform method for nodal analysis to solve for the nodal voltages (v_1, v_2, and v_3) for the circuit in Figure 6.68.

Figure 6.68

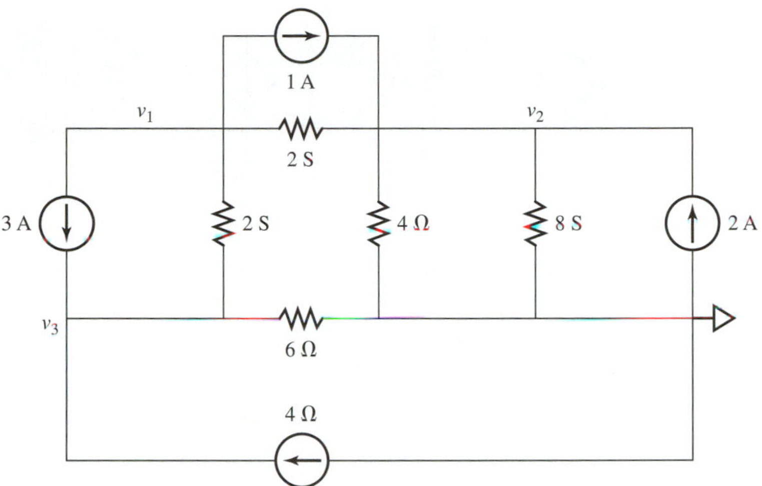

19. Use the uniform method for nodal analysis to solve for the nodal voltages (v_1, v_2, v_3, and v_4) in determinant form for the circuit in Figure 6.69.

Figure 6.69

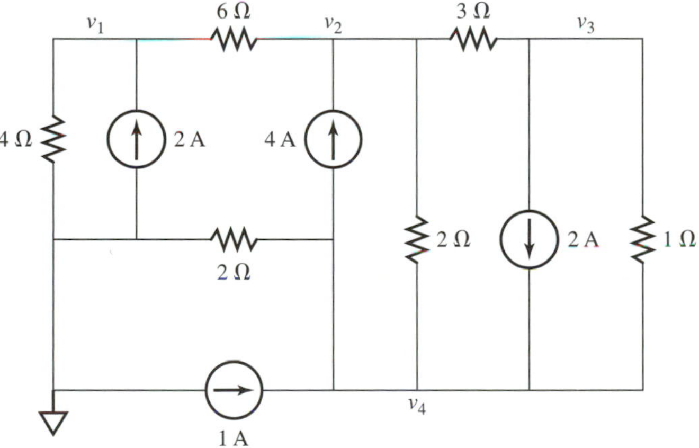

20. Use the mesh-analysis method to solve for the mesh current for the circuit in Figure 6.70.

Figure 6.70

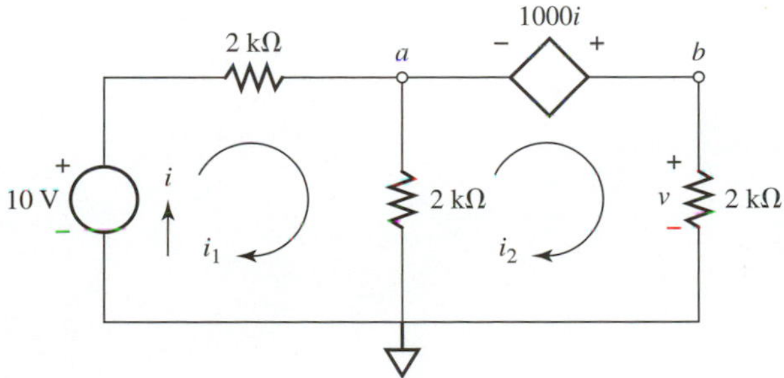

21. Use the nodal-current method to solve for v in Figure 6.71.

Figure 6.71

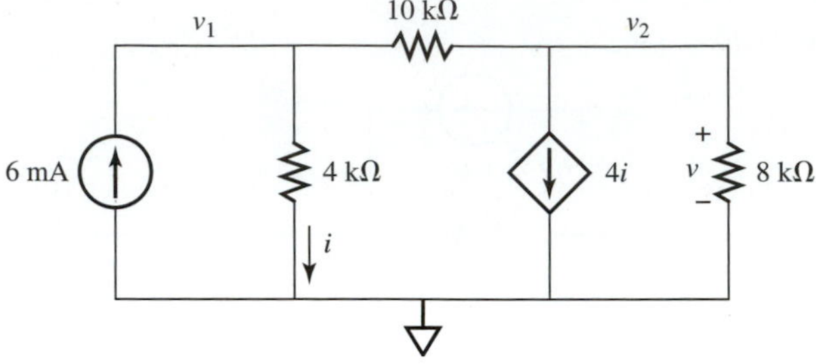

22. Solve Problem 21 using the mesh-analysis method.

CHAPTER

7

Network Theorems

After completing this chapter the student should be able to

* apply the superposition method to circuits with multiple sources
* apply Thevenin's theorem to analyze circuits
* apply Norton's theorem to analyze circuits
* use the maximum power transfer theorem to determine loads for maximum power transfer

The circuit-analysis methods presented in this chapter make use of what is traditionally called the network theorems. These theorems and related analysis methods allow us to simplify circuits for analysis purposes and thus solve complicated circuits with less effort than normally required.

7.1 SUPERPOSITION METHOD

The superposition method is based on the linearity of the basic circuit elements and allows us to determine the current through or the voltage across an electrical element by separately considering the contributions to the current or voltage from each current and voltage source in the circuit. When considering the contribution from each source, all other sources in the circuit are replaced by their internal resistances. That is, the voltage sources are replaced by short circuits and the current sources are replaced by open circuits.

Consider the circuit in Figure 7.1, where v_1, i_1, R_1, R_2, and R_3 are known

Figure 7.1 Superposition method.

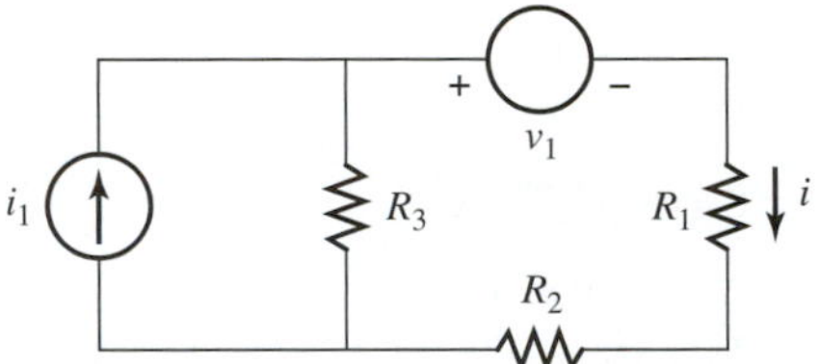

and we wish to determine the current i. The *first step* in the superposition method is to draw a separate circuit for each source in the circuit with all the other sources replaced by their internal resistances. Because the circuit in Figure 7.1 contains two sources (i_1 and v_1), two circuits are drawn, as shown in Figure 7.2.

Figure 7.2 Sources replaced by internal resistances.

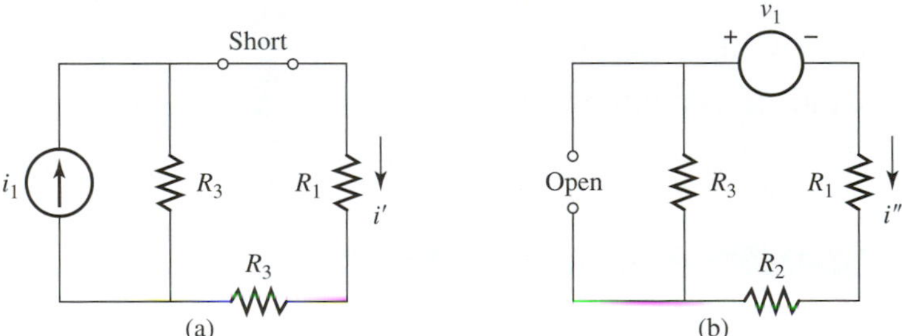

In Figure 7.2(a) voltage source v_1 is replaced by a short and in Figure 7.2(b) current source i_1 is replaced by an open. The current i' is the contribution to i from the current source i_1, and the current i'' is the contribution to i from the voltage source v_1.

The *second step* in the superposition method is to solve the circuits obtained in step 1 for the contributed (primed) currents or voltages. For the circuits in Figure 7.2, the contributed (primed) currents are i' and i''. The current i' can be determined from Equation (4.1) for current division as

$$i' = \left(\frac{R_3}{R_1 + R_2 + R_3}\right)(i_1)$$

and the current i'' can be determined from Ohm's law as

$$i'' = \frac{-v_1}{R_1 + R_2 + R_3}$$

The *third step* in the superposition method is to algebraically sum the contributed (primed) currents or voltages obtained in step two. For the circuit in Figure 7.1,

$$i = i' + i''$$

$$= \frac{R_3 i_1 - v_1}{R_1 + R_2 + R_3}$$

The three-step procedure for the superposition method is summarized here:

1. Draw a circuit for each current and voltage source by replacing all of the other current and voltage sources in the circuit under analysis by their internal resistances. The number of circuits drawn will equal the number of sources in the circuit under analysis.
2. Solve each circuit in step 1 for the current or voltage contributed by each source.
3. Algebraically sum the contributed currents or voltages obtained in step 2 to determine the desired current or voltage.

EXAMPLE 7.1 Determine the voltage v for the circuit in Figure 7.3. ■

Figure 7.3

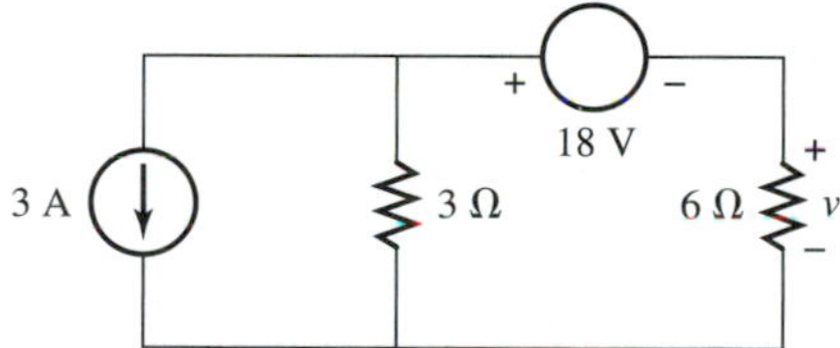

SOLUTION Because there are two sources in the circuit, there are two circuits to consider, as shown in Figures 7.4 and 7.5.

Figure 7.4

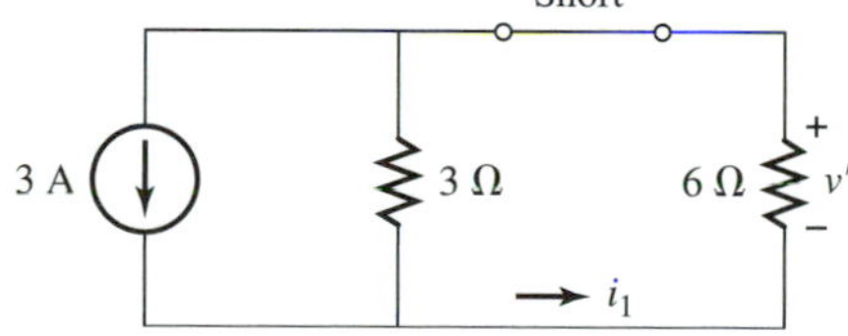

Figure 7.5

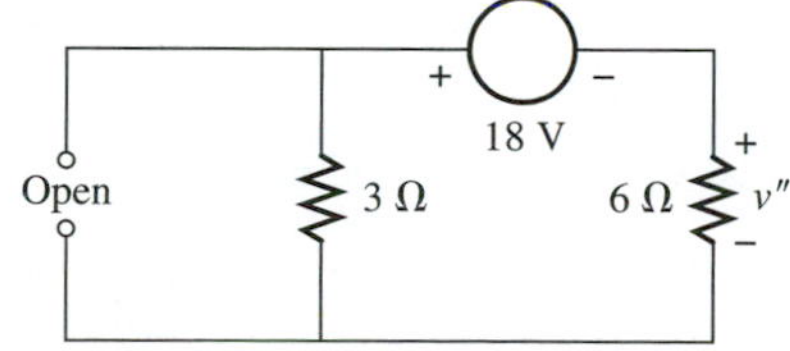

For the circuit in Figure 7.4, voltage v' is the contribution to v from the 3-A current source and can be determined from the current-division formula and Ohm's law. From Equation (4.1) for current division,

$$i_1 = \left(\frac{3}{6 + 3}\right)(3) = 1 \text{ A}$$

and from Ohm's law,

$$v' = -6i_1 = -6 \text{ V}$$

For the circuit in Figure 7.5, voltage v'' is the contribution to v from the 18-V voltage source and can be determined from Equation (4.3) for voltage division as follows.

$$v'' = \left(\frac{6}{6+3}\right)(-18) = -12 \text{ V}$$

Voltage v is equal to the algebraic sum of the contributed voltages v' and v''.

$$v = v' + v'' = -6 - 12$$
$$= -18 \text{ V}$$

EXAMPLE 7.2 Determine the current i for the circuit in Figure 7.6. ■

Figure 7.6

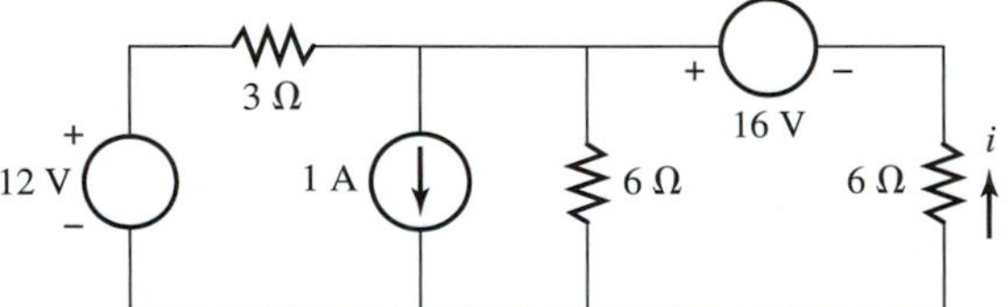

SOLUTION Because there are three sources in the circuit, there are three circuits to consider, as shown in Figures 7.7, 7.8, and 7.9.

Figure 7.7

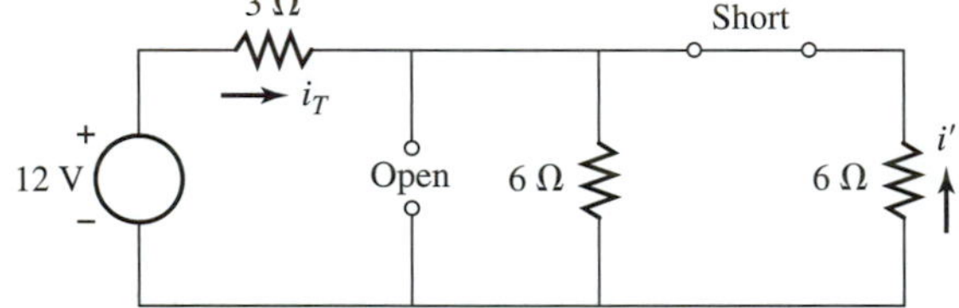

Figure 7.8

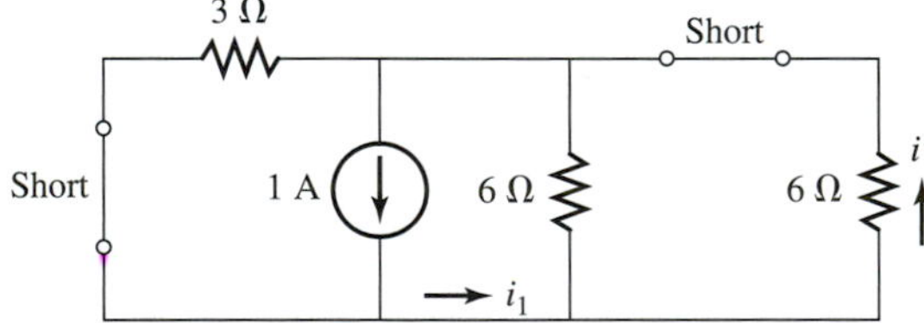

Figure 7.9

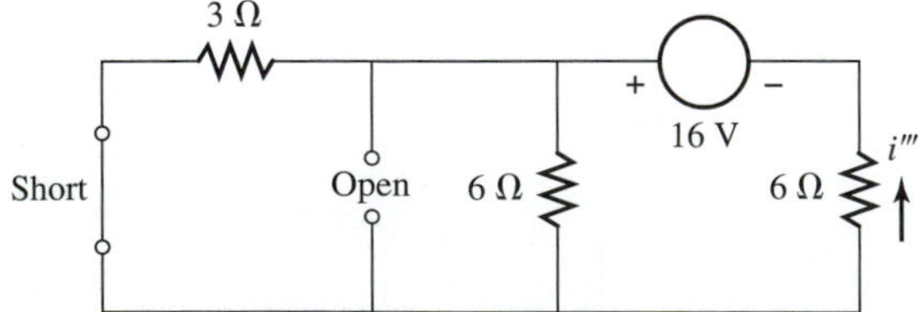

For the circuit in Figure 7.7, current i' is the contribution to i from the 12-V source and can be determined from Ohm's law and Equation (4.1) for current division. The total equivalent resistance R_T across the 12-V source is

$$R_T = 3 + (6\|6) = 6\ \Omega$$

and from Ohm's law the current through the 3-Ω resistor is

$$i_T = \frac{12}{R_T} = 2 \text{ A}$$

Using the current-division formula

$$i' = \left(\frac{6}{6 + 6}\right)(-i_T) = -1 \text{ A}$$

For the circuit in Figure 7.8, the current i'' is the contribution to i from the 1-A current source and can be determined by applying the current-division equation twice:

$$i' = \left\{\frac{3}{3 + (6\|6)}\right\}(1) = \frac{1}{2} \text{ A}$$

$$i'' = \left(\frac{6}{6 + 6}\right)(i_1)$$

$$i'' = \left(\frac{6}{6 + 6}\right)\left(\frac{1}{2}\right) = \frac{1}{4} \text{ A}$$

For the circuit in Figure 7.9, current i''' is the contribution to i from the 16-V voltage source and can be determined from Ohm's law. Using Ohm's law

$$i''' = \frac{16}{6 + (3\|6)} = 2 \text{ A}$$

Algebraically summing the contributed currents results in

$$i = i' + i'' + i'''$$

$$= -1 + \frac{1}{4} + 2 = \frac{5}{4} = 1.25 \text{ A}$$

EXAMPLE 7.3 Determine the voltages v_1 and v_2 for the circuit in Figure 7.10 using the superposition theorem. ■

Figure 7.10

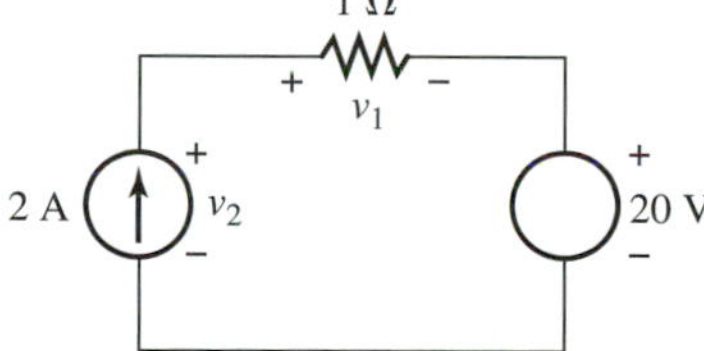

SOLUTION Because there are two sources in the circuit, there are two circuits to consider (Figures 7.11 and 7.12). For the circuit in Figure 7.11, v_1' is the

Figure 7.11

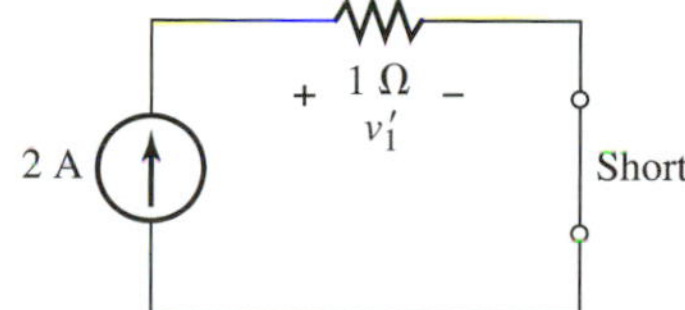

Figure 7.12

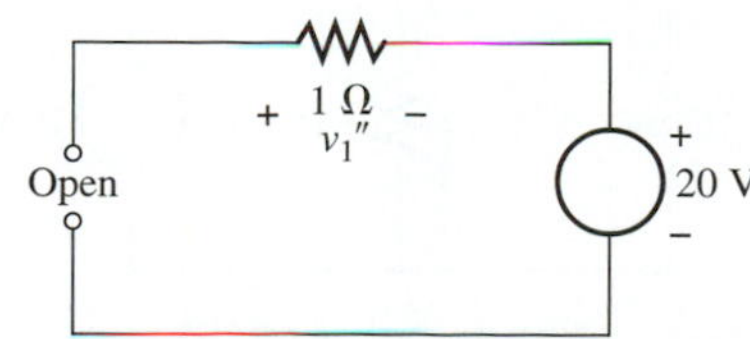

contribution to v_1 from the 2-A current source. From Ohm's law,

$$v_1' = (2)(1) = 2 \text{ V}$$

For the circuit in Figure 7.12, v_1'' is the contribution to v_1 from the 20-V voltage source. Because no current flows in the circuit,

$$v_1'' = 0$$

Voltage v_1 is equal to the algebraic sum of the contributed voltages:

$$\begin{aligned} v_1 &= v_1' + v_1'' = 2 + 0 \\ &= 2 \text{ V} \end{aligned}$$

Applying Kirchhoff's voltage law to the original circuit results in

$$v_2 - v_1 - 20 = 0$$

and the voltage across the current source is

$$\begin{aligned} v_2 &= 20 + v_1 \\ &= 22 \text{ V} \end{aligned}$$

EXAMPLE 7.4 Determine the current i_1 for the circuit in Figure 7.13 using the superposition theorem. ■

Figure 7.13

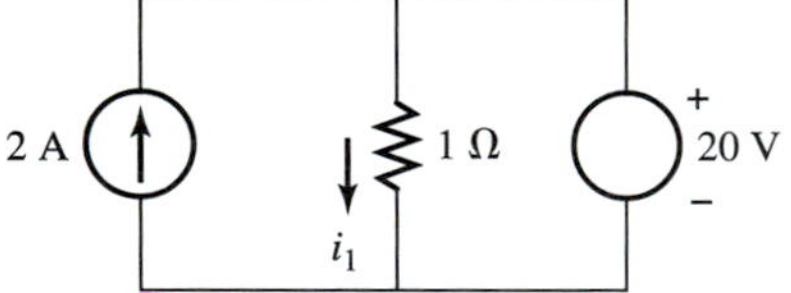

SOLUTION Because there are two sources in the circuit, there are two circuits to consider (Figures 7.14 and 7.15). For the circuit in Figure 7.14, i_1' is the

Figure 7.14

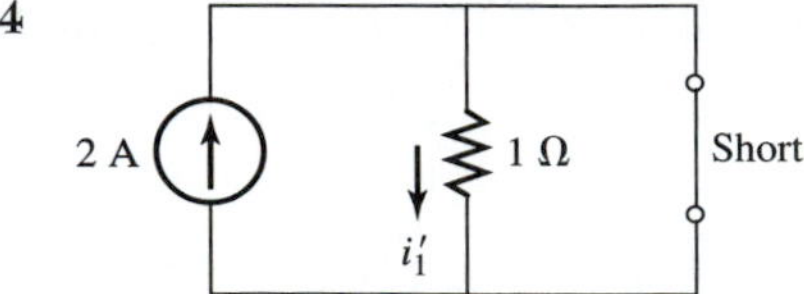

Figure 7.15

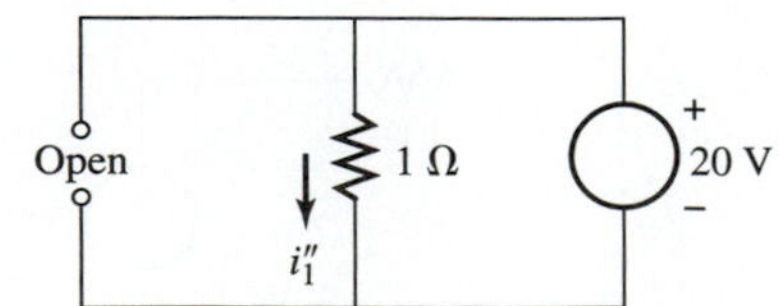

contribution to i_1 from the 2-A current source. Because all the current from the 2-A current source flows through the short,

$$i_1' = 0$$

This result can be arrived at from Equation (4.1) for current division as follows:

$$i_1' = \left(\frac{0}{0 + 1}\right)(2) = 0$$

For the circuit in Figure 7.15, i_1'' is the contribution to i_1 from the 20-V voltage source. From Ohm's law

$$i_1'' = \frac{20}{1} = 20 \text{ A}$$

Current i_1 is equal to the algebraic sum of the contributed currents.

$$\begin{aligned} i_1 = i_1' + i_1'' &= 0 + 20 \\ &= 20 \text{ A} \end{aligned}$$

7.2 THEVENIN'S THEOREM

A very useful theorem for circuit analysis was presented by Charles Thevenin, a French engineer, in 1883. It is known as Thevenin's theorem and can be stated as follows for resistive circuits.

> **THEVENIN'S THEOREM** A resistive circuit connected across two terminals *a-b* can be replaced by a voltage source in series with a resistance.
>
> The series combination of voltage source and resistance is known as a Thevenin equivalent circuit with respect to terminals *a-b*.
>
> The series resistance, known as the Thevenin resistance and designated by R_T, is the resistance seen across terminals *a-b* when all voltage and current sources in the circuit are replaced by their internal resistances (a short for voltage sources and an open for current sources).
>
> The series voltage source, known as the Thevenin voltage source and designated by v_T, is equal to the open-circuited voltage across terminals *a-b*.

The *first step* in applying Thevenin's theorem to a circuit is to isolate the two terminals in the circuit connected to the portion of the circuit that is to be replaced by a Thevenin equivalent. Consider the circuit in Figure 7.16, where it is desired to replace the circuit to the left of terminals *a-b* by a

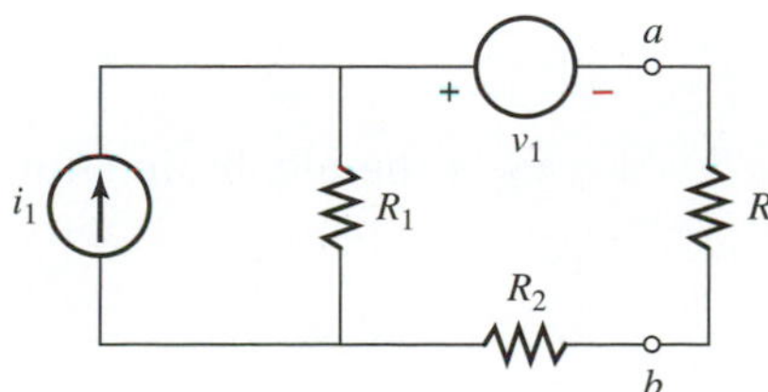

Figure 7.16 Thevenin's theorem.

Thevenin equivalent. All the electrical elements connected to terminals *a*-*b* are removed (disconnected) except for the circuit that is to be replaced by the Thevenin equivalent. Resistance R is therefore removed, and the resulting circuit with terminals *a*-*b* isolated is as shown in Figure 7.17.

Figure 7.17 Terminals *a*-*b* isolated.

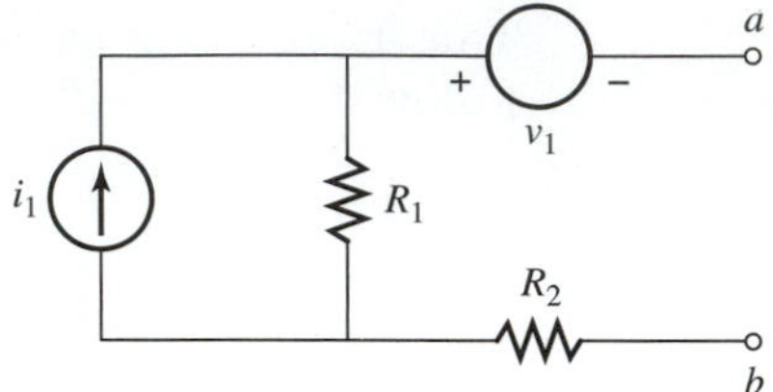

Using Thevenin's theorem, the circuit to the left of terminals *a*-*b* in Figure 7.17 can be replaced by the Thevenin equivalent in Figure 7.18.

Figure 7.18 Thevenin equivalent.

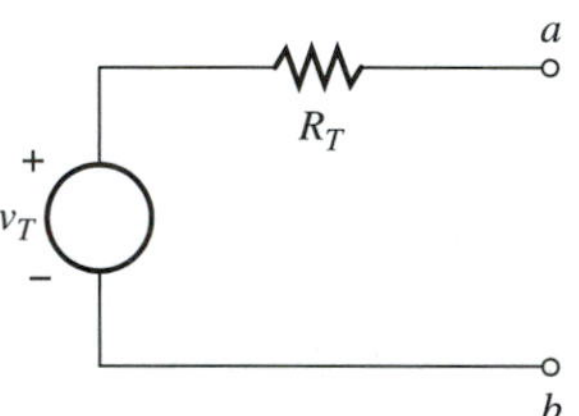

The *second step* in applying Thevenin's theorem is to determine the Thevenin voltage v_T for the Thevenin equivalent. Any analysis method or laboratory procedure can be used to determine the Thevenin voltage v_T.

We shall apply the superposition method to determine v_T for the circuit in Figure 7.17. Because there are two sources in the circuit, there are two circuits to consider, as shown in Figure 7.19: v'_T is the contribution to v_T from the current source i_1 and v''_T is the contribution to v_T from the voltage source v_1.

Figure 7.19 Thevenin circuits.

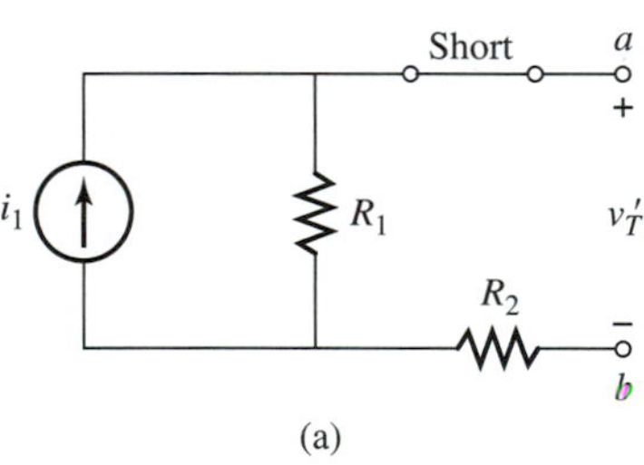

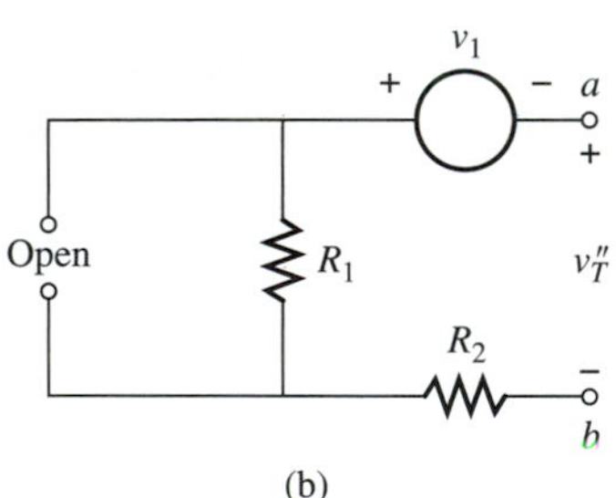

For the circuit in Figure 7.19(a), all the current i_1 flows through R_1; from Ohm's law,

$$v'_T = i_1 R_1$$

For the circuit in Figure 7.19(b), no current flows in the circuit and

$$v''_T = -v_1$$

v_T is the algebraic sum of the contributed voltages v'_T and v''_T.

$$v_T = v'_T + v''_T$$

$$v_T = i_1 R_1 - v_1$$

The *third step* in applying Thevenin's theorem is to determine the Thevenin resistance R_T for the Thevenin equivalent. The circuit in Figure 7.17 is shown in Figure 7.20 with all sources replaced by their internal resistances, as required by Thevenin's theorem for determining R_T. The Thevenin resistance R_T is the resistance across terminals *a-b* in Figure 7.20; from Equation (3.1)

Figure 7.20 Thevenin's circuit for R_T.

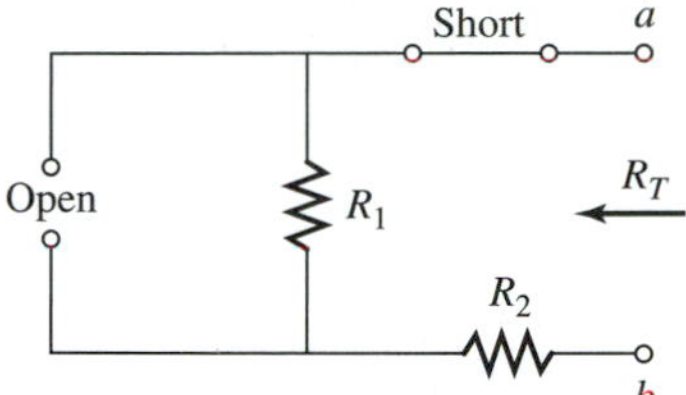

$$R_T = R_1 + R_2$$

The *fourth step* in applying Thevenin's theorem is to form the Thevenin equivalent circuit using v_T and R_T. For the circuit to the left of terminals *a-b* in Figure 7.16, the Thevenin equivalent circuit is shown in Figure 7.21.

Figure 7.21 Thevenin equivalent.

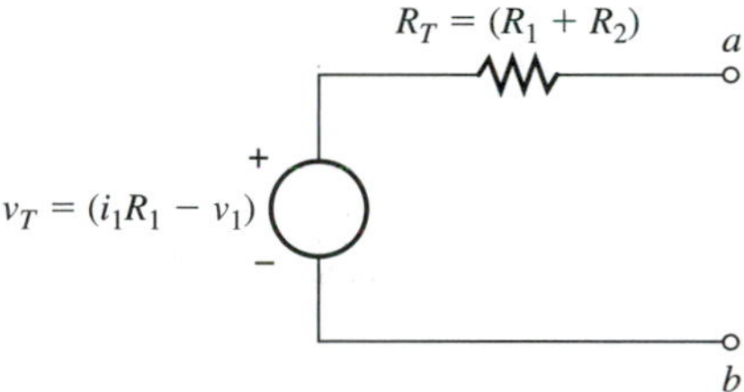

The four-step procedure for applying Thevenin's theorem is summarized as follows:

1. Isolate the two terminals connected to the portion of the circuit to be replaced by a Thevenin equivalent.
2. Determine the Thevenin voltage v_T.
3. Determine the Thevenin resistance R_T.
4. Use v_T and R_T to form the Thevenin equivalent.

EXAMPLE 7.5 Determine a Thevenin equivalent for the circuit to the left of terminals *a-b* in Figure 7.22. ■

Figure 7.22

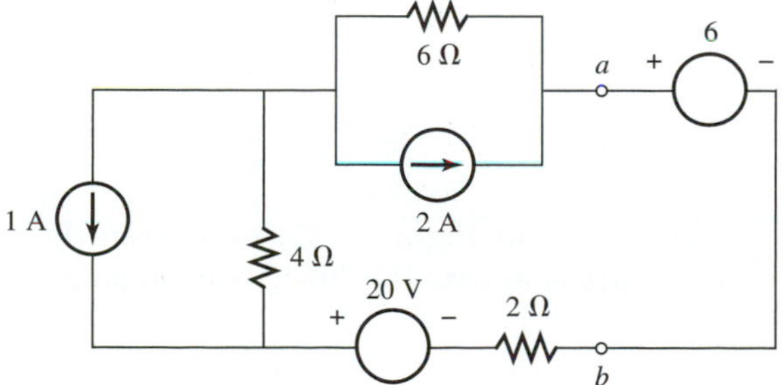

SOLUTION The circuit to the left of terminals *a-b* is shown in Figure 7.23. We shall use the superposition method to determine v_T. Because there are three

Figure 7.23

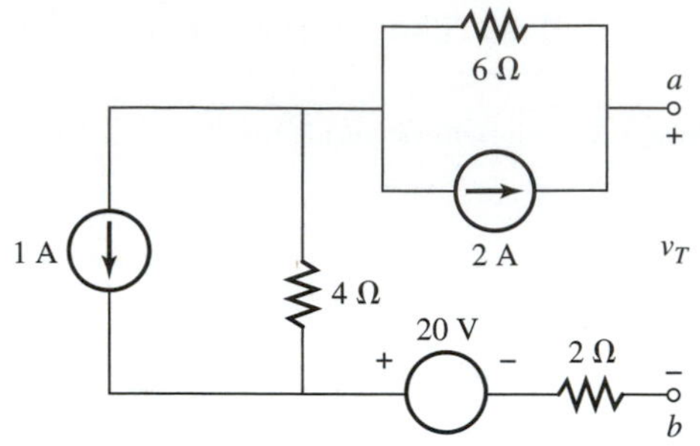

sources, three circuits are considered, as shown in Figures 7.24, 7.25, and 7.26.

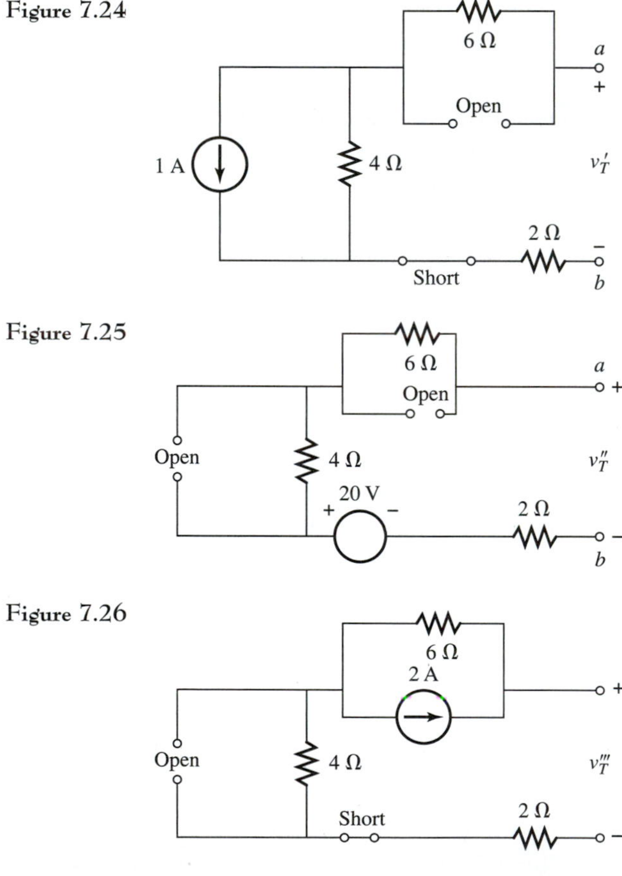

Figure 7.24

Figure 7.25

Figure 7.26

v'_T is the contribution to v_T from the 1-A current source. v''_T is the contribution to v_T from the 20-V voltage source. v'''_T is the contribution to v_T from the 2-A current source.

For the circuit in Figure 7.24, all of the current from the 1-A current source flows through the 4-Ω resistance and from Ohm's law

$$v'_T = -(4)(1) = -4 \text{ V}$$

For the circuit in Figure 7.25, no current flows in the circuit and v''_T is equal to the voltage across the 20-V voltage source and

$$v''_T = 20 \text{ V}$$

For the circuit in Figure 7.26, all of the current from the 2-A current source flows through the 6-Ω resistance and from Ohm's law

$$v_T''' = (6)(2) = 12 \text{ V}$$

v_T is the algebraic sum of the contributed voltages and

$$\begin{aligned} v_T &= v_T' + v_T'' + v_T''' \\ &= -4 + 20 + 12 \\ &= 28 \text{ V} \end{aligned}$$

The circuit to the left of terminals *a*-*b* is shown in Figure 7.27 with all of the sources replaced by their internal resistances. The Thevenin resistance R_T is determined from this circuit.

Figure 7.27

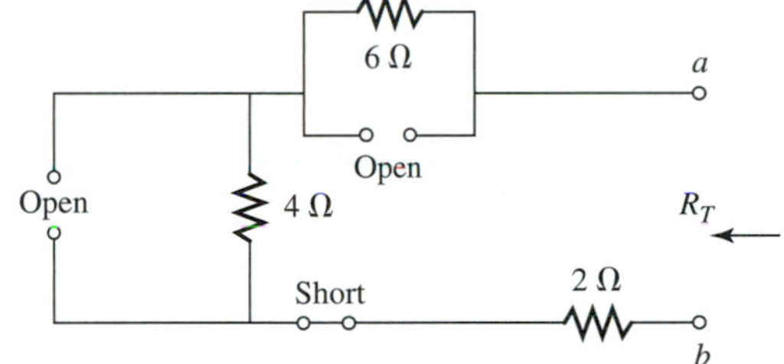

From Equation (3.1)

$$\begin{aligned} R_T &= 6 + 4 + 2 \\ &= 12 \ \Omega \end{aligned}$$

The Thevenin equivalent for the circuit to the left of terminals *a*-*b* is shown in Figure 7.28.

Figure 7.28

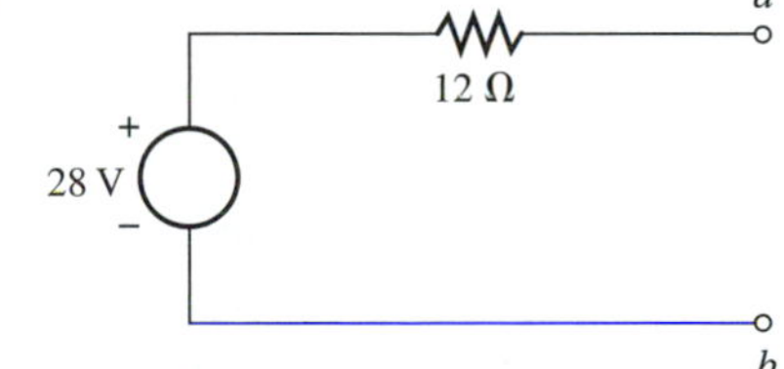

EXAMPLE 7.6 For the circuit in Example 7.5, determine the current *i* flowing through the 6-V voltage source. The circuit is redrawn in Figure 7.29. ■

Figure 7.29

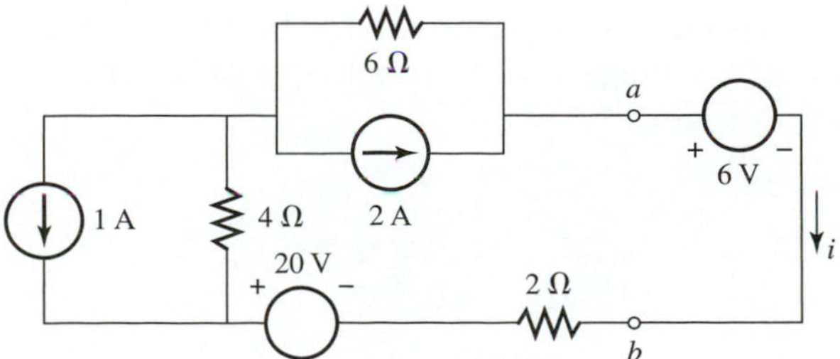

SOLUTION The circuit to the left of terminals a-b can be replaced by the Thevenin equivalent determined in Example 7.5, as shown in Figure 7.30. Employing Kirchhoff's voltage law results in

Figure 7.30

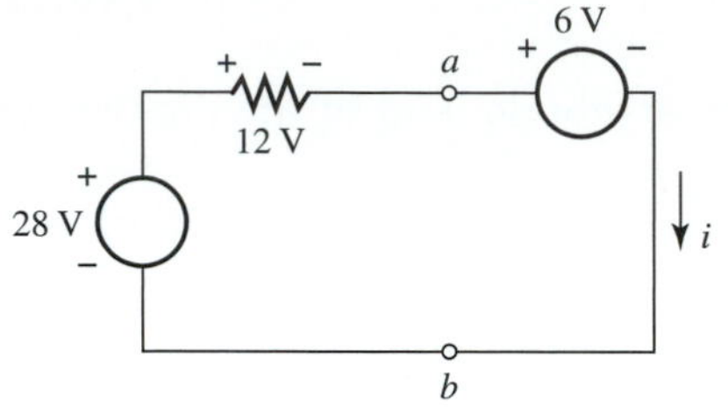

$$28 - 12i - 6 = 0$$

$$i = \frac{28 - 6}{12}$$

$$= \frac{11}{6}\ \text{A}$$

EXAMPLE 7.7 Determine the current i for the circuit shown in Figure 7.31. ■

Figure 7.31

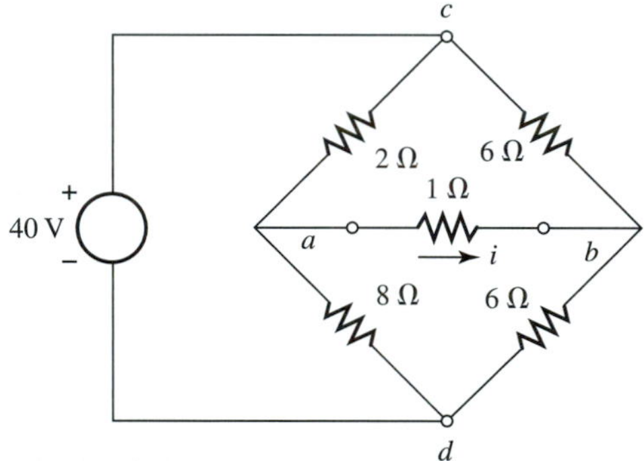

SOLUTION We shall remove the 1-Ω resistor, as shown in Figure 7.32, and determine a Thevenin equivalent with respect to terminals a-b. The Thevenin

Figure 7.32

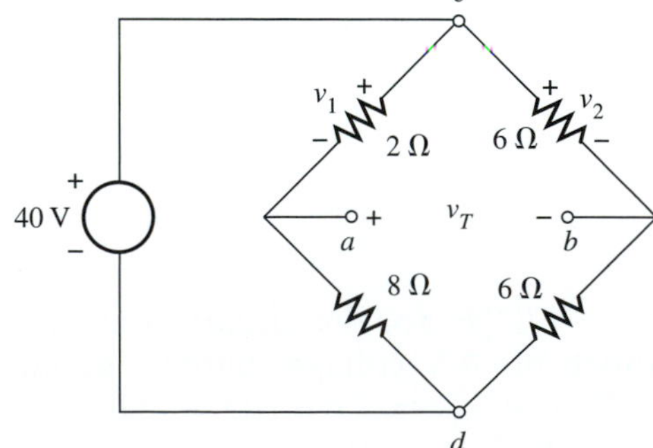

voltage v_T can be determined by finding voltages v_1 and v_2 and applying Kirchhoff's voltage law to the loop containing v_1, v_2, and terminals a-b, as shown in Figure 7.33.

Figure 7.33

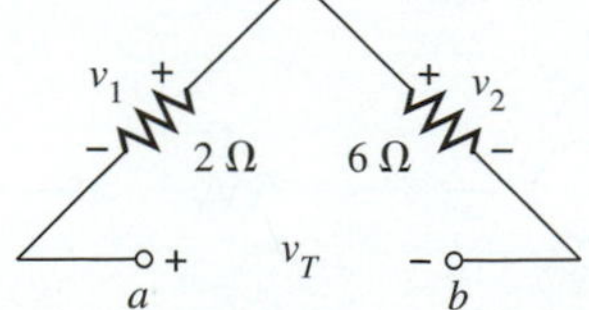

From Kirchhoff's voltage law,

$$v_T + v_1 - v_2 = 0$$
$$v_T = v_2 - v_1$$

Referring to the circuit in Figure 7.32, voltages v_1 and v_2 can be determined from Equation (4.3) for voltage division as

$$v_1 = \left(\frac{2}{2+8}\right)(40)$$
$$= 8 \text{ V}$$
$$v_2 = \left(\frac{6}{6+6}\right)(40)$$
$$= 20 \text{ V}$$

Because

$$v_T = v_2 - v_1,$$
$$v_T = 12 \text{ V}$$

The Thevenin resistance R_T is determined by replacing the 40-V voltage source with a short and finding the resulting resistance across terminals *a*-*b*. The circuit used to determine R_T is shown in Figure 7.34. Notice that shorting the 40-V voltage source is the same as connecting terminals *c* and *d*.

Figure 7.34

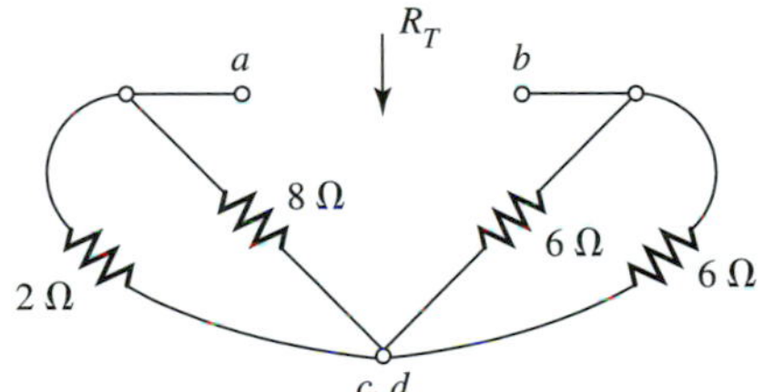

From Equations (3.1) and (3.5), the Thevenin resistance R_T is

$$R_T = (2\|8) + (6\|6) = \frac{(2)(8)}{2+8} + \frac{(6)(6)}{6+6}$$
$$= 1.6 + 3$$
$$= 4.6 \ \Omega$$

The Thevenin equivalent for the circuit across terminals *a*-*b* with the 1-Ω resistance reconnected is shown in Figure 7.35.

Figure 7.35

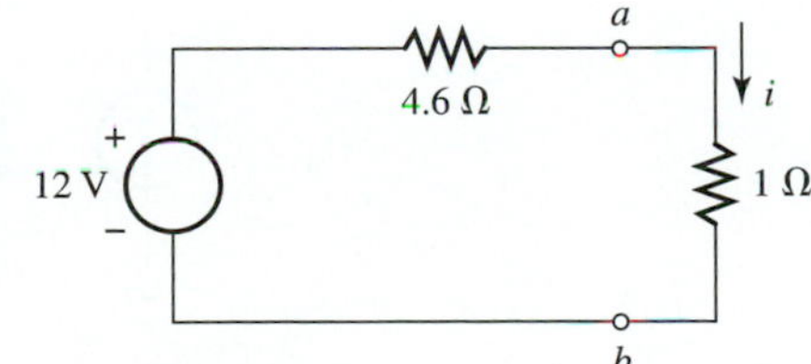

From Ohm's law

$$i = \frac{12}{4.6 + 1}$$

$$= 2.1 \text{ A}$$

7.3 NORTON'S THEOREM

Another very useful theorem for circuit analysis was presented by E. L. Norton, an American engineer, in 1933. The theorem is known as Norton's theorem and can be stated as follows for resistive circuits.

> **NORTON'S THEOREM** The resistive circuit connected across any two terminals *a-b* can be replaced by a current source in parallel with a resistance.
>
> The parallel combination of the current source and the resistance is known as a Norton equivalent with respect to terminals *a-b*.
>
> The parallel resistance, known as the Norton resistance and designated by R_N, is the resistance across terminals *a-b* with all voltage and current sources in the circuit replaced by their internal resistances (a short for voltage sources and an open for current sources).
>
> The parallel current source, known as the Norton current source and designated by i_N, is equal to the current flowing through a short connected across terminals *a-b*.

The *first step* in applying Norton's theorem to a circuit is to isolate the two terminals in the circuit connected to the portion of the circuit that is to be replaced by a Norton equivalent. Consider the circuit in Figure 7.16 (redrawn in Figure 7.36), where it is desired to replace the circuit to the left of terminals *a-b* by a Norton equivalent. The resistance R is removed as shown in Figure 7.37 to isolate terminals *a-b*, as required by Norton's theorem.

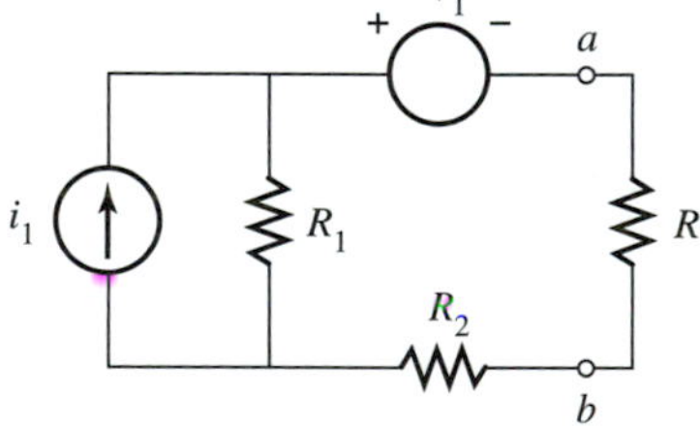

Figure 7.36 Norton's theorem.

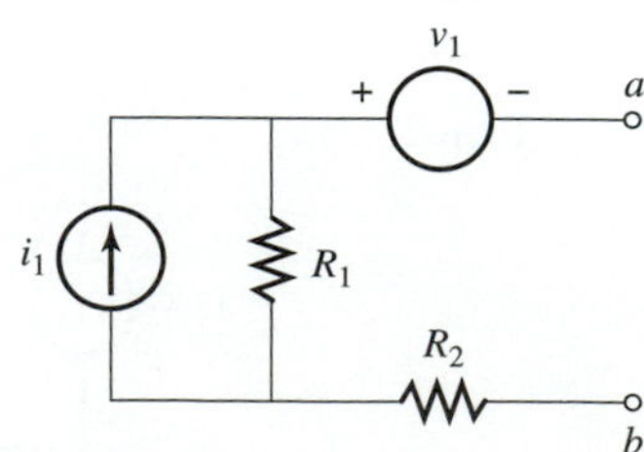

Figure 7.37 Terminals *a-b* isolated.

Using Norton's theorem, the circuit in Figure 7.37 can be replaced by the Norton equivalent in Figure 7.38.

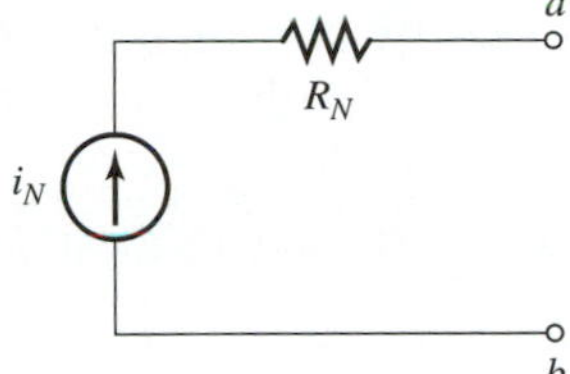

Figure 7.38 Norton equivalent.

The *second step* in applying Norton's theorem is to determine the Norton current i_N for the Norton equivalent. Because i_N is the current through a short across terminals *a*-*b*, the circuit in Figure 7.37 with terminals *a*-*b* shorted is shown in Figure 7.39. Any analysis method or laboratory procedure can be used to determine the Norton current, i_N.

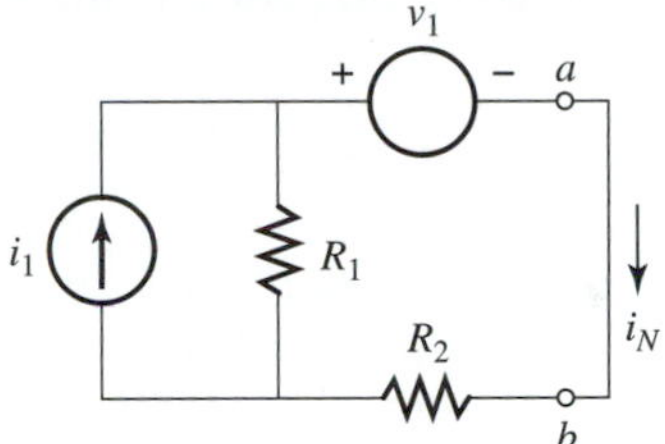

Figure 7.39 Terminals *a*-*b* shorted.

We shall apply the superposition method to determine i_N for the circuit in Figure 7.39. Because there are two sources in the circuit, there are two circuits to consider, as shown in Figure 7.40: i'_N is the contribution to i_N from the current source i_1 and i''_N is the contribution to i_N from the voltage source v_1.

Figure 7.40 Norton circuits.

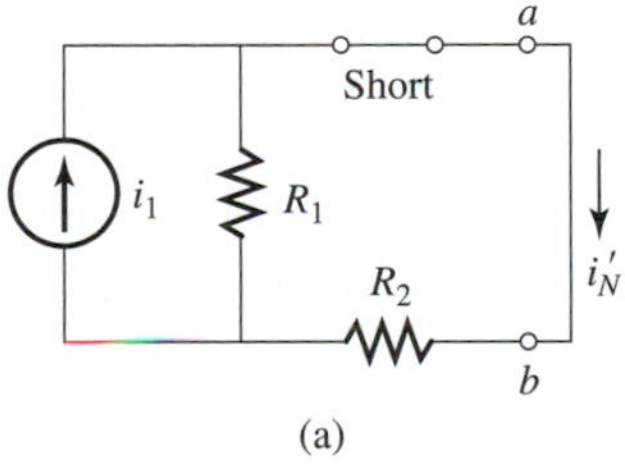

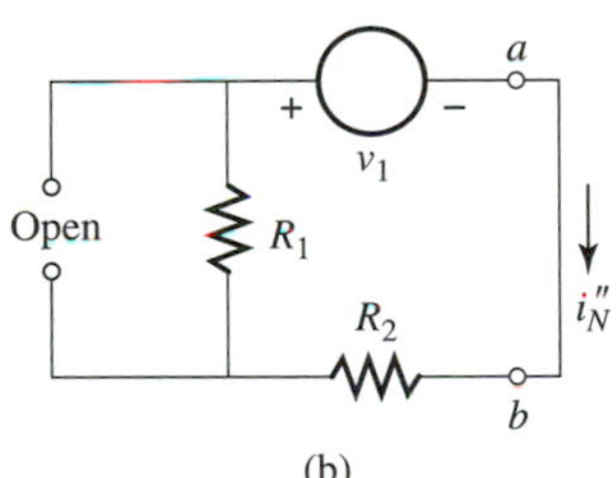

For the circuit in Figure 7.40(a), i'_N can be determined from Equation (4.1) for current division.

$$i'_N = \left(\frac{R_1}{R_1 + R_2}\right)(i_1)$$

For the circuit in Figure 7.40(b), i''_N can be determined from Ohm's law:

$$i''_N = \left(\frac{-v_1}{R_1 + R_2}\right)$$

i_N is the algebraic sum of the contributed currents i'_N and i''_N and is

$$\begin{aligned} i_N &= i'_N + i''_N \\ &= \frac{R_1 i_1 - v_1}{R_1 + R_2} \end{aligned}$$

The *third step* in applying Norton's theorem is to determine the Norton resistance R_N for the Norton equivalent. The circuit in Figure 7.36 is shown in Figure 7.41 with all sources replaced by their internal resistances, as required by Norton's theorem for determining R_N. Notice that the Norton resistance R_N is identical to the Thevenin resistance R_T.

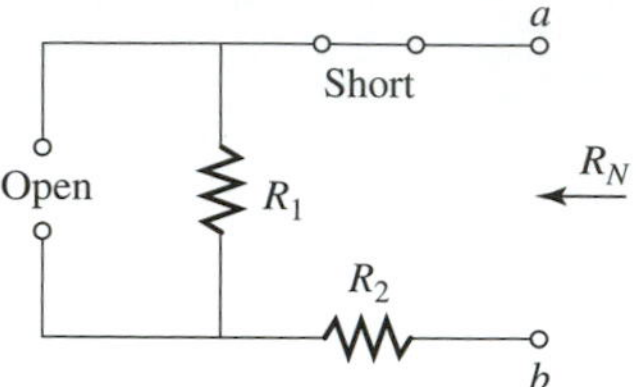

Figure 7.41 Norton's circuit for R_N.

The Norton resistance R_N is the resistance across terminals *a-b* in Figure 7.41, and from Equation (3.1)

$$R_N = R_1 + R_2$$

The *fourth step* in applying Norton's theorem is to form the Norton equivalent circuit using i_N and R_N determined in steps 2 and 3. For the circuit to the left of terminals *a-b* in Figure 7.36, the Norton equivalent circuit is shown in Figure 7.42.

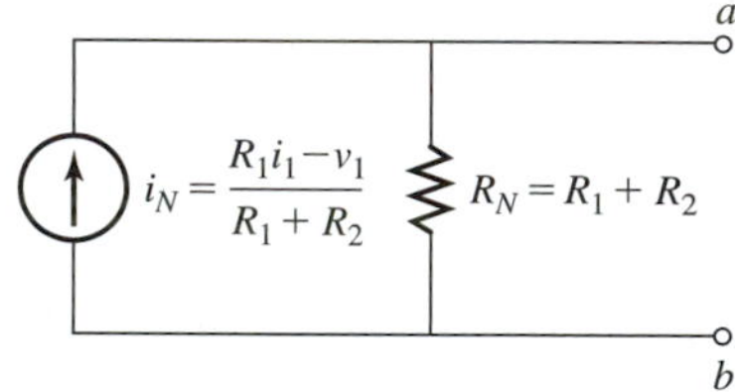

Figure 7.42 Norton equivalent.

The four-step procedure for applying Norton's theorem is summarized as follows:

1. Isolate the two terminals connected to the portion of the circuit to be replaced by a Norton equivalent.
2. Determine the Norton current i_N.
3. Determine the Norton resistance R_N.
4. Use i_N and R_N to form the Norton equivalent.

EXAMPLE 7.8 Determine a Norton equivalent for the circuit to the left of terminals *a-b* in Figure 7.43. ■

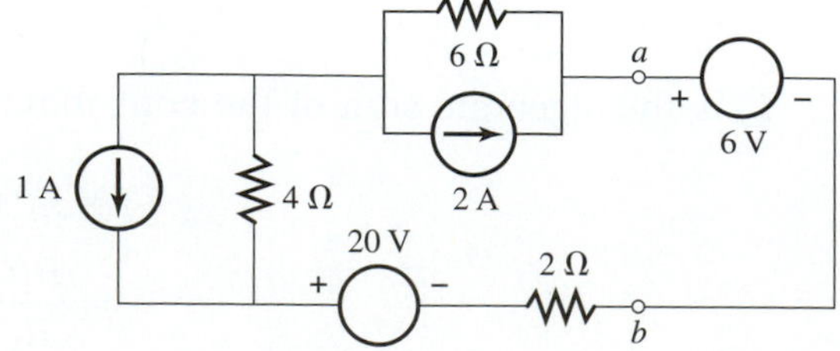

Figure 7.43

SOLUTION The circuit to the left of terminals *a-b* is shown in Figure 7.44 with terminals *a-b* shorted. We shall use the superposition method to determine i_N. Because there are three sources, three circuits are considered, as shown in Figures 7.45, 7.46, and 7.47.

Figure 7.44

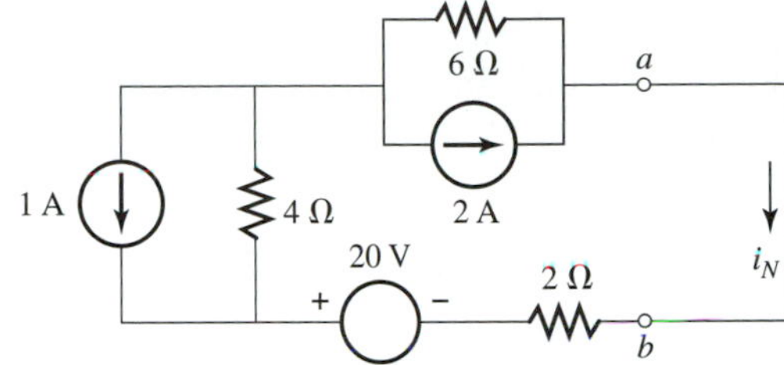

Figure 7.45

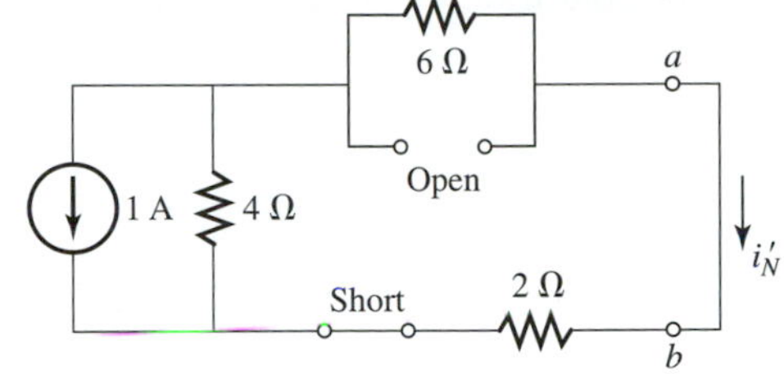

Figure 7.46

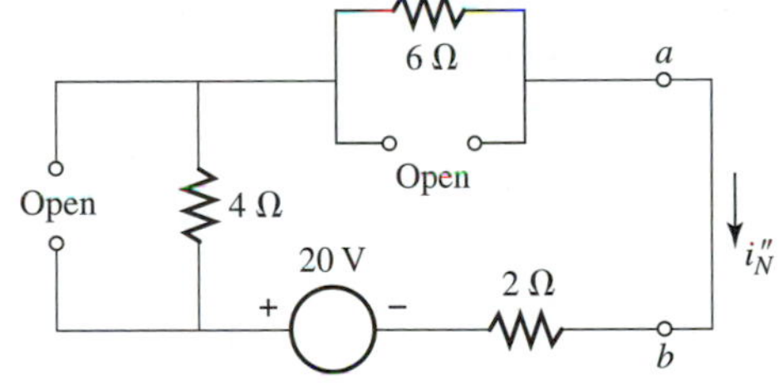

Figure 7.47

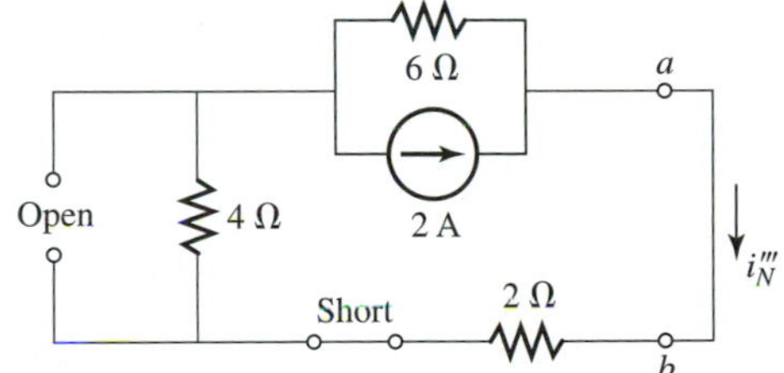

i'_N is the contribution to i_N from the 1-A current source. i''_N is the contribution to i_N from the 20-V voltage source. i'''_N is the contribution to i_N from the 2-A current source.

For the circuit in Figure 7.45, i'_N is determined from Equation (4.1) for current division:

$$i'_N = \left(\frac{4}{4 + 6 + 2}\right)(-1)$$

$$= -\frac{1}{3}\text{ A}$$

For the circuit in Figure 7.46, i''_N is determined by Ohm's law:

$$i''_N = \frac{20}{6 + 2 + 4}$$
$$= \frac{5}{3}\text{ A}$$

For the circuit in Figure 7.47, i'''_N is determined from Equation (4.1) for current division:

$$i'''_N = \left(\frac{6}{6 + 4 + 2}\right)(2)$$
$$= 1\text{ A}$$

i_N is the algebraic sum of the contributed voltages.

$$i_N = i'_N + i''_N + i'''_N$$
$$= -\frac{1}{3} + \frac{5}{3} + 1$$
$$= \frac{7}{3}\text{ A}$$

The circuit to the left of terminals *a-b* is shown in Figure 7.48, with all the sources replaced by their internal resistances. The Norton resistance R_N is determined from this circuit.

Figure 7.48

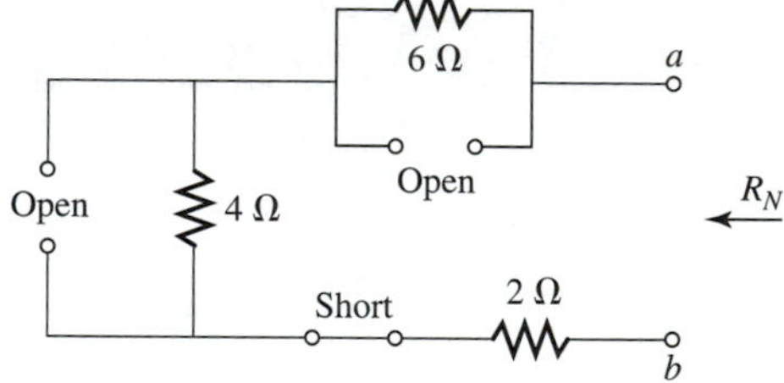

From Equation (3.1)

$$R_N = 6 + 4 + 2$$
$$= 12\ \Omega$$

The Norton equivalent for the circuit to the left of terminals *a-b* is shown in Figure 7.49.

Figure 7.49

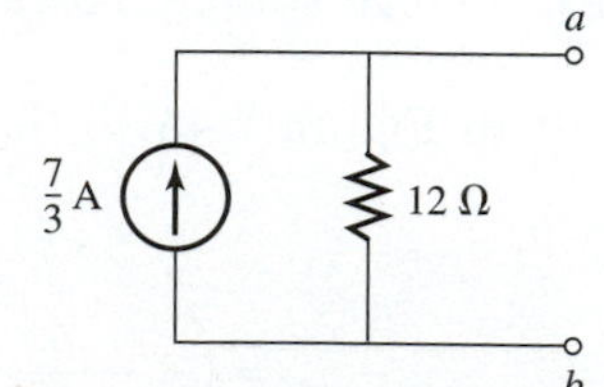

EXAMPLE 7.9 For the circuit in Example 7.8, determine the current *i* flowing through the 6-V voltage source. ■

SOLUTION The circuit is redrawn in Figure 7.50. The circuit to the left of

Figure 7.50

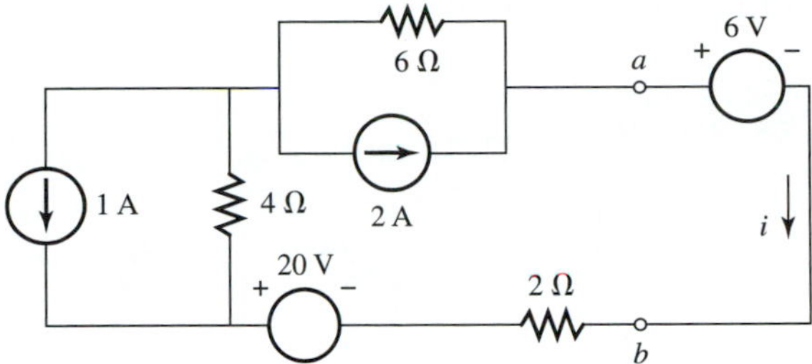

terminals a-b can be replaced by the Norton equivalent determined in Example 7.8, as shown in Figure 7.51.

Figure 7.51

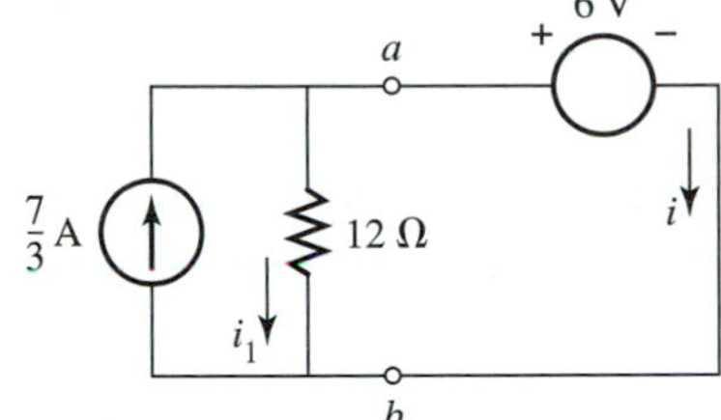

From Ohm's law,

$$i_1 = \frac{6}{12} = \frac{1}{2}\text{ A}$$

Applying Kirchhoff's current law to node a,

$$\frac{7}{3} - i_1 - i = 0$$

$$i = \frac{7}{3} - \frac{1}{2}$$

$$= \frac{11}{6}\text{ A}$$

EXAMPLE 7.10 Determine the current i for the circuit shown in Figure 7.52. ■

Figure 7.52

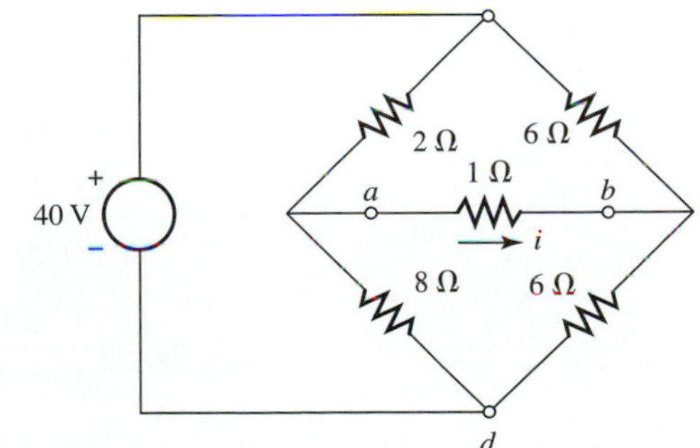

SOLUTION We first remove the 1-Ω resistor and determine a Norton equivalent with respect to terminals *a*-*b*. In accordance with Norton's theorem, terminals *a*-*b* are shorted, as shown in Figure 7.53. The Norton current i_N is the current through the short, as indicated.

Figure 7.53

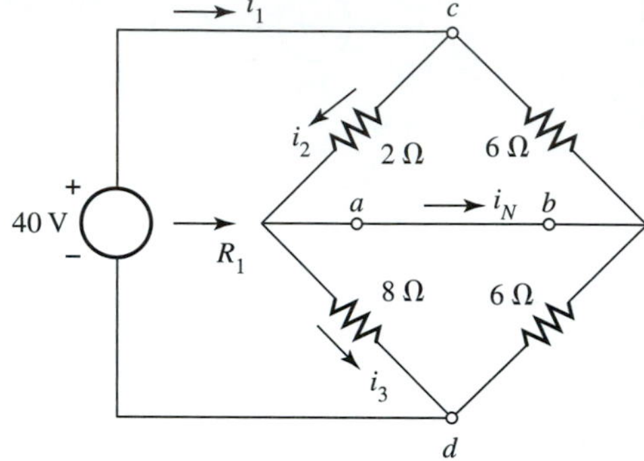

The circuit is redrawn in Figure 7.54 to indicate the parallel configuration of the resistors when terminals *a*-*b* are shorted.

Figure 7.54

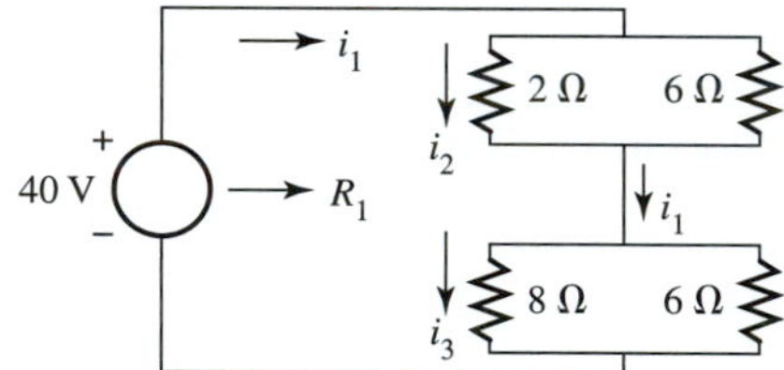

Using Equations (3.1) and (3.5) to combine resistors results in

$$\begin{aligned} R_1 &= (2 \| 6) + (8 \| 6) \\ &= \frac{(2)(6)}{2+6} + \frac{(8)(6)}{8+6} \\ &= 4.9\ \Omega \end{aligned}$$

where R_1 is the resistance seen by the 40-V source. From Ohm's law

$$\begin{aligned} i_1 &= \frac{40}{R_1} = \frac{40}{4.9} \\ &= 8.2\ \text{A} \end{aligned}$$

From Equation (4.1) for current division,

$$\begin{aligned} i_2 &= \left(\frac{6}{6+2}\right)(i_1) \\ &= 6.1\ \text{A} \end{aligned}$$

$$\begin{aligned} i_3 &= \left(\frac{6}{6+8}\right)(i_1) \\ &= 3.5\ \text{A} \end{aligned}$$

Let us refer back to Figure 7.53 with terminals a-b shorted and the Norton current i_N indicated. Applying Kirchhoff's current law at node a results in

$$i_2 - i_N - i_3 = 0$$
$$i_N = i_2 - i_3$$
$$= 6.1 - 3.5$$
$$= 2.6 \text{ A}$$

The Norton resistance R_N is determined by replacing the 40-V voltage source with a short and finding the resulting resistance across terminals a-b. The circuit used to determine R_N is shown in Figure 7.55. Notice that shorting the 40-V voltage source is the same as connecting terminals c and d.

Figure 7.55

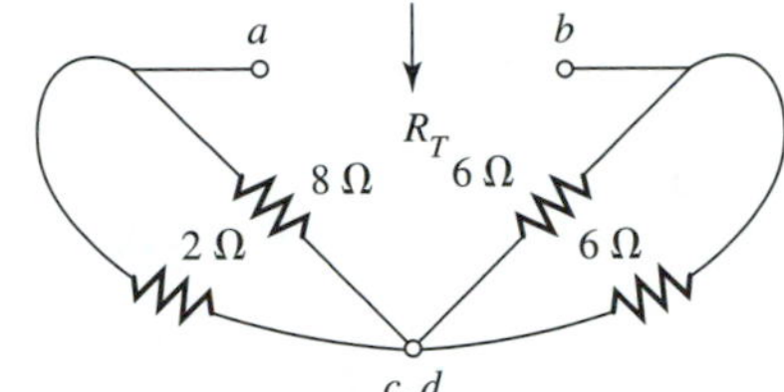

From Equations (3.1) and (3.5), the Norton resistance R_N is

$$R_N = (2\|8) + (6\|6)$$
$$= \left(\frac{(2)(8)}{2 + 8}\right) + \frac{(6)(6)}{6 + 6}$$
$$= 1.6 + 3$$
$$= 4.6 \ \Omega$$

The Norton equivalent for the circuit across terminals a-b with the 1-Ω resistance removed is shown in Figure 7.56 with the 1-Ω resistance reconnected.

Figure 7.56

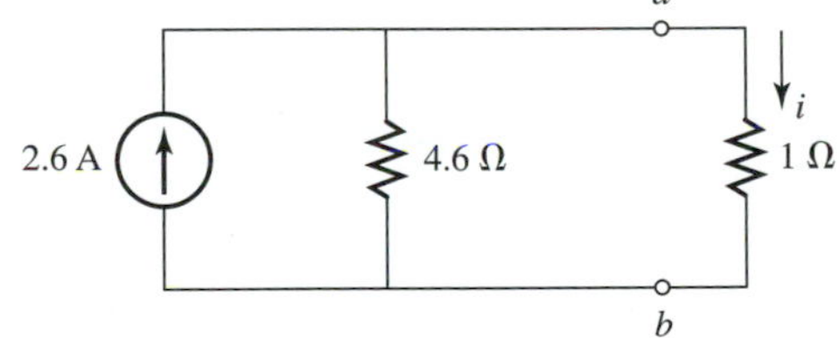

From Equation (4.1) for current division,

$$i = \left(\frac{4.6}{1 + 4.6}\right)(2.6)$$
$$= 2.1 \text{ A}$$

The Thevenin and Norton resistances are identical for a given circuit and a given pair of terminals. This is easily seen from the Thevenin and Norton theorems. Because R_N equals R_T for a given circuit, the choice of applying Norton's theorem or Thevenin's theorem depends on which is easiest to determine, the Norton current or the Thevenin voltage.

Using the rules presented in Section 4.6 for transforming imperfect sources, we can transform a Thevenin equivalent to a Norton equivalent and a Norton equivalent to a Thevenin equivalent, as shown in Figure 7.57. Notice that the transformation rules are easily verified by applying Thevenin's and Norton's theorems to the sources to be transformed.

Figure 7.57 Transformations for Thevenin and Norton circuits.

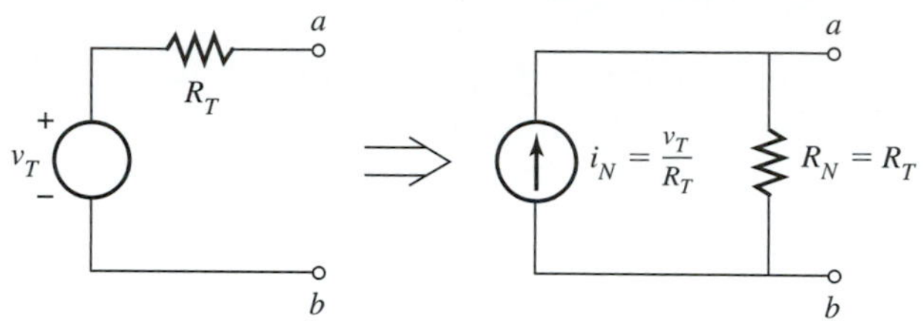

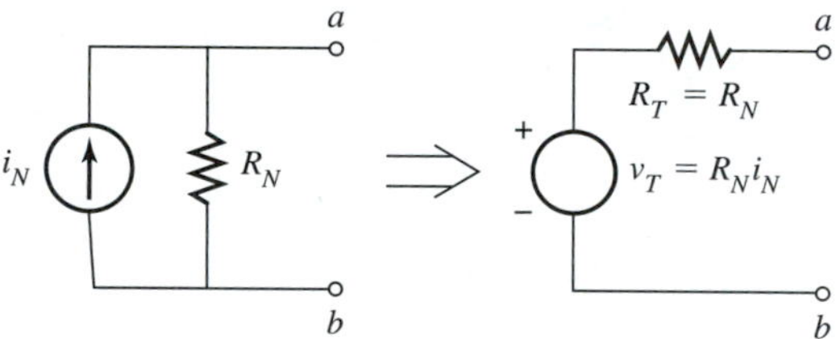

7.4 ALTERNATIVE METHODS FOR DETERMINING THEVENIN AND NORTON RESISTANCES

Consider a Thevenin equivalent with the terminals shorted, as shown in Figure 7.58. The current through the short is the Norton current i_N for the circuit across terminals *a*-*b* represented by the Thevenin equivalent. From Ohm's law,

$$R_T = \frac{v_T}{i_N} \tag{7.1}$$

and, because R_N is identical to R_T,

$$R_N = \frac{v_T}{i_N} \tag{7.2}$$

Equations (7.1) and (7.2) give us a second method for determining the Thevenin and Norton resistances. Consider the following example.

Figure 7.58 Thevenin equivalent.

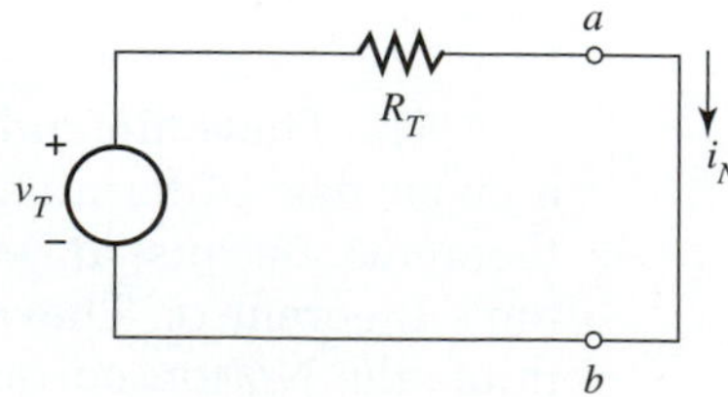

EXAMPLE 7.11 Determine Norton and Thevenin equivalents with respect to terminals *a-b* for the circuit in Figure 7.59. ■

Figure 7.59

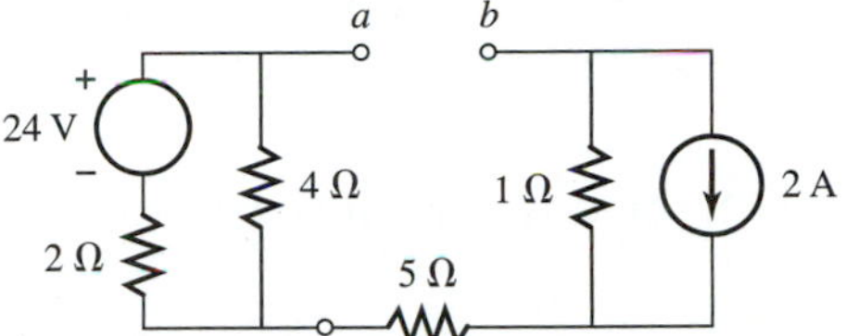

SOLUTION Using the conventional method for determining Thevenin and Norton resistances, R_T and R_N for terminals *a-b* can easily be determined by shorting the 24-V voltage source, opening the 2-A current source, and determining the resulting resistance across the terminals as

$$R_T = R_N = (2\|4) + 5 + 1$$
$$= 7.3\ \Omega$$

However, we shall determine the Thevenin and Norton resistances using i_N and v_T to demonstrate another method.

Figure 7.60

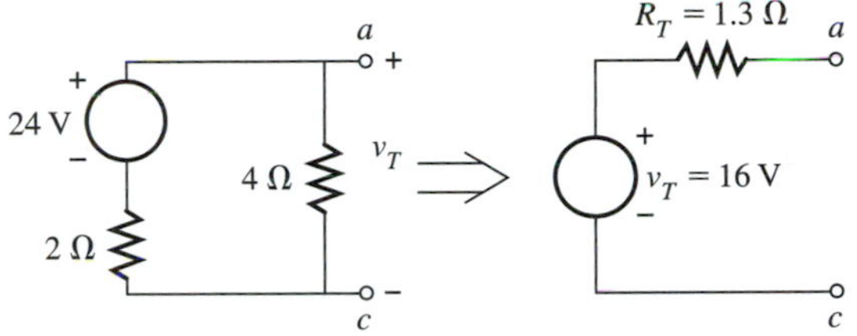

The circuit to the left of terminals *a-c* can be replaced by a Thevenin equivalent, as shown in Figure 7.60. From Equation (4.3) for voltage division,

$$v_T = \left(\frac{4}{4+2}\right)(24)$$
$$= 16\ \text{V}$$

Replacing the 24-V voltage source by a short and using Equation (3.5) results in

$$R_T = (4)\|(2)$$
$$= \frac{(2)(4)}{2+4} = \frac{8}{6}$$
$$= 1.3\ \Omega$$

as shown in Figure 7.60.

The circuit to the right of terminals *b-c* can be replaced by a Thevenin equivalent, as shown in Figure 7.61.

Figure 7.61

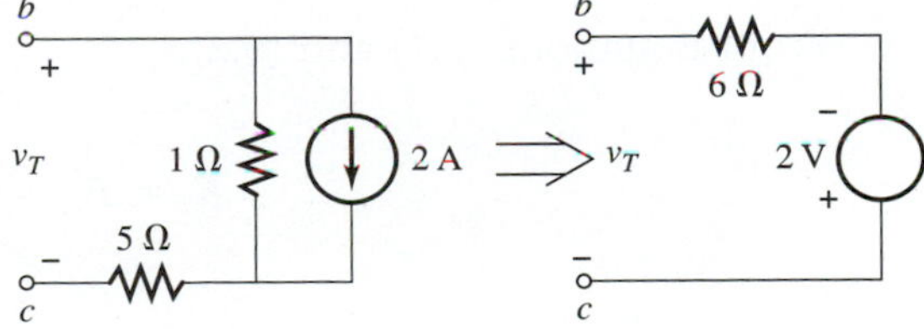

From Ohm's law

$$v_T = -(2)(1)$$
$$= -2 \text{ V}$$

and with the 2-A current source replaced by an open, Equation (3.1) results in

$$R_T = 1 + 5$$
$$= 6\ \Omega$$

as shown in Figure 7.61.

Replacing the circuits to the left of terminals *a*-*c* and to the right of terminals *b*-*c* by their Thevenin equivalents results in the circuit shown in Figure 7.62.

Figure 7.62

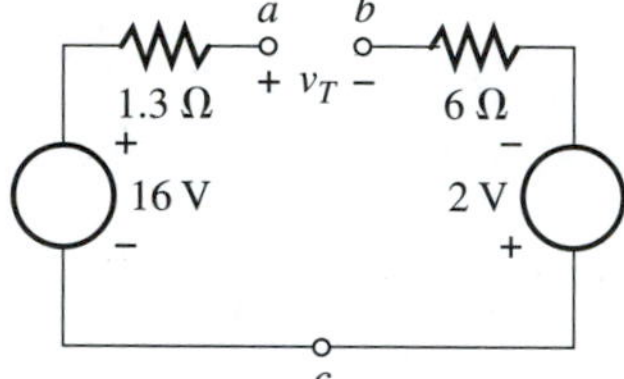

Because no current flows in the circuit, Kirchhoff's voltage law results in

$$v_T = 16 + 2$$
$$= 18 \text{ V}$$

Redrawing the circuit with terminals *a*-*b* shorted results in the circuit in Figure 7.63, with the Norton current i_N indicated.

Figure 7.63

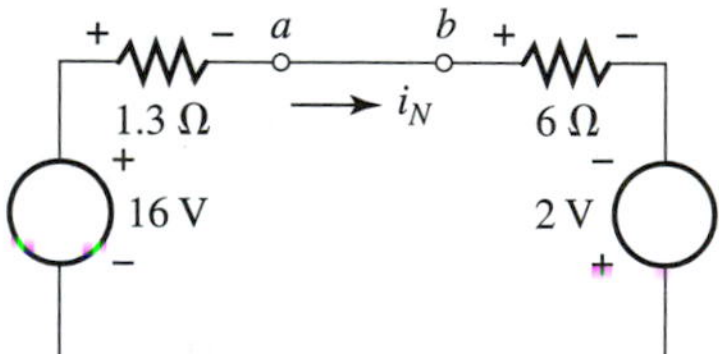

From Kirchhoff's voltage law,

$$16 - 1.3i_N - 6i_N + 2 = 0$$
$$i_N = \frac{16 + 2}{6 + 1.3}$$
$$= 2.5 \text{ A}$$

From Equations (7.1) and (7.2),

$$R_T = R_N = \frac{v_T}{i_N} = \frac{18}{2.5}$$
$$= 7.3\ \Omega$$

The Norton and Thevenin equivalents of the original circuit with respect to terminals *a-b* are shown in Figure 7.64.

Figure 7.64

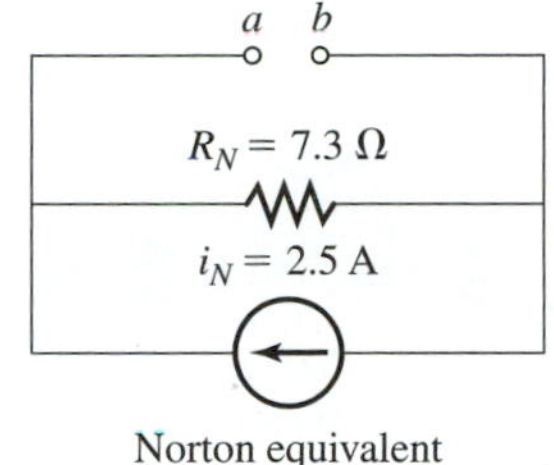

Norton equivalent

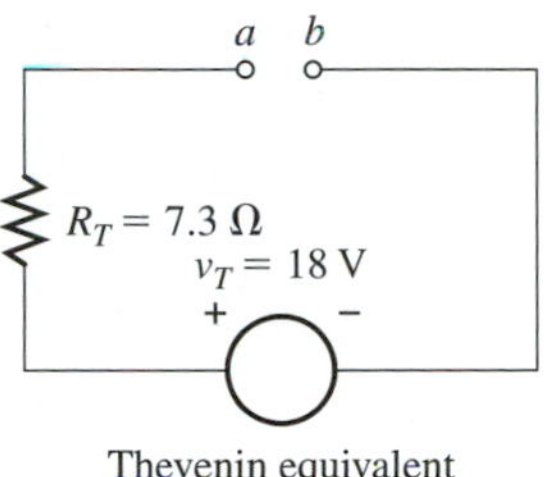

Thevenin equivalent

Another method for determining R_N and R_T involves removing the independent sources and applying a voltage or current source to the circuit, as shown in Figure 7.65. This method is especially useful when dependent sources are involved.

Figure 7.65 Method for determining R_T and R_N.

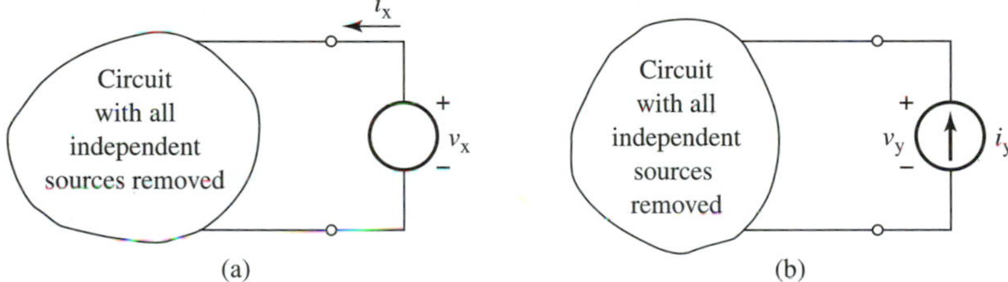

A voltage source v_x is applied to the circuit, as in Figure 7.65(a), with all independent sources removed, and i_x is determined; or a current source i_y is applied to the circuit, as in Figure 7.65(b), with all independent sources removed, and v_y is determined. Either of the ratios v_x/i_x or v_y/i_y can be used to determine R_N and R_T.

$$R_N = R_T = \frac{v_x}{i_x}$$

or

$$R_N = R_T = \frac{v_y}{i_y}$$

EXAMPLE 7.12 Determine R_T and R_N for the circuit in Example 7.11. ■

SOLUTION The circuit is shown in Figure 7.66 with the independent sources removed and a voltage source applied across terminals a-b. From Ohm's law

Figure 7.66

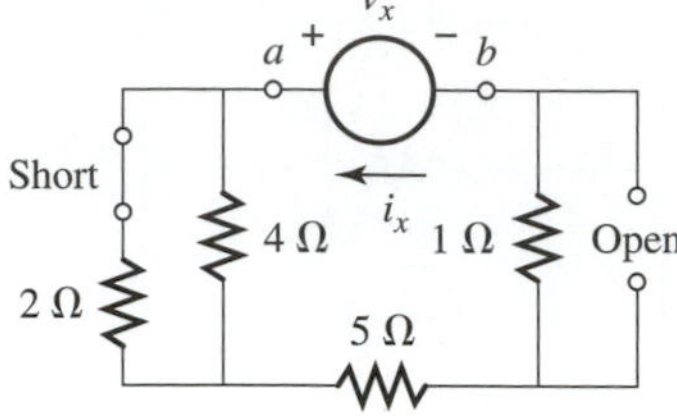

and Equations (3.1) and (3.5) for combining resistors in series and parallel,

$$i_x = \frac{v_x}{(2\|4) + 5 + 1}$$

and

$$R_N = R_T = \frac{v_x}{i_x} = \{(2\|4) + 5 + 1\}$$
$$= 7.3\ \Omega$$

7.5 SUPERPOSITION METHOD WITH DEPENDENT SOURCES

The superposition method is applicable to circuits containing dependent sources. However, the method as previously presented for analyzing circuits with only independent sources must be altered to accommodate the presence of dependent sources. Because the value of a dependent source depends on a voltage or current in another part of the circuit, the dependent source cannot be removed from the circuit or separated from its controlling variable when applying the superposition method.

Consider the following example.

EXAMPLE 7.13 Determine the current i for the circuit in Figure 7.67. ■

Figure 7.67

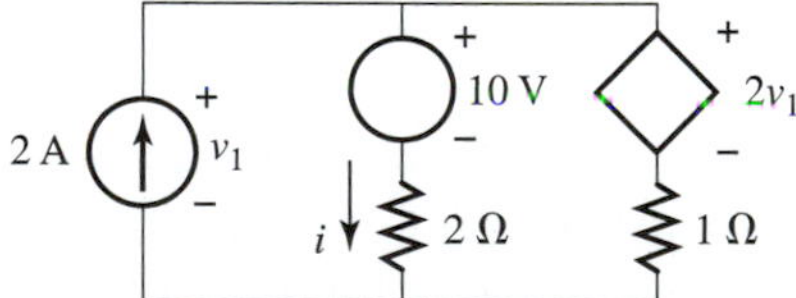

SOLUTION Because there are two independent sources, there are two circuits to consider for the superposition method. Notice that the dependent voltage source $2v_1$ depends on the voltage across the 2-A current source. The circuit shown in Figure 7.68 with the 10-V voltage source shorted is used to determine the current i', which is the contribution to i from the 2-A current source. Notice that the dependent source remains in the circuit.

Figure 7.68

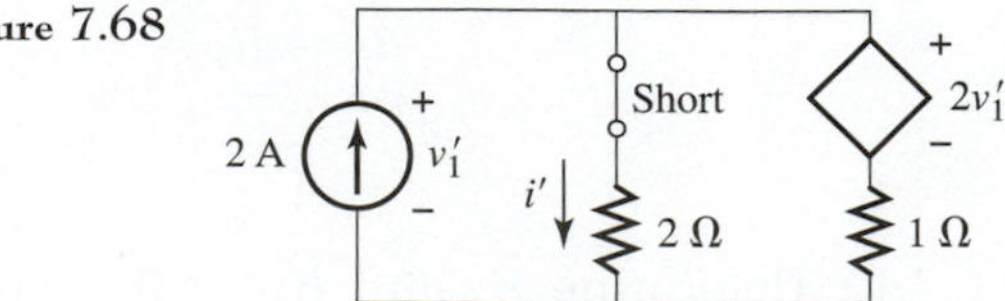

Converting the voltage source $2v_1'$ and the 1-Ω series resistance to a current source results in the circuit shown in Figure 7.69.

Figure 7.69

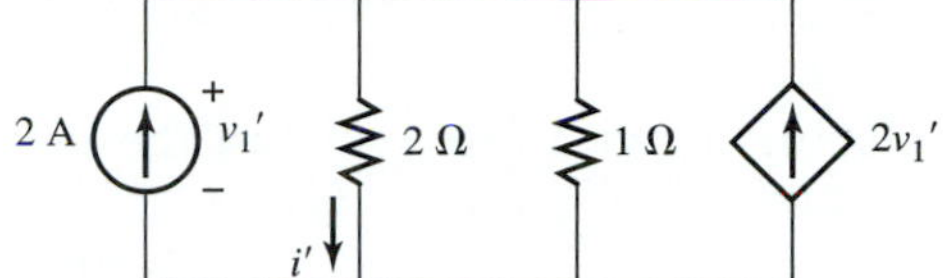

Combining the parallel current sources results in the circuit shown in Figure 7.70.

Figure 7.70

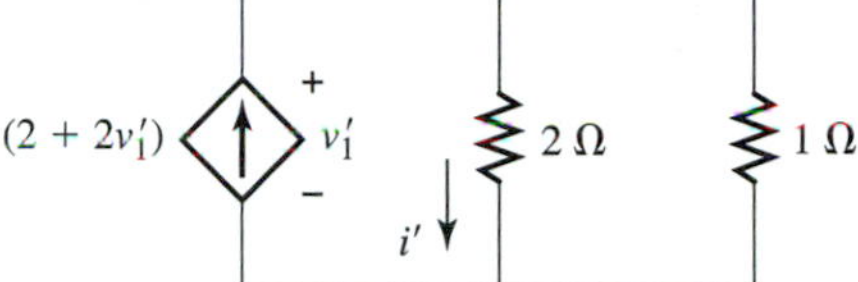

From Equation (4.1) for current division,

$$i' = \left(\frac{1}{1+2}\right)(2 + 2v_1')$$

Recognizing that v_1' is the voltage across the 2-Ω resistance in Figure 7.68 and using Ohm's law results in

$$v_1' = 2i'$$

Replacing v_1' with $2i'$ in the previous equation for i' results in

$$\begin{aligned} i' &= \frac{2}{3} + \left(\frac{4}{3}\right)i' \\ &= -2 \text{ A} \end{aligned}$$

The circuit in Figure 7.71 with the 2-A current source replaced by an open is used to determine the contribution to i from the 10-V source. i'' is the contribution to i from the 10-V voltage source. Notice that the dependent voltage source remains in the circuit.

Figure 7.71

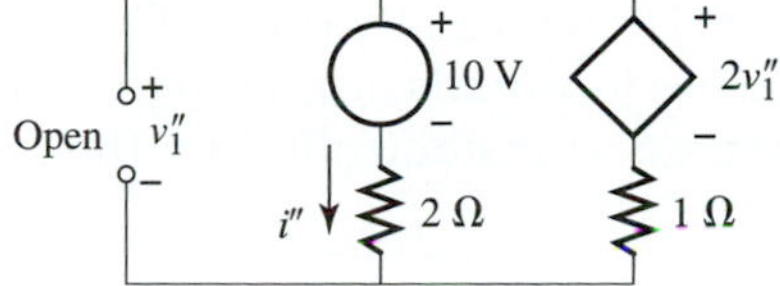

Combining the voltage sources and resistors in series and employing Ohm's law results in

$$\begin{aligned} i'' &= \frac{2v_1'' - 10}{2+1} \\ &= \frac{2v_1''}{3} - \frac{10}{3} \end{aligned}$$

Equating the voltage across the open to the voltage across the branch containing the 10-V voltage source and the 2-Ω resistor results in

$$v_1'' = 10 + 2i''$$

Replacing v'' with $(10 + 2i'')$ in the previous equation results in

$$\begin{aligned} i'' &= \frac{2(10 + 2i'')}{3} - \frac{10}{3} \\ &= -10 \text{ A} \end{aligned}$$

The current i is the algebraic sum of i' and i''.

$$\begin{aligned} i &= i' + i'' \\ &= -2 - 10 \\ &= -12 \text{ A} \end{aligned}$$

7.6 THEVENIN AND NORTON EQUIVALENTS WITH DEPENDENT SOURCES

The Norton and Thevenin theorems can be applied to circuits with dependent sources. However, the procedure presented previously for applying the Thevenin and Norton theorems to circuits with only independent sources must be slightly modified to accommodate the presence of the dependent sources as illustrated in the following examples.

EXAMPLE 7.14 Determine a Thevenin equivalent for the circuit to the left of terminals a-b in Figure 7.72. Use Equation (7.1) to determine R_T. ■

Figure 7.72

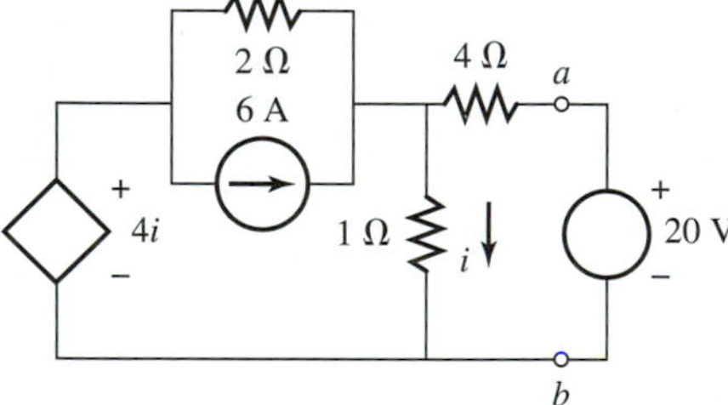

SOLUTION The 20-V voltage source is removed to isolate terminals a-b as required by Thevenin's theorem. The Thevenin voltage v_T is the open-circuited voltage across terminals a-b, as shown in Figure 7.73. Notice that the dependent voltage source $4i$ is controlled by the current i through the 1-Ω resistor.

Figure 7.73

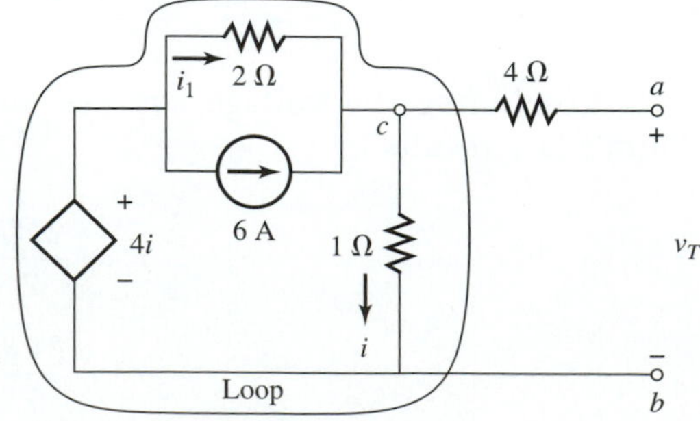

Applying Kirchhoff's current law to node c results in

$$i_1 + 6 - i = 0$$
$$i_1 = i - 6$$

Applying Kirchhoff's voltage law to the loop indicated in the preceding circuit results in

$$4i - 2i_1 - 1i = 0$$

Substituting $(i - 6)$ for i_1 results in

$$4i - 2(i - 6) - i = 0$$
$$i = -12 \text{ A}$$

The Thevenin voltage v_T is the voltage across the 1-Ω resistor, because no current flows through the 4-Ω resistor.

From Ohm's law

$$v_T = 1i$$
$$= -12 \text{ V}$$

Terminals a-b are now shorted to determine the Norton current i_N, as shown in Figure 7.74.

Figure 7.74

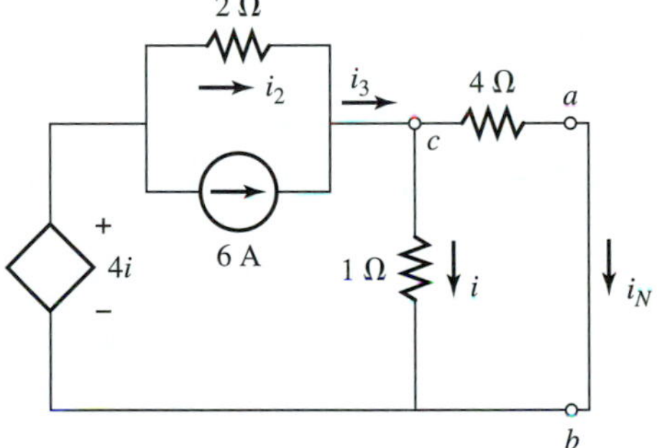

From Kirchhoff's current law

$$i_3 = i_2 + 6$$

From Equation (4.1) for current division,

$$i_N = \left(\frac{1}{1 + 4}\right)(i_3)$$
$$= \left(\frac{1}{5}\right)(i_2 + 6)$$

and

$$i = \left(\frac{4}{1 + 4}\right)(i_3)$$
$$= \left(\frac{4}{5}\right)(i_2 + 6)$$

Solving for $(i_2 + 6)$ in each of these equations results in

$$(i_2 + 6) = 5i_N$$

$$(i_2 + 6) = \frac{5}{4}i$$

It follows that

$$i = 4i_N$$

and

$$i_2 = 5i_N - 6$$

Applying Kirchhoff's voltage law to the loop indicated in the circuit in Figure 7.75 results in

$$4i - 2i_2 - 1i = 0$$

Figure 7.75

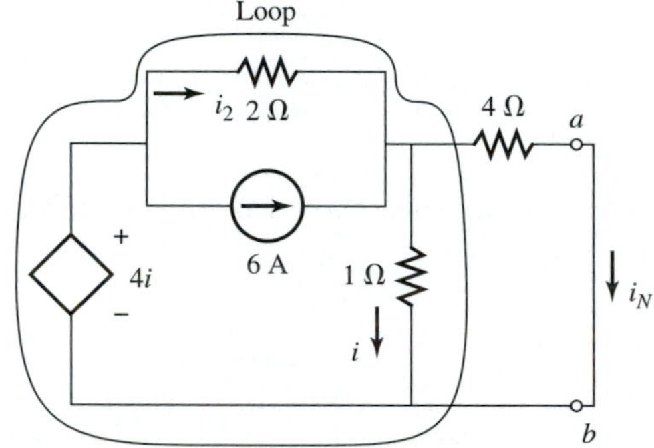

Substituting $(4i_N)$ for i and $(5i_N - 6)$ for i_2 results in

$$4(4i_N) - 2(5i_N - 6) - 4i_N = 0$$
$$2i_N = -12$$
$$i_N = -6 \text{ A}$$

From Equation (7.1)

$$R_T = \frac{v_T}{i_N}$$
$$= \frac{-12}{-6}$$
$$= 2\ \Omega$$

The Thevenin equivalent for the circuit to the left of terminals *a-b* is shown in Figure 7.76.

Figure 7.76

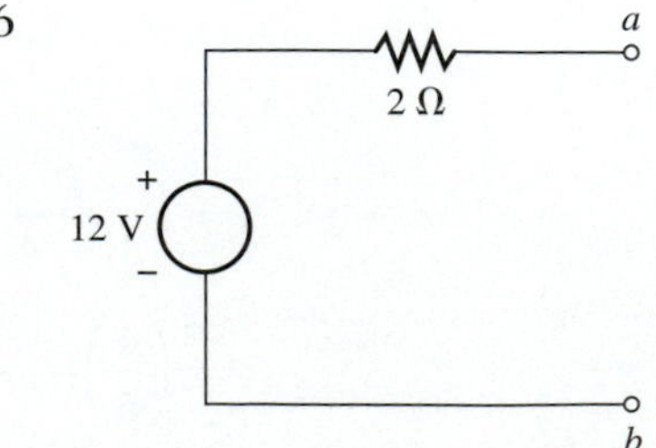

EXAMPLE 7.15 Determine the Norton resistance for the circuit in the previous example by applying a voltage source to terminals a-b. ■

SOLUTION The circuit with the independent current source removed and the voltage source v_x applied to terminals a-b is shown in Figure 7.77.

Figure 7.77

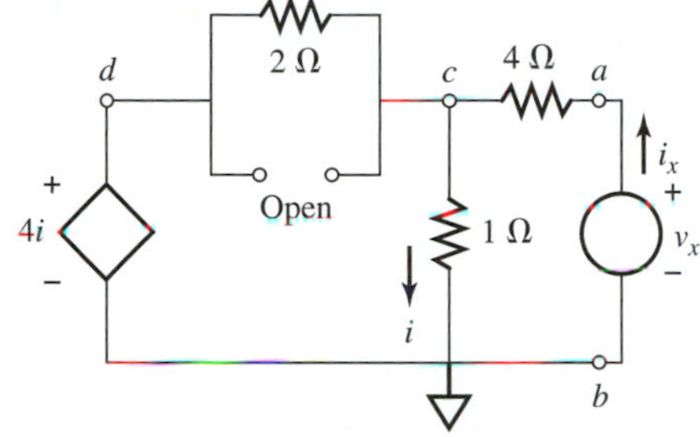

Employing Equation (6.8) with b as a reference node and applying Kirchhoff's current law at node c results in

$$\frac{v_d - v_c}{2} + \frac{v_a - v_c}{4} = i$$

and because $v_d = 4i$ and $v_a = v_x$,

$$\frac{4i - v_c}{2} + \frac{v_x - v_c}{4} = i$$

From Equation (6.8)

$$i = \frac{v_c - 0}{1}$$

$$= v_c$$

Substituting v_c for i in the preceding equation results in

$$\frac{4v_c - v_c}{2} + \frac{v_x - v_c}{4} = v_c$$

and

$$v_c = -v_x$$

From Equation (6.8)

$$i_x = \frac{v_a - v_c}{4}$$

and because $v_a = v_x$ and $v_c = -v_x$

$$i_x = \frac{v_x - (-v_x)}{4}$$

$$i_x = \frac{v_x + v_x}{4}$$

$$2v_x = 4i_x$$

$$\frac{v_x}{i_x} = 2$$

From Equation (7.2)

$$R_N = \frac{v_x}{i_x} = 2\ \Omega$$

7.7 SUCCESSIVE SOURCE-CONVERSION METHOD

Simply converting imperfect current sources to imperfect voltage sources and vice versa often reduces the work necessary to determine a particular current or voltage in an electric circuit. The successive source-conversion method involves reducing the circuit by successive source conversions to an equivalent circuit that can be easily solved for the desired current or voltage. The method is best described by an example and is demonstrated in Example 7.16.

Let us first review the process for converting imperfect sources, as shown in Figure 7.78. An imperfect voltage source is converted to an imperfect current source in Figure 7.78(a), and an imperfect current source is converted to an imperfect voltage source in Figure 7.78(b). Recall that an imperfect voltage source is a perfect voltage source in series with a resistance and an imperfect current source is a perfect current source in parallel with a resistance.

Figure 7.78 Source conversions.

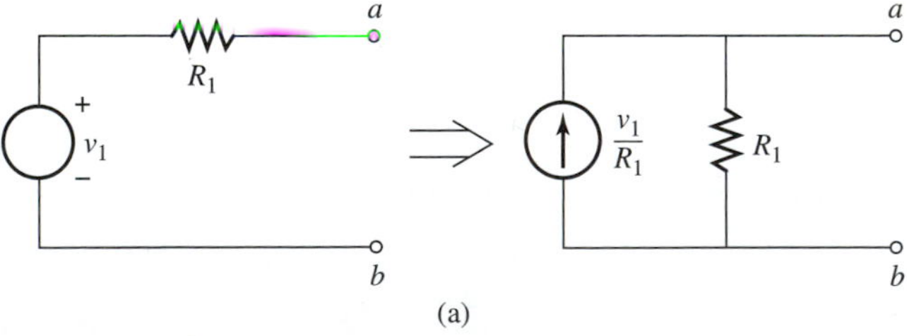

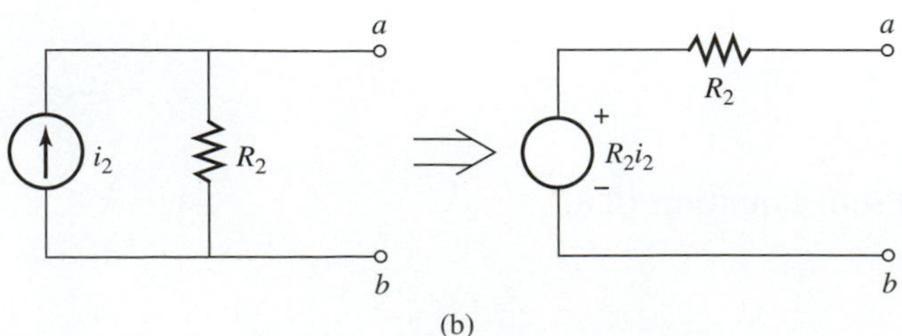

EXAMPLE 7.16 Determine i for the circuit in Figure 7.79. ■

Figure 7.79

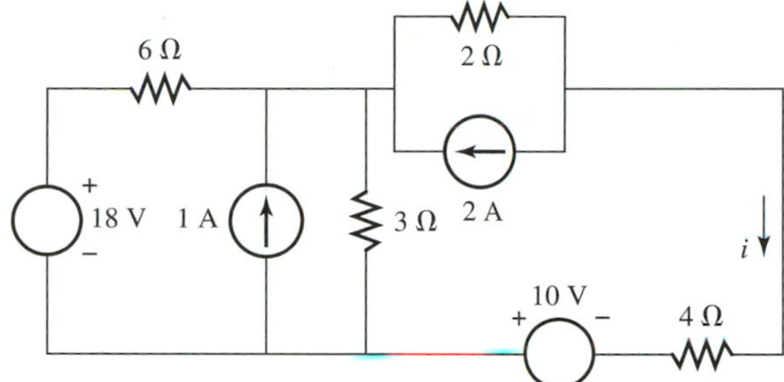

SOLUTION First convert the 18-V voltage source with the 6-Ω series resistance to a current source, as shown in Figure 7.80.

Figure 7.80

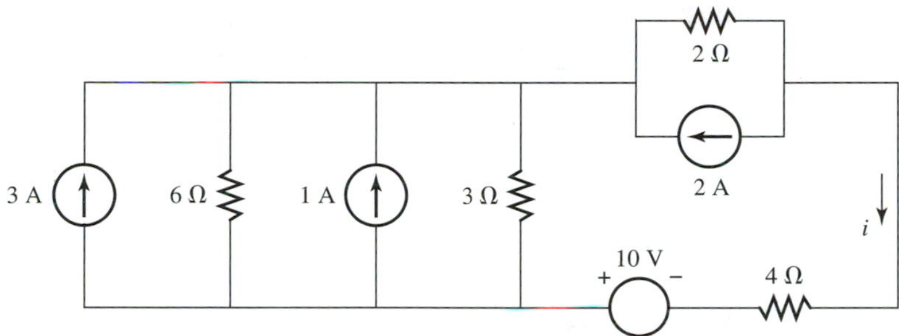

Next add the 3-A and 1-A current sources in parallel and the 6-Ω and 3-Ω resistors in parallel, as shown in Figure 7.81.

Figure 7.81

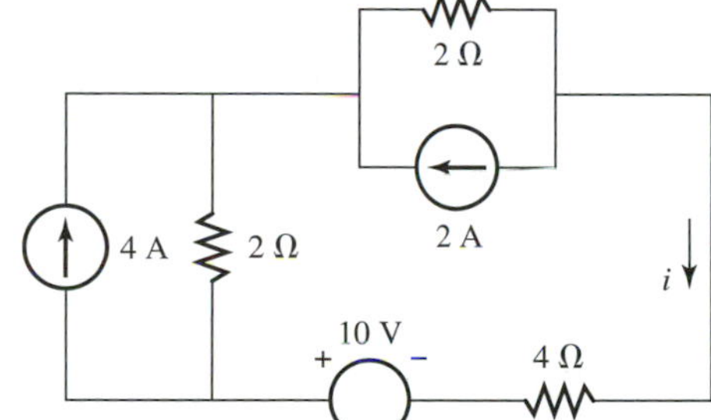

Then convert the 2-A current source with the 2-Ω parallel resistance and the 4-A current source with the 2-Ω parallel resistance to voltage sources, as shown in Figure 7.82.

Figure 7.82

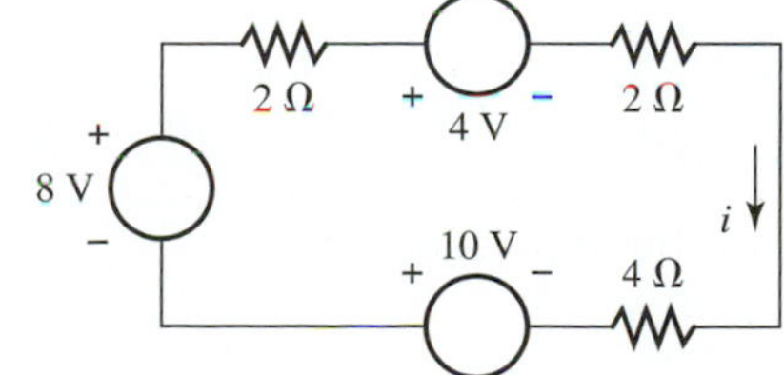

Finally, combining the voltage sources and resistances in series results in the circuit shown in Figure 7.83.

Figure 7.83

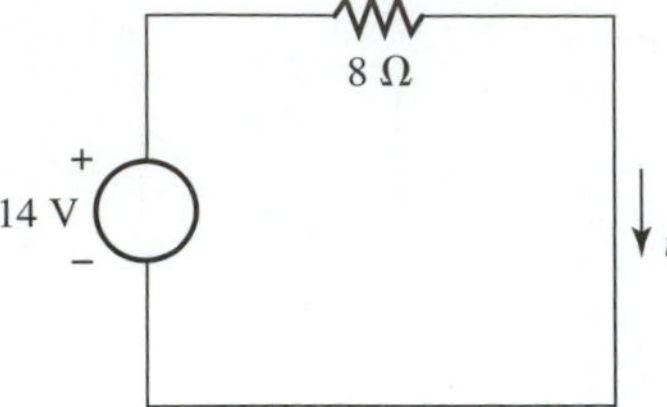

From Ohm's law

$$i = \frac{14}{8}$$
$$= 1.8 \text{ A}$$

7.8 MAXIMUM POWER-TRANSFER THEOREM

In electric circuit applications it is often required to determine the maximum power that a circuit can deliver to a given load. Consider the circuit in Figure 7.84, where it is desired to determine the value of R_L that results in maximum power delivered to R_L by the circuit to the left of terminals *a-b*.

Figure 7.84 Maximum power transfer.

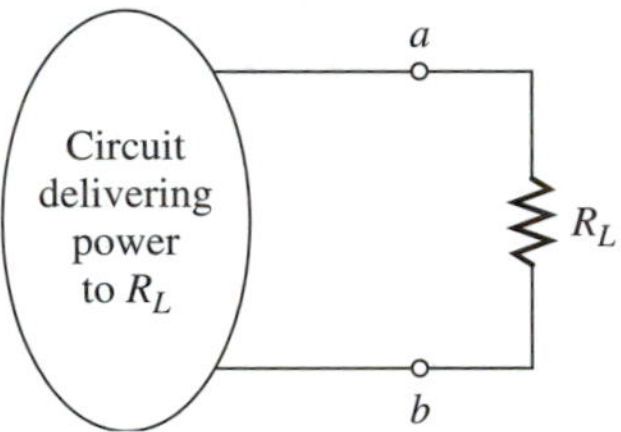

The circuit to the left of terminals *a-b* can be replaced by a Thevenin equivalent, as shown in Figure 7.85.

Figure 7.85 Thevenin equivalent.

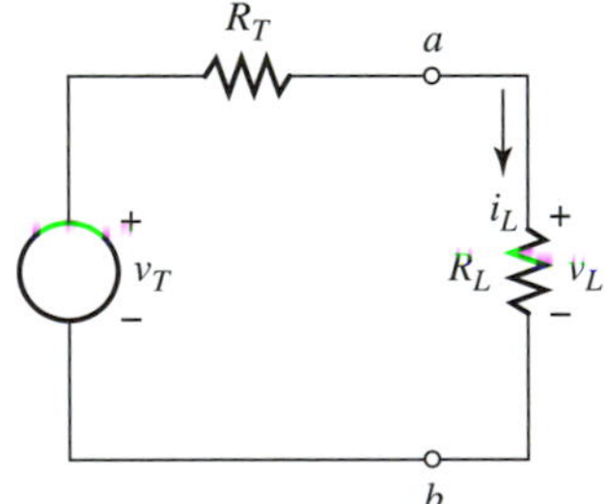

Recall from Equation (2.7) that the power delivered to R_L in Figure 7.85 is

$$p_L = i_L^2 R_L$$

From Ohm's law

$$i_L = \frac{v_T}{R_T + R_L}$$

and p_L can be expressed as

$$p_L = \frac{(v_T)^2 (R_L)}{(R_T + R_L)^2}$$

p_L varies with changes in R_L, as shown in the sketch in Figure 7.86. The sketch can be obtained by assigning values to R_L and determining the corresponding values of p_L from the preceding equation. Notice that the maximum power point occurs when R_L is equal to the Thevenin resistance R_T of the circuit furnishing the power.

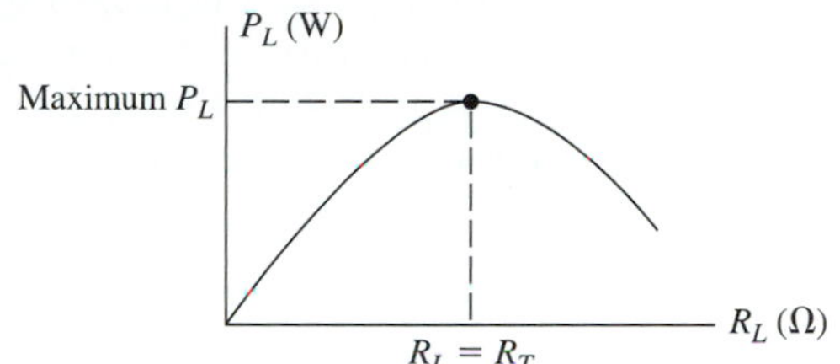

Figure 7.86 Maximum power.

The fact that p_L is maximum when $R_L = R_T$ is arrived at through a mathematical process known as differentiation, which is discussed in Chapter 9. The preceding expression, p_L, is differentiated with respect to R_L and the resulting equation is set equal to zero. The solution of the resulting equation results in $R_L = R_T$ for maximum power transfer. The process for determining the maximum value is presented in Chapter 9.

MAXIMUM POWER–TRANSFER THEOREM A resistive circuit will deliver maximum power to a resistive load R_L when R_L is set equal to the Thevenin resistance R_T of the circuit delivering the power.

From the maximum power–transfer theorem, maximum power is delivered to R_L for the circuit in Figure 7.85 when

$$R_L = R_T \tag{7.3}$$

Using Ohm's law for the circuit in Figure 7.85,

$$i_L = \frac{v_T}{R_L + R_T}$$

and

$$i_L = \frac{v_T}{2R_T} \tag{7.4}$$

when maximum power is delivered to R_L.

Using Equation (4.3) for voltage division for the circuit in Figure 7.85,

$$v_L = \left(\frac{R_L}{R_L + R_T}\right)v_T$$

and

$$v_L = \frac{v_T}{2} \tag{7.5}$$

when maximum power is delivered to R_L.

From Equation (2.7) the power delivered to R_L for the circuit in Figure 7.85 is

$$p_L = i_L^2 R_L$$

Because $R_L = R_T$ and $i_L = v_T/(2R_T)$ for maximum power transfer,

$$p_{max} = \frac{v_T^2}{4R_T} \tag{7.6}$$

where p_{max} is the maximum power delivered to R_L.

The circuit delivering power in Figure 7.84 can also be replaced by a Norton equivalent. Using the relationships between the Norton equivalent and the Thevenin equivalent ($R_N = R_T$ and $i_N = v_T/R_T$), Equations (7.4), (7.5), and (7.6) can be written as

$$i_L = \frac{i_N}{2} \tag{7.7}$$

$$v_L = \frac{R_N i_N}{2} \tag{7.8}$$

$$p_{max} = \frac{i_N^2 R_N}{4} \tag{7.9}$$

Consider the following example.

EXAMPLE 7.17 Determine the value of R_L for maximum power transfer to R_L and find the maximum power transferred for the circuit in Figure 7.87. ■

Figure 7.87

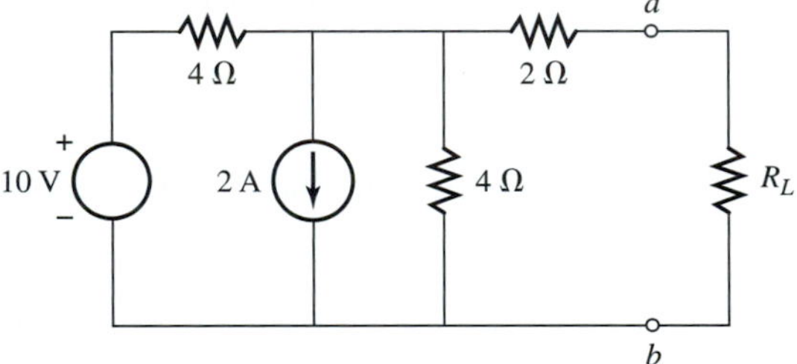

SOLUTION From the maximum power–transfer theorem

$$R_L = R_T$$

for maximum power transfer to R_L from the circuit to the left of terminals *a*-*b*. R_T is the Thevenin resistance for the circuit to the left of terminals *a*-*b*.

The circuit in Figure 7.88 is used to determine R_T. All sources are replaced by their internal resistances and R_L is disconnected, as required by Thevenin's theorem.

Figure 7.88

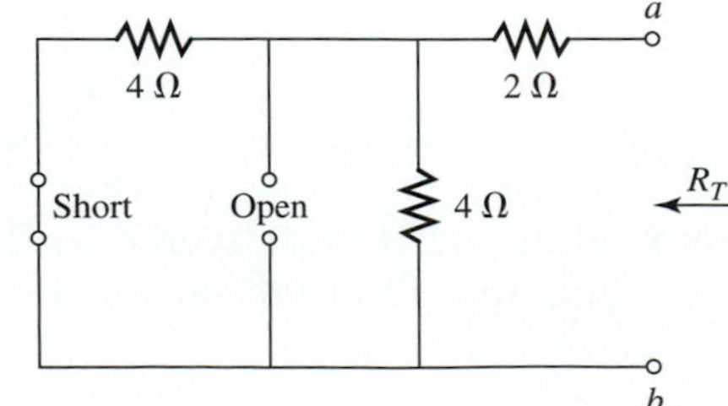

Using Equations (3.1) and (3.5) to combine the resistors in series and parallel results in

$$R_T = (4\|4) + 2$$
$$= 4\ \Omega$$

From Equation (7.3) maximum power is transferred to R_L when

$$R_L = R_T = 4\ \Omega$$

The maximum power transferred to R_L is given by Equation (7.6) as

$$p_{\max} = \frac{v_T^2}{4R_T}$$
$$= \frac{v_T^2}{16}$$

The circuit in Figure 7.89 is used to determine v_T. R_L is disconnected as required by Thevenin's theorem, and v_T is the open-circuited voltage across terminals *a*-*b*, as indicated.

Figure 7.89

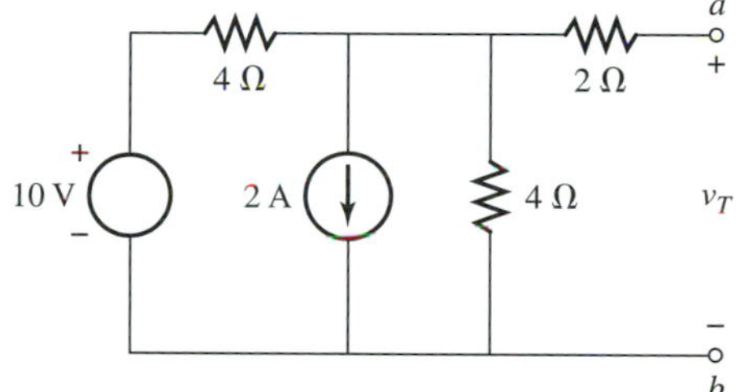

We shall use the superposition method to determine v_T. For the circuits in Figures 7.90 and 7.91, v_T' is the contribution to v_T from the 10-V voltage source and v_T'' is the contribution to v_T from the 2-A current source.

Figure 7.90

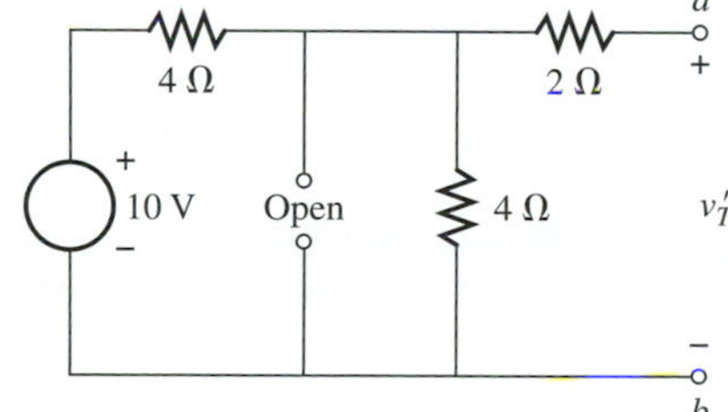

Figure 7.91

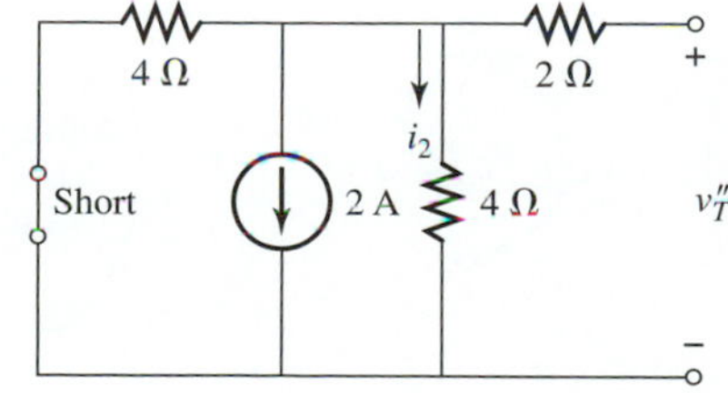

Because no current flows through the 2-Ω resistance in Figure 7.90, v'_T is the voltage across the 4-Ω resistance and can be determined from Equation (4.7) for voltage division as

$$v'_T = \left(\frac{4}{4+4}\right)(10)$$
$$= 5 \text{ V}$$

From Equation (4.1) for current division, i_2 in Figure 7.91 is

$$i_2 = \left(\frac{4}{4+4}\right)(-2)$$
$$= -1 \text{ A}$$

Because no current flows through the 2-Ω resistance, v''_T is the voltage across the 4-Ω resistance and can be determined from Ohm's law.

$$v''_T = 4i_2 = (4)(-1)$$
$$= -4 \text{ V}$$

The Thevenin voltage v_T is

$$v_T = v'_T + v''_T$$
$$= 5 - 4$$
$$= 1 \text{ V}$$

From Equation (6.6)

$$p_{\max} = \frac{(v_T)^2}{4R_T}$$
$$= \frac{(1)^2}{(4)(4)}$$
$$= \frac{1}{16} \text{ W}$$

7.9 LABORATORY METHOD FOR MAXIMUM POWER TRANSFER

A simple laboratory procedure using a voltmeter and a variable resistor can also be used to determine R_L for maximum power transfer. Consider the circuits in Figure 7.92.

Figure 7.92 Maximum power transfer.

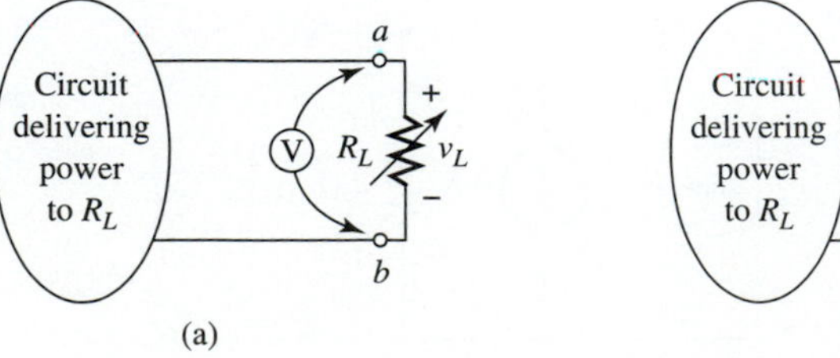

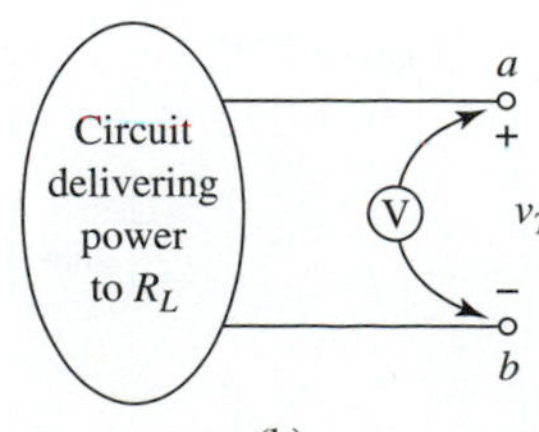

The circuit to the left of terminals *a-b* in Figure 7.92(a) is delivering power to the variable resistor R_L. The voltmeter is in the circuit to measure the voltage v_L across terminals *a-b*. In Figure 7.92(b) R_L is removed and the voltmeter measures the open-circuited voltage across terminals *a-b*, which is the Thevenin voltage v_T for the circuit to the left of terminals *a-b*. From Equation (7.5), $v_L = v_T/2$ for the circuit in Figure 7.92(a) when maximum power is delivered to R_L. R_L is, therefore, varied until the voltmeter reads $v_T/2$. At this value of R_L the circuit to the left of terminals *a-b* is delivering maximum power to R_L.

7.10 EXERCISES

1. Use the superposition method to solve for v_1 and v_2 in Figure 7.93.

Figure 7.93

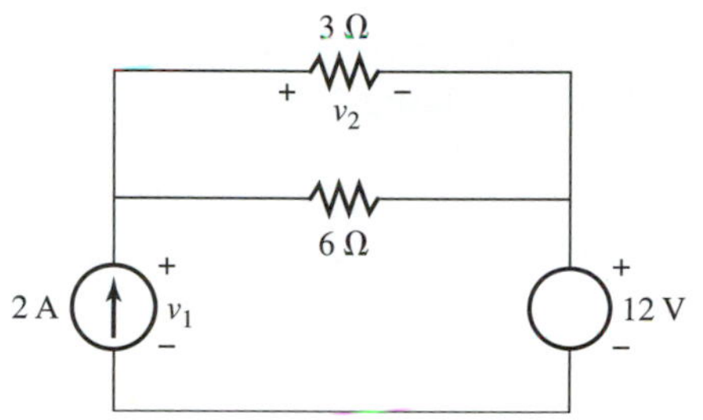

2. Use the superposition method to solve for i_1, i_2, and i_3 in Figure 7.94.

Figure 7.94

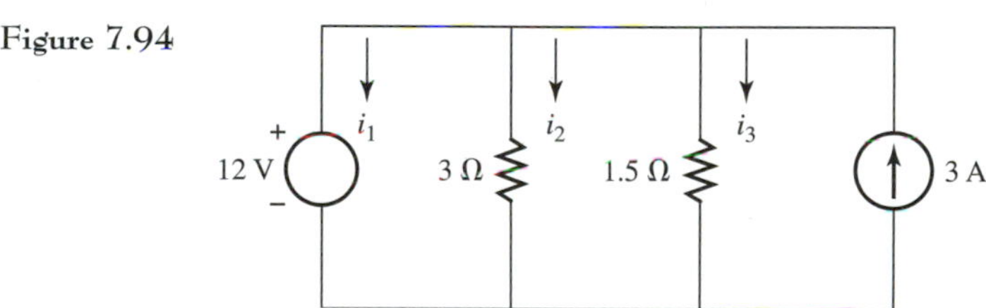

3. Use the superposition method to solve for i in Figure 7.95.

Figure 7.95

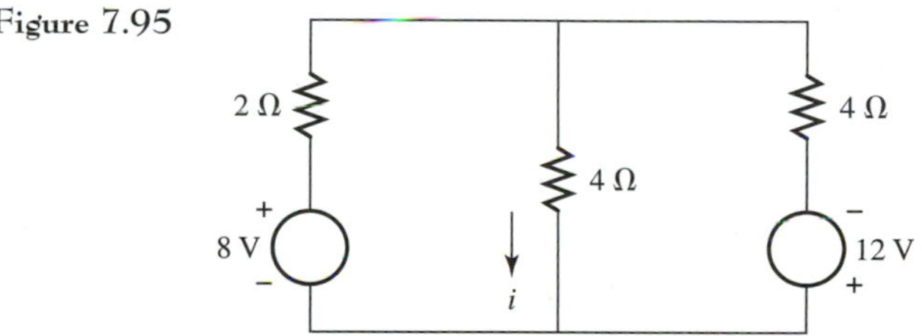

4. Use the superposition method to solve for i in Figure 7.96.

Figure 7.96

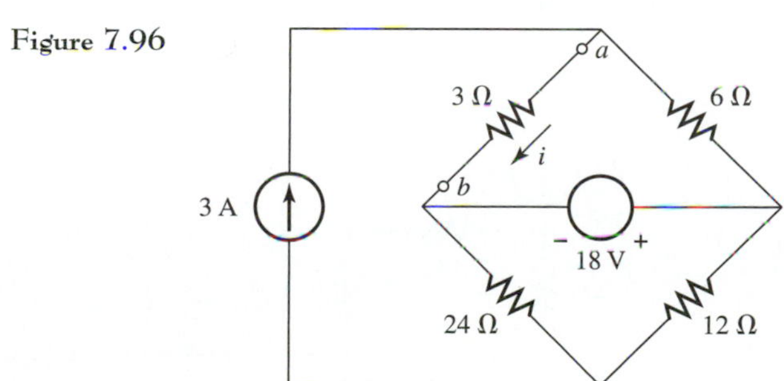

5. Use the superposition method to solve for v and i in Figure 7.97.

Figure 7.97

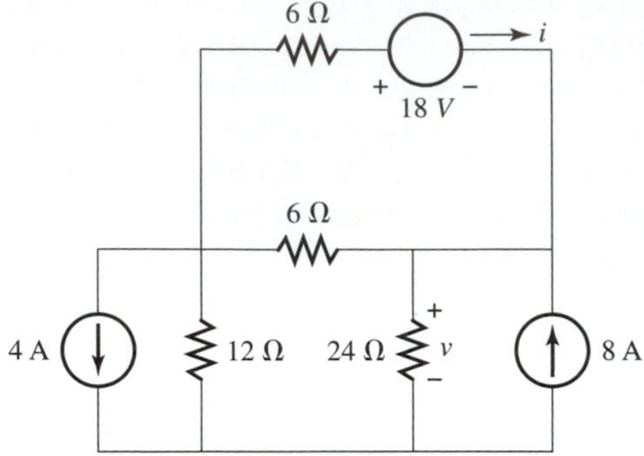

6. Replace the circuit to the left of terminals a-b in Figure 7.98 with a Thevenin equivalent.

Figure 7.98

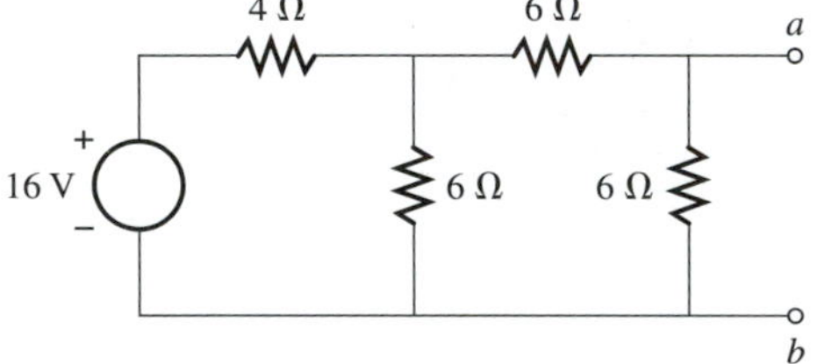

7. A 10-V source and a 4-Ω resistor are connected in series across terminals a-b in Figure 7.98. Find the current flowing from a to b.

8. Remove the 3-Ω resistor from the circuit in Problem 4 and determine a Thevenin equivalent with respect to terminals a-b for the remaining circuit. Reconnect the 3-Ω resistor to terminals a-b and find i.

9. Replace the circuit to the left of terminals a-b in Figure 7.99 with a Thevenin equivalent.

Figure 7.99

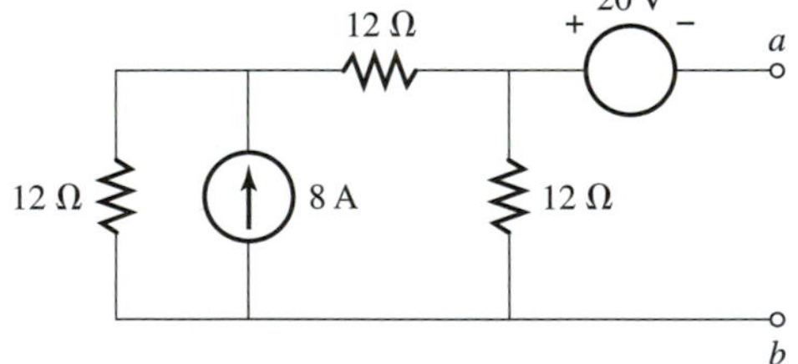

10. Replace the circuit to the left of terminals a-b in Figure 7.100 with a Thevenin equivalent.

Figure 7.100

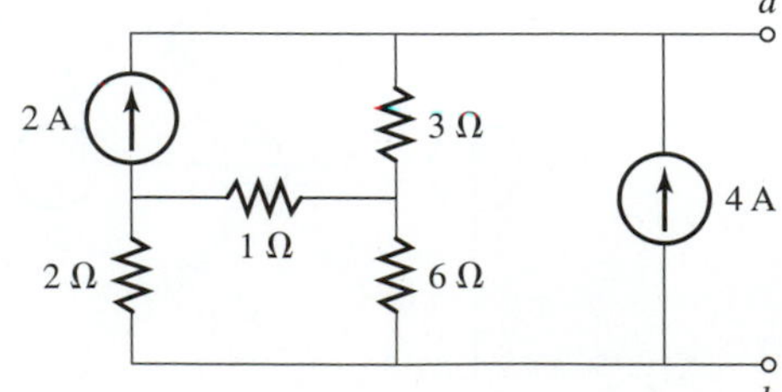

11. Replace the circuit to the left of terminals *a-b* in Figure 7.101 with a Norton equivalent.

Figure 7.101

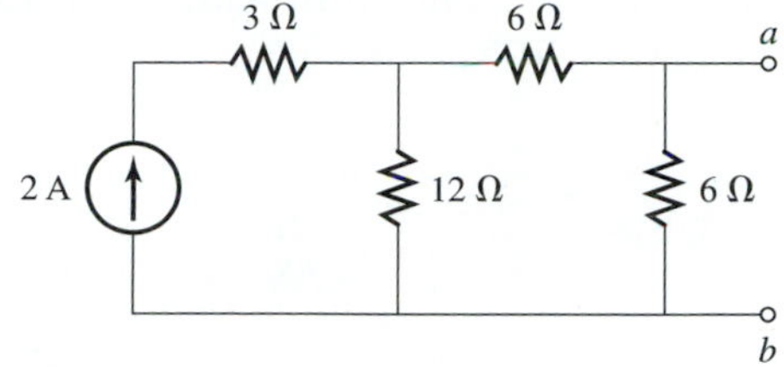

12. Replace the circuit to the left of terminals *a-b* in Figure 7.98 with a Norton equivalent.

13. Replace the circuit to the left of terminals *a-b* in Figure 7.99 with a Norton equivalent.

14. Replace the circuit to the left of terminals *a-b* in Figure 7.100 with a Norton equivalent.

15. Remove the 3-Ω resistor from the circuit in Problem 4 and replace the resulting circuit across terminals *a-b* with a Norton equivalent.

16. Determine a Norton equivalent for the circuit in Figure 7.102 with respect to terminals *a-b*.

Figure 7.102

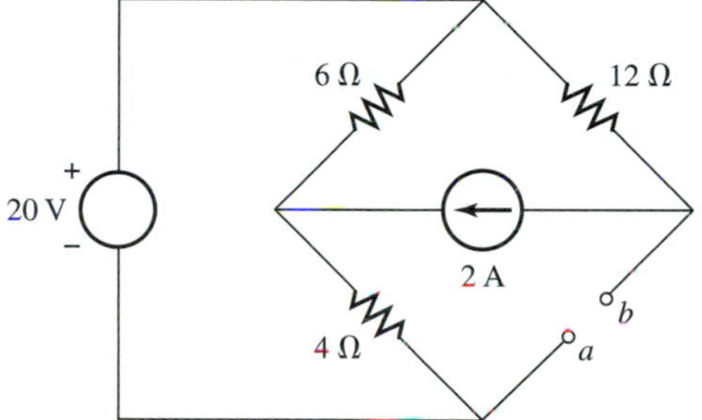

17. Determine a Norton equivalent for the circuit in Figure 7.103 with respect to terminals *a-b*.

Figure 7.103

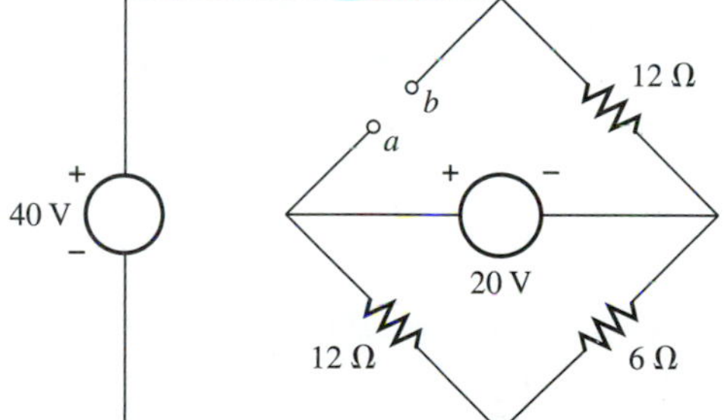

18. Determine a Thevenin equivalent circuit for the circuit to the left of terminals *a-b* in Figure 7.104.

Figure 7.104

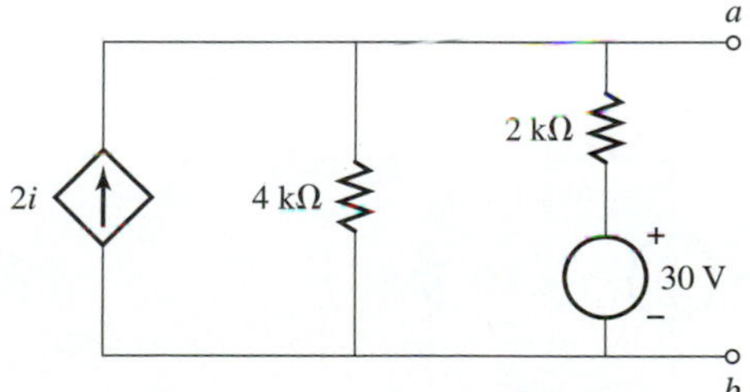

19. Determine a Norton equivalent circuit for the circuit to the left of terminals a-b in Problem 18.

20. Determine a Thevenin equivalent for the circuit to the left of terminals a-b in Figure 7.105.

Figure 7.105

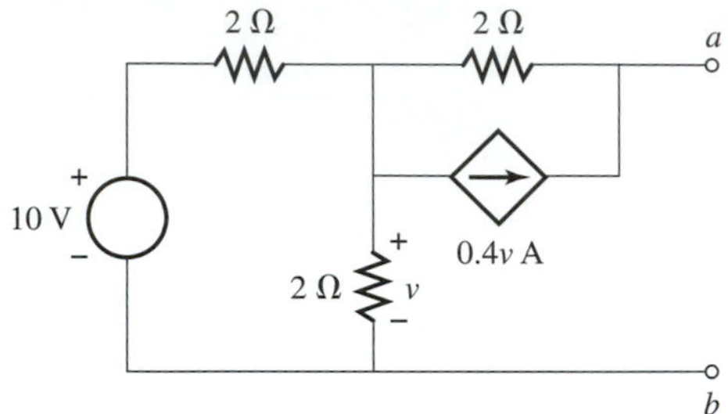

21. Determine a Norton equivalent for the circuit to the left of terminals a-b in Figure 7.105.

22. Determine a Thevenin resistance for Problem 6 by applying a voltage source v_x and finding the current i_x flowing through the voltage source. Recall that $R_T = v_x/i_x$.

23. Determine a Thevenin resistance for Problem 9 by applying a voltage source v_x and finding the current i_x flowing through the voltage source. Recall that $R_T = v_x/i_x$.

24. Use the successive source-transformation method to determine i in Figure 7.106.

Figure 7.106

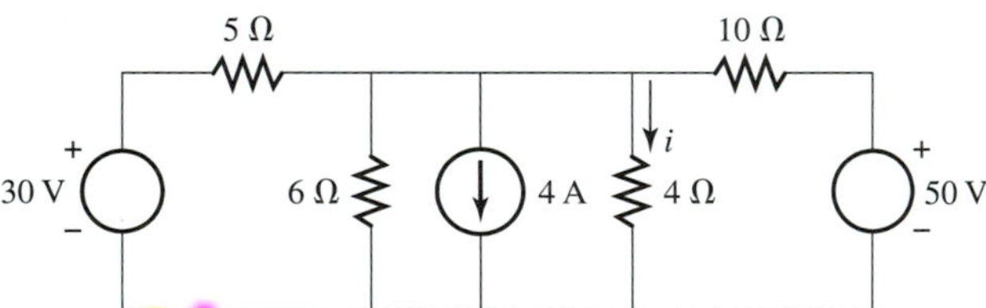

25. Use the successive source-transformation method to determine i in Figure 7.107.

Figure 7.107

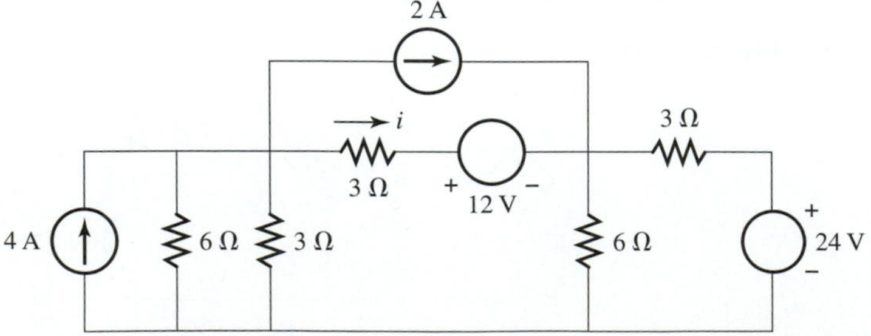

26. Use the successive source-transformation method to determine *i* in Figure 7.108.

Figure 7.108

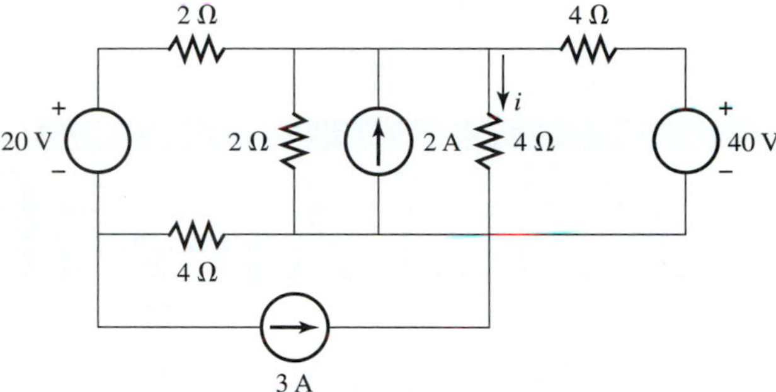

27. Determine the load resistor R_L to connect across terminals *a*-*b* in Problem 10 for maximum power transfer to R_L.

28. Determine the load resistor R_L to connect across terminals *a*-*b* in Problem 18 for maximum power transfer to R_L.

29. Determine the load resistor R_L to connect across terminals *a*-*b* in Problem 20 for maximum power transfer to R_L.

CHAPTER

8

The Operational Amplifier and Basic Measuring Devices

After completing this chapter the student should be able to

* analyze basic circuits containing operational amplifiers
* design basic dc measuring devices

Prior to this chapter our main focus has been on analytical methods for determining voltage and current. It seems appropriate at this point to introduce several basic devices that can be utilized to measure voltages and currents under laboratory or field conditions. Simple devices capable of making these measurements are presented in this chapter.

Another basic device, called the operational amplifier, is often encountered in electric circuits and is also presented in this chapter. Although it is not considered as significant as the five basic electrical elements introduced in Chapter 1, the operational amplifier plays an important role in electric circuits and is worthy of our consideration.

8.1 OPERATIONAL AMPLIFIERS

The operational amplifier (commonly called an op amp) is an electronic device that has many applications in electric circuits. It is the heart of the analog computer and has myriad uses in electronics, control, and computer applications. Although the operational amplifier is composed of rather complicated electronic circuitry, we shall consider it only as an electric circuit element, confining our interest to its external characteristics. The symbol for a basic operational amplifier is given in Figure 8.1, where v_o is the output voltage, v_i is the input voltage, and i_i is the input current. The input terminal marked with a negative sign is called the inverting terminal.

Figure 8.1 Operational amplifier.

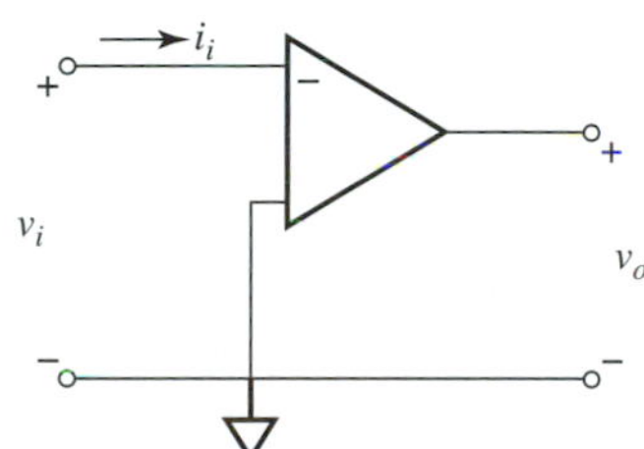

The operational amplifier has a very high input resistance (ratio v_i/i_i), which allows us to approximate the input current i_i as zero. The operational amplifier also has a very high voltage gain (ratio v_o/v_i), which allows us to approximate the input voltage v_i as zero. These approximations do not mean that i_i and v_i are equal to zero but that i_i and v_i are very small compared to other currents and voltages of interest in the circuit containing the operational amplifier. The operational amplifier is seldom used as a single electrical element, as shown in Figure 8.1. It is usually used in conjunction with other circuit elements, as shown in Figure 8.2. In this chapter we consider resistive circuits that contain operational amplifiers as circuit elements; in a later chapter we consider the use of capacitors and inductors with operational amplifiers.

The operational amplifier has more external terminals than indicated in Figure 8.1. The terminals not shown are for the power supply and other applications that are of no concern to us at this time, because our interest is in the external characteristics of the operational amplifier used as a basic circuit element. However, the terminals not shown in Figure 8.1 do affect the analysis of op-amp circuits regarding the use of Kirchhoff's current law. Because the terminals not shown in Figure 8.1 are usually connected to the reference node, there are currents at this node of which we are unaware when considering the operational amplifier as a basic circuit element. Therefore, we cannot apply Kirchhoff's current law at the reference node of the operational amplifier in our analysis procedures.

The operational amplifier is seldom used without all or a portion of the output signal fed back to the input. This mode of operation is called feedback and is almost always present in op-amp circuits. Feedback is accomplished by providing a path from an output terminal to the inverting input terminal (marked with a negative sign) of the operational amplifier. The path usually consists of one or more resistors, capacitors, or inductors. If more than one path is provided from the output to the input terminals of the operational amplifier, at least one of the paths must be to the inverting terminal for feedback to exist. The circuits in the following sections illustrate some of the basic operational amplifier configurations.

8.2 INVERTER

Consider the circuit in Figure 8.2. Voltages v_e, v_c, and v_a are all with respect

Figure 8.2 Inverter.

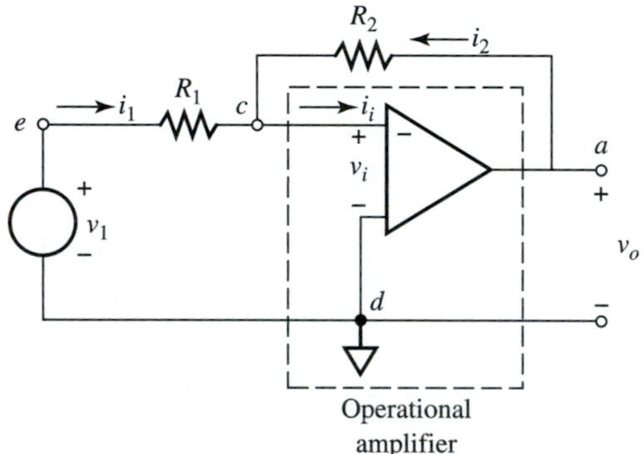

to the reference node d. Voltage v_i is the voltage of the top input terminal of the op amp with respect to the bottom input terminal, and in this particular circuit v_i equals v_c. Voltage v_a is equal to the output voltage v_o. Notice that the op amp is in the feedback mode, because R_2 connects the output of the op amp to the inverting input terminal.

Using Equation (6.8) to determine i_1 and i_2 and recognizing that $v_e = v_1$ and $v_a = v_o$ results in

$$i_1 = \frac{v_e - v_c}{R_1}$$

and

$$i_2 = \frac{v_a - v_c}{R_2}$$

Because v_i can be assumed equal to zero and $v_c = v_i$,

$$i_1 = \frac{v_1}{R_1}$$

and

$$i_2 = \frac{v_o}{R_2}$$

Applying Kirchhoff's current law to node c results in

$$i_1 + i_2 - i_i = 0$$

Because i_i can be assumed equal to zero,

$$\left(\frac{v_1}{R_1}\right) + \left(\frac{v_o}{R_2}\right) = 0$$

and

$$v_o = -\left(\frac{R_2}{R_1}\right)(v_1) \tag{8.1}$$

The circuit in Figure 8.2 is called an inverter, because the output voltage v_o is proportional to the negative of the input voltage with proportionality constant (R_2/R_1). Notice that if $R_1 = R_2$, $v_o = -v_1$.

EXAMPLE 8.1 Determine v_o for the inverter circuit shown in Figure 8.3. ■

Figure 8.3

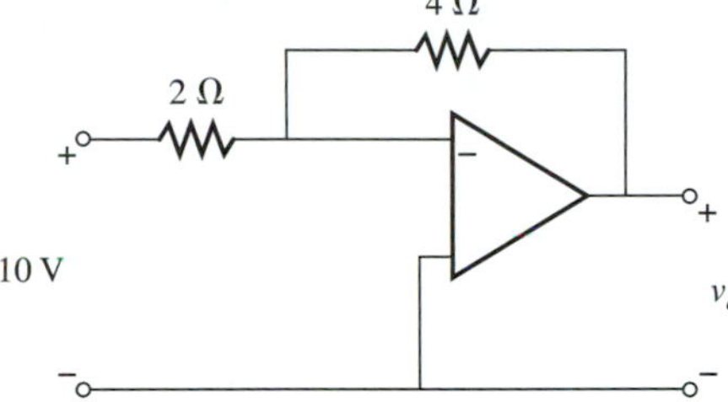

SOLUTION From Equation (8.1)

$$\begin{aligned} v_o &= -\left(\frac{R_2}{R_1}\right)(v_1) \\ &= -\left(\frac{4}{2}\right)(10) \\ &= -20 \text{ V} \end{aligned}$$

8.3 SUMMER

Consider the circuit in Figure 8.4. Voltages v_b, v_c, v_e, and v_f are all with respect to the reference node d. Voltage v_i is the voltage of the top input terminal with respect to the bottom input terminal of the op amp, and in this particular op-amp circuit, $v_i = v_c$. Voltage v_b is equal to the output voltage v_o. Notice that the op amp is in the feedback mode, because R_f connects the output of the op amp to the inverting input terminal.

Figure 8.4 Summer.

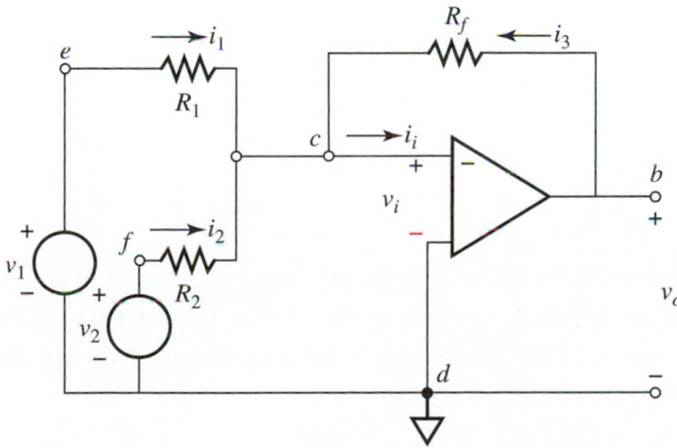

Using Equation (6.8) for i_1, i_2, and i_3 and recognizing that $v_c = v_i$, $v_e = v_1$, $v_f = v_2$, and $v_b = v_o$ results in

$$i_1 = \frac{v_e - v_c}{R_1}$$

$$= \frac{v_1 - v_i}{R_1}$$

$$i_2 = \frac{v_f - v_c}{R_2}$$

$$= \frac{v_o - v_i}{R_2}$$

and

$$i_3 = \frac{v_b - v_c}{R_f}$$

$$= \frac{v_o - v_i}{R_f}$$

Because v_i can be assumed equal to zero

$$i_1 = \frac{v_1}{R_1}$$

$$i_2 = \frac{v_2}{R_2}$$

and

$$i_3 = \frac{v_o}{R_f}$$

Applying Kirchhoff's current law to node c results in

$$i_1 + i_2 + i_3 - i_i = 0$$

Because i_i can be assumed equal to zero,

$$\frac{v_1}{R_1} + \frac{v_2}{R_2} + \frac{v_o}{R_f} = 0$$

and

$$v_o = -\left(\frac{R_f}{R_1}\right)v_1 - \left(\frac{R_f}{R_2}\right)v_2 \tag{8.2}$$

The circuit in Figure 8.4 is called a summer, because the output voltage v_o equals the negative of the sum of the input voltages v_1 and v_2 if $R_1 = R_2 = R_f$. The circuit in Figure 8.4 can be extended to sum more than two voltages, as illustrated in Example 8.3.

EXAMPLE 8.2 Determine v_o for the summer circuit shown in Figure 8.5. ■

Figure 8.5

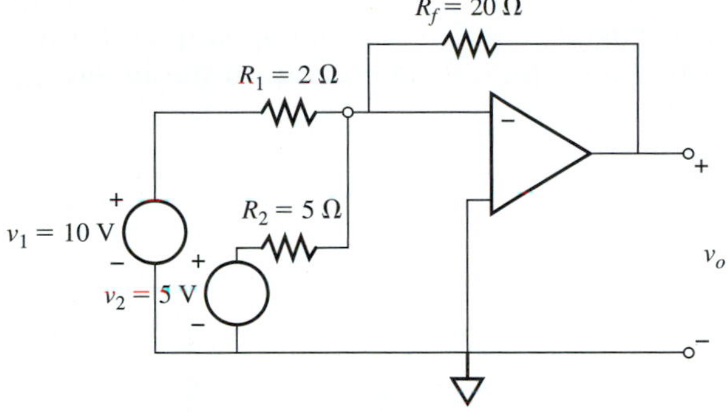

SOLUTION From Equation (8.2)

$$v_o = -\left(\frac{R_f}{R_1}\right)v_1 - \left(\frac{R_f}{R_2}\right)v_2$$

$$v_o = -\left(\frac{20}{2}\right)(10) - \left(\frac{20}{5}\right)(5)$$

$$= -120 \text{ V}$$

EXAMPLE 8.3 Determine v_o for the summer circuit shown in Figure 8.6. ■

Figure 8.6

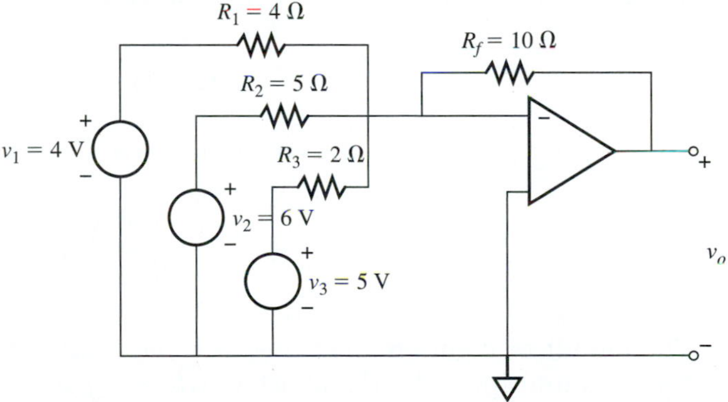

SOLUTION Extending Equation (8.2),

$$v_o = -\left(\frac{R_f}{R_1}\right)v_1 - \left(\frac{R_f}{R_2}\right)v_2 - \left(\frac{R_f}{R_3}\right)v_3$$

$$v_o = -\left(\frac{10}{4}\right)(4) - \left(\frac{10}{5}\right)(6) - \left(\frac{10}{2}\right)5$$

$$= -47 \text{ V}$$

8.4 AMPLIFIER

Consider the circuit in Figure 8.7. Voltages v_a, v_b, and v_c are all with respect to the reference node d. Voltage v_i is the voltage of the top input terminal with respect to the bottom input terminal of the op amp. Voltage v_a is equal to the output voltage v_o. Notice that the op amp is in the feedback mode, because R_2 connects the output of the op amp to the inverting input terminal.

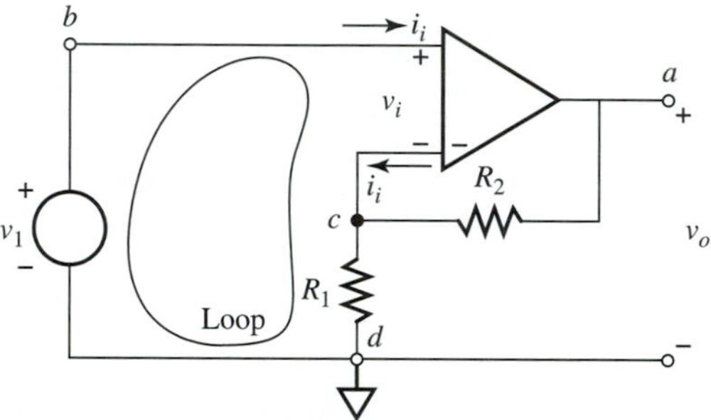

Figure 8.7 Amplifier.

Because i_i can be assumed equal to zero, resistors R_1 and R_2 are effectively in series and v_o is the voltage across the series combination.

From Equation (4.3) for voltage division,

$$v_c = \left(\frac{R_1}{R_1 + R_2}\right)v_o$$

From Kirchhoff's voltage law for the loop indicated in Figure 8.7,

$$v_1 - v_i - v_c = 0$$

and because v_i can be assumed equal to zero,

$$v_c = v_1$$

Substituting $v_c = v_1$ into the equation above results in

$$v_o = \left(\frac{R_1 + R_2}{R_1}\right)v_1 \tag{8.3}$$

The circuit in Figure 8.7 is called a voltage amplifier, because the input voltage v_1 is multiplied by the constant $(R_1 + R_2)/R_1$.

EXAMPLE 8.4 Determine v_o for the circuit shown in Figure 8.8. ■

Figure 8.8

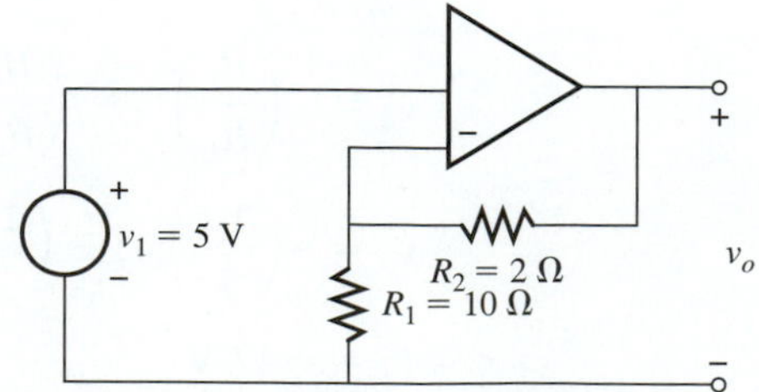

SOLUTION From Equation (8.3)

$$v_o = \left(\frac{R_1 + R_2}{R_2}\right)v_1$$

$$v_o = \left(\frac{2 + 10}{10}\right)(5)$$

$$= 6 \text{ V}$$

8.5 MISCELLANEOUS OP-AMP CIRCUITS

The following examples illustrate the analysis procedures previously presented to analyze circuits containing operational amplifiers.

EXAMPLE 8.5 Determine i for the circuit shown in Figure 8.9. ■

Figure 8.9

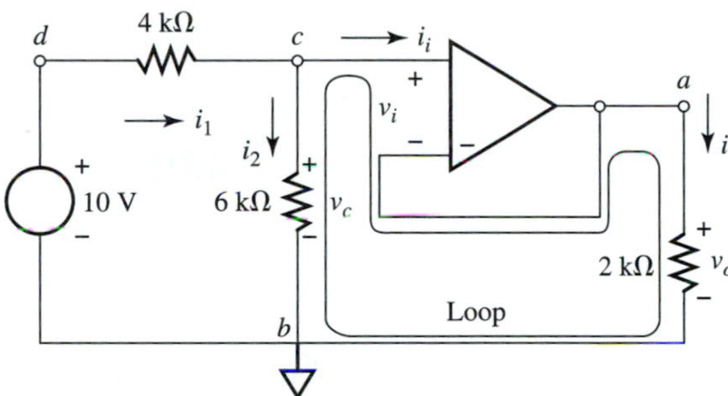

SOLUTION Voltages v_d, v_c, and v_a are all with respect to the reference node b. From Equation (6.8)

$$i_1 = \frac{v_d - v_c}{4 \text{ k}\Omega}$$

and

$$i_2 = \frac{v_c}{6 \text{ k}\Omega}$$

Applying Kirchhoff's voltage law to the loop indicated in Figure 8.9 results in

$$v_c - v_i - v_o = 0$$

Because v_i can be assumed equal to zero,

$$v_c = v_o$$

Because $v_d = 10$ and $v_c = v_o$,

$$i_1 = \frac{10 - v_o}{4 \text{ k}\Omega}$$

and

$$i_2 = \frac{v_o}{6 \text{ k}\Omega}$$

Applying Kirchhoff's current law at node c results in

$$i_1 - i_2 - i_i = 0$$

Because i_i can be assumed equal to zero,

$$i_1 = i_2$$

and

$$\frac{10 - v_o}{4\ \text{k}\Omega} = \frac{v_o}{6\ \text{k}\Omega}$$

$$v_o = 6\ \text{V}$$

From Ohm's law

$$i = \frac{v_o}{2\ \text{k}\Omega} = \frac{6}{2\ \text{k}\Omega} = 3\ \text{mA}$$

EXAMPLE 8.6 Determine i_2 for the circuit shown in Figure 8.10. ■

Figure 8.10

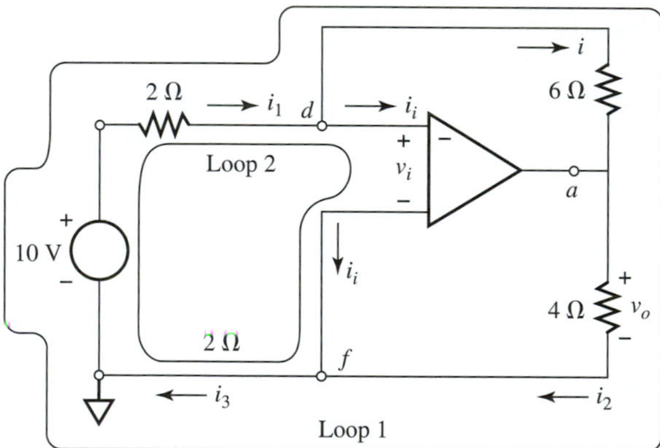

SOLUTION Applying Kirchhoff's voltage law to loop 1 results in

$$10 - 6i - 4i_2 - 2i_1 = 0$$

Applying Kirchhoff's current law at node d and recognizing that $i_i = 0$ results in $i = i_1$.

Substituting i_1 for i in the preceding equation results in

$$10 - 8i_1 - 4i_2 = 0$$

Applying Kirchhoff's voltage law to loop 2 results in

$$10 - 2i_1 - v_i = 0$$

and since $i_1 = 5$.

Substituting $i_1 = 5$ and $i = 5$ in the equation from loop 1 $(10 - 6i - 4i_2 - 2i_1 = 0)$ results in

$$i_2 = -7.5 \text{ A}$$

An alternative approach is to recognize the circuit as an inverter, so

$$v_a = v_o = -\left(\frac{6}{2}\right)10 = -30 \text{ V}$$

$$i_2 = \frac{v_o}{4} = -\frac{30}{4} = -7.5 \text{ A}$$

EXAMPLE 8.7 Determine v_o for the circuit shown in Figure 8.11. ■

Figure 8.11

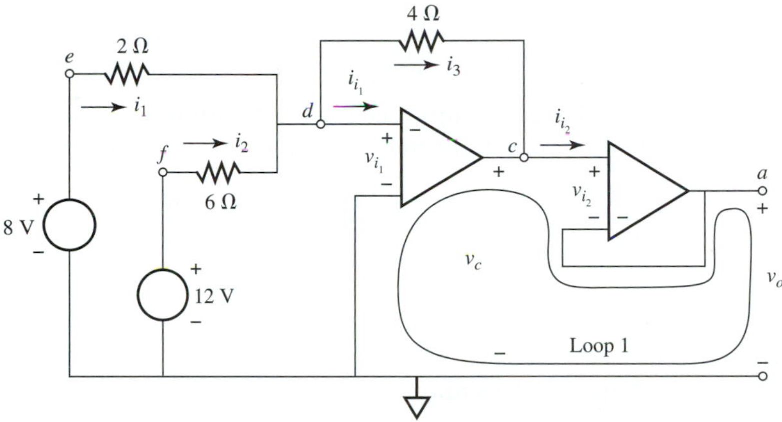

SOLUTION Voltages v_e, v_f, v_d, v_c, and v_a are all with respect to the reference node. Applying Kirchhoff's voltage law to loop 1 results in

$$v_c - v_{i_2} - v_o = 0$$

Because v_{i_2} can be assumed equal to zero,

$$v_c = v_o$$

Because $v_d = v_{i_1}$ and v_{i_1} can be assumed equal to zero,

$$v_d = 0$$

From Equation (6.8)

$$i_1 = \frac{v_e - v_d}{2}$$

$$i_2 = \frac{v_f - v_d}{6}$$

and

$$i_3 = \frac{v_d - v_c}{4}$$

Applying Kirchhoff's current law at node d results in

$$i_1 + i_2 - i_3 - i_{i_1} = 0$$

Recalling that i_{i_1} can be assumed equal to zero and substituting the expressions determined here for i_1, i_2, and i_3 results in

$$\frac{v_e - v_d}{2} + \frac{v_f - v_d}{6} - \frac{v_d - v_c}{4} = 0$$

Substituting $v_c = v_o$, $v_d = 0$, $v_e = 8$, and $v_f = 12$ into the preceding equation results in

$$\frac{8}{2} + \frac{12}{6} - \frac{-v_o}{4} = 0$$

$$v_o = -24 \text{ V}$$

An alternative approach for determining v_o involves recognizing that the first operational amplifier is a summer with v_c as the output. It follows that

$$v_c = -\left(\frac{4}{2}\right)(8) - \frac{4}{6}(12)$$

$$= -24$$

Applying Kirchhoff's voltage law to loop 1 results in

$$v_c - v_{i_2} - v_o = 0$$

and because v_{i_2} can be assumed equal to zero,

$$v_o = v_c$$

$$= -24 \text{ V}$$

8.6 DC MEASURING DEVICES

Measuring devices are central to the analysis and study of electrical systems under laboratory and field conditions. In this section we present some of the basic measuring devices for measuring dc voltage, dc current, and resistance.

The heart of the measuring devices considered in this section is the D'Arsonval meter movement, which is basically an electric coil under the influence of a magnetic field from a permanent magnet. Current through the coil and the magnetic field cause forces on the coil, and the resulting rotation of the coil is proportional to the current. A pointer is attached to the coil to allow observation and measurement of the rotation. The amount of rotation is calibrated to indicate the value of dc voltage, dc current, or resistance measured.

The D'Arsonval meter movement can be represented as shown in Figure 8.12, where R_M is the internal resistance of the meter.

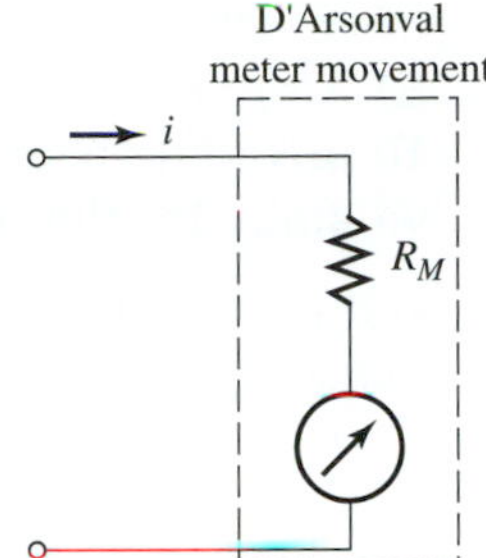

Figure 8.12 D'Arsonval meter movement.

8.7 DC AMMETER

The circuit in Figure 8.12 is a dc ammeter, and the reading of the meter is proportional to the current i. However, the D'Arsonval meter movement is limited to small currents and will deflect full scale at the limiting value of i. The circuit in Figure 8.12 is modified to allow for the measurement of larger currents, as shown in Figure 8.13.

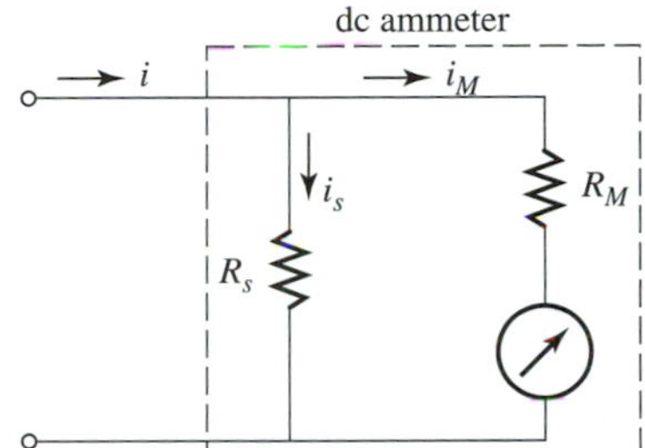

Figure 8.13 Ammeter.

We shall designate the current through the meter movement as i_M, the current through the shunting (parallel) resistor R_s as i_s, and the measured current as i.

From Equation (4.1) for current division,

$$i_M = \left(\frac{R_s}{R_s + R_M}\right)i$$

and solving for i results in

$$i = \left(\frac{R_s + R_M}{R_s}\right)i_M \tag{8.4}$$

Because i_M is proportional to the meter reading, i is also proportional to the meter reading and is determined from Equation (8.4). In Equation (8.4) the quantity $\{(R_s + R_M)/(R_s)\}$ is the proportionality constant relating i to i_M.

The maximum current, i_{max}, that can be measured by the ammeter in Figure 8.13 with a meter movement that can tolerate a maximum current $i_{M_{max}}$ depends on the shunting resistor R_s. Solving Equation (8.4) for R_s with i set equal to i_{max} and i_M set equal to $i_{M_{max}}$ results in

$$R_s = \left(\frac{R_M}{i_{max} - i_{M_{max}}}\right)(i_{M_{max}}) \tag{8.5}$$

When R_s in Figure 8.13 is equal to the value determined in Equation (8.5) and i is equal to its maximum value i_{max}, the meter will read full scale.

Consider the following example.

EXAMPLE 8.8 A D'Arsonval meter movement has a full-scale current rating of 10 mA and an internal resistance of 20 Ω. Use the meter movement to design an ammeter that can measure dc currents up to 100 mA. ■

SOLUTION Consider the circuit in Figure 8.14. Using Equation (8.5) to determine R_s results in

Figure 8.14

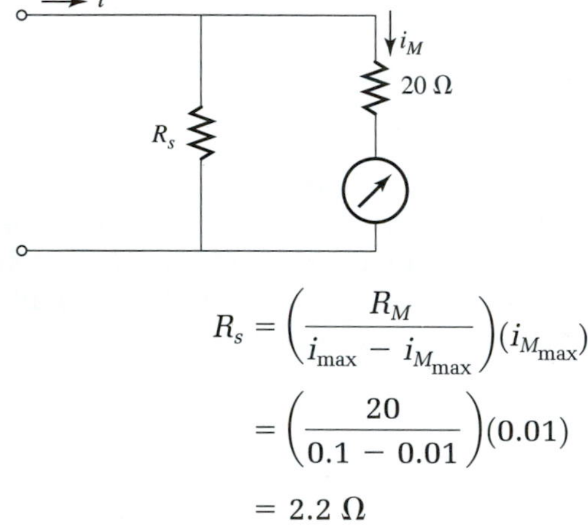

$$R_s = \left(\frac{R_M}{i_{max} - i_{M_{max}}}\right)(i_{M_{max}})$$

$$= \left(\frac{20}{0.1 - 0.01}\right)(0.01)$$

$$= 2.2\ \Omega$$

The dc ammeter shown in Figure 8.15 will measure a current i up to 100 mA. The meter will deflect full scale when i = 100 mA.

Figure 8.15

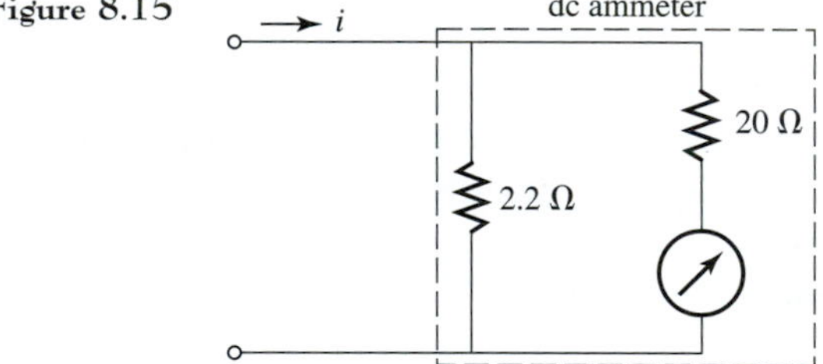

8.8 DC VOLTMETER

We now consider the basic D'Arsonval meter movement, as connected in Figure 8.16, to measure dc voltages. R_M is the internal resistance of the meter movement, R_s is a series resistance external to the meter movement, and v is the measured voltage.

Figure 8.16 Voltmeter.

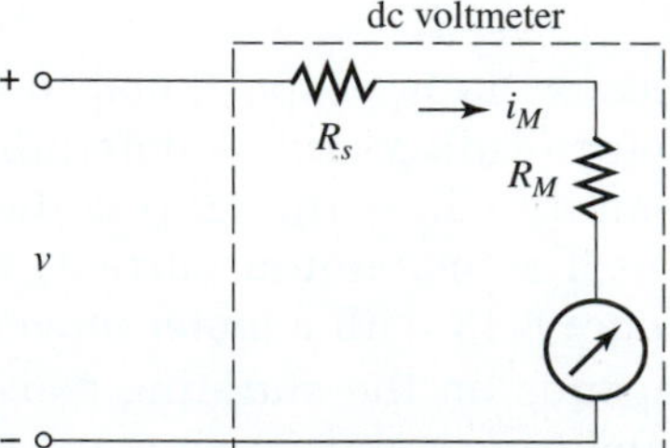

Applying Kirchhoff's voltage law to the circuit in Figure 8.16 results in

$$v - R_s i_M - R_M i_M = 0$$

and

$$v = (R_s + R_M)i_M \tag{8.6}$$

Because i_M is proportional to the meter reading, v is also proportional to the meter reading. The constant of proportionality relating v to i_M is $(R_s + R_M)$. For a given value of i_M, a corresponding value of v is determined from Equation (8.6).

If it is desired to measure voltage up to a maximum value v_{max}, we choose a value of R_s that will cause a maximum deflection of the meter when v equals v_{max}. This value of R_s is obtained from Equation (8.6) by setting v equal to v_{max} and i_M equal to $i_{M_{max}}$, where $i_{M_{max}}$ is the value of i_M that causes a full-scale meter deflection. Solving Equation (8.6) for R_s when $i_M = i_{M_{max}}$ and $v = v_{max}$ results in

$$R_s = \frac{v_{max} - R_M i_{M_{max}}}{i_{M_{max}}} \tag{8.7}$$

EXAMPLE 8.9 A D'Arsonval meter has a full-scale current rating of 10 mA and an internal resistance of 20 Ω. Use the meter movement to design a voltmeter that can measure dc voltages up to 200 V. ■

SOLUTION Consider the circuit in Figure 8.17. Using Equation (8.7) to determine R_s results in

$$R_s = \frac{v_{max} - R_M i_{M_{max}}}{i_{M_{max}}}$$

$$= \frac{200 - (20)(0.01)}{0.01}$$

$$= 19.98 \text{ k}\Omega$$

Figure 8.17

dc voltmeter

+ − v R_s i_M R_M

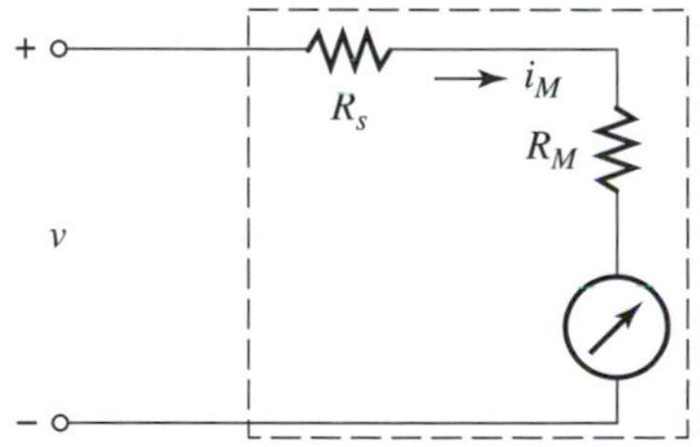

The dc voltmeter in Figure 8.18 will measure a voltage v up to 200 V. The meter will deflect full scale when $v = 200$ V.

Figure 8.18

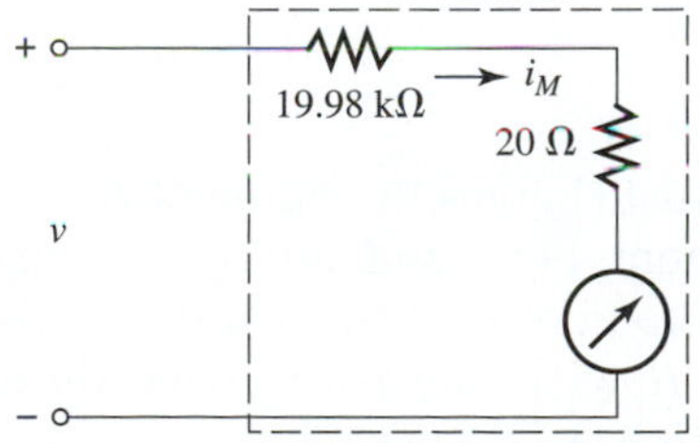

dc voltmeter

+ − v 19.98 kΩ i_M 20 Ω

8.9 OHMMETER

We now consider the basic D'Arsonval meter movement as connected in Figure 8.19 to measure resistance. R_M is the internal resistance of the meter movement, R_s is a series resistance external to the meter movement, v is a dc voltage source external to the meter movement, and R is the measured resistance.

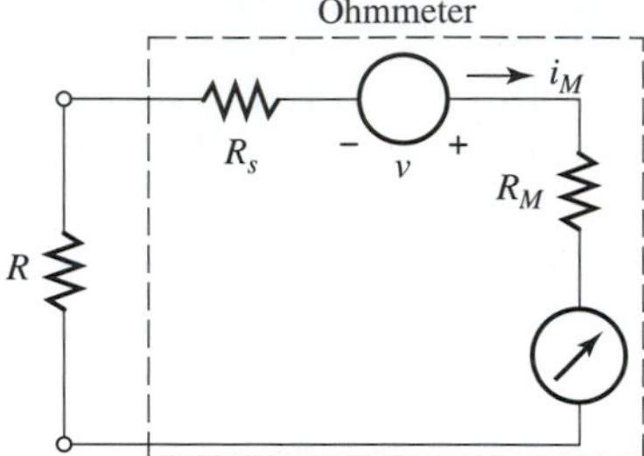

Figure 8.19 Ohmmeter.

Applying Kirchhoff's voltage law to the circuit results in

$$-Ri_M - R_s i_M + v - R_M i_M = 0$$

Solving for R results in

$$R = \frac{v - R_M i_M - R_s i_M}{i_M} \tag{8.8}$$

We select a value for v that will cause i_M to equal $i_{M\max}$ when $R = 0$. This will ensure full-scale meter deflection when $R = 0$. Solving Equation (8.8) for v and R_s when $R = 0$ and $i_M = i_{M\max}$ results in

$$v = (R_M + R_s)i_{M\max} \tag{8.9}$$

and

$$R_s = \frac{v - R_M i_{M\max}}{i_{M\max}} \tag{8.10}$$

The numerical value of v is usually known and can be substituted into Equation (8.10) to determine R_s.

Substituting the expression for v given in Equation (8.9) into Equation (8.8) results in

$$\begin{aligned} R &= \frac{(R_M + R_s)i_{M\max} - R_M i_M - R_s i_M}{i_M} \\ &= \left\{\frac{(R_M + R_s)(i_{M\max})}{i_M}\right\} - (R_M + R_s) \\ &= (R_M + R_s)\left\{\left(\frac{i_{M\max}}{i_M}\right) - 1\right\} \end{aligned} \tag{8.11}$$

Equation (8.11) gives the value of R in terms of the current i_M through the meter movement, because R and i_M are the only variables in the equation. A value of R equal to zero will cause a full-scale deflection of the meter, and all other values of R will cause a downscale deflection with readings, as given by Equation (8.11).

Consider the following example.

EXAMPLE 8.10 A D'Arsonval meter has a full-scale current rating of 10 mA and an internal resistance of 20 Ω. Use the meter movement with a 20-V voltage source to design an ohmmeter. ■

SOLUTION Consider the circuit in Figure 8.20, where $v = 20$ V and $R_M = 20\ \Omega$.

Figure 8.20

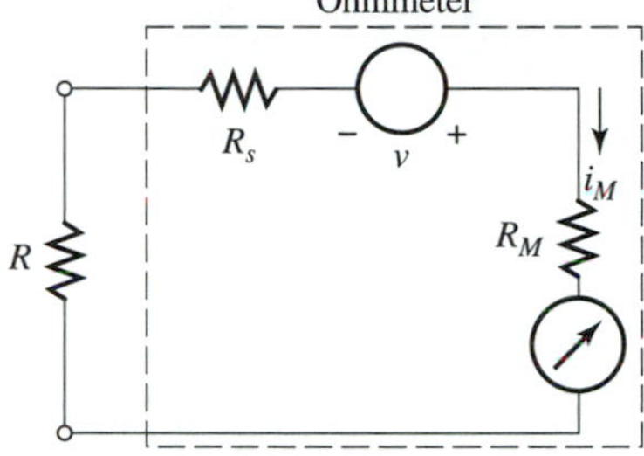

From Equation (8.10)

$$\begin{aligned} R_s &= \frac{v - R_M i_{M_{\max}}}{i_{M_{\max}}} \\ &= \frac{20 - 20(0.01)}{0.01} \\ &= 1980\ \Omega \end{aligned}$$

The ohmmeter circuit shown in Figure 8.21 will measure resistance R from 0 (short circuit) to infinity (open circuit) in ohms.

Figure 8.21

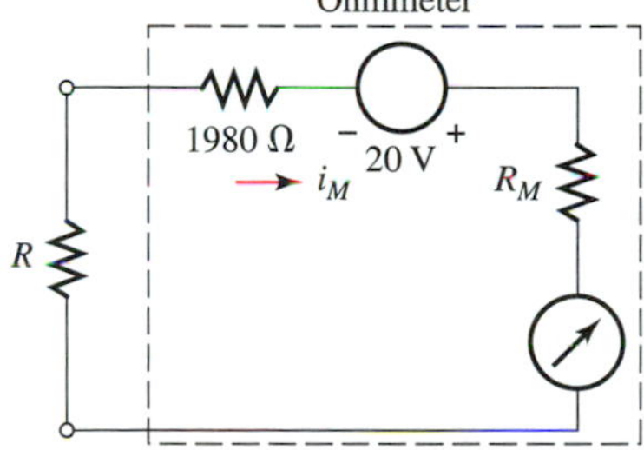

A full-scale reading will correspond to 0 Ω, and all readings will be governed by the equation

$$R = \left(\frac{20}{i_M}\right) - 2000$$

This equation results when $i_{M_{\max}} = 10$ mA, $R_M = 20\ \Omega$, $v = 20$ V, and $R_s = 1980\ \Omega$ are substituted into Equation (8.11).

8.10 WHEATSTONE BRIDGE

Another measuring device of interest to us at this point is the Wheatstone bridge, which can be used to measure resistance. The Wheatstone bridge is composed of high-precision components and can measure resistance to a high degree of accuracy. The essential elements of a Wheatstone bridge are a voltage source, three variable resistors, and a galvanometer. A galvanometer is a very sensitive current-measuring device that can measure current flowing through it in either direction. The basic Wheatstone bridge circuit is given in Figure 8.22.

Figure 8.22 Wheatstone bridge.

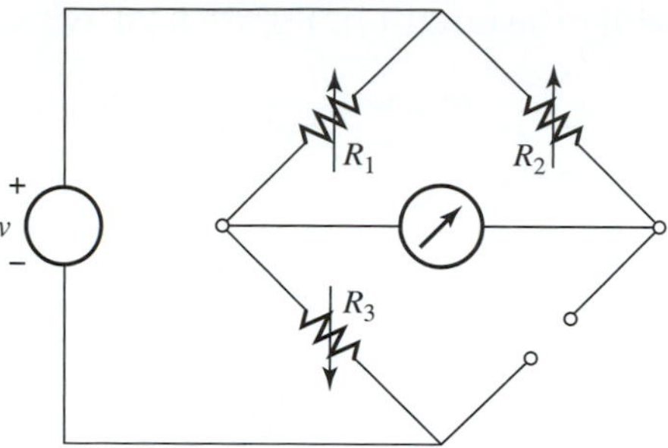

The unknown resistance R to be measured is placed across the open terminals shown in Figure 8.23. The variable resistors R_1, R_2, and R_3 are adjusted for a zero galvanometer reading ($i = 0$).

Figure 8.23 Wheatstone bridge.

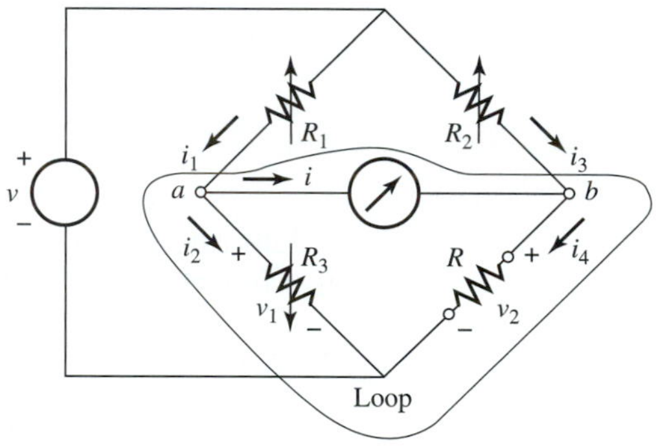

Applying Kirchhoff's current law at nodes a and b for the condition $i = 0$ results in

$$i_1 = i_2$$

and

$$i_3 = i_4$$

Because $i_1 = i_2$, R_1 is effectively in series with R_3; and because $i_3 = i_4$, R_2 is effectively in series with R.

Applying Equation (4.3) for voltage division to the series combinations results in

$$v_1 = \left(\frac{R_3}{R_1 + R_3}\right)(v)$$

and

$$v_2 = \left(\frac{R}{R + R_2}\right)(v)$$

Applying Kirchhoff's voltage law to the loop indicated in Figure 8.23 results in

$$v_1 + v_{ba} - v_2 = 0$$

and because $v_{ba} = 0$,

$$v_1 = v_2$$

Equating the expressions for v_1 and v_2 results in

$$\left(\frac{R_3}{R_1 + R_3}\right)(v) = \left(\frac{R}{R + R_2}\right)(v)$$

$$\frac{R_3}{R_1 + R_3} = \frac{R}{R + R_2}$$

$$R_3R + R_3R_2 = RR_1 + RR_3$$

and

$$R = \frac{R_2R_3}{R_1} \tag{8.12}$$

The value of the unknown resistance R can be determined from Equation (8.12) when the galvanometer reading is zero. The Wheatstone bridge is usually constructed to allow for a direct reading of R from the dials used to set R_1, R_2, and R_3 for zero galvanometer current.

EXAMPLE 8.11 An unknown resistance R is placed in the Wheatstone bridge circuit in Figure 8.24. The galvanometer reading is zero. Determine the unknown resistance. ■

Figure 8.24

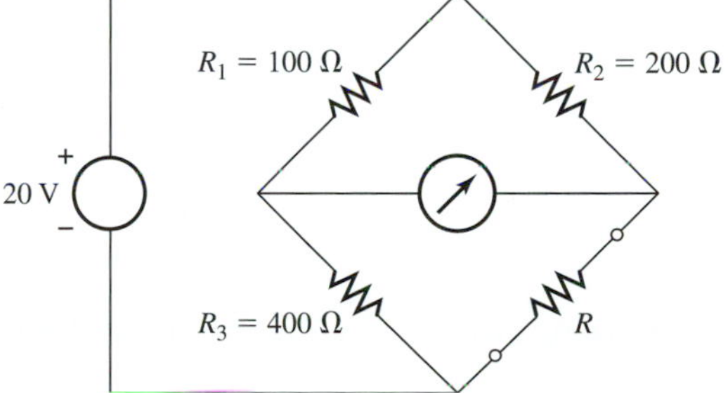

SOLUTION From Equation (8.12)

$$R = \frac{R_2R_3}{R_1}$$

$$= \frac{(200)(400)}{100}$$

$$= 800\ \Omega$$

8.11 EXERCISES

1. Determine i, i_1, and v_o for the circuit in Figure 8.25.

Figure 8.25

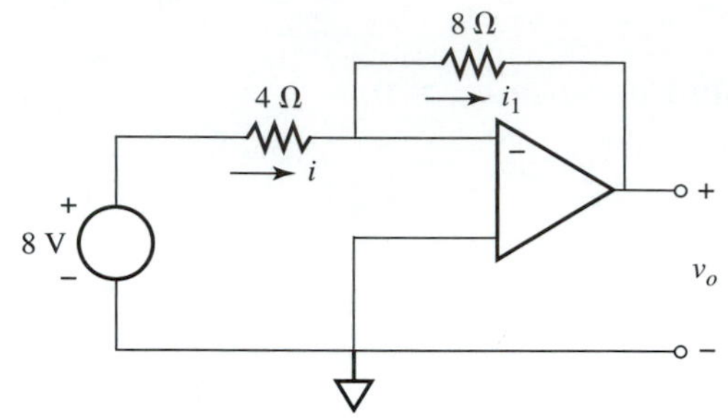

2. Determine i, i_1, i_2, and v_o for the circuit in Figure 8.26.

Figure 8.26

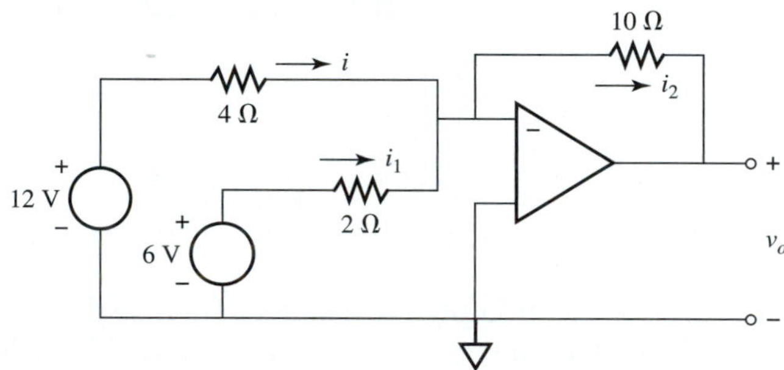

3. Determine i_1, i_2, and v_o for the circuit in Figure 8.27.

Figure 8.27

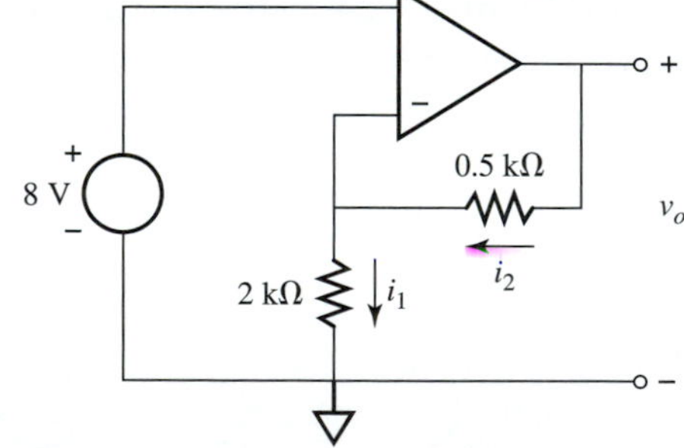

4. Determine v_o for the circuit in Figure 8.28.

Figure 8.28

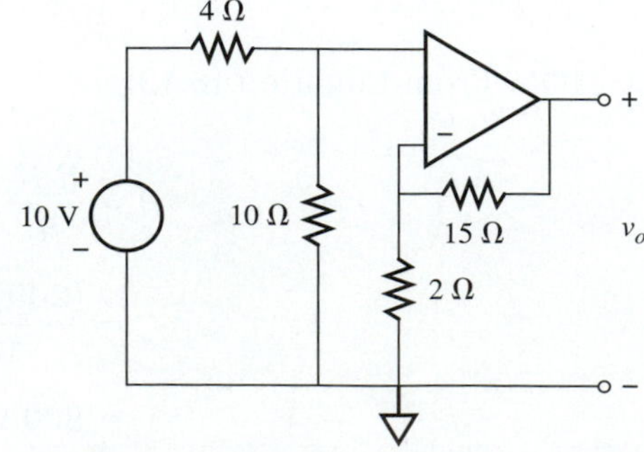

5. Determine a value for R in Figure 8.29 that will result in $v_o = 20$ V.

Figure 8.29

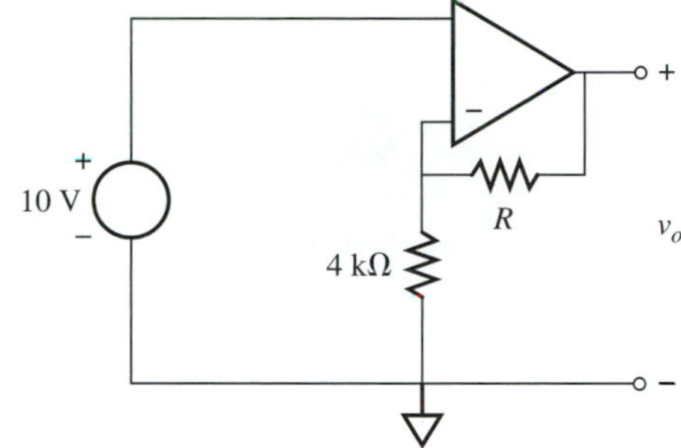

6. Determine a value for v in Figure 8.30 that will result in $v_o = 20$ V.

Figure 8.30

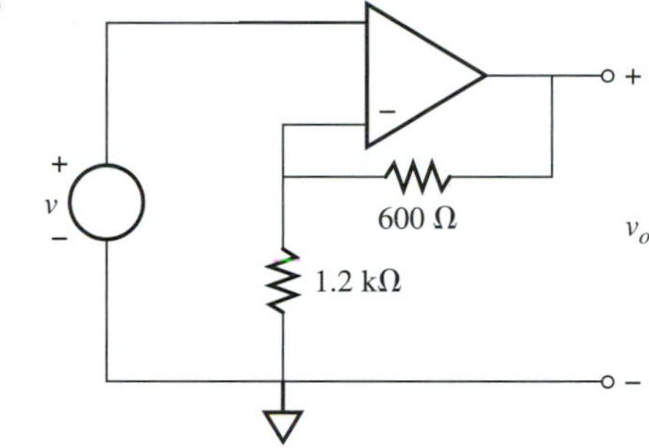

7. Determine v_o for the circuit in Figure 8.31.

Figure 8.31

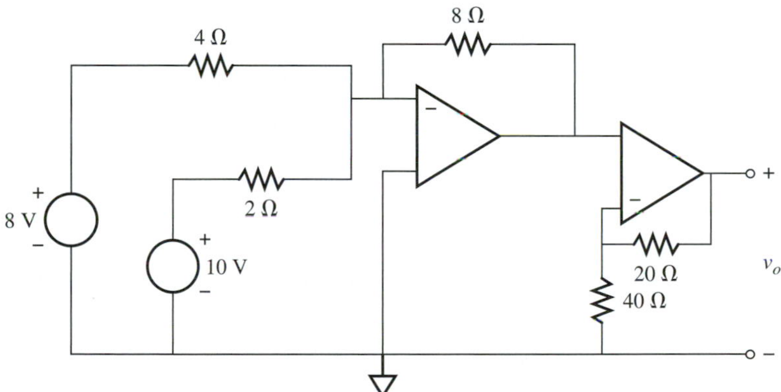

8. Determine v_o for the circuit in Figure 8.32.

Figure 8.32

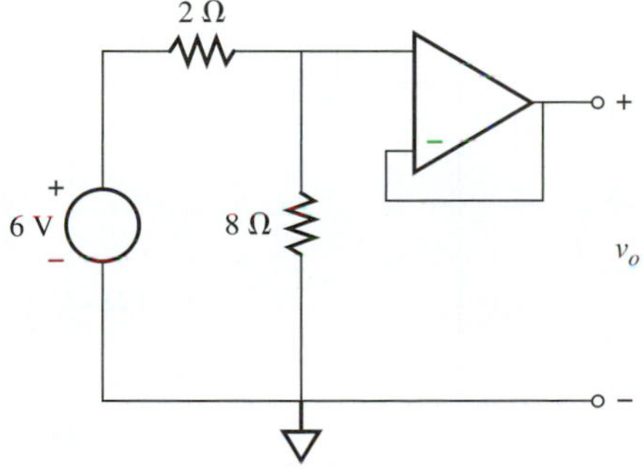

9. Determine i for the circuit in Figure 8.33.

Figure 8.33

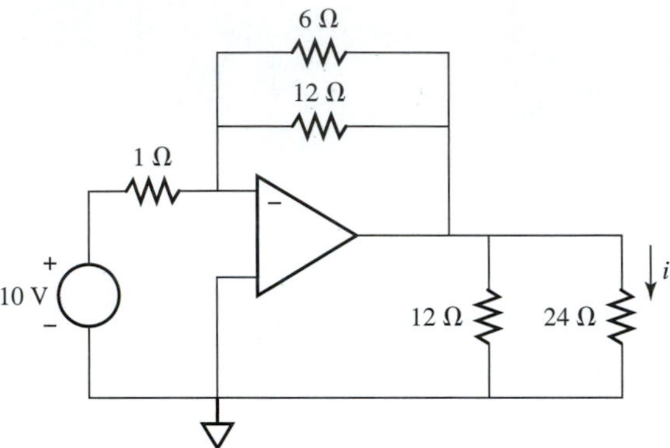

10. Determine v_o for the circuit in Figure 8.34.

Figure 8.34

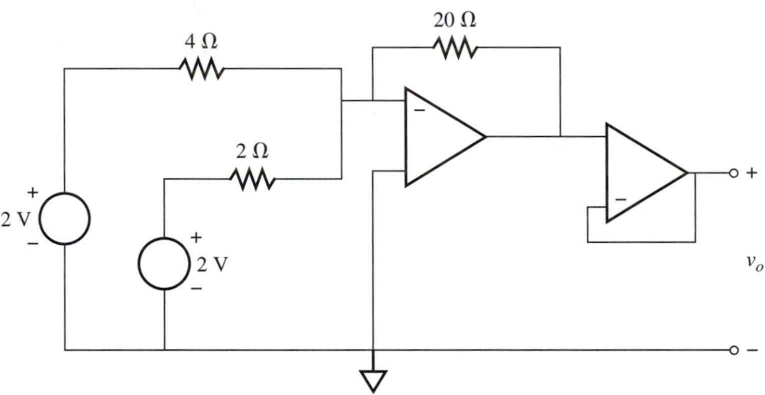

11. Determine v_o, i_1, and i_2 for the circuit in Figure 8.35.

Figure 8.35

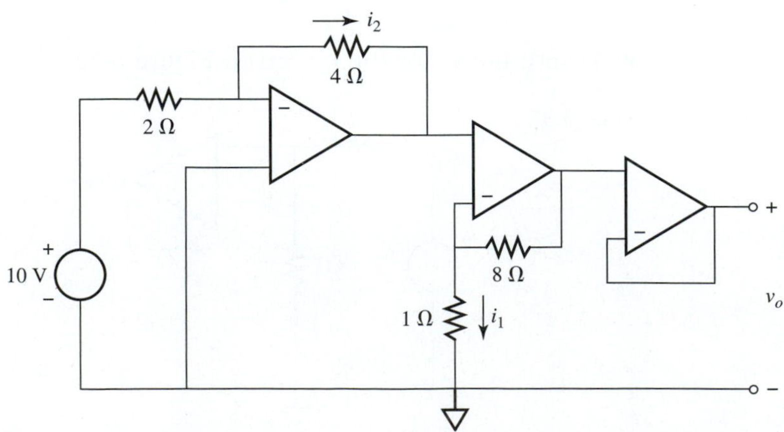

12. Determine v_o, i_1, and i_2 for the circuit in Figure 8.36.

Figure 8.36

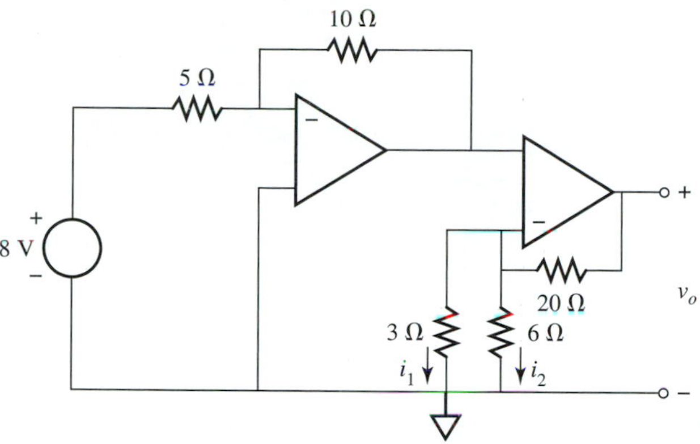

13. Determine v_o, i_1, and i_2 for the circuit in Figure 8.37.

Figure 8.37

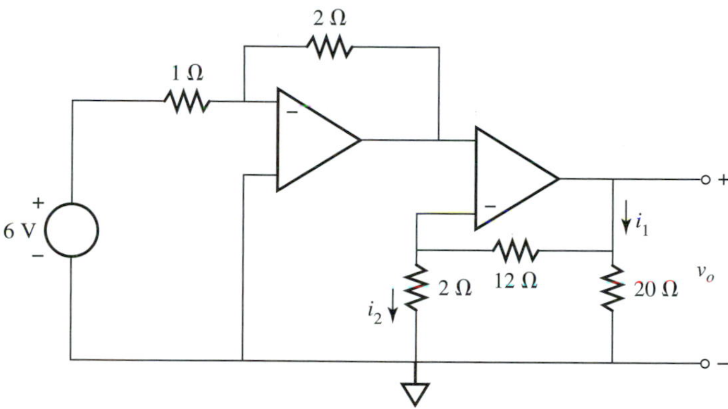

14. Determine v_o, i_1, and i_2 for the circuit in Figure 8.38.

Figure 8.38

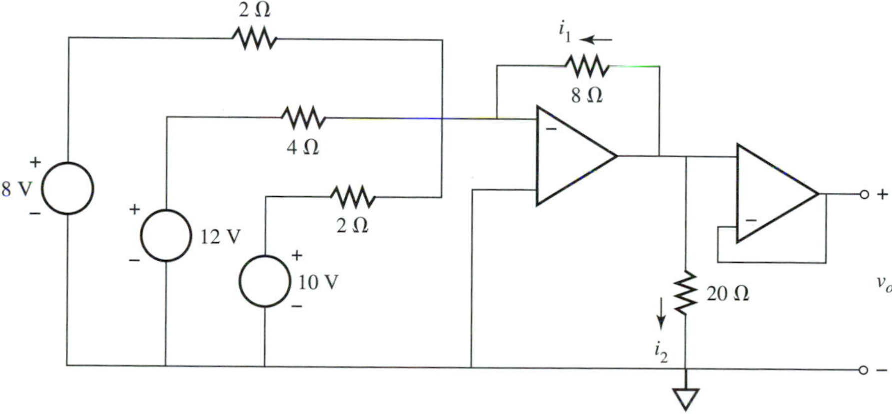

15. A D'Arsonval meter movement has a full scale current rating of 20 mA and an internal resistance of 30 Ω. Use the movement to design an ammeter to measure dc currents up to 120 mA.

16. Using the D'Arsonval meter movement given in Problem 15, design a dc voltmeter to measure voltages up to 150 V.

17. Using the D'Arsonval meter movement given in Problem 15 and a 12 V voltage source, design an ohmmeter.

18. A Wheatstone bridge is shown in Figure 8.39. What is the value of R if the galvanometer reading is zero?

Figure 8.39

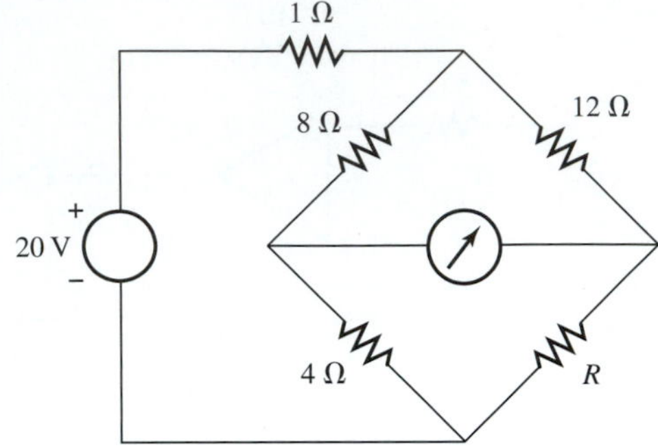

19. In Problem 18 what is the current through the resistor R when the galvanometer reading is zero?

20. In Problem 18 what is the voltage across the 8 Ω resistor when the galvanometer reading is zero?

21. If the resistor R is removed from the circuit, What is the galvanometer reading in Figure 8.39?

22. If the resistor R is replaced by a short in Figure 8.39, What is the galvanometer reading?

CHAPTER

9

Basic Mathematical Concepts

After completing this chapter the student should be able to

* differentiate a function
* determine the indefinite integral of a function
* determine the definite integral of a function

Most of the circuits considered in the previous chapters consist of resistors, current sources, and voltage sources. These circuits are called resistive circuits and their associated mathematical models consist of one or more algebraic equations. Unlike resistive circuits, circuits containing inductors and/or capacitors are described by mathematical models that involve derivatives and integrals. An introduction to derivatives and integrals, along with related mathematical concepts, is presented in this chapter preparatory to considering circuits containing inductors and capacitors.

9.1 FUNCTIONS

A *function* can be defined as a relationship between two variables. An example is

$$y(t) = t^2 + 2$$

which is read as "y function of t equals $(t^2 + 2)$." The variable t is called the *independent variable,* and the variable y is called the *dependent variable.* For a given value of t, the rule for obtaining y is "square the value of t and add two to the result." For example, if $t = 3$

$$y(3) = (3)^2 + 2 = 11$$

which is read as "$y = 11$ when $t = 3$" or "y function of t evaluated at $t = 3$ is equal to 11." The symbol $y(t)$ is usually written as y, where it is understood that y is a function of t from the expression. Using this notation, the preceding expression can be written as

$$y = t^2 + 2$$

Some of the more important measurable quantities that characterize electric circuits are voltage, current, energy, power, electric charge, and magnetic flux, which are functions of time for a given circuit. In the analysis of electric circuits, time is the independent variable, and the measurable quantities are the dependent variables.

Consider the electric circuit in Figure 9.1 as an example. The current provided by the current source varies exponentially with time. The measurable quantities v_1 and i_1 are dependent variables, and the time t is the independent variable.

Figure 9.1 Circuit example.

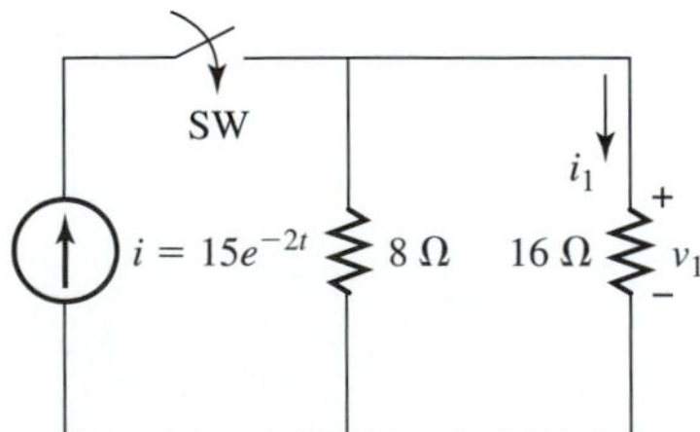

Applying Equation (4.1) for current division to the circuit with the switch closed results in

$$\begin{aligned} i_1 &= \left(\frac{8}{16 + 8}\right)(i) \\ &= \left(\frac{8}{16 + 8}\right)\{15e^{-2t}\} \\ &= 5e^{-2t}\text{ A} \end{aligned}$$

Using Ohm's law, v_1 can be expressed as a function of time as

$$\begin{aligned} v_1 &= 16i_1 \\ &= 80e^{-2t}\ \text{V} \end{aligned}$$

If the switch in Figure 9.1 is closed at $t = 0$ (starting time), then i_1 and v_1 can be determined for all values of t after the switch is closed by simply substituting the value of t into the preceding equations. For example, $\frac{1}{2}$ s after the switch is closed,

$$\begin{aligned} v_1 &= 90e^{-2(t)} \\ &= 90e^{-2(1/2)} \\ &= \frac{90}{e} \\ &= 54.6\ \text{V} \end{aligned}$$

and

$$\begin{aligned} i_1 &= 5e^{-2t} \\ &= 5e^{-2(1/2)} \\ &= \frac{5}{e} \\ &= 1.84\ \text{A} \end{aligned}$$

9.2 DIFFERENTIATION

When a mathematical operation is performed on a function, the result can be another function with a different rule relating the independent and dependent variables. For example, if

$$y = t^2 + 2$$

and y is multiplied by $2t$ to form a new function x, then

$$\begin{aligned} x &= 2t(t^2 + 2) \\ &= 2t^3 + 4t \end{aligned}$$

The rule for obtaining y for a given value of t obviously differs from the rule for obtaining x.

The *derivative* of a function is the new function that results from performing a particular mathematical operation on a function. The particular mathematical operation performed to obtain a derivative is called differentiation. All functions that are differentiable can be differentiated using the definition of a derivative. However, since differentiation using the difinition of a derivative is often a long and tedious process, tables and rules for differentiation derived from the definition can be used to obtain derivatives in problem-solving situations.

We shall present rules for differentiating the functions that generally occur in basic electric circuit analysis problems. The notation for the derivative used in this text is dy/dt, which is read as "the derivative of y with respect to t." Sometimes we use the expression for y in the derivative symbol. For example, if $y = 3t^2$, we can write

$$\frac{dy}{dt} \quad \text{as} \quad \frac{d(3t^2)}{dt}$$

which is read as "the derivative of $3t^2$ with respect to t."

The following rules allow us to obtain the derivatives of functions with respect to the independent variables.

RULE D1 The *power function* is $y = t^n$, where n is a nonzero constant, t is the independent variable and y is the dependent variable.
The *rule* for differentiating y with respect to t is

$$\frac{dy}{dt} = nt^{n-1}$$

EXAMPLE 9.1 Determine $\frac{dy}{dt}$ if $y = t^8$. ■

SOLUTION From rule D1

$$\frac{dy}{dt} = \frac{d(t^8)}{dt}$$
$$= (8)t^{8-1}$$
$$= 8t^7$$

EXAMPLE 9.2 Determine $\frac{dy}{dt}$ if $y = t^{-4}$. ■

SOLUTION From rule D1

$$\frac{dy}{dt} = \frac{d(t^{-4})}{dt}$$
$$= (-4)t^{-4-1}$$
$$= -4t^{-5}$$

RULE D2 The *constant function* is $y = K$, where K is a constant. The derivative of a constant is zero.

$$\frac{dy}{dt} = \frac{d\{K\}}{dt} = 0$$

EXAMPLE 9.3 Determine $\frac{dy}{dt}$ if $y = 17$. ■

SOLUTION From rule D2

$$\frac{dy}{dt} = \frac{d(17)}{dt}$$
$$= 0$$

RULE D3 The *exponential function* is $y = e^{nt}$, where n is a constant, e is the base of the natural logarithms, t is the independent variable, and y is the dependent variable.

The *rule* for differentiating y with respect to t is

$$\frac{dy}{dt} = ne^{nt}$$

EXAMPLE 9.4 Determine $\frac{dy}{dt}$ if $y = e^{-2t}$. ■

SOLUTION From rule D3

$$\frac{dy}{dt} = \frac{d(e^{-2t})}{dt}$$

$$= -2e^{-2t}$$

RULE D4 The *sine function* is $y = \sin(\omega t + \theta)$, where ω and θ are constants, t is the independent variable, and y is the dependent variable.

The *rule* for differentiating y with respect to t is

$$\frac{dy}{dt} = \omega \cos(\omega t + \theta)$$

EXAMPLE 9.5 Determine $\frac{dy}{dt}$ if $y = \sin\left(2t + \frac{\pi}{8}\right)$. ■

SOLUTION From rule D4

$$\frac{dy}{dt} = \frac{d\{\sin(2t + \pi/8)\}}{dt}$$

$$= 2 \cos\left(2t + \frac{\pi}{8}\right)$$

EXAMPLE 9.6 Determine $\frac{dy}{dt}$ if $y = \sin(4t)$. ■

SOLUTION From rule D4

$$\frac{dy}{dt} = \frac{d\{\sin(4t)\}}{dt}$$

$$= 4 \cos(4t)$$

RULE D5 The *cosine function* is $y = \cos(\omega t + \theta)$, where ω and θ are constants, t is the independent variable, and y is the dependent variable.

The *rule* for differentiating y with respect to t is

$$\frac{dy}{dt} = -\omega \sin(\omega t + \theta)$$

EXAMPLE 9.7 Determine $\frac{dy}{dt}$ if $y = \cos\left(4t + \frac{\pi}{6}\right)$. ■

SOLUTION From rule D5

$$\frac{dy}{dt} = \frac{d\{\cos(4t + \pi/6)\}}{dt}$$

$$= -4\sin\left(4t + \frac{\pi}{6}\right)$$

RULE D6 The *derivative of a constant times a function* is equal to the constant times the derivative of the function.

The *rule* for differentiation is

$$\frac{d\{Ky\}}{dt} = K\frac{dy}{dt}$$

where K is a constant, t is the independent variable, and y is the dependent variable.

EXAMPLE 9.8 Determine the derivative of $12y$ if $y = -3t^3$. ■

SOLUTION From rule D6

$$\frac{d(12y)}{dt} = 12\frac{dy}{dt} = 12\frac{d\{-3t^3\}}{dt}$$

Applying rule D6 a second time results in

$$\frac{d(12y)}{dt} = (12)(-3)\frac{d(t^3)}{dt}$$

From rule D1

$$\frac{d(12y)}{dt} = (-36)(3t^2)$$

$$= -108t^2$$

RULE D7 The *derivative of the sum of two functions* equals the sum of the derivatives of the functions.

The *rule* for differentiation is

$$\frac{d(y_1 + y_2)}{dt} = \frac{dy_1}{dt} + \frac{dy_2}{dt}$$

where t is the independent variable and y_1 and y_2 are dependent variables.

EXAMPLE 9.9 Determine $\frac{dy}{dt}$ if $y = 6t^4 + 2e^{-2t}$. ■

SOLUTION From rule D7

$$\frac{dy}{dt} = \frac{d(6t^4 + 2e^{-2t})}{dt}$$

$$= \frac{d(6t^4)}{dt} + \frac{d(2e^{-2t})}{dt}$$

From rule D6

$$= 6\frac{d(t^4)}{dt} + 2\frac{d(e^{-2t})}{dt}$$

From rules D1 and D3

$$\frac{dy}{dt} = (6)(4)t^{(4-3)} + (2)(-2)e^{-2t}$$

$$= 24t^3 - 4e^{-2t}$$

RULE D8 The *derivative of the difference of two functions* equals the difference of the derivatives of the two functions.

The *rule* for differentiation is

$$\frac{d(y_1 - y_2)}{dt} = \frac{dy_1}{dt} - \frac{dy_2}{dt}$$

where t is the independent variable and y_1 and y_2 are dependent variables.

Rules D7 and D8 can be extended and/or combined to include more than two functions, as illustrated in the following example.

EXAMPLE 9.10 Determine $\frac{dy}{dt}$ if $y = 10 - 8\sin(4t) + t^2$. ■

SOLUTION From rules D7 and D8

$$\frac{dy}{dt} = \frac{d\{10 - 8\sin(4t) + t^2\}}{dt}$$

$$= \frac{d(10)}{dt} - \frac{d\{8\sin(4t)\}}{dt} + \frac{d\{t^2\}}{dt}$$

From rules D1, D2, D4, and D6

$$\frac{dy}{dt} = 0 - (8)\{(4)\cos(4t)\} + 2t$$

$$= -32\cos(4t) + 2t$$

RULE D9 The *derivative of the product of two functions* equals the first function times the derivative of the second function plus the second function times the derivative of the first function.

The rule for differentiation is

$$\frac{d(y_1 y_2)}{dt} = y_1 \frac{dy_2}{dt} + y_2 \frac{dy_1}{dt}$$

where t is the independent variable and y_1 and y_2 are dependent variables.

EXAMPLE 9.11 Determine $\frac{dy}{dt}$ if $y = -2e^{-t}\cos(4t)$. ■

SOLUTION Let $y_1 = -2e^{-t}$ and $y_2 = \cos(4t)$. From rule D9

$$\begin{aligned}\frac{d(y_1 y_2)}{dt} &= y_1 \frac{dy_2}{dt} + y_2 \frac{dy_1}{dt} \\ &= (-2e^{-2t})\frac{dy_2}{dt} + (\cos 4t)\frac{dy_1}{dt} \\ &= (-2e^{-2t})\frac{d(\cos 4t)}{dt} + (\cos 4t)\frac{d(-2e^{-2t})}{dt}\end{aligned}$$

From rule D6

$$\frac{d(y_1 y_2)}{dt} = (-2e^{-2t})\frac{d(\cos 4t)}{dt} + (\cos 4t)(-2)\frac{d(e^{-2t})}{dt}$$

From rules D3 and D5

$$\begin{aligned}\frac{d(y_1 y_2)}{dt} &= (-2e^{-2t})(-4\sin 4t) + (\cos 4t)(-2)(-2e^{-2t}) \\ &= 8e^{-2t}\sin 4t + 4e^{-2t}\cos 4t\end{aligned}$$

RULE D10 The *derivative of the quotient of two functions* equals the denominator times the derivative of the numerator minus the numerator times the derivative of the denominator, all divided by the denominator squared.

The *rule* for differentiation is

$$\frac{d\left\{\frac{y_1}{y_2}\right\}}{dt} = \frac{(y_2)\frac{dy_1}{dt} - (y_1)\frac{dy_2}{dt}}{(y_2)^2}$$

where t is the independent variable and y_1 and y_2 are dependent variables.

EXAMPLE 9.12 Determine $\frac{d\{y_1/y_2\}}{dt}$ where $y_1 = 4$ and $y_2 = t^2$. ■

SOLUTION From rule D10

$$\begin{aligned}\frac{d\{y_1/y_2\}}{dt} &= \frac{y_2\frac{dy_1}{dt} - y_1\frac{dy_2}{dt}}{(y_2)^2} \\ &= \frac{t^2\left\{\frac{d(4)}{dt}\right\} - 4\left\{\frac{d(t^2)}{dt}\right\}}{(t^2)^2}\end{aligned}$$

From rules D2 and D3

$$\frac{d(y_1/y_2)}{dt} = \frac{t^2(0) - (4)(2t)}{t^4}$$

$$= \frac{-8t}{t^4}$$

$$= -8t^{-3}$$

An alternative method for determining $d\{y_1/y_2\}/dt$ is to recognize that

$$\frac{y_1}{y_2} = \frac{4}{t^2}$$

$$= 4t^{-2}$$

and from rules D3 and D6,

$$\frac{d(4t^{-2})}{dt} = (4)(-2t^{-2-1})$$

$$= -8t^{-3}$$

9.3 HIGHER DERIVATIVES

When we differentiate y with respect to t, we obtain another function called the derivative of y with respect to t, which is symbolized as dy/dt. If we differentiate the function dy/dt with respect to t the result is yet another function called the *second derivative* of y with respect to t, and it is symbolized as d^2y/dt^2. If we differentiate the function d^2y/dt^2, we obtain the *third derivative* of y with respect to t which is written as d^3y/dt^3. This process can be continued, but only the first and second derivatives are of primary concern in basic circuit analysis work.

EXAMPLE 9.13 Determine $\frac{dy}{dt}, \frac{d^2y}{dt^2}$, and $\frac{d^3y}{dt^3}$ if $y = -4t^6 + e^{-t}$. ■

SOLUTION From rules D1, D3, D6, and D7

$$\frac{dy}{dt} = \frac{d\{-4t^6 + e^{-t}\}}{dt}$$

$$= \frac{d(-4t^6)}{dt} + \frac{d(e^{-t})}{dt}$$

$$= -4\frac{d(t^6)}{dt} + \frac{d(e^{-t})}{dt}$$

$$= -4(6)t^{6-1} + (-1)e^{-t}$$

$$= -24t^5 - e^{-t}$$

$$\frac{d^2y}{dt^2} = \frac{d\{-24t^5 - e^{-t}\}}{dt}$$

$$= \frac{d(-24t^5)}{dt} - \frac{d(e^{-t})}{dt}$$

$$= -24\frac{d(t^5)}{dt} - \frac{d(e^{-t})}{dt}$$

$$= (-24)(5)t^{5-1} - (-1)(e^{-t})$$

$$= -120t^4 + e^{-t}$$

$$\frac{d^3y}{dt^3} = \frac{d\{-120t^4 + e^{-t}\}}{dt}$$

$$= \frac{d\{-120t^4\}}{dt} + \frac{d\{e^{-t}\}}{dt}$$

$$= -120\frac{d(t^4)}{dt} + \frac{d(e^{-t})}{dt}$$

$$= (-120)(4)t^{4-1} + (-1)(e^{-t})$$

$$= -480t^3 - e^{-t}$$

9.4 RATE OF CHANGE WITH RESPECT TO TIME

The first derivative of any function y with respect to time represents the rate of change in y with respect to time. For example, if $y = 3t^2$, then y is changing at the rate $6t$ with respect to time, because $dy/dt = 6t$. If t is measured in seconds and y equals the number of gallons of water in a tank at any time t, then the amount of water in the tank is changing at the rate of 12 gal/s at the instant $t = 2$ s.

Because the measurable quantities associated with electric circuits can be expressed mathematically as functions of time, differentiation with respect to time allows us to determine the rate of change for the measurable electrical quantities at any instant in time. This concept is most important when analyzing electric circuits containing inductors and/or capacitors as illustrated in a later chapter.

9.5 SLOPE OF A TANGENT LINE RELATED TO DIFFERENTIATION

The first derivative of a function y with respect to time represents the slope of a line tangent to the y-versus-t curve, as shown in Figure 9.2. For the curve

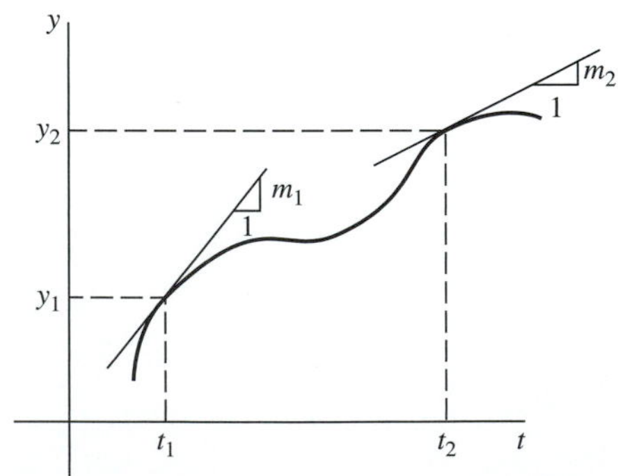

Figure 9.2 Differentiation related to tangent slope.

in Figure 9.2, $dy(t_1)/dt$, which is read as "the derivative of y with respect to t evaluated at $t = t_1$," is equal to the slope of the tangent line drawn at point (t_1, y_1). That is, $dy(t_1)/dt = m_1$; in a similar manner, $dy(t_2)/dt = m_2$.

For example, if $y = 4t^2$, then from rules D1 and D6

$$\frac{dy}{dt} = 8t$$

and the slope m of a tangent line drawn at $t = 2$ as shown in Figure 9.3 is

$$m = \frac{dy(2)}{dt} = 8(2)$$
$$= 16$$

Figure 9.3 Slope of a tangent line.

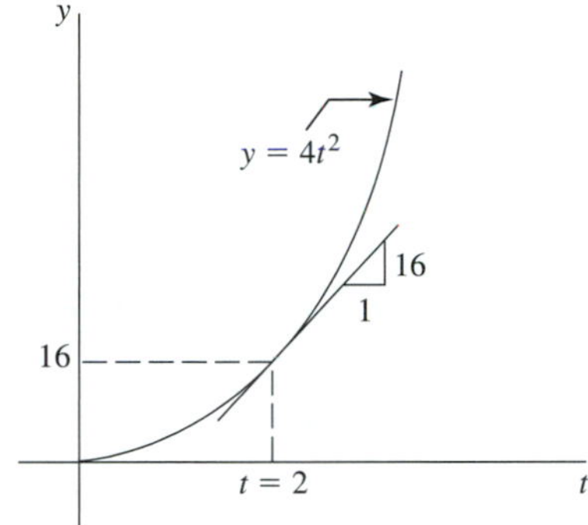

In electric circuits containing inductors and/or capacitors, it is possible to determine voltages and currents using the tangent slope property, as illustrated in a later chapter.

9.6 MAXIMUM AND MINIMUM VALUES

Consider the graph of y versus t shown in Figure 9.4. The tangent line drawn at (t_1, y_1) has zero slope, and therefore $d\{y(t_1)\}/dt$ (derivative of y with respect to t evaluated at t_1) is equal to zero. An inspection of the y-versus-t plot in Figure 9.4 reveals that y_1 is a maximum value of y relative to nearby values of y.

Figure 9.4 Maximum power.

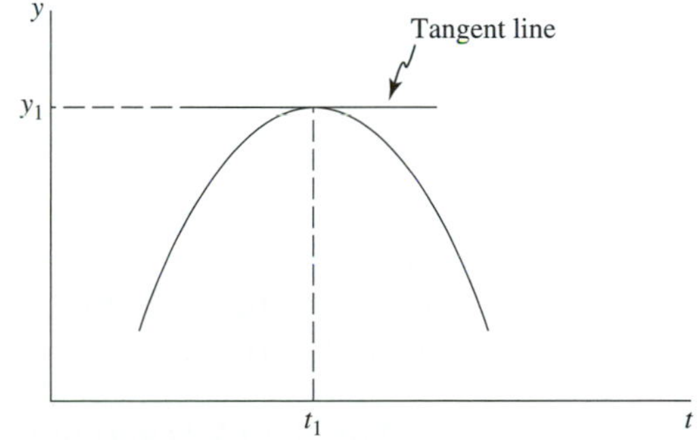

It follows that if we differentiate y with respect to t and set the derivative equal to zero, the solution of the resulting equation will yield the value of t where y is a maximum. Substituting the resulting value of t into $y(t)$ will usually result in the maximum value of y.

Consider the following example.

EXAMPLE 9.14 y is given as $y = -t^2 + 9t - 18$. Determine the maximum value of y. ■

SOLUTION Using rules D1, D2, D7, and D8

$$\frac{dy}{dt} = -2t + 9$$

Setting $\frac{dy}{dt} = 0$ results in

$$-2t + 9 = 0$$
$$t = 4.5$$

and the maximum value of y occurs at $t = 4.5$. Substituting $t = 4.5$ into the expression for y results in

$$y_{max} = -(4.5)^2 + 9(4.5) - 18$$
$$= 2.25$$

A sketch of y versus t is shown in Figure 9.5.

Figure 9.5

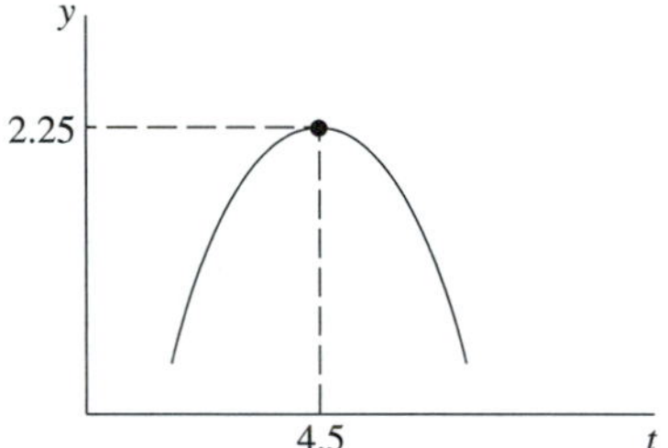

Notice that the slope of a tangent line drawn to the y-versus-t plot at $t = 4.5$ has a slope equal to zero.

The procedure just presented for determining an extreme point that is the maximum value of a function can also yield a minimum value or an extreme point that is neither a maximum nor a minimum value. Although there are methods for determining the nature of an extreme point, we shall not investigate them because we are only minimally interested in the determination of extreme points in basic electric circuit analysis.

9.7 ANTIDERIVATIVES

In circuit analysis we also encounter a mathematical quantity called an antiderivative. The antiderivative of a function y with respect to t is another function g such that the derivative of g with respect to t is equal to y. That is, if g is an antiderivative of y, then $dg/dt = y$.

EXAMPLE 9.15 Determine the antiderivative of y if $y = 2t$. ■

SOLUTION Because the derivative of t^2 equals $2t$,

$$g = t^2$$

is an antiderivative of y. Using rules D2 and D7, we observe that $(t^2 + 10)$, $(t^2 + 101)$, and, in general, $(t^2 + K)$, where K is any constant, are all antiderivatives of y because $dg/dt = y$ in each case. This occurs because the derivative of a constant is equal to zero.

9.8 INDEFINITE INTEGRALS

The mathematical process of determining an antiderivative of a function is called *integration* and is symbolized as

$$\int y\,dt = g + K$$

which is read "the indefinite integral of the function y with respect to t equals g (the antiderivative of y) plus an arbitrary constant K." It follows from the definition of the antiderivative that

$$\frac{d(g + K)}{dt} = y$$

or

$$\frac{d\{\int y\,dt\}}{dt} = y \tag{9.1}$$

The quantity

$$\int y\,dt$$

is called the *indefinite integral* of the function y with respect to t, and y is called the *integrand* of the indefinite integral. In determining an indefinite integral, we must find the antiderivative of the integrand.

Consider the following examples.

EXAMPLE 9.16 Determine the indefinite integral of the function $y = 20$ with respect to t. ■

SOLUTION

$$\int y\,dt = \int 20\,dt = 20t + K$$

because

$$\frac{d(20t + K)}{dt} = 20$$

EXAMPLE 9.17 Determine the indefinite integral of the function $y = -6t^3$ with respect to t. ■

SOLUTION

$$\begin{aligned}\int y\,dt &= \int (-6t^3)\,dt \\ &= -(\tfrac{3}{2})t^4 + K\end{aligned}$$

because

$$\frac{d\{-(\frac{3}{2})t^4 + K\}}{dt} = -6t^3$$

9.9 INTEGRATION

The process of obtaining an indefinite integral is called *integration*. There are rules for integration, just as there are rules for differentiation. Rules for integrating the functions that generally occur in basic electric circuit analysis are listed next.

RULE I1 The *power function* is $y = t^n$, where n is a constant, t is the independent variable and y is the dependent variable.

The *rule* for integrating t^n with respect to t is

$$\int t^n \, dt = \left\{\frac{1}{n+1}\right\}(t^{n+1}) + K_1$$

where K_1 is an arbitrary constant.

EXAMPLE 9.18 Determine the indefinite integral of $y = t^4$ with respect to t. ■

SOLUTION From rule I1

$$\begin{aligned}\int y \, dt &= \int (t^4) \, dt \\ &= \left(\frac{1}{4+1}\right)(t^{4+1}) + K_1 \\ &= \left(\frac{1}{5}\right) t^5 + K_1\end{aligned}$$

Notice that

$$\frac{d\{(\frac{1}{5})t^5 + K_1\}}{dt} = t^4$$

RULE I2 The *constant function* is $y = K$, where K is a constant.

The *rule* for integrating K with respect to t is

$$\int K \, dt = Kt + K_1$$

where K_1 is an arbitrary constant.

EXAMPLE 9.19 Determine the indefinite integral of $y = 17$ with respect to t. ■

SOLUTION From rule I2

$$\int 17\,dt = 17t + K_1$$

Notice that

$$\frac{d\{17t + K_1\}}{dt} = 17$$

RULE I3 The *exponential function* is $y = e^{nt}$, where n is a constant, e is the base of the natural logarithms, t is the independent variable, and y is the dependent variable.

The *rule* for integrating e^{nt} with respect to t is

$$\int e^{nt} = \left(\frac{1}{n}\right)e^{nt} + K_1$$

where K_1 is an arbitrary constant.

EXAMPLE 9.20 Determine the indefinite integral of $y = e^{-2t}$ with respect to t. ■

SOLUTION From rule I3

$$\int e^{-2t}\,dt = \left(-\frac{1}{2}\right)e^{-2t} + K_1$$

$$= \left(-\frac{1}{2}\right)e^{-2t} + K_1$$

Notice that

$$\frac{d\{(-\frac{1}{2})\,e^{-2t} + K_1\}}{dt} = e^{-2t}$$

RULE I4 The *sine function* is $y = \sin(\omega t + \theta)$, where ω and θ are constants, t is the independent variable, and y is the dependent variable.

The *rule* for integrating $\sin(\omega t + \theta)$ with respect to t is

$$\int \sin\,(\omega t + \theta)\,dt = -\left\{\left(\frac{1}{\omega}\right)\cos(\omega t + \theta)\right\} + K_1$$

where K_1 is an arbitrary constant.

EXAMPLE 9.21 Determine the indefinite integral of $y = \sin(6t + \pi/2)$ with respect to t. ■

SOLUTION From rule I4

$$\int \sin\left(6t + \frac{\pi}{2}\right) dt = -\left\{\left(\frac{1}{6}\right)\cos\left(6t + \frac{\pi}{2}\right)\right\} + K_1$$
$$= \left(-\frac{1}{6}\right)\cos\left(6t + \frac{\pi}{2}\right) + K_1$$

Notice that

$$\frac{d\{(-\frac{1}{6})\cos(6t + \pi/2) + K_1\}}{dt} = \sin\left(6t + \frac{\pi}{2}\right)$$

RULE I5 The *cosine function* is $y = \cos(\omega t + \theta)$, where ω and θ are constants, t is the independent variable, and y is the dependent variable.

The *rule* for integrating $\cos(\omega t + \theta)$ with respect to t is

$$\int \cos(\omega t + \theta)\, dt = \left(\frac{1}{\omega}\right)\sin(\omega t + \theta) + K_1$$

where K_1 is an arbitrary constant.

EXAMPLE 9.22 Determine the indefinite integral of $y = \cos(8t + \pi)$ with respect to t. ■

SOLUTION From rule I5

$$\int \cos(8t + \pi)\, dt = \frac{1}{8}\sin(8t + \pi) + K_1$$

Notice that

$$\frac{d\{(\frac{1}{8})\sin(8t + \pi) + K_1\}}{dt} = \cos(8t + \pi)$$

RULE I6 The *indefinite integral of a constant times a function* is equal to the constant times the indefinite integral of the function.

The *rule* for integration is

$$\int K_y\, dt = K\int y\, dt$$

where t is the independent variable, y is the dependent variable, and K is a constant.

EXAMPLE 9.23 Determine the indefinite integral of $y = 6e^{-7t}$ with respect to t. ■

SOLUTION From rule I6

$$\int 6e^{-7t}\, dt = 6\int e^{-7t}\, dt$$

and from rule I3

$$\int y\,dt = 6\int e^{-7t}\,dt$$
$$= 6\left\{\left(-\frac{1}{7}\right)e^{-7t}\right\}$$
$$= -\left(\frac{6}{7}\right)e^{-7t}$$

Notice that

$$\frac{d\{-(\frac{6}{7})e^{-7t}\}}{dt} = 6e^{-7t}$$

RULE I7 The *indefinite integral of the sum or difference of two functions* equals the sum or difference of the indefinite integrals of the functions.

The *rule* for integration is

$$\int (y_1 \pm y_2)\,dt = \int y_1\,dt \pm \int y_2\,dt + K_1$$

where t is the independent variable, y_1 and y_2 are dependent variables, and K_1 is an arbitrary constant.

EXAMPLE 9.24 Determine the indefinite integral of $(16 + 2t^3)$ with respect to t. ■

SOLUTION From rule I7

$$\int (16 + 2t^3)\,dt = \int 16\,dt + \int 2t^3\,dt + K_1$$

From rules I1, I2, and I6

$$\int (16 + t^3)\,dt = 16t + (2)\left(\frac{1}{3+1}\right)t^{3+1} + K_1$$
$$= 16t + \left(\frac{1}{2}\right)t^4 + K_1$$

Notice that

$$\frac{d\left\{16t + \left(\frac{1}{2}\right)t^4 + K_1\right\}}{dt} = 16 + 2t^3$$

9.10 DEFINITE INTEGRALS

As previously stated, the indefinite integral of a function y with respect to t can be written as

$$\int y(t)\,dt = g(t) + K$$

where $g(t)$ is the antiderivative of $y(t)$ and K is an arbitrary constant. Recall that $y(t)$ and $g(t)$ are read as "y function of t" and "g function of t," respectively.

The quantity

$$g(t_2) - g(t_1)$$

where $t_2 \geq t_1$, is called the *definite integral* of y with respect to t between t_1 and t_2 and is symbolized as

$$\int_{t_1}^{t_2} y(t)\, dt = g(t_2) - g(t_1) \tag{9.2}$$

Recall that $g(t_2)$ and $g(t_1)$ are read as "function g evaluated at time t_2" and "function g evaluated at time t_1," respectively.

EXAMPLE 9.25 Determine the definite integral of $y = 2t^3$ with respect to t between $t = 1$ and $t = 2$. ■

SOLUTION From rules I1 and I6, the antiderivative of $2t^3$ is

$$g = \left(\frac{1}{2}\right)t^4$$

From Equation (9.2)

$$\begin{aligned}\int_1^2 y\, dt &= g(2) - g(1) \\ &= \left(\frac{1}{2}\right)t^4\Big|_{t=2} - \left(\frac{1}{2}\right)t^4\Big|_{t=1} \\ &= \left(\frac{1}{2}\right)(2)^4 - \left(\frac{1}{2}\right)(1)^4 \\ &= \frac{15}{2}\end{aligned}$$

9.11 AREA RELATED TO INTEGRATION

Consider a function $y(t)$ in graphical form, as shown in Figure 9.6. The definite integral of y with respect to t from t_1 to t_2 is equal to the area under the y curve between t_1 and t_2, as indicated by the cross-hatching in Figure 9.6. That is,

$$\int_{t_1}^{t_2} y\, dt = A$$

Figure 9.6 Area under a curve.

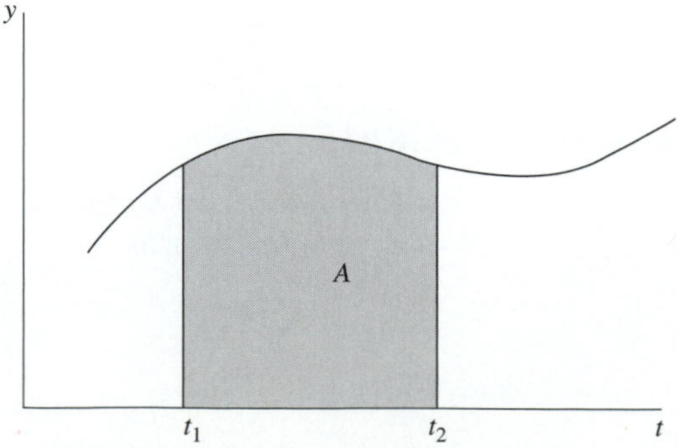

If the curve extends below the t axis, as indicated in Figure 9.7, the definite integral of y with respect to t from t_1 to t_2 is equal to the net area between

Figure 9.7 Negative area under a curve.

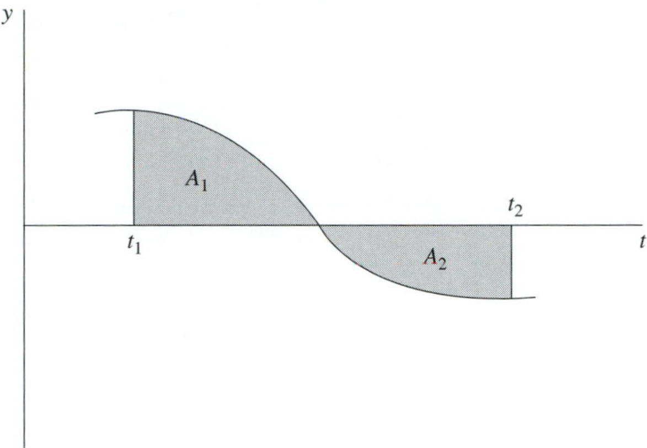

the curve and the t axis, with area below the t axis considered as negative area. That is,

$$\int_{t_1}^{t_2} y\,dt = A_1 - A_2$$

for the curve in Figure 9.7.

EXAMPLE 9.26 Determine the area A shown in Figure 9.8. ■

Figure 9.8

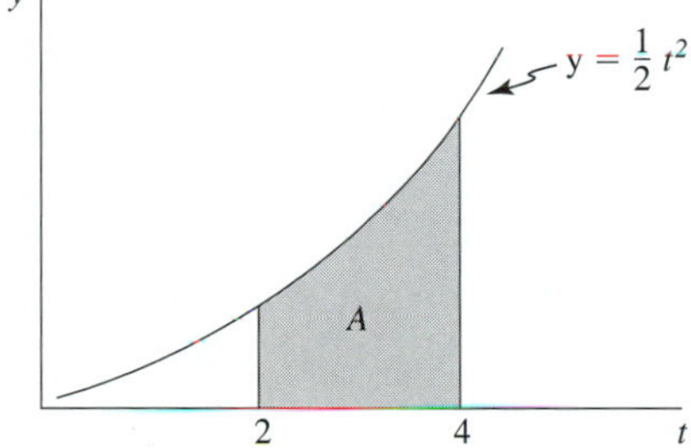

SOLUTION From rules I1 and I6, the antiderivative of $\frac{1}{2}t^2$ is

$$\left(\frac{1}{6}\right)t^3$$

and

$$\begin{aligned}\text{area } A &= \int_2^4 \left(\frac{1}{2}\right)t^2\,dt \\ &= \left(\frac{1}{6}\right)t^3\bigg|_{t=4} - \left(\frac{1}{6}\right)t^3\Big|_{t=2} \\ &= \frac{64}{6} - \frac{8}{6} \\ &= \frac{28}{3}\end{aligned}$$

Associating the area under a curve with the definite integral is useful for determining current, voltage, power, and energy in electric circuits, as illustrated in later chapters.

9.12 EXERCISES

1. Determine the derivative of y with respect to t for the following functions.
(a) $y = t^{-4}$ (b) $y = 8t^3$ (c) $y = -4t^{-6}$ (d) $y = -15t$

2. Determine the derivative of y with respect to t for the following functions.
(a) $y = -50$ (b) $y = 4e^{-t}$ (c) $y = -10e^{-2t}$ (d) $y = 6e^{4t}$

3. Determine the derivative of y with respect to t for the following functions.
(a) $y = 14e^{-30t}$ (b) $y = -2\sin(20t + 30°)$
(c) $y = 40\sin(t - 45°)$ (d) $y = 8\sin(t)$

4. Determine the derivative of y with respect to t for the following functions.
(a) $y = -100\cos(4t + 20°)$ (b) $y = 9\cos(t + 180°)$
(c) $y = -12\cos(t)$

5. Determine the derivative of y with respect to t for the following functions.
(a) $y = 6t^3 + e^{-t}$ (b) $y = 16 - t^{-4}$
(c) $y = 12e^{2t} - 100$

6. Determine the derivative of y with respect to t for the following functions.
(a) $y = -4e^{2t} + 16e^{t}$ (b) $y = 40 + 20t - 7e^{-2t}$
(c) $y = 20\sin(3t) - e^{-4t} + 100$

7. Determine the derivative of y with respect to t for the following functions.
(a) $y = 6e^{-t}\sin(t)$ (b) $y = -t^2\cos(2t + 40°)$
(c) $y = 10\sin(3t)\cos(t + 20°)$

8. Determine the derivative of y with respect to t for the following functions.
(a) $y = (2e^{-4t})/\{\cos(t)\}$ (b) $y = -20/(4t^2)$
(c) $y = (14t^{-4})/4e^{3t}$

9. An object travels along a line so that its position x from the starting point is $x = t^2 + 2t + 4$ feet after t seconds. What is the instantaneous velocity of the object after 4 s.

10. If the position x from the starting point is changed to $x = -t^3 + 8t^2 + 5t + 6$ in Problem 3, find the instantaneous velocity of the object after 2 s.

11. Determine the slope of a tangent line to the curve $y = x^2 - 2x + 1$ at the points $x = 0$, $x = 1$, and $x = 4$.

12. Determine the slope of a tangent line to the curve $y = 2x^3 - x^2 + 2x + 6$ at $x = 0$, $x = 2$, and $x = 4$.

13. Determine the indefinite integrals.
(a) $\int 10\,dt$ (b) $\int 6t^6\,dt$
(c) $\int 3\sin(2t + 60°)\,dt$ (d) $\int -40\sin(5t)\,dt$

14. Determine the indefinite integrals.
(a) $\int -20t^{-2}\,dt$ (b) $\int 16e^{-2t}\,dt$
(c) $\int 7e^{4t}\,dt$ (d) $\int 14\cos(4t - 30°)\,dt$
(e) $\int -4\cos(t)\,dt$ (f) $\int 5\sin(t + 20°)\,dt$

15. Determine the indefinite integrals.

(a) $\int \{6e^{-2t} + 20\sin(t)\}dt$ (b) $\int \{-4\cos(4t) + t^2\}dt$

16. Determine the indefinite integrals.

(a) $\int \{5 - t^5\}dt$ (b) $\int \{t^4 + \cos(2t) + 3e^{-8t}\}dt$

(c) $\int \{-20 + 6\sin(3t - 40°)\}dt$

17. Determine the definite integrals.

(a) $\int_0^4 12\,dt$ (b) $\int_{-2}^{8} -4e^{-2t}dt$

(c) $\int_0^{10} 7\sin(3t - 40°)dt$

18. Determine the definite integrals.

(a) $\int_{-8}^{-6} -t^4\,dt$ (b) $\int_1^4 -20 - 6\sin(2t + 40°)dt$

19. Determine the area under the curve $y = 2x + 4$ from $x = 0$ to $x = 4$.

20. Determine the area under the curve $y = 4x^2 + 2$ from $x = 1$ to $x = 4$.

CHAPTER 10

Capacitors

After completing this chapter the student should be able to

* describe a basic capacitor
* describe the charge-voltage relationship for capacitors
* describe the current-voltage relationship for capacitors
* combine capacitors in series and/or parallel
* use the current and voltage division formulas to determine currents and voltages in capacitor circuits

10.1 THE BASIC CAPACITOR

A basic *capacitor* consists of two conductors separated by a dielectric. A *dielectric* is a material (such as rubber, air, glass, mica) that is a nonconductor of electric charge and acts as an insulator. A basic capacitor is illustrated in Figure 10.1.

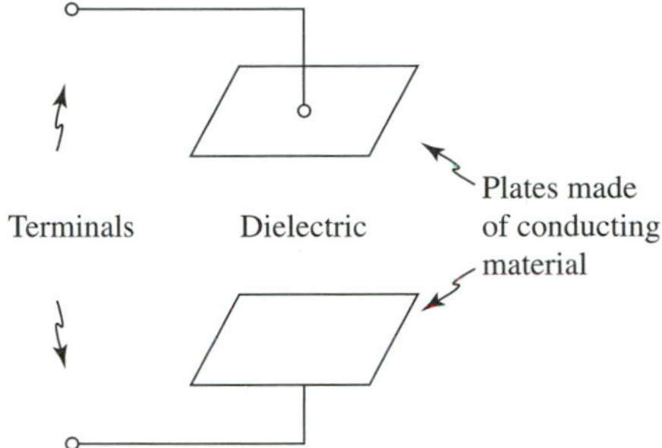

Figure 10.1 Basic capacitor.

The measure of the ability of a device to perform as a capacitor is called capacitance. *Capacitance* exists whenever two conductors are separated by a dielectric. If the capacitance is unintended and occurs inadvertently because of the physical structure of the circuit, it is called stray capacitance. If a device is deliberately constructed as a circuit element with the capacitance intended, it is called a capacitor. The capacitance of a capacitor is a measure of the capacitor's ability to store energy in the form of an electric field.

The voltage source applied across the terminals of the capacitor, as shown in Figure 10.2, has the potential for moving charge from one plate of the capacitor to the other. Because electric charge cannot flow from plate to plate through an ideal dielectric, it must flow through the circuitry that is connected externally to the capacitor. In this case the external circuit is the voltage source.

Consider the capacitor in Figure 10.2, where a constant voltage v is applied across the two plates, as shown. The voltage source v will attract free electrons from the top plate and cause them to flow through the voltage source

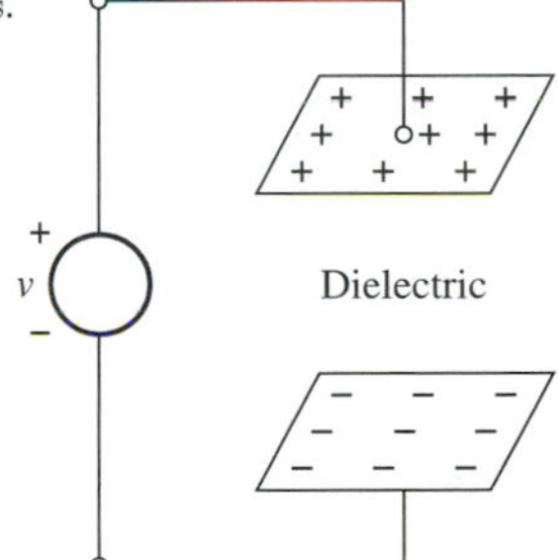

Figure 10.2 Charge on capacitor plates.

to the bottom plate. The top plate will become positively charged (absence of electrons) and the bottom plate will become negatively charged (excess of electrons). The accumulation of charge on the plates creates a voltage across the plates that is opposite in polarity to the applied voltage v. The charge flow will continue until the voltage caused by the accumulation of charge on the plates equals v. The total charge on both plates is always zero, because the charge deposited on one plate comes from the other plate.

10.2 ELECTRIC FIELD

The accumulated charge on the capacitor plates causes an electric field to exist between the plates. The electric field is distinguishable by the force that it causes on a charged particle placed in its presence. If a positive charge q_T is placed between the capacitor plates, a force F is exerted on the charge by the electric field, as shown in Figure 10.3.

Figure 10.3 Electric field.

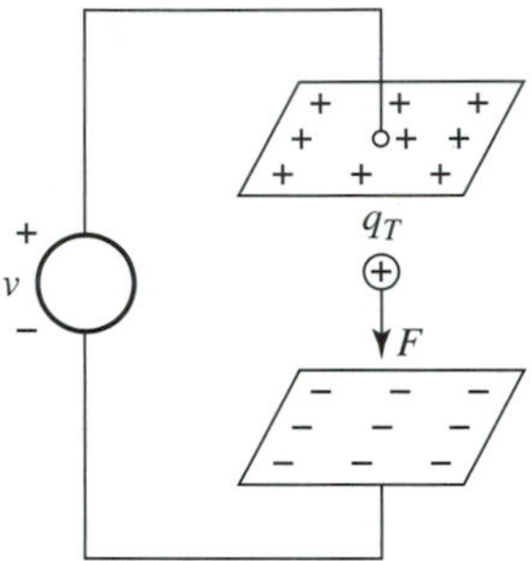

The *electric field strength* (also called the electric field intensity) is defined as the force per unit of charge and can be written as

$$\mathscr{E} = \frac{F}{q_T}$$

where $\mathscr{E}$ (Greek letter epsilon) is the symbol for electric field strength and has units of newtons per coulomb. The direction of the electric field is the direction of the force on a positive charge placed between the capacitor plates. Because a positive charge placed between the capacitor plates will be repelled by the positively charged top plate and attracted by the negatively charged bottom plate, the electric field strength $\mathscr{E}$ is directed downward, as shown in Figure 10.4.

Figure 10.4 Electric field direction.

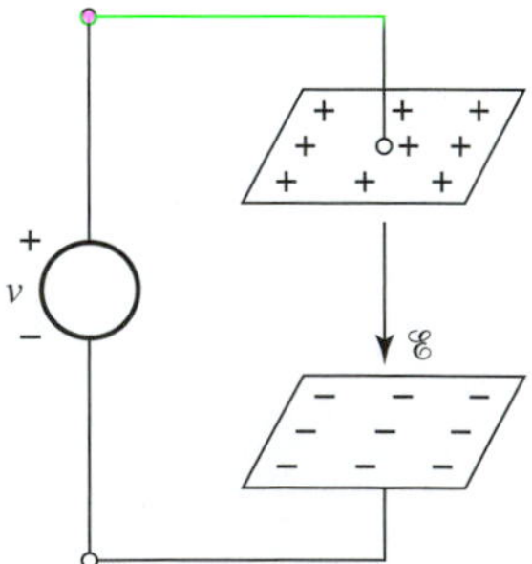

10.3 ELECTRIC FLUX

The charged plates of a capacitor give rise to lines of electric flux, which emanate at the positive charges on the top plate and terminate on the negative charges on the bottom plate, as shown in Figure 10.5.

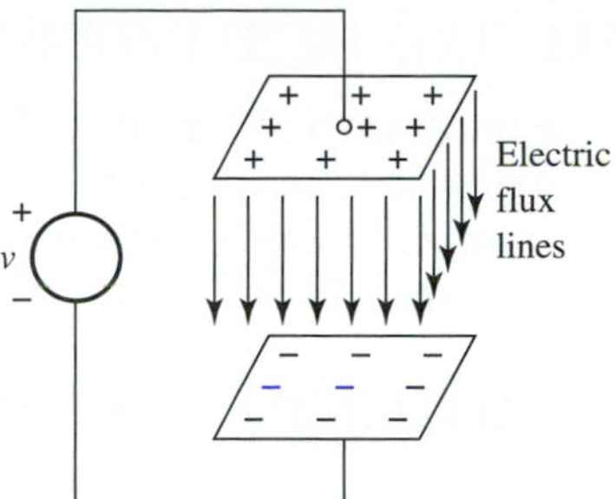

Figure 10.5 Flux lines.

Electric flux lines are used to characterize and help describe the presence of an electric field. The combined lines of flux between the plates is called the electric flux and is equal to the charge causing the flux. The electric flux is symbolized by the Greek letter psi (ψ) and is equal to the charge on each plate of the capacitor. That is, if the charge on each capacitor plate is q,

$$\psi = q$$

where q and ψ are both expressed in coulombs.

10.4 FLUX DENSITY

The electric flux between the plates of the capacitor divided by the cross-sectional area through which the electric flux passes is called the *flux density* and can be written as

$$D = \frac{\psi}{A}$$

where D has units of coulombs per square meter, ψ is measured in coulombs, and A (the plate area) is measured in square meters. Because $\psi = q$, D can also be expressed as

$$D = \frac{q}{A}$$

The electric field strength $\mathcal{E}$ and the flux density D both increase as the charge on the capacitor plates increases, and the two are proportional. The proportionality can be written as

$$D \propto \mathcal{E}$$

10.5 PERMITTIVITY

The constant of proportionality relating D and ε is called the permittivity of the dielectric material between the capacitor plates and is designated by the lowercase Greek letter epsilon (ϵ). Employing the proportionality constant, D can be written as

$$D = \epsilon\mathcal{E}$$

10.6 UNIFORMITY OF ELECTRIC FIELD INTENSITY

For a constant voltage, the electric field intensity is the same at all points in the space between the capacitor plates and is said to be uniform. If the voltage across the capacitor varies with time, then q and $\mathscr{E}$ will vary with time, but $\mathscr{E}$ is uniform at any instant.

10.7 RELATING ELECTRIC FIELD INTENSITY TO VOLTAGE

Consider the movement of a positive charge q_T from the positively charged plate to the negatively charged plate of a capacitor with electric field strength $\mathscr{E}$. constant voltage v, and plates separated by d meters, as shown in Figure 10.6.

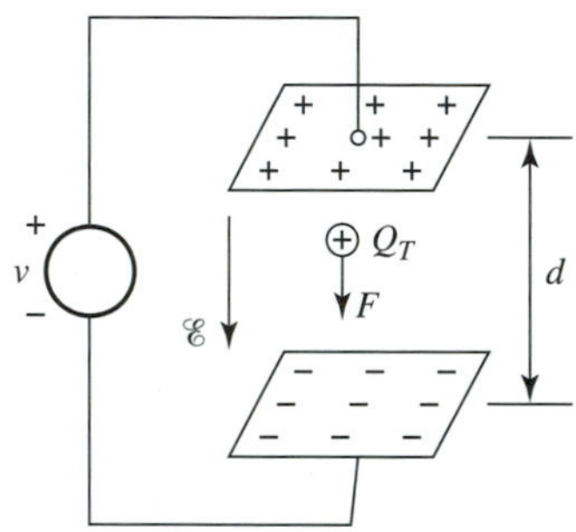

Figure 10.6 Electric field intensity.

The electric field between the plates is uniform and will cause a constant force F to act on q_T. The work done on the charge by the electric field $\mathscr{E}$ in moving the charge from plate to plate is

$$w = Fd$$

and because $F = \mathscr{E}q_T$,

$$w = \mathscr{E}q_T d$$

From the definition of the voltage between two points, 1 V of voltage across the capacitor plates causes 1 J (newton-meter) of energy to act on each unit of positive charge while it is moving from the top plate to the bottom plate. Because v volts are applied across the capacitor plates, the work done on q_T by the constant voltage source v while q_T is moving from the top to the bottom plate can be expressed as

$$w = vq_T$$

Equating the equal but different expressions for the work done on the positive charge q_T while it moves between the plates of the capacitor results in

$$vq_T = \mathscr{E}q_T d$$

and

$$\mathscr{E} = \frac{v}{d}$$

where v is in volts, d is in meters, and $\mathscr{E}$ is in volts per meter.

10.8 CHARGE-VOLTAGE RELATIONSHIP

The charge q on the plates of a capacitor is proportional to the voltage v applied across the plates and can be written as

$$q \propto v$$

The constant of proportionality is the capacitance of the capacitor and depends on the physical aspects of the capacitor (plate area, separation of plates, and the permittivity of the dielectric). The symbol for capacitance is C and the preceding equation in terms of the proportionality constant can be written as

$$q = Cv \tag{10.1}$$

where the charge q is in coulombs, v is in volts, and the capacitance C is in farads. One *farad* is equal to 1 C/V and is named for the British physicist Michael Faraday (1741–1867). Because a 1-F capacitor is physically much larger than the capacitors that we normally encounter, capacitance is usually expressed in microfarads or millifarads or in terms of another reduction prefix.

10.9 PERMITTIVITY AND CAPACITANCE

Consider the parallel-plate capacitance in Figure 10.1. Because $\epsilon = D/\mathscr{E}$, $D = q/A$, $\mathscr{E} = v/d$, and $q = Cv$, as developed previously, it follows that

$$\epsilon = \frac{D}{\mathscr{E}} = \frac{q/A}{v/d} = \frac{qd}{vA} = \frac{Cvd}{vA} = \frac{Cd}{A}$$

and

$$C = \epsilon\frac{A}{d}$$

where ϵ is the permittivity of the capacitor dielectric, A is the area of the plates, and d is the distance between the plates. Since the units for C, A, and d are farads, square meters, and meters, respectively, the units for ϵ are farads per meter.

10.10 RELATIVE PERMITTIVITY

The permittivity of a vacuum is denoted by ϵ_o and is equal to 8.85×10^{-12} F/m (farads/meter). A dielectric with permittivity ϵ has a relative permittivity ϵ_R with respect to the permittivity of a vacuum. Relative permittivity is defined as

$$\epsilon_R = \frac{\epsilon}{\epsilon_o}$$

and permittivity can be written as

$$\epsilon = \epsilon_R\epsilon_o$$

where ϵ and ϵ_o have units of farads per meter and ϵ_R is unitless. Because $C = \epsilon(A/d)$ and $\epsilon = \epsilon_R\epsilon_o$, the capacitance of a capacitor can be written as

$$C = \epsilon_R\epsilon_o\left(\frac{A}{d}\right)$$

Because $C = \epsilon(A/d)$, the capacitance C_o of a given capacitor with a vacuum between the plates is

$$C_o = \epsilon_o\left(\frac{A}{d}\right)$$

and if a dielectric with permittivity ϵ is placed between the plates of the same capacitor,

$$C = \epsilon\left(\frac{A}{d}\right)$$

Because the quantity (A/d) is the same in both equations,

$$\frac{C_o}{\epsilon_o} = \frac{C}{\epsilon}$$

and

$$C = \left(\frac{\epsilon}{\epsilon_o}\right)C_o$$

$$C = \epsilon_R C_o$$

This equation indicates that the capacitance of a given capacitor is increased by inserting a dielectric with an increased relative permittivity between the plates.

Consider the capacitor in Figure 10.7 to arrive at an explanation to support the preceding equation ($C = \epsilon_R C_o$). Although a dielectric is without free electrons, the charge on the capacitor plates causes the dielectric to polarize.

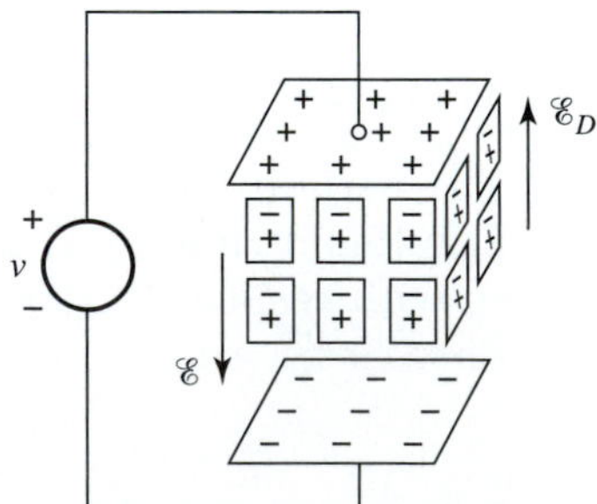

Figure 10.7 Polarized dielectric.

The protons and electrons of each atom of the dielectric are aligned and form dipoles, as illustrated. The polarized dielectric tends to reduce the electric field $\mathscr{E}$ between the plates by establishing an opposing electric field $\mathscr{E}_D$, as illustrated in Figure 10.7. However, electric field strength must remain constant for a given applied voltage and a given plate separation ($\mathscr{E} = v/d$), regardless of the dielectric. Therefore, if a dielectric with increased permittivity is inserted between the plates of a capacitor, the charge q on the plates must increase to maintain a constant electric field. Because $C = q/v$ and v is constant, the capacitance C is increased if a dielectric with increased permittivity is placed between the plates. Table 10.1 includes the relative permittivity for several well-known dielectrics.

TABLE 10.1

Dielectric	ϵ_R
Vacuum	1.0
Air	1.0006
Teflon	2.0
Paraffined paper	2.5
Rubber	3.0
Mica	5.0
Porcelain	6.0
Bakelite	7.0
Glass	7.5

EXAMPLE 10.1 A voltage of 40 V is applied across a capacitor, as shown in Figure 10.8. Determine the electric field strength between the plates, the charge on the plates, and the capacitance if the dielectric is air and if the dielectric is mica. ■

Figure 10.8

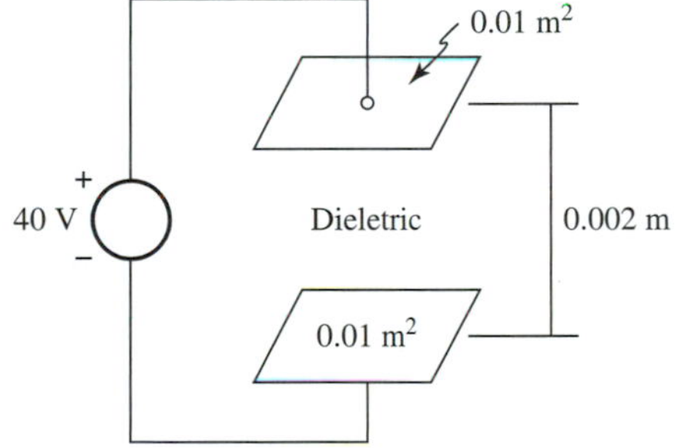

SOLUTION Because the electric field strength $\mathscr{E} = v/d$,

$$\mathscr{E} = \frac{40}{0.002}$$
$$= 20(10)^3 \text{ V/m}$$

for both dielectrics.

If a vacuum is used as a dielectric,

$$C_o = \epsilon_o\left(\frac{A}{d}\right)$$
$$= (8.85)(10^{-12})\left\{\frac{0.01}{0.002}\right\}$$
$$= 44.25(10^{-12})$$
$$= 44.25 \text{ pF}$$

Because $C = \epsilon_R C_o$ and $\epsilon_R = 1.0006$ for air and 5 for mica,

$$\begin{aligned} C &= \epsilon_R C_o \\ &= (1.0006)(44.25)(10^{-12}) \\ &= 44.3(10^{-12}) \\ &= 44.30 \text{ pF} \end{aligned}$$

for air as a dielectric and

$$\begin{aligned} C &= \epsilon_R C_o \\ &= 5(44.25)(10^{-12}) \\ &= 221(10^{-12}) \\ &= 221 \text{ pF} \end{aligned}$$

for mica as a dielectric.

From Equation (10.1)

$$\begin{aligned} q &= Cv \\ &= 44.3(10^{-12})40 \\ &= 1772(10^{-12}) \\ &= 1772 \text{ pc} \end{aligned}$$

for air as a dielectric and

$$\begin{aligned} q &= 221(10^{-6})40 \\ &= 8840(10^{-12}) \\ &= 8840 \text{ pc} \end{aligned}$$

for mica as a dielectric.

10.11 DIELECTRIC STRENGTH

The value of the voltage across a capacitor that causes the dielectric to break down and allow current to flow through the dielectric is called the capacitor *breakdown voltage.* The dielectric changes in behavior from an insulator to a partial conductor at the breakdown voltage and the capacitor ceases to behave as a pure capacitor.

10.12 TYPES OF CAPACITORS

Capacitors are constructed as *fixed* or *variable.* The capacitance of a fixed capacitor is constant and the capacitance of a variable capacitor can be varied by mechanical adjustment, which is usually an integral part of the capacitor. Regardless of the materials used to construct a capacitor, the arrangement of the materials is always the same—conductors separated by dielectrics.

Variable capacitors are often called air capacitors, because air is usually used as the dielectric. A typical variable capacitor consists of a series of multiple parallel conducting plates, with every other plate attached to a common shaft running perpendicular to the parallel plates. When the shaft is rotated, the distance between the plates and the effective plate area is altered, and the capacitance is varied in that manner.

Most *fixed* capacitors are constructed by alternating sheets of dielectric material with sheets of conducting material, rolling the sheets into cylindrical, tubular, or other shapes, and encapsulating the resulting capacitor in a wax, plastic, or other protective material. The dielectrics used for fixed capacitors are paper, mylar, plastic, mica, ceramic material, aluminum dioxide, or tantalum. The conductors used for fixed capacitors are metallic foil, metallized paper, or metal film. Some fixed capacitors are constructed for use in dc circuits only and are called *electrolytic* capacitors. When a dc voltage is applied across an electrolytic capacitor, the positive-marked terminal must be maintained positive with respect to the other terminal; otherwise capacitor damage could result.

10.13 CURRENT-VOLTAGE RELATIONSHIP FOR CAPACITORS

In arriving at the capacitor equations in the previous section, we assumed a constant voltage across the capacitor for convenience. However, the voltage across a capacitor often varies, and Equation (10.1) ($q = Cv$) indicates that if the voltage across a capacitor varies with time, then the charge on the capacitor plates varies with time.

If we differentiate both sides of the expression in Equation (10.1) with respect to time, the result is

$$\frac{d\{q\}}{dt} = \frac{d\{Cv\}}{dt}$$

and because C is a constant for any given capacitor,

$$\frac{dq}{dt} = C\frac{dv}{dt}$$

Recalling that electric current is the rate of flow of electric charge, electric current is the first derivative of q with respect to time and can be written as

$$i = \frac{dq}{dt}$$

Equation (10.1) can then be written as

$$i = C\frac{dv}{dt} \tag{10.2}$$

where i is in amperes, v is in volts, t is in seconds, and C is in farads. Because $dv/dt = (1/C)i$, v is the antiderivative of $(1/C)i$ and can be written in terms of the indefinite integral as

$$v = \frac{1}{C}\int i\, dt \tag{10.3}$$

Consider the following example.

EXAMPLE 10.2 If $v = 100e^{-6t}$ V and $C = \frac{1}{100}$ F for the circuit in Figure 10.9, determine the current i and the charge q on the capacitor plates. ■

Figure 10.9

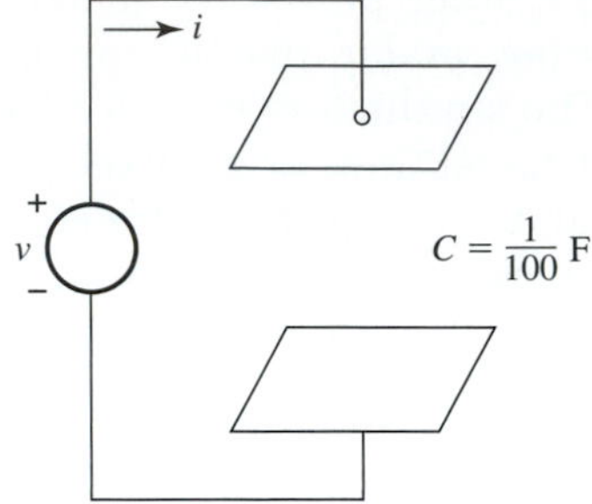

SOLUTION From Equation (10.2) and differentiation rules D3 and D6,

$$\begin{aligned} i &= C\frac{dv}{dt} \\ &= \left(\frac{1}{100}\right)\frac{d\{100e^{-6t}\}}{dt} \\ &= \left(\frac{1}{100}\right)(100)(-6)e^{-6t} \\ &= -6e^{-6t}\text{ A} \end{aligned}$$

From Equation (10.1)

$$\begin{aligned} q &= Cv \\ &= \left(\frac{1}{100}\right)100e^{-6t} \\ &= e^{-6t}\text{ C} \end{aligned}$$

EXAMPLE 10.3 If $v = 100$ V for the circuit in Example 10.2, determine the current i and the charge q on the capacitor plates. ■

SOLUTION From Equation (10.1), the charge on each plate is

$$\begin{aligned} q &= Cv \\ &= \left(\frac{1}{100}\right)(100) \\ &= 1\text{ C} \end{aligned}$$

From Equation (10.2)

$$\begin{aligned} i &= C\frac{dv}{dt} \\ &= \left(\frac{1}{100}\right)(0) \\ &= 0 \end{aligned}$$

because the voltage v is constant.

The symbol for a capacitor, as given in Figure 1.1(c), is illustrated in Figure 10.10. The relationship between the voltage v across the capacitor and the current i flowing through the capacitor is given by Equations (10.2) and (10.3). As previously noted, charge does not flow through a capacitor but flows through the external circuitry connected to the capacitor. However, in applying Equations (10.2) and (10.3), we shall use the terminology in common use and refer to current flow through a capacitor.

Figure 10.10 Symbol for a capacitor.

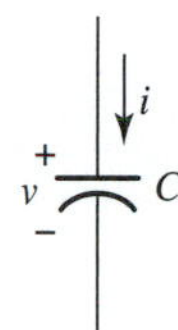

When employing the relationship between current and voltage given by Equation (10.2), the direction of current as it relates to the voltage polarity must be considered. Consider the capacitor in Figure 10.11. The conventional current flowing from point a to point b is designated by i_{ab}, and the conventional current flowing from point b to point a is designated by i_{ba}, where

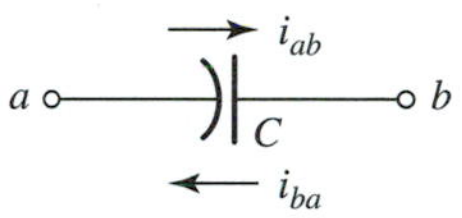

Figure 10.11 Current flow for a capacitor.

$$i_{ab} = -i_{ba}$$

Applying Equation (10.2) to the capacitor in Figure 10.11,

$$i_{ab} = C\frac{d(v_{ab})}{dt}$$

and

$$i_{ba} = C\frac{d(v_{ba})}{dt}$$

where v_{ab} is the voltage of point a with respect to b, v_{ba} is the voltage of point b with respect to a, and

$$v_{ab} = -v_{ba}$$

Consider the following examples.

EXAMPLE 10.4 Determine i_{ab} and i_{ba} for the capacitor shown in Figure 10.12 if $v_{ab} = 2e^{-t}$ volts. ■

Figure 10.12

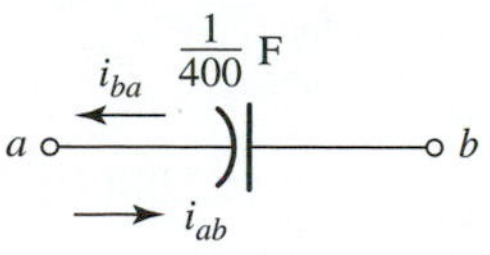

SOLUTION From Equation (10.2) and differentiation rules D3 and D6,

$$
\begin{aligned}
i_{ab} &= C\frac{d(v_{ab})}{dt} \\
&= \left(\frac{1}{400}\right)\frac{d\{2e^{-t}\}}{dt} \\
&= \left(\frac{1}{400}\right)(-2e^{-t}) \\
&= -\left(\frac{1}{200}\right)e^{-t}\text{ A}
\end{aligned}
$$

Because $v_{ba} = -v_{ab}$, Equation (10.2) results in

$$
\begin{aligned}
i_{ba} &= C\frac{d(v_{ba})}{dt} \\
&= \left(\frac{1}{400}\right)\frac{d\{-2e^{-t}\}}{dt} \\
&= \left(\frac{1}{400}\right)(2e^{-t}) \\
&= \left(\frac{1}{200}\right)e^{-t}\text{ A}
\end{aligned}
$$

Alternatively, because $i_{ba} = -i_{ab}$,

$$
\begin{aligned}
i_{ba} &= -\left\{-\left(\frac{1}{200}\right)e^{-t}\right\} \\
&= \left(\frac{1}{200}\right)e^{-t}\text{ A}
\end{aligned}
$$

EXAMPLE 10.5 Determine i_{ab} and i_{ba} for the capacitor shown in Figure 10.13 if $v_{ab} = -\sin(6t)$ volts. ■

Figure 10.13

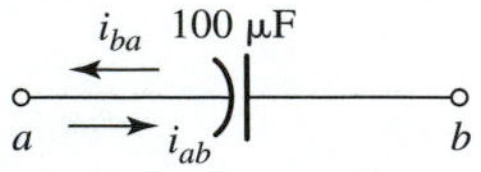

i_{ba} 100 μF

a i_{ab} b

SOLUTION From Equation (10.2) and differentiation rules D4 and D6

$$
\begin{aligned}
i_{ab} &= C\frac{dv_{ab}}{dt} \\
&= 100(10)^{-6}\frac{d\{-\sin(6t)\}}{dt}
\end{aligned}
$$

$$= 100(10)^{-6}(-6)\cos(6t)$$
$$= -600(10)^{-6}\cos(6t)\text{ A}$$
$$= -600\cos(6t)\ \mu\text{A}$$

Because $v_{ba} = -v_{ab}$, Equation (10.2) results in

$$V_{ba} = -\{-\sin(6t)\}$$
$$= \sin(6t)$$
$$i_{ba} = 100(10)^{-6}\frac{d\{\sin(6t)\}}{dt}$$
$$= 100(10)^{-6}(6)\cos(6t)$$
$$= 600(10)^{-6}\cos(6t)\text{ A}$$
$$= 600\cos(6t)\ \mu\text{A}$$

Alternatively, because $i_{ba} = -i_{ab}$,

$$i_{ba} = -\{-600(10)^{-6}\cos(6t)\}$$
$$= 600(10)^{-6}\cos(6t)\text{ A}$$
$$= 600\cos(6t)\ \mu\text{A}$$

Although the procedure for labeling terminals with letters to designate voltages and currents is convenient, the polarity notation, as presented in Chapter 2 for resistors, is more commonly used. Consider the capacitor in Figure 10.14, where v is the voltage of the positive-marked terminal with respect to the negative-marked terminal.

Figure 10.14 Voltage-current relationship for a capacitor.

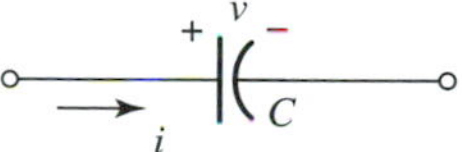

If the current i is expressed as flowing from the positive-marked terminal to the negative-marked terminal in Figure 10.14, then v must be expressed as the voltage of the positive-marked terminal with respect to the negative-marked terminal when applying Equation (10.2) to the capacitor in Figure 10.14.

The positive and negative markings on the terminals of the capacitor do not necessarily correspond to the actual polarities of the terminals. For instance, if $v = -10$ V in Figure 10.14, the positive-marked terminal is 10 V negative with respect to the negative-marked terminal. Similarly, if i is negative, then the actual current flows opposite to the direction assigned.

Consider the following examples.

EXAMPLE 10.6 Determine the current i if $v = 2\sin(t)$ volts, as shown in Figure 10.15. ■

Figure 10.15

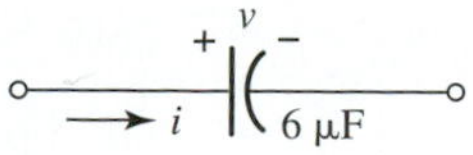

SOLUTION From Equation (10.2) and differentiation rules D4 and D6,

$$
\begin{aligned}
i &= C\frac{dv}{dt} \\
&= \{6(10)^{-6}\}\frac{d\{2\sin(t)\}}{dt} \\
&= 6(10)^{-6}(2)(-\cos t) \\
&= -12(10)^{-6}\cos(t)\text{ A} \\
&= -12\cos(t)\ \mu\text{A}
\end{aligned}
$$

EXAMPLE 10.7 Determine the current i if $v = -e^{-2t}$ volts, as shown in Figure 10.16. ■

Figure 10.16

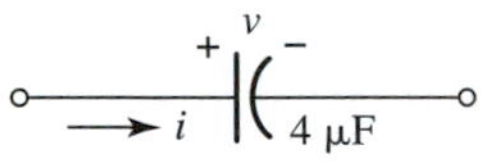

$+\ v\ -$, i, $4\ \mu\text{F}$

SOLUTION From Equation (10.2) and differentiation rules D3 and D6,

$$
\begin{aligned}
i &= C\frac{dv}{dt} \\
&= \{4(10)^{-6}\}\frac{d\{-e^{-2t}\}}{dt} \\
&= 4(10)^{-6}(-1)(-2)e^{-2t} \\
&= 8(10)^{-6}e^{-2t}\text{ A} \\
&= 8e^{-2t}\ \mu\text{A}
\end{aligned}
$$

EXAMPLE 10.8 Determine the current i in Example 10.7 if i is assigned in the opposite direction, as shown in Figure 10.17. ■

Figure 10.17

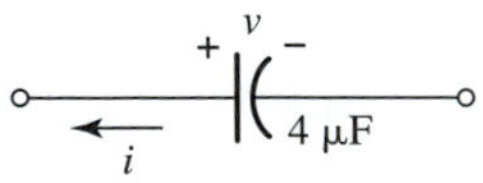

$+\ v\ -$, i, $4\ \mu\text{F}$

SOLUTION From Equation (10.2) and differentiation rules D3 and D6,

$$
\begin{aligned}
i &= C\frac{d\{-v\}}{dt} \\
&= \{4(10)^{-6}\}\frac{d\{-(-e^{-2t})\}}{dt} \\
&= 4(10)^{-6}\{-2e^{-2t}\} \\
&= -8(10)^{-6}e^{-2t}\text{ A} \\
&= -8e^{-2t}\ \mu\text{A}
\end{aligned}
$$

Notice that the current in Example 10.8 is the negative of the current in Example 10.7.

EXAMPLE 10.9 Determine the current i if $v = 6$ V, as shown in Figure 10.18. ■

Figure 10.18

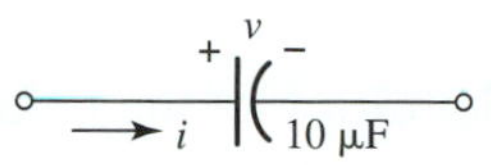

SOLUTION From Equation (10.2) and differentiation rule D2

$$\begin{aligned} i &= C\frac{dv}{dt} \\ &= \{10(10)^{-6}\}\frac{d(6)}{dt} \\ &= 0 \end{aligned}$$

The rate of flow of charge is equal to zero, because $dv/dt = 0$. However, the net charge on each plate of the capacitor, from Equation (10.1), is

$$\begin{aligned} q &= Cv \\ &= 10(10)^{-6}(6) \\ &= 60(10)^{-6} \\ &= 60\ \mu\text{c} \end{aligned}$$

10.14 ENERGY STORAGE IN CAPACITORS

An ideal capacitor does not dissipate energy. However, if the dielectric material in a capacitor contains free electrons, a current will flow through the dielectric, and energy is dissipated in this manner. The dielectric acts as a conductor, and the current flowing through the dielectric is called *leakage current*. The energy loss is very small for a high-quality capacitor, and we shall assume for analysis purposes that all capacitors used in this text are ideal unless otherwise noted.

Consider the capacitor in Figure 10.14 to determine the energy stored in a capacitor. From Equation (2.6) for the instantaneous power p and Equation (10.2),

$$\begin{aligned} p &= vi \\ &= v\left\{C\frac{dv}{dt}\right\} \\ &= Cv\frac{dv}{dt} \end{aligned}$$

Because power is the rate of energy change

$$\begin{aligned} \frac{dw}{dt} &= p \\ &= Cv\frac{dv}{dt} \end{aligned}$$

From the definitions of the antiderivative and the indefinite integral, w is the antiderivative of $\left\{Cv\frac{dv}{dt}\right\}$ with respect to time, so

$$w = \int \left\{Cv\frac{dv}{dt}\right\} dt$$

Also recalling the definition of the definite integral and again recognizing that w is the antiderivative of $\left\{Cv\frac{dv}{dt}\right\}$, we can write the expression

$$\int_0^t \left\{Cv\frac{dv}{dt}\right\} dt = w(t) - w(0)$$

where t is any time greater than $t = 0$ and $w(t)$ and $w(0)$ are the values of w at any time t and at $t = 0$, respectively. It follows that

$$w = \int_0^t \left\{Cv\frac{dv}{dt}\right\} dt + w(0)$$

Consider the expression $\{Cv^2/2\}$, which can be written as $\{(C/2)(v)(v)\}$. Recognizing that $(C/2)$ is a constant and employing rules D6 and D9 to differentiate the expression results in

$$\frac{d\{(C/2)(v)(v)\}}{dt} = \frac{C}{2}\left\{v\frac{dv}{dt} + \frac{dv}{dt}v\right\}$$

$$= Cv\frac{dv}{dt}$$

It follows that $C\frac{v^2}{2}$ is the antiderivative of $\left\{Cv\frac{dv}{dt}\right\}$, and therefore the definite integral for w can be written as

$$w = C\frac{v^2}{2}\bigg|_0^t + w(0)$$

If the initial energy stored and the capacitor voltage are both zero at the beginning of the capacitor life, then $w(0) = 0$ and $v(0) = 0$; substituting the limits of integration (0 and t) into the expression $\left\{C\frac{v^2}{2}\right\}$ results in

$$w = C\frac{v^2}{2}\bigg|_t - C\frac{v^2}{2}\bigg|_{t=0} + w(0)$$

$$w = C\frac{v^2}{2} \tag{10.4}$$

where C is in farads, v is the voltage across the capacitor in volts, and w is the energy stored in joules. The energy stored is constant if v is constant and varies with time if the voltage across the capacitor varies with time.

Equation (10.4) can be arrived at more easily by examining the previous equation,

$$w = \int_0^t \left\{ Cv \frac{dv}{dt} \right\} dt + w(0)$$

Although we have not discussed algebraic operations involving the differential quantity dt, it is possible to cancel out dt from this equation, as follows:

$$w = \int_0^t \left\{ Cv \frac{dv}{\cancel{dt}} \right\} \cancel{dt} + w(0)$$

$$= \int_0^t \{Cv\}\, dv + w(0)$$

The integration is now with respect to v, and using rule I1 from Chapter 9,

$$w = C\frac{v^2}{2}\bigg|_0^t + w(0)$$

as previously determined.

EXAMPLE 10.10 Determine the energy stored in the capacitor shown in Figure 10.19. ■

Figure 10.19

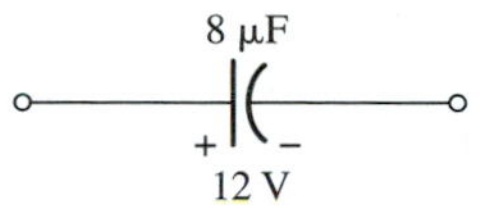

8 μF, + 12 V −

SOLUTION From Equation (10.4)

$$w = \left(\frac{1}{2}\right)Cv^2$$

$$= \left(\frac{1}{2}\right)(8)(10)^{-6}(12)^2$$

$$= 576(10)^{-6}\text{ J}$$

$$= 576\ \mu\text{J}$$

EXAMPLE 10.11 Determine the energy stored in the capacitor in Figure 10.20 when $v = 6e^{-t}$ V. ■

Figure 10.20

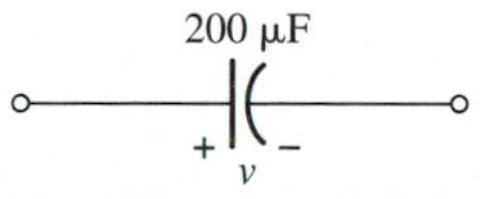

200 μF, + v −

SOLUTION From Equation (10.4)

$$w = \left(\frac{1}{2}\right)Cv^2$$

$$= \left(\frac{1}{2}\right)(200)(10)^{-6}(6e^{-t})^2$$

$$= 3600(10)^{-6}e^{-2t}$$

$$= 3600e^{-2t}\ \mu\text{J}$$

Notice that the energy stored in the capacitor varies with time, because the voltage across the capacitor varies with time.

10.15 CAPACITORS IN SERIES

Consider the series combination of capacitors in Figure 10.21(a). From the equivalency concept previously presented, the series combination (C_1 and C_2) in Figure 10.21(a) is equivalent to the single capacitor (C) in Figure 10.21(b) when the voltage-current relationship for terminals a-b is the same for both circuits. That is, if the voltage source v is the same in both circuits, then the current i will also be the same in both circuits when the single equivalent capacitor replaces the series combination.

Figure 10.21 Capacitors in series.

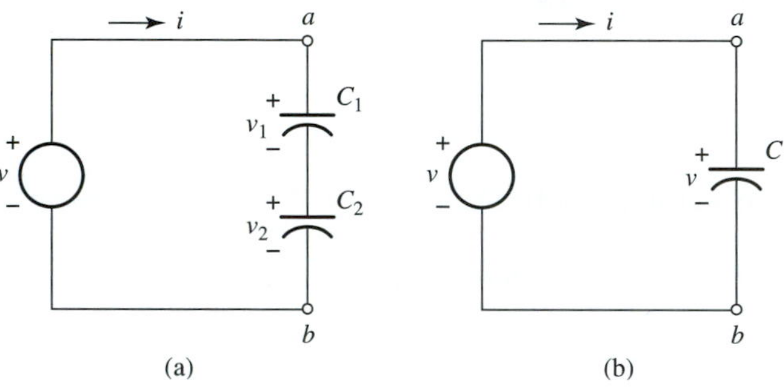

Because the current i is the same for both capacitors (C_1 and C_2) and is the rate of flow of charge, then the charge on C_1 must equal the charge on C_2. That is,

$$q_1 = q_2$$

where q_1 is the charge on C_1 and q_2 is the charge on C_2. And if capacitor C is equivalent to the series combination (C_1 and C_2), then the current i flowing through C is equal to the current i flowing through C_1 and C_2. It follows that

$$q = q_1 = q_2$$

where q is the charge on C.

Applying Kirchhoff's voltage law to the circuit in Figure 10.21(a) results in

$$v = v_1 + v_2$$

Recalling that v is also the voltage across the capacitor C in Figure 10.21(b) and employing Equation (10.1) results in

$$\frac{q}{C} = \frac{q_1}{C_1} + \frac{q_2}{C_2}$$

Since $q = q_1 = q_2$,

$$\frac{q}{C} = \frac{q}{C_1} + \frac{q}{C_2}$$

Dividing both sides of the equation by q results in

$$\frac{1}{C} = \frac{1}{C_1} + \frac{1}{C_2}$$

and

$$C = \frac{1}{1/C_1 + 1/C_2}$$

The preceding equation can be extended to include any number of capacitors in series.

In general, n capacitors in series can be combined as indicated in Figure 10.22. The equation for combining n capacitors in series is

$$C = \frac{1}{1/C_1 + 1/C_2 + 1/C_3 + \cdots + 1/C_n} \tag{10.5}$$

Figure 10.22 Equivalent capacitor.

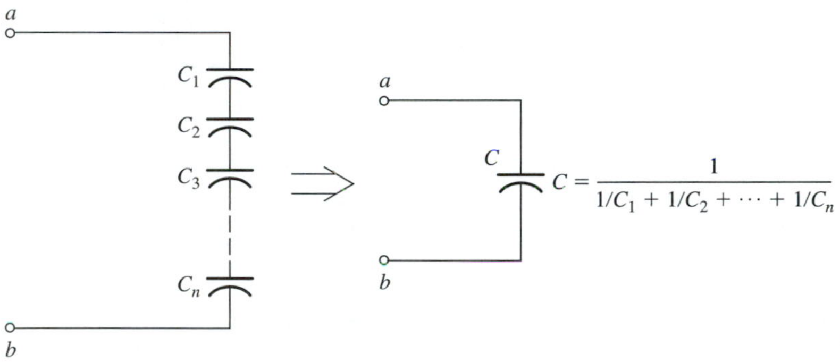

Notice the similarity between Equation (10.5) and Equation (3.4) for combining resistors in parallel. Notice also that for two capacitors in series, Equation (10.5) can be written as

$$C = \frac{C_1 C_2}{C_1 + C_2}$$

Consider the following examples.

EXAMPLE 10.12 Replace the two capacitors in Figure 10.23 by a single capacitor. ■

Figure 10.23

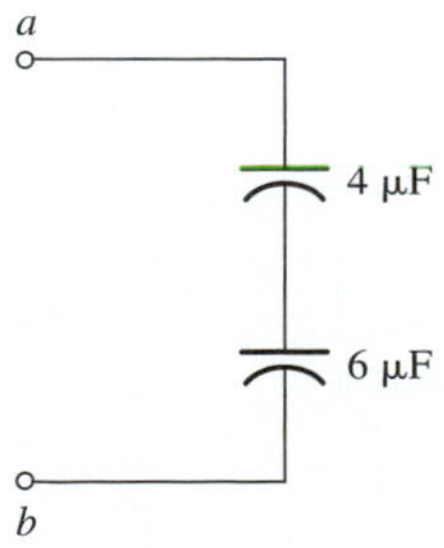

a

4 µF

6 µF

b

SOLUTION From Equation (10.5)

$$C = \frac{\{4(10)^{-6}\}\{6(10)^{-6}\}}{4(10)^{-6} + 6(10)^{-6}}$$
$$= 2.4(10)^{-6}$$
$$= 2.4\ \mu F$$

The result is shown in Figure 10.24.

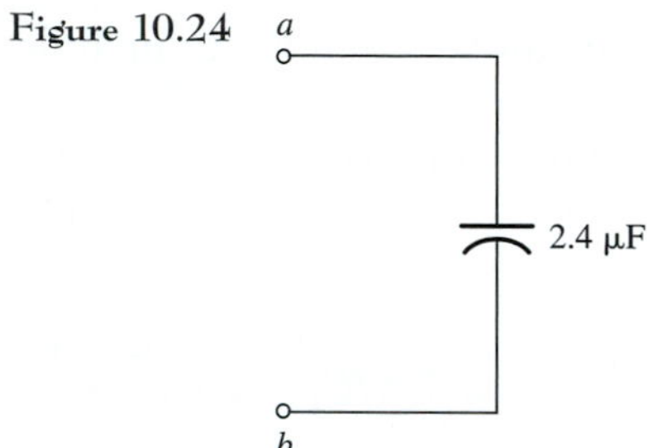

Figure 10.24

a
2.4 μF
b

EXAMPLE 10.13 Replace the three capacitors in Figure 10.25 by a single capacitor. ■

Figure 10.25

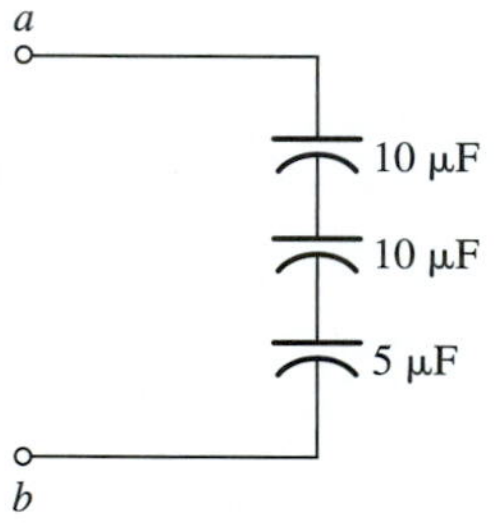

a
10 μF
10 μF
5 μF
b

SOLUTION From Equation (10.5)

$$C = \frac{1}{\frac{1}{10} + \frac{1}{10} + \frac{1}{5}}$$
$$= 2.5\ \mu F$$

The result is shown in Figure 10.26.

Figure 10.26

a
2.5 μF
b

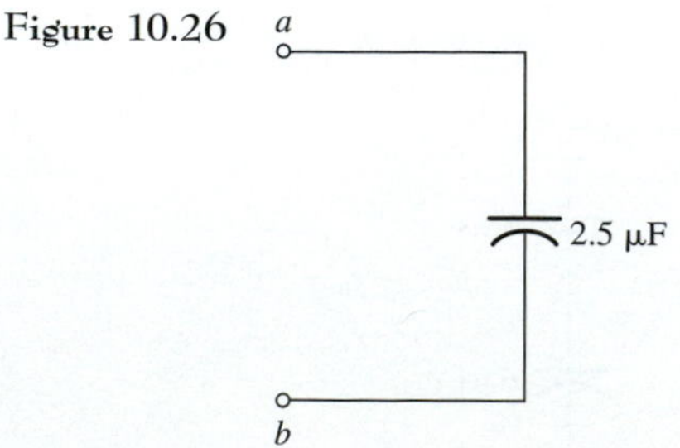

10.16 CAPACITORS IN PARALLEL

Consider the parallel combination of capacitors in Figure 10.27(a). From the equivalency concept previously presented, the parallel combination (C_1 and

Figure 10.27 Capacitors in parallel.

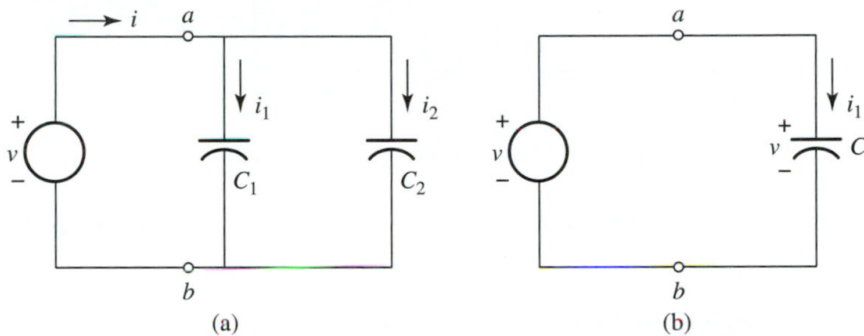

C_2) in Figure 10.27(a) is equivalent to the single capacitor (C) in Figure 10.27(b) when the voltage-current relationship for terminals *a*-*b* is the same for both circuits. That is, if the voltage source v is the same in both circuits, then the current i will be the same in both circuits when the single equivalent capacitor replaces the parallel combination.

Applying Kirchhoff's current law to the circuit in Figure 10.27(a) results in

$$i = i_1 + i_2$$

Recognizing that v is the voltage across each of the capacitors C_1 and C_2 and employing Equation (10.2) along with Kirchhoff's current law results in

$$i = C_1 \frac{dv}{dt} + C_2 \frac{dv}{dt}$$

Because v also appears across capacitor C in Figure 10.27(b) and $i = C\dfrac{dv}{dt}$

$$C\frac{dv}{dt} = C_1 \frac{dv}{dt} + C_2 \frac{dv}{dt}$$

Dividing both sides of the equation by dv/dt results in

$$C = C_1 + C_2$$

This equation can be extended to include any number of capacitors in parallel.

In general, n capacitors in parallel can be combined as indicated in Figure 10.28. The equation for combining n capacitors in parallel is

$$C = C_1 + C_2 + C_3 + \cdots + C_n \tag{10.6}$$

Figure 10.28 Equivalent capacitors.

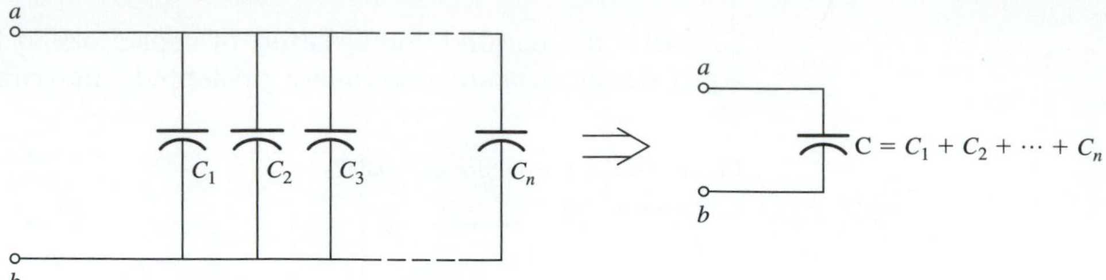

Consider the following example.

EXAMPLE 10.14 Replace the four capacitors in Figure 10.29 with a single capacitor. ■

Figure 10.29

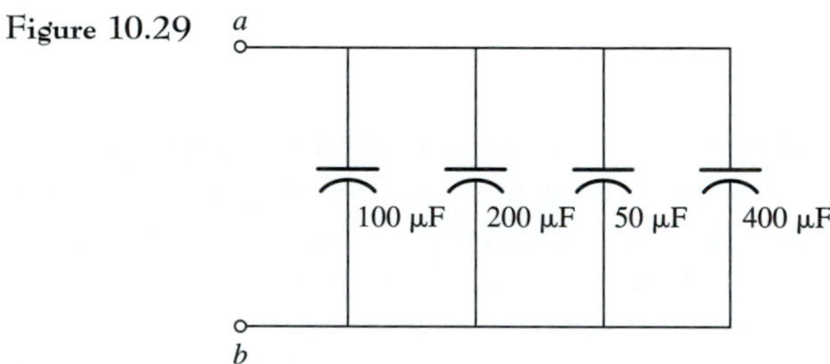

SOLUTION From Equation (10.6)

$$C = 100 + 200 + 50 + 400$$
$$= 750\ \mu\text{F}$$

The result is shown in Figure 10.30.

Figure 10.30

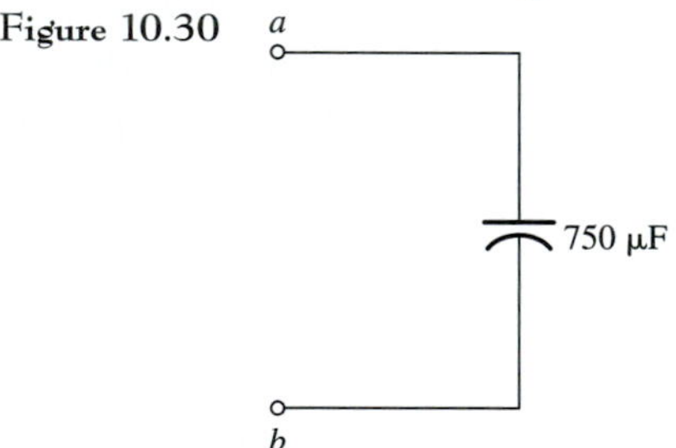

10.17 SERIES-PARALLEL COMBINATIONS

We shall now consider capacitors connected in series-parallel combinations. It is advantageous to use a concise notation, as follows:

$(C_1 \oplus C_2)$ is read as "C_1 and C_2 are in series."
$(C_1 \| C_2)$ is read as "C_1 and C_2 are in parallel."

As an example

$$(C_1 \oplus C_2 \oplus C_3)\|(C_4 \oplus C_5)$$

is read as "the series combination of C_1, C_2, and C_3 is in parallel with the series combination of C_4 and C_5." The notation is illustrated in the following examples.

EXAMPLE 10.15 Determine a single equivalent capacitor for terminals *a-b* in Figure 10.31. ■

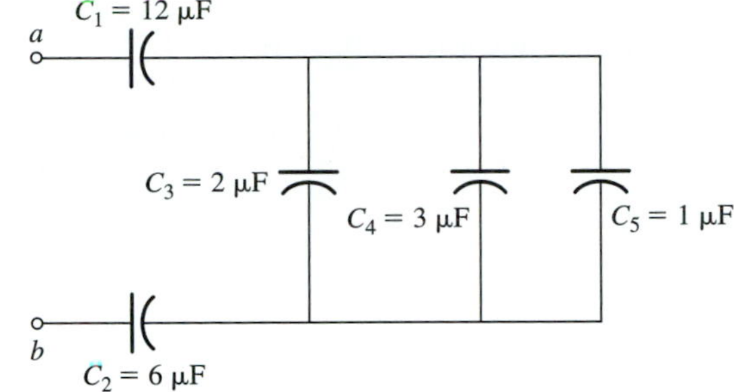

Figure 10.31

SOLUTION The equivalent capacitance C_{EQ} for terminals *a-b* is

$$C_{EQ} = (C_1 \oplus C_2) \oplus (C_3 \| C_4 \| C_5)$$

which is read as

"C_3, C_4, and C_5 are in parallel and the parallel combination is in series with C_1 and C_2."

From Equations (10.5) and (10.6),

$$\begin{aligned} C_1 \oplus C_2 &= \frac{(12)(6)}{12 + 6} \\ &= 4\ \mu F \\ C_3 \| C_4 \| C_5 &= 2 + 3 + 1 \\ &= 6\ \mu F \\ C_{EQ} &= \frac{(4)(6)}{4 + 6} \\ &= 2.4\ \mu F \end{aligned}$$

EXAMPLE 10.16 Determine a single equivalent capacitor for terminals *a-b* in Figure 10.32. ■

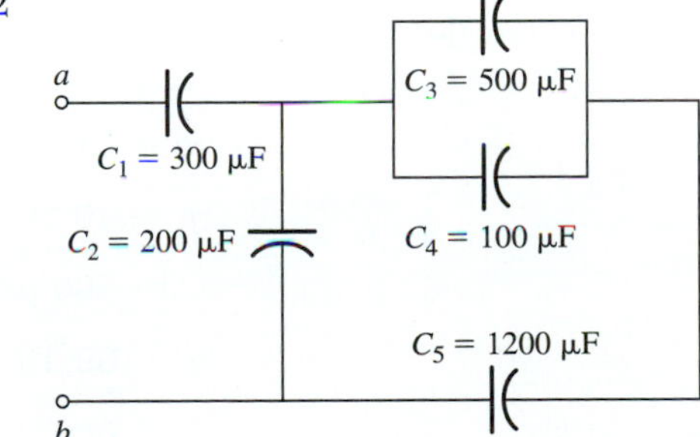

Figure 10.32

SOLUTION The equivalent capacitance C_{EQ} for terminals a-b is

$$C_{EQ} = \{\{(C_3 \| C_4) \oplus C_5\} \| (C_2)\} \oplus C_1$$

From Equations (10.5) and (10.6),

$$C_3 \| C_4 = 500 + 100$$
$$= 600\ \mu\text{F}$$
$$(C_3 \| C_4) \oplus (C_5) = \frac{(600)(1200)}{600 + 1200}$$
$$= 400\ \mu\text{F}$$
$$\{(C_3 \| C_4) \oplus C_5\} \| C_2 = 400 + 200$$
$$= 600\ \mu\text{F}$$
$$\{\{(C_3 \| C_4) \oplus C_5\} \| C_2\} \oplus C_1 = \frac{(300)(600)}{300 + 600}$$

It follows that

$$C_{EQ} = 200\ \mu\text{F}$$

10.18 CHARGE ON SERIES-PARALLEL CAPACITORS

It is often necessary to determine the charge on and the voltage across capacitors in circuits containing several capacitors in series and parallel. A procedure is illustrated in the following examples.

EXAMPLE 10.17 Determine the charge on each capacitor and the voltage across each capacitor for the circuit shown in Figure 10.33. ■

Figure 10.33

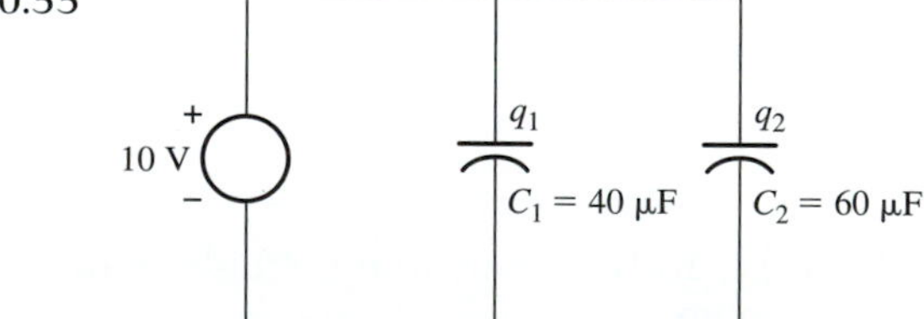

SOLUTION The voltage across C_1 and C_2 is equal to 10 V, because C_1 and C_2 are each in parallel with the voltage source. From Equation (10.1)

$$q = Cv$$
$$q_1 = 40(10^{-6})(10)$$
$$= 400\ \mu c$$
$$q_2 = 60(10^{-6})(10)$$
$$= 600\ \mu c$$

EXAMPLE 10.18 Determine the voltage across and the charge on each capacitor in the circuit shown in Figure 10.34. ■

Figure 10.34

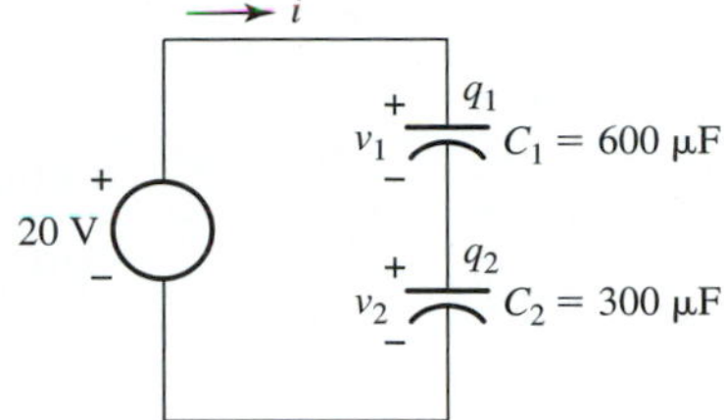

SOLUTION For capacitors in series, the same current flows through each capacitor; therefore, the same charge appears on each capacitor. For the circuit in Figure 10.34 the current i flows through C_1 and C_2, so

$$q_1 = q_2$$

From Equation (10.5) an equivalent capacitor for C_1 and C_2 in series is

$$C_{EQ} = \frac{(C_1)(C_2)}{C_1 + C_2}$$

$$= \frac{(600)(300)}{600 + 300}$$

$$= 200\ \mu F$$

As shown in Figure 10.35.

Figure 10.35

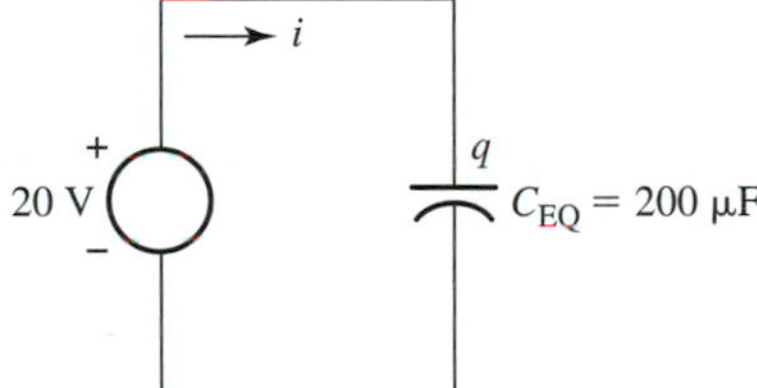

Because the same current i that flows through C_{EQ} also flows through C_1 and C_2,

$$q = q_1 = q_2$$

From Equation (10.1)

$$q = C_{EQ}v$$

$$= 200(20)$$

$$= 4000\ \mu c$$

and since $q = q_1 = q_2$,

$$q_1 = 4000\ \mu c$$

$$q_2 = 4000\ \mu c$$

It follows that

$$v_1 = \frac{q_1}{C_1} = \frac{4000(10)^{-6}}{600(10)^{-6}} = \frac{20}{3}\text{ V}$$

$$v_2 = \frac{q_2}{C_2} = \frac{4000(10)^{-6}}{300(10)^{-6}} = \frac{40}{3}\text{ V}$$

EXAMPLE 10.19 Determine the voltage across and the charge on each capacitor in the circuit shown in Figure 10.36. ■

Figure 10.36

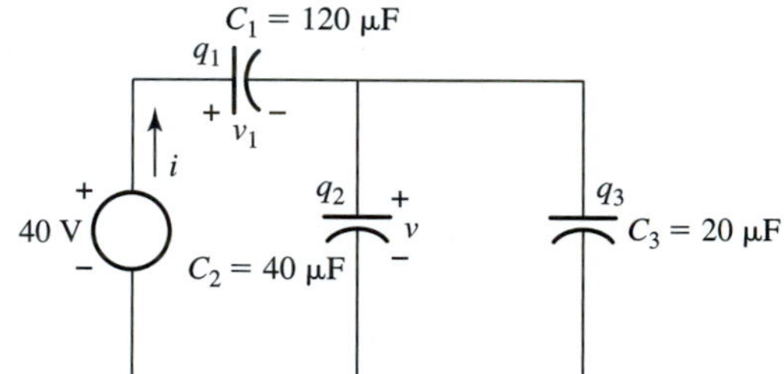

SOLUTION An equivalent capacitor for C_1, C_2, and C_3 is

$$C_{EQ} = \{C_2 \| C_3\} \oplus (C_1)$$

From Equations (10.5) and (10.6),

$$C_{EQ} = \frac{(C_2 + C_3)(C_1)}{C_1 + (C_2 + C_3)} = \frac{(40 + 20)(120)}{120 + (40 + 20)} = 40\ \mu F$$

as shown in Figure 10.37.

Figure 10.37

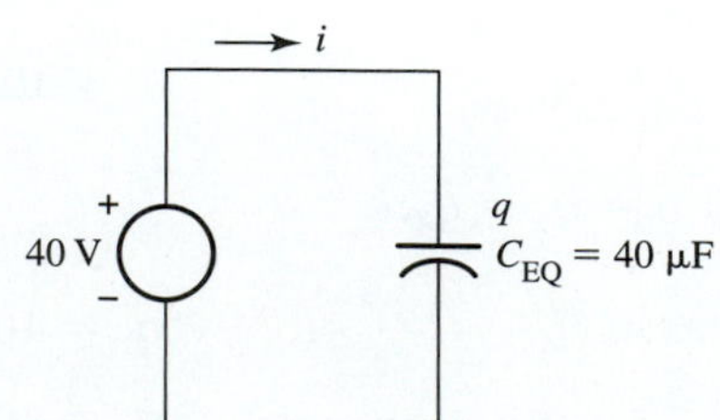

Because the same current i that flows through C_{EQ} also flows through C_1,

$$q_1 = q$$

From Equation (10.1)

$$\begin{aligned} q &= C_{EQ}v \\ &= 40(10^{-6})40 \\ &= 1600\ \mu c \end{aligned}$$

and

$$\begin{aligned} q_1 &= 1600\ \mu c \\ v_1 &= \frac{q_1}{C_1} \\ &= \frac{1600(10)^{-6}}{120(10)^{-6}} \\ &= \frac{40}{3}\ \text{V} \end{aligned}$$

From Equation (10.6) an equivalent capacitor, C'_{EQ}, for C_2 and C_3 is

$$\begin{aligned} C'_{EQ} &= C_2 \| C_3 \\ &= C_2 + C_3 \\ &= 40 + 20 \\ &= 60\ \mu\text{F} \end{aligned}$$

Because the same current i flows through C'_{EQ} as flows through C_1 in Figure 10.38,

$$\begin{aligned} q' &= q_1 \\ &= 1600\ \mu c \end{aligned}$$

Figure 10.38

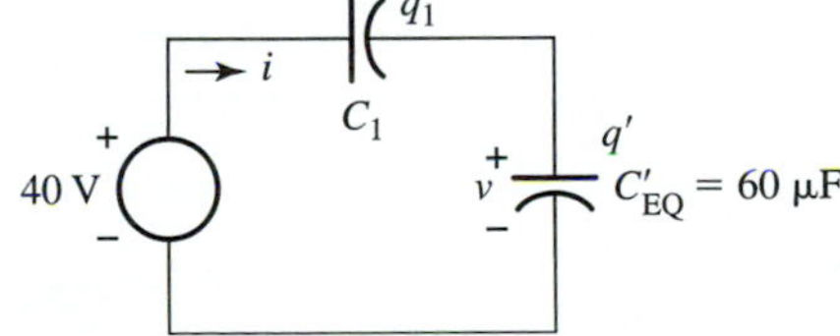

From Equation (10.1)

$$\begin{aligned} v &= \frac{q'}{C'_{EQ}} \\ &= \frac{1600(10)^{-6}}{60(10)^{-6}} \\ &= \frac{80}{3}\ \text{V} \end{aligned}$$

and
$$q_2 = C_2 v$$
$$= 40(10^{-6})(80/3)$$
$$= \frac{3200}{3}\ \mu C$$
$$q_3 = C_3 v$$
$$= 20(10^{-6})\left(\frac{80}{3}\right)$$
$$= \frac{1600}{3}\ \mu C$$

An alternative approach after v_1 is determined is to find v in the original circuit using Kirchhoff's voltage law.

$$40 - v_1 - v = 0$$
$$40 - \frac{40}{3} - v = 0$$
$$v = \frac{80}{3}\ \text{V}$$

Because v appears across C_2 and C_3, Equation (10.1) results in

$$q_2 = C_2 v$$
$$= 40(10)^{-6}\left(\frac{80}{3}\right)$$
$$= \frac{3200}{3}\ \mu C$$

and
$$q_3 = C_3 v$$
$$= 20(10)^{-6}\left(\frac{80}{3}\right)$$
$$= \frac{1600}{3}\ \mu C$$

10.19 CURRENT DIVISION

Recall that the current-division formula for resistors in parallel allows us to determine the division of current without determining the voltage across the resistors that the divided currents flow through. We shall now develop a current-division formula for capacitors in parallel. It should be observed that current division and voltage division (presented in the next section) enable us to more easily solve the example problems in the last section.

Figure 10.39 Capacitors in parallel.

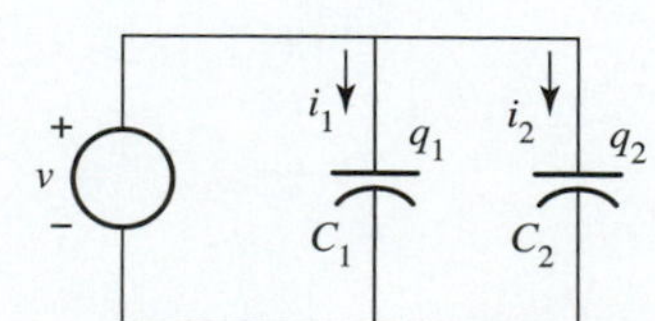

Consider the circuit in Figure 10.39. The voltage source v appears across C_1 and C_2. From Equation (10.1) the charge on each capacitor is

$$q_1 = C_1 v$$
$$q_2 = C_2 v$$

From Equation (10.6) the equivalent capacitor for the parallel combination of C_1 and C_2 is

$$C_{EQ} = C_1 \| C_2$$
$$= C_1 + C_2$$

and from the equivalent circuit in Figure 10.40,

$$q = v(C_1 + C_2)$$

Figure 10.40 Capacitor equivalent.

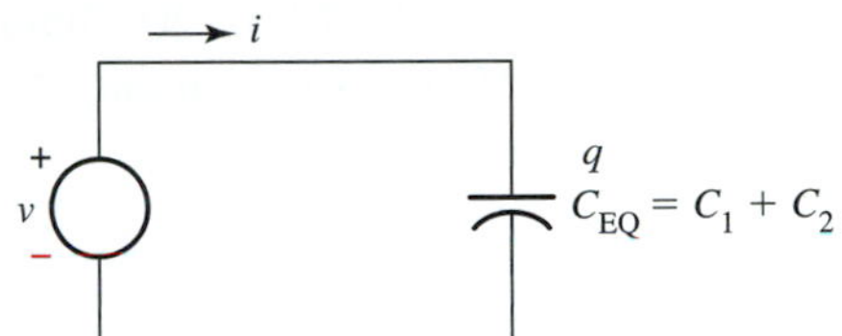

From the preceding equations for q, q_1, and q_2,

$$v = \frac{q}{C_1 + C_2}$$
$$q_1 = C_1 v$$
$$= \left\{\frac{C_1}{C_1 + C_2}\right\} q$$
$$q_2 = C_2 v$$
$$= \left\{\frac{C_2}{C_1 + C_2}\right\} q$$

Differentiating both sides of the preceding equations for q_1 and q_2 with respect to t results in

$$\frac{dq_1}{dt} = \left\{\frac{C_1}{C_1 + C_2}\right\} \frac{dq}{dt}$$
$$\frac{dq_2}{dt} = \left\{\frac{C_2}{C_1 + C_2}\right\} \frac{dq}{dt}$$

and since $i = dq/dt$, $i_1 = dq_1/dt$, and $i_2 = dq_2/dt$,

$$i_1 = \left\{\frac{C_1}{C_1 + C_2}\right\} i$$
$$i_2 = \left\{\frac{C_2}{C_1 + C_2}\right\} i$$

These equations can be extended to n capacitors in parallel, as shown in Figure 10.41. The expression for i_x is

$$i_x = \left\{\frac{C_x}{C_{EQ}}\right\}(i) \tag{10.7}$$

Figure 10.41 Capacitors in parallel.

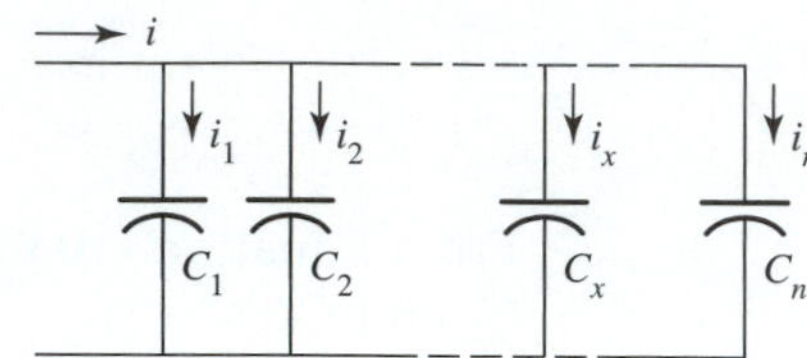

where C_x is the capacitance through which i_x flows and

$$C_{EQ} = C_1 + C_2 + \cdots + C_n$$

Consider the following examples.

EXAMPLE 10.20 Determine the currents i_1 and i_2 for the circuit shown in Figure 10.42 at the instant the current is as shown. ■

Figure 10.42

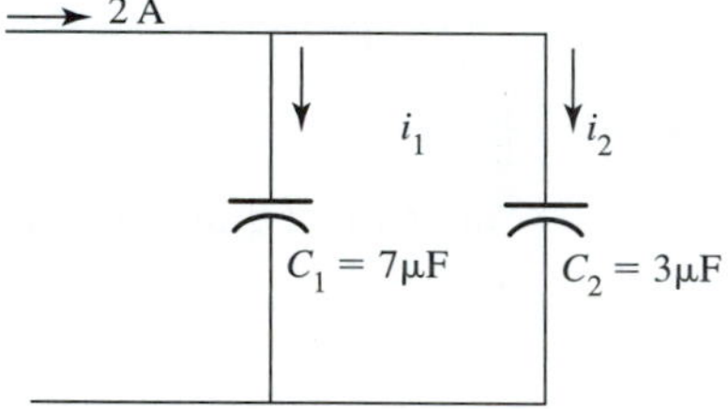

SOLUTION From Equation (10.7)

$$i_1 = \left\{\frac{C_1}{C_1 + C_2}\right\}(i)$$

$$= \left\{\frac{7}{7 + 3}\right\}(2)$$

$$= 1.4 \text{ A}$$

$$i_2 = \left\{\frac{C_2}{C_1 + C_2}\right\}(i)$$

$$= \left\{\frac{3}{7 + 3}\right\}(2)$$

$$= 0.6 \text{ A}$$

EXAMPLE 10.21 Determine the currents i_1, i_2, and i_3 for the circuit shown in Figure 10.43 at the instant the current is as shown. ■

Figure 10.43

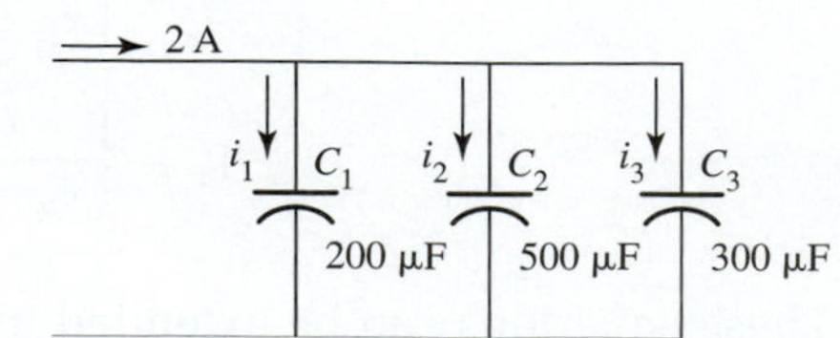

SOLUTION From Equation (10.7)

$$i_1 = \left\{\frac{C_1}{C_{EQ}}\right\}(i)$$

$$= \left\{\frac{C_1}{C_1 + C_2 + C_3}\right\}(i)$$

$$= \left\{\frac{200}{200 + 500 + 300}\right\}(2)$$

$$= 0.4 \text{ A}$$

$$i_2 = \left\{\frac{C_2}{C_1 + C_2 + C_3}\right\}(i)$$

$$= \left\{\frac{500}{200 + 500 + 300}\right\}(2)$$

$$= 1 \text{ A}$$

$$i_3 = \left\{\frac{C_3}{C_1 + C_2 + C_3}\right\}(i)$$

$$= \left\{\frac{300}{200 + 500 + 300}\right\}(2)$$

$$= 0.6 \text{ A}$$

EXAMPLE 10.22 Determine the currents i_1, i_2, i_3, and i_4 for the circuit shown in Figure 10.44 at the instant the current is as shown. ■

Figure 10.44

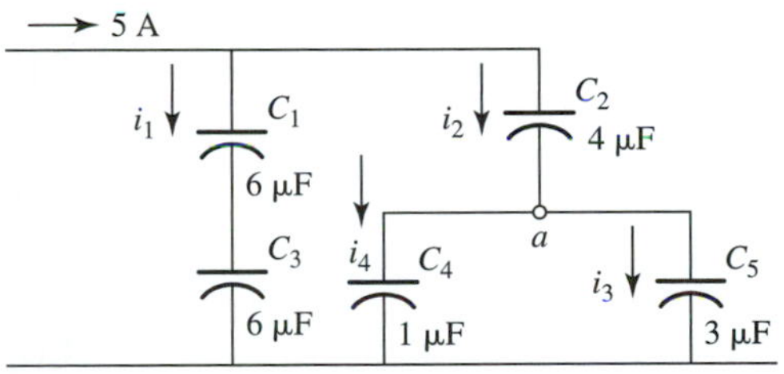

SOLUTION Using Equations (10.5) and (10.6) to combine the two 6-μF capacitors in series and the 1-μF and 3-μF capacitors in parallel results in Figure 10.45.

Figure 10.45

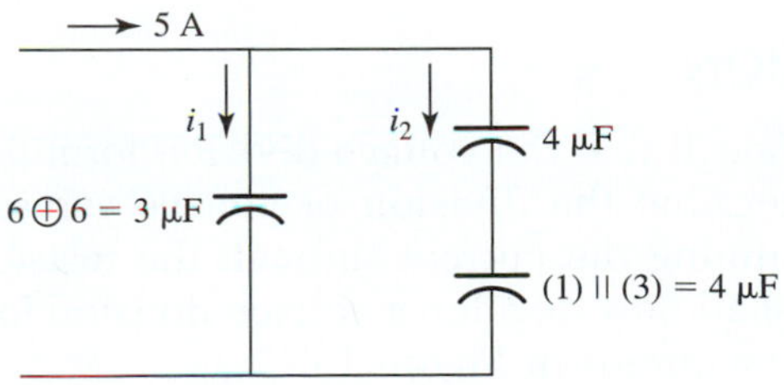

Using Equation (10.5) to combine the two 4-μF capacitors in Figure 10.45 results in the circuit in Figure 10.46. From Equation (10.7)

Figure 10.46

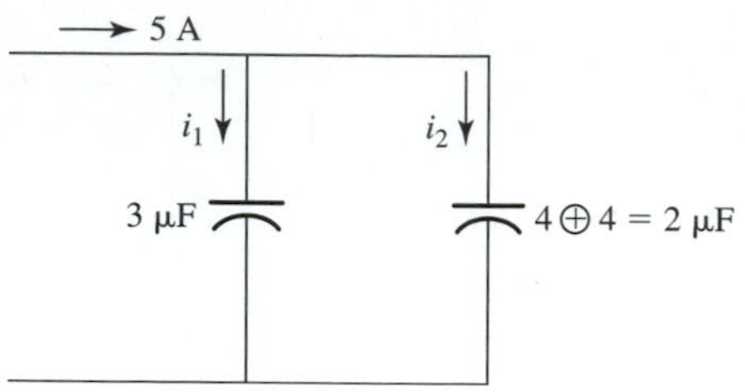

$$i_1 = \left\{\frac{3}{3+2}\right\}(i)$$
$$= \left(\frac{3}{5}\right)(5)$$
$$= 3 \text{ A}$$
$$i_2 = \left\{\frac{2}{3+2}\right\}(5)$$
$$= 2 \text{ A}$$

Applying Equation (10.7) to node a in Figure 10.44,

$$i_4 = \left\{\frac{C_4}{C_4 + C_5}\right\}(i_2)$$
$$= \left\{\frac{1}{1+3}\right\}(2)$$
$$= \frac{1}{2} \text{ A}$$
$$i_3 = \left\{\frac{C_5}{C_4 + C_5}\right\}(i_2)$$
$$= \left\{\frac{3}{1+3}\right\}(2)$$
$$= \frac{3}{2} \text{ A}$$

10.20 VOLTAGE DIVISION

Recall that the voltage-division formula for resistors in series allows us to determine the division of voltage across two or more resistors without determining the current through the resistors that the voltage divides across. We shall now develop a voltage-division formula for capacitors in series. Consider the circuit in Figure 10.47.

Figure 10.47 Capacitors in series.

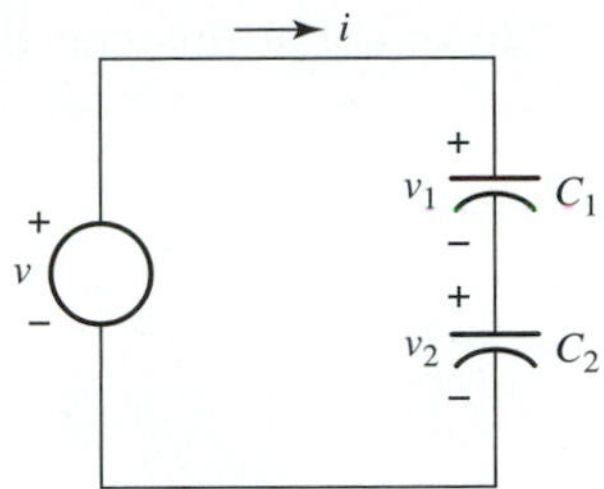

Because the current i flows through C_1 and C_2, the charge on both capacitors is the same. From Equation (10.1)

$$q = C_1 v_1$$
$$q = C_2 v_2$$

where q is the charge on each capacitor.

From Equation (10.6) the equivalent capacitor for C_1 and C_2 in series is

$$C_{EQ} = \frac{C_1 C_2}{C_1 + C_2}$$

and from the equivalent circuit in Figure 10.48 and Equation (10.1),

$$q = \left\{\frac{C_1 C_2}{C_1 + C_2}\right\}(v)$$

where q is the same for C_1, C_2, and C_{EQ} because the same current flows through the three capacitors.

Figure 10.48 Capacitor equivalent.

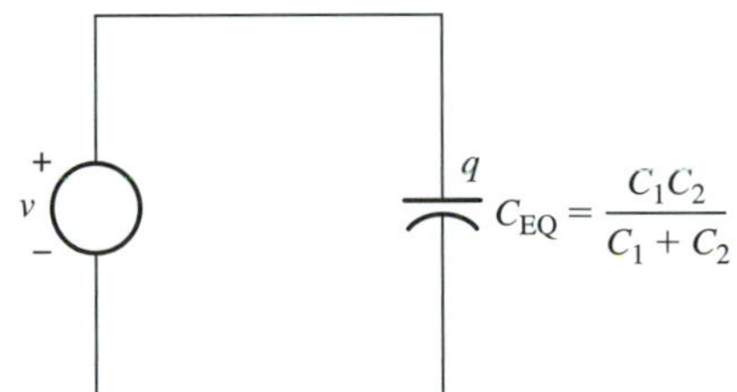

From the preceding equations: $q = C_1 v_1$, $q = C_2 v_2$, and $q = \{C_1 C_2/(C_1 + C_2)\}(v)$.

$$\left\{\frac{C_1 C_2}{C_1 + C_2}\right\} v = C_1 v_1$$

$$\left\{\frac{C_1 C_2}{C_1 + C_2}\right\} v = C_2 v_2$$

$$v_1 = \left\{\frac{C_2}{C_1 + C_2}\right\}(v)$$

$$v_2 = \left\{\frac{C_1}{C_1 + C_2}\right\}(v)$$

These equations can be written as

$$v_1 = \left\{\frac{C_{EQ}}{C_1}\right\}v$$

$$v_2 = \left\{\frac{C_{EQ}}{C_2}\right\}v$$

where

$$C_{EQ} = \frac{C_1C_2}{C_1 + C_2}$$

Voltage division can be extended to n capacitors in series, as shown in Figure 10.49. The expression for v_x for the circuit in Figure 10.49 is

$$v_x = \left\{\frac{C_{EQ}}{C_x}\right\}(v) \tag{10.8}$$

Figure 10.49 Capacitors in series.

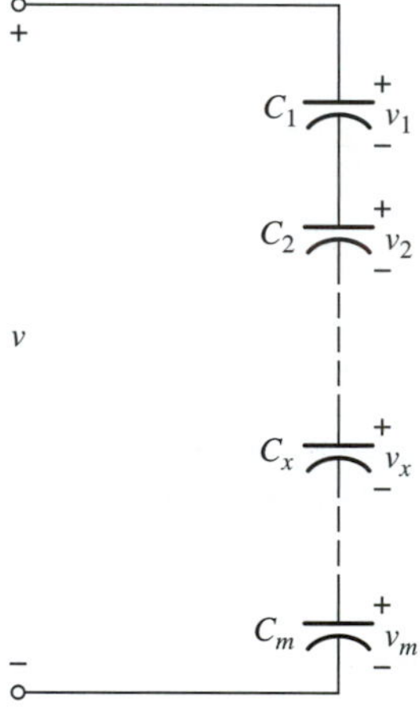

where C_x is the capacitor across which v_x appears and

$$C_{EQ} = \frac{1}{1/C_1 + 1/C_2 + \cdots + 1/C_n}$$

Consider the following examples.

EXAMPLE 10.23 Determine the voltages v_1 and v_2 for the circuit in Figure 10.50 at the instant that the voltage is as shown. ■

Figure 10.50

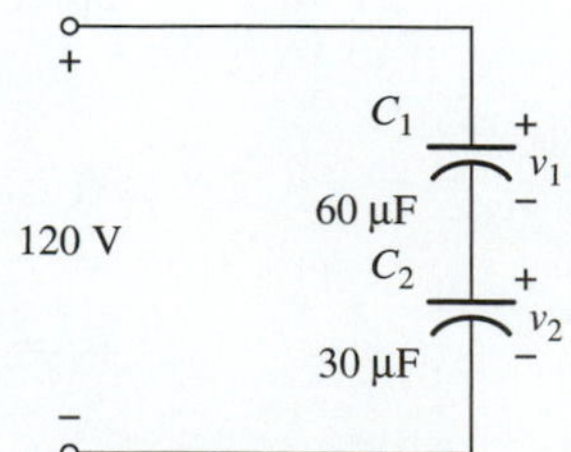

SOLUTION From Equation (10.8)

$$v_1 = \left\{\frac{C_2}{C_1 + C_2}\right\}(v)$$

$$= \left\{\frac{30}{60 + 30}\right\}(120)$$

$$= 40 \text{ V}$$

$$v_2 = \left\{\frac{C_1}{C_1 + C_2}\right\}(v)$$

$$= \left\{\frac{60}{60 + 30}\right\}(120)$$

$$= 80 \text{ V}$$

EXAMPLE 10.24 Determine the voltages v_1, v_2, and v_3 for the circuit in Figure 10.51 at the instant the voltage is as shown. ■

Figure 10.51

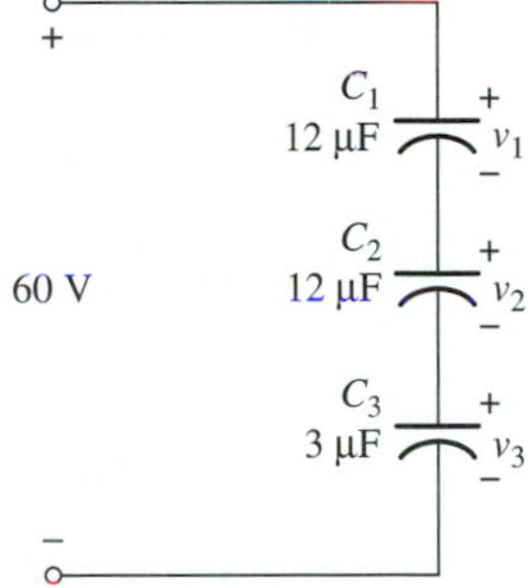

SOLUTION From Equation (10.5)

$$C_{EQ} = \frac{1}{1/C_1 + 1/C_2 + 1/C_3}$$

$$= \frac{1}{\frac{1}{2} + \frac{1}{12} + \frac{1}{12}}$$

$$= 2 \text{ μF}$$

and from Equation (10.8)

$$v_1 = \left\{\frac{C_{EQ}}{C_1}\right\}(v)$$

$$= \left\{\frac{2}{12}\right\}(60)$$

$$= 10 \text{ V}$$

$$v_2 = \left\{\frac{C_{EQ}}{C_2}\right\}(v)$$

$$= \left\{\frac{2}{12}\right\}(60)$$

$$= 10 \text{ V}$$

$$v_3 = \left\{\frac{C_{EQ}}{C_3}\right\}(v)$$

$$= \left\{\frac{2}{3}\right\}(60)$$

$$= 40\,\text{V}$$

EXAMPLE 10.25 Determine the voltages v_1, v_2, v_3, and v_4 for the circuit in Figure 10.52 at the instant the voltage is as shown. ■

Figure 10.52

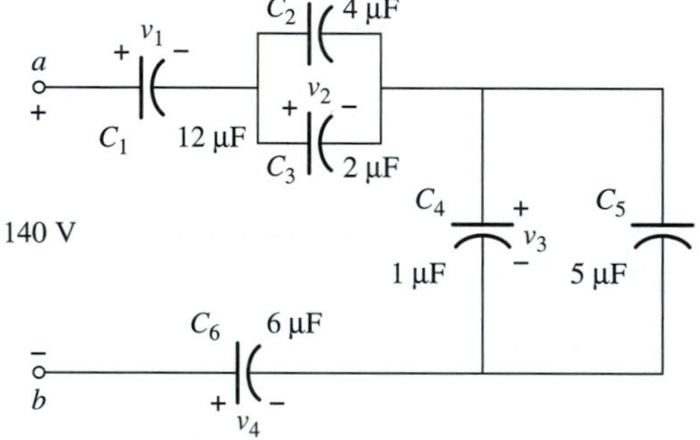

SOLUTION Using Equations (10.5) and (10.6) for the circuit to the right of terminals *a-b*

$$C_{EQ} = C_1 \oplus (C_2 \| C_3) \oplus (C_4 \| C_5) \oplus C_6$$

$$= 12 \oplus 6 \oplus 6 \oplus 6$$

$$= \frac{1}{\frac{1}{12} + \frac{1}{6} + \frac{1}{6} + \frac{1}{6}}$$

$$= \frac{12}{7}\,\mu\text{F}$$

Let C' and C'' be as follows.

$$C' = C_2 \| C_3 = 4 + 2$$

$$= 6\,\mu\text{F}$$

$$C'' = C_4 \| C_5 = 1 + 5$$

$$= 6\,\mu\text{F}$$

From Equation (10.8)

$$v_1 = \left\{\frac{C_{EQ}}{C_1}\right\}(v)$$

$$= \left\{\frac{\frac{12}{7}}{12}\right\}(140)$$

$$= 20\,\text{V}$$

$$v_2 = \left\{\frac{C_{EQ}}{C'}\right\}(v)$$

$$= \left\{\frac{\frac{12}{7}}{6}\right\}(140)$$

$$= 40 \text{ V}$$

$$v_3 = \left\{\frac{C_{EQ}}{C''}\right\}(v)$$

$$= \left\{\frac{\frac{12}{7}}{6}\right\}(140)$$

$$= 40 \text{ V}$$

$$v_4 = \left\{\frac{C_{EQ}}{C_6}\right\}(v)$$

$$= \left\{\frac{\frac{12}{7}}{6}\right\}(140)$$

$$= 40 \text{ V}$$

10.21 GRAPHICAL DETERMINATION OF VOLTAGE AND CURRENT

The basic relationship between current and voltage for a capacitor allows us to graphically determine the waveform of one from the waveform of the other. Consider the capacitor in Figure 10.14 and recall from Equation (10.2) that

$$i = C\frac{dv}{dt}$$

Recall from Chapter 9 that the value of a derivative at a point is equal to the slope of a tangent line drawn at the point. This allows us to graphically evaluate i if a plot of v versus t is given. Because the slope of a tangent line to the v-versus-t plot at any time t_1 is equal to dv/dt evaluated at $t = t_1$, it follows that $i(t_1)$ is equal to C multiplied by the slope of the tangent to the v-versus-t plot at $t = t_1$. Although the graphical procedure presented here is valid for most v-versus-t plots, it is easiest to apply when the v-versus-t plot consists of straight lines as illustrated in the following example.

EXAMPLE 10.26 The voltage across a 20-μF capacitor is given in Figure 10.53. Determine the current i for values of t from $t = 0$ to $t = 5$ s. ■

Figure 10.53

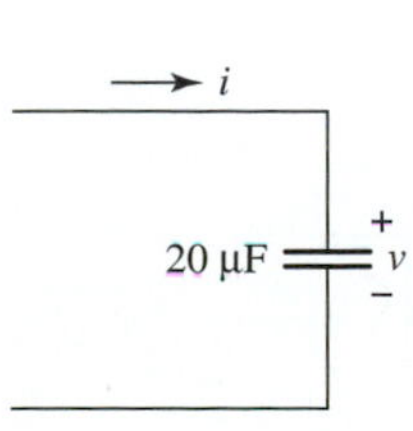

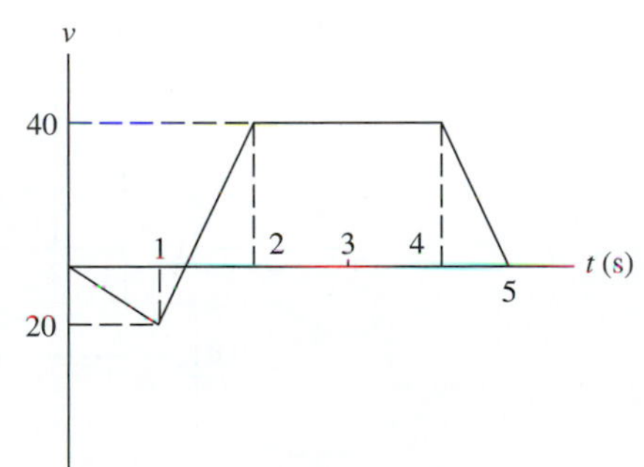

SOLUTION Because the slope of a tangent to v versus t for any time from $t = 0$ to $t = 1$ equals (-20), the current for this time period is

$$
\begin{aligned}
i &= C\frac{dv}{dt} \\
&= \{(20 \times 10^{-6})\}(-20) \\
&= (-400)10^{-6} \\
&= -0.4 \text{ mA}
\end{aligned}
$$

Because the slope of a tangent to v versus t for any time from $t = 1$ to $t = 2$ equals 60, the current for this period of time is

$$
\begin{aligned}
i &= C\frac{dv}{dt} \\
&= \{(20 \times 10^{-6})\}(60) \\
&= 1200 \times 10^{-6} \\
&= 1.2 \text{ mA}
\end{aligned}
$$

Because the slope of a tangent to v versus t for any time from $t = 2$ to $t = 4$ equals 0, the current for this period of time is

$$
\begin{aligned}
i &= C\frac{dv}{dt} \\
&= \{(20 \times 10^{-6})\}(0) \\
&= 0
\end{aligned}
$$

Because the slope of a tangent to v versus t for any time from $t = 4$ to $t = 5$ equals (-40), the current for this period of time is

$$
\begin{aligned}
i &= C\frac{dv}{dt} \\
&= \{(20 \times 10^{-6})\}(-40) \\
&= -800 \times 10^{-6} \\
&= -0.8 \text{ mA}
\end{aligned}
$$

A plot of i versus t from $t = 0$ to $t = 5$ is shown in Figure 10.54.

Figure 10.54

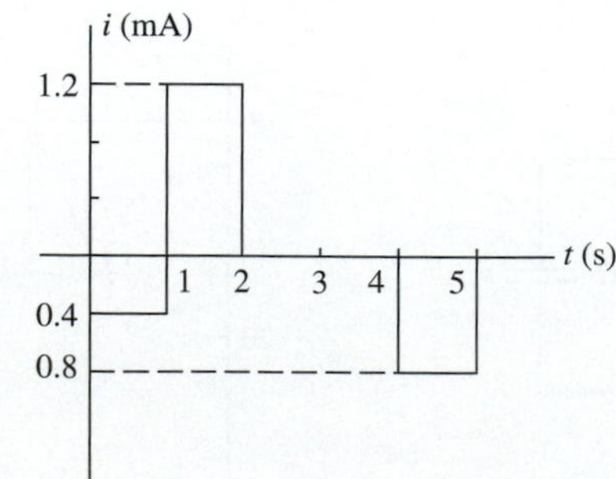

Recalling the definitions of the antiderivative and the indefinite integral and recognizing that $dv/dt = (1/C)i$ from Equation (10.2), we conclude that v is the antiderivative of $(1/C)i$ and can be expressed in terms of the indefinite integral as

$$v = \int\left(\frac{1}{C}\right)i\,dt$$

Because C is a constant, v can also be expressed as

$$v = \frac{1}{C}\int i\,dt$$

Also recalling the definition of the definite integral and again recognizing that v is the antiderivative of $(1/C)i$, we can write the following expression

$$\frac{1}{C}\int_{t_0}^{t} i\,dt = v(t) - v(t_0)$$

where t is any time greater than time t_0 and $v(t)$ and $v(t_0)$ are the values of v at any time t and at $t = t_0$, respectively. It follows that

$$v = \frac{1}{C}\int_{t_0}^{t} i\,dt + v(t_0)$$

and if $t_0 = 0$,

$$v = \frac{1}{C}\int_{0}^{t} i\,dt + v(0)$$

Recalling the area property of the definite integral, the quantity $\int_{t_0}^{t} i\,dt$ is the area under the i-versus-t waveform from time t_0 to any time t. It follows that the voltage at any given time t_1 can be written as

$$v(t_1) = \left(\frac{1}{C}\right)\begin{Bmatrix}\text{area under } i\text{-versus-}t \\ \text{plot from } t_0 \text{ to } t_1\end{Bmatrix} + v(t_0)$$

Although the graphical procedure presented here is valid for most i-versus-t plots, it is easiest to apply when the i-versus-t plot consists of straight lines, as illustrated in the following example.

EXAMPLE 10.27 The current flowing through a 0.2-μF capacitor is given in Figure 10.55. Determine the voltage v for t = 2, 4, and 5 s. The capacitor is initially uncharged. ■

Figure 10.55

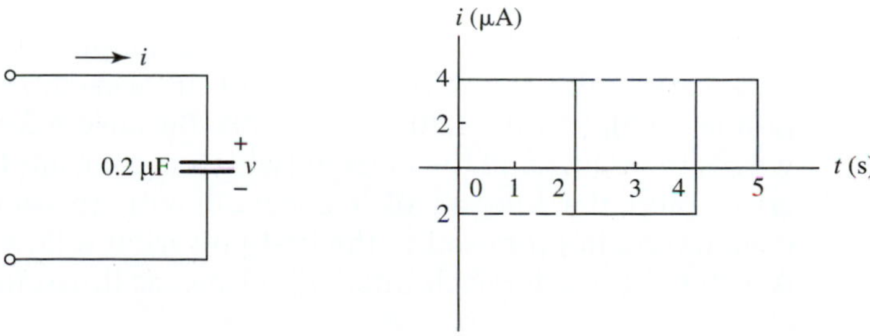

SOLUTION The areas under the i-versus-t curve are shown in Figure 10.56.

Figure 10.56

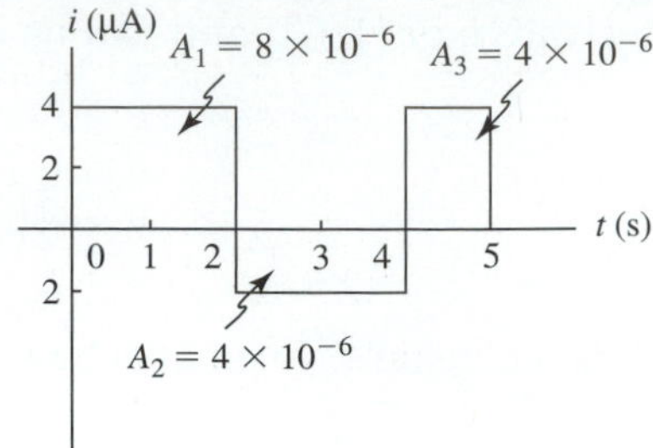

Because the area under i versus t from $t = 0$ to $t = 2$ equals 8×10^{-6} A and $v(0) = 0$, the equation for $v(2)$ results in

$$\begin{aligned} v(2) &= \left(\frac{1}{C}\right)A_1 + v(0) \\ &= \left(\frac{1}{0.2 \times 10^{-6}}\right)(8 \times 10^{-6}) + 0 \\ &= 40 \text{ V} \end{aligned}$$

which is read as "v evaluated at 2 s equals 40 V."

Because the area under i versus t from $t = 2$ to $t = 4$ equals (-4×10^{-6}) and $v(2) = 40$, from before, the equation for $v(4)$ results in

$$\begin{aligned} v(4) &= \left(\frac{1}{C}\right)A_2 + v(2) \\ &= \left(\frac{1}{0.2 \times 10^{-6}}\right)(-2 \times 10^{-6}) + 40 \\ &= -20 + 40 \\ &= 20 \text{ V} \end{aligned}$$

which is read as "v evaluated at 4 s equals 20 V."

Because the area under i versus t from $t = 4$ to $t = 5$ equals (4×10^{-6}) and $v(4) = 20$, from before, the equation for $v(5)$ results in

$$\begin{aligned} v(5) &= \left(\frac{1}{C}\right)A_3 + v(4) \\ &= \left(\frac{1}{0.2 \times 10^{-6}}\right)(4 \times 10^{-6}) + 20 \\ &= 40 \text{ V} \end{aligned}$$

which is read as "v evaluated at 5 s equals 40 V."

If it is desired to plot v versus t in Example 10.27, further investigation is necessary to determine the portion of the waveforms that connect the known points, (0,0), (2,40), (4,20), and (5,40). Because v for each interval of time involves the integral of the current (which is constant for each interval of time as given) and the integral of a constant with respect to t is equal to a first-degree term in t (t raised to the first power), it follows that the expression for v is a straight line for each interval of time, as illustrated in Example 10.28.

EXAMPLE 10.28 Draw a plot of v versus t for the capacitor in Example 10.27.■

SOLUTION From Example 10.27 the known points are (0,0), (2,40), (4,20), and (5,40). Because i is a constant for each interval of time, the known points can be connected by straight lines to form v versus t, as shown in Figure 10.57.

Figure 10.57

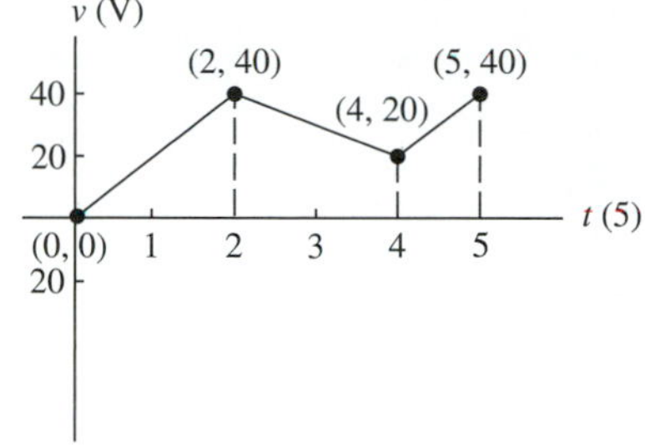

10.22 EXERCISES

1. Determine i for each of the capacitors in Figure 10.58.

Figure 10.58

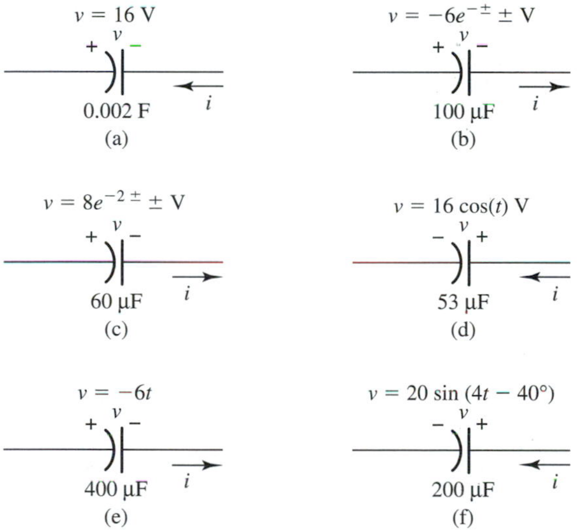

(a) $v = 16$ V, 0.002 F

(b) $v = -6e^{-\pm}\,\pm$ V, 100 μF

(c) $v = 8e^{-2\pm}\,\pm$ V, 60 μF

(d) $v = 16\cos(t)$ V, 53 μF

(e) $v = -6t$, 400 μF

(f) $v = 20\sin(4t - 40°)$, 200 μF

2. Determine the energy associated with each capacitor in Figure 10.58.

3. Determine a single capacitor equivalent with respect to terminals a-b for each circuit in Figure 10.59.

Figure 10.59

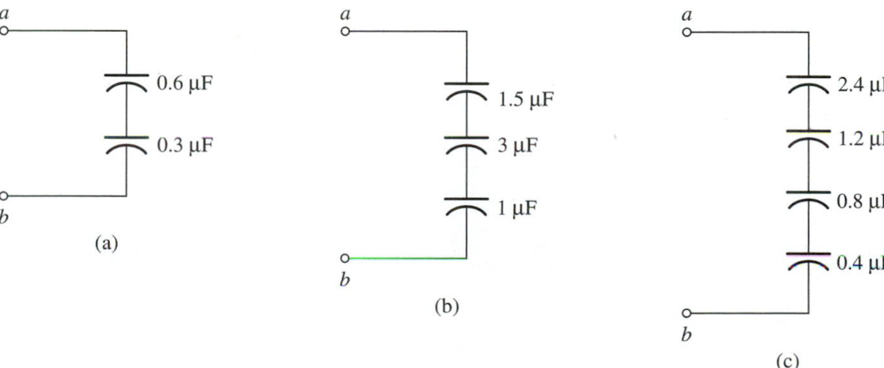

4. Determine a single capacitor equivalent with respect to terminals *a*-*b* for each circuit in Figure 10.60.

Figure 10.60

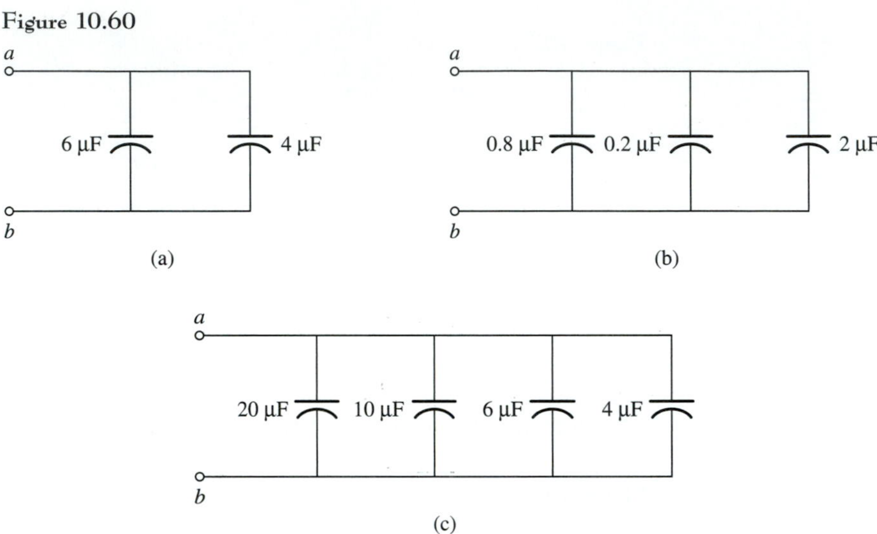

5. Determine a single capacitor equivalent with respect to terminals *a*-*b* for each circuit in Figure 10.61.

Figure 10.61

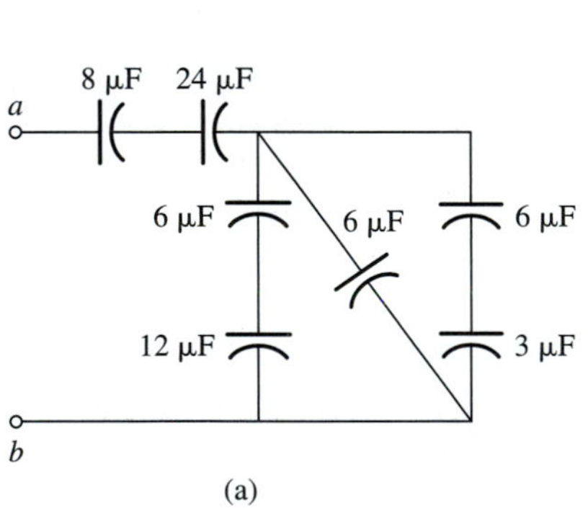

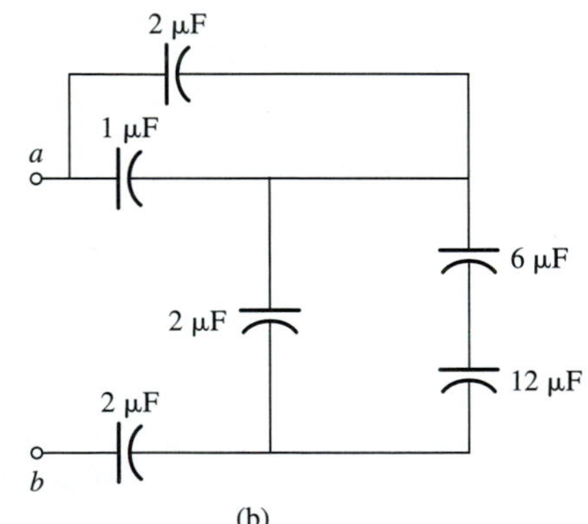

6. Determine a single capacitor equivalent with respect to terminals *a-b* for each circuit in Figure 10.62.

Figure 10.62

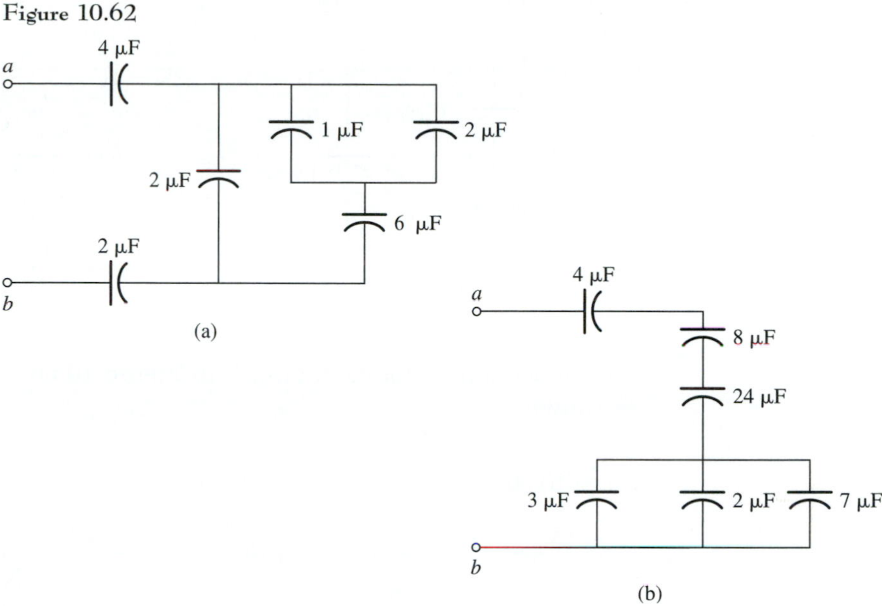

7. Determine the charge on capacitors C_1 and C_2 for each circuit in Figure 10.63.

Figure 10.63

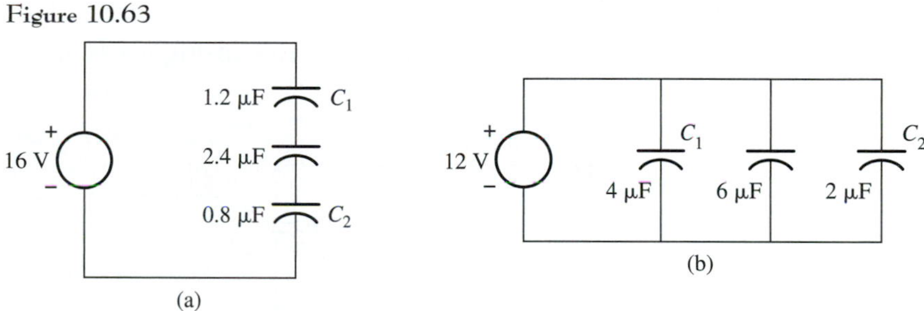

8. Determine the charge on capacitors C_1 and C_2 for each circuit in Figure 10.64.

Figure 10.64

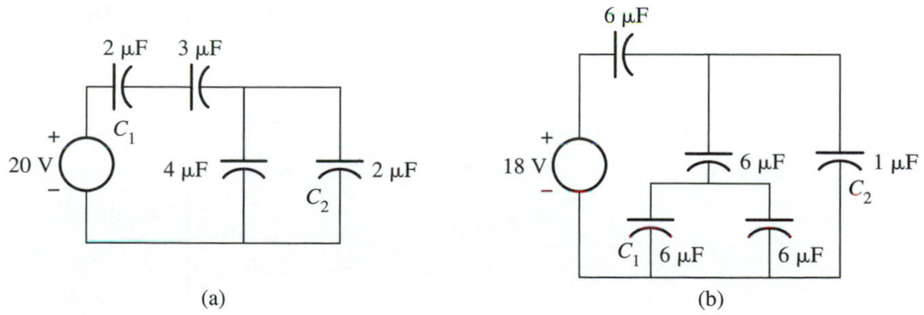

2 μF 3 μF
C_1
+
20 V
−
4 μF
2 μF
C_2
(a)

6 μF
+
18 V
−
6 μF
1 μF
C_2
C_1 6 μF
6 μF
(b)

9. Find i_1 and i_2 for each circuit in Figure 10.65 at the instant the current is as shown.

Figure 10.65

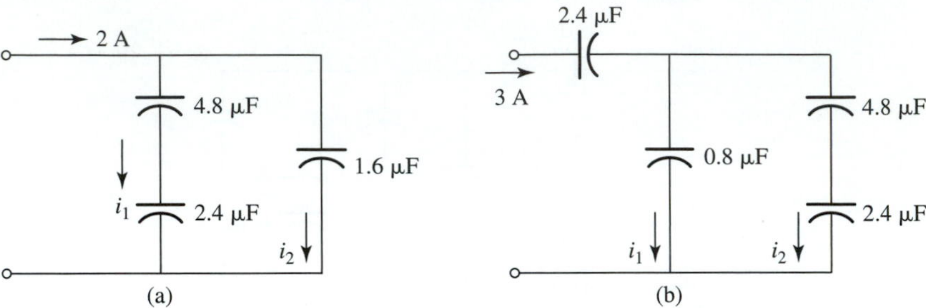

10. Find i_1 and i_2 for each circuit in Figure 10.66 at the instant the current is as shown.

Figure 10.66

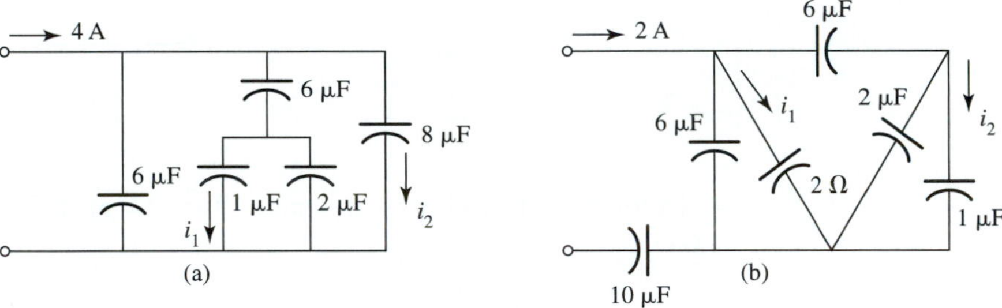

11. Find v_1 and v_2 for the circuit in Figure 10.67 at the instant the voltage is as shown.

Figure 10.67

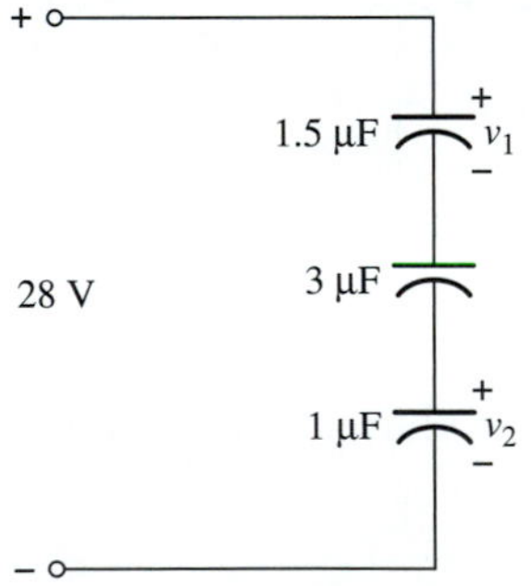

28 V

1.5 µF

3 µF

1 µF

12. Find v_1 and v_2 for the circuit in Figure 10.68 at the instant the voltage is as shown.

Figure 10.68

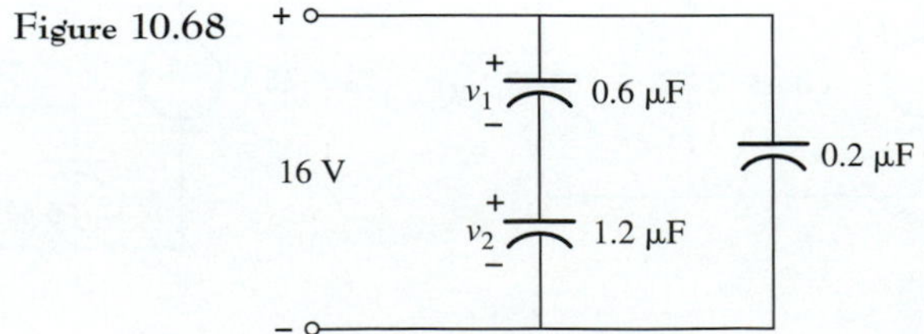

13. Find v_1 and v_2 for the circuit in Figure 10.69 at the instant the voltage is as shown.

Figure 10.69

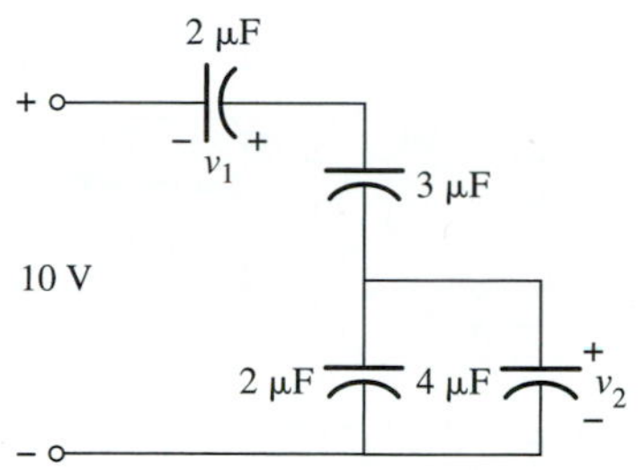

14. Find v_1 and v_2 for the circuit in Figure 10.70 at the instant the voltage is as shown.

Figure 10.70

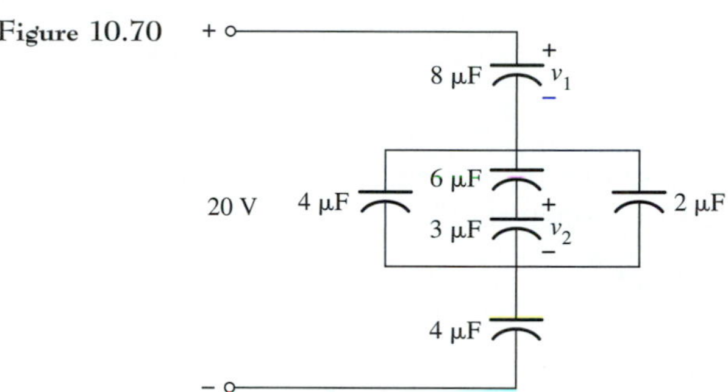

15. The equivalent capacitor for terminals *a-b* is given for the circuit in Figure 10.71. Find *C* for the circuit.

Figure 10.71

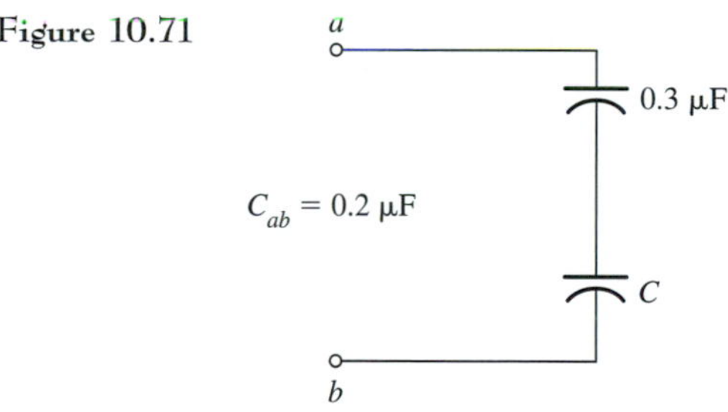

16. The equivalent capacitor for terminals *a-b* is given for the circuit in Figure 10.72. Find *C* for the circuit.

Figure 10.72

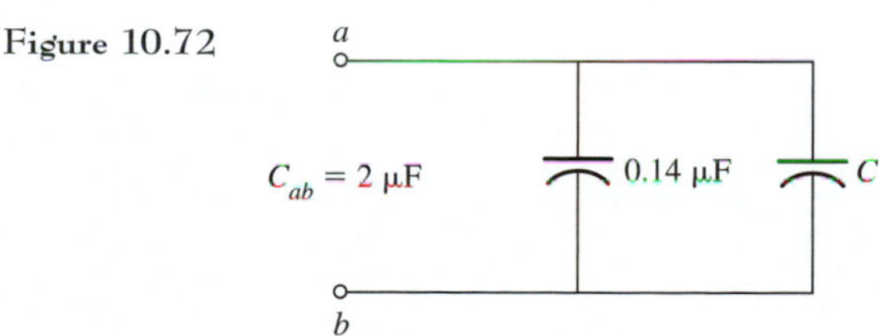

17. The equivalent capacitor for terminals *a-b* is given for the circuit in Figure 10.73. Find *C* for the circuit.

Figure 10.73

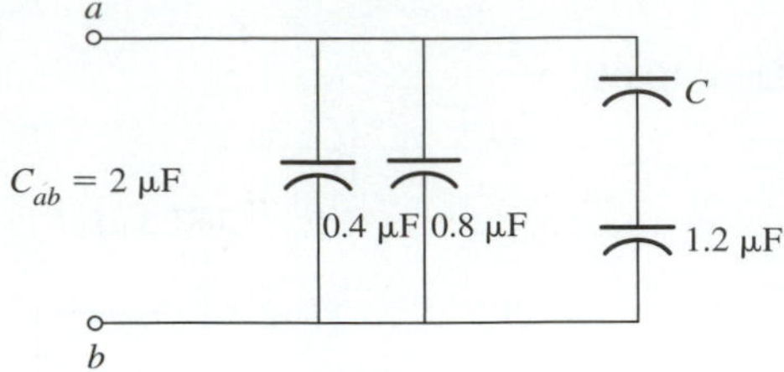

18. The equivalent capacitor for terminals *a-b* is given for the circuit in Figure 10.74. Find *C* for the circuit.

Figure 10.74

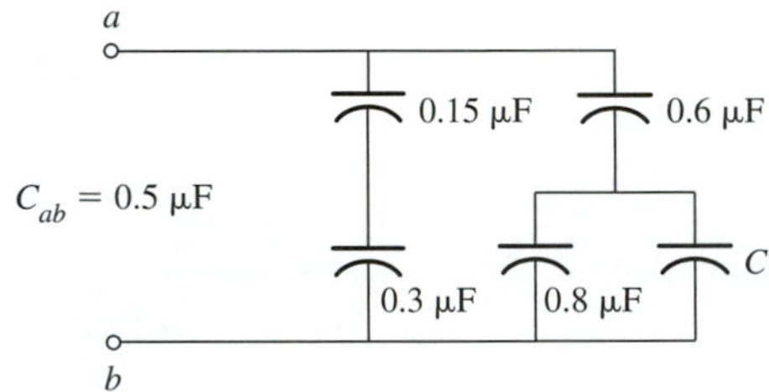

19. A plot of *v* (volts) versus *t* (seconds) for a 200-μF capacitor is given in Figure 10.75. Construct the *i* versus *t* plot.

Figure 10.75

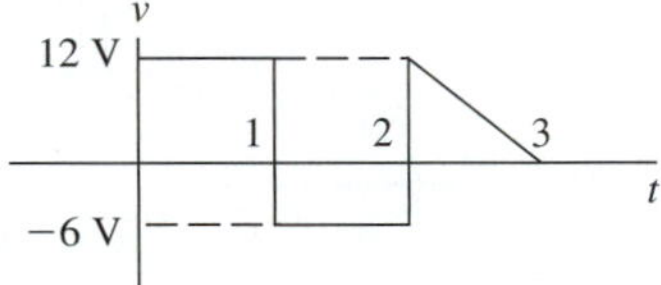

20. A plot of *i* (amperes) versus *t* (seconds) for a 100-μF capacitor is given in Figure 10.76. Construct the *v* versus *t* plot.

Figure 10.76

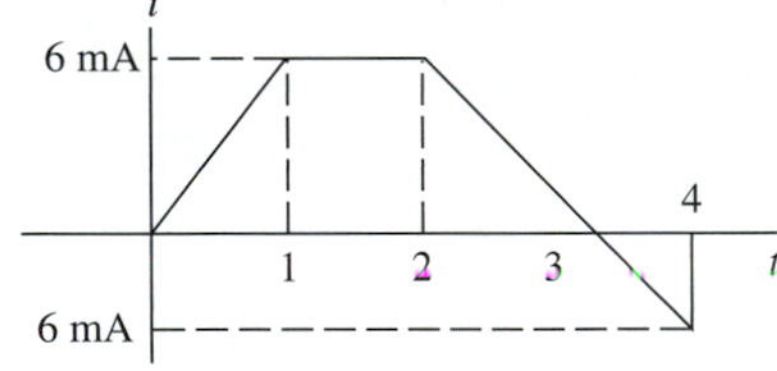

CHAPTER

11

Inductors

After completing this chapter the student should be able to

* describe a basic inductor
* describe the flux-current relationship for inductors
* describe the current-voltage relationship for inductors
* combine inductors in series and/or parallel
* use the current and voltage division formulas to determine currents and voltages for inductor circuits

11.1 BASIC INDUCTOR

A basic inductor consists of a coiled conductor (winding) wound around a magnetic core (such as iron) or a nonmagnetic core (such as air), as illustrated in Figure 11.1.

The measure of the ability of a device to perform as an inductor is called inductance. If the inductance is unintended and occurs inadvertently because of the physical structure of the circuit, it is called *stray inductance*. If a device is deliberately constructed as a circuit element with the inductance intended, it is called an *inductor*. Inductance is a measure of an inductor's ability to store energy in the form of a magnetic field.

Figure 11.1 Basic inductor.

11.2 MAGNETIC FIELDS

Magnetic fields are characterized by the presence of magnetic flux lines. The magnetic flux forms continuous lines (called flux lines) around a conductor when a current flows through the conductor. The magnetic flux is assigned the symbol ϕ (Greek letter phi), and the flux lines are illustrated in Figure 11.2.

Figure 11.2 Lines of flux.

The magnetic field is distinguishable by the force that it causes on an isolated north pole placed in the presence of the field. Magnetic flux is measured in webers (Wb) in honor of the German physicist Wilheim Weber (1804–1891). The flux lines per unit area is called flux density, which is measured in teslas (T), where one weber per square meter is equal to a flux density of one tesla.

When a conductor is in the presence of a magnetic field and there is relative motion of one with respect to the other, a voltage is induced across the conductor. As an example, if the flux lines in Figure 11.3 are stationary and the conductor moves across the flux lines as shown, a voltage is induced across the conductor. In a similar manner, if the conductor in Figure 11.3 is stationary and the current establishing the magnetic field is varied, causing the flux lines to change position with respect to the conductor, a voltage will be induced across the conductor.

Figure 11.3 Voltage induced across a conductor.

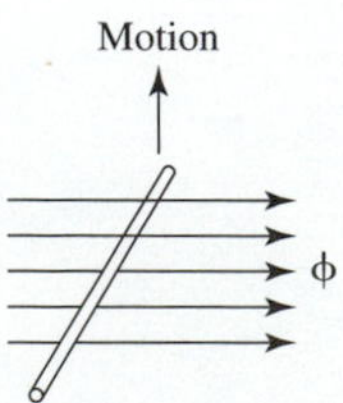

When a current flows through an inductor, the flux lines encircle each coil, as shown in Figure 11.4. The flux in each coil adds to form the total flux

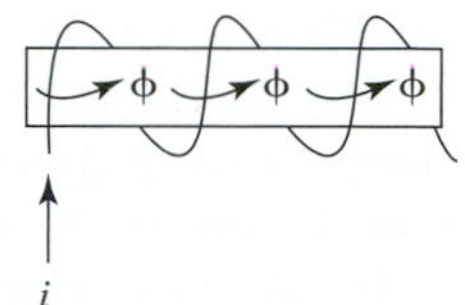

Figure 11.4 Flux per coil.

for the inductor, as shown in Figure 11.5. The total flux is given the symbol of ϕ_T and is equal to $N\phi$, where N is the number of coils in the inductor.

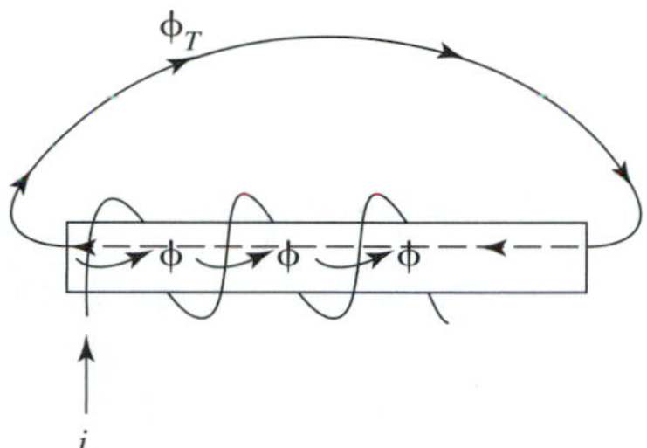

Figure 11.5 Total flux.

Flux direction is related to the direction of the current causing the flux. A rule known as the *right-hand rule* allows us to determine the direction of flux relative to the current direction.

> If the fingers of the right hand are wrapped around the coils of an inductor in the direction of current flow, the thumb will point in the direction of ϕ_T, as seen in Figure 11.5.
>
> If the thumb of the right hand is placed in the direction of current for a single conductor, the fingers will indicate the direction of flux lines encircling the conductor, as seen in Figure 11.2.

11.3 FLUX-CURRENT RELATIONSHIP

Consider the inductor in Figure 11.6, where ϕ_T is the total flux associated with the inductor, as shown. The flux direction is determined by the right-hand rule.

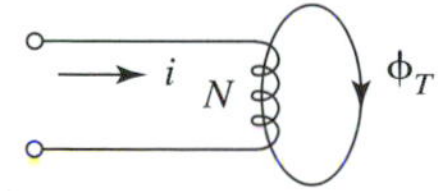

Figure 11.6 Total flux.

The total flux ϕ_T is proportional to the current causing the flux.

$$\phi_T \propto i$$

That proportionality constant is called inductance and is assigned the symbol L. The preceding equation can be written as

$$\phi_T = Li \tag{11.1}$$

where ϕ_T is in webers, i is in amperes, and L is in henrys. The henry is named in honor of American physicist Joseph Henry (1797–1878), and one henry is equal to one weber per ampere.

The inductance L is a constant for a given inductor and depends on the length, area, permeability of the magnetic core material, and the number of turns in the winding of the inductor. The *permeability* of a magnetic core material indicates the material's ability to conduct magnetic flux. Materials that allow flux to flow easily are called magnetic materials and have relatively high values of permeability. Materials that tend to restrict the flow of flux are called diamagnetic and have relatively low values of permeability.

Most inductors are formed by coiling a conducting wire around a core material such as air, iron, or ferrite and are constructed as fixed inductors or variable inductors. The inductance of a variable inductor is usually varied by moving the magnetic core material in or out of the coiled winding by mechanical adjustment to vary the permeability of the inductor core. Inductors are commercially available in sizes ranging from a few microhenrys to the neighborhood of 100 H.

11.4 CURRENT-VOLTAGE RELATIONSHIP

If the current through an inductor varies with time, then the associated total flux, ϕ_T, also varies with time, as seen in Equation (11.1).

If we differentiate both sides of Equation (11.1) with respect to time, the result is

$$\frac{d\{\phi_T\}}{dt} = \frac{d\{Li\}}{dt}$$

and because L is a constant for any given inductor,

$$\frac{d\{\phi_T\}}{dt} = L\frac{di}{dt}$$

The law of electromagnetic induction presented by British physicist Michael Faraday (1791–1867) states that the voltage induced across an inductor caused by a changing magnetic flux is given by

$$v = \frac{d\phi_T}{dt} \tag{11.2}$$

Combining Equation (11.2) with the preceding equation,

$$\frac{d\phi_T}{dt} = L\frac{di}{dt}$$

results in

$$v = L\frac{di}{dt} \tag{11.3}$$

where v is in volts, i is in amperes, t is in seconds, and L is in henrys. Because $di/dt = (1/L)v$, i is the antiderivative of $\{(1/L)v\}$ and can also be written in terms of the indefinite integral as

$$i = \frac{1}{L}\int v\,dt$$

The induced voltage v_{ab} described in Equation (11.3) is shown in Figure 11.7. We arrive at the polarity for v_{ab} in Figure 11.7 by applying a basic law

Figure 11.7 Induced voltage.

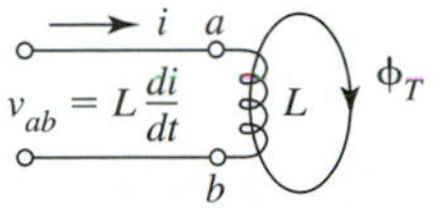

known as Lenz's law, formulated by the German scientist Heinrich Lenz around 1834. The law states that the polarity of the voltage induced across an inductor by a changing flux is in a direction that opposes the current causing the change in flux. That is, if the current is increasing, the polarity of the induced voltage will be in a direction that tends to decrease the current; and if the current is decreasing, the polarity of the inducted voltage will tend to increase the current.

Consider the following example.

EXAMPLE 11.1 If $i = 16e^{-4t}$ A and $L = 2$ H for the circuit in Figure 11.8, determine v_{ab}. ■

Figure 11.8

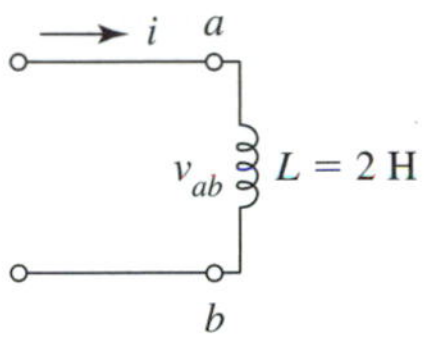

SOLUTION From Equation (11.3) and differentiation rules D3 and D6,

$$
\begin{aligned}
v_{ab} &= L\frac{di}{dt} \\
&= (2)\left\{\frac{d\{16e^{-4t}\}}{dt}\right\} \\
&= (2)(16)(-4)e^{-4t} \\
&= -(128)e^{-4t}\text{ V}
\end{aligned}
$$

Notice that because the current is decreasing, v_{ab} is negative. That is, point b is positive with respect to point a, and the induced voltage tends to increase the current.

The symbol for the inductor as a circuit element is given in Figure 1.3(b) and is redrawn in Figure 11.9. The relationship between the voltage v across the inductor and the current i flowing through the inductor is given by Equation (11.3).

Figure 11.9 Inductor symbol.

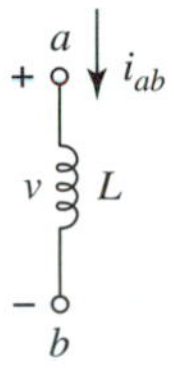

The conventional current flowing from point a to point b is designated by i_{ab}, and the conventional current flowing from point b to point a is designated by i_{ba}, where

$$i_{ab} = -i_{ba}$$

Applying Equation (11.3) to the inductor in Figure 11.9 results in

$$v_{ab} = L\frac{di_{ab}}{dt}$$

and

$$v_{ba} = L\frac{di_{ba}}{dt}$$

where v_{ab} is the voltage of point a with respect to b, v_{ba} is the voltage of point b with respect to a, and

$$v_{ab} = -v_{ba}$$

Consider the following examples.

EXAMPLE 11.2 Determine v_{ab} and v_{ba} for the inductor in Figure 11.10 if $i_{ab} = e^{-t}$ amperes. ■

Figure 11.10

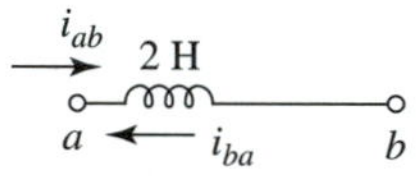

i_{ab} 2 H

a i_{ba} b

SOLUTION From Equation (11.3) and differentiation rule D3,

$$\begin{aligned} v_{ab} &= L\frac{di_{ab}}{dt} \\ &= 2\frac{d\{e^{-t}\}}{dt} \\ &= -2e^{-t} \end{aligned}$$

Because $i_{ba} = -i_{ab}$ and from Equation (11.3),

$$\begin{aligned} v_{ba} &= L\frac{di_{ba}}{dt} \\ &= 2\frac{d\{-e^{-t}\}}{dt} \\ &= 2e^{-t} \end{aligned}$$

EXAMPLE 11.3 Determine v_{ab} and v_{ba} for the inductor in Figure 11.11 if $i_{ab} = 2 \cos 4t$ amperes. ■

Figure 11.11

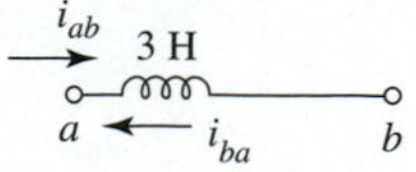

i_{ab} 3 H

a i_{ba} b

SOLUTION From Equation (11.3) and differentiation rules D5 and D6,

$$\begin{aligned} v_{ab} &= L\frac{di_{ab}}{dt} \\ &= 3\frac{d\{2\cos 4t\}}{dt} \\ &= (3)(2)(4)(-\sin 4t) \\ &= -24\sin 4t \text{ V} \end{aligned}$$

Because $i_{ba} = -i_{ab}$ and from Equation (11.3),

$$\begin{aligned} v_{ba} &= L\frac{di_{ba}}{dt} \\ &= 3\frac{d\{-2\cos 4t\}}{dt} \\ &= (3)(-2)(4)(-\sin 4t) \\ &= 24\sin 4t \text{ V} \end{aligned}$$

Alternatively, because $v_{ba} = -v_{ab}$,

$$\begin{aligned} v_{ba} &= -\{-24\sin 4t\} \\ &= 24\sin 4t \text{ V} \end{aligned}$$

Although the procedure for labeling terminals with letters to designate voltages and currents is convenient, we shall also present the more commonly used polarity notation, as given previously for resistors and capacitors. Consider the inductor in Figure 11.12, where v is the voltage of the positive-marked terminal with respect to the negative-marked terminal.

Figure 11.12 v–i relationship.

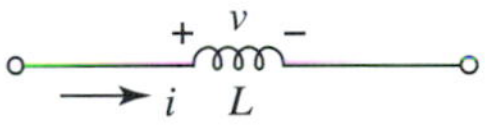

When applying Equation (11.3), if the current i is expressed as flowing from the positive-marked terminal to the negative-marked terminal, as shown, then v must be expressed as the voltage of the positive-marked terminal with respect to the negative-marked terminal.

The positive and negative markings on the terminals of the inductor do not necessarily correspond to the actual polarities of the terminals. For instance, if $v = -10$ volts in Figure 11.12, the positive-marked terminal is 10 volts negative with respect to the negative-marked terminal. Similarly, if i is negative, then the actual current flows opposite to the direction assigned.

Consider the following examples.

EXAMPLE 11.4 Determine the voltage v if $i = 6e^{-8t}$ amperes, as shown in Figure 11.13. ■

Figure 11.13

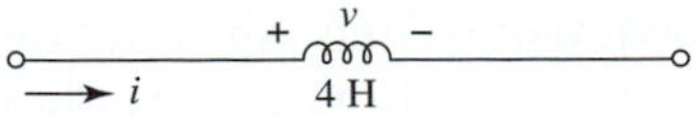

SOLUTION From Equation (11.3) and differentiation rules D3 and D6,

$$\begin{aligned} v &= L\frac{di}{dt} \\ &= (4)\frac{d\{6e^{-8t}\}}{dt} \\ &= (4)(6)(-8)(e^{-8t}) \\ &= -192e^{-8t}\text{ V} \end{aligned}$$

EXAMPLE 11.5 Determine the voltage v if $i = -2\cos 4t$ amperes, as shown in Figure 11.14. ■

Figure 11.14

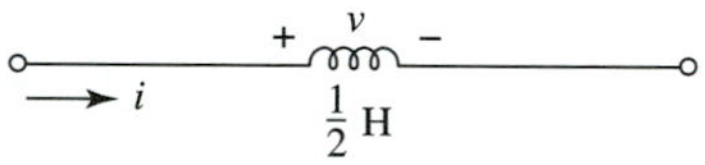

SOLUTION From Equation (11.3) and differentiation rules D5 and D6,

$$\begin{aligned} v &= L\frac{di}{dt} \\ &= \left(\frac{1}{2}\right)\frac{d\{-2\cos 4t\}}{dt} \\ &= \left(\frac{1}{2}\right)(-2)(4)(-\sin 4t) \\ &= 4\sin 4t\text{ V} \end{aligned}$$

EXAMPLE 11.6 Determine the voltage v in Example 11.5 if i is in the opposite direction. ■

SOLUTION

$$\begin{aligned} v &= L\frac{d\{-i\}}{dt} \\ &= \left(\frac{1}{2}\right)\frac{d\{-(-2\cos 4t)\}}{dt} \\ &= \left(\frac{1}{2}\right)(-1)(-2)(4)(-\sin 4t) \\ &= -4\sin 4t\text{ V} \end{aligned}$$

11.5 ENERGY STORAGE

An ideal inductor has zero resistance and does not dissipate energy. However, all realizable inductors have resistance and dissipate a small amount of energy, but the energy loss is small for high-quality inductors. All inductors used in this text are considered ideal for analysis purposes unless otherwise indicated.

As previously stated, an inductor stores energy in a magnetic field. The energy stored at any time t depends on the current flowing through the inductor at that time. Consider the inductor in Figure 11.12. From Equation (2.7) for the instantaneous power $\{p = vi\}$ and Equation (11.3), $\left\{v = L\dfrac{di}{dt}\right\}$,

$$\begin{aligned} p &= vi \\ &= \left\{L\frac{di}{dt}\right\}i \\ &= Li\frac{di}{dt} \end{aligned}$$

Because power is the rate of energy change,

$$\begin{aligned} \frac{dw}{dt} &= p \\ &= Li\frac{di}{dt} \end{aligned}$$

From the definitions of the antiderivative and the indefinite integral, w is the antiderivative of $\left\{Li\dfrac{di}{dt}\right\}$ with respect to time and can be written as

$$w = \int\left\{Li\frac{di}{dt}\right\}dt$$

Also recalling the definition of the definite integral and again recognizing that w is the antiderivative of $\left\{Li\dfrac{di}{dt}\right\}$, we can write the following expression:

$$\int_0^t\left\{Li\frac{di}{dt}\right\}dt = w(t) - w(0)$$

where t is any time greater than $t = 0$ and $w(t)$ and $w(0)$ are the values of w at any time t and at $t = 0$, respectively. It follows that

$$w = \int_0^t\left\{Li\frac{di}{dt}\right\}dt + w(0)$$

Consider the expression $\{L(i^2/2)\}$, which can be written as $\{(L/2)(i)(i)\}$. Recognizing that $(L/2)$ is a constant and employing rules D6 and D9 to differentiate the expression with respect to time results in

$$\begin{aligned} \frac{d\{(L/2)(i)(i)\}}{dt} &= \left(\frac{L}{2}\right)\left\{i\frac{di}{dt} + \frac{di}{dt}i\right\} \\ &= Li\frac{di}{dt} \end{aligned}$$

It follows that $\{L(i^2/2)\}$ is the antiderivative of $\left\{Li\dfrac{di}{dt}\right\}$, so the definite integral for w can be written as

$$w = L\frac{i^2}{2}\bigg|_0^t + w(0)$$

If the initial energy and the inductor current are both zero at the beginning of the inductor life, then $w(0) = 0$, $i(0) = 0$, and substitution of the limits of integration (0 and t) into the expression $\{L(i^2/2)\}$ results in

$$w = L\frac{i^2}{2}\bigg|_t - L\frac{i^2}{2}\bigg|_0 + w(0)$$

$$w = L\frac{i^2}{2} \tag{11.4}$$

where L is in henrys, i is the current flowing through the inductor in amperes, and w is the energy stored in joules. The energy stored varies with time if the current varies with time.

Equation (11.4) can be arrived at more easily by examining the previous equation,

$$w = \int_0^t \left\{Li\frac{di}{dt}\right\} dt + w(0)$$

Although we have not discussed algebraic operations involving the differential quantity dt, it is possible to cancel dt from the preceding equation as follows.

$$w = \int_0^t \left\{Li\frac{di}{\cancel{dt}}\right\} \cancel{dt} + w(0)$$

$$= \int_0^t \{Li\}\, di + w(0)$$

The integration is now with respect to i, and using rule I1 from Chapter 9,

$$w = L\frac{i^2}{2}\bigg|_0^t + w(0)$$

as previously determined.

EXAMPLE 11.7 Determine the energy stored in the inductor in Figure 11.15. ■

Figure 11.15

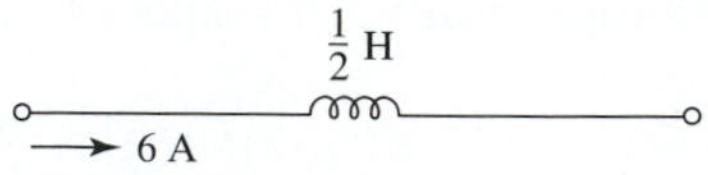

SOLUTION From Equation (11.4)

$$w = \left(\frac{1}{2}\right)Li^2$$

$$= \left(\frac{1}{2}\right)\left(\frac{1}{2}\right)(6)^2$$

$$= 9 \text{ J}$$

EXAMPLE 11.8 Determine the energy stored in the inductor in Figure 11.16, where $i = e^{-2t}$ amperes. ■

Figure 11.16 (2 H inductor, current i)

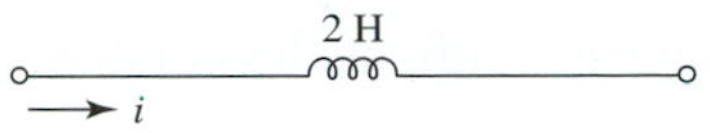

SOLUTION From Equation (11.4)

$$w = \left(\frac{1}{2}\right)Li^2$$

$$= \left(\frac{1}{2}\right)(2)(e^{-2t})^2$$

$$= e^{-4t}\text{ J}$$

Notice that the energy stored in the inductor varies with time, because the current flowing through the inductor varies with time.

11.6 INDUCTORS IN SERIES

Consider the two circuits in Figure 11.17. From the equivalency concept previously presented, the series combination (L_1 and L_2) in Figure 11.17(a) is equivalent to the single inductor (L_{EQ}) in Figure 11.17(b) if the voltage-current relationship for terminals a-b is the same for both circuits. That is, if the voltage source v is the same in both circuits, then the current i is the same in both circuits when the single inductor is equivalent to the series combination.

Figure 11.17 Inductors in series.

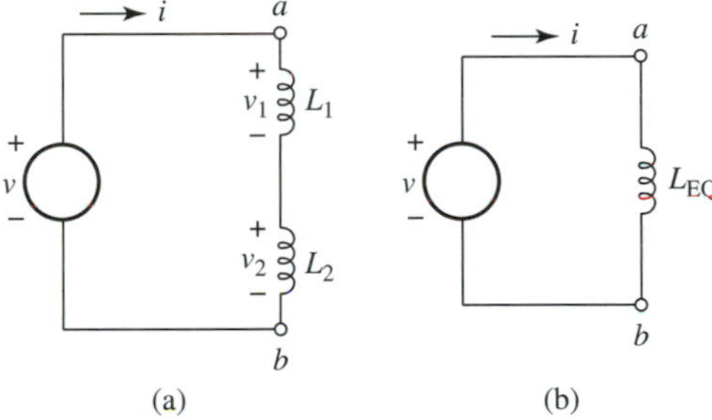

Applying Equation (11.3) and noting that the current i is the same for both inductors for the circuit in Figure 11.17(a) results in

$$v_1 = L_1\frac{di}{dt}$$

$$v_2 = L_2\frac{di}{dt}$$

Applying Equation (11.3) to the circuit in Figure 11.17(b) results in

$$v = L_{EQ}\frac{di}{dt}$$

Applying Kirchhoff's voltage law to the circuit in Figure 11.17(a) results in

$$v = v_1 + v_2$$

Because v is the same for both circuits

$$L_{EQ}\frac{di}{dt} = L_1\frac{di}{dt} + L_2\frac{di}{dt}$$

and dividing each side of the equation by di/dt results in

$$L_{EQ} = L_1 + L_2$$

This equation can be extended to include any number of inductors in series.

In general, n inductors in series can be combined as indicated in Figure 11.18. The equation for combining n inductors in series is

$$L_{EQ} = L_1 + L_2 + L_3 + \cdots + L_n \tag{11.5}$$

Figure 11.18 Inductors in series.

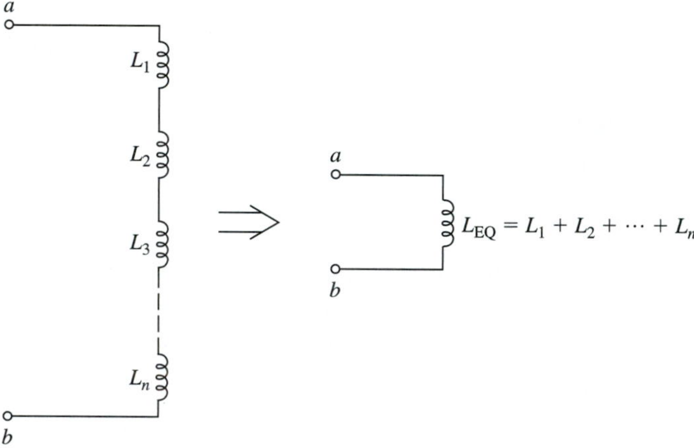

Notice the similarity between Equation (11.5) and Equation (3.1) for combining resistors in series.

Consider the following example.

EXAMPLE 11.9 Replace the three series inductors in Figure 11.19 by a single inductor. ■

Figure 11.19

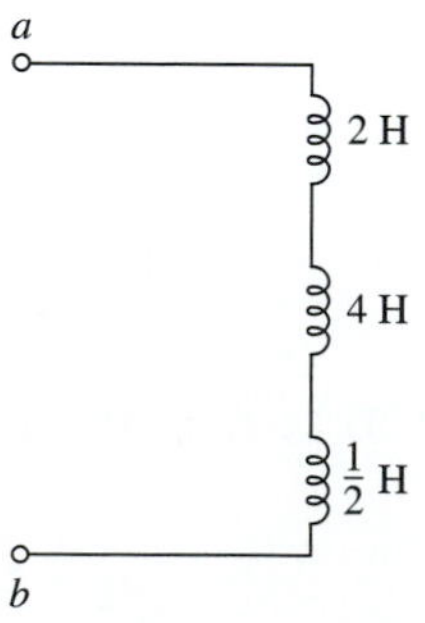

SOLUTION From Equation (11.5)

$$L_{EQ} = 2 + 4 + \left(\frac{1}{2}\right)$$

$$= 6.5 \text{ H}$$

See Figure 11.20.

Figure 11.20

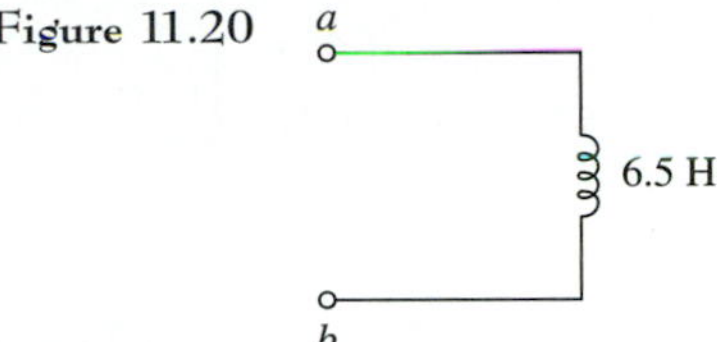

11.7 INDUCTORS IN PARALLEL

Consider the parallel combination of inductors in Figure 11.21. From the equivalency concept previously presented the parallel combination (L_1 and L_2) in Figure 11.21(a) is equivalent to the single inductor (L_{EQ}) in Figure 11.21(b) when the voltage-current relationship for terminals *a-b* is the same for both circuits. That is, if the voltage source v is the same in both circuits, then the current i is the same in both circuits when the single inductor is equivalent to the parallel combination.

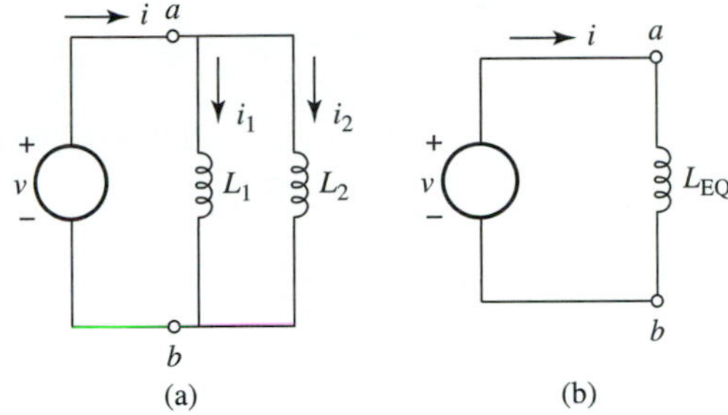

Figure 11.21 Inductors in parallel.

Applying Kirchhoff's current law to the circuit in Figure 11.21(a) results in

$$i = i_1 + i_2$$

where i is the same for both circuits in Figure 11.21. Recognizing that v is the voltage across each of the inductors L_1, L_2, and L_{EQ} and applying Equation (11.3) to the circuits in Figure 11.21(a) and (b) results in

$$v = (L_1)\frac{di_1}{dt}$$

$$v = (L_2)\frac{di_2}{dt}$$

$$v = (L_{EQ})\frac{di}{dt}$$

and it follows that

$$\frac{di_1}{dt} = \left(\frac{1}{L_1}\right)v$$

$$\frac{di_2}{dt} = \left(\frac{1}{L_2}\right)v$$

$$\frac{di}{dt} = \left(\frac{1}{L_{EQ}}\right)v$$

From the definition of the indefinite integral,

$$i = \int \left(\frac{1}{L_{EQ}}\right) v\, dt$$

and from rule I6,

$$i = \left(\frac{1}{L_{EQ}}\right) \int v\, dt$$

In a similar manner,

$$i_1 = \left(\frac{1}{L_1}\right) \int v\, dt$$

$$i_2 = \left(\frac{1}{L_2}\right) \int v\, dt$$

Substituting these expressions into the preceding equation ($i = i_1 + i_2$) results in

$$\left(\frac{1}{L_{EQ}}\right) \int v\, dt = \left(\frac{1}{L_1}\right) \int v\, dt + \left(\frac{1}{L_2}\right) \int v\, dt$$

Dividing both sides of the equation by $\int v\, dt$ results in

$$\frac{1}{L_{EQ}} = \left(\frac{1}{L_1}\right) + \left(\frac{1}{L_2}\right)$$

and

$$L_{EQ} = \frac{1}{(1/L_1) + (1/L_2)}$$

This equation can be extended to include any number of inductors in parallel.

In general, n inductors in parallel can be combined as indicated in Figure 11.22. The equation for combining n inductors in parallel is

$$L_{EQ} = \frac{1}{1/L_1 + 1/L_2 + 1/L_3 + \cdots + 1/L_n} \tag{11.6}$$

Figure 11.22 Inductors in parallel.

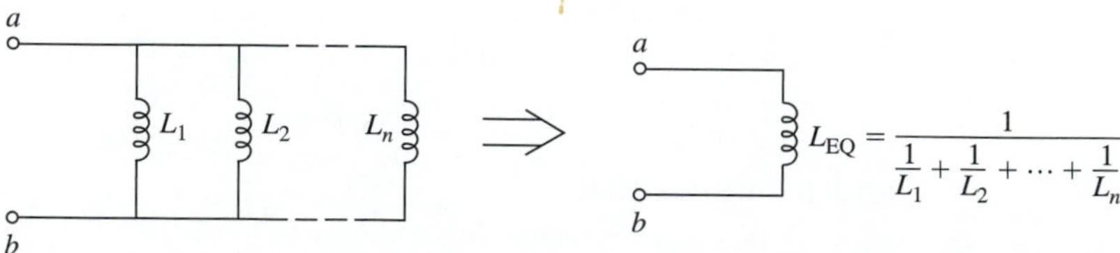

Notice the similarity between Equation (11.6) and Equation (3.4) for combining resistors in parallel. Notice also that for two inductors in parallel, Equation (11.6) can be written as

$$L_{EQ} = \frac{L_1 L_2}{L_1 + L_2}$$

Consider the following examples.

EXAMPLE 11.10 Replace the two inductors in Figure 11.23 by a single inductor. ■

Figure 11.23

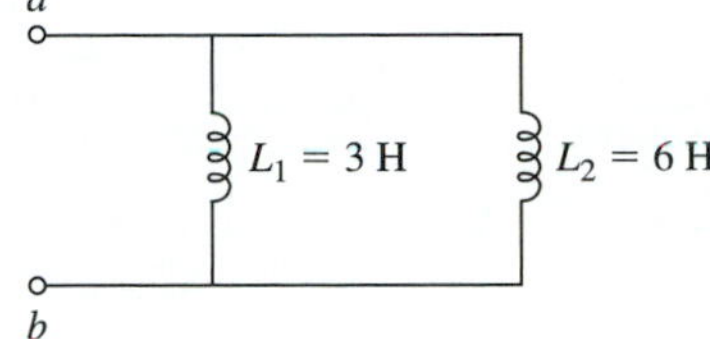

SOLUTION From Equation (11.6)

$$L_{EQ} = \frac{(3)(6)}{3 + 6} = 2 \text{ H}$$

See Figure 11.24.

Figure 11.24

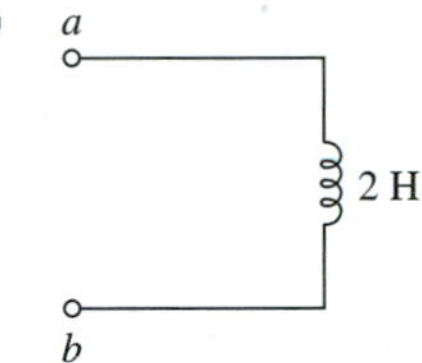

EXAMPLE 11.11 Replace the three inductors in Figure 11.25 by a single inductor. ■

Figure 11.25

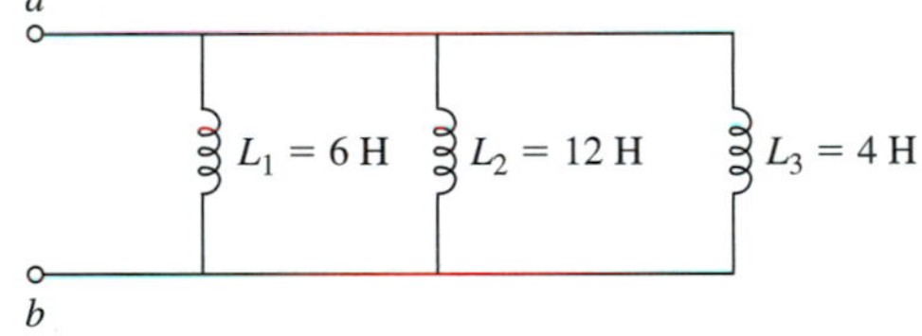

SOLUTION From Equation (11.6)

$$L_{EQ} = \frac{1}{1/L_1 + 1/L_2 + 1/L_3} = \frac{1}{\frac{1}{6} + \frac{1}{12} + \frac{1}{4}} = \frac{1}{\frac{6}{12}} = 2 \text{ H}$$

See Figure 11.26.

Figure 11.26

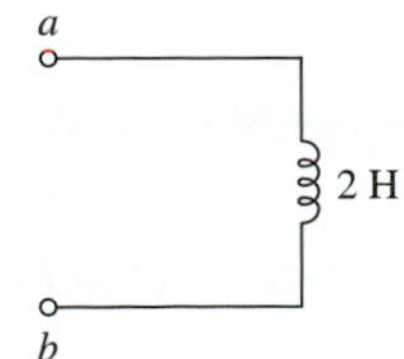

Alternatively, using Equation (11.6) for two inductors twice,

$$L'_{EQ} = \frac{L_1 L_2}{L_1 + L_2} = \frac{(6)(12)}{6 + 12} = 4 \text{ H}$$

and

$$L_{EQ} = \frac{L'_{EQ} L_3}{L'_{EQ} + L_3} = \frac{(4)(4)}{4 + 4} = 2 \text{ H}$$

11.8 SERIES-PARALLEL COMBINATIONS

We shall now consider inductors connected in series-parallel combinations. It is advantageous to use a concise notation as we did for representing resistors and capacitors in parallel and series.

$(L_1 + L_2)$ is read as "L_1 and L_2 are in series."
$(L_1 \| L_2)$ is read as "L_1 and L_2 are in parallel."

As an example,

$$(L_1 + L_2) \| (L_1 + L_2 + L_3)$$

is read as "the series combination of L_1 and L_2 is in parallel with the series combination of L_1, L_2, and L_3." The notation is illustrated in the following examples.

EXAMPLE 11.12 Determine a single equivalent inductor for terminals a-b in Figure 11.27. ■

Figure 11.27

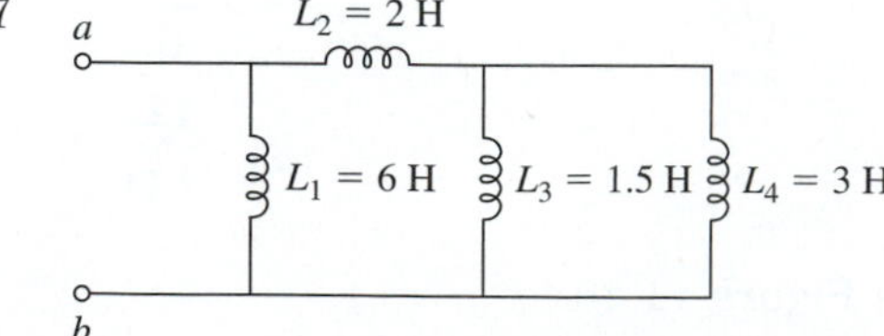

SOLUTION The single equivalent inductor L_{EQ} for terminals a-b is

$$L_{EQ} = \{(L_3 \| L_4) + L_2\} \| (L_1)$$

From Equations (11.5) and (11.6),

$$
\begin{aligned}
L_3 \| L_4 &= \frac{L_3 L_4}{L_3 + L_4} \\
&= \frac{(1.5)(3)}{1.5 + 3} \\
&= 1\text{ H} \\
(L_3 \| L_4) + L_2 &= 1 + L_2 \\
&= 1 + 2 \\
&= 3\text{ H} \\
L_{EQ} &= \{(L_3 \| L_4) + L_2\} \| L_1 \\
&= \frac{3L_1}{3 + L_1} \\
&= \frac{3(6)}{3 + 6} \\
&= 2\text{ H}
\end{aligned}
$$

EXAMPLE 11.13 Determine a single equivalent inductor for terminals a-b in Figure 11.28. ■

Figure 11.28

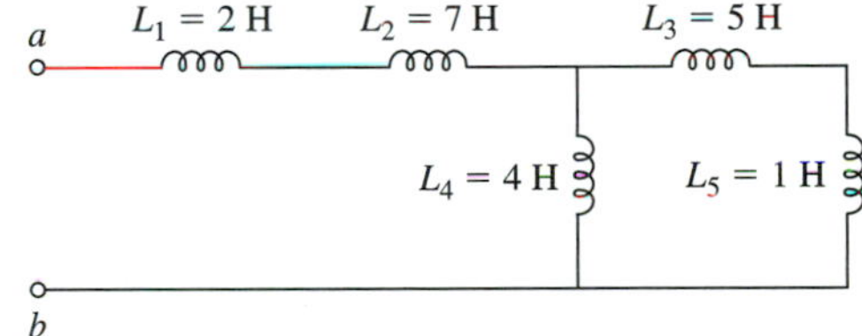

SOLUTION The single equivalent inductor for terminals a-b is

$$L_{EQ} = (L_1 + L_2) + \{(L_3 + L_5) \| (L_4)\}$$

From Equations (11.5) and (11.6),

$$
\begin{aligned}
L_3 + L_5 &= 5 + 1 \\
&= 6\text{ H} \\
(L_3 + L_5) \| (L_4) &= \frac{6(L_4)}{6 + L_4} \\
&= \frac{6(4)}{6 + 4} \\
&= 2.4\text{ H}
\end{aligned}
$$

$$L_1 + L_2 = 2 + 7$$
$$= 9\text{ H}$$
$$L_{EQ} = (L_1 + L_2) + \{(L_3 + L_5)\|(L_4)\}$$
$$= 9 + 2.4$$
$$= 11.4\text{ H}$$

11.9 MAGNETIC FLUX ASSOCIATED WITH SERIES-PARALLEL INDUCTORS

It is often necessary to determine the total magnetic flux ϕ_T associated with inductors in a circuit. The procedure is illustrated in the following examples.

EXAMPLE 11.14 Determine the total magnetic flux associated with each inductor for the circuit in Figure 11.29. ■

Figure 11.29

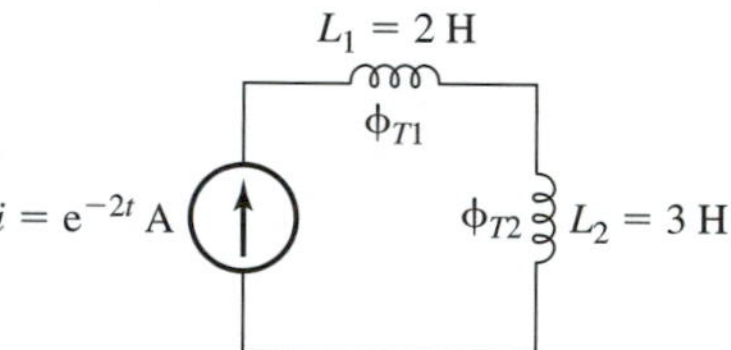

SOLUTION Realizing that the same current flows through L_1 and L_2, and using Equation (11.1) results in

$$\phi_{T1} = L_1 i$$
$$= 2e^{-2t}\text{ Wb}$$
$$\phi_{T2} = L_2 i$$
$$= 3e^{-2t}\text{ Wb}$$

The total magnetic flux varies with time, because the current varies with time.

EXAMPLE 11.15 Determine the current and the total magnetic flux associated with each inductor for the circuit in Figure 11.30. ■

Figure 11.30

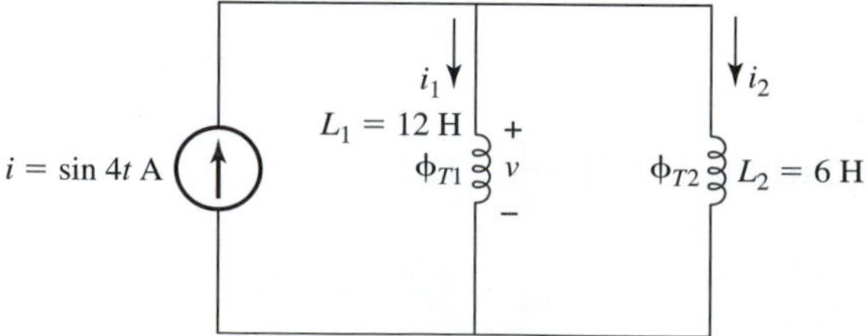

SOLUTION From Equation (11.6) an equivalent inductor for L_1 and L_3 in parallel is

$$L_{EQ} = \frac{(12)(6)}{12 + 6}$$
$$= 4\text{ H}$$

See Figure 11.31.

Figure 11.31

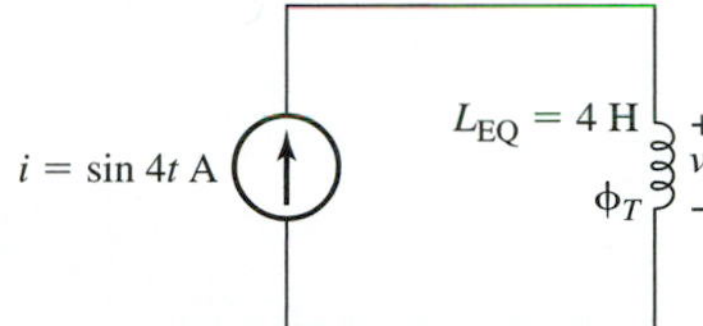

Because the same voltage v appears across L_1, L_2, and L_{EQ}, employing Equation (11.3) results in

$$v = \frac{d\phi_{T1}}{dt} = \frac{d\phi_{T2}}{dt} = \frac{d\phi_T}{dt}$$

Because the derivatives of ϕ_T, ϕ_{T1}, and ϕ_{T2} with respect to time are equal, we can assume

$$\phi_{T1} = \phi_{T2} = \phi_T$$

From Equation (11.1)

$$\begin{aligned} \phi_T &= L_{EQ} i \\ &= 4 \sin 4t \end{aligned}$$

and therefore

$$\phi_{T1} = 4 \sin 4t$$
$$\phi_{T2} = 4 \sin 4t$$

From Equation (11.1)

$$\begin{aligned} i_1 &= \frac{\phi_{T1}}{L_1} \\ &= \left(\frac{1}{3}\right)\sin(4t) \text{ A} \\ i_2 &= \frac{\phi_{T2}}{L_2} \\ &= \left(\frac{2}{3}\right)\sin(4t) \text{ A} \end{aligned}$$

EXAMPLE 11.16 Determine the currents i_2 and i_3 and the total magnetic flux associated with each inductor in the circuit in Figure 11.32. ■

Figure 11.32

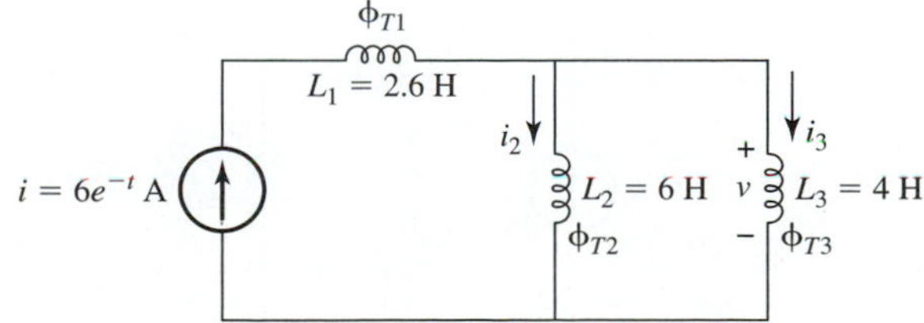

SOLUTION From Equations (11.5) and (11.6), an equivalent inductor for L_1, L_2, and L_3 is

$$\begin{aligned} L_{EQ} &= (L_2 \| L_3) + L_1 \\ &= \left\{ \frac{L_2 L_3}{L_2 + L_3} \right\} + L_1 \\ &= \frac{(6)(4)}{6 + 4} + 2.6 \\ &= 5 \text{ H} \end{aligned}$$

L_1, L_2, and L_3 are replaced by L_{EQ} in the circuit shown in Figure 11.33.

Figure 11.33

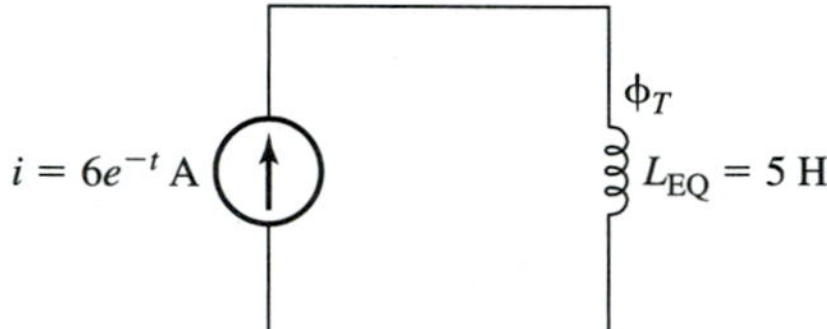

From Equation (11.1)

$$\begin{aligned} \phi_T &= L_{EQ} i \\ &= (5)6e^{-t} \\ &= 30e^{-t} \text{ Wb} \end{aligned}$$

Because the same current flows through L_{EQ} and L_1

$$\begin{aligned} \phi_{T1} &= L_1 i \\ &= (2.6)6e^{-t} \\ &= 15.6e^{-t} \text{ Wb} \end{aligned}$$

From Equation (11.6) an equivalent inductor for L_2 and L_3 in parallel is

$$\begin{aligned} L'_{EQ} &= \frac{L_2 L_3}{L_2 + L_3} \\ &= \frac{(6)(4)}{6 + 4} \\ &= 2.4 \text{ H} \end{aligned}$$

Replacing L_2 and L_3 by L'_{EQ} results in the circuit in Figure 11.34.

Figure 11.34

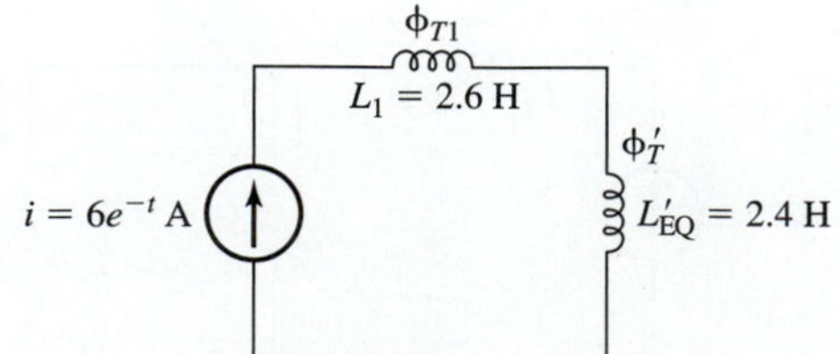

From Equation (11.1)

$$\begin{aligned}\phi'_T &= L'_{EQ} i \\ &= (2.4)6e^{-t} \\ &= 14.4e^{-t}\ \text{Wb}\end{aligned}$$

Because the same voltage v appears across L_2, L_3, and L'_{EQ},

$$v = \frac{d\phi_{T2}}{dt} = \frac{d\phi_{T3}}{dt} = \frac{d\phi'_T}{dt}$$

Therefore,

$$\phi_{T2} = \phi_{T3} = \phi'_T$$

Because $\phi'_T = 14.4e^{-t}$,

$$\begin{aligned}\phi_{T2} &= 14.4e^{-t}\ \text{Wb} \\ \phi_{T3} &= 14.4e^{-t}\ \text{Wb}\end{aligned}$$

From Equation (11.1)

$$\begin{aligned}i_2 &= \frac{\phi_{T2}}{L_2} \\ &= \frac{14.4e^{-t}}{6} \\ &= 2.4e^{-t}\ \text{A} \\ i_3 &= \frac{\phi_{T3}}{L_3} \\ &= \frac{14.4e^{-t}}{4} \\ &= 3.6e^{-t}\ \text{A}\end{aligned}$$

11.10 CURRENT DIVISION

The current-division formula for inductors allows us to determine the division of a current at a node without determining the voltage across the inductors through which the divided currents flow. It should be observed that current division and voltage division (presented in the next section) enable us to more easily solve the example problems in the last section.

Consider the circuit in Figure 11.35. From Equation (11.6) the equivalent inductor for the parallel combination of L_1 and L_2 is

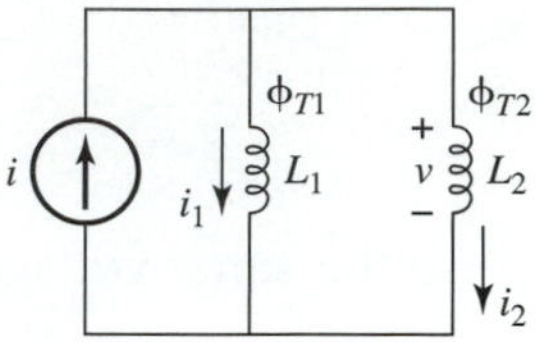

Figure 11.35 Current division.

$$L_{EQ} = \frac{L_1 L_2}{L_1 + L_2}$$

as shown in Figure 11.36.

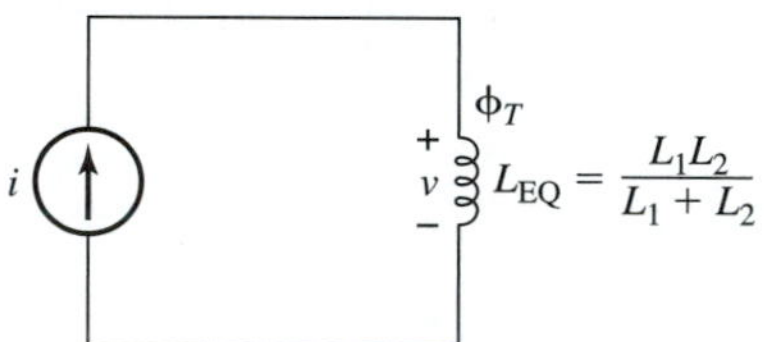

Figure 11.36 Equivalent inductor.

Recognizing that the same voltage v appears across L_1, L_2, and L_{EQ} and applying Equation (11.3) results in

$$v = \frac{d\phi_T}{dt} = \frac{d\phi_{T1}}{dt} = \frac{d\phi_{T2}}{dt}$$

Because the derivatives of ϕ_T, ϕ_{T1}, and ϕ_{T2} with respect to time are equal,

$$\phi_T = \phi_{T1} = \phi_{T2}$$

From Equation (11.1)

$$\begin{aligned}\phi_T &= L_{EQ} i \\ &= \left\{\frac{L_1 L_2}{L_1 + L_2}\right\} i\end{aligned}$$

and therefore

$$\phi_{T1} = \left\{\frac{L_1 L_2}{L_1 + L_2}\right\} i$$

$$\phi_{T2} = \left\{\frac{L_1 L_2}{L_1 + L_2}\right\} i$$

It follows from Equation (11.1) that

$$i_1 = \frac{\phi_{T1}}{L_1} = \left\{\frac{L_2}{L_1 + L_2}\right\} i$$

$$i_2 = \frac{\phi_{T2}}{L_2} = \left\{\frac{L_1}{L_1 + L_2}\right\} i$$

The preceding equations can be written as

$$i_1 = \left\{\frac{L_{\text{EQ}}}{L_1}\right\} i$$

$$i_2 = \left\{\frac{L_{\text{EQ}}}{L_2}\right\} i$$

These equations can be extended to n inductors in parallel, as shown in Figure 11.37.

Figure 11.37 Current division.

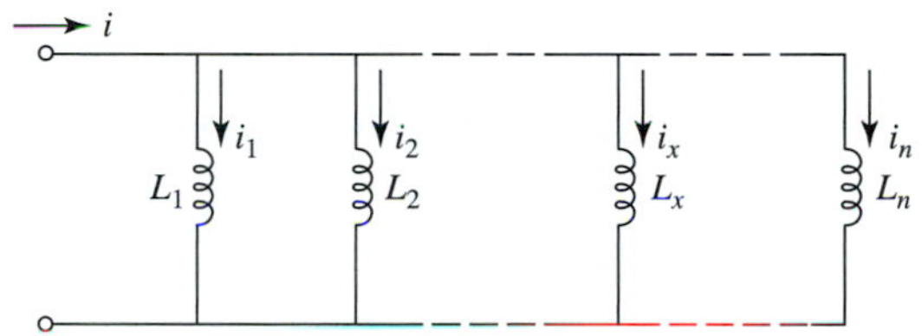

The expression for i_x is

$$i_x = \left\{\frac{L_{\text{EQ}}}{L_x}\right\}(i) \qquad (11.7)$$

where L_x is the inductance through which i_x flows and

$$L_{\text{EQ}} = \frac{1}{1/L_1 + 1/L_2 + \cdots + 1/L_n}$$

Consider the following examples.

EXAMPLE 11.17 Determine the currents i_1 and i_2 for the circuit in Figure 11.38. ■

Figure 11.38

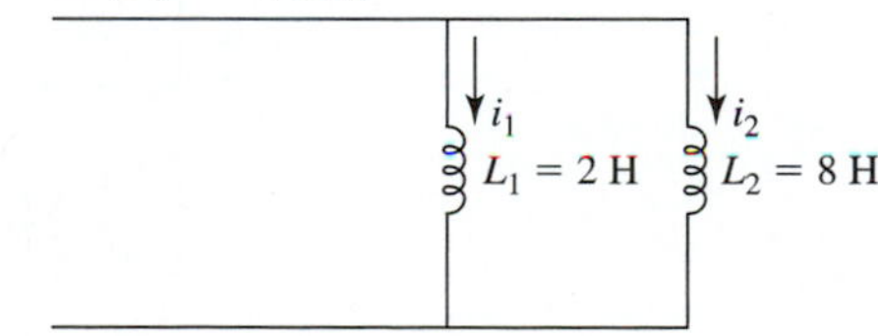

SOLUTION From Equation (11.7)

$$i_1 = \left\{\frac{8}{2+8}\right\}\{-4\cos(t)\}$$
$$= -3.2\cos t$$
$$i_2 = \left\{\frac{2}{2+8}\right\}\{-4\cos(t)\}$$
$$= -0.8\cos t$$

EXAMPLE 11.18 Determine the currents i_1, i_2, and i_3 for the circuit in Figure 11.39. ■

Figure 11.39

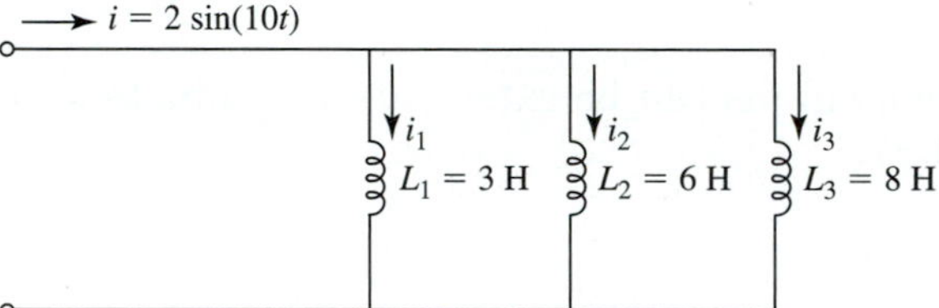

SOLUTION From Equation (11.7)

$$L_{EQ} = \frac{1}{1/L_1 + 1/L_2 + 1/L_3}$$
$$= \frac{1}{\frac{1}{3} + \frac{1}{6} + \frac{1}{8}}$$
$$= 1.6 \text{ H}$$
$$i_1 = \left\{\frac{L_{EQ}}{L_1}\right\}(i)$$
$$= \left\{\frac{1.6}{3}\right\}\{2\sin(10t)\}$$
$$= 1.07\sin(10t)\text{ A}$$
$$i_2 = \left\{\frac{L_{EQ}}{L_2}\right\}(i)$$
$$= \left\{\frac{1.6}{6}\right\}\{2\sin(10t)\}$$
$$= 0.53\sin(10t)\text{ A}$$
$$i_3 = \left\{\frac{L_{EQ}}{L_3}\right\}(i)$$
$$= \left\{\frac{1.6}{8}\right\}\{2\sin(10t)\}$$
$$= 0.4\sin(10t)\text{ A}$$

11.11 VOLTAGE DIVISION

The voltage-division formula for inductors allows us to determine the division of a voltage across two or more inductors in series. Consider the circuit in Figure 11.40. From Equation (11.5) an equivalent inductor for L_1 and L_2 in series is

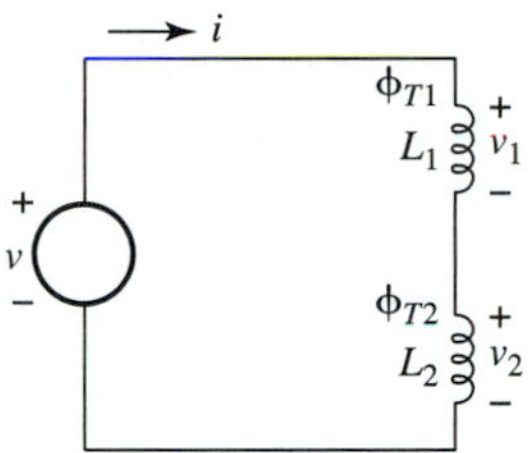

Figure 11.40 Voltage division.

$$L_{EQ} = L_1 + L_2$$

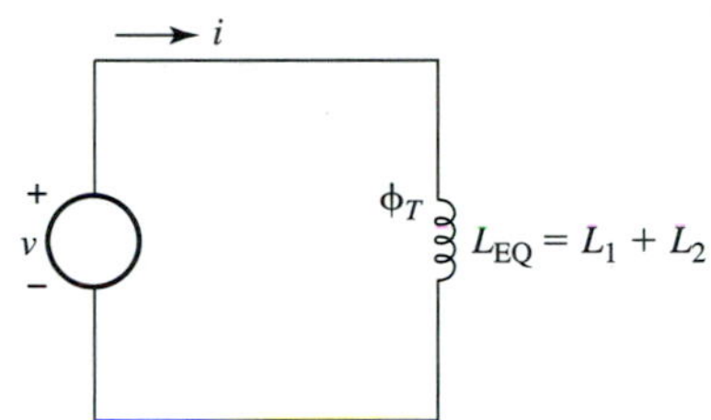

Figure 11.41 Equivalent inductor.

as shown in Figure 11.41.

From Equation (11.1)

$$\phi_T = (L_1 + L_2)i$$

and

$$i = \frac{\phi_T}{L_1 + L_2}$$

for the circuits in Figures 11.40 and 11.41.

Because the same current i flows through L_{EQ}, L_1, and L_2,

$$\phi_{T1} = L_1 i$$
$$= \left\{\frac{L_1}{L_1 + L_2}\right\}\phi_T$$
$$\phi_{T2} = L_2 i$$
$$= \left\{\frac{L_2}{L_1 + L_2}\right\}\phi_T$$

Differentiating both of these equations with respect to time results in

$$\frac{d\phi_{T1}}{dt} = \left\{\frac{L_1}{L_1 + L_2}\right\}\frac{d\phi_T}{dt}$$
$$\frac{d\phi_{T2}}{dt} = \left\{\frac{L_2}{L_1 + L_2}\right\}\frac{d\phi_T}{dt}$$

From Equation (11.3)

$$\frac{d\phi_T}{dt} = v$$

$$\frac{d\phi_{T1}}{dt} = v_1$$

$$\frac{d\phi_{T2}}{dt} = v_2$$

and therefore

$$v_1 = \left\{\frac{L_1}{L_1 + L_2}\right\}(v)$$

$$v_2 = \left\{\frac{L_2}{L_1 + L_2}\right\}(v)$$

These equations can be extended to n inductors in series, as shown in Figure 11.42. The expression for v_x for the circuit in Figure 11.42 is

$$v_x = \left\{\frac{L_x}{L_{EQ}}\right\}v \qquad (11.8)$$

Figure 11.42 Voltage division.

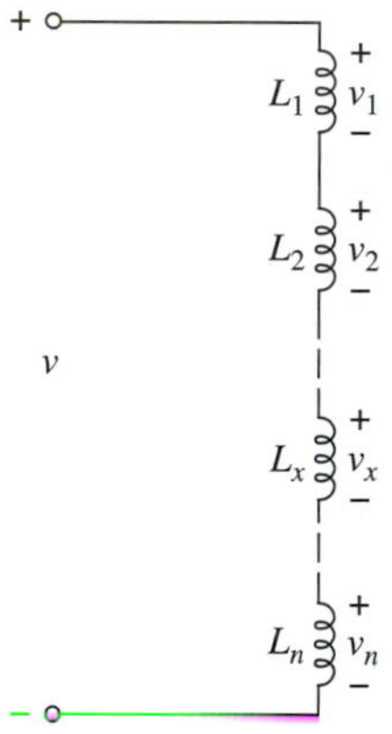

where L_x is the inductor across which v_x appears and

$$L_{EQ} = L_1 + L_2 + \cdots + L_n$$

Consider the following examples.

EXAMPLE 11.19 Determine the voltages v_1 and v_2 for the circuit in Figure 11.43. ■

Figure 11.43

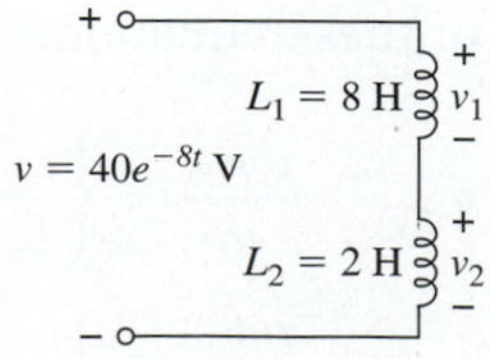

SOLUTION From Equation (11.8)

$$v_1 = \left\{\frac{8}{8+2}\right\}(40e^{-8t})$$
$$= 32\{e^{-8t}\}\text{ V}$$
$$v_2 = \left\{\frac{2}{8+2}\right\}(40e^{-8t})$$
$$= 8\{e^{-8t}\}\text{ V}$$

EXAMPLE 11.20 Determine the voltages v_1, v_2, and v_3 for the circuit in Figure 11.44. ■

Figure 11.44

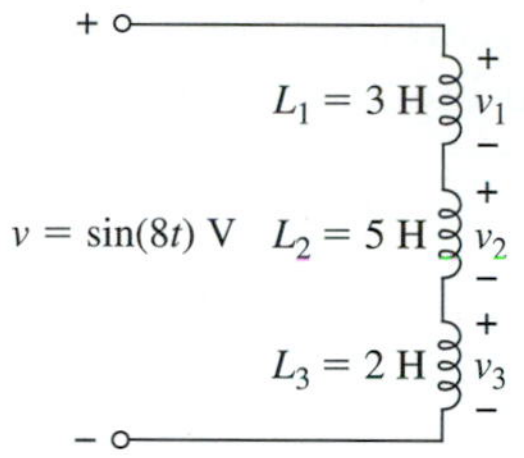

+ ○
$L_1 = 3$ H, + v_1 −
$v = \sin(8t)$ V, $L_2 = 5$ H, + v_2 −
$L_3 = 2$ H, + v_3 −
− ○

SOLUTION From Equation (11.8)

$$L_{EQ} = L_1 + L_2 + L_3$$
$$= 3 + 5 + 2 = 10\text{ H}$$
$$v_1 = \left\{\frac{L_1}{L_{EQ}}\right\}(v)$$
$$= \left\{\frac{3}{10}\right\}(\sin 8t)$$
$$= 0.3\sin(8t)\text{ V}$$
$$v_2 = \left\{\frac{L_2}{L_{EQ}}\right\}(v)$$
$$= \left\{\frac{5}{10}\right\}(\sin 8t)$$
$$= 0.5\sin(8t)\text{ V}$$
$$v_3 = \left\{\frac{L_3}{L_{EQ}}\right\}(v)$$
$$= \left\{\frac{2}{10}\right\}(\sin 8t)$$
$$= 0.2\sin(8t)\text{ V}$$

EXAMPLE 11.21 Determine the voltages v_1, v_2, v_3, and v_4 for the circuit in Figure 11.45. ■

Figure 11.45

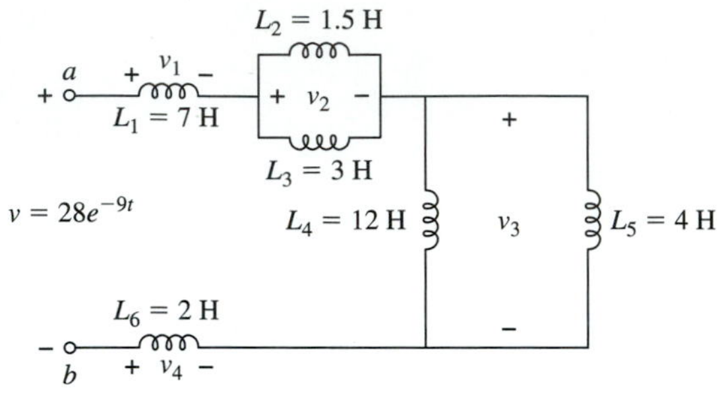

SOLUTION Let L'_{EQ} and L''_{EQ} be the equivalent inductances for $L_2 \| L_3$ and $L_4 \| L_5$, respectively. From Equation (11.6)

$$L'_{EQ} = L_2 \| L_3 = \frac{(1.5)(3)}{1.5 + 3}$$

$$= 1 \text{ H}$$

$$L''_{EQ} = L_4 \| L_5 = \frac{(12)(6)}{12 + 6}$$

$$= 4 \text{ H}$$

From Equation (11.5), L_{EQ} for the circuit to the right of terminals a-b is

$$L_{EQ} = L_1 + L'_{EQ} + L''_{EQ} + L_6$$

$$= 7 + 1 + 4 + 2$$

$$= 14 \text{ H}$$

From Equation (11.8)

$$v_1 = \left\{\frac{L_1}{L_{EQ}}\right\}(v)$$

$$= \left(\frac{7}{14}\right)(28e^{-9t})$$

$$= 14\{e^{-9t}\} \text{ V}$$

$$v_2 = \left\{\frac{L'}{L_{EQ}}\right\}(v)$$

$$= \left(\frac{1}{14}\right)(28e^{-9t})$$

$$= 2e^{-9t} \text{ V}$$

$$v_3 = \left\{\frac{L''}{L_{EQ}}\right\}(v)$$

$$= \left(\frac{4}{14}\right)(28e^{-9t})$$

$$= 8e^{-9t}\text{ V}$$

$$v_4 = \left\{\frac{L_6}{L_{EQ}}\right\}(v)$$

$$= \left(\frac{2}{14}\right)(28e^{-9t})$$

$$= 4e^{-9t}\text{ V}$$

11.12 GRAPHICAL DETERMINATION OF VOLTAGE AND CURRENT

The basic relationship between current and voltage for an inductor allows us to graphically determine the waveform of one from the waveform of the other. Recall from Equation (11.3) that the voltage across an inductor is

$$v = L\frac{di}{dt}$$

Recalling that the value of a derivative at a point is equal to the slope of a tangent line drawn to the point allows us to evaluate v graphically if a plot of i versus t is given. Because the slope of a tangent line to the i-versus-t waveform at any time t_1 is equal to di/dt evaluated at $t = t_1$, it follows that $v(t_1)$ is equal to L multiplied by the slope of the tangent to i versus t at $t = t_1$. Although the graphical procedure presented here is valid for any i-versus-t waveform, it is easiest to apply when the i-versus-t waveform consists of straight lines, as illustrated in the following example.

EXAMPLE 11.22 The current across a 40-mH inductor is given in Figure 11.46. Determine the voltage v from $t = 0$ to $t = 5$ s. ■

Figure 11.46

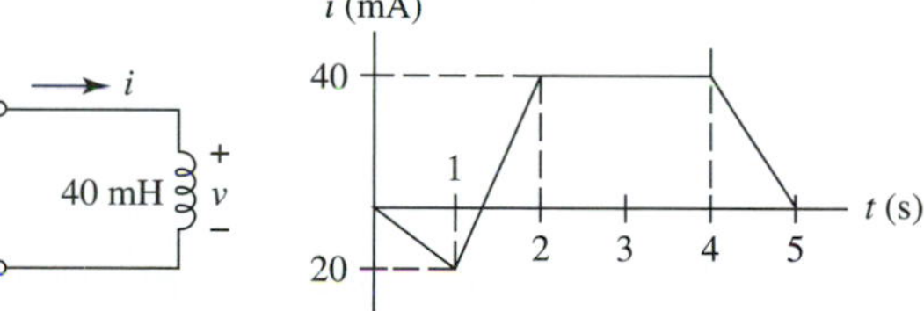

SOLUTION Because the slope of a tangent to i versus t for any time from $t = 0$ to $t = 1$ equals (-20×10^{-3}), the voltage for this time period is

$$v = L\frac{di}{dt}$$

$$= (40 \times 10^{-3})(-20 \times 10^{-3})$$

$$= (-800)10^{-6}$$

$$= -0.8\text{ mV}$$

Because the slope of a tangent to i versus t for any time from $t = 1$ to $t = 2$ equals 60×10^{-3}, the voltage for this time period is

$$\begin{aligned} v &= L\frac{di}{dt} \\ &= (40 \times 10^{-3})(60 \times 10^{-3}) \\ &= 2400 \times 10^{-6} \\ &= 2.4 \text{ mV} \end{aligned}$$

Because the slope of a tangent to i versus t for any time from $t = 2$ to $t = 4$ equals 0, the voltage for this time period is

$$\begin{aligned} v &= L\frac{di}{dt} \\ &= (40 \times 10^{-3})(0) \\ &= 0 \end{aligned}$$

Because the slope of a tangent to i versus t for any time from $t = 4$ to $t = 5$ equals (-40), the voltage for this time period is

$$\begin{aligned} v &= L\frac{di}{dt} \\ &= (40 \times 10^{-3})(-40 \times 10^{-3}) \\ &= -1600 \times 10^{-6} \\ &= -1.6 \text{ mV} \end{aligned}$$

A plot of i versus t from $t = 0$ to $t = 5$ is shown in Figure 11.47.

Figure 11.47

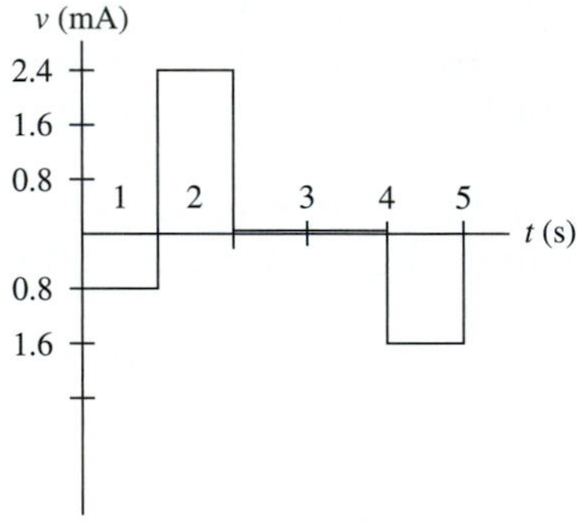

Recalling the definition of the antiderivative and the indefinite integral and recognizing from Equation (11.3) that $di/dt = (1/L)v$, we conclude that i is the antiderivative of $(1/L)v$ and can be expressed in terms of the indefinite integral as

$$i = \int \left(\frac{1}{L}\right) v \, dt$$

Because L is a constant, i can also be expressed as

$$i = \frac{1}{L}\int v \, dt$$

Also recalling the definition of the definite integral and again recognizing that i is the antiderivative of $(1/L)v$, we can write the following expression:

$$\frac{1}{L}\int_{t_0}^{t} v\,dt = i(t) - i(t_0)$$

where t is any time greater than time t_0 and $i(t)$ and $i(t_0)$ are the values of i at any time t and at $t = t_0$, respectively. It follows that

$$i = \frac{1}{L}\int_{t_0}^{t} v\,dt + i(t_0)$$

and if $t_0 = 0$,

$$i = \frac{1}{L}\int_{0}^{t} v\,dt + i(0)$$

Recalling the area property of the definite integral, the quantity $\int_{t_0}^{t} v\,dt$ is the area under the v-versus-t waveform from time t_0 to any time t. It follows that the current at any given time t_1 can be written as

$$i(t_1) = \left(\frac{1}{L}\right)\left\{\begin{matrix}\text{area under } v\text{-versus-}t \\ \text{plot from } t_0 \text{ to } t_1\end{matrix}\right\} + i(t_0)$$

Although the graphical procedure presented here is valid for any v-versus-t waveform, it is easiest to apply when the v-versus-t waveform consists of straight lines, as illustrated in the following example.

EXAMPLE 11.23 The voltage across an 800-mH inductor is given in Figure 11.48. Determine the current i for t = 2, 4, and 5 s. The current through the inductor is initially zero.

Figure 11.48

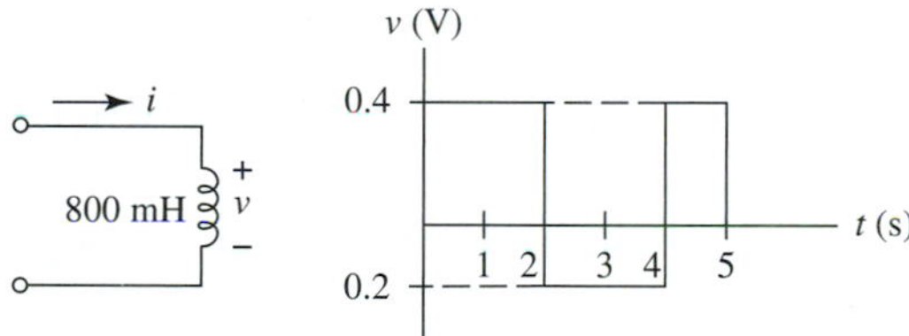

SOLUTION The areas under the v-versus-t curve are shown in Figure 11.49.

Figure 11.49

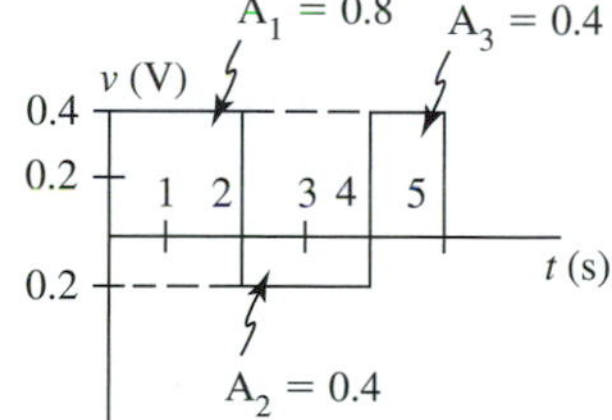

Because the area under v versus t from $t = 0$ to $t = 2$ equals 0.8 and $i(0) = 0$, the equation for $i(2)$ results in

$$\begin{aligned} i(2) &= \left(\frac{1}{L}\right)A_1 + i(0) \\ &= \left(\frac{1}{800 \times 10^{-3}}\right)(0.8) + 0 \\ &= 1\text{ A} \end{aligned}$$

which is read as "i evaluated at 2 s equals 1 A."

Because the area under v versus t from $t = 2$ to $t = 4$ equals (-0.4) and $i(2) = 1$ from before, the equation for $i(4)$ results in

$$\begin{aligned} i(4) &= \left(\frac{1}{L}\right)A_2 + i(2) \\ &= \left(\frac{1}{800 \times 10^{-3}}\right)(-0.4) + 1 \\ &= -0.5 + 1 \\ &= 0.5 \text{ A} \end{aligned}$$

which is read as "i evaluated at 4 s equals 0.5 A."

Because the area under v versus t from $t = 4$ to $t = 5$ equals (0.4) and $i(4) = 0.5$ from before, the equation for $i(5)$ results in

$$\begin{aligned} i(5) &= \left(\frac{1}{L}\right)A_3 + i(4) \\ &= \left(\frac{1}{800 \times 10^{-3}}\right)(0.4) + 0.5 \\ &= 0.5 + 0.5 \\ &= 1 \text{ A} \end{aligned}$$

If it is desired to plot i versus t in Example 11.23, further investigation is necessary to determine the portion of the waveforms that connect the known points (0, 0), (2, 1), (4, 0.5), and (5, 1). Because i for each interval of time involves the integral of the voltage (which is constant for each interval of time as given) and the integral of a constant with respect to t is equal to a first-degree term in t (t raised to the first power), it follows that the expression for i is a straight line for each interval of time, as illustrated in Example 11.24.

EXAMPLE 11.24 Draw a plot of i versus t for the inductor in Example 11.23 for the period of time from 0 to 5 s. ■

SOLUTION From Example 11.23 the known points are (0, 0), (2, 1), (4, 0.5), and (5, 1). Because v is a constant for each interval of time, the known points can be connected by straight lines to form the i-versus-t plot, as shown in Figure 11.50.

Figure 11.50

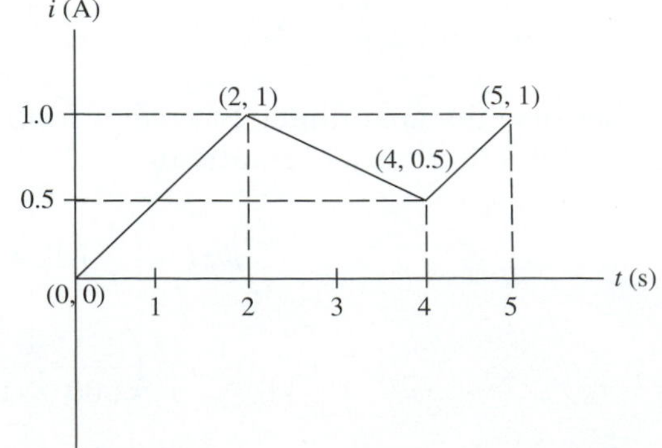

11.13 EXERCISES

1. Determine v for each of the inductors in Figure 11.51.

Figure 11.51

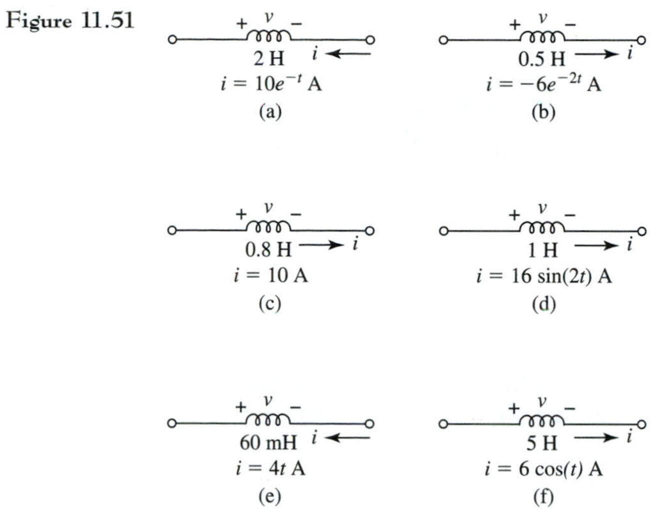

2. Determine the energy associated with each inductor in Figure 11.51.

3. Determine a single inductor equivalent with respect to terminals a-b for each circuit in Figure 11.52.

Figure 11.52

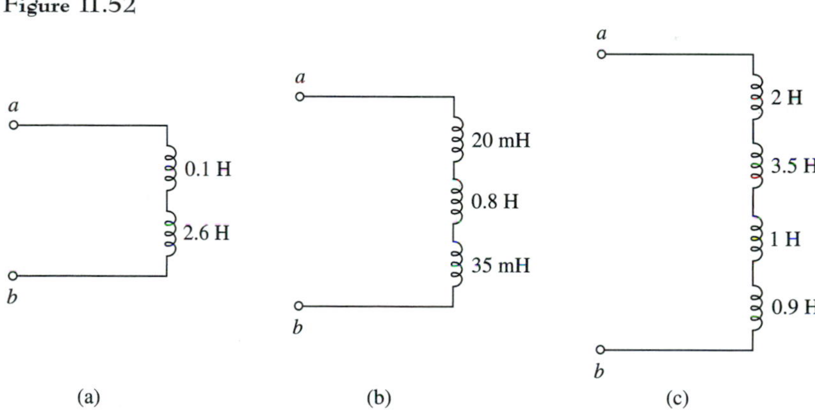

4. Determine a single inductor equivalent with respect to terminals a-b for each circuit in Figure 11.53.

Figure 11.53

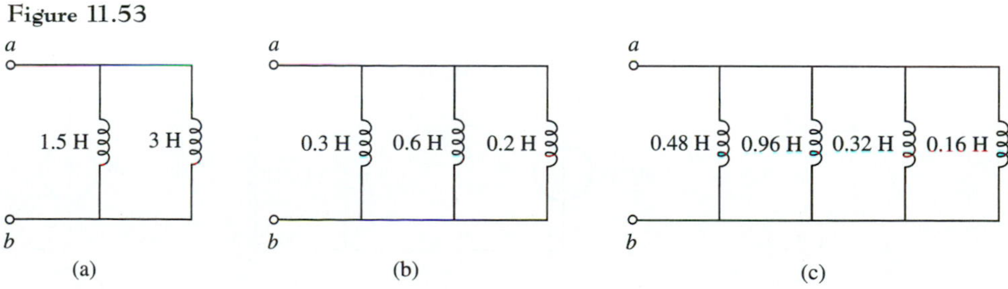

5. Determine a single inductor equivalent with respect to terminals *a-b* for each circuit in Figure 11.54.

Figure 11.54

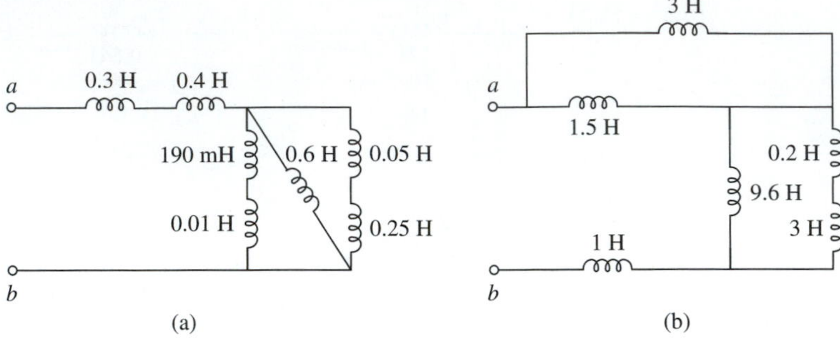

6. Determine a single inductor equivalent with respect to terminals *a-b* for each circuit in Figure 11.55.

Figure 11.55

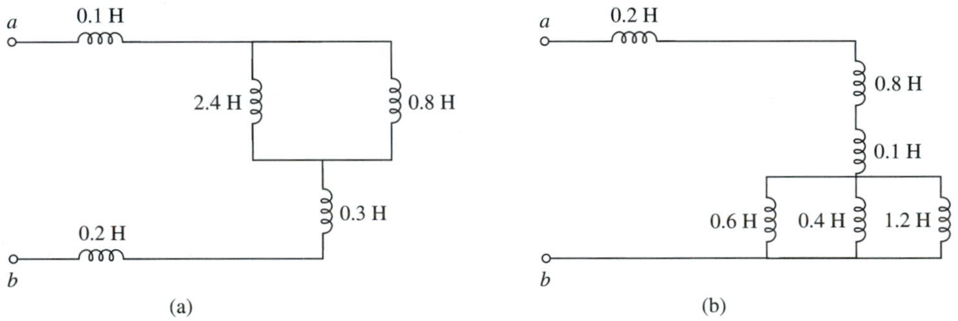

7. Determine the total flux (ϕ_T) associated with inductors L_1 and L_2 in Figure 11.56.

Figure 11.56

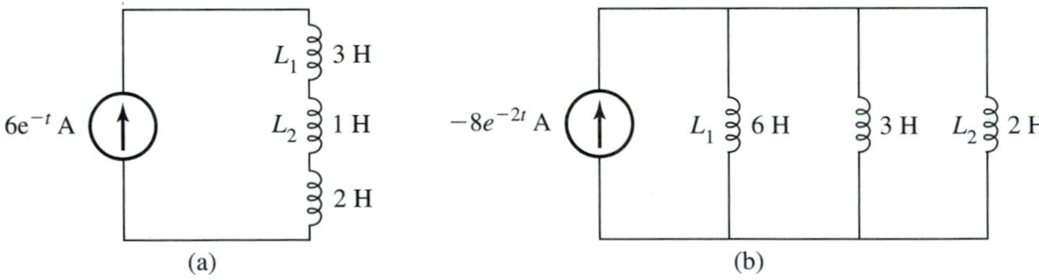

8. Determine the total flux (ϕ_T) associated with inductors L_1 and L_2 in Figure 11.57.

Figure 11.57

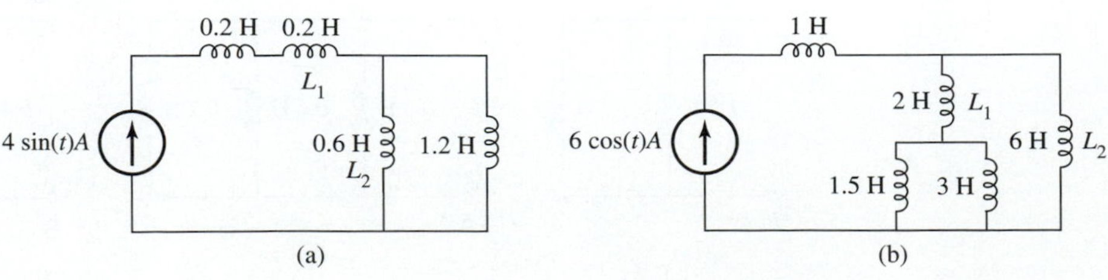

9. Find i_1 and i_2 for each circuit in Figure 11.58 at the instant the current is as shown.

Figure 11.58

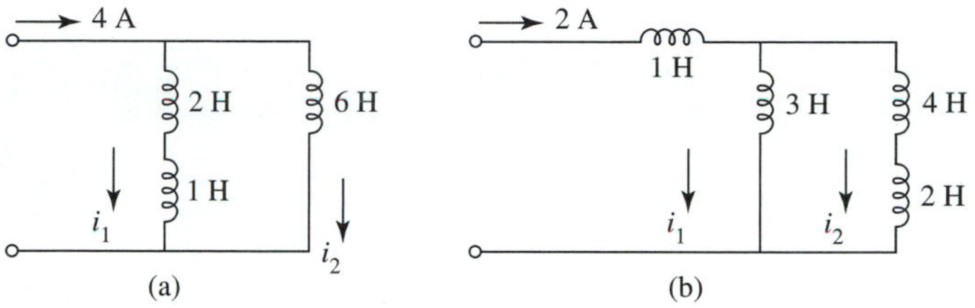

10. Find i_1 and i_2 for each circuit in Figure 11.59 at the instant the current is as shown.

Figure 11.59

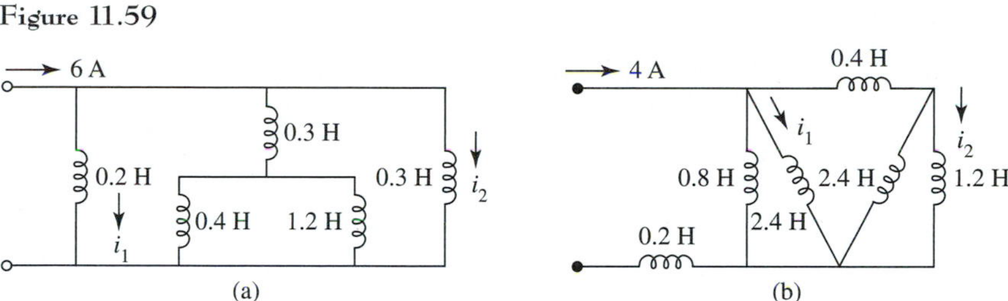

11. Find v_1 and v_2 for the circuit in Figure 11.60 at the instant the voltage is as shown.

Figure 11.60

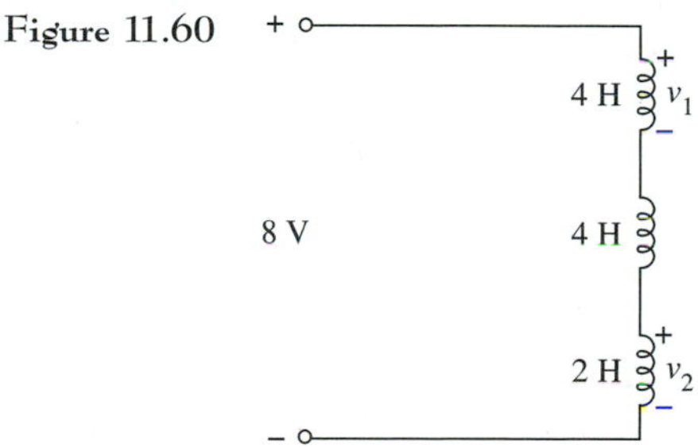

12. Find v_1 and v_2 for the circuit in Figure 11.61 at the instant the voltage is as shown.

Figure 11.61

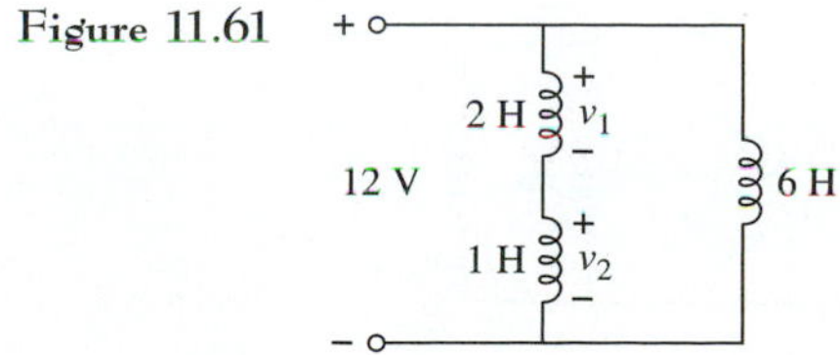

13. Find v_1 and v_2 for the circuit in Figure 11.62 at the instant the voltage is as shown.

Figure 11.62

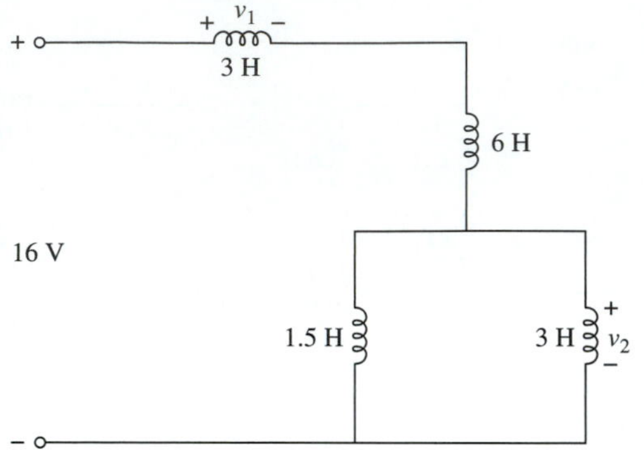

14. Find v_1 and v_2 for the circuit in Figure 11.63 at the instant the voltage is as shown.

Figure 11.63

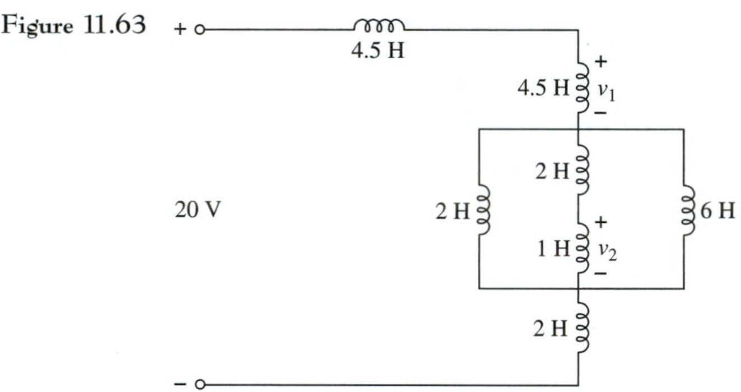

15. The equivalent inductor for terminals *a*-*b* is given for the circuit in Figure 11.64. Find L for the circuit.

Figure 11.64

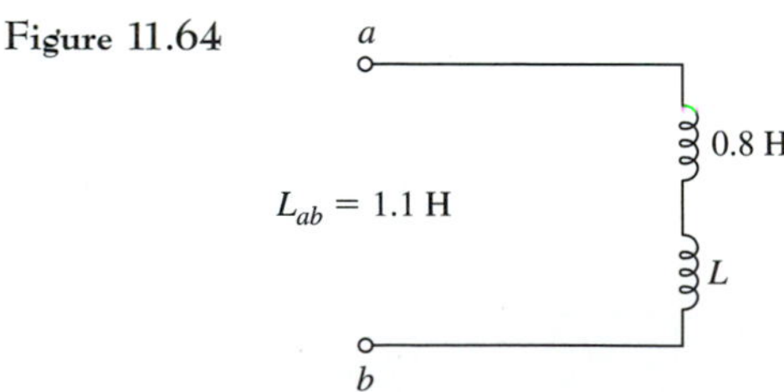

16. The equivalent inductor for terminals *a*-*b* is given for the circuit in Figure 11.65. Find L for the circuit.

Figure 11.65

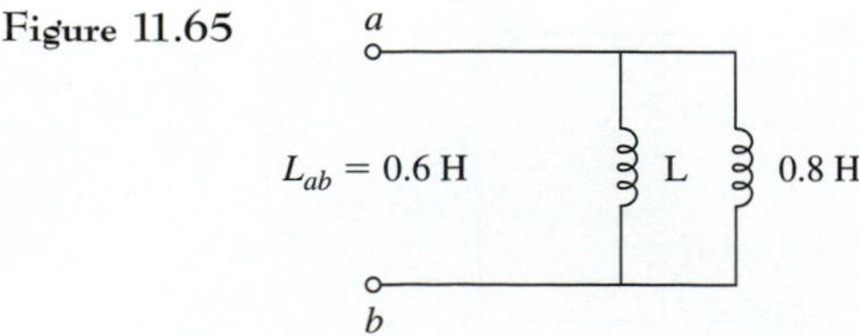

17. The equivalent inductor for terminals a-b is given for the circuit in Figure 11.66. Find L for the circuit.

Figure 11.66

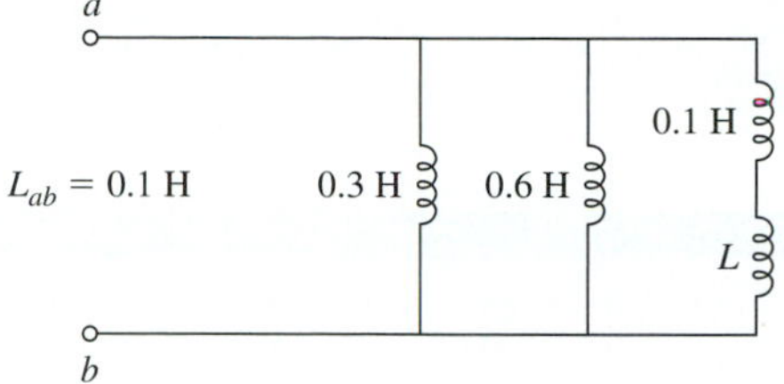

18. The equivalent inductor for terminals a-b is given for the circuit in Figure 11.67. Find L for the circuit.

Figure 11.67

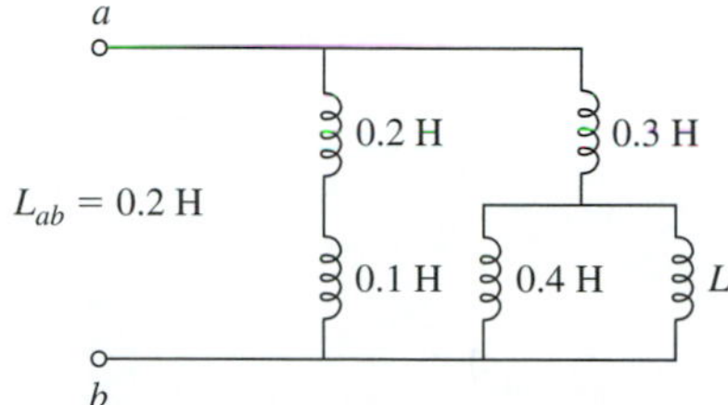

19. A plot of i (amperes) versus t (seconds) for a 2-H inductor is given in Figure 11.68. Construct the v-versus-t plot.

Figure 11.68

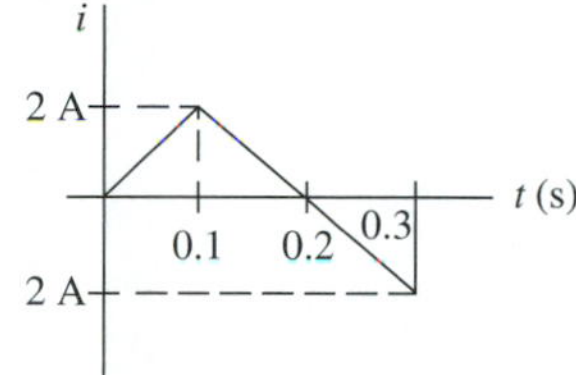

20. A plot of v (volts) versus t (seconds) for a 1-H inductor is given in Figure 11.69. Construct the i-versus-t plot.

Figure 11.69

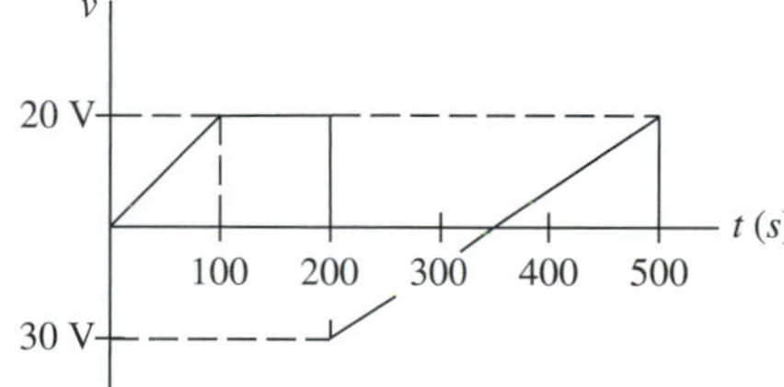

CHAPTER

12

Differential Equations

After completing this chapter the student should be able to

* solve first- and second-order homogeneous and nonhomogeneous differential equations by the classical method
* solve first- and second-order homogeneous and nonhomogeneous differential equations by the LaPlace transform method

12.1 DIFFERENTIAL EQUATIONS

The resistive circuits discussed in previous chapters are described by a mathematical model involving only algebraic equations. However, when analyzing electric circuits containing inductors and/or capacitors, the mathematical models encountered often involve equations containing derivatives. Such equations are called differential equations and the methods of solving them are discussed in this chapter.

Differential equations are classified as *linear* and *nonlinear*. In a linear differential equation, none of the dependent variables or their derivatives are multiplied by each other or raised to a power greater than 1, as is the case with nonlinear differential equations. Examples of linear and nonlinear equations are shown here, where the dependent variable is y and the independent variable is t.

The following are examples of linear differential equations:

$$5\frac{dy}{dt} + 6y = t$$

$$\frac{dy}{dt} + 4y = \sin(t) + t^2$$

Notice that the coefficients of the dependent variable terms are all constants, but they could involve the independent variable without making the differential equation nonlinear. In the following differential equation, the t^2y term does not make the equation nonlinear, because t is the independent variable.

$$\frac{dy}{dt} + t^2y = 6e^t$$

The following are examples of nonlinear differential equations:

$$\frac{dy}{dt} + y^2 = 2t$$

$$y\frac{dy}{dt} + 6y = \sin(4t)$$

The first equation is nonlinear because of the y^2 term and the second equation, because of the $y\dfrac{dy}{dt}$ term.

In basic electric circuits the differential equations that we encounter are linear with constant coefficients, and in this text our discussion is limited to this type of differential equation.

Consider the following differential equation:

$$\frac{dy}{dt} + 2y = 0$$

In this equation, t is the independent variable and y is the dependent variable. The functional expression for y in terms of t that yields an equivalency when substituted into the differential equation is called a solution to the differential equation.

12.2 SOLUTION TO A DIFFERENTIAL EQUATION

A *solution* to the differential equation

$$\frac{dy}{dt} + 2y = 0$$

is

$$y = 6e^{-2t}$$

because an equivalency results when $6e^{-2t}$ is substituted for y, as shown here. From rules D3 and D6 in Chapter 9,

$$\frac{d\{6e^{-2t}\}}{dt} + 2\{6e^{-2t}\} = 0$$

$$-(2)(6e^{-2t}) + 12e^{-2t} = 0$$

$$-12e^{-2t} + 12e^{-2t} = 0$$

$$0 = 0$$

A *solution* to the differential equation

$$\frac{dy}{dt} + 6y = 4$$

is

$$y = 4e^{-6t} + \left(\frac{2}{3}\right)$$

because an equivalency results when $\left(4e^{-6t} + \frac{2}{3}\right)$ is substituted for y, as shown here. From rules D2, D3, D6, and D7 in Chapter 9,

$$\frac{d\left\{4e^{-6t} + \frac{2}{3}\right\}}{dt} + 6\left\{4e^{-6t} + \left(\frac{2}{3}\right)\right\} = 4$$

$$\frac{d\{4e^{-6t}\}}{dt} + \frac{d\left(\frac{3}{2}\right)}{dt} + 24^{-6t} + 4 = 4$$

$$-24e^{-6t} + 0 + 24e^{-6t} + 4 = 4$$

$$4 = 4$$

12.3 ORDER OF A DIFFERENTIAL EQUATION

The highest derivative in a differential equation determines the *order* of the differential equation. The two differential equations previously considered are both first order, because the highest derivative in each equation is a first derivative. The following differential equations are second and third order, respectively, because the highest derivatives are second and third derivatives.

$$\frac{d^2y}{dt^2} + 3\frac{dy}{dt} + 2y = 0$$

$$\frac{d^3y}{dt^3} + 5\frac{d^2y}{dt^2} + 6\frac{dy}{dt} = 0$$

12.4 FORCING FUNCTION

Consider the following differential equation:

$$\frac{d^2y}{dt} + 3\frac{dy}{dt} + y = t - 2$$

Notice that all the dependent variable terms (terms containing y or its derivatives) are on the left side of the equal sign and all the independent variable terms (terms containing t or constants) are on the right side of the equal sign. When the terms of the differential equation are arranged in this manner, the differential equation is said to be in *proper form*. When the differential equation is in proper form, the expression on the right side of the equal sign is called the *forcing function*. The forcing function never contains terms involving the dependent variable. The forcing function for the differential equation considered here is $(t - 2)$.

12.5 HOMOGENEOUS AND NONHOMOGENEOUS EQUATIONS

If the forcing function is zero, the differential equation is called a *homogeneous* differential equation. An example of a second-order homogeneous equation is

$$\frac{d^2y}{dt^2} + 6\frac{dy}{dt} + 14y = 0$$

If the forcing function is not zero, the differential equation is called a *nonhomogeneous* differential equation. An example of a first-order nonhomogeneous differential equation with a forcing function of $\{6t^2 + \sin(t)\}$ is

$$\frac{dy}{dt} + 4y = 6t^2 + \sin(t)$$

12.6 GENERAL SOLUTION

Consider the following differential equation:

$$\frac{dy}{dt} + 4y = 8t$$

The solution is

$$y = Ke^{-4t} + 2t - \frac{1}{2}$$

where K is any constant. Whenever a solution contains arbitrary constants such as K, the solution is called a *general solution*. The solution can be verified by substituting it into the differential equation:

$$\frac{d\left\{Ke^{-4t} + 2t - \frac{1}{2}\right\}}{dt} + 4\left\{Ke^{-4t} + 2t - \frac{1}{2}\right\} = 8t$$

$$-4Ke^{-4t} + 2 + 4Ke^{-4t} + 8t - 2 = 8t$$

$$8t = 8t$$

The general solution will have as many arbitrary constants as the order of the differential equation.

12.7 COMPLETE, TRANSIENT, AND STEADY-STATE SOLUTIONS

Consider the same nonhomogeneous differential equation

$$\frac{dy}{dt} + 4y = 8t$$

with solution

$$y = Ke^{-4t} + 2t - \frac{1}{2}$$

Notice that if we set the forcing function ($8t$) equal to zero, the resulting homogeneous differential equation

$$\frac{dy}{dt} + 4y = 0$$

has the solution

$$y = Ke^{-4t}$$

which is part of the solution to the original nonhomogeneous differential equation. Notice also that the other part of the solution, $\left(2t - \frac{1}{2}\right)$, is itself a solution to the original nonhomogeneous differential equation because

$$\frac{d\left\{2t - \frac{1}{2}\right\}}{dt} + 4\left\{2t - \frac{1}{2}\right\} = 8t$$

$$2 + 8t - 2 = 8t$$

$$8t = 8t$$

Because $y = Ke^{-4t}$ yields zero when substituted into the expression

$$\frac{dy}{dt} + 4y$$

and $y = \left(2t - \frac{1}{2}\right)$ yields $8t$ when substituted into the expression

$$\frac{dy}{dt} + 4y$$

it follows that

$$y = Ke^{-4t} + 2t - \frac{1}{2}$$

yields $8t$ when substituted into the expression

$$\frac{dy}{dt} + 4y$$

and is therefore a solution to the nonhomogeneous differential equation

$$\frac{dy}{dt} + 4y = 8t$$

as previously demonstrated.

The *complete solution* to a nonhomogeneous differential equation consists of two parts. One part of the complete solution, called the *transient solution,* is the solution to the resulting homogeneous differential equation when the forcing function of the nonhomogeneous differential equation is set to zero (Ke^{-4t} for the differential equation just considered). The other part of the complete solution, called the *steady-state solution,* is the contribution to the complete solution by the forcing function (($2t - 1/2$) for the differential equation just considered). Considering a nonhomogeneous differential equation in proper form, the transient solution is the contribution to the complete solution from the left side of the equation (the dependent-variable terms), and the steady-state solution is the contribution to the complete solution from the forcing function (the independent-variable terms on the right side of the equation). In electric circuit analysis the transient solution comes from the passive circuit elements (resistors, inductors, and capacitors), and the steady-state solution comes from the active circuit elements (voltage and current sources). The transient solution is appropriately named, because it usually decreases and disappears with the passage of time in the life of an electric circuit.

Consider the nonhomogeneous differential equation presented previously:

$$\frac{dy}{dt} + 4y = 8t$$

The general transient solution is

$$y_T = Ke^{-4t}$$

The steady-state solution is

$$y_{ss} = 2t - \frac{1}{2}$$

and the general complete solution y is the sum of y_T and y_{ss}.

$$\begin{aligned} y &= y_T + y_{ss} \\ &= Ke^{-4t} + 2t - \frac{1}{2} \end{aligned}$$

Notice that $y = Ke^{-4t} + 2t - \frac{1}{2}$ is a general complete solution, because it contains an arbitrary constant K.

Because the forcing function is equal to zero for a homogeneous differential equation, the steady-state solution is equal to zero and the general complete solution is equal to the transient solution. Consider the following first-order homogeneous differential equation:

$$\frac{dy}{dt} + 10y = 0$$

The general transient solution is

$$y_T = Ke^{-10t}$$

The steady-state solution is

$$y_{ss} = 0$$

and the general complete solution is

$$\begin{aligned} y &= y_T + y_{ss} \\ &= Ke^{-10t} \end{aligned}$$

We now consider methods for determining solutions for first- and second-order, linear, constant-coefficient, homogeneous differential equations. The order of a differential equation in basic electric circuit analysis is seldom higher than second. However, the methods presented here can easily be extended to higher-order differential equations.

12.8 SOLVING FIRST-ORDER HOMOGENEOUS EQUATIONS

Consider the following first-order homogeneous differential equation.

$$\frac{dy}{dt} + 8y = 0$$

Replace dy/dt by p and y by unity. This results in

$$p + 8 = 0$$

which is called the *characteristic equation*. p is called a dummy variable, and it is only the value of the roots of the characteristic equation that are significant.

Solving the characteristic equation for p results in

$$p = -8$$

The general complete solution to the differential equation can than be written as

$$y = Ke^{pt} = Ke^{-8t}$$

where K is any constant. The solution can be verified by substituting it into the differential equation, as follows:

$$\begin{aligned} \frac{d\{Ke^{-8t}\}}{dt} + 8\{Ke^{-8t}\} &= 0 \\ -8Ke^{-8t} + 8Ke^{-8t} &= 0 \\ 0 &= 0 \end{aligned}$$

Recall that the complete solution and the transient solution are one and the same for a homogeneous differential equation, because the steady-state solution is zero.

The following procedure can be used to determine the general complete solution for a first-order homogeneous differential equation with dependent variable y and independent variable t.

1. Replace $\dfrac{dy}{dt}$ by p and y by unity. The resulting algebraic equation is the characteristic equation.

2. Solve for the root, p, of the characteristic equation.
3. Write the general complete solution to the homogeneous differential equation as

$$y = Ke^{pt}$$

where K is any constant.

Consider the following example.

EXAMPLE 12.1 Determine the general complete solution for the following first-order homogeneous differential equation:

$$\frac{dy}{dt} + 6y = 0$$

■

SOLUTION The characteristic equation is

$$p + 6 = 0$$

The root of the characteristic equation is

$$p = -6$$

The general complete solution of the differential equation is

$$y = Ke^{pt}$$
$$= Ke^{-6t}$$

12.9 SOLVING SECOND-ORDER HOMOGENEOUS EQUATIONS

For a second-order homogeneous differential equation, the degree of the characteristic equation is 2, and therefore the number of roots of the characteristic equation is equal to 2. When the characteristic equation has only one root, the root is a real number, and the solution to the associated first-order differential equation is obtained as demonstrated above. When considering second-order differential equations, the two roots of the associated characteristic equation can be real and unequal, real and equal, or complex. The solution to a second-order differential equation depends on the nature of the roots of the associated characteristic equation.

The following procedure can be used to determine the general complete solution for a second-order homogeneous differential equation with dependent variable y and independent variable t. The first and second steps in the process are as follows.

1. Replace $\frac{dy}{dt}$ by p, $\frac{d^2y}{dt^2}$ by p^2 and y by unity. The resulting algebraic equation is the characteristic equation.
2. Solve for the roots (p_1 and p_2) of the characteristic equation.

If p_1 and p_2 are real and unequal, write the general solution to the homogeneous differential equation as

$$y = K_1e^{p_1t} + K_2e^{p_2t}$$

where K_1 and K_2 are any constants.

Consider the following example.

EXAMPLE 12.2 Determine the general complete solution for the following second-order homogeneous differential equation.

$$\frac{d^2y}{dt^2} + 3\frac{dy}{dt} + 2y = 0$$

■

SOLUTION The characterisitic equation is

$$p^2 + 3p + 2 = 0$$

From factoring, the roots of the characteristic equation are

$$(p + 2)(p + 1) = 0$$
$$p_1 = -2$$
$$P_2 = -1$$

The general complete solution of the differential equation is

$$y = K_1e^{p_1t} + K_2e^{p_2t}$$
$$= K_1e^{-2t} + K_2e^{-1t}$$

If $p_1 = p_2$, write the general complete solution of the homogeneous differential equation as

$$y = K_1e^{pt} + K_2te^{pt}$$

where K_1 and K_2 are any constants and $p = p_1 = p_2$.

Consider the following example.

EXAMPLE 12.3 Determine the general complete solution for the following second-order homogeneous differential equation:

$$\frac{d^2y}{dt^2} + 4\frac{dy}{dt} + 4y = 0$$

■

SOLUTION The characteristic equation is

$$p^2 + 4p + 4 = 0$$

From factoring, the roots of the characteristic equation are

$$(p + 2)(p + 2) = 0$$
$$p_1 = -2$$
$$p_2 = -2$$

The general complete solution of the differential equation is

$$y = K_1e^{pt} + K_2te^{pt}$$
$$= K_1e^{-2t} + K_2te^{-2t}$$

Recall from algebra that the complex roots for a second-degree algebraic equation are complex conjugates; that is, the roots have equal real parts and the imaginary part of one root is the negative of the imaginary part of the other root. The roots can be written as $p_1 = \alpha + j\beta$ and $p_2 = \alpha - j\beta$, where α and β are, respectively, the real and imaginary parts of the roots.

If p_1 and p_2 are complex conjugates, write the general complete solution for the homogeneous differential equation as

$$y = K_1 e^{\alpha t} \sin(\beta t + \theta)$$

where K_1 and θ are any constants.

EXAMPLE 12.4 Determine the general complete solution for the following second-order homogeneous differential equation:

$$\frac{d^2y}{dt^2} + 4\frac{dy}{dt} + 20y = 0$$ ■

SOLUTION The characteristic equation is

$$p^2 + 4p + 20 = 0$$

The quadratic formula states that the roots for a second-degree algebraic equation of the form

$$p^2 + bp + c = 0$$

are

$$p = \frac{-b \pm \sqrt{b^2 - 4c}}{2}$$

Applying the quadratic formula to the characteristic equation

$$p^2 + 4p + 20 = 0$$

results in

$$p = \frac{-4 \pm \sqrt{(4)^2 - (4)(20)}}{2}$$

$$= -2 \pm \frac{\sqrt{-64}}{2}$$

$$= -2 \pm \frac{\sqrt{(-1)(64)}}{2}$$

$$= -2 \pm \frac{(\sqrt{-1})(\sqrt{64})}{2}$$

Recalling from algebra that $\sqrt{-1}$ is the imaginary number j,

$$p = -2 \pm j4$$

The roots of the characteristic equation are

$$p_1 = -2 + j4$$
$$p_2 = -2 - j4$$

and

$$\alpha = -2$$
$$\beta = 4$$

The general complete solution of the differential equation is

$$y = K_1 e^{-2t} \sin(4t + \theta)$$

The roots of a characteristic equation for a second-order homogeneous differential equation can be obtained by factoring or by the quadratic formula, as demonstrated in the previous examples. In general, the characteristic equation for a second-order differential equation is a second-degree algebraic equation of the form

$$p^2 + bp + c = 0$$

with roots

$$p = -\frac{b}{2} \pm \frac{\sqrt{b^2 - 4c}}{2}$$

If $b^2 > 4c$, the expression under the radical is positive and the roots of the characteristic equation are real and unequal.

If $b^2 = 4c$, the expression under the radical is equal to zero and the roots of the characteristic equation are real and equal.

If $b^2 < 4c$, the expression under the radical is negative and the roots of the characteristic equation are complex conjugates.

12.10 PARTICULAR SOLUTION

When the complete solution to a homogeneous or nonhomogeneous differential equation includes arbitrary constants (the Ks and θs in the methods previously presented), the complete solution is called a *general complete solution.* The number of arbitrary constants always equals the order of the differential equation. When the arbitrary constants are evaluated, the resulting solution is called a *particular complete solution.* In order to evaluate the arbitrary constants, initial conditions for the dependent variable are required. There must be as many initial conditions as the order of the differential equation. An *initial condition* is the value of the dependent variable or a derivative of the dependent variable at time $t = 0$. In an electric circuit, an initial condition is usually the value of voltage or current at the instant a switch is opened or closed.

Consider a first-order differential equation with dependent variable y and independent variable t. In order to obtain a particular complete solution from the general complete solution, $y(0)$ (which is read as "y evaluated at $t = 0$") must be known. In order to obtain a particular complete solution from the general complete solution for a second-order differential equation with dependent variable y and independent variable t, $y(0)$ and $dy(0)/dt$, which is read as "the derivative of y with respect to t evaluated at $t = 0$," must be known. Extending this concept to an nth-order differential equation,

$$y(0), \frac{dy(0)}{dt}, \frac{d^2y(0)}{dt^2}, \ldots, \frac{d^{n-1}y(0)}{dt^n}$$

must be known to obtain a particular complete solution from the general complete solution. A particular complete solution is always with respect to the given initial conditions.

Consider the following examples.

EXAMPLE 12.5 Determine the particular complete solution for the following first-order homogeneous differential equation if $y(0) = 4$:

$$\frac{dy}{dt} + 2y = 0$$

■

SOLUTION The characteristic equation is

$$p + 2 = 0$$

The root of the characteristic equation is

$$p = -2$$

The general complete solution is

$$\begin{aligned} y &= Ke^{pt} \\ &= Ke^{-2t} \end{aligned}$$

Because $y(0) = 4$,

$$\begin{aligned} 4 &= Ke^{-2(0)} \\ &= Ke^{0} \\ &= K \end{aligned}$$

Because $K = 4$, the particular complete solution for $y(0) = 4$ is

$$y = 4e^{-2t}$$

EXAMPLE 12.6 Determine the particular complete solution for the following second-order homogeneous differential equation if $y(0) = 1$ and $\frac{dy(0)}{dt} = -4$:

$$\frac{d^2y}{dt^2} + 4\frac{dy}{dt} + 3y = 0$$

■

SOLUTION The characteristic equation is

$$p^2 + 4p + 3 = 0$$

From factoring, the two roots are

$$(p + 1)(p + 3) = 0$$

$$p_1 = -1$$

$$p_2 = -3$$

The general complete solution is

$$y = K_1 e^{p_1 t} + K_2 e^{p_2 t}$$

$$= K_1 e^{-t} + K_2 e^{-3t}$$

Because $y(0) = 1$

$$1 = K_1 e^{-(0)} + K_2 e^{-3(0)}$$

$$= K_1 + K_2$$

Differentiating the general solution with respect to t results in

$$\frac{dy}{dt} = -K_1 e^{-t} - 3K_2 e^{-3t}$$

Because $\dfrac{dy(0)}{dt} = -4,$

$$-4 = -K_1 e^{-(0)} - 3K_2 e^{-3(0)}$$

$$= -K_1 - 3K_2$$

Solving the two simultaneous equations

$$K_1 + K_2 = 1$$

$$-K_1 - 3K_2 = -4$$

results in

$$K_1 = -\frac{1}{2}$$

$$K_2 = \frac{3}{2}$$

The particular complete solution for $y(0) = 1$ and $\dfrac{dy(0)}{dt} = -4$ is

$$y = \left(-\frac{1}{2}\right)e^{-t} + \left(\frac{3}{2}\right)e^{-3t}$$

EXAMPLE 12.7 Determine the particular complete solution for the following second-order homogeneous differential equation if $y(0) = 1$ and $\dfrac{dy(0)}{dt} = 2$:

$$\frac{d^2y}{dt^2} + 2\frac{dy}{dt} + y = 0$$

■

SOLUTION The characteristic equation is

$$p^2 + 2p + 1 = 0$$

From factoring, the roots are

$$(p + 1)(p + 1) = 0$$
$$p_1 = -1$$
$$p_2 = -1$$

The general complete solution is

$$y = K_1e^{-t} + K_2te^{-t}$$

Because $y(0) = 1$

$$1 = K_1e^{-(0)} + K_2(0)e^{-(0)}$$

and

$$K_1 = 1$$

Differentiating the general complete solution with respect to t results in

$$\frac{dy}{dt} = -K_1e^{-t} + K_2t(-e^{-t}) + K_2e^{-t}$$
$$= (-K_1 + K_2)e^{-t} - K_2te^{-t}$$

Because

$$\frac{dy(0)}{dt} = 2$$
$$2 = (-K_1 + K_2)e^{-(0)} - K_2(0)e^{-(0)}$$
$$= -K_1 + K_2$$

Because $K_1 = 1$,

$$K_2 = 3$$

The particular complete solution for $y(0) = 1$ and $\frac{dy(0)}{dt} = 2$ is

$$y = 1e^{-t} + 3te^{-t}$$

EXAMPLE 12.8 Determine the particular complete solution for the second-order homogeneous differential equation given in Example 12.4 if $y(0) = 1$ and $dy(0)/dt = 2$. ■

SOLUTION The differential equation given in Example 12.4 is

$$\frac{d^2y}{dt^2} + 4\frac{dy}{dt} + 20y = 0$$

and the general complete solution as determined in Example 12.4 is

$$y = K_1 e^{-2t}\sin(4t + \theta)$$

where K_1 and θ are arbitrary constants.

Because $y(0) = 1$,

$$1 = K_1 e^{-2(0)}\sin\{4(0) + \theta\}$$
$$1 = K_1 \sin(\theta)$$

Using rules D3 and D9 in Chapter 9 to differentiate the general complete solution with respect to t and using the initial condition $\dfrac{dy(0)}{dt} = 2$ results in

$$\frac{dy}{dt} = 4K_1 e^{-2t}\cos(4t + \theta) - 2K_1 e^{-2t}\sin(4t + \theta)$$
$$2 = 4K_1 e^{-2(0)}\cos\{4(0) + \theta\} - 2K_1 e^{-2(0)}\sin\{4(0) + \theta\}$$
$$2 = 4K_1 \cos(\theta) - 2K_1 \sin(\theta)$$

Substituting unity for (K_1 sin θ) and 1/(sin θ) for K_1, as determined earlier results in

$$2 = 4\left\{\frac{1}{\sin\theta}\right\}(\cos\,\theta) - 2(1)$$

This expression can be rewritten as

$$\frac{\sin\,\theta}{\cos\,\theta} = 1$$

Because $(\sin\theta)/(\cos\theta) = \tan\theta$,

$$\tan\,\theta = 1$$

and

$$\theta = \frac{\pi}{4}\text{ rad}$$

Because $K_1 \sin\theta = 1$, from before,

$$K_1 \sin\left(\frac{\pi}{4}\right) = 1$$

and because $\sin(\pi/4) = 1/\sqrt{2}$,

$$K_1\left(\frac{1}{\sqrt{2}}\right) = 1$$

and

$$K_1 = \sqrt{2}$$

The particular complete solution for $y(0) = 1$ and $\dfrac{dy(0)}{dt} = 2$ is

$$y = \sqrt{2}\, e^{-2t} \sin\left(4t + \frac{\pi}{4}\right)$$

12.11 SOLVING NONHOMOGENEOUS EQUATIONS

We now consider a method for determining the steady-state solution for nonhomogeneous differential equations. Recall that the complete solution to a nonhomogeneous differential equation consists of two parts, the transient solution and the steady-state solution. Also recall the transient solution is the solution to the resulting homogeneous differential equation when the forcing function of the nonhomogeneous differential equation is set equal to zero. The previous examples well illustrate the method for determining the transient part of the complete solution for a nonhomogeneous differential equation.

The steady-state solution can be determined by assuming the steady-state solution to consist of the forcing function plus all the derivatives of the forcing function, with each term multiplied by an undetermined constant. The assumed solution is then substituted into the differential equation to determine the values of the undetermined constants. Do not confuse the undetermined constants associated with determining a steady-state solution with the arbitrary constants contained in a general solution.

Consider the following differential equation:

$$\frac{dy}{dt} + 2y = 2t$$

The steady-state solution is assumed as

$$y_{ss} = At + B$$

because the t term appears in the forcing function and the derivative of the t term is a constant. A and B are undetermined constants.

Substituting y_{ss} into the differential equation results in

$$\frac{d\{At + B\}}{dt} + 2\{At + B\} = 2t$$

$$A + 2At + 2B = 2t$$

$$(2A)t + (A + 2B) = 2t$$

Equating the coefficients on both sides of the equation results in

$$2A = 2$$

$$A + 2B = 0$$

and

$$A = 1$$

$$B = -\frac{1}{2}$$

It follows that

$$y_{ss} = 1t - \frac{1}{2}$$

Notice that the assumed steady-state solution consists of the derivative of the forcing function plus the forcing function, but the coefficients of the forcing function terms are disregarded. If the forcing function were $(8t)$ or $(50t)$ instead of $(2t)$ for the differential equation considered before, the assumed steady-state solution would remain $(At + B)$. Table 12.1 is a table of assumed steady-state solutions for basic forcing functions usually encountered in basic electric circuits.

TABLE 12.1

Entry	Forcing Function	Assumed Steady-State Solution
1	constant	A
2	t	$At + B$
3	t^2	$At^2 + Bt + C$
4	e^{pt}	Ae^{pt}
5	$\sin(\omega t)$	$A\sin(\omega t) + B\cos(\omega t)$
6	$\cos(\omega t)$	$A\sin(\omega t) + B\cos(\omega t)$

The undetermined constants A, B, and C in Table 12.1 are determined by substituting the assumed steady-state solution into the differential equation, as illustrated. Consider the following examples.

EXAMPLE 12.9 Determine the steady-state solution for the following first-order nonhomogeneous differential equation:

$$\frac{dy}{dt} + 4y = 2t^2$$ ■

SOLUTION From Entry 3 of Table 12.1 we assume the steady-state solution as

$$y_{ss} = At^2 + Bt + C$$

Substituting y_{ss} into the differential equation results in

$$\frac{d\{At^2 + Bt + C\}}{dt} + 4\{At^2 + Bt + C\} = 2t^2$$

$$2At + B + 4At^2 + 4Bt + 4C = 2t^2$$

$$(4A)t^2 + (2A + 4B)t + (B + 4C) = 2t^2$$

Equating coefficients results in

$$4A = 2$$

$$2A + 4B = 0$$

$$B + 4C = 0$$

and

$$A = \frac{1}{2}$$

$$B = -\frac{1}{4}$$

$$C = \frac{1}{16}$$

It follows that

$$y_{ss} = \frac{1}{2}t^2 - \frac{1}{4}t + \frac{1}{16}$$

EXAMPLE 12.10 Determine the particular complete solution for the differential equation ($dy/dt + 4y = 2t^2$) in Example 12.9 for the initial condition $y(0) = -1$. ■

SOLUTION From Example 12.9 the steady-state solution is

$$y_{ss} = \frac{1}{2}t^2 - \frac{1}{4}t + \frac{1}{16}$$

Recall that the transient part of the complete solution for a nonhomogeneous differential equation is determined by setting the forcing function equal to zero and solving the resulting homogeneous differential equation. Using the method presented previously for solving homogeneous differential equations results in

$$y_T = Ke^{-4t}$$

The general complete solution to the nonhomogeneous differential equation is then

$$\begin{aligned} y &= y_T + y_{ss} \\ &= Ke^{-4t} + \frac{1}{2}t^2 - \frac{1}{4}t + \frac{1}{16} \end{aligned}$$

The particular complete solution to the nonhomogeneous differential equation is obtained by using the given initial condition to solve for the constant K by setting $y = -1$ and $t = 0$ in the preceding equation.

$$-1 = Ke^{-4(0)} + \left(\frac{1}{2}\right)(0)^2 - \left(\frac{1}{4}\right)(0) + \frac{1}{16}$$

$$K = -\frac{17}{16}$$

The particular complete solution for the initial condition $y(0) = -1$ is

$$y = -\frac{17}{16}e^{-4t} + \frac{1}{2}t^2 - \frac{1}{4}t + \frac{1}{16}$$

EXAMPLE 12.11 Determine the transient solution, the steady-state solution, the general complete solution, and the particular complete solution for the following first-order nonhomogeneous differential equation and initial condition:

$$\frac{dy}{dt} + 4y = 2e^{-t}$$

$$y(0) = 2$$

■

SOLUTION The transient solution is determined by setting the forcing function equal to zero and solving the resulting homogeneous differential equation.

$$\frac{dy}{dt} + 4y = 0$$

$$p + 4 = 0$$

$$p = -4$$

$$y_T = Ke^{-4t}$$

The steady-state solution is determined by assuming a solution from entry 4 of Table 12.1 and solving for the undetermined coefficients:

$$y_{ss} = Ae^{-t}$$

$$\frac{d\{Ae^{-t}\}}{dt} + 4\{Ae^{-t}\} = 2e^{-t}$$

$$-Ae^{-t} + 4Ae^{-t} = 2e^{-t}$$

$$(-A + 4A)e^{-t} = 2e^{-t}$$

$$-A + 4A = 2$$

$$A = \frac{2}{3}$$

$$y_{ss} = \left(\frac{2}{3}\right)e^{-t}$$

The general complete solution is

$$y = y_T + y_{ss}$$

$$= Ke^{-4t} + \left(\frac{2}{3}\right)e^{-t}$$

The particular complete solution is determined as follows. Because $y(0) = 2$, we set $y = 2$ and $t = 0$. The result is

$$2 = Ke^{-4(0)} + \left(\frac{2}{3}\right)e^{-(0)}$$

$$K = \frac{4}{3}$$

and the particular complete solution for the initial condition given is

$$y = \left(\frac{4}{3}\right)e^{-4t} + \left(\frac{2}{3}\right)e^{-t}$$

EXAMPLE 12.12 Determine the particular complete solution for the following second-order nonhomogeneous differential equation with initial conditions $y(0) = 0$ and $dy(0)/dt = 0$:

$$\frac{d^2y}{dt^2} + 5\frac{dy}{dt} + 4y = t + e^{-2t}$$

■

SOLUTION The transient solution is determined by setting the forcing function to zero and solving the resulting homogeneous differential equation.

$$\frac{d^2y}{dt^2} + 5\frac{dy}{dt} + 4y = 0$$

The characteristic equation is

$$p^2 + 5p + 4 = 0$$

The roots are

$$(p + 4)(p + 1) = 0$$

$$p_1 = -4$$

$$p_2 = -1$$

and the general transient solution is

$$y_T = K_1e^{-4t} + K_2e^{-t}$$

The steady-state solution is determined by assuming a solution based on the forcing function. From entries 2 and 4 of Table 12.1, y_{ss} is assumed as

$$y_{ss} = At + B + Ce^{-2t}$$

A, B, and C are evaluated by substituting y_{ss} into the differential equation:

$$\frac{d^2(y_{ss})}{dt^2} + 5\frac{d(y_{ss})}{dt} + 4y_{ss} = t + e^{-2t}$$

$$4Ce^{-2t} + 5\{A - 2Ce^{-2t}\} + 4\{At + B + Ce^{-2t}\} = t + e^{-2t}$$

$$(4A)t + (4C - 10C + 4C)e^{-2t} + (5A + 4B) = t + e^{-2t}$$

Equating coefficients results in

$$4A = 1$$

$$-2C = 1$$

$$5A + 4B = 0$$

$$A = \frac{1}{4}$$

$$C = -\frac{1}{2}$$

$$B = -\frac{5}{16}$$

and

$$y_{ss} = \frac{1}{4}t - \frac{5}{16} - \frac{1}{2}e^{-2t}$$

The general complete solution is

$$y = y_T + y_{ss}$$

$$y = K_1e^{-4t} + K_2e^{-t} + \frac{1}{4}t - \frac{5}{16} - \frac{1}{2}e^{-2t}$$

The particular complete solution is determined by using the initial conditions, $y(0) = 0$ and $dy(0)/dy = 0$, to evaluate K_1 and K_2.

Because $y(0) = 0$,

$$0 = K_1e^{-4(0)} + K_2e^{-1(0)} + \left(\frac{1}{4}\right)(0) - \frac{5}{16} - \frac{1}{2}e^{-2(0)}$$

which can be written as

$$K_1 + K_2 = \frac{13}{16}$$

In order to apply the initial condition $dy(0)/dt = 0$, we first differentiate the general complete solution and then set $dy/dt = 0$ and $t = 0$ as follows:

$$\frac{dy}{dt} = -4K_1e^{-4t} - K_2e^{-t} + \frac{1}{4} + e^{-2t}$$

$$0 = -4K_1e^{-4(0)} - K_2e^{-1(0)} + \frac{1}{4} + e^{-2(0)}$$

$$4K_1 + K_2 = \frac{5}{4}$$

Solving the two algebraic equations

$$K_1 + K_2 = \frac{13}{16}$$

$$4K_1 + K_2 = \frac{5}{4}$$

for a simultaneous solution results in

$$K_1 = \frac{7}{48}$$

$$K_2 = \frac{2}{3}$$

The particular complete solution is

$$y = \frac{7}{48}e^{-4t} + \frac{2}{3}e^{-t} + \frac{1}{4}t - \frac{1}{2}e^{-2t} - \frac{5}{16}$$

As a review, let us describe the differential equation considered in this example by using all the descriptive terms for differential equations previously presented.

The differential equation

$$\frac{d^2y}{dt^2} + 5\frac{dy}{dt} + 4y = t + e^{-2t}$$

can be described as follows.

Dependent variable is y.
Independent variable is t.
Order is second.
Nonhomogeneous.
Forcing function is $(t + e^{-2t})$.
Characteristic equation is $(p^2 + 5p + 4 = 0)$.
General transient solution is

$$y_T = K_1e^{-4t} + K_2e^{-t}$$

Steady-state solution is

$$y_{ss} = \frac{1}{4}t - \frac{1}{2}e^{-2t} - \frac{5}{16}$$

General complete solution is

$$y = K_1e^{-4t} + K_2e^{-t} + \frac{1}{4}t - \frac{1}{2}e^{-2t} - \frac{5}{16}$$

Initial conditions are

$$y(0) = 0 \quad \text{and} \quad \frac{dy(0)}{dt} = 0$$

Particular complete solution for the given initial conditions is

$$y = \frac{7}{48}e^{-4t} + \frac{2}{3}e^{-t} + \frac{1}{4}t - \frac{1}{2}e^{-2t} - \frac{5}{16}$$

12.12 LAPLACE TRANSFORMS

The Laplace transform method, named for Pierre-Simon de Laplace (1749–1827), a French mathematician, is an alternative approach for solving differential equations. It is presented here because it is systematic and lends itself to the solution of basic electric circuits.

The Laplace transform method is called a transform method because the differential equation is transformed into an algebraic equation in the process of arriving at a solution. The method yields a particular complete solution and can be used to solve homogeneous or nonhomogeneous differential equations.

All the functions usually encountered in basic electric circuit analysis can be transformed from the *t domain* to the *s domain* by applying the definition of the Laplace transform. A function in the t domain is simply an expression of the function in terms of t, and a function in the s domain is an expression of the Laplace transform of the function in terms of the Laplace transform variable s. For example, the function $y = e^{-6t}$ is e^{-6t} in the t domain and is $1/(s + 6)$ in the s domain. The s domain expression $1/(s + 6)$ is obtained by applying the definition of the Laplace transform to the function $y = e^{-6t}$. The quantity $1/(s + 6)$ is called the Laplace transform of (e^{-6t}) and can be written

$$\mathcal{L}\{e^{-6t}\} = \frac{1}{s + 6}$$

which is read as "the Laplace transform of e^{-6t} is $1/(s + 6)$." The quantity (e^{-6t}) is called the inverse Laplace transform of $1/(s + 6)$ and can be written as

$$\mathcal{L}^{-1}\left\{\frac{1}{s + 6}\right\} = e^{-6t}$$

which is read as "the inverse Laplace transform of $1/(s + 6)$ is (e^{-6t})."

The Laplace transform of the function $y = e^{-6t}$ can also be written as

$$\mathbf{Y} = \frac{1}{s + 6}$$

which is read as "the Laplace transform of the function y is $1/(s + 6)$ where $\mathbf{Y}$ symbolizes the Laplace transform of y and is identical to the symbol $\mathcal{L}\{y\}$. The capital letter represents the Laplace transform of the function and the associated lowercase letter represents the function in the t domain. For example, $\mathbf{G}$ represents the Laplace transform of g and can be written as

$$\mathbf{G} = \mathcal{L}\{g\}$$

If $g = t^2$

$$\mathbf{G} = \mathcal{L}\{t^2\}$$

and

$$\begin{aligned} g &= \mathcal{L}^{-1}\{\mathbf{G}\} \\ &= t^2 \end{aligned}$$

Because the determination of the Laplace transform of a function is often a long and tedious process using the definition of a Laplace transform, tables of Laplace transforms derived from the definition are usually used to determine Laplace transforms in problem solving situations. We shall present the Laplace transforms of the functions usually encountered in basic electric circuit analysis and the rules necessary for solving differential equations using the Laplace transform method.

12.13 DEFINITION OF THE LAPLACE TRANSFORM

Although we shall always refer to Table 12.2 to obtain Laplace transforms when analyzing electric circuits, we now illustrate the determination of a Laplace transform of a function from the definition of the Laplace transform.

The Laplace transform of a function $f(t)$ is, by definition,

$$\mathscr{L}\{f(t)\} = \int_0^\infty f(t)e^{-st}\,dt$$

This integration is performed with respect to t and with s considered as a constant. The integral can be determined by replacing the upper integration limit (infinity) with a dummy variable a, performing the definite integration between the limits 0 and a, and then determining the resulting expression as a becomes infinitely large.

We shall determine the Laplace transform of a constant (entry 1 in Table 12.2) as an example. If $f(t) = K$, then

$$\mathscr{L}\{f(t)\} = \int_0^\infty f(t)e^{-st}\,dt$$

$$= \int_0^\infty Ke^{-st}\,dt$$

From rule I6 in Chapter 9,

$$\mathscr{L}\{K\} = K\int_0^\infty e^{-st}\,dt$$

Replacing infinity (∞) with a results in

$$\mathscr{L}\{K\} = K\int_0^a e^{-st}\,dt$$

From rule I3 and the procedure for definite integration in Chapter 9,

$$\mathscr{L}\{K\} = K\left\{\frac{e^{-st}}{-s}\right\}\Bigg|_0^a$$

$$= K\left\{\frac{e^{-sa}}{-s} - \frac{e^{-s(0)}}{-s}\right\}$$

$$= K\left\{-\frac{1}{se^{as}} + \frac{1}{s}\right\}$$

If a becomes infinitely large, then $\{-1/se^{as}\}$ goes to zero and

$$\mathscr{L}\{K\} = \frac{K}{s}$$

as shown in entry 1 of Table 12.2. From the illustration just given, it is easily seen that the derivation of the Laplace transform of a function is a lengthy and tedious process. Principally for this reason, we use tables to determine Laplace transforms, as illustrated next.

12.14 LAPLACE TRANSFORM TABLE

Table 12.2 includes the Laplace transform of the functions usually encountered in basic electric circuits analysis. In Table 12.2, a, K, and ω are constants and s is the Laplace transform variable. y is the function and $\mathbf{Y}$ is the Laplace transform of the function.

TABLE 12.2

Entry	y	Y
1	K (constant)	$\frac{K}{s}$
2	e^{-at}	$\frac{1}{s+a}$
3	t	$\frac{1}{s^2}$
4	t^2	$\frac{2}{s^3}$
5	$\sin(\omega t)$	$\frac{\omega}{s^2+\omega^2}$
6	$\cos(\omega t)$	$\frac{s}{s^2+\omega^2}$
7	$e^{-at}\sin(\omega t)$	$\frac{\omega}{(s+a)+\omega^2}$
8	$e^{-at}\cos(\omega t)$	$\frac{s+a}{(s+a)^2+\omega^2}$
9	te^{-at}	$\frac{1}{(s+a)^2}$
10	$\sin(\omega t+\theta)$	$\frac{s\sin\theta+\omega\cos\theta}{s^2+\omega^2}$
11	$\cos(\omega t+\theta)$	$\frac{s\cos\theta-\omega\sin\theta}{s^2+\omega^2}$

Consider the following examples.

EXAMPLE 12.13 Determine the Laplace transform of y if $y = 42$. ■

SOLUTION From entry 1 of Table 12.2,

$$\mathbf{Y} = \frac{42}{s}$$

or stated equivalently,

$$\mathcal{L}\{y\} = \frac{42}{s}$$

EXAMPLE 12.14 Determine the Laplace transforms of y and z if $y = e^{-4t}$ and $z = e^{2t}$. ■

SOLUTION From entry 2 of Table 12.2,

$$\mathbf{Y} = \frac{1}{s+4}$$

$$\mathbf{Z} = \frac{1}{s-2}$$

EXAMPLE 12.15 Determine the Laplace transforms of y and z if $y = t$ and $z = t^2$. ■

SOLUTION From entries 3 and 4 of Table 12.2,

$$\mathbf{Y} = \frac{1}{s^2}$$

$$\mathbf{Z} = \frac{2}{s^3}$$

EXAMPLE 12.16 Determine the Laplace transforms y and g if $y = \sin(20t)$ and $g = \cos(3t)$. ■

SOLUTION From entries 5 and 6 of Table 12.2,

$$\mathbf{Y} = \frac{20}{s^2 + (20)^2}$$

$$= \frac{20}{s^2 + 400}$$

$$\mathbf{G} = \frac{s}{s^2 + (3)^2}$$

$$= \frac{s}{s^2 + 9}$$

EXAMPLE 12.17 Determine the inverse Laplace transform of

$$\left\{\frac{1}{s + 15}\right\}$$ ■

SOLUTION From entry 2 of Table 12.2,

$$\mathcal{L}^{-1}\left\{\frac{1}{s + 15}\right\} = e^{-15t}$$

EXAMPLE 12.18 Determine the inverse Laplace transform of

$$\left\{\frac{12}{s}\right\}$$ ■

SOLUTION From entry 1 of Table 12.2,

$$\mathcal{L}^{-1}\left\{\frac{12}{s}\right\} = 12$$

EXAMPLE 12.19 Determine the inverse Laplace transform of

$$\left\{\frac{6}{s^2 + 36}\right\}$$ ■

SOLUTION From entry 5 of Table 12.2,

$$\mathcal{L}^{-1}\left\{\frac{6}{s^2+36}\right\} = \sin(6t)$$

12.15 TRANSFORMATION RULES

The rules presented here provide the tools necessary for solving differential equations by the Laplace transform method.

RULE L1 *The Laplace transform of the first derivative* is

$$\mathcal{L}\left\{\frac{dy}{dt}\right\} = s\mathbf{Y} - y(0)$$

where $\mathbf{Y}$ is the Laplace transform of y and $y(0)$ is the value of y at $t = 0$.

EXAMPLE 12.20 Determine the Laplace transform of dy/dt if $y = e^{-4t}$. ■

SOLUTION From rule L1,

$$\mathcal{L}\left\{\frac{dy}{dt}\right\} = s\mathbf{Y} - y(0)$$

Because $\mathbf{Y} = \dfrac{1}{(s+4)}$ from entry 2 of Table 12.2 and $y(0) = e^{-4(0)} = 1$,

$$\begin{aligned}\mathcal{L}\left\{\frac{dy}{dt}\right\} &= s\left\{\frac{1}{s+4}\right\} - 1\\ &= \frac{s}{s+4} - 1\\ &= \frac{s-(s+4)}{s+4}\\ &= \frac{-4}{s+4}\end{aligned}$$

RULE L2 *The Laplace transform of the second derivative* is

$$\mathcal{L}\left\{\frac{d^2y}{dt^2}\right\} = s^2\mathbf{Y} - sy(0) - \frac{dy(0)}{dt}$$

where $\mathbf{Y}$ is the Laplace transform of y, $dy(0)/dt$ is the value of dy/dt at $t = 0$, and $y(0)$ is the value of y at $t = 0$.

EXAMPLE 12.21 Determine the Laplace transform of d^2y/dt^2 if $y = t^2$. ■

SOLUTION From rule L2,

$$\mathcal{L}\left\{\frac{d^2y}{dt^2}\right\} = s^2\mathbf{Y} - sy(0) - \frac{dy(0)}{dt}$$

From entry 3 of Table 12.2,

$$\mathbf{Y} = \frac{2}{s^3}$$

Because $dy/dt = \dfrac{d(t^2)}{dt} = 2t$,

$$\frac{dy(0)}{dt} = 0$$

From the function $y = t^2$,

$$y(0) = 0$$

It follows that

$$\begin{aligned}\mathcal{L}\left\{\frac{d^2y}{dt^2}\right\} &= s^2\left\{\frac{2}{s^3}\right\} - s(0) - 0\\ &= \frac{2}{s}\end{aligned}$$

Alternatively, because $d^2y/dt^2 = 2$ for $y = t^2$ and $\mathcal{L}\{2\} = 2/s$ from entry 1 of Table 12.2,

$$\mathcal{L}\left\{\frac{d^2y}{dt^2}\right\} = \frac{2}{s}$$

RULE L3(a) *The Laplace transform of a constant times a function* is equal to the constant times the Laplace transform of the function.

$$\mathcal{L}\{Ky\} = (K)\mathcal{L}\{y\}$$

where K is a constant and y is a function of t.

RULE L3(b) *The inverse Laplace transform of a constant times a function* is equal to the constant times the inverse Laplace transform of the function.

$$\mathcal{L}^{-1}\{K\mathbf{Y}\} = K\mathcal{L}^{-1}\{\mathbf{Y}\}$$

where K is a constant and $\mathbf{Y}$ is a function of s.

EXAMPLE 12.22 Determine the Laplace transform of $16y$ if $y = e^{-8t}$ and the inverse Laplace transform is $10/(s + 3)$. ■

SOLUTION From rule L3(a) and entry 2 of Table 12.2,

$$\begin{aligned}\mathcal{L}\{16y\} &= (16)\mathcal{L}\{y\}\\ &= (16)\mathcal{L}\{e^{-8t}\}\\ &= (16)\left\{\frac{1}{s+8}\right\}\\ &= \frac{16}{s+8}\end{aligned}$$

From rule L3(b) and entry 2 of Table 12.2,

$$\mathcal{L}^{-1}\left\{\frac{10}{s+3}\right\} = 10\mathcal{L}^{-1}\left\{\frac{1}{s+3}\right\}$$
$$= 10e^{-3t}$$

RULE L4(a) *The Laplace transform of the sum or difference of two functions is equal to the sum or difference of the Laplace transforms of the two functions.*

$$\mathcal{L}\{y_1 \pm y_2\} = \mathcal{L}\{y_1\} \pm \mathcal{L}\{y_2\}$$

where y_1 and y_2 are functions of t.

RULE L4(b) *The inverse Laplace transform of the sum or difference of two functions is equal to the sum or difference of the inverse Laplace transforms of the two functions.*

$$\mathcal{L}^{-1}\{\mathbf{Y}_1 \pm \mathbf{Y}_2\} = \mathcal{L}^{-1}\{\mathbf{Y}_1\} \pm \mathcal{L}^{-1}\{\mathbf{Y}_2\}$$

where $\mathbf{Y}_1$ and $\mathbf{Y}_2$ are functions of s.

EXAMPLE 12.23 Determine the Laplace transform of $y = e^{-6t} + t$. ■

SOLUTION From rule L4(a) and entries 2 and 3 of Table 12.2,

$$\mathcal{L}\{y\} = \mathcal{L}\{e^{-6t} + t\}$$
$$= \mathcal{L}\{e^{-6t}\} + \mathcal{L}\{t\}$$
$$= \frac{1}{s+6} + \frac{1}{s^2}$$
$$= \frac{s^2+s+6}{s^2(s+6)}$$
$$= \frac{s^2+s+6}{s^3+6s^2}$$

EXAMPLE 12.24 Determine the inverse Laplace transform of

$$40\left\{\frac{1}{s+4}\right\}$$ ■

SOLUTION From entry 2 of Table 12.2 and rule L3(b),

$$\mathcal{L}^{-1}\left\{40\left\{\frac{1}{s+4}\right\}\right\} = 40\mathcal{L}^{-1}\left\{\frac{1}{s+4}\right\}$$
$$= 40e^{-4t}$$

EXAMPLE 12.25 Determine the inverse Laplace transform of

$$\left\{\frac{6}{s+1} + \frac{14}{s}\right\}$$ ■

SOLUTION From entries 1 and 2 of Table 12.2 and rules L3(b) and L4(b),

$$\mathcal{L}^{-1}\left\{\frac{6}{s+1}+\frac{14}{s}\right\} = \mathcal{L}^{-1}\left\{\frac{6}{s+1}\right\} + \mathcal{L}^{-1}\left\{\frac{14}{s}\right\}$$

$$= 6\mathcal{L}^{-1}\left\{\frac{1}{s+1}\right\} + \mathcal{L}^{-1}\left\{\frac{14}{s}\right\}$$

$$= 6e^{-t} + 14$$

EXAMPLE 12.26 Determine the inverse Laplace transform of

$$\left\{\frac{4}{s^2+64}\right\}$$ ■

SOLUTION The Laplace transform must be put in the form of entry 5, Table 12.2, which is $\omega/(s^2+\omega^2)$:

$$\frac{4}{s^2+64} = \frac{4}{s^2+8^2}$$

$$= \frac{(4)}{(8)}\left\{\frac{8}{s^2+8^2}\right\}$$

$$= \left(\frac{1}{2}\right)\left\{\frac{8}{s^2+8^2}\right\}$$

Notice that $\{8/(s^2+8^2)\}$ is in the same form as entry 5 of Table 12.2, where $\omega = 8$. From entry 5 and rule L3(b),

$$\mathcal{L}^{-1}\left\{\frac{4}{s^2+64}\right\} = \mathcal{L}^{-1}\left\{\left(\frac{1}{2}\right)\left\{\frac{8}{s^2+8^2}\right\}\right\}$$

$$= \left(\frac{1}{2}\right)\mathcal{L}^{-1}\left\{\frac{8}{s^2+8^2}\right\}$$

$$= \left(\frac{1}{2}\right)\sin(8t)$$

12.16 PARTIAL FRACTION EXPANSION

As illustrated previously, the Laplace transform must be expressed in the exact form as the table entry to determine the inverse Laplace transform from a table of Laplace transforms. Algebraic manipulation of the Laplace transform is usually required, as illustrated in Example 12.26. The use of a larger table of transforms (many are commercially available) greatly reduces the algebraic manipulation necessary to determine an inverse transform.

The expansion of Laplace transform expressions into partial fractions allows us to make more effective use of Laplace transform tables. The rules presented here are valid only for expressions that are proper fractions. That is, the degree of the denominator must be at least 1 greater than the degree of the numerator to expand an expression into partial fractions. If the degree of the denominator is not greater than the degree of the numerator, division must be performed to form a proper fraction prior to expanding into partial fractions.

The denominator of a proper fraction must be factored prior to expanding the proper fraction into partial fractions. We shall consider denominator factors s^n, $(s + a)^n$, and $(s^2 + bs + c)$, where n is any positive integer and is called the *multiplicity* of the factor. We shall consider the quadratic factor $(s^2 + bs + c)$ only where it has a multiplicity of 1. It is noted that quadratic factors can be factored and the factors written in the form $(s + e)(s + f)$.

For *each factor of the form* $(s)^n$ in the denominator of a proper fraction, n terms of the form

$$\frac{K_n}{(s)^n} + \frac{K_{n-1}}{(s)^{n-1}} + \cdots + \frac{K_1}{s}$$

are required in the partial fraction expansion of the proper fraction.

For *each factor of the form* $(s + a)^n$ in the denominator of a proper fraction, n terms of the form

$$\frac{K_n}{(s + a)^n} + \frac{K_{n-1}}{(s + a)^{n-1}} + \cdots + \frac{K_1}{(s + a)}$$

are required in the partial fraction expansion of the proper fraction. For example, the proper fraction

$$\frac{s - 1}{(s + 2)^3(s + 4)^2(s + 1)(s)^2}$$

can be represented in terms of partial fractions as

$$\begin{aligned}\frac{s - 1}{(s + 2)^3(s + 4)^2(s + 1)(s)^2} &= \frac{K_1}{(s + 2)^3} + \frac{K_2}{(s + 2)^2} + \frac{K_3}{(s + 2)} \\ &+ \frac{K_4}{(s + 4)^2} + \frac{K_5}{(s + 4)} \\ &+ \frac{K_6}{(s + 1)} + \frac{K_7}{(s)^2} + \frac{K_8}{(s)}\end{aligned}$$

where the Ks are constants to be determined. Notice that there are three terms in the expansion for the factor $(s + 2)^3$, two terms for $(s + 4)^2$, one term for $(s + 1)$, and two terms for $(s)^2$.

For *each factor of the form* $(s^2 + bs + c)$ in the denominator of a proper fraction, a single term of the form

$$\frac{K_1 s + K_2}{(s^2 + bs + c)}$$

is required in the partial fraction expansion of the proper fraction. For example, the proper fraction

$$\frac{s + 4}{(s^2 + 2s + 2)(s + 8)(s + 6)(s)}$$

can be represented in terms of partial fractions as

$$\frac{s + 4}{(s^2 + 3s + 2)(s + 8)(s + 6)(s)} = \frac{K_1 s + K_2}{(s^2 + 3s + 2)} + \frac{K_3}{(s + 8)} + \frac{K_4}{(s + 6)} + \frac{K_5}{s}$$

where the Ks are constants to be determined.

EXAMPLE 12.27 Determine the form of the partial fraction expansion for the proper fraction

$$\frac{s-1}{(s+9)^2(s+4)(s^2+3s+2)(s+7)^2}$$ ■

SOLUTION From the preceding rules for expanding partial fractions,

$$\frac{s-1}{(s+9)^2(s+4)(s^2+3s+2)(s+7)^2} = \frac{K_1}{(s+9)^2} + \frac{K_2}{(s+9)} + \frac{K_3}{(s+4)} + \frac{K_4 s + K_5}{(s^2+3s+2)} + \frac{K_6}{(s+7)^2} + \frac{K_7}{s+7}$$

Alternatively, $(s^2 + 3s + 2)$ can be factored and written as $(s + 2)(s + 1)$, and the resulting partial fraction expansion can be written as

$$\frac{s-1}{(s+9)^2(s+4)(s+2)(s+1)(s+7)^2} = \frac{K_1}{(s+9)^2} + \frac{K_2}{(s+9)} + \frac{K_3}{(s+4)} + \frac{K_4}{(s+2)} + \frac{K_5}{(s+1)} + \frac{K_6}{(s+7)^2} + \frac{K_7}{(s+7)}$$

EXAMPLE 12.28 Determine the form for a partial fraction expansion for the improper fraction

$$\frac{6s^3 + 100s^2 + 85s + 52}{s^3 + 7s^2 + 14s + 8}$$ ■

SOLUTION Because the expression is not a proper fraction, it cannot be expanded into partial fractions as written. However, if the denominator is divided into the numerator, part of the expression becomes a proper fraction, and that part can be expanded.

$$\begin{array}{r|l} & 6 \\ \hline s^3 + 7s^2 + 14s + 8 & 6s^3 + 100s^2 + 85s + 52 \\ & 6s^3 + 42s^2 + 84s + 48 \\ \hline & 58s^2 + s + 4 \end{array}$$

$$\frac{6s^3 + 100s^2 + 85s + 52}{s^3 + 7s^2 + 14s + 8} = 6 + \frac{58s^2 + s + 4}{s^3 + 7s^2 + 14s + 8}$$

Factoring $(s^3 + 7s^2 + 14s + 18)$ results in

$$\frac{6s^3 + 100s^2 + 85s + 52}{s^3 + 7s^2 + 14s + 8} = 6 + \frac{58s^2 + s + 4}{(s+1)(s+2)(s+4)}$$

From the preceding rules for expanding partial fractions,

$$\frac{6s^3 + 100s^2 + 85s + 52}{s^3 + 7s^2 + 14s + 8} = 6 + \frac{K_1}{(s+1)} + \frac{K_2}{(s+2)} + \frac{K_3}{(s+3)}$$

We shall now consider the evaluation of the constants (Ks) associated with partial fraction expansions. The examples that follow demonstrate methods for reducing Laplace transforms to simpler algebraic forms by partial fraction expansion to facilitate locating them in Laplace transform tables.

EXAMPLE 12.29 Determine the inverse Laplace transform of

$$\left\{\frac{1}{(s)(s+2)}\right\}$$ ■

SOLUTION The Laplace transform

$$\left\{\frac{1}{(s)(s+2)}\right\}$$

can be expressed in a form that is compatible with entries in Table 12.2 by expanding the fraction

$$\left\{\frac{1}{(s)(s+2)}\right\}$$

into partial fractions:

$$\frac{1}{(s)(s+2)} = \frac{K_1}{(s)} + \frac{K_2}{(s+2)}$$

K_1 can be determined by first multiplying both sides of the equation by s (the denominator of K_1):

$$\cancel{s}\left\{\frac{1}{\cancel{s}(s+2)}\right\} = \cancel{s}\left\{\frac{K_1}{\cancel{s}}\right\} + (s)\left\{\frac{K_2}{s+2}\right\}$$

$$\frac{1}{s+2} = K_1 + s\left\{\frac{K_2}{s+2}\right\}$$

Because the equation is valid for all values of s, we set $s = 0$ to eliminate the term containing K_2:

$$\frac{1}{0+2} = K_1 + 0\left\{\frac{K_2}{0+2}\right\}$$

and

$$K_1 = \frac{1}{2}$$

K_2 can be determined by first multiplying both sides of the equation by $(s + 2)$, the denominator of K_2, as follows:

$$\cancel{(s+2)}\left\{\frac{1}{s\cancel{(s+2)}}\right\} = (s+2)\left\{\frac{K_1}{s}\right\} + \cancel{(s+2)}\left\{\frac{K_2}{\cancel{s+2}}\right\}$$

$$\frac{1}{s} = (s+2)\left\{\frac{K_1}{s}\right\} + K_2$$

Setting $s = -2$ to eliminate the K_1 term results in

$$\frac{1}{(-2)} = (-2+2)\left\{\frac{K_1}{-2}\right\} + K_2$$

and $$K_2 = -\frac{1}{2}$$

The Laplace transform

$$\left\{\frac{1}{(s)(s+2)}\right\}$$

can now be written as

$$\left\{\frac{\left(\frac{1}{2}\right)}{s} - \frac{\left(\frac{1}{2}\right)}{s+2}\right\}$$

and $$\mathcal{L}^{-1}\left\{\frac{1}{(s)(s+2)}\right\} = \mathcal{L}^{-1}\left\{\frac{\left(\frac{1}{2}\right)}{s} - \frac{\left(\frac{1}{2}\right)}{s+2}\right\}$$

From rules L3(b) and L4(b) and entries 1 and 2 of Table 12.2,

$$\mathcal{L}^{-1}\left\{\frac{1}{(s)(s+2)}\right\} = \mathcal{L}^{-1}\left\{\frac{\left(\frac{1}{2}\right)}{s}\right\} - \left(\frac{1}{2}\right)\mathcal{L}^{-1}\left\{\frac{1}{s+2}\right\}$$

$$= \frac{1}{2} - \left(\frac{1}{2}\right)e^{-2t}$$

Notice that K_1 and K_2 can also be determined by putting

$$\frac{K_1}{s} + \frac{K_2}{s+2}$$

on a common denominator and equating the numerator coefficient:

$$\frac{1}{(s)(s+2)} = \frac{K_1(s+2) + K_2(s)}{(s)(s+2)}$$

Equating the numerators results in

$$1 = K_1(s+2) + K_2(s)$$

$$1 = (K_1 + K_2)s + 2K_1$$

Equating the coefficients on each side of the equation results in

$$2K_1 = 1$$

$$K_1 = \frac{1}{2}$$

$$K_1 + K_2 = 0$$

$$K_2 = K_1$$

$$K_2 = -\frac{1}{2}$$

EXAMPLE 12.30 Determine the inverse Laplace transform of

$$\left\{\frac{1}{(s+2)(s+4)}\right\}$$

SOLUTION The expression can be expanded into partial fractions as

$$\frac{1}{(s+2)(s+4)} = \frac{K_1}{s+2} + \frac{K_2}{s+4}$$

Multiplying both sides of the equation by $(s + 2)$ and setting $s = -2$ to determine K_1 results in

$$\cancel{(s+2)}\left\{\frac{1}{\cancel{(s+2)}(s+4)}\right\} = \cancel{(s+2)}\left\{\frac{K_1}{\cancel{s+2}}\right\} + (s+2)\left\{\frac{K_2}{s+4}\right\}$$

$$\frac{1}{s+4} = K_1 + (s+2)\left\{\frac{K_2}{s+4}\right\}$$

$$\frac{1}{(-2+4)} = K_1 + (-2+2)\left\{\frac{K_2}{-2+4}\right\}$$

$$K_1 = \frac{1}{2}$$

Multiplying both sides of the equation by $(s + 4)$ and setting $s = -4$ to determine K_2 results in

$$\cancel{(s+4)}\left\{\frac{1}{(s+2)\cancel{(s+4)}}\right\} = (s+4)\left\{\frac{K_1}{s+2}\right\} + \cancel{(s+4)}\left\{\frac{K_2}{\cancel{(s+4)}}\right\}$$

$$\frac{1}{s+2} = (s+4)\left\{\frac{K_1}{s+2}\right\} + K_2$$

$$\frac{1}{(-4+2)} = (-4+4)\left\{\frac{K_1}{-4+2}\right\} + K_2$$

$$K_2 = -\frac{1}{2}$$

Because $K_1 = \frac{1}{2}$ and $K_2 = -\frac{1}{2}$,

$$\frac{1}{(s+2)(s+4)} = \frac{\left(\frac{1}{2}\right)}{(s+2)} - \frac{\left(\frac{1}{2}\right)}{(s+4)}$$

From rules L3(b) and L4(b) and entry 2 of Table 12.2,

$$\mathcal{L}^{-1}\left\{\frac{1}{(s+2)(s+4)}\right\} = \mathcal{L}^{-1}\left\{\frac{\left(\frac{1}{2}\right)}{s+2} - \frac{\left(\frac{1}{2}\right)}{s+4}\right\}$$

$$= \mathcal{L}^{-1}\left\{\frac{\left(\frac{1}{2}\right)}{s+2}\right\} - \mathcal{L}^{-1}\left\{\frac{\left(\frac{1}{2}\right)}{s+4}\right\}$$

$$= \left(\frac{1}{2}\right)\mathcal{L}^{-1}\left\{\frac{1}{s+2}\right\} - \left(\frac{1}{2}\right)\mathcal{L}^{-1}\left\{\frac{1}{s+4}\right\}$$

$$= \left(\frac{1}{2}\right)e^{-2t} - \left(\frac{1}{2}\right)e^{-4t}$$

EXAMPLE 12.31 Determine the inverse Laplace transform of **Y**, where

$$\mathbf{Y} = \frac{s}{(s+1)^2(s+4)}$$ ■

SOLUTION From the rules given before for expanding proper fractions

$$\frac{s}{(s+1)^2(s+4)} = \frac{K_1}{(s+1)^2} + \frac{K_2}{(s+1)} + \frac{K_3}{(s+4)}$$

Multiplying both sides of the equation by $(s+1)^2$ to determine K_1 results in

$$\cancel{(s+1)^2}\left\{\frac{s}{\cancel{(s+1)^2}(s+4)}\right\} = \cancel{(s+1)^2}\left\{\frac{K_1}{\cancel{(s+1)^2}}\right\} + (s+1)^{\cancel{2}}\left\{\frac{K_2}{\cancel{(s+1)}}\right\} + (s+1)^2\left\{\frac{K_3}{(s+4)}\right\}$$

and setting $s = -1$,

$$\left\{\frac{-1}{-1+4}\right\} = K_1 + (-1+1)K_2 + (-1+1)^2\frac{K_3}{(-1+4)}$$

$$-\frac{1}{3} = K_1 + 0 + 0$$

$$K_1 = -\frac{1}{3}$$

Multiplying both sides of the equation by $(s+4)$ to determine K_3 results in

$$\cancel{(s+4)}\left\{\frac{s}{(s+1)^2\cancel{(s+4)}}\right\} = (s+4)\left\{\frac{K_1}{(s+1)^2}\right\} + (s+4)\left\{\frac{K_2}{(s+1)}\right\} + \cancel{(s+4)}\left\{\frac{K_3}{\cancel{(s+4)}}\right\}$$

and setting $s = -4$

$$\frac{-4}{(-4+1)^2} = (-4+4)\left\{\frac{K_1}{(-4+1)^2}\right\} + (-4+4)\left\{\frac{K_2}{(-4+1)}\right\} + K_3$$

$$-\frac{4}{9} = 0 + 0 + K_3$$

$$K_3 = -\frac{4}{9}$$

We can determine K_2 by putting the right side of the equation on a common denominator and equating numerators:

$$\frac{s}{(s+1)^2(s+4)} = \frac{K_1(s+4) + K_2(s+1)(s+4) + K_3(s+1)^2}{(s+1)^2(s+4)}$$

Since $K_1 = -\frac{1}{3}$ and $K_3 = -\frac{4}{9}$,

$$\frac{s}{(s+1)^2(s+4)} = \frac{\left(-\frac{1}{3}\right)(s+4) + K_2(s+1)(s+4) - \left(\frac{4}{9}\right)(s+1)^2}{(s+1)^2(s+4)}$$

$$= \frac{\left(-\frac{1}{3}\right)(s+4) + K_2(s^2+5s+4) + \left(-\frac{4}{9}\right)(s^2+2s+1)}{(s+1)^2(s+4)}$$

$$= \frac{\left(-\frac{1}{3} + 5K_2 - \frac{8}{9}\right)s + \left(K_2 - \frac{4}{9}\right)s^2 + \left(-\frac{4}{3} + 4K_2 - \frac{4}{9}\right)}{(s+1)^2(s+4)}$$

Equating numerators results in

$$s = \left(-\frac{11}{9} + 5K_2\right)s + \left(K_2 - \frac{4}{9}\right)s^2 + \left(-\frac{16}{9} + 4K_2\right)$$

Equating coefficients results in

$$5K_2 - \frac{11}{9} = 1$$

$$K_2 = \frac{4}{9}$$

or

$$K_2 - \frac{4}{9} = 0$$

$$K_2 = \frac{4}{9}$$

or

$$-\frac{16}{9} + 4K_2 = 0$$

$$K_2 = \frac{4}{9}$$

The expanded transform can be written as

$$\mathbf{Y} = \frac{s}{(s+1)^2(s+4)} = \frac{\left(-\frac{1}{3}\right)}{(s+1)^2} + \frac{\left(\frac{4}{9}\right)}{(s+1)} + \frac{\left(-\frac{4}{9}\right)}{(s+4)}$$

and

$$\mathcal{L}^{-1}\{\mathbf{Y}\} = \mathcal{L}^{-1}\left\{\frac{-\frac{1}{3}}{(s+1)^2}\ \frac{\frac{4}{9}}{(s+1)}\ \frac{-\frac{4}{9}}{(s+4)}\right\}$$

From rules L3(b) and L4(b),

$$y = \left(-\frac{1}{3}\right)\mathcal{L}^{-1}\left\{\frac{1}{(s+1)^2}\right\} + \left(\frac{4}{9}\right)\mathcal{L}^{-1}\left\{\frac{1}{(s+1)}\right\} - \left(\frac{4}{9}\right)\mathcal{L}^{-1}\left\{\frac{1}{(s+4)}\right\}$$

From entries 2 and 9 in Table 12.2,

$$y = -\left(\frac{1}{3}\right)te^{-t} + \left(\frac{4}{9}\right)e^{-t} - \left(\frac{4}{9}\right)e^{-4t}$$

Notice that K_2 in Example (12.31) cannot be determined by multiplying both sides of the expanded equation by the denominator of K_2 and then setting $s = -1$, as would be the usual procedure. If this procedure is attempted, division by zero results and K_2 cannot be determined.

Although we shall not generally use differentiation to determine the constants associated with partial fraction expansions, it is noted that K_2 in Example (12.31) can also be determined by multiplying both sides of the expanded equation by the denominator of K_1, differentiating both sides of the resulting equation with respect to s, and finally setting $s = -1$ to arrive at the value of K_2. The determination of K_2 in this manner is left as a student exercise.

EXAMPLE 12.32 Expand the following proper fraction into partial fractions:

$$\mathbf{Y} = \frac{1}{(s^2+3s+2)(s+4)}$$ ■

SOLUTION From the rules given above for expanding proper fractions,

$$\frac{1}{(s^2+3s+2)(s+4)} = \frac{K_1 s + K_2}{(s^2+3s+4)} + \frac{K_3}{(s+4)}$$

Multiplying both sides of the equation by $(s+4)$ to determine K_3 results in

$$\cancel{(s+4)}\left\{\frac{1}{(s^2+3s+2)\cancel{(s+4)}}\right\} = (s+4)\left\{\frac{K_1 s + K_2}{(s^2+3s+2)}\right\} + \cancel{(s+4)}\left\{\frac{K_3}{\cancel{(s+4)}}\right\}$$

and setting $s = -4$,

$$\frac{1}{6} = 0 + K_3$$

$$K_3 = \frac{1}{6}$$

Putting the right-hand side of the expanded equation over a common denominator to determine K_1 and K_2 results in

$$\frac{1}{(s^2 + 3s + 2)(s + 4)} = \frac{K_1 s + K_2}{(s^2 + 3s + 4)} + \frac{\frac{1}{6}}{(s + 4)}$$

$$= \frac{K_1 s^2 + 4sK_1 + K_2 s + 4K_2 + \left(\frac{1}{6}\right)s^2 + \left(\frac{1}{2}\right)s + \frac{1}{3}}{(s^2 + 3s + 2)(s + 4)}$$

$$= \frac{\left(K_1 + \frac{1}{6}\right)s^2 + \left(4K_1 + K_2 + \frac{1}{2}\right)s + \left(4K_2 + \frac{1}{3}\right)}{(s^2 + 3s + 2)(s + 4)}$$

Equating numerators results in

$$1 = \left(K_1 + \frac{1}{6}\right)s^2 + \left(4K_1 + K_2 + \frac{1}{2}\right)s + \left(4K_2 + \frac{1}{3}\right)$$

Equating coefficients results in

$$K_1 + \frac{1}{6} = 0$$

$$4K_1 + K_2 + \frac{1}{2} = 0$$

$$4K_2 + \frac{1}{3} = 1$$

From the first and third equations

$$K_1 = -\frac{1}{6}$$

$$K_2 = \frac{1}{6}$$

Notice that the second equation is also satisfied by $K_1 = -\frac{1}{6}$ and $K_2 = \frac{1}{6}$.

The required partial fraction expansion is

$$\mathbf{Y} = \frac{\left(-\frac{1}{6}\right)s + \left(\frac{1}{6}\right)}{(s^2 + 3s + 2)} + \frac{\frac{1}{6}}{(s + 4)}$$

Alternatively, the second-degree term of the denominator of the proper fraction can be factored and the proper fraction written as

$$\mathbf{Y} = \frac{1}{(s+2)(s+1)(s+4)}$$

The partial fraction expansion is then

$$\mathbf{Y} = \frac{K_1}{(s+2)} + \frac{K_2}{(s+1)} + \frac{K_3}{(s+4)}$$

and using the method previously described to evaluate the constants,

$$\mathbf{Y} = \frac{\left(-\frac{1}{2}\right)}{(s+2)} + \frac{\left(\frac{1}{3}\right)}{(s+1)} + \frac{\frac{1}{6}}{(s+4)}$$

which is equivalent to the previous partial fraction expansion

$$\mathbf{Y} = \frac{\left(-\frac{1}{6}\right)s + \left(\frac{1}{6}\right)}{(s^2+3s+2)} + \frac{\frac{1}{6}}{(s+4)}$$

12.17 SOLVING DIFFERENTIAL EQUATIONS BY THE LAPLACE TRANSFORM METHOD

We shall now present the Laplace transform method for solving differential equations. All the terms of the differential equations that we encounter in basic electric circuit analysis are usually Laplace transformable; that is, the Laplace transform for each term can be determined. Consider the differential equation

$$\frac{dy}{dt} + 6y = 0$$

with initial condition $y(0) = 2$. In this differential equation, the terms dy/dt, $6y$, and 0 each have a Laplace transform.

Since the left side of the preceding differential equation equals the right side, the Laplace transform of the left side equals the Laplace transform of the right side and

$$\mathscr{L}\left\{\frac{dy}{dt} + 6y\right\} = \mathscr{L}\{0\}$$

From rule L4

$$\mathscr{L}\left\{\frac{dy}{dt}\right\} + \mathscr{L}\{6y\} = \mathscr{L}\{0\}$$

From rule L3

$$\mathscr{L}\left\{\frac{dy}{dt}\right\} + 6\,\mathscr{L}\{y\} = \mathscr{L}\{0\}$$

From entry 1 of Table 12.2 and rule L1

$$\{s\mathbf{Y} - y(0)\} + 6\mathbf{Y} = \frac{0}{s}$$

Since $y(0)$ is given as 2,

$$s\mathbf{Y} - 2 + 6\mathbf{Y} = 0$$
$$(s + 6)\mathbf{Y} = 2$$

and

$$\mathbf{Y} = \frac{2}{s + 6}$$

Y is the Laplace transform of the particular complete solution of the differential equation for initial condition $y(0) = 2$.

From entry 2 of Table 12.2 and rule L3(b)

$$y = \mathcal{L}^{-1}\left\{\frac{2}{s + 6}\right\}$$
$$= 2\,\mathcal{L}^{-1}\left\{\frac{1}{s + 6}\right\}$$
$$= 2e^{-6t}$$

is the particular complete solution.

This procedure is demonstrated in the following examples.

EXAMPLE 12.33 Determine the particular complete solution for the differential equation

$$\frac{dy}{dt} + 6y = e^{-t}$$

with initial condition $y(0) = 0$. ■

SOLUTION Taking the Laplace transform of each side of the differential equation results in

$$\mathcal{L}\left\{\frac{dy}{dt} + 6y\right\} = \mathcal{L}\{e^{-t}\}$$

From rule L4

$$\mathcal{L}\left\{\frac{dy}{dt}\right\} + \mathcal{L}\{6y\} = \mathcal{L}\{e^{-t}\}$$

From rules L1 and L3 and entry 2 of Table 12.2,

$$\{s\mathbf{Y} - y(0)\} + 6\mathcal{L}\{y\} = \mathcal{L}\{e^{-t}\}$$
$$s\mathbf{Y} - y(0) + 6\mathbf{Y} = \frac{1}{s + 1}$$

Because $y(0) = 0$

$$s\mathbf{Y} + 6\mathbf{Y} = \frac{1}{s + 1}$$

$$(s + 6)\mathbf{Y} = \frac{1}{s + 1}$$

$$\mathbf{Y} = \frac{1}{(s + 1)(s + 6)}$$

Since the Laplace transform

$$\left\{\frac{1}{(s + 1)(s + 6)}\right\}$$

does not appear in Table 12.2, it must be expanded to a form that appears in Table 12.2 using the method previously demonstrated for partial fraction expansion.

$$\frac{1}{(s + 1)(s + 6)} = \frac{K_1}{s + 1} + \frac{K_2}{s + 6}$$

$$K_1 = \frac{1}{5}$$

$$K_2 = -\frac{1}{5}$$

Substituting the values for K_1 and K_2 into the expression for **Y** results in

$$\mathbf{Y} = \frac{\frac{1}{5}}{s + 1} - \frac{\frac{1}{5}}{s + 6}$$

Taking the inverse of both sides of the preceding equation results in

$$y = \mathcal{L}^{-1}\left\{\frac{\frac{1}{5}}{s + 1} - \frac{\frac{1}{5}}{s + 6}\right\}$$

From rule L4(b)

$$y = \mathcal{L}^{-1}\left\{\frac{\frac{1}{5}}{s + 1}\right\} - \mathcal{L}^{-1}\left\{\frac{\frac{1}{5}}{s + 6}\right\}$$

From rule L3(b)

$$y = \left(\frac{1}{5}\right)\mathcal{L}^{-1}\left\{\frac{1}{s + 1}\right\} - \left(\frac{1}{5}\right)\mathcal{L}^{-1}\left\{\frac{1}{s + 6}\right\}$$

From entry 2 of Table 12.2,

$$y = \left(\frac{1}{5}\right)e^{-t} - \left(\frac{1}{5}\right)e^{-6t}$$

EXAMPLE 12.34 Determine the particular complete solution for the differential equation

$$\frac{d^2y}{dt^2} + 3\frac{dy}{dt} + 2y = 0$$

with initial conditions $y(0) = 1$ and $\dfrac{dy(0)}{dt} = 0$. ■

SOLUTION Taking the Laplace transform of each side of the differential equation results in

$$\mathcal{L}\left\{\frac{d^2y}{dt^2} + 3\frac{dy}{dt} + 2y\right\} = \mathcal{L}\{0\}$$

From rules L3 and L4

$$\mathcal{L}\left\{\frac{d^2y}{dt^2}\right\} + 3\mathcal{L}\left\{\frac{dy}{dt}\right\} + 2\mathcal{L}\{y\} = \mathcal{L}\{0\}$$

From rules L1 and L2 and entry 1 of Table 12.2,

$$\left\{s^2\mathbf{Y} - sy(0) - \frac{d(0)}{dt}\right\} + 3\{s\mathbf{Y} - y(0)\} + 2\mathbf{Y} = \frac{0}{s}$$

Substituting the given initial conditions for $y(0)$ and $\dfrac{dy(0)}{dt}$ results in

$$s^2\mathbf{Y} - s + 3s\mathbf{Y} - 3 + 2\mathbf{Y} = 0$$

$$(s^2 + 3s + 2)\mathbf{Y} = s + 3$$

$$\mathbf{Y} = \frac{s + 3}{s^2 + 3s + 2}$$

$$= \frac{s + 3}{(s + 2)(s + 1)}$$

Because the Laplace transform

$$\left\{\frac{s + 3}{s^2 + 3s + 2}\right\}$$

does not appear in Table 12.2, it must be expanded to a form that is in the table. Using the method previously demonstrated for partial fraction expansion results in

$$\frac{s+3}{(s+1)(s+2)} = \frac{K_1}{s+2} + \frac{K_2}{s+1}$$

$$K_1 = -1$$

$$K_2 = 2$$

Substituting the values for K_1 and K_2 into the expression for **Y** results in

$$\mathbf{Y} = \frac{-1}{s+2} + \frac{2}{s+1}$$

Taking the inverse of both sides of this equation results in

$$y = \mathcal{L}^{-1}\left\{\frac{-1}{s+2} + \frac{2}{s+1}\right\}$$

From rules L3(b) and L4(b)

$$y = -\mathcal{L}^{-1}\left\{\frac{1}{s+2}\right\} + 2\mathcal{L}^{-1}\left\{\frac{1}{s+1}\right\}$$

From entry 2 of Table 12.2

$$y = -e^{-2t} + 2e^{-t}$$

EXAMPLE 12.35 Determine the particular complete solution for the differential equation

$$\frac{dy}{dt} + 10y = 4 - e^{-2t}$$

with initial condition $y(0) = 0$. ■

SOLUTION Taking the Laplace transform of each side of the differential equation results in

$$\mathcal{L}\left\{\frac{dy}{dt} + 10y\right\} = \mathcal{L}\{4 - e^{-2t}\}$$

From rules L3 and L4,

$$\mathcal{L}\left\{\frac{dy}{dt}\right\} + 10\,\mathcal{L}\{y\} = \mathcal{L}\{4\} - \mathcal{L}\{e^{-2t}\}$$

From rule L1 and entries 1 and 2 of Table 12.2,

$$\{s\mathbf{Y} - y(0)\} + 10\mathbf{Y} = \frac{4}{s} - \frac{1}{s+2}$$

Because $y(0) = 0$,

$$s\mathbf{Y} + 10\mathbf{Y} = \frac{4}{s} - \frac{1}{s+2}$$

$$(s+10)\mathbf{Y} = \frac{4(s+2) - s}{(s)(s+2)}$$

$$(s+10)\mathbf{Y} = \frac{3s+8}{s(s+2)}$$

$$\mathbf{Y} = \frac{3s+8}{s(s+2)(s+10)}$$

Using the method previously demonstrated for partial fraction expansion to put $\mathbf{Y}$ in a form that appears in Table 12.2 results in

$$\frac{3s+8}{s(s+2)(s+10)} = \frac{K_1}{s} + \frac{K_2}{s+2} + \frac{K_3}{s+10}$$

$$K_1 = \frac{2}{5}$$

$$K_2 = -\frac{1}{8}$$

$$K_3 = -\frac{11}{40}$$

and

$$\mathbf{Y} = \frac{\frac{2}{5}}{s} - \frac{\frac{1}{8}}{s+2} - \frac{\frac{11}{40}}{s+10}$$

Taking the inverse of both sides of this equation results in

$$y = \mathcal{L}^{-1}\left\{\frac{\frac{2}{5}}{s} - \frac{\frac{1}{8}}{s+2} - \frac{\frac{11}{40}}{s+10}\right\}$$

From rules L3(b) and L4(b),

$$y = \left(\frac{2}{5}\right)\mathcal{L}^{-1}\left\{\frac{1}{s}\right\} - \left(\frac{1}{8}\right)\mathcal{L}^{-1}\left\{\frac{1}{s+2}\right\} - \left(\frac{11}{40}\right)\mathcal{L}^{-1}\left\{\frac{1}{s+10}\right\}$$

From entries 1 and 2 of Table 12.2

$$y = \frac{2}{5} - \left(\frac{1}{8}\right)e^{-2t} - \left(\frac{11}{40}\right)e^{-10t}$$

12.18 COMPARISON OF METHODS

The classical method for solving differential equations presented previously is usually comparable in degree of difficulty to the Laplace transform method.

Both methods are easy to apply to the linear constant coefficient differential equations arising in basic electric circuit analysis. Although the student will enjoy the systematic aspects of the Laplace transform method, the classical method should also be mastered, because it gives the beginner a "feel" and understanding for differential equations.

12.19 EXERCISES

1. Define each of the following terms as they apply to differential equations.

(a) Dependent variable
(b) Independent variable
(c) Order
(d) Homogeneous
(e) Nonhomogeneous
(f) Forcing function
(g) Characteristic equation
(h) General transient solution
(i) Steady-state solution
(j) General complete solution
(k) Initial condition
(l) Particular complete solution

2. Determine a particular solution for each of the following homogeneous differential equations using the classical method.

(a) $\frac{dy}{dt} + 6y = 0, \quad y(0) = -2$ (b) $\frac{dy}{dt} - 2y = 0, \quad y(0) = 6$

3. Determine a particular solution for each of the following homogeneous differential equations using the classical method.

(a) $-\frac{dy}{dt} + 6y = 0, \quad y(0) = -3$

(b) $\frac{d^2y}{dt^2} + 4\frac{dy}{dt} + 3y = 0, y(0) = -4 \quad \frac{dy(0)}{dt} = 0$

(c) $\frac{d^2y}{dt^2} + 6\frac{dy}{dt} + 8y = 0, y(0) = 1 \quad \frac{dy(0)}{dt} = -2$

4. Determine a particular solution for each of the following homogeneous differential equations using the classical method.

(a) $\frac{d^2y}{dt^2} + 4\frac{dy}{dt} + 4y = 0, y(0) = -4 \quad \frac{dy(0)}{dt} = 1$

(b) $\frac{d^2y}{dt^2} + 8\frac{dy}{dt} + 16y = 0, y(0) = 0 \quad \frac{dy(0)}{dt} = -1$

(c) $\frac{d^2y}{dt^2} + 16y = 0, y(0) = 4 \quad \frac{dy(0)}{dt} = 0$

5. Determine a particular complete solution for each of the following nonhomogeneous differential equations using the classical method.

(a) $\frac{dy}{dt} + 4y = 10, \quad y(0) = 4$ (b) $\frac{dy}{dt} + 2y = 6e^{-4t}, \quad y(0) = -2$

6. Determine a particular complete solution for each of the following nonhomogeneous differential equations using the classical method.

(a) $\frac{dy}{dt} + 8y = -e^{-2t} + 12, \quad y(0) = 0$

(b) $\frac{d^2y}{dt^2} + 5\frac{dy}{dt} + 4y = -8, \quad y(0) = -1 \quad \frac{dy(0)}{dt} = 0$

7. Determine a particular complete solution for each of the following nonhomogeneous differential equations using the classical method.

(a) $\frac{d^2y}{dt^2} + 6\frac{dy}{dt} + 9y = 4, \quad y(0) = -2 \quad \frac{dy(0)}{dt} = 1$

(b) $\frac{d^2y}{dt^2} + 10\frac{dy}{dt} + 16y = 10e^{-4t}, \quad y(0) = 0 \quad \frac{dy(0)}{dt} = 0$

8. Expand each of the Laplace transform expressions into partial fractions.

(a) $\dfrac{s + 1}{s^2 + 10s + 16}$ (b) $\dfrac{-1}{(s + 4)^2(s + 1)}$

9. Expand each of the Laplace transform expressions into partial fractions.

(a) $\dfrac{2s + 6}{(s^2 + 6s + 8)(s + 8)}$ (b) $\dfrac{s^2}{s(s + 2)(s + 5)}$

10. Expand each of the Laplace transform expressions into partial fractions.

(a) $\dfrac{s^2 + 6s}{(s + 1)^2(s + 4)}$ (b) $\dfrac{10}{(s + 1)^2 (s + 4)^2(s + 6)}$

11. Expand each of the Laplace transform expressions into partial fractions.

(a) $\dfrac{-4s}{(s - 1)^2 (s + 5)}$ (b) $\dfrac{2}{(s^2 + 4) (s + 1)}$

12. Find the inverse Laplace transform of each Laplace transform expression in Problem 8.

13. Find the inverse Laplace transform of each Laplace transform expression in Problem 9.

14. Find the inverse Laplace transform of each Laplace transform expression in Problem 10.

15. Find the inverse Laplace transform of each Laplace transform expression in Problem 11.

16. Solve each differential equation in Problem 2 using the Laplace transform method.

17. Solve each differential equation in Problem 3 using the Laplace transform method.

18. Solve each differential equation in Problem 4 using the Laplace transform method.

19. Solve each differential equation in Problem 5 using the Laplace transform method.

20. Solve each differential equation in Problem 6 using the Laplace transform method.

21. Solve each differential equation in Problem 7 using the Laplace transform method.

CHAPTER

13

Source-Free Circuits

After completing this chapter the student should be able to

* determine currents and voltages in source-free *RC* circuits
* determine currents and voltages in source-free *RL* circuits
* determine currents and voltages in source-free *RLC* circuits

Circuits containing only passive electrical elements (resistors, inductors, and capacitors) are called source-free circuits. Source-free circuits are energized by the initial voltages or currents associated with capacitors and/or inductors prior to their connection in the circuit. The voltages and currents in source-free circuits containing inductors and/or capacitors are usually described by homogeneous differential equations.

13.1 *RC* CIRCUIT

A circuit containing only resistors and capacitors is called an *RC* circuit. Consider the simple *RC* circuit in Figure 13.1. The circuit in Figure 13.1(a) is energized by the initial voltage across the capacitor when the switch is closed at

Figure 13.1 *RC* circuit.

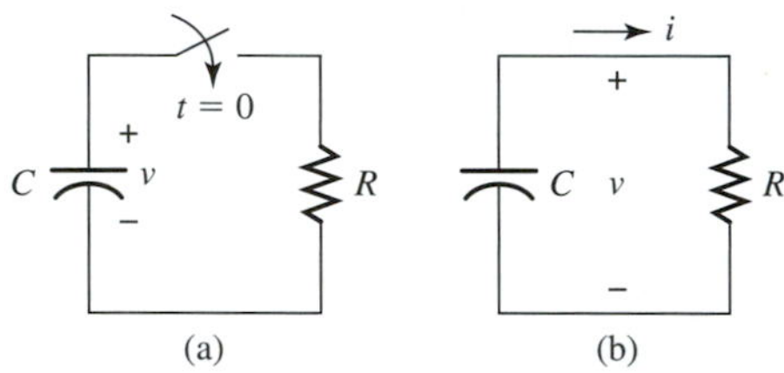

time $t = 0$, and from that time forward the circuit appears as shown in Figure 13.1(b). The measuring of voltages and currents for the circuit in Figure 13.1(b) begins at time $t = 0$, when the switch is closed. Time t is the independent variable and v and i are the dependent variables for the expressions that we shall develop for voltage and current.

It is useful at this point to introduce some concise notation regarding the switching of electrical elements in and out of a circuit.

> $v(0^-)$ and $i(0^-)$ represent the values of v and i an instant before a switch is closed or opened.
>
> $v(0)$ and $i(0)$ represent the values of v and i at the instant a switch is closed or opened.
>
> $v(0^+)$ and $i(0^+)$ represent the values of v and i an instant after a switch is closed or opened.

It is important to recognize that the voltage across a capacitor cannot change instantaneously. That is, when a capacitor is switched into or out of a circuit,

$$v(0^-) = v(0) = v(0^+)$$

where v is the voltage across the capacitor. This follows from the voltage-current relationship for a capacitor, which is

$$i = C\frac{dv}{dt}$$

If v could change instantaneously (instantaneously means without using any time), the quantity dv/dt would approach infinity, and from the preceding equation the current i would also approach infinity. Because $i = dq/dt$, an infinitely large current would furnish charge at an infinitely large rate, which would result in an infinite amount of charge. Because it is physically impossible to accumulate an infinite amount of charge, it follows that the voltage across a capacitor cannot change instantaneously.

From Equation (10.2) the current i flowing through the capacitor in Figure 13.1(b) is

$$i = C\frac{d(-v)}{dt}$$

$$= -C\frac{dv}{dt}$$

Notice that the assigned direction for i and the polarity of v necessitates introducing a minus sign in this equation.

From Ohm's law the current flowing through the resistor is

$$i = \frac{v}{R}$$

Because the same current flows through the resistor and capacitor,

$$-C\frac{dv}{dt} = \frac{v}{R}$$

and

$$C\frac{dv}{dt} + \frac{v}{R} = 0$$

Note that the mathematical model describing v is a first-order linear homogeneous differential equation with dependent variable v and independent variable t.

Dividing both sides of this equation by C results in

$$\frac{dv}{dt} + \frac{v}{RC} = 0$$

Using the classical method for solving first-order homogeneous differential equations presented in Chapter 12, the characteristic equation is

$$p + \frac{1}{RC} = 0$$

with root

$$p = -\frac{1}{RC}$$

and the general complete solution is then

$$v = Ke^{-(1/RC)t}$$

Recall that the steady-state solution is zero and the complete solution is equal to the transient solution for a homogeneous differential equation. Substituting the initial condition for v into the preceding equation results in

$$v(0) = Ke^{-(1/RC)(0)}$$

and

$$K = v(0)$$

The expression for the voltage v in Figure 13.1(b) is then

$$v = v(0)e^{-(1/RC)t} \tag{13.1}$$

where $v(0) = v(0^-)$, because the voltage across a capacitor cannot change instantaneously. $v(0^-)$ is the voltage across the capacitor prior to closing the switch.

From Ohm's law

$$\begin{aligned} i &= \frac{v}{R} \\ &= \left\{\frac{v(0)}{R}\right\}e^{-(1/RC)t} \end{aligned} \tag{13.2}$$

or alternatively from Equation (10.2),

$$\begin{aligned} i &= C\frac{d(-v)}{dt} \\ &= -C\frac{dv}{dt} \\ &= -C\frac{d\{v(0)e^{-(1/RC)t}\}}{dt} \\ &= -C\{v(0)\}\left(-\frac{1}{RC}\right)e^{-(1/RC)t} \\ &= \left\{\frac{v(0)}{R}\right\}e^{-(1/RC)t} \end{aligned}$$

Consider the following example.

EXAMPLE 13.1 Determine v and i after the switch is closed for the circuit in Figure 13.2. ■

Figure 13.2

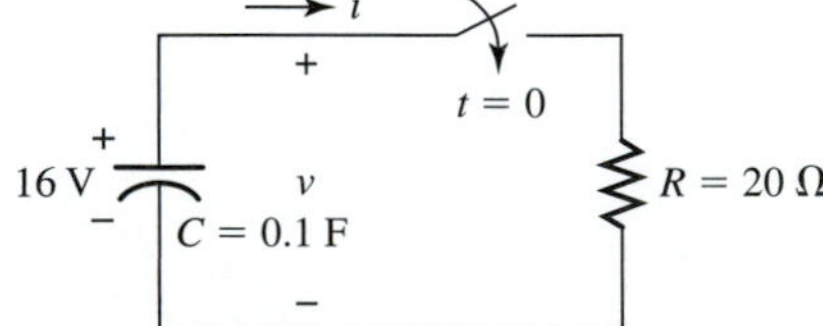

SOLUTION After the switch is closed, the circuit appears as shown in Figure 13.3.

Figure 13.3

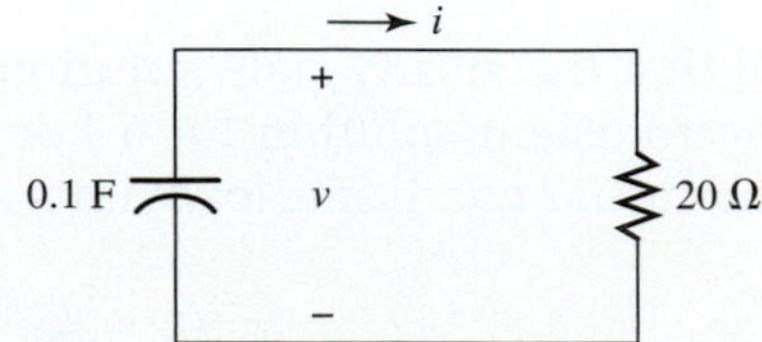

Because the voltage across a capacitor cannot change instantaneously and $v(0^-) = 16$,

$$\begin{aligned} v(0) &= v(0^-) \\ &= 16 \text{ V} \end{aligned}$$

From Equation (13.1)

$$\begin{aligned} v &= v(0)e^{-(1/RC)t} \\ &= 16e^{-\{1/(20)(0.1)\}(t)} \\ &= 16e^{-(1/2)t} \text{ V} \end{aligned}$$

From Equation (13.2)

$$\begin{aligned} i &= \left(\frac{v(0)}{R}\right)e^{-(1/RC)t} \\ &= \left(\frac{16}{20}\right)e^{-\{1/(20)(0.1)\}t} \\ &= \left(\frac{4}{5}\right)e^{-(1/2)t} \end{aligned}$$

Alternatively, from Ohm's law,

$$\begin{aligned} i &= \frac{v}{R} \\ &= \frac{\{16e^{-(1/2)t}\}}{20} \\ &= \left(\frac{4}{5}\right)e^{-(1/2)t} \end{aligned}$$

13.2 *RC* TIME CONSTANT

Graphs of v versus t and i versus t for the *RC* circuit in Figure 13.1 are shown in Figure 13.4. The graphs are described by Equations (13.1) and (13.2), respectively.

Figure 13.4 v, i versus t.

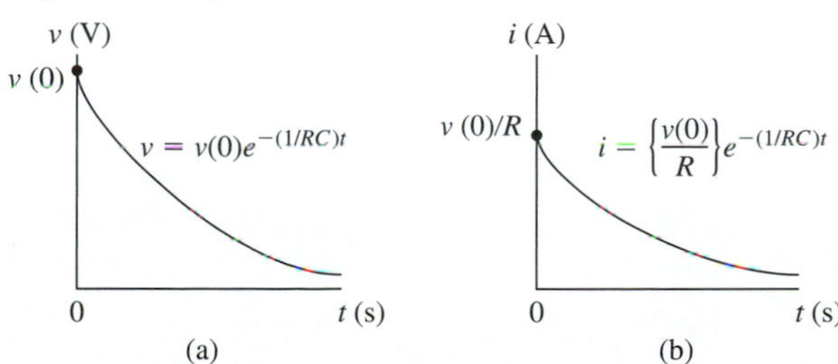

Notice that v begins at $v(0)$ (initial voltage across the capacitor at $t = 0$) and continues to decrease exponentially as time increases. Notice also that i begins at $v(0)/R$ amperes and continues to decrease exponentially as time increases. Theoretically, v and i never reach zero because they are asymtotic to the t axis. The time required for v and i to decrease to a value close enough to zero to approximate them as equal to zero depends on the product RC. The graphs of v for different values of RC are shown in Figure 13.5, where

$$R_3C_3 > R_2C_2 > R_1C_1$$

The higher the value of RC, the slower the voltage decreases with time. The current i begins at $v(0)/R$ and decreases in a manner similar to that shown in Figure 13.5 with regard to the different values of RC.

Figure 13.5 v versus t.

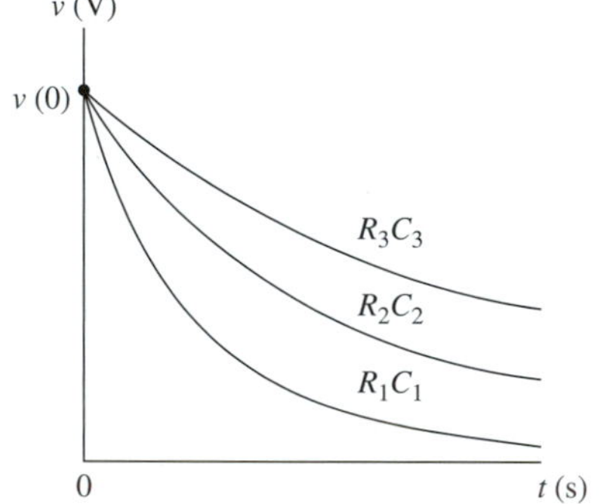

The time required for the voltage or current in Figure 13.4 to decrease from its initial value at $t = 0$ to a value of $1/e$ (e is the base of the natural logarithms and is approximately equal to 2.72) times the initial value of the voltage or current is called the *time constant* for the circuit. The time constant for an RC circuit is designated by the Greek letter tau (τ) and is equal to RC:

$$\tau = RC \tag{13.3}$$

where τ is in seconds, R is in ohms, and C is in farads. Equation (13.1) easily establishes that at time $t = \tau$, the voltage v is equal to $(1/e)$ times its initial value. From Equation (13.1)

$$v = v(0)e^{-(1/RC)t}$$
$$v(\tau) = v(0)e^{-(1/RC)\tau}$$

and if $\tau = RC$,

$$v(\tau) = v(0)e^{-(1/RC)(RC)}$$
$$= \frac{v(0)}{e}$$

The time constant is illustrated in Figure 13.6 and applies to current as well as voltage. In most instances, current and voltage are assumed to equal zero

Figure 13.6 Time constant.

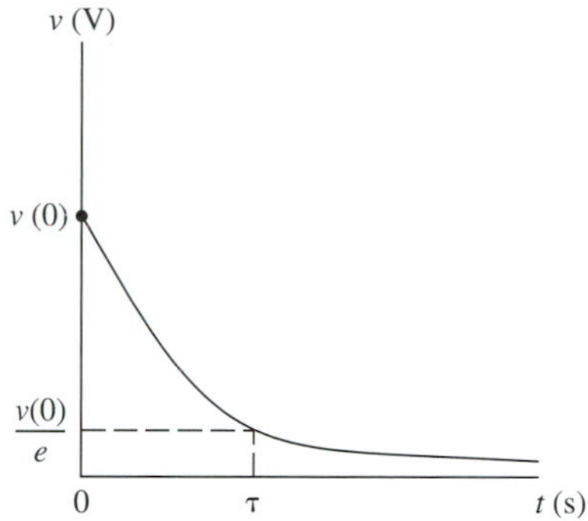

after six or more time constants, but this rule varies with the precision required for the circuit under consideration.

EXAMPLE 13.2 Determine τ and v after the switch is closed for the circuit in Figure 13.7. ■

Figure 13.7

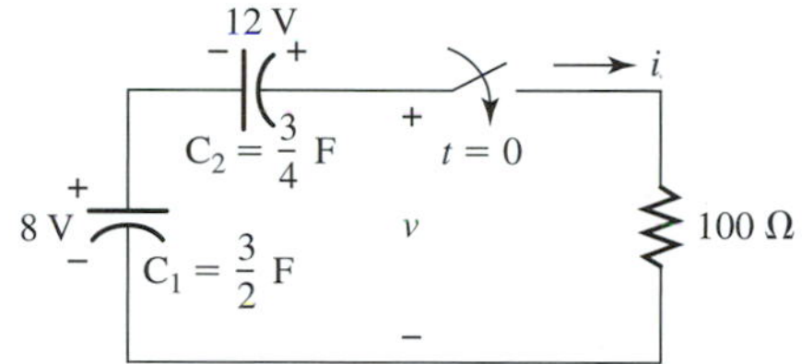

SOLUTION Prior to applying Equation (13.2), the circuit must be put in the same form as the circuit in Figure 13.1.

From Equation (10.5) the equivalent capacitor for C_1 and C_2 in series is

$$C_{EQ} = \frac{C_1 C_2}{C_1 + C_2}$$

$$= \frac{\left(\frac{3}{2}\right)\left(\frac{3}{4}\right)}{\left(\frac{3}{2}\right) + \left(\frac{3}{4}\right)}$$

$$= \frac{1}{2} \text{ F}$$

Prior to closing the switch, the voltage across the series combination of C_1 and C_2 is

$$v(0^-) = 12 + 8$$

$$= 20 \text{ V}$$

Because the voltage across a capacitor cannot change instantaneously,

$$v(0) = v(0^-)$$

$$= 20 \text{ V}$$

The resulting circuit after closing the switch is shown in Figure 13.8.

Figure 13.8

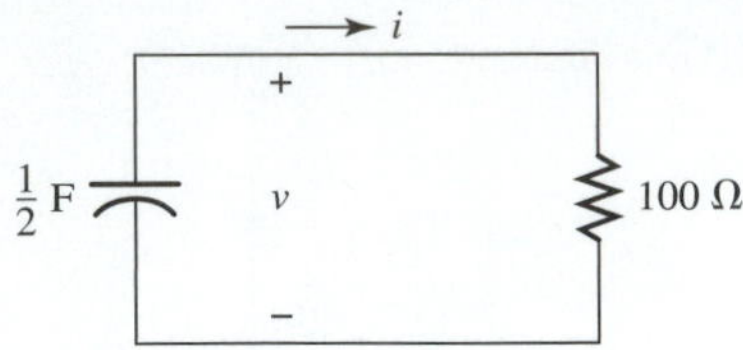

From Equation (13.1)

$$v = v(0)e^{-(1/RC)t}$$
$$= 20e^{-\{1/(100)(1/2)\}(t)}$$
$$= 20e^{-(1/50)t}\ \mathrm{V}$$

From Equation (13.3) the time constant is

$$\tau = RC$$
$$= (100)\left(\frac{1}{2}\right)$$
$$= 50\ \mathrm{s}$$

13.3 CAPACITOR BEHAVIOR AS AN OPEN

If a capacitor is connected in a circuit with a constant voltage source, as shown in Figure 13.9, current will reduce to zero after enough time elapses.

Figure 13.9 Constant voltage source.

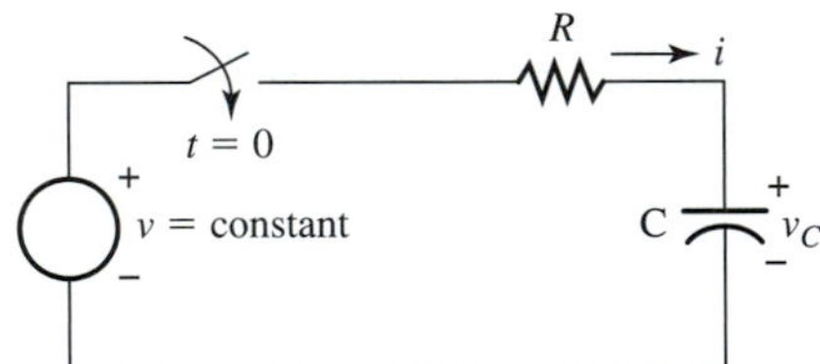

Consider the circuit in Figure 13.9. After the switch is closed at $t = 0$, the current will continue to charge the capacitor until $v_C = v$. After the time required to charge the capacitor to v volts, the voltage across the resistor will equal zero and the current i will cease. The capacitor will then appear as an open circuit, as shown in Figure 13.10. Recall that an open circuit is a break or gap in a circuit path through which no current flows.

Figure 13.10 Constant voltage source.

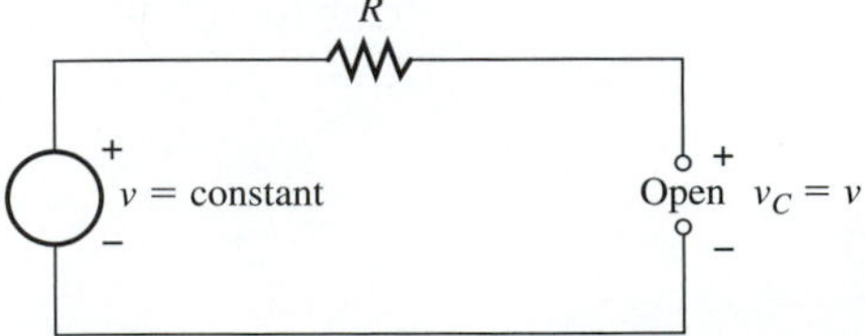

13.4 MISCELLANEOUS *RC* CIRCUITS

The following examples contain constant voltage sources, but they are used only to establish initial voltages across capacitors and are then switched out of the circuit. The resulting circuits are source-free circuits.

EXAMPLE 13.3 For the circuit in Figure 13.11, the switch remains closed for a long period of time and is opened at $t = 0$. Determine τ, v, i, i_2, and i_3 for $t \geq 0$. ■

Figure 13.11

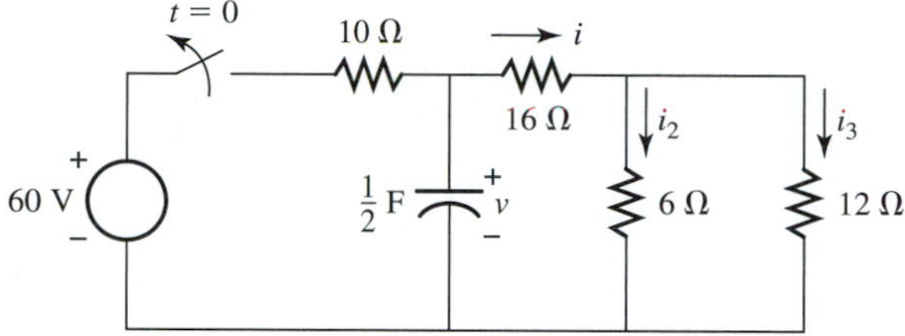

SOLUTION Because the switch is in the closed position for a long period of time and the source is a constant voltage source, the capacitor appears as an open circuit at the instant prior to the opening of the switch, as shown in Figure 13.12. $v(0^-)$ is the voltage across the capacitor an instant before the switch is opened.

Figure 13.12

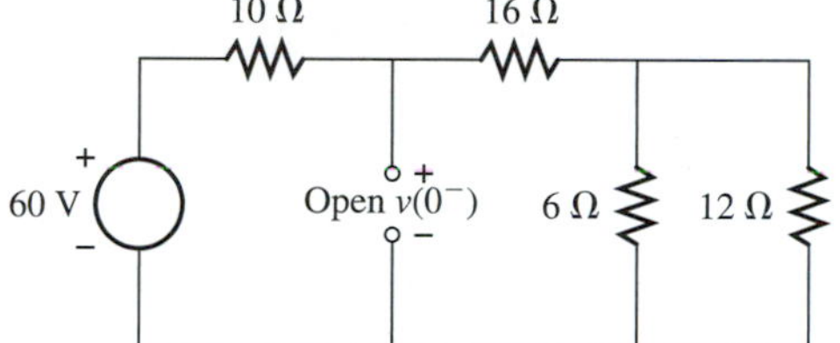

Using Equations (3.1) and (3.5) to combine the 16-Ω, 6-Ω and 12-Ω resistors results in

$$R = 16 + (6 \| 12)$$
$$= 16 + \frac{(6)(12)}{6 + 12}$$
$$= 20\ \Omega$$

and the circuit can be redrawn as shown in Figure 13.13.

Figure 13.13

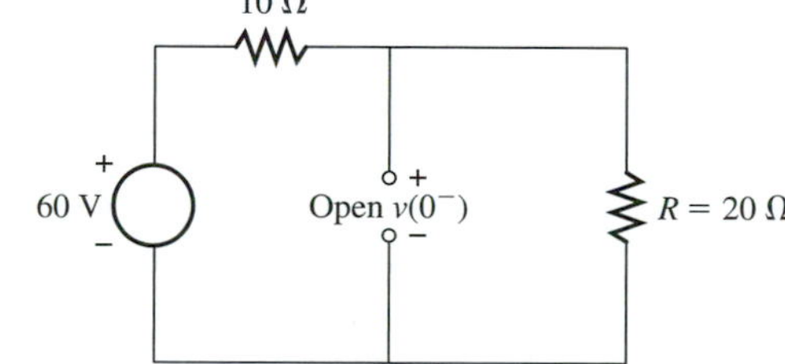

Recognizing that $v(0^-)$ is the voltage across R and using Equation (4.3) for voltage division to determine $v(0^-)$ results in

$$v(0^-) = \left\{\frac{20}{20 + 10}\right\}(60)$$
$$= 40\ \text{V}$$

Because the 60-V voltage source and the 10-Ω resistor are eliminated from the original circuit when the switch is opened, the circuit appears as shown in Figure 13.14 after the switch is opened.

Figure 13.14

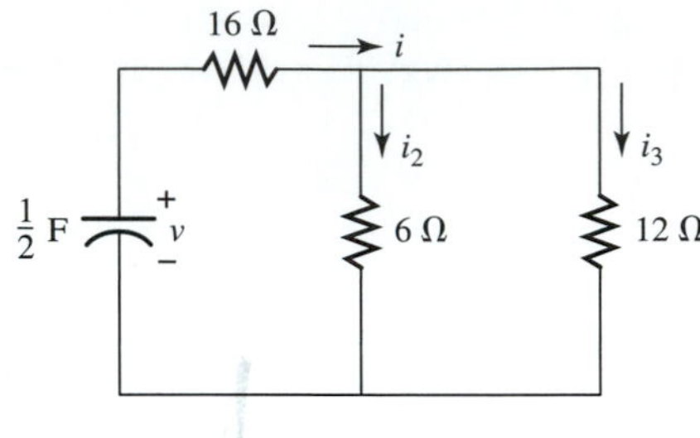

Putting the circuit in the same form as the circuit in Figure 13.1 results in the circuit in Figure 13.15.

Figure 13.15

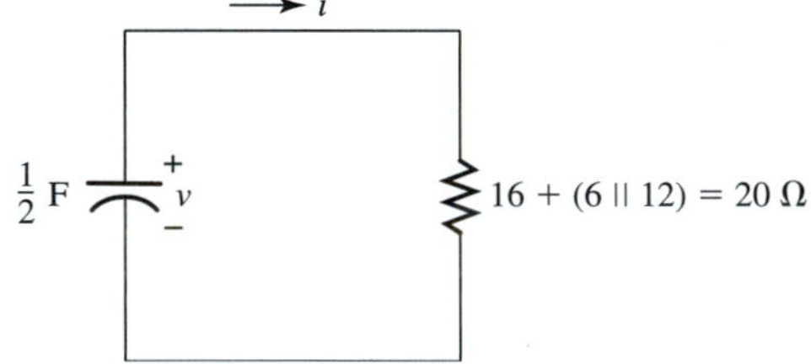

Applying Equation (13.1) and recalling that

$$
\begin{aligned}
v(0) &= v(0^-) \\
&= 40 \text{ V}
\end{aligned}
$$

results in

$$
\begin{aligned}
v &= v(0)e^{-(1/RC)t} \\
&= 40e^{-\{1/(20)(1/2)\}(t)} \\
&= 40e^{-(1/10)t} \text{ V}
\end{aligned}
$$

Because v appears across the 20-Ω resistor, Ohm's law results in

$$
\begin{aligned}
i &= \frac{v}{R_{EQ}} \\
&= \frac{\{40e^{-(1/10)t}\}}{20} \\
&= 2e^{-(1/10)t} \text{ A}
\end{aligned}
$$

Using Equation (4.1) for the division of the current i in Figure 13.14 results in

$$i_2 = \left(\frac{12}{6 + 12}\right)(i)$$
$$= \left(\frac{2}{3}\right)(2e^{-(1/10)t})$$
$$= \left(\frac{4}{3}\right)e^{-(1/10)t}\text{ A}$$

and

$$i_3 = \left(\frac{6}{6 + 12}\right)(i)$$
$$= \left(\frac{1}{3}\right)(2e^{-(1/10)t})$$
$$= \left(\frac{2}{3}\right)e^{-(1/10)t}\text{ A}$$

From Equation (13.3)

$$\tau = RC$$
$$= 20\left(\frac{1}{2}\right)$$
$$= 10\text{ s}$$

EXAMPLE 13.4 For the circuit shown in Figure 13.16, the switch remains closed for a long period of time and is opened at $t = 0$. Determine v and i for $t \geq 0$. ■

Figure 13.16

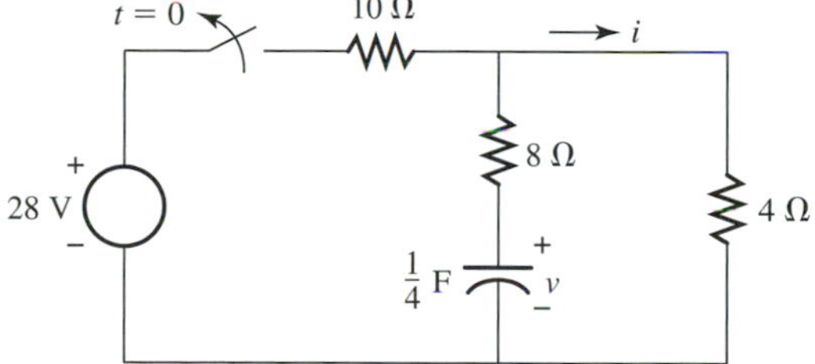

SOLUTION Because the switch is in the closed position for a long period of time and the source is a constant voltage source, the capacitor appears as an open circuit at the instant prior to the opening of the switch, as shown in Figure 13.17.

Figure 13.17

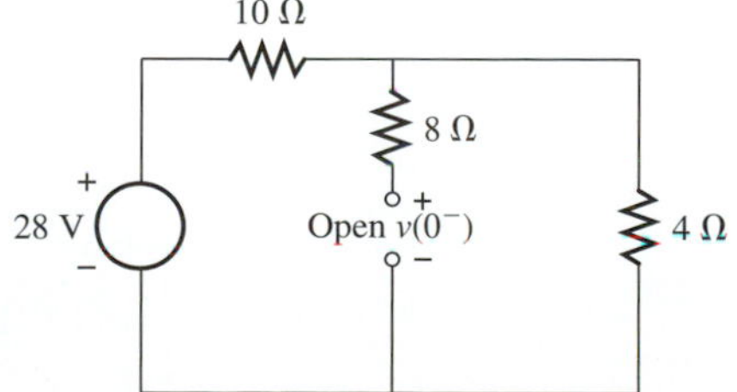

$v(0^-)$ is the voltage across the capacitor an instant before the switch is opened. Because no current flows through the 8-Ω resistor, $v(0^-)$ is also the voltage across the 4-Ω resistor. Using Equation (4.3) for voltage division to determine $v(0^-)$ results in

$$v(0^-) = \left\{\frac{4}{4 + 10}\right\}(28)$$
$$= 8 \text{ V}$$

Because the 28-V voltage source and the 10-Ω resistor are eliminated from the original circuit when the switch is opened, the circuit appears as shown in Figure 13.18 after the switch is opened.

Figure 13.18

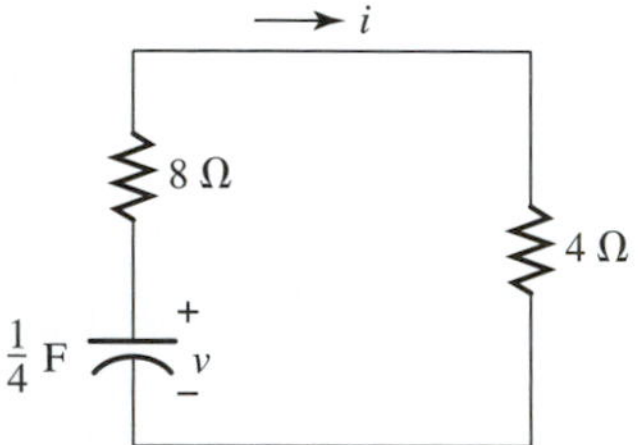

Putting the circuit in the same form as the circuit in Figure 13.1 results in the circuit in Figure 13.19. Applying Equation (13.1) and recalling that

$$v(0) = v(0^-)$$
$$= 8 \text{ V}$$

Figure 13.19

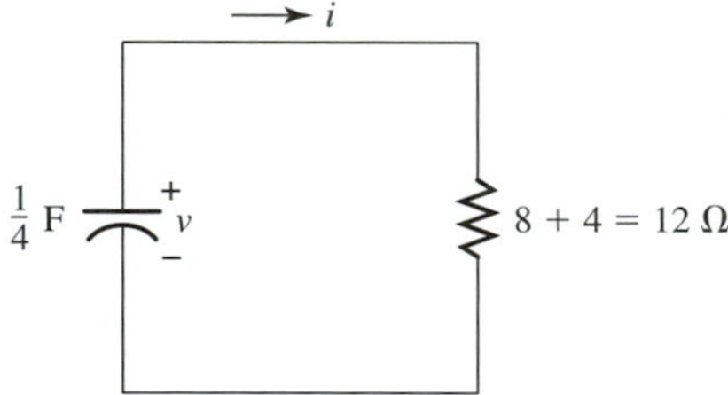

results in

$$v = v(0)e^{-(1/RC)t}$$
$$= 8e^{-\{1/(12)(1/4)\}(t)}$$
$$= 8e^{-(1/3)t} \text{ V}$$

Because v appears across the 12-Ω resistor, Ohm's law results in

$$i = \frac{v}{12}$$
$$= \left(\frac{8}{12}\right)e^{-(1/3)t}$$
$$= \left(\frac{2}{3}\right)e^{-(1/3)t} \text{ V}$$

EXAMPLE 13.5 For the circuit in Figure 13.20, the switch remains closed for a long period of time and is opened at $t = 0$. Determine v_1, v_2, and i for $t \geq 0$. ■

Figure 13.20

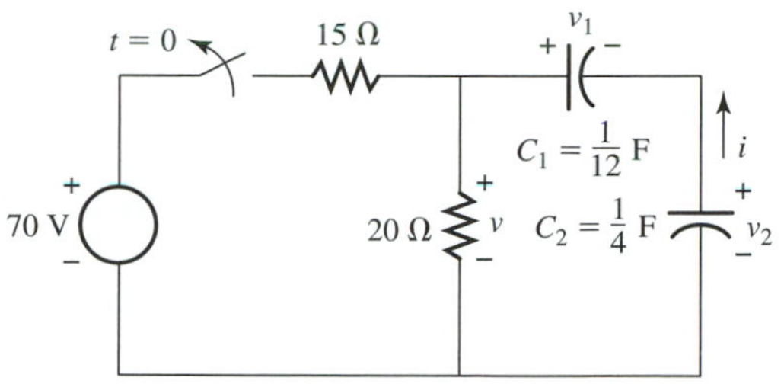

SOLUTION Because the switch is in the closed position for a long period of time and the source is a constant voltage source, the capacitors appear as open circuits at the instant prior to the opening of the switch, as shown in Figure 13.21.

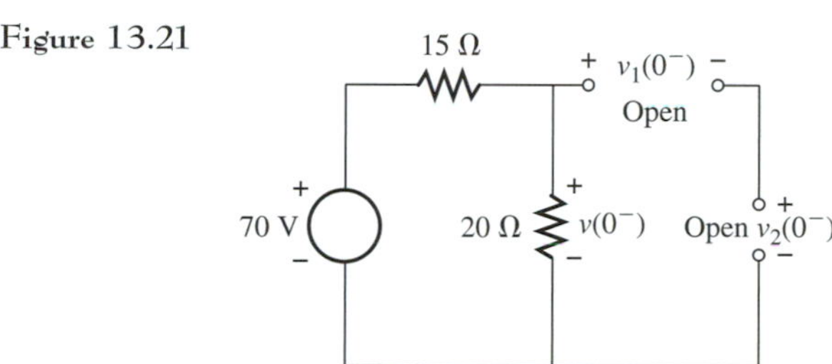

Figure 13.21

$v_1(0^-)$ is the voltage across the $\frac{1}{12}$-F capacitor and $v_2(0^-)$ is the voltage across the $\frac{1}{4}$-F capacitor an instant before the switch is opened. $v(0^-)$ is the voltage across the series combination of the capacitors and is equal to the voltage across the 20-Ω resistor in Figure 13.21. Using Equation (4.3) for voltage division to determine $v(0^-)$ results in

$$v(0^-) = \left\{\frac{20}{20 + 15}\right\}(70)$$
$$= 40 \text{ V}$$

Using Equation (10.5) to combine capacitors C_1 and C_2 in series to form C_{EQ}, replacing C_1 and C_2 in Figure 13.20 by C_{EQ}, and recognizing that the 70-V voltage source and the 15-Ω resistor are eliminated from the circuit when the switch is opened results in the circuit in Figure 13.22.

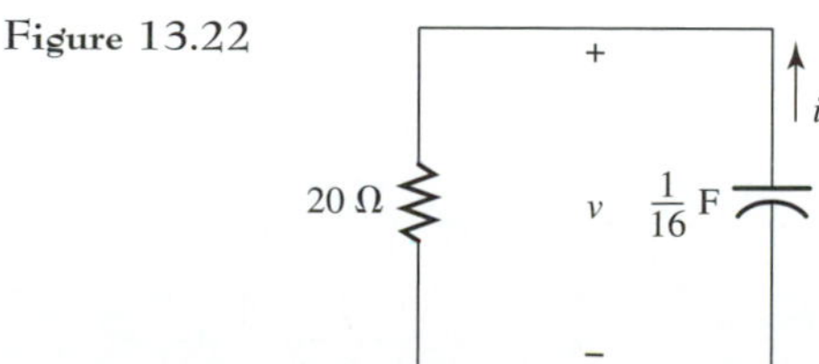

Figure 13.22

$$C_{EQ} = \frac{C_1 C_2}{C_1 + C_2}$$
$$= \frac{\left(\frac{1}{12}\right)\left(\frac{1}{4}\right)}{\frac{1}{12} + \frac{1}{4}}$$
$$= \frac{1}{16}\ \text{F}$$

Applying Equation (13.1) and recalling that

$$v(0) = v(0^-)$$
$$= 40\ \text{V}$$

results in

$$v = v(0)e^{-(1/RC)t}$$
$$= 40e^{-\{1/(20)(1/16)\}(t)}$$
$$= 40e^{-(4/5)t}\ \text{V}$$

From Ohm's law

$$i = \frac{v}{20}$$
$$= \frac{40e^{-(4/5)t}}{20}$$
$$= 2e^{-(4/5)t}\ \text{A}$$

Using Equation (10.8) for voltage division in Figure 13.23 results in

Figure 13.23

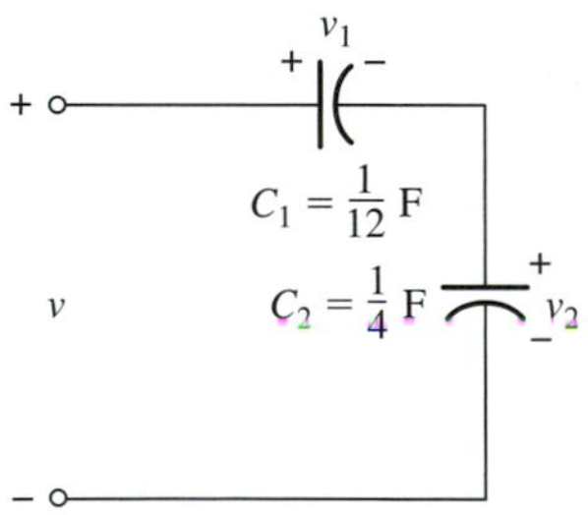

$$v_1 = \left\{\frac{C_2}{C_1 + C_2}\right\}(v)$$
$$= \left\{\frac{\frac{1}{4}}{\frac{1}{12} + \frac{1}{4}}\right\}(40e^{-(4/5)t})$$
$$= 30e^{-(4/5)t}\ \text{V}$$
$$v_2 = \left\{\frac{C_1}{C_1 + C_2}\right\}v$$

$$= \left\{ \frac{\frac{1}{12}}{\frac{1}{12} + \frac{1}{4}} \right\} (40e^{-(4/5)t})$$

$$= 10e^{-(4/5)t} \text{ V}$$

EXAMPLE 13.6 For the circuit in Figure 13.1, establish Equation (13.1) using the Laplace transform method for solving the differential equation. ■

SOLUTION The circuit in Figure 13.1(b) for $t \geq 0$ is shown in Figure 13.24.

Figure 13.24

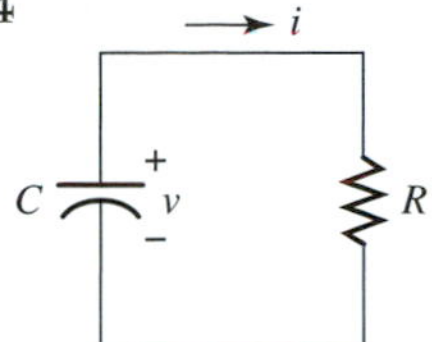

From Equation (10.2) the current i flowing through the capacitor is

$$i = C\frac{d(-v)}{dt}$$

$$= -C\frac{dv}{dt}$$

Notice that the assigned direction for i and the assumed polarity of v necessitates introducing the minus sign in the preceding equation.

From Ohm's law the current flowing through the resistor is

$$i = \frac{v}{R}$$

Because the same current flows through the resistor and capacitor,

$$-C\frac{dv}{dt} = \frac{v}{R}$$

$$C\frac{dv}{dt} + \frac{v}{R} = 0$$

and

$$\frac{dv}{dt} + \frac{v}{RC} = 0$$

Using rules L1 and L3 in Chapter 12 for Laplace transforms results in

$$\mathscr{L}\left\{\frac{dv}{dt} + \frac{v}{RC}\right\} = \mathscr{L}\{0\}$$

$$\mathscr{L}\left\{\frac{dv}{dt}\right\} + \mathscr{L}\left\{\frac{v}{RC}\right\} = 0$$

$$\{s\mathbf{V} - v(0)\} + \left(\frac{1}{RC}\right)\mathbf{V} = 0$$

$$\mathbf{V} = \frac{v(0)}{s + (1/RC)}$$

and from rule L4 in Chapter 12,

$$v = \mathcal{L}^{-1}\left\{\frac{v(0)}{s + (1/RC)}\right\}$$

$$= \{v(0)\}\mathcal{L}^{-1}\left\{\frac{1}{s + (1/RC)}\right\}$$

From entry 2 of Table 12.2,

$$v = v(0)e^{-(1/RC)t}$$

where $v(0) = v(0^-)$, because the voltage across a capacitor cannot change instantaneously.

13.5 *RL* CIRCUIT

A circuit containing only resistors and inductors is called an *RL* circuit. Consider the simple *RL* circuit in Figure 13.25. The circuit in Figure 13.25(a) is energized by an initial current existing in the inductor when the switch is closed at $t = 0$, and from that time forward the circuit appears as shown in

Figure 13.25 *RL* circuit.

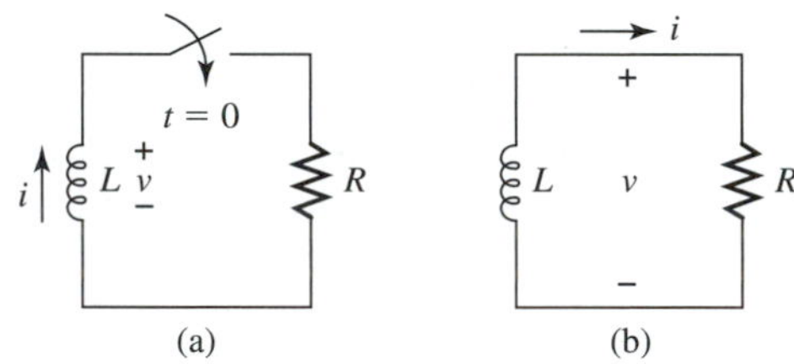

Figure 13.25(b). The measuring of voltages and currents for the circuit in Figure 13.25(b) begins at time $t = 0$ when the switch is closed. Time t is the independent variable, and v and i are the dependent variables in the expressions that we shall develop for voltage and current.

Because current cannot flow unless there is a closed path, the inductor in Figure 13.25(a) must be connected to an energizing circuit prior to closing the switch in order to provide an initial current for the inductor at the instant the switch is closed. The energizing circuit is not shown in Figure 13.25(a) and is switched out of the circuit at the instant the switch is closed. The resulting circuit is as shown in Figure 13.25(b). Switching arrangements to accomplish these actions are considered later in this chapter.

It is important to recognize that the current flowing through an inductor cannot change instantaneously. That is, when an inductor is switched into or out of a circuit,

$$i(0^-) = i(0) = i(0^+)$$

where $i(0^-)$, $i(0)$, and $i(0^+)$ are, respectively, the currents flowing through the inductor an instant before switching, at the instant of switching, and an instant after switching. This follows from the voltage-current relationship for an inductor, which is

$$v = L\frac{di}{dt}$$

If i could change instantaneously (instantaneously means without using time), the quantity di/dt would approach infinity, and from the preceding equation the voltage v would also approach infinity. Because $v = d\phi_T/dt$, an infinitely large voltage would produce total magnetic flux ϕ_T at an infinitely large rate, which would result in an infinite amount of magnetic flux. Because it is physically impossible to accumulate an infinite amount of magnetic flux, it follows that the current flowing through an inductor cannot change instantaneously.

From Equation (11.3) the voltage v across the inductor in Figure 13.25(b) is

$$v = L\frac{d(-i)}{dt}$$

$$= -L\frac{di}{dt}$$

Notice that the assigned direction for i and the assumed polarity of v necessitates the introduction of the minus sign in this equation.

From Ohm's law the voltage v across the resistor is

$$v = Ri$$

Because the same voltage appears across the resistor and inductor

$$-L\frac{di}{dt} = Ri$$

and

$$L\frac{di}{dt} + Ri = 0$$

It is noted that the mathematical model describing v is a first-order linear homogeneous differential equation with dependent variable i and independent variable t.

Dividing both sides of the preceding equation by L results in

$$\frac{di}{dt} + \frac{R}{L}i = 0$$

Using the classical method for solving first-order homogeneous differential equations presented in Chapter 12, the characteristic equation is

$$p + \frac{R}{L} = 0$$

with root

$$p = -\frac{R}{L}$$

The general, complete solution is then

$$i = Ke^{-(R/L)t}$$

Recall that the steady-state solution is zero and the complete solution is equal to the transient solution for a homogeneous differential equation. Substituting the initial conditions for i into the preceding equation results in

$$i(0) = Ke^{-(R/L)(0)}$$

and

$$K = i(0)$$

The expression for i in Figure 13.25(b) is then

$$i = i(0)e^{-(R/L)t} \tag{13.4}$$

where $i(0) = i(0^-)$, because the current through an inductor cannot change instantaneously. $i(0^-)$ is the current through the inductor prior to closing the switch.

From Ohm's law

$$\begin{aligned} v &= Ri \\ &= Ri(0)e^{-(R/L)t} \end{aligned} \tag{13.5}$$

or, alternatively,

$$\begin{aligned} v &= L\frac{d(-i)}{dt} \\ &= -L\frac{di}{dt} \\ &= -L\frac{d\{i(0)e^{-(R/L)t}\}}{dt} \\ &= -L\{i(0)\}(-R/L)e^{-(R/L)t} \\ &= Ri(0)e^{-(R/L)t} \end{aligned}$$

Consider the following example.

EXAMPLE 13.7 Determine v and i after the switch is closed for the circuit in Figure 13.26. The circuit providing the current $i(0^-)$ flowing through the inductor prior to closing the switch is not shown and is eliminated from the circuit at the instant the switch is closed (at $t = 0$). ■

Figure 13.26

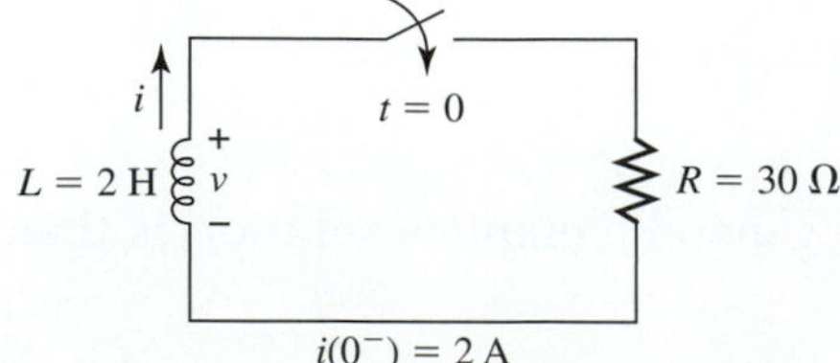

SOLUTION After the switch is closed, the circuit appears as shown in Figure 13.27. Because the current flowing through an inductor cannot change instantaneously and $i(0^-) = 2$ A,

$$\begin{aligned} i(0) &= i(0^-) \\ &= 2 \text{ A} \end{aligned}$$

Figure 13.27

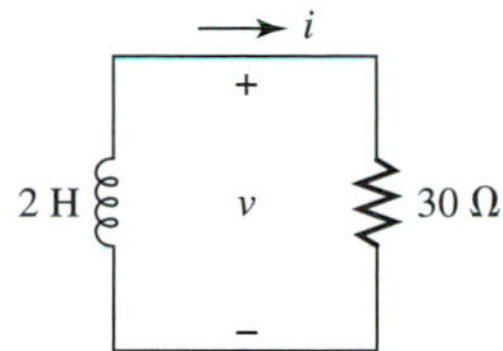

From Equation (13.4)

$$\begin{aligned} i &= i(0)e^{-(R/L)t} \\ &= 2e^{-(30/2)t} \\ &= 2e^{-15t} \text{ A} \end{aligned}$$

From Equation (13.5)

$$\begin{aligned} v &= Ri(0)e^{-(R/L)t} \\ &= (30)(2)e^{-(30/2)t} \\ &= 60e^{-15t} \text{ V} \end{aligned}$$

13.6 *RL* TIME CONSTANT

Graphs of v versus t and i versus t for the RL circuit in Figure 13.25 are shown in Figure 13.28. The graphs are described by Equations (13.4) and (13.5). Notice that i begins at $i(0)$ (initial current through the inductor at $t = 0$) and continues to decrease exponentially as time increases. Notice also that v begins at a value of $Ri(0)$ volts and continues to decrease exponentially as time increases. Theoretically, v and i never reach zero because they are asymtotic to the t axis.

Figure 13.28 v, i versus t.

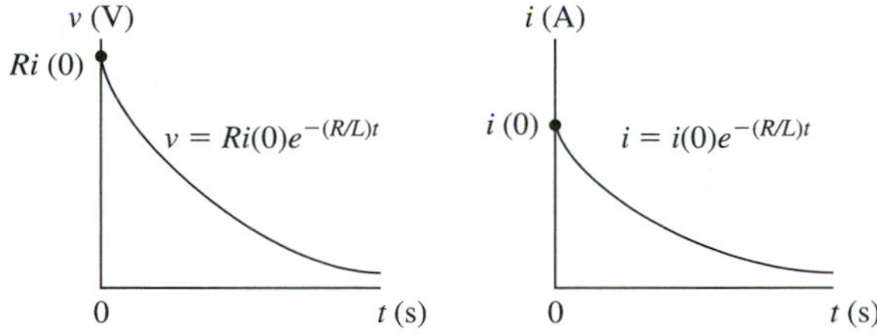

The time required for v and i to decrease to a value close enough to zero to approximate them as equal to zero depends on the quotient L/R. The graphs of i for different values of L/R are shown in Figure 13.29, where

$$\frac{L_3}{R_3} > \frac{L_2}{R_2} > \frac{L_1}{R_1}$$

The higher the value of L/R, the slower the current decreases with time.

Figure 13.29 i versus t.

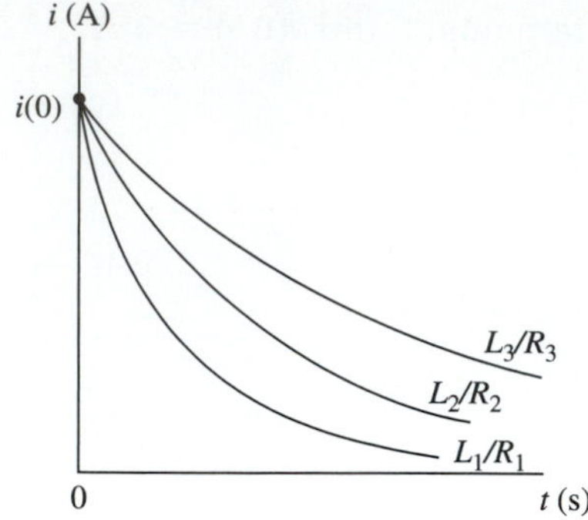

The voltage v begins at $Ri(0)$ and decreases in a manner similar to that shown in Figure 13.29 with regard to the value of L/R.

The time required for the voltage or current in Figure 13.28 to decrease from its initial value at $t = 0$ to a value $1/e$ (e is the base of the natural logarithms and is approximately equal 2.72) times the initial value of the voltage or current is called the *time constant* for the circuit. The time constant for an RL circuit is designated by the Greek letter tau (τ) and is

$$\tau = \frac{L}{R} \tag{13.6}$$

where τ is in seconds, R is in ohms, and L is in henrys. Equation (13.4) easily establishes that at time $t = \tau$, the current i is equal to $(1/e)$ times its initial value.

From Equation (13.4)

$$i = i(0)e^{-(R/L)t}$$

$$i(\tau) = i(0)e^{-(R/L)\tau}$$

and if $\tau = L/R$,

$$i(\tau) = i(0)e^{-(R/L)(L/R)}$$

$$= \frac{i(0)}{e}$$

The time constant is illustrated in Figure 13.30 and applies to voltage as well as current. In most instances, current and voltage are assumed to equal

Figure 13.30 Time constant.

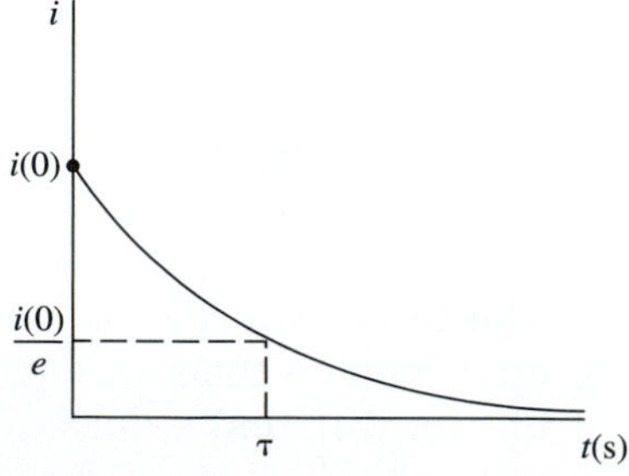

zero after six time constants, but this rule varies with the precision required for the circuit under consideration.

13.7 INDUCTOR BEHAVIOR AS A SHORT

If an inductor is connected to a circuit with only a constant voltage source, the voltage across the inductor will reduce to zero after enough time elapses. The constant source tends to suppress changes in current through the inductor and cause the voltage across the inductor to reduce to zero.

Consider the circuit shown in Figure 13.31. The inductor in Figure 13.31(a) can be replaced by a short, as shown in Figure 13.31(b), after enough time elapses. The current i flowing through the inductor when it appears as a short is v/R. The energized inductor can be switched into another circuit and used as an energy source, as illustrated in the following examples.

Figure 13.31 Constant voltage source.

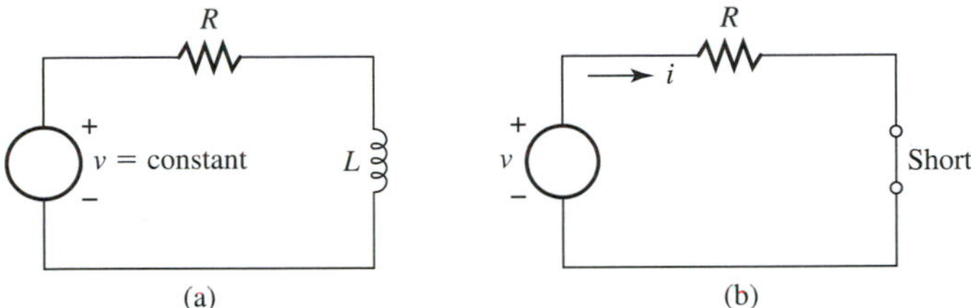

13.8 MISCELLANEOUS *RL* CIRCUITS

The following examples contain constant voltage sources, but they are used only to establish initial currents through inductors and are then switched out of the circuit. The resulting circuits are source-free circuits.

EXAMPLE 13.8 For the circuit in Figure 13.32, the switch remains closed for a long period of time and is opened at $t = 0$. Determine τ, v, i_1, i_2, and i_3 for $t \geq 0$. ■

Figure 13.32

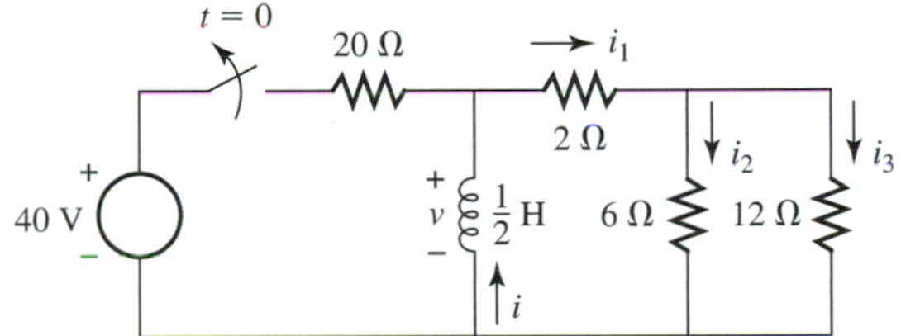

SOLUTION Because the switch is in the closed position for a long period of time and the source is a constant voltage source, the inductor appears as a short circuit at the instant prior to the opening of the switch, as shown in Figure 13.33.

Figure 13.33

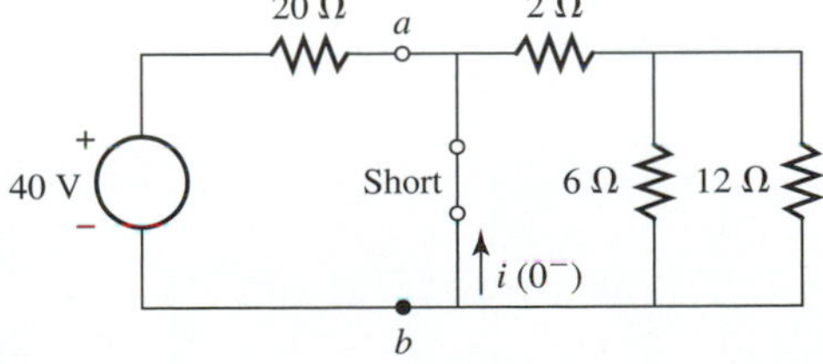

$i(0^-)$ is the current flowing through the inductor an instant before the switch is opened. Since the equivalent resistance of the circuit to the right of a-b is zero because of the short, the circuit can be redrawn as shown in Figure 13.34.

Figure 13.34

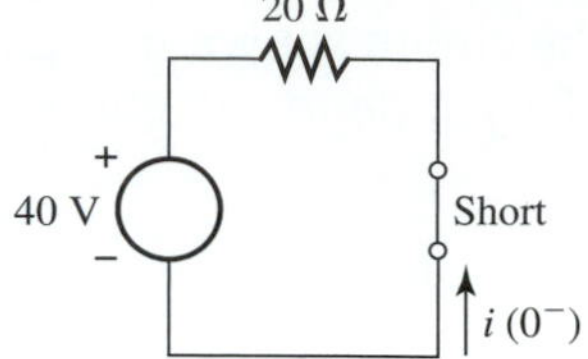

From Ohm's law

$$i(0^-) = -\frac{40}{20}$$
$$= -2 \text{ A}$$

Because the 40-V voltage source and the 20-Ω resistor are eliminated from the original circuit when the switch is opened, the circuit appears as shown in Figure 13.35 after the switch is opened.

Figure 13.35

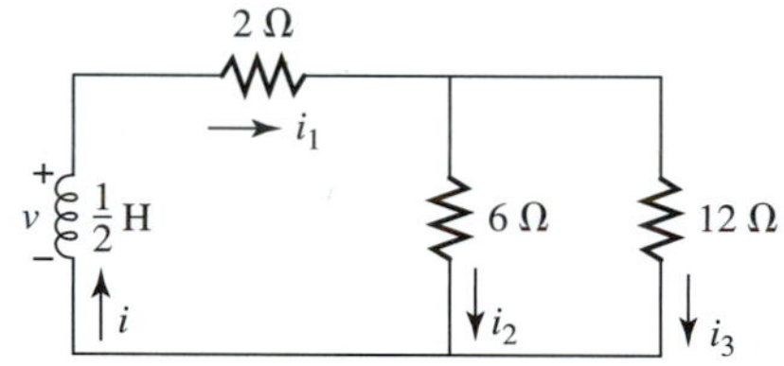

Putting the circuit in the same form as the circuit in Figure 13.25 by using Equations (3.1) and (3.4) to combine resistances results in the circuit shown in Figure 13.36.

Figure 13.36

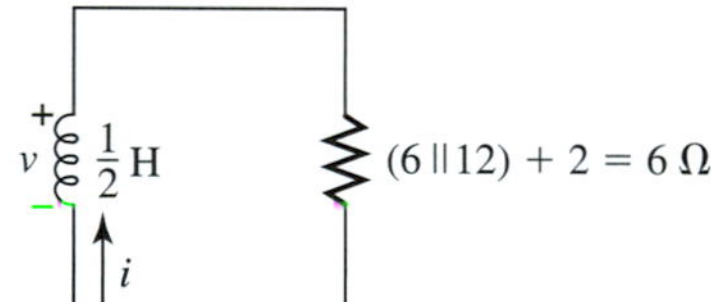

Applying Equation (13.4) and recalling that

$$i(0) = i(0^-)$$
$$= -2$$

results in

$$i = i(0)e^{-(R/L)t}$$
$$= -2e^{-\{6/(1/2)\}t}$$
$$= -2e^{-12t} \text{ A}$$

From Ohm's law

$$\begin{aligned} v &= Ri \\ &= 6\{-2e^{-12t}\} \\ &= -12e^{-12t}\ \text{V} \end{aligned}$$

From an inspection of the circuit in Figure 13.32,

$$\begin{aligned} i_1 &= i \\ &= -2e^{-12t}\ \text{A} \end{aligned}$$

when the switch is opened.

Using Equation (4.1) for current division results in

$$\begin{aligned} i_2 &= \left\{\frac{12}{6+12}\right\} i_1 \\ &= \left\{\frac{12}{6+12}\right\}(-2e^{-12t}) \\ &= -\left(\frac{4}{3}\right)e^{-12t}\ \text{A} \\ i_3 &= \left\{\frac{6}{6+12}\right\} i_1 \\ &= \left\{\frac{6}{6+12}\right\}(-2e^{-12t}) \\ &= -\left(\frac{2}{3}\right)e^{-12t}\ \text{A} \end{aligned}$$

From Equation (13.6)

$$\begin{aligned} \tau &= \frac{L}{R} \\ &= \frac{\frac{1}{2}}{6} \\ &= \frac{1}{12}\ \text{s} \end{aligned}$$

EXAMPLE 13.9 For the circuit in Figure 13.37 the switch remains closed for a long period of time and is opened at $t = 0$. Determine v and i for $t \geq 0$. ■

Figure 13.37

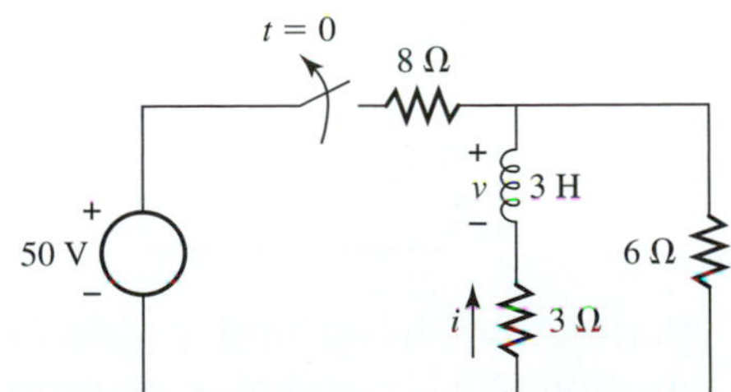

SOLUTION Because the switch is in the closed position for a long period of time and the source is a constant voltage source, the inductor appears as a short circuit at the instant prior to the opening of the switch, as shown in Figure 13.38.

Figure 13.38

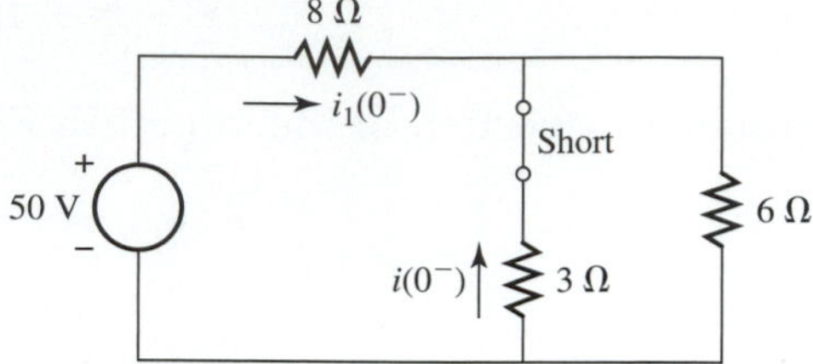

$i(0^-)$ is the current flowing through the inductor an instant before the switch is opened. Using Equations (3.1) and (3.4) to combine the 3-Ω and 6-Ω resistors in parallel with the 8-Ω resistor in series results in the circuit in Figure 13.39.

Figure 13.39

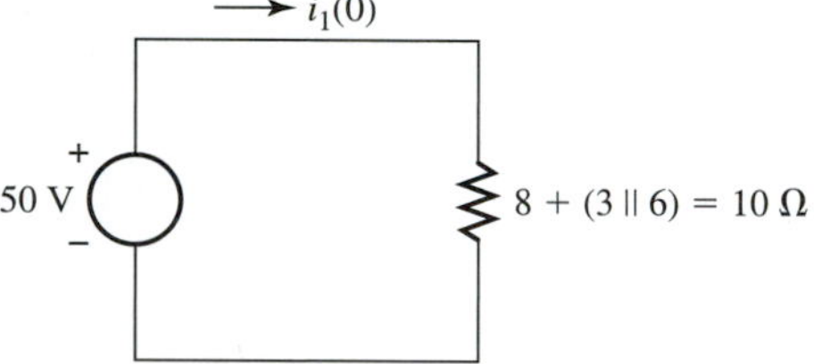

From Ohm's law

$$i_1(0^-) = \frac{50}{10}$$

$$= 5 \text{ A}$$

Using Equation (4.1) for current division results in

$$i(0^-) = -\left\{\frac{6}{6+3}\right\} i_1(0^-)$$

$$= -\frac{10}{3} \text{ A}$$

Because the 50-V voltage source and the 8-Ω resistor are eliminated from the original circuit when the switch is opened, the circuit appears as shown in Figure 13.40 after the switch is opened.

Figure 13.40

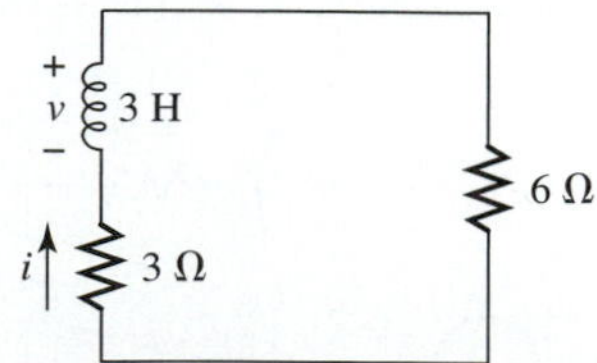

Putting the circuit in the same form as the circuit in Figure 13.7 by using Equation (3.1) to combine resistors results in the circuit in Figure 13.41.

Figure 13.41

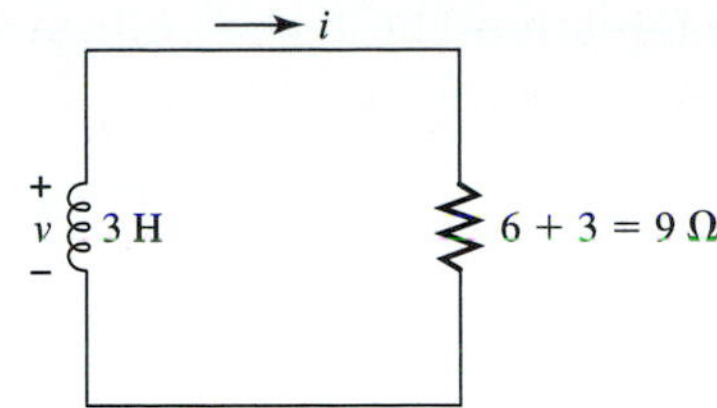

Applying Equation (13.4) and recalling that

$$\begin{aligned} i(0) &= i(0^-) \\ &= -\frac{10}{3}\text{ A} \end{aligned}$$

results in

$$\begin{aligned} i &= i(0)e^{-(R/L)t} \\ &= -\left(\frac{10}{3}\right)e^{-(9/3)t} \\ &= -\left(\frac{10}{3}\right)e^{-3t}\text{ A} \end{aligned}$$

From Ohm's law

$$\begin{aligned} v &= 9i \\ &= 9\left\{-\left(\frac{10}{3}\right)e^{-3t}\right\} \\ &= -30e^{-3t}\text{ V} \end{aligned}$$

or, alternatively, from Equation (10.3)

$$\begin{aligned} v &= L\frac{d(-i)}{dt} \\ &= 3\,\frac{d\left\{-\left(-\frac{10}{3}e^{-3t}\right)\right\}}{dt} \\ &= (3)(-1)\left(-\frac{10}{3}\right)(-3)\,e^{-3t} \\ &= -30e^{-3t}\text{ V} \end{aligned}$$

EXAMPLE 13.10 For the circuit in Figure 13.25, establish Equation (13.4) using the Laplace transform method for solving differential equations. ■

SOLUTION The circuit in Figure 13.25 for $t \geq 0$ is shown in Figure 13.42.

Figure 13.42

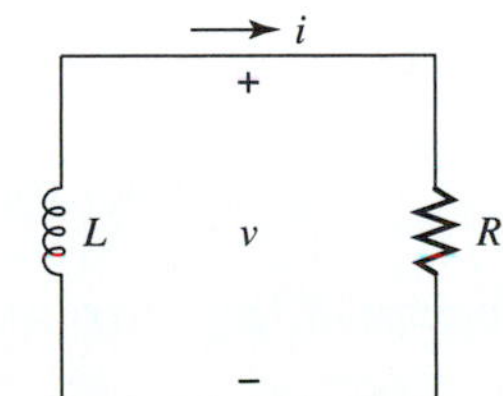

From Equation (11.3) the voltage v across the inductor is

$$v = L\frac{d(-i)}{dt}$$

$$= -L\frac{di}{dt}$$

Notice that the assigned direction for i and the polarity of v necessitates the introduction of the minus sign in the preceding equation.

From Ohm's law the voltage v across the resistor is

$$v = Ri$$

Because the same voltage appears across the resistor and inductor,

$$-L\frac{di}{dt} = Ri$$

$$L\frac{di}{dt} + Ri = 0$$

and

$$\frac{di}{dt} + \frac{Ri}{L} = 0$$

Using rules L1, L3, and L4 presented in Chapter 12 for Laplace transforms results in

$$\mathcal{L}\left\{\frac{di}{dt} + \frac{R}{L}i\right\} = 0$$

$$\mathcal{L}\left\{\frac{di}{dt}\right\} + \mathcal{L}\left\{\frac{R}{L}i\right\} = 0$$

$$\{s\mathbf{I} - i(0)\} + \left(\frac{R}{L}\right)\mathbf{I} = 0$$

$$\mathbf{I} = \frac{i(0)}{s + (R/L)}$$

and from rule L3

$$i = \mathcal{L}^{-1}\left\{\frac{i(0)}{s + (R/L)}\right\}$$

$$= \{i(0)\}\mathcal{L}^{-1}\left\{\frac{1}{s + (R/L)}\right\}$$

From entry 2 of Table 12.2,

$$i = i(0)e^{-(R/L)t}$$

where $i(0) = i(0^-)$, because the current through an inductor cannot change instantaneously.

13.9 *RLC* SERIES CIRCUIT

An *RLC* circuit contains resistors, inductors, and capacitors. A simple *RLC* series circuit is shown in Figure 13.43. The circuit is energized by the initial voltage across the capacitor and/or the initial current flowing through the inductor prior to closing the switch at $t = 0$. From that time forward, the circuit appears as shown in Figure 13.43(b). The measurement of voltages and currents for the circuit in Figure 13.43(b) begins at time $t = 0$ when the switch is closed. Time t is the independent variable and v and i are the dependent variables for the expressions that we shall develop for voltages and current. The circuit furnishing the initial current $i(0^-)$ is not shown and is switched out of the original circuit when the switch is closed at $t = 0$.

Figure 13.43 *RLC* series circuit.

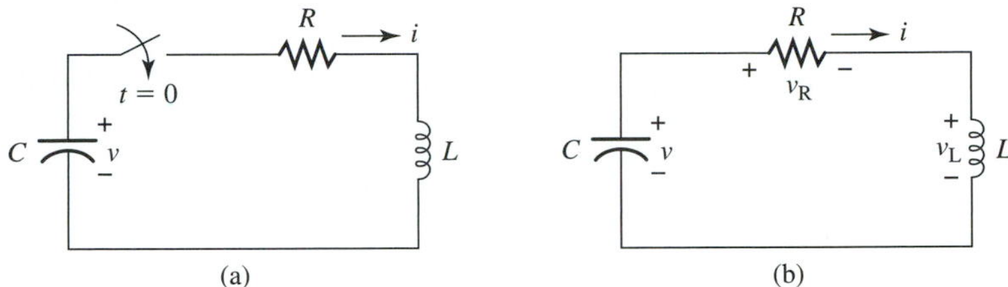

From Equation (10.2)

$$i = C\frac{d(-v)}{dt}$$

$$= -C\frac{dv}{dt}$$

and

$$\frac{dv}{dt} = -\left(\frac{1}{C}\right)i$$

Notice that the assigned direction for i and the assumed polarity for v across the capacitor necessitates the introduction of the minus sign in the above equation for i.

From the preceding equation and the definition of an antiderivative given in Chapter 9, v is the antiderivative of $-(1/C)i$. From the definition of an indefinite integral given in Chapter 9,

$$v = \int -\left(\frac{1}{C}\right)i\, dt$$

and from rule I6 in Chapter 9,

$$v = -\left(\frac{1}{C}\right)\int i\, dt$$

From Ohm's law and Equation (11.3),

$$v_R = Ri$$

and

$$v_L = L\frac{di}{dt}$$

Employing Kirchhoff's voltage law for the circuit in Figure 13.12(b) results in

$$v - v_R - v_L = 0$$

$$-\frac{1}{C}\int i\,dt - Ri - L\frac{di}{dt} = 0$$

$$L\frac{di}{dt} + Ri + \frac{1}{C}\int i\,dt = 0$$

Differentiating both sides of this equation and recognizing that the derivative of the indefinite integral is equal to the integrand of the integral results in

$$\frac{d\left\{L\frac{di}{dt} + Ri + \frac{1}{C}\int i\,dt\right\}}{dt} = \frac{d\{0\}}{dt}$$

$$\frac{d\left\{L\frac{di}{dt}\right\}}{dt} + \frac{d\{Ri\}}{dt} + \frac{d\left\{\frac{1}{C}\int i\,dt\right\}}{dt} = 0$$

From rule D6 in Chapter 9,

$$L\frac{d^2i}{dt^2} + R\frac{di}{dt} + \left(\frac{1}{C}\right)i = 0$$

$$\frac{d^2i}{dt^2} + \left(\frac{R}{L}\right)\frac{di}{dt} + \left(\frac{1}{LC}\right)i = 0$$

The preceding equation is a second-order homogeneous differential equation with dependent variable i and independent variable t. Because the differential equation is homogeneous, the complete solution is equal to the transient solution. The steady-state solution is equal to zero.

Because the differential equation is second order, two initial conditions are required to determine a particular complete solution. The required initial conditions are $i(0)$ (i evaluated at $t = 0$) and $di(0)/dt$ (the derivative of i with respect to t evaluated at $t = 0$).

Because the current through an inductor cannot change instantaneously, the first initial condition is

$$i(0) = i(0^-)$$

where $i(0^-)$ is the current flowing through the inductor prior to closing the switch.

The second initial condition, $di(0)/dt$, can be determined by applying Kirchhoff's voltage law to the circuit in Figure 13.43(b) at $t = 0$. The circuit at the instant $t = 0$ appears in Figure 13.44.

Figure 13.44 Initial conditions.

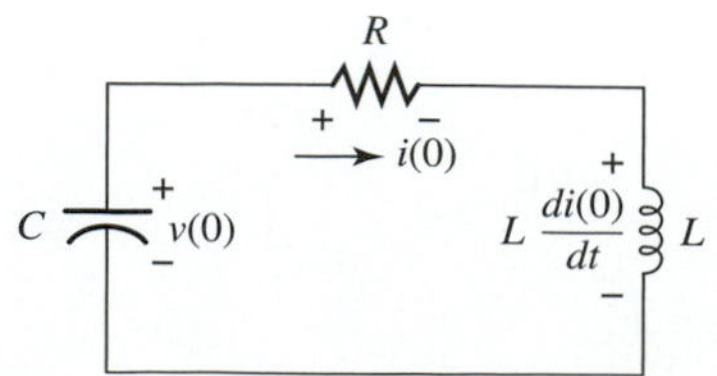

Employing Kirchhoff's voltage law results in

$$v(0) - L\frac{di(0)}{dt} - Ri(0) = 0$$

and

$$\frac{di(0)}{dt} = \frac{v(0) - Ri(0)}{L}$$

Because neither the current flowing through an inductor nor the voltage across a capacitor can change instantaneously,

$$i(0) = i(0^-)$$
$$v(0) = v(0^-)$$

in the equation for the second initial condition.

Using the procedure for solving second-order homogeneous differential equations presented in Chapter 12, the characteristic equation is

$$p^2 + \left(\frac{R}{L}\right)p + \frac{1}{LC} = 0$$

Using the quadratic equation to determine the roots of the characteristic equation results in

$$p_{1,2} = \frac{-(R/L) \pm \sqrt{(R/L)^2 - (4/LC)}}{2}$$
$$= -\left(\frac{R}{2L}\right) \pm \sqrt{\left(\frac{R}{2L}\right)^2 - \left(\frac{1}{LC}\right)}$$

The roots p_1 and p_2 can be real and unequal, real and equal, or complex conjugates. The nature of the roots is determined by the quantity under the radical sign.

Case 1 If $(R/2L)^2 > (1/LC)$, the roots are real and unequal and are

$$p_1 = -\left(\frac{R}{2L}\right) + \sqrt{\left(\frac{R}{2L}\right)^2 - \left(\frac{1}{LC}\right)}$$
$$p_2 = -\left(\frac{R}{2L}\right) - \sqrt{\left(\frac{R}{2L}\right)^2 - \left(\frac{1}{LC}\right)}$$

From Chapter 12, the general complete solution to the differential equation is

$$i = K_1e^{p_1t} + K_2e^{p_2t} \tag{13.7}$$

where K_1 and K_2 are arbitrary constants determined by the initial conditions.

Case 2 If $(R/2L)^2 = (1/LC)$, the roots are real and equal and are

$$p_1 = -\left(\frac{R}{2L}\right)$$

$$p_2 = -\left(\frac{R}{2L}\right)$$

From Chapter 12, the general complete solution to the differential equation is

$$i = K_1 e^{pt} + K_2 t e^{pt} \tag{13.8}$$

where $p = -(R/2L)$ and K_1 and K_2 are arbitrary constants determined by the initial conditions.

Case 3 If $(1/LC) > (R/2L)^2$, the roots are complex conjugates and are

$$p_1 = -\left(\frac{R}{2L}\right) + j\sqrt{\frac{1}{LC} - \left(\frac{R}{2L}\right)^2}$$

$$p_2 = -\left(\frac{R}{2L}\right) - j\sqrt{\frac{1}{LC} - \left(\frac{R}{2L}\right)^2}$$

In accordance with the procedure presented in Chapter 12,

$$\alpha = -\left(\frac{R}{2L}\right)$$

$$\beta = \sqrt{\frac{1}{LC} - \left(\frac{R}{2L}\right)^2} \tag{13.9}$$

and the general complete solution to the differential equation is

$$i = K_1 e^{\alpha t} \sin(\beta t + \theta) \tag{13.10}$$

where K_1 and θ are arbitrary constants determined by the initial conditions.

In each of these three possible cases, the initial conditions $i(0) = i(0^-)$ and $di(0)/dt = \{v(0) - Ri(0)\}/L$, as previously determined, are used to evaluate the arbitrary constants and determine a particular complete solution.

Consider the following three examples, which illustrate each of these cases.

EXAMPLE 13.11 Determine i for $t \geq 0$ if $i(0^-) = 0$ in Figure 13.45. ■

Figure 13.45

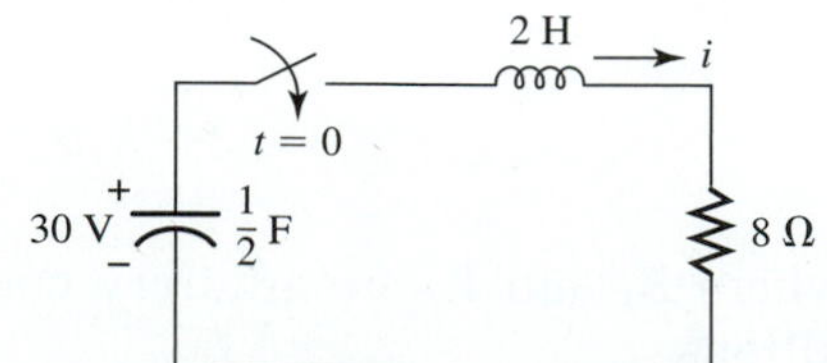

SOLUTION The circuit for $t \geq 0$ appears as shown in Figure 13.46.

Figure 13.46

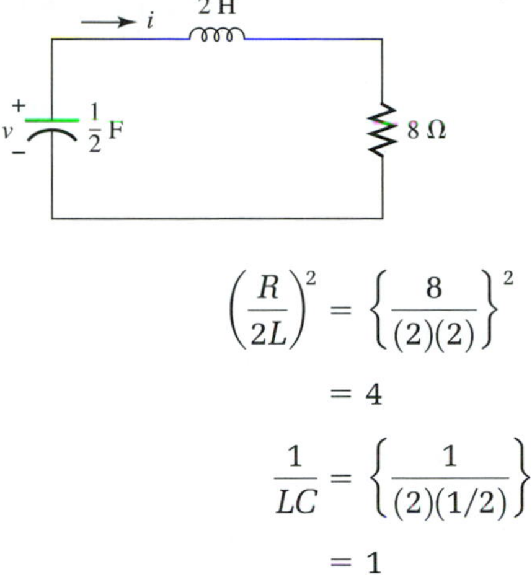

$$\left(\frac{R}{2L}\right)^2 = \left\{\frac{8}{(2)(2)}\right\}^2 = 4$$

$$\frac{1}{LC} = \left\{\frac{1}{(2)(1/2)}\right\} = 1$$

Because $(R/2L)^2 > (1/LC)$, case 1 applies and

$$p_1 = -\left(\frac{R}{2L}\right) + \sqrt{\left(\frac{R}{2L}\right)^2 - \left(\frac{1}{LC}\right)}$$
$$= -\left(\frac{8}{(2)(2)}\right) + \sqrt{\left\{\frac{8}{(2)(2)}\right\}^2 - \left\{\frac{1}{(2)(1/2)}\right\}}$$
$$= -2 + \sqrt{3}$$
$$= -0.27$$

$$p_2 = -\left(\frac{R}{2L}\right) - \sqrt{\left(\frac{R}{2L}\right)^2 - \left(\frac{1}{LC}\right)}$$
$$= -\left(\frac{8}{(2)(2)}\right) - \sqrt{\left\{\frac{8}{(2)(2)}\right\}^2 - \left\{\frac{1}{(2)(1/2)}\right\}}$$
$$= -2 - \sqrt{3}$$
$$= -3.73$$

From Equation (13.7) the general solution for i is

$$i = K_1 e^{p_1 t} + K_2 e^{p_2 t}$$
$$= K_1 e^{-0.27t} + K_2 e^{-3.73t}$$

Because $i(0) = i(0^-) = 0$,

$$0 = K_1 + K_2$$

Differentiating i with respect to t results in

$$\frac{di}{dt} = -0.27K_1 e^{-0.27t} - 3.73K_2 e^{-3.73t}$$

$$\frac{di(0)}{dt} = -0.27K_1 - 3.73K_2$$

Because $i(0) = i(0^-) = 0$ and $v(0) = v(0^-) = 30$ V,

$$\frac{di(0)}{dt} = \frac{v(0) - Ri(0)}{L} = \frac{30}{2} = 15$$

and

$$15 = -0.27K_1 - 3.73K_2$$

Solving the two algebraic equations

$$K_1 + K_2 = 0$$
$$-0.27K_1 - 3.73K_2 = 15$$

for K_1 and K_2 results in

$$K_1 = 4.34$$
$$K_2 = -4.34$$

and

$$i = 4.34e^{-0.27t} - 4.34e^{-3.73t} \text{ A}$$

EXAMPLE 13.12 Determine i for $t \geq 0$ if $i(0^-) = 0$ in Figure 13.47. ■

Figure 13.47

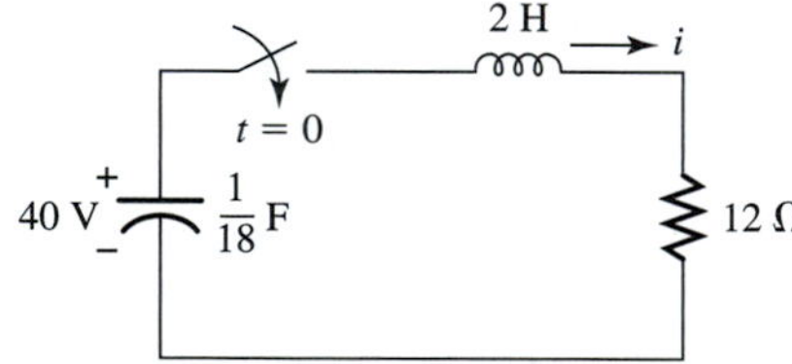

SOLUTION The circuit for $t \geq 0$ appears in Figure 13.48.

Figure 13.48

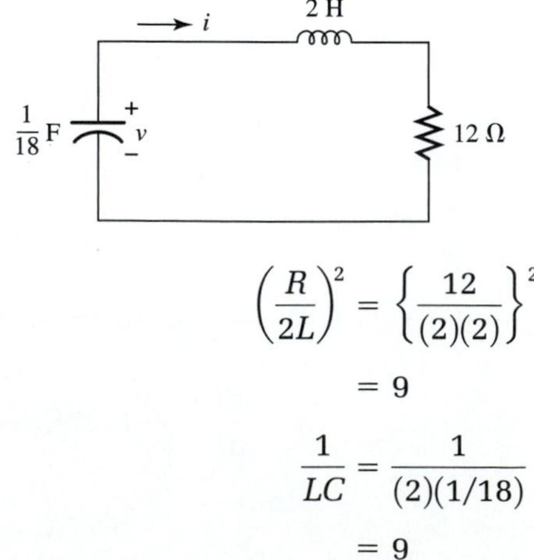

$$\left(\frac{R}{2L}\right)^2 = \left\{\frac{12}{(2)(2)}\right\}^2 = 9$$

$$\frac{1}{LC} = \frac{1}{(2)(1/18)} = 9$$

Because $(R/2L)^2 = (1/LC)$, case 2 applies and the two roots of the characteristic equation are equal.

$$\begin{aligned} p &= p_1 = p_2 \\ &= -\left(\frac{R}{2L}\right) \pm \sqrt{\left(\frac{R}{2L}\right)^2 - \left(\frac{1}{LC}\right)} \\ &= -\left\{\frac{12}{(2)(2)}\right\} \pm \sqrt{\left\{\frac{12}{(2)(2)}\right\}^2 - \left\{\frac{1}{(2)(1/18)}\right\}} \\ &= -3 \pm 0 \\ &= -3 \end{aligned}$$

From Equation (13.8) the general complete solution for i is

$$\begin{aligned} i &= K_1 e^{pt} + K_2 t e^{pt} \\ &= K_1 e^{-3t} + K_2 t e^{-3t} \end{aligned}$$

Because $i(0) = i(0^-) = 0$,

$$\begin{aligned} 0 &= K_1 + (K_2)(0) \\ K_1 &= 0 \end{aligned}$$

Differentiating i with respect to t and employing rule D9 from Chapter 9 results in

$$\begin{aligned} \frac{di}{dt} &= -3K_1 e^{-3t} + K_2 e^{-3t} - 3K_2 t e^{-3t} \\ \frac{di(0)}{dt} &= -3K_1 + K_2 \end{aligned}$$

Because $i(0) = i(0^-) = 0$ and $v(0) = v(0^-) = 40$,

$$\begin{aligned} \frac{di(0)}{dt} &= \frac{v(0) - Ri(0)}{L} \\ \frac{di(0)}{dt} &= \frac{40}{2} \\ &= 20 \end{aligned}$$

and

$$20 = -3K_1 + K_2$$

Because $K_1 = 0$

$$K_2 = 20$$

and

$$i = 20te^{-3t}\text{ A}$$

EXAMPLE 13.13 Determine i for $t \geq 0$ if $i(0^-) = 0$ in Figure 13.49. ■

Figure 13.49

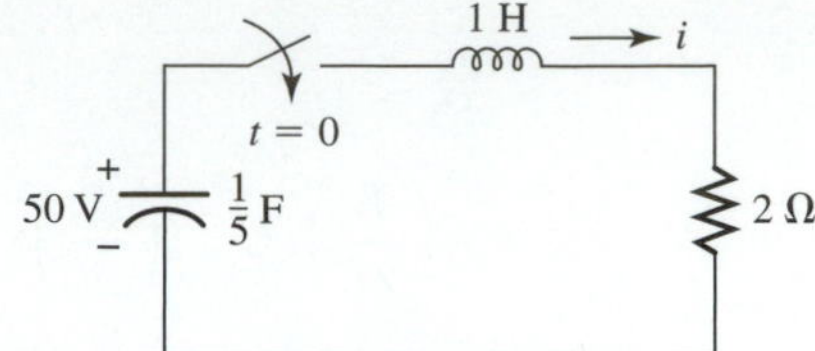

SOLUTION The circuit for $t \geq 0$ is shown in Figure 13.50.

Figure 13.50

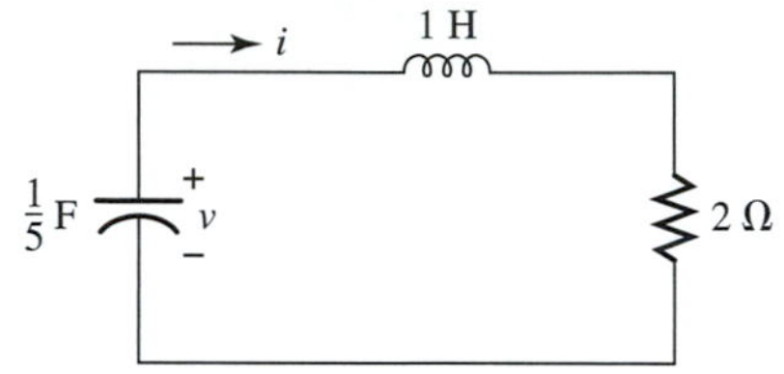

$$\left(\frac{R}{2L}\right)^2 = \left\{\frac{2}{(2)(1)}\right\}^2$$
$$= 1$$
$$\frac{1}{LC} = \frac{1}{(1)\left(\frac{1}{5}\right)}$$
$$= 5$$

Because $(1/LC) > (R/2L)^2$, case 3 applies, and

$$p_1 = -\left(\frac{R}{2L}\right) + j\sqrt{\frac{1}{LC} - \left(\frac{R}{2L}\right)^2}$$
$$= -\frac{2}{(2)(1)} + j\sqrt{\frac{1}{(1)\left(\frac{1}{5}\right)} - \frac{2}{(2)(1)}}$$
$$= -1 + j2$$
$$p_2 = -\left(\frac{R}{2L}\right) - j\sqrt{\frac{1}{LC} - \left(\frac{R}{2L}\right)^2}$$
$$= -\frac{2}{(2)(1)} - j\sqrt{\frac{1}{(1)\left(\frac{1}{5}\right)} - \frac{2}{(2)(1)}}$$
$$= -1 - j2$$
$$\alpha = -\left(\frac{R}{2L}\right)$$
$$= -1$$
$$\beta = \sqrt{\frac{1}{LC} - \left(\frac{R}{2L}\right)^2}$$
$$= 2$$

From Equation (13.9) the general complete solution for i is

$$\begin{aligned} i &= K_1 e^{\alpha t}\sin(\beta t + \theta) \\ &= K_1 e^{-t}\sin(2t + \theta) \end{aligned}$$

Because $i(0) = i(0^-) = 0$,

$$0 = K_1 \sin(\theta)$$

Using rule D9 of Chapter 9 to differentiate i with respect to t results in

$$\begin{aligned} \frac{di}{dt} &= -K_1 e^{-t}\sin(2t + \theta) \\ &\quad + K_1 e^{-t}(2)\cos(2t + \theta) \\ \frac{di(0)}{dt} &= -K_1 \sin\theta + 2K_1 \cos\theta \end{aligned}$$

Because $i(0) = i(0^-) = 0$ and $v(0) = v(0^-) = 50$,

$$\begin{aligned} \frac{di(0)}{dt} &= \frac{v(0) - Ri(0)}{L} \\ &= \frac{50}{1} \\ &= 50 \end{aligned}$$

and

$$50 = -K_1 \sin(\theta) + 2K_1 \cos(\theta)$$

The trigonometric equations

$$\begin{aligned} K_1 \sin(\theta) &= 0 \\ -K_1 \sin(\theta) + 2K_1 \cos(\theta) &= 50 \end{aligned}$$

must be solved for K_1 and θ.

Because $K_1 \sin(\theta) = 0$ and K_1 cannot equal zero, as seen from the general complete solution for i, it follows that

$$\sin\theta = 0$$

and

$$\theta = 0$$

Substituting $\theta = 0$ into the second equation results in

$$\begin{aligned} -K_1 \sin(0) + 2K_1 \cos(0) &= 50 \\ 2K_1 &= 50 \\ K_1 &= 25 \end{aligned}$$

The expression for the current i is

$$i = 25e^{-t}\sin(2t) \text{ A}$$

13.10 *RLC* PARALLEL CIRCUIT

Consider the simple parallel *RLC* circuit shown in Figure 13.51. The circuit is energized by the initial voltage across the capacitor and/or the initial current flowing through the inductor when the switch is closed at $t = 0$, and from that

Figure 13.51 *RLC* parallel circuit.

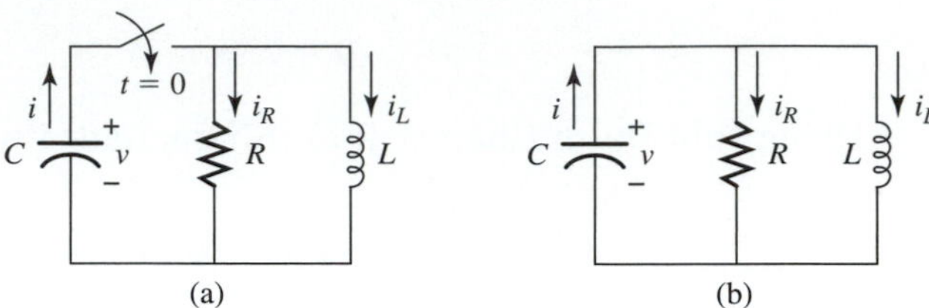

time forward the circuit appears as shown in Figure 13.51(b). The measurement of voltages and currents for the circuit in Figure 13.51(b) begins at time $t = 0$ when the switch is closed. Time t is the independent variable and v and i are the dependent variables for the expressions that we shall develop for voltages and currents. The circuit furnishing the initial current $i_L(0^-)$ prior to closing the switch is not shown and is switched out of the original circuit when the switch is closed at $t = 0$.

From Equation (11.3)

$$v = L\frac{di_L}{dt}$$

and

$$\frac{di_L}{dt} = \frac{v}{L}$$

From this equation and the definition of an antiderivative given in Chapter 9, i_L is the antiderivative of v/L.

From the definition of an indefinite integral given in Chapter 9,

$$i_L = \int \left(\frac{1}{L}\right) v\, dt$$

and from rule I6 in Chapter 9,

$$i_L = \left(\frac{1}{L}\right) \int v\, dt$$

From Equation (10.2) and Ohm's law, respectively,

$$\begin{aligned} i &= C\frac{d(-v)}{dt} \\ &= -C\frac{dv}{dt} \end{aligned}$$

and

$$i_R = \frac{v}{R}$$

Notice that the assigned direction for i and the assumed polarity of v necessitate the introduction of the minus sign in the preceding equation for i.

Employing Kirchhoff's current law for the circuit in Figure 13.51(b) results in

$$i = i_R + i_L$$

$$-C\frac{dv}{dt} = \frac{v}{R} + \left(\frac{1}{L}\right)\int v\,dt$$

$$C\frac{dv}{dt} + \frac{v}{R} + \left(\frac{1}{L}\right)\int v\,dt = 0$$

Because the derivative of the indefinite integral is equal to the integrand of the integral, differentiation of both sides of the preceding equation with respect to t rids the equation of the integral. Differentiating both sides of the equation and employing rule D7 of Chapter 9 results in

$$\frac{d\{C\,dv/dt + (1/R)v + (1/L)\int v\,dt\}}{dt} = \frac{d(0)}{dt}$$

$$\frac{d\{C\,dv/dt\}}{dt} + \frac{d\{(1/R)v\}}{dt} + \frac{d\{(1/L)\int v\,dt\}}{dt} = 0$$

and from rule D6 in Chapter 9,

$$C\frac{d^2v}{dt^2} + \left(\frac{1}{R}\right)\frac{dv}{dt} + \left(\frac{1}{L}\right)v = 0$$

$$\frac{d^2v}{dt^2} + \left(\frac{1}{RC}\right)\frac{dv}{dt} + \left(\frac{1}{LC}\right)v = 0$$

This differential equation is a second-order homogenous differential equation with dependent variable v and independent variable t. Because the differential equation is homogeneous, the complete solution is equal to the transient solution, and the steady-state solution is equal to zero.

Because the differential equation is second order, two initial conditions are required to determine a particular complete solution. The required initial conditions are $v(0)$ (v evaluated at $t = 0$) and $dv(0)/dt$ (the derivative of v with respect to t evaluated at $t = 0$). Because the voltage across the capacitor cannot change instantaneously

$$v(0) = v(0^-)$$

which is the first initial condition.

The second initial condition can be determined by applying Kirchhoff's current law to the circuit in Figure 13.51(b) at $t = 0$. The circuit at the instant $t = 0$ appears in Figure 13.52.

Figure 13.52 Initial conditions.

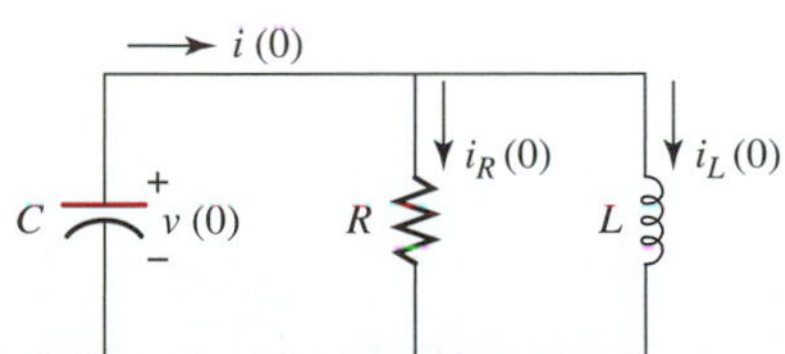

Employing Kirchhoff's current law results in

$$i(0) = i_R(0) + i_L(0)$$

From Equation (10.2)

$$i = C\frac{d(-v)}{dt}$$

$$= -C\frac{dv}{dt}$$

and

$$i(0) = -C\frac{dv(0)}{dt}$$

Substituting $-C\dfrac{dv(0)}{dt}$ for $i(0)$ in the preceding equation results in

$$-C\frac{dv(0)}{dt} = i_R(0) + i_L(0)$$

From Ohm's law

$$i_R(0) = \frac{v(0)}{R}$$

and the second initial condition is

$$\frac{dv(0)}{dt} = -\left\{\frac{v(0)}{RC}\right\} - \left\{\frac{i_L(0)}{C}\right\}$$

Because neither the initial current flowing through an inductor nor the initial voltage across a capacitor can change instantaneously,

$$i_L(0) = i_L(0^-)$$

$$v(0) = v(0^-)$$

where $i_L(0^-)$ and $v(0^-)$ are the current and voltage prior to closing the switch in Figure 13.51(a). We shall use the classical method presented in Chapter 12 to solve the second-order homogeneous differential equation

$$\frac{d^2v}{dt^2} + \left(\frac{1}{RC}\right)\frac{dv}{dt} + \left(\frac{1}{LC}\right)v = 0$$

with initial conditions

$$v(0) = v(0^-)$$

and

$$\frac{dv(0)}{dt} = -\left\{\frac{v(0)}{RC}\right\} - \left\{\frac{i_L(0)}{C}\right\}$$

for a particular complete solution.

Using the procedure for solving second-order homogeneous differential equations presented in Chapter 12, the characteristic equation is

$$p^2 + \left(\frac{1}{RC}\right)p + \frac{1}{LC} = 0$$

Using the quadratic equation to determine the roots of the characteristic equation results in

$$p_{1,2} = \frac{-(1/RC) \pm \sqrt{(1/RC)^2 - (4/LC)}}{2}$$

$$= -\left(\frac{1}{2RC}\right) \pm \sqrt{\left(\frac{1}{2RC}\right)^2 - \frac{1}{LC}}$$

The roots p_1 and p_2 can be real and unequal, real and equal, or complex conjugates.

Case 1 If $(1/2RC)^2 > (1/LC)$, the roots are real and unequal and are

$$p_1 = -\left(\frac{1}{2RC}\right) + \sqrt{\left(\frac{1}{2RC}\right)^2 - \frac{1}{LC}}$$

$$p_2 = -\left(\frac{1}{2RC}\right) - \sqrt{\left(\frac{1}{2RC}\right)^2 - \frac{1}{LC}}$$

From Chapter 12, the general complete solution to the differential equation is

$$v = K_1 e^{p_1 t} + K_2 e^{p_2 t} \qquad (13.11)$$

where K_1 and K_2 are arbitrary constants determined by the initial conditions.

Case 2 If $(1/2RC)^2 = (1/LC)$, the roots are real and equal and are

$$p_1 = -\left(\frac{1}{2RC}\right)$$

$$p_2 = -\left(\frac{1}{2RC}\right)$$

From Chapter 12, the general complete solution to the differential equation is

$$v = K_1 e^{pt} + K_2 t e^{pt} \qquad (13.12)$$

where $p = -(1/2RC)$ and K_1 and K_2 are arbitrary constants determined by the initial conditions.

Case 3 If $(1/LC) > (1/2RC)^2$, the roots are complex conjugates and are

$$p_1 = -\left(\frac{1}{2RC}\right) + j\sqrt{\frac{1}{LC} - \left(\frac{1}{2RC}\right)^2}$$

$$p_2 = -\left(\frac{1}{2RC}\right) - j\sqrt{\frac{1}{LC} - \left(\frac{1}{2RC}\right)^2}$$

In accordance with the procedure presented in Chapter 12,

$$\alpha = -\left(\frac{1}{2RC}\right)$$
$$\beta = \sqrt{\frac{1}{LC} - \left(\frac{1}{2RC}\right)^2} \tag{13.13}$$

and the general complete solution is

$$v = K_1 e^{\alpha t} \sin(\beta\, t + \theta) \tag{13.14}$$

where K_1 and θ are arbitrary constants determined by the initial conditions.

In each of the three possible cases, the initial conditions $v(0) = v(0^-)$ and

$$\frac{dv(0)}{dt} = -\left\{\frac{v(0)}{RC}\right\} - \left\{\frac{i_L(0)}{C}\right\}$$

are used to evaluate the arbitrary constants and determine a particular complete solution. $v(0) = v(0^-)$ and $i_L(0) = i_L(0^-)$, because the voltage across a capacitor and the current through an inductor cannot change instantaneously.

Consider the following three examples, which illustrate the preceding cases.

EXAMPLE 13.14 Determine v, i, i_R, and i_L for $t \geq 0$ if $i_L(0^-) = 0$ in Figure 13.53. ■

Figure 13.53

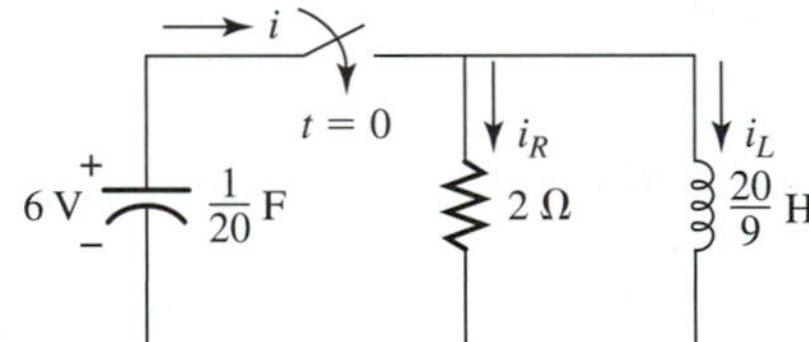

SOLUTION The circuit for $t \geq 0$ appears as shown in Figure 13.54.

Figure 13.54

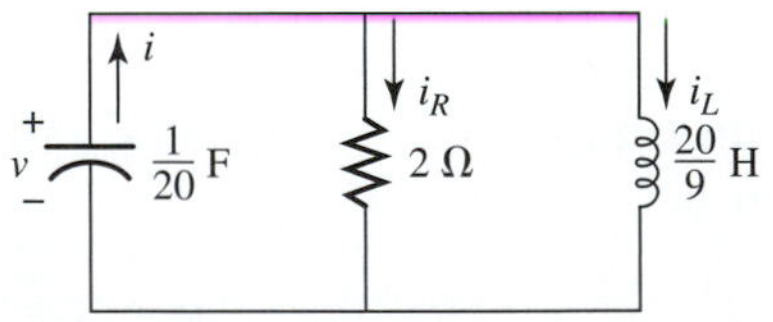

$$\left(\frac{1}{2RC}\right)^2 = \left\{\frac{1}{2(2)\left(\frac{1}{20}\right)}\right\}^2$$
$$= 25$$
$$\left(\frac{1}{LC}\right) = \frac{1}{\left(\frac{20}{9}\right)\left(\frac{1}{20}\right)}$$
$$= 9$$

Because $(1/2RC)^2 > (1/LC)$, case 1 applies and

$$\begin{aligned} p_1 &= -\left(\frac{1}{2RC}\right) + \sqrt{\left(\frac{1}{2RC}\right)^2 - \frac{1}{LC}} \\ &= \frac{-1}{(2)(2)(1/20)} + \sqrt{\left\{\frac{1}{(2)(2)\left(\frac{1}{20}\right)}\right\}^2 - \frac{1}{\left(\frac{20}{9}\right)\left(\frac{1}{20}\right)}} \\ &= -1 \end{aligned}$$

$$\begin{aligned} p_2 &= -\left(\frac{1}{2RC}\right) - \sqrt{\left(\frac{1}{2RC}\right)^2 - \frac{1}{LC}} \\ &= \frac{-1}{(2)(2)\left(\frac{1}{20}\right)} + \sqrt{\left\{\frac{1}{(2)(2)\left(\frac{1}{20}\right)}\right\}^2 - \frac{1}{\left(\frac{20}{9}\right)\left(\frac{1}{20}\right)}} \\ &= -9 \end{aligned}$$

From Equation (13.11) the general complete solution for v is

$$\begin{aligned} v &= K_1 e^{p_1 t} + K_2 e^{p_2 t} \\ &= K_1 e^{-t} + K_2 e^{-9t} \end{aligned}$$

Because $v(0^-) = 6$ V, the first initial condition is

$$\begin{aligned} v(0) &= v(0^-) \\ &= 6 \text{ V} \end{aligned}$$

and setting $t = 0$ and $v = 6$ in the equation for v results in

$$6 = K_1 + K_2$$

Differentiating this equation for v with respect to t results in

$$\frac{dv}{dt} = -K_1 e^{-t} - 9K_2 e^{-9t}$$

Because $v(0^-) = 6$ and $i_L(0^-) = 0$,

$$\begin{aligned} v(0) &= v(0^-) = 6 \\ i_L(0) &= i_L(0^-) = 0 \end{aligned}$$

and the second initial condition is

$$\begin{aligned} \frac{dv(0)}{dt} &= -\left(\frac{1}{RC}\right)v(0^-) - \left(\frac{1}{C}\right)i_L(0^-) \\ &= -\left\{\frac{1}{(2)\left(\frac{1}{20}\right)}\right\}(6) - \frac{1}{\left(\frac{1}{20}\right)}(0) \\ &= -60 \end{aligned}$$

Setting $t = 0$ and $dv/dt = -60$ in the preceding equation for dv/dt results in

$$-60 = -K_1 - 9K_2$$

Solving the two algebraic equations

$$K_1 + K_2 = 6$$
$$-K_1 - 9K_2 = -60$$

for K_1 and K_2 results in

$$K_2 = \frac{27}{4}$$

$$K_1 = -\frac{3}{4}$$

and

$$v = \left(-\frac{3}{4}\right)e^{-t} + \left(\frac{27}{4}\right)e^{-9t}\ \text{V}$$

Because $i_R = v/R$,

$$i_R = \frac{\left(-\frac{3}{4}\right)e^{-t} + \left(\frac{27}{4}\right)e^{-9t}}{2}$$
$$= -\left(\frac{3}{8}\right)e^{-t} + \left(\frac{27}{8}\right)e^{-9t}\ \text{A}$$

Since $i = -C\dfrac{dv}{dt}$,

$$i = \left(-\frac{1}{20}\right)\frac{d\left\{\left(-\frac{3}{4}\right)e^{-t} + \left(\frac{27}{4}\right)e^{-9t}\right\}}{dt}$$
$$= \left(-\frac{1}{20}\right)\left\{\left\{(-1)\left(-\frac{3}{4}\right)e^{-t}\right\} + \left\{(-9)\left(\frac{27}{4}\right)e^{-9t}\right\}\right\}$$
$$= -\left(\frac{3}{80}\right)e^{-t} + \left(\frac{243}{80}\right)e^{-9t}\ \text{A}$$

From Kirchhoff's current law,

$$i_L = i - i_R$$
$$= -\left(\frac{3}{80}\right)e^{-t} + \left(\frac{243}{80}\right)e^{-9t} + \left(\frac{3}{8}\right)e^{-t} - \left(\frac{27}{8}\right)e^{-9t}$$
$$= \left(\frac{27}{80}\right)e^{-t} - \left(\frac{27}{80}\right)e^{-9t}\ \text{A}$$

EXAMPLE 13.15 Determine v, i, i_R, and i_L for $t \geq 0$ if $i_L(0^-) = 0$ in Figure 13.55. ■

Figure 13.55

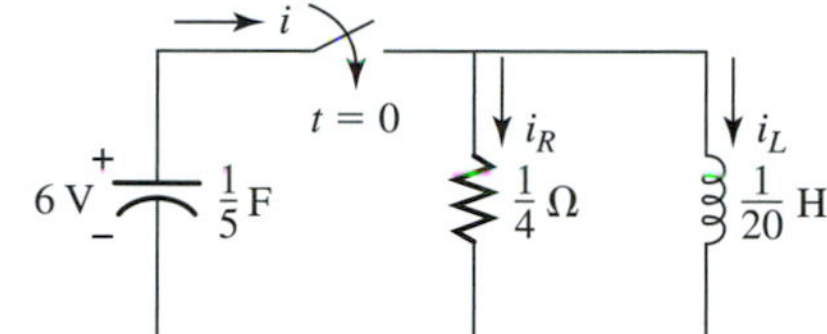

SOLUTION The circuit for $t \geq 0$ appears as shown in Figure 13.56.

Figure 13.56

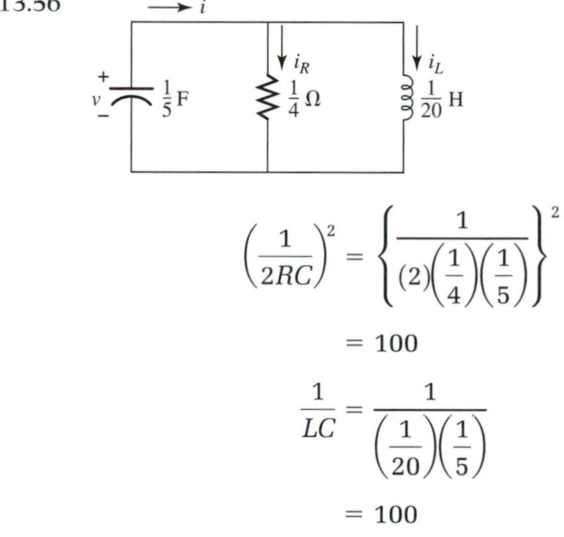

$$\left(\frac{1}{2RC}\right)^2 = \left\{\frac{1}{(2)\left(\frac{1}{4}\right)\left(\frac{1}{5}\right)}\right\}^2$$

$$= 100$$

$$\frac{1}{LC} = \frac{1}{\left(\frac{1}{20}\right)\left(\frac{1}{5}\right)}$$

$$= 100$$

Because $(1/2RC)^2 = (1/LC)$, case 2 applies and

$$p_1 = p_2 = -\left(\frac{1}{2RC}\right)$$

$$= -\frac{1}{(2)\left(\frac{1}{4}\right)\left(\frac{1}{5}\right)}$$

$$= -10$$

From Equation (13.12) the general complete solution for v is

$$v = K_1 e^{pt} + K_2 t e^{pt}$$

where $p = p_1 = p_2$ and

$$v = K_1 e^{-10t} + K_2 t e^{-10t} \text{ V}$$

Because $v(0^-) = 6$ V, the first initial condition is

$$v(0) = v(0^-)$$

$$= 6 \text{ V}$$

Setting $t = 0$ and $v = 6$ in the preceding equation for v results in

$$K_1 = 6$$

Employing rule D9 in Chapter 9 to differentiating the preceding equation for v with respect to t results in

$$\frac{dv}{dt} = -10K_1e^{-10t} + K_2e^{-10t} - 10K_2te^{-10t}$$

Because $v(0^-) = 6$ and $i_L(0^-) = 0$,

$$v(0) = v(0^-) = 6$$
$$i_L(0) = i_L(0^-) = 0$$

and the second initial condition is

$$\begin{aligned}\frac{dv(0)}{dt} &= -\left(\frac{1}{RC}\right)v(0) - \left(\frac{1}{C}\right)i_L(0)\\ &= -\left\{\frac{1}{\left(\frac{1}{4}\right)\left(\frac{1}{5}\right)}\right\}6\\ &= -120\end{aligned}$$

Setting $t = 0$ and $dv/dt = -120$ in the preceding equation for dv/dt results in

$$-120 = -10K_1 + K_2$$

Solving the two algebraic equations

$$K_1 = 6$$
$$-120 = -10K_1 + K_2$$

for K_1 and K_2 results in

$$K_1 = 6$$
$$K_2 = 60$$

and
$$v = 6e^{-10t} - 60te^{-10t}\text{ V}$$

Because $i_R = v/R$,

$$\begin{aligned}i_R &= \{6e^{-10t} - 120te^{-10t}\}/(1/4)\\ &= 24e^{-10t} - 240te^{-10t}\text{ A}\end{aligned}$$

Because $i = -C\dfrac{dv}{dt}$,

$$\begin{aligned}i &= -\left(\frac{1}{5}\right)\frac{d\{6e^{-10t} - 60te^{-10t}\}}{dt}\\ &= -\left(\frac{1}{5}\right)\{-60e^{-10t} - 60e^{-10t} + 600te^{-10t}\}\\ &= 24e^{-10t} - 120te^{-10t}\text{ A}\end{aligned}$$

From Kirchhoff's current law,

$$\begin{aligned} i_L &= i - i_R \\ &= 24e^{-10t} - 120te^{-10t} - 24e^{-10t} + 240te^{-10t} \\ &= 120te^{-10t}\text{ A} \end{aligned}$$

EXAMPLE 13.16 Determine v for $t \geq 0$ if $i_L(0^-) = 0$ (Figure 13.57). ■

Figure 13.57

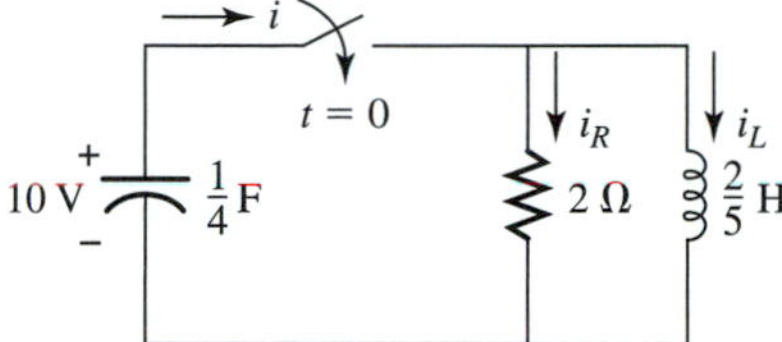

SOLUTION The circuit for $t \geq 0$ appears as shown in Figure 13.58.

Figure 13.58

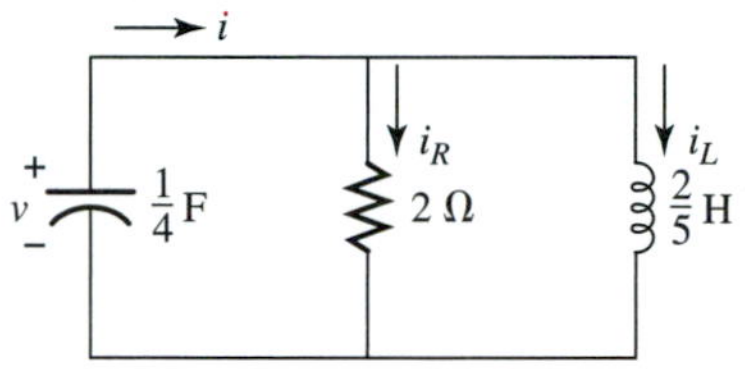

$$\begin{aligned} \left(\frac{1}{2RC}\right)^2 &= \left\{\frac{1}{(2)(2)\left(\frac{1}{4}\right)}\right\}^2 \\ &= 1 \end{aligned}$$

$$\begin{aligned} \frac{1}{LC} &= \left\{\frac{1}{\left(\frac{2}{5}\right)\left(\frac{1}{4}\right)}\right\} \\ &= 10 \end{aligned}$$

Because $(1/LC) > (1/2RC)^2$, case 3 applies, and from Equation (13.13)

$$\begin{aligned} \alpha &= -\left(\frac{1}{2RC}\right) \\ &= -\frac{1}{(2)(2)\left(\frac{1}{4}\right)} \\ &= -1 \end{aligned}$$

$$\begin{aligned} \beta &= \sqrt{\frac{1}{LC} - \left(\frac{1}{2RC}\right)^2} \\ &= \sqrt{10 - 1} \\ &= 3 \end{aligned}$$

From Equation (13.14) the general solution for v is

$$\begin{aligned} v &= K_1 e^{\alpha t}\sin(\beta t + \theta) \\ &= K_1 e^{-t}\sin(3t + \theta) \end{aligned}$$

Because $v(0^-) = 10$ V, the first initial condition is

$$\begin{aligned} v(0) &= v(0^-) \\ &= 10 \text{ V} \end{aligned}$$

and setting $t = 0$ and $v = 10$ in the preceding equation for v results in

$$10 = K_1 \sin\theta$$

Differentiating this equation for v with respect to t results in

$$\frac{dv}{dt} = 3K_1e^{-t}\cos(3t + \theta) - K_1e^{-t}\sin(3t + \theta)$$

Because $v(0^-) = 10$ and $i_L(0^-) = 0$

$$\begin{aligned} v(0) &= v(0^-) = 10 \\ i_L(0) &= i_L(0^-) = 0 \end{aligned}$$

and the second initial condition is

$$\begin{aligned} \frac{dv(0)}{dt} &= -\left(\frac{1}{RC}\right)v(0) - \left(\frac{1}{C}\right)i_L(0) \\ &= -20 \end{aligned}$$

Setting $t = 0$ and $dv/dt = -20$ in the preceding equation for dv/dt results in

$$-20 = 3K_1\cos(\theta) - K_1\sin(\theta)$$

The two trigonometric equations

$$\begin{aligned} K_1\sin(\theta) &= 10 \\ 3K_1\cos(\theta) - K_1\sin(\theta) &= -20 \end{aligned}$$

must be solved for K_1 and θ. From the first equation,

$$K_1 = \frac{10}{\sin(\theta)}$$

Substituting $(10/\sin(\theta))$ for K_1 and 10 for $K_1\sin(\theta)$ into the second equation results in

$$\begin{aligned} 3\left\{\frac{10}{\sin(\theta)}\right\}(\cos(\theta)) - 10 &= -20 \\ 30\left\{\frac{\cos(\theta)}{\sin(\theta)}\right\} &= -10 \\ \frac{\sin(\theta)}{\cos(\theta)} &= \frac{30}{-10} \\ \tan(\theta) &= -3 \\ \theta &= -71.6^\circ \end{aligned}$$

Since $K_1 \sin(\theta) = 10$,

$$K_1 = \frac{10}{\sin(\theta)} = \frac{10}{\sin(-71.6°)} = -10.5$$

The particular complete solution for v is

$$v = K_1 e^{\alpha t} \sin(\beta t + \theta) = -10.5e^{-t} \sin(3t - 71.6°) \text{ V}$$

Because $i_R = v/R$, $i = -C\dfrac{dv}{dt}$, and $i_L = (i - i_R)$, i_R, i, and i_L can be determined from the previous expression for v.

13.11 EXERCISES

1. For each circuit in Figure 13.59 determine v and i for $t \geq 0$ after the switch is closed at $t = 0$.

Figure 13.59

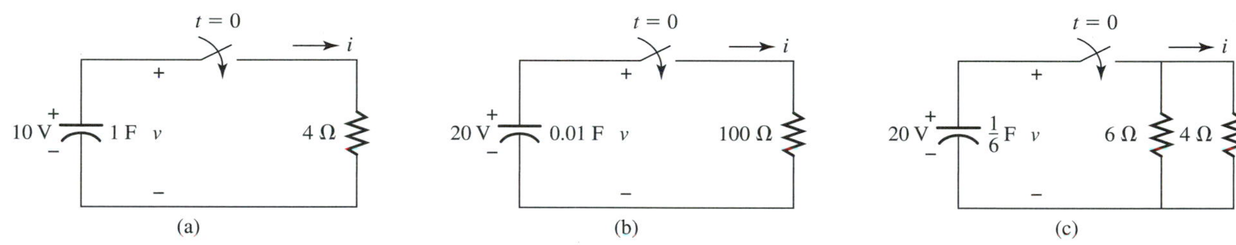

2. For each circuit in Figure 13.60 determine v and i for $t \geq 0$ after the switch is closed at $t = 0$.

Figure 13.60

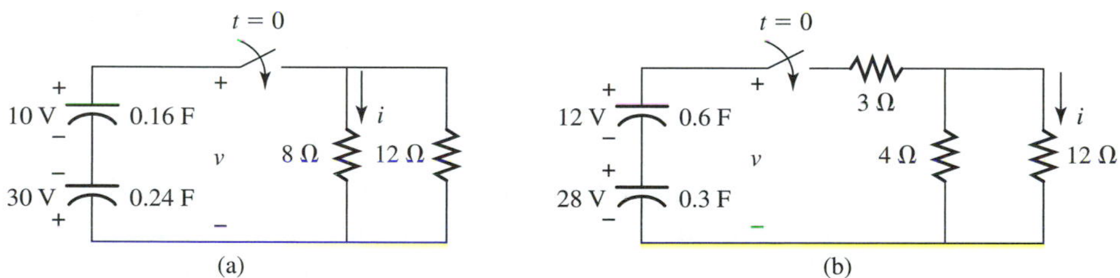

3. For each circuit in Figure 13.61 determine v and i for $t \geq 0$ after the switch is closed at $t = 0$.

Figure 13.61

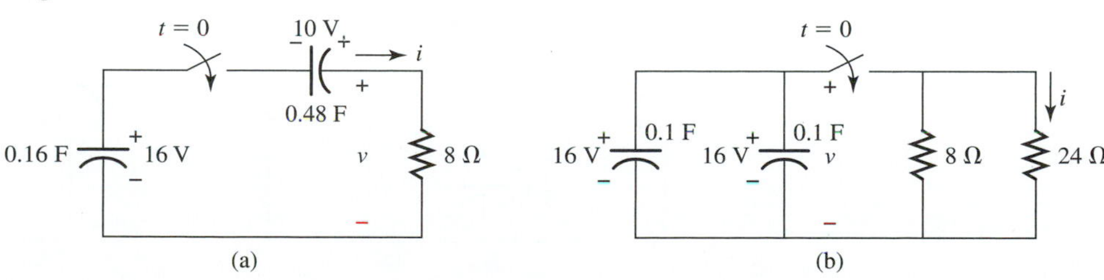

4. Determine the time constant (τ) for each circuit in Problem 1.

5. Determine i_1 and i_2 for the circuit in Figure 13.62 if the switch is in the closed position for a long period of time.

Figure 13.62

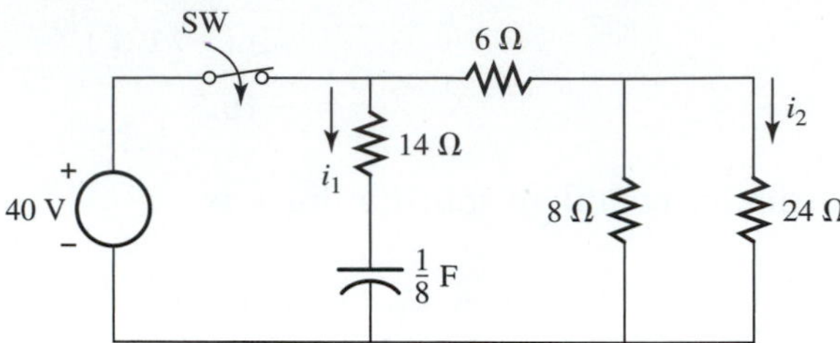

6. Determine i_1 and i_2 for the circuit in Figure 13.63 if the switch is in the closed position for a long period of time.

Figure 13.63

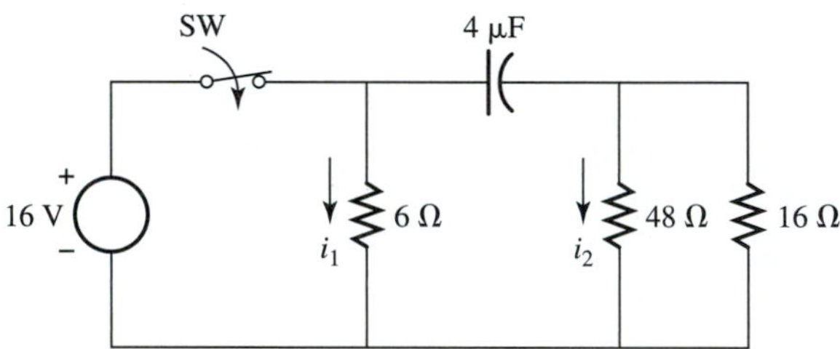

7. Determine i_1 and i_2 for the circuit in Figure 13.64 if the switch is in the closed position for a long period of time.

Figure 13.64

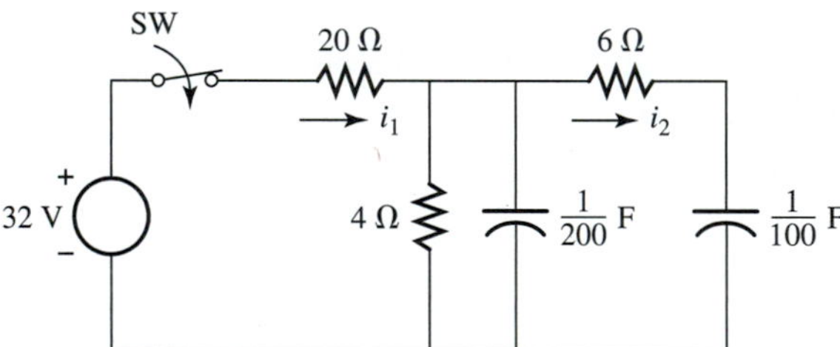

8. For each circuit in Figure 13.65 determine v, i_1, and i_2 for $t \geq 0$. The switch remains closed for a long period of time and is opened at $t = 0$.

Figure 13.65

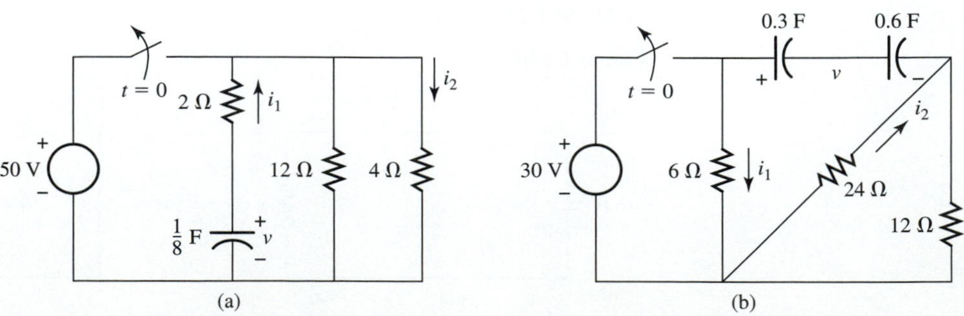

9. For each circuit in Figure 13.66 determine v, i_1, and i_2 for $t \geq 0$. The switch remains closed for a long period of time and is opened at $t = 0$.

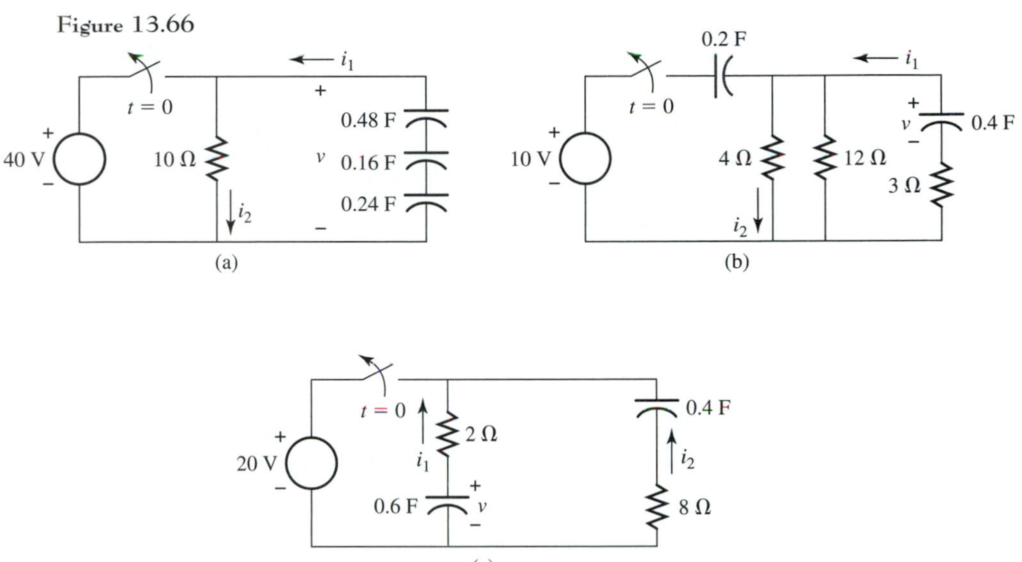

Figure 13.66

10. For each circuit in Figure 13.67 determine v and i for $t \geq 0$ after the switch is closed at $t = 0$.

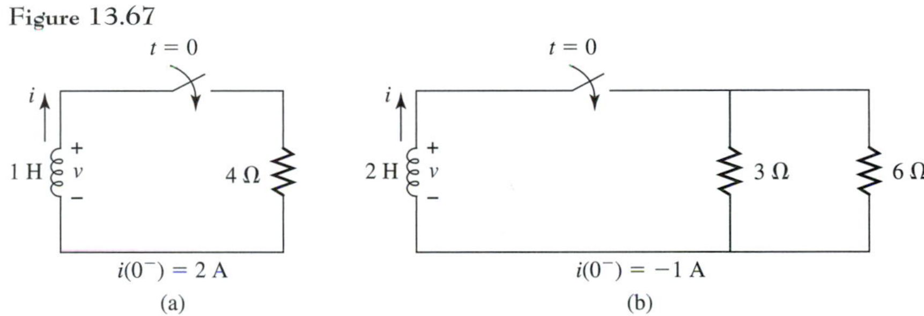

Figure 13.67

11. For each circuit in Figure 13.68 determine v and i for $t \geq 0$ after the switch is closed at $t = 0$.

Figure 13.68

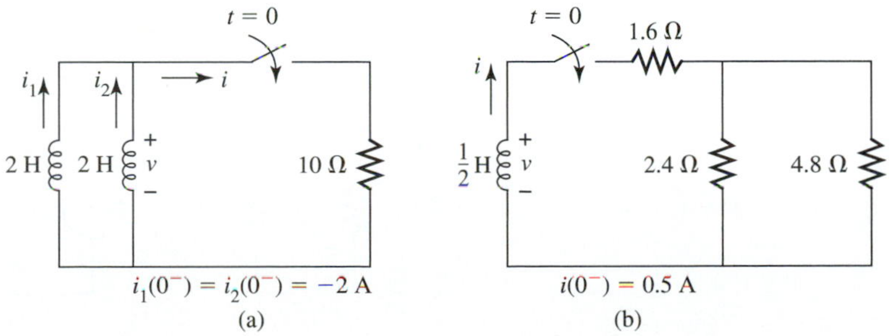

t = 0
2 H 2 H
10 Ω
$i_1(0^-) = i_2(0^-) = -2$ A
(a)

t = 0
1.6 Ω
$\frac{1}{2}$ H
2.4 Ω
4.8 Ω
$i(0^-) = 0.5$ A
(b)

12. For each circuit in Figure 13.69 determine v and i and $t \geq 0$ after the switch is closed at $t = 0$.

Figure 13.69

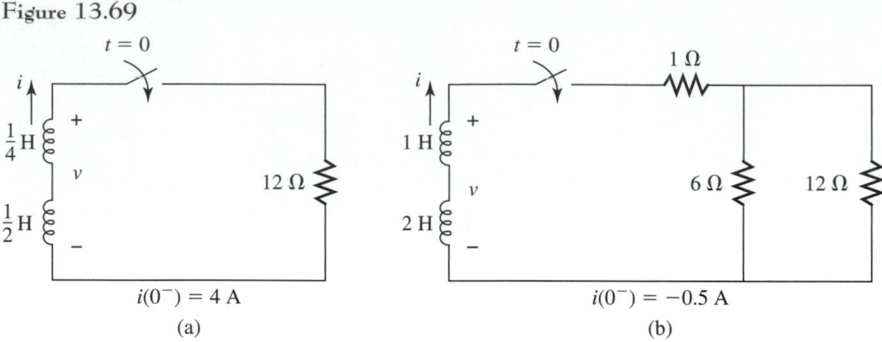

13. Determine the time constant (τ) for each circuit in Problem 10.

14. Determine i_1 and i_2 for each circuit in Figure 13.70 if the switch is in the closed position for a long period of time.

Figure 13.70

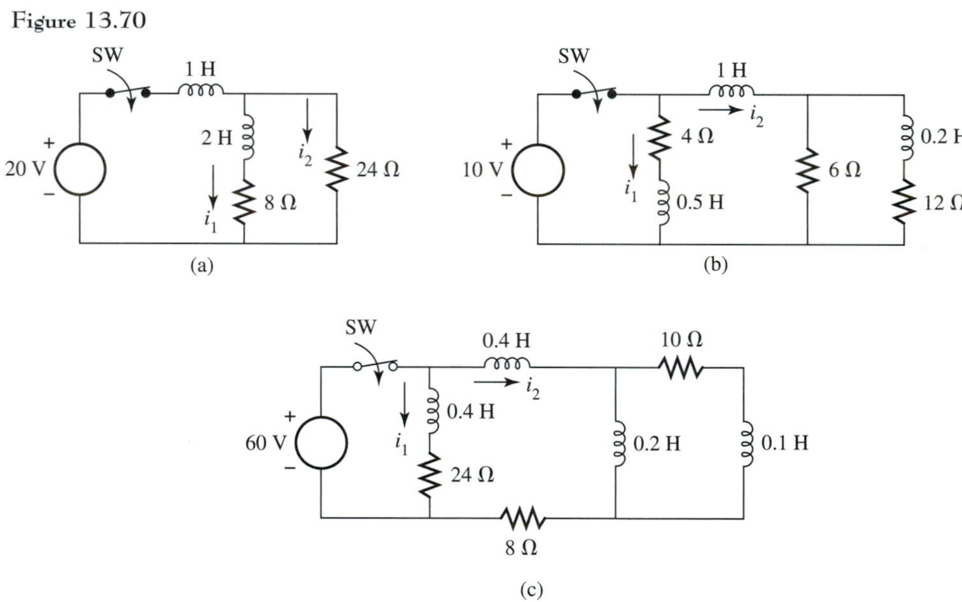

15. For each circuit in Figure 13.71 determine v, i_1, and i_2 for $t \geq 0$. The switch remains closed for a long period of time and is opened at $t = 0$.

Figure 13.71

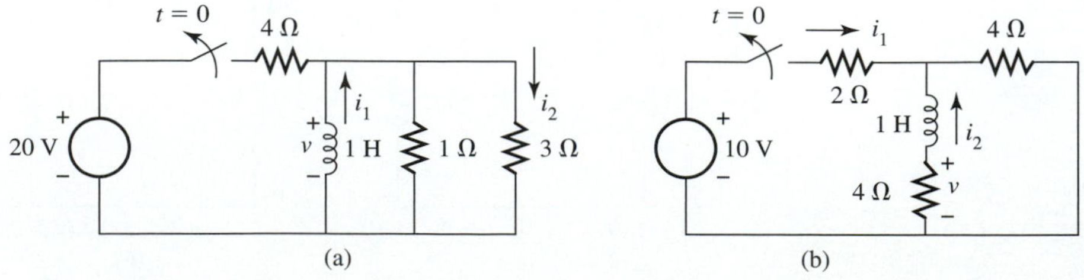

(a)

(b)

16. For each circuit in Figure 13.72 determine v, i_1, and i_2 for $t \geq 0$. The switch remains closed for a long period of time and is opened at $t = 0$.

Figure 13.72

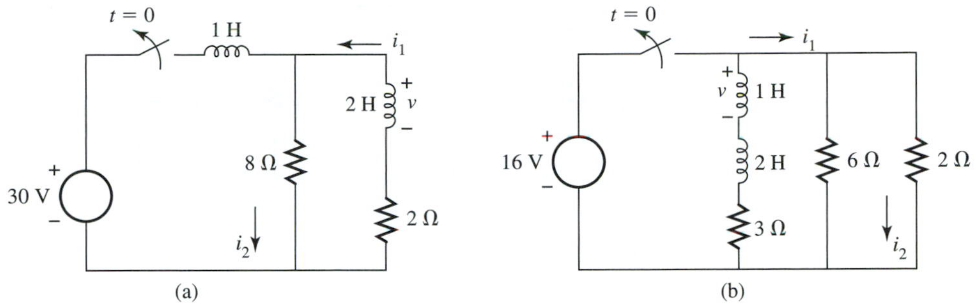

(a) (b)

17. For each circuit in Figure 13.73 determine v, i_1, and i_2 for $t \geq 0$. The switch remains closed for a long period of time and is opened at $t = 0$.

Figure 13.73

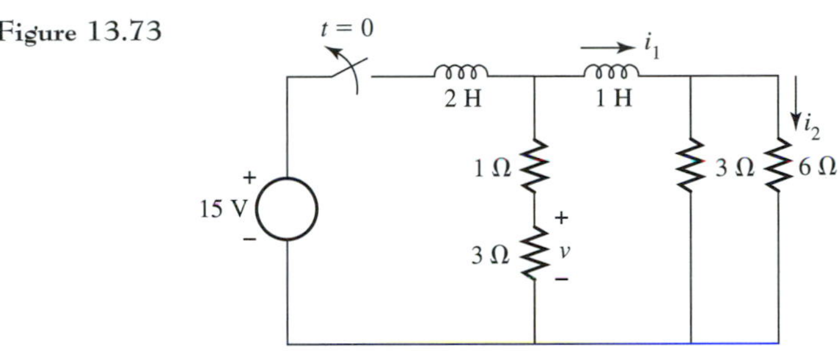

18. Determine v and i for $t \geq 0$ for each circuit in Figure 13.74. The switch is closed at $t = 0$.

Figure 13.74

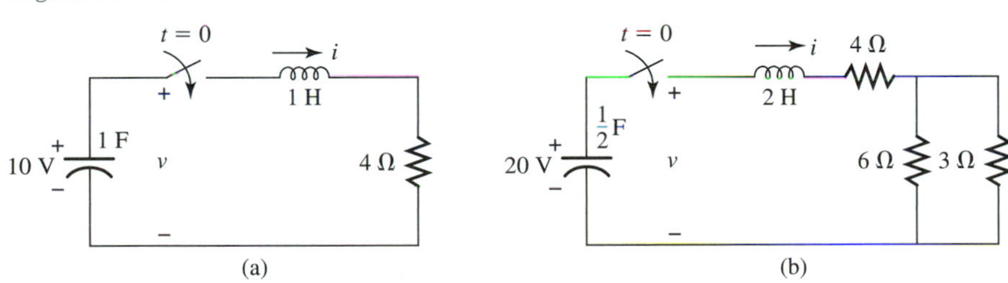

19. Determine v and i for $t \geq 0$ for each circuit in Figure 13.75. The switch is closed at $t = 0$.

Figure 13.75

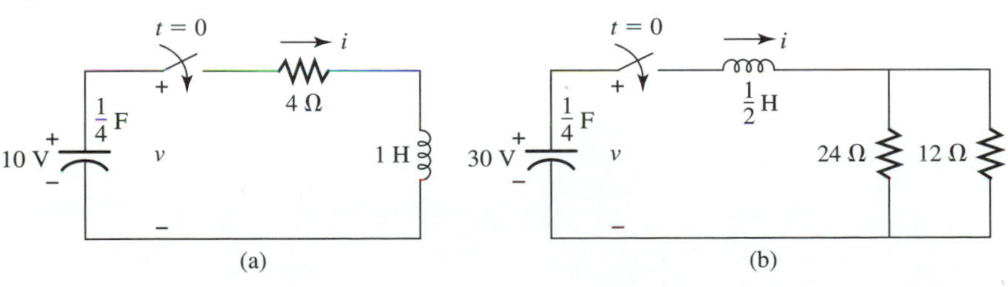

20. Determine v and i for $t \geq 0$ for each circuit in Figure 13.76. The switch is closed at $t = 0$.

Figure 13.76

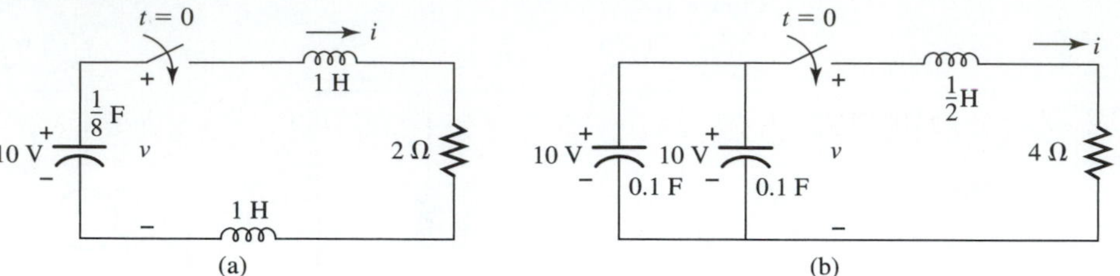

21. Determine v, i, i_L, and i_R for $t \geq 0$ for each of the circuits in Figure 13.77. The switch is closed at $t = 0$.

Figure 13.77

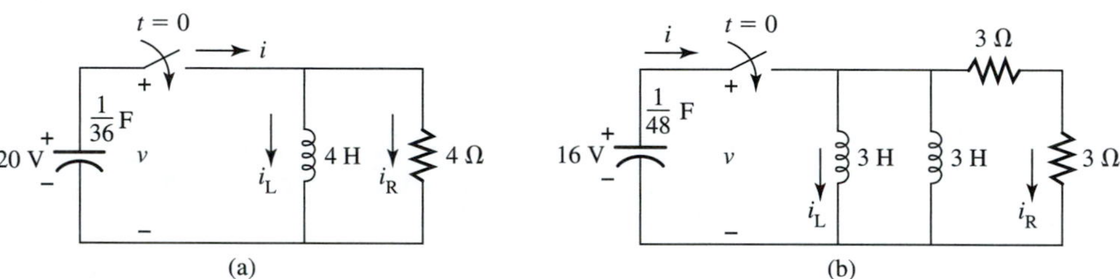

22. Determine v, i, i_L, and i_R for $t \geq 0$ for each of the circuits in Figure 13.78. The switch is closed at $t = 0$.

Figure 13.78

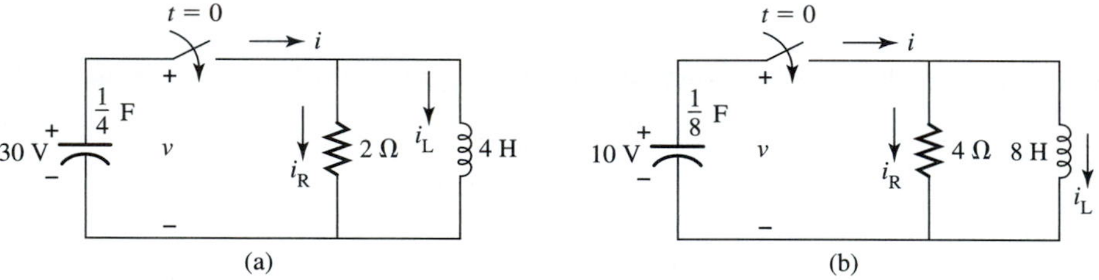

23. Determine v, i, i_L, and i_R for $t \geq 0$ for each of the circuits in Figure 13.79. The switch is closed at $t = 0$.

Figure 13.79

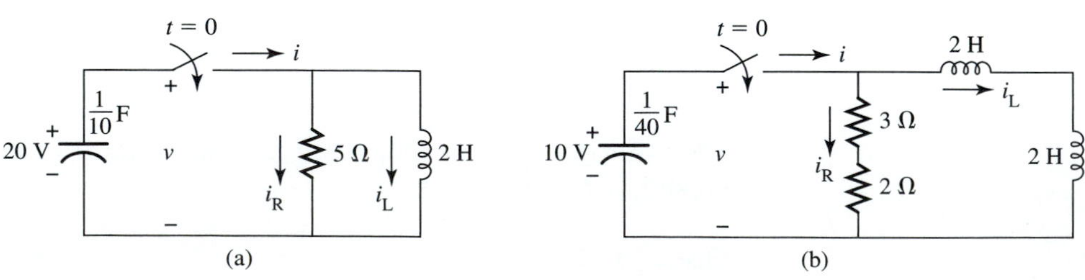

CHAPTER

14

Source-Driven Circuits

After completing this chapter the student should be able to

* determine currents and voltages in *RL*, *RC*, and *RLC* circuits driven by a constant voltage source
* analyze circuits driven by general driving sources

A circuit containing inductors and/or capacitors that is driven by an independent voltage or current source is described by a nonhomogeneous differential equation. Because the describing differential equation is nonhomogeneous, the complete solution consists of a transient component and a steady-state component. The steady-state component of the solution is related to the driving source, and the transient component is determined by the passive circuit elements.

Unlike the source-free circuits discussed in the last chapter, source-driven circuits may or may not contain initially stored energy in the inductors and capacitors. Source-free circuits require initially stored energy to energize the circuit.

The procedure for analyzing source-driven circuits is similar to the procedure for analyzing source-free circuits. The former involves solving nonhomogeneous differential equations, and the latter involves solving homogeneous differential equations.

14.1 *RL* CIRCUIT DRIVEN BY A CONSTANT-VOLTAGE SOURCE

Consider the source-driven RL circuit in Figure 14.1, where v is a constant-voltage source and $i(0^-)$ is the current flowing through the inductor prior to closing the switch.

Figure 14.1 *RL* circuit.

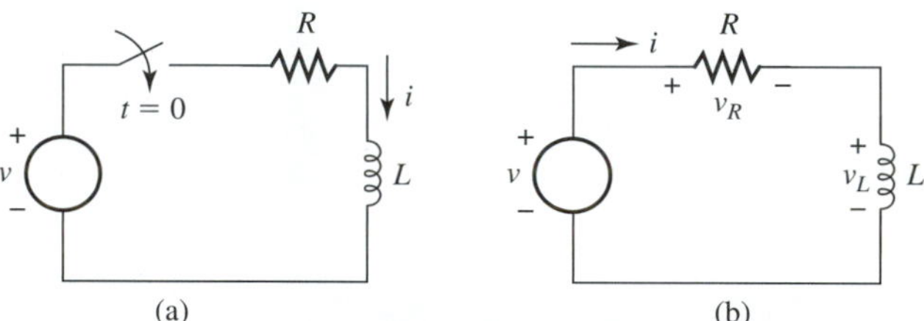

The circuit is energized by the constant-voltage source and the initial current flowing through the inductor. The switch is closed at $t = 0$, and from that time forward the circuit appears as shown in Figure 14.1(b). Time t is the independent variable, and v_R, v_L, and i are the dependent variables in the expressions that we shall develop for the voltages and current in the circuit. The circuit furnishing the initial current $i(0^-)$ through the inductor is not shown and is switched out of the circuit when the switch is closed.

From Ohm's law and Equation (11.3)

$$v_R = Ri$$

$$v_L = L\frac{di}{dt}$$

and employing Kirchhoff's voltage law for the circuit in Figure 14.1(b) results in

$$v_L + v_R = v$$

$$L\frac{di}{dt} + Ri = v$$

$$\frac{di}{dt} + \left(\frac{R}{L}\right)i = \frac{v}{L}$$

The differential equation is first order and nonhomogeneous. The independent variable is t, the dependent variable is i, and the forcing function is v/L. Because the differential equation is first order, one initial condition is required to determine a particular complete solution. The initial condition is $i(0)$ {i evaluated at $t = 0$}, and because the current flowing through an inductor cannot change instantaneously, $i(0) = i(0^-)$. We shall use the classical method presented in Chapter 12 to solve the differential equation.

Using the method for solving first-order nonhomogeneous differential equations presented in Chapter 12, the transient solution is obtained by setting the forcing function to zero and solving the resulting homogeneous differential equation, which is

$$\frac{di}{dt} + \left(\frac{R}{L}\right)i = 0$$

The characteristic equation is

$$p + \frac{R}{L} = 0$$

$$p = -\frac{R}{L}$$

and the general transient solution is

$$i_T = Ke^{-(R/L)t}$$

where K is an arbitrary constant.

Using entry 1 of Table 12.1 and following the procedure for determining steady-state solutions in Chapter 12, the steady-state solution is obtained by assuming

$$i_{ss} = A$$

where A is a constant to be determined. We assume $i_{ss} = A$, because the forcing function v/L is a constant. Substituting the assumed steady-state solution into the nonhomogeneous differential equation results in

$$\frac{d\{i_{ss}\}}{dt} + \left(\frac{R}{L}\right)i_{ss} = \frac{v}{L}$$

$$\frac{d\{A\}}{dt} + \left(\frac{R}{L}\right)A = \frac{v}{L}$$

$$0 + \left(\frac{R}{L}\right)A = \frac{v}{L}$$

$$A = \left(\frac{v}{L}\right)\left(\frac{L}{R}\right)$$

$$= \frac{v}{R}$$

and the steady-state solution is

$$i_{ss} = \frac{v}{R}$$

The general complete solution is

$$\begin{aligned} i &= i_T + i_{ss} \\ &= Ke^{-(R/L)t} + \frac{V}{R} \end{aligned}$$

The particular complete solution is determined by using the initial condition to evaluate K. This results in

$$\begin{aligned} i(0) &= Ke^{-(R/L)(0)} + \frac{V}{R} \\ &= K + \frac{V}{R} \\ K &= i(0) - \frac{V}{R} \end{aligned}$$

Substituting $(i(0) - v/R)$ for K in the general solution results in the particular complete solution

$$i = \left(i(0) - \frac{V}{R}\right)e^{-(R/L)t} + \frac{V}{R} \tag{14.1}$$

where $i(0) = i(0^-)$.

Alternatively, because the only independent source contained in the circuit in Figure 14.1 is a constant voltage source, the inductor can be replaced by a short, as shown in Figure 14.2, and the resulting circuit can be solved for

Figure 14.2 Steady-state circuit.

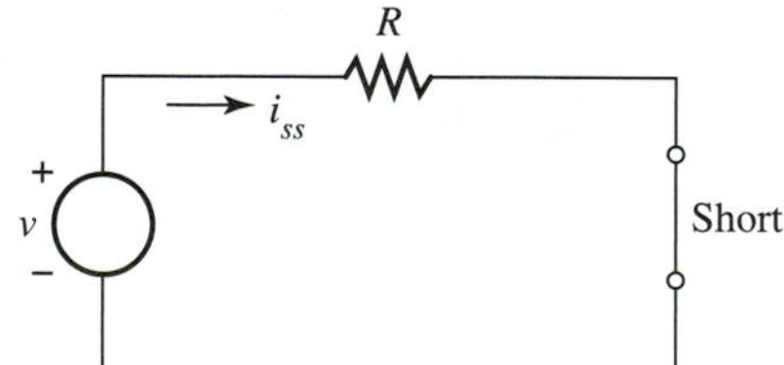

the steady-state current i. This procedure is presented in Chapter 13. Applying Ohm's law to the circuit

$$i_{ss} = \frac{V}{R}$$

as previously determined.

For the circuit in Figure 14.1(b) the current i is given by Equation (14.1), and because $v_R = Ri$, from Ohm's law

$$v_R = \{Ri(0) - v\}e^{-(R/L)t} + v \tag{14.2}$$

Because $v_L = L\dfrac{di}{dt}$

$$v_L = L\frac{d\{(i(0) - v/R)e^{-(R/L)t} + (v/R)\}}{dt}$$

$$= \{v - Ri(0)\}e^{-(R/L)t} \tag{14.3}$$

where $i(0) = i(0^-)$.

If the initial current through the inductor in Figure 14.1 is equal to zero, Equations (14.1), (14.2), and (14.3) become

$$i = -\left(\frac{v}{R}\right)e^{-(R/L)t} + \frac{v}{R} \tag{14.4}$$

$$v_R = -ve^{-(R/L)t} + v \tag{14.5}$$

$$v_L = ve^{-(R/L)t} \tag{14.6}$$

Consider the following example.

EXAMPLE 14.1 Determine i, v_R, and v_L after the switch is closed for the circuit in Figure 14.3 if $i(0^-) = 0$. ■

Figure 14.3

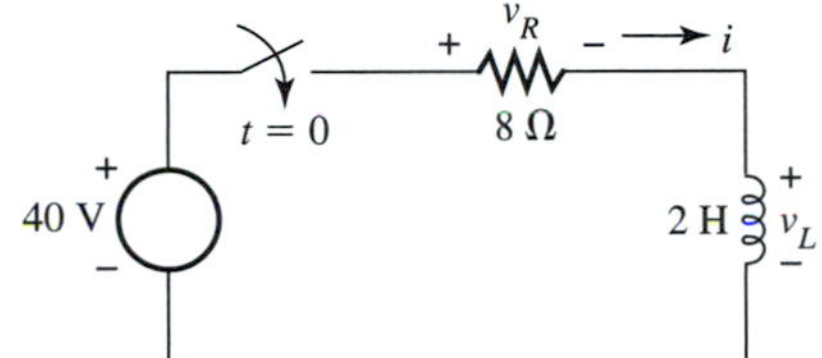

SOLUTION From Equation (14.1)

$$i = \left\{i(0) - \frac{v}{R}\right\}e^{-(R/L)t} + \frac{v}{R}$$

$$= \left(\frac{40}{8}\right)\{1 - e^{-(8/2)t}\}$$

$$= 5 - 5e^{-4t}\text{ A}$$

From Equation (14.2)

$$v_R = \{Ri(0) - v\}e^{-(R/L)t+v}$$

$$= 40\{1 - e^{-(8/2)t}\}$$

$$= 40 - 40e^{-4t}\text{ V}$$

From Equation (14.3)

$$v_L = \{v - Ri(0)\}e^{-(R/L)t}$$

$$= 40e^{-(8/2)t}$$

$$= 40e^{-4t}\text{ V}$$

14.2 *RC* CIRCUIT DRIVEN BY A CONSTANT-VOLTAGE SOURCE

Consider the source driven *RC* circuit in Figure 14.4, where v is a constant voltage and the voltage across the capacitor prior to closing the switch is $v_C(0^-)$.

Figure 14.4 *RC* circuit.

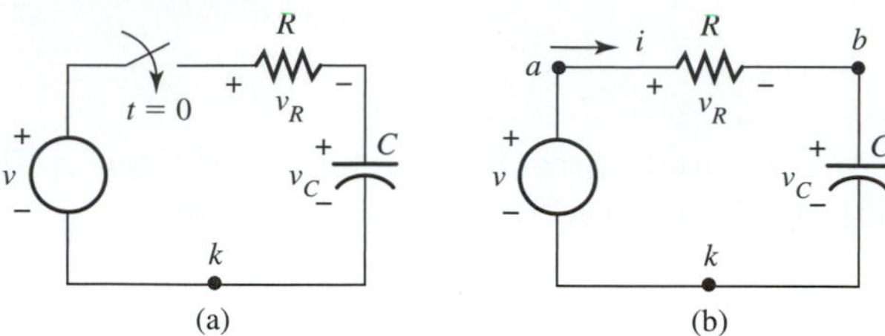

The circuit is energized by the constant voltage source and the initial voltage across the capacitor. The switch is closed at $t = 0$, and from that time forward the circuit appears as shown in Figure 14.4(b). Time t is the independent variable and v_R, v_C, and i are the dependent variables in the expression that we shall develop for the voltages and current in the circuit.

For the circuit in Figure 14.4(b), voltages v_a and v_b are written with respect to reference point k and v_C is the voltage across the capacitor. From Equation (6.8)

$$i = \frac{v_a - v_b}{R}$$

and from Equation (10.2)

$$i = C\frac{dv_C}{dt}$$

Because $v_b = v_C$ and $v_a = v$,

$$i = \frac{v - v_C}{R}$$

Combining the above equations results in

$$C\frac{dv_C}{dt} = \frac{v}{R} - \frac{v_C}{R}$$

$$\frac{dv_C}{dt} + \left(\frac{1}{RC}\right)v_C = \left(\frac{1}{RC}\right)v$$

The differential equation describing the voltage across the capacitor for the circuit in Figure 14.4(b) is first order and nonhomogeneous. The independent variable is t, the dependent variable is v_C, and the forcing function is v/RC. Because the differential equation is first order, one initial condition is required to determine a particular complete solution; because the voltage across a capacitor cannot change instantaneously, the initial condition is

$$v_C(0) = v_C(0^-)$$

We use the classical method in chapter 12 to solve the first-order nonhomogeneous differential equation

$$\frac{dv_C}{dt} + \left(\frac{1}{RC}\right)v_C = \frac{v}{RC}$$

with initial condition $v_C(0) = v_C(0^-)$. The transient solution is obtained by setting the forcing function equal to zero and solving the resulting homogeneous differential equation, which is

$$\frac{dv_C}{dt} + \left(\frac{1}{RC}\right)v_C = 0$$

The characteristic equation is

$$p + \left(\frac{1}{RC}\right) = 0$$

$$p = -\left(\frac{1}{RC}\right)$$

and the transient solution is

$$v_{C_T} = Ke^{-(1/RC)t}$$

where K is an arbitrary constant.

Using entry 1 of Table 12.1 and following the procedure for determining steady-state solutions in Chapter 12, the steady-state solution is obtained by assuming

$$v_{C_{ss}} = A$$

where A is a constant to be determined. The steady-state solution for v_C is assumed constant because the forcing function (v/RC) is a constant. Substituting the assumed steady-state solution into the nonhomogeneous differential equation results in

$$\frac{d\{v_{C_{ss}}\}}{dt} + \left(\frac{1}{RC}\right)v = \frac{v}{RC}$$

$$\frac{d(A)}{dt} + \left(\frac{1}{RC}\right)A = \frac{v}{RC}$$

$$0 + \left(\frac{1}{RC}\right)A = \frac{v}{RC}$$

$$A = \left(\frac{v}{RC}\right)(RC)$$

$$= v$$

and the steady-state solution is

$$v_{C_{ss}} = v$$

The general complete solution for v_C is

$$v_C = v_{C_T} + v_{C_{ss}}$$
$$= Ke^{-(1/RC)t} + v$$

The particular complete solution is determined by using the initial condition to evaluate K. This results in

$$v_C(0) = Ke^{-(1/RC)(0)} + v$$
$$K = v_C(0) - v$$

Substituting $(v_C(0) - v)$ for K in the general complete solution results in the following particular complete solution for v_C for the circuit in Figure 14.4(b):

$$v_C = \{v_C(0) - v\}e^{-(1/RC)t} + v \tag{14.7}$$

where $v_C(0) = v_C(0^-)$.

Alternatively, because the only independent source contained in the circuit in Figure 14.4 is a constant voltage source, the capacitor can be replaced by an open, as shown in Figure 14.5, and the resulting circuit can be solved for the steady-state voltage across the open ($v_{C_{ss}}$). This method is presented in Chapter 13.

Figure 14.5 Steady-state circuit.

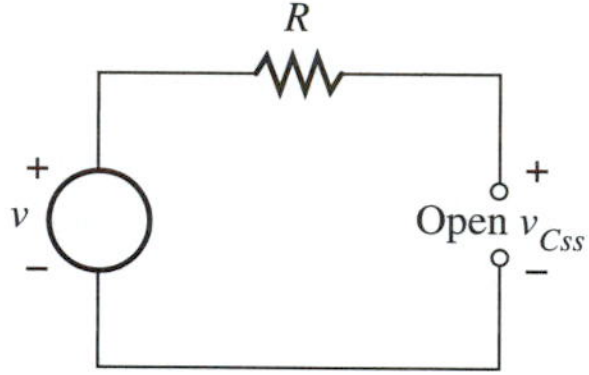

Because no current flows in this circuit,

$$v_{C_{ss}} = v$$

as previously determined.

Because $i = C\dfrac{dv_C}{dt}$,

$$i = C\frac{d\{(v_C(0) - v)\,e^{-(1/RC)t} + v\}}{dt}$$
$$= \left\{\frac{v - v_C(0)}{R}\right\}e^{-(1/RC)t} \tag{14.8}$$

and because $v_R = Ri$,

$$v_R = (R)\left\{\frac{v - v_C(0)}{R}\right\}e^{-(1/RC)t}$$
$$= \{v - v_C(0)\}e^{-(1/RC)t} \tag{14.9}$$

where $v_C(0) = v_C(0^-)$.

If the initial voltage $v_C(0)$ across the capacitor in Figure 14.4 is equal to zero, Equations (14.7), (14.8), and (14.9) become

$$v_C = v\{1 - e^{-(1/RC)t}\} \tag{14.10}$$

$$i = \left(\frac{v}{R}\right)e^{-(1/RC)t} \tag{14.11}$$

$$v_R = ve^{-(1/RC)t} \tag{14.12}$$

respectively.

Consider the following example.

EXAMPLE 14.2 Determine i, v_R, and v_C after the switch is closed for the circuit in Figure 14.6. The voltage v_C across the capacitor is 20 V prior to closing the switch. ■

Figure 14.6

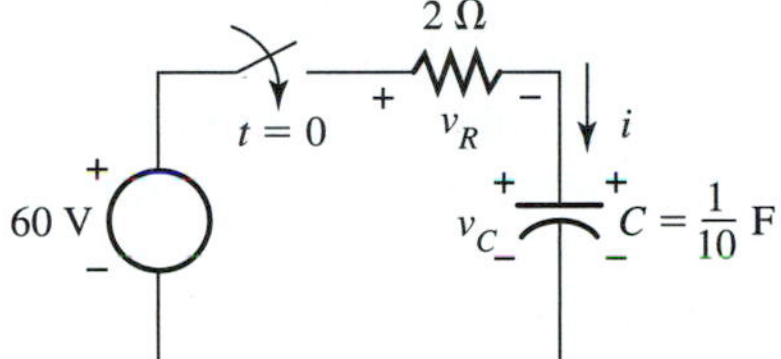

SOLUTION From Equation (14.7)

$$\begin{aligned} v_C &= \{v_C(0) - v\}e^{-(1/RC)t} + v \\ &= \{20 - 60\}e^{-\{(1/(2)(1/10)\}t} + 60 \\ &= -40e^{-5t} + 60 \text{ V} \end{aligned}$$

From Equation (14.8)

$$\begin{aligned} i &= \left\{\frac{v - v_C(0)}{R}\right\}e^{-(1/RC)t} \\ &= \left\{\frac{60 - 20}{2}\right\}e^{-5t} \\ &= 20e^{-5t} \text{ A} \end{aligned}$$

From Equation (14.9)

$$\begin{aligned} v_R &= \{v - v_C(0)\}e^{-(1/RC)t} \\ &= (60 - 20)e^{-5t} \\ &= 40e^{-5t} \text{ V} \end{aligned}$$

14.3 *RLC* CIRCUIT DRIVEN BY A CONSTANT-VOLTAGE SOURCE

Consider the source-driven *RLC* circuit in Figure 14.7, where v is a constant voltage source, $i(0^-)$ is the current flowing through the inductor an instant prior to closing the switch, and $v_C(0^-)$ is the voltage across the capacitor an instant prior to closing the switch.

Figure 14.7 *RLC* series circuit.

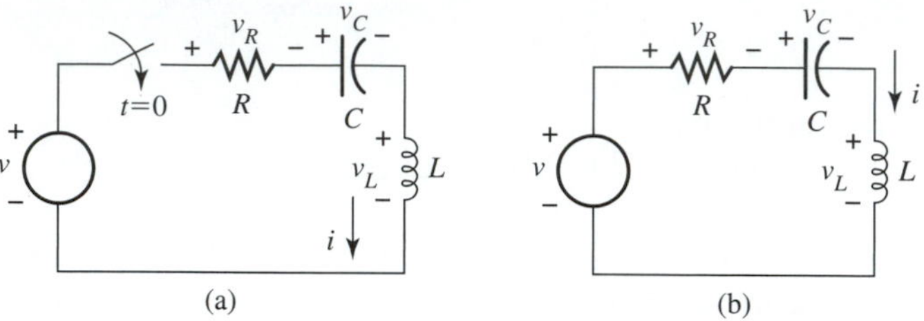

The circuit is energized by the constant-voltage source, the initial voltage across the capacitor, and the initial current flowing through the inductor. The switch is closed at $t = 0$, and from that time forward the circuit appears as shown in Figure 14.7(b). Time t is the independent variable and v_R, v_C, v_L, and i are the dependent variables in the equations that we shall develop for the voltages and current in the circuit. The circuit furnishing the initial current $i(0^-)$ through the inductor is not shown and is switched out of the circuit in Figure 14.7 when the switch is closed.

From Equation (10.2) the current for the circuit in Figure 14.7(b) can be expressed as

$$i = C\frac{dv_C}{dt}$$

and it follows that

$$\frac{dv_C}{dt} = \left(\frac{1}{C}\right)i$$

From this equation and the definition for an antiderivative given in Chapter 9, v_C is the antiderivative of $(1/C)i$ and can be written as

$$v_C = \int\left(\frac{1}{C}\right)i\,dt$$

From rule I6 in Chapter 9

$$v_C = \left(\frac{1}{C}\right)\int i\,dt$$

From Ohm's law and Equation (11.3),

$$v_R = Ri$$

$$v_L = L\frac{di}{dt}$$

Employing Kirchhoff's voltage law for the circuit in Figure 14.7(b) results in

$$v - v_R - v_C - v_L = 0$$

$$v - Ri - \left(\frac{1}{C}\right)\int i\,dt - L\frac{di}{dt} = 0$$

$$L\frac{di}{dt} + Ri + \frac{1}{C}\int i\,dt = v$$

Because the derivative of the indefinite integral is equal to the integrand of the integral, as demonstrated in Chapter 9, differentiation of both sides of the preceding equation rids the equation of the integral. Recalling also that the derivative of a sum equals the sum of the derivatives results in

$$\frac{d}{dt}\left\{L\frac{di}{dt} + Ri + \frac{1}{C}\int i\,dt\right\} = \frac{dv}{dt}$$

$$\frac{d\left\{L\frac{di}{dt}\right\}}{dt} + \frac{d\{Ri\}}{dt} + \frac{d\left\{\frac{1}{C}\int i\,dt\right\}}{dt} = \frac{d(v)}{dt}$$

Using rule D6 in Chapter 9 and recognizing that v is a constant results in

$$L\frac{d^2i}{dt^2} + R\frac{di}{dt} + \left(\frac{1}{C}\right)i = 0$$

$$\frac{d^2i}{dt} + \left(\frac{R}{L}\right)\frac{di}{dt} + \left(\frac{1}{LC}\right)i = 0$$

The preceding equation is a second-order linear homogeneous differential equation with dependent variable i and independent variable t.

Because the second-order differential equation describing the circuit,

$$\frac{d^2i}{dt} + \left(\frac{R}{L}\right)\frac{di}{dt} + \left(\frac{1}{LC}\right)i = 0$$

is identical to the equation describing the *RLC* source-free series circuit in Chapter 13, one may question the effect of the driving voltage source v. The driving voltage source v appears in the initial conditions for i and affects the solution to the differential equation in that manner.

Because the differential equation is second order, two initial conditions are required to determine a particular complete solution. The required initial conditions are $i(0)$, i evaluated at $t = 0$, and $di(0)/dt$, the derivative or i with respect to t, evaluated at $t = 0$.

The first initial condition is

$$i(0) = i(0^-)$$

because the current through an inductor cannot change instantaneously.

The second initial condition, $di(0)/dt$, can be determined by applying

Figure 14.8 Initial conditions.

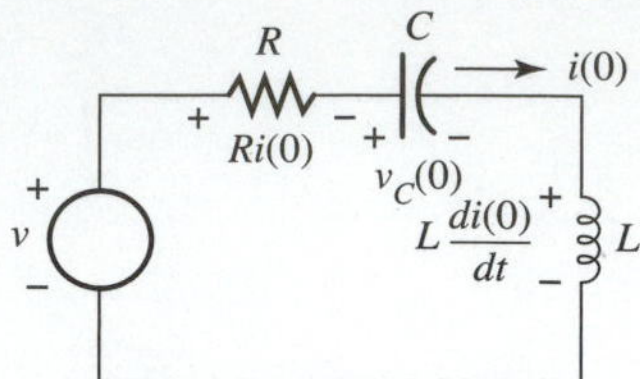

Kirchhoff's voltage law to the representation of the circuit at $t = 0$, as shown in Figure 14.8. Employing Kirchhoff's voltage law results in

$$v - Ri(0) - v_C(0) - L\frac{di(0)}{dt} = 0$$

and

$$\frac{di(0)}{dt} = \frac{v - Ri(0) - v_C(0)}{L}$$

Because neither the current flowing through an inductor nor the voltage across a capacitor can change instantaneously,

$$i(0) = i(0^-)$$

and

$$v_C(0) = v_C(0^-)$$

in the equation for the second initial condition.

We shall use the classical method presented in Chapter 12 to solve the second-order homogeneous differential equation

$$\frac{d^2i}{dt^2} + \left(\frac{R}{L}\right)\frac{di}{dt} + \left(\frac{1}{LC}\right)i = 0$$

with initial conditions

$$i(0) = i(0^-)$$

and

$$\frac{di(0)}{dt} = \frac{v - Ri(0) - v_C(0)}{L}$$

Using the procedure for solving second-order homogeneous differential equations presented in Chapter 12, the characteristic equation is

$$p^2 + \left(\frac{R}{L}\right)p + \frac{1}{LC} = 0$$

Using the quadratic equation to determine the roots of the characteristic equation results in

$$p_{1,2} = \frac{-(R/L) \pm \sqrt{(R/L)^2 - (4/LC)}}{2}$$

$$= -\left(\frac{R}{2L}\right) \pm \sqrt{\left(\frac{R}{2L}\right)^2 - \frac{1}{LC}}$$

The roots p_1 and p_2 can be real and unequal, real and equal, or complex conjugates. The nature of the roots is determined by the quantity under the radical sign.

Case 1 If $(R/2L)^2 > (1/LC)$, the roots are real and unequal and are

$$p_1 = -\left(\frac{R}{2L}\right) + \sqrt{\left(\frac{R}{2L}\right)^2 - \frac{1}{LC}}$$

$$p_2 = -\left(\frac{R}{2L}\right) - \sqrt{\left(\frac{R}{2L}\right)^2 - \frac{1}{LC}}$$

From Chapter 12, the general complete solution to the differential equation is

$$i = K_1 e^{p_1 t} + K_2 e^{p_2 t} \tag{14.13}$$

where K_1 and K_2 are arbitrary constants.

Case 2 If $(R/2L)^2 = (1/LC)$, the roots are real and equal and are

$$p_1 = -\left(\frac{R}{2L}\right)$$

$$p_2 = -\left(\frac{R}{2L}\right)$$

From Chapter 12, the general complete solution to the differential equation is

$$i = K_1 e^{pt} + K_2 t e^{pt} \tag{14.14}$$

where $p = -(R/2L)$ and K_1 and K_2 are arbitrary constants.

Case 3 If $(1/LC) > (R/2L)^2$, the roots are complex conjugates and are

$$p_1 = -\left(\frac{R}{2L}\right) + j\sqrt{\frac{1}{LC} - \left(\frac{R}{2L}\right)^2}$$

$$p_2 = -\left(\frac{R}{2L}\right) - j\sqrt{\frac{1}{LC} - \left(\frac{R}{2L}\right)^2}$$

In accordance with the procedure presented in Chapter 12,

$$\left.\begin{aligned} \alpha &= -\left(\frac{R}{2L}\right) \\ \beta &= \sqrt{\frac{1}{LC} - \left(\frac{R}{2L}\right)^2} \end{aligned}\right\} \tag{14.15}$$

and the general complete solution to the differential equation is

$$i = K_1 e^{\alpha t} \sin(\beta t + \theta) \tag{14.16}$$

where K_1 and θ are arbitrary constants.

In each of these three possible cases, the initial conditions $i(0) = i(0^-)$ and $di(0)/dt = \{v - Ri(0) - v_C(0)\}/L$, as previously determined, are used to evaluate the arbitrary constants and determine a particular complete solution. An example of each of the three cases is given next.

EXAMPLE 14.3 Determine i for the circuit in Figure 14.9 for $t \geq 0$ if $i(0^-) = 0$ and $v_C(0^-) = -4$ V. ■

Figure 14.9

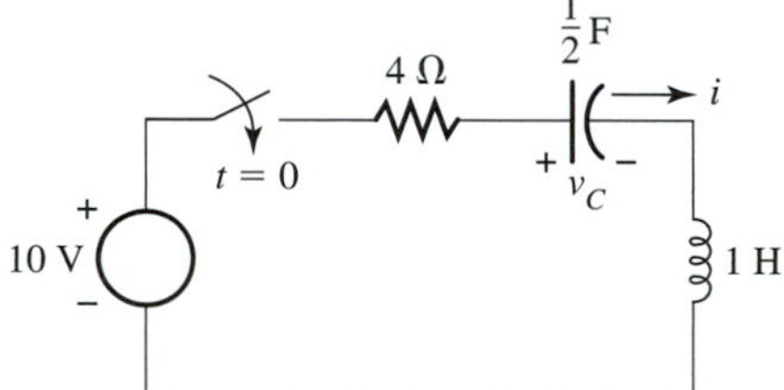

SOLUTION The circuit for $t \geq 0$ appears as shown in Figure 14.10.

Figure 14.10

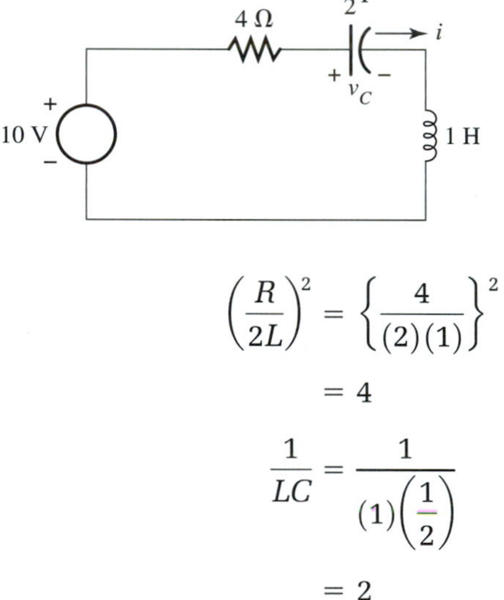

$$\left(\frac{R}{2L}\right)^2 = \left\{\frac{4}{(2)(1)}\right\}^2$$
$$= 4$$

$$\frac{1}{LC} = \frac{1}{(1)\left(\frac{1}{2}\right)}$$
$$= 2$$

Because $(R/2L)^2 > (1/LC)$, case 1 applies and

$$p_1 = -\left(\frac{R}{2L}\right) + \sqrt{\left(\frac{R}{2L}\right)^2 - \frac{1}{LC}}$$
$$= \frac{4}{(2)(1)} + \sqrt{\left\{\frac{4}{(2)(1)}\right\}^2 - \frac{1}{(1)\left(\frac{1}{2}\right)}}$$
$$= -2 + \sqrt{2}$$
$$= -0.59$$

$$\begin{aligned}
p_2 &= -\left(\frac{R}{2L}\right) - \sqrt{\left(\frac{R}{2L}\right)^2 - \frac{1}{LC}} \\
&= -\left\{\frac{4}{(2)(1)}\right\} - \sqrt{\left\{\frac{4}{(2)(1)}\right\}^2 - \frac{1}{(1)(\frac{1}{2})}} \\
&= -2 - \sqrt{2} \\
&= -3.41
\end{aligned}$$

From Equation (14.13) the general solution for i is

$$\begin{aligned}
i &= K_1e^{p_1t} + K_2e^{p_2t} \\
&= K_1e^{-0.59t} + K_2e^{-3.41t}
\end{aligned}$$

Because $i(0) = i(0^-) = 0$

$$0 = K_1 + K_2$$

Differentiating i with respect to t results in

$$\frac{di}{dt} = -0.59K_1e^{-0.59t} - 3.14K_2e^{-3.41t}$$

$$\frac{di(0)}{dt} = -0.59K_1 - 3.41K_2$$

Because $i(0) = i(0^-) = 0$ and $v_C(0) = v_C(0^-) = -4$ V,

$$\begin{aligned}
\frac{di(0)}{dt} &= \frac{v}{L} - \frac{Ri(0)}{L} - \frac{v_C(0)}{L} \\
&= \frac{10}{1} - \frac{(4)(0)}{1} - \left(-\frac{4}{1}\right) \\
&= 14
\end{aligned}$$

and

$$14 = -0.59K_1 - 3.41K_2$$

Solving the two algebraic equations

$$\begin{aligned}
K_1 + K_2 &= 0 \\
-0.59K_1 - 3.41K_2 &= 14
\end{aligned}$$

for K_1 and K_2 results in

$$\begin{aligned}
K_1 &= 4.98 \\
K_2 &= -4.98
\end{aligned}$$

and

$$i = 4.98e^{-0.59t} - 4.98e^{-3.41t}$$

EXAMPLE 14.4 Determine i for the circuit in Figure 14.11 for $t \geq 0$ if $i(0^-) = 0$ and $v_C(0^-) = -4$ V. ■

Figure 14.11

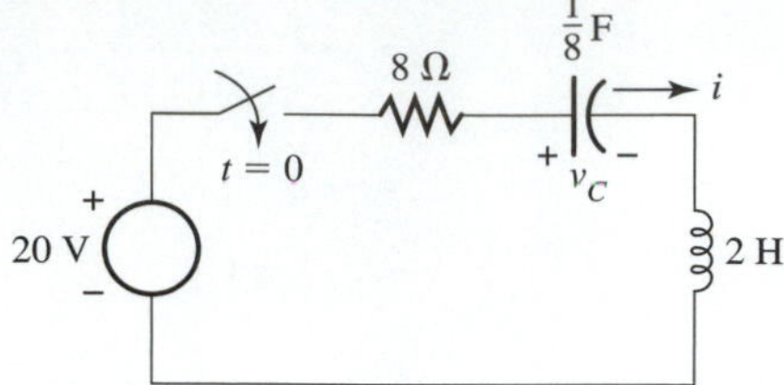

SOLUTION The circuit for $t \geq 0$ appears as shown in Figure 14.12.

Figure 14.12

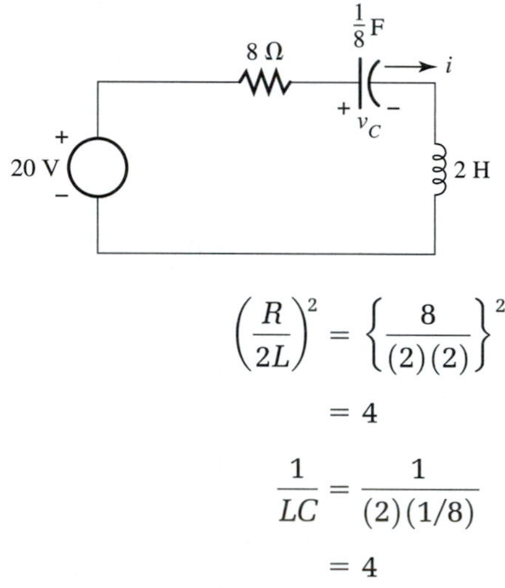

$$\left(\frac{R}{2L}\right)^2 = \left\{\frac{8}{(2)(2)}\right\}^2$$

$$= 4$$

$$\frac{1}{LC} = \frac{1}{(2)(1/8)}$$

$$= 4$$

Because $(R/2L)^2 = (1/LC)$, case 2 applies, and the two roots of the characteristic equation are equal.

$$p = p_1 = p_2$$

$$= -\left(\frac{R}{2L}\right) \pm \sqrt{\left(\frac{R}{2L}\right)^2 - \frac{1}{LC}}$$

$$= -\left\{\frac{8}{(2)(2)}\right\} \pm \sqrt{4 - 4}$$

$$= -2$$

From Equation (14.14) the general complete solution for i is

$$i = K_1 e^{pt} + K_2 t e^{pt}$$

$$= K_1 e^{-2t} + K_2 t e^{-2t}$$

Because $i(0) = i(0^-) = 0$,

$$0 = K_1 + (K_2)(0)$$

$$K_1 = 0$$

Differentiating i with respect to t results in

$$\frac{di}{dt} = -2K_1 e^{-2t} + K_2 e^{-2t} - 2K_2 t e^{-2t}$$

$$\frac{di(0)}{dt} = -2K_1 + K_2$$

Because $i(0) = i(0^-) = 0$ and $v_C(0) = v_C(0^-) = -4$ V,

$$\begin{aligned}\frac{di(0)}{dt} &= \frac{v}{L} - \frac{Ri(0)}{L} - \frac{v_C(0)}{L}\\ &= \left(\frac{20}{2}\right) - \left\{\frac{8(0)}{2}\right\} - \left(-\frac{4}{2}\right)\\ &= 12\end{aligned}$$

and

$$12 = -2K_1 + K_2$$

Because $K_1 = 0$

$$K_2 = 12$$

and

$$i = 12te^{-2t}$$

EXAMPLE 14.5 Determine i for the circuit in Figure 14.13 for $t \geq 0$ if $i(0^-) = 0$ and $v_C(0^-) = -10$ V. ■

Figure 14.13

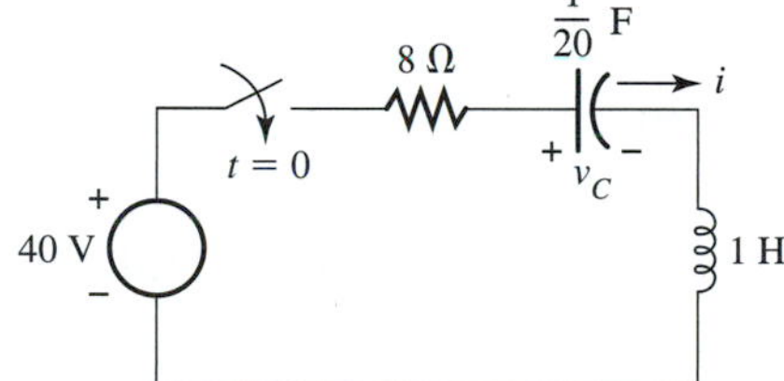

SOLUTION The circuit for $t \geq 0$ appears as shown in Figure 14.14.

Figure 14.14

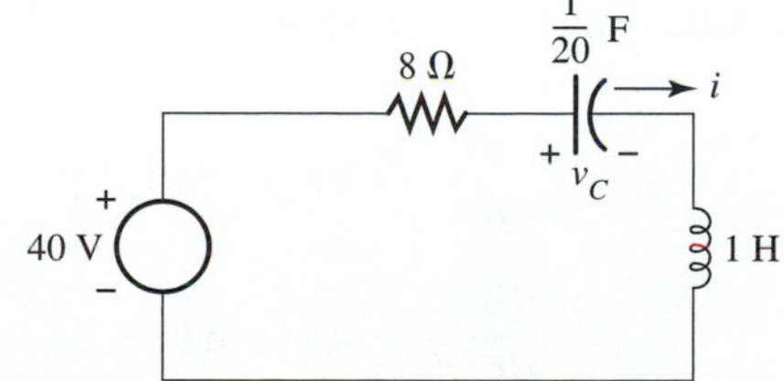

$$\left(\frac{R}{2L}\right)^2 = \left\{\frac{8}{(2)(1)}\right\}^2$$
$$= 16$$
$$\left(\frac{1}{LC}\right) = \frac{1}{(1)(1/20)}$$
$$= 20$$

Because $(1/LC) > (R/2L)^2$, case 3 applies and

$$p_1 = -\left(\frac{R}{2L}\right) + j\sqrt{\frac{1}{LC} - \left(\frac{R}{2L}\right)^2}$$
$$= -\left\{\frac{8}{(2)(1)}\right\} + j\sqrt{20 - 16}$$
$$= -4 + j2$$
$$p_2 = -\left(\frac{R}{2L}\right) - j\sqrt{\frac{1}{LC} - \left(\frac{R}{2L}\right)^2}$$
$$= -\left\{\frac{8}{(2)(1)}\right\} - j\sqrt{20 - 16}$$
$$= -4 - j2$$

From Equation (14.15)

$$\alpha = -\left(\frac{R}{2L}\right)$$
$$= -4$$
$$\beta = \sqrt{\frac{1}{LC} - \left(\frac{R}{2L}\right)^2}$$
$$= 2$$

From Equation (14.16) the general complete solution for i is

$$i = K_1 e^{\alpha t} \sin(\beta t + \theta)$$
$$= K_1 e^{-4t} \sin(2t + \theta)$$

Because $i(0) = i(0^-) = 0$

$$0 = K_1 \sin(\theta)$$

Differentiating i with respect to t results in

$$\frac{di}{dt} = -4K_1 e^{-4t} \sin(2t + \theta) + 2K_1 e^{-4t} \cos(2t + \theta)$$
$$\frac{di(0)}{dt} = -4K_1 \sin\theta + 2K_1 \cos\theta$$

where

$$K_1 = \frac{5}{2}$$

$$K_2 = -\frac{5}{2}$$

and

$$\mathbf{I} = \frac{\frac{5}{2}}{(s+1)} + \frac{-\frac{5}{2}}{(s+5)}$$

Taking the inverse transform of both sides of the equation results in

$$\mathcal{L}^{-1}\{\mathbf{I}\} = \mathcal{L}^{-1}\left\{\frac{\frac{5}{2}}{(s+1)} + \frac{-\frac{5}{2}}{(s+5)}\right\}$$

From rules L3(b) and L4(b) and entry 2 in Table 12.2 of Chapter 12,

$$\mathcal{L}^{-1}\{\mathbf{I}\} = \mathcal{L}^{-1}\left\{\frac{\frac{5}{2}}{(s+1)}\right\} + \mathcal{L}^{-1}\left\{\frac{-\frac{5}{2}}{(s+5)}\right\}$$

$$\mathcal{L}^{-1}\{\mathbf{I}\} = \frac{5}{2}\mathcal{L}^{-1}\left\{\frac{1}{s+1}\right\} - \frac{5}{2}\mathcal{L}^{-1}\left\{\frac{1}{s+5}\right\}$$

$$i = \left(\frac{5}{2}\right)e^{-t} - \left(\frac{5}{2}\right)e^{-5t}$$

EXAMPLE 14.7 Determine the voltage v_C in the circuit in Figure 14.17 for $t \geq 0$ if the switch is closed at $t = 0$. ■

Figure 14.17

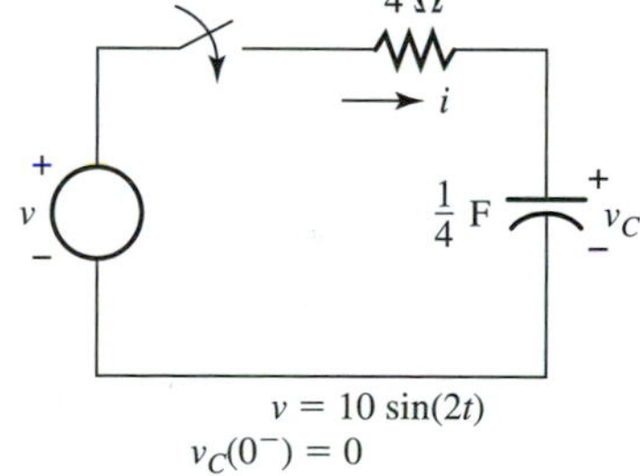

SOLUTION The circuit for $t \geq 0$ is shown in Figure 14.18.

Figure 14.18

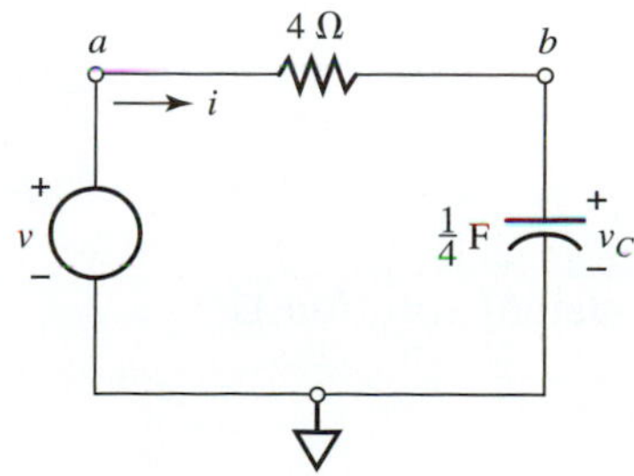

With respect to the indicated reference,

$$v_a = v$$
$$v_b = v_C$$

From Equation (6.8)

$$i = \frac{v_a - v_b}{4}$$
$$= \frac{v - v_C}{4}$$

From Equation (10.2)

$$i = C\frac{dv_C}{dt}$$
$$= \left(\frac{1}{4}\right)\frac{dv_C}{dt}$$

Equating the expressions for i results in

$$\frac{v - v_C}{4} = \left(\frac{1}{4}\right)\frac{dv_C}{dt}$$
$$\frac{dv_C}{dt} + v_C = v$$
$$\frac{dv_C}{dt} + v_C = 10\sin(2t)$$

because the voltage across a capacitor cannot change instantaneously,

$$v_C(0) = v_C(0^-) = 0$$

We shall use the classical method presented in Chapter 12 to solve the differential equation

$$\frac{dv_C}{dt} + v_C = 10\sin(2t)$$

for v_C. The characteristic equation is

$$p + 1 = 0$$
$$p = -1$$

and the transient solution is

$$v_{C_T} = Ke^{-t}$$

Using entry 5 of Table 12.1, the steady-state solution is assumed as

$$v_{C_{ss}} = A\sin(2t) + B\cos(2t)$$

and substituting $v_{C_{ss}}$ into the differential equation results in

$$\frac{d\{A\sin 2t + B\cos 2t\}}{dt} + A\sin 2t + B\cos 2t = 10\sin 2t$$

$$2A\cos 2t - 2B\sin 2t + A\sin 2t + B\cos 2t = 10\sin 2t$$

$$(2A + B)\cos 2t + (-2B + A)\sin 2t = 10\sin 2t$$

Equating coefficients results in

$$2A + B = 0$$

$$-2B + A = 10$$

$$B = -4$$

$$A = 2$$

and

$$v_{C_{ss}} = 2\sin(2t) - 4\cos(2t)$$

The general complete solution is

$$\begin{aligned} v_C &= v_{C_T} + v_{C_{ss}} \\ &= Ke^{-t} + 2\sin(2t) - 4\cos(2t) \end{aligned}$$

Because $v_C(0) = v_C(0^-) = 0$, K can be determined as follows.

$$0 = Ke^{-(0)} + 2\sin\{(2)(0)\} - 4\cos\{(2)(0)\}$$

$$K = 4$$

The particular complete solution for initial condition $v(0) = 0$ is as follows.

$$v_C = 4e^{-t} + 2\sin(2t) - 4\cos(2t)$$

14.5 THE INTEGRATOR AND DIFFERENTIATOR

Although the integrator is not a source-driven passive circuit of the type that we are considering in this chapter, it is, nevertheless, an important source-driven circuit. The integrator is an integral part of the analog computer and is an important component in control systems. The name of the circuit derives from the fact that the output of the circuit is proportional to the integral of the input (the driving source). The basic integrator circuit consists of a resistor, a capacitor and an operational amplifier (an active device), which is presented in Chapter 8. The circuit is shown in Figure 14.19.

Figure 14.19 The integrator.

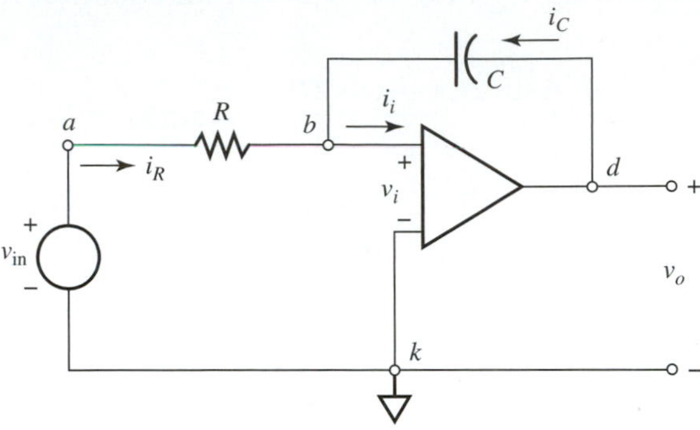

For the circuit in Figure 14.19, nodal voltages v_a, v_b, and v_d are written with respect to the reference node k. From Equation (6.8)

$$i_R = \frac{v_a - v_b}{R}$$

and from Equations (6.7) and (10.2), the voltage across the capacitor is $(v_d - v_b)$, and the current i_C is

$$i_C = C\frac{d(v_d - v_b)}{dt}$$

Because $v_a = v_{in}$, $v_b = v_i$, v_i is assumed to equal zero, and $v_d = v_o$,

$$i_R = \frac{v_{in}}{R}$$

$$i_C = C\frac{dv_o}{dt}$$

Applying Kirchhoff's current law at node b results in

$$i_R + i_C = i_i$$

and because i_i is assumed to equal zero,

$$i_R + i_C = 0$$

$$\frac{v_{in}}{R} + C\frac{dv_o}{dt} = 0$$

$$\frac{dv_o}{dt} = -\left(\frac{1}{RC}\right)v_{in}$$

From the definitions of the antiderivative and the indefinite integral presented in Chapter 9, if

$$\frac{dv_o}{dt} = -\left(\frac{1}{RC}\right)v_{in}$$

then v_o is the antiderivative of $\{-(1/RC)v_{\text{in}}\}$ and

$$v_o = \int\left(-\frac{1}{RC}\right)v_{\text{in}}\,dt \tag{14.17}$$

However, the op amp does not directly yield an analytical expression for the antiderivative of $\{(-1/RC)v_{\text{in}}\}$. The op amp yields instantaneous values of the antiderivative for each value of t as time elapses, starting with $t = 0$ when the circuit is energized. Equation (14.17) is best written in terms of the definite integral from time $t = 0$ to time t (any instant after the circuit is energized):

$$\int_0^t\left(-\frac{1}{RC}\right)v_{\text{in}}\,dt$$

Because v_0 is the antiderivative of $\{(-1/RC)v_{\text{in}}\}$, applying Equation (9.2) results in

$$\int_0^t\left(-\frac{1}{RC}\right)v_{\text{in}}\,dt = v_o(t) - v_o(0)$$

and from rule I6 in Chapter 9,

$$v_o = -\left(\frac{1}{RC}\right)\int_o^t v_{\text{in}}\,dt - v_o(0) \tag{14.18}$$

The integrator in Figure 14.19 is redrawn in Figure 14.20 with the loop *kbdk* indicated. Applying Kirchhoff's voltage law to the loop results in

$$v_i + v_C - v_o = 0$$

Figure 14.20 The integrator redrawn.

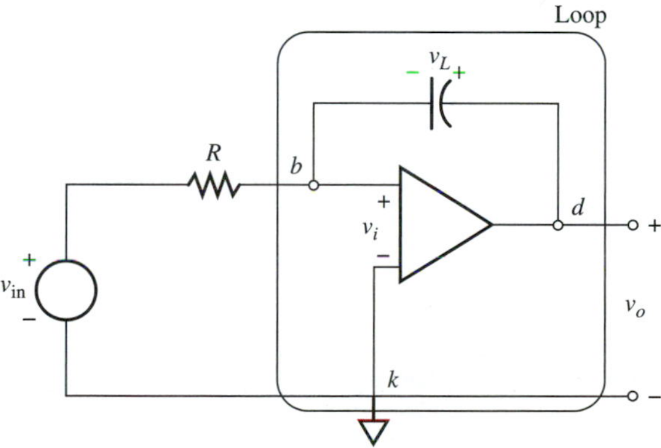

and at $t = 0$ (the instant the circuit is energized)

$$v_i(0) + v_C(0) - v_o(0) = 0$$

If the capacitor is initially uncharged ($v_C(0) = 0$), then

$$v_o(0) = 0$$

because v_i is assumed to equal zero.

Equation (14.18) can then be written as

$$v_o = -\left(\frac{1}{RC}\right)\int_0^t v_{in}\,dt \tag{14.19}$$

If R is set equal to $1/C$,

$$v_o = -\int_0^t v_{in}\,dt$$

and the output voltage v_o is equal to the negative of the integral of the input voltage with respect to time. A graph of v_o versus t is usually obtained by connecting the output of the integrator to a ploting device.

EXAMPLE 14.8 Determine v_o for the circuit in Figure 14.21 if the capacitor is initially uncharged. ■

Figure 14.21

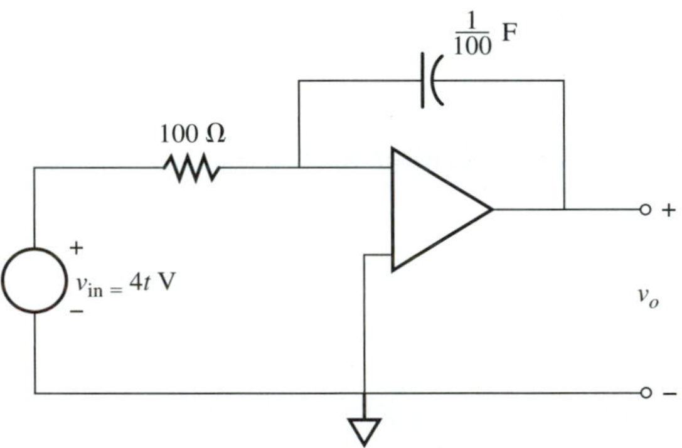

SOLUTION From Equation (14.19)

$$\begin{aligned} v_o &= -\left(\frac{1}{RC}\right)\int_0^t v_{in}\,dt \\ &= -\left\{\frac{1}{(100)\left(\frac{1}{100}\right)}\right\}\int_0^t 4t\,dt \\ &= -\int_0^t 4t\,dt \end{aligned}$$

From rule I1 in Chapter 9, the antiderivative of $4t$ is $2t^2$, and from Equation (9.2)

$$\begin{aligned} \int_0^t 4t\,dt &= 2t^2\Big|_t - 2t^2\Big|_0 \\ &= 2t^2 - 2(0)^2 \\ &= 2t^2 \end{aligned}$$

Therefore,

$$v_o = -2t^2$$

Sketches of the output and input voltages versus time are shown in Figure 14.22.

Figure 14.22

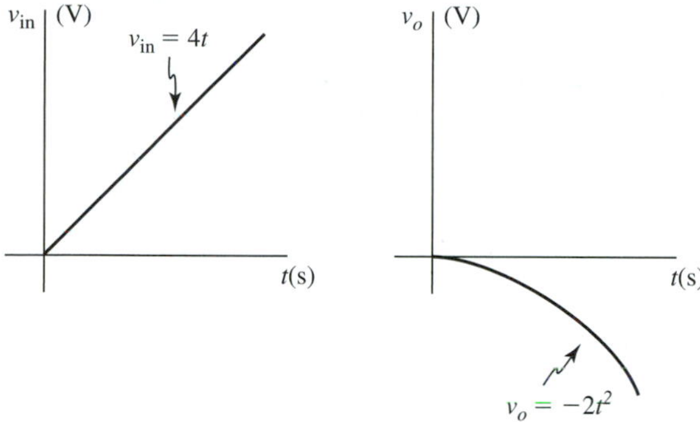

Recall the area property of the definite integral from Chapter 9. For the v_o-versus-t and v_{in}-versus-t curves shown in Figure 14.22,

$$
\begin{aligned}
v_o(t_1) &= -(\text{area under the } v_{in}\text{-versus-}t \text{ curve from } t = 0 \text{ to } t = t_1) \\
&= -A
\end{aligned}
$$

as shown in Figure 14.23, where t_1 is any instant of time after the circuit is energized.

Figure 14.23

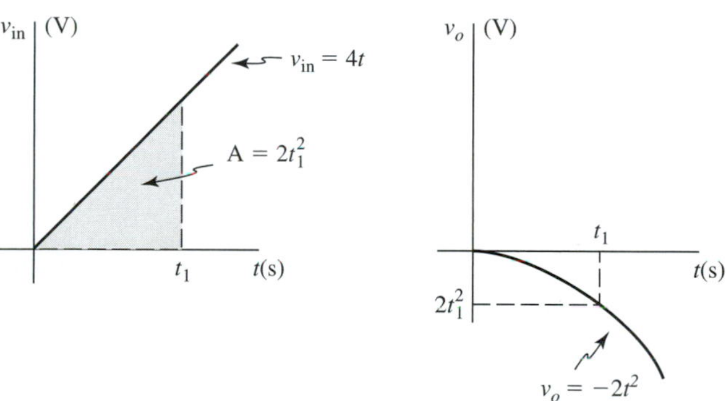

The integrator circuit can be transformed to a differentiator by interchanging the capacitor and resistor in Figure 14.19. The resulting circuit is

Figure 14.24

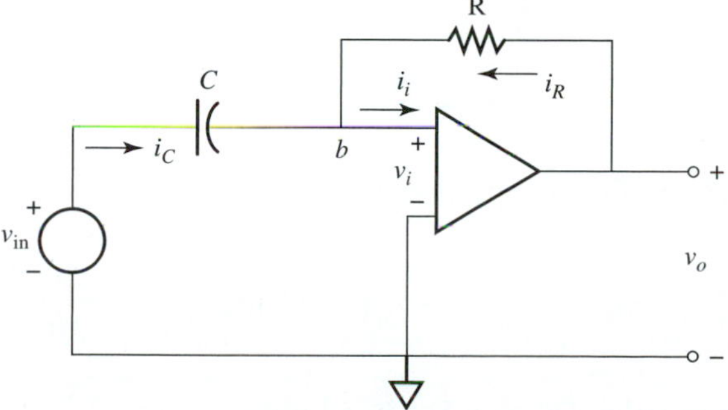

shown in Figure 14.24. The currents i_R and i_C are determined by using Equations (6.8) and (10.2), respectively,

$$i_R = \frac{v_o - v_i}{R}$$

$$i_C = C\frac{d(v_{in} - v_i)}{dt}$$

and recognizing that $v_i = 0$:

$$i_R = \frac{v_o}{R}$$

$$i_C = C\frac{dv_{in}}{dt}$$

Employing Kirchhoff's current law at node b results in

$$i_R + i_C = i_i$$

Because $i_i = 0$,

$$(C)\frac{dv_{in}}{dt} = \frac{-v_o}{R}$$

$$v_o = -(RC)\frac{dv_{in}}{dt} \tag{14.20}$$

and if $RC = 1$,

$$v_o = -\frac{dv_{in}}{dt}$$

EXAMPLE 14.9 Determine v_o for the circuit in Figure 14.25 if the capacitor is initially uncharged. ■

Figure 14.25

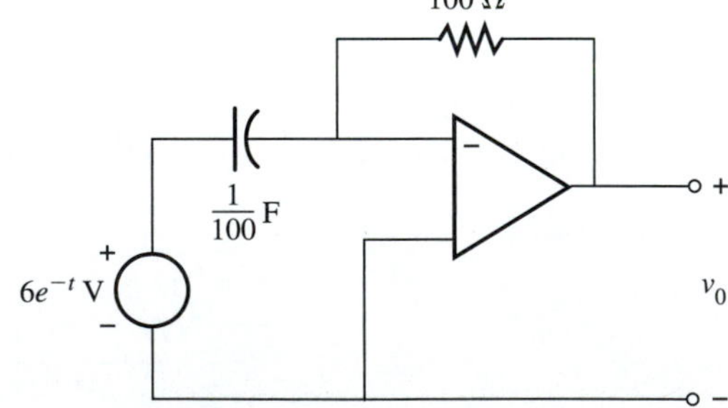

SOLUTION From Equation (14.20)

$$v_o = -(100)\left(\frac{1}{100}\right)\frac{d(6e^{-t})}{dt}$$

and from rules D6 and D3 in Chapter 9,

$$v_o = 6e^{-t}$$

14.6 EXERCISES

1. Determine v_C and i for the circuit in Figure 14.26 for $t \geq 0$. The switch is closed at $t = 0$.

Figure 14.26

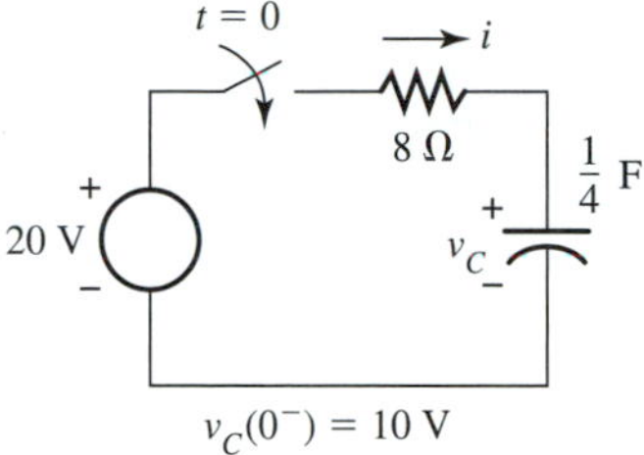

2. Determine v_C and i for the circuit in Figure 14.27 for $t \geq 0$. The switch is closed at $t = 0$.

Figure 14.27

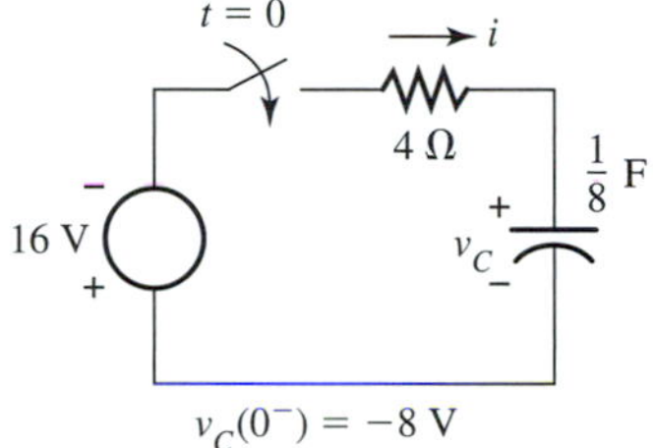

3. Determine v_C and i for the circuit in Figure 14.28 for $t \geq 0$. The switch is closed at $t = 0$.

Figure 14.28

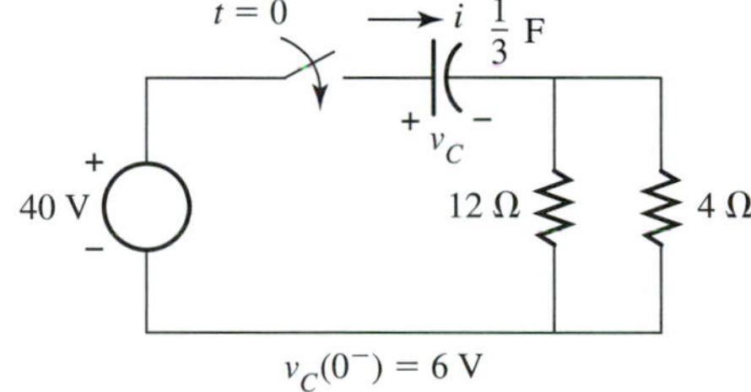

4. Determine v_C and i for the circuit in Figure 14.29 for $t \geq 0$. The switch is closed at $t = 0$.

Figure 14.29

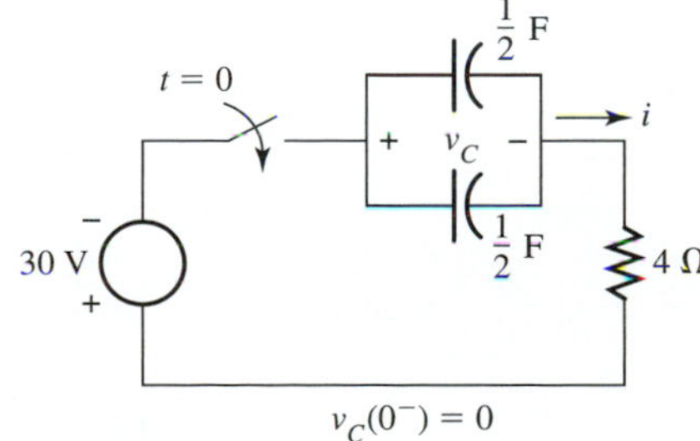

5. Determine v_C and i for the circuit in Figure 14.30 for $t \geq 0$. The switch is closed at $t = 0$.

Figure 14.30

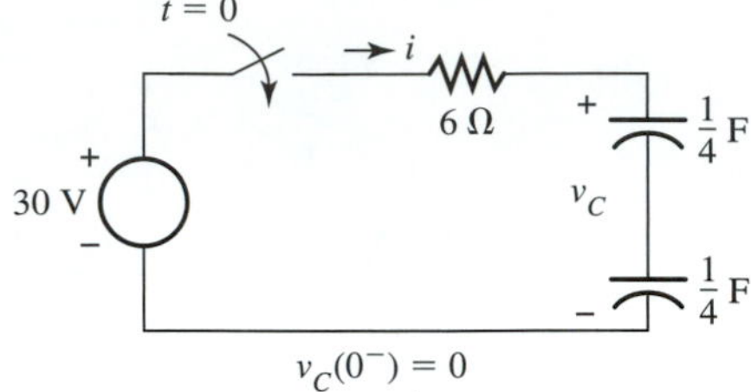

6. Determine v and i for the circuit in Figure 14.31 for $t \geq 0$. The switch is closed at $t = 0$.

Figure 14.31

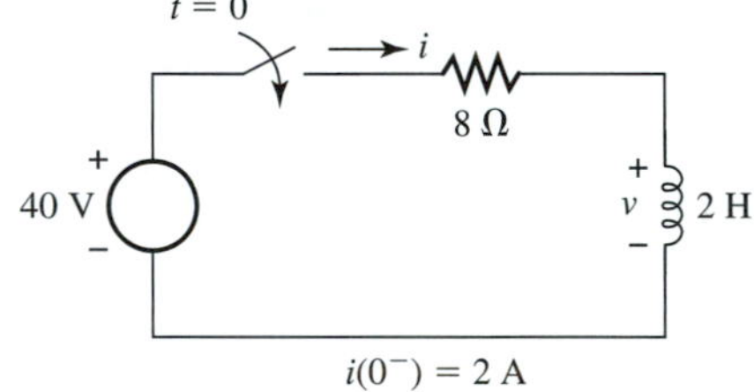

7. Determine v and i for the circuit in Figure 14.32 for $t \geq 0$. The switch is closed at $t = 0$.

Figure 14.32

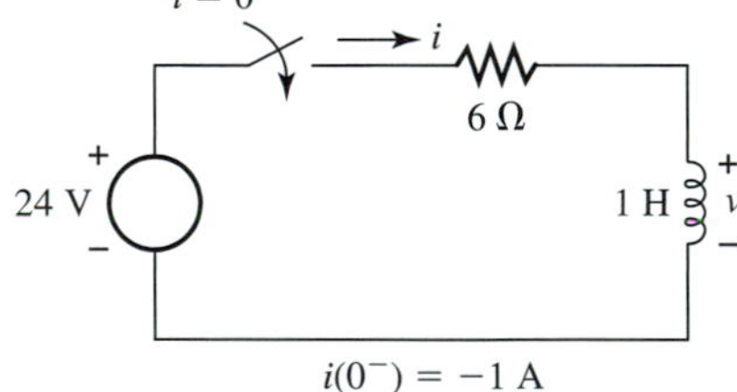

8. Determine v and i for the circuit in Figure 14.33 for $t \geq 0$. The switch is closed at $t = 0$.

Figure 14.33

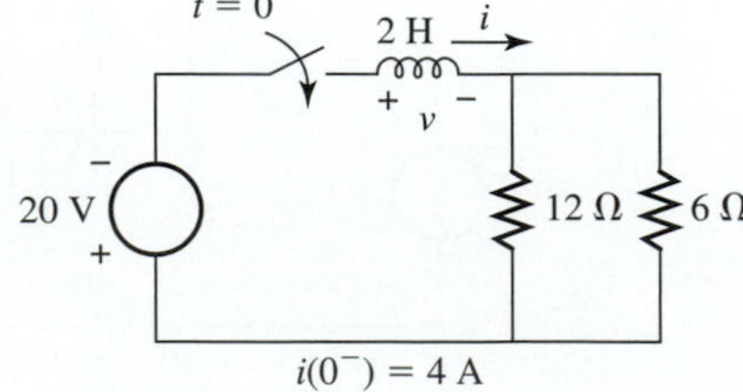

9. Determine v and i for the circuit in Figure 14.34 for $t \geq 0$. The switch is closed at $t = 0$.

Figure 14.34

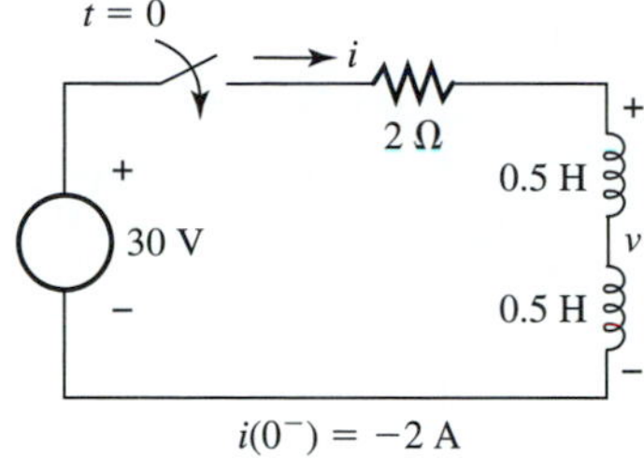

10. Determine v and i for the circuit in Figure 14.35 for $t \geq 0$. The switch is closed at $t = 0$.

Figure 14.35

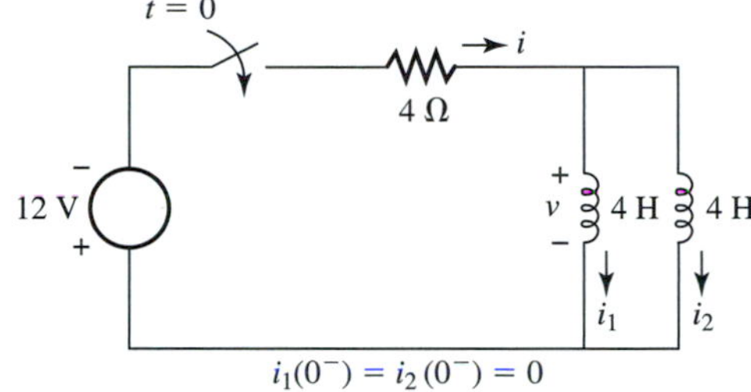

11. Determine v_L, v_C, and i for the circuit in Figure 14.36 for $t \geq 0$. The switch is closed at $t = 0$.

Figure 14.36

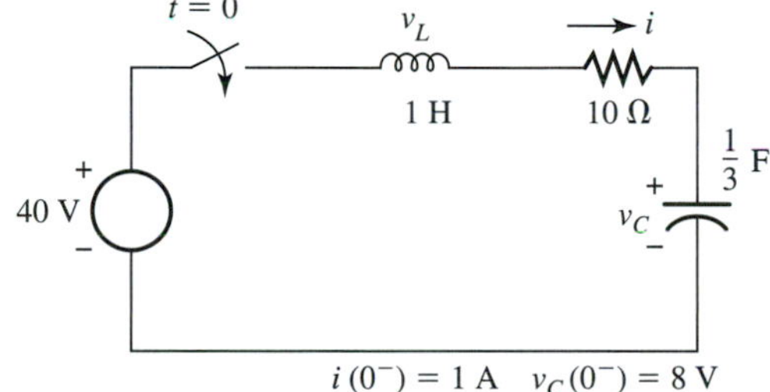

12. Determine v_L, v_C, and i for the circuit in Figure 14.37 for $t \geq 0$. The switch is closed at $t = 0$.

Figure 14.37

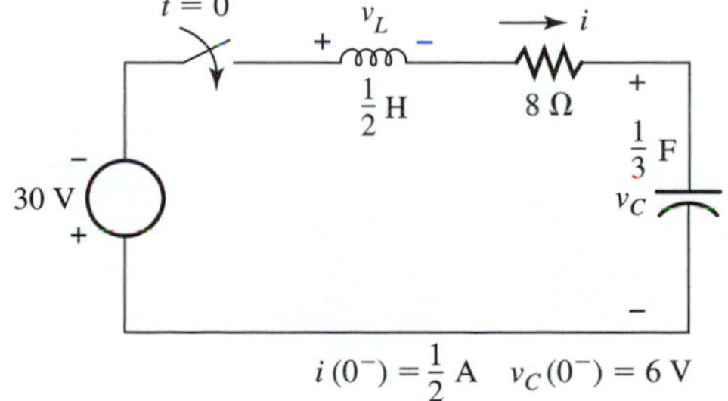

13. Determine v_L, v_C, and i for the circuit in Figure 14.38 for $t \geq 0$. The switch is closed at $t = 0$.

Figure 14.38

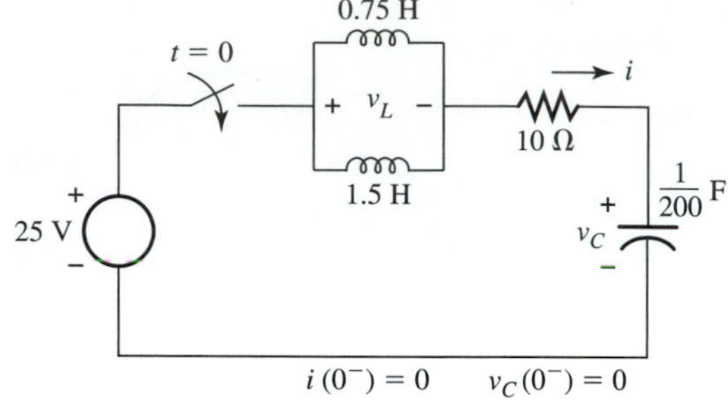

14. Determine v_L, v_C, and i for the circuit in Figure 14.39 for $t \geq 0$. The switch is closed at $t = 0$.

Figure 14.39

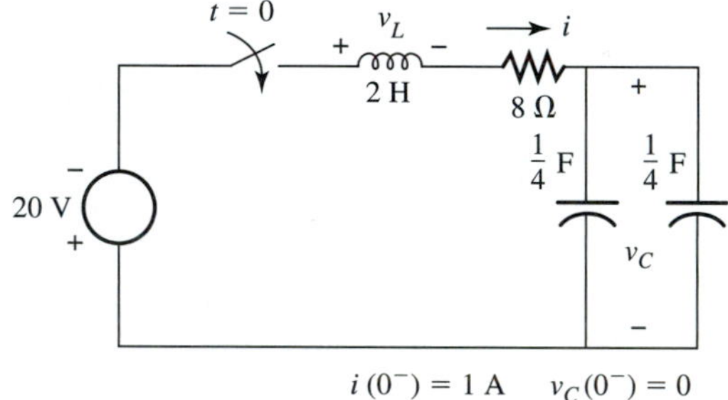

15. Determine v_L, v_C, and i for the circuit in Figure 14.40 for $t \geq 0$. The switch is closed at $t = 0$.

Figure 14.40

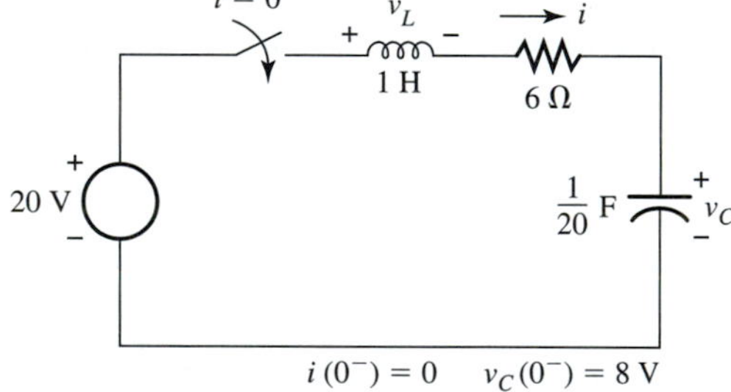

16. Determine v and i for $t \geq 0$ for each circuit in Figure 14.41. The switch is closed at $t = 0$.

Figure 14.41

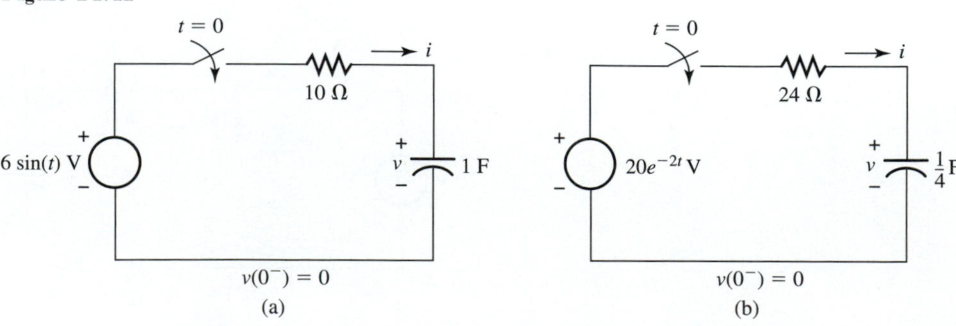

17. Determine v and i for $t \geq 0$ for each circuit in Figure 14.42. The switch is closed at $t = 0$.

Figure 14.42

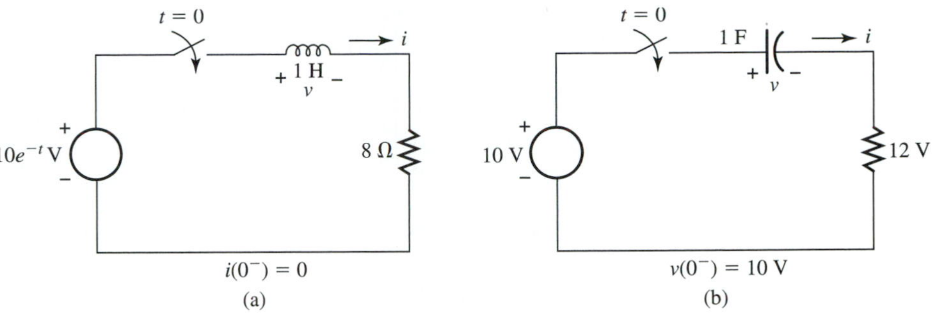

18. Determine v and i for $t \geq 0$ for each circuit in Figure 14.43. The switch is closed at $t = 0$.

Figure 14.43

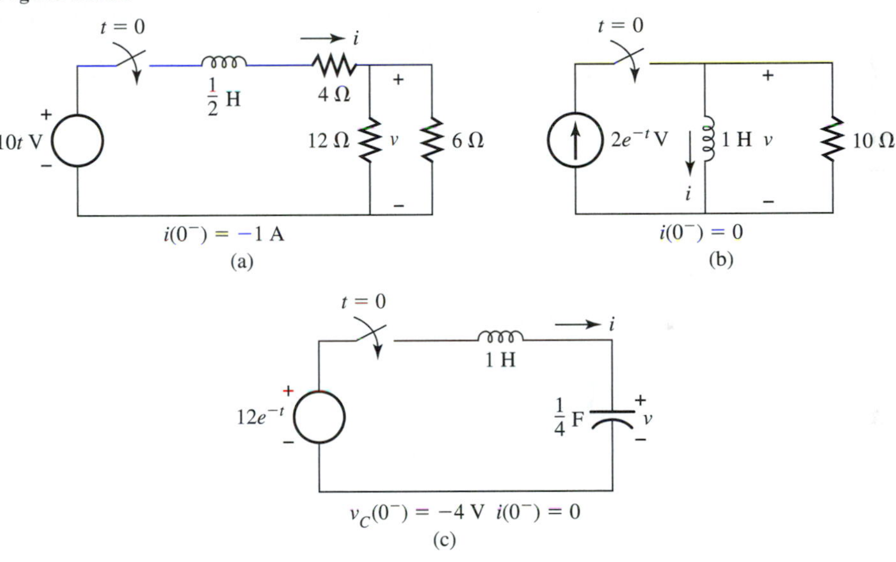

19. Determine v_o for $t \geq 0$ for each circuit in Figure 14.44. The capacitor is initially uncharged.

Figure 14.44

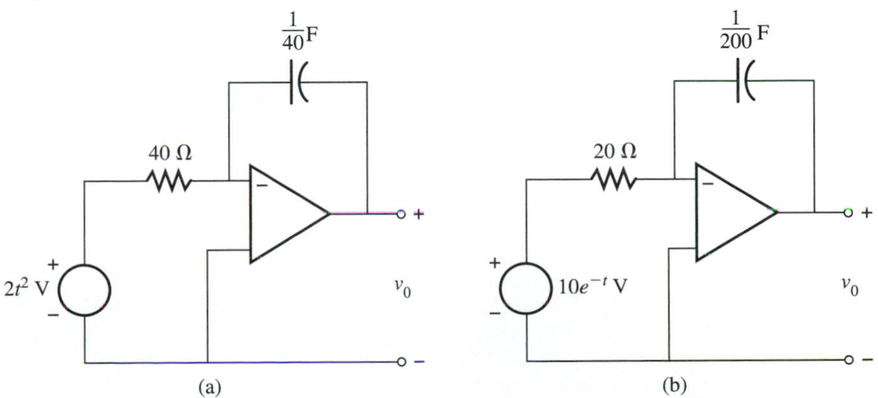

20. Determine v_o for $t \geq 0$ for each circuit in Figure 14.45. The capacitor is initially uncharged.

Figure 14.45

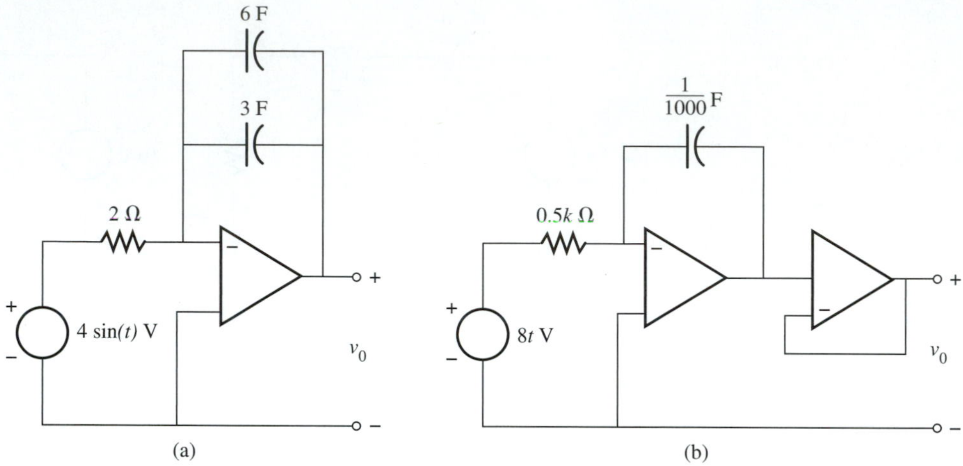

21. Determine v_o for $t \geq 0$ for each circuit in Figure 14.46. The capacitor is initially uncharged.

Figure 14.46

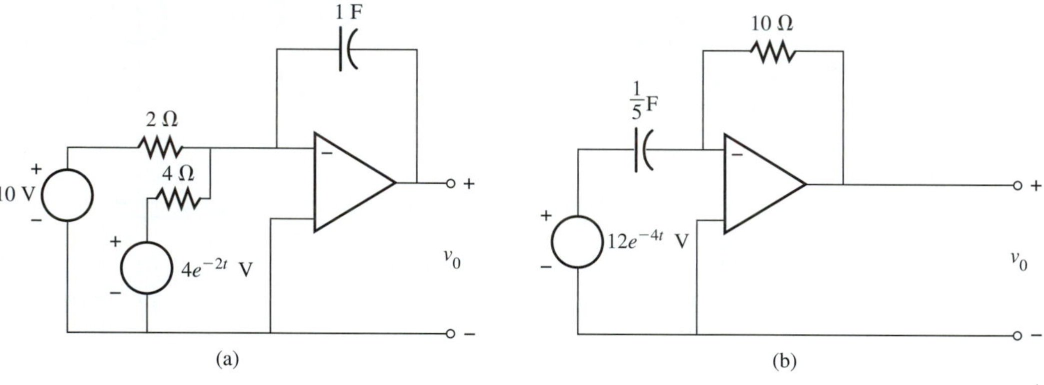

CHAPTER

15

Laplace Transform Methods

After completing this chapter the student should be able to

* transform circuits from the time (t) domain to the LaPlace (s) domain
* use LaPlace transform methods to analyze electric circuits in the LaPlace (s) domain

In Chapter 12 we presented methods for solving differential equations and in Chapters 13 and 14 we used the methods to analyze circuits described by differential equations. The general procedure employed in Chapters 13 and 14 is to describe the circuit by a mathematics model (differential equation) and then solve the differential equation using the classical method or the Laplace transform method. In this chapter we extend the analysis of electric circuits by the use of Laplace transforms.

15.1 TRANSFER FUNCTION

A *transfer function* (often called a network function) in the s (Laplace) domain relates the Laplace transforms of the input and output of a circuit. It is defined as the ratio of the Laplace transform of the output divided by the Laplace transform of the input and is designated $\mathbf{H}(s)$, where

$$\mathbf{H}(s) = \frac{\mathscr{L}\{\text{output}\}}{\mathscr{L}\{\text{input}\}} \tag{15.1}$$

$\mathbf{H}(s)$ is read as "**H** function of s" and is usually written as **H**, where it is understood that **H** is a function of s.

The circuit input is usually the voltage or current source driving the circuit and the circuit output is any current or voltage appearing in the circuit.

Consider the following example.

EXAMPLE 15.1 Determine the transfer function relating i (output) to v (input) for the circuit in Figure 15.1 after the switch is closed at $t = 0$. ■

Figure 15.1

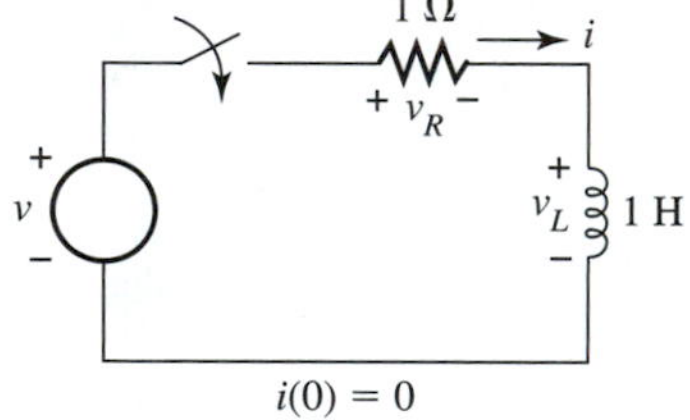

SOLUTION For $t \geq 0$ the circuit appears as shown in Figure 15.2. From Kirchhoff's voltage law

$$v = v_L + v_R$$

Figure 15.2

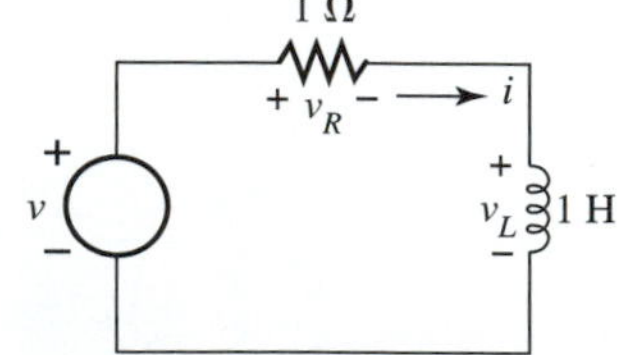

and from Ohm's law and Equation (11.3),

$$L\frac{di}{dt} + Ri = v$$

Taking the Laplace transform of both sides of this equation and using rules L1, L3(a), and L4(a) from Chapter 12 results in

$$\mathcal{L}\left\{L\frac{di}{dt} + Ri\right\} = \mathcal{L}\{v\}$$

$$L\mathcal{L}\left\{\frac{di}{dt}\right\} + R\mathcal{L}\{i\} = \mathcal{L}\{v\}$$

$$L\{s\mathbf{I} - i(0)\} + R\mathbf{I} = \mathbf{V}$$

$$\mathbf{I} = \left\{\frac{1}{sL + R}\right\}\mathbf{V}$$

where **I** and **V** are the Laplace transforms of the output and input, respectively.

From Equation (15.1) the transfer function relating i and v is

$$\mathbf{H} = \frac{\mathbf{I}}{\mathbf{V}}$$

$$= \frac{1}{sL + R}$$

If the transfer function and the input are known, the Laplace transform of the output can be determined; and if the transfer function and the output are known, the Laplace transform of the input can be determined using Equation (15.1). The output function obtained directly from a transfer function is always for zero initial conditions and contains the transient and steady-state components of the output. That is, the output obtained from a transfer function is the particular complete solution of the differential equation describing the output for zero initial conditions.

Consider the following example.

EXAMPLE 15.2 The transfer function relating the output voltage v_o and the input voltage v_i for the circuit in Figure 15.3 is given as

$$\mathbf{H} = \frac{2}{(s + 1)(s + 2)}$$ ■

Figure 15.3

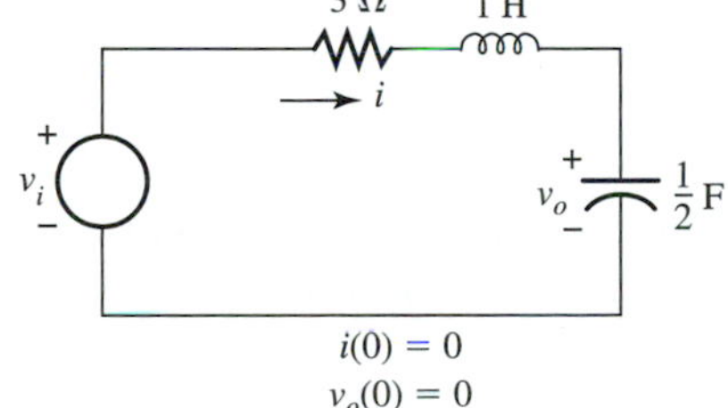

Determine v_o for $v_i = 20$ V.

SOLUTION From entry 1 of Table 12.2,

$$\mathbf{V}_i = \mathcal{L}\{20\} = \frac{20}{s}$$

for input voltage $v_i = 20$.

From Equation (15.1)

$$\begin{aligned}\mathbf{V}_o &= \mathbf{H}\mathbf{V}_i \\ &= \left\{\frac{2}{(s+1)(s+2)}\right\}\left\{\frac{20}{s}\right\} \\ &= \frac{40}{(s)(s+1)(s+2)}\end{aligned}$$

Using the methods in Chapter 12, the partial fraction expansion for $\mathbf{V}_o$ is

$$\mathbf{V}_o = \frac{K_1}{s} + \frac{K_2}{s+1} + \frac{K_3}{s+2}$$

$$K_1 = 20$$

$$K_2 = -40$$

$$K_3 = 20$$

and

$$\mathbf{V}_o = \frac{20}{s} + \frac{-40}{s+1} + \frac{20}{s+2}$$

Taking the inverse Laplace transform of each side of this equation and using rules L3(b) and L4(b) of Chapter 12 and entries 1 and 2 of Table 12.2 results in

$$\begin{aligned}\mathcal{L}^{-1}\{\mathbf{V}_o\} &= \mathcal{L}^{-1}\left\{\frac{20}{s} + \frac{-40}{s+1} + \frac{20}{s+2}\right\} \\ &= 20\mathcal{L}^{-1}\left\{\frac{1}{s}\right\} - 40\mathcal{L}^{-1}\left\{\frac{1}{s+1}\right\} + 20\mathcal{L}^{-1}\left\{\frac{1}{s+2}\right\} \\ v_o &= 20 - 40e^{-t} + 2e^{-2t}\end{aligned}$$

which is the output voltage v_o for the input $v_i = 20$ V.

15.2 LAPLACE IMPEDANCE

From Ohm's law and Equations (11.3) and (10.2), the current-voltage relationships for resistors, inductors, and capacitors are, respectively,

$$v = Ri$$

$$v = L\frac{di}{dt}$$

and

$$i = C\frac{dv}{dt}$$

Using rules L1 and L3(a) of Chapter 12 and taking the Laplace transform of both sides of each of these equations results in

$$\mathbf{V} = R\mathbf{I}$$

$$\mathbf{V} = L\{s\mathbf{I} - i(0)\}$$

$$\mathbf{I} = C\{s\mathbf{V} - v(0)\}$$

where **V** and **I** are the Laplace transforms for voltage and current, respectively. Setting the initial conditions equal to zero results in

$$\mathbf{V} = R\mathbf{I}$$

$$\mathbf{V} = Ls\mathbf{I}$$

$$\mathbf{I} = Cs\mathbf{V}$$

The Laplace impedance $\mathbf{Z}(s)$ (read as "**Z** function of s") for a pair of terminals is defined as the Laplace transform of the voltage across the terminals divided by the Laplace transform of the current at the terminals. That is, $\mathbf{Z}(s) = \mathbf{V}(s)/\mathbf{I}(s)$, which is written as

$$\mathbf{Z} = \frac{\mathbf{V}}{\mathbf{I}} \tag{15.2}$$

where it is understood that **Z**, **V**, and **I** are functions of s.

Solving these equations for **V**/**I**, the Laplace impedances for resistors, inductors, and capacitors are, respectively,

$$\mathbf{Z}_R = R \tag{15.3}$$

$$\mathbf{Z}_L = sL \tag{15.4}$$

$$\mathbf{Z}_c = \frac{1}{sC} \tag{15.5}$$

Notice that the Laplace impedances are derived for zero initial conditions—that is, zero initial current through the inductor and zero initial voltage across the capacitor. Notice also that the Laplace impedance is a particular transfer function relating input voltage and input current for a pair of terminals.

15.3 CIRCUIT IN THE *s* DOMAIN

A circuit is transformed to the s (Laplace) domain by using Equations (15.3), (15.4), and (15.5) for the passive elements and Table 12.2 to transform the sources. It can be shown that Laplace impedances can be combined in series and parallel using the same rules for combining resistors given in Chapter 3; that is,

$$\mathbf{Z}(s) = \mathbf{Z}_1(s) + \mathbf{Z}_2(s) + \cdots + \mathbf{Z}_n(s) \tag{15.6}$$

for Laplace impedances in series and

$$\mathbf{Z}(s) = \frac{1}{\dfrac{1}{\mathbf{Z}_1(s)} + \dfrac{1}{\mathbf{Z}_2(s)} + \cdots + \dfrac{1}{\mathbf{Z}_n(s)}} \tag{15.7}$$

for Laplace impedances in parallel.

Consider the following examples.

EXAMPLE 15.3 Transform the circuit in Figure 15.4 to the s domain and determine the Laplace impedance $\mathbf{Z}$ as indicated. ■

Figure 15.4

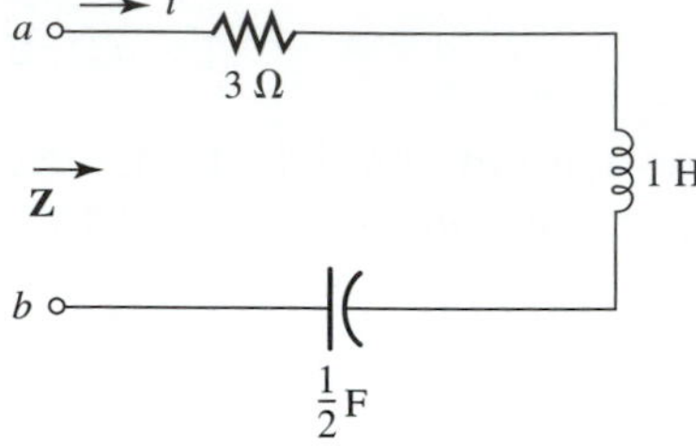

SOLUTION From Equations (15.3), (15.4), and (15.5), the transformed circuit is as shown in Figure 15.5.

Figure 15.5

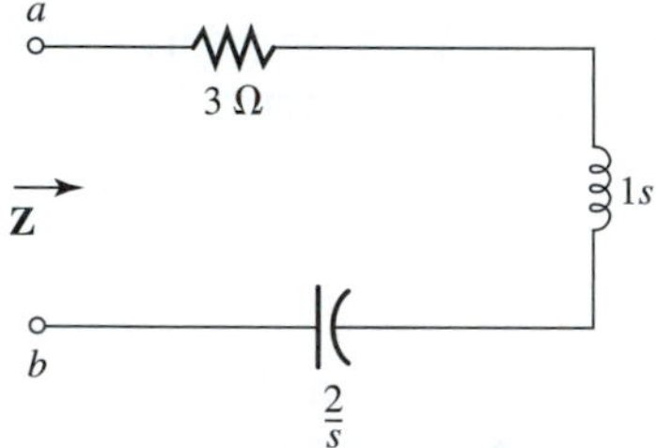

Using Equation (15.6)

$$\mathbf{Z} = 3 + s + (2/s)$$

$$= \frac{s^2 + 3s + 2}{s}$$

EXAMPLE 15.4 Transform the circuit in Figure 15.6 to the s domain and determine the Laplace impedance as indicated. ■

Figure 15.6

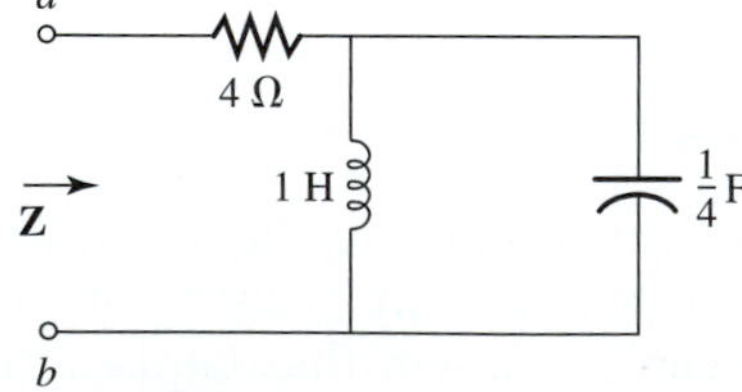

SOLUTION From Equations (15.3), (15.4), and (15.5), the transformed circuit is as shown in Figure 15.7. Using Equations (15.6) and (15.7) to combine the elements in series and parallel results in

Figure 15.7

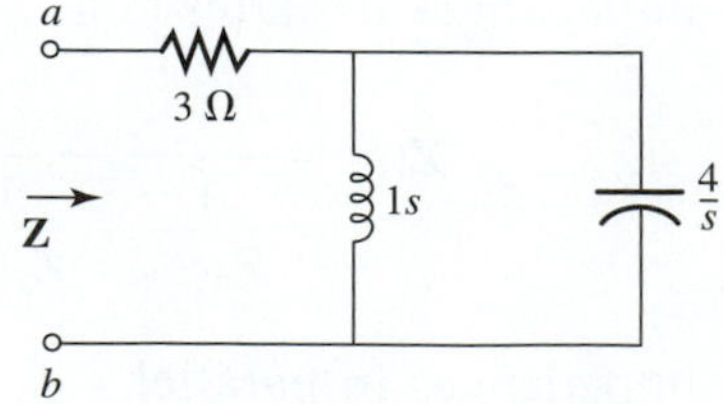

$$\mathbf{Z} = 3 + \left\{(s) \| \left(\frac{4}{s}\right)\right\}$$

$$= 3 + \frac{s\left(\frac{4}{5}\right)}{s + (4/s)}$$

$$= 3 + \frac{4}{(s^2 + 4)/s}$$

$$= 3 + \frac{4s}{s^2 + 4}$$

$$= \frac{3s^2 + 4s + 12}{(s^2 + 4)}$$

Because the Laplace impedance for a pair of terminals is defined as $\mathbf{Z} = \mathbf{V}/\mathbf{I}$ (Equation 15.2), the Laplace impedance can be used to determine $\mathbf{V}$ and $\mathbf{I}$. That is, if $\mathbf{Z}$ and $\mathbf{V}$ are known for a pair of terminals, then $\mathbf{I}$ can be determined; and if $\mathbf{Z}$ and $\mathbf{I}$ are known for a pair of terminals then $\mathbf{V}$ can be determined.

Consider the following example.

EXAMPLE 15.5 For the circuit in Example 15.3, determine the current i if the circuit is driven by a voltage source as shown in Figure 15.8. The initial value of the voltage across the capacitor and the initial current through the inductor are both zero. ■

Figure 15.8

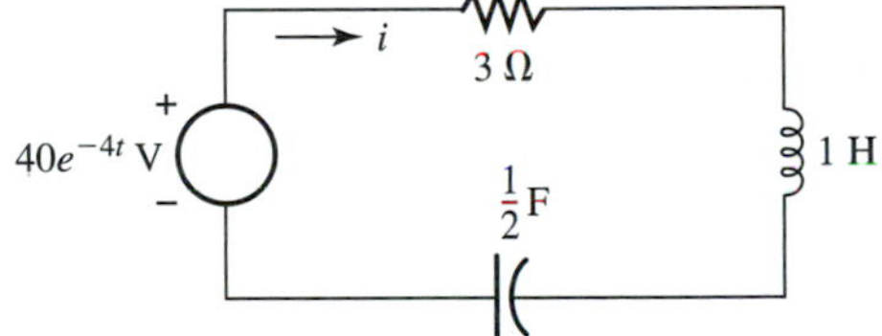

SOLUTION From Example 15.3 and entry 2 in Table 12.2, the transformed

Figure 15.9

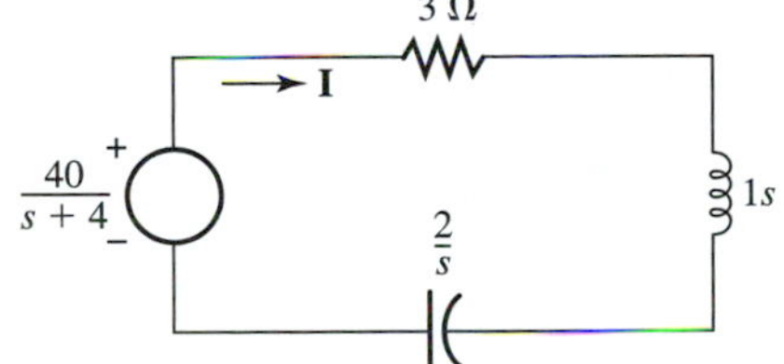

circuit is as shown in Figure 15.9, and the total Laplace impedance across the voltage source from Equation (15.6) is

$$\mathbf{Z} = 3 + s + \frac{2}{s}$$

$$\mathbf{Z} = \frac{s^2 + 3s + 2}{s}$$

From Equation (15.2)

$$\mathbf{Z} = \frac{\mathbf{V}}{\mathbf{I}}$$

$$\mathbf{I} = \frac{\mathbf{V}}{\mathbf{Z}}$$

$$= \frac{40/(s + 4)}{(s^2 + 3s + 2)/s}$$

$$= \frac{40s}{(s^2 + 3s + 2)(s + 4)}$$

$$= \frac{40s}{(s + 2)(s + 1)(s + 4)}$$

Taking the inverse Laplace transform of both sides of the equation results in

$$i = \mathcal{L}^{-1}\left\{\frac{40s}{(s + 2)(s + 1)(s + 4)}\right\}$$

Applying the procedure presented in Chapter 12 for expanding into partial fractions results in

$$i = \mathcal{L}^{-1}\left\{\frac{K_1}{(s + 2)} + \frac{K_2}{(s + 1)} + \frac{K_3}{(s + 4)}\right\}$$

$$= \mathcal{L}^{-1}\left\{\frac{40}{(s + 2)} + \frac{\left(-\frac{40}{3}\right)}{(s + 1)} + \frac{\left(-\frac{80}{3}\right)}{(s + 4)}\right\}$$

From rules L3(b) and L4(b) of Chapter 12 and entry 2 of Table 12.2

$$i = 40\mathcal{L}^{-1}\left\{\frac{1}{s + 2}\right\} - \left(\frac{40}{3}\right)\mathcal{L}^{-1}\left\{\frac{1}{s + 1}\right\} - \left(\frac{80}{3}\right)\mathcal{L}^{-1}\left\{\frac{1}{s + 4}\right\}$$

$$= 40e^{-2t} - \left(\frac{40}{3}\right)e^{-t} - \left(\frac{80}{3}\right)e^{-4t}$$

Recall that transfer functions and Laplace impedances involve zero initial conditions. We now present methods for transforming circuits to the s domain when nonzero initial conditions are present.

15.4 INDUCTORS WITH INITIAL CONDITIONS

Consider the inductor in Figure 15.10 where the initial current is $i(0)$. From Equation (11.3)

$$v = L\frac{di}{dt}$$

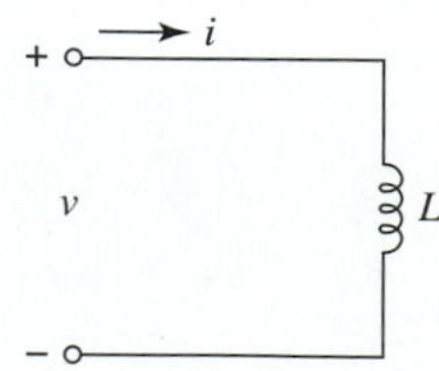

Figure 15.10 Inductor circuit.

Taking the Laplace transform of both sides of this equation and using rules L1 and L3(a) of Chapter 12 results in

$$\mathcal{L}\{v\} = \mathcal{L}\left\{L\frac{di}{dt}\right\}$$

$$\mathbf{V} = L\mathcal{L}\left\{\frac{di}{dt}\right\}$$

$$= Ls\mathbf{I} - Li(0)$$

The circuit in Figure 15.11 is a valid transformation of the circuit in Figure 15.10 to the s domain, because both circuits are described by the same equation in the s domain, as shown next.

Kirchhoff's voltage law and the manner of assigning voltage polarities to passive elements based on current flow apply to circuits in the s domain in the same manner that they apply to circuits in the time domain. Applying Kirchhoff's voltage law to the circuit in Figure 15.11 and using Equation (15.4) to write the voltage across the inductor as $(Ls)\mathbf{I}$ results in

$$\mathbf{V} - (Ls)\mathbf{I} + Li(0) = 0$$

and

$$\mathbf{V} = Ls\mathbf{I} - Li(0)$$

which is the same equation previously arrived at for $\mathbf{V}$ for the circuit in Figure 15.10.

Figure 15.11 Inductor with initial conditions.

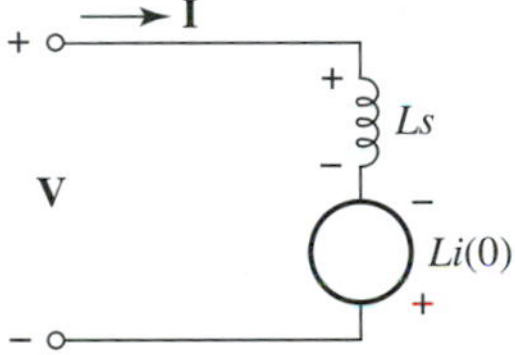

The rules in Figure 4.10 for converting sources in resistive circuits also hold in the s (Laplace) domain. In Figure 15.12 the voltage source conversion rule in Figure 4.10(a) is restated for the s domain.

Figure 15.12 Voltage source transformation.

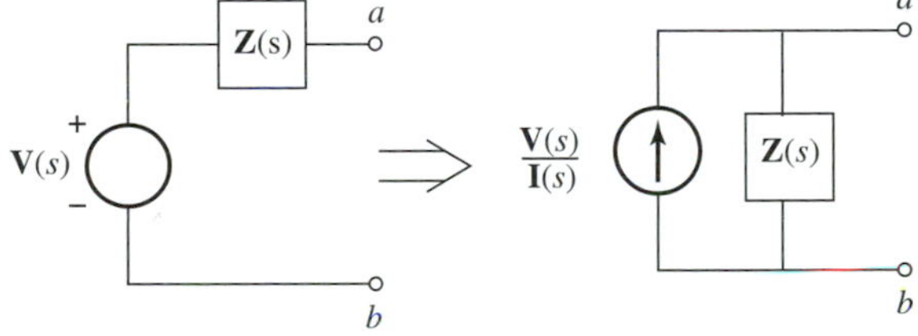

From Figure 15.12 the circuit in Figure 15.11 can be drawn as shown in Figure 15.13.

Figure 15.13 Current source equivalent.

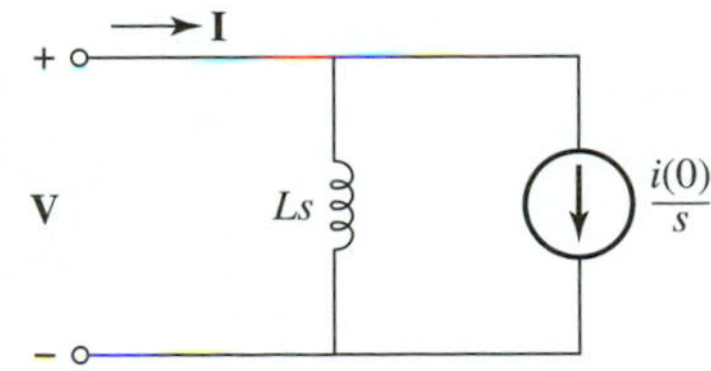

Notice that applying Kirchhoff's current law to the top node in Figure 15.13 and using Equation (15.2) to write the current through the inductor results in

$$\mathbf{I} - \frac{\mathbf{V}}{Ls} - \frac{i(0)}{s} = 0$$

and

$$\mathbf{V} = Ls\mathbf{I} - Li(0)$$

which is the same equation that describes the circuit in Figure 15.11.

In summary, the inductor with initial current $i(0)$ can be represented in the s domain in either of the two ways shown in Figure 15.14.

Figure 15.14 Inductor with initial conditions.

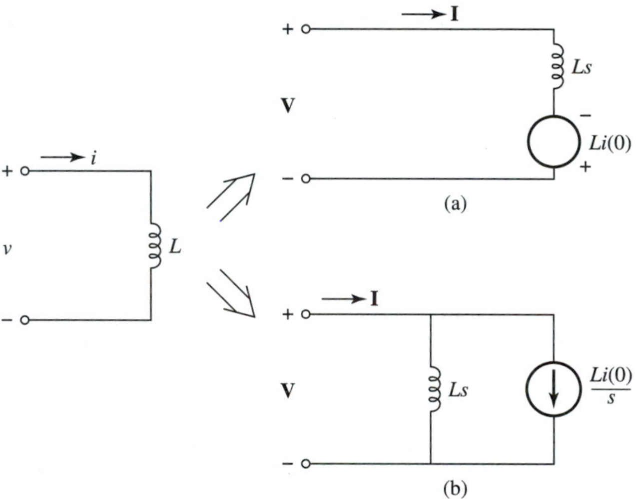

Consider the following examples.

EXAMPLE 15.6 Determine the current i for $t \geq 0$ if $i(0) = 1$ A for the circuit in Figure 15.15. ■

Figure 15.15

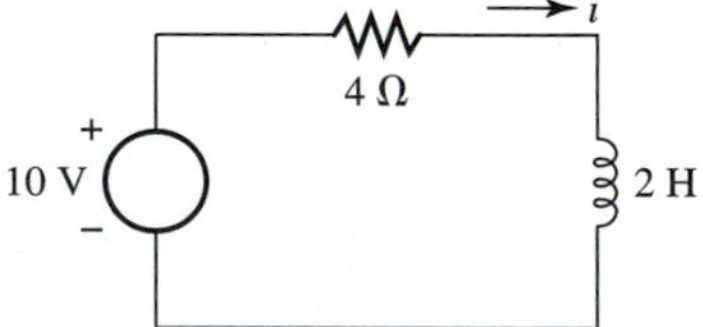

SOLUTION Applying Equations (15.3) and (15.4), entry 1 of Table 12.2, and the rule in Figure 15.4(a) for representing inductors with initial current in the s domain results in the circuit in Figure 15.16.

Figure 15.16

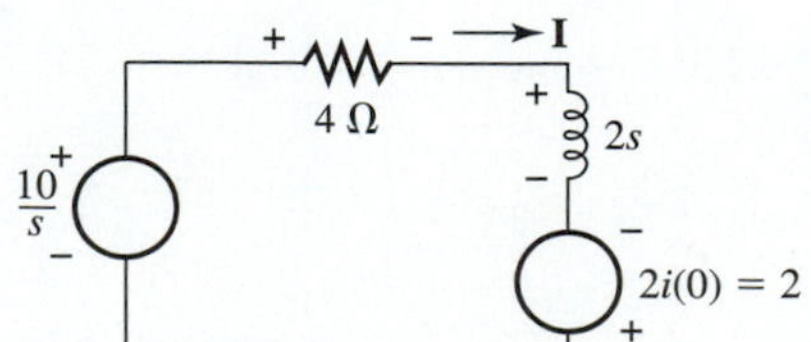

We use Equation (15.2) to determine the voltages across the resistor and inductor as (4**I**) and (2s**I**), respectively. We then assign polarities as shown to the passive elements based on the current flow and apply Kirchhoff's voltage law to the loop, which results in

$$\frac{10}{s} - 4\mathbf{I} - 2s(\mathbf{I}) + 2 = 0$$

$$10 - 4s\mathbf{I} - 2s^2(\mathbf{I}) + 2s = 0$$

$$\mathbf{I} = \frac{10 + 2s}{2s(s+2)}$$

$$\mathbf{I} = \frac{5 + s}{s(s+2)}$$

Using the methods of Chapter 12 for partial fraction expansion results in

$$\mathbf{I} = \frac{K_1}{s} + \frac{K_2}{s+2}$$

$$K_1 = \frac{5}{2}$$

$$K_2 = -\frac{3}{2}$$

$$\mathbf{I} = \frac{\frac{5}{2}}{s} + \frac{-\frac{3}{2}}{s+2}$$

From rules L3(b) and L4(b) of Chapter 12,

$$i = \mathcal{L}^{-1}\left\{\frac{\frac{5}{2}}{s} + \frac{\left(-\frac{3}{2}\right)}{s+2}\right\}$$

$$= \frac{5}{2}\mathcal{L}^{-1}\left\{\frac{1}{s}\right\} - \left(\frac{3}{2}\right)\mathcal{L}^{-1}\left\{\frac{1}{s+2}\right\}$$

and from entries 1 and 2 of Table 12.2,

$$i = \left(\frac{5}{2}\right) - \left(\frac{3}{2}\right)e^{-2t}$$

Alternatively, the inductor initial condition can be represented by a current source according to the transformation rule in Figure 15.14(b), as shown in Figure 15.17. The current source representation is shown to the right of terminals a-b. $\mathbf{I}_1$ and $\mathbf{I}_2$ are mesh currents.

Figure 15.17

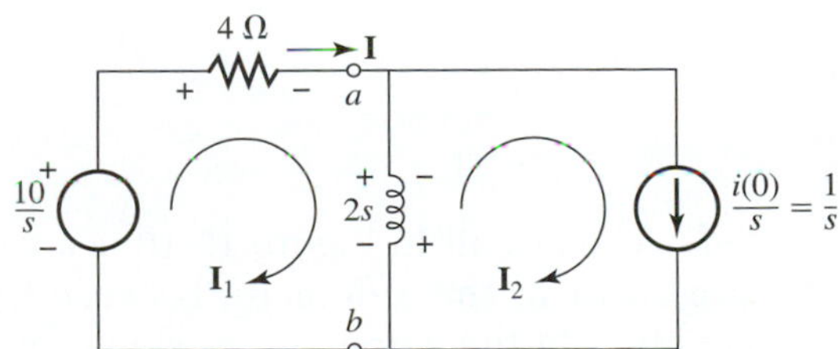

From an inspection of the circuit,

$$\mathbf{I}_2 = \frac{1}{s}$$

and applying Kirchhoff's voltage law to mesh 1 results in

$$\frac{10}{s} - 4\mathbf{I}_1 - 2s\mathbf{I}_1 + 2s\mathbf{I}_2 = 0$$

Because $\mathbf{I}_1 = \mathbf{I}$ and $\mathbf{I}_2 = 1/s$,

$$\frac{10}{s} - 4\mathbf{I} - 2s\mathbf{I} + 2s(1/s) = 0$$

$$10 - 4s\mathbf{I} - 2s^2\mathbf{I} + 2s = 0$$

$$\mathbf{I} = \frac{5 + s}{s(s + 2)}$$

and

$$i = \left(\frac{5}{2}\right) - \left(\frac{3}{2}\right)e^{-2t}$$

as previously determined.

15.5 CAPACITORS WITH INITIAL CONDITIONS

Consider the capacitor in Figure 15.18, where the initial voltage across the capacitor is $v(0)$. From Equation (10.2)

$$i = C\frac{dv}{dt}$$

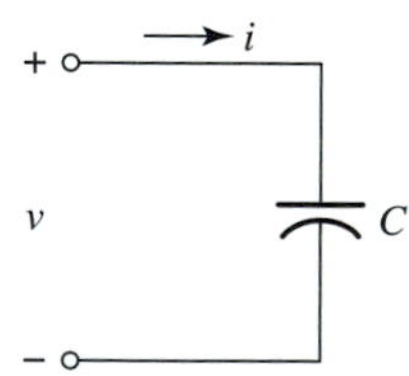

Figure 15.18 Capacitor circuit.

Taking the Laplace transform of both sides of this equation and using rules L1 and L3(a) of Chapter 12 results in

$$\mathscr{L}\{i\} = \mathscr{L}\left\{C\frac{dv}{dt}\right\}$$

$$\mathbf{I} = C\mathscr{L}\left\{\frac{dv}{dt}\right\}$$

$$= Cs\mathbf{V} - Cv(0)$$

The circuit in Figure 15.19 is a valid transformation of the circuit in Figure 15.18 in the s domain, because both circuits are described by the same equation in the s domain, as shown next.

Kirchhoff's current law applies to circuits in the s domain in the same manner that it applies to circuits in the t domain. Applying Kirchhoff's current law to the circuit in Figure 15.19 and using Equation (15.2) to obtain the capacitor current ($\mathbf{V}/(1/Cs)$) results in

$$\mathbf{I} - \frac{\mathbf{V}}{(1/Cs)} + Cv(0) = 0$$

$$\mathbf{I} = Cs\mathbf{V} - Cv(0)$$

which is the equation previously arrived at for **I** for the circuit in Figure 15.18.

Figure 15.19 Capacitor with initial condition.

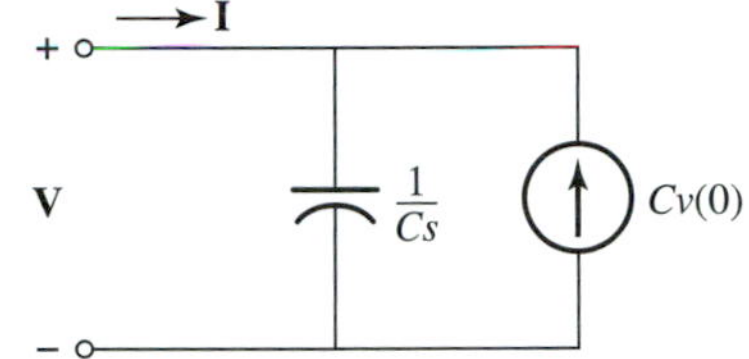

The rules in Figure 4.10 for converting sources in resistive circuits also hold in the s (Laplace) domain. In Figure 15.20 the current source conversion rule in Figure 4.10(b) is restated for the s domain.

Figure 15.20 Current source transformation.

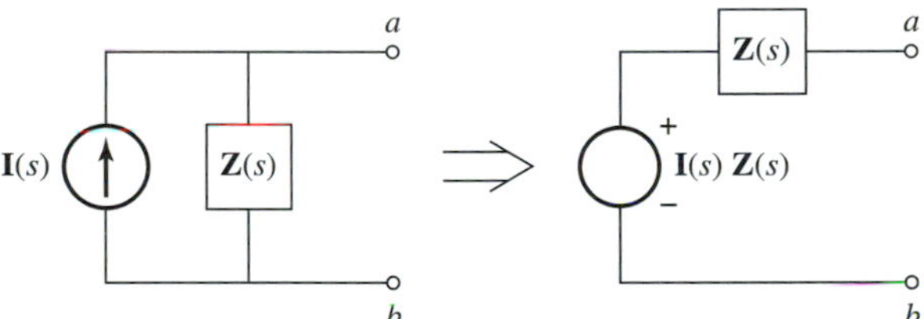

From Figure 15.20, the circuit in Figure 15.19 can be redrawn as shown in Figure 15.21. Notice that applying Kirchhoff's voltage law to the circuit in

Figure 15.21 Voltage source equivalent.

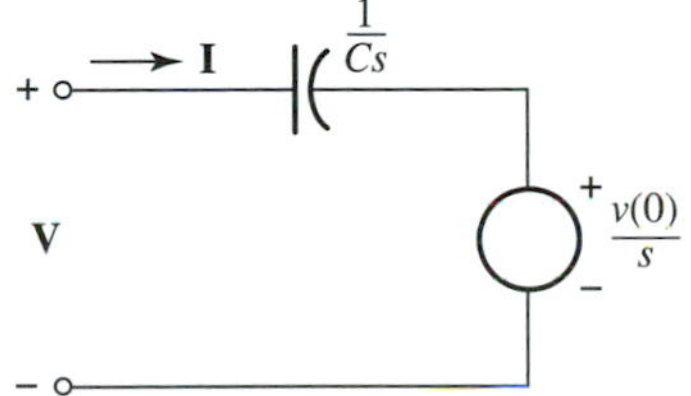

Figure 15.21 and using Equation (15.2) to write the voltage across the capacitor as $(\mathbf{I})(1/Cs)$ results in

$$\mathbf{V} - \left\{\frac{1}{Cs}\right\}\mathbf{I} - \frac{v(0)}{s} = 0$$

and

$$\mathbf{I} = Cs\mathbf{V} - Cv(0)$$

which is the same equation that describes the circuit in Figure 15.18.

In summary, the capacitor with initial voltage $v(0)$ can be represented in the s domain in either of the two ways shown in Figure 15.22.

Figure 15.22 Capacitor with initial condition.

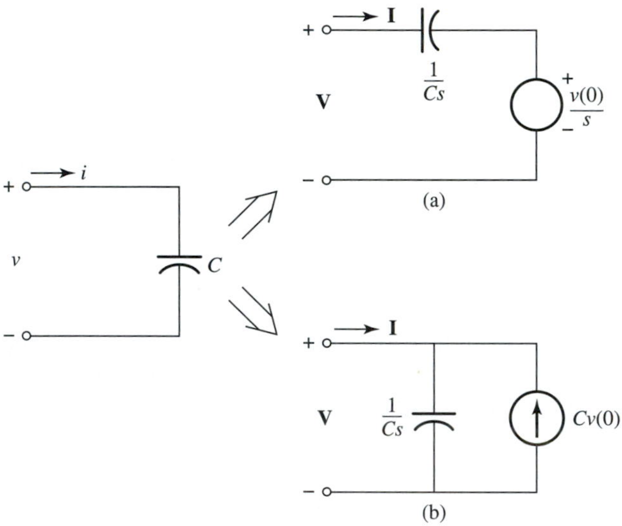

Consider the following example.

EXAMPLE 15.7 Determine the current i for $t \geq 0$ if $v_c(0) = 4$ V for the circuit in Figure 15.23. ■

Figure 15.23

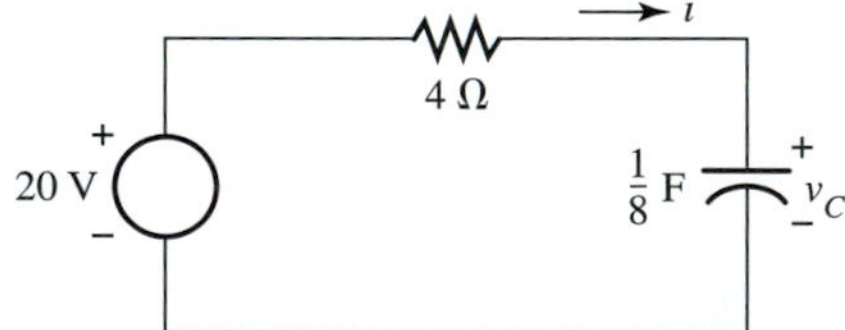

SOLUTION From Equations (15.3) and (15.5), entry 1 of Table 12.2, and the rule in Figure 15.22 for representing capacitors with initial conditions in the s domain, the preceding circuit in the s domain is shown in Figure 15.24.

Figure 15.24

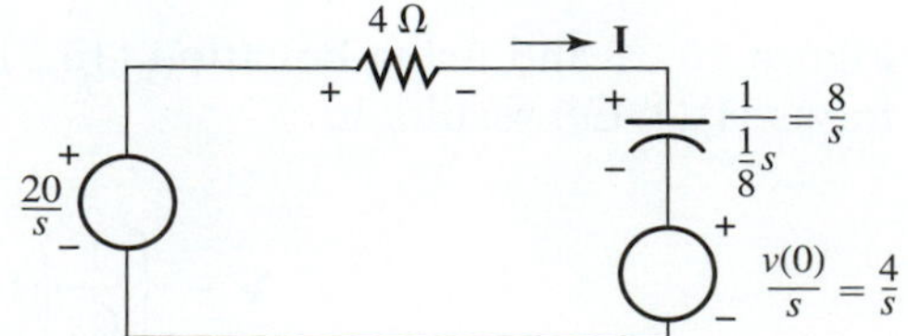

We use Equation (15.2) to determine the voltages across the resistor and capacitor as (4**I**) and {(8/s)**I**}, respectively. We then assign polarities as shown to the passive elements based on the assigned current-flow direction and apply Kirchhoff's voltage law to the loop, which results in

$$\frac{20}{s} - 4\mathbf{I} - \left(\frac{8}{s}\right)\mathbf{I} - \frac{4}{s} = 0$$

$$20 - 4s\mathbf{I} - 8\mathbf{I} - 4 = 0$$

$$\mathbf{I} = \frac{16}{4s + 8} = \frac{4}{s + 2}$$

From rule L3(b) of Chapter 12 and entry 2 of Table 12.2,

$$i = \mathscr{L}^{-1}\left\{\frac{4}{s + 2}\right\} = 4\mathscr{L}^{-1}\left\{\frac{1}{s + 2}\right\} = 4e^{-2t}$$

Alternatively, the capacitor with initial condition can be represented by a current source according to the transformation rule in Figure 15.22(b), as shown in Figure 15.25. The current representation is shown to the right of terminals *a-b*. $\mathbf{I}_1$ and $\mathbf{I}_2$ are mesh currents.

Figure 15.25

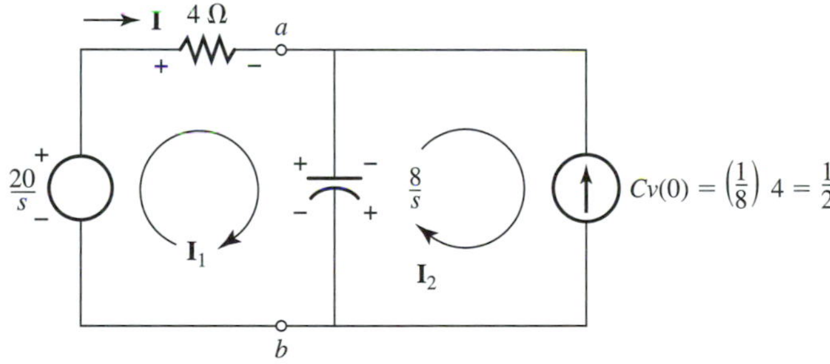

From an inspection of the circuit

$$\mathbf{I}_2 = -\frac{1}{2}$$

and applying Kirchhoff's voltage law to mesh 1 results in

$$\frac{20}{s} - 4\mathbf{I}_1 - \left(\frac{8}{s}\right)\mathbf{I}_1 + \left(\frac{8}{s}\right)\left(-\frac{1}{2}\right) = 0$$

Because $\mathbf{I}_1 = \mathbf{I}$ and $\mathbf{I}_2 = -\frac{1}{2}$,

$$\frac{20}{s} - 4\mathbf{I} - \left(\frac{8}{s}\right)\mathbf{I} + \left(\frac{8}{s}\right)\left(-\frac{1}{2}\right) = 0$$

$$20 - 4s\mathbf{I} - 8\mathbf{I} - 4 = 0$$

$$\mathbf{I} = \frac{4}{s+2}$$

and

$$i = 4e^{-2t}$$

as previously determined.

15.6 CONVERSION OF SOURCES IN THE *s* DOMAIN

The rules for the conversion of sources in the *t* domain given in Figure 4.10 for resistors also holds in the *s* domain, as previously shown and repeated in Figure 15.26.

Figure 15.26 Source transformations.

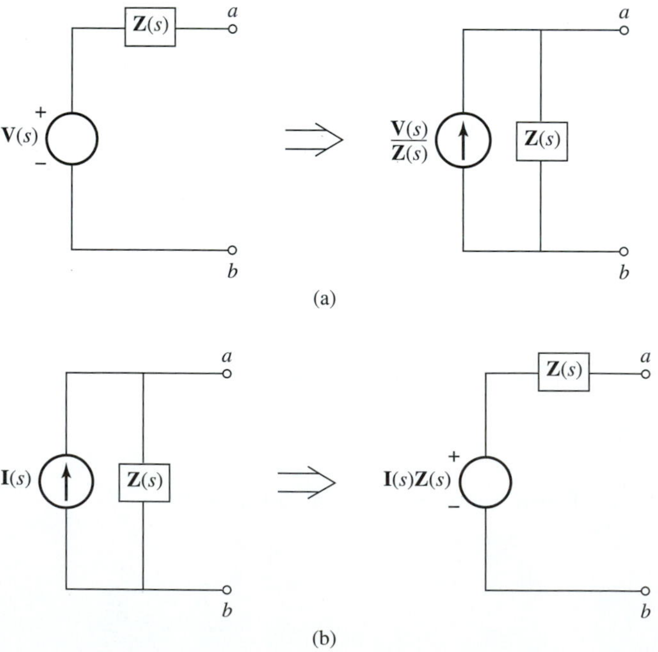

Consider the following examples.

EXAMPLE 15.8 Convert the current source in Figure 15.27 to a voltage source in the s domain. ■

Figure 15.27

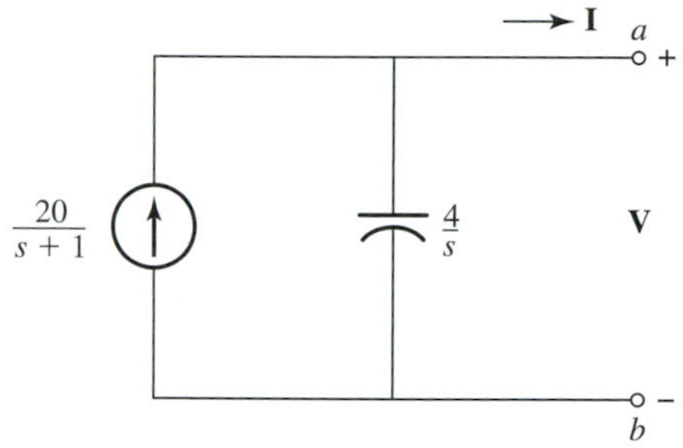

SOLUTION Applying the rule in Figure 15.26(b) to the circuit in the s domain results in the circuit shown in Figure 15.28.

Figure 15.28

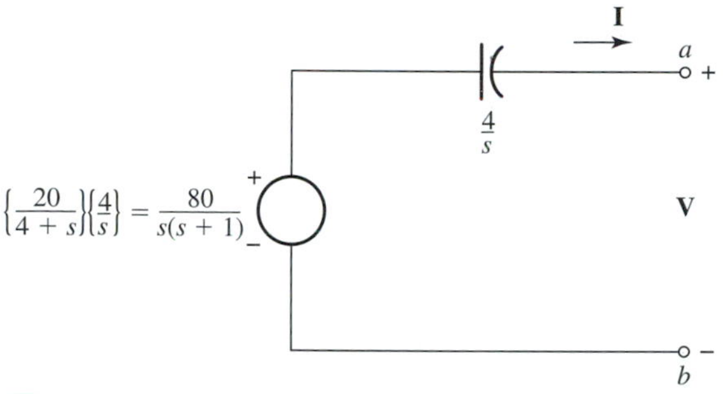

$$\left\{\frac{20}{4+s}\right\}\left\{\frac{4}{s}\right\} = \frac{80}{s(s+1)}$$

EXAMPLE 15.9 Convert the voltage source in Figure 15.29 to a current source in the s domain. ■

Figure 15.29

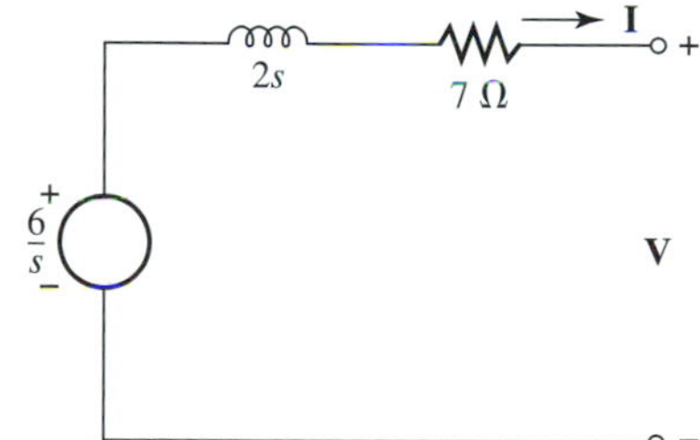

SOLUTION Applying the rule in Figure 15.26(a) to a circuit in the s domain results in the circuit shown in Figure 15.30.

Figure 15.30

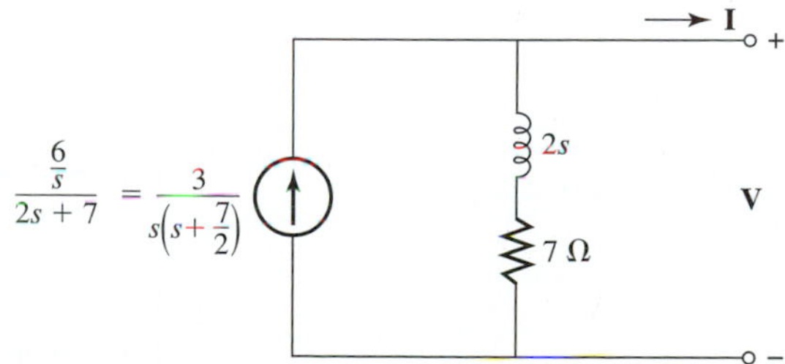

$$\frac{\frac{6}{s}}{2s+7} = \frac{3}{s\left(s+\frac{7}{2}\right)}$$

15.7 VOLTAGE DIVISION IN THE *s* DOMAIN

The voltage-division formula (Equation (4.3) and Figure 4.11) developed for resistors in Chapter 4 also holds in the *s* domain. In the *s* domain the voltage-division formula is

$$\mathbf{V}_x = \left\{\frac{\mathbf{Z}_x}{\mathbf{Z}_{\text{EQ}}}\right\}\mathbf{V} \tag{15.8}$$

where $\mathbf{Z}_{\text{EQ}}$ is the equivalent Laplace transform impedance for terminals *a-b*, $\mathbf{Z}_x$ is the Laplace transform impedance of the element across which $\mathbf{V}_x$ appears, and $\mathbf{V}$ is the Laplace transform of the voltage across terminals *a-b*.

The voltage-division formula is illustrated in the following example.

EXAMPLE 15.10 Determine v_o for the circuit in Figure 15.31. ■

Figure 15.31

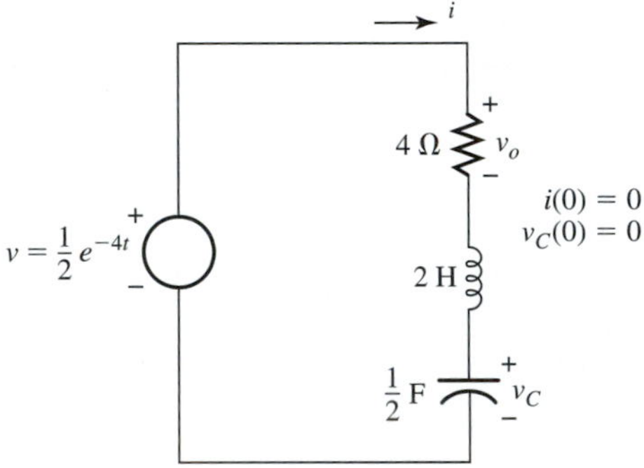

SOLUTION From Equations (15.3), (15.4), (15.5), and entry 2 of Table 12.2, the circuit in Figure 15.31 in the *s* domain is as shown in Figure 15.32.

Figure 15.32

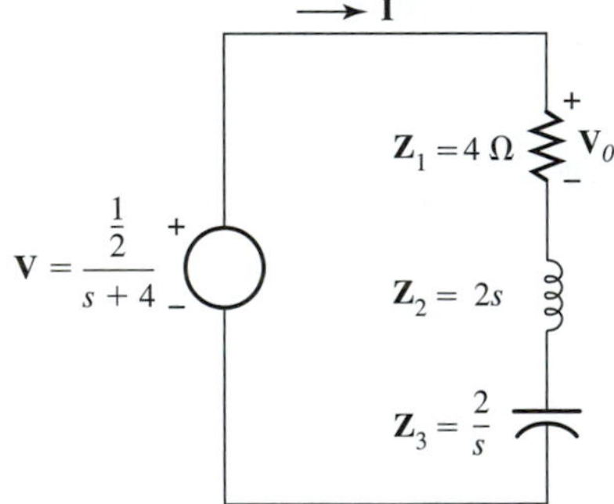

Applying Equation (15.8) for voltage division,

$$\mathbf{Z}_{\text{EQ}} = \mathbf{Z}_1 + \mathbf{Z}_2 + \mathbf{Z}_3$$

$$\begin{aligned}\mathbf{V}_o &= \left\{\frac{\mathbf{Z}_1}{\mathbf{Z}_1 + \mathbf{Z}_2 + \mathbf{Z}_3}\right\}\mathbf{V} \\ &= \left\{\frac{4}{2s + 4 + 2/s}\right\}\left\{\frac{\frac{1}{2}}{s+4}\right\} \\ &= \left\{\frac{2}{s + 2 + 1/s}\right\}\left\{\frac{\frac{1}{2}}{s+4}\right\}\end{aligned}$$

$$= \left\{\frac{2s}{s^2 + 2s + 1}\right\}\left\{\frac{\frac{1}{2}}{s + 4}\right\}$$

$$= \frac{s}{(s + 4)(s^2 + 2s + 1)}$$

$$= \frac{s}{(s + 4)(s + 1)^2}$$

Using the procedure presented in Chapter 12 to expand into partial fractions results in

$$\mathbf{V}_o = \frac{K_1}{(s + 1)^2} + \frac{K_2}{s + 1} + \frac{K_3}{s + 4}$$

$$K_1 = -\frac{1}{3}$$

$$K_2 = \frac{4}{9}$$

$$K_3 = -\frac{4}{9}$$

$$\mathbf{V}_o = \frac{\left(-\frac{1}{3}\right)}{(s + 1)^2} + \frac{\frac{4}{9}}{s + 1} + \frac{\left(-\frac{4}{9}\right)}{s + 4}$$

Taking the inverse Laplace transform of both sides of the equation results in

$$\mathcal{L}^{-1}\{\mathbf{V}_o\} = \mathcal{L}^{-1}\left\{\frac{\left(-\frac{1}{3}\right)}{(s + 1)^2} + \frac{\frac{4}{9}}{s + 1} + \frac{\left(-\frac{4}{9}\right)}{s + 4}\right\}$$

From rules L3(b) and L4(b)

$$v_o = -\frac{1}{3}\mathcal{L}^{-1}\left\{\frac{1}{(s + 1)^2}\right\} + \frac{4}{9}\mathcal{L}^{-1}\left\{\frac{1}{(s + 1)}\right\} - \frac{4}{9}\mathcal{L}^{-1}\left\{\frac{1}{s + 4}\right\}$$

Using entries 2 and 9 in Table 12.2 results in

$$v_o = \left(-\frac{1}{3}\right)te^{-t} + \left(\frac{4}{9}\right)e^{-t} - \left(\frac{4}{9}\right)e^{-4t}$$

15.8 CURRENT DIVISION IN THE *s* DOMAIN

The current-division formula (Equation (4.1) and Figure 4.1) developed for resistors in Chapter 4 also holds in the *s* domain. The current-division formula is

$$\mathbf{I}_x = \left\{\frac{\mathbf{Z}_{\text{EQ}}}{\mathbf{Z}_x}\right\}\mathbf{I} \tag{15.9}$$

where $\mathbf{Z}_{\text{EQ}}$ is the equivalent Laplace transform impedance for terminals *a*-*b*, $\mathbf{Z}_x$ is the Laplace transform impedance that $\mathbf{I}_x$ flows through, and $\mathbf{I}$ is the Laplace transform of the current entering the node.

The current-division formula is illustrated in the following example.

EXAMPLE 15.11 Determine i_1, i_2, v, and v_1 for the circuit in Figure 15.33. ■

Figure 15.33

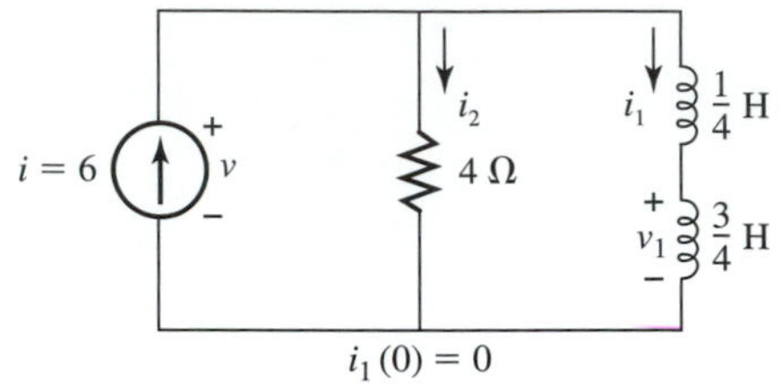

SOLUTION From Equations (15.3) and (15.4) and entry 1 of Table 12.2, the

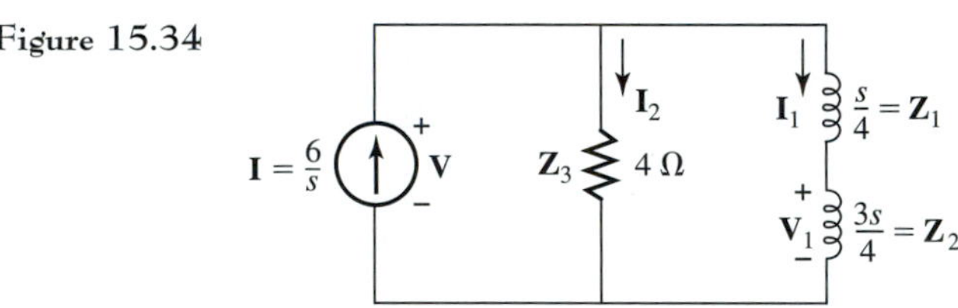

Figure 15.34

circuit (Figure 15.33) in the s domain is as shown in Figure 15.34. Applying Equation (15.9) for current division to the circuit in the s domain,

$$\mathbf{Z}_{\text{EQ}} = 4 \| \left(\frac{s}{4} + \frac{3s}{4}\right)$$

$$\mathbf{I}_1 = \left\{\frac{\mathbf{Z}_{\text{EQ}}}{\mathbf{Z}_1 + \mathbf{Z}_2}\right\}\left(\frac{6}{5}\right)$$

$$= \frac{24}{s(s + 4)}$$

Using the procedure presented in Chapter 12 to expand into partial fractions results in

$$\mathbf{I}_1 = \frac{K_1}{s} + \frac{K_2}{s + 4}$$

$$K_1 = 6$$

$$K_2 = -6$$

$$\mathbf{I}_1 = \frac{6}{s} + \frac{(-6)}{s + 4}$$

Taking the inverse Laplace transform of both sides of the equation results in

$$\mathcal{L}^{-1}\{\mathbf{I}_1\} = \mathcal{L}^{-1}\left\{\frac{6}{s} + \frac{(-6)}{s + 4}\right\}$$

From rules L3(b) and L4(b),

$$i_1 = 6\mathcal{L}^{-1}\left\{\frac{1}{s}\right\} - 6\mathcal{L}^{-1}\left\{\frac{1}{s+4}\right\}$$

From entries 1 and 2 of Table 12.2,

$$i_1 = 6 - 6e^{-4t}$$

From Kirchhoff's current law,

$$\begin{aligned}
\frac{6}{s} &= \mathbf{I}_2 + \mathbf{I}_1 \\
&= \mathbf{I}_2 + \frac{24}{s(s+4)} \\
\mathbf{I}_2 &= \frac{6}{s} - \frac{24}{s(s+4)} \\
&= \frac{6s + 24 - 24}{s(s+4)} \\
&= \frac{6}{s+4}
\end{aligned}$$

and from Equation (15.2),

$$\begin{aligned}
\mathbf{V} &= 4\mathbf{I}_2 \\
&= 4\left\{\frac{6}{s+4}\right\} \\
&= \frac{24}{s+4} \\
\mathbf{V}_1 &= \left\{\frac{3s}{4}\right\}\mathbf{I}_1 \\
&= \left\{\frac{3s}{4}\right\}\left\{\frac{24}{s(s+4)}\right\} \\
&= \frac{18}{s+4}
\end{aligned}$$

From rule L3(b) and entry 2 of Table 12.2,

$$\begin{aligned}
\mathcal{L}^{-1}\{\mathbf{I}_2\} &= \mathcal{L}^{-1}\left\{\frac{6}{s+4}\right\} \\
i_2 &= 6\mathcal{L}^{-1}\left\{\frac{1}{s+4}\right\} \\
&= 6e^{-4t}
\end{aligned}$$

$$\mathcal{L}^{-1}\{\mathbf{V}\} = \mathcal{L}^{-1}\left\{\frac{24}{s+4}\right\}$$

$$v = 24\mathcal{L}^{-1}\left\{\frac{1}{s+4}\right\}$$

$$= 24e^{-4t}$$

$$\mathcal{L}^{-1}\{\mathbf{V}_1\} = \mathcal{L}^{-1}\left\{\frac{18}{s+4}\right\}$$

$$v_1 = 18\mathcal{L}^{-1}\left\{\frac{1}{s+4}\right\}$$

$$= 18e^{-4t}$$

Alternatively, applying Equations (15.6) and (15.7) to circuits in the s domain results in

$$\mathbf{Z} = (4)\|\left\{\frac{s}{4} + \frac{3s}{4}\right\}$$

$$= 4\|s$$

$$= \frac{4s}{s+4}$$

which is the Laplace impedance seen by the current source. From Equation (15.2)

$$\mathbf{Z} = \frac{\mathbf{V}}{(6/s)}$$

$$\mathbf{V} = \mathbf{Z}\left(\frac{6}{s}\right)$$

$$= \left\{\frac{4s}{s+4}\right\}\left\{\frac{6}{s}\right\}$$

$$= \frac{24}{s+4}$$

$$\mathbf{I}_2 = \frac{\mathbf{V}}{4}$$

$$= \frac{24/(s+4)}{4}$$

$$= \frac{6}{s+4}$$

$$\mathbf{I}_1 = \frac{\mathbf{V}}{s/4 + 3s/4}$$

$$= \frac{24/(s+4)}{s}$$

$$= \frac{24}{s(s+4)}$$

$$\mathbf{V}_1 = \left\{\frac{3s}{4}\right\}\mathbf{I}_1$$
$$= \left\{\frac{3s}{4}\right\}\left\{\frac{24}{s(s+4)}\right\}$$
$$= \frac{18}{s+4}$$

The inverses of $\mathbf{V}$, $\mathbf{I}_2$, $\mathbf{I}_1$, and $\mathbf{V}_1$ can be determined as demonstrated previously.

15.9 MESH ANALYSIS IN THE *s* DOMAIN

Mesh analysis in the s domain is very similar to mesh analysis in the t domain, as presented in Chapter 6. The following example illustrates mesh analysis in the s domain.

EXAMPLE 15.12 Determine the voltage v for the circuit in Figure 15.35. ■

Figure 15.35

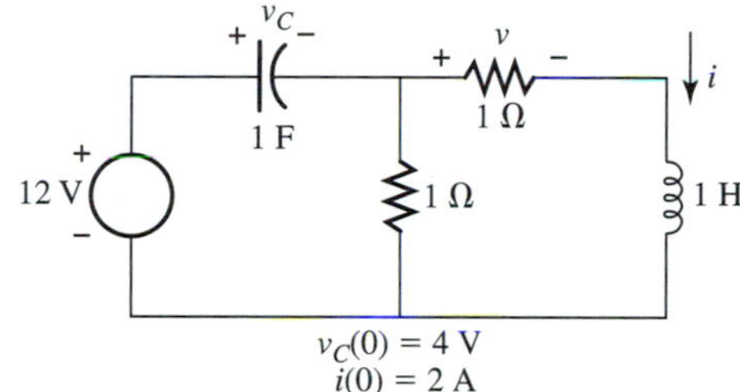

SOLUTION Employing Equations (15.3), (15.4), and (15.5), entry 1 in Table 12.2, and the rules in Figures 15.14(a) and 15.22(a) to transform this circuit (Figure 15.35) to the s domain results in the circuit in Figure 15.36.

Figure 15.36

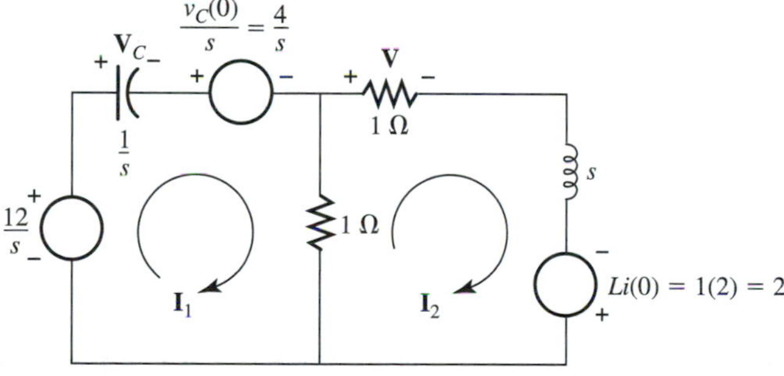

Applying the uniform approach to the mesh analysis method, as presented in Chapter 6, to this circuit (Figure 15.36) in the s domain results in

$$\mathbf{I}_2 = \frac{\begin{vmatrix} 1+\dfrac{1}{s} & \dfrac{12}{s}-\dfrac{4}{s} \\ -1 & 2 \end{vmatrix}}{\begin{vmatrix} 1+\dfrac{1}{s} & -1 \\ -1 & 1+1+s \end{vmatrix}}$$

Using the rule for evaluating a 2×2 determinant given in Figure 5.2 results in

$$\begin{aligned}
\mathbf{I}_2 &= \frac{2\left(1 + \frac{1}{s}\right) + \left(\frac{8}{s}\right)}{\left(1 + \frac{1}{s}\right)(2 + s) - 1} \\
&= \frac{2\left(\frac{s + 1}{s}\right) + \left(\frac{8}{s}\right)}{\left(\frac{s + 1}{s}\right)(2 + s) - 1} \\
&= \frac{2s + 2 + 8}{(s + 1)(s + 2) - s} \\
&= \frac{2s + 10}{s^2 + 3s + 2 - s} \\
&= \frac{2s + 10}{s^2 + 2s + 2}
\end{aligned}$$

From Equations (15.2) and (15.3),

$$\begin{aligned}
\mathbf{V} &= R\mathbf{I}_2 \\
&= (1)\left\{\frac{2s + 10}{s^2 + 2s + 2}\right\} \\
&= \frac{2s + 10}{s^2 + 2s + 2}
\end{aligned}$$

Algebraically changing the form of the expression for $\mathbf{V}$ to compare with entries 7 and 8 of Table 12.2 results in

$$\begin{aligned}
\mathbf{V} &= \frac{2s + 2 + 8}{s^2 + 2s + 2} \\
&= 2\left\{\frac{s + 1}{s^2 + 2s + 2}\right\} + 8\left\{\frac{1}{s^2 + 2s + 2}\right\} \\
&= 2\left\{\frac{s + 1}{(s + 1)^2 + 1}\right\} + 8\left\{\frac{1}{(s + 1)^2 + 1}\right\}
\end{aligned}$$

Taking the inverse Laplace transform of both sides of the equation using rule L3(b) of Chapter 12 and entries 7 and 8 from Table 12.2 results in

$$\begin{aligned}
\mathscr{L}^{-1}\{\mathbf{V}\} &= \mathscr{L}^{-1}\left\{2\left\{\frac{s + 1}{(s + 1)^2 + 1}\right\}\right\} + \mathscr{L}^{-1}\left\{8\left\{\frac{1}{(s + 1)^2 + 1}\right\}\right\} \\
v &= 2\mathscr{L}^{-1}\left\{\frac{s + 1}{(s + 1)^2 + 1}\right\} + 8\mathscr{L}^{-1}\left\{\frac{1}{(s + 1)^2 + 1}\right\} \\
&= 2e^{-t}\cos t + 8e^{-t}\sin t
\end{aligned}$$

15.10 NODAL ANALYSIS IN THE *s* DOMAIN

Nodal analysis in the s domain is very similar to nodal analysis in the t domain (Chapter 6). The following example illustrates nodal analysis in the s domain.

EXAMPLE 15.13 Determine the voltage v for the circuit in Figure 15.37. ■

Figure 15.37

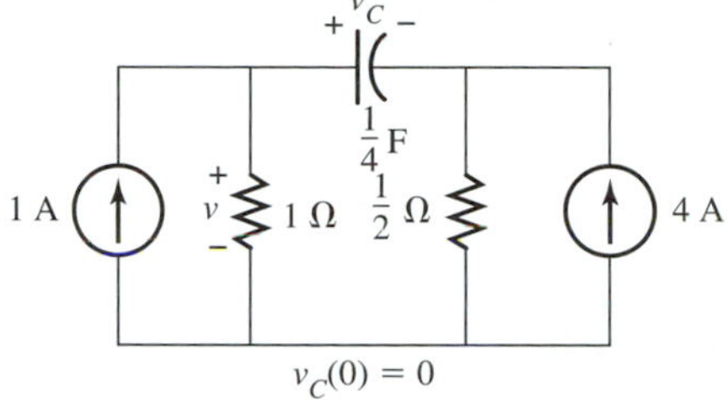

SOLUTION Employing Equations (15.3) and (15.5) and entry 1 in Table 12.2 to transform this circuit (Figure 15.37) to the s domain results in the circuit in Figure 15.38.

Figure 15.38

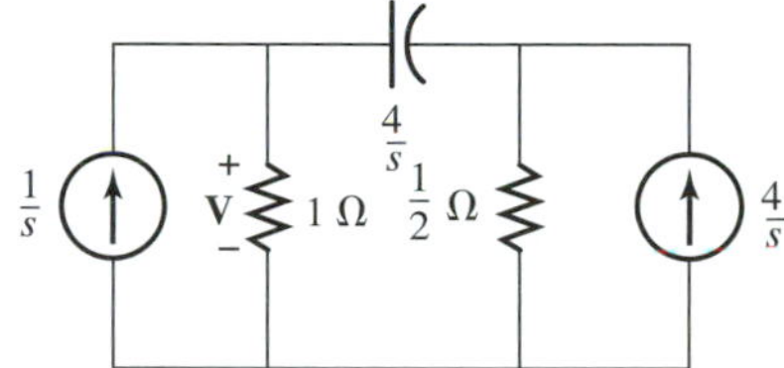

Because the Laplace impedance is given by Equation (15.2) as $\mathbf{Z} = \mathbf{V}/\mathbf{I}$, it follows that the Laplace admittance is

$$\mathbf{Y} = \frac{1}{\mathbf{Z}}$$

From Equation (15.3) for the resistor,

$$\mathbf{Y}_R = \frac{1}{\mathbf{Z}_R} = \frac{1}{R}$$

and from Equation (15.5) for the capacitor,

$$\mathbf{Y}_C = \frac{1}{\mathbf{Z}_C} = \frac{1}{1/sC} = sC$$

Replacing the Laplace impedances for R and C with Laplace admittances and designating the nodes, as required by the uniform approach for nodal analysis presented in Chapter 6, results in the circuit in Figure 15.39.

Figure 15.39

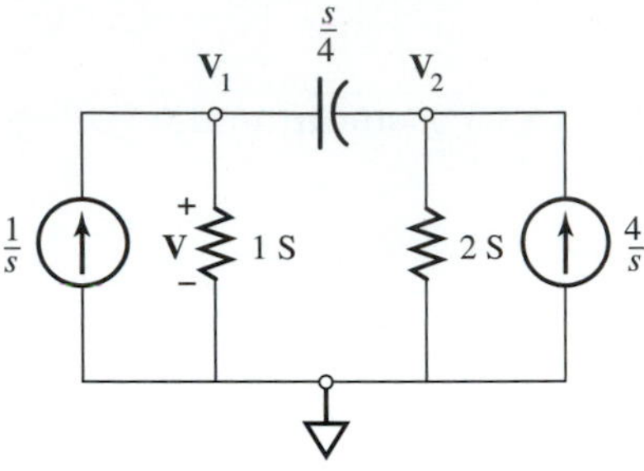

Applying the method of uniform approach for nodal analysis presented in Chapter 6 to the circuit in Figure 15.39 results in

$$\mathbf{V}_1 = \frac{\begin{vmatrix} \frac{1}{s} & -\frac{s}{4} \\ \frac{4}{s} & 2+\frac{s}{4} \end{vmatrix}}{\begin{vmatrix} 1+\frac{s}{4} & -\frac{s}{4} \\ -\frac{s}{4} & 2+\frac{s}{4} \end{vmatrix}}$$

Using the rule for evaluating a 2×2 determinant in Figure 5.2 results in

$$\begin{aligned}\mathbf{V}_1 &= \frac{(1/s)\{2+(s/4)\} - (4/s)(-s/4)}{\{1+(s/4)\}\{2+(s/4)\} - (-s/4)(-s/4)} \\ &= \frac{(2/s)+(1/4)+1}{(s^2/16)+(s/2)+2+(s/4)-(s^2/16)} \\ &= \frac{5s+8}{s(3s+8)} \\ &= \frac{(5/3)s+(8/3)}{s(s+8/3)}\end{aligned}$$

Using the procedure presented in Chapter 12 to expand into partial fractions results in

$$\mathbf{V}_1 = \frac{K_1}{s} + \frac{K_2}{s+\left(\frac{8}{3}\right)}$$

$$K_1 = 1$$

$$K_2 = \frac{2}{3}$$

$$\mathbf{V}_1 = \frac{1}{s} + \frac{\frac{2}{3}}{\left(s+\frac{8}{3}\right)}$$

Taking the inverse Laplace transform of each side of the equation results in

$$\mathcal{L}^{-1}\{\mathbf{V}_1\} = \mathcal{L}^{-1}\left\{\frac{1}{s} + \frac{\frac{2}{3}}{\left(s + \frac{8}{3}\right)}\right\}$$

From rules L3(b) and L4(b)

$$v_1 = \mathcal{L}^{-1}\left\{\frac{1}{s}\right\} + \left(\frac{2}{3}\right)\mathcal{L}^{-1}\left\{\frac{1}{s + 8/3}\right\}$$

From entries 1 and 2 of Table 12.2,

$$v_1 = 1 + \left(\frac{2}{3}\right)e^{-(8/3)t}$$

and because $v = v_1$,

$$v = 1 + \left(\frac{2}{3}\right)e^{-(8/3)t}$$

15.11 THEVENIN'S THEOREM IN THE *s* DOMAIN

Thevenin's theorem as presented in Chapter 7 also holds for circuits in the *s* domain. The following example illustrates the application of Thevenin's theorem in the *s* domain.

EXAMPLE 15.14 Determine the voltage *v* for the circuit in Example 15.13 using Thevenin's theorem. ■

SOLUTION We begin with the transformed circuit from Example 15.13 with the passive elements expressed as Laplace impedances as shown in Figure 15.40.

Figure 15.40

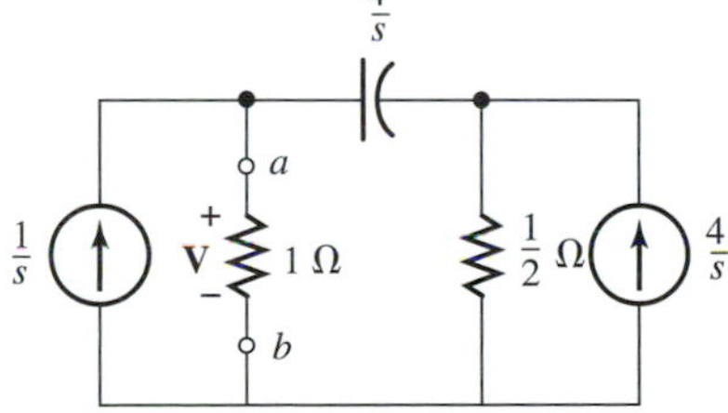

The application of Thevenin's theorem from Chapter 7 to this circuit (Figure 15.40) in the *s* domain requires the removal of the 1-Ω resistor, as shown in Figure 15.41, and the determination of the resulting open-circuited voltage.

Figure 15.41

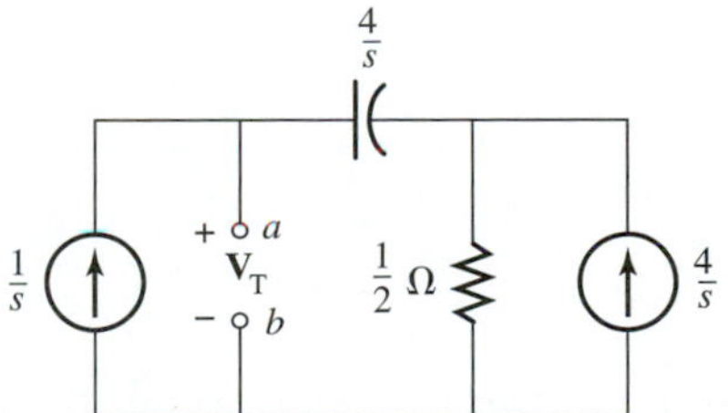

Applying the superposition method results in the circuits in Figures 15.42 and 15.43, where $\mathbf{V}'_T$ and $\mathbf{V}''_T$ are the contributions to $\mathbf{V}_T$ from the Laplace-transformed current sources $(1/s)$ and $(4/s)$, respectively.

Figure 15.42

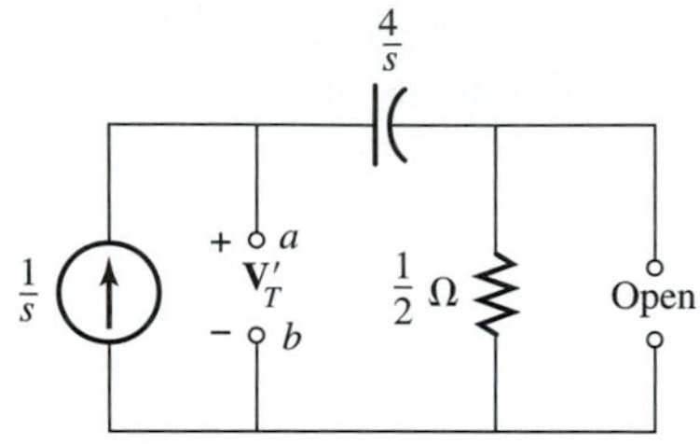

Figure 15.43

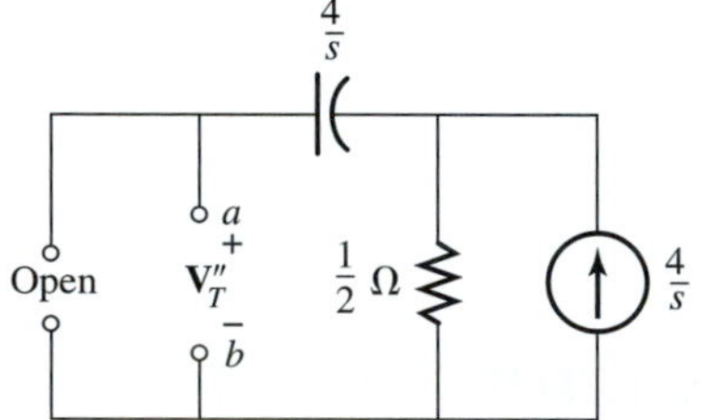

Applying Equations (15.6) and (15.2) to the circuit in Figure 15.42 results in

$$\begin{aligned}\mathbf{V}'_T &= \mathbf{IZ} \\ &= \left(\frac{1}{s}\right)\left\{\frac{4}{s} + \frac{1}{2}\right\} \\ &= \frac{8 + s}{s(2s)} \\ &= \frac{8 + s}{2s^2}\end{aligned}$$

Applying Equation (15.2) to the circuit in Figure 15.43 results in

$$\begin{aligned}\mathbf{V}''_T &= \mathbf{IZ} \\ &= \left(\frac{4}{s}\right)\left(\frac{1}{2}\right) \\ &= \frac{2}{s}\end{aligned}$$

because no current flows through the capacitor.

From the superposition method

$$\begin{aligned}\mathbf{V}_T &= \mathbf{V}'_T + \mathbf{V}''_T \\ &= \frac{8+s}{2s^2} + \frac{2}{s} \\ &= \frac{8+s+4s}{2s^2} \\ &= \frac{5s+8}{2s^2}\end{aligned}$$

Replacing both current sources by opens, as required by Thevenin's theorem to determine the Thevenin impedance, results in Figure 15.44.

Figure 15.44

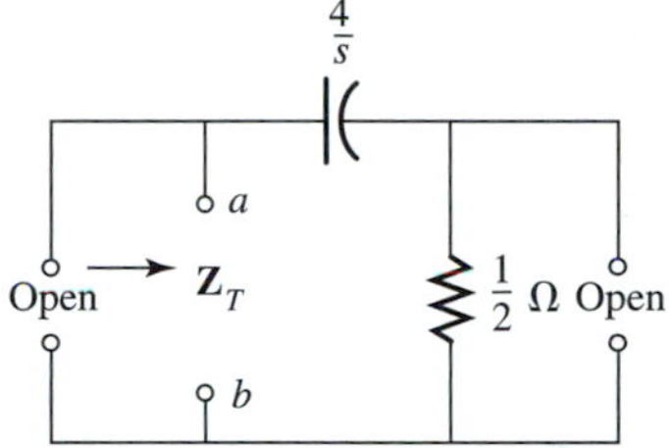

Applying Equation (15.6) to the circuit in the s domain

$$\begin{aligned}\mathbf{Z}_T &= \frac{4}{s} + \frac{1}{2} \\ &= \frac{s+8}{2s}\end{aligned}$$

and the Thevenin equivalent circuit for terminals a-b is as shown in Figure 15.45.

Figure 15.45

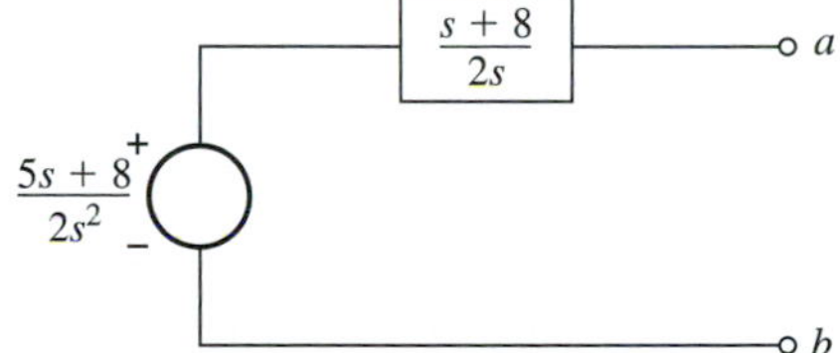

If the 1-Ω resistor is reconnected across terminals a-b, then $\mathbf{V}$ can be determined by applying Equation (15.8) for voltage division to the circuit in Figure 15.46.

Figure 15.46

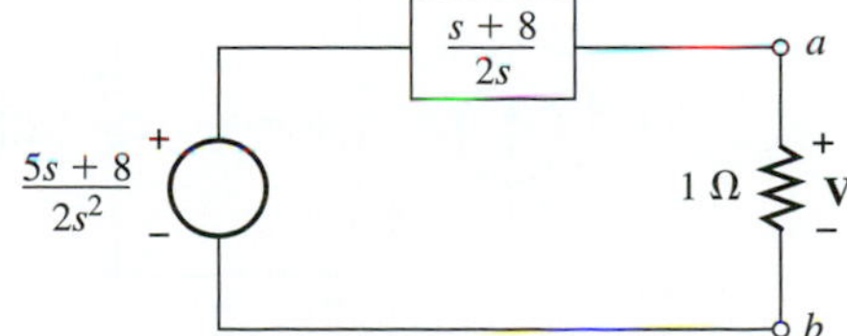

$$\mathbf{V} = \left\{\frac{1}{1 + \dfrac{s+8}{2s}}\right\}\left\{\frac{5s+8}{2s^2}\right\}$$

$$= \left\{\frac{2s}{3s+8}\right\}\left\{\frac{5s+8}{2s^2}\right\}$$

$$= \frac{5s+8}{s(3s+8)}$$

The inverse Laplace transform of $\mathbf{V}$ is

$$v = 1 + \left(\frac{2}{3}\right)e^{-(8/3)t}$$

as determined in Example 15.13.

15.12 NORTON'S THEOREM IN THE *s* DOMAIN

Norton's theorem as presented in Chapter 7 also holds for circuits in the s domain. The following example illustrates the application of Norton's theorem in the s domain.

EXAMPLE 15.15 Determine the voltage v for the circuit in Example 15.13 using Norton's theorem. ■

SOLUTION We begin with the transformed circuit from Example 15.13 with the passive elements expressed as Laplace impedances, as shown in Figure 15.47.

Figure 15.47

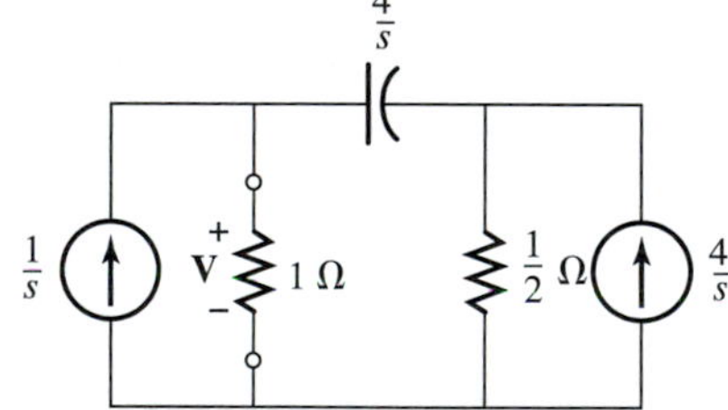

The application of Norton's theorem from Chapter 7 to this circuit (Figure 15.47) in the s domain requires the removal of the 1-Ω resistor, as shown in Figure 15.48, and the determination of the resulting short-circuited current.

Figure 15.48

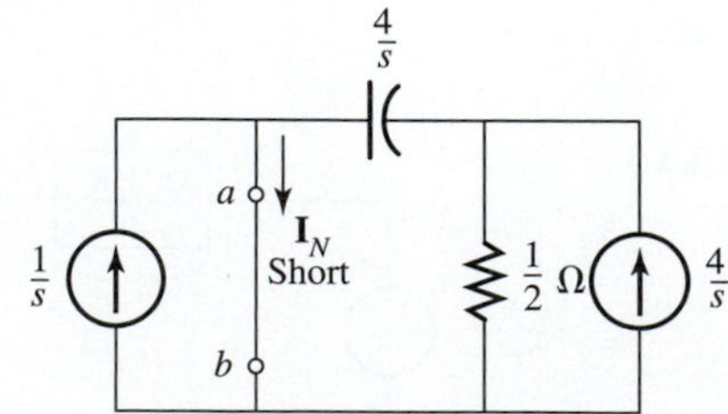

Applying the superposition method from Chapter 7 results in the circuits in Figures 15.49 and 15.50, where $\mathbf{I}_N'$ and $\mathbf{I}_N''$ are the contributions to $\mathbf{I}_N$ from the Laplace-transformed current sources $(1/s)$ and $(4/s)$, respectively.

Figure 15.49

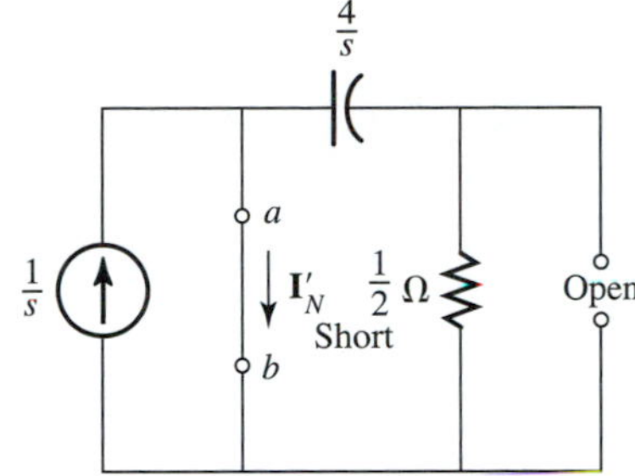

Figure 15.50

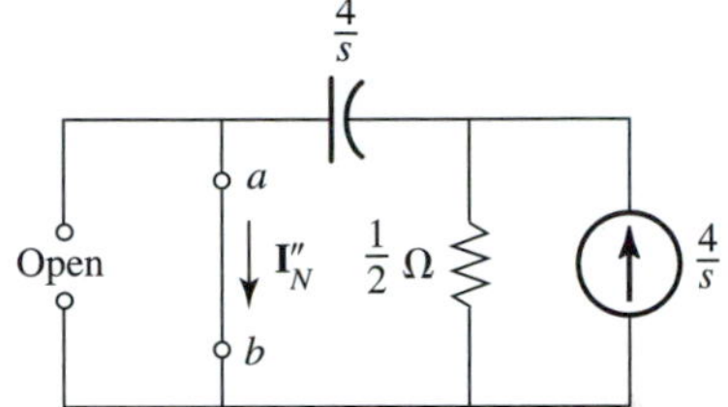

From an inspection of the circuit in Figure 15.49,

$$\mathbf{I}_N' = \frac{1}{s}$$

Applying Equation (15.9) for current division to the circuit in Figure 15.49 results in

$$\mathbf{I}_N'' = \left\{\frac{\frac{1}{2}}{\left(\frac{1}{2}\right) + (4/s)}\right\}\left\{\frac{4}{s}\right\}$$

$$= \frac{4}{s+8}$$

From the superposition method

$$\mathbf{I}_N = \mathbf{I}_N' + \mathbf{I}_N''$$

$$= \frac{1}{s} + \frac{4}{s+8}$$

$$= \frac{5s+8}{s(s+8)}$$

Because Thevenin and Norton impedances are equal,

$$\mathbf{Z}_N = \frac{s+8}{2s}$$

from Example 15.14. The Norton equivalent circuit for terminals a-b is as shown in Figure 15.51.

Figure 15.51

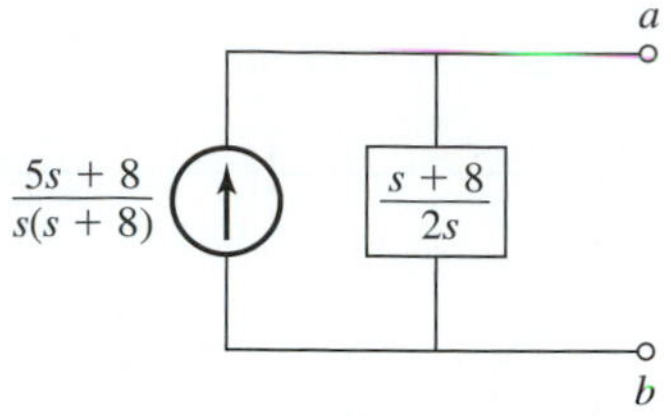

If the 1-Ω resistor is reconnected across terminals a-b, the voltage $\mathbf{V}$ can be determined by applying Equations (15.2) and (15.7) to the circuit in Figure 15.52.

Figure 15.52

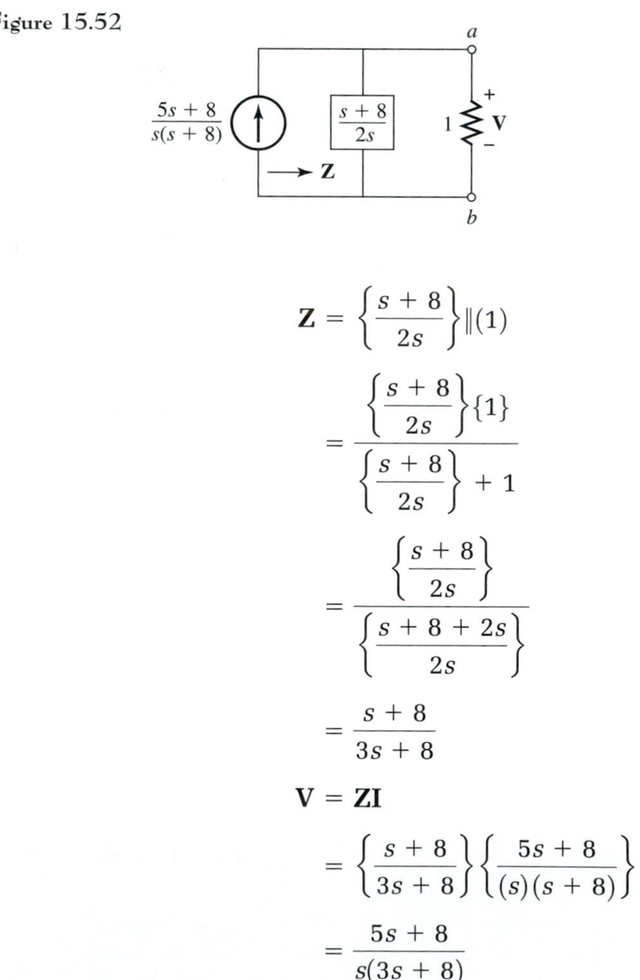

$$\mathbf{Z} = \left\{\frac{s+8}{2s}\right\} \| (1)$$

$$= \frac{\left\{\dfrac{s+8}{2s}\right\}\{1\}}{\left\{\dfrac{s+8}{2s}\right\} + 1}$$

$$= \frac{\left\{\dfrac{s+8}{2s}\right\}}{\left\{\dfrac{s+8+2s}{2s}\right\}}$$

$$= \frac{s+8}{3s+8}$$

$$\mathbf{V} = \mathbf{ZI}$$

$$= \left\{\frac{s+8}{3s+8}\right\}\left\{\frac{5s+8}{(s)(s+8)}\right\}$$

$$= \frac{5s+8}{s(3s+8)}$$

The inverse Laplace transform of $\mathbf{V}$ is

$$v = 1 + \left(\frac{2}{3}\right)e^{-(8/3)t}$$

as determined in Example 15.13.

Alternatively, the Laplace transform voltage **V** can be determined using current division. Applying Equation (15.9) to the Norton equivalent circuit in Figure 15.52 results in

$$\mathbf{I}_{ab} = \left\{ \frac{\left\{ \dfrac{s+8}{2s} \right\}}{1 + \left\{ \dfrac{s+8}{2s} \right\}} \right\} \left\{ \frac{5s+8}{s(s+8)} \right\}$$

$$= \left\{ \frac{s+8}{3s+8} \right\} \left\{ \frac{5s+8}{s(s+8)} \right\}$$

$$= \frac{5s+8}{s(3s+8)}$$

and from Equation (15.2)

$$\mathbf{V} = (1)\mathbf{I}_{ab}$$

$$= \frac{5s+8}{s(3s+8)}$$

as previously determined.

15.13 EXERCISES

1. Determine the transfer function relating v (input) to i (output) for each of the circuits in Figure 15.53.

Figure 15.53

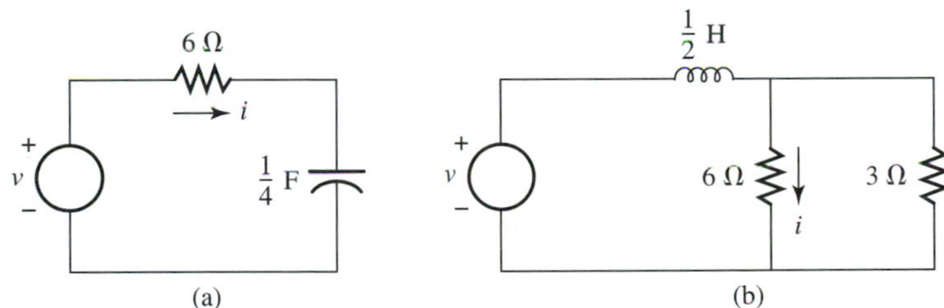

2. Determine the transfer function relating v (input) to i (output) for each of the circuits in Figure 15.54.

Figure 15.54

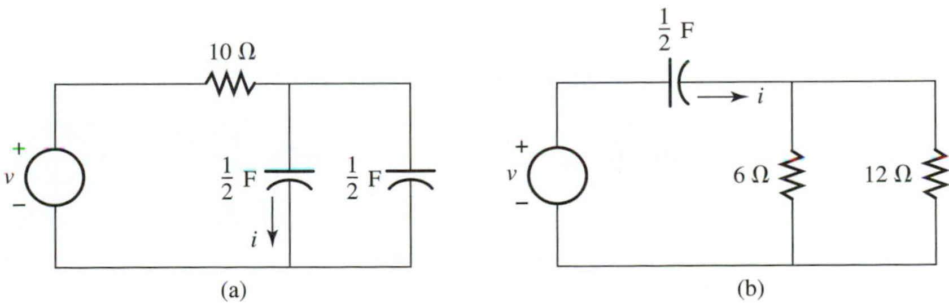

3. Determine the Laplace impedance for terminals a-b for each circuit in Figure 15.55.

Figure 15.55

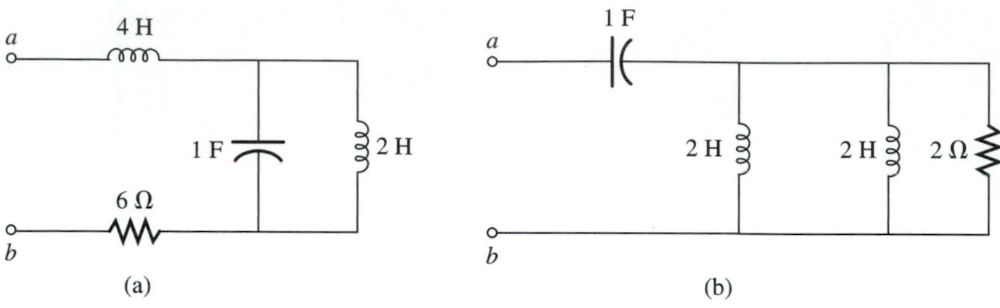

4. Determine the Laplace impedance for terminals a-b for each circuit in Figure 15.56.

Figure 15.56

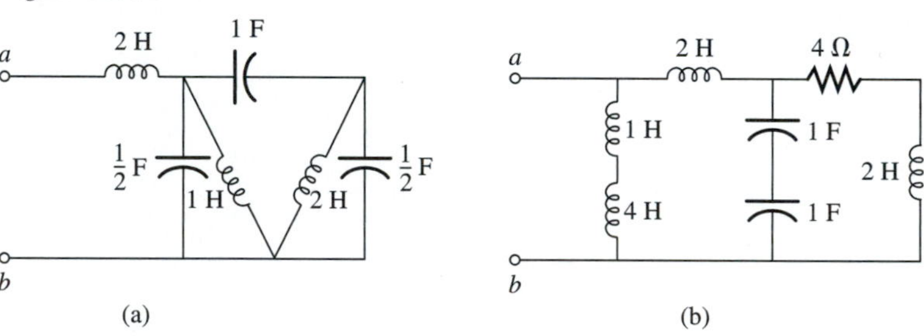

5. Transform each of the circuits in Figure 15.57 to the s domain after the switch is closed at $t = 0$.

Figure 15.57

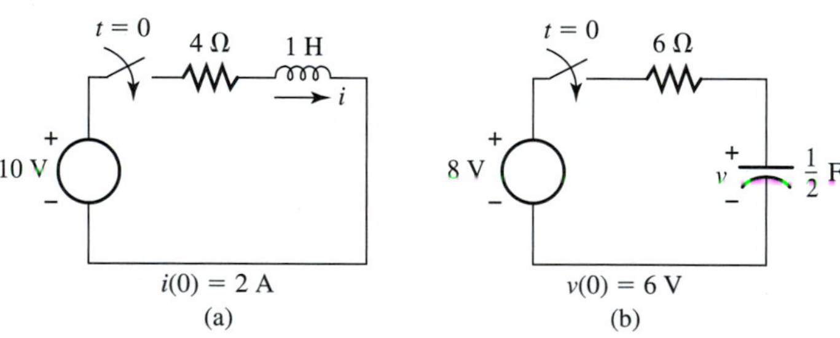

6. Transform each of the circuits in Figure 15.58 to the s domain after the switch is closed at $t = 0$.

Figure 15.58

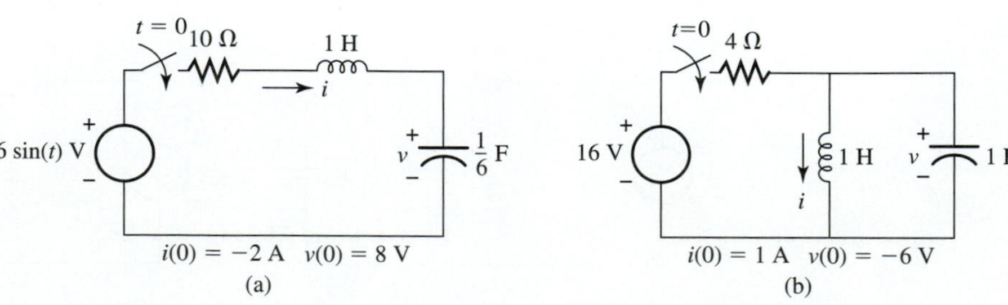

7. Transform each of the circuits in Figure 15.59 to the s domain after the switch is closed at $t = 0$.

Figure 15.59

$t = 0$, 20 V, 4 Ω, $\frac{1}{2}$ F, 1 H, v, i

$v(0) = 0$ $i(0) = -1$ A

(a)

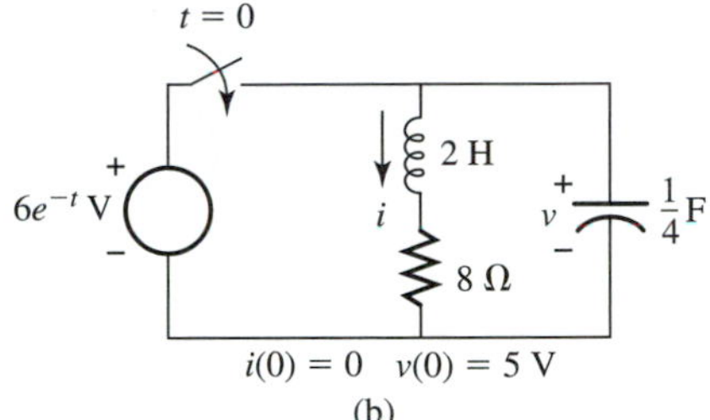

$i(0) = 0$ $v(0) = 5$ V

(b)

8. Transform each of the circuits in Figure 15.60 to the s domain and solve for $\mathbf{I}(s)$ and $\mathbf{V}(s)$ after the switch is closed at $t = 0$.

Figure 15.60

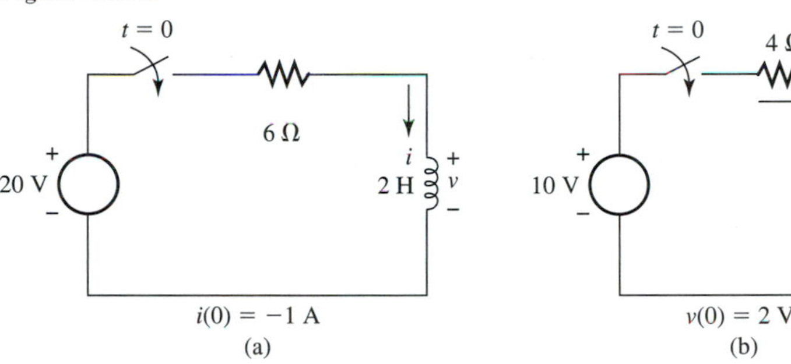

$i(0) = -1$ A

(a)

$v(0) = 2$ V

(b)

9. Transform each of the circuits in Figure 15.61 to the s domain and solve for $\mathbf{I}(s)$ and $\mathbf{V}(s)$ after the switch is closed at $t = 0$.

Figure 15.61

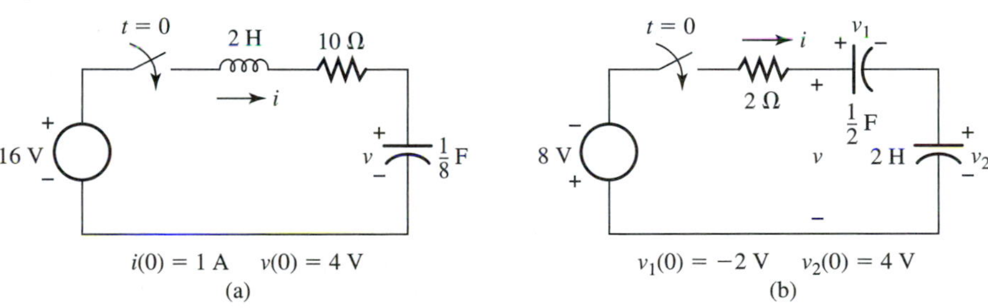

$i(0) = 1$ A $v(0) = 4$ V

(a)

$v_1(0) = -2$ V $v_2(0) = 4$ V

(b)

10. Determine v_0 for each circuit in Figure 15.62 by first converting each circuit to the s domain and finding $\mathbf{V}_0(s)$.

Figure 15.62

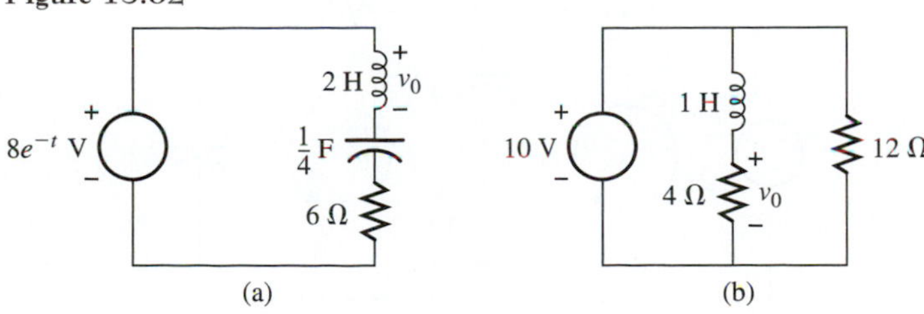

11. Determine v_0 for each circuit in Figure 15.63 by first converting each circuit to the s domain and finding $\mathbf{V}_0(s)$.

Figure 15.63

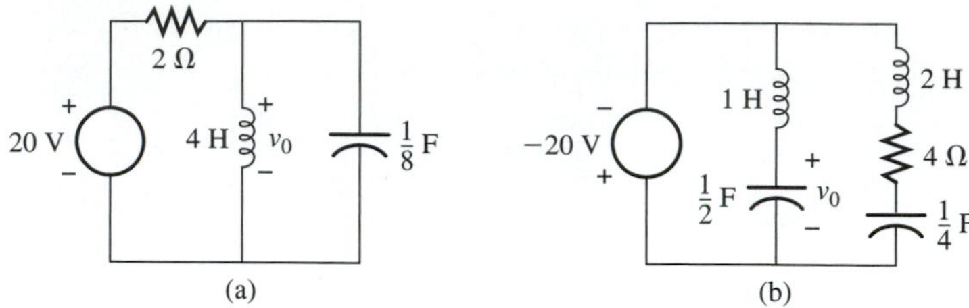

12. Determine i for the circuit in Figure 15.64 by first converting the circuit to the s domain and finding $\mathbf{I}(s)$.

Figure 15.64

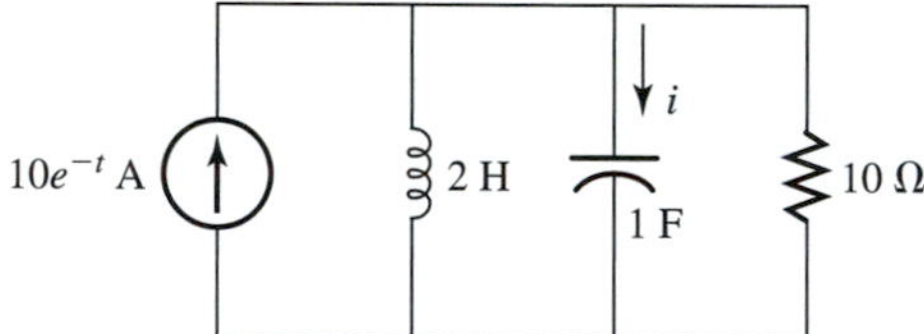

13. Determine i for the circuit in Figure 15.65 by first converting the circuit to the s domain and finding $\mathbf{I}(s)$.

Figure 15.65

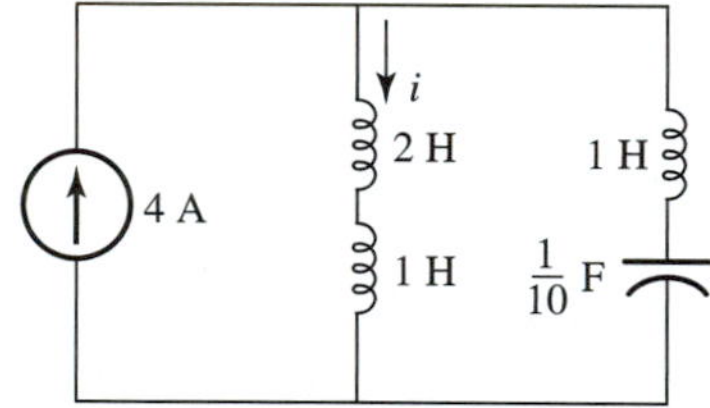

14. Determine i for the circuit in Figure 15.66 by first converting the circuit to the s domain and finding $\mathbf{I}(s)$.

Figure 15.66

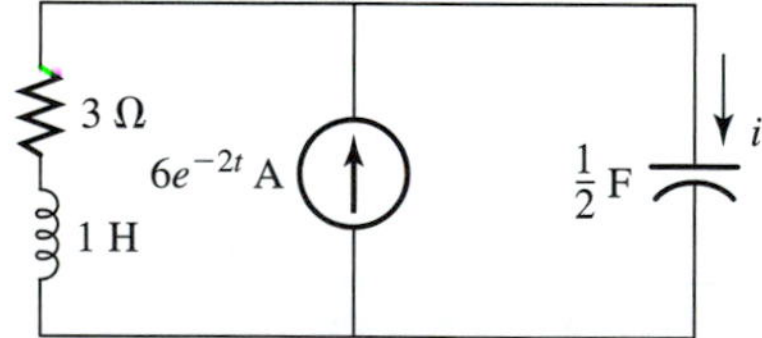

15. Determine the current i for the circuit in Figure 15.67 by first converting the circuit to the s domain and finding $\mathbf{I}(s)$. (Use the Mesh method).

Figure 15.67

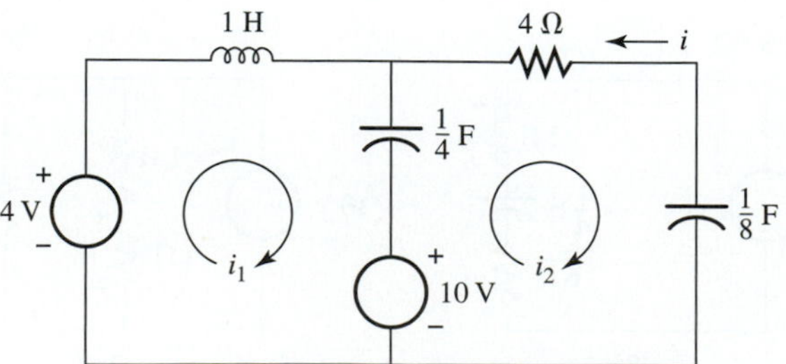

16. Determine the current i for the circuit in Figure 15.68 by first converting the circuit to the s domain and finding $\mathbf{I}(s)$. (Use the Mesh method).

Figure 15.68

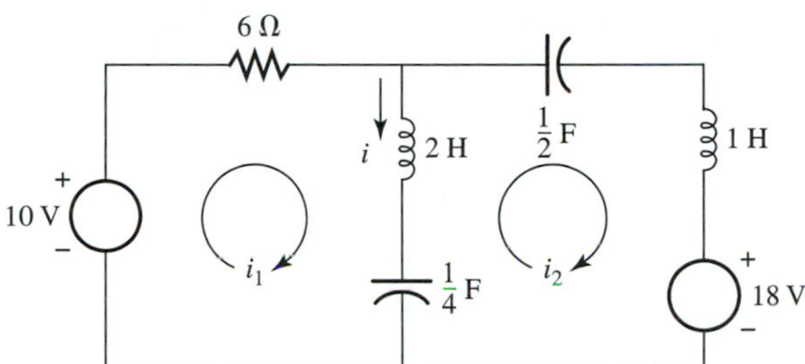

17. Determine the current i for the circuit in Figure 15.69 by first converting the circuit to the s domain and finding $\mathbf{I}(s)$. (Use the Mesh method).

Figure 15.69

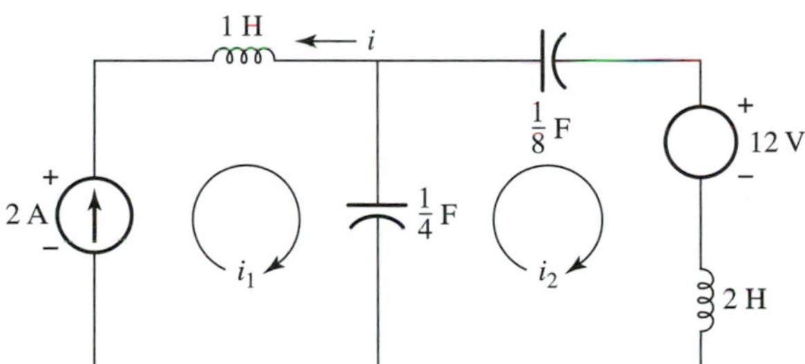

18. Determine the voltage v for the circuit in Figure 15.70 by first converting the circuit to the s domain and finding $\mathbf{V}(s)$. (Use the Nodal method).

Figure 15.70

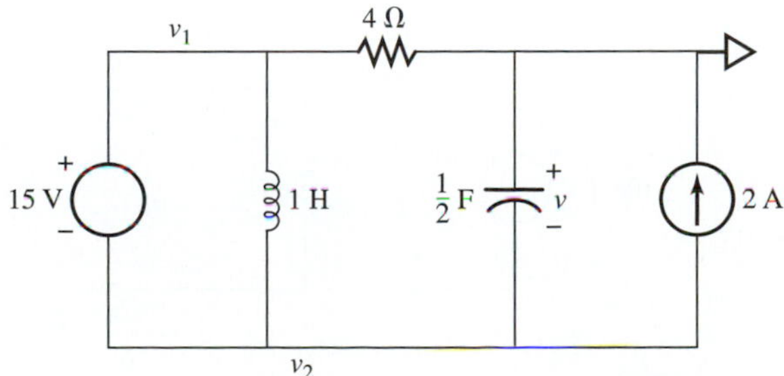

19. Determine the voltage v for the circuit in Figure 15.71 by first converting the circuit to the s domain and finding $\mathbf{V}(s)$. (Use the Nodal method).

Figure 15.71

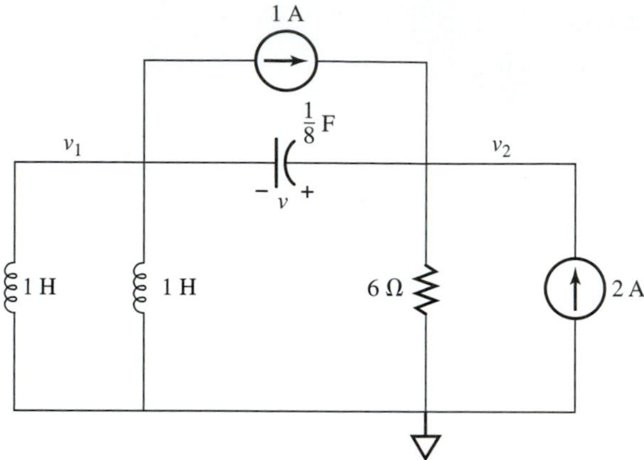

20. Determine the voltage v for the circuit in Figure 15.72 by first converting the circuit to the s domain and finding $\mathbf{V}(s)$. (Use the Nodal method).

Figure 15.72

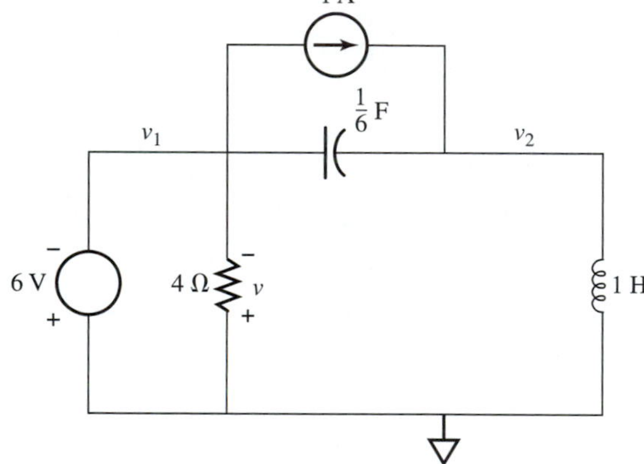

21. Determine the Thevenin equivalent circuit for terminals a-b in the s domain for the circuit in Figure 15.73. (Use Thevenin's theorem).

Figure 15.73

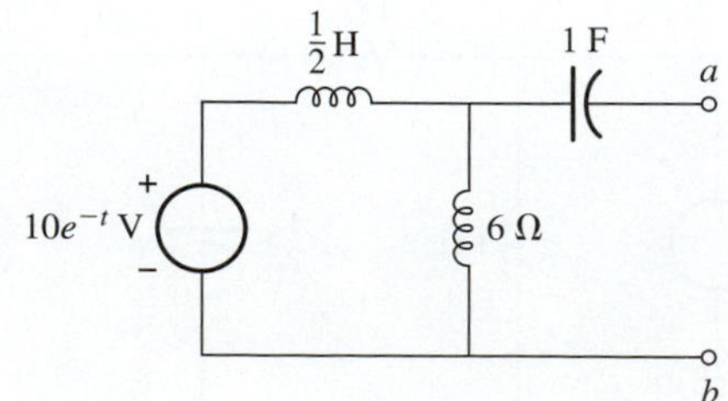

22. Determine the Thevenin equivalent circuit for terminals *a-b* in the *s* domain for the circuit in Figure 15.74. (Use Thevenin's theorem).

Figure 15.74

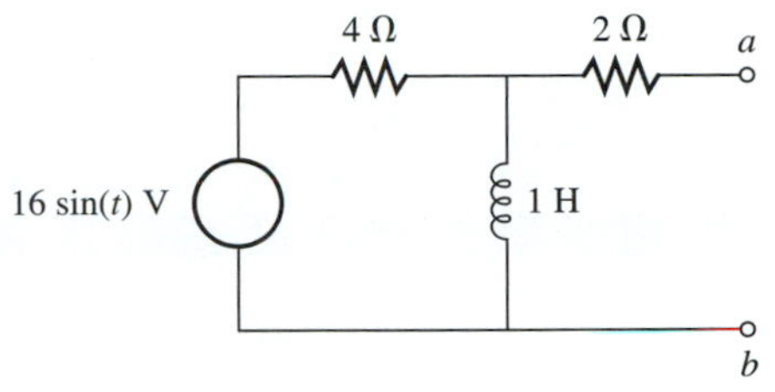

23. Determine the Thevenin equivalent circuit for terminals *a-b* in the *s* domain for the circuit in Figure 15.75. (Use Thevenin's theorem).

Figure 15.75

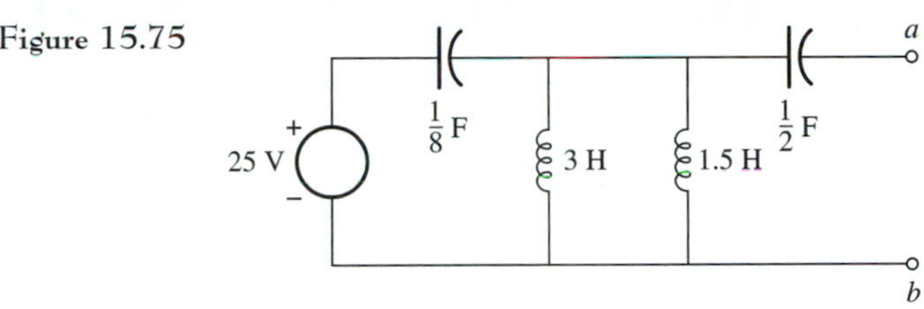

24. Determine a Norton equivalent circuit for terminals *a-b* in the *s* domain for the circuit in Figure 15.73. (Use Norton's theorem).

25. Determine a Norton equivalent circuit for terminals *a-b* in the *s* domain for the circuit in Figure 15.74. (Use Norton's theorem).

26. Determine a Norton equivalent circuit for terminals *a-b* in the *s* domain for the circuit in Figure 15.75. (Use Norton's theorem).

CHAPTER

16

Mathematics for AC Circuits

After completing this chapter the student should be able to

* express sinusoidal voltages and currents in complex form
* perform basic arithmetic operations on complex numbers

Recall from Chapter 12 that the complete solution of a differential equation describing a voltage or current in a circuit consists of two parts—the transient solution and the steady-state solution. The steady-state part of the solution is due to the forcing function, which is directly related to the voltage or current sources driving the circuit. If a circuit is driven by a sinusoidal voltage or current source, the steady-state voltages or currents appearing in the circuit are themselves sinusoidal. In this chapter we present the mathematics necessary for determining steady-state solutions for circuits driven by sinusoidal sources.

16.1 SINE AND COSINE WAVEFORMS

Sine and cosine waveforms are illustrated in Figure 16.1.The waveforms in Figure 16.1(a) and (b), respectively, are described as

$$y = Y_M \sin(\alpha)$$

$$z = Z_M \cos(\alpha)$$

Figure 16.1 Sine and cosine waveforms.

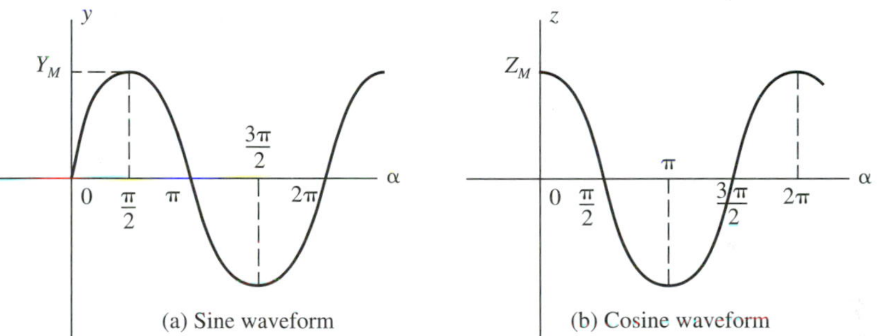

where α (Greek letter alpha) is the independent variable measured in radians or degrees, y and z are the dependent variables, and Y_M and Z_M are the maximum values of the waveform. The cosine and sine waveforms are both sinusoids and are related by the following trigonometric identities:

$$\left.\begin{aligned} \cos(\alpha) &= \sin\left(\alpha + \frac{\pi}{2}\right) \\ \sin(\alpha) &= \cos\left(\alpha - \frac{\pi}{2}\right) \end{aligned}\right\} \tag{16.1}$$

Other trigonometric identities that are useful in the study of ac circuits are as follows.

$$\begin{aligned} \sin(-\alpha) &= -\sin(\alpha) \\ \cos(-\alpha) &= \cos(\alpha) \\ -\sin(\alpha) &= \sin(\alpha \pm \pi) \\ -\cos(\alpha) &= \cos(\alpha \pm \pi) \\ \sin(\alpha + \beta) &= \sin\alpha\cos\beta + \cos\alpha\sin\beta \\ \cos(\alpha + \beta) &= \cos\alpha\cos\beta - \sin\alpha\sin\beta \end{aligned}$$

$$\sin^2(\alpha) + \cos^2(\alpha) = 1$$

$$2\sin^2(\alpha) = 1 - \cos(2\alpha)$$

$$2\cos^2(\alpha) = 1 + \cos(2\alpha)$$

The quantity in parentheses following the sin or cos term is called the *argument.* The argument is the quantity on which the sine, cosine, or other trigonometric operation is performed. The argument is an angle measured in *degrees* or *radians.* The relationship between degrees and radians is

$$\pi \text{ radians} = 180 \text{ degrees}$$

and is written symbolically as

$$\pi \text{ rad} = 180°$$

The number π is irrational and is approximated by 3.14 in most basic circuit analysis work.

> To convert from degrees to radians, multiply degrees by $(\pi/180)$, and to convert from radians to degrees, multiply radians by $(180/\pi)$.

EXAMPLE 16.1 Convert 30° to radians and $\pi/3$ rad to degrees. ■

SOLUTION

$$\{30°\}\left\{\frac{\pi}{180}\right\} = \pi/6 \text{ rad}$$

$$30° = \frac{\pi}{6} \text{ rad}$$

$$\left\{\frac{\pi}{3} \text{ rad}\right\}\left\{\frac{180}{\pi}\right\} = 60°$$

$$\frac{\pi}{3} \text{ rad} = 60°$$

16.2 PHASORS

We shall now consider the generation of a sinusoid by the rotation of a line of constant length at a constant angular velocity. Consider the rotating line and the associated sine wave in Figure 16.2.

Figure 16.2 Phasor-generated sine wave.

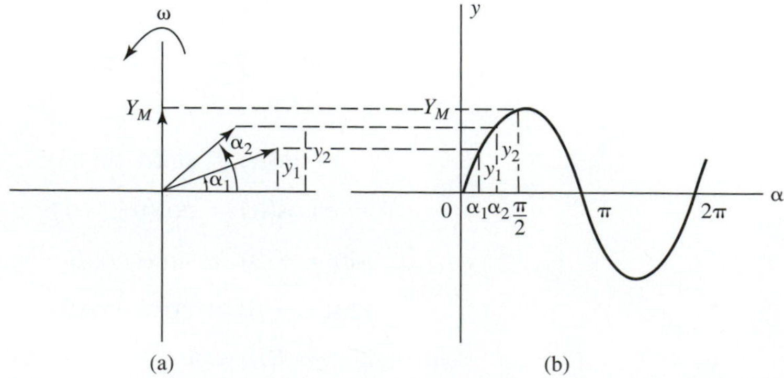

The rotating line segment of length Y_M is called a *phasor,* and it rotates in a counterclockwise direction at a constant angular velocity ω (Greek letter omega) radians/second. The angle α is measured from the positive horizontal axis in Figure 16.2(a) and forms the abscissa for the waveform in Figure 16.2(b). Both α and ω are positive for counterclockwise rotation. Because ω is constant,

$$\alpha = \omega t$$

The vertical distances ($y_1, y_2, \ldots$) from the horizontal axis to the tip of the phasor are projected across for each value of α to form a sine wave, as shown in Figure 16.2. The projection for each value of α can be expressed as

$$y = Y_M \sin(\alpha)$$

as illustrated for α_1 and α_2 in Figure 16.3.

Figure 16.3 Phasors.

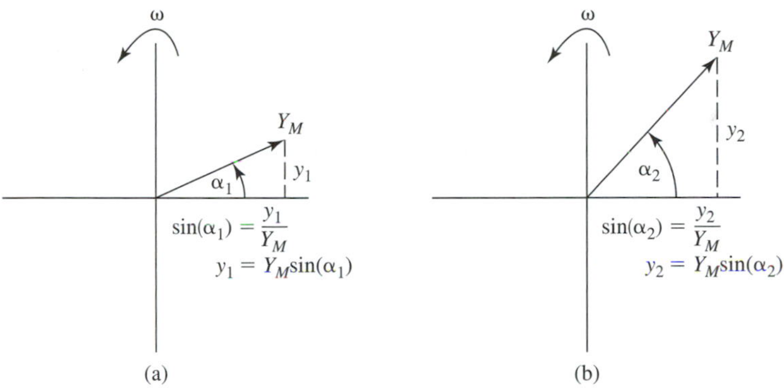

Because the value of y for any α is $Y_M \sin(\alpha)$, the expression for the resulting sine wave generated by the rotating phasor is

$$y = Y_M \sin(\alpha)$$

Because $\alpha = \omega t$,

$$y = Y_M \sin(\omega t)$$

where ω is the angular velocity of the rotating phasor expressed in radians/second, y is the dependent variable, t is the independent variable, and Y_M is the maximum value (amplitude) of the sine wave.

The sine wave in Figure 16.1(a) can be represented as shown in Figure 16.4 using ωt instead of α as the abscissa. The angles π, 2π, 3π, . . . are

Figure 16.4 Sine wave.

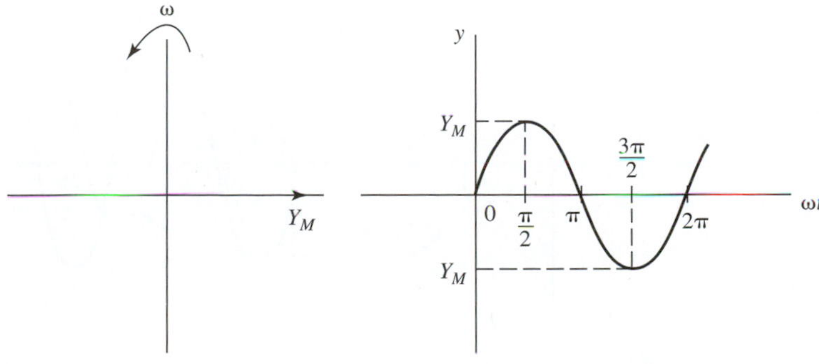

the points where the waveform crosses the horizontal axis, because the sine of each of these angles is equal to zero.

16.3 PERIOD AND FREQUENCY

The time required for the phasor in Figure 16.4 to rotate 2π rad (360°) is called the *period* of the sinusoidal waveform and is given the symbol T. The time T is also the time required for the sinusoidal waveform in Figure 16.4 to complete one cycle and begin to repeat itself. For this reason the sine and cosine waveforms are said to be periodic. Because ω is the constant angular velocity of the phasor, the angle turned through by the phasor in time t is

$$\alpha = \omega t$$

and because $\alpha = 2\pi$ rad when $t = T$ seconds,

$$2\pi = \omega T$$

$$T = \frac{2\pi}{\omega} \tag{16.2}$$

Because T is the time required for the sinusoidal waveform to complete one cycle, the frequency of the waveform is

$$f = \frac{1}{T} \tag{16.3}$$

where f is measured in cycles/second and T is measured in seconds. Substituting $1/f$ for T in Equation (16.2) results in

$$\omega = 2\pi f \tag{16.4}$$

Equation (16.4) relates the angular velocity of the phasor to the frequency of the generated sine wave. The frequency f is measured in hertz (Hz) in honor of the German scientist Heinrich Hertz (1857–1894). One hertz equals one cycle per second.

16.4 PHASE ANGLE

The *phase angle* of a sinusoidal waveform is the angular location of the generating phasor at time $t = 0$. The phase angle is included in the argument of

Figure 16.5 Phase angle.

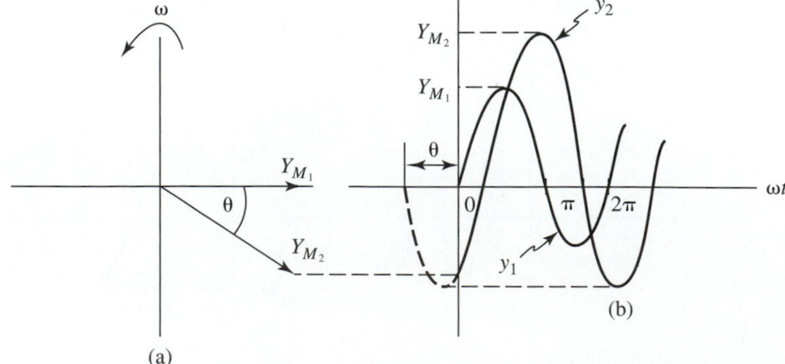

the expression for a sinusoidal waveform, as illustrated for the waveforms in Figure 16.5. In Figure 16.5, y_1 can be expressed as

$$y_1 = Y_{M_1} \sin(\omega t)$$

and y_2 can be expressed as

$$y_2 = Y_{M_2} \sin(\omega t - \theta)$$

θ is always measured from the positive horizontal axis and is positive if counterclockwise and negative if clockwise. Notice that the argument equals the angle locating the phasor at $t = 0$. For y_1, the argument is zero at $t = 0$, and for y_2, the argument is $(-\theta)$ at $t = 0$. Notice also that a change in the phase angle shifts the waveform along the horizontal axis.

Consider the phasors and associated sinusoids in Figure 16.6. An example is given for a phasor located in each of the four quadrants at $t = 0$. The length of the phasor is equal to the maximum value of the sinusoid, the angular velocity of the phasor determines the frequency and period of the sinusoid, and the angular position of the phasor at $t = 0$ determines the phase angle of the sinusoid. The angles θ_1 and θ_2 in Figure 16.6 locate the phasor at $t = 0$; that is, they determine the starting point for the phasor. Angle θ_1 is measured in a positive (counterclockwise direction), and angle θ_2 is measured in a negative (clockwise direction); both can be used to locate the phasor, as illustrated in Figure 16.6.

Figure 16.6 Phase angles.

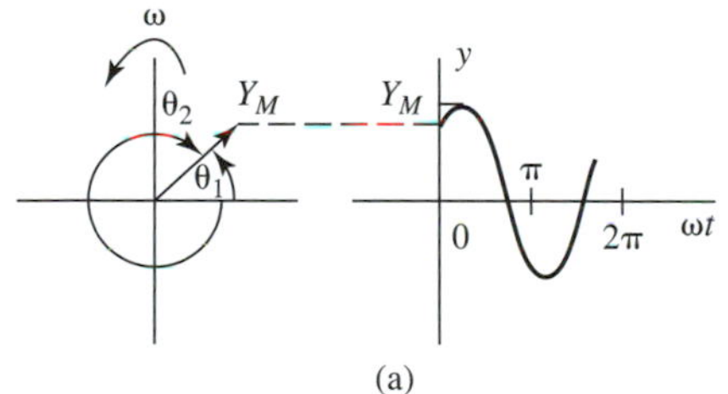

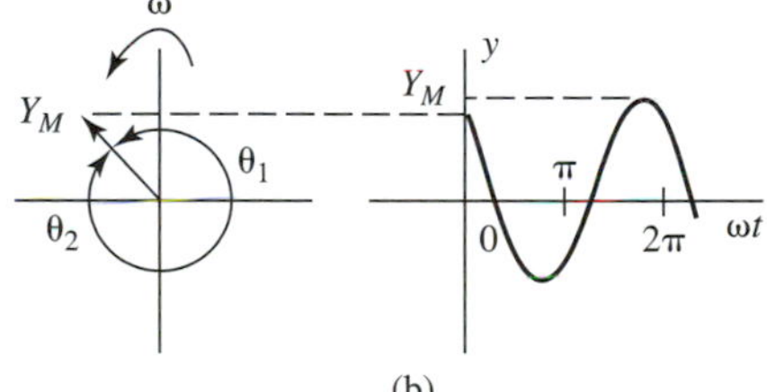

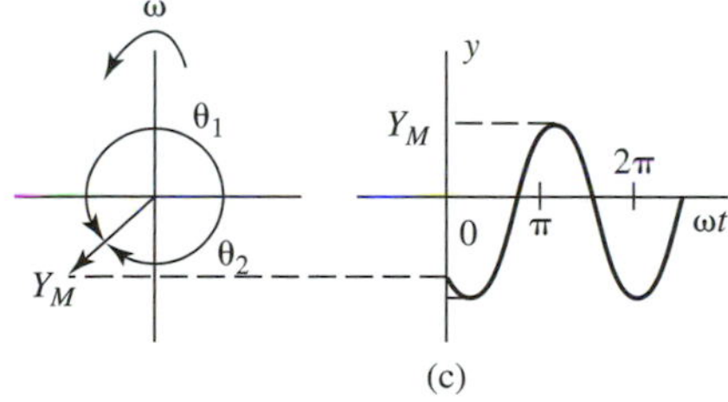

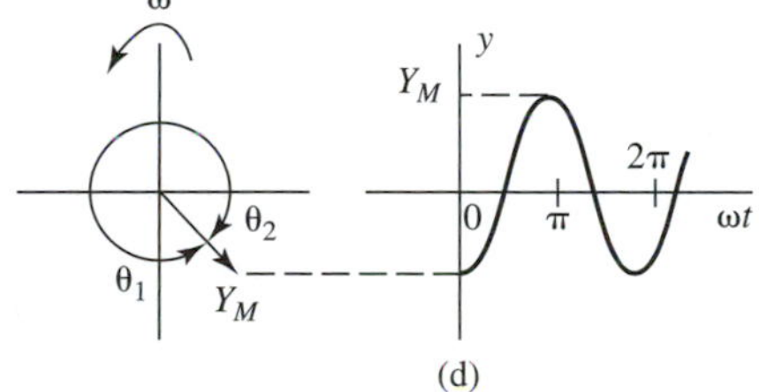

Each sinusoid in Figure 16.6 can be expressed as

$$y = Y_M \sin(\omega t + \theta_1)$$

or

$$y = Y_M \sin(\omega t - \theta_2)$$

with the appropriate values of θ_1 or θ_2, Y_M, and ω inserted.

16.5 GENERAL SINE EXPRESSION

The general expression for the sine wave is

$$y = Y_M \sin(\omega t + \theta) \tag{16.5}$$

where Y_M is the length of the generating phasor and the maximum value of the generated waveform; ω is the angular velocity of the generating phasor; and the phase angle θ is the angular location of the generating phasor at $t = 0$. θ is measured from the positive horizontal axis of the phasor diagram with counterclockwise measurement positive and clockwise measurement negative.

Consider the following example.

EXAMPLE 16.2 Determine an expression for the sine wave generated by the phasor in Figure 16.7. ■

Figure 16.7

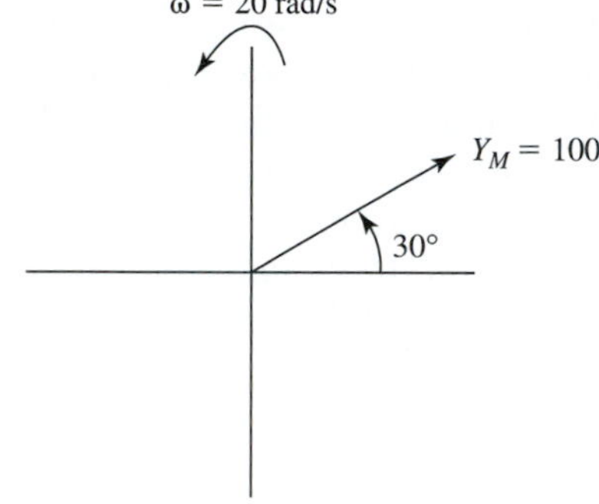

SOLUTION If y is designated as the dependent variable, the phasor has the following characteristics:

$$Y_M = 100$$

$$\omega = 20 \text{ rad/s}$$

$$\theta = 30° \text{ or } -330°$$

and from Equation (16.5)

$$y = 100 \sin(20t + 30°)$$

or

$$y = 100 \sin(20t - 330°)$$

The dimensional units of Y_M determine the units of y, because $\sin(20t + \theta)$ is unitless.

Consider the argument $(\omega t + \theta)$ in the expression

$$y = Y_M \sin(\omega + \theta) \tag{16.6}$$

The quantity ωt is measured in radians, because ω has units of radians/second and t has units of seconds. For dimensional integrity θ must have the same units as ωt. For convenience, we often express the phase angle θ in degrees, as in Example 16.2. However, to preserve dimensional integrity prior to evaluating y for a particular value of t, ωt and θ must be expressed in the same units.

EXAMPLE 16.3 Determine the phasor associated with the generation of the following sine wave:

$$y = 60 \sin(40t - 300°)$$ ■

SOLUTION The phasor characteristics are

$$Y_M = 60$$

$$\omega = 40 \text{ rad/s}$$

$$\theta = -300°$$

and the phasor diagram is as shown in Figure 16.8. If it is desired to deter-

Figure 16.8

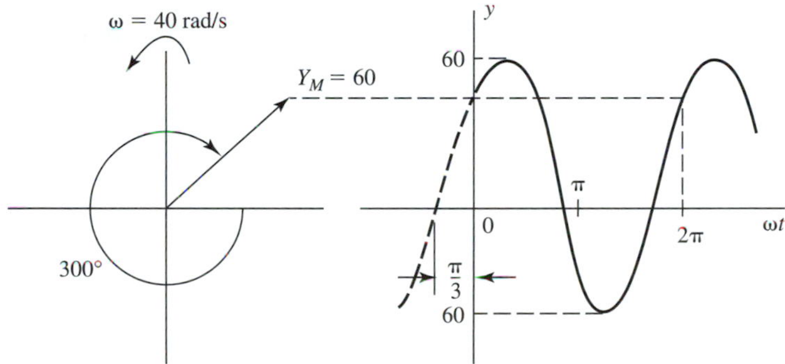

mine y for a particular value of t, 300° must be converted to radians or 40 ± rad must be converted to degrees before performing the calculation.

The units for the horizontal axis in Figure 16.9 are radians and the units for the vertical axis are the units assigned to the dependent variable, y. Be-

Figure 16.9 Sine wave.

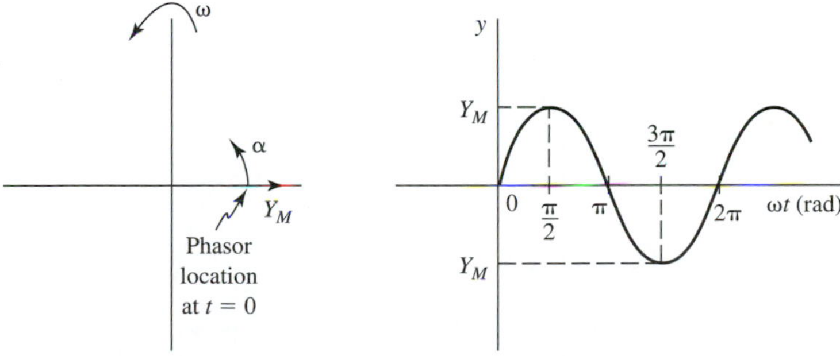

cause $\alpha = \omega t$, each value of α on the horizontal axis has a corresponding value of t, which is given by

$$t = \frac{\alpha}{\omega}$$

The sine wave in Figure 16.7 can be represented with t replacing α along the horizontal axis, as shown in Figure 16.10. Time t on the horizontal axis of

Figure 16.10 Sine wave.

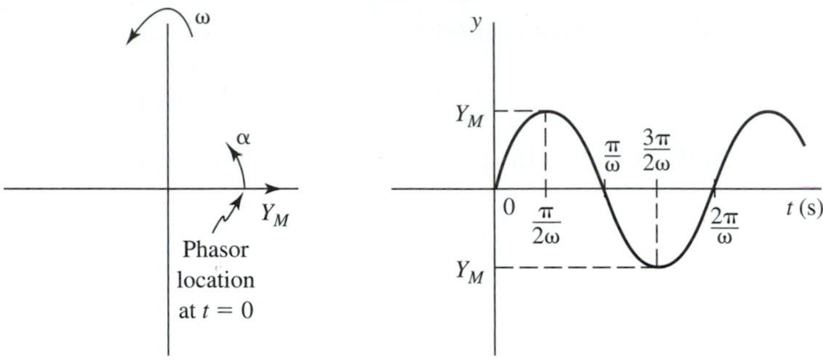

the sine wave corresponds to the time that has elapsed since the start of the phasor at $t = 0$. The time t on the horizontal axis can also be determined by dividing the period T into fractions, as indicated in Example 16.4.

EXAMPLE 16.4 Sketch the sine wave given here using time as the measurement along the horizontal axis.

$$y = 100 \sin\left(\pi t + \frac{\pi}{2}\right)$$

■

SOLUTION

$$Y_M = 100$$

$$\omega = \pi \text{ rad/s}$$

$$\theta = \frac{\pi}{2} \text{ rad}$$

The sine wave and phasor are shown in Figure 16.11 with period T divided into fractions.

Figure 16.11

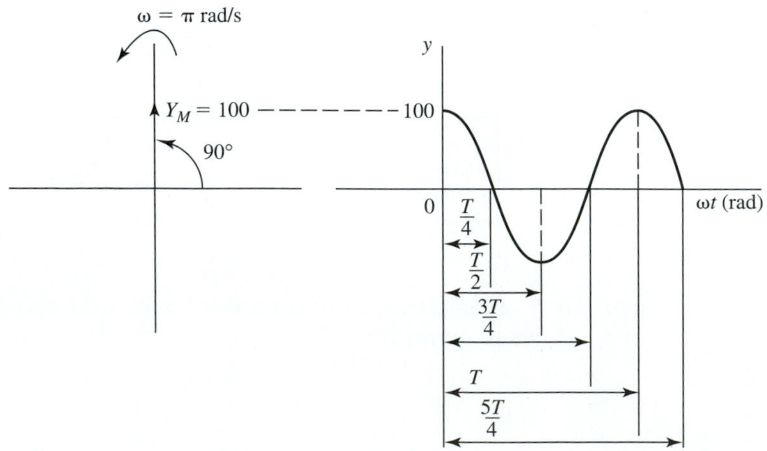

Because $T = 2\pi/\omega$,

$$T = \frac{2\pi}{\pi} = 2 \text{ s}$$

and it follows that

$$\frac{T}{4} = \frac{2}{4} = \frac{1}{2} \text{ s}$$

$$\frac{T}{2} = \frac{2}{2} = 1 \text{ s}$$

$$\frac{3T}{4} = \frac{(3)(2)}{4} = \frac{3}{2} \text{ s}$$

$$\frac{5T}{4} = \frac{(5)(2)}{4} = \frac{5}{2} \text{ s}$$

The sine wave with time as the measurement along the horizontal axis is shown in Figure 16.12.

Figure 16.12

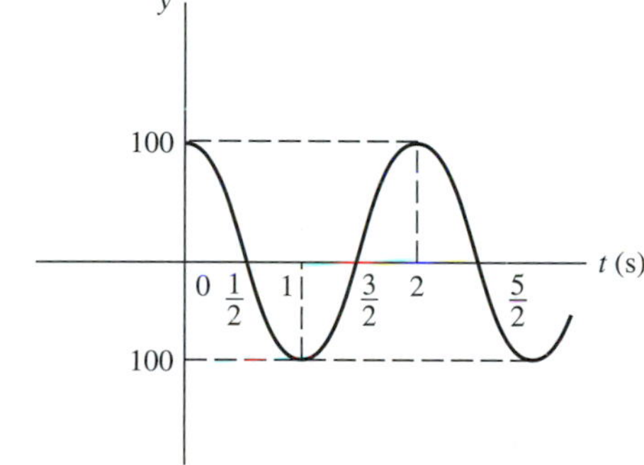

16.6 PHASE DIFFERENCE

Consider two sine waves with the following frequencies:

$$y = Y_M \sin(\omega t + \theta)$$

$$z = Z_M \sin(\omega t + \beta)$$

The phasors for y and z are as indicated on the phasor diagram in Figure 16.13.

Figure 16.13 Phase difference.

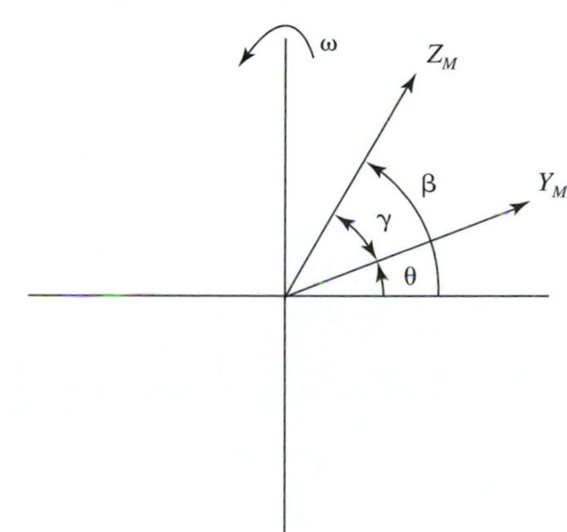

The *phase difference* between two phasors is the smaller of the two angles that separate the two phasors. In Figure 16.13, the phase difference γ (Greek letter gamma) is

$$\gamma = \beta - \theta$$

The sine wave z is said to *lead* the sine wave y by the angle γ; equivalently, sine wave y is said to *lag* sine wave z by the angle γ.

EXAMPLE 16.5 Determine the phase difference and sketch the two sine waves given here.

$$y = 100 \sin(3t - 30°)$$
$$z = 200 \sin(3t + 60°)$$ ■

SOLUTION The phasor diagram is shown in Figure 16.14.

Figure 16.14

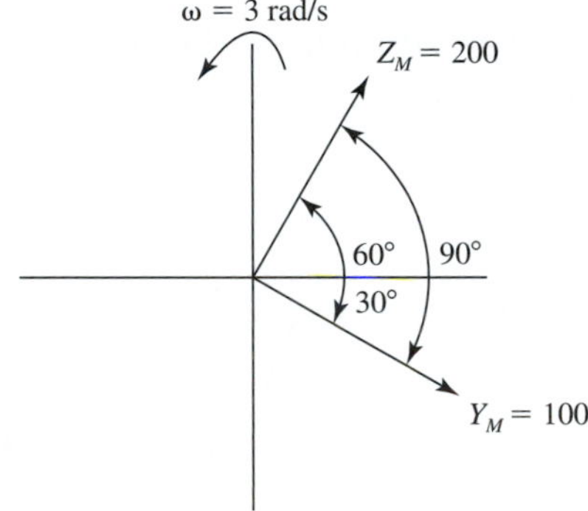

The phase difference is

$$\gamma = 90°$$

with sine wave z leading sine wave y by 90° or sine wave y lagging sine wave z by 90°. The sine waves are sketched in Figure 16.15.

Figure 16.15

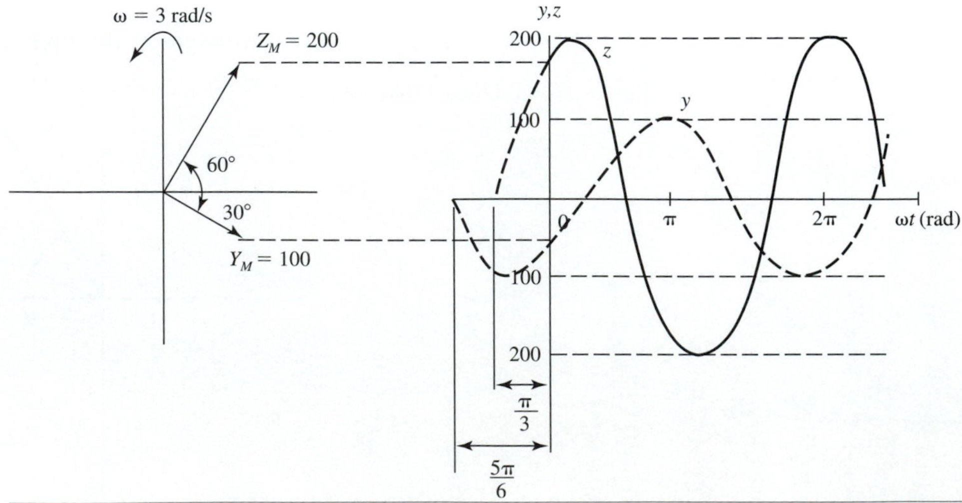

EXAMPLE 16.6 Determine the phase difference and sketch the two sine waves given here.

$$y = 50 \sin(t - 260°)$$
$$z = 25 \sin(t + 100°)$$

SOLUTION The phasor diagram is shown in Figure 16.16.

Figure 16.16

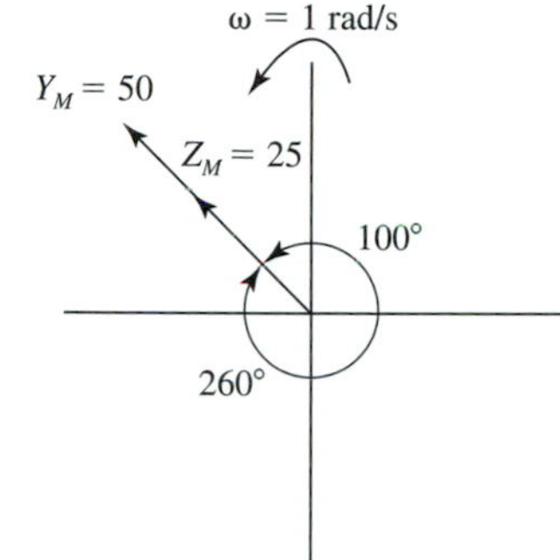

The phase difference is

$$\gamma = 0$$

and sine waves y and z are said to be in phase. The sine waves are sketched in Figure 16.17. Sine waves y and z are said to be in phase, because $\gamma = 0$.

Figure 16.17

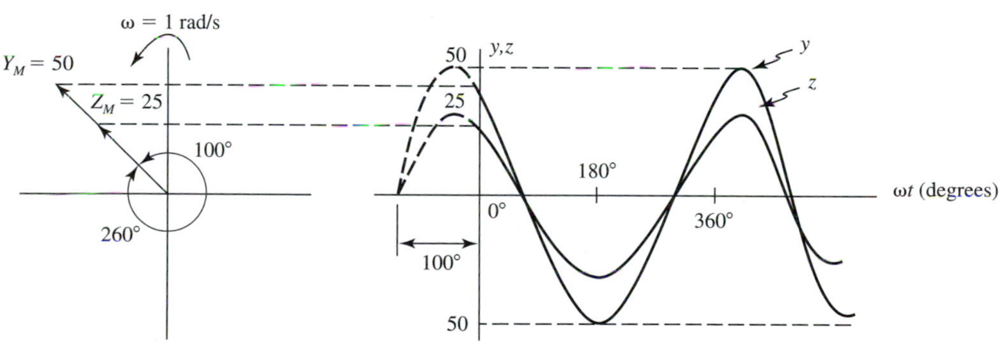

16.7 INVERSE SINE

The symbol $\sin^{-1}\{A\}$ is read as "the angle whose sine is equal to A" and is called the inverse sine of A. For example,

$$\sin^{-1}\left\{\frac{1}{2}\right\} = 30°$$

because $\sin 30° = \frac{1}{2}$. The inverse sine is multivalued, because $\sin^{-1}\left(\frac{1}{2}\right)$ also equals 150°, 390°, 410°, etc.

16.8 GENERAL COSINE EXPRESSION

Consider the generation of the cosine waveform by rotating a phasor of length Y_M counterclockwise at ω rad/s with the phasor located as shown in Figure 16.18 at $t = 0$. The vertical distances ($y_1, y_2, \ldots$) measured from the horizontal axis to the tip of the phasor as the phasor rotates are projected across for

Figure 16.18 Cosine waveform.

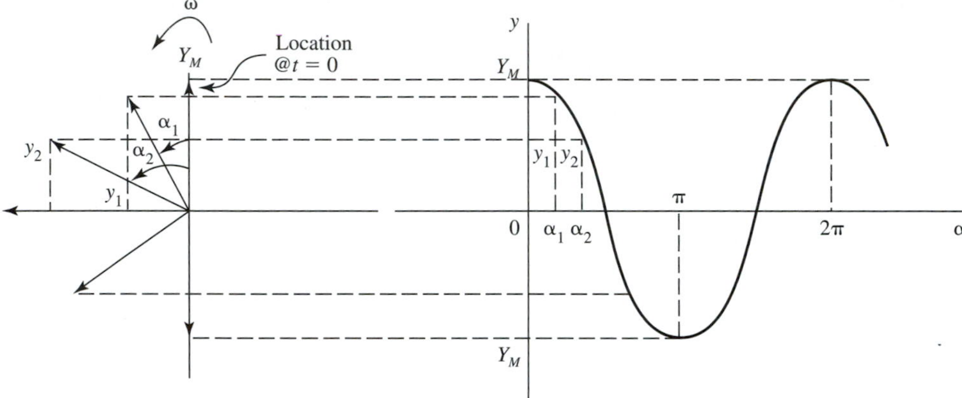

each value of α to form the cosine wave, as shown in Figure 16.18. The projection of y for each value of α is

$$y = Y_M \cos(\alpha)$$

as illustrated in Figure 16.19 for y_1 and y_2.

Figure 16.19 Phasors.

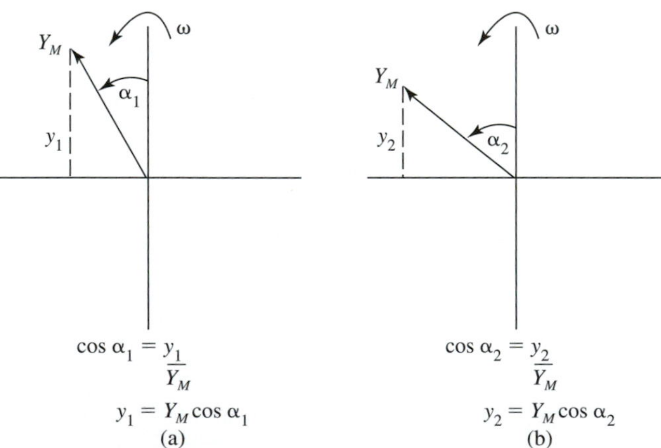

The waveform resulting from the rotating phasor in Figure 16.18 is

$$y = Y_M \cos(\alpha) \tag{16.6}$$

Because $\alpha = \omega t$

$$y = Y_M \cos(\omega t)$$

where ω is the angular velocity of the phasor expressed in radians/second, y is the dependent variable, and Y_M is the maximum value (amplitude) of the cosine wave. Notice that the angle α is measured from the positive vertical axis for the cosine wave and is measured from the positive horizontal axis for the sine wave.

The cosine wave in Figure 16.18 can be represented as shown in Figure 16.20 using ωt instead of α as the abscissa. The angles $\pi/2$, $3\pi/2$, . . . are the

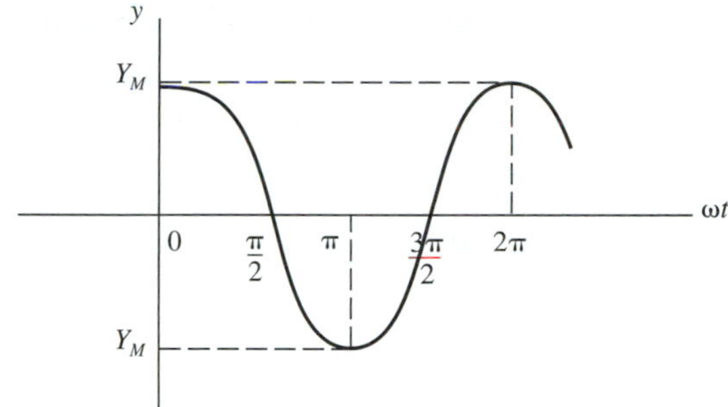

Figure 16.20 Cosine wave.

points where the waveform crosses the horizontal axis, because the cosine of each of these angles is equal to zero.

The general expression for the cosine wave is

$$y = Y_M \cos(\omega t + \theta) \tag{16.7}$$

where Y_M is the length of the generating phasor and the maximum value of the generated waveform, ω is the angular velocity of the generating phasor, and the phase angle θ is the angular location of the generating phasor at $t = 0$. θ is measured from the positive vertical axis of the phasor diagram with counterclockwise measurement positive and clockwise measurement negative.

EXAMPLE 16.7 Determine an expression for the cosine wave generated by the phasor in Figure 16.21. ■

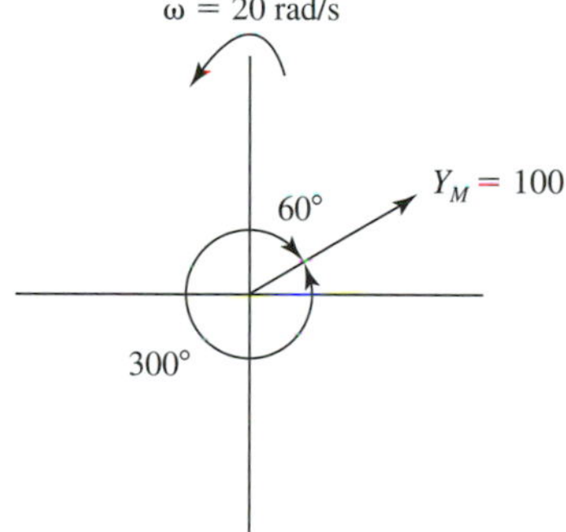

Figure 16.21

SOLUTION If y is designated as the dependent variable, the phasor has the following characteristics:

$$Y_M = 100$$
$$\omega = 20 \text{ rad/s}$$
$$\theta = -60^\circ$$

and from Equation (16.7),

$$y = 100 \cos(20t - 60^\circ)$$

or

$$y = 100 \cos(20t + 300^\circ)$$

The dimensional units for Y_M determine the units for y, because $\cos(20t - 60°)$ is unitless.

Notice that the location of the phasor at $t = 0$ is 30° positive when measured from the positive horizontal axis and the equivalent sine expression is

$$y = 100 \sin(20t + 30°)$$

or

$$y = 100 \sin(20t - 330°)$$

EXAMPLE 16.8 Determine the phasor associated with the generation of the following cosine wave:

$$y = 60 \cos(40t - 30°)$$ ■

SOLUTION The phasor characteristics are

$$Y_M = 60$$
$$\omega = 40 \text{ rad/s}$$
$$\theta = -30°$$

and the phasor is located on the phasor diagram as shown in Figure 16.22.

Figure 16.22

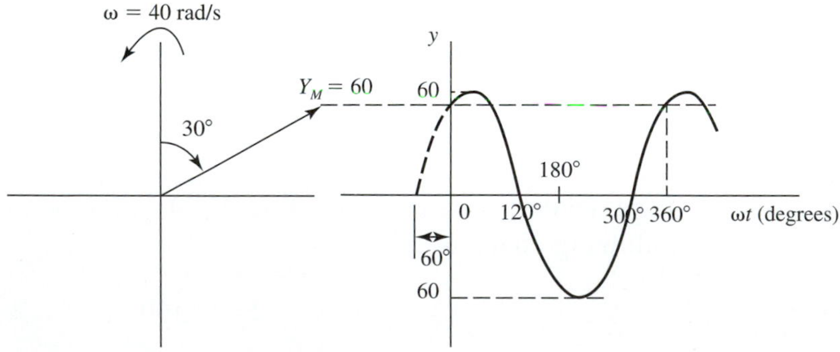

Notice that the generated waveform can also be expressed as

$$y = 60 \sin(40t + 60°)$$

because the phasor is located 60° from the positive horizontal axis at $t = 0$.

EXAMPLE 16.9 Determine sine and cosine expressions for the waveform generated by the phasor in Figure 16.23. ■

Figure 16.23

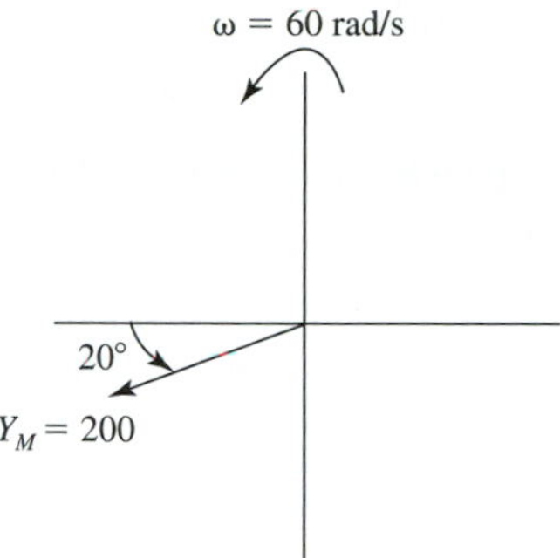

SOLUTION Recalling that the phase angle is measured from the positive horizontal axis for sine-wave generation and employing Equation (16.5) results in (Figure 16.24)

$$y = 200 \sin(60t + 200°)$$

or

$$y = 200 \sin(60t - 160°)$$

Figure 16.24

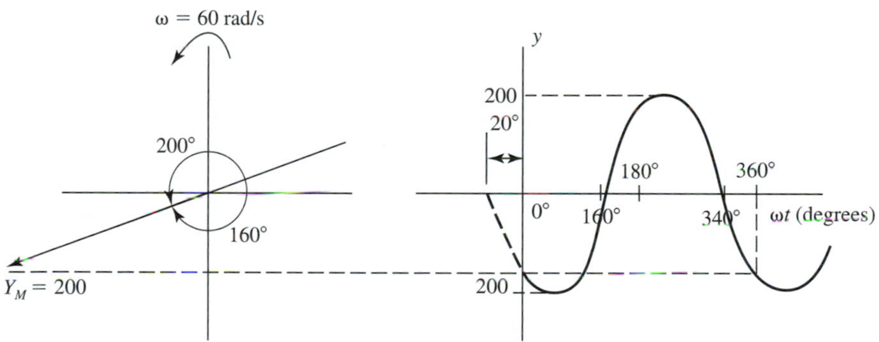

Recalling that the phase angle is measured from the positive vertical axis for the cosine wave generation and employing Equation (16.7) results in (Figure 16.25)

$$y = 200 \cos(60t + 110°)$$

Figure 16.25

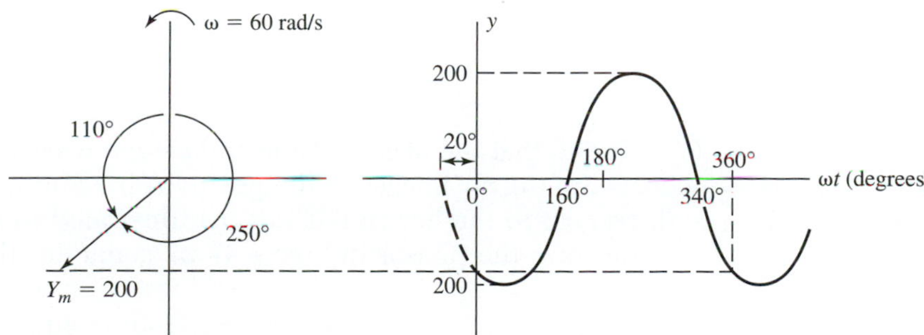

or

$$y = 200\cos(60t - 250°)$$

EXAMPLE 16.10 Determine the phasor associated with the generation of the following sine wave:

$$y = -200\sin(100t + 40°)$$ ■

SOLUTION From the trigonometric identities on page 543,

$$\sin(\theta) = -\sin(\theta \pm 180°)$$

$$\cos(\theta) = -\cos(\theta \pm 180°)$$

the negative of a phasor generates a negative sinusoid, as shown in Figure 16.26. The negative of a phasor is another phasor of equal length separated from the original phasor by 180°.

Figure 16.26

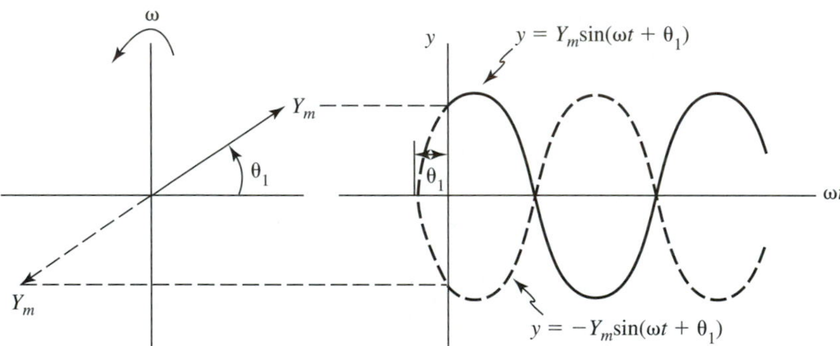

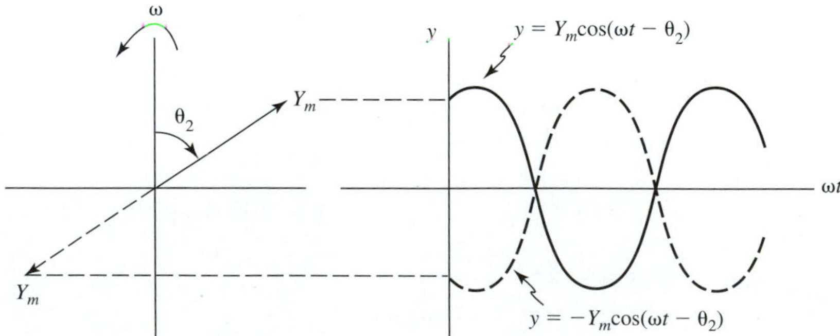

Notice that the negative sine and cosine waves generated by the negative phasors are mirror images of the positive sine and cosine waves, respectively, with respect to the horizontal axis, as illustrated in Figure 16.26.

Because the phasor in Figure 16.27 generates the sine wave

$$y = 200\sin(100t + 40°)$$

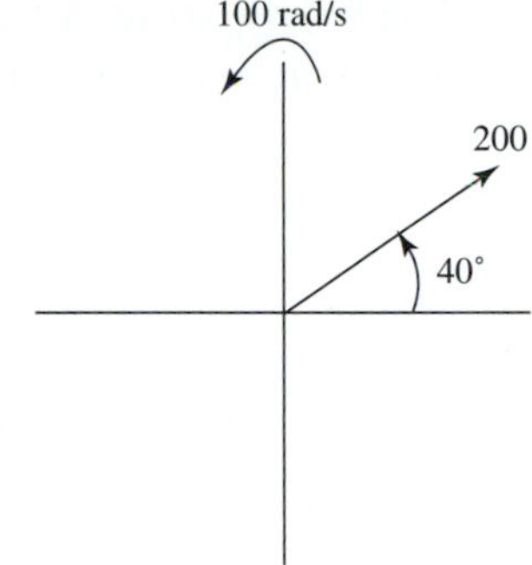

Figure 16.27

the phasor in Figure 16.28 generates the negative sine wave

$$y = -200 \sin(100t + 40°)$$

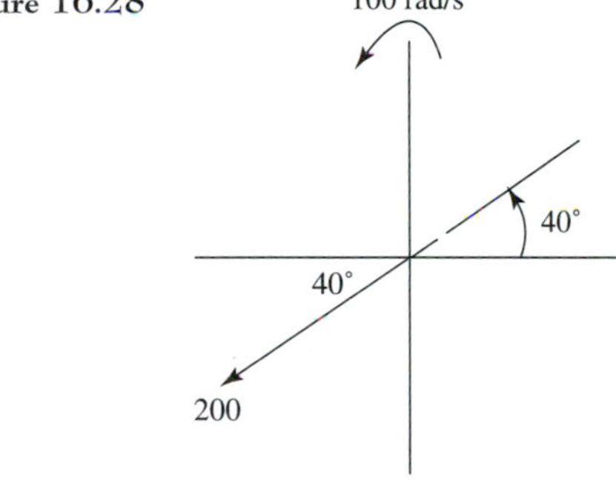

Figure 16.28

16.9 COMPLEX NUMBERS

We now represent the phasors used to generate sinusoids by complex numbers. This approach allows us to utilize the algebra of complex numbers to perform mathematical operations on phasors and therefore on the sinusoidal waveforms generated by the phasors.

A review of complex numbers is in order at this point. A complex number consists of a real part and an imaginary part and can be geometrically represented in the complex plane, where the horizontal axis is the real axis and the vertical axis is the imaginary axis (also called the j axis). Complex number $\mathbf{A}$ can be written as

$$\mathbf{A} = a + jb$$

where a is the real part of $\mathbf{A}$, b is the imaginary part of $\mathbf{A}$, and $j = \sqrt{-1}$. We can express a and b as

$$a = \text{Re}\,\mathbf{A}$$

$$b = \text{Im}\,\mathbf{A}$$

where Re **A** is read as "real part of **A**" and Im **A** is read as "imaginary part of **A**." The complex number **A** is represented in the complex plane in Figure 16.29.

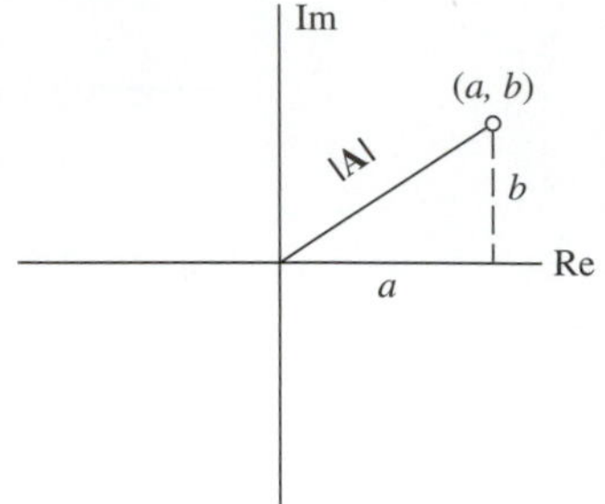

Figure 16.29 Complex number.

The complex number **A** is designated in the complex plane by the ordered pair (a, b). A (or $|\mathbf{A}|$) is the magnitude of the complex number and is the distance from the origin of the complex plane to the ordered pair (a, b); that is,

$$A = |\mathbf{A}| = \sqrt{a^2 + b^2}$$

The complex number is always symbolized in boldface type, whereas the magnitude of the complex number is designated by an italic letter. For example, the magnitude of the complex number **B** is designated by B.

16.10 RECTANGULAR FORM

The rectangular form of the complex number **A** is as previously given. That is, complex number **A** in rectangular form is

$$\mathbf{A} = \text{Re}\{\mathbf{A}\} + j\,\text{Im}\{\mathbf{A}\}$$

and it appears in the complex plane as shown in Figure 16.30.

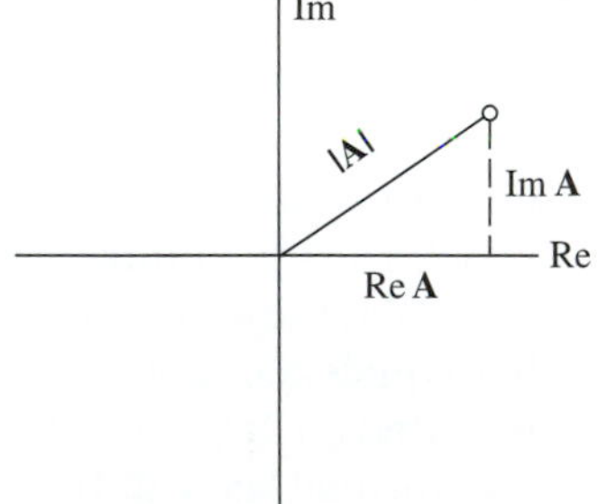

Figure 16.30 Rectangular form.

EXAMPLE 16.11 Locate the following complex numbers in the complex plane:

$$\mathbf{A} = 2 + j3$$

$$\mathbf{B} = -3 + j4$$

$$\mathbf{C} = -4 - j3$$

$$\mathbf{D} = 6 - j2$$

■

SOLUTION See Figure 16.31.

Figure 16.31

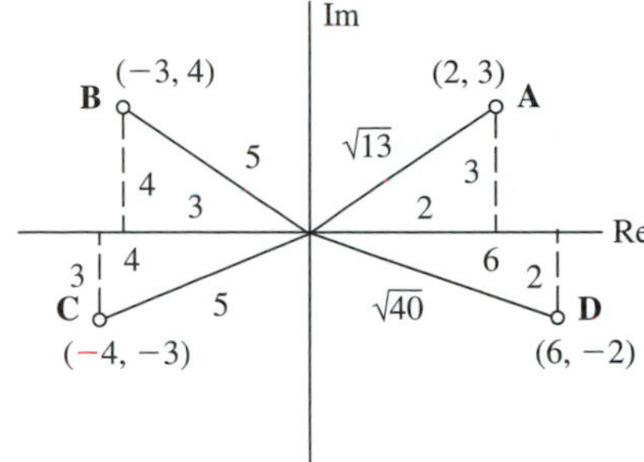

16.11 POLAR FORM

In rectangular form the distances along the real and imaginary axes are given to determine a complex number. In polar form the distance from the origin to the complex number and the angle measured from the positive real axis of the complex plane are used to determine a complex number, as shown in Figure 16.32. That is, complex number **A** in polar form is

$$\mathbf{A} = |\mathbf{A}|\underline{/\theta}$$

or

$$\mathbf{A} = A\underline{/\theta}$$

where $|\mathbf{A}|$ is the distance from the origin and θ is the angle measured from the positive real axis, as shown in Figure 16.32. Counterclockwise measurement is positive and clockwise measurement is negative.

Figure 16.32 Polar form.

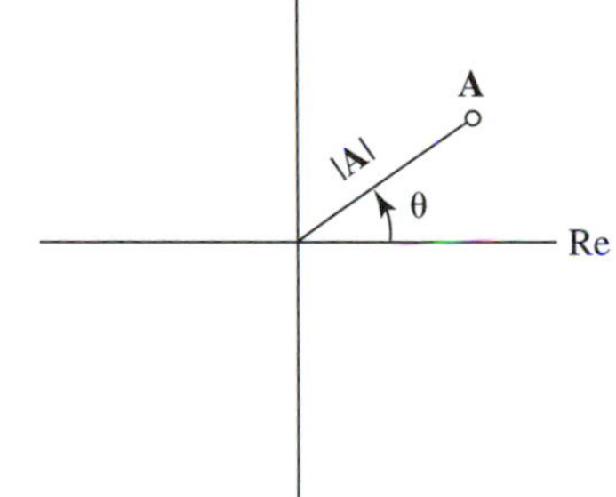

EXAMPLE 16.12 Locate the following complex numbers in the complex plane.

$$\mathbf{A} = 20\underline{/30^\circ}$$
$$\mathbf{B} = 15\underline{/-210^\circ}$$
$$\mathbf{C} = 40\underline{/210^\circ}$$

$$\mathbf{D} = 60\angle 300^\circ$$
$$\mathbf{E} = -12\angle 45^\circ$$
$$\mathbf{F} = -20\angle -60^\circ$$

■

SOLUTION See Figure 16.33.

Figure 16.33

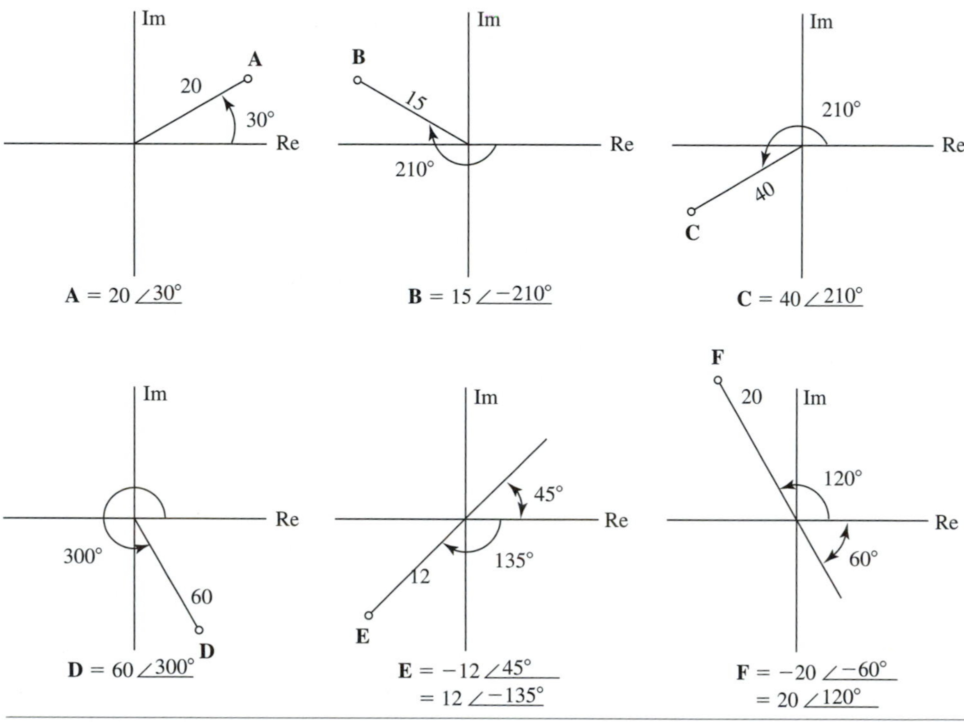

16.12 RECTANGULAR TO POLAR FORM

The conversion of a complex number from rectangular to polar form is illustrated in Figure 16.34.

Figure 16.34 Rectangular to polar form.

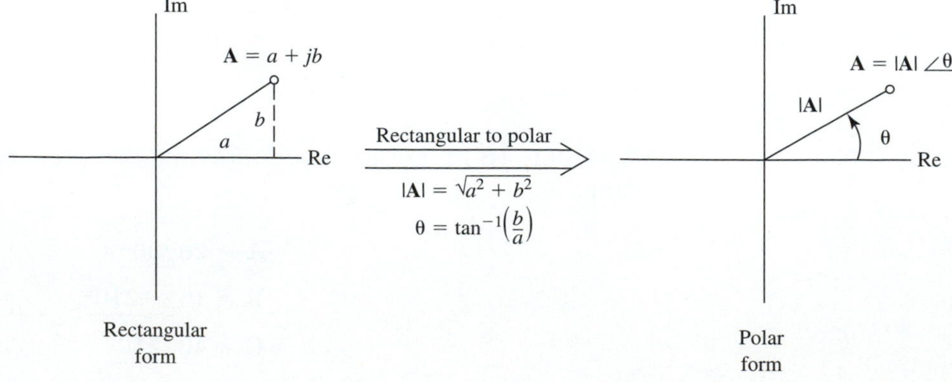

16.13 POLAR TO RECTANGULAR FORM

The conversion of a complex number from polar to rectangular form is illustrated in Figure 16.35.

Figure 16.35 Polar to rectangular form.

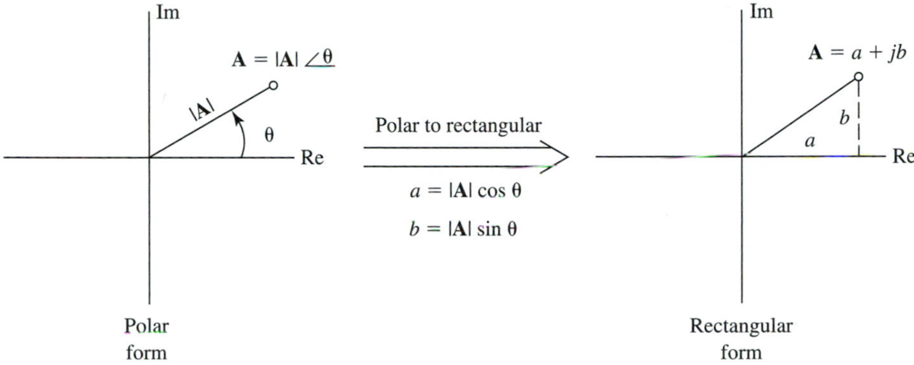

EXAMPLE 16.13 Convert the following complex numbers to polar form.

$$\mathbf{A} = 3 + j4$$

$$\mathbf{B} = -3 - j4$$

$$\mathbf{C} = -3 + j4$$

$$\mathbf{D} = 3 - j4$$

■

SOLUTION Using Figure 16.34, the result is Figure 16.36.

Figure 16.36

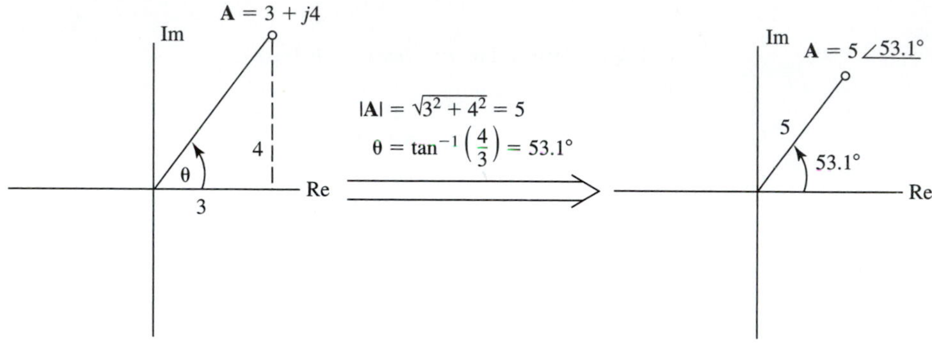

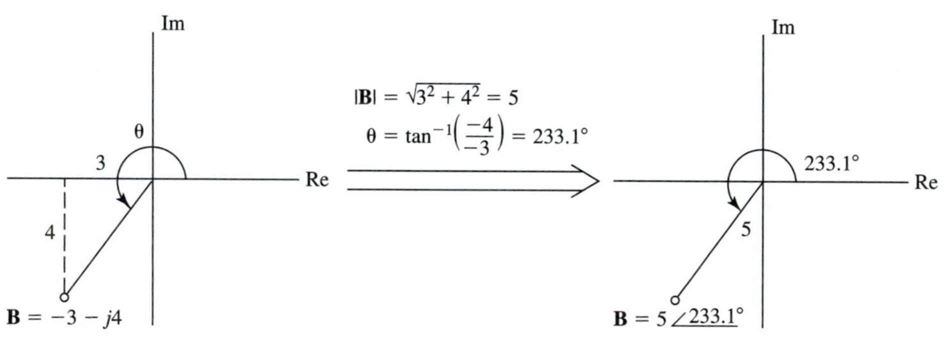

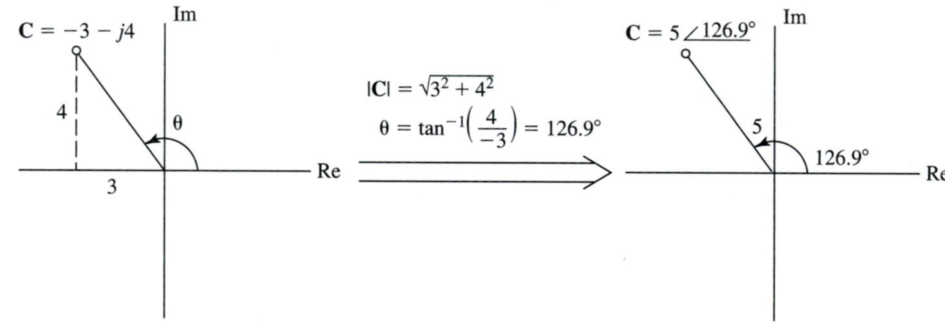

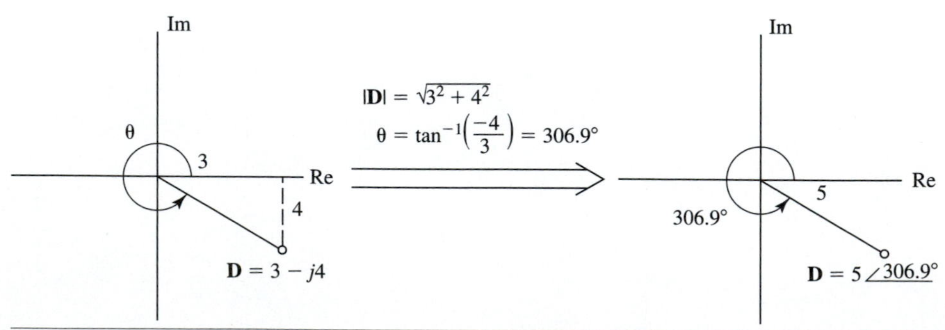

EXAMPLE 16.14 Convert the following complex numbers to rectangular form.

$$\mathbf{A} = 10\angle 53.1^\circ$$
$$\mathbf{B} = \sqrt{2}\angle -45^\circ$$
$$\mathbf{C} = 6\angle 90^\circ$$
$$\mathbf{D} = 4\angle 180^\circ$$

■

SOLUTION Using Figure 16.35, the result is Figure 16.37.

Figure 16.37

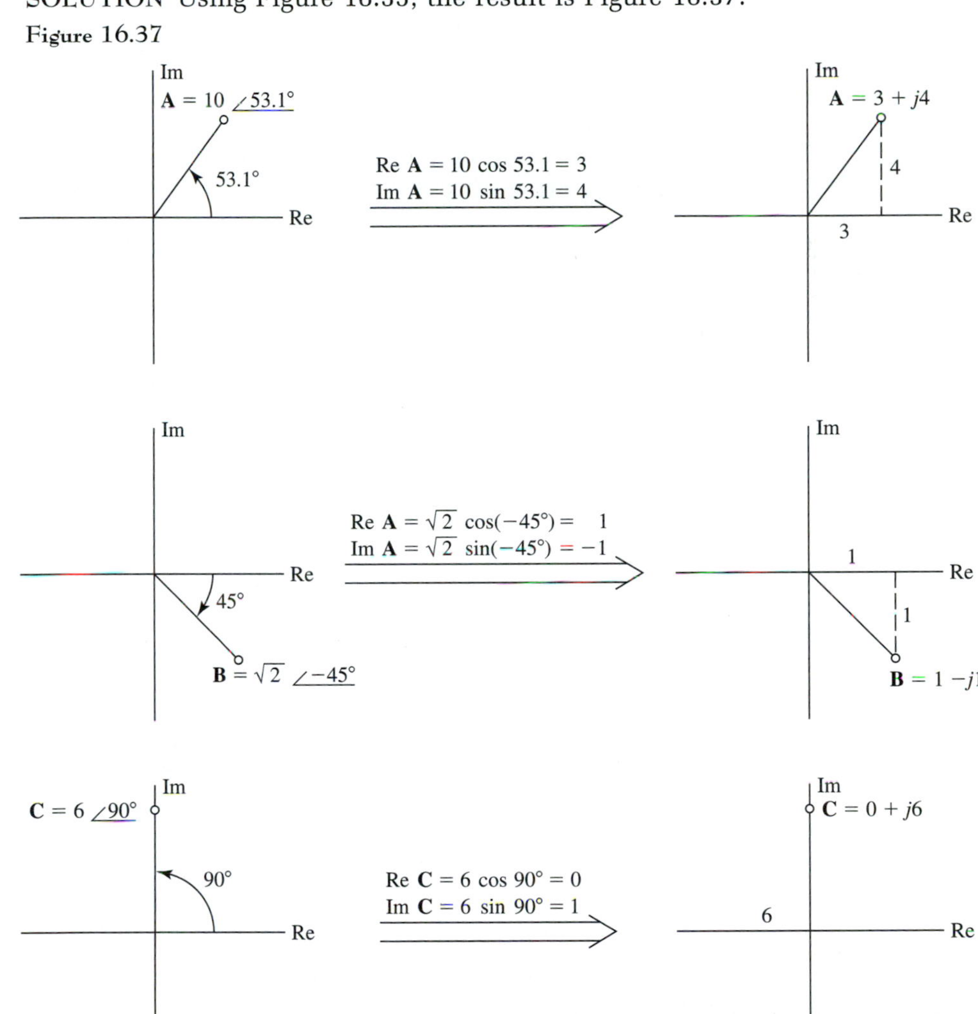

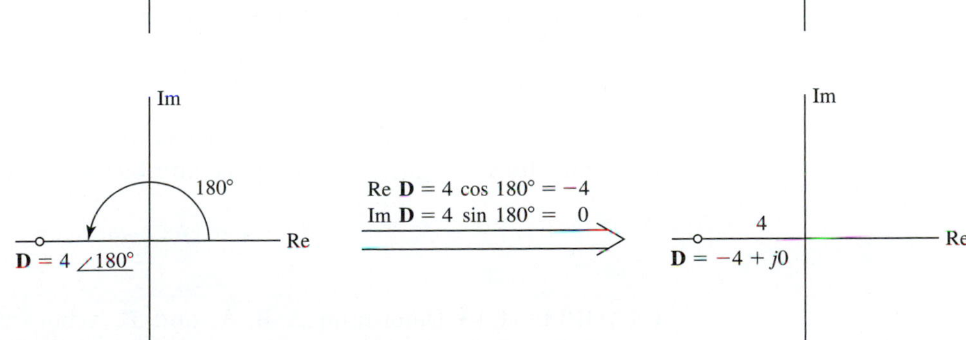

16.14 THE COMPLEX NUMBER j

In analyzing physical systems, equations that have no real-number solution are often encountered. An example of an equation that has no real-number solution is

$$y^2 = -25$$

Although there is no real value of y that satisfies this equation, the complex number system allows us to determine a solution. Because the complex number $j = \sqrt{-1}$ and $j^2 = -1$,

$$y = j5$$

is a solution to the equation $y^2 = -25$.

The complex number j is treated as an algebraic quantity and has the following properties:

$$j = \sqrt{-1}$$
$$j^2 = -1$$
$$j^3 = -j$$
$$j^4 = 1$$
$$j^5 = j$$

If we continue to raise j to higher powers, this pattern is cyclic and repeats itself beginning with j^5.

The reciprocal of j is

$$\frac{1}{j} = -j$$

This is determined by multiplying the numerator and denominator of $1/j$ by j:

$$\frac{1}{j} = \left(\frac{1}{j}\right)\left(\frac{j}{j}\right) = \frac{j}{j^2} = \frac{j}{-1} = -j$$

16.15 COMPLEX CONJUGATES

The conjugate of a complex number $\mathbf{A} = a + jb$ is defined as

$$\tilde{\mathbf{A}} = a - jb$$

A wavy line over the complex number is the symbol for the conjugate. In polar form, the conjugate of a complex number $\mathbf{A} = |\mathbf{A}|\angle\theta$ is

$$\tilde{\mathbf{A}} = |\mathbf{A}|\angle-\theta$$

EXAMPLE 16.15 Determine $\tilde{\mathbf{A}}$, $\tilde{\mathbf{B}}$, $\tilde{\tilde{\mathbf{A}}}$, and $\tilde{\tilde{\mathbf{B}}}$, where $\mathbf{A} = -6 - j12$ and $\mathbf{B} = 40\angle-80°$ and the double wavy line symbolizes the conjugate of the conjugate. ■

SOLUTION

$$\mathbf{A} = -6 - j12$$
$$\tilde{\mathbf{A}} = -6 + j12$$
$$\tilde{\tilde{\mathbf{A}}} = -6 - j12$$
$$\mathbf{B} = 40\angle{-80^\circ}$$
$$\tilde{\mathbf{B}} = 40\angle{-(-80^\circ)}$$
$$= 40\angle{80^\circ}$$
$$\tilde{\tilde{\mathbf{B}}} = 40\angle{-80^\circ}$$

Notice that $\tilde{\tilde{\mathbf{A}}} = \mathbf{A}$ and $\tilde{\tilde{\mathbf{B}}} = \mathbf{B}$.

16.16 ADDITION AND SUBTRACTION

Addition and subtraction of complex numbers is illustrated next. If $\mathbf{A} = a + jb$ and $\mathbf{B} = c + jd$, the sum $\mathbf{A} + \mathbf{B}$ is

$$\mathbf{A} + \mathbf{B} = (a + jb) + (c + jd)$$
$$= (a + c) + j(b + d)$$

and the difference $\mathbf{A} - \mathbf{B}$ is

$$\mathbf{A} - \mathbf{B} = (a + jb) - (c + jd)$$
$$= (a - c) + j(b - d)$$

The real parts of the complex number are combined and the imaginary parts are combined, as illustrated.

Consider the following example.

EXAMPLE 16.16 Determine $(\mathbf{A} - \mathbf{B})$, $(\mathbf{A} + \mathbf{B})$, and $(\mathbf{A} + \mathbf{B} + \mathbf{C})$ if

$$\mathbf{A} = -6 - j2$$
$$\mathbf{B} = 1 - j8$$
$$\mathbf{C} = 2 + j4$$

■

SOLUTION

$$\mathbf{A} - \mathbf{B} = (-6 - j2) - (1 - j8)$$
$$= -6 - j2 - 1 + j8$$
$$= -7 + j6$$
$$\mathbf{A} + \mathbf{B} = (-6 - j2) + (1 - j8)$$
$$= -6 - j2 + 1 - j8$$
$$= -5 - j10$$

$$\begin{aligned}\mathbf{A}+\mathbf{B}+\mathbf{C} &= (-6 - j2) + (1 - j8) + (2 + j4)\\ &= -6 - j2 + 1 - j8 + 2 + j4\\ &= -3 - j6\end{aligned}$$

16.17 MULTIPLICATION

In rectangular form, complex numbers are multiplied as shown next. The complex number j is treated as an algebraic quantity, as indicated. If $\mathbf{A} = a + jb$ and $\mathbf{B} = c + jd$,

$$\begin{aligned}(\mathbf{A})(\mathbf{B}) &= (a + jb)(c + jd)\\ &= ac + jad + jbc + j^2bd\\ &= ac + jad + jbc - bd\\ &= (ac - bd) + j(ad + bc)\end{aligned}$$

In polar form, complex numbers are multiplied as follows. If $\mathbf{A} = |\mathbf{A}|\underline{/\theta}$ and $\mathbf{B} = |\mathbf{B}|\underline{/\beta}$,

$$\begin{aligned}(\mathbf{A})(\mathbf{B}) &= \{|\mathbf{A}|\underline{/\theta}\}\{|\mathbf{B}|\underline{/\beta}\}\\ &= |\mathbf{A}||\mathbf{B}|\underline{/\theta + \beta}\end{aligned}$$

EXAMPLE 16.17 Determine the products **AB**, **CD**, **BC**, and **CDE** if

$$\begin{aligned}\mathbf{A} &= 2 + j4\\ \mathbf{B} &= -6 - j8\\ \mathbf{C} &= 40\underline{/30^\circ}\\ \mathbf{D} &= 8\underline{/120^\circ}\\ \mathbf{E} &= 2\underline{/-20^\circ}\end{aligned}$$

Express the answers in polar form. ■

SOLUTION

$$\begin{aligned}\mathbf{AB} &= (1 + j4)(-6 - j8)\\ &= -12 - j16 - j24 + 32\\ &= 20 - j40\\ &= 44.7\underline{/-63.4^\circ}\\ \mathbf{CD} &= \{40\underline{/30^\circ}\}\{8\underline{/120^\circ}\}\\ &= 320\underline{/150^\circ}\\ \mathbf{CDE} &= \{40\underline{/30^\circ}\}\{8\underline{/120^\circ}\}\{2\underline{/-20^\circ}\}\\ &= (40)(8)(2)\underline{/30^\circ + 120^\circ - 20^\circ}\\ &= 640\underline{/130^\circ}\end{aligned}$$

The determination of **BC** can be approached in two ways. Either convert $(-6 - j8)$ to polar form or convert $40\angle 30^\circ$ to rectangular form and then pro-

Figure 16.38

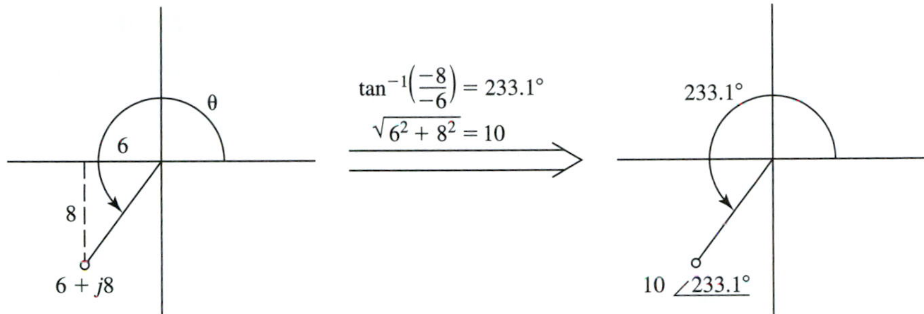

ceed with the multiplication. The result of converting $(-6 - j8)$ to polar form is shown in Figure 16.38, and

$$\begin{aligned}\mathbf{BC} &= (-6 - j8)(40\angle 30^\circ)\\ &= \{10\angle 233.1^\circ\}\{40\angle 30^\circ\}\\ &= 400\angle 233.1^\circ + 30^\circ\\ &= 400\angle 263.1^\circ\end{aligned}$$

The result of converting $40\angle 30^\circ$ to rectangular form is shown in Figure 16.39

Figure 16.39

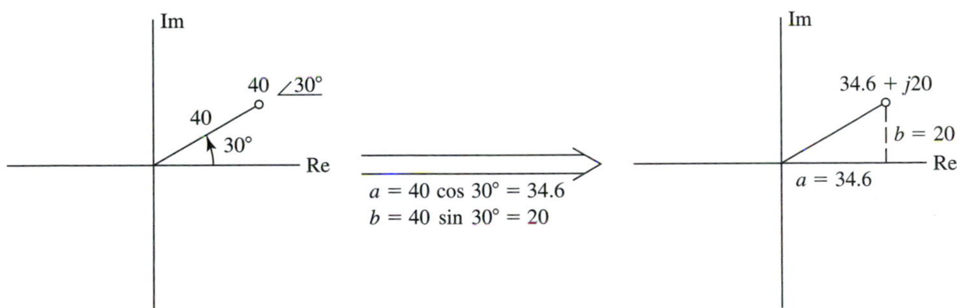

and

$$\begin{aligned}\mathbf{BC} &= (-6 - j8)(40\angle 30^\circ)\\ &= (-6 - j8)(34.6 + j20)\\ &= -207.6 - j120 - j276.8 + 160\\ &= -47.6 - j396.8\\ &= 400\angle 263.1^\circ\end{aligned}$$

16.18 DIVISION

Using the rule for multiplying complex numbers and the definition of a conjugate, the multiplication of a complex number by its conjugate results in a real number. If $\mathbf{A} = a + jb$,

$$\begin{aligned}(\tilde{\mathbf{A}})(\mathbf{A}) &= (a - jb)(a + jb)\\ &= a^2 + jab - jab - j^2b^2\\ &= a^2 + b^2\end{aligned}$$

The process of multiplying a complex number by its conjugate is called *rationalizing the complex number.*

Division is accomplished for complex numbers in *rectangular form* by rationalizing the denominator (divisor) of the expression. Consider the quotient $\mathbf{A}/\mathbf{C}$, where $\mathbf{A} = a + jb$ and $\mathbf{C} = c + jd$. The denominator $(c + jd)$ can be rationalized by multiplying it by its conjugate $(c - jd)$, but we must also multiply the numerator by $(c - jd)$ to preserve the equality. This results in

$$\begin{aligned}\frac{\mathbf{A}}{\mathbf{C}} &= \frac{a + jb}{c + jd}\\ &= \frac{(a + jb)(c - jd)}{(c + jd)(c + jd)}\\ &= \frac{ac - jad + jbc + bd}{c^2 + d^2}\\ &= \left\{\frac{ac + bd}{c^2 + d^2}\right\} + j\left\{\frac{bc - ad}{c^2 + d^2}\right\}\end{aligned}$$

Notice that

$$\mathrm{Re}\left\{\frac{\mathbf{A}}{\mathbf{C}}\right\} = \frac{ac + bd}{c^2 + d^2}$$

and

$$\mathrm{Im}\left\{\frac{\mathbf{A}}{\mathbf{C}}\right\} = \frac{bc - ad}{c^2 + d^2}$$

In *polar form*, complex numbers are divided by using the magnitudes and angles. If $\mathbf{A} = |\mathbf{A}|\underline{/\alpha}$ and $\mathbf{B} = |\mathbf{B}|\underline{/\beta}$,

$$\begin{aligned}\frac{\mathbf{A}}{\mathbf{B}} &= \frac{|\mathbf{A}|\underline{/\alpha}}{|\mathbf{B}|\underline{/\beta}}\\ &= \left(\frac{|\mathbf{A}|}{|\mathbf{B}|}\right)\underline{/\alpha - \beta}\end{aligned}$$

Consider the following example.

EXAMPLE 16.18 Determine the quotients **A**/**B**, **C**/**D**, and **CD**/**E** if

$$\mathbf{A} = 12 - j3$$
$$\mathbf{B} = 4 - j2$$
$$\mathbf{C} = 10\angle 45^\circ$$
$$\mathbf{D} = 5\angle -20^\circ$$
$$\mathbf{E} = 25\angle 10^\circ$$

Express the answers in polar form. ■

SOLUTION

$$\frac{\mathbf{A}}{\mathbf{B}} = \frac{12 - j3}{4 - j2}$$
$$= \left\{\frac{12 - j3}{4 - j2}\right\}\left\{\frac{4 + j2}{4 + j2}\right\}$$
$$= \frac{48 + j24 - j12 + 6}{16 + 4}$$
$$= \frac{52 + j12}{20}$$
$$= \left(\frac{52}{20}\right) + j\left(\frac{12}{20}\right)$$
$$= \left(\frac{13}{5}\right) + j\left(\frac{3}{5}\right)$$
$$= 2.7\angle 13^\circ$$

$$\frac{\mathbf{C}}{\mathbf{D}} = \frac{10\angle 45^\circ}{5\angle -20^\circ}$$
$$= (10/5)\angle 45^\circ - (-20^\circ)$$
$$= 2\angle 65^\circ$$

$$\frac{\mathbf{CD}}{\mathbf{E}} = \frac{(10\angle 45^\circ)(5\angle -20^\circ)}{(25\angle 10^\circ)}$$
$$= \left\{\frac{(10)(5)}{25}\right\}\angle 45^\circ + (-20^\circ) - 10^\circ$$
$$= 2\angle 15^\circ$$

16.19 PHASORS IN THE COMPLEX PLANE

If we consider the plane of the rotating phasor (phasor diagram) as the complex plane, the location of a phasor at $t = 0$ can be expressed as a complex number. A phasor and the generated sine wave are illustrated in Figure 16.40,

Figure 16.40 Phasor.

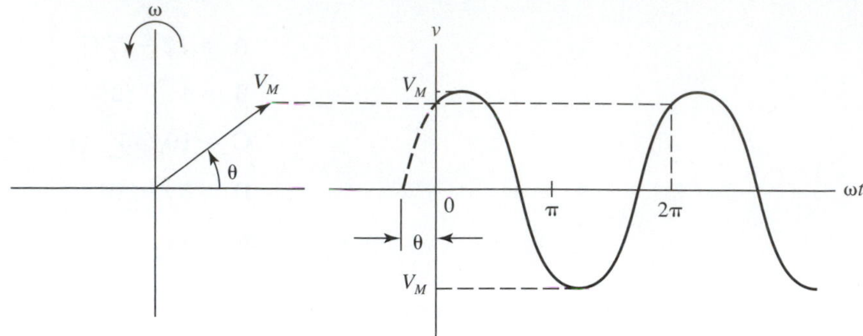

where the plane of the phasor diagram is considered the complex plane. The phasor used to generate the waveform v in Figure 16.40 can be expressed in polar form as

$$\mathbf{V} = V_M \angle \theta$$

where $\mathbf{V} = V_M \angle \theta$ is the location of the phasor in the complex plane at $t = 0$.

16.20 SUM OF TWO PHASORS IN COMPLEX FORM

Two phasors

$$\mathbf{Y} = Y_M \angle \theta_y$$
$$\mathbf{Z} = Z_M \angle \theta_z$$

generate the sine waves

$$y = Y_M \sin(\omega t + \theta_y)$$
$$z = Z_M \sin(\omega t + \theta_z)$$

Consider a phasor **X** where **X** generates a sine wave x such that

$$x = y + z$$

The phasor **X** is determined by using complex-number algebra to add phasors **Y** and **Z.**

$$\mathbf{X} = \mathbf{Y} + \mathbf{Z}$$
$$= Y_M \angle \theta_y + Z_M \angle \theta_z$$

Using the method given in Figure 16.35 to convert from polar to rectangular form,

$$\begin{aligned}\mathbf{X} &= Y_M \cos(\theta_y) + jY_M \sin(\theta_y) + Z_M \cos(\theta_z) + jZ_M \sin(\theta_z)\\ &= \{Y_M \cos(\theta_y) + Z_M \cos(\theta_z)\} + j\{Y_M \sin(\theta_y) + Z_M \sin(\theta_z)\}\\ &= X_M \angle \theta_x\end{aligned}$$

where

$$X_M = \sqrt{\{Y_M \cos(\theta_y) + Z_M \cos(\theta_z)\}^2 + \{Y_M \sin(\theta_y) + Z_M \sin(\theta_z)\}^2}$$

$$\theta_x = \tan^{-1}\left\{\frac{Y_M \sin(\theta_y) + Z_M \sin(\theta_z)}{Y_M \cos(\theta_y) + Z_M \cos(\theta_z)}\right\}$$

Because $\mathbf{X} = X_M \angle \theta_x$,

$$x = X_M \sin(\omega t + \theta_x)$$

Consider the following example.

EXAMPLE 16.19 Determine x if

$$x = y + z$$

and

$$y = 10 \sin(\omega t + 30°)$$
$$z = 20 \sin(\omega t + 60°)$$

■

SOLUTION The phasors **Y** and **Z** that generate the sinusoids y and z are

$$\mathbf{Y} = 10\angle 30°$$
$$\mathbf{Z} = 20\angle 60°$$

In order to add **Y** and **Z**, we convert to rectangular form:

$$\mathbf{Y} = 10 \cos 30° + j10 \sin 30°$$
$$= 8.7 + j5$$
$$\mathbf{Z} = 20 \cos 60° + j20 \sin 60°$$
$$= 10 + j17.3$$

Because $x = y + z$,

$$\mathbf{X} = \mathbf{Y} + \mathbf{Z}$$
$$= 8.7 + j5 + 10 + j17.3$$
$$= 18.7 + j22.3$$

Converting **X** to polar form results in

$$\mathbf{X} = \sqrt{18.7^2 + 22.3^2} \angle \tan^{-1}(22.3/28.7)$$
$$= 29.1\angle 50°$$

The phasor **X** generates the sine wave

$$x = 29.1 \sin(\omega t + 50°)$$

EXAMPLE 16.20 Consider the series circuit with voltages as shown in Figure 16.41.

Figure 16.41

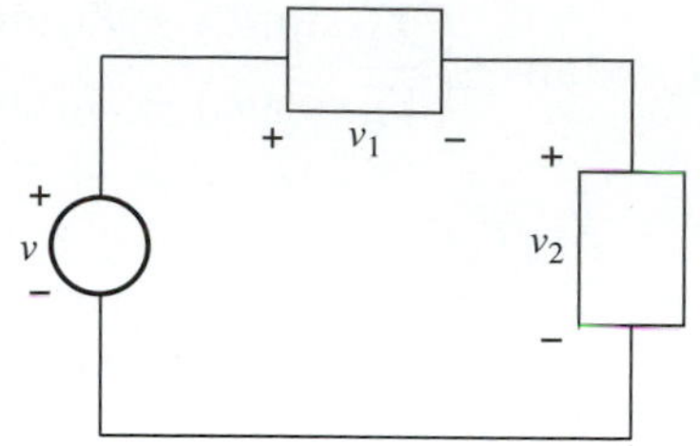

$$v = 40 \sin(2t - 30°)$$
$$v_2 = 60 \sin(2t + 15°)$$

Determine v_1, and draw a phasor diagram that includes generating phasors for v, v_1, and v_2. ■

SOLUTION From Kirchhoff's voltage law

$$v = v_1 + v_2$$

and, therefore,

$$\mathbf{V} = \mathbf{V}_1 + \mathbf{V}_2$$
$$\mathbf{V}_1 = \mathbf{V} - \mathbf{V}_2$$

From the expression for v and v_2,

$$\begin{aligned}\mathbf{V} &= 40\angle -30° \\ &= 40 \cos(-30°) + j40 \sin(-30°) \\ &= 34.6 - j20 \\ \mathbf{V}_2 &= 60\angle 15° \\ &= 60 \cos(15°) + j60 \sin(15°) \\ &= 58 + j15.5\end{aligned}$$

and from Kirchhoff's voltage law,

$$\begin{aligned}\mathbf{V}_1 &= \mathbf{V} - \mathbf{V}_2 \\ &= 34.6 - j20 - 58 - j15.5 \\ &= -23.4 - j35.5 \\ &= 42.5\angle -123.4°\end{aligned}$$

The phasor voltage $\mathbf{V}_1$ generates the voltage waveform

$$v_1 = 42.5 \sin(2t - 123.4°)$$

The phasor diagram is shown in Figure 16.42.

Figure 16.42

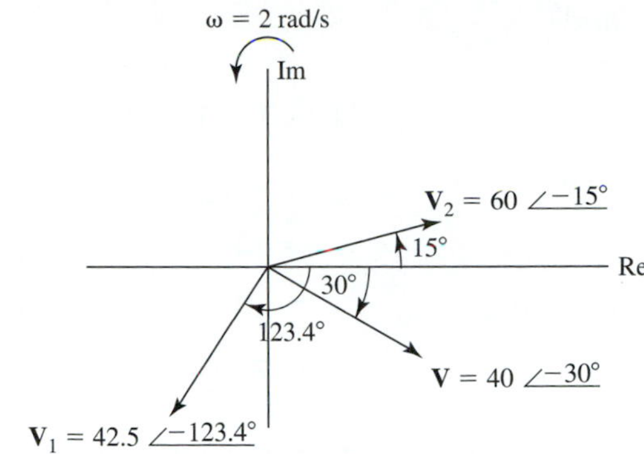

The procedure illustrated in Example 16.20 for combining two phasors is valid for combining any number of phasors. Consider the following example, where the addition of three phasors is required.

EXAMPLE 16.21 Consider the parallel circuit with currents shown in Figure 16.43.

Figure 16.43

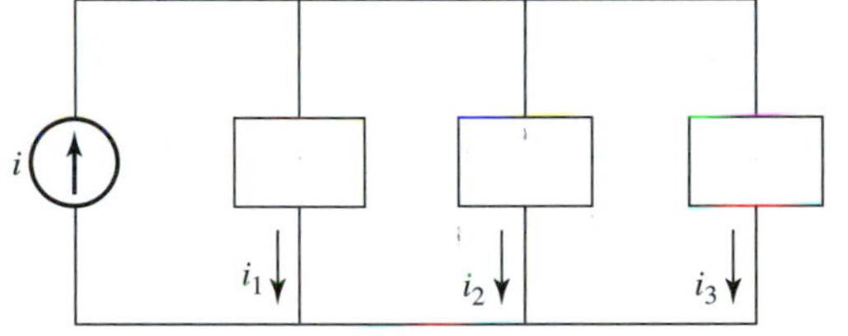

$$i = 10 \sin(\omega t)$$
$$i_1 = 4 \cos(\omega t)$$
$$i_3 = 5 \sin(\omega t - 53°)$$

Determine i_2 and draw a phasor diagram that includes the generating phasors for i, i_1, i_2, and i_3. ■

SOLUTION From Kirchhoff's current law,

$$i = i_1 + i_2 + i_3$$

and, therefore,

$$\mathbf{I} = \mathbf{I}_1 + \mathbf{I}_2 + \mathbf{I}_3$$
$$\mathbf{I}_2 = \mathbf{I} - \mathbf{I}_1 - \mathbf{I}_3$$

From the expressions for i, i_1, and i_3,

$$\mathbf{I} = 10\angle 0°$$
$$\mathbf{I}_1 = 4\angle 90°$$
$$\mathbf{I}_3 = 5\angle -53°$$

and

$$
\begin{aligned}
\mathbf{I}_2 &= \mathbf{I} - \mathbf{I}_1 - \mathbf{I}_2 \\
&= 10\underline{/0^\circ} - 4\underline{/90^\circ} - 5\underline{/-53^\circ} \\
&= 10 - j4 - 5\cos(-53^\circ) - j5\sin(-53^\circ) \\
&= 10 - j4 - 3 + j4 \\
&= 7\underline{/0^\circ}
\end{aligned}
$$

The phasor current $\mathbf{I}_2$ generates the current waveform

$$i_2 = 7\sin(\omega t)$$

The phasor diagram is as shown in Figure 16.44.

Figure 16.44

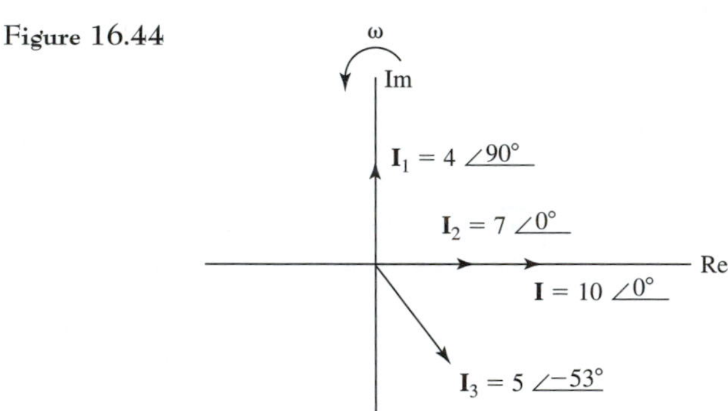

16.21 PARALLELOGRAM LAW

Phasors can also be added using a law known as the *parallelogram law*. Preparatory to presenting the parallelogram law, let us recall (from trigonometry) the law of sines and the law of cosines, as illustrated in Figure 16.45(a) and (b), respectively.

Figure 16.45 Laws of sines and cosines.

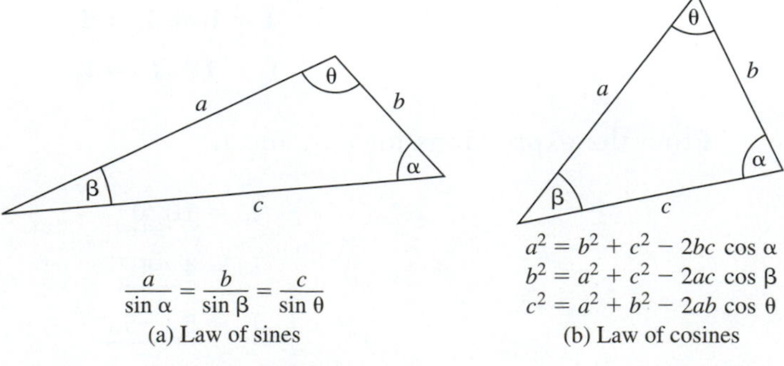

The parallelogram law states that if two phasors to be added are arranged to form a parallelogram with the two phasors as sides, then the major diagonal of the resulting parallelogram will be the phasor sum of the two phasors, as illustrated in Figure 16.46.

Figure 16.46 Parallelogram law.

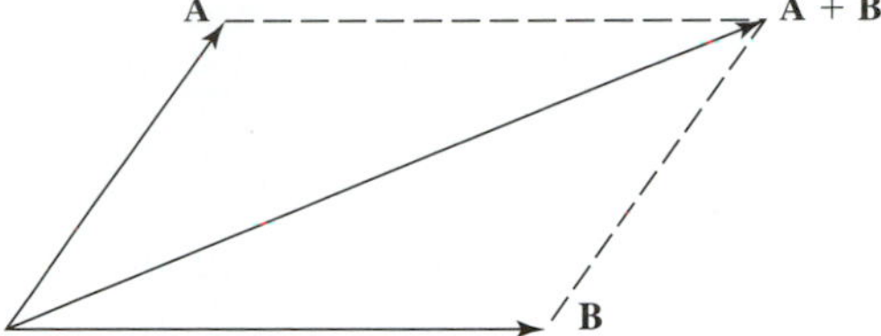

Consider the following example.

EXAMPLE 16.22 Determine the sum of phasors **A** and **B**, where

$$\mathbf{A} = 100\angle 60^\circ$$

and

$$\mathbf{B} = 100\angle 20^\circ$$ ■

SOLUTION From the parallelogram law, the phasor sum **A** + **B** is shown in Figure 16.47.

Figure 16.47

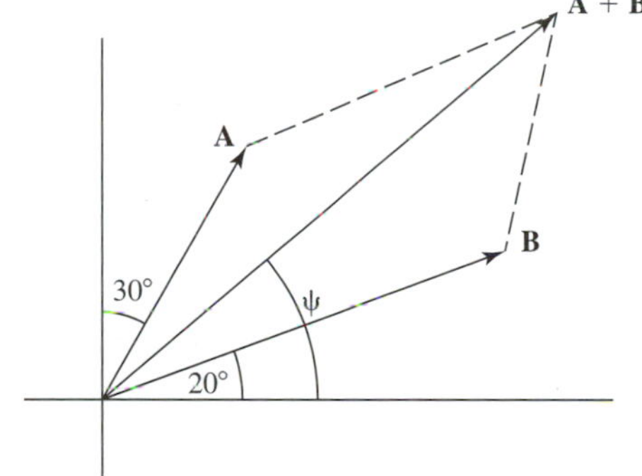

The relationship between the angles associated with a parallelogram are illustrated in Figure 16.48.

Figure 16.48

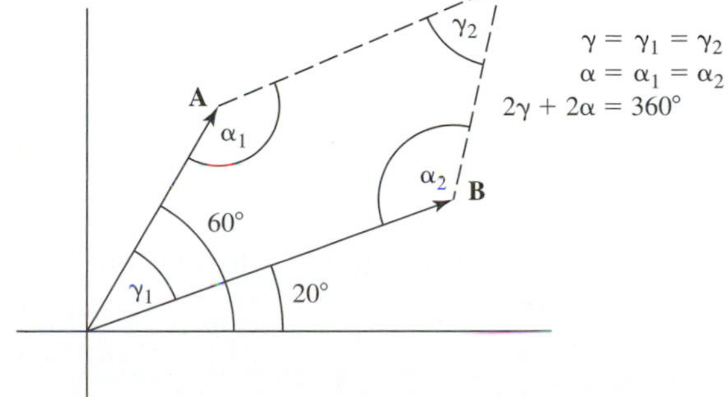

The angles labeled γ_1 and γ_2 are equal from the geometric symmetry of the parallelogram, and for the parallelogram under consideration,

$$\gamma = 60^\circ - 20^\circ = 40^\circ$$

The angles labeled α_1 and α_2 are also equal from geometric symmetry, and because the sum of the interior angles of a four-sided figure equals 360°,

$$2\alpha + 2\gamma = 360°$$
$$2\alpha = 360 - 2(40°)$$
$$\alpha = 140°$$

for the parallelogram under consideration.

The original parallelogram can be redrawn as shown in Figure 16.49. Letting $\mathbf{R} = \mathbf{A} + \mathbf{B}$ and R, A, and B be the magnitudes of $\mathbf{R}$, $\mathbf{A}$, and $\mathbf{B}$, respectively, the lower triangle of the parallelogram can be drawn as shown in Figure 16.50.

Figure 16.49

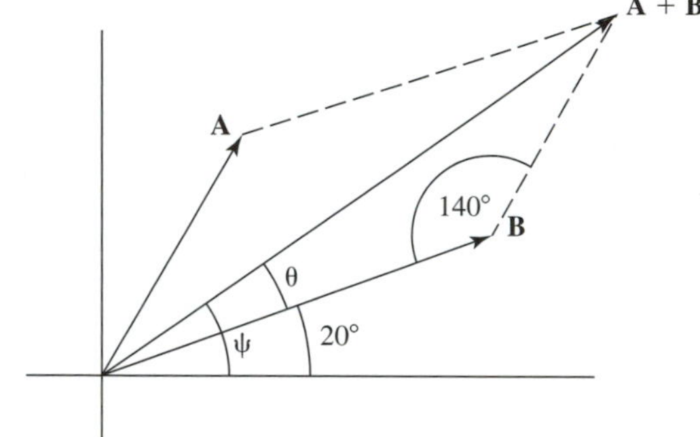

Figure 16.50

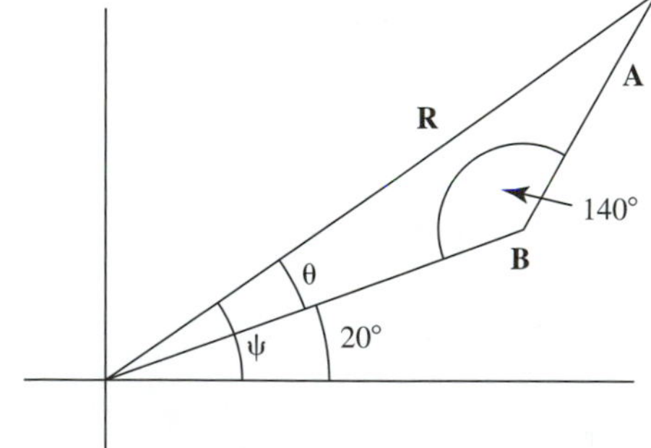

From the given information, $A = 100$ and $B = 100$, and from the law of cosines,

$$R^2 = A^2 + B^2 - 2AB\cos(140°)$$
$$R = \sqrt{100^2 + 100^2 - 2(100)(100)\cos(140°)}$$
$$= 188$$

From the law of sines,

$$\frac{R}{\sin(140°)} = \frac{A}{\sin\theta}$$

$$\frac{188}{\sin(140°)} = \frac{100}{\sin\theta}$$

$$\sin\theta = (100)\left\{\frac{\sin(140°)}{188}\right\}$$

$$= 0.342$$

$$\theta = \sin^{-1}(0.342)$$

$$= 20°$$

and the angle ψ that **R** makes with the positive horizontal axis is

$$\psi = \theta + 20° = 20° + 20°$$

$$= 40°$$

The phasor **R** (where **R** = **A** + **B**) can be written as

$$\mathbf{R} = 188\underline{/40°}$$

The parallelogram law allows us to express easily the sum of a sine wave and a cosine wave with zero phase angles and the same frequency as a single sinusoid. Consider the sine wave and the cosine wave shown in Figure 16.51.

Figure 16.51 Generation of sine and cosine waves.

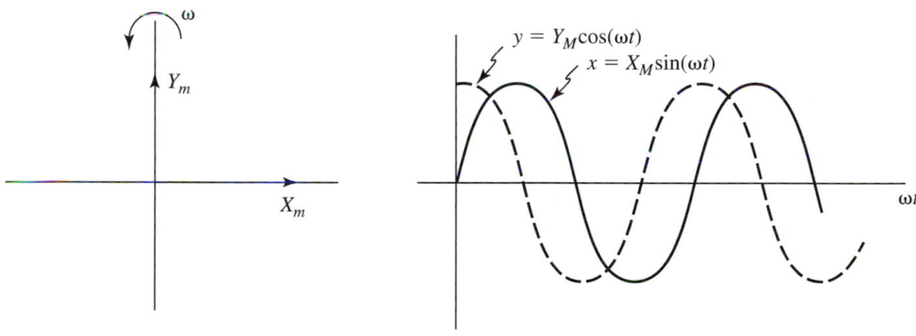

The phasor representation in complex form for x and y is

$$\mathbf{X} = X_M\underline{/0°}$$

$$\mathbf{Y} = Y_M\underline{/90°}$$

From the parallelogram law, the sum of **X** + **Y** is shown in Figure 16.52, where **X** + **Y** is designated by **R** and the waveform generated by **R** is called *r*.

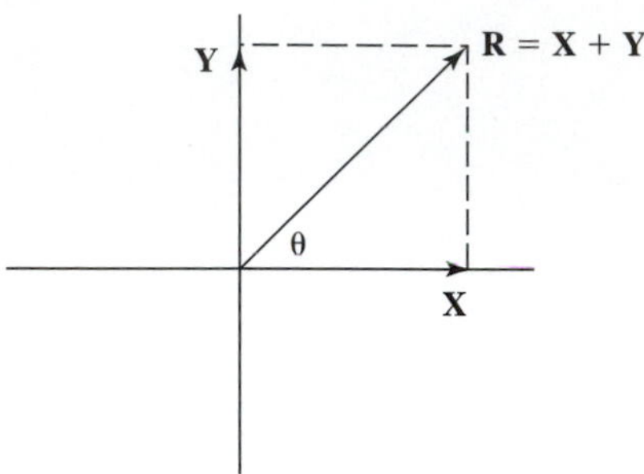

Figure 16.52 Parallelogram law.

Because the angle between **X** and **Y** is 90°, we can use the Pythagorean theorem instead of the law of cosines to determine the magnitude of **R**; the result is

$$R_M = \sqrt{X_M^2 + Y_M^2}$$

and

$$\theta = \tan^{-1}\left\{\frac{Y_M}{X_M}\right\}$$

where R_M is the maximum value of r and θ is the phase angle.

The phasor **R** can be expressed in complex form as

$$\mathbf{R} = R_M\angle\theta$$

and in the time domain r is

$$r = R_M \sin(\omega t + \theta)$$

where $r = x + y$.

EXAMPLE 16.23 Express z as a sine wave, where

$$z = x + y$$

and

$$x = 40 \sin(2t)$$

$$y = 30 \cos(2t)$$

■

SOLUTION The phasors for x and y are

$$\mathbf{X} = 40\angle 0^\circ$$

$$\mathbf{Y} = 30\angle 90^\circ$$

From the parallelogram law, the phasor diagram is as shown in Figure 16.53.

Figure 16.53

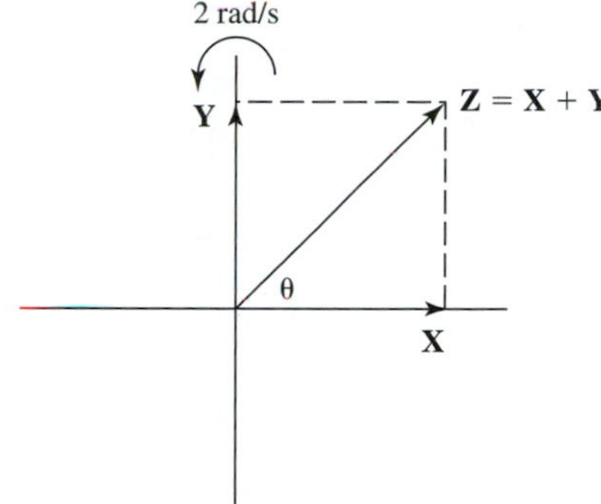

Because $X_M = 40$ and $Y_M = 30$

$$Z_M = \sqrt{X_M^2 + Y_M^2}$$
$$= \sqrt{40^2 + 30^2}$$
$$= 50$$
$$\theta = \tan^{-1}\left\{\frac{Y_M}{X_M}\right\}$$
$$= \tan^{-1}\left\{\frac{30}{40}\right\}$$
$$= 36.9°$$

The phasor **Z** is

$$\mathbf{Z} = Z_M\angle\theta$$
$$= 50\angle 36.9°$$

and

$$z = 50\sin(2t + 36.9°)$$

16.22 EXERCISES

1. Given the angular velocity of the generating phasor, find the frequency of the generated wave form.

(a) π rad/s (b) 88 rad/s (c) 4 π rad/s (d) $\frac{\pi}{2}$ rad/s
(e) 50 rad/s (f) 16 rad/s

2. Given the frequency of the generated waveform, find the angular velocity of the generating phasor.

(a) 60 Hz (b) $\frac{1}{4}$ Hz (c) 5 Hz (d) 80 Hz
(e) 0.5 Hz (f) 2000 Hz

3. Given the frequencies in Problem 2, find the periods.

4. Determine the period and frequency for each waveform in Figure 16.54.

Figure 16.54

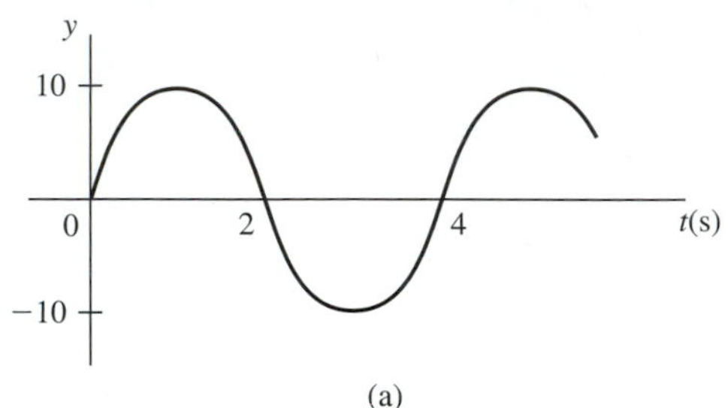

(a)

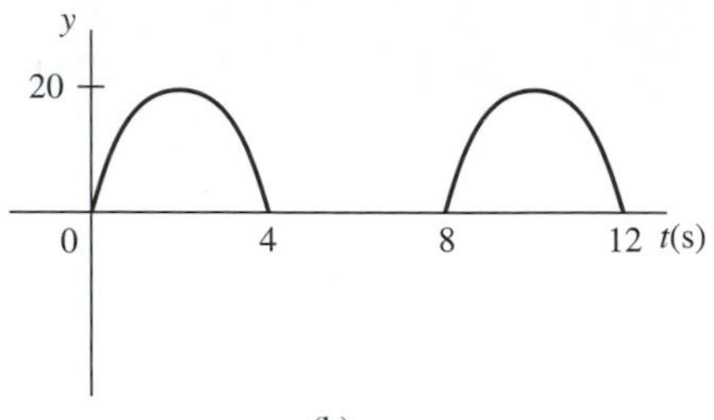

(b)

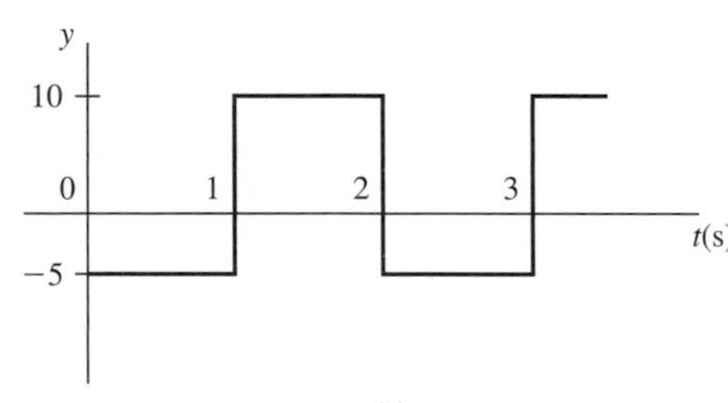

(c)

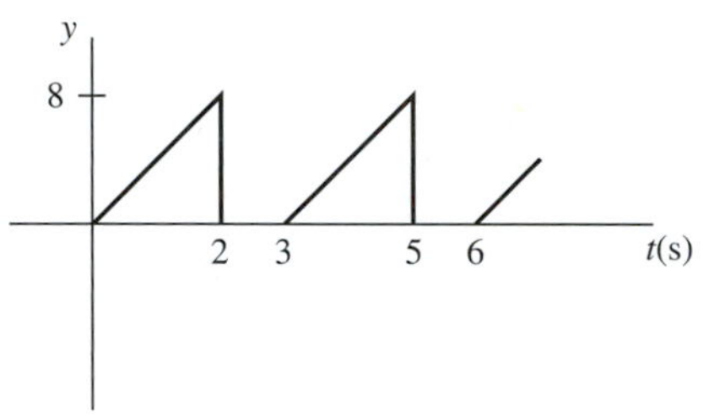

(d)

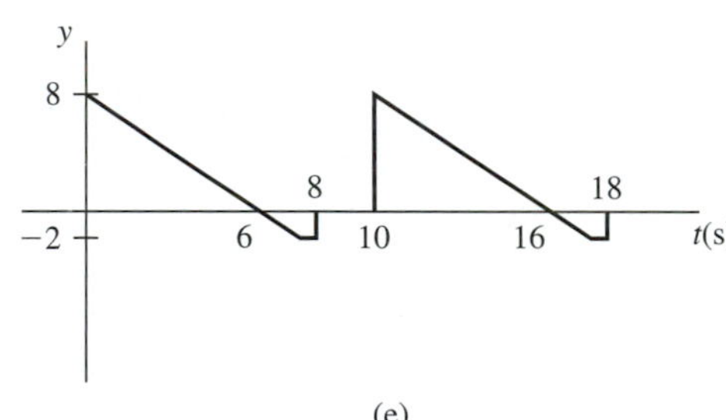

(e)

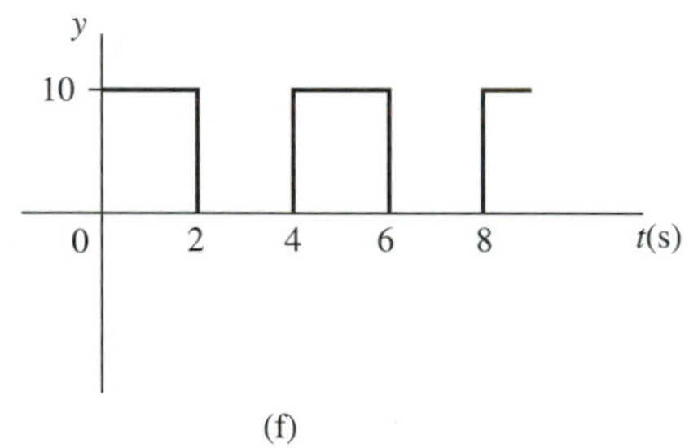

(f)

5. Find the period and frequency of a periodic waveform that completes 20 cycles in 4 s.

6. Find the frequency for each periodic waveform with the given period.

(a) 100 ms (b) $\frac{1}{10}$ s (c) 50 μs (d) 1 ms
(e) 80 s (f) 10 ms (g) 5 s (h) 10 μs

7. Find the period for each periodic waveform with the given frequency.

(a) 10 Hz (b) 1 Hz (c) 60 Hz (d) 15 Hz
(e) $\frac{1}{2}$ Hz (f) 1000 Hz (g) 0.1 Hz (h) 7 Hz

8. Convert from radians to degrees.

(a) π rad (b) 2 rad (c) 1 rad (d) 2π rad
(e) $\frac{\pi}{2}$ rad (f) $\frac{2\pi}{3}$ rad (g) 5 rad (h) 15 rad

9. Convert from degrees to radians.

(a) 30° (b) 20° (c) 60° (d) 180°
(e) 270° (f) 45° (g) 360° (h) 95°

10. Determine the phase angle for each sinusoid in Figure 16.55.

Figure 16.55

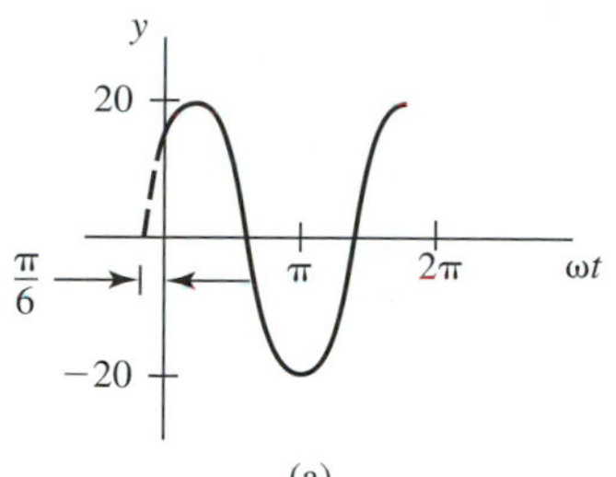

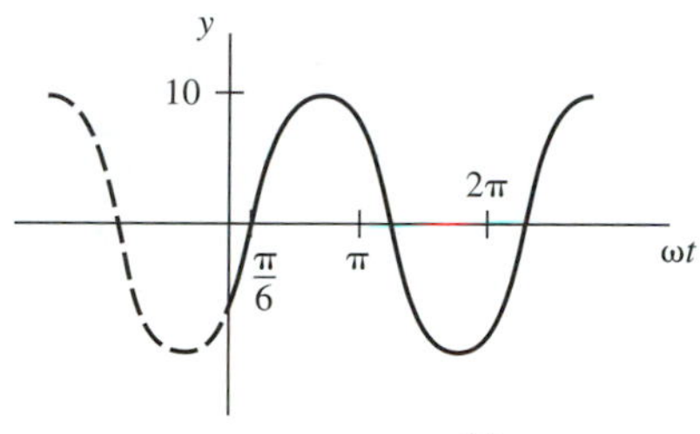

11. Determine the expression for the sine wave generated by each of the following phasors.

(a) $Y_M = 100$, $\omega = 20$ rad/s, $\theta = 40°$
(b) $Y_M = 50$, $\omega = 100$ rad/s, $\theta = -200°$
(c) $Y_M = 100$, $\omega = 10$ rad/s, $\theta = \pi$ rad
(d) $Y_M = 200$, $\omega = 40$ rad/s, $\theta = -100°$

12. Sketch the sine waves generated for each phasor in Problem 11.

13. Given the following sine waves, determine the generating phasors.

(a) $y = 40\sin(30t + 30°)$ (b) $y = -80\sin(2t - 40°)$
(c) $y = 10\sin(4t + 270°)$ (d) $y = 50\sin(10t - 80°)$
(e) $y = -10\sin(2t - 60°)$ (f) $y = -20\sin(8t)$

14. Sketch the following waveforms using time (t) as the horizontal axis.

(a) $y = 50\sin\left(\pi t + \dfrac{3\pi}{2}\right)$ (b) $y = 100\sin(2\pi t)$
(c) $y = -10\sin\left(\dfrac{3\pi}{2t} + \dfrac{\pi}{2}\right)$ (d) $y = -80\sin(4t - 180°)$

15. Determine the phase difference for y_1 and y_2.

(a) $y_1 = 2\sin(\omega t - 40°)$, $y_2 = 6\sin(\omega t + 40°)$
(b) $y_1 = 4\sin(\omega t + 60°)$, $y_2 = 10\sin(\omega t + 150°)$
(c) $y_1 = -50\sin(\omega t - 20°)$, $y_2 = 10\sin(\omega t + 95°)$
(d) $y_1 = -30\sin(\omega t - 100°)$, $y_2 = -40\sin(\omega t + 180°)$

16. Write a sinusoidal expression for each of the waveforms in Figure 16.56.

Figure 16.56

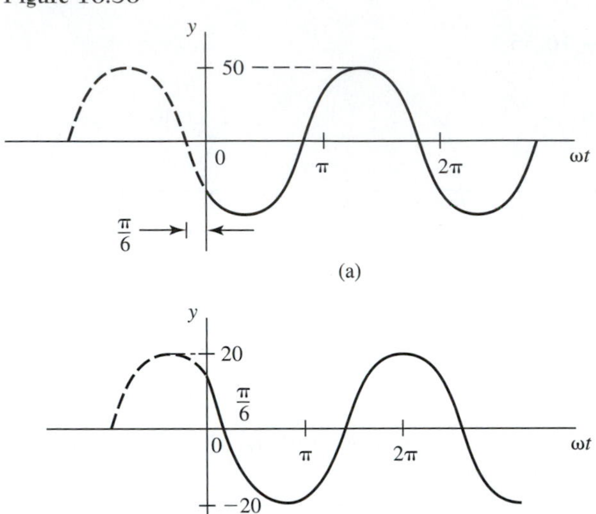

(a)

(b)

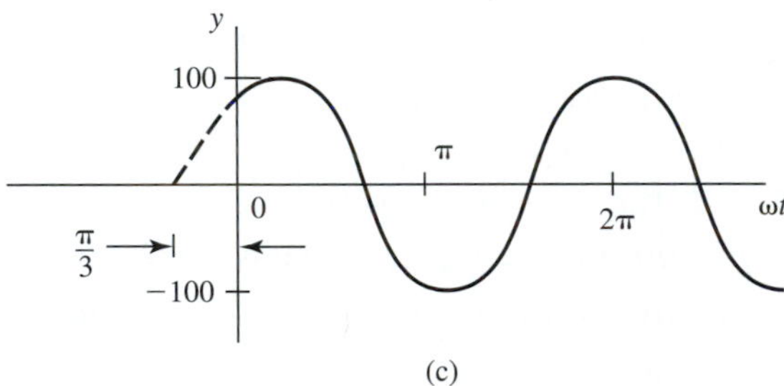

(c)

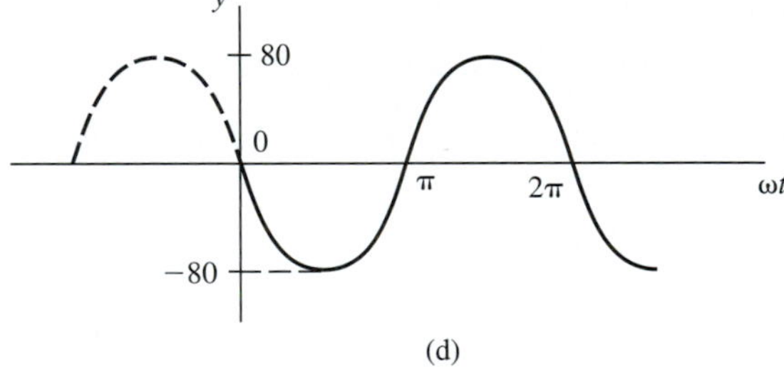

(d)

17. Find the smallest value of t if $y = 25$ in the waveform $y = 50 \sin(2\pi t)$.

18. Determine the expressions for the cosine waves generated by the phasors in Problem 11.

19. Sketch the cosine waves determined in Problem 11.

20. Given the following cosine waves, determine the generating phasors.

(a) $y = 50 \cos(40t + \pi)$ (b) $y = -\cos(3t - 150°)$
(c) $y = 20 \cos(4t + 270°)$ (d) $y = \cos(2t - 200°)$
(e) $y = -40 \cos(3t - 100°)$ (f) $y = -\cos(6t)$

21. Write a cosine expression for each of the waveforms in Problem 16.

22. Define a complex number.

23. Plot the following complex numbers in the complex plane.

(a) $6 - j2$ (b) $3 + j4$ (c) $-8 + j6$
(d) $-5 - j5$ (e) $700 - j450$ (f) $0.2 - j0.4$
(g) $30 + j40$ (h) $-60 - j80$

24. Determine the magnitude of each complex number in Problem 23.

25. Plot the following complex numbers in the complex plane.
(a) $4\angle 30^\circ$ (b) $-6\angle 60^\circ$ (c) $10\angle -140^\circ$
(d) $-10\angle -30^\circ$ (e) $-600\angle 270^\circ$ (f) $0.15\angle 300^\circ$
(g) $4\angle 180^\circ$ (h) $-10\angle 360^\circ$

26. Convert the following complex numbers from rectangular to polar form.
(a) $3 - j4$ (b) $4 - j3$ (c) $10 + j10$
(d) $-10 + j10$ (e) $8 + j20$ (f) $-6 - j6$
(g) $30 + j40$ (h) $-300 + j500$

27. Convert the following complex numbers from polar to rectangular form.
(a) $20\angle 45^\circ$ (b) $10\angle -30^\circ$ (c) $9 \times 10^{-6}\angle 270^\circ$
(d) $50\angle -180^\circ$ (e) $15\angle 150^\circ$ (f) $10 \times 10^{-2}\angle -270^\circ$
(g) $-30\angle 360^\circ$ (h) $-4\angle 45^\circ$

28. Determine the value of **Y** in each of the following expressions.
(a) $(j)^3 - j + \mathbf{Y} = 0$ (b) $j - \sqrt{-1} - \mathbf{Y} = 0$
(c) $\frac{6}{j} + j2 - \mathbf{Y} = 0$ (d) $j^5 + j + 6\mathbf{Y} = 0$
(e) $(j)4\mathbf{Y} - 16 = 0$ (f) $\frac{1}{(j)^3} + \mathbf{Y} = 0$

29. Determine the conjugates of the following complex numbers.
(a) $6 - j4$ (b) $40\angle -50^\circ$ (c) $-30\angle 20^\circ$
(d) $-10\angle -90^\circ$ (e) $12 + j10$ (f) $-8 + j20$
(g) $-150 - j100$ (h) $-30\angle -360^\circ$

30. Perform the indicated operations and express the result in rectangular form.
(a) $(6 + j9) + (3 - j5)$ (b) $(6.8) + (2 + j2)$
(c) $(3 - j6) - (-10 + j6)$ (d) $30\angle 40^\circ + 16\angle 90^\circ$
(e) $(40 + j20) + (10\angle 90^\circ)$ (f) $j14 - 10\angle 90^\circ$
(g) $(300 + j4000) - 500\angle 90^\circ$ (h) $(j)^2 80 + j80$

31. Perform the indicated operations and express the result in polar form.
(a) $(j2)(4 + j8)$ (b) $(6 - j8)(4 + j4)$
(c) $(-j)(6 - j10)$ (d) $(j + 2)(8\angle 40^\circ)$
(e) $(10 + j6)(2 - j2)(4\angle 90^\circ)$ (f) $(6\angle 90^\circ - j)(12\angle 45^\circ)$
(g) $(-6\angle 60^\circ)(4\angle 90^\circ)(8\angle -180^\circ)$ (h) $(-4 - j6)(j6)(-4\angle -210^\circ)$

32. Perform the indicated operations and express the result in polar form.
(a) $\frac{6 + j8}{2 - j^2}$ (b) $\frac{4 + j4}{j^2}$
(c) $\left(\frac{2}{j^2}\right)(3 + j)$ (d) $\left(\frac{1}{j}\right)(-9 - j9)$
(e) $\frac{(40\angle 30^\circ)(20\angle 45^\circ)}{1 + j1}$ (f) $\frac{3 + j4}{6 - j8}$
(g) $\frac{6 + j8}{2 - j2}$ (h) $\frac{-8 - j2}{6 - j8}$

33. State the advantage of using complex numbers to represent phasors.

34. Express the following waveforms in phasor form using complex numbers in polar form to represent the phasors.
(a) $70 \sin(\omega t - 20^\circ)$ (b) $-10 \sin(\omega t + 30^\circ)$
(c) $4 \cos(\omega t - 200^\circ)$ (d) $-60 \cos(\omega t - 45^\circ)$
(e) $-10 \sin(\omega t + 300^\circ)$ (f) $16 \cos(\omega t + 270^\circ)$
(g) $-7 \cos(\omega t - 360^\circ)$ (h) $-100 \sin(\omega t - 180^\circ)$

35. Determine the current phasor $\mathbf{I}_2$ in polar form and the current i_2 for the circuit in Figure 16.57.

Figure 16.57

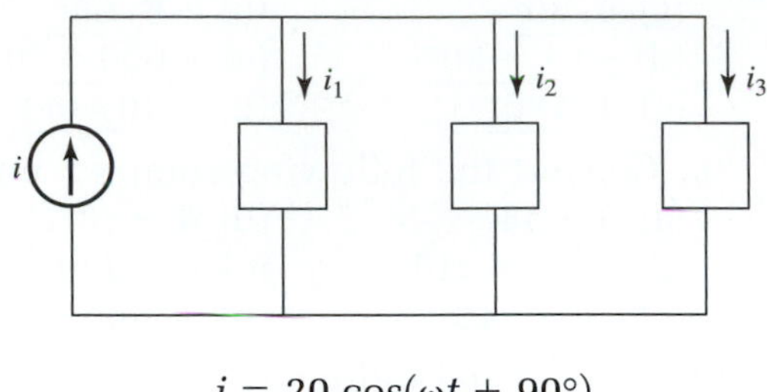

$$i = 20\cos(\omega t + 90°)$$

$$i_1 = -10\sin(\omega t)$$

$$i_3 = 10\sin(\omega t - 90°)$$

36. Determine the voltage phasor $\mathbf{V}_3$ in polar form and the voltage v_3 for the circuit in Figure 16.58.

Figure 16.58

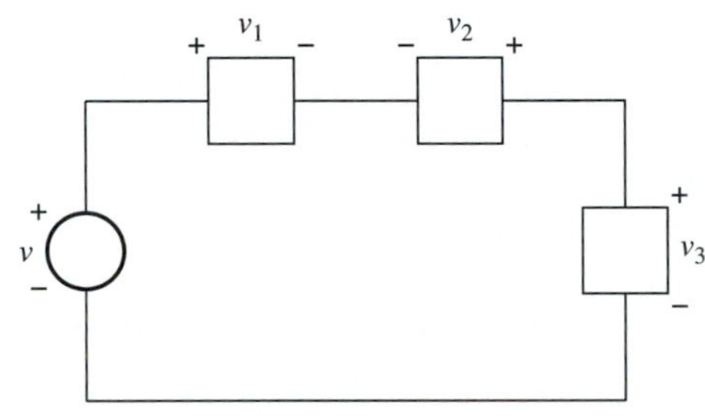

$$v = 20\cos(\omega t - 180°)$$

$$v_1 = 20\sin(\omega t - 90°)$$

$$v_2 = -30\sin(\omega t)$$

37. Determine the sum of phasors **A** and **B** using complex number algebra. Express the result in polar form.

(a) $\mathbf{A} = -40\angle 20°$
$\mathbf{B} = 20\angle 180°$

(b) $\mathbf{A} = 30 + j30$
$\mathbf{B} = 20 - j10$

(c) $\mathbf{A} = -80\angle -110°$
$\mathbf{B} = 50\angle 30°$

(d) $\mathbf{A} = 30\angle -180°$
$\mathbf{B} = -20 - j20$

38. Determine the sum of phasors **A** and **B** using the parallelogram law. Express the result in polar form.

$$\mathbf{A} = 200\angle 30°$$

$$\mathbf{B} = 100\angle 60°$$

CHAPTER

17

Basic AC Circuits

After completing this chapter the student should be able to

* determine ac impedance for a pair of terminals
* determine ac admittance for a pair of terminals
* analyze ac circuits using the impedance and admittance concepts

As presented previously, the current-voltage relationships for the resistor, inductor, and capacitor, respectively, are

$$v = Ri$$

$$v = L\frac{di}{dt}$$

$$i = C\frac{dv}{dt}$$

in derivative form and

$$i = \frac{v}{R}$$

$$i = \frac{1}{L}\int v\,dt$$

$$v = \frac{1}{C}\int i\,dt$$

in integral form.

Determining a current from a voltage or a voltage from a current involves multiplication by a constant, differentiation, or integration, as indicated by the preceding equations. Because multiplying a sinusoidal function by a constant, differentiating a sinusoidal function, or integrating a sinusoidal function all result in another sinusoidal function of the same frequency, the steady-state currents and voltages in an electric circuit driven by sinusoidal voltage and current sources are themselves sinusoidal waveforms. We shall use the impedance and admittance concepts to relate the phasor voltage to the phasor current for any pair of circuit terminals. The impedance and admittance concepts allow us to determine the steady-state sinusoidal current flowing through an electrical element if we know the sinusoidal voltage across the element and vice versa.

17.1 THE IMPEDANCE CONCEPT

The impedance of a pair of terminals for ac voltages and currents is defined as the phasor voltage divided by the phasor current, as illustrated in Figure 17.1. The symbol **Z** is the symbol for ac impedance. Impedance **Z** is a complex quantity and can be expressed in polar or rectangular form. The symbol Z is used for the magnitude of **Z**. That is, if $\mathbf{Z} = 5\underline{/30^\circ}$, then $Z = 5$. The same terminology applies to current and voltage phasors. If $\mathbf{V} = 100\underline{/30^\circ}$ and $\mathbf{I} = 10\underline{/-20^\circ}$, the magnitudes of **V** and **I** are written as $V = 100$ and $I = 10$, where 100 and 10 are, respectively, the maximum values of the sinusoidal voltage and current generated by phasors **V** and **I**. We sometimes subscript V and I with the letter M to emphasize the maximum value. For example, if $\mathbf{V} = 100\underline{/30^\circ}$, the magnitude of **V** can be written as $V = 100$ or $V_M = 100$.

Figure 17.1 Impedance.

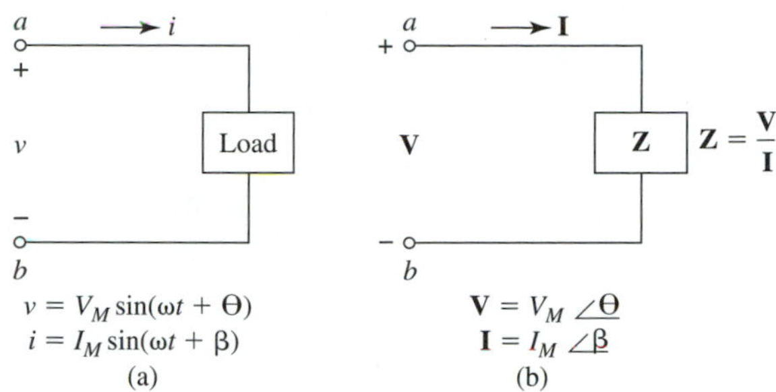

In Figure 17.1(a) the circuit is represented in the time domain, and in Figure 17.1(b) the currents and voltages are represented as phasors. In Figure 17.1(b) the impedance **Z** "looking into terminals *ab*" is

$$\mathbf{Z} = \frac{\mathbf{V}}{\mathbf{I}} \tag{17.1}$$

From the given current and voltage phasor expressions,

$$\begin{aligned}\mathbf{Z} &= \frac{V_M\angle\theta}{I_M\angle\beta} \\ &= \left(\frac{V_M}{I_M}\right)\angle\theta - \beta\end{aligned}$$

and the magnitude of **Z** is

$$Z = \frac{V_M}{I_M} \tag{17.2}$$

Because **Z** has units of volts/ampere, the units for impedance are ohms. If the impedance for a pair of terminals is known, the voltage can be obtained from the current expression and the current can be obtained from the voltage expression, as illustrated in the examples that follow.

We shall now consider the impedance of the three basic passive elements—resistors, inductors, and capacitors.

17.2 IMPEDANCE OF RESISTORS

Consider the resistor in Figure 17.2, with current and voltage as indicated. From the given voltage expression, the phasor for v is

$$\mathbf{V} = V_M\angle\theta$$

Figure 17.2 Resistor circuit.

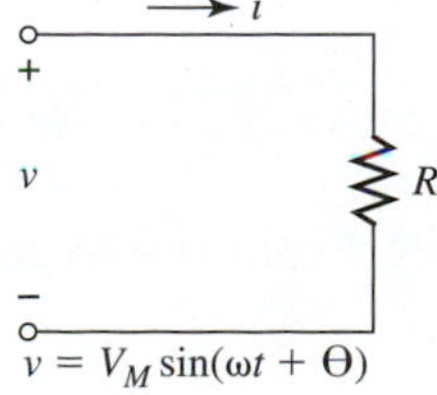

From Ohm's law

$$i = \frac{v}{R}$$
$$= \left(\frac{V_M}{R}\right)\sin(\omega t + \theta)$$

The phasor for i is

$$\mathbf{I} = \left(\frac{V_M}{R}\right)\angle\theta$$

From Equation (17.1)

$$\mathbf{Z}_R = \frac{\mathbf{V}}{\mathbf{I}}$$
$$= \frac{V_M\angle\theta}{(V_M/R)\angle\theta}$$
$$= R\angle\theta - \theta$$
$$= R\angle 0^\circ$$

and in rectangular form,

$$\mathbf{Z}_R = R\angle 0^\circ = R + j0 \tag{17.3}$$

Because $\mathbf{I} = (V_M/R)\angle\theta$ and $i = (V_M/R)\sin(\omega t + \theta)$, it follows that

$$I_M = \frac{V_M}{R}$$

and

$$R = \frac{V_M}{I_M} \tag{17.4}$$

Because **V** and **I** have the same phase angle the current and voltage are in phase for a pure resistor.

Consider the following examples.

EXAMPLE 17.1 Determine the current i for the circuit in Figure 17.3. ■

Figure 17.3

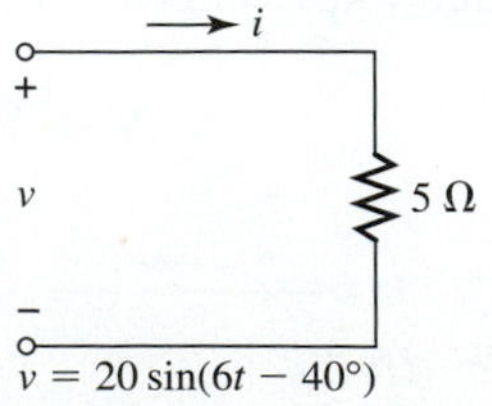

$v = 20\sin(6t - 40^\circ)$

SOLUTION From the expression for v

$$\mathbf{V} = 20\angle -40^\circ$$

From Equation (17.3)

$$\mathbf{Z}_R = R\angle 0^\circ$$
$$= 5\angle 0^\circ$$

and from Equation (17.1)

$$\mathbf{Z}_R = \frac{\mathbf{V}}{\mathbf{I}}$$

$$\mathbf{I} = \frac{\mathbf{V}}{\mathbf{Z}_R}$$
$$= \frac{20\angle -40^\circ}{5\angle 0^\circ}$$
$$= \left(\frac{20}{5}\right)\angle -40^\circ - 0^\circ$$
$$= 4\angle -40^\circ$$

The current i generated by the phasor $4\angle -40^\circ$ is

$$i = 4\sin(6t - 40^\circ)$$

Notice that the voltage v and the current i are in phase. That is, **V** and **I** have the same phase angle.

EXAMPLE 17.2 Determine the voltage v for the circuit in Figure 17.4. ■

Figure 17.4

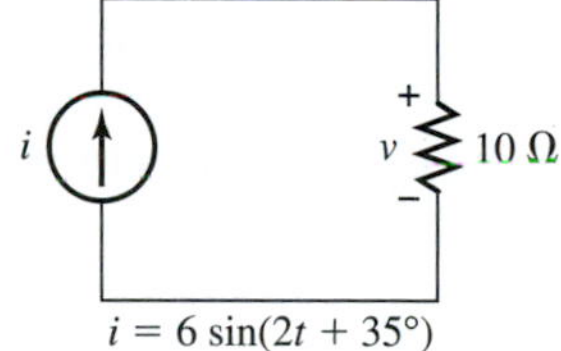

SOLUTION From the expression given, the phasor **I** is

$$\mathbf{I} = 6\angle 35^\circ$$

From Equation (17.3)

$$\mathbf{Z}_R = R\angle 0^\circ$$
$$= 10\angle 0^\circ$$

and from Equation (17.1)

$$\mathbf{Z}_R = \frac{\mathbf{V}}{\mathbf{I}}$$

$$\mathbf{V} = \mathbf{Z}_R\mathbf{I}$$
$$= (10\angle 0^\circ)(6\angle 35^\circ)$$

$$= \{(10)(6)\}\underline{/0^\circ + 35^\circ}$$
$$= 60\underline{/35^\circ}$$

The voltage v generated by the phasor $60\underline{/35^\circ}$is

$$v = 60\sin(2t + 35^\circ)$$

Notice that the voltage v and the current i are in phase.

17.3 IMPEDANCE OF INDUCTORS

Consider the inductor in Figure 17.5 with current as indicated. From the given current expression the phasor representing i is

$$\mathbf{I} = I_M\underline{/\beta}$$

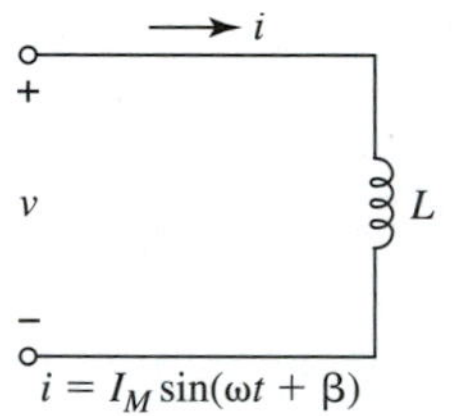

Figure 17.5 Inductor circuit.

Because $v = L\dfrac{di}{dt}$ (Equation 11.3) for an inductor,

$$v = L\left\{\frac{d\{I_M\sin(\omega t + \beta)\}}{dt}\right\}$$

From rules D4 and D6 in Chapter 12 for differentiation,

$$v = (\omega L I_M)\cos(\omega t + \beta)$$

and the phasor necessary to generate v is

$$\mathbf{V} = (\omega L I_M)\underline{/\beta + 90^\circ}$$

It follows that

$$V_M = \omega L I_M$$

and

$$\frac{V_M}{I_M} = \omega L \tag{17.5}$$

A comparison of phasors V and I reveals that *voltage leads current by* 90° *for a pure inductor.*

From Equation (17.1)

$$\mathbf{Z}_L = \frac{\mathbf{V}}{\mathbf{I}}$$

$$= \frac{(\omega L\mathbf{I}_M)\underline{/\beta + 90^\circ}}{I_M\underline{/\beta}}$$

$$= \{(\omega L I_M)/I_M\}\underline{/\beta + 90° - \beta}$$

$$\mathbf{Z}_L = \omega L\underline{/90°} = j\omega L \qquad (17.6)$$

The magnitude of $\mathbf{Z}_L$ is called *inductive reactance* and is designated by the symbol X_L:

$$X_L = \omega L \qquad (17.7)$$

In terms of inductive reactance, $\mathbf{Z}_L$ can be expressed as

$$\mathbf{Z}_L = X_L\underline{/90°} = jX_L \qquad (17.8)$$

The units for X_L are ohms, because $\mathbf{Z}_L$ has units of ohms.

EXAMPLE 17.3 Determine the current i for the circuit in Figure 17.6. ■

Figure 17.6

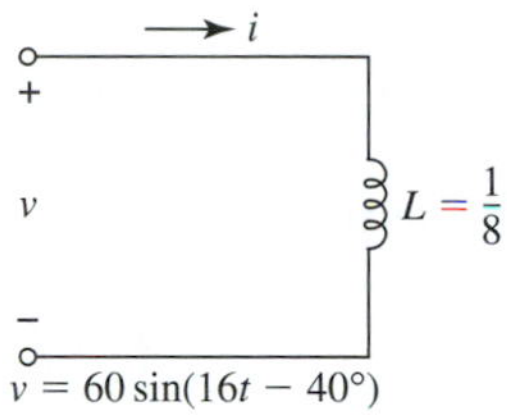

i
+
v
−
$L = \frac{1}{8}$ H
$v = 60 \sin(16t - 40°)$

SOLUTION From the expression for v,

$$\mathbf{V} = 60\underline{/-40°}$$

From Equations (17.7) and (17.8),

$$X_L = \omega L$$

$$= (16)\left(\frac{1}{8}\right)$$

$$= 2\ \Omega$$

and

$$\mathbf{Z}_L = X_L\underline{/90°}$$

$$= 2\underline{/90°}$$

From Equation (17.1)

$$\mathbf{Z}_L = \frac{\mathbf{V}}{\mathbf{I}}$$

and

$$\mathbf{I} = \frac{\mathbf{V}}{\mathbf{Z}_L}$$

$$= \frac{60\underline{/-40°}}{2\underline{/90°}}$$

$$= \frac{60}{2}\underline{/-40^\circ - 90^\circ}$$

$$= 30\underline{/-130^\circ}$$

The current i generated by the phasor $30\underline{/-130^\circ}$ is

$$i = 30 \sin(16t - 130^\circ)$$

Notice that the voltage v leads the current i by 90°.

EXAMPLE 17.4 Determine the voltage v for the circuit in Figure 17.7. ■

Figure 17.7

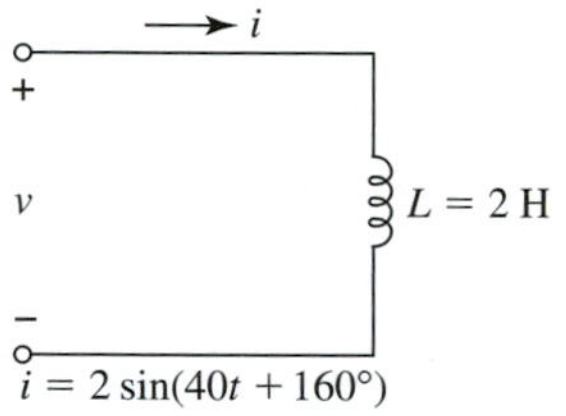

i, +, v, −, $L = 2$ H, $i = 2 \sin(40t + 160^\circ)$

SOLUTION From the expression for i,

$$\mathbf{I} = 2\underline{/160^\circ}$$

From Equations (17.7) and (17.8),

$$X_L = \omega L$$
$$= (40)(2)$$
$$= 80\ \Omega$$

and

$$\mathbf{Z}_L = X_L\underline{/90^\circ}$$
$$= 80\underline{/90^\circ}$$

From Equation (17.1)

$$\mathbf{Z}_L = \frac{\mathbf{V}}{\mathbf{I}}$$

and

$$\mathbf{V} = (\mathbf{Z}_L)(\mathbf{I})$$
$$= (80\underline{/90^\circ})(2\underline{/160^\circ})$$
$$= 160\underline{/250^\circ}$$

The voltage v generated by the phasor $160\underline{/250^\circ}$ is

$$v = 160 \sin(40t + 250^\circ)$$

Notice that the voltage v leads the current i by 90°.

17.4 IMPEDANCE OF CAPACITORS

Consider the capacitor in Figure 17.8 with the voltage as indicated. From the given voltage expression, the phasor representing v is

$$\mathbf{V} = V_M \angle \theta$$

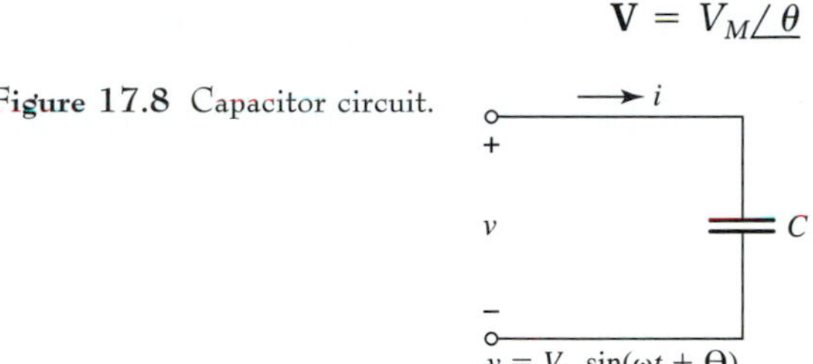

Figure 17.8 Capacitor circuit.

Because $i = C\dfrac{dv}{dt}$ (Equation 10.2) for a capacitor,

$$i = C\left\{\frac{d\{V_M \sin(\omega t + \theta)\}}{dt}\right\}$$

From rules D4 and D6 in Chapter 12 for differentiation,

$$i = \omega C V_M \cos(\omega t + \theta)$$

and the phasor necessary to generate i is

$$\mathbf{I} = (\omega C V_M) \angle \theta + 90^\circ$$

It follows that

$$I_M = \omega C V_M$$

and

$$\frac{V_M}{I_M} = \frac{1}{\omega C} \qquad (17.9)$$

A comparison of phasors **V** and **I** reveals that *current leads voltage by* 90° *for a pure capacitor.*

From Equation (17.1)

$$\mathbf{Z}_C = \frac{\mathbf{V}}{\mathbf{I}}$$

$$= \frac{V_M \angle \theta}{(\omega C V_M) \angle \theta + 90^\circ}$$

$$= \left\{\frac{1}{\omega C}\right\} \angle \theta - (\theta + 90^\circ)$$

$$\mathbf{Z}_C = \left\{\frac{1}{\omega C}\right\} \angle -90^\circ = -j\left\{\frac{1}{\omega C}\right\} \qquad (17.10)$$

The magnitude of $\mathbf{Z}_C$ is called *capacitive reactance* and is designated by the symbol X_C:

$$X_C = \frac{1}{\omega C} \qquad (17.11)$$

In terms of capacitive reactance, $\mathbf{Z}_C$ can be expressed as

$$\mathbf{Z}_C = X_C\underline{/-90^\circ} = -jX_C \tag{17.12}$$

The units for X_C are ohms, because $\mathbf{Z}_C$ has units of ohms.

EXAMPLE 17.5 Determine the current i for the circuit in Figure 17.9. ■

Figure 17.9

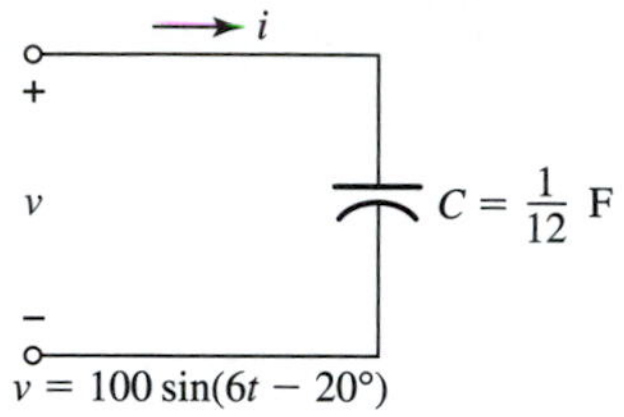

i, +, v, −, $C = \frac{1}{12}$ F

$v = 100 \sin(6t - 20^\circ)$

SOLUTION From the expression for v,

$$\mathbf{V} = 100\underline{/-20^\circ}$$

From Equations (17.11) and (17.12),

$$X_C = \frac{1}{\omega C}$$

$$= \frac{1}{(6)\left(\frac{1}{12}\right)}$$

$$= 2\ \Omega$$

and

$$\mathbf{Z}_C = X_C\underline{/-90^\circ}$$

$$= 2\underline{/-90^\circ}$$

From Equation (17.1)

$$\mathbf{Z}_C = \frac{\mathbf{V}}{\mathbf{I}}$$

and

$$\mathbf{I} = \frac{\mathbf{V}}{\mathbf{Z}_C}$$

$$= \frac{100\underline{/-20^\circ}}{2\underline{/-90^\circ}}$$

$$= \frac{100}{2}\underline{/-20^\circ - (-90^\circ)}$$

$$= 50\underline{/70^\circ}$$

The current i generated by the phasor $50\angle 70°$ is

$$i = 50\ \sin(6t + 70°)$$

Notice that the current i leads the voltage v by 90°.

EXAMPLE 17.6 Determine the voltage v for the circuit in Figure 17.10. ■

Figure 17.10

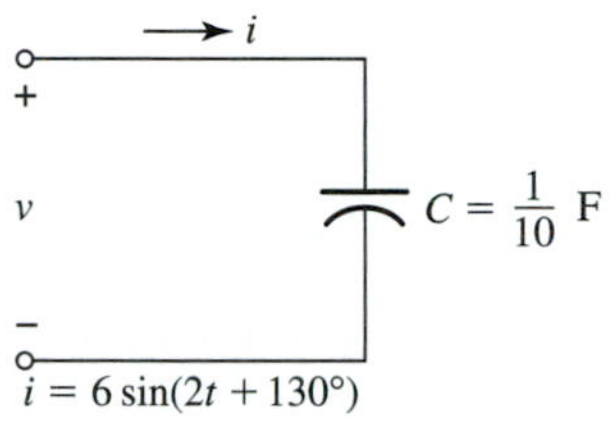

SOLUTION From the expression for i,

$$\mathbf{I} = 6\angle 130°$$

From Equations (17.11) and (17.12),

$$\begin{aligned} X_C &= \frac{1}{\omega C} \\ &= \frac{1}{(2)\left(\frac{1}{10}\right)} \\ &= 5\ \Omega \end{aligned}$$

and

$$\begin{aligned} \mathbf{Z}_C &= X_C\angle -90° \\ &= 5\angle -90° \end{aligned}$$

From Equation (17.1)

$$\begin{aligned} \mathbf{V} &= (\mathbf{Z}_C)(\mathbf{I}) \\ &= (5\angle -90°)(6\angle 130°) \\ &= 30\angle 40° \end{aligned}$$

The voltage v generated by the phasor $30\angle 40°$ is

$$v = 30\ \sin(2t + 40°)$$

Notice that the current i leads the voltage v by 90°.

17.5 THE ADMITTANCE CONCEPT

The admittance of a pair of terminals for ac voltages and currents is defined as the phasor current divided by the phasor voltage, as illustrated in Figure 17.11. The symbol **Y** is the symbol for ac admittance, which is a complex quantity. The symbol Y is used to represent the magnitude of **Y**. That is, if $\mathbf{Y} = 2\underline{/60^\circ}$,then $Y = 2$.

Figure 17.11 Admittance.

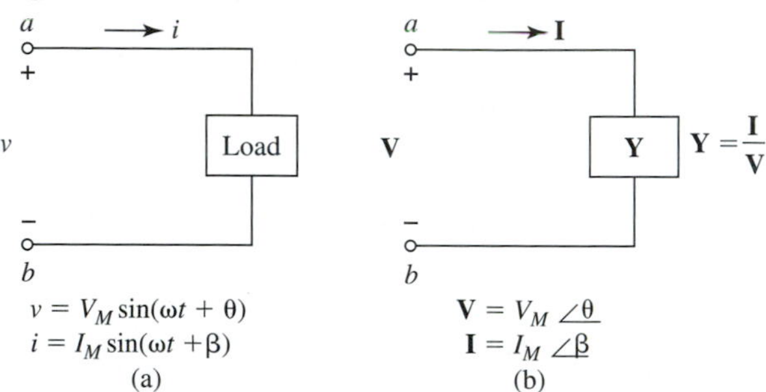

In Figure 17.11(a) the circuit is represented in the time domain, and in Figure 17.11(b) the currents and voltages are represented as phasors. In Figure 17.11(b) the admittance "looking into terminals *ab*" is

$$\mathbf{Y} = \frac{\mathbf{I}}{\mathbf{V}} \tag{17.13}$$

From the current and voltage phasor expressions,

$$\mathbf{Y} = \frac{I_M\underline{/\beta}}{V_M\underline{/\theta}}$$

$$= \left(\frac{I_M}{V_M}\right)\underline{/\beta - \theta}$$

and the magnitude of **Y** is

$$Y = \frac{I_M}{V_M} \tag{17.14}$$

The units for admittance are *siemens*, where one siemen equals one amp per volt and the symbol for siemens is S. If the admittance for a pair of terminals is known, the voltage can be obtained from the current expression and the current can be obtained from the voltage expression.

A comparison of Equations (17.1) and (17.13) reveals that admittance and impedance are reciprocals of each other, as indicated by the following equations.

$$\left.\begin{aligned} \mathbf{Y} &= \frac{1}{\mathbf{Z}} \\ \mathbf{Z} &= \frac{1}{\mathbf{Y}} \end{aligned}\right\} \tag{17.15}$$

Because $\mathbf{Z}_R$, $\mathbf{Z}_L$ and $\mathbf{Z}_C$ have been determined in the previous section, we shall use Equation (17.15) to determine the admittance for resistors, inductors, and capacitors.

17.6 ADMITTANCE OF RESISTORS

From Equation (17.3)

$$\mathbf{Z}_R = R\angle 0^\circ$$

and from Equation (17.15)

$$\mathbf{Y}_R = \frac{1}{\mathbf{Z}_R}$$

$$= \frac{1}{R\angle 0^\circ}$$

$$\mathbf{Y}_R = \left(\frac{1}{R}\right)\angle 0^\circ = \frac{1}{R} + j0 \tag{17.16}$$

The quantity $1/R$ is called *conductance* and is given the symbol G. Because the units for $\mathbf{Y}_R$ are siemens, the units for G are siemens, and $\mathbf{Y}_R$ can be written as

$$\mathbf{Y}_R = G\angle 0^\circ = G + j0 \tag{17.17}$$

where

$$G = \frac{1}{R} \tag{17.18}$$

17.7 ADMITTANCE OF INDUCTORS

From Equation (17.6)

$$\mathbf{Z}_L = (\omega L)\angle 90^\circ$$

and from Equation (17.15)

$$\mathbf{Y}_L = \frac{1}{\mathbf{Z}_L}$$

$$= \frac{1}{(\omega L)\angle 90^\circ}$$

$$\mathbf{Y}_L = \left\{\frac{1}{\omega L}\right\}\angle -90^\circ = -j\left\{\frac{1}{\omega L}\right\} \tag{17.19}$$

The quantity $1/(\omega L)$ is called *inductive susceptance* and is given the symbol B_L. Because the units for $\mathbf{Y}_L$ are siemens, the units for B_L are siemens, and $\mathbf{Y}_L$ can be written as

$$\mathbf{Y}_L = B_L\angle -90^\circ = -jB_L \tag{17.20}$$

where

$$B_L = \frac{1}{\omega L} \tag{17.21}$$

A comparison of Equations (17.7) and (17.21) reveals that

$$B_L = \frac{1}{X_L} \tag{17.22}$$

17.8 ADMITTANCE OF CAPACITORS

From Equation (17.10)

$$\mathbf{Z}_C = \left\{\frac{1}{\omega C}\right\}\underline{/-90^\circ}$$

and from Equation (17.15)

$$\mathbf{Y}_C = \frac{1}{\mathbf{Z}_C}$$

$$= \frac{1}{1/(\omega C)\underline{/-90^\circ}}$$

$$\mathbf{Y}_C = (\omega C)\underline{/90^\circ} = j\omega C \tag{17.23}$$

The quantity ωC is called *capacitive susceptance* and is given the symbol B_C. Because the units for $\mathbf{Y}_C$ are siemens, the units for B_C are siemens, and $\mathbf{Y}_C$ can be written as

$$\mathbf{Y}_C = B_C\underline{/90^\circ} = jB_C \tag{17.24}$$

where

$$B_C = \omega C \tag{17.25}$$

A comparison of Equations (17.11) and (17.25) reveals that

$$B_C = \frac{1}{X_C} \tag{17.26}$$

Consider the following examples.

EXAMPLE 17.7 Determine the voltage v for the circuit in Figure 17.12. ■

Figure 17.12

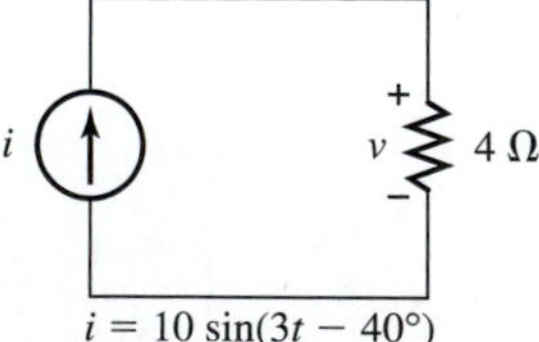

SOLUTION From the expression for i,

$$\mathbf{I} = 10\underline{/-40^\circ}$$

From Equation (17.16)

$$\mathbf{Y}_R = \left(\frac{1}{R}\right)\underline{/0^\circ}$$

$$= \left(\frac{1}{4}\right)\underline{/0^\circ}$$

and from Equation (17.13)

$$\mathbf{Y}_R = \frac{\mathbf{I}}{\mathbf{V}}$$

$$\mathbf{V} = \frac{\mathbf{I}}{\mathbf{Y}_R}$$

$$= \frac{(10\angle -40^\circ)}{\left(\frac{1}{4}\right)\angle 0^\circ}$$

$$= \left(\frac{10}{\frac{1}{4}}\right)\angle -40^\circ - 0^\circ$$

$$= 40\angle -40^\circ$$

The voltage v generated by the phasor $40\angle -40^\circ$ is

$$v = 40\sin(3t - 40^\circ)$$

Notice that the voltage v and the current i are in phase.

EXAMPLE 17.8 Determine the current i for the circuit in Figure 17.13. ■

Figure 17.13

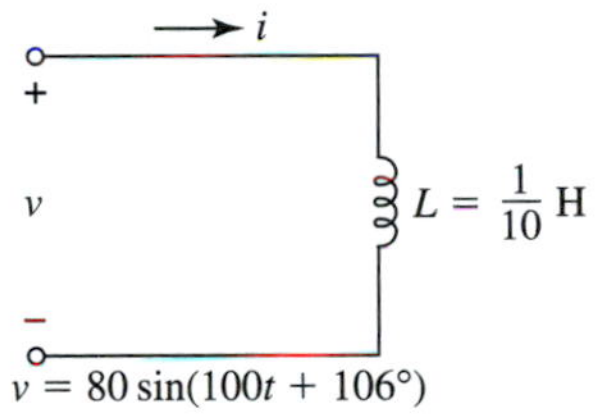

SOLUTION From the expression for v,

$$\mathbf{V} = 80\angle 106^\circ$$

From Equations (17.20) and (17.21),

$$B_L = \frac{1}{\omega L}$$

$$= \frac{1}{(100)\left(\frac{1}{10}\right)}$$

$$= \frac{1}{10}\ \text{S}$$

$$\mathbf{Y}_L = B_L\angle -90^\circ$$

$$= \left(\frac{1}{10}\right)\angle -90^\circ$$

From Equation (17.13)

$$\mathbf{Y}_L = \frac{\mathbf{I}}{\mathbf{V}}$$

$$\begin{aligned}\mathbf{I} &= \mathbf{V}\mathbf{Y}_L \\ &= \{80\underline{/106^\circ}\}\left\{\left(\frac{1}{10}\right)\underline{/-90^\circ}\right\} \\ &= \left\{(80)\left(\frac{1}{10}\right)\right\}\underline{/106^\circ + (-90^\circ)} \\ &= 8\underline{/16^\circ}\end{aligned}$$

The current i generated by the phasor $10\underline{/16^\circ}$ is

$$i = 8\sin(100t + 16^\circ)$$

Notice that the voltage v leads the current i by 90°.

EXAMPLE 17.9 Determine the voltage v for the circuit in Figure 17.14. ■

Figure 17.14

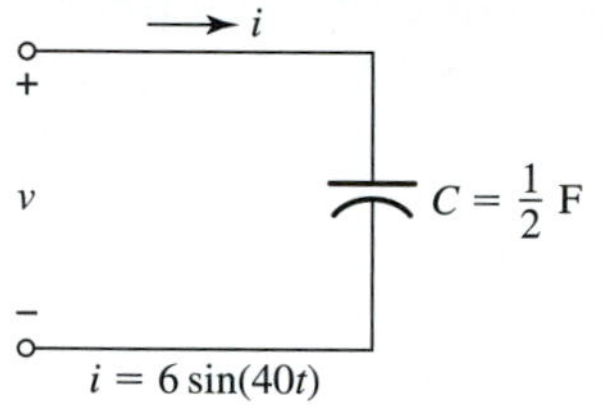

SOLUTION From the expression for i,

$$\mathbf{I} = 6\underline{/0^\circ}$$

From Equations (17.24) and (17.25),

$$\begin{aligned}B_C &= \omega C \\ &= (40)\left(\frac{1}{2}\right) \\ &= 20\text{ S} \\ \mathbf{Y}_C &= B_C\underline{/90^\circ} \\ &= 20\underline{/90^\circ}\end{aligned}$$

From Equation (17.13)

$$\begin{aligned}\mathbf{Y}_C &= \frac{\mathbf{I}}{\mathbf{V}} \\ \mathbf{V} &= \frac{\mathbf{I}}{\mathbf{Y}_C} \\ &= \frac{6\underline{/0^\circ}}{20\underline{/90^\circ}} \\ &= \left(\frac{6}{20}\right)\underline{/0^\circ - 90^\circ} \\ &= \left(\frac{3}{10}\right)\underline{/-90^\circ}\end{aligned}$$

The voltage v generated by the phasor $\left(\frac{3}{10}\right)\underline{/\ 90°}$ is

$$v = 0.3\sin(40t - 90°)$$

Notice that the current i leads voltage v by 90°.

The impedance and admittance concepts are similar. In both concepts, the voltage for a pair of terminals can be obtained if the current is known and the current can be obtained if the voltage is known. Because the degree of difficulty for obtaining impedance or admittance varies with the circuit and can favor impedance or admittance, both concepts are important. It is usually easier to work with impedance for series circuits and admittance for parallel circuits.

17.9 KIRCHHOFF'S VOLTAGE LAW

Kirchhoff's voltage law is valid for circuits in the phasor domain, just as it is for circuits in the t domain. *Kirchhoff's voltage law* for phasor voltages can be stated as follows.

> The algebraic sum of the phasor voltages around a closed path is equal to zero.

Consider the circuit in Figure 17.15 as an example. We shall use Kirchhoff's voltage law to determine the current flow in the circuit. We first assume a direction for the current, as shown in Figure 17.16(a), and redraw the circuit

Figure 17.15 Kirchhoff's voltage law.

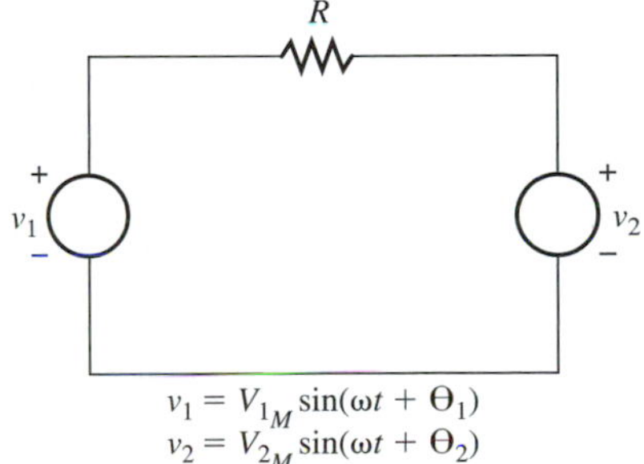

in phasor form, as shown in Figure 17.16(b). Notice that the assigned voltage polarity for the resistance is based on the assumed current, as indicated in Figure 17.16. Finally, we apply Kirchhoff's voltage law to the circuit in

Figure 17.16 Kirchhoff's voltage law.

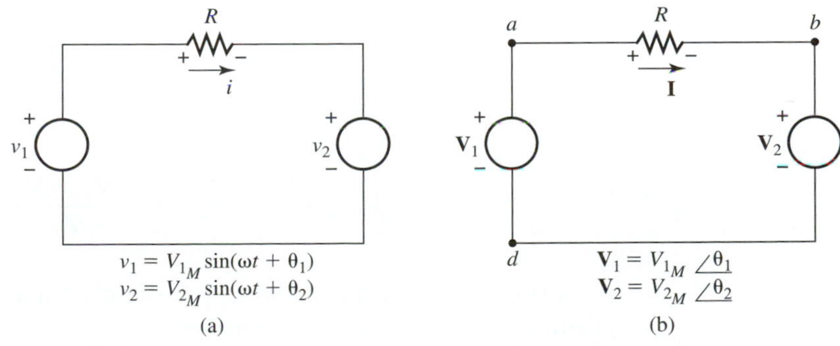

Figure 17.16(b). Using subscripted voltage notation, Kirchhoff's voltage law results in

$$\mathbf{V}_{ad} + \mathbf{V}_{ba} + \mathbf{V}_{db} = 0$$

From an inspection of the circuit $\mathbf{V}_{ad} = \mathbf{V}_1$ and $\mathbf{V}_{db} = -\mathbf{V}_2$, and from Equation (17.1) $\mathbf{V}_{ba} = -\mathbf{Z}_R\mathbf{I}$. It follows that

$$\mathbf{V}_1 - \mathbf{Z}_R\mathbf{I} - \mathbf{V}_2 = 0$$

and

$$\mathbf{I} = \frac{\mathbf{V}_1 - \mathbf{V}_2}{\mathbf{Z}_R}$$

From Equation (17.3) $\mathbf{Z}_R = R\underline{/0^\circ}$ and $\mathbf{I}$ can be written as

$$\mathbf{I} = \frac{V_1\underline{/\theta_1} - V_2\underline{/\theta_2}}{(R\underline{/0^\circ})}$$

For given values of V_1, V_2, θ_1, θ_2, and R, the expression for $\mathbf{I}$ can be reduced to

$$\mathbf{I} = I_M\underline{/\beta}$$

and the current i can then be expressed as

$$i = I_M \sin(\omega t + \beta)$$

Another commonly used procedure of applying Kirchhoff's voltage law involves the use of voltage rises and drops as previously presented.

A voltage rise occurs when we travel from − to + through an electrical element.

A voltage drop occurs when we travel from + to − through an electrical element.

A voltage rise is assigned a positive sign and a voltage drop is assigned a negative sign.

The algebraic sum of the phasor voltage rises and drops around a closed path is equal to zero.

Applying Kirchhoff's voltage law to the circuit in Figure 17.16(b) by algebraically summing rises and drops as we travel around the closed path in a clockwise direction results in

$$\mathbf{V}_1 - \mathbf{Z}_R\mathbf{I} - \mathbf{V}_2 = 0$$

and

$$\mathbf{I} = \frac{\mathbf{V}_1 - \mathbf{V}_2}{\mathbf{Z}_R}$$

which is the same result arrived at previously for $\mathbf{I}$.

Consider the following examples.

EXAMPLE 17.10 Determine the current i in the circuit in Figure 17.17. ■

Figure 17.17

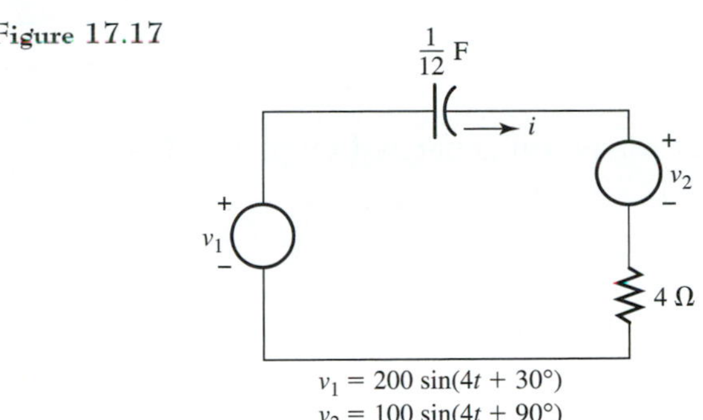

SOLUTION We first assign voltage polarities to the passive elements based on the assumed current and represent the circuit in phasor form, as shown in Figure 17.18.

Figure 17.18

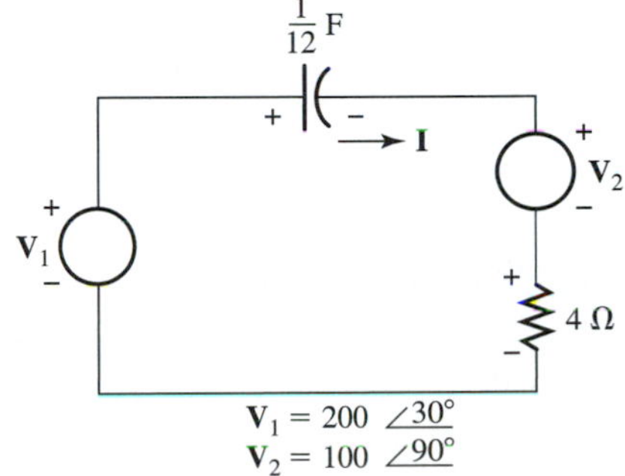

From Equations (17.3), (17.11), and (17.12),

$$\mathbf{Z}_R = 4\angle 0°$$

$$X_C = \frac{1}{\omega C} = \frac{1}{(4)\left(\frac{1}{12}\right)} = 3\ \Omega$$

$$\mathbf{Z}_C = 3\angle -90°$$

From Equation (17.1) the voltages across the capacitor and resistor are $\mathbf{Z}_C\mathbf{I}$ and $\mathbf{Z}_R\mathbf{I}$, respectively. Applying Kirchhoff's voltage law results in

$$\mathbf{V}_1 - \mathbf{Z}_C\mathbf{I} - \mathbf{V}_2 - \mathbf{Z}_R\mathbf{I} = 0$$

$$200\angle 30° - (3\angle -90°)\mathbf{I} - 100\angle 90° - (4\angle 0°)\mathbf{I} = 0$$

$$\mathbf{I} = \frac{200\angle 30° - 100\angle 90°}{3\angle -90° + 4\angle 0°} = \frac{173.2 + j100 - j100}{4 - j3}$$

$$= \frac{173.2\angle 0^\circ}{5\angle -36.9^\circ}$$

$$= 34.6\angle 36.9$$

and the waveform generated by phasor **I** is

$$i = 34.6 \sin(4t + 36.9)$$

EXAMPLE 17.11 Determine the voltages v_1 and v_2 for the circuit in Figure 17.19. ■

Figure 17.19

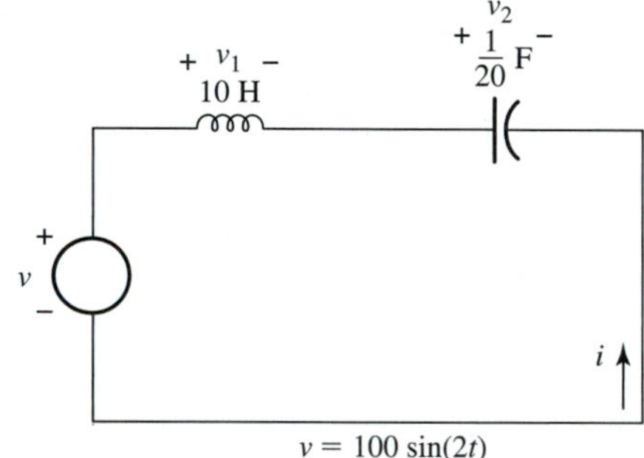

SOLUTION We first assign voltage polarities to the passive elements based on the current direction and represent the circuit in phasor form as shown in Figure 17.20.

Figure 17.20

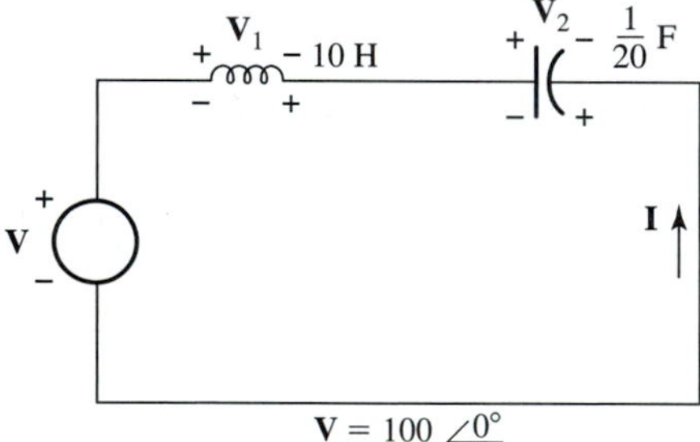

From Equations (17.7), (17.8), (17.11), and (17.12)

$$X_L = \omega L$$

$$= (2)(10)$$

$$= 20\ \Omega$$

$$\mathbf{Z}_L = 20\angle 90^\circ$$

$$X_C = \frac{1}{\omega C}$$

$$= \frac{1}{(2)\left(\frac{1}{20}\right)}$$

$$= 10\ \Omega$$

$$\mathbf{Z}_C = 10\angle -90^\circ$$

From Equation (17.1) the voltages across the inductor and capacitor are $\mathbf{Z}_C\mathbf{I}$ and $\mathbf{Z}_L\mathbf{I}$, respectively. When applying Kirchhoff's voltage law, we use the polarities associated with the assumed current flow and not the polarity markings associated with $\mathbf{V}_1$ and $\mathbf{V}_2$. This results in

$$\mathbf{V} + \mathbf{Z}_L\mathbf{I} + \mathbf{Z}_C\mathbf{I} = 0$$

$$100\underline{/0^\circ} + (20\underline{/90^\circ})\mathbf{I} + (10\underline{/-90^\circ})\mathbf{I} = 0$$

$$\begin{aligned} \mathbf{I} &= \frac{-100\underline{/0^\circ}}{20\underline{/90^\circ} + 10\underline{/-90^\circ}} \\ &= \frac{100\underline{/180^\circ}}{j20 - j10} \\ &= \frac{(100\underline{/180^\circ})}{(10\underline{/90^\circ})} \\ &= 10\underline{/90^\circ} \end{aligned}$$

and the waveform generated by phasor $\mathbf{I}$ is

$$i = 10\sin(2t + 90^\circ)$$

From Equation (17.1)

$$\begin{aligned} \mathbf{V}_1 &= \mathbf{Z}_C(-\mathbf{I}) \\ &= (10\underline{/-90^\circ})(-10\underline{/90^\circ}) \\ &= (10\underline{/-90^\circ})(10\underline{/-90^\circ}) \\ &= 100\underline{/-180^\circ} \\ \mathbf{V}_2 &= \mathbf{Z}_L(-\mathbf{I}) \\ &= (20\underline{/90^\circ})(-10\underline{/90^\circ}) \\ &= (20\underline{/90^\circ})(10\underline{/-90^\circ}) \\ &= 200\underline{/0^\circ} \end{aligned}$$

and the waveforms generated by phasors $\mathbf{V}_1$ and $\mathbf{V}_2$ are

$$\begin{aligned} v_1 &= 100\sin(2t - 180^\circ) \\ v_2 &= 200\sin(2t) \end{aligned}$$

Alternatively, we can use the polarities associated with $\mathbf{V}_1$ and $\mathbf{V}_2$ and write, from Kirchhoff's voltage law,

$$\mathbf{V} - \mathbf{V}_1 - \mathbf{V}_2 = 0$$

Substituting $\mathbf{V}_2 = \mathbf{Z}_C(-\mathbf{I})$ and $\mathbf{V}_1 = \mathbf{Z}_L(-\mathbf{I})$ into these equations yields the previous results for $\mathbf{I}$, $\mathbf{V}_1$, and $\mathbf{V}_2$.

EXAMPLE 17.12 Determine the voltages v_1 and v_2 in the circuit in Figure 17.21. ■

Figure 17.21

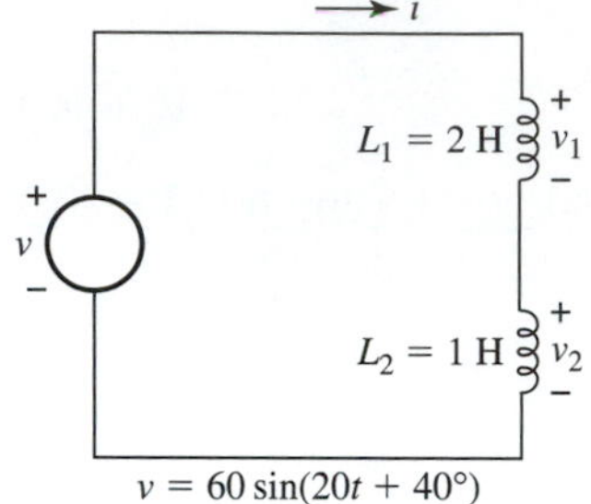

SOLUTION We first assign voltage polarities to the passive elements based on the current direction and represent the circuit in phasor form, as shown in Figure 17.22.

Figure 17.22

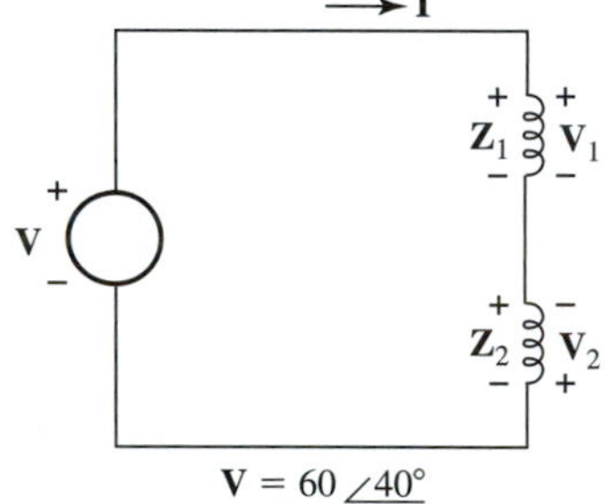

From Equations (17.7) and (17.8),

$$\begin{aligned} X_1 &= \omega L_1 \\ &= (20)(2) \\ &= 40\ \Omega \\ \mathbf{Z}_1 &= 40\angle 90° \\ X_2 &= \omega L_2 \\ &= (20)(1) \\ &= 20\ \Omega \\ \mathbf{Z}_2 &= 20\angle 90° \end{aligned}$$

From Equation (17.1) the voltages across the inductors are $\mathbf{Z}_1\mathbf{I}$ and $\mathbf{Z}_2\mathbf{I}$.

When applying Kirchhoff's voltage law, we use the polarities associated with the assumed current flow and not the polarity markings associated with the $\mathbf{V}_1$ and $\mathbf{V}_2$ labels. This results in

$$\mathbf{V} - \mathbf{Z}_1\mathbf{I} - \mathbf{Z}_2\mathbf{I} = 0$$

$$60\angle 40° - (40\angle 90°)\mathbf{I} - (20\angle 90°)\mathbf{I} = 0$$

$$\begin{aligned}\mathbf{I} &= \frac{60\angle 40^\circ}{40\angle 90^\circ + 20\angle 90^\circ} \\ &= \frac{60\angle 40^\circ}{60\angle 90^\circ} \\ &= 1\angle -50^\circ\end{aligned}$$

From Equation (17.1)

$$\begin{aligned}\mathbf{V}_1 &= \mathbf{Z}_1\mathbf{I} \\ &= (40\angle 90^\circ)(1\angle -50^\circ) \\ &= 40\angle 40^\circ \\ \mathbf{V}_2 &= \mathbf{Z}_2(-\mathbf{I}) \\ &= (20\angle 90^\circ)(-1\angle -50^\circ) \\ &= (20\angle 90^\circ)(1\angle 130^\circ) \\ &= 20\angle 220^\circ\end{aligned}$$

The waveforms generated by $\mathbf{I}$, $\mathbf{V}_1$, and $\mathbf{V}_2$ are

$$\begin{aligned}i &= 1\sin(20t - 50^\circ) \\ v_1 &= 40\sin(20t + 40^\circ) \\ v_2 &= 20\sin(20t + 220^\circ)\end{aligned}$$

17.10 KIRCHHOFF'S CURRENT LAW

Kirchhoff's current law is valid for circuits in the phasor domain, just as it is for circuits in the t domain. *Kirchhoff's current law* for phasor currents can be stated as follows.

> The sum of the phasor currents entering a node equals the sum of the phasor currents leaving the node.

Consider the partial circuit in Figure 17.23. The sum of the phasor currents entering node a equals

$$\mathbf{I}_1 + \mathbf{I}_3$$

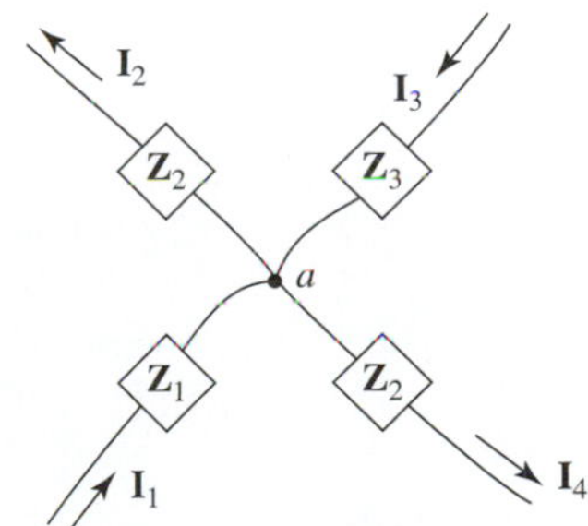

Figure 17.23 Kirchhoff's current law.

and the sum of the phasor currents leaving node *a* equals

$$\mathbf{I}_2 + \mathbf{I}_4$$

From Kirchhoff's current law

$$\mathbf{I}_1 + \mathbf{I}_3 = \mathbf{I}_2 + \mathbf{I}_4$$

Alternatively, we can assign a positive sign to phasor currents entering node *a* and a negative sign to phasor currents leaving node *a* and algebraically sum the phasor currents at node *a*. This results in

$$\mathbf{I}_1 + \mathbf{I}_3 - \mathbf{I}_2 - \mathbf{I}_4 = 0$$

and

$$\mathbf{I}_1 + \mathbf{I}_3 = \mathbf{I}_2 + \mathbf{I}_4$$

which is the same as the result obtained previously by equating the phasor currents entering the node to the phasor currents leaving.

Consider the following example.

EXAMPLE 17.13 Determine the phasor current **I** for the partial circuit in Figure 17.24. ■

Figure 17.24

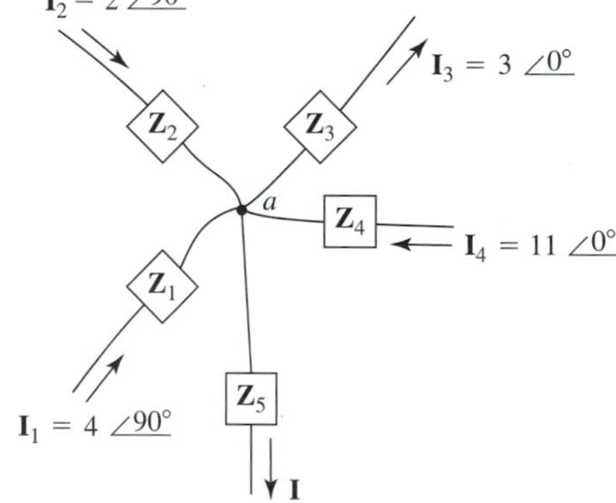

SOLUTION Assigning positive signs to the phasor currents entering node *a* and negative signs to the phasor currents leaving node *a* and applying Kirchhoff's current law results in

$$\mathbf{I}_2 + \mathbf{I}_1 - \mathbf{I}_3 + \mathbf{I}_4 - \mathbf{I} = 0$$

$$2\underline{/90^\circ} + 4\underline{/90^\circ} - 3\underline{/0^\circ} + 11\underline{/0^\circ} - \mathbf{I} = 0$$

$$\mathbf{I} = 8\underline{/0^\circ} + 6\underline{/90^\circ}$$

$$= 8 + j6$$

$$= 10\underline{/36.9^\circ}$$

17.11 GENERAL IMPEDANCE AND ADMITTANCE EXPRESSIONS

The general expression for impedance for a pair of terminals from Equation (17.1) is

$$\mathbf{Z} = \frac{\mathbf{V}}{\mathbf{I}}$$

It can be expressed in rectangular form as

$$\mathbf{Z} = R \pm jX \tag{17.27}$$

where R is resistance and X is reactance. The positive sign is used for inductive circuits and the negative sign is used for capacitive circuits. When considering pure resistors, inductors, and capacitors, $\mathbf{Z}$ is equal to R, jX_L, and $-jX_C$, respectively, as given by Equations (17.3), (17.8), and (17.12); but when combining impedances for different elements in series or parallel, $\mathbf{Z}$ is as given in Equation (17.27).

The general expression for admittance for a pair of terminals from Equation (17.13) is

$$\mathbf{Y} = \frac{\mathbf{I}}{\mathbf{V}}$$

It can be expressed in rectangular form as

$$\mathbf{Y} = G \pm jB \tag{17.28}$$

where G is conductance and B is susceptance. The negative sign is used for inductive circuits and the positive sign is used for capacitive circuits. When considering pure resistors, inductors, and capacitors, $\mathbf{Y}$ is equal to G, $-jB_L$, and jB_C, respectively, as given by Equations (17.15), (17.20), and (17.24); but when combining admittances for different elements in series or parallel, $\mathbf{Y}$ is as given in Equation (17.28).

Consider the following example.

EXAMPLE 17.14 Determine $\mathbf{Z}$, $\mathbf{Y}$, R, G, X, and B for the circuit in Figure 17.25. ■

Figure 17.25

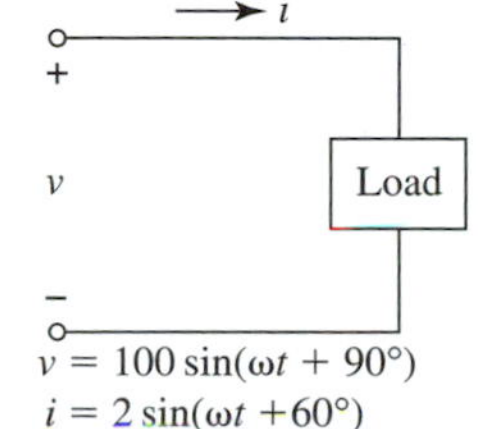

SOLUTION From the expressions for v and i

$$\mathbf{V} = 100\angle 90^\circ$$

$$\mathbf{I} = 2\angle 60^\circ$$

From Equation (17.1)

$$\begin{aligned}\mathbf{Z} &= \frac{\mathbf{V}}{\mathbf{I}} \\ &= \frac{100\angle 90^\circ}{2\angle 60^\circ} \\ &= 50\angle 90^\circ - 60^\circ \\ &= 50\angle 30^\circ \\ &= 50\cos 30^\circ + j50\sin 30^\circ \\ &= 43.3 + j25\end{aligned}$$

From Equation (17.27)

$$\begin{aligned}R &= 43.3\ \Omega \\ X &= 25\ \Omega\end{aligned}$$

and the circuit is inductive, because the imaginary part of **Z** is positive.

From Equation (17.13)

$$\begin{aligned}\mathbf{Y} &= \frac{\mathbf{I}}{\mathbf{V}} \\ &= \frac{2\angle 60^\circ}{100\angle 90^\circ} \\ &= \left(\frac{1}{50}\right)\angle 60^\circ - 90^\circ \\ &= \left(\frac{1}{50}\right)\angle -30^\circ \\ &= \left(\frac{1}{50}\right)\cos(-30^\circ) + j\left(\frac{1}{50}\right)\sin(-30^\circ) \\ &= \left(\frac{1}{50}\right)\cos(30^\circ) - j\left(\frac{1}{50}\right)\sin(30^\circ) \\ &= 0.0173 - j0.01\end{aligned}$$

From Equation (17.28)

$$\begin{aligned}G &= 0.0173\ \text{S} \\ B &= 0.01\ \text{S}\end{aligned}$$

and the circuit is inductive, because the imaginary part of **Y** is negative.

If either **Y** or **Z** is known, the other can be determined from Equation (17.15) as follows. From Equation (17.15), if $\mathbf{Z} = R \pm jX$ is known, then

$$\begin{aligned}\mathbf{Y} &= \frac{1}{\mathbf{Z}} \\ &= \frac{1}{R \pm jX}\end{aligned}$$

Using the procedure in Chapter 16 to rationalize the denominator,

$$\mathbf{Y} = \left\{\frac{1}{R \pm jX}\right\}\left\{\frac{R \mp jX}{R \mp jX}\right\}$$

$$= \frac{R \mp jX}{R^2 + X^2}$$

$$= \left\{\frac{R}{R^2 + X^2}\right\} \mp j\left\{\frac{X}{R^2 + X^2}\right\}$$

$$\mathbf{Y} = G \mp jB \tag{17.29}$$

where

$$G = \frac{R}{R^2 + X^2} \tag{17.30}$$

and

$$B = \frac{X}{R^2 + X^2} \tag{17.31}$$

A negative sign for jB in Equation (17.29) indicates an inductive circuit, and a positive sign for jB indicates a capacitive circuit.

From Equation (17.15), if $\mathbf{Y} = G \pm jB$ is known, then

$$\mathbf{Z} = \frac{1}{\mathbf{Y}}$$

$$= \frac{1}{G \mp jB}$$

$$= \left\{\frac{1}{G \mp jB}\right\}\left\{\frac{G \pm jB}{G \pm jB}\right\}$$

$$= \frac{G \pm jB}{G^2 + B^2}$$

$$= \left\{\frac{G}{G^2 + B^2}\right\} \pm j\left\{\frac{B}{G^2 + B^2}\right\}$$

$$\mathbf{Z} = R \pm jX \tag{17.32}$$

where

$$R = \frac{G}{G^2 + B^2} \tag{17.33}$$

and

$$X = \frac{B}{G^2 + B^2} \tag{17.34}$$

A negative sign for jX in Equation (17.32) indicates a capacitive circuit, and a positive sign for jX indicates an inductive circuit.

17.12 SERIES IMPEDANCE

Consider the two impedances in series in Figure 17.26(a). Applying Kirchhoff's voltage law to the phasor voltages in Figure 17.26(a) results in

$$\mathbf{V} = \mathbf{V}_1 + \mathbf{V}_2$$

Figure 17.26 Series impedance.

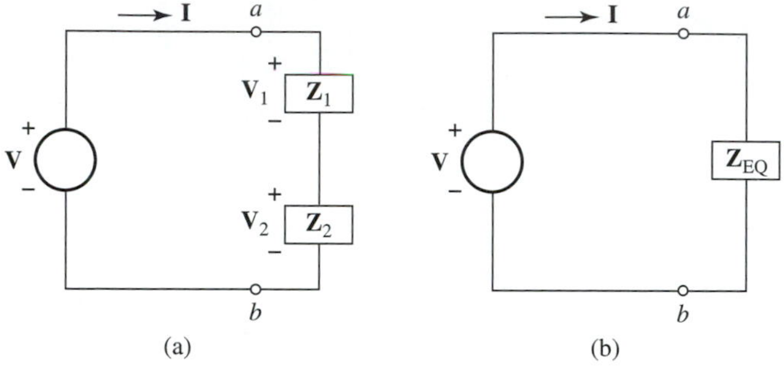

Applying Equation (17.1) to the circuit in Figure 17.26(a) results in

$$\mathbf{V}_1 = \mathbf{Z}_1\mathbf{I}$$
$$\mathbf{V}_2 = \mathbf{Z}_2\mathbf{I}$$

and from the preceding equation for **V**,

$$\begin{aligned}\mathbf{V} &= \mathbf{Z}_1\mathbf{I} + \mathbf{Z}_2\mathbf{I} \\ &= (\mathbf{Z}_1 + \mathbf{Z}_2)\mathbf{I}\end{aligned}$$

Applying Equation (17.1) to the circuit in Figure 17.26(b) results in

$$\mathbf{V} = (\mathbf{Z}_{EQ})\mathbf{I}$$

If the current-voltage relationships with respect to terminals *a-b* for the circuits in Figure 17.26(a) and (b) are to be equivalent, then

$$\mathbf{Z}_{EQ} = \mathbf{Z}_1 + \mathbf{Z}_2$$

where $\mathbf{Z}_{EQ}$ is called the equivalent impedance for terminals *a-b*.

In general, *n* impedances in series can be combined using Equation (17.35), as indicated in Figure 17.27.

$$\mathbf{Z}_{EQ} = \mathbf{Z}_1 + \mathbf{Z}_2 + \cdots + \mathbf{Z}_n \qquad (17.35)$$

Figure 17.27 Series impedances.

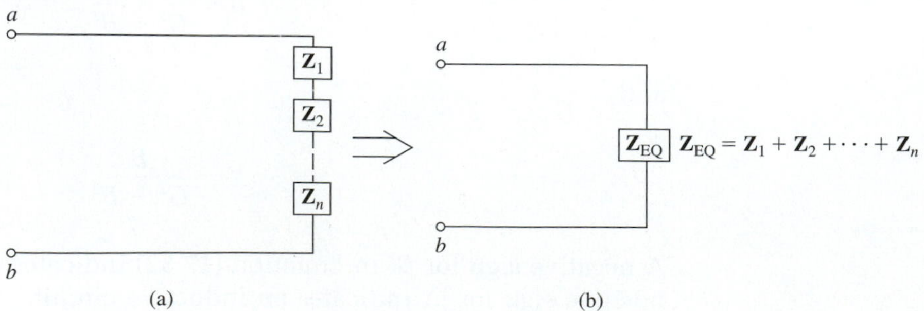

EXAMPLE 17.15 Determine the impedance for terminals *a*-*b* in polar form in Figure 17.28. ■

Figure 17.28

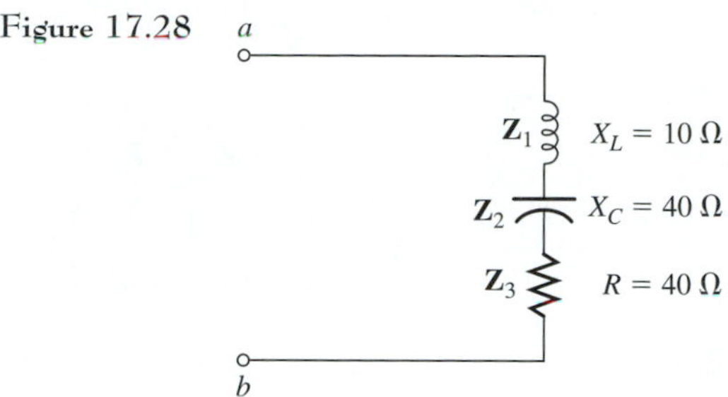

SOLUTION From Equations (17.8), (17.12), and (17.3), respectively,

$$\mathbf{Z}_1 = j10$$
$$\mathbf{Z}_2 = -j40$$
$$\mathbf{Z}_3 = 40$$

and from Equation (17.35)

$$\begin{aligned}\mathbf{Z}_{ab} &= \mathbf{Z}_1 + \mathbf{Z}_2 + \mathbf{Z}_3 \\ &= j10 - j40 + 40 \\ &= 40 - j30 \\ &= 50\angle -36.9^\circ\end{aligned}$$

17.13 PARALLEL IMPEDANCE

Consider the two parallel impedances in Figure 17.29(a). Applying Kirchhoff's current law to the phasor currents in Figure 17.29(a) results in

$$\mathbf{I} = \mathbf{I}_1 + \mathbf{I}_2$$

Figure 17.29 Parallel impedances.

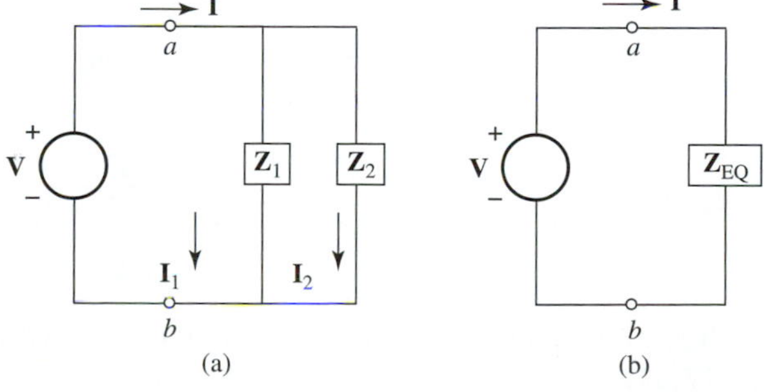

Applying Equation (17.1) to the circuit in Figure 17.29(a) results in

$$\mathbf{I}_1 = \frac{\mathbf{V}}{\mathbf{Z}_1}$$

$$\mathbf{I}_2 = \frac{\mathbf{V}}{\mathbf{Z}_2}$$

and from the preceding equation for **I**,

$$\mathbf{I} = \left(\frac{\mathbf{V}}{\mathbf{Z}_1}\right) + \left(\frac{\mathbf{V}}{\mathbf{Z}_2}\right)$$

$$= (\mathbf{V})\left\{\frac{1}{\mathbf{Z}_1} + \frac{1}{\mathbf{Z}_2}\right\}$$

$$\mathbf{V} = \left\{\frac{1}{1/\mathbf{Z}_1 + 1/\mathbf{Z}_2}\right\}\mathbf{I}$$

Applying Equation (17.1) to the circuit in Figure 17.29(b) results in

$$\mathbf{V} = \mathbf{Z}_{\text{EQ}}\mathbf{I}$$

If the current-voltage relationships with respect to terminals *a-b* for the circuits in Figure 17.29(a) and (b) are to be equivalent, then

$$\mathbf{Z}_{\text{EQ}} = \frac{1}{1/\mathbf{Z}_1 + 1/\mathbf{Z}_2}$$

where $\mathbf{Z}_{\text{EQ}}$ is called the equivalent impedance for terminals *a-b*.

In general, *n* impedances in parallel can be combined using Equation (17.36), as indicated in Figure 17.30.

$$\mathbf{Z}_{\text{EQ}} = \frac{1}{1/\mathbf{Z}_1 + 1/\mathbf{Z}_2 + \cdots + 1/\mathbf{Z}_n} \tag{17.36}$$

Figure 17.30 Parallel impedances.

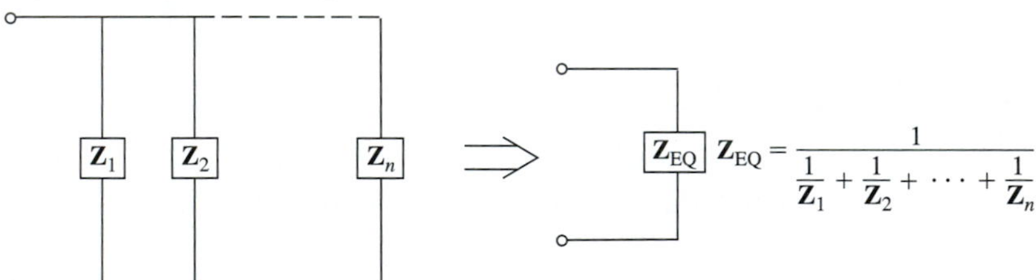

For two impedances in parallel, Equation (17.36) can be written as

$$\mathbf{Z}_{\text{EQ}} = \frac{1}{1/\mathbf{Z}_1 + 1/\mathbf{Z}_2}$$

$$= \frac{\mathbf{Z}_1\mathbf{Z}_2}{\mathbf{Z}_1 + \mathbf{Z}_2} \tag{17.37}$$

EXAMPLE 17.16 Determine the impedance for terminals *a-b* for the circuit in Figure 17.31. ■

Figure 17.31

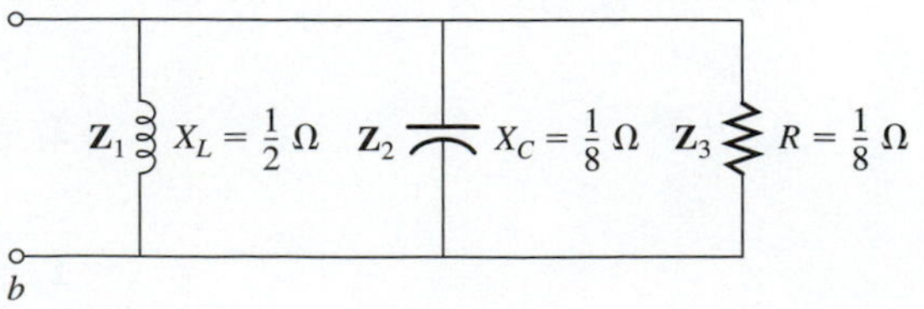

a

$\mathbf{Z}_1$ $X_L = \frac{1}{2}\ \Omega$ $\mathbf{Z}_2$ $X_C = \frac{1}{8}\ \Omega$ $\mathbf{Z}_3$ $R = \frac{1}{8}\ \Omega$

b

SOLUTION From Equations (17.3), (17.8), and (17.12), respectively,

$$Z_1 = j\frac{1}{2}\,\Omega$$

$$Z_2 = -j\frac{1}{8}\,\Omega$$

$$Z_3 = \frac{1}{8}\,\Omega$$

and from Equation (17.36)

$$\begin{aligned}\mathbf{Z}_{ab} &= \frac{1}{1/\mathbf{Z}_1 + 1/\mathbf{Z}_2 + 1/\mathbf{Z}_3} \\ &= \frac{1}{1/\left(j\frac{1}{2}\right) + 1/\left(-j\frac{1}{8}\right) + 1/\left(\frac{1}{8}\right)} \\ &= \frac{1}{-j2 + j8 + 8} \\ &= \frac{1}{8 + j6} \\ &= \frac{1}{10\angle 36.9^\circ} \\ &= \left(\frac{1}{10}\right)\angle -36.9^\circ \\ &= 0.08 - j0.06\ \Omega\end{aligned}$$

17.14 SERIES ADMITTANCE

Consider the two series admittances in Figure 17.32(a). Applying Kirchhoff's voltage law to the phasor voltages in Figure 17.32(a) results in

$$\mathbf{V} = \mathbf{V}_1 + \mathbf{V}_2$$

Figure 17.32 Series admittances.

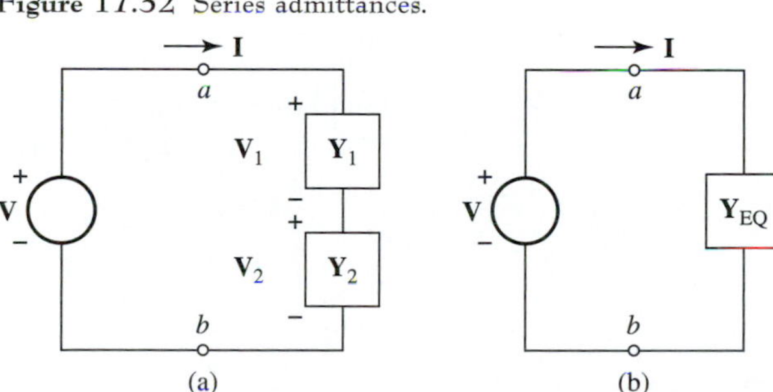

Applying Equation (17.13) to the circuit in Figure 17.32(a) results in

$$\mathbf{V}_1 = \frac{\mathbf{I}}{\mathbf{Y}_1}$$

$$\mathbf{V}_2 = \frac{\mathbf{I}}{\mathbf{Y}_2}$$

and from the preceding equation for **V**,

$$\mathbf{V} = \frac{\mathbf{I}}{\mathbf{Y}_1} + \frac{\mathbf{I}}{\mathbf{Y}_2}$$

$$\mathbf{V} = \left\{\frac{1}{\mathbf{Y}_1} + \frac{1}{\mathbf{Y}_2}\right\}\mathbf{I}$$

$$\mathbf{I} = \left\{\frac{1}{1/\mathbf{Y}_1 + 1/\mathbf{Y}_2}\right\}\mathbf{V}$$

Applying Equation (17.13) to the circuit in Figure 17.32(b) results in

$$\mathbf{I} = (\mathbf{Y}_{EQ})\mathbf{V}$$

If the current-voltage relationships for the circuits in Figure 17.32(a) and (b) are to be equivalent with respect to terminals *a-b*, then

$$\mathbf{Y}_{EQ} = \frac{1}{1/\mathbf{Y}_1 + 1/\mathbf{Y}_2}$$

where $\mathbf{Y}_{EQ}$ is called the equivalent admittance for terminals *a-b*.

In general, *n* admittances in series can be combined using Equation (17.38), as indicated in Figure 17.33.

$$\mathbf{Y}_{EQ} = \frac{1}{1/\mathbf{Y}_1 + 1/\mathbf{Y}_2 + \cdots + 1/\mathbf{Y}_n} \tag{17.38}$$

Figure 17.33 Series admittances.

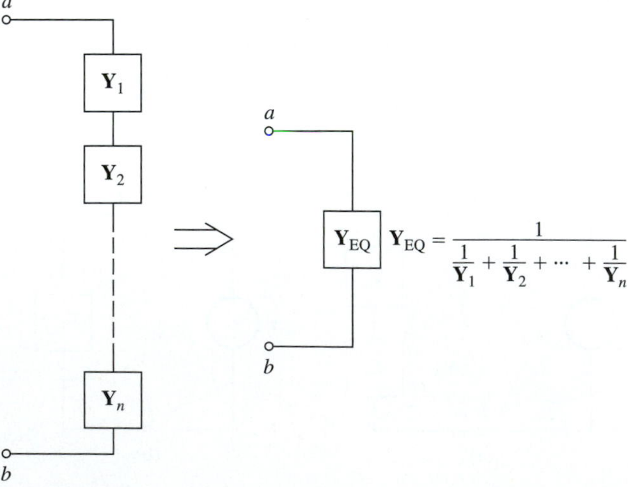

For two admittances in series Equation (17.38) can be written as

$$\mathbf{Y}_{EQ} = \frac{\mathbf{Y}_1\mathbf{Y}_2}{\mathbf{Y}_1 + \mathbf{Y}_2} \tag{17.39}$$

EXAMPLE 17.17 Determine the admittance for terminals *a*-*b* in polar and rectangular form for the circuit in Figure 17.34. ■

Figure 17.34

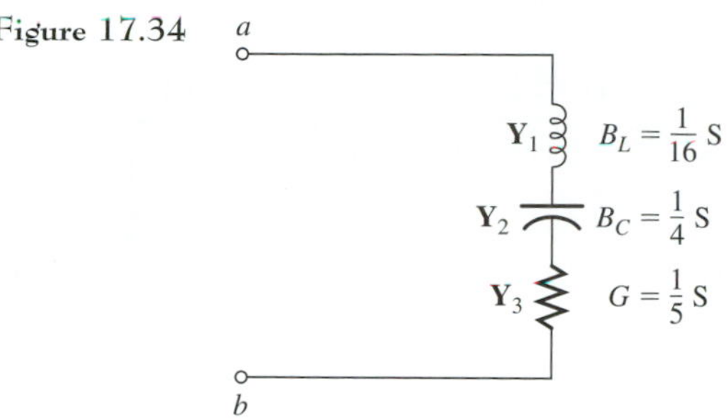

SOLUTION From Equations (17.20), (17.24), and (17.16), respectively,

$$\mathbf{Y}_1 = -j\frac{1}{16} \text{ S}$$

$$\mathbf{Y}_2 = j\frac{1}{4} \text{ S}$$

$$\mathbf{Y}_3 = \frac{1}{5} \text{ S}$$

and from Equation (17.38)

$$\begin{aligned}
\mathbf{Y}_{ab} &= \frac{1}{1/\mathbf{Y}_1 + 1/\mathbf{Y}_2 + 1/\mathbf{Y}_3} \\
&= \frac{1}{\left\{1/\left(-j\frac{1}{16}\right)\right\} + \left\{1/\left(j\frac{1}{4}\right)\right\} + \left\{1/\frac{1}{5}\right\}} \\
&= \frac{1}{j16 - j4 + 5} \\
&= \frac{1}{5 + j12} \\
&= \frac{1}{13\angle 67.4^\circ} \\
&= \left(\frac{1}{13}\right)\angle -67.4^\circ \\
&= 0.03 - j0.07 \text{ S}
\end{aligned}$$

17.15 PARALLEL ADMITTANCE

Consider the two parallel admittances in Figure 17.35(a). Applying Kirchhoff's current law to the phasor currents in Figure 17.35(a) results in

$$\mathbf{I} = \mathbf{I}_1 + \mathbf{I}_2$$

Figure 17.35 Parallel admittances.

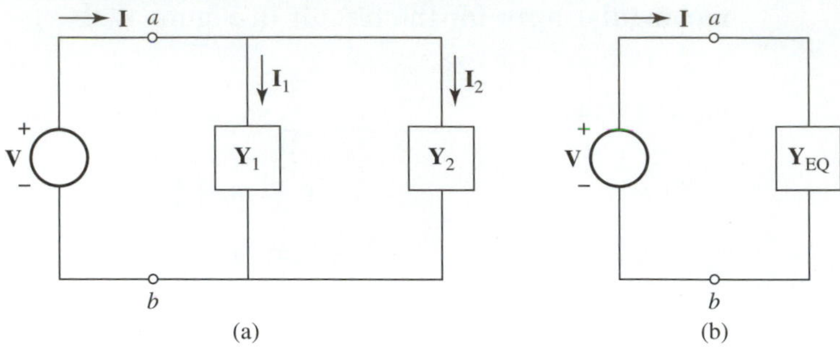

Applying Equation (17.13) to $\mathbf{Y}_1$ and $\mathbf{Y}_2$ in Figure 17.35(a) results in

$$\mathbf{I}_1 = \mathbf{V}\mathbf{Y}_1$$
$$\mathbf{I}_2 = \mathbf{V}\mathbf{Y}_2$$

and from the preceding equation for $\mathbf{I}$,

$$\mathbf{I} = \mathbf{V}\mathbf{Y}_1 + \mathbf{V}\mathbf{Y}_2$$
$$= (\mathbf{Y}_1 + \mathbf{Y}_2)\mathbf{V}$$

Applying Equation (17.13) to the circuit in Figure 17.35(b) results in

$$\mathbf{I} = (\mathbf{Y}_{EQ})\mathbf{V}$$

If the current-voltage relationship for the circuits in Figure 17.35(a) and (b) are to be equivalent with respect to terminals *a-b*, then

$$\mathbf{Y}_{EQ} = \mathbf{Y}_1 + \mathbf{Y}_2$$

where $\mathbf{Y}_{EQ}$ is called the equivalent admittance for terminals *a-b*.

In general, *n* admittances in parallel can be combined using Equation (17.40), as indicated in Figure 17.36.

$$\mathbf{Y}_{EQ} = \mathbf{Y}_1 + \mathbf{Y}_2 + \cdots + \mathbf{Y}_n \qquad (17.40)$$

Figure 17.36 Parallel admittances

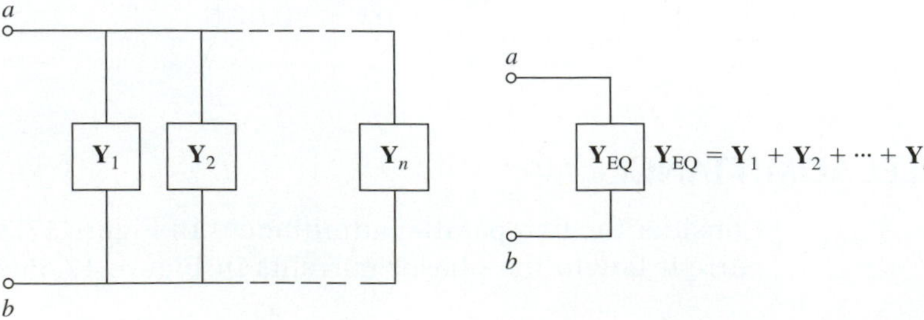

EXAMPLE 17.18 Determine the admittance for terminals *a-b* in polar and rectangular form for the circuit in Figure 17.37. ■

Figure 17.37

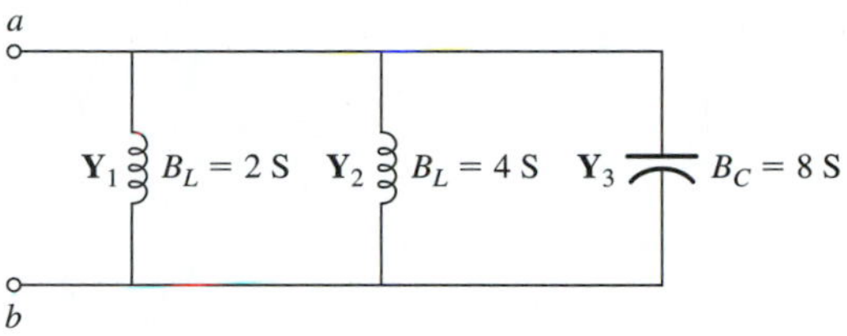

SOLUTION From Equations (17.20) and (17.24),

$$\mathbf{Y}_1 = -j2$$
$$\mathbf{Y}_2 = -j4$$
$$\mathbf{Y}_3 = j8$$

and from Equation (17.40)

$$\begin{aligned}\mathbf{Y}_{ab} &= \mathbf{Y}_1 + \mathbf{Y}_2 + \mathbf{Y}_3 \\ &= -j2 - j4 + j8 \\ &= j2 \\ &= 2\angle 90^\circ\end{aligned}$$

17.16 SERIES-PARALLEL CIRCUITS

Most electric circuits consist of electrical elements connected in series and in parallel and not just in one or the other, as previously presented. The following examples illustrate the use of the concepts previously presented to reduce series-parallel combinations to single equivalent impedances and admittances.

EXAMPLE 17.19 Determine a single equivalent impedance and a single equivalent admittance for terminals *a-b* in the circuit in Figure 17.38. ■

Figure 17.38

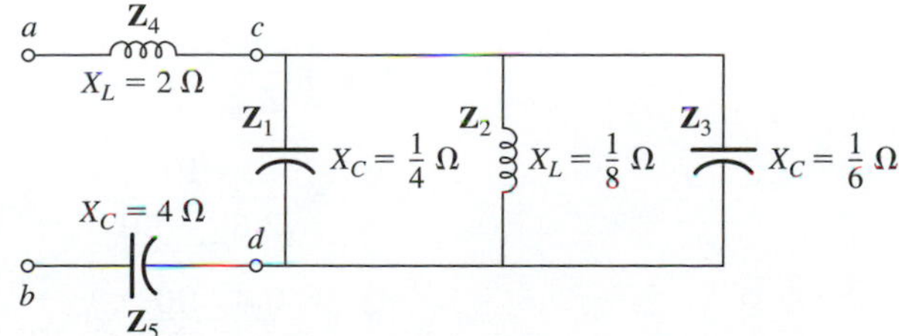

SOLUTION A comparison of Equations (17.36) and (17.40) reveals that it is easier to add admittances in parallel than it is to add impedances in parallel.

Because $\mathbf{Z}_1$, $\mathbf{Z}_2$, and $\mathbf{Z}_3$ are in parallel, we convert the given impedances to admittances and employ Equation (17.40) to combine the resulting admittances.

$$\mathbf{Y}_1 = \frac{1}{\mathbf{Z}_1} = \frac{1}{-j\frac{1}{4}} = j4$$

$$\mathbf{Y}_2 = \frac{1}{\mathbf{Z}_2} = \frac{1}{j\frac{1}{8}} = -j8$$

$$\mathbf{Y}_3 = \frac{1}{\mathbf{Z}_3} = \frac{1}{-j\frac{1}{6}} = j6$$

The admittance for the circuit to the right of terminals *c-d* is

$$\begin{aligned}\mathbf{Y}_{cd} &= \mathbf{Y}_1 + \mathbf{Y}_2 + \mathbf{Y}_3 \\ &= j4 - j8 + j6 \\ &= j2\end{aligned}$$

A comparison of Equations (17.35) and (17.38) reveals that it is easier to add impedances in series than it is to add admittances in series. Because $\mathbf{Y}_{cd}$ is in series with $\mathbf{Z}_4$ and $\mathbf{Z}_5$, we convert $\mathbf{Y}_{cd}$ to impedance and apply Equation (17.35) to combine $\mathbf{Z}_{cd}$, $\mathbf{Z}_4$, and $\mathbf{Z}_5$.

$$\begin{aligned}\mathbf{Z}_{cd} &= \frac{1}{\mathbf{Y}_{cd}} \\ &= \frac{1}{j2} \\ &= -j\left(\frac{1}{2}\right)\end{aligned}$$

$$\begin{aligned}\mathbf{Z}_{ab} &= \mathbf{Z}_4 + \mathbf{Z}_5 + \mathbf{Z}_{cd} \\ &= j2 - j4 - j\left(\frac{1}{2}\right) \\ &= -j(2.5) \\ &= 2.5\underline{/-90^\circ}\end{aligned}$$

From Equation 17.15

$$\begin{aligned}\mathbf{Y}_{ab} &= \frac{1}{\mathbf{Z}_{ab}} \\ &= \frac{1}{2.5\underline{/-90^\circ}} \\ &= 0.4\underline{/90^\circ}\end{aligned}$$

EXAMPLE 17.20 Determine a single equivalent impedance for terminals a-b in polar and rectangular form for the circuit in Figure 17.39. ■

Figure 17.39

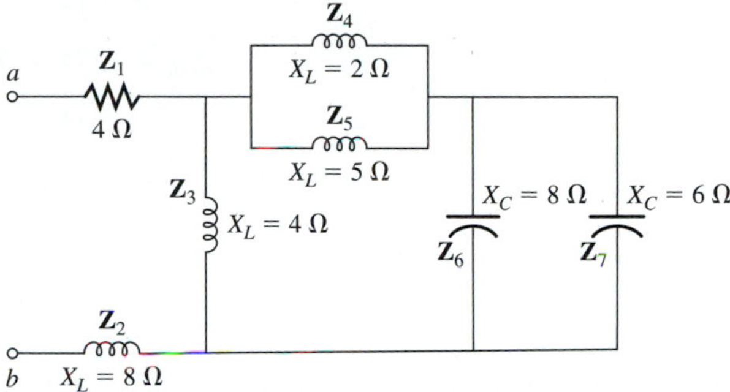

SOLUTION The impedance $\mathbf{Z}_{ab}$ is

$$\mathbf{Z}_{ab} = \{\{(\mathbf{Z}_6 \| \mathbf{Z}_7) + (\mathbf{Z}_4 \| \mathbf{Z}_5)\} \| \{\mathbf{Z}_3\}\} + \mathbf{Z}_1 + \mathbf{Z}_2$$

From Equations (17.35) and (17.36),

$$\begin{aligned}
\mathbf{Z}_6 \| \mathbf{Z}_7 &= \frac{\mathbf{Z}_6\mathbf{Z}_7}{\mathbf{Z}_6 + \mathbf{Z}_7} \\
&= \frac{(-j8)(-j6)}{-j8 - j6} \\
&= \frac{-48}{-j14} \\
&= -j\left(\frac{24}{7}\right) \\
\mathbf{Z}_4 \| \mathbf{Z}_5 &= \frac{\mathbf{Z}_4\mathbf{Z}_5}{\mathbf{Z}_4 + \mathbf{Z}_5} \\
&= \frac{(j2)(j5)}{j2 + j5} \\
&= j\left(\frac{10}{7}\right) \\
(\mathbf{Z}_6 \| \mathbf{Z}_7) + (\mathbf{Z}_4 \| \mathbf{Z}_5) &= -j\left(\frac{24}{7}\right) + j\left(\frac{10}{7}\right) \\
&= -j\left(\frac{14}{7}\right) \\
&= -j2
\end{aligned}$$

$$\{(\mathbf{Z}_6 \| \mathbf{Z}_7) + (\mathbf{Z}_4 \| \mathbf{Z}_5)\} \| \{\mathbf{Z}_3\} = \{-j2\} \| \{j4\}$$
$$= \frac{(-j2)(j4)}{(-j2) + (j4)}$$
$$= \frac{8}{j2}$$
$$= -j4$$
$$\mathbf{Z}_{ab} = -j4 + \mathbf{Z}_1 + \mathbf{Z}_2$$
$$= -j4 + 4 + j8$$
$$= 4 + j4$$
$$= 5.66\underline{/45^\circ}$$

EXAMPLE 17.21 Determine a simple series combination of electrical elements to connect across terminals *a-b* in Figure 17.40 that satisfies the given current and voltage expressions. (Simple implies the smallest number of elements.) ■

Figure 17.40

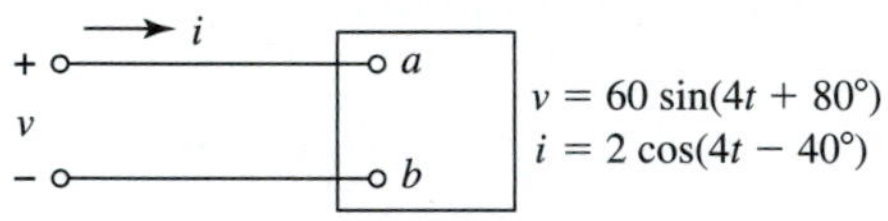

$v = 60 \sin(4t + 80^\circ)$
$i = 2 \cos(4t - 40^\circ)$

SOLUTION v and i in phasor form are

$$\mathbf{V} = 60\underline{/80^\circ}$$
$$\mathbf{I} = 2\underline{/50^\circ}$$

From Equation (17.1)

$$\mathbf{Z} = \frac{\mathbf{V}}{\mathbf{I}}$$
$$= \frac{60\underline{/80^\circ}}{2\underline{/50^\circ}}$$
$$= 30\underline{/30^\circ}$$
$$= 26 + j15$$

If we consider the real and the imaginary parts of $\mathbf{Z}$ as separate impedances, then $\mathbf{Z}$ can be thought of as a 26-Ω resistor in series with a 15-Ω inductor.

Because $X_L = \omega L$ (Equation 17.7),

$$L = \frac{X_L}{\omega}$$
$$= \frac{15}{4} \text{ H}$$

A simple series representation for the circuit in Figure 17.40 is shown in Figure 17.41.

Figure 17.41

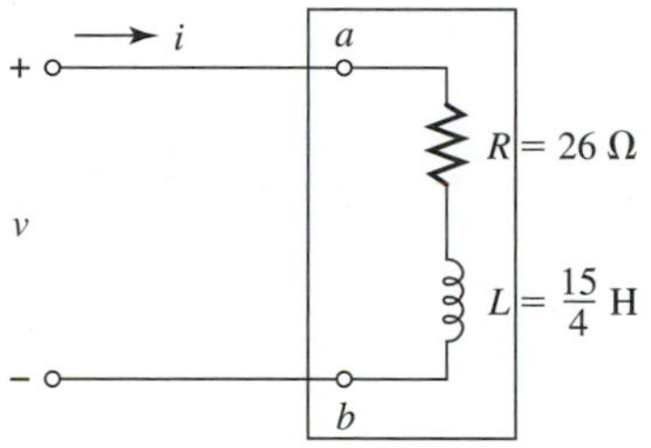

EXAMPLE 17.22 Determine a simple parallel combination of electrical elements to connect across terminals *a-b* in Figure 17.42 that satisfies the given current and voltage expressions. (Simple implies the smallest number of elements.) ■

Figure 17.42

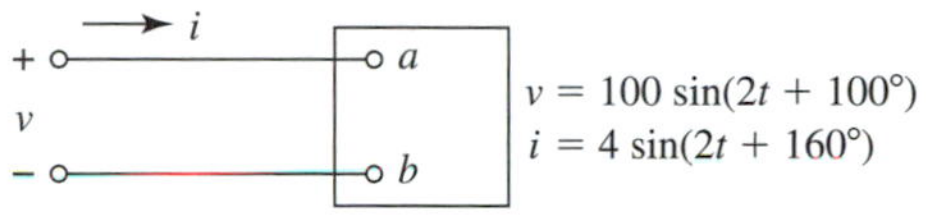

SOLUTION In phasor form v and i are

$$\mathbf{V} = 100\underline{/100°}$$

$$\mathbf{I} = 4\underline{/160°}$$

From Equation (17.13)

$$\begin{aligned} \mathbf{Y} &= \frac{\mathbf{I}}{\mathbf{V}} \\ &= \frac{4\underline{/160°}}{100\underline{/100°}} \\ &= \left(\frac{1}{25}\right)\underline{/160° - 100°} \\ &= \left(\frac{1}{25}\right)\underline{/60°} \\ &= 0.02 + j0.035 \end{aligned}$$

If we consider the real and the imaginary parts of **Y** as separate admittances, then **Y** can be thought of as a 0.025-S conductance in parallel with a 0.035-S capacitance.

Because $B_C = \omega C$ (Equation (2.25),

$$\begin{aligned} C &= \frac{B_C}{\omega} \\ &= \frac{0.035}{2} \\ &= 0.0175 \text{ F} \end{aligned}$$

A simple parallel representation for the circuit in Figure 17.42 is shown in Figure 17.43.

Figure 17.43

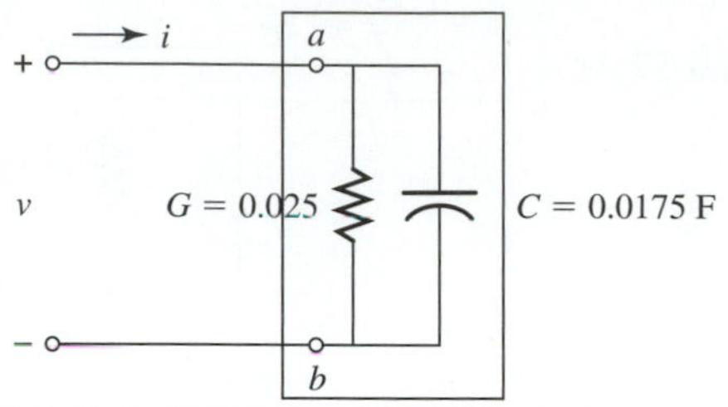

i, a, b, $+$, v, $-$, $G = 0.025$, $C = 0.0175$ F

17.17 CURRENT-DIVISION FORMULA FOR IMPEDANCE

The current-division formula allows us to determine the division of a current at a node without determining the voltage across the elements through which the divided currents flow. Consider the circuit in Figure 17.44.

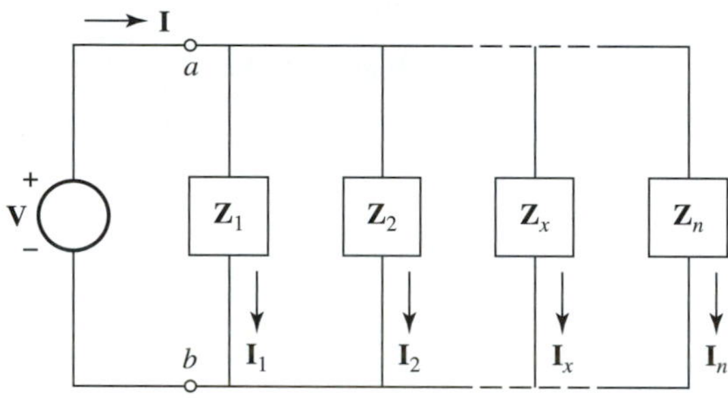

Figure 17.44 Current division.

From Equation (17.36), the equivalent impedance for terminals a-b is

$$\mathbf{Z}_{\text{EQ}} = \frac{1}{1/\mathbf{Z}_1 + 1/\mathbf{Z}_2 + \cdots + \mathbf{Z}_n}$$

and the circuit can be redrawn as shown in Figure 17.45. From Equation (17.1)

$$\mathbf{V} = \mathbf{Z}_{\text{EQ}}\mathbf{I}$$

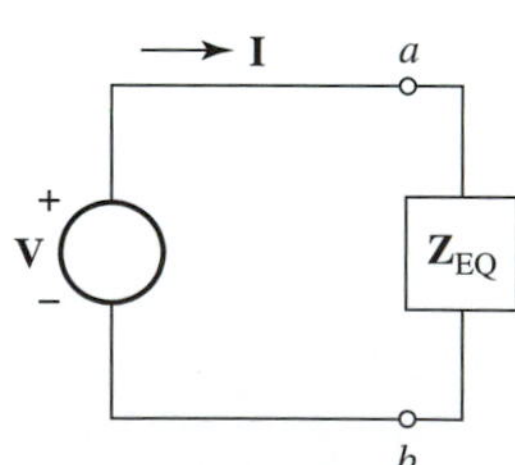

Figure 17.45 Equivalent impedance.

and because $\mathbf{V}$ appears across each impedance in Figure 17.44,

$$\mathbf{I}_1 = \frac{\mathbf{V}}{\mathbf{Z}_1}$$

$$\mathbf{I}_2 = \frac{\mathbf{V}}{\mathbf{Z}_2}$$

$$\vdots$$

$$\mathbf{I}_n = \frac{\mathbf{V}}{\mathbf{Z}_n}$$

Substituting $\mathbf{V} = \mathbf{Z}_{EQ}\mathbf{I}$ into each equation results in

$$\mathbf{I}_1 = \left(\frac{\mathbf{Z}_{EQ}}{\mathbf{Z}_1}\right)\mathbf{I}$$

$$\mathbf{I}_2 = \left(\frac{\mathbf{Z}_{EQ}}{\mathbf{Z}_2}\right)\mathbf{I}$$

$$\vdots$$

$$\mathbf{I}_n = \left(\frac{\mathbf{Z}_{EQ}}{\mathbf{Z}_n}\right)\mathbf{I}$$

In general, the current $\mathbf{I}_x$ through an impedance $\mathbf{Z}_x$ in Figure 17.44 is

$$\mathbf{I}_x = \left(\frac{\mathbf{Z}_{EQ}}{\mathbf{Z}_x}\right)\mathbf{I} \tag{17.41}$$

For two impedances, $\mathbf{Z}_1$ and $\mathbf{Z}_2$, in parallel, Equation (17.41) can be written as

$$\left.\begin{aligned} \mathbf{I}_1 &= \left\{\frac{\mathbf{Z}_2}{\mathbf{Z}_1 + \mathbf{Z}_2}\right\}\mathbf{I} \\ \mathbf{I}_2 &= \left\{\frac{\mathbf{Z}_1}{\mathbf{Z}_1 + \mathbf{Z}_2}\right\}\mathbf{I} \end{aligned}\right\} \tag{17.42}$$

17.18 CURRENT-DIVISION FORMULA FOR ADMITTANCE

Consider the circuit in Figure 17.46. From Equation (17.40), the equivalent admittance for terminals *a-b* is

$$\mathbf{Y}_{EQ} = \mathbf{Y}_1 + \mathbf{Y}_2 + \cdots + \mathbf{Y}_n$$

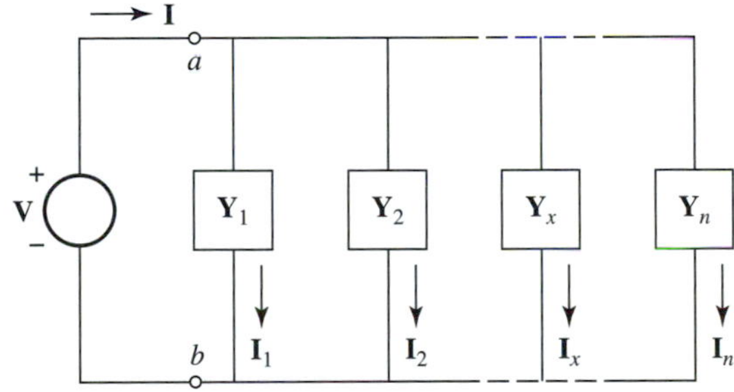

Figure 17.46 Current division.

and the circuit can be redrawn as shown in Figure 17.47.

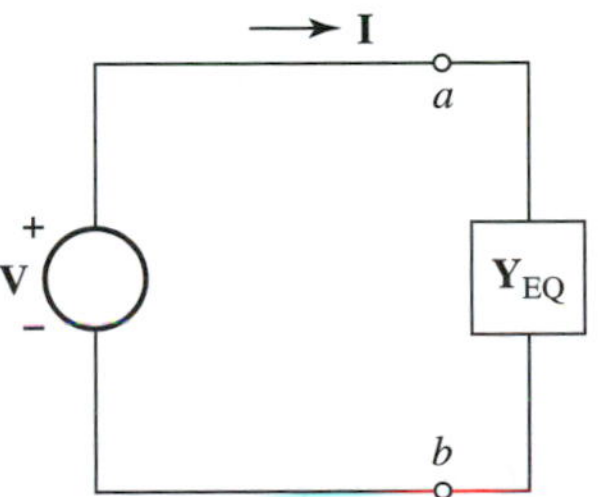

Figure 17.47 Equivalent admittance.

From Equation (17.13)

$$\mathbf{V} = \frac{\mathbf{I}}{\mathbf{Y}_{EQ}}$$

and because $\mathbf{V}$ appears across each admittance in Figure 17.46,

$$\mathbf{I}_1 = \mathbf{V}\,\mathbf{Y}_1$$
$$\mathbf{I}_2 = \mathbf{V}\,\mathbf{Y}_2$$
$$\vdots$$
$$\mathbf{I}_n = \mathbf{V}\,\mathbf{Y}_n$$

Substituting $\mathbf{V} = \mathbf{I}/\mathbf{Y}_{EQ}$ into each equation results in

$$\mathbf{I}_1 = \left(\frac{\mathbf{Y}_1}{\mathbf{Y}_{EQ}}\right)\mathbf{I}$$
$$\mathbf{I}_2 = \left(\frac{\mathbf{Y}_2}{\mathbf{Y}_{EQ}}\right)\mathbf{I}$$
$$\vdots$$
$$\mathbf{I}_n = \left(\frac{\mathbf{Y}_n}{\mathbf{Y}_{EQ}}\right)\mathbf{I}$$

In general, the current $\mathbf{I}_x$ through an admittance $\mathbf{Y}_x$ in Figure 17.46 is

$$\mathbf{I}_x = \left(\frac{\mathbf{Y}_x}{\mathbf{Y}_{EQ}}\right)\mathbf{I} \tag{17.43}$$

For two admittances $\mathbf{Y}_1$ and $\mathbf{Y}_2$ in parallel, Equation (17.43) can be written as

$$\left.\begin{aligned} \mathbf{I}_1 &= \left\{\frac{\mathbf{Y}_1}{\mathbf{Y}_1 + \mathbf{Y}_2}\right\}\mathbf{I} \\ \mathbf{I}_2 &= \left\{\frac{\mathbf{Y}_2}{\mathbf{Y}_1 + \mathbf{Y}_2}\right\}\mathbf{I} \end{aligned}\right\} \tag{17.44}$$

Because an equivalent admittance $\mathbf{Y}_{EQ}$ is easier to determine than an equivalent impedance $\mathbf{Z}_{EQ}$ for parallel elements, it is usually advantageous to employ Equation (17.43) rather than Equation (17.41) when using current division.

Consider the following examples.

EXAMPLE 17.23 Determine currents i_1, i_2, and i_3 for the circuit in Figure 17.48 using Equation (17.41) for current division. ■

Figure 17.48

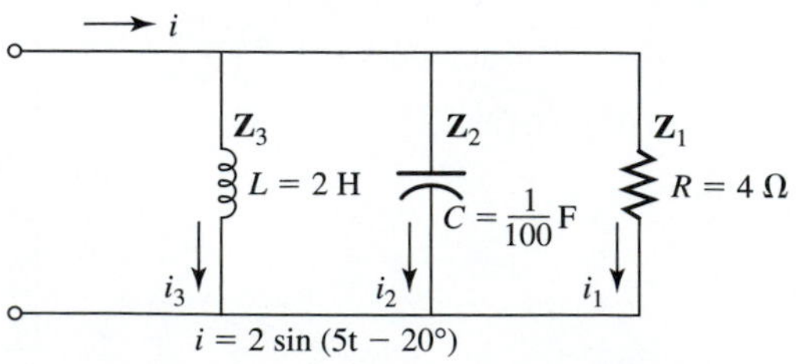

SOLUTION From the expression for i,

$$\mathbf{I} = 2\angle{-20°}$$

From Equation (17.3)

$$\mathbf{Z}_1 = 4\underline{/\ 0^\circ}$$

From Equations (17.7) and (17.8),

$$\begin{aligned} X_L &= \omega L \\ &= (5)(2) \\ &= 10 \\ \mathbf{Z}_3 &= X_L\underline{/\ 90^\circ} \\ &= 10\underline{/\ 90^\circ} \\ &= j10 \end{aligned}$$

From Equations (17.11) and (17.12),

$$\begin{aligned} X_C &= \frac{1}{\omega C} \\ &= \frac{1}{(5)\left(\frac{1}{100}\right)} \\ &= 20 \\ \mathbf{Z}_2 &= X_c\underline{/\ -90^\circ} \\ &= 20\underline{/\ -90^\circ} \\ &= -j20 \end{aligned}$$

From Equation (17.35)

$$\begin{aligned} \mathbf{Z}_{\mathrm{EQ}} &= \frac{1}{1/\mathbf{Z}_1 + 1/\mathbf{Z}_3 + 1/\mathbf{Z}_2} \\ &= \frac{1}{\{1/4\} + \{1/j10\} + \{1/(-j20)\}} \\ &= \frac{1}{0.25 - j(0.1) + j(0.05)} \\ &= \frac{1}{0.25 - j0.05} \\ &= \frac{1}{0.26\underline{/\ -11.3^\circ}} \\ &= 3.9\underline{/\ 11.3^\circ} \end{aligned}$$

From Equation (17.41)

$$\begin{aligned} \mathbf{I}_1 &= \left\{\frac{\mathbf{Z}_{\mathrm{EQ}}}{\mathbf{Z}_1}\right\}\mathbf{I} \\ &= \left\{\frac{(3.9\underline{/\ 11.3^\circ})}{(4\underline{/\ 0^\circ})}\right\}(2\underline{/\ -20^\circ}) \\ &= 1.9\underline{/\ -8.7^\circ} \end{aligned}$$

$$\mathbf{I}_2 = \left\{\frac{\mathbf{Z}_{EQ}}{\mathbf{Z}_3}\right\}\mathbf{I}$$

$$= \left\{\frac{(3.9\angle 11.3^\circ)}{(20\angle -90^\circ)}\right\}(2\angle -20^\circ)$$

$$= 0.39\angle 81.3^\circ$$

$$\mathbf{I}_3 = \left\{\frac{\mathbf{Z}_{EQ}}{\mathbf{Z}_3}\right\}\mathbf{I}$$

$$= \left\{\frac{(3.9\angle 11.3^\circ)}{(10\angle 90^\circ)}\right\}(2\angle -20^\circ)$$

$$= 0.78\angle -98.7^\circ$$

From current phasors $\mathbf{I}_1$, $\mathbf{I}_2$, and $\mathbf{I}_3$, currents i_1, i_2, and i_3 in the time domain are

$$i_1 = 1.9\sin(5t - 8.7^\circ)$$

$$i_2 = 0.39\sin(5t + 81.3^\circ)$$

$$i_3 = 0.78\sin(5t - 98.7^\circ)$$

EXAMPLE 17.24 Determine currents i_1, i_2, and i_3 in Example 17.23 using Equation (17.43) for current division. ■

SOLUTION Referring to the circuit in Example 17.23,

$$\mathbf{I} = 2\angle -20^\circ$$

From Equation (17.16)

$$\mathbf{Y}_1 = \left(\frac{1}{R}\right)\angle 0^\circ$$

$$= \left(\frac{1}{4}\right)\angle 0^\circ$$

From Equations (17.20) and (17.21),

$$B_L = \frac{1}{\omega L}$$

$$= \frac{1}{(5)(2)}$$

$$= \frac{1}{10}$$

$$\mathbf{Y}_3 = B_L\angle -90^\circ$$

$$= \left(\frac{1}{10}\right)\angle -90^\circ$$

$$= -j\left(\frac{1}{10}\right)$$

From Equations (17.24) and (17.25),

$$B_c = \omega C$$
$$= (5)\left(\frac{1}{100}\right)$$
$$= \frac{1}{20}$$
$$\mathbf{Y}_2 = B_c\angle 90^\circ$$
$$= \left(\frac{1}{20}\right)\angle 90^\circ$$
$$= j\left(\frac{1}{20}\right)$$

From Equation (17.40)

$$\mathbf{Y}_{EQ} = \mathbf{Y}_1 + \mathbf{Y}_2 + \mathbf{Y}_3$$
$$= \left(\frac{1}{4}\right) + j\left(\frac{1}{20}\right) - j\left(\frac{1}{10}\right)$$
$$= \left(\frac{1}{4}\right) - j\left(\frac{1}{20}\right)$$
$$= 0.25\angle 11.3^\circ$$

From Equation (17.43)

$$\mathbf{I}_1 = \left\{\frac{\mathbf{Y}_1}{\mathbf{Y}_{EQ}}\right\}\mathbf{I}$$
$$= \left\{\frac{0.25\angle 0^\circ}{0.25\angle 11.3^\circ}\right\}(2\angle -20^\circ)$$
$$= 1.9\angle -8.7^\circ$$
$$\mathbf{I}_2 = \left\{\frac{\mathbf{Y}_2}{\mathbf{Y}_{EQ}}\right\}\mathbf{I}$$
$$= \left\{\frac{0.05\angle 90^\circ}{0.25\angle 11.3^\circ}\right\}(2\angle -20^\circ)$$
$$= 0.39\angle 81.3^\circ$$
$$\mathbf{I}_3 = \left\{\frac{\mathbf{Y}_3}{\mathbf{Y}_{EQ}}\right\}\mathbf{I}$$
$$= \left\{\frac{0.1\angle -90^\circ}{0.25\angle 11.3^\circ}\right\}(2\angle -20^\circ)$$
$$= 0.78\angle -98.7^\circ$$

From current phasors $\mathbf{I}_1$, $\mathbf{I}_2$, and $\mathbf{I}_3$, the currents i_1, i_2, and i_3 in the time domain are

$$i_1 = 1.9 \sin(5t - 8.7^\circ)$$
$$i_2 = 0.39 \sin(5t + 81.3^\circ)$$
$$i_3 = 0.78 \sin(5t - 98.7^\circ)$$

17.19 VOLTAGE-DIVISION FORMULA FOR IMPEDANCE

The voltage-division formula allows us to determine the division of a voltage appearing across several electrical elements in series without determining the current through the elements. Consider the circuit in Figure 17.49, where voltage $\mathbf{V}$ divides across impedances $\mathbf{Z}_1, \mathbf{Z}_2, \ldots, \mathbf{Z}_n$.

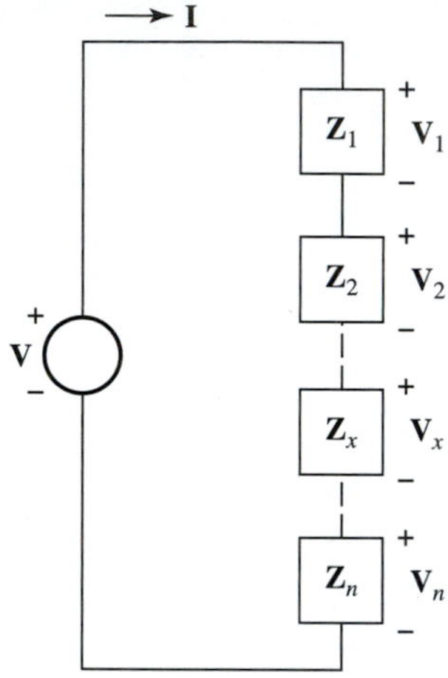

Figure 17.49 Voltage division.

From Equation (17.35) the equivalent impedance for terminals *a-b* is

$$\mathbf{Z}_{\text{EQ}} = \mathbf{Z}_1 + \mathbf{Z}_2 + \cdots + \mathbf{Z}_n$$

and the circuit in Figure 17.49 can be redrawn as shown in Figure 17.50.

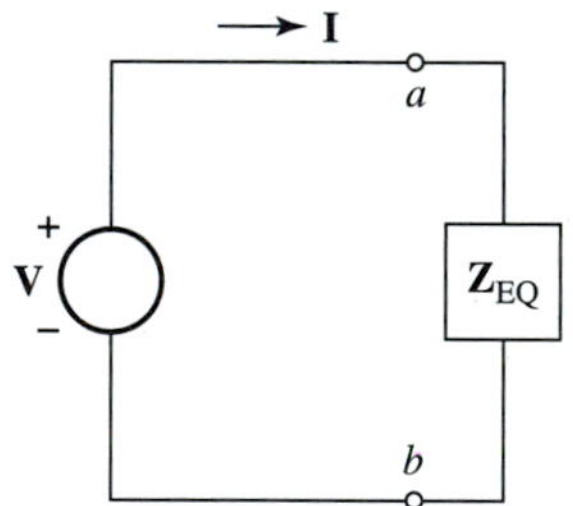

Figure 17.50 Equivalent impedance.

From Equation (17.1)

$$\mathbf{V} = \mathbf{Z}_{\text{EQ}}\mathbf{I}$$

$$\mathbf{I} = \frac{\mathbf{V}}{\mathbf{Z}_{\text{EQ}}}$$

and because $\mathbf{I}$ flows through each impedance in Figure 17.49,

$$\mathbf{V}_1 = \mathbf{Z}_1\mathbf{I}$$

$$\mathbf{V}_2 = \mathbf{Z}_2\mathbf{I}$$

$$\vdots$$

$$\mathbf{V}_n = \mathbf{Z}_n\mathbf{I}$$

Substituting $\mathbf{I} = \mathbf{V}/\mathbf{Z}_{EQ}$ into each equation results in

$$\mathbf{V}_1 = \left(\frac{\mathbf{Z}_1}{\mathbf{Z}_{EQ}}\right)\mathbf{V}$$

$$\mathbf{V}_2 = \left(\frac{\mathbf{Z}_2}{\mathbf{Z}_{EQ}}\right)\mathbf{V}$$

$$\vdots$$

$$\mathbf{V}_n = \left(\frac{\mathbf{Z}_n}{\mathbf{Z}_{EQ}}\right)\mathbf{V}$$

In general, the voltage $\mathbf{V}_x$ across an impedance $\mathbf{Z}_x$ in Figure 17.49 is

$$\mathbf{V}_x = \left(\frac{\mathbf{Z}_x}{\mathbf{Z}_{EQ}}\right)\mathbf{V} \tag{17.45}$$

For two series impedances $\mathbf{Z}_1$ and $\mathbf{Z}_2$, Equation (17.45) can be written as

$$\left.\begin{aligned} \mathbf{V}_1 &= \left\{\frac{\mathbf{Z}_1}{\mathbf{Z}_1 + \mathbf{Z}_2}\right\}\mathbf{V} \\ \mathbf{V}_2 &= \left\{\frac{\mathbf{Z}_2}{\mathbf{Z}_1 + \mathbf{Z}_2}\right\}\mathbf{V} \end{aligned}\right\} \tag{17.46}$$

17.20 VOLTAGE-DIVISION FORMULA FOR ADMITTANCE

Consider the circuit in Figure 17.51. From Equation (17.38), the equivalent admittance for terminals a-b is

$$\mathbf{Y}_{EQ} = \frac{1}{1/\mathbf{Y}_1 + 1/\mathbf{Y}_2 + \cdots + 1/\mathbf{Y}_n}$$

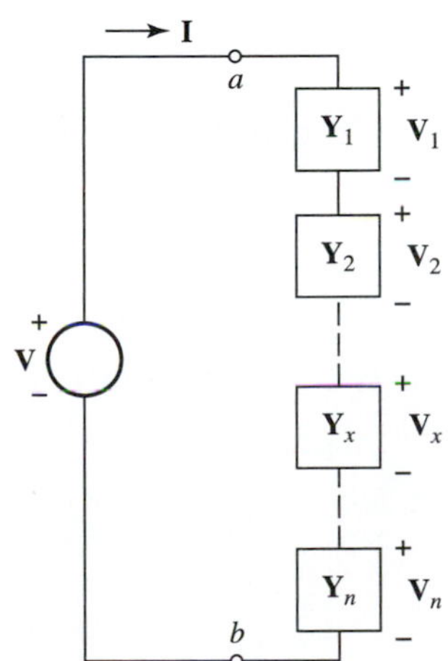

Figure 17.51 Voltage division.

and the circuit can be redrawn as shown in Figure 17.52.

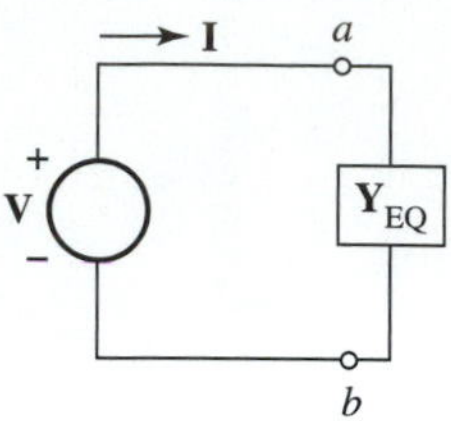

Figure 17.52 Equivalent admittance.

From Equation (17.13)

$$\mathbf{V} = \frac{\mathbf{I}}{\mathbf{Y}_{EQ}}$$

$$\mathbf{I} = \mathbf{V}\,\mathbf{Y}_{EQ}$$

and because **I** flows through each admittance in Figure 17.51,

$$\mathbf{V}_1 = \frac{\mathbf{I}}{\mathbf{Y}_1}$$

$$\mathbf{V}_2 = \frac{\mathbf{I}}{\mathbf{Y}_2}$$

$$\vdots$$

$$\mathbf{V}_n = \frac{\mathbf{I}}{\mathbf{Y}_n}$$

Substituting $\mathbf{I} = \mathbf{V}\mathbf{Y}_{EQ}$ into each equation results in

$$\mathbf{V}_1 = \left(\frac{\mathbf{Y}_{EQ}}{\mathbf{Y}_1}\right)\mathbf{V}$$

$$\mathbf{V}_2 = \left(\frac{\mathbf{Y}_{EQ}}{\mathbf{Y}_2}\right)\mathbf{V}$$

$$\vdots$$

$$\mathbf{V}_n = \left(\frac{\mathbf{Y}_{EQ}}{\mathbf{Y}_n}\right)\mathbf{V}$$

In general, the voltage $\mathbf{V}_x$ across an admittance $\mathbf{Y}_x$ in Figure 17.51 is

$$\mathbf{V}_x = \left(\frac{\mathbf{Y}_{EQ}}{\mathbf{Y}_x}\right)\mathbf{V} \tag{17.47}$$

For two series admittances, $\mathbf{Y}_1$ and $\mathbf{Y}_2$, Equation (17.47) can be written as

$$\mathbf{V}_1 = \left\{\frac{\mathbf{Y}_2}{\mathbf{Y}_1 + \mathbf{Y}_2}\right\}\mathbf{V}$$

$$\mathbf{V}_2 = \left\{\frac{\mathbf{Y}_1}{\mathbf{Y}_1 + \mathbf{Y}_2}\right\}\mathbf{V} \tag{17.48}$$

Because an equivalent impedance $\mathbf{Z}_{EQ}$ is easier to determine than an equivalent admittance $\mathbf{Y}_{EQ}$ for series elements, it is usually advantageous to employ Equation (17.45) rather than Equation (17.47) when using voltage division.

Consider the following examples.

EXAMPLE 17.25 Determine phasor voltages $\mathbf{V}_1$ and $\mathbf{V}_2$ for the following circuit in Figure 17.53 using Equation (17.46). ■

Figure 17.53

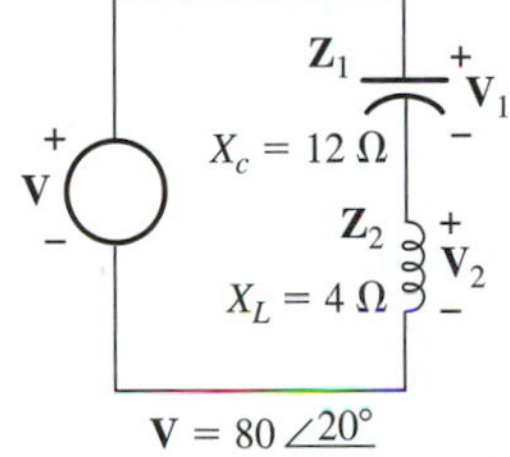

SOLUTION From Equation (17.46)

$$\mathbf{V}_1 = \left\{\frac{\mathbf{Z}_1}{\mathbf{Z}_1 + \mathbf{Z}_2}\right\}\mathbf{V}$$

$$= \left\{\frac{-j12}{-j12 + j4}\right\}/(80\underline{/20^\circ})$$

$$= \left\{\frac{-j12}{-j8}\right\}(80\underline{/20^\circ})$$

$$= 120\underline{/20^\circ}$$

$$\mathbf{V}_2 = \left\{\frac{\mathbf{Z}_2}{\mathbf{Z}_1 + \mathbf{Z}_2}\right\}\mathbf{V}$$

$$= \left\{\frac{j4}{-j12 + j4}\right\}(80\underline{/20^\circ})$$

$$= 40\underline{/200^\circ}$$

EXAMPLE 17.26 Determine phasor voltages $\mathbf{V}_1$ and $\mathbf{V}_2$ in Example 17.25 using Equation (17.48). ■

SOLUTION From Equation (17.48)

$$\mathbf{V}_1 = \left\{\frac{\mathbf{Y}_2}{\mathbf{Y}_1 + \mathbf{Y}_2}\right\}\mathbf{V}$$

$$= \left\{\left(-j\frac{1}{4}\right)/\left(-j\frac{1}{4} + j\frac{1}{12}\right)\right\}(80\underline{/20^\circ})$$

$$= \left\{\left(-j\frac{1}{4}\right)/\left(-j\frac{1}{6}\right)\right\}(80\underline{/20^\circ})$$

$$= \left(\frac{3}{2}\right)(80\underline{/20^\circ})$$

$$= 120\underline{/20^\circ}$$

$$\mathbf{V}_2 = \left\{\frac{\mathbf{Y}_1}{\mathbf{Y}_1 + \mathbf{Y}_2}\right\}\mathbf{V}$$

$$= \left\{\left(j\frac{1}{2}\right)/\left(-j\frac{1}{4} + j\frac{1}{12}\right)\right\}(80\angle 20^\circ)$$

$$= \left\{\left(j\frac{1}{12}\right)/\left(-j\frac{1}{6}\right)\right\}(80\angle 20^\circ)$$

$$= 40\angle 200^\circ$$

EXAMPLE 17.27 Determine voltages v_1, v_2, v_3, and v_4 for the circuit in Figure 17.54. ■

Figure 17.54

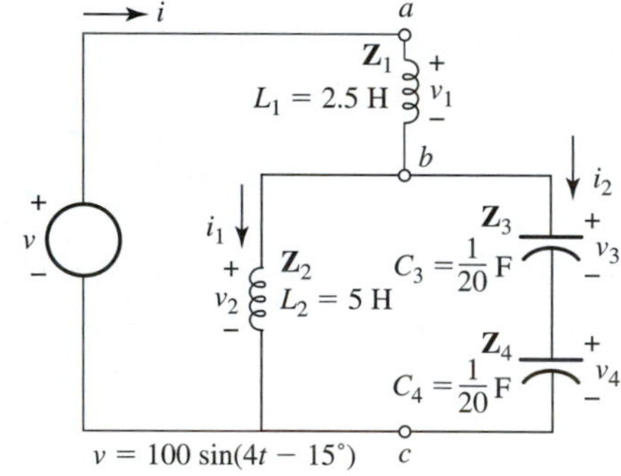

SOLUTION From the given expression for the voltage v,

$$\mathbf{V} = 100\angle -15^\circ$$

Combining $\mathbf{Z}_2$, $\mathbf{Z}_3$, and $\mathbf{Z}_4$ results in

$$\mathbf{Z}_{bc} = \{\mathbf{Z}_3 + \mathbf{Z}_4\} \| \{\mathbf{Z}_2\}$$

$$\mathbf{Z}_3 + \mathbf{Z}_4 = \left\{-j\left(\frac{1}{\omega C_3}\right) - j\left(\frac{1}{\omega C_4}\right)\right\}$$

$$= \left\{-j\frac{1}{(4)\left(\frac{1}{20}\right)} - j\frac{1}{(4)\left(\frac{1}{20}\right)}\right\}$$

$$= -j10$$

$$\mathbf{Z}_{bc} = \{-j10\} \| \{\mathbf{Z}_2\}$$

$$= \frac{(-j10)(j\omega L_2)}{-j10 + j\omega L_2}$$

$$= \frac{\{(-j10)\}\{j(4)(5)\}}{-j10 + j(4)(5)}$$

$$= -j20$$

From Equation (17.48)

$$\mathbf{V}_1 = \left\{\frac{\mathbf{Z}_1}{\mathbf{Z}_1 + \mathbf{Z}_{bc}}\right\}\mathbf{V}$$

$$= \left\{\frac{j\omega L_1}{j\omega L_1 - j20}\right\}(100\underline{/-15^\circ})$$

$$= \left\{\frac{j(4)(2.5)}{j(4)(2.5) - j20}\right\}(100\underline{/-15^\circ})$$

$$= \left\{\frac{j10}{-j10}\right\}(100\underline{/-15^\circ})$$

$$= (1\underline{/180^\circ})(100\underline{/-15^\circ})$$

$$= 100\underline{/165^\circ}$$

$$\mathbf{V}_{bc} = \left\{\frac{\mathbf{Z}_{bc}}{\mathbf{Z}_1 + \mathbf{Z}_{bc}}\right\}\mathbf{V}$$

$$= \left\{\frac{-j20}{j\omega L - j20}\right\}(100\underline{/-15^\circ})$$

$$= \left\{\frac{-j20}{j(4)(2.5) - j20}\right\}(100\underline{/-15^\circ})$$

$$= \left\{\frac{-j20}{-j10}\right\}(100\underline{/-15^\circ})$$

$$= 200\underline{/-15^\circ}$$

Because $\mathbf{V}_2 = \mathbf{V}_{bc}$,

$$\mathbf{V}_2 = 200\underline{/-15^\circ}$$

Applying Equation (17.46) to the voltage across the series combination $\mathbf{Z}_3$ and $\mathbf{Z}_4$ results in

$$\mathbf{V}_3 = \left\{\frac{\mathbf{Z}_3}{\mathbf{Z}_3 + \mathbf{Z}_4}\right\}\mathbf{V}_{bc}$$

$$= \left\{\frac{1/j\omega C_3}{1/(j\omega C_3) - j5}\right\}(200\underline{/-15^\circ})$$

$$= \{(-j5)/(-j5 - j5)\}(200\underline{/-15^\circ})$$

$$= 100\underline{/-15^\circ}$$

$$\mathbf{V}_4 = \left\{\frac{\mathbf{Z}_4}{\mathbf{Z}_3 + \mathbf{Z}_4}\right\}\mathbf{V}_{bc}$$

$$= \left\{\frac{1/j\omega C_4}{1/(j\omega C_4) - j5}\right\}(200\underline{/-15^\circ})$$

$$= \left\{\frac{-j5}{-j5 - j5}\right\}(200\underline{/-15^\circ})$$

$$= 100\underline{/-15^\circ}$$

The phasor voltages $\mathbf{V}_1$, $\mathbf{V}_2$, $\mathbf{V}_3$, and $\mathbf{V}_4$ generate the following waveforms:

$$v_1 = 100 \sin(4t + 165°)$$
$$v_2 = 200 \sin(4t - 15°)$$
$$v_3 = 100 \sin(4t - 15°)$$
$$v_4 = 100 \sin(4t - 15°)$$

EXAMPLE 17.28 Determine the currents i, i_1, and i_2 for the circuit in Example 17.27 and draw a phasor diagram that includes $\mathbf{V}$, $\mathbf{V}_1$, $\mathbf{V}_2$, $\mathbf{V}_3$, $\mathbf{V}_4$, $\mathbf{I}$, $\mathbf{I}_1$, and $\mathbf{I}_2$.■

SOLUTION The circuit is redrawn in Figure 17.55 with the impedances indicated as determined in Example 17.27.

Figure 17.55

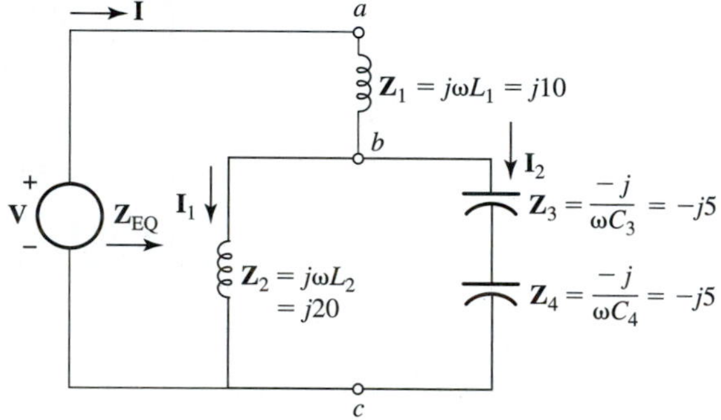

$$\mathbf{Z}_{EQ} = \mathbf{Z}_1 + \{\mathbf{Z}_2 \| \{\mathbf{Z}_3 + \mathbf{Z}_4\}\}$$

From Example 17.27

$$\mathbf{Z}_2 \| \{\mathbf{Z}_3 + \mathbf{Z}_4\} = -j20$$

$$\begin{aligned} \mathbf{Z}_{EQ} &= \mathbf{Z}_1 + \{\mathbf{Z}_2 \| (\mathbf{Z}_3 + \mathbf{Z}_4)\} \\ &= j10 + \{-j20\} \\ &= -j10 \\ &= 10\underline{/-90°} \end{aligned}$$

From Equation (17.1)

$$\begin{aligned} \mathbf{I} &= \frac{\mathbf{V}}{\mathbf{Z}_{EQ}} \\ &= \frac{100\underline{/-15°}}{10\underline{/-90°}} \\ &= 10\underline{/75°} \end{aligned}$$

Alternatively, because $\mathbf{V}_1$ is known from Example 17.27, $\mathbf{I}$ can be determined from $\mathbf{I} = \mathbf{V}_1/\mathbf{Z}_1$ as

$$\mathbf{I} = \frac{100\angle 165^\circ}{10\angle 90^\circ} = 10\angle 75^\circ$$

Current $\mathbf{I}$ divides into $\mathbf{I}_1$ and $\mathbf{I}_2$ at node b, as shown in Figure 17.56.

Figure 17.56

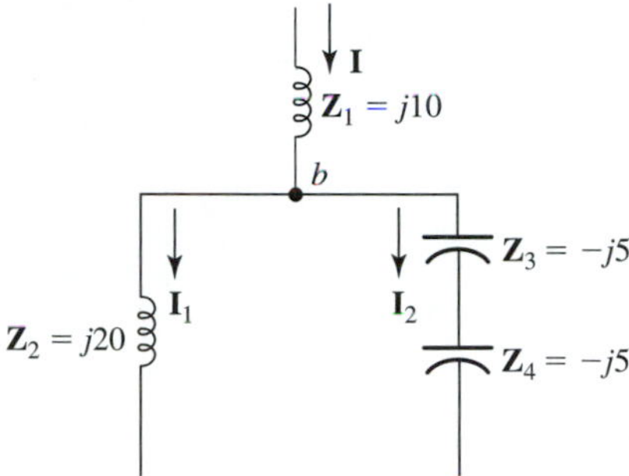

From Equation (17.35)

$$\mathbf{Z}_3 + \mathbf{Z}_4 = -j5 - j5 = -j10$$

From Equation (17.42)

$$\mathbf{I}_1 = \left\{\frac{-j10}{-j10 + j20}\right\}\mathbf{I} = \left\{\frac{-j10}{j10}\right\}(10\angle 75^\circ) = 10\angle -105^\circ$$

and

$$\mathbf{I}_2 = \left\{\frac{j20}{-j10 + j20}\right\}\mathbf{I} = \left\{\frac{j20}{j10}\right\}(10\angle 75^\circ) = 20\angle 75^\circ$$

The phasor currents $\mathbf{I}$, $\mathbf{I}_1$, and $\mathbf{I}_2$ generate the following waveforms.

$$i = 10\sin(4t + 75^\circ)$$
$$i_1 = 10\sin(4t - 105^\circ)$$
$$i_2 = 20\sin(4t + 75^\circ)$$

The phasor diagram for $\mathbf{V}$, $\mathbf{V}_1$, $\mathbf{V}_2$, $\mathbf{V}_3$, $\mathbf{V}_4$, $\mathbf{I}$, $\mathbf{I}_1$ and $\mathbf{I}_2$ is shown in Figure 17.57.

Figure 17.57

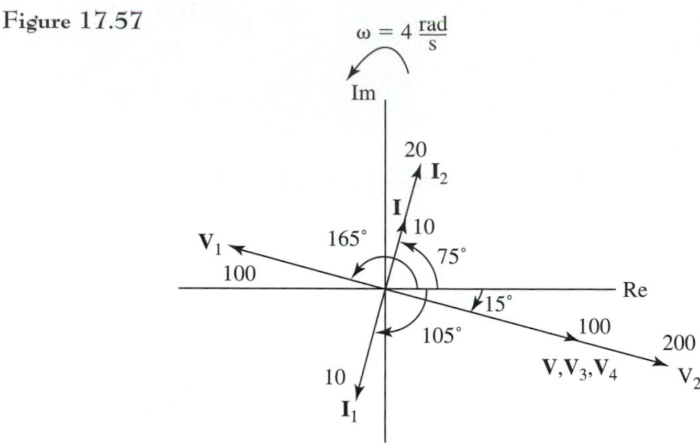

17.21 CONVERSION OF AC VOLTAGE AND CURRENT SOURCES

Perfect (ideal) and imperfect (physically realizable) ac voltage and current sources in phasor form are shown in Figure 17.58. The conversion of a current source to a voltage source (or vice versa) is useful in ac circuit analysis. Source conversion can greatly simplify an otherwise complicated circuit for analysis purposes. It can also alter the circuit in a manner that allows us to employ circuit-analysis procedures that cannot be applied to the circuit prior to the source conversion. Perfect voltage and current sources are primal electrical elements

Figure 17.58 AC sources.

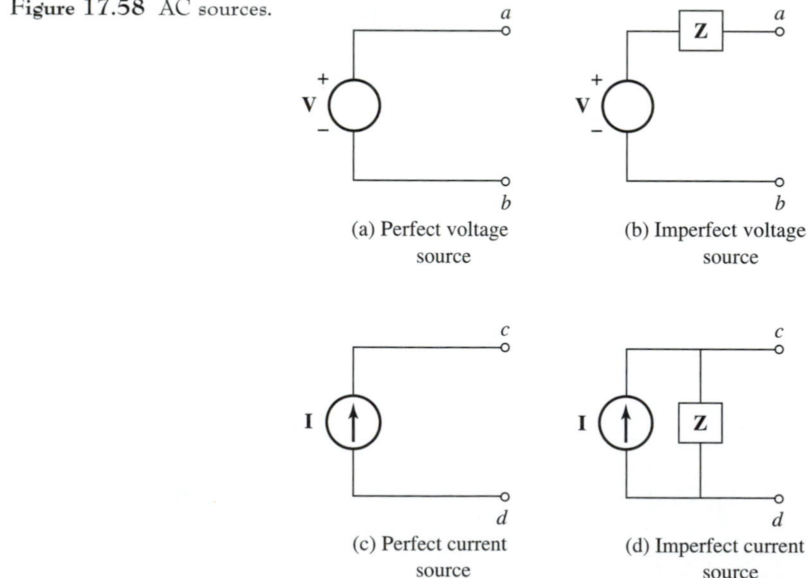

and cannot be converted. Imperfect voltage and current sources can be converted as illustrated in Figure 17.59.

Figure 17.59 Source conversion.

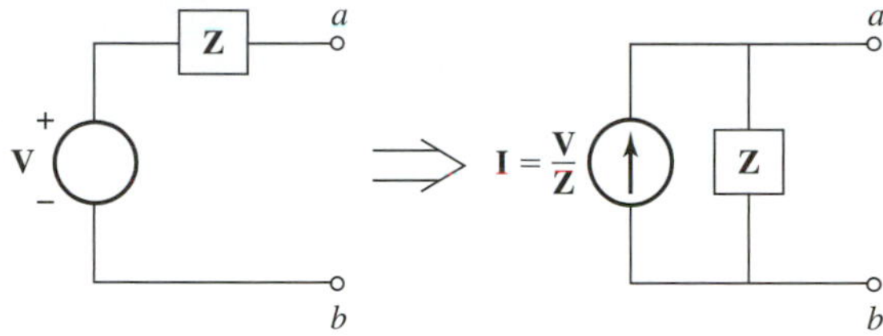

(a) Conversion of a voltage source to a current source

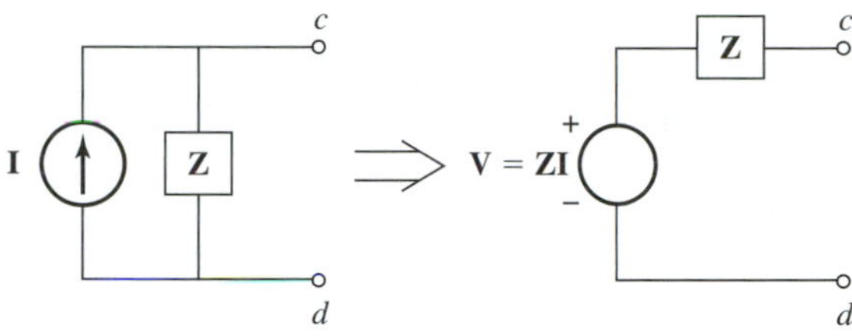

(b) Conversion of a current source to a voltage source

The conversion of sources is always with respect to a pair of terminals. In Figure 17.59(a), the current source equivalent is with respect to terminals *a-b*; in Figure 17.59(b), the voltage source equivalent is with respect to terminals *c-d*. Thevenin's or Norton's theorems—presented in Chapter 19 for ac circuits—allow us easily to verify the equivalences in Figure 17.59.

Consider the following examples.

EXAMPLE 17.29 Convert the voltage source to a current source for the circuit in Figure 17.60. ■

Figure 17.60

$X_L = 30\ \Omega$ a

$\mathbf{V} = 60\angle 40^\circ$ + −

b

SOLUTION From Equation (17.8)

$$\mathbf{Z} = 30\angle 90^\circ$$

and from the conversion rule given in Figure 17.59(a), the equivalent current source is as shown in Figure 17.61, where

$$\mathbf{I} = \frac{\mathbf{V}}{\mathbf{Z}}$$

$$= \frac{60\underline{/40^\circ}}{30\underline{/90^\circ}}$$

$$= 2\underline{/-50^\circ}$$

Figure 17.61

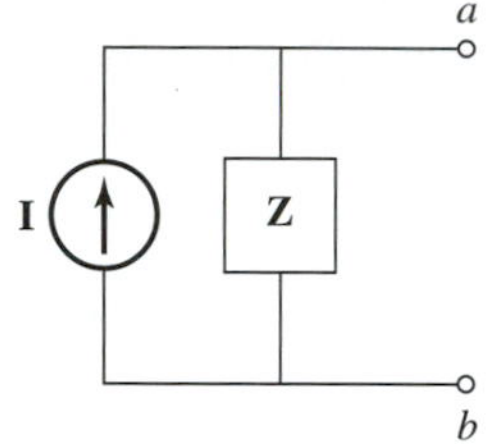

The current source equivalent is shown in Figure 17.62.

Figure 17.62

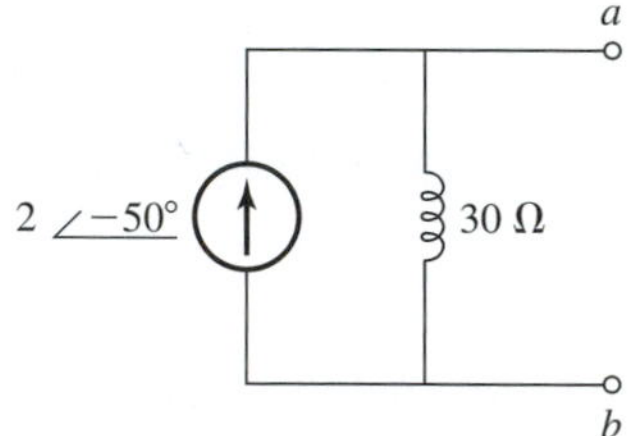

EXAMPLE 17.30 Convert the current source to a voltage source for the circuit in Figure 17.63. ■

Figure 17.63

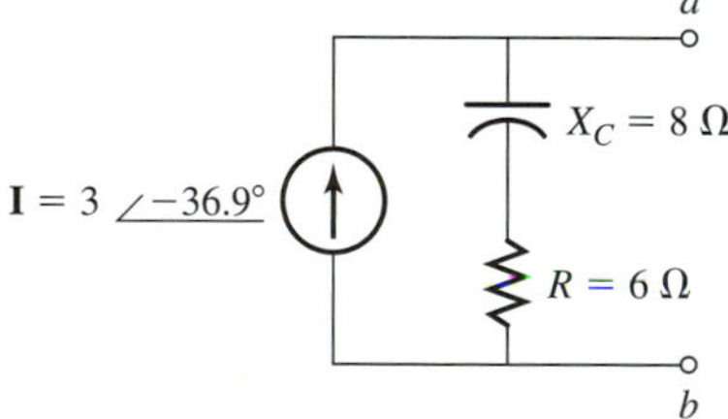

SOLUTION From Equations (17.3), (17.12), and (17.35),

$$\mathbf{Z} = 6 - j8$$

$$= 10\underline{/-53.1^\circ}$$

and from the conversion rule given in Figure 17.59(b), the equivalent voltage source is as shown in Figure 17.64, where

$$\mathbf{V} = \mathbf{IZ}$$

$$= (3\underline{/-36.9^\circ})(10\underline{/-53.1^\circ})$$

$$= 30\underline{/-90^\circ}$$

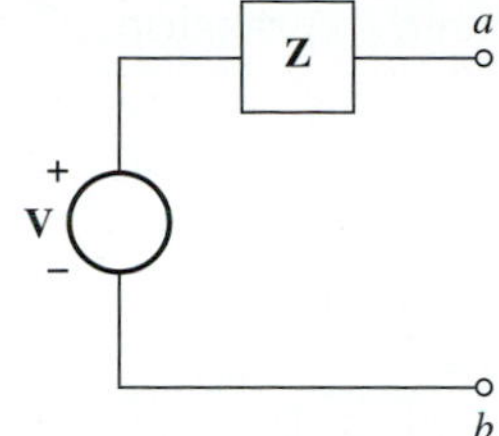

Figure 17.64

The voltage source equivalent can be drawn as shown in Figure 17.65.

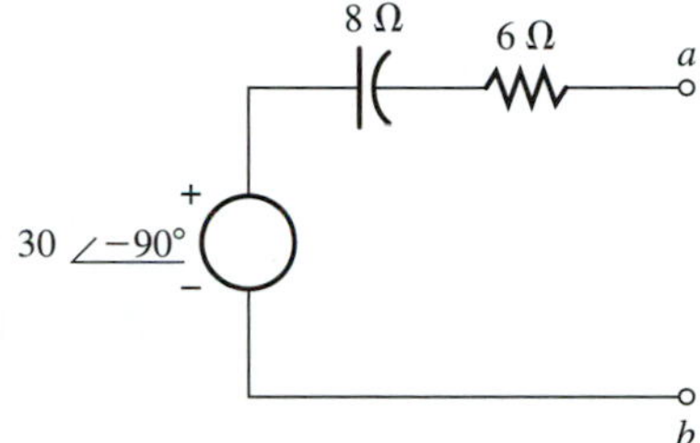

Figure 17.65

EXAMPLE 17.31 Convert the voltage source to a current source for the circuit in Figure 17.66. ■

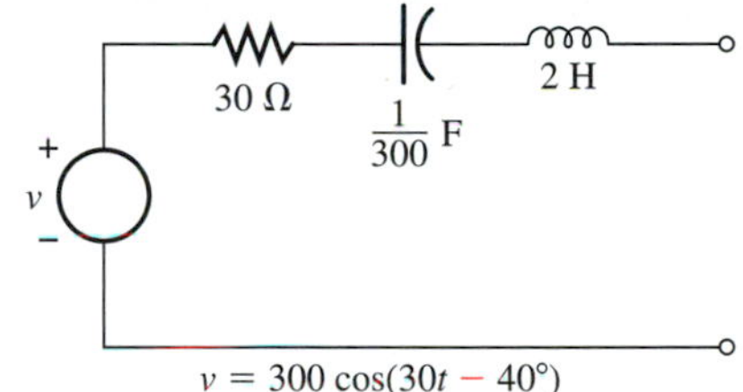

Figure 17.66

SOLUTION From Equations (17.3), (17.7), (17.8), (17.11), (17.12), and (17.35),

$$\mathbf{Z}_R = 30\,\Omega$$

$$X_L = \omega L$$

$$= (30)(2)$$

$$= 60\,\Omega$$

$$\mathbf{Z}_L = j60$$

$$X_C = \frac{1}{\omega C}$$

$$= \frac{1}{(30)/\left(\frac{1}{300}\right)}$$

$$= 10\,\Omega$$

$$\mathbf{Z}_C = -j10$$

$$\mathbf{Z} = \mathbf{Z}_C + \mathbf{Z}_L + \mathbf{Z}_R$$

$$= 30 + j50$$

$$= 58.3\angle 59^\circ$$

From the given voltage expression,

$$\mathbf{V} = 300\underline{/50^\circ}$$

The voltage source in phasor form is shown in Figure 17.67,

Figure 17.67

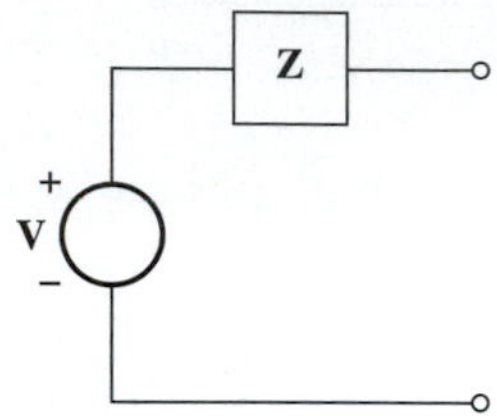

where

$$\mathbf{Z} = 58.3\underline{/50^\circ}$$

and

$$\mathbf{V} = 300\underline{/50^\circ}$$

From the conversion rule given in Figure 17.59(a), the equivalent current source in phasor form is as shown in Figure 17.68, where

$$\mathbf{I} = \frac{\mathbf{V}}{\mathbf{Z}}$$

$$= \frac{300\underline{/50^\circ}}{58.3\underline{/59^\circ}}$$

$$= 5.2\underline{/-9^\circ}$$

Figure 17.68

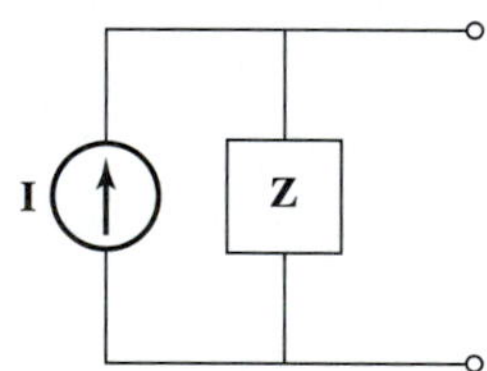

and

$$\mathbf{Z} = 58.3\underline{/59^\circ}$$

In the time domain the equivalent current source is as shown in Figure 17.69, where

$$i = 5.2\ \sin(30t - 9^\circ)$$

Figure 17.69

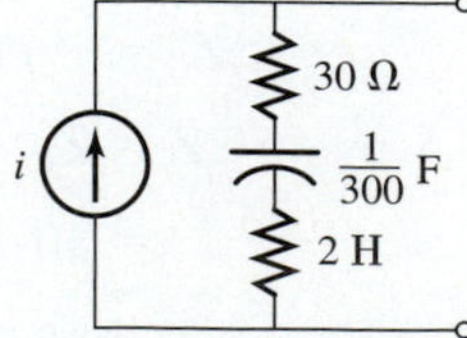

EXAMPLE 17.32 Convert the current source to a voltage source for the circuit in Figure 17.70. ■

Figure 17.70

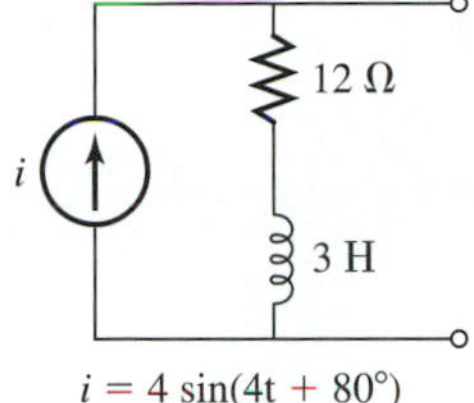

SOLUTION From Equations (17.3), (17.7), (17.8), and (17.35),

$$\begin{aligned}
\mathbf{Z}_R &= 12\,\Omega \\
X_L &= \omega L \\
&= (4)(3) \\
&= 12\,\Omega \\
\mathbf{Z}_L &= j12\,\Omega \\
\mathbf{Z} &= \mathbf{Z}_R + \mathbf{Z}_L \\
&= 12 + j12 \\
&= 17\angle 45^\circ\ \Omega
\end{aligned}$$

From the given current expression,

$$\mathbf{I} = 4\angle 80^\circ$$

The given current source in phasor form is as shown in Figure 17.71, where

$$\mathbf{Z} = 17\angle 45^\circ$$

Figure 17.71

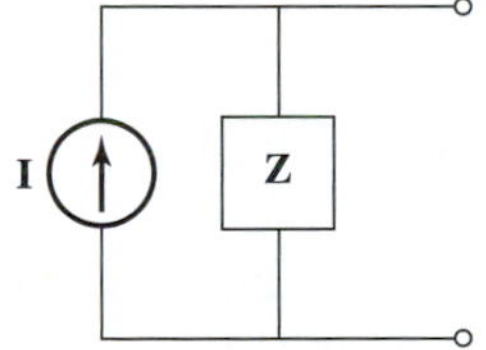

and

$$\mathbf{I} = 4\angle 80^\circ$$

From the conversion rule given in Figure 17.59(b), the equivalent voltage source in phasor form is as shown in Figure 17.72, where

$$\begin{aligned}
\mathbf{V} &= \mathbf{IZ} \\
&= (4\angle 80^\circ)(17\angle 45^\circ) \\
&= 68\angle 125^\circ
\end{aligned}$$

Figure 17.72

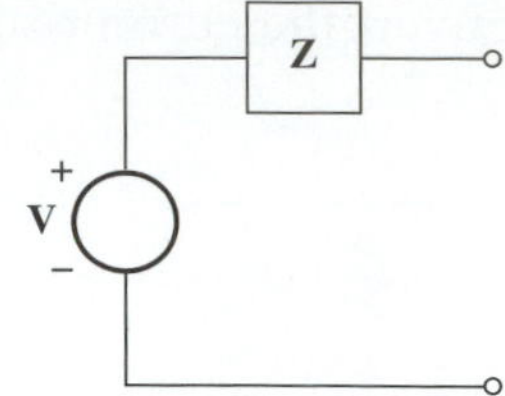

and

$$\mathbf{Z} = 17\angle 45^\circ$$

In the time domain the equivalent voltage source is as shown in Figure 17.73, where

$$v = 68\ \sin(4t + 125^\circ)$$

Figure 17.73

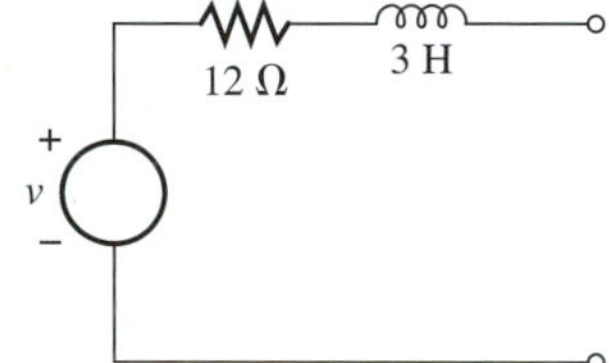

17.22 EXERCISES

1. Define the ac impedance for a pair of terminals.

2. If the current and voltage for a pair of terminals are $i = 60\ \sin(\omega t + 50^\circ)$ and $v = 90\ \sin(\omega t + 20^\circ)$, determine the impedance $\mathbf{Z}$ in polar form.

3. Repeat Problem 2 for the following currents and voltages.

(a) $v = 10\ \sin(\omega t - 40^\circ)$
$i = 20\ \sin(\omega t + 20^\circ)$

(b) $v = 40\ \cos(\omega t + 40^\circ)$
$i = -10\ \sin(\omega t - 170^\circ)$

(c) $v = 400\ \cos(\omega t)$
$i = 50\ \sin(\omega t + 120^\circ)$

(d) $v = -50\ \sin(\omega t)$
$i = 50\ \cos(\omega t)$

(e) $v = -100\ \cos(\omega t + 90^\circ)$
$i = 2\ \cos(\omega t - 60^\circ)$

(f) $v = -100\ \sin(\omega t + 180^\circ)$
$i = 5\ \sin(\omega t + 45^\circ)$

4. The current through a 20-Ω resistor is as follows. Determine $\mathbf{Z}_R$ and the voltage across the resistor.

(a) $i = 4\ \sin(\omega t + 30^\circ)$
(b) $i = 6\ \cos(\omega t)$
(c) $i = -2\ \cos(\omega t + 40^\circ)$
(d) $i = 5\ \sin(\omega t - 170^\circ)$
(e) $i = -2\ \sin(\omega t - 180^\circ)$
(f) $i = 4\ \sin(\omega t + 270^\circ)$

5. The voltage across a 40-Ω resistor is as follows. Determine $\mathbf{Z}_R$ and the current i.

(a) $v = 100\ \sin(\omega t + 60^\circ)$
(b) $v = -100\ \sin(\omega t)$
(c) $v = 60\ \cos(\omega t + 120^\circ)$
(d) $v = 40\ \cos(\omega t + 60^\circ)$
(e) $v = -80\ \cos(\omega t + 20^\circ)$
(f) $v = -100\ \sin(\omega t - 180^\circ)$

6. Determine the inductive reactance of a $\frac{1}{2}$-H coil for the following frequencies.

(a) 60 Hz
(b) 6 Hz
(c) 4 Hz
(d) 2500 Hz
(e) 1000 kHz
(f) 600 Hz

7. Determine the inductance of a coil with an inductive reactance of
(a) 600 Ω at 100 Hz (b) 50 Ω at 100 Hz
(c) 20 Ω at 20 rad/s (d) 40 Ω at 2 rad/s
(e) 1000 Ω at 60 Hz (f) 2 Ω at 1000 Hz

8. Determine the current through a 2-H inductance for the following applied voltages.
(a) $v = 50\cos(2t + 40°)$ (b) $v = -100\sin(4t)$
(c) $v = 200\sin(8t + 60°)$ (d) $v = -50\cos(6t)$
(e) $v = -40\sin(t - 180°)$ (f) $v = -60\cos(4t - 270°)$

9. Determine the voltage across a $\frac{1}{2}$-H inductance for the following currents.
(a) $i = -6\sin(4t + 60°)$ (b) $i = 2\cos(2t - 40°)$
(c) $i = 10\sin(t + 40°)$ (d) $i = -2\cos(8t + 80°)$
(e) $i = -5\cos(60t - 270°)$ (f) $i = -4\sin(40t - 180°)$

10. Determine the impedance $\mathbf{Z}_L$ of the inductors in Problem 8.

11. Determine the impedance $\mathbf{Z}_L$ of the inductors in Problem 9.

12. Determine the capacitive reactance of a 100-μF capacitor for the following frequencies.
(a) 2000 Hz (b) 40 Hz
(c) 600 rad/s (d) 15,000 Hz
(e) 6 Hz (f) 1000 rad/s

13. Determine the capacitance of a capacitor with a capacitive reactance of
(a) 1000 Ω at 200 Hz (b) 550 Ω at 120 Hz
(c) 60 Ω at 30 rad/s (d) 2 Ω at 10 rad/s
(e) 10 Ω at 100 rad/s (f) 40 Ω at 1000 rad/s

14. Determine the current through a 100-μF capacitor for the following applied voltages.
(a) $v = 600\sin(300t - 10°)$ (b) $v = -100\cos(400t)$
(c) $v = -1000\sin(500t)$ (d) $v = 30\sin(1000t + 40°)$
(e) $v = -40\sin(200t - 90°)$ (f) $v = -60\cos(400t + 90°)$

15. Determine the voltage across a $\frac{1}{2}$-mF capacitor for the following currents.
(a) $i = 0.5\sin(4t)$ (b) $i = -6\cos(t - 40°)$
(c) $i = 4\sin(2t - 90°)$ (d) $i = 2\sin(t + 270°)$
(e) $i = -2\sin(3t - 90°)$ (f) $i = -4\cos(2t + 30°)$

16. Determine the impedance $\mathbf{Z}_c$ of the capacitors in Problem 12.

17. Determine the impedance $\mathbf{Z}_c$ of the capacitors in Problem 13.

18. Define the ac admittance for a pair of terminals.

19. If the current and voltage for a pair of terminals is $i = 40\sin(\omega t - 80°)$ and $v = 100\sin(\omega t - 50°)$, determine the admittance $\mathbf{Y}$.

20. Repeat Problem 19 for the following currents and voltages.
(a) $v = 100\cos(\omega t)$, $i = 2\sin(\omega t + 60°)$ (b) $v = -50\sin(\omega t)$, $i = 4\cos(\omega t + 60°)$
(c) $v = 60\sin(\omega t + 60°)$, $i = 2\sin(\omega t + 30°)$ (d) $v = 100\sin(\omega t)$, $i = 5\cos(\omega t - 120°)$

21. Determine the admittance $\mathbf{Y}$ for the following resistors.
(a) 400 Ω (b) $\frac{1}{2}$ Ω
(c) 4000×10^6 Ω (d) 0.002 Ω

22. The current or voltage is given for a 20-Ω resistor. In each case find the voltage or current using the admittance concept.
(a) $v = -20\sin(\omega t + 90°)$ (b) $v = 2\cos(\omega t + 20°)$
(c) $i = 8\sin(\omega t)$ (d) $i = -2\sin(\omega t - 40°)$

23. Determine the inductive susceptance in each part of Problem 6.

24. Determine the inductance of a coil with an inductive susceptance of

(a) $\frac{1}{100}$ S at 20 Hz (b) 20 S at 1000 Hz
(c) 5 S at 10 Hz (d) 200 S at 400 Hz

25. Determine the admittance $\mathbf{Y}_L$ for inductors with the following impedances.

(a) $\mathbf{Z}_L = j6$ (b) $\mathbf{Z}_L = j\frac{1}{1000}$
(c) $\mathbf{Z}_L = j(0.002)$ (d) $\mathbf{Z}_L = j400$

26. Determine the capacitive susceptance in each part of Problem 12.

27. Determine the capacitance of a capacitor with a capacitive susceptance of

(a) 100 S at 40 Hz (b) 0.60 S at 500 Hz
(c) 2 S at 60 Hz (d) $\frac{1}{200}$ S at 2000 Hz

28. Determine the admittance **Y** for the following impedances.

(a) $\mathbf{Z} = -j40$ (b) $\mathbf{Z} = -j(0.004)$
(c) $\mathbf{Z} = -j1000$ (d) $\mathbf{Z} = -j\frac{1}{8}$

29. Consider the circuit in Figure 17.74.

Figure 17.74

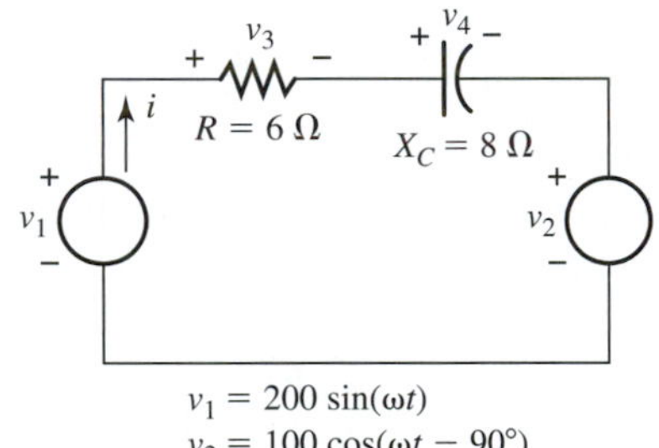

(a) Find the phasor current **I**. (b) Find i.
(c) Find the phasor voltage $\mathbf{V}_3$. (d) Find v_3.
(e) Find the phasor voltage $\mathbf{V}_4$. (f) Find v_4.

30. Consider the circuit in Figure 17.75.

Figure 17.75

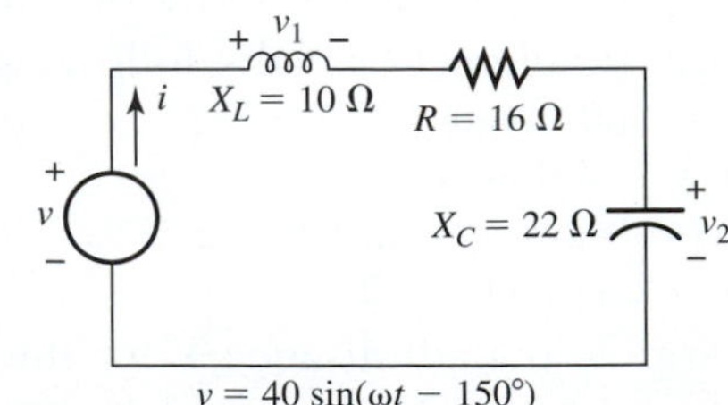

(a) Find the phasor current **I**. (b) Find i.
(c) Find the phasor voltage $\mathbf{V}_1$. (d) Find v_1.
(e) Find the phasor voltage $\mathbf{V}_2$. (f) Find v_2.

31. Consider the circuit in Figure 17.76.

Figure 17.76

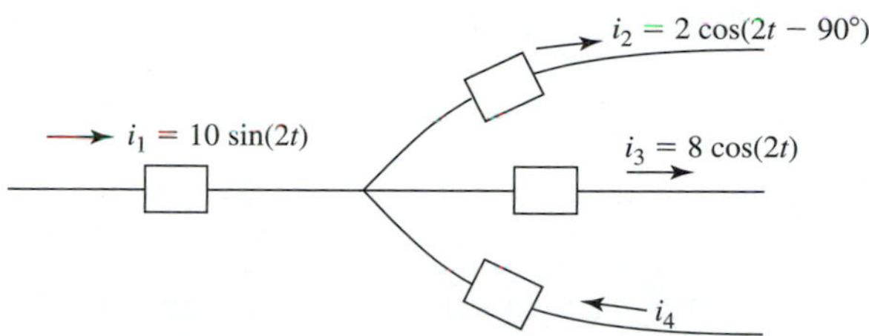

(a) Find the phasor current $\mathbf{I}_4$. (b) Find i_4.

32. A load is connected across a pair of terminals with the voltage and current as given in Figure 17.77.

Figure 17.77

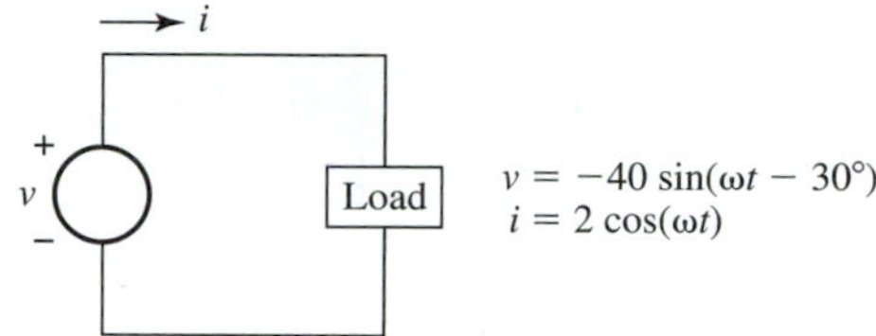

(a) Find R for the load. (b) Find G for the load.
(c) Find X for the load. (d) Find B for the load.
(e) Find $\mathbf{Z}$ for the load. (f) Find $\mathbf{Y}$ for the load.

33. Find $\mathbf{Z}$ in rectangular form if $\mathbf{Y} = 0.06 - j0.08$.

34. Find $\mathbf{Y}$ in rectangular form if $\mathbf{Z} = 32 + j24$.

35. Determine the impedance for terminals a-b in polar form for each circuit in Figure 17.78.

Figure 17.78

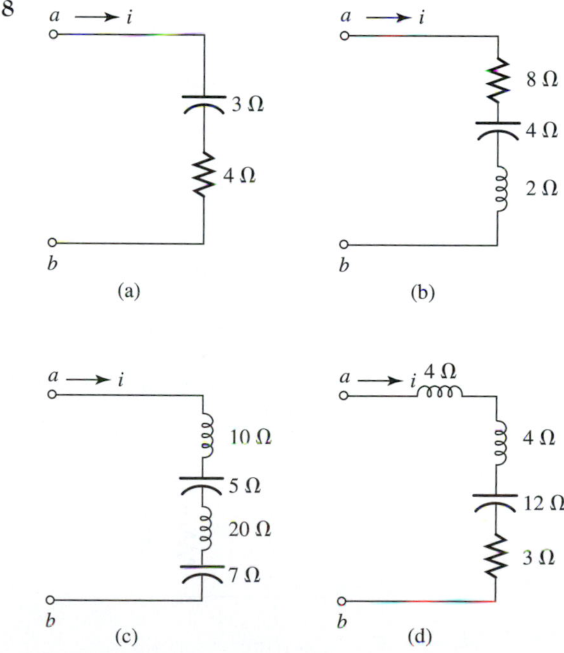

36. If $v_{ab} = 200\sin(\omega t)$ for each circuit in Problem 35, find the phasor current $\mathbf{I}$ in polar form and i for each circuit.

37. Determine the impedance for terminals *a*-*b* in polar form for each circuit in Figure 17.79.

Figure 17.79

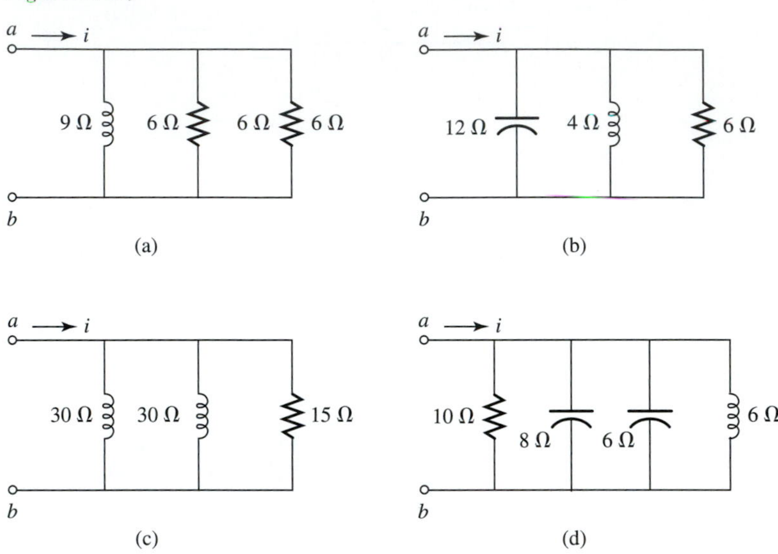

38. If $v_{ab} = 100 \sin(\omega t + 30°)$ for each circuit in Problem 37, find the phasor current **I** in polar form and *i* for each circuit.

39. Determine the admittance for terminals *a*-*b* in polar form for each circuit in Figure 17.80.

Figure 17.80

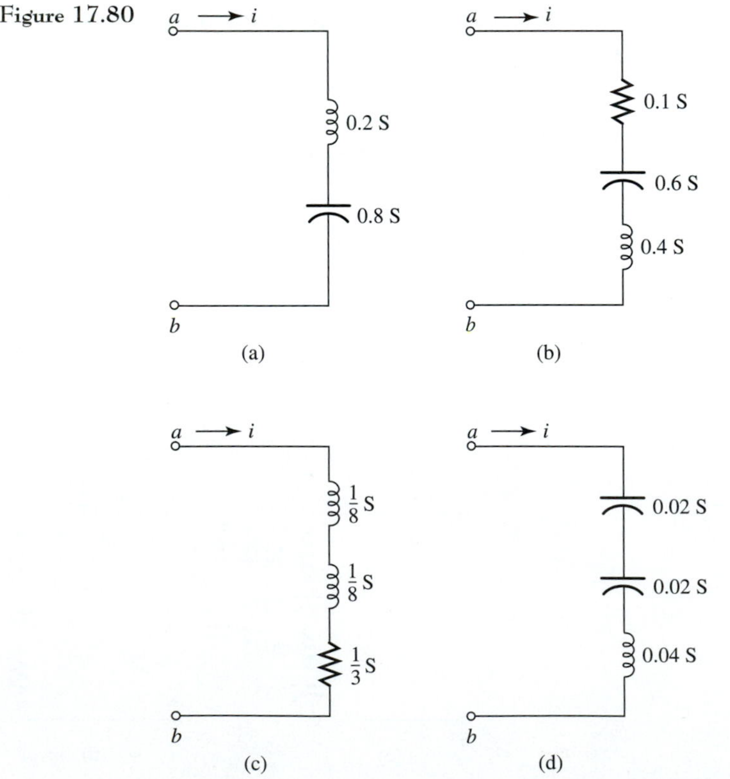

40. If $v_{ab} = 150 \sin(\omega t + 45°)$ for each circuit in Problem 39, find the phasor current **I** in polar form and *i* for each circuit.

41. Determine the admittance for terminals *a*-*b* in polar form for each circuit in Figure 17.81.

Figure 17.81

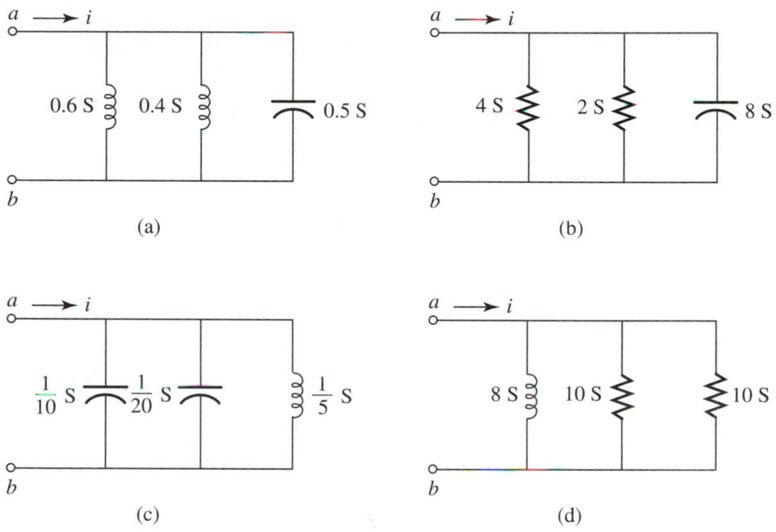

42. If $v_{ab} = 40 \sin(\omega t + 90°)$ for each circuit in Problem 41, find the phasor current **I** in polar form and *i* for each circuit.

43. Determine the impedance $\mathbf{Z}_{ab}$ and the admittance $\mathbf{Y}_{ab}$ for terminals *a*-*b* for each circuit in Figure 17.82.

Figure 17.82

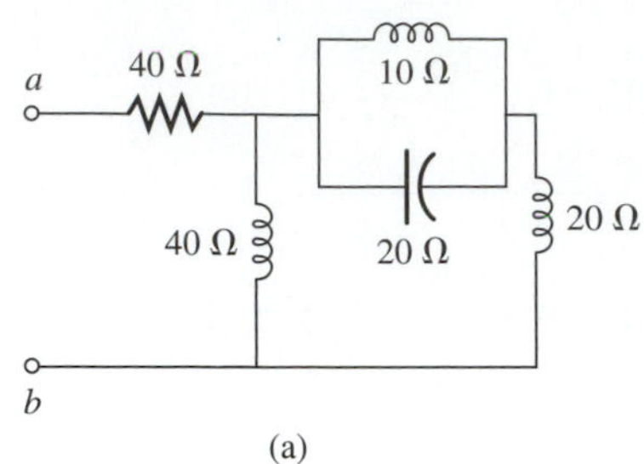

(a)

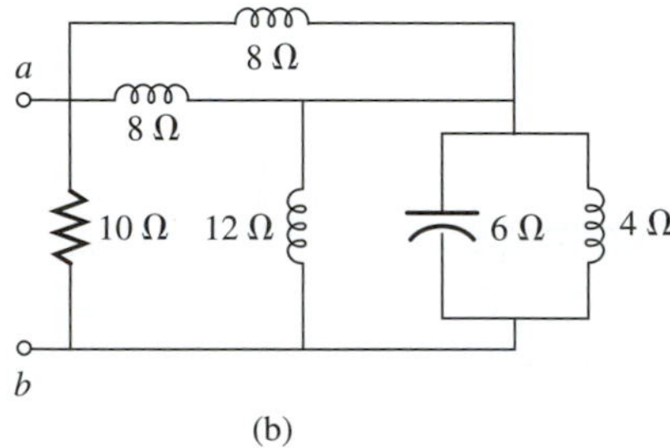

(b)

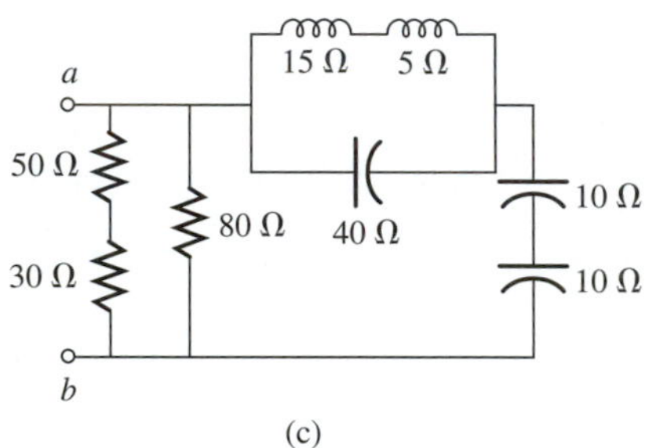

(c)

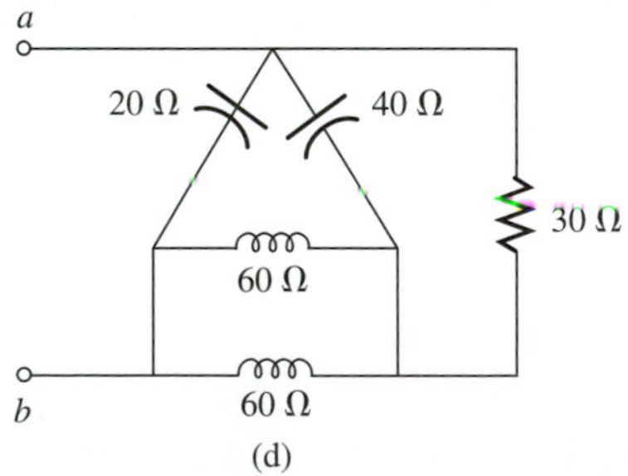

(d)

44. Consider the circuit in Figure 17.83.

Figure 17.83

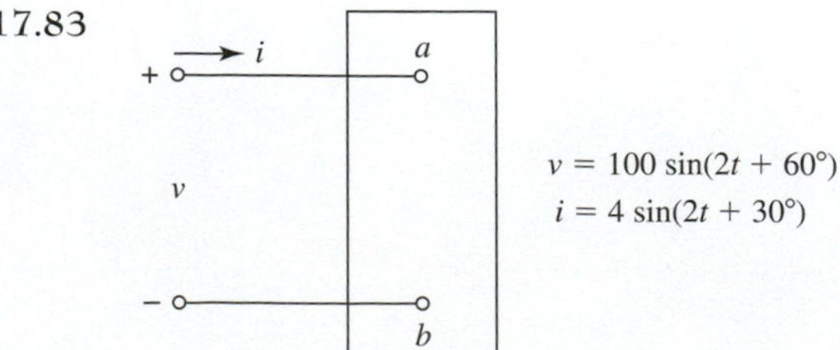

(a) Determine a simple series combination of electrical elements to connect across terminals *a*-*b* that satisfy the given current and voltage expressions. (Simple implies the smallest number of elements).
(b) Determine a simple parallel combination of electrical elements to connect across terminals *a*-*b* that satisfy the given current and voltage expressions. (Simple implies the smallest number of elements.)
(c) Determine $\mathbf{Z}_{ab}$.
(d) Determine $\mathbf{Y}_{ab}$.
(e) Determine R.
(f) Determine X.
(g) Determine G.
(h) Determine B.

45. Determine the impedance $\mathbf{Z}_{ab}$, the phasor current $\mathbf{I}_1$, and i_1 for each circuit in Figure 17.84 if $i = 2\cos(\omega t)$.

Figure 17.84

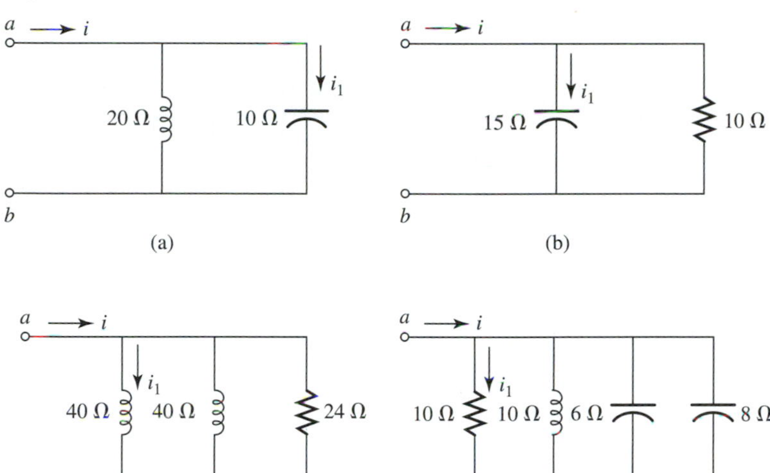

46. Determine the admittance $\mathbf{Y}_{ab}$, the phasor current $\mathbf{I}_1$, and i_1 for each circuit in Figure 17.85 if $i = 6\sin(\omega t + 45°)$.

Figure 17.85

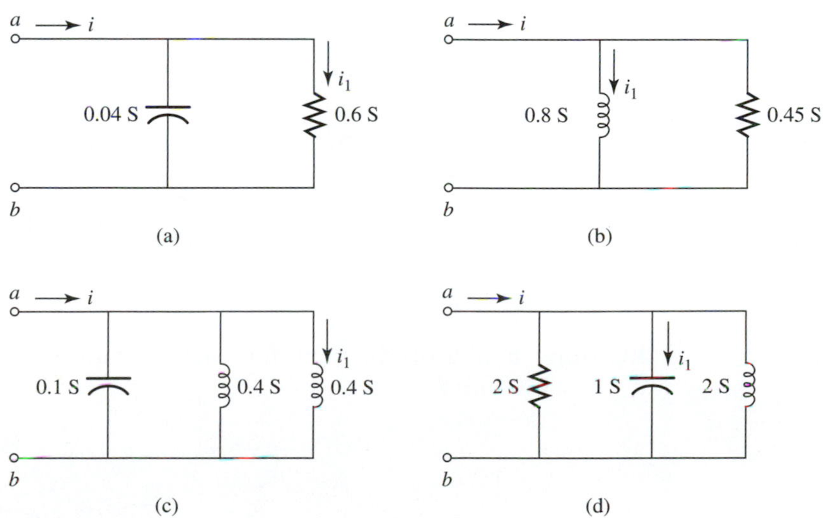

47. Determine the phasor voltages $\mathbf{V}_1$ and $\mathbf{V}_2$ in polar form and v_1 and v_2 for each circuit in Figure 17.86 if $v_{ab} = 120 \sin(\omega t)$.

Figure 17.86

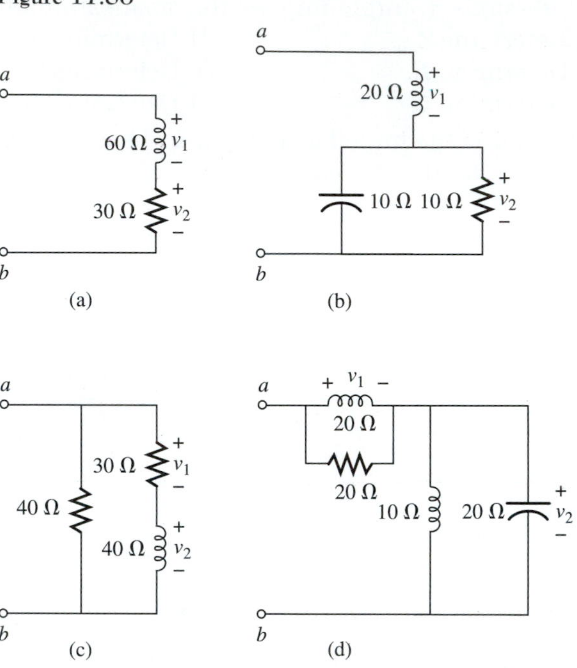

48. Determine phasor voltages $\mathbf{V}_1$, $\mathbf{V}_2$, $\mathbf{V}_3$ in polar form and v_1, v_2, and v_3 for each circuit in Figure 17.87 if $v_{ab} = 8 \cos(\omega t + 40°)$.

Figure 17.87

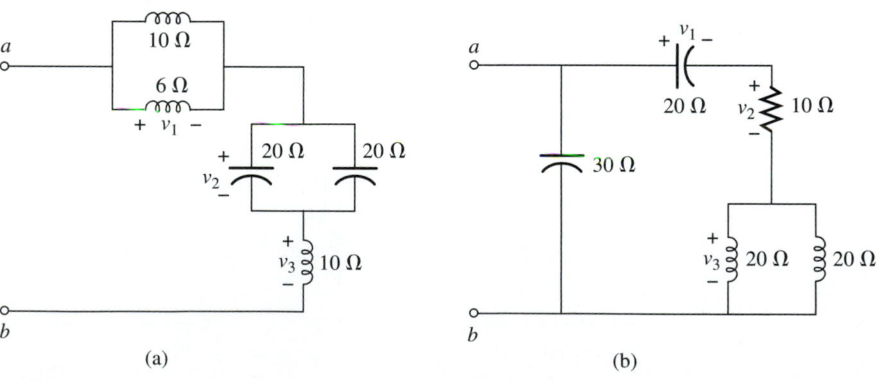

49. Draw a phasor diagram for each circuit in Problem 48 that includes $\mathbf{V}_{ab}$, $\mathbf{V}_1$, $\mathbf{V}_2$, and $\mathbf{V}_3$.

50. Determine the phasor voltages $\mathbf{V}_1$ and $\mathbf{V}_2$ in polar form and v_1 and v_2 for each circuit in Figure 17.88 if $v_{ab} = 50\sin(\omega t)$.

Figure 17.88

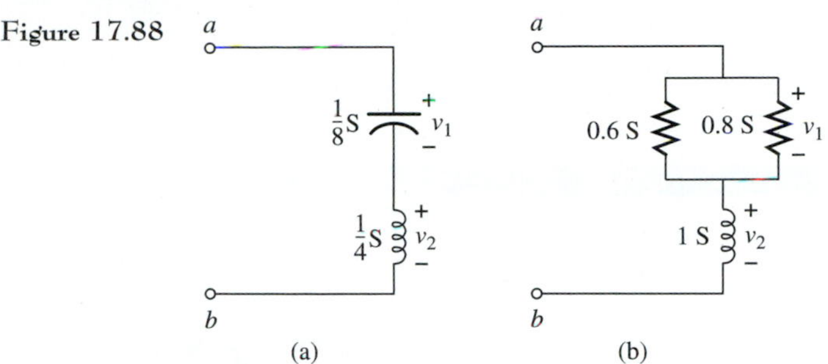

51. Convert the voltage source across terminals *a*-*b* to a current source for each circuit in Figure 17.89.

Figure 17.89

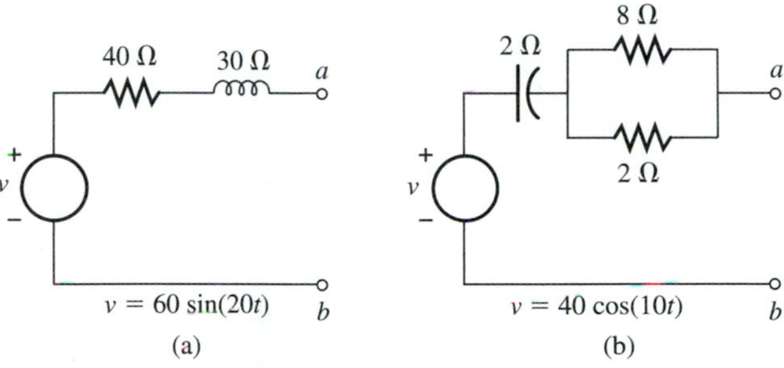

52. Convert the current source across terminals *a*-*b* to a voltage source for each circuit in Figure 17.90.

Figure 17.90

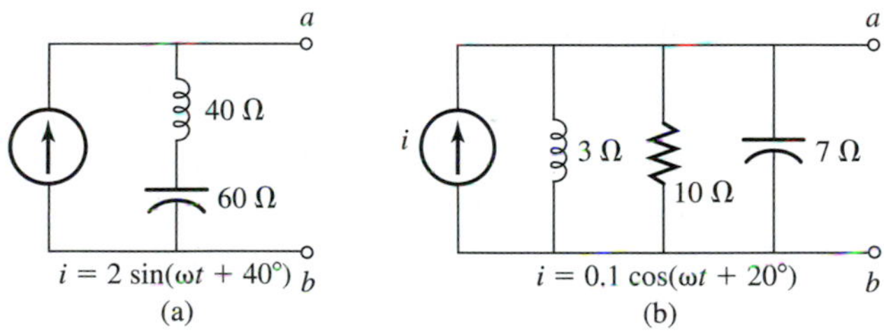

CHAPTER

18

Multiple-Loop AC Circuits

After completing this chapter the student should be able to

* use the mesh method of analysis to solve multiple-loop ac circuits
* use the nodal method of analysis to solve multiple-loop ac circuits

The methods for analyzing multiple-loop dc circuits presented in Chapter 6 are also applicable for the analysis of multiple-loop ac circuits. The only variation in the methods is the use of phasors and complex numbers for the analysis of ac circuits. We shall demonstrate the analysis of multiple-loop ac circuits using the procedures outlined in Chapter 6.

18.1 GENERAL APPROACH TO MESH ANALYSIS

Consider the ac circuit in phasor form given in Figure 18.1, where it is desired to analyze the circuit using the mesh analysis method. Following the procedure in Chapter 6, the *first step* in the mesh-analysis method is to identify the meshes, assign mesh currents, and assign voltage polarities to the passive circuit elements based on the assigned mesh currents. The resulting circuit is shown in Figure 18.2.

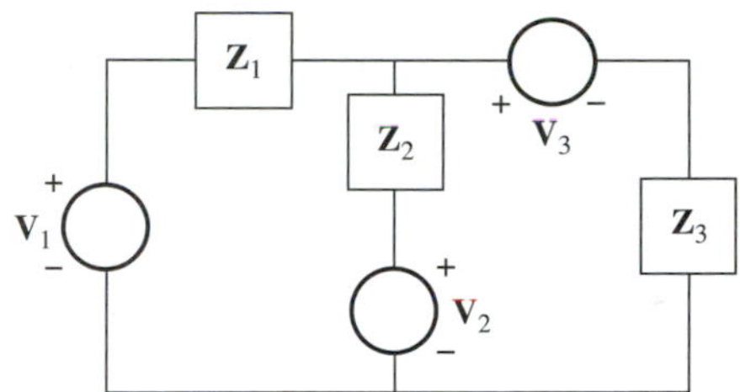

Figure 18.1 Mesh analysis.

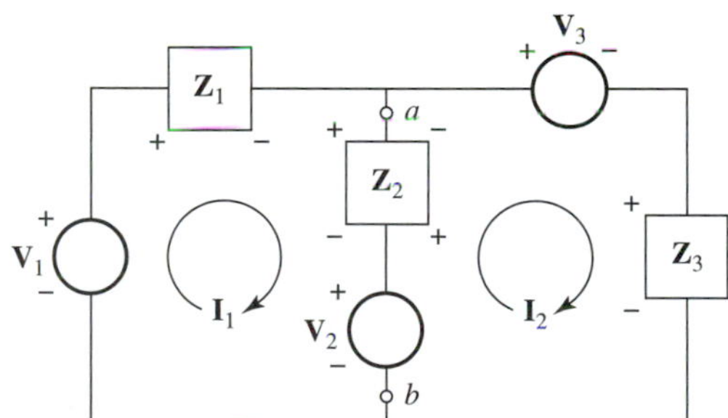

Figure 18.2 Mesh analysis.

Notice that element $\mathbf{Z}_2$ has two sets of voltage polarities, because two mesh currents flow through the element. Recall that the actual phasor current flowing through elements $\mathbf{V}_1$ and $\mathbf{Z}_1$ is mesh current $\mathbf{I}_1$ and the actual phasor current flowing through elements $\mathbf{V}_3$ and $\mathbf{Z}_3$ is mesh current $\mathbf{I}_2$. The actual phasor current flowing from point *a* to point *b* is equal to $(\mathbf{I}_1 - \mathbf{I}_2)$, and the actual phasor current flowing from point *b* to point *a* is equal to $(\mathbf{I}_2 - \mathbf{I}_1)$.

The *second step* in the mesh method is to apply Kirchhoff's voltage law to each mesh, which results in

$$\text{mesh 1:}\quad \mathbf{V}_1 - \mathbf{Z}_1\mathbf{I}_1 - \mathbf{Z}_2\mathbf{I}_1 + \mathbf{Z}_2\mathbf{I}_2 - \mathbf{V}_2 = 0$$

$$\text{mesh 2:}\quad \mathbf{V}_2 - \mathbf{Z}_2\mathbf{I}_2 + \mathbf{Z}_1\mathbf{I}_1 - \mathbf{V}_3 - \mathbf{Z}_3\mathbf{I}_2 = 0$$

These equations are arrived at by traversing the meshes in a clockwise direction and algebraically summing voltage rises and drops. Voltage rises (− to +) are assigned positive signs and voltage drops (+ to −) are assigned negative signs. Equation (17.1) is used to write the phasor voltages across the passive elements in terms of the mesh currents.

The *final step* in the mesh-analysis method is to solve the mesh equations that describe the circuit. These mesh equations can be written as

$$\begin{aligned} (\mathbf{Z}_1 + \mathbf{Z}_2)\mathbf{I}_1 - \qquad (\mathbf{Z}_2)\mathbf{I}_2 &= \mathbf{V}_1 - \mathbf{V}_2 \\ -(\mathbf{Z}_2)\mathbf{I}_1 - (\mathbf{Z}_2 + \mathbf{Z}_3)\mathbf{I}_2 &= \mathbf{V}_2 - \mathbf{V}_3 \end{aligned}$$

Employing Cramer's rule from Chapter 5 results in

$$\mathbf{I}_1 = \frac{\begin{vmatrix} (\mathbf{V}_1 - \mathbf{V}_2) & -(\mathbf{Z}_2) \\ (\mathbf{V}_2 - \mathbf{V}_3) & (\mathbf{Z}_2 + \mathbf{Z}_3) \end{vmatrix}}{\begin{vmatrix} (\mathbf{Z}_1 + \mathbf{Z}_2) & -(\mathbf{Z}_2) \\ (-\mathbf{Z}_2) & (\mathbf{Z}_2 + \mathbf{Z}_3) \end{vmatrix}}$$

and

$$\mathbf{I}_2 = \frac{\begin{vmatrix} (\mathbf{Z}_1 + \mathbf{Z}_2) & (\mathbf{V}_1 - \mathbf{V}_2) \\ -(\mathbf{Z}_2) & (\mathbf{V}_2 - \mathbf{V}_3) \end{vmatrix}}{\begin{vmatrix} (\mathbf{Z}_1 + \mathbf{Z}_2) & -(\mathbf{Z}_2) \\ -(\mathbf{Z}_2) & (\mathbf{Z}_2 + \mathbf{Z}_3) \end{vmatrix}}$$

Methods for evaluating 2×2 and higher-order determinants are described in Chapter 5.

Consider the following examples.

EXAMPLE 18.1 Determine $\mathbf{I}_{ab}$, $\mathbf{I}_{dc}$, $\mathbf{I}_{bd}$, and $\mathbf{I}_{db}$ for the circuit in Figure 18.3. ■

Figure 18.3

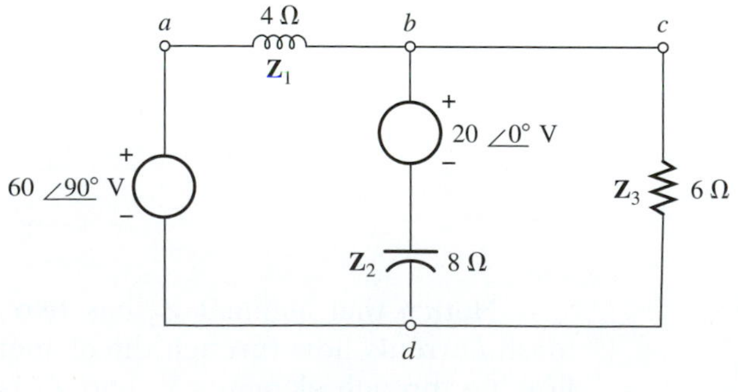

SOLUTION The circuit with assigned mesh currents and associated voltage polarities for the passive elements is shown in Figure 18.4. From Equations (17.8), (17.12), and (17.3),

$$\begin{aligned} \mathbf{Z}_1 &= j4 \\ \mathbf{Z}_2 &= -j8 \\ \mathbf{Z}_3 &= 6 \end{aligned}$$

Figure 18.4

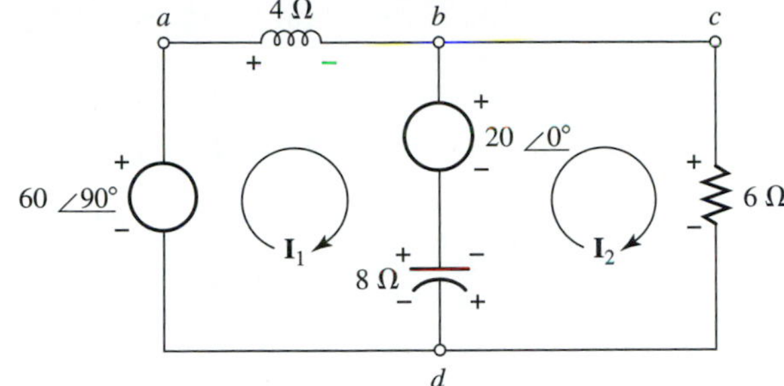

Applying Kirchhoff's voltage law to meshes 1 and 2 and using Equation (17.1) to write the voltages across the passive elements in terms of the mesh currents results in

$$\text{mesh 1:} \quad 6\underline{/90^\circ} - (j4)\mathbf{I}_1 - 20\underline{/0^\circ} - (-j8)\mathbf{I}_1 + (-j8)\mathbf{I}_2 = 0$$

$$\text{mesh 2:} \quad -(-j8)\mathbf{I}_2 + (-j8)\mathbf{I}_1 + 20\underline{/0^\circ} - (6)\mathbf{I}_2 = 0$$

The mesh equations can be written as

$$(j4 - j8)\mathbf{I}_1 + (j8)\mathbf{I}_2 = j60 - 20$$

$$(j8)\mathbf{I}_1 + (6 - j8)\mathbf{I}_2 = 20$$

From Cramer's rule, the rule for evaluating 2 × 2 determinants (Figure 5.2), and the rules for complex numbers in Chapter 16,

$$\mathbf{I}_1 = \frac{\begin{vmatrix} -20 + j60 & j8 \\ 20 & 6 - j8 \end{vmatrix}}{\begin{vmatrix} -j4 & j8 \\ j8 & 6 - j8 \end{vmatrix}}$$

$$= \frac{(-20 + j60)(6 - j8) - (20)(j8)}{(-j4)(6 - j8) - (j8)(j8)}$$

$$= \frac{360 + j360}{32 - j24} = \frac{509.1\underline{/45^\circ}}{40.6\underline{/-36.9^\circ}}$$

$$= 12.7\underline{/81.9^\circ}$$

$$\mathbf{I}_2 = \frac{\begin{vmatrix} -j4 & -20 + j60 \\ j8 & 20 \end{vmatrix}}{\begin{vmatrix} -j4 & j8 \\ j8 & 6 - j8 \end{vmatrix}}$$

$$= \frac{(-j4)(20) - (-20 + j60)(j8)}{(-j4)(6 - j8) - (j8)(j8)}$$

$$= \frac{480 + j80}{32 - j24} = \frac{486.6\underline{/9.5^\circ}}{40.0\underline{/-36.9^\circ}}$$

$$= 12.2\underline{/46.4^\circ}$$

$$\mathbf{I}_{ab} = \mathbf{I}_1 = 12.7\underline{/81.9^\circ}$$

$$\mathbf{I}_{dc} = -\mathbf{I}_2 = -12.2\underline{/46.4^\circ}$$
$$= 12.2\underline{/226.3^\circ}$$
$$\mathbf{I}_{bd} = \mathbf{I}_1 - \mathbf{I}_2$$
$$= 12.7\underline{/81.9^\circ} - 12.2\underline{/46.4^\circ}$$
$$= 1.8 + j12.6 - \{8.4 + j8.8\}$$
$$= -6.6 + j3.8$$
$$= 7.6\underline{/150.5^\circ}$$
$$\mathbf{I}_{db} = \mathbf{I}_2 - \mathbf{I}_1 = -\mathbf{I}_{bd}$$
$$= -7.6\underline{/150.5^\circ}$$
$$= 7.6\underline{/-29.5^\circ}$$

EXAMPLE 18.2 Determine i_a, i_b, and i_c for the circuit in Figure 18.5. ■

Figure 18.5

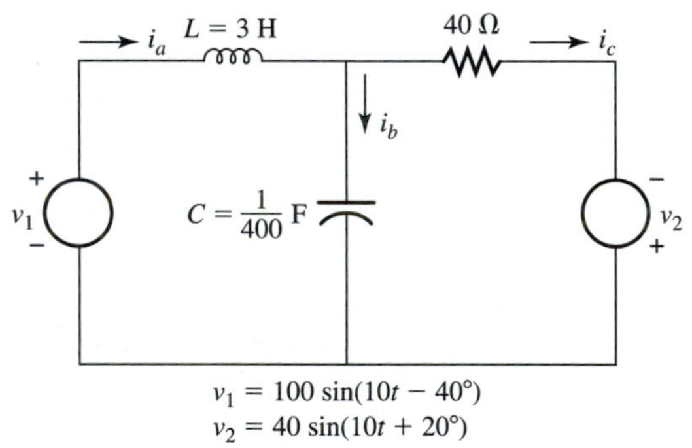

SOLUTION The circuit with assigned mesh currents and associated voltage polarities for the passive elements is shown in Figure 18.6 in phasor form. From an inspection of the circuit in Figure 18.5, phasors for v_1 and v_2 are

$$\mathbf{V}_1 = 100\underline{/-40^\circ}$$
$$\mathbf{V}_2 = 40\underline{/20^\circ}$$

From Equations (17.4), (17.7), (17.8), (17.11), and (17.12),

$$\mathbf{Z}_R = 40\ \Omega$$
$$X_L = \omega L = (10)(3)$$
$$= 30\ \Omega$$
$$\mathbf{Z}_L = j30$$
$$X_C = \frac{1}{\omega C} = \frac{1}{(10)\left(\frac{1}{40}\right)}$$
$$= 40\ \Omega$$
$$\mathbf{Z}_C = -j40$$

Figure 18.6

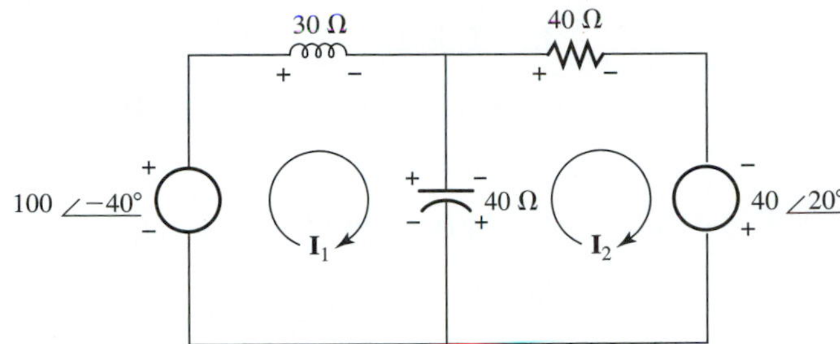

Using Equation (17.1) to write the voltages across the passive elements in terms of the mesh currents and applying Kirchhoff's voltage law to each mesh results in

$$\text{mesh 1:}\quad 100\angle -40^\circ - (j30)\mathbf{I}_1 - (-j40)\mathbf{I}_1 + (-j40)\mathbf{I}_2 = 0$$

$$\text{mesh 2:}\quad (-j40)\mathbf{I}_1 - (-j40)\mathbf{I}_2 - 40\mathbf{I}_2 + 40\angle 20^\circ = 0$$

The mesh equations can be written as

$$(-j10)\mathbf{I}_1 + (j40)\mathbf{I}_2 = 100\angle -40^\circ$$

$$(j40)\mathbf{I}_1 + (40 - j40)\mathbf{I}_2 = 40\angle 20^\circ$$

From Cramer's rule, the rule for evaluating 2 × 2 determinants (Figure 5.2), and the rules for complex numbers in Chapter 16,

$$\mathbf{I}_1 = \frac{\begin{vmatrix} 100\angle -40^\circ & j40 \\ 40\angle 20^\circ & 40 - j40 \end{vmatrix}}{\begin{vmatrix} -j10 & j40 \\ j40 & 40 - j40 \end{vmatrix}}$$

$$= \frac{(100\angle -40^\circ)(40 - j40) - (40\angle 20^\circ)(j40)}{(-j10)(40 - j40) - (j40)(j40)}$$

$$= \frac{1040.3 - j7138.8}{1200.0 - j400.0} = \frac{7214.2\angle -81.7^\circ}{1264.9\angle -18.4^\circ}$$

$$= 5.7\angle -63.3^\circ$$

$$\mathbf{I}_2 = \frac{\begin{vmatrix} -j10 & 100\angle -40^\circ \\ j40 & 40\angle 20^\circ \end{vmatrix}}{\begin{vmatrix} -j10 & j40 \\ j40 & 40 - j40 \end{vmatrix}}$$

$$= \frac{(-j10)(40\angle 20^\circ) - (j40)(100\angle -40^\circ)}{(-j10)(40 - j40) - (j40)(j40)}$$

$$= \frac{-2371.2 - j3464.2}{1200 - j400} = \frac{4198.0\angle -124.4^\circ}{1264.9\angle -18.4^\circ}$$

$$= 3.3\angle -106.8^\circ$$

$$\mathbf{I}_a = \mathbf{I}_1 = 5.7\underline{/-63.3^\circ}$$

$$\begin{aligned}\mathbf{I}_b &= \mathbf{I}_1 - \mathbf{I}_2 \\ &= 5.7\underline{/-63.3^\circ} - 3.3\underline{/-106.8^\circ} \\ &= 3.5 - j1.9 \\ &= 4.0\underline{/-28.8^\circ}\end{aligned}$$

$$\mathbf{I}_C = \mathbf{I}_2 = 3.3\underline{/-106.8^\circ}$$

From the phasor-current expressions, the time-domain currents are

$$i_a = 5.7\sin(10t - 63.3^\circ)$$

$$i_b = 4.0\sin(10t - 28.8^\circ)$$

$$i_C = 3.3\sin(10t - 106.8^\circ)$$

18.2 UNIFORM APPROACH TO MESH ANALYSIS

We shall now demonstrate the uniform approach to the mesh-analysis method for analyzing ac circuits. Recall from Chapter 6 that the uniform approach to mesh analysis allows us to write the mesh currents in determinant form by a simple inspection of the circuit.

Consider the circuit in Figure 18.1. The circuit with the mesh currents assigned is redrawn for convenience in Figure 18.7. Recall that the method presented in Chapter 6 requires that all assigned mesh currents be clockwise.

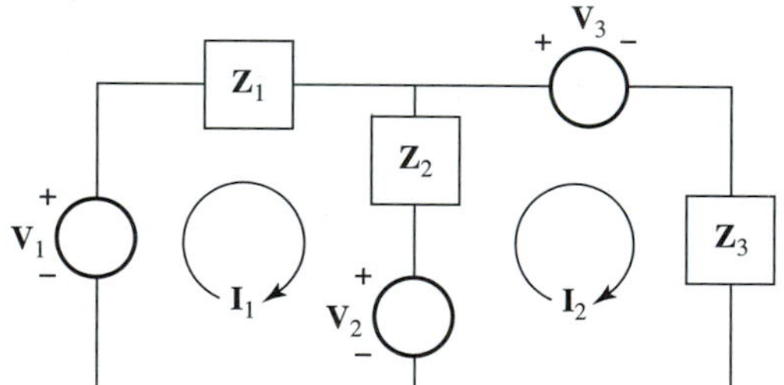

Figure 18.7 Mesh analysis.

Because there are two meshes, the numerator and denominator determinants for $\mathbf{I}_1$ and $\mathbf{I}_2$ consist of two rows and two columns. The denominator determinant is

$$\begin{vmatrix} \mathbf{Z}_{11} & -\mathbf{Z}_{12} \\ -\mathbf{Z}_{21} & \mathbf{Z}_{22} \end{vmatrix}$$

where

$$\begin{aligned}\mathbf{Z}_{11} &= \text{total impedance in mesh 1} \\ \mathbf{Z}_{12} &= \text{impedance common to meshes 1 and 2} \\ \mathbf{Z}_{21} &= \text{impedance common to meshes 2 and 1} \\ \mathbf{Z}_{22} &= \text{total impedance in mesh 2}\end{aligned}$$

The denominator determinant for the circuit in Figure 18.7 can be written as

$$\begin{vmatrix} \mathbf{Z}_1 + \mathbf{Z}_2 & -\mathbf{Z}_2 \\ -\mathbf{Z}_2 & \mathbf{Z}_2 + \mathbf{Z}_3 \end{vmatrix}$$

The voltage summation column is

$$\begin{pmatrix} \mathbf{V}_a \\ \mathbf{V}_b \end{pmatrix}$$

where

$\mathbf{V}_a$ = algebraic sum of the phasor voltage sources in mesh 1 obtained by proceeding around mesh 1 in the direction of phasor mesh current $\mathbf{I}_1$

$\mathbf{V}_b$ = algebraic sum of the phasor voltage sources in mesh 2 obtained by proceeding around mesh 2 in the direction of phasor mesh current $\mathbf{I}_2$

The voltage summation column for the circuit in Figure 18.7 is

$$\begin{pmatrix} \mathbf{V}_1 - \mathbf{V}_2 \\ \mathbf{V}_2 - \mathbf{V}_3 \end{pmatrix}$$

The phasor currents $\mathbf{I}_1$ and $\mathbf{I}_2$ are then

$$\mathbf{I}_1 = \frac{\begin{vmatrix} \mathbf{V}_1 - \mathbf{V}_2 & -\mathbf{Z}_2 \\ \mathbf{V}_2 - \mathbf{V}_3 & \mathbf{Z}_2 + \mathbf{Z}_3 \end{vmatrix}}{\begin{vmatrix} \mathbf{Z}_1 + \mathbf{Z}_2 & -\mathbf{Z}_2 \\ -\mathbf{Z}_2 & \mathbf{Z}_2 + \mathbf{Z}_3 \end{vmatrix}}$$

$$\mathbf{I}_2 = \frac{\begin{vmatrix} \mathbf{Z}_1 + \mathbf{Z}_2 & \mathbf{V}_1 - \mathbf{V}_2 \\ -\mathbf{Z}_2 & \mathbf{V}_2 - \mathbf{V}_3 \end{vmatrix}}{\begin{vmatrix} \mathbf{Z}_1 + \mathbf{Z}_2 & -\mathbf{Z}_2 \\ -\mathbf{Z}_2 & \mathbf{Z}_2 + \mathbf{Z}_3 \end{vmatrix}}$$

Notice that the voltage summation column replaces the first and second columns of the denominator determinant to form the numerator determinants for $\mathbf{I}_1$ and $\mathbf{I}_2$, respectively.

Consider the following example.

EXAMPLE 18.3 Determine mesh currents $\mathbf{I}_1$ and $\mathbf{I}_2$ for the circuit in Figure 18.8. ■

Figure 18.8

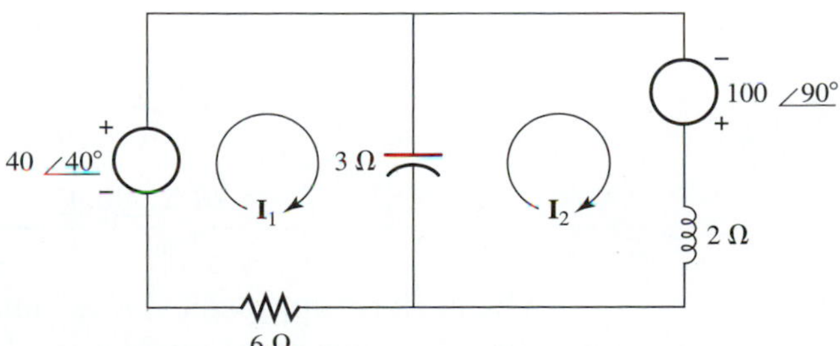

SOLUTION Using the uniform approach to mesh analysis, the entries for the denominator determinant are

$$\mathbf{Z}_{11} = 6 - j3$$
$$\mathbf{Z}_{22} = -j3 + j2 = -j1$$
$$\mathbf{Z}_{12} = -j3$$
$$\mathbf{Z}_{21} = -j3$$

and the denominator determinant is

$$\begin{vmatrix} 6 - j3 & -(-j3) \\ -(-j3) & -j1 \end{vmatrix}$$

The voltage summation column is

$$\begin{pmatrix} 40\angle 40^\circ \\ 100\angle 90^\circ \end{pmatrix}$$

The phasor currents are

$$\mathbf{I}_1 = \frac{\begin{vmatrix} 40\angle 40^\circ & j3 \\ 100\angle 90^\circ & -j1 \end{vmatrix}}{\begin{vmatrix} 6 - j3 & j3 \\ j3 & -j1 \end{vmatrix}}$$
$$= \frac{(40\angle 40^\circ)(-j1) - (100\angle 90^\circ)(j3)}{(6 - j3)(-j1) - (j3)(j3)}$$
$$= \frac{325.7 - j30.6}{6 - j6}$$
$$= 38.5\angle 39.6^\circ$$

and

$$\mathbf{I}_2 = \frac{\begin{vmatrix} 6 - j3 & 40\angle 40^\circ \\ j3 & 100\angle 90^\circ \end{vmatrix}}{\begin{vmatrix} 6 - j3 & j3 \\ j3 & -j1 \end{vmatrix}}$$
$$= \frac{(6 - j3)(100\angle 90^\circ) - (j3)(40\angle 40^\circ)}{(6 - j3)(-j1) - (j3)(j3)}$$
$$= \frac{377.1 + j508.1}{6 - j6}$$
$$= 74.5\angle 98.4^\circ$$

The uniform approach to mesh analysis is easily extended to circuits with more than two meshes. Consider the following example.

EXAMPLE 18.4 Determine i for the circuit in Figure 18.9.

Figure 18.9

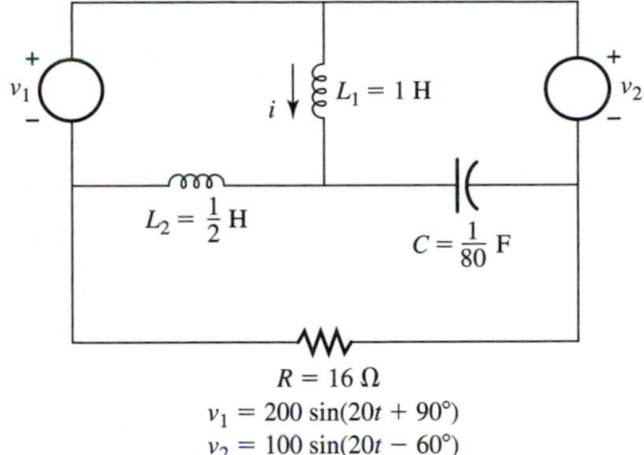

SOLUTION From an inspection of the circuit, the phasors for v_1 and v_2 are

$$\mathbf{V}_1 = 200\angle 90°$$

$$\mathbf{V}_2 = 100\angle -60°$$

From Equations (17.4), (17.7), (17.8), (17.11), and (17.12),

$$\mathbf{Z}_R = 16\ \Omega$$

$$X_{L_1} = \omega L_1 = 20(1)$$

$$= 20\ \Omega$$

$$\mathbf{Z}_{L_1} = j20\ \Omega$$

$$X_{L_2} = \omega L_2 = 20\left(\frac{1}{2}\right)$$

$$= 10\ \Omega$$

$$\mathbf{Z}_{L_2} = j10\ \Omega$$

$$X_c = \frac{1}{\omega C} = \frac{1}{(20)\left(\frac{1}{80}\right)}$$

$$= 4\ \Omega$$

$$\mathbf{Z}_c = -j4\ \Omega$$

The circuit is redrawn in phasor form in Figure 18.10.

Figure 18.10

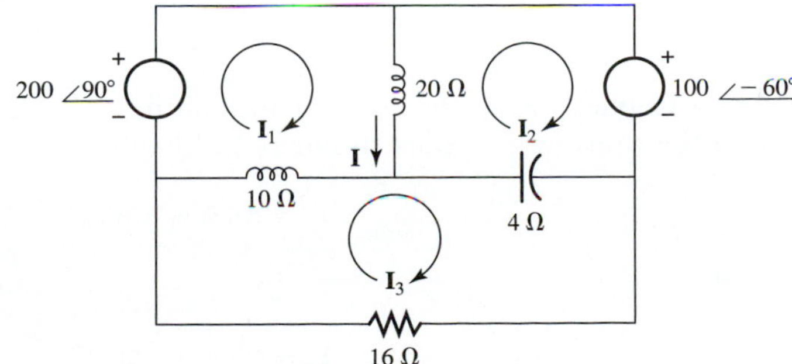

Using the uniform approach to mesh analysis, the entries for the denominator determinant are

$$\mathbf{Z}_{11} = j20 + j10$$
$$= j30$$
$$\mathbf{Z}_{22} = j20 - j4$$
$$= j16$$
$$\mathbf{Z}_{33} = j10 - j4 + 16$$
$$= 16 + j6$$
$$\mathbf{Z}_{12} = \mathbf{Z}_{21} = j20$$
$$\mathbf{Z}_{13} = \mathbf{Z}_{31} = j10$$
$$\mathbf{Z}_{23} = \mathbf{Z}_{32} = -j4$$

and the denominator determinant is

$$\begin{vmatrix} j30 & -j20 & -j10 \\ -j20 & j16 & j4 \\ -j10 & j4 & 16 + j6 \end{vmatrix}$$

The voltage summation column is

$$\begin{pmatrix} 200\angle 90^\circ \\ -100\angle -60^\circ \\ 0 \end{pmatrix}$$

The phasor currents in determinant form are

$$\mathbf{I}_1 = \frac{\begin{vmatrix} 200\angle 90^\circ & -j20 & -j10 \\ -100\angle -60^\circ & j16 & j4 \\ 0 & j4 & 16 + j6 \end{vmatrix}}{\begin{vmatrix} j30 & -j20 & -j10 \\ -j20 & j16 & j4 \\ -j10 & j4 & 16 + j6 \end{vmatrix}}$$

$$\mathbf{I}_2 = \frac{\begin{vmatrix} j30 & 200\angle 90^\circ & -j10 \\ -j20 & -100\angle -60^\circ & j4 \\ -j10 & 0 & 10 + j6 \end{vmatrix}}{\begin{vmatrix} j30 & -j20 & -j10 \\ -j20 & j16 & j4 \\ -j10 & j4 & 16 + j6 \end{vmatrix}}$$

From Cramer's rule, the rule for evaluating 3×3 determinants (Figure 5.3), and the rules for complex numbers in Chapter 16,

$$\mathbf{I}_1 = 58.9 + j30.4$$

and

$$\mathbf{I}_2 = 79.4 + j36.7$$

From the phasor form of the circuit given previously,

$$\begin{aligned} \mathbf{I} &= \mathbf{I}_1 - \mathbf{I}_2 \\ &= 58.9 + j30.4 - 79.4 - j36.7 \\ &= -20.5 - j6.3 \\ &= 21.9\underline{/-163.4^\circ} \end{aligned}$$

and the current generated by phasor **I** is

$$i = 21.9 \sin(20t - 163.4^\circ)$$

18.3 GENERAL APPROACH TO NODAL ANALYSIS

We demonstrate the nodal-analysis method for ac circuits by analyzing the simple three-node ac circuit in Figure 18.11. The procedure for the nodal analysis of ac circuits parallels the procedure for dc circuits presented in Chapter 6.

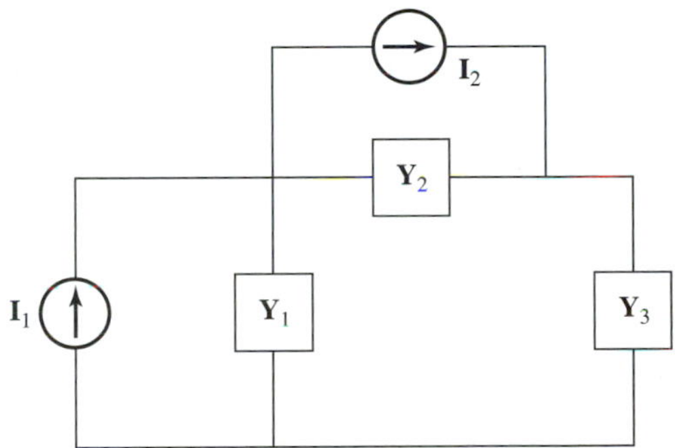

Figure 18.11 Nodal analysis.

The *first step* in the nodal-analysis method is to identify the nodes of the circuit, choose a reference node, and label the remaining nodes. The resulting circuit is shown in Figure 18.12, where $\mathbf{V}_a$ and $\mathbf{V}_b$ are the phasor nodal voltages of nodes a and b, respectively.

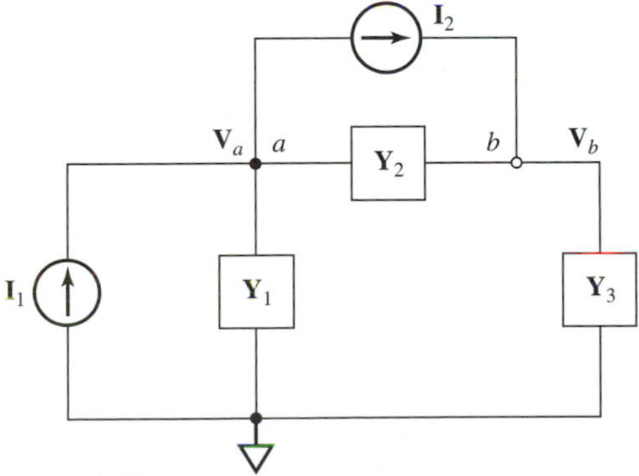

Figure 18.12 Nodal analysis.

Recall that all nodal voltages are with respect to the reference node. Recall also from Equation (6.7) that $v_{ab} = v_a - v_b$. It follows that the phasor voltage of point a with respect to point b and the phasor voltage of point b with respect to point a are $(\mathbf{V}_a - \mathbf{V}_b)$ and $(\mathbf{V}_b - \mathbf{V}_a)$, respectively. In addition, recall from Equation (6.9) that $i_{ab} = (v_a - v_b)G$. It follows that the phasor current flowing from a to b through $\mathbf{V}_2$ can be expressed as $(\mathbf{V}_a - \mathbf{V}_b)\mathbf{Y}_2$, and the phasor current flowing from b to a can be expressed as $(\mathbf{V}_b - \mathbf{V}_a)\mathbf{Y}_2$ for the circuit in Figure 18.12.

The *second step* is to apply Kirchhoff's current law to each labeled node, which results in

$$\begin{aligned} \text{node } a&: \quad \mathbf{I}_1 - \mathbf{Y}_1\mathbf{V}_a - \mathbf{I}_2 - (\mathbf{Y}_2)(\mathbf{V}_a - \mathbf{V}_b) = 0 \\ \text{node } b&: \quad \mathbf{I}_2 - (\mathbf{Y}_2)(\mathbf{V}_b - \mathbf{V}_a) - (\mathbf{Y}_3)(\mathbf{V}_b) = 0 \end{aligned}$$

These nodal equations are arrived at by calling the phasor currents entering the node positive and the phasor currents leaving the node negative.

The *final step* in the nodal-analysis method is to solve the nodal equations that describe the circuit. The preceding nodal equations can be written as

$$\begin{aligned} (\mathbf{Y}_1 + \mathbf{Y}_2)\mathbf{V}_a - (\mathbf{Y}_2)\mathbf{V}_b &= \mathbf{I}_1 - \mathbf{I}_2 \\ (-\mathbf{Y}_2)\mathbf{V}_a + (\mathbf{Y}_2 + \mathbf{Y}_3)\mathbf{V}_b &= \mathbf{I}_2 \end{aligned}$$

and employing Cramer's rule from Chapter 5 results in

$$\mathbf{V}_a = \frac{\begin{vmatrix} (\mathbf{I}_1 - \mathbf{I}_2) & -(\mathbf{Y}_2) \\ (\mathbf{I}_2) & (\mathbf{Y}_2 + \mathbf{Y}_3) \end{vmatrix}}{\begin{vmatrix} (\mathbf{Y}_1 + \mathbf{Y}_2) & -(\mathbf{Y}_2) \\ -(\mathbf{Y}_2) & (\mathbf{Y}_2 + \mathbf{Y}_3) \end{vmatrix}}$$

$$\mathbf{V}_b = \frac{\begin{vmatrix} (\mathbf{Y}_1 + \mathbf{Y}_2) & (\mathbf{I}_1 - \mathbf{I}_2) \\ -(\mathbf{Y}_2) & (\mathbf{I}_2) \end{vmatrix}}{\begin{vmatrix} (\mathbf{Y}_1 + \mathbf{Y}_2) & -(\mathbf{Y}_2) \\ -(\mathbf{Y}_2) & (\mathbf{Y}_2 + \mathbf{Y}_3) \end{vmatrix}}$$

The evaluation of 2×2 and higher-order determinants is described in Chapter 5.

Consider the following examples.

EXAMPLE 18.5 Determine $\mathbf{V}_a$, $\mathbf{V}_b$, $\mathbf{I}_1$, $\mathbf{I}_2$, and $\mathbf{I}_3$ for the circuit in phasor form in Figure 18.13. Use the reference node indicated. ■

Figure 18.13

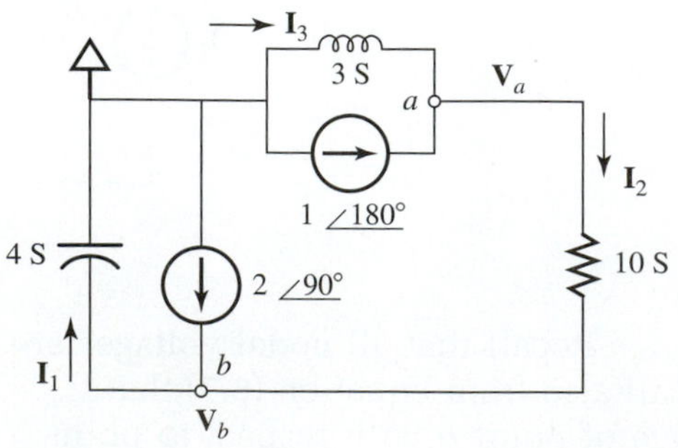

SOLUTION Applying Kirchhoff's current law to nodes a and b results in

$$\text{node } a: \quad 1\angle 180^\circ - (-j3)\mathbf{V}_a - (10)(\mathbf{V}_a - \mathbf{V}_b) = 0$$

$$\text{node } b: \quad 2\angle 90^\circ - (j4)\mathbf{V}_b - 10(\mathbf{V}_b - \mathbf{V}_a) = 0$$

The nodal equations can be written as

$$(10 - j3)\mathbf{V}_a - (10)\mathbf{V}_b = 1\angle 180^\circ$$

$$-(10)\mathbf{V}_a + (10 + j4)\mathbf{V}_b = 2\angle 90^\circ$$

From Cramer's rule, the rule for evaluating 2×2 determinants in Chapter 5, and the rules for complex numbers in Chapter 16,

$$\mathbf{V}_a = \frac{\begin{vmatrix} 1\angle 180^\circ & -10 \\ 2\angle 90^\circ & 10 + j4 \end{vmatrix}}{\begin{vmatrix} 10 - j3 & -10 \\ -10 & 10 + j4 \end{vmatrix}}$$

$$= \frac{(1\angle 180^\circ)(10 + j4) - (2\angle 90^\circ)(-10)}{(10 - j3)(10 + j4) - (-10)(-10)}$$

$$= \frac{-10.0 + j16.0}{12.0 + j10.0}$$

$$= \frac{18.9\angle 122.9^\circ}{15.6\angle 39.8^\circ}$$

$$= 0.2 + j1.2$$

$$= 1.2\angle 82.2^\circ$$

$$\mathbf{V}_b = \frac{\begin{vmatrix} 10 - j3 & 1\angle 180^\circ \\ -10 & 2\angle 90^\circ \end{vmatrix}}{\begin{vmatrix} 10 - j3 & -10 \\ -10 & 10 + j4 \end{vmatrix}}$$

$$= \frac{(10 - j3)(2\angle 90^\circ) - (-10)(1\angle 180^\circ)}{(10 - j3)(10 + j4) - (-10)(-10)}$$

$$= \frac{-4.0 + j20.0}{12.0 + j10.0}$$

$$= \frac{20.4\angle 101.3^\circ}{15.6\angle 39.8^\circ}$$

$$= 0.6 + j1.1$$

$$= 1.3\angle 61.4^\circ$$

$$\mathbf{I}_1 = (j4)(\mathbf{V}_b)$$

$$= (j4)(1.3\angle 61.4^\circ)$$

$$= -4.6 + j2.5$$

$$= 5.2\angle 151.4^\circ$$

$$\begin{aligned}\mathbf{I}_2 &= (10)(\mathbf{V}_a - \mathbf{V}_b)\\ &= (10)\{1.2\underline{/82.2^\circ} - 1.3\underline{/61.4^\circ}\}\\ &= 4.6\underline{/174.1^\circ}\\ \mathbf{I}_3 &= (-j3)(-\mathbf{V}_a)\\ &= (-j3)(-1.2\underline{/82.2^\circ})\\ &= 3.6\underline{/172.2^\circ}\end{aligned}$$

EXAMPLE 18.6 Determine the nodal voltages v_a and v_b and the current i for the circuit in Figure 18.14. Use the reference node indicated. ■

Figure 18.14

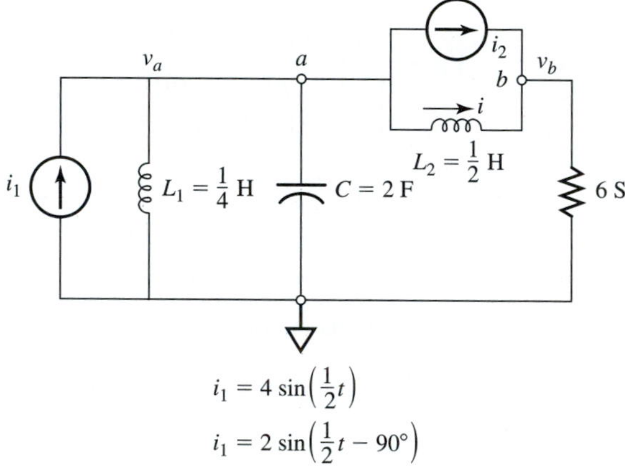

$i_1 = 4\sin\left(\frac{1}{2}t\right)$

$i_1 = 2\sin\left(\frac{1}{2}t - 90^\circ\right)$

SOLUTION From Equations (17.17), (17.19), and (17.23),

$$\mathbf{Y}_R = 6\underline{/0^\circ}$$

$$\begin{aligned}\mathbf{Y}_{L_1} &= -j\left(\frac{1}{\omega L_1}\right) = -j\left\{\frac{1}{\left(\frac{1}{2}\right)\left(\frac{1}{4}\right)}\right\}\\ &= -j8\end{aligned}$$

$$\begin{aligned}\mathbf{Y}_{L_2} &= -j\left(\frac{1}{\omega L_2}\right) = -j\left\{\frac{1}{\left(\frac{1}{2}\right)\left(\frac{1}{2}\right)}\right\}\\ &= -j4\end{aligned}$$

$$\begin{aligned}\mathbf{Y}_C &= j\omega C = j\left(\frac{1}{2}\right)(2)\\ &= j1\end{aligned}$$

From an inspection of the current equations, the phasors generating i_1 and i_2 are

$$\begin{aligned}\mathbf{I}_1 &= 4\underline{/0^\circ}\\ \mathbf{I}_2 &= 2\underline{/-90^\circ}\end{aligned}$$

The circuit in phasor form is shown in Figure 18.15.

Figure 18.15

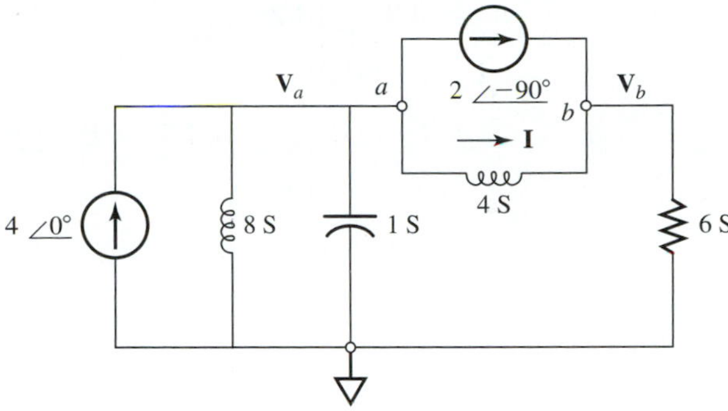

Applying Kirchhoff's current law to nodes a and b results in

$$\text{node } a\text{:} \quad 4\angle 0^\circ - (-j8)\mathbf{V}_a - (j1)\mathbf{V}_a - 2\angle -90^\circ - (-j4)(\mathbf{V}_a - \mathbf{V}_b) = 0$$

$$\text{node } b\text{:} \quad 2\angle -90^\circ - (-j4)(\mathbf{V}_b - \mathbf{V}_a) - 6\mathbf{V}_b = 0$$

The nodal equations can be written as

$$(-j11)\mathbf{V}_a + (j4)\mathbf{V}_b = 4\angle 0^\circ - 2\angle -90^\circ$$
$$(j4)\mathbf{V}_a + (6 - j4)\mathbf{V}_b = 2\angle -90^\circ$$

From Cramer's rule, the rule for evaluating 2×2 determinants in Chapter 5, and the rules for complex numbers in Chapter 16,

$$\mathbf{V}_a = \frac{\begin{vmatrix} (4\angle 0^\circ - 2\angle -90^\circ) & j4 \\ 2\angle -90^\circ & 6 - j4 \end{vmatrix}}{\begin{vmatrix} -j11 & j4 \\ j4 & 6 - j4 \end{vmatrix}}$$

$$= \frac{(4\angle 0^\circ - 2\angle -90^\circ)(6 - j4) - (2\angle -90^\circ)(j4)}{(-j11)(6 - j4) - (j4)(j4)}$$

$$= \frac{24.0 - j4.0}{-28.0 - j66.0}$$

$$= \frac{24.3\angle -9.5^\circ}{71.7\angle -113.0^\circ}$$

$$= 0.34\angle 103.5^\circ$$

$$\mathbf{V}_b = \frac{\begin{vmatrix} -j11 & (4\angle 0^\circ - 2\angle -90^\circ) \\ j4 & 2\angle -90^\circ \end{vmatrix}}{\begin{vmatrix} -j11 & j4 \\ j4 & 6 - j4 \end{vmatrix}}$$

$$= \frac{(-j11)(2\angle -90^\circ) - (j4)(4\angle 0^\circ - 2\angle -90^\circ)}{(-j11)(6 - j4) - (j4)(j4)}$$

$$= \frac{-14.0 - j16}{-28.0 - j66.0}$$

$$= \frac{21.3\angle -131.2^\circ}{71.7\angle -113.0^\circ}$$

$$= 0.3\angle -18.2^\circ$$

$$\mathbf{I} = (-j4)(\mathbf{V}_a - \mathbf{V}_b)$$

$$= (-j4)(0.34\angle 103.5^\circ - 0.3\angle -18.2^\circ)$$

$$= 1.7 + j1.5$$

$$= 2.2\angle 40.7^\circ$$

From the phasor representations of v_a, v_b, and i,

$$v_a = 0.34 \sin\left(\frac{1}{2}t + 103.5^\circ\right)$$

$$v_b = 0.3 \sin\left(\frac{1}{2}t - 18.2^\circ\right)$$

$$i = 2.2 \sin\left(\frac{1}{2}t + 40.7^\circ\right)$$

18.4 UNIFORM APPROACH TO NODAL ANALYSIS

We now demonstrate the uniform approach to the nodal analysis method for analyzing ac circuits. Recall from Chapter 6 that the uniform approach to nodal analysis allows us to write the nodal voltages in determinant form by a simple inspection of the circuit.

Consider the circuit in Figure 18.12, which is redrawn in Figure 18.16 with the nodes labeled 1 and 2 and the reference node as indicated. Because

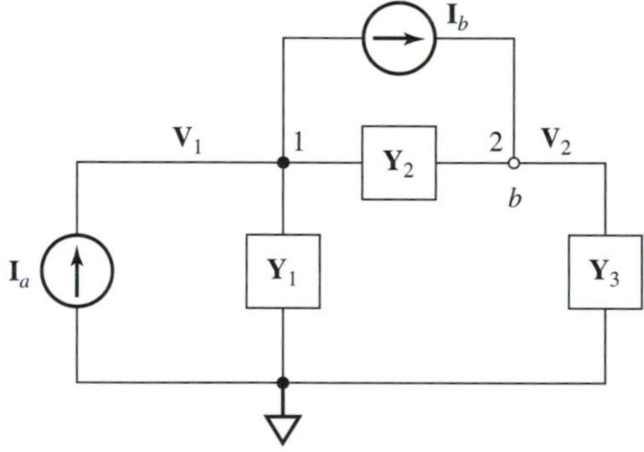

Figure 18.16 Nodal analysis.

there are two nodes (other than the reference node), the numerator and denominator determinants for $\mathbf{V}_1$ and $\mathbf{V}_2$ each consist of two rows and two columns. The denominator determinant is

$$\begin{vmatrix} \mathbf{Y}_{11} & -\mathbf{Y}_{12} \\ -\mathbf{Y}_{21} & \mathbf{Y}_{22} \end{vmatrix}$$

where

$\mathbf{Y}_{11}$ = total admittance connected to node 1

$\mathbf{Y}_{12}$ = admittance common to nodes 1 and 2

$\mathbf{Y}_{22}$ = total admittance connected to node 2

$\mathbf{Y}_{21}$ = admittance common to nodes 2 and 1

The current source summation column is

$$\begin{pmatrix} \mathbf{I}_1 \\ \mathbf{I}_2 \end{pmatrix}$$

where

$\mathbf{I}_1$ = algebraic sum of phasor currents from current sources entering node 1

$\mathbf{I}_2$ = algebraic sum of phasor currents from current sources entering node 2

For the circuit in Figure 18.16, the denominator determinant and the current source summation column are

$$\begin{vmatrix} \mathbf{Y}_1 + \mathbf{Y}_2 & -\mathbf{Y}_2 \\ -\mathbf{Y}_2 & \mathbf{Y}_2 + \mathbf{Y}_3 \end{vmatrix}$$

$$\begin{pmatrix} \mathbf{I}_a - \mathbf{I}_b \\ \mathbf{I}_b \end{pmatrix}$$

and

$$\mathbf{V}_1 = \frac{\begin{vmatrix} \mathbf{I}_a - \mathbf{I}_b & -\mathbf{Y}_2 \\ \mathbf{I}_b & \mathbf{Y}_2 + \mathbf{Y}_3 \end{vmatrix}}{\begin{vmatrix} \mathbf{Y}_1 + \mathbf{Y}_2 & -\mathbf{Y}_2 \\ -\mathbf{Y}_2 & \mathbf{Y}_2 + \mathbf{Y}_3 \end{vmatrix}}$$

$$\mathbf{V}_2 = \frac{\begin{vmatrix} \mathbf{Y}_1 + \mathbf{Y}_2 & \mathbf{I}_a - \mathbf{I}_b \\ -\mathbf{Y}_2 & \mathbf{I}_b \end{vmatrix}}{\begin{vmatrix} \mathbf{Y}_1 + \mathbf{Y}_2 & -\mathbf{Y}_2 \\ -\mathbf{Y}_2 & \mathbf{Y}_2 + \mathbf{Y}_3 \end{vmatrix}}$$

Notice that the current source summation column replaces column 1 or 2 of the denominator determinant to form the numerator determinants for nodal voltages $\mathbf{V}_1$ and $\mathbf{V}_2$, respectively.

Consider the following example.

EXAMPLE 18.7 Determine the nodal voltages $\mathbf{V}_1$ and $\mathbf{V}_2$ for the circuit in Figure 18.17. ■

Figure 18.17

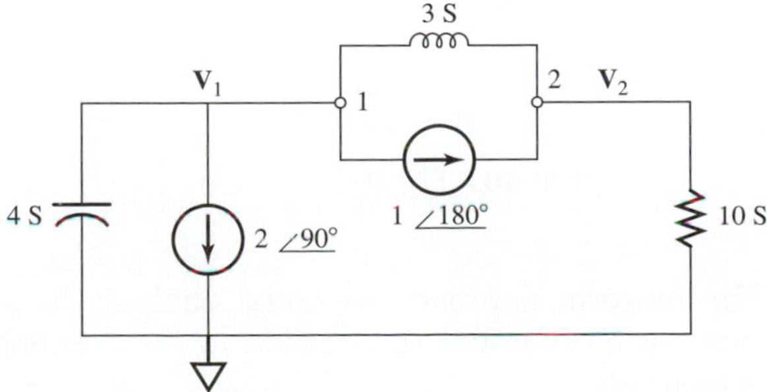

SOLUTION The elements of the denominator determinant are

$$\mathbf{Y}_{11} = -j3 + j4$$
$$= j1$$
$$\mathbf{Y}_{22} = 10 - j3$$
$$\mathbf{Y}_{12} = \mathbf{Y}_{21} = -j3$$

and the denominator determinant is

$$\begin{vmatrix} j1 & -(-j3) \\ -(-j3) & 10 - j3 \end{vmatrix}$$

The current source summation column is

$$\begin{pmatrix} -2\angle 90^\circ - 1\angle 180^\circ \\ 1\angle 180^\circ \end{pmatrix}$$

The phasor nodal voltages are

$$\mathbf{V}_1 = \frac{\begin{vmatrix} (-2\angle 90^\circ - 1\angle 180^\circ) & j3 \\ 1\angle 180^\circ & 10 - j3 \end{vmatrix}}{\begin{vmatrix} j1 & j3 \\ j3 & 10 - j3 \end{vmatrix}}$$

$$= \frac{(-2\angle 90^\circ - 1\angle 180^\circ)(10 - j3) - (1\angle 180^\circ)(j3)}{(j1)(10 - j3) - (j3)(j3)}$$

$$= \frac{4 - j20}{12 + j10} - \frac{20.4\angle -78.6^\circ}{15.6\angle 39.8^\circ}$$

$$= -0.62 - j1.15$$

$$= 1.31\angle -118.5^\circ$$

$$\mathbf{V}_2 = \frac{\begin{vmatrix} j1 & (-2\angle 90^\circ - 1\angle 180^\circ) \\ j3 & 1\angle 180^\circ \end{vmatrix}}{\begin{vmatrix} j1 & j3 \\ j3 & 10 - j3 \end{vmatrix}}$$

$$= \frac{(j1)(1\angle 180^\circ) - (j3)(-2\angle 90^\circ - 1\angle 180^\circ)}{(j1)(10 - j3) - (j3)(j3)}$$

$$= \frac{-6 - j4}{12 + j10} = \frac{7.2\angle -146.3^\circ}{15.6\angle 39.8^\circ}$$

$$= -0.46 + j0.05$$

$$= 0.46\angle 173.9^\circ$$

The uniform approach to nodal analysis is easily extended to circuits with three or more nodes in addition to the reference node. Consider the following example.

EXAMPLE 18.8 Determine the phasor nodal voltages $\mathbf{V}_1$, $\mathbf{V}_2$, and $\mathbf{V}_3$ in determinant form for the circuit in Figure 18.18 with the reference node as indicated. ■

Figure 18.18

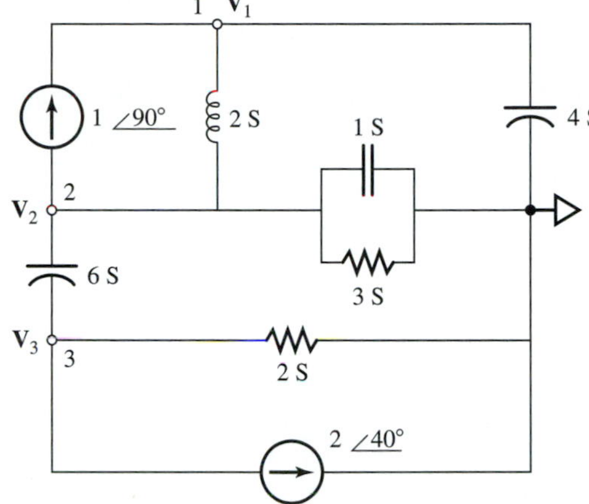

SOLUTION The elements of the denominator determinant are

$$\begin{aligned}
\mathbf{Y}_{11} &= -j2 + j4 \\
&= j2 \\
\mathbf{Y}_{22} &= -j2 + j1 + 3 + j6 \\
&= 3 + j5 \\
\mathbf{Y}_{33} &= 2 + j6 \\
\mathbf{Y}_{12} &= \mathbf{Y}_{21} = -j2 \\
\mathbf{Y}_{13} &= \mathbf{Y}_{31} = 0 \\
\mathbf{Y}_{23} &= \mathbf{Y}_{32} = j6
\end{aligned}$$

and the denominator determinant is

$$\begin{vmatrix} j2 & -(-j2) & 0 \\ -(-j2) & 3 + j5 & -j6 \\ 0 & -j6 & 2 + j6 \end{vmatrix}$$

The current source summation column is

$$\begin{pmatrix} 1\angle 90^\circ \\ -1\angle 90^\circ \\ -2\angle 40^\circ \end{pmatrix}$$

The phasor nodal voltages in determinant form are

$$\mathbf{V}_1 = \frac{\begin{vmatrix} 1\angle 90^\circ & j2 & 0 \\ -1\angle 90^\circ & 3 + j5 & -j6 \\ -2\angle 40^\circ & -j6 & 2 + j6 \end{vmatrix}}{\begin{vmatrix} j2 & j2 & 0 \\ j2 & 3 + j5 & -j6 \\ 0 & -j6 & 2 + j6 \end{vmatrix}}$$

$$\mathbf{V}_2 = \frac{\begin{vmatrix} j2 & 1\angle 90^\circ & 0 \\ j2 & -1\angle 90^\circ & -j6 \\ 0 & -2\angle 40^\circ & 2+j6 \end{vmatrix}}{\begin{vmatrix} j2 & j2 & 0 \\ j2 & 3+j5 & -j6 \\ 0 & -j6 & 2+j6 \end{vmatrix}}$$

$$\mathbf{V}_3 = \frac{\begin{vmatrix} j2 & j2 & 1\angle 90^\circ \\ j2 & 3+j5 & -1\angle 90^\circ \\ 0 & -j6 & -2\angle 40^\circ \end{vmatrix}}{\begin{vmatrix} j2 & j2 & 0 \\ j2 & 3+j5 & -j6 \\ 0 & -j6 & 2+j6 \end{vmatrix}}$$

18.5 DEPENDENT SOURCES

In general, the mesh and nodal methods of analysis can be applied to multiple-loop ac circuits with dependent sources. However, the more systematic uniform-approach methods are generally restricted to circuits with independent sources.

Multiple-loop methods with dependent sources are considered in Chapter 6. Except for the use of phasors and complex numbers, the methods presented in Chapter 6 are not altered when applied to ac circuits. Consider the following examples.

EXAMPLE 18.9 Use the mesh-analysis method to determine the phasor current $\mathbf{I}_a$ for the circuit in Figure 18.19 with dependent current source $\mathbf{I}$. ■

Figure 18.19

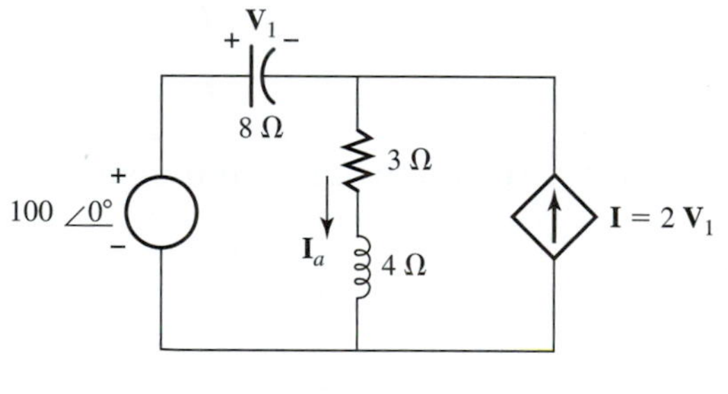

SOLUTION The circuit is redrawn in Figure 18.20 with the phasor mesh currents assigned and the associated voltage polarities for the passive elements indicated.

Figure 18.20

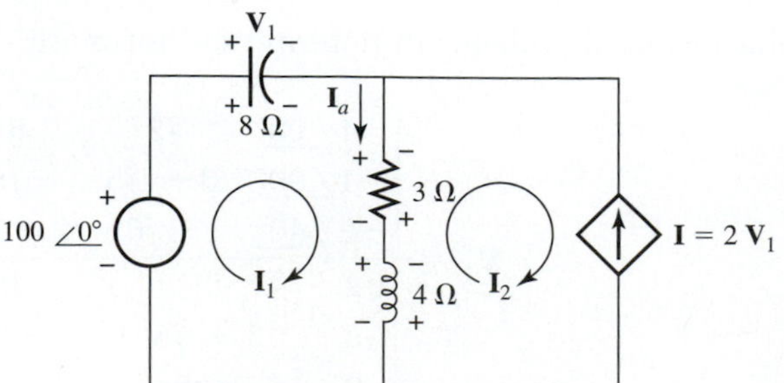

An inspection of the circuit reveals that

$$\begin{aligned}\mathbf{I}_2 &= -\mathbf{I}\\ &= -2\mathbf{V}_1\end{aligned}$$

and from Equation (17.1)

$$\mathbf{V}_1 = (-j8)\mathbf{I}_1$$

It follows that

$$\begin{aligned}\mathbf{I}_2 &= -2(-j8\mathbf{I}_1)\\ &= j16\mathbf{I}_1\end{aligned}$$

Applying Kirchhoff's voltage law to mesh 1 results in

$$\begin{aligned}100\underline{/0^\circ} - (-j8)\mathbf{I}_1 - (3 + j4)\mathbf{I}_1 + (3 + j4)\mathbf{I}_2 &= 0\\ (3 - j4)\mathbf{I}_1 - (3 + j4)\mathbf{I}_2 &= 100\underline{/0^\circ}\end{aligned}$$

Substituting $(j16\mathbf{I}_1)$ for $\mathbf{I}_2$ results in

$$\begin{aligned}(3 - j4)\mathbf{I}_1 - (3 + j4)(j16\mathbf{I}_1) &= 100\underline{/0^\circ}\\ (67 - j52)\mathbf{I}_1 &= 100\underline{/0^\circ}\\ \mathbf{I}_1 &= \frac{100\underline{/0^\circ}}{67 - j52}\\ &= 1.18\underline{/37.8^\circ}\\ \mathbf{I}_a &= \mathbf{I}_1 - \mathbf{I}_2\\ &= \mathbf{I}_1 - j16\mathbf{I}_1\\ &= (1 - j16)\mathbf{I}_1\\ &= (1 - j16)(1.18\underline{/37.82^\circ})\\ &= 12.5 - j14.2\\ &= 18.9\underline{/-48.6^\circ}\end{aligned}$$

EXAMPLE 18.10 Use the nodal-analysis method to determine the phasor current $\mathbf{I}_a$ for the circuit in Figure 18.21 with dependent current source $\mathbf{I}$. Use the reference node indicated. ■

Figure 18.21

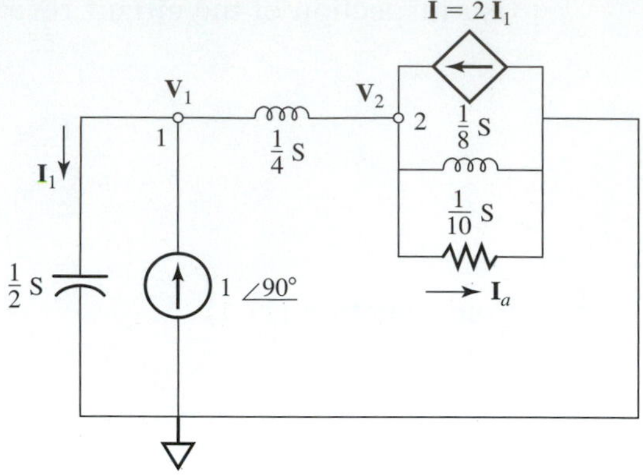

SOLUTION From Equation (17.13)

$$\mathbf{I}_1 = \left(j\frac{1}{2}\right)\mathbf{V}_1$$

and from the relationship of the dependent source to $\mathbf{I}_1$,

$$\begin{aligned}\mathbf{I} &= 2\mathbf{I}_1 \\ &= 2\left(j\frac{1}{2}\right)\mathbf{V}_1 \\ &= j\mathbf{V}_1\end{aligned}$$

Applying Kirchhoff's current law to nodes 1 and 2 results in

$$\text{node 1:} \qquad -\left(j\frac{1}{2}\right)\mathbf{V}_1 + 1\underline{/\,90^\circ} - \left(-j\frac{1}{4}\right)(\mathbf{V}_1 - \mathbf{V}_2) = 0$$

$$\text{node 2:} \quad \mathbf{I} - \left(-j\frac{1}{8}\right)\mathbf{V}_2 - \left(\frac{1}{10}\right)\mathbf{V}_2 - \left(-j\frac{1}{4}\right)(\mathbf{V}_2 - \mathbf{V}_1) = 0$$

Combining terms and substituting $(j\mathbf{V}_1)$ for $\mathbf{I}$ in the second equation results in

$$\begin{aligned}\left(j\frac{1}{4}\right)\mathbf{V}_1 + \left(j\frac{1}{4}\right)\mathbf{V}_2 &= 1\underline{/\,90^\circ} \\ \left(-j\frac{3}{4}\right)\mathbf{V}_1 + \left(\frac{1}{10} - j\frac{3}{8}\right)\mathbf{V}_2 &= 0\end{aligned}$$

Using Cramer's rule to determine $\mathbf{V}_2$ results in

$$\mathbf{V}_2 = \frac{\begin{vmatrix} j\frac{1}{4} & 1\underline{/\,90^\circ} \\ -j\frac{3}{4} & 0 \end{vmatrix}}{\begin{vmatrix} j\frac{1}{4} & j\frac{1}{4} \\ -j\frac{3}{4} & \frac{1}{10} - j\frac{3}{8} \end{vmatrix}}$$

$$= 7.5 + j2.0$$

$$= 7.7\underline{/\,14.9^\circ}$$

and from Equation (17.13)

$$\mathbf{I}_a = \left(\frac{1}{10}\right)\mathbf{V}_2$$

$$= \left(\frac{1}{10}\right)(7.7\angle 14.9^\circ)$$

$$= 0.77\angle 14.9^\circ$$

18.6 DELTA AND WYE CIRCUITS

The delta and wye conversion formulas presented in Chapter 4 for resistive circuits apply directly to ac circuits, where the resistances are replaced by general impedances. The conversion formulas given in Figure 4.78 are reproduced in Figure 18.22 in terms of ac impedances.

Figure 18.22 Delta and wye conversions.

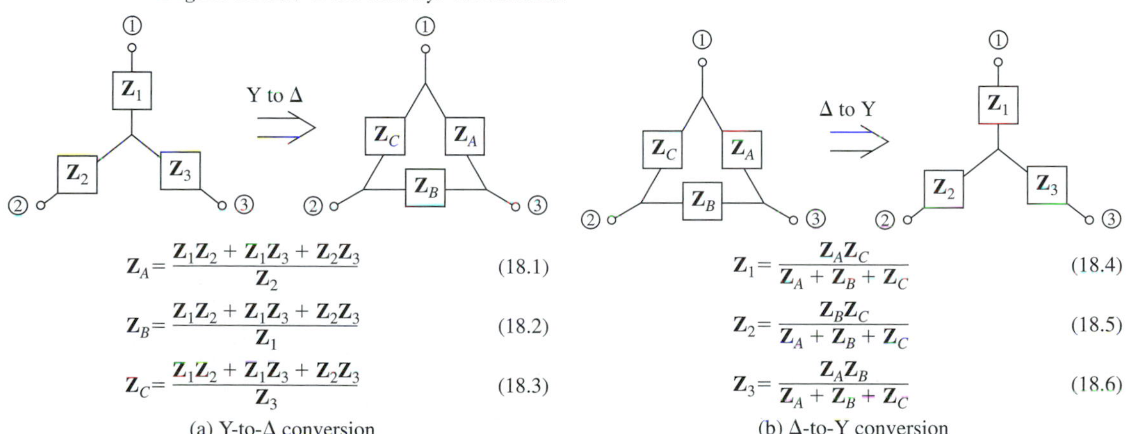

Examples of circuits containing delta and wye connections are illustrated in Figure 18.23. Notice that neither of the circuits in Figure 18.23 can be simplified by combining the impedances in series or in parallel. The delta and wye conversion formulas allow us to convert the circuits to equivalent circuits, where the impedances can be combined in series and/or parallel.

Figure 18.23 Delta and wye conversions.

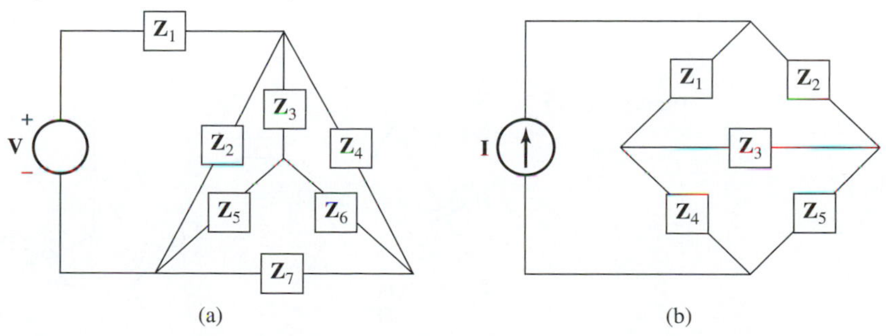

Consider the following examples.

EXAMPLE 18.11 Determine $\mathbf{Z}_{ab}$ for the circuit in Figure 18.24. ■

Figure 18.24

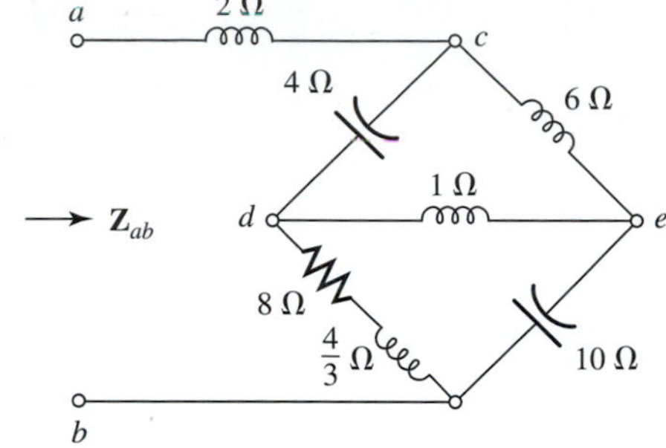

SOLUTION The Δ consisting of the 4-Ω capacitor and the 1-Ω and 6-Ω inductors is converted to an equivalent Y, as shown in Figure 18.25.

From Equations (18.4), (18.5), and (18.6)

$$\begin{aligned}\mathbf{Z}_1 &= \frac{(-j4)(j6)}{-j4 + j6 + j1} \\ &= \frac{24}{j3} \\ &= -j8\end{aligned}$$

$$\begin{aligned}\mathbf{Z}_2 &= \frac{(-j4)(j1)}{-j4 + j6 + j1} \\ &= \frac{4}{j3} \\ &= -j\frac{4}{3}\end{aligned}$$

$$\begin{aligned}\mathbf{Z}_3 &= \frac{(j1)(j6)}{-j4 + j6 + j1} \\ &= \frac{-6}{j3} \\ &= j2\end{aligned}$$

Figure 18.25

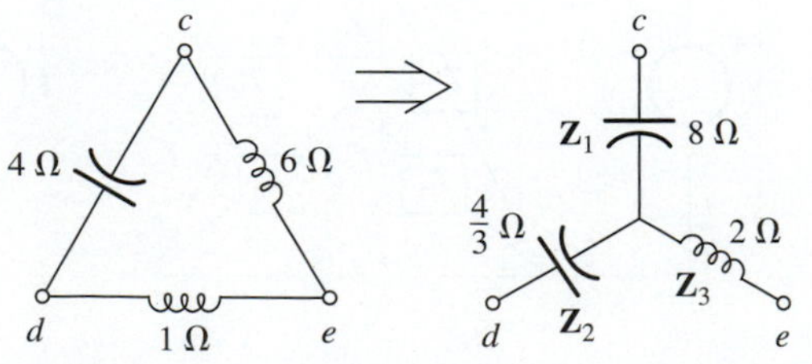

c
4 Ω
6 Ω
d
1 Ω
e
⟹
c
$\mathbf{Z}_1$
8 Ω
$\frac{4}{3}$ Ω
2 Ω
$\mathbf{Z}_3$
d
$\mathbf{Z}_2$
e

Replacing the Δ by the equivalent Y results in the circuit in Figure 18.26.

Figure 18.26

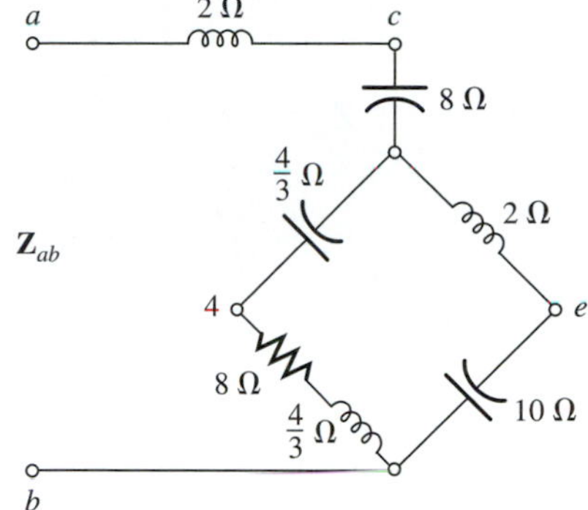

and

$$\mathbf{Z}_{ab} = (j2) + (-j8) + \left\{\left(8 - j\frac{4}{3} + j\frac{4}{3}\right)\|(j2 - j10)\right\}$$

$$= -j6 + \{(8)\|(-j8)\}$$

$$= -j6 + \frac{8(-j8)}{8 - j8}$$

$$= 4 - j10$$

$$= 10.8\angle -68.2°$$

EXAMPLE 18.12 Determine $\mathbf{Z}_{ab}$ for the circuit in Figure 18.27. ■

Figure 18.27

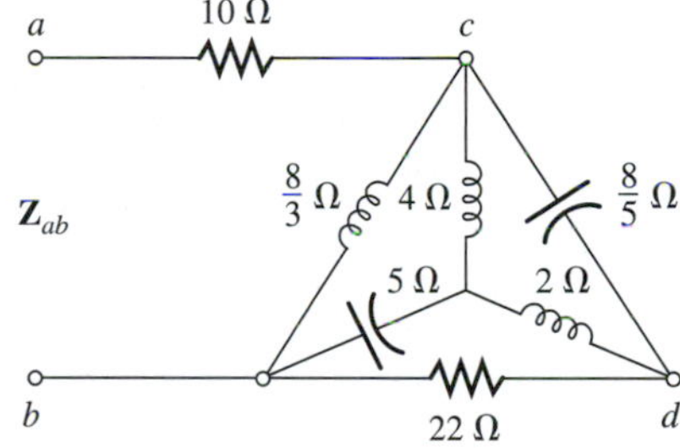

SOLUTION The Y can be converted to an equivalent Δ, as shown in Figure 18.28.

From Equations (18.1), (18.2), and (18.3),

$$\mathbf{Z}_A = \frac{(j4)(-j5) + (j4)(j2) + (-j5)(j2)}{-j5}$$

$$= \frac{22}{-j5}$$

$$= j\frac{22}{5}$$

$$\mathbf{Z}_B = \frac{(j4)(-j5) + (j4)(j2) + (-j5)(j2)}{j1}$$

$$= \frac{22}{j1}$$

$$= -j22$$

$$\mathbf{Z}_C = \frac{(j4)(-j5) + (j4)(j2) + (-j5)(j2)}{j3}$$

$$= \frac{22}{j3}$$

$$= -j\frac{22}{3}$$

Figure 18.28

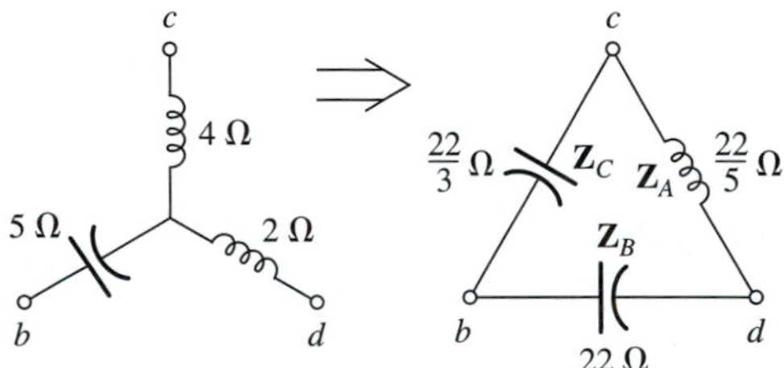

Replacing the Y by an equivalent Δ results in the circuit in Figure 18.29.

Figure 18.29

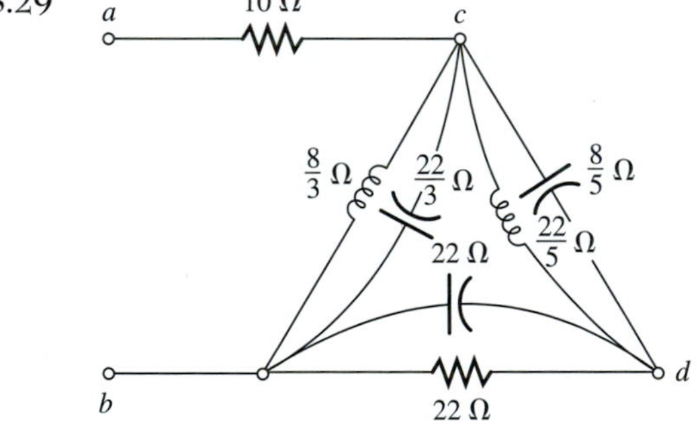

$$\mathbf{Z}_{ab} = 10 + \left[\left\{\left\{\left(j\frac{22}{5}\right)\middle\|\left(-j\frac{8}{5}\right)\right\} + \{(-j22)\|(22)\}\right\}\middle\|\left\{\left(j\frac{8}{3}\right)\middle\|\left(-j\frac{22}{3}\right)\right\}\right]$$

$$= 10 + \left[\left\{\left\{\frac{\left(j\frac{22}{5}\right)\left(-j\frac{8}{5}\right)}{\left(j\frac{22}{5} - j\frac{8}{5}\right)}\right\} + \left\{\frac{(-j22)(22)}{(22 - j22)}\right\}\right\}\middle\|\left\{\frac{\left(j\frac{8}{3}\right)\left(-j\frac{22}{3}\right)}{\left(j\frac{8}{3} - j\frac{22}{3}\right)}\right\}\right]$$

$$= 10 + \{[11 - j13.5]\|[j4.2]\}$$

$$= 10 + \frac{(11 - j13.5)(j4.2)}{11 - j13.5 + j4.2}$$

$$= 10.9 + j4.9$$

$$= 11.9\underline{/24.2^\circ}$$

18.7 EXERCISES

1. Use the general approach method to mesh analysis for the circuit in Figure 18.30 to determine $\mathbf{I}_1$, $\mathbf{I}_2$, $\mathbf{I}_3$, $\mathbf{V}_3$, $\mathbf{V}_4$, and $\mathbf{V}_5$ in polar form.

Figure 18.30

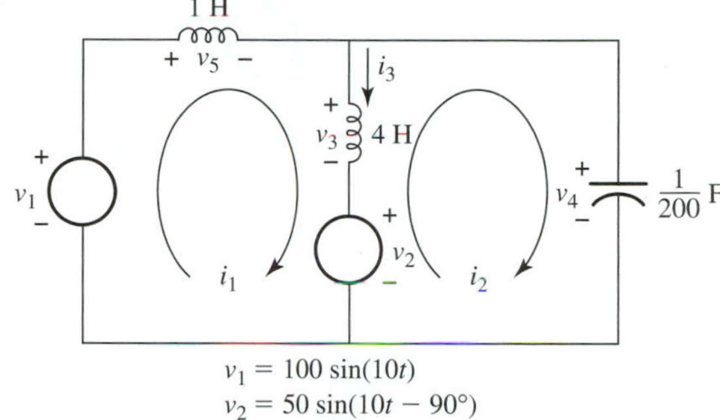

2. Use the general approach method to mesh analysis for the circuit in Figure 18.31 to determine the phasor mesh currents $\mathbf{I}_1$ and $\mathbf{I}_2$ and the phanor voltage $\mathbf{V}_C$ in polar form.

Figure 18.31

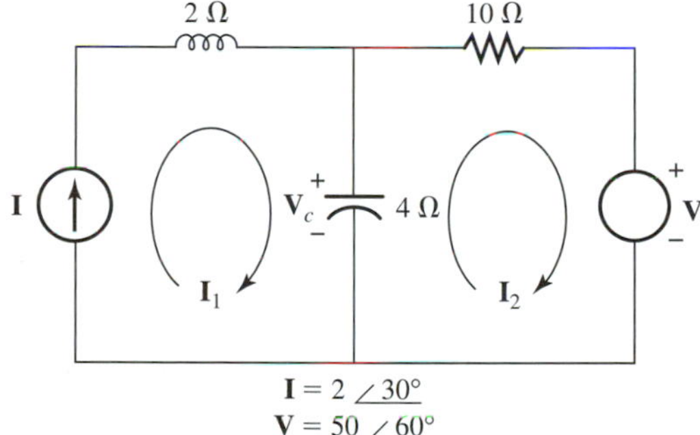

3. Use the general approach method to mesh analysis for the circuit in Figure 18.32 to write the mesh equations for the circuit in phasor form, and determine $\mathbf{I}_1$, $\mathbf{I}_2$, $\mathbf{I}_3$, and $\mathbf{V}_{ab}$.

Figure 18.32

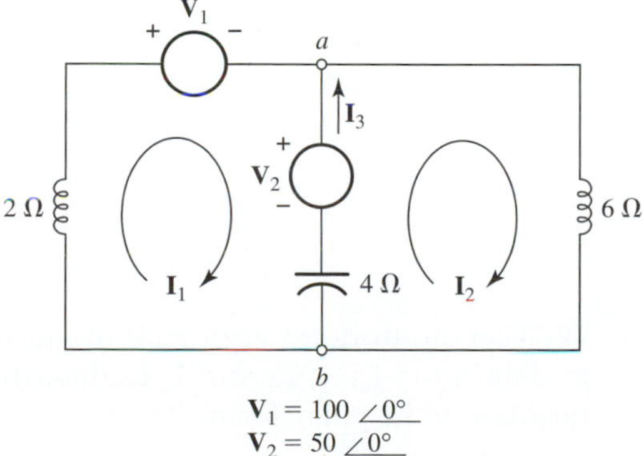

4. Use the general approach method to mesh analysis for the circuit in Figure 18.33 to determine $\mathbf{I}_1$, $\mathbf{I}_2$, $\mathbf{I}_3$, and $\mathbf{I}$ in polar form.

Figure 18.33

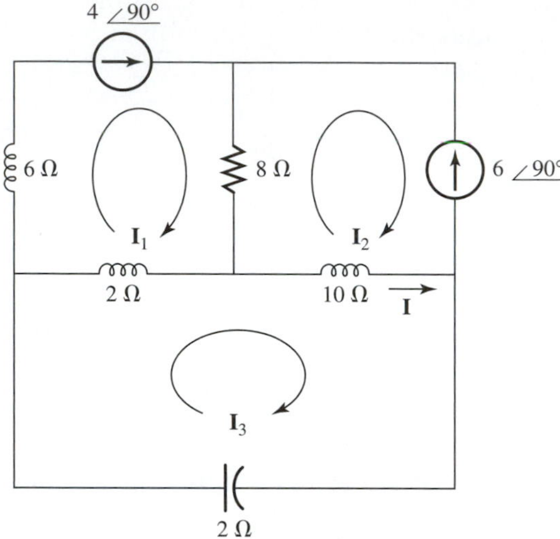

5. Compare the uniform approach method to the general approach method for mesh analysis. State the advantages and disadvantages of each method.

6. Use the uniform approach to the mesh analysis method for the circuit in Figure 18.34 to determine $\mathbf{I}_1$, $\mathbf{I}_2$, $\mathbf{I}_3$, and $\mathbf{I}_{ab}$ in polar form.

Figure 18.34

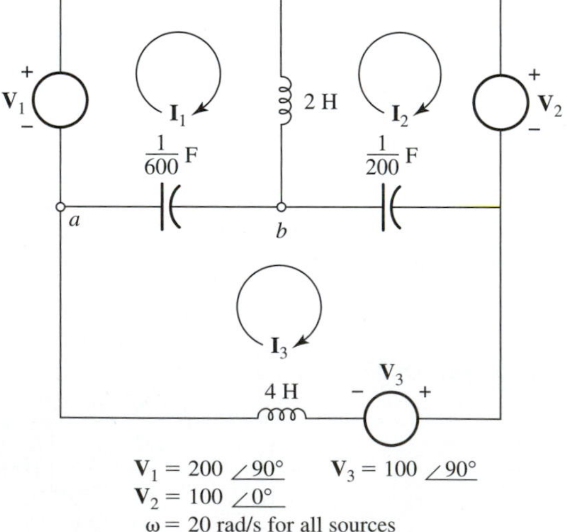

7. Use the uniform approach to mesh analysis for the circuit in Figure 18.35 to determine $\mathbf{I}_1$, $\mathbf{I}_2$, $\mathbf{I}_3$, and $\mathbf{I}_4$ in determinant form with each element of the determinants in polar form.

Figure 18.35

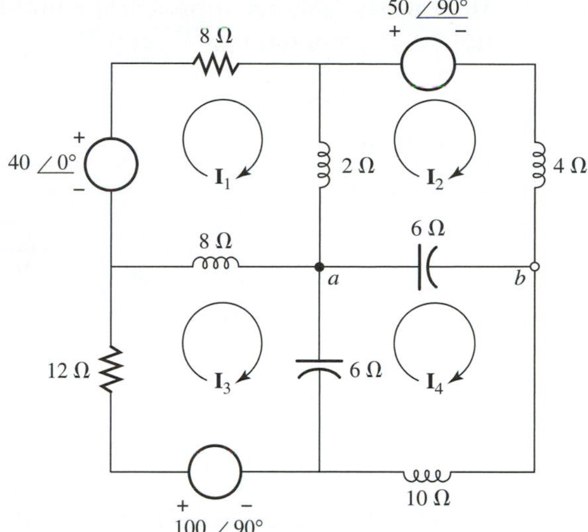

8. Use the general approach method for nodal analysis for the circuit in Figure 18.36 to determine voltages $\mathbf{V}_1$, $\mathbf{V}_2$, $\mathbf{V}_s$ and currents $\mathbf{I}_1$, $\mathbf{I}_2$, $\mathbf{I}_3$, and $\mathbf{I}_4$.

Figure 18.36

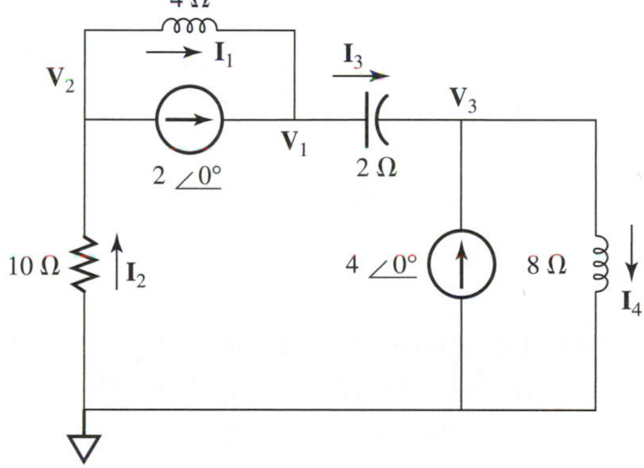

9. The reference node in Problem 8 has been changed as shown in Figure 18.37. Determine $\mathbf{V}_1$, $\mathbf{V}_2$, $\mathbf{V}_3$, $\mathbf{I}_1$, $\mathbf{I}_2$, $\mathbf{I}_3$, and $\mathbf{I}_4$. Compare $\mathbf{V}_1$, $\mathbf{V}_2$, $\mathbf{V}_3$, $\mathbf{I}_1$, $\mathbf{I}_2$, $\mathbf{I}_3$, and $\mathbf{I}_4$ to the voltages and currents determined in Problem 8. What is your conclusion based on the comparison?

Figure 18.37

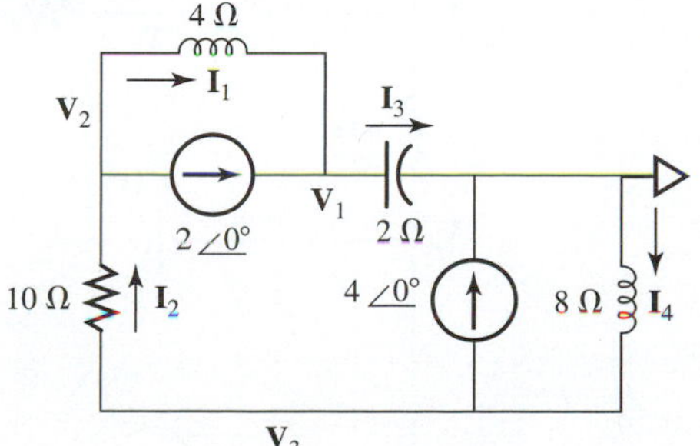

10. Use the general approach method for nodal analysis for the circuit in Figure 18.38 to determine $\mathbf{V}_1$ and $\mathbf{V}_2$.

Figure 18.38

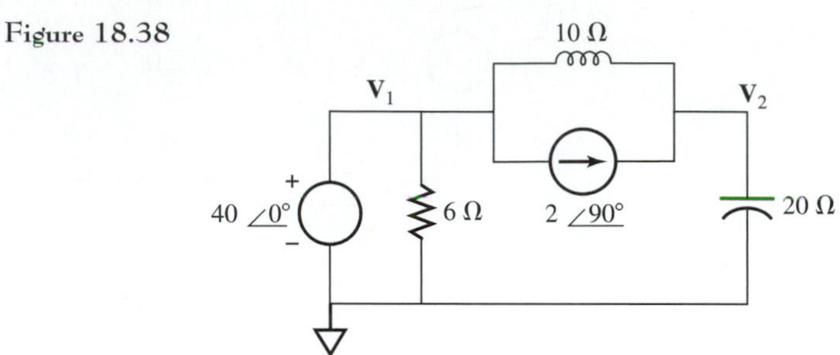

11. Use the general approach method for nodal analysis for the circuit in Figure 18.39 to determine $\mathbf{V}_1$, $\mathbf{V}_2$, $\mathbf{I}_1$, and $\mathbf{I}_2$,

Figure 18.39

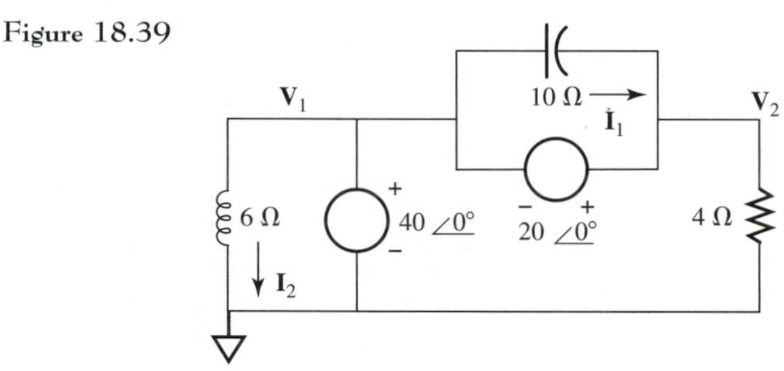

12. Compare the uniform approach method to the general approach method for nodal analysis. State the advantages and disadvantages of each method.

13. Use the uniform approach method to nodal analysis for the circuit in Figure 18.40 to determine $\mathbf{V}_1$, $\mathbf{V}_2$, $\mathbf{V}_3$, $\mathbf{V}_4$, and $\mathbf{I}$ in determinant form.

Figure 18.40

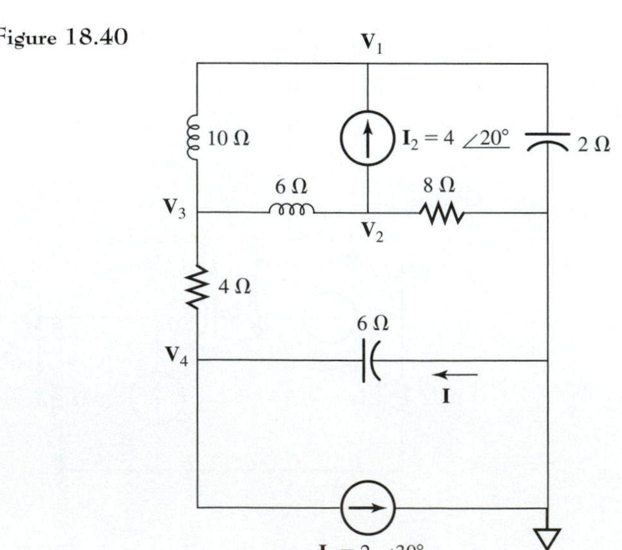

14. Use the mesh analysis method for the circuit in Figure 18.41 to determine the phasor currents $\mathbf{I}_1$ and $\mathbf{I}_2$ in polar form.

Figure 18.41

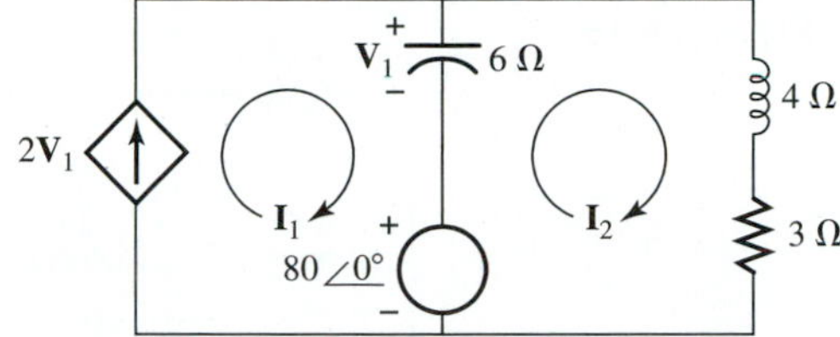

15. Use the nodal analysis method for the circuit in Figure 18.42 to determine the phasor nodal voltages $\mathbf{V}_1$ and $\mathbf{V}_2$ in polar form.

Figure 18.42

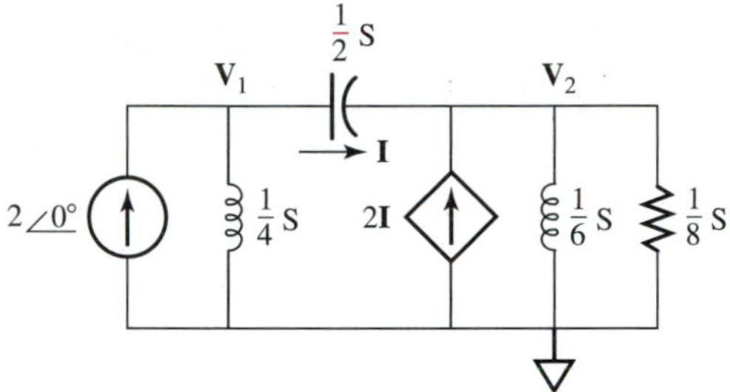

16. Determine $\mathbf{Z}_{ab}$ and $\mathbf{Y}_{ab}$ for the circuit in Figure 18.43.

Figure 18.43

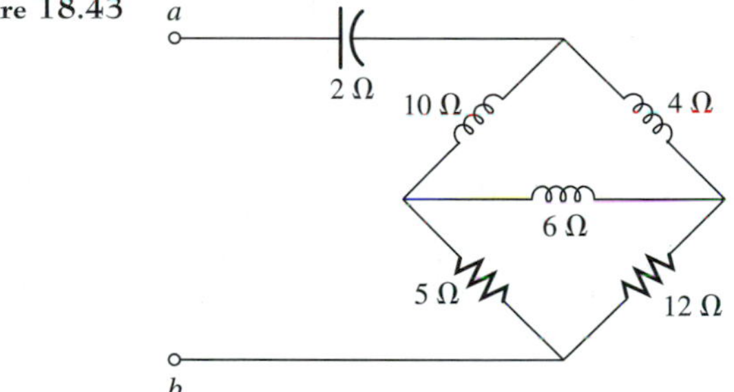

17. Find R_{ab}, X_{ab}, G_{ab}, and B_{ab} for the circuit in Figure Problem 16.

18. Determine $\mathbf{Z}_{ab}$ and $\mathbf{Y}_{ab}$ for the circuit in Figure 18.44.

Figure 18.44

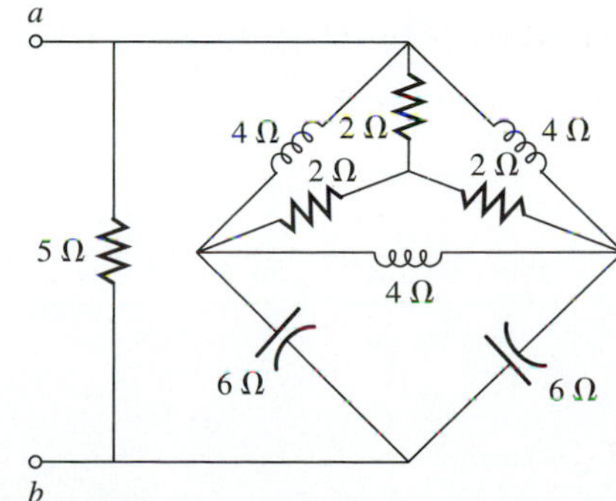

19. Find R_{ab}, X_{ab}, G_{ab}, and R_{ab} for the circuit in Problem 18.

20. Determine $\mathbf{Z}_{ab}$ and $\mathbf{Y}_{ab}$ for the circuit in Figure 18.45.

Figure 18.45

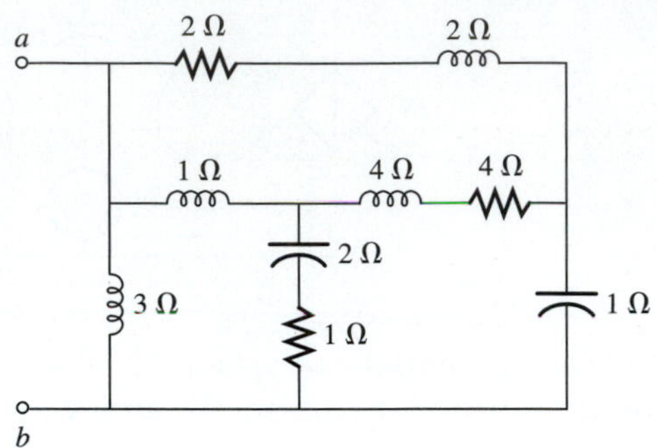

21. Find R_{ab}, G_{ab}, X_{ab}, and B_{ab} for the circuit in Problem 20.

CHAPTER

19

Network Theorems Applied to AC Circuits

After completing this chapter the student should be able to

* apply the superposition method to ac circuits with multiple sources
* apply Thevenin's theorem to analyze ac circuits
* apply Norton's theorem to analyze ac circuits
* use the maximum power transfer theorem to determine ac loads for maximum power transfer

The network theorems presented in Chapter 7 for resistive circuits are also applicable to the analysis of ac circuits. The only variation in applying the network theorems to ac circuits is the use of phasors and complex numbers instead of real numbers. The network theorems allow us to simplify circuits for analysis purposes and thus solve complicated electric circuits with less effort than normally required.

19.1 SUPERPOSITION METHOD

The superposition method for ac circuits allows us to determine the current through or the voltage across an electrical element by separately considering the contribution to the current or voltage from each current and voltage source in the circuit. When considering the contribution from each source, all other sources in the circuit are replaced by their internal impedances. That is, the voltage sources are replaced by short circuits and the current sources are replaced by open circuits.

Consider the circuit in Figure 19.1, where $\mathbf{V}_1$, $\mathbf{I}_1$, $\mathbf{Z}_1$, $\mathbf{Z}_2$, and $\mathbf{Z}_3$ are known and we wish to determine the current $\mathbf{I}$.

Figure 19.1 Superposition method.

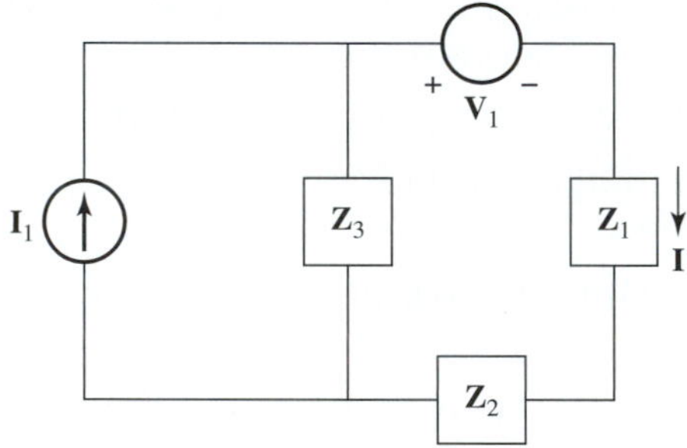

The *first step* in the superposition method is to draw a separate circuit for each source in the circuit, with all the other sources replaced by their internal impedances. Because the circuit in Figure 19.1 contains two sources ($\mathbf{I}_1$ and $\mathbf{V}_1$), two circuits are drawn as shown in Figure 19.2.

Figure 19.2 Superposition method.

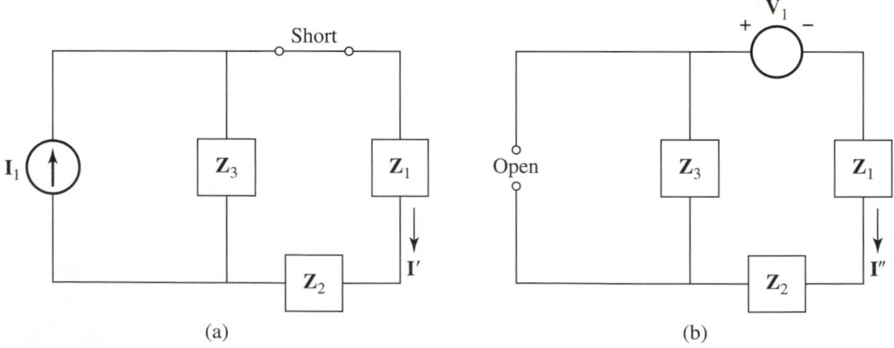

In Figure 19.2(a) the voltage source $\mathbf{V}_1$ is replaced by a short, and in Figure 19.2(b) the current source $\mathbf{I}_1$ is replaced by an open. The current $\mathbf{I}'$ is the contribution to $\mathbf{I}$ from the current source $\mathbf{I}_1$ and the current $\mathbf{I}''$ is the contribution to $\mathbf{I}$ from the voltage source $\mathbf{V}_1$.

The *second step* in the superposition method is to solve the circuits obtained in step 1 for the contributed (primed) currents or voltages. For the circuits in Figure 19.2, the contributed (primed) currents are **I**′ and **I**″. From Equation (17.41) for current division,

$$\mathbf{I}' = \left\{\frac{\mathbf{Z}_3}{\mathbf{Z}_1 + \mathbf{Z}_2 + \mathbf{Z}_3}\right\}(\mathbf{I}_1) = \frac{\mathbf{Z}_3\mathbf{I}_1}{\mathbf{Z}_1 + \mathbf{Z}_2 + \mathbf{Z}_3}$$

From Equations (17.1) and (17.35),

$$\mathbf{I}'' = -\left\{\frac{\mathbf{V}_1}{\mathbf{Z}_1 + \mathbf{Z}_2 + \mathbf{Z}_3}\right\}$$

The *third step* in the superposition method is to algebraically sum the contributed (primed) currents or voltages obtained in step 2. For the circuit in Figure 19.1,

$$\mathbf{I} = \mathbf{I}' + \mathbf{I}'' = \frac{\mathbf{Z}_3\mathbf{I}_1 - \mathbf{V}_1}{\mathbf{Z}_1 + \mathbf{Z}_2 + \mathbf{Z}_3}$$

The three-step procedure for the superposition method is summarized next.

1. Draw a circuit for each current and voltage source by replacing all the other current and voltage sources in the circuit under analysis by their internal impedances. The number of circuits drawn will equal the number of sources in the circuit under analysis.
2. Solve each circuit in step 1 for the current or voltage contributed by each source.
3. Algebraically sum the contributed currents or voltages obtained in step 2 to determine the desired current or voltage.

EXAMPLE 19.1 Determine the voltage **V** for the circuit in Figure 19.3. ■

Figure 19.3

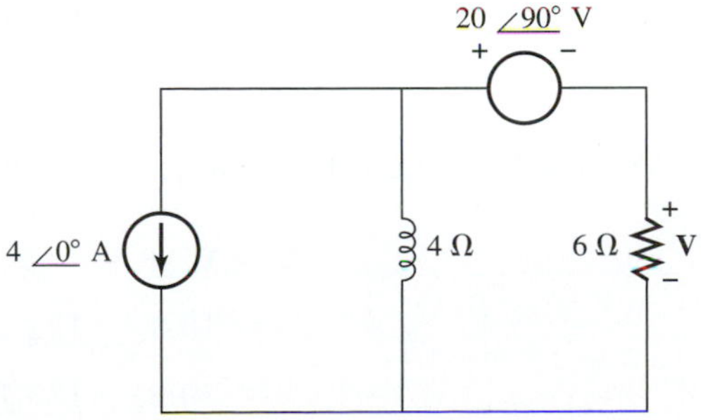

SOLUTION Because there are two sources in the circuit, there are two circuits to consider, as shown in Figures 19.4 and 19.5.

Figure 19.4

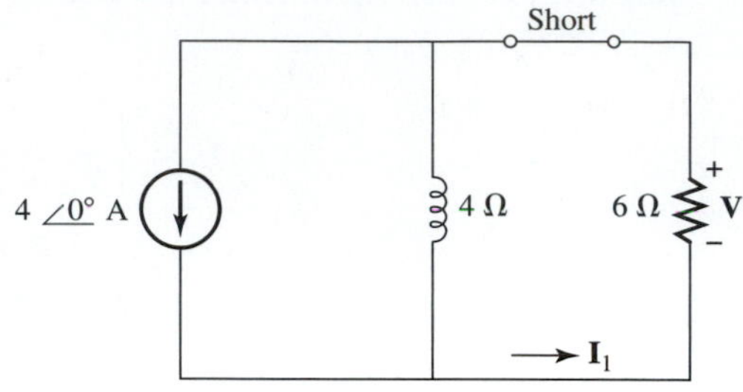

Figure 19.5

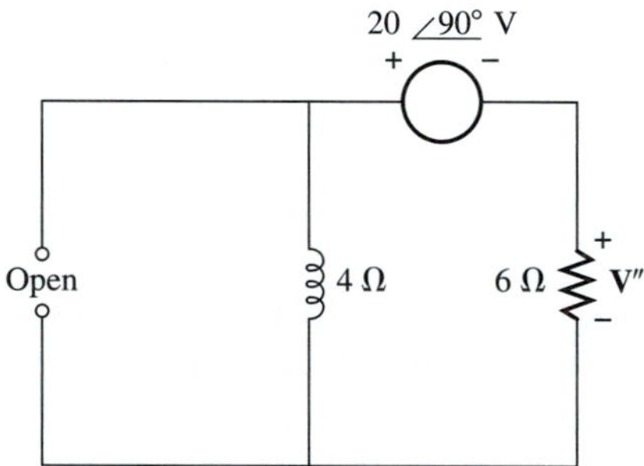

In the first circuit, voltage $\mathbf{V}'$ is the contribution to $\mathbf{V}$ from the current source and can be determined as follows. From Equation (17.42) for current division,

$$\begin{aligned}\mathbf{I}_1 &= \{(j4)/(j4 + 6)\}\,(4\angle 0^\circ)\\ &= 2.2\angle 56.3^\circ\end{aligned}$$

and from Equation (17.1)

$$\begin{aligned}\mathbf{V}' &= 6(-\mathbf{I}_1)\\ &= 6(-2.2\angle 56.3^\circ)\\ &= 13.3\angle -123.7^\circ\end{aligned}$$

In the second circuit, voltage $\mathbf{V}''$ is the contribution to $\mathbf{V}$ from the voltage source, and it can be determined from Equation (17.45) for voltage division:

$$\begin{aligned}\mathbf{V}'' &= \left\{\frac{6}{6 + j4}\right\}(-20\angle 90^\circ)\\ &= 16.6\angle -123.7^\circ\end{aligned}$$

Voltage $\mathbf{V}$ is equal to the algebraic sum of the contributed voltages $\mathbf{V}'$ and $\mathbf{V}''$.

$$\begin{aligned}\mathbf{V} &= \mathbf{V}' + \mathbf{V}''\\ &= 13.3\angle -123.7^\circ + 16.6\angle -123.7^\circ\\ &= 29.9\angle -123.7^\circ\end{aligned}$$

EXAMPLE 19.2 Determine the current **I** for the circuit in Figure 19.6. ■

Figure 19.6

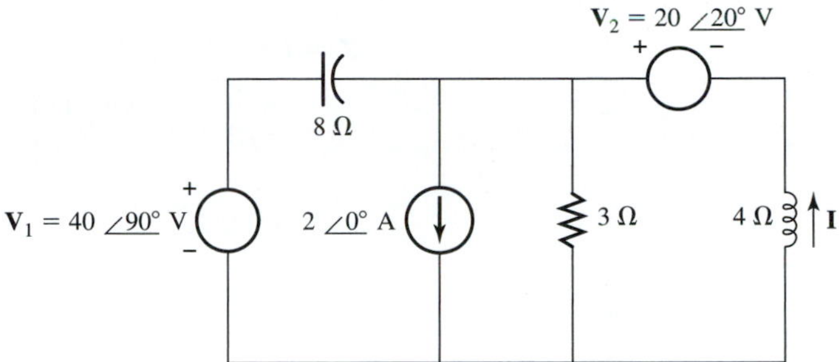

SOLUTION There are three sources in the circuit, so there are three circuits to consider, as shown in Figures 19.7, 19.8, and 19.9.

Figure 19.7

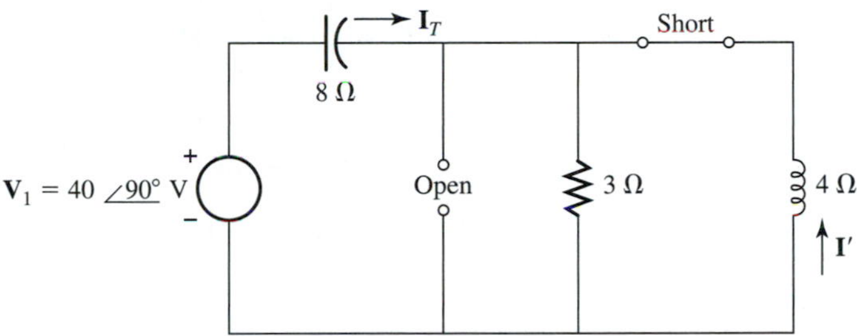

Figure 19.8

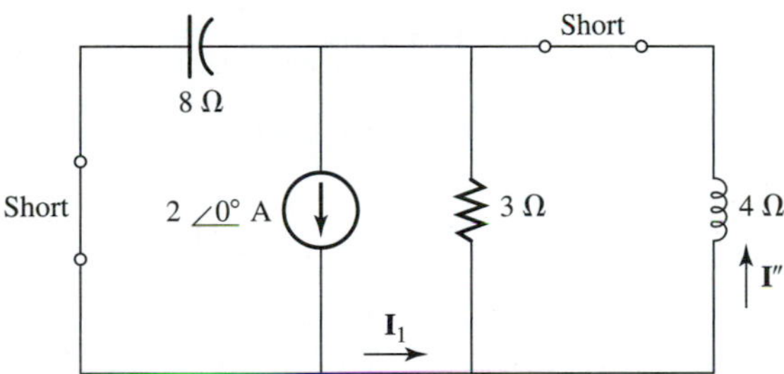

Figure 19.9

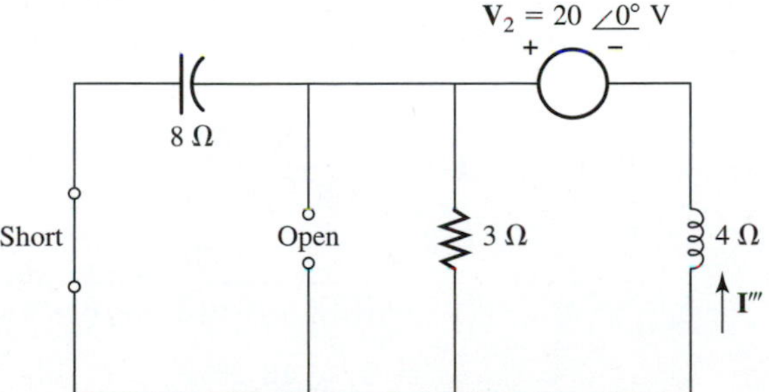

For the circuit in Figure 19.7, the current $\mathbf{I}'$ is the contribution to $\mathbf{I}$ from the voltage source $\mathbf{V}_1$ and can be determined from Equation (17.41) for current division after $\mathbf{I}_T$ is determined. The total impedance $\mathbf{Z}$ across voltage source $\mathbf{V}_1$ is

$$\begin{aligned}\mathbf{Z} &= (-j8) + (3\,\|\,j4)\\ &= -j8 + \left\{\frac{(3)(j4)}{3 + j4}\right\}\\ &= 6.8\underline{/-73.7^\circ}\ \Omega\end{aligned}$$

and from Equation (17.1) the current $\mathbf{I}_T$ is

$$\begin{aligned}\mathbf{I}_T &= \frac{\mathbf{V}_1}{\mathbf{Z}} = \frac{40\underline{/90^\circ}}{6.8\underline{/-73/7^\circ}}\\ &= 5.9\underline{/163.7^\circ}\ \text{A}\end{aligned}$$

Using Equation (17.41) for current division,

$$\begin{aligned}\mathbf{I}' &= \left\{\frac{3}{3 + j4}\right\}(-\mathbf{I}_T)\\ &= \left\{\frac{3}{3 + j4}\right\}(-5.9\underline{/163.7^\circ})\\ &= 3.5\underline{/-69.4^\circ}\ \text{A}\end{aligned}$$

For the circuit in Figure 19.8, the current $\mathbf{I}''$ is the contribution to $\mathbf{I}$ from the current source and can be determined by applying the current division equation twice, as follows:

$$\begin{aligned}\mathbf{I}_1 &= \frac{-j8}{(-j8) + (3\,\|\,j4)}(2\underline{/0^\circ})\\ &= 2.3\underline{/-16.3^\circ}\\ \mathbf{I}'' &= \left\{\frac{3}{3 + j4}\right\}\mathbf{I}_1\\ &= \left\{\frac{3}{3 + j4}\right\}(2.3\underline{/-16.3^\circ})\\ &= 1.4\underline{/-69.4^\circ}\end{aligned}$$

For the circuit in Figure 19.9, current $\mathbf{I}'''$ is the contribution to $\mathbf{I}$ from voltage source $\mathbf{V}_2$ and can be determined from Equation (17.1).

$$\begin{aligned}\mathbf{I}''' &= \frac{20\underline{/0^\circ}}{(3\,\|(-j8)) + j4}\\ &= \frac{20\underline{/0^\circ}}{\{(3)(-j8)/(3 - j8)\} + j4}\\ &= 5\underline{/-48.9^\circ}\end{aligned}$$

Algebraically summing the contributed currents results in

$$\begin{aligned}\mathbf{I} &= \mathbf{I}' + \mathbf{I}'' + \mathbf{I}''' \\ &= 3.5\angle -69.4^\circ + 1.4\angle -69.4^\circ + 5\angle -48.9^\circ \\ &= 5.01 - j8.35 \\ &= 9.7\angle -59.1^\circ\end{aligned}$$

EXAMPLE 19.3 Determine the voltages $\mathbf{V}_1$ and $\mathbf{V}_2$ for the circuit in Figure 19.10. ■

Figure 19.10

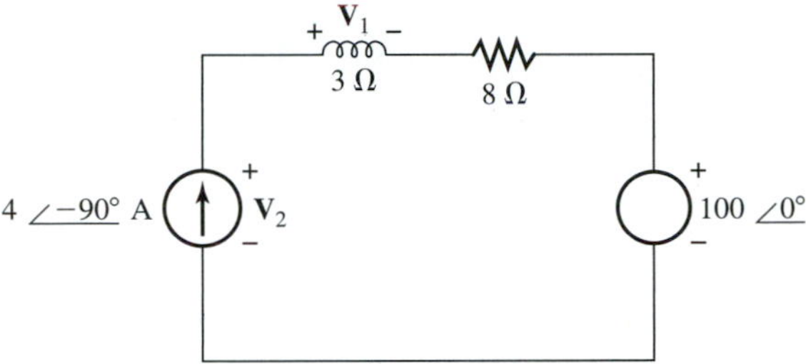

SOLUTION Because there are two sources in the circuit, there are two circuits to consider, as shown in Figure 19.11. For the circuit in Figure 19.11(a), $\mathbf{V}_1'$

Figure 19.11

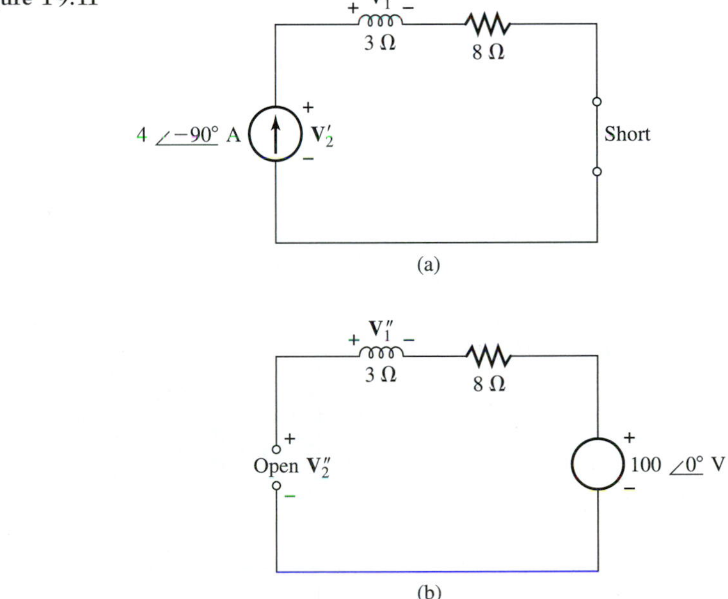

and $\mathbf{V}_2'$ are the contributions to $\mathbf{V}_1$ and $\mathbf{V}_2$ from the current source. From Equation (17.1)

$$\begin{aligned} \mathbf{V}_1' &= (j3)(4\angle -90^\circ) \\ &= 12\angle 0^\circ \\ \mathbf{V}_2' &= (8 + j3)(4\angle -90^\circ) \\ &= 34.2\angle -69.4^\circ \end{aligned}$$

For the circuit in Figure 19.11(b), $\mathbf{V}_1''$ and $\mathbf{V}_2''$ are the contributions to $\mathbf{V}_1$ and $\mathbf{V}_2$ from the voltage source. Because no current flows in the circuit,

$$\mathbf{V}_1'' = 0$$

and because the voltage source $100\angle 0^\circ$ appears across the open,

$$\mathbf{V}_2'' = 100\angle 0^\circ$$

Voltages $\mathbf{V}_1$ and $\mathbf{V}_2$ are equal to the algebraic sum of the contributed voltages:

$$\begin{aligned} \mathbf{V}_1 &= \mathbf{V}_1' + \mathbf{V}_1'' \\ &= 12\angle 0^\circ \\ \mathbf{V}_2 &= \mathbf{V}_2' + \mathbf{V}_2'' \\ &= 34.2\angle -69.4^\circ + 100\angle 0^\circ \\ &= 116.5\angle -15.9^\circ \end{aligned}$$

EXAMPLE 19.4 Determine the current $\mathbf{I}_1$ for the circuit in Figure 19.12. ■

Figure 19.12

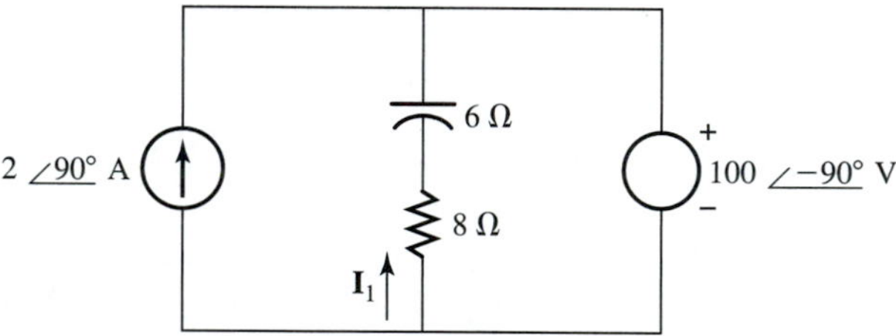

SOLUTION Because there are two sources in the circuit, there are two circuits to consider, as shown in Figure 19.13.

Figure 19.13

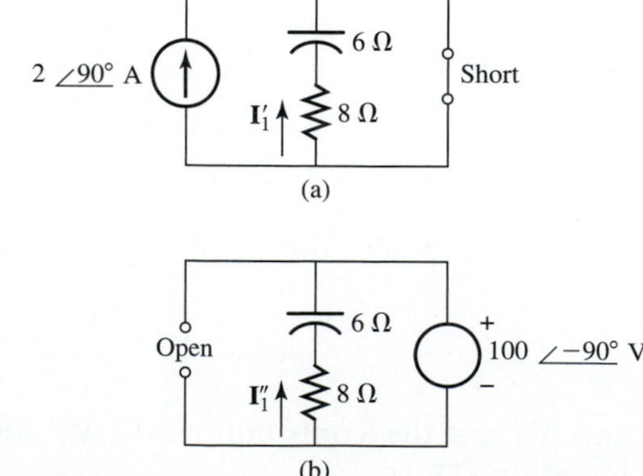

For the circuit in Figure 19.13(a), $\mathbf{I}_1'$ is the contribution to $\mathbf{I}_1$ from the current source. Because all the current from the current source flows through the short,

$$\mathbf{I}_1' = 0$$

This result can also be arrived at from Equation (17.41) for current division:

$$\begin{aligned}\mathbf{I}_1' &= \left\{\frac{0}{0 + 1 - j6}\right\}(2\underline{/90^\circ}) \\ &= 0\end{aligned}$$

For the circuit in Figure 19.13(b), $\mathbf{I}_1''$ is the contribution to $\mathbf{I}_1$ from the voltage source. From Equation (17.1)

$$\begin{aligned}\mathbf{I}_1'' &= (-100\underline{/-90^\circ})/(8 - j6) \\ &= 10\underline{/126.9^\circ}\end{aligned}$$

Current $\mathbf{I}_1$ is equal to the algebraic sum of the contributed currents.

$$\begin{aligned}\mathbf{I}_1 &= \mathbf{I}_1' + \mathbf{I}_1'' \\ &= 0 + 10\underline{/126.9^\circ} \\ &= 10\underline{/126.9^\circ}\end{aligned}$$

19.2 THEVENIN'S THEOREM

A very useful theorem for circuit analysis presented by Charles Thevenin, a French engineer, in 1883 is known as Thevenin's theorem and can be applied to ac circuits in the same manner that we applied it to resistive circuits in Chapter 7.

THEVENIN'S THEOREM A circuit connected across two terminals *a-b* can be replaced by a voltage source in series with an impedance.

The series combination of voltage source and impedance is known as a Thevenin equivalent circuit with respect to terminals *a-b*.

The series impedance, known as the Thevenin impedance and designated $\mathbf{Z}_T$, is the impedance across terminals *a-b* when all voltage and current sources in the circuit are replaced by their internal impedances (a short for voltage sources and an open for current sources).

The series voltage source, known as the Thevenin voltage source and designated $\mathbf{V}_T$ in phasor form, is equal to the open-circuited phasor voltage across terminals *a-b*.

The *first step* in applying Thevenin's theorem is to isolate the two terminals connected to the portion of the circuit that is to be replaced by a Thevenin equivalent. Consider the circuit in Figure 19.14, where it is desired

Figure 19.14 Thevenin's theorem.

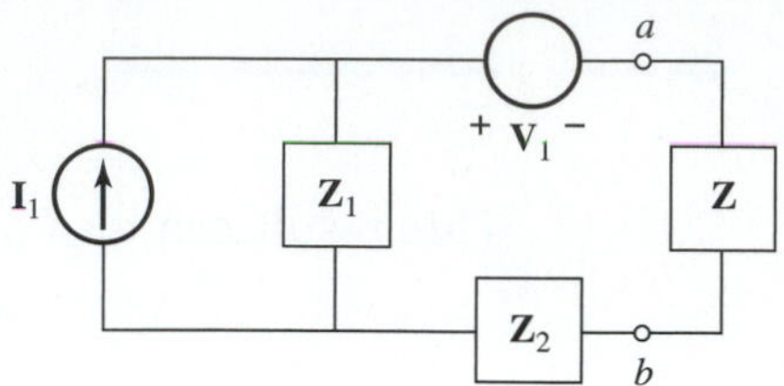

to replace the circuit to the left of terminals *a-b* by a Thevenin equivalent. All the electrical elements connected to terminals *a-b* are removed except for the circuit that is to be replaced by the Thevenin equivalent. Impedance **Z** is therefore removed, and the resulting circuit with terminals *a-b* isolated is as shown in Figure 19.15.

Figure 19.15 Thevenin's theorem.

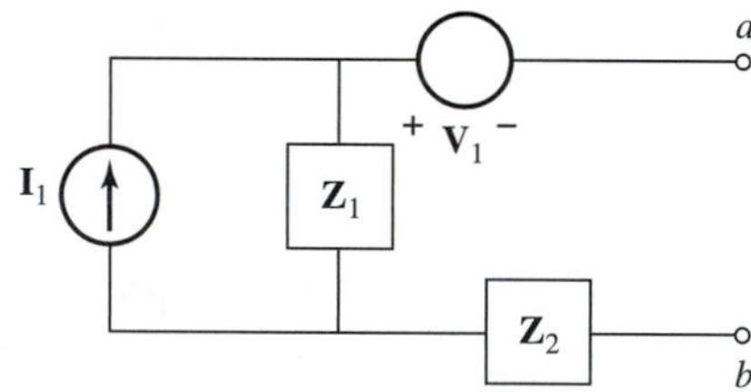

Using Thevenin's theorem, the circuit to the left of terminals *a-b* in Figure 19.15 can be replaced by the Thevenin equivalent in Figure 19.16.

Figure 19.16 Thevenin's theorem.

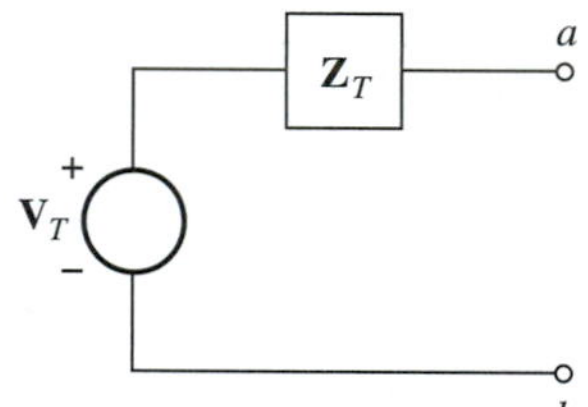

The *second step* in applying Thevenin's theorem is to determine the Thevenin voltage $\mathbf{V}_T$ for the Thevenin equivalent. Any analysis method or laboratory procedure can be used to determine the Thevenin voltage $\mathbf{V}_T$.

We apply the superposition method to determine $\mathbf{V}_T$ for the circuit in Figure 19.15. Because there are two sources in the circuit, there are two circuits to consider, as shown in Figure 19.17. $\mathbf{V}_T'$ is the contribution to $\mathbf{V}_T$ from the current source $\mathbf{I}_1$, and $\mathbf{V}_T''$ is the contribution to $\mathbf{V}_T$ from the voltage source $\mathbf{V}_1$.

Figure 19.17 Superposition.

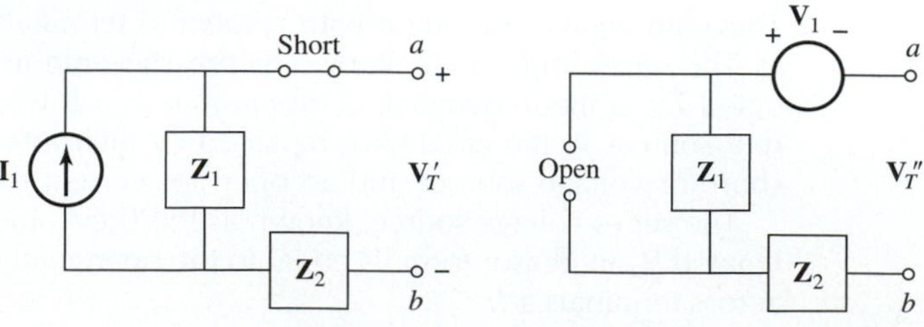

For the circuit in Figure 19.17(a), all the current $\mathbf{I}_1$ flows through $\mathbf{Z}_1$ and from Equation (17.1)

$$\mathbf{V}'_T = \mathbf{Z}_1\mathbf{I}_1$$

For the circuit in Figure 19.17(b), no current flows in the circuit and

$$\mathbf{V}''_T = -\mathbf{V}_1$$

$\mathbf{V}_T$ is the algebraic sum of the contributed voltages $\mathbf{V}'_T$ and $\mathbf{V}''_T$.

$$\begin{aligned}\mathbf{V}_T &= \mathbf{V}'_T + \mathbf{V}''_T \\ &= \mathbf{Z}_1\mathbf{I}_1 - \mathbf{V}_1\end{aligned}$$

The *third step* in applying Thevenin's theorem is to determine the Thevenin impedance $\mathbf{Z}_T$ for the Thevenin equivalent. The circuit in Figure 19.15 is shown in Figure 19.18, with all the sources replaced by their internal

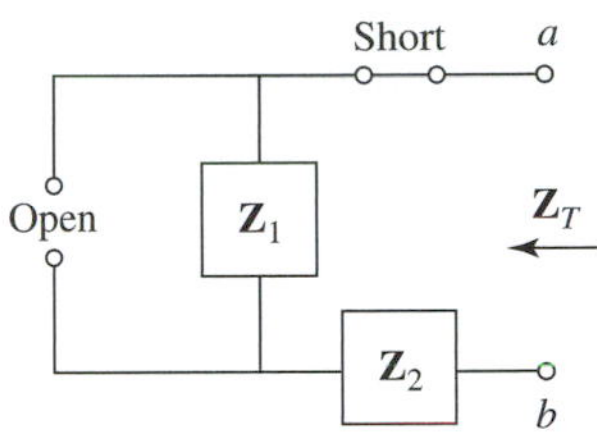

Figure 19.18 Thevenin's impedance.

impedances, as required by Thevenin's theorem for determining $\mathbf{Z}_T$. The Thevenin impedance $\mathbf{Z}_T$ is the impedance across terminals *a-b* in Figure 19.18 and is

$$\mathbf{Z}_T = \mathbf{Z}_1 + \mathbf{Z}_2$$

The *fourth step* in applying Thevenin's theorem is to form the Thevenin equivalent circuit using $\mathbf{V}_T$ and $\mathbf{Z}_T$. For the circuit to the left of terminals *a-b* in Figure 19.14, the Thevenin equivalent circuit is shown in Figure 19.19.

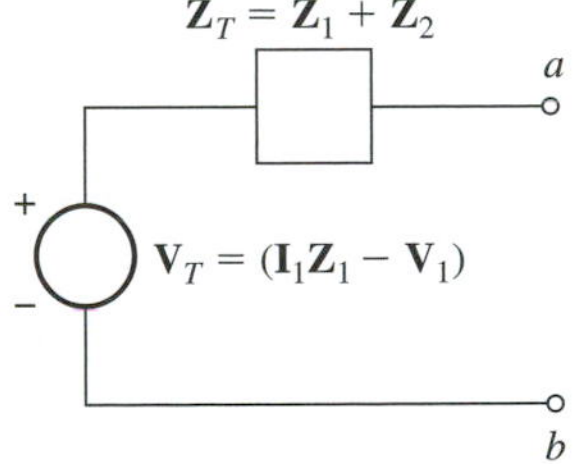

Figure 19.19 Thevenin equivalent.

The four-step procedure for applying Thevenin's theorem is summarized here.

1. Isolate the two terminals connected to the portion of the circuit to be replaced by a Thevenin equivalent.
2. Determine the Thevenin voltage $\mathbf{V}_T$.
3. Determine the Thevenin impedance $\mathbf{Z}_T$.
4. Use $\mathbf{V}_T$ and $\mathbf{Z}_T$ to form the Thevenin equivalent.

Consider the following examples.

EXAMPLE 19.5 Determine a Thevenin equivalent for the circuit to the left of terminals *a*-*b* in Figure 19.20. ■

Figure 19.20

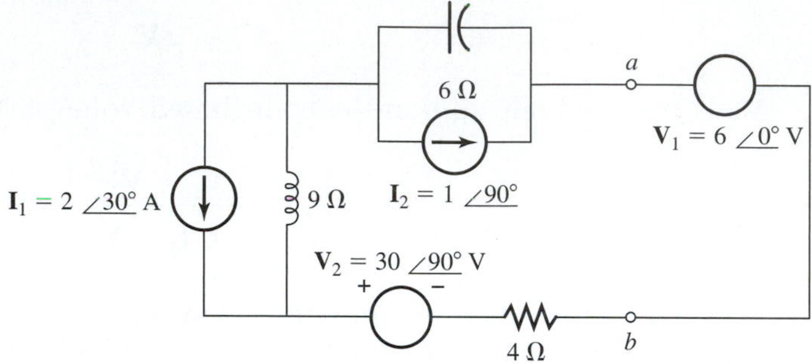

SOLUTION The circuit with terminals *a*-*b* isolated is shown in Figure 19.21.

Figure 19.21

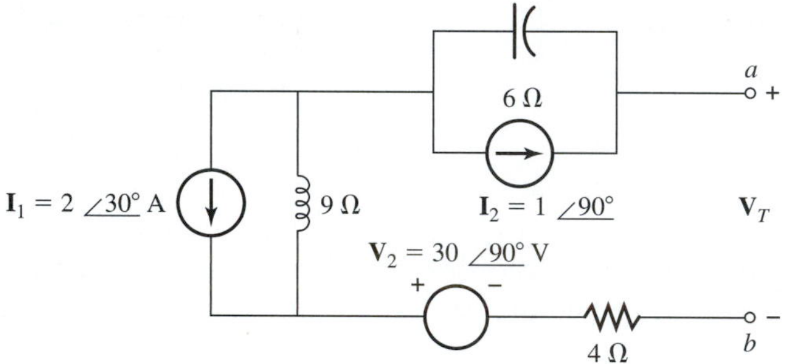

We use the superposition method to determine $\mathbf{V}_T$. Because there are three sources, three circuits are considered, as shown in Figures 19.22, 19.23, and 19.24.

Figure 19.22

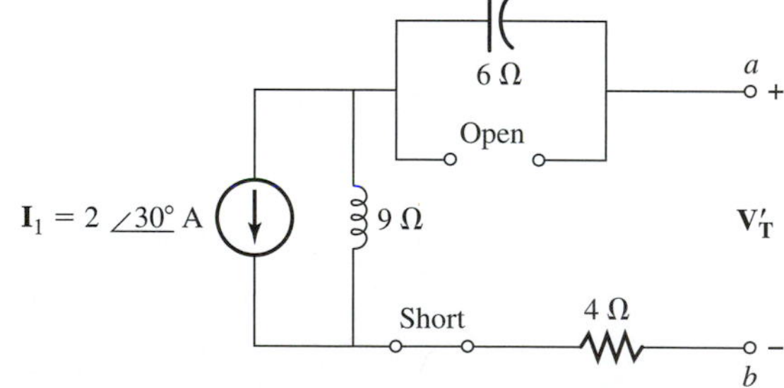

Figure 19.23

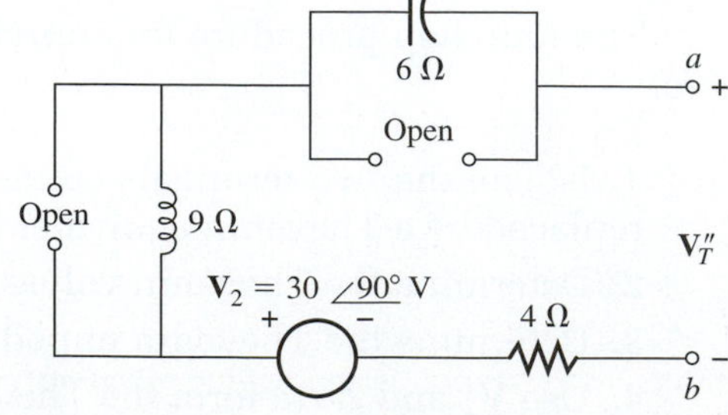

Figure 19.24

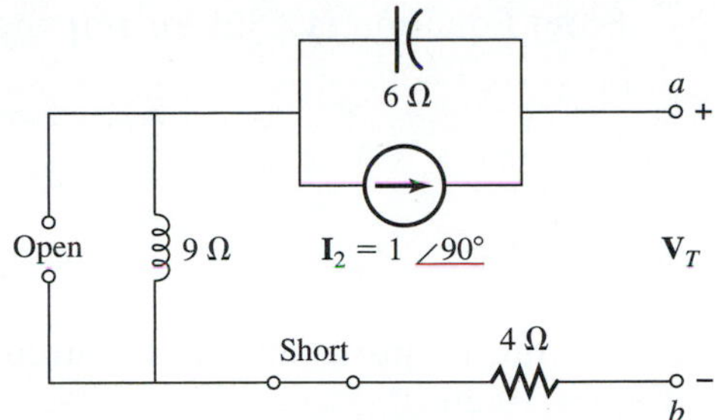

$\mathbf{V}'_T$ is the contribution to $\mathbf{V}_T$ from current source $\mathbf{I}_1$. $\mathbf{V}''_T$ is the contribution to $\mathbf{V}_T$ from the voltage source $\mathbf{V}_2$. $\mathbf{V}'''_T$ is the contribution to $\mathbf{V}_T$ from current source $\mathbf{I}_2$. For the circuit in Figure 19.22, all the current from source $\mathbf{I}_1$ flows through the 9-Ω inductance and from Equation (17.1)

$$\begin{aligned} \mathbf{V}'_T &= -(j9)(2\angle 30^\circ) \\ &= 18\angle -60^\circ \end{aligned}$$

For the circuit in Figure 19.23, no current flows in the circuit, and $\mathbf{V}''_T$ is equal to the voltage across the voltage source $\mathbf{V}_2$.

$$\mathbf{V}''_T = 30\angle 90^\circ$$

For the circuit in Figure 19.24, all the current from current source $\mathbf{I}_2$ flows through the capacitor, and from Equation (17.1)

$$\begin{aligned} \mathbf{V}'''_T &= (-j6)(1\angle 90^\circ) \\ &= 6\angle 0^\circ \end{aligned}$$

$\mathbf{V}_T$ is the algebraic sum of the contributed voltages, and

$$\begin{aligned} \mathbf{V}_T &= \mathbf{V}'_T + \mathbf{V}''_T + \mathbf{V}'''_T \\ &= 18\angle -60^\circ + 30\angle 90^\circ + 6\angle 0^\circ \\ &= 15 + j14.4 \\ &= 20.8\angle 43.8^\circ \end{aligned}$$

The circuit to the left of terminals *a-b* is shown in Figure 19.25, with all the sources replaced by their internal impedances. The Thevenin impedance $\mathbf{Z}_T$ is determined from this circuit.

Figure 19.25

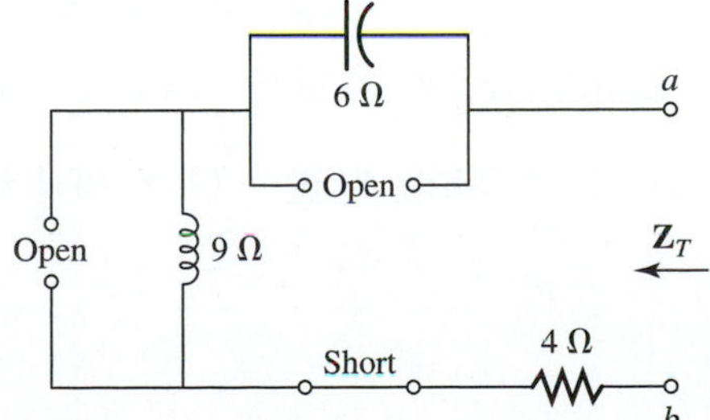

From Equation (17.35) for impedances in series,

$$\mathbf{Z}_T = 4 + j9 - j6$$
$$= 4 + j3$$
$$= 5\angle 36.9^\circ$$

The Thevenin equivalent for the circuit to the left of terminals *a*-*b* is shown in Figure 19.26.

Figure 19.26

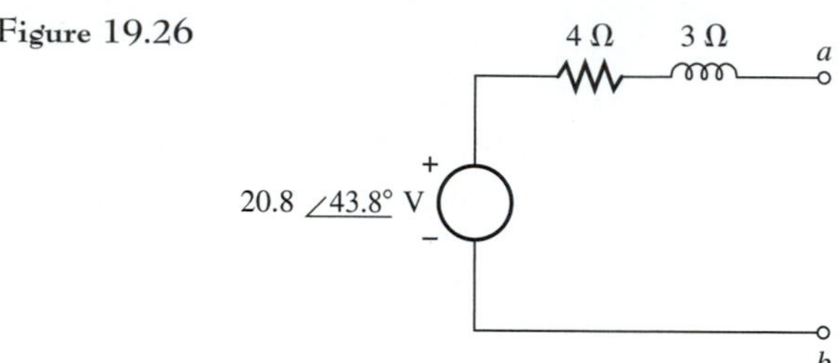

EXAMPLE 19.6 For the circuit in Example 19.5, determine the current **I** flowing through the voltage source $\mathbf{V}_1$. The circuit is redrawn in Figure 19.27. ■

Figure 19.27

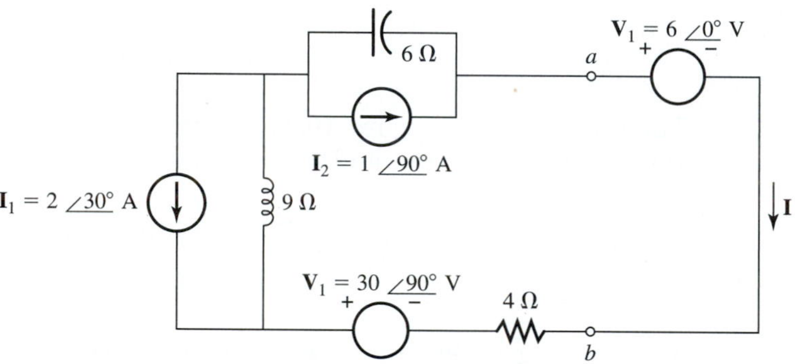

SOLUTION The circuit to the left of terminals *a*-*b* can be replaced by the Thevenin equivalent determined in Example 19.5, as shown in Figure 19.28.

Figure 19.28

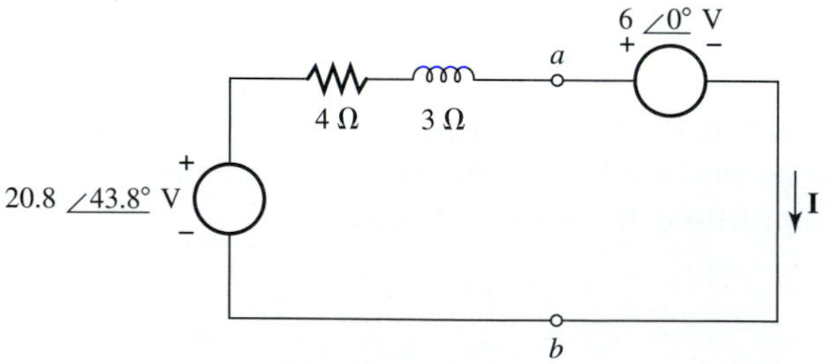

Employing Kirchhoff's voltage law results in

$$20.8\angle 43.8^\circ - (4 + j3)\mathbf{I} - 6\angle 0^\circ = 0$$

$$\mathbf{I} = \frac{20.8\angle 43.8^\circ - 6}{4 + j3}$$
$$= 3.4\angle 21.1^\circ$$

EXAMPLE 19.7 Determine the current **I** for the circuit in Figure 19.29. ■

Figure 19.29

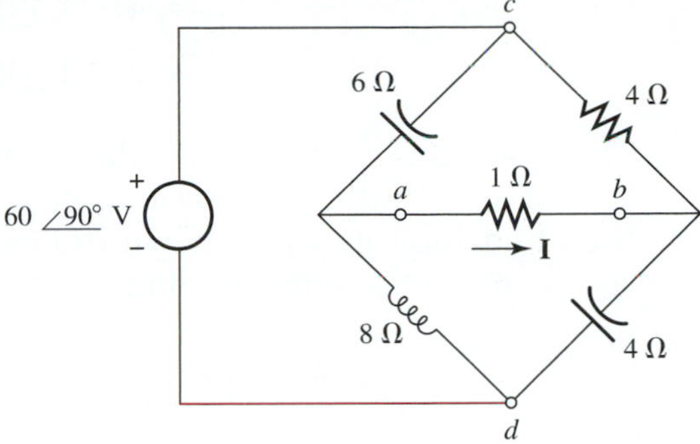

SOLUTION We shall remove the 1-Ω resistor as shown in Figure 19.30 and determine a Thevenin equivalent for the rest of the circuit with respect to terminals *a-b* (Figure 19.30).

Figure 19.30

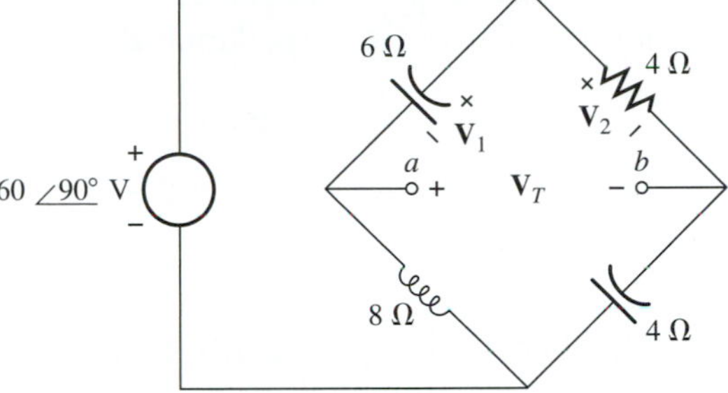

The Thevenin voltage $\mathbf{V}_T$ can be determined by finding voltages $\mathbf{V}_1$ and $\mathbf{V}_2$ and applying Kirchhoff's voltage law to the loop containing $\mathbf{V}_1$, $\mathbf{V}_2$, and terminals *a-b* in Figure 19.31.

Figure 19.31

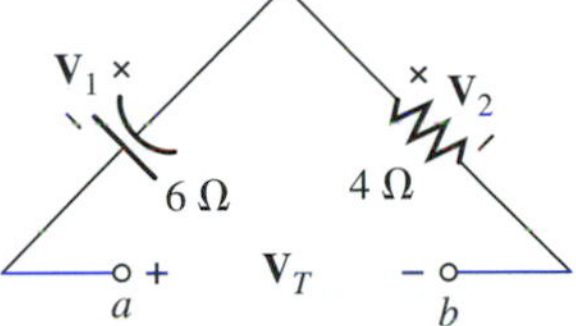

From Kirchhoff's voltage law,

$$\mathbf{V}_T + \mathbf{V}_1 - \mathbf{V}_2 = 0$$

$$\mathbf{V}_T = \mathbf{V}_2 - \mathbf{V}_1$$

Referring to the circuit in Figure 19.30, voltages $\mathbf{V}_1$ and $\mathbf{V}_2$ can be determined from Equation (17.45) for voltage division as

$$\mathbf{V}_1 = \left\{\frac{-j6}{-j6 + j8}\right\}(60\angle 90^\circ)$$

$$= 180\angle -90^\circ$$

$$\mathbf{V}_2 = \left\{\frac{4}{4 - j4}\right\}(60\angle 90^\circ)$$

$$= 42.4\angle 135^\circ$$

and because $\mathbf{V}_T = \mathbf{V}_2 - \mathbf{V}_1$,

$$\mathbf{V}_T = 42.4\angle 135^\circ - 180\angle -90^\circ$$
$$= 212.1\angle 98.1^\circ$$

The Thevenin resistance $\mathbf{Z}_T$ is determined by replacing the voltage source with a short and finding the resulting impedance across terminals *a*-*b*. The circuit used to determine $\mathbf{Z}_T$ is shown in Figure 19.32. Notice that shorting the voltage source is the same as connecting terminals *c* and *d*.

Figure 19.32

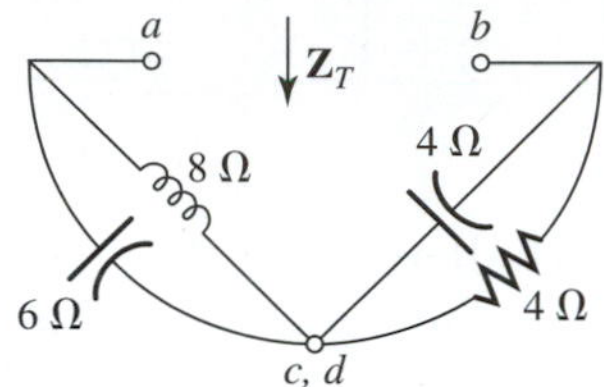

From Equations (17.35) and (17.36) for impedances in series and parallel, the Thevenin impedance $\mathbf{Z}_T$ is

$$\mathbf{Z}_T = \{(-j6)\|(j8)\} + \{(-j4)\|(4)\}$$
$$= 2 - j26$$
$$= 26.1\angle -85.6^\circ$$

The Thevenin equivalent for the circuit across terminals *a*-*b* with the 1-Ω resistance reconnected is shown in Figure 19.33.

Figure 19.33

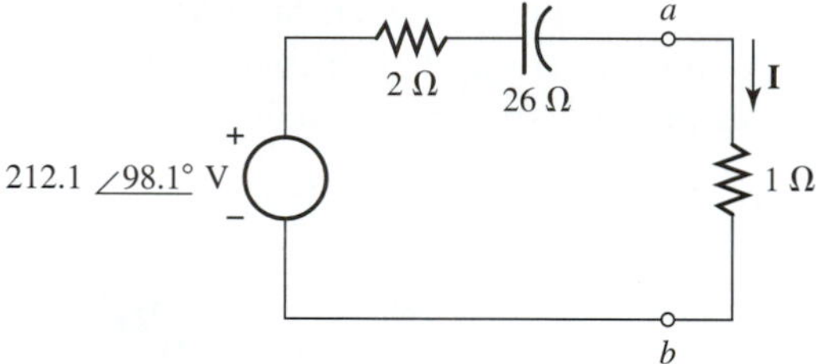

From Equation (17.1)

$$\mathbf{I} = \frac{(212\angle 98.1^\circ)}{3 - j26}$$
$$= 8.11\angle -178.5^\circ$$

19.3 NORTON'S THEOREM

Another very useful theorem for circuit analysis presented by E. L. Norton, an American engineer, in 1933 is known as Norton's theorem and can be applied to ac circuits in the same manner that we applied it to resistive circuits in Chapter 7.

NORTON'S THEOREM A circuit connected across two terminals *a*-*b* can be replaced by a current source in parallel with an impedance.

The parallel combination of the current source and the impedance is known as a Norton equivalent with respect to terminals *a*-*b*.

The parallel impedance, known as the Norton impedance and designated $\mathbf{Z}_N$, is the impedance across terminals *a-b* with all voltage and current sources in the circuit replaced by their internal impedances (a short for voltage sources and an open for current sources).

The parallel current source, known as the Norton current source and designated $\mathbf{I}_N$ in phasor form, is equal to the current flowing through a short connected across terminals *a-b*.

The *first step* in applying Norton's theorem is to isolate the two terminals connected to the portion of the circuit that is to be replaced by a Norton equivalent. Consider the circuit in Figure 19.14 (redrawn in Figure 19.34), where it is desired to replace the circuit to the left of terminals *a-b* by a Norton equivalent.

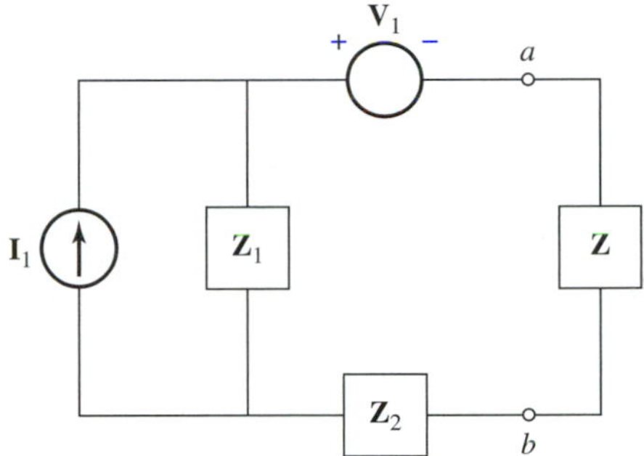

Figure 19.34 Norton's theorem.

The impedance **Z** is removed as shown in Figure 19.35 to isolate terminals *a-b* as required by Norton's theorem. Using Norton's theorem, the circuit

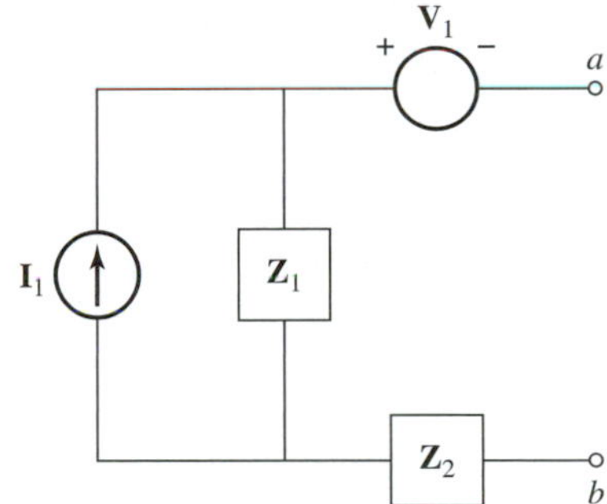

Figure 19.35 Norton's theorem.

in Figure 19.35 can be replaced by the Norton equivalent in Figure 19.36.

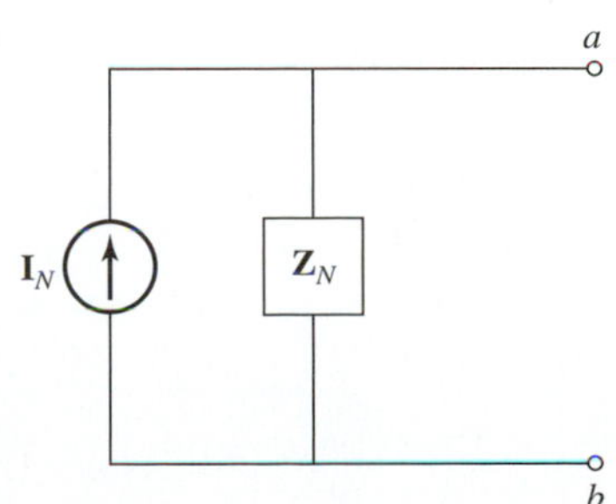

Figure 19.36 Norton's equivalent.

The *second step* in applying Norton's theorem is to determine the Norton current $\mathbf{I}_N$ for the Norton equivalent. Because $\mathbf{I}_N$ is the current through a short across terminals *a-b*, the circuit in Figure 19.35 with terminals *a-b* shorted is shown in Figure 19.37.

Figure 19.37 Norton's theorem.

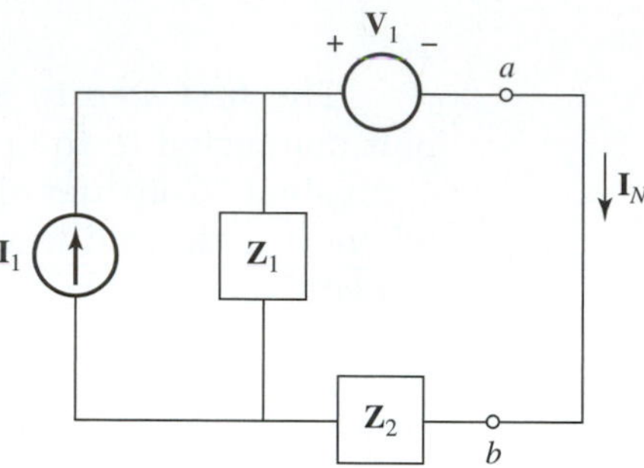

Any analysis method or laboratory procedure that applies can be used to determine the Norton current $\mathbf{I}_N$. We apply the superposition method to determine $\mathbf{I}_N$ for the circuit in Figure 19.37. Because there are two sources in the circuit there are two circuits to consider, as shown in Figure 19.38. $\mathbf{I}'_N$ is the contribution to $\mathbf{I}_N$ from the current source $\mathbf{I}_1$, and $\mathbf{I}''_N$ is the contribution to $\mathbf{I}_N$ from the voltage source $\mathbf{V}_1$.

Figure 19.38 Superposition.

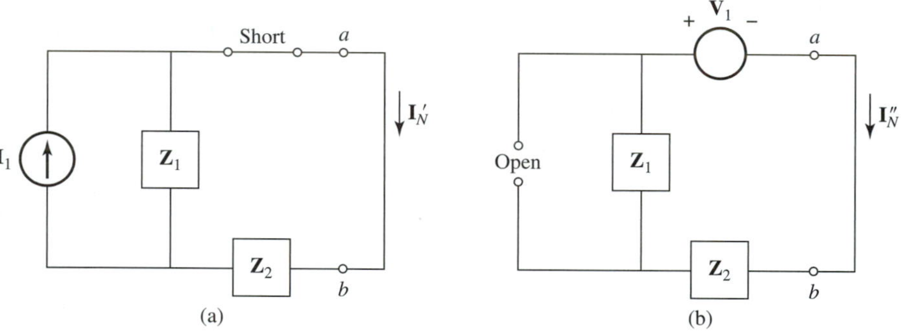

For the circuit in Figure 19.38(a), $\mathbf{I}'_N$ can be determined from Equation (17.41) for current division as

$$\mathbf{I}'_N = \left\{\frac{\mathbf{Z}_1}{\mathbf{Z}_1 + \mathbf{Z}_2}\right\}(\mathbf{I}_1)$$

For the circuit in Figure 19.38(b), $\mathbf{I}''_N$ can be determined from Equations (17.1) and (17.35) as

$$\mathbf{I}''_N = \frac{-\mathbf{V}_1}{\mathbf{Z}_1 + \mathbf{Z}_2}$$

$\mathbf{I}_N$ is the algebraic sum of the contributed currents $\mathbf{I}'_N$ and $\mathbf{I}''_N$.

$$\mathbf{I}_N = \mathbf{I}'_N + \mathbf{I}''_N$$

$$\mathbf{I}_N = \frac{\mathbf{Z}_1\mathbf{I}_1 - \mathbf{V}_1}{\mathbf{Z}_1 + \mathbf{Z}_2}$$

The *third step* in applying Norton's theorem is to determine the Norton impedance $\mathbf{Z}_N$ for the Norton equivalent circuit. The circuit in Figure 19.34 is shown in Figure 19.39 with all sources replaced by their internal impedances, as required by Norton's theorem for determining $\mathbf{Z}_N$.

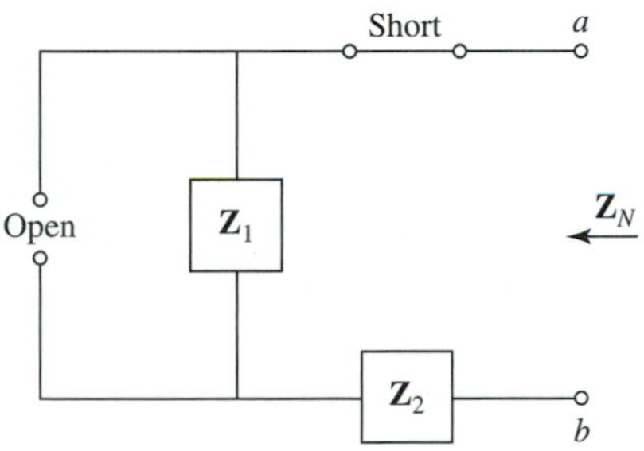

Figure 19.39 Norton impedance.

The Norton impedance $\mathbf{Z}_N$ is the impedance across terminals *a-b* in Figure 19.39. From Equation (17.35)

$$\mathbf{Z}_N = \mathbf{Z}_1 + \mathbf{Z}_2$$

The *fourth step* in applying Norton's theorem is to form the Norton equivalent circuit using $\mathbf{I}_N$ and $\mathbf{Z}_N$ determined in steps 2 and 3. For the circuit to the left of terminals *a-b* in Figure 19.34, the Norton equivalent circuit is shown in Figure 19.40.

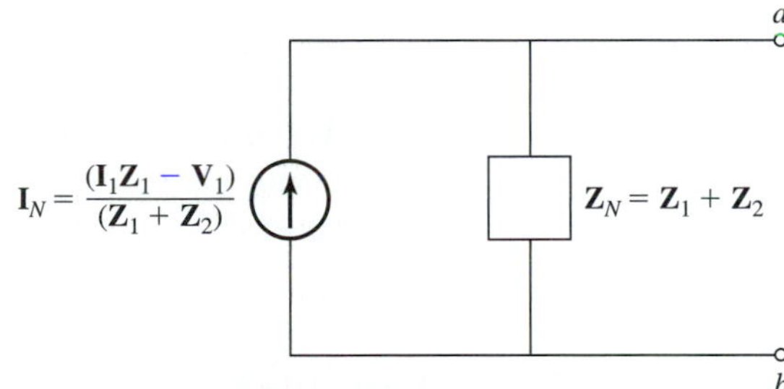

Figure 19.40 Norton equivalent.

The four-step procedure for applying Norton's theorem to ac circuits is summarized next.

1. Isolate the two terminals connected to the portion of the circuit to be replaced by a Norton equivalent.
2. Determine the Norton current $\mathbf{I}_N$.

3. Determine the Norton impedance $\mathbf{Z}_N$.
4. Use $\mathbf{I}_N$ and $\mathbf{Z}_N$ to form the Norton equivalent.

EXAMPLE 19.8 Determine a Norton equivalent for the circuit to the left of terminals *a*-*b* in Figure 19.41. ■

Figure 19.41

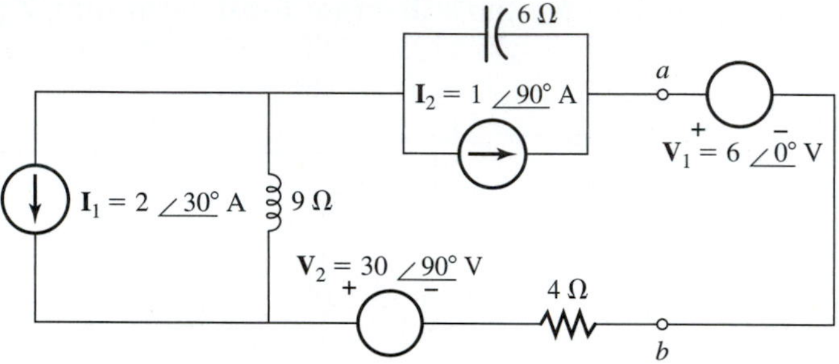

SOLUTION The circuit with terminals *a*-*b* isolated is shown in Figure 19.42.

Figure 19.42

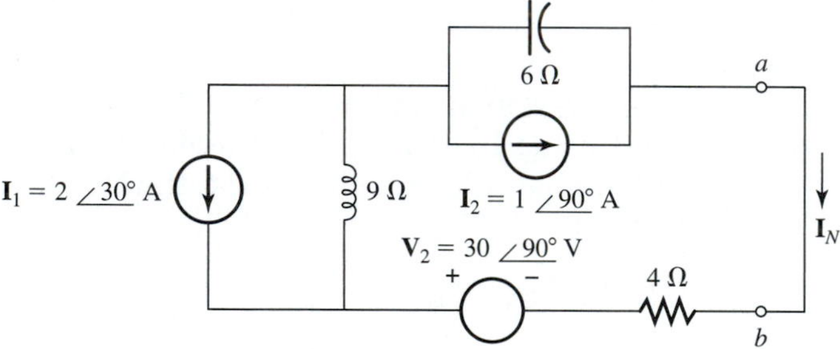

We shall use the superposition method to determine $\mathbf{I}_N$. Because there are three sources, three circuits are considered, as shown in Figures 19.43, 19.44, and 19.45.

Figure 19.43

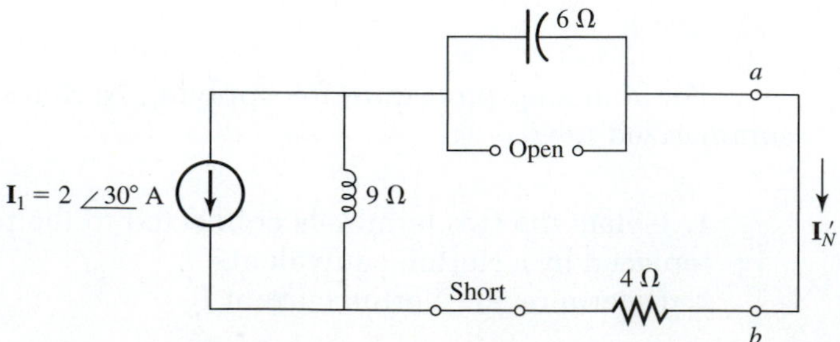

Figure 19.44

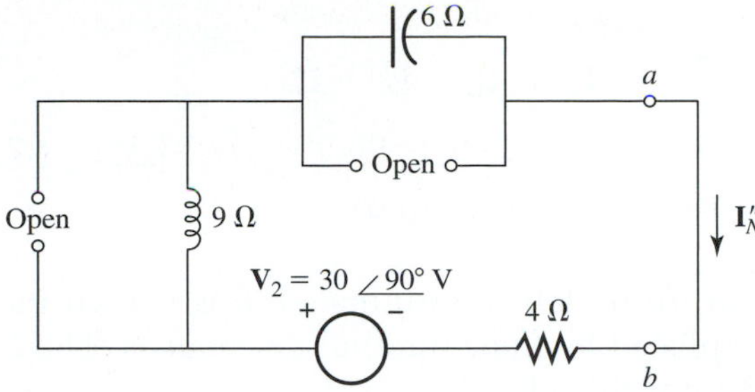

Figure 19.45

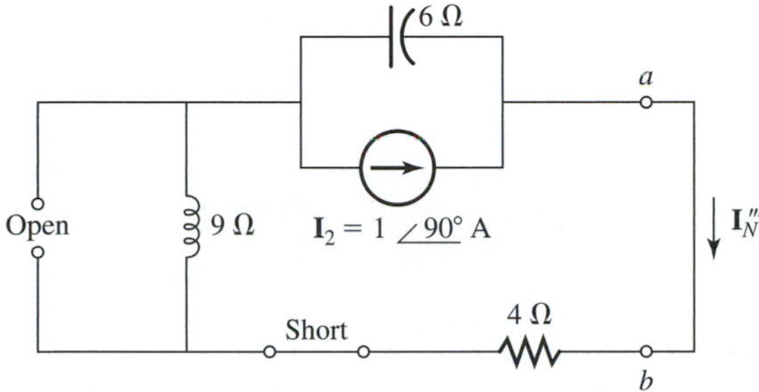

$\mathbf{I}'_N$ is the contribution to $\mathbf{I}_N$ from the current source $\mathbf{I}_1$. $\mathbf{I}''_N$ is the contribution to $\mathbf{I}_N$ from the voltage source $\mathbf{V}_2$. $\mathbf{I}'''_N$ is the contribution to $\mathbf{I}_N$ from the current source $\mathbf{I}_2$. For the circuit in Figure 19.43, $\mathbf{I}'_N$ is determined from Equation (17.41) for current division.

$$\begin{aligned}\mathbf{I}'_N &= -\left\{\frac{j9}{4 - j6 + j9}\right\}(2\angle 30°)\\ &= -0.43 - j3.6\\ &= 3.6\angle -96.9°\end{aligned}$$

For the circuit in Figure 19.44, $\mathbf{I}''_N$ is determined from Equations (17.1) and (17.35).

$$\begin{aligned}\mathbf{I}''_N &= \frac{30\angle 90°}{4 + j9 - j6}\\ &= 3.6 + j4.8\\ &= 6\angle 53/1°\end{aligned}$$

For the circuit in Figure 19.45, $\mathbf{I}'''_N$ is determined from Equation (17.41) for current division.

$$\begin{aligned}\mathbf{I}'''_N &= \left\{\frac{-j6}{4 + j9 - j6}\right\}(1\angle 90°)\\ &= 0.96 - j0.72\\ &= 1.2\angle -36.9°\end{aligned}$$

$\mathbf{I}_N$ is the algebraic sum of the contributed currents.

$$\begin{aligned}\mathbf{I}_N &= \mathbf{I}'_N + \mathbf{I}''_N + \mathbf{I}'''_N \\ &= 3.6\underline{/-96.9^\circ} + 6\underline{/53.1^\circ} + 1.2\underline{/-36.9^\circ} \\ &= 4.2\underline{/6.9^\circ}\end{aligned}$$

The circuit to the left of terminals *a-b* is shown in Figure 19.46, with all the sources replaced by their internal resistances. The Norton resistance $\mathbf{Z}_N$ is determined from this circuit.

Figure 19.46

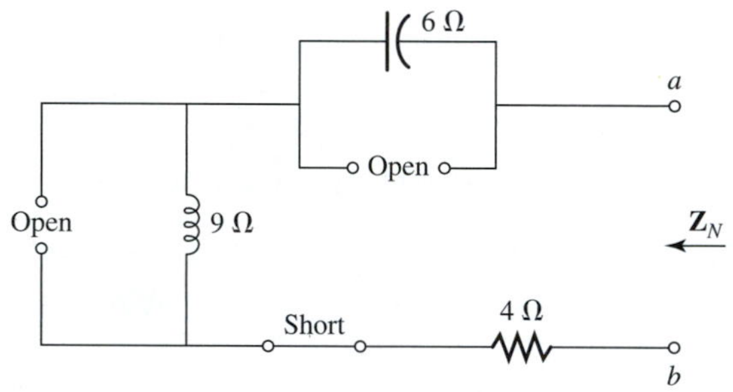

From Equation (17.35)

$$\begin{aligned}\mathbf{Z}_N &= 4 + j9 - j6 \\ &= 4 + j3 \\ &= 5\underline{/36.9^\circ}\end{aligned}$$

The Norton equivalent for the circuit to the left of terminals *a-b* is shown in Figure 19.47.

Figure 19.47

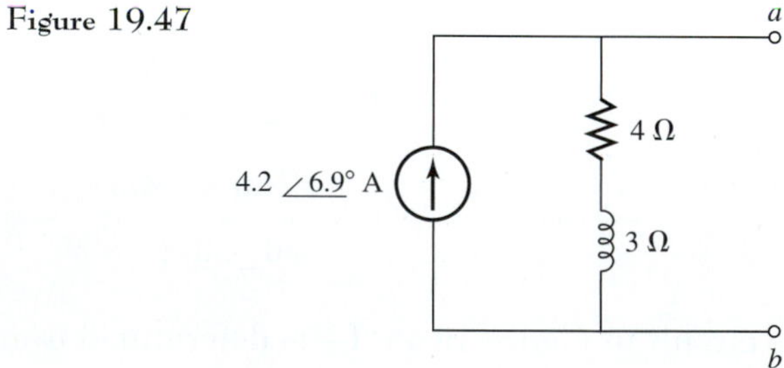

Notice that the Norton equivalent determined here is equivalent to the Thevenin equivalent determined in Example 19.5 for the same circuit.

EXAMPLE 19.9 For the circuit in Example 19.8, determine the current **I** flowing through the voltage source $\mathbf{V}_1$. The circuit is redrawn in Figure 19.48. ■

Figure 19.48

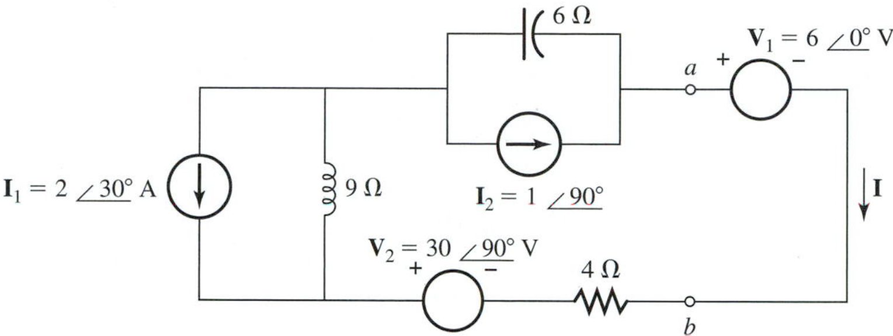

SOLUTION The circuit to the left of terminals *a-b* can be replaced by the Norton equivalent determined in Example 19.8, as shown in Figure 19.49.

Figure 19.49

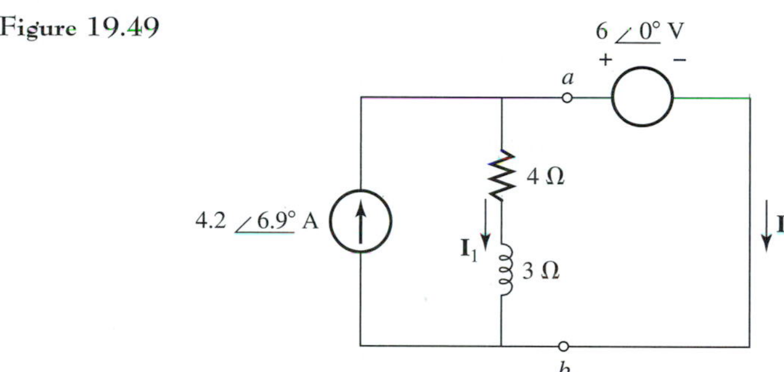

Because the voltage source ($6\angle 0°$) appears directly across the branch with impedance ($4 + j3$), applying Equation (17.1) to the branch results in

$$\begin{aligned}\mathbf{I}_1 &= \frac{6\angle 0°}{4 + j3} \\ &= 0.96 - j0.72 \\ &= 1.2\angle -36.9°\end{aligned}$$

Applying Kirchhoff's current law to node *a* results in

$$\begin{aligned}\mathbf{I} &= 4.2\angle 6.9° - 1.2\angle -36.9° \\ &= 3.2 + j1.2 \\ &= 3.4\angle 21.1°\end{aligned}$$

EXAMPLE 19.10 Determine the current **I** for the circuit in Figure 19.50. ■

Figure 19.50

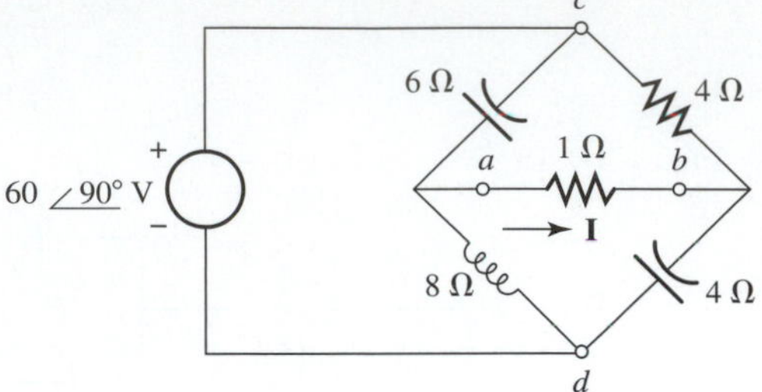

SOLUTION We shall remove the 1-Ω resistor and determine a Norton equivalent with respect to terminals *a*-*b*. In accordance with Norton's theorem, terminals *a*-*b* are shorted as shown in Figure 19.51. The Norton current $\mathbf{I}_N$ is the current through the short, as indicated.

Figure 19.51

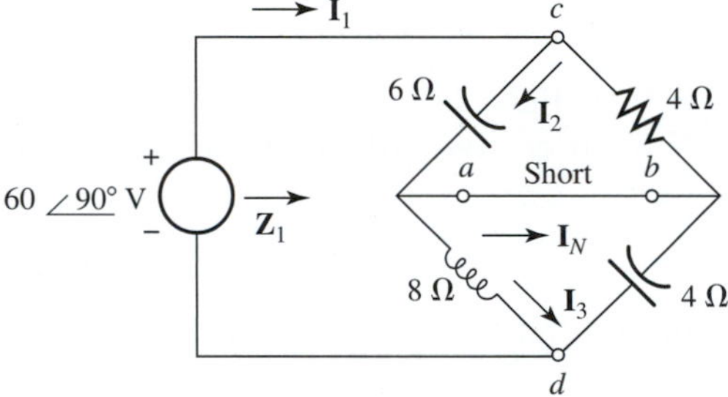

The circuit is redrawn in Figure 19.52 to indicate the parallel configuration of the elements when terminals *a*-*b* are shorted. Using Equations (17.35) and (17.36) to combine impedances results in

Figure 19.52

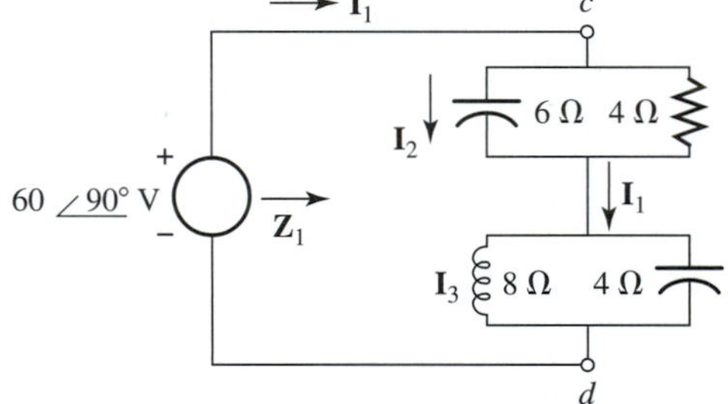

$$\begin{aligned}\mathbf{Z}_1 &= \{(-j6)\|4\} + \{(-j4)\|(j8)\}\\ &= \left\{\frac{(-j6)(4)}{-j6+4}\right\} + \left\{\frac{(-j4)(j8)}{-j4+j8}\right\}\\ &= 2.8 - j1.8 - 8\\ &= 10.2\underline{/-74.3^\circ}\end{aligned}$$

where $\mathbf{Z}_1$ is the impedance seen by the voltage source. From Equation (17.1)

$$\begin{aligned}\mathbf{I}_1 &= \frac{60\underline{/90^\circ}}{10.2\underline{/-74.3^\circ}}\\ &= 5.9\underline{/164.3^\circ}\end{aligned}$$

From Equation (17.41) for current division,

$$\mathbf{I}_2 = \left\{\frac{4}{4 - j6}\right\}\mathbf{I}_1$$
$$= \left\{\frac{4}{4 - j6}\right\}(5.9\underline{/164.3^\circ})$$
$$= 3.3\underline{/-139.4^\circ}$$
$$\mathbf{I}_3 = \left\{\frac{-j4}{-j4 + j8}\right\}\mathbf{I}_1$$
$$= \left\{\frac{-j4}{-j4 + j8}\right\}(5.9\underline{/164.3^\circ})$$
$$= 5.9\underline{/-15.7^\circ}$$

Let us refer back to the circuit in Figure 19.51 with terminals *a-b* shorted and the Norton current $\mathbf{I}_N$ indicated. Applying Kirchhoff's current law at node *a* results in

$$\mathbf{I}_2 - \mathbf{I}_N - \mathbf{I}_3 = 0$$
$$\mathbf{I}_N = \mathbf{I}_2 - \mathbf{I}_3$$
$$= 3.3\underline{/-139.4^\circ} - 5.9\underline{/-15.7^\circ}$$
$$= 8.14\underline{/-176.3^\circ}$$

The Norton resistance $\mathbf{Z}_N$ is determined by replacing the voltage source with a short and finding the resulting impedance across terminals *a-b*. The circuit used to determine $\mathbf{Z}_N$ is shown in Figure 19.53. Notice that shorting the voltage source is the same as connecting terminals *c* and *d*.

Figure 19.53

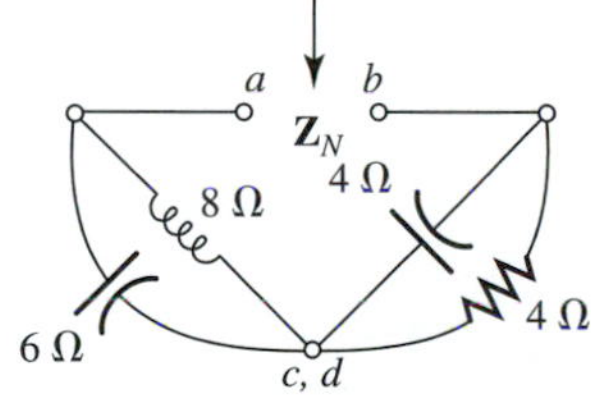

From Equations (17.35) and (17.37), the Norton resistance $\mathbf{Z}_N$ is

$$\mathbf{Z}_N = \{(-j6)\|(j8)\} + \{(-j4)\|(4)\}$$
$$= (-j24) + (2.83\underline{/-45^\circ})$$
$$= 2 - j26$$
$$= 26.08\underline{/-85.6^\circ}$$

The Norton equivalent for the circuit across terminals *a-b* with the 1-Ω resistance removed is shown in Figure 19.54, with the 1-Ω resistance reconnected.

Figure 19.54

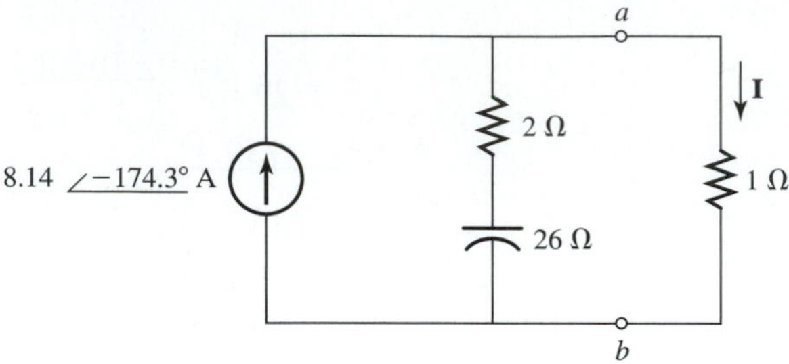

From Equation (17.41) for current division,

$$\mathbf{I} = \left\{\frac{2 - j26}{2 - j26 + 1}\right\}(8.14\angle -176.3°)$$

$$= 8.11\angle -178.5°$$

From the Thevenin and Norton theorems, we see that the Thevenin and Norton impedances are identical for any given circuit. Because $\mathbf{Z}_N$ equals $\mathbf{Z}_T$ for a given circuit, the choice of applying Norton's theorem or Thevenin's theorem depends on which is easiest to determine, the Norton current or the Thevenin voltage.

It is sometimes useful in the analysis of ac circuits to transform a Thevenin equivalent circuit to a Norton equivalent circuit, and vice versa. Recalling the rules for transforming imperfect sources in Chapter 17, we can transform a Thevenin equivalent to a Norton equivalent and a Norton equivalent to a Thevenin equivalent for ac circuits, as shown here.

Figure 19.55 Transformation of sources.

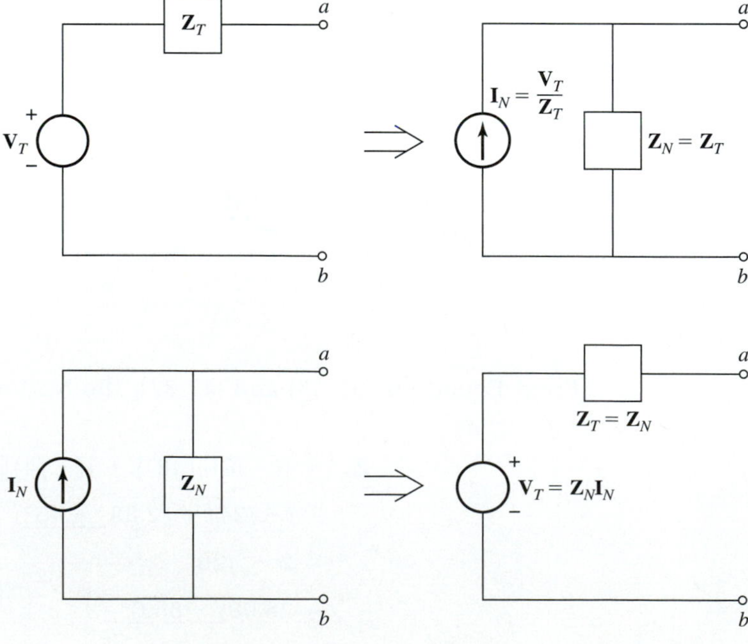

The transformation rules in Figure 19.55 can be verified by simply applying Norton's or Thevenin's theorem to the circuits.

19.4 SUCCESSIVE-SOURCE-CONVERSION METHOD

It is often possible to determine a current or voltage in an ac circuit by simply converting imperfect current sources to imperfect voltage sources, and vice versa. The successive-source-conversion method involves reducing the circuit by successive source conversions to an equivalent circuit that can be easily solved for the desired current or voltage. The method is best described by an example and is demonstrated in Example 19.11.

EXAMPLE 19.11 Determine **I** for the circuit in Figure 19.56. ■

Figure 19.56

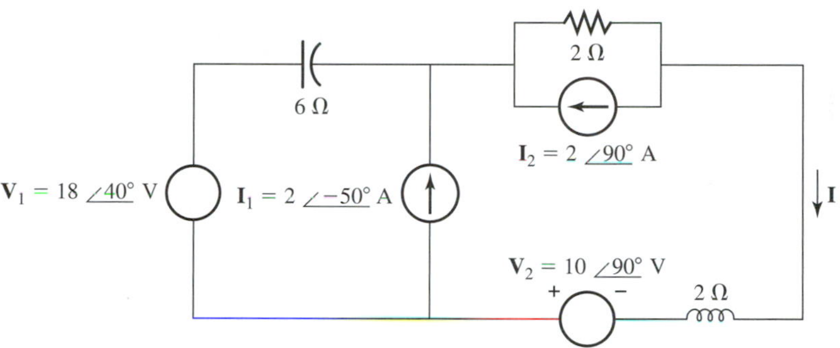

SOLUTION We first convert voltage source $\mathbf{V}_1$ with the 6-Ω capacitor in series to a current source, as shown in Figure 19.57.

Figure 19.57

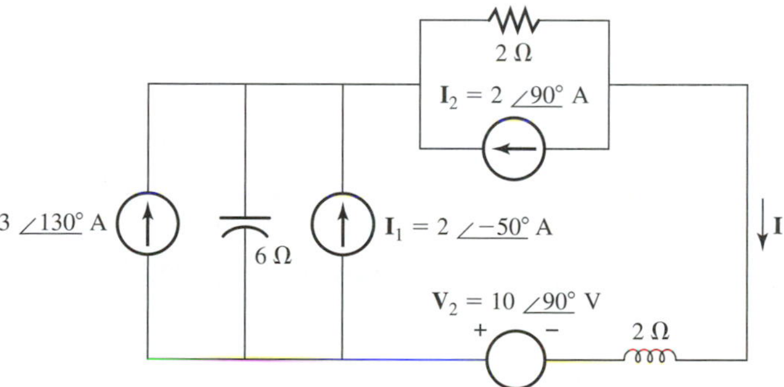

We next combine the parallel current sources $3\angle 130°$ and $\mathbf{I}_1$, which results in

$$3\angle 130° + 2\angle -50° = 1\angle 130° \text{ A}$$

The resulting circuit is shown in Figure 19.58.

Figure 19.58

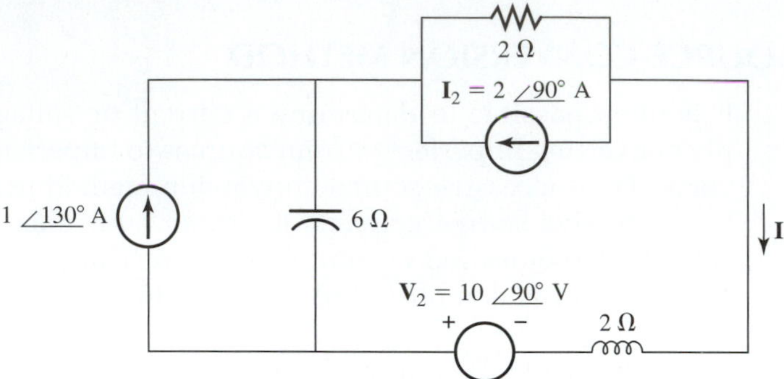

Next we convert the two current sources to voltage sources, which results in

$$(1\angle 130°)(-j6) = 6\angle 40°\ \text{V}$$

$$(2\angle 90°)(2) = 4\angle 90°\ \text{V}$$

The resulting circuit is shown in Figure 19.59.

Figure 19.59

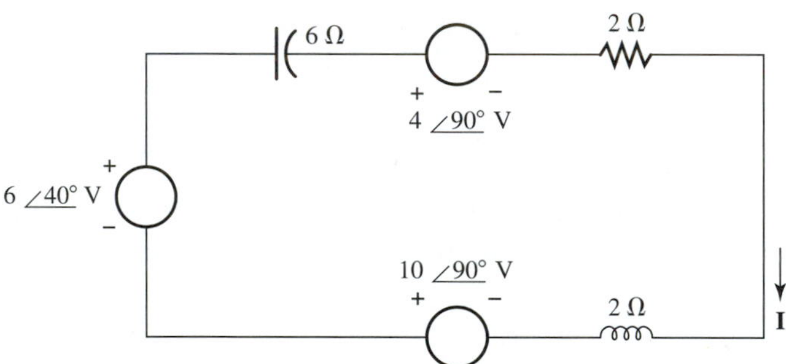

Finally, combining the voltage sources and using Equation (17.35) to combine impedances in series results in

$$6\angle 40° - 4\angle 90° + 10\angle 90° = 10.87\angle 65°\ \text{V}$$

$$-j6 + 2 + j2 = 4.47\angle -63.4°\ \Omega$$

The resulting circuit is shown in Figure 19.60.

Figure 19.60

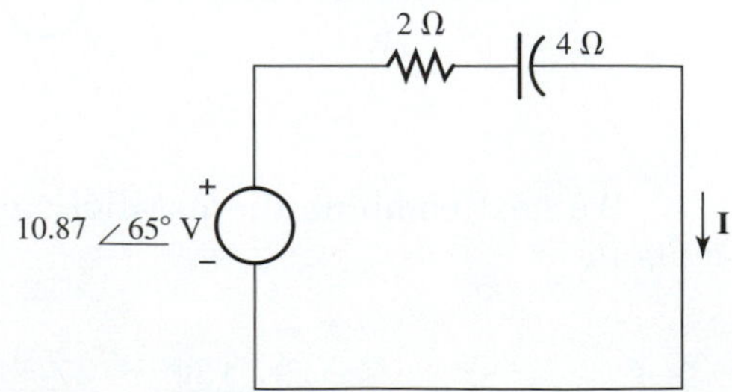

From Equation (17.1)

$$\mathbf{I} = \frac{10.87\angle 65^\circ}{4.47\angle -63.4^\circ}$$

$$= 2.43\angle 128.4^\circ \text{ A}$$

19.5 THE SUPERPOSITION METHOD AND DEPENDENT SOURCES

The superposition method is applicable to ac circuits containing dependent sources. However, the method as previously presented for analyzing ac circuits with only independent sources must be altered to accommodate the presence of the dependent sources. Because the value of a dependent source depends on a voltage or current in another part of the circuit, the dependent source cannot be removed when applying the superposition method. When determining the voltage or current contributed by each independent source for the superposition method, the dependent sources are left in the circuit.

Consider the following example.

EXAMPLE 19.12 Determine the current **I** for the circuit in Figure 19.61. ■

Figure 19.61

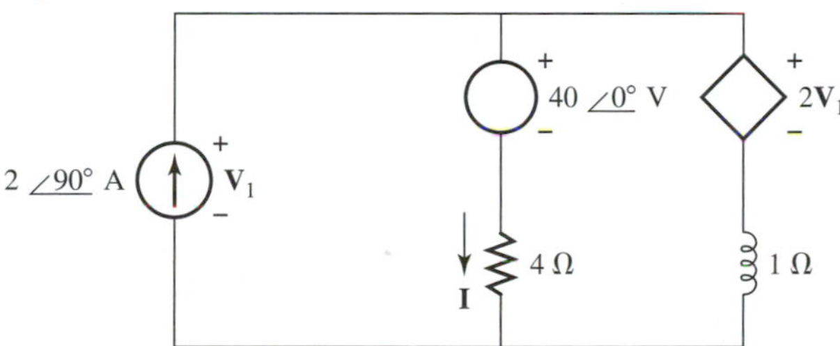

SOLUTION Because there are two independent sources, there are two circuits to consider for the superposition method. Notice that the dependent voltage source ($2\mathbf{V}_1$) depends on the voltage across the independent current source.

The circuit shown in Figure 19.62 is used to determine $\mathbf{I}'$, the contribution to **I** from the independent voltage source.

Figure 19.62

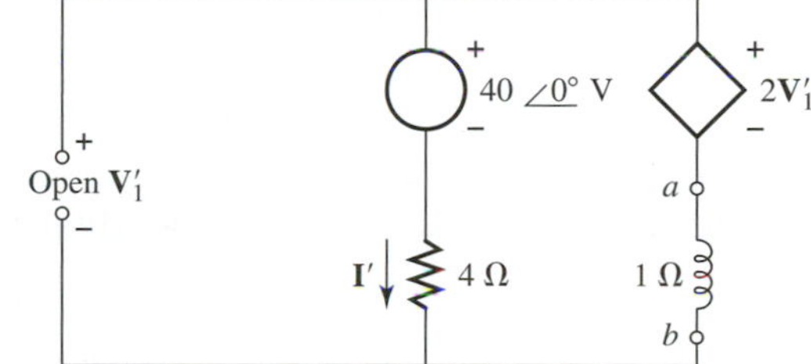

From an inspection of the circuit in Figure 19.62 and Equation (17.3), the voltage across the open is

$$\mathbf{V}_1' = 40\angle 0^\circ + 4\mathbf{I}'$$

and the voltage across the dependent voltage source is

$$\begin{aligned} 2\mathbf{V}_1' &= 2(40\angle 0^\circ + 4\mathbf{I}') \\ &= 80\angle 0^\circ + 8\mathbf{I}' \end{aligned}$$

From Equations (17.1) and (17.8), the voltage across the 1-Ω inductor is

$$\mathbf{V}_{ab} = -(j1)\mathbf{I}'$$

Because the voltage across the open is equal to the voltage across the branch containing the dependent source,

$$\begin{aligned} \mathbf{V}_1' &= 2\mathbf{V}_1' + \mathbf{V}_{ab} \\ 40\angle 0^\circ + 4\mathbf{I}' &= 80\angle 0^\circ + 8\mathbf{I}' - (j1)\mathbf{I}' \\ \mathbf{I}' &= -\frac{40}{4 - j} \end{aligned}$$

The circuit shown in Figure 19.63 is used to determine $\mathbf{I}''$, the contribution to **I** from the independent current source.

Figure 19.63

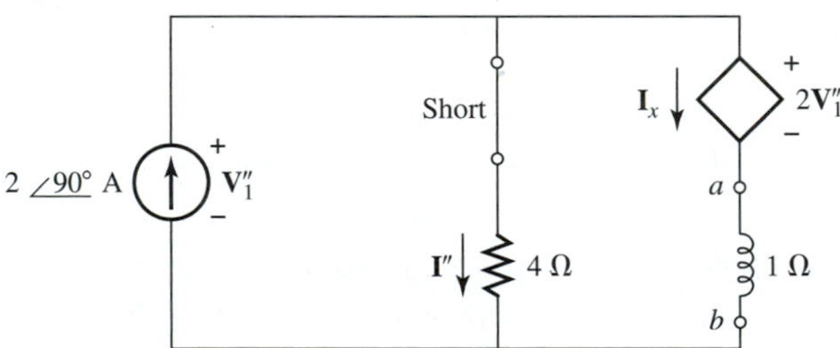

From an inspection of the circuit and Equation (17.3),

$$\mathbf{V}_1'' = 4\mathbf{I}''$$

and the voltage across the dependent voltage source is

$$2\mathbf{V}_1'' = 8\mathbf{I}''$$

From Kirchhoff's current law,

$$\begin{aligned} \mathbf{I}_x &= 2\angle 90^\circ - \mathbf{I}'' \\ &= j2 - \mathbf{I}'' \end{aligned}$$

and from Equation (17.8),

$$\begin{aligned}\mathbf{V}_{ab} &= (j1)\mathbf{I}_x \\ &= (j1)(j2 - \mathbf{I}'')\end{aligned}$$

Because the voltage across the current source is equal to the voltage across the branch containing the dependent source,

$$\begin{aligned}\mathbf{V}''_1 &= 2\mathbf{V}''_1 + \mathbf{V}_{ab} \\ 4\mathbf{I}'' &= 8\mathbf{I}'' + j(j2 - \mathbf{I}'') \\ \mathbf{I}'' &= \frac{2}{4 - j}\end{aligned}$$

From the superposition method

$$\begin{aligned}\mathbf{I} &= \mathbf{I}' + \mathbf{I}'' \\ &= \left\{-\frac{40}{4 - j}\right\} + \left\{\frac{2}{4 - j}\right\} \\ &= -\frac{38}{4 - j} \\ &= 9.22\underline{/-165.9^\circ}\end{aligned}$$

19.6 THEVENIN AND NORTON EQUIVALENT CIRCUITS AND DEPENDENT SOURCES

The Norton and Thevenin theorems can be applied to circuits with dependent sources. However, the procedures presented previously for applying the Thevenin and Norton theorems to circuits with only independent ac sources must be slightly modified to accommodate the presence of the dependent sources.

The modification necessary to accommodate dependent sources when applying Thevenin's and Norton's theorems involves the determination of $\mathbf{Z}_T$ and $\mathbf{Z}_N$. If the circuit to be replaced by a Thevenin or Norton equivalent contains dependent sources, the dependent sources cannot be removed from the circuit as independent sources are removed to determine $\mathbf{Z}_T$ and $\mathbf{Z}_N$. The alternative methods presented in Chapter 7 for determining Thevenin and Norton resistances for resistive circuits can be used to determine Thevenin and Norton impedances when dependent sources are present in ac circuits. The two alternative methods presented in Chapter 7 for determining $\mathbf{R}_T$ and $\mathbf{R}_N$ when dependent sources are present can be modified to determine $\mathbf{Z}_T$ and $\mathbf{Z}_N$ for ac circuits. The two methods are as follows.

1. Determine $\mathbf{V}_T$ and $\mathbf{I}_N$ and then set $\mathbf{Z}_T$ to $\mathbf{V}_T/\mathbf{I}_N$ or $\mathbf{Z}_N = \mathbf{V}_T/\mathbf{I}_N$.
2. Apply a current source $\mathbf{I}_X$ and determine the voltage $\mathbf{V}_X$ across the source; or apply a voltage source $\mathbf{V}_X$ and determine the current $\mathbf{I}_X$ flowing through the source. $\mathbf{Z}_T$ and $\mathbf{Z}_N$ are then set equal to $\mathbf{V}_X/\mathbf{I}_X$.

These methods for determining Thevenin and Norton impedances for ac circuits with dependent sources are demonstrated in the following examples.

EXAMPLE 19.13 Determine a Thevenin equivalent for the circuit to the left of terminals *a*-*b* in Figure 19.64. ■

Figure 19.64

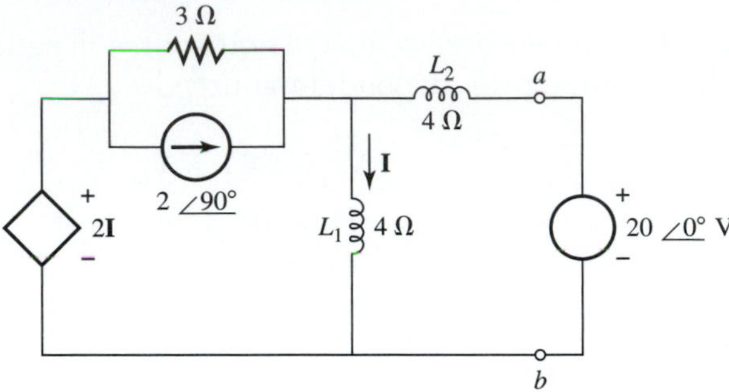

SOLUTION The voltage source across terminals *a*-*b* is removed to isolate the terminals, as required by Thevenin's theorem. The Thevenin voltage $\mathbf{V}_T$ is the open-circuited voltage across terminals *a*-*b*, as shown in Figure 19.65. Notice that the dependent voltage source (2**I**) is controlled by the current **I** through the inductor L_1, and both the dependent source and the controlling element are in the circuit to the left of terminals *a*-*b*.

Figure 19.65

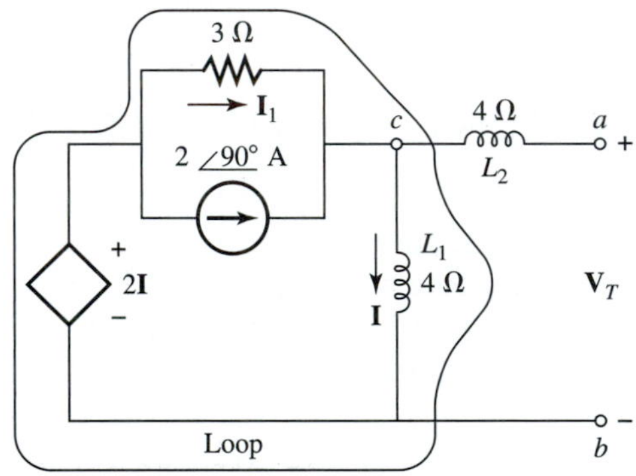

Applying Kirchhoff's current law to node *c* results in

$$\mathbf{I}_1 + 2\angle 90^\circ - \mathbf{I} = 0$$

$$\mathbf{I}_1 = \mathbf{I} - j2$$

Applying Kirchhoff's voltage law to the loop indicated in Figure 19.65 results in

$$2\mathbf{I} - 3\mathbf{I}_1 - j4\mathbf{I} = 0$$

Substituting $(\mathbf{I} - j2)$ for $\mathbf{I}_1$ results in

$$2\mathbf{I} - 3(\mathbf{I} - j2) - j4\mathbf{I} = 0$$

$$\mathbf{I} = \frac{j6}{1 + j4}$$

Because no current flows through L_2, $\mathbf{V}_T$ is the voltage across L_1, and Equation (17.1) results in

$$\begin{aligned}\mathbf{V}_T &= (j4)\mathbf{I}\\ &= \frac{(j4)(j6)}{1 + j4}\\ &= 5.82\underline{/\ 104.03^\circ}\end{aligned}$$

We shall determine $\mathbf{I}_N$ in order to arrive at $\mathbf{Z}_T$, because $\mathbf{Z}_T = \mathbf{V}_T/\mathbf{I}_N$. Terminals *a-b* are shorted, as required by Norton's theorem, to determine the Norton current $\mathbf{I}_N$, as shown in Figure 19.66.

Figure 19.66

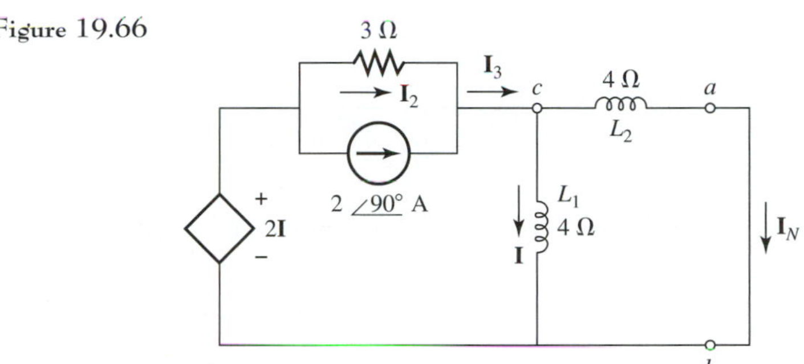

From Kirchhoff's current law

$$\mathbf{I}_3 = \mathbf{I}_2 + 2\underline{/\ 90^\circ}$$

From Equation (17.41) for current division,

$$\begin{aligned}\mathbf{I}_N &= \left\{\frac{j4}{j4 + j4}\right\}\mathbf{I}_3\\ &= \frac{\mathbf{I}_2 + j2}{2}\end{aligned}$$

and

$$\begin{aligned}\mathbf{I} &= \left\{\frac{j4}{j4 + j4}\right\}\mathbf{I}_3\\ &= \frac{\mathbf{I}_3}{2}\end{aligned}$$

From these equations

$$\mathbf{I} = \mathbf{I}_N$$

and

$$\mathbf{I}_2 = 2\mathbf{I}_N - j2$$

Applying Kirchhoff's voltage law to the loop indicated in the circuit shown in Figure 19.67 results in

$$2\mathbf{I} - 3\mathbf{I}_2 - j4\mathbf{I} = 0$$

Figure 19.67

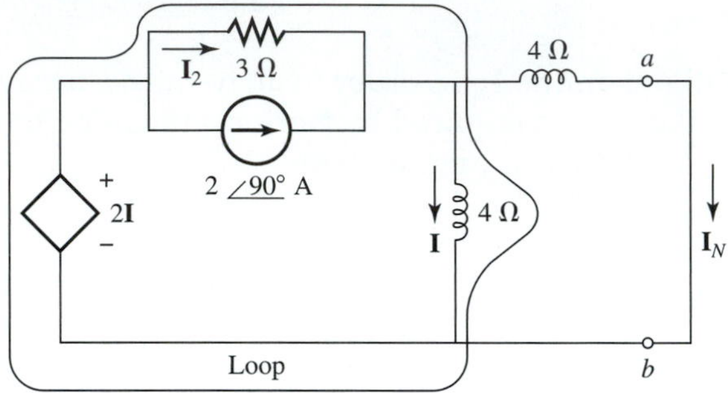

Substituting ($\mathbf{I}_N$) for $\mathbf{I}$ and ($2\mathbf{I}_N - j2$) for $\mathbf{I}_2$ results in

$$2\mathbf{I}_N - 3(2\mathbf{I}_N - j2) - j4\mathbf{I}_N = 0$$

$$(-4 - j4)\mathbf{I}_N = -j6$$

$$\mathbf{I}_N = \frac{j6}{4 + j4}$$

$$= 1.061\angle 45^\circ$$

It follows that

$$\mathbf{Z}_T = \frac{\mathbf{V}_T}{\mathbf{I}_N}$$

$$= \frac{5.82\angle 104.03^\circ}{1.061\angle 45^\circ}$$

$$= 5.49\angle 59.03^\circ$$

$$= 2.82 + j4.70$$

The Thevenin equivalent for the original circuit to the left of terminals *a*-*b* is shown in Figure 19.68.

Figure 19.68

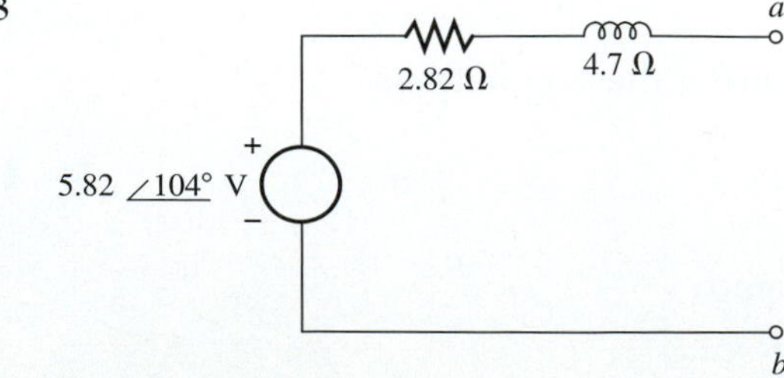

EXAMPLE 19.14 Determine the Thevenin impedance $\mathbf{Z}_T$ for the circuit to the left of terminals *a-b* in Example 19.13 by applying an external voltage source to the circuit. ■

SOLUTION The circuit with an applied voltage source $\mathbf{V}_y$ and the independent source removed is shown in Figure 19.69.

Figure 19.69

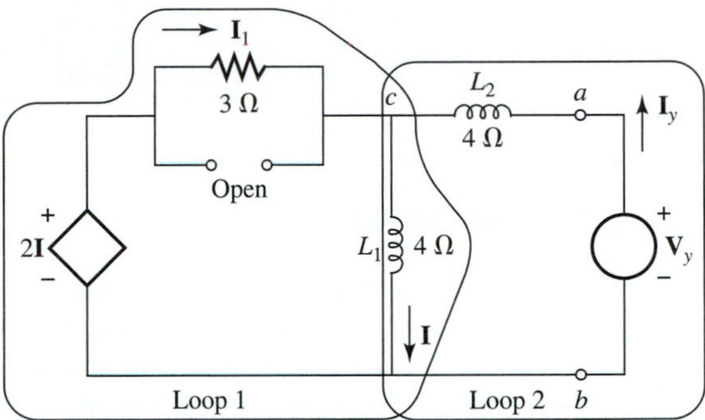

Applying Kirchhoff's current law to node *c* results in

$$\mathbf{I}_1 + \mathbf{I}_y = \mathbf{I}$$
$$\mathbf{I}_1 = \mathbf{I} - \mathbf{I}_y$$

Applying Kirchhoff's voltage law to loop 1 in Figure 19.69 results in

$$2\mathbf{I} - 3\mathbf{I}_1 - j4\mathbf{I} = 0$$

Substituting $(\mathbf{I} - \mathbf{I}_y)$ for $\mathbf{I}_1$ into the preceding equation results in

$$2\mathbf{I} - 3(\mathbf{I} - \mathbf{I}_y) - j4\mathbf{I} = 0$$
$$(1 + j4)\mathbf{I} = 3\mathbf{I}_y$$
$$\mathbf{I} = \frac{3\mathbf{I}_y}{1 + j4}$$

Applying Kirchhoff's voltage law to loop 2 indicated in Figure 19.69 results in

$$(j4)\mathbf{I} + (j4)\mathbf{I}_y = \mathbf{V}_y$$

Substituting $\{(3\mathbf{I}_y)/(1 + j4)\}$ for $\mathbf{I}$ into the preceding equation results in

$$(j4)\left\{\frac{(3\mathbf{I}_y)}{1 + j4}\right\} + (j4)\mathbf{I}_y = \mathbf{V}_y$$
$$\left\{\frac{j12}{1 + j4} + j4\right\}\mathbf{I}_y = \mathbf{V}_y$$
$$\mathbf{V}_y/\mathbf{I}_y = \left\{\frac{j12}{1 + j4}\right\} + j4$$
$$= \frac{j12 + j4 - 16}{1 + j4}$$

$$= \frac{-16 + j16}{1 + j4}$$
$$= 5.49\angle 59.03^\circ$$
$$= 2.82 + j4.70$$

It follows that the Thevenin impedance is

$$\mathbf{Z}_T = 2.82 + j4.70$$

19.7 MAXIMUM-POWER-TRANSFER THEOREM

In ac electric circuit applications it is often required to determine maximum power transfer from a circuit to a given load. Consider the circuit in Figure 19.70, where it is desired to determine the value of $\mathbf{Z}_L$ that results in maximum power delivered to $\mathbf{Z}_L$ by the circuit to the left of terminals *a-b*.

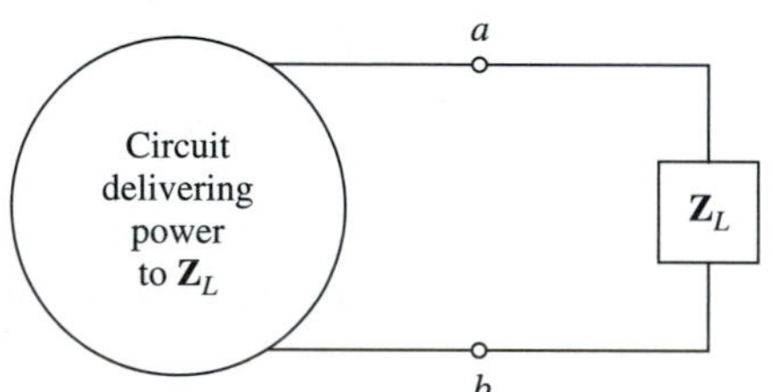

Figure 19.70 Maximum power transfer.

The circuit to the left of terminals *a-b* can be replaced by a Thevenin equivalent, as shown in Figure 19.71.

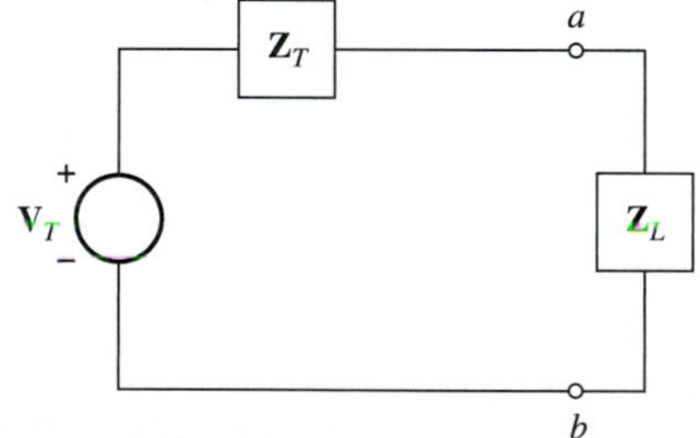

Figure 19.71 Maximum power transfer.

The maximum power transfer theorem for ac circuits states that maximum power will be delivered to $\mathbf{Z}_L$ by the circuit to the left of terminals *a-b* in Figure 19.71 when $\mathbf{Z}_L$ is equal to the conjugate of the Thevenin impedance $\mathbf{Z}_T$. That is, for maximum power transfer to $\mathbf{Z}_L$,

$$\mathbf{Z}_L = \tilde{\mathbf{Z}}_T$$

where $\tilde{\mathbf{Z}}_T$ is the conjugate of $\mathbf{Z}_T$.

In rectangular form if $\mathbf{Z}_T = R_T + jX_T$, then

$$\tilde{\mathbf{Z}}_L = R_T - jX_T \tag{19.1}$$

for maximum power transfer.

In polar form if $\mathbf{Z}_T = Z_T \angle \theta_T$, then

$$\tilde{\mathbf{Z}}_L = Z_T \angle -\theta_T \tag{19.2}$$

for maximum power transfer. In Equation (19.2), Z_T is the magnitude of the Thevenin impedance and θ_T is the angle associated with the Thevenin impedance.

EXAMPLE 19.15 Determine the value of the impedance **Z** for maximum power transfer to **Z** by the circuit to the left of terminals *a-b* in Figure 19.72.

Figure 19.72

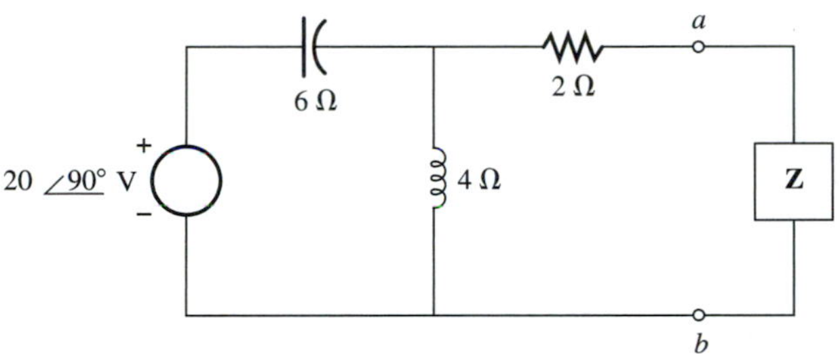

SOLUTION The Thevenin impedance for the circuit to the left of terminals *a-b* can be determined from the circuit in Figure 19.73.

Figure 19.73

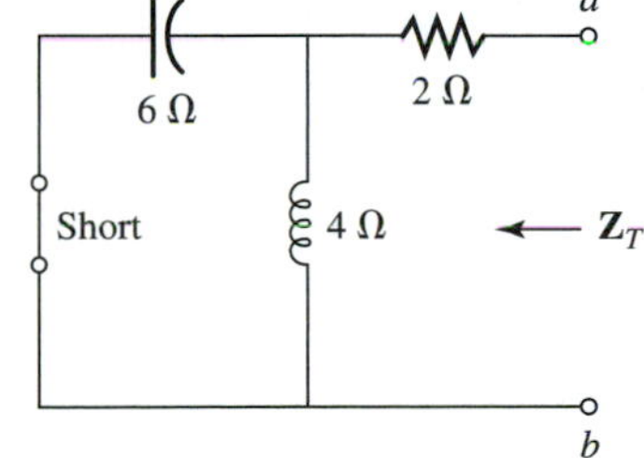

From Equations (17.35) and (17.36),

$$\begin{aligned}\mathbf{Z}_T &= \{(-j6)\|(j4)\} + 2\\ &= \left\{\frac{(-j6)(j4)}{-j6 + j4}\right\} + 2\\ &= 2 + j12\\ &= 12.17\angle 80.5^\circ\end{aligned}$$

For maximum power transfer to impedance **Z** from the circuit to the left of terminals a-b,

$$\mathbf{Z} = \tilde{\mathbf{Z}}_T = \widetilde{2 + j12} = 2 - j12$$

as shown in Figure 19.74.

Figure 19.74

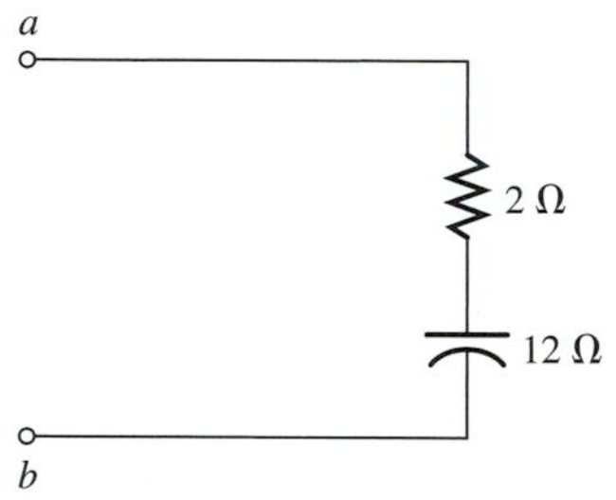

a

2 Ω

12 Ω

b

19.8 EXERCISES

1. Determine the phasor current **I** for the circuit in Figure 19.75. (Use the superposition method.)

Figure 19.75

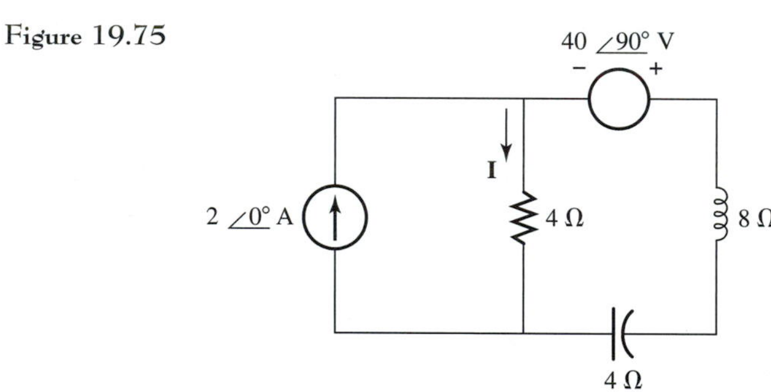

2. Determine the phasor current **I** for the circuit in Figure 19.76. (Use the superposition method.)

Figure 19.76

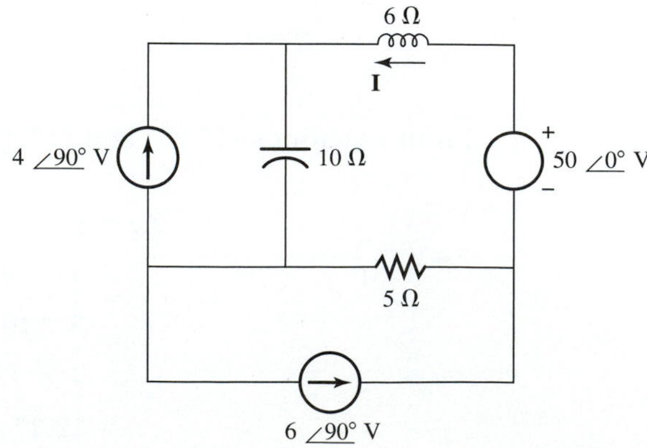

3. Determine the phasor voltages $\mathbf{V}_1$ and $\mathbf{V}_2$ for the circuit in Figure 19.77. (Use the superposition method.)

Figure 19.77

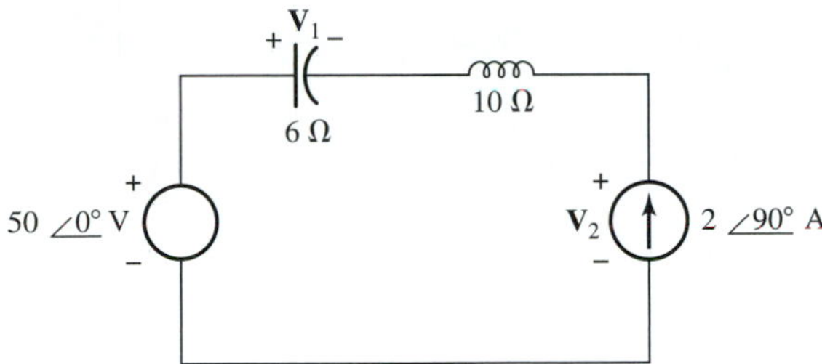

4. Determine the phasor current **I** for the circuit in Figure 19.78. (Use the superposition method.)

Figure 19.78

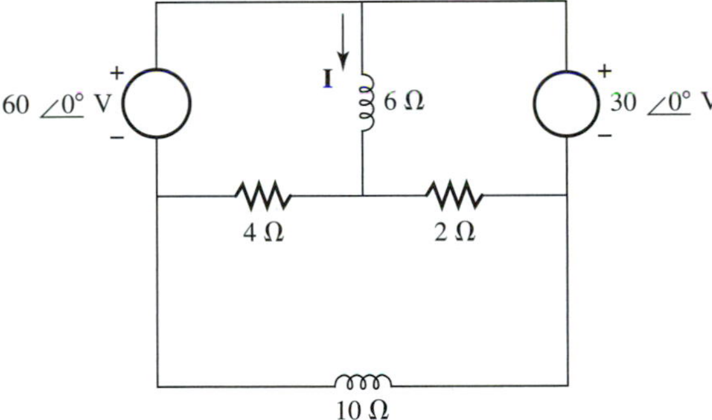

5. Determine a Thevenin equivalent circuit for terminals *a*-*b* for the circuit in Figure 19.79.

Figure 19.79

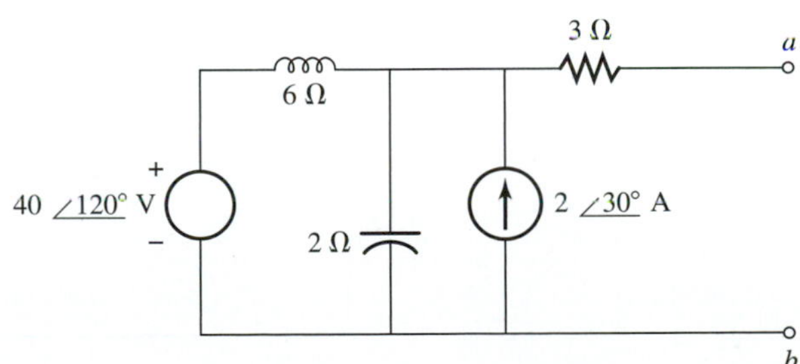

6. Determine a Thevenin equivalent circuit for terminals *a*-*b* for the circuits in Figure 19.80.

Figure 19.80

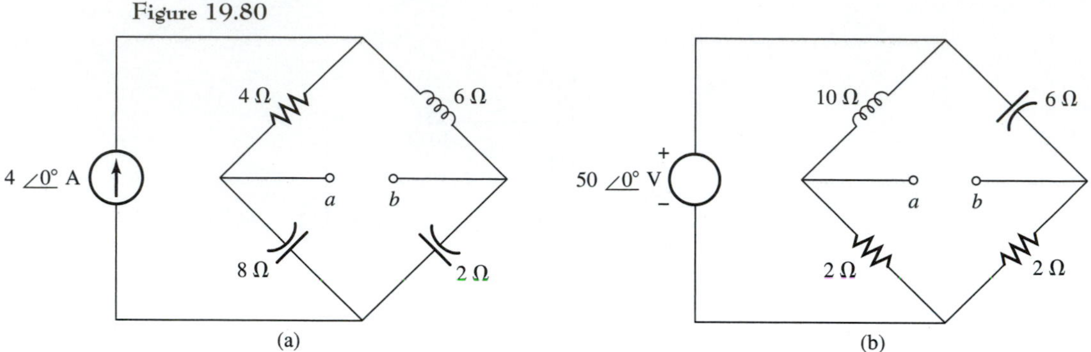

7. If a 2-Ω inductor is placed across terminals *a*-*b* in Problem 6, what is the phasor current for *a* to *b* for each circuit in Figure 19.80?

8. Determine a Thevenin equivalent circuit for terminals *a*-*b* for the circuit in Figure 19.81.

Figure 19.81

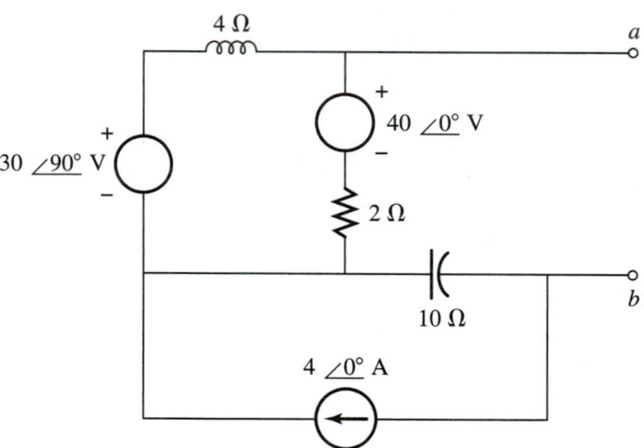

9. Determine a Norton equivalent circuit for terminals *a*-*b* for the circuit in Figure 19.82.

Figure 19.82

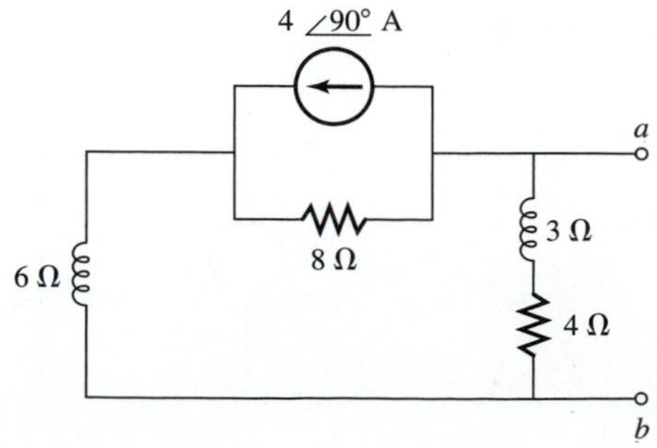

4 ∠90° A

8 Ω

6 Ω

3 Ω

4 Ω

a

b

10. Determine a Norton equivalent circuit for terminals *a-b* for the circuit in Figure 19.83.

Figure 19.83

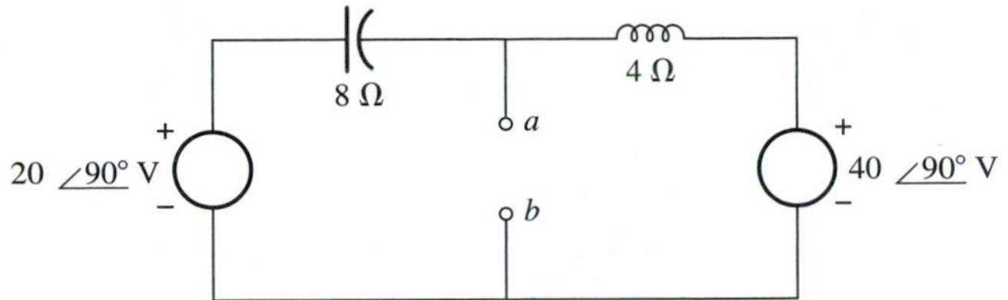

11. If a 10-Ω inductor is placed across terminals *a-b* in Problem 10, determine the phasor current $\mathbf{I}_{ab}$.

12. Determine a Norton equivalent circuit for terminals *a-b* for the circuit in Figure 19.84.

Figure 19.84

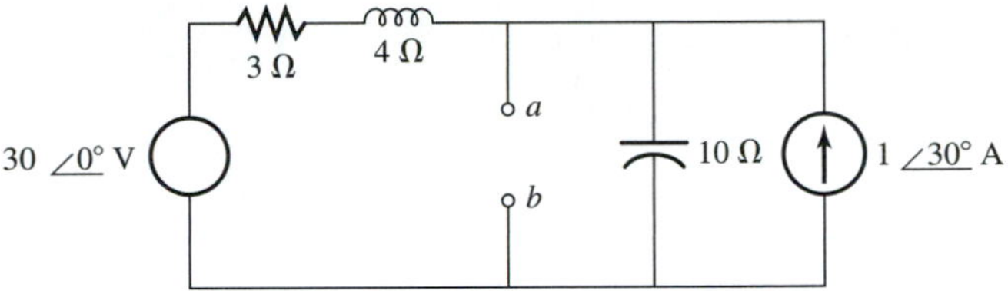

13. Determine a Norton equivalent circuit for terminals *a-b* for each circuit in Problem 6 by applying Norton's theorem to the circuit.

14. Determine the phasor current **I** for the circuit in Figure 19.85. (Use the source-conversion method.)

Figure 19.85

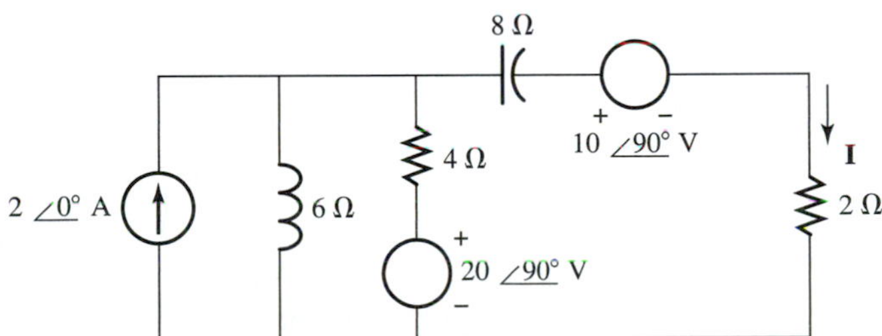

15. Determine the phasor current **I** for the circuit in Figure 19.86. (Use the source-conversion method.)

Figure 19.86

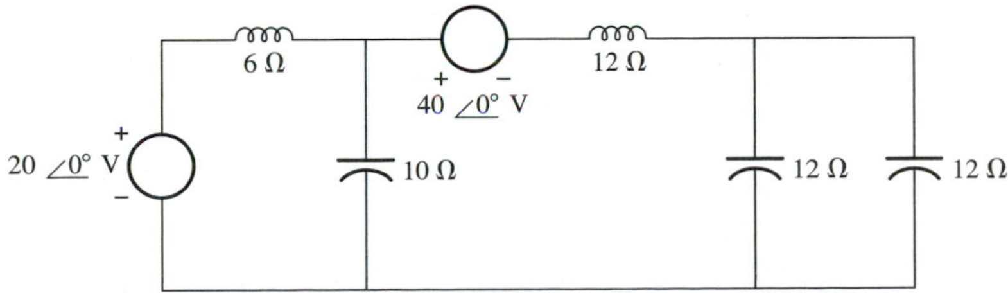

16. Determine the phasor current **I** for the circuit in Problem 1 by applying the successive-source-conversion method.

17. Determine the phasor current **I** for the circuit in Problem 2 by applying the successive-source-conversion method.

18. Determine the phasor voltage **V** for the circuit in Figure 19.87.

Figure 19.87

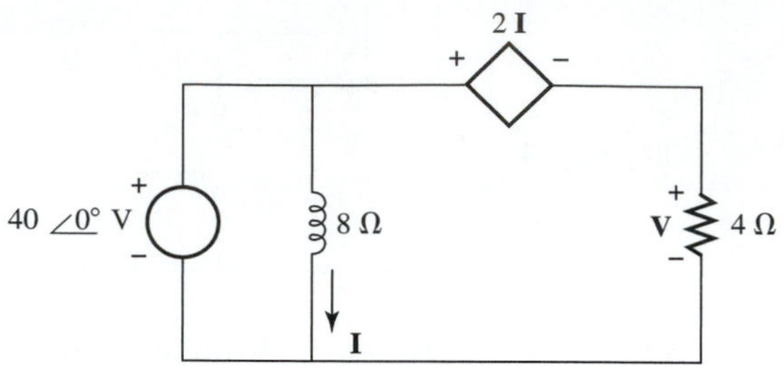

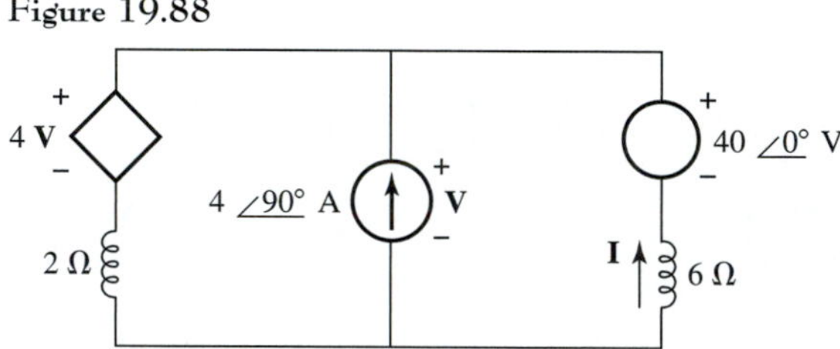

19. Determine the phasor current **I** for the circuit in Figure 19.88.

Figure 19.88

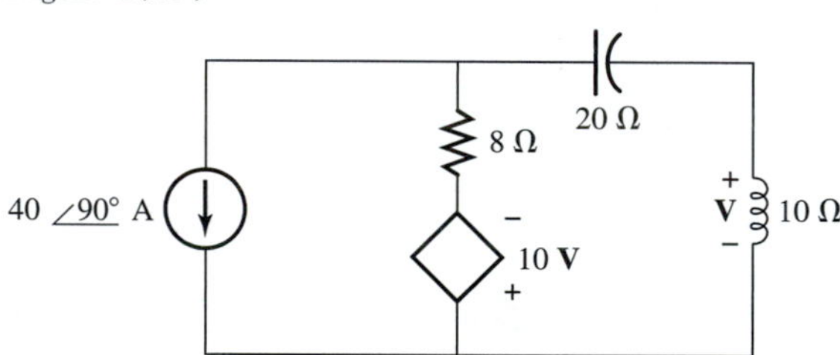

20. Determine the phasor voltage **V** for the circuit in Figure 19.89.

Figure 19.89

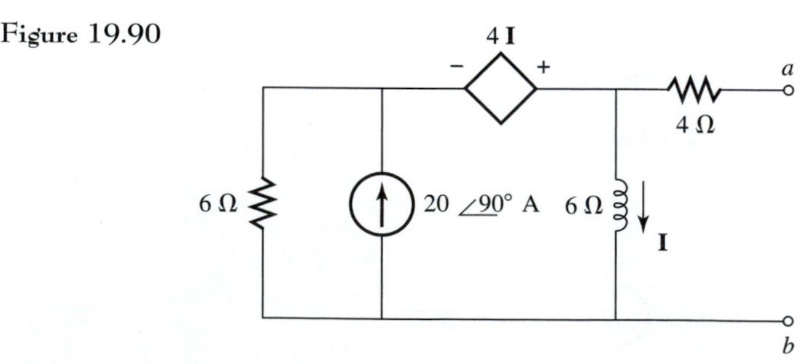

21. Determine a Thevenin equivalent circuit for terminals *a-b* for the circuit in Figure 19.90 by applying Thevenin's theorem.

Figure 19.90

22. Determine a Norton equivalent circuit for terminals *a-b* in Figure 19.90 by applying Norton's theorem.

23. Determine the Thevenin impedance $\mathbf{Z}_T$ for terminals *a-b* for the circuit in Figure 19.90 by applying an external voltage to the terminals.

24. Determine a Thevenin voltage $\mathbf{V}_T$ and a Norton current $\mathbf{I}_N$ for terminals *a*-*b* for the circuit in Figure 19.91.

Figure 19.91

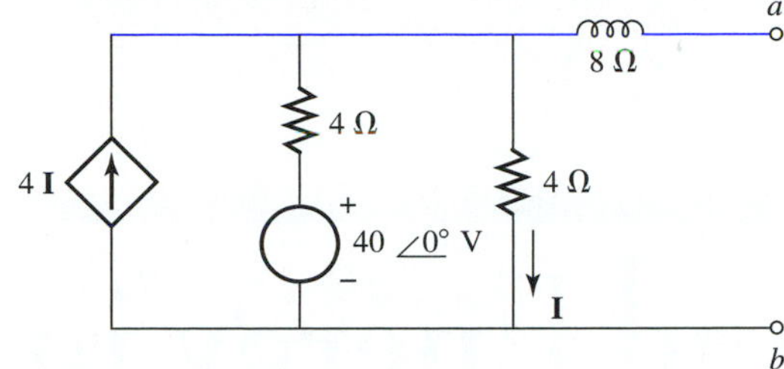

25. Determine the Norton impedance for the circuit in Figure 19.91 from $\mathbf{V}_T$ and $\mathbf{I}_N$ determined in Problem 24.

26. Determine the load impedance $\mathbf{Z}_L$ for maximum power transfer to the load for the circuits in Figure 19.92.

Figure 19.92

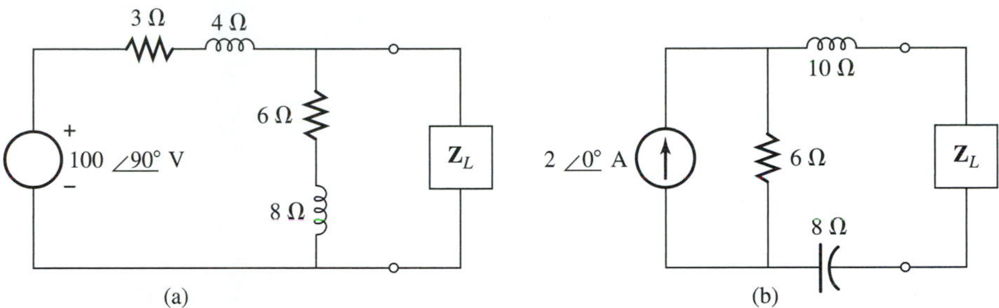

27. Determine the load impedance $\mathbf{Z}_L$ for maximum power transfer to the load for the circuit in Figure 19.93. What are the phasor voltage $\mathbf{V}_L$ and the phasor current $\mathbf{I}_L$ at maximum power transfer?

Figure 19.93

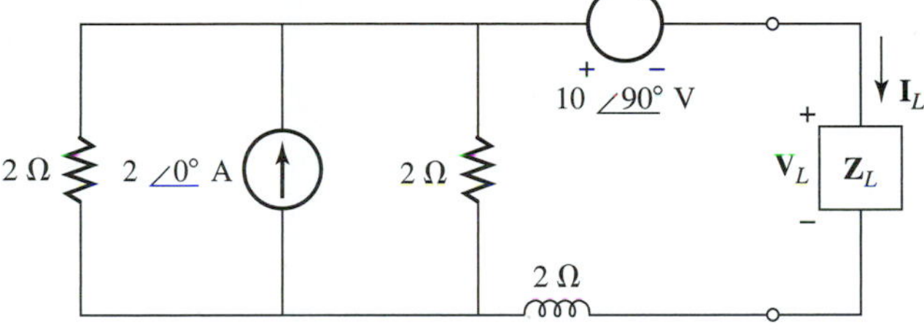

CHAPTER

20

Power and Energy in AC Circuits

After completing this chapter the student should be able to

* determine the effective value of a sinusoidal voltage or current waveform
* determine the average (real) power for an ac circuit
* determine the apparent power for an ac circuit
* determine the reactive power for an ac circuit
* determine the energy associated with the passive elements in an ac circuit

The energy and power associated with ac circuits is a primary concern for the circuit analyst. In this chapter we consider the power and energy for circuits driven by sinusoidal sources.

20.1 AVERAGE VALUE

The average value of a waveform is important when considering power and energy for ac circuits. Consider the sinusoidal waveform and the generating phasor in Figure 20.1.

Figure 20.1 Average value of a sine wave.

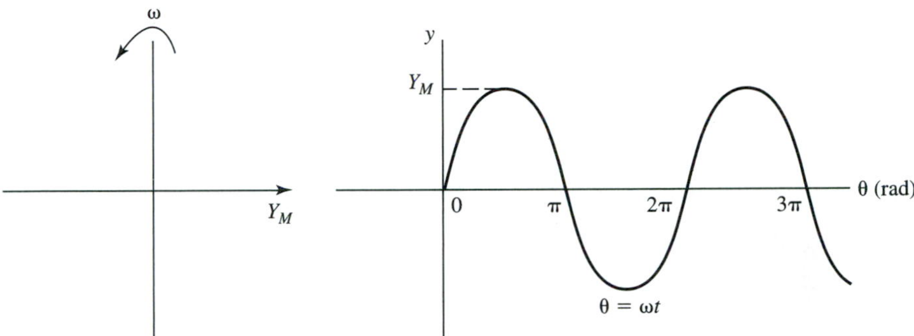

The average value of y for any full period of the waveform is zero, because the area above the horizontal axis is equal to the area below the horizontal axis for a full period.

The average value of y for half of a sine wave, as shown in Figure 20.2, is useful in many calculations and is determined here.

Figure 20.2 Average value for half of a sine wave.

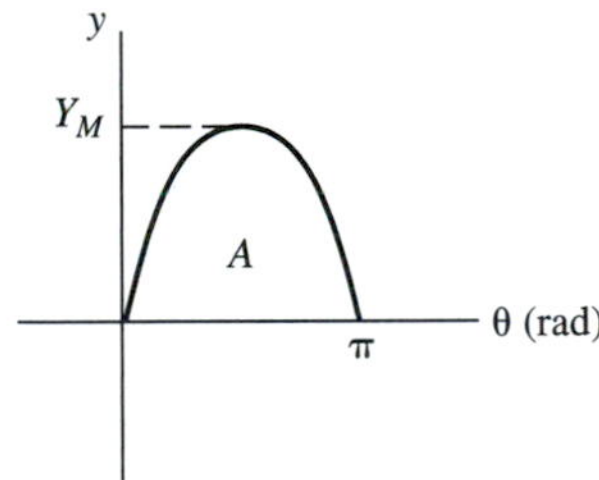

The sinusoidal waveform in Figure 20.2 can be described as

$$y = Y_M \sin\theta$$

where $0 \le \theta \le \pi$. The average value of y for the interval 0 to π is

$$y_{\text{avg}} = \frac{\text{net area under the waveform}}{\text{base}}$$

$$= \frac{A}{\pi}$$

The area A in Figure 20.2 can be determined by definite integration, as demonstrated in Chapter 9.

$$A = \int_0^{\pi} y\, d\theta$$
$$= \int_0^{\pi} Y_M \sin\theta\, d\theta$$

From rules I4 and I6 and the definition of a definite integral in Chapter 9,

$$A = Y_M \int_0^{\pi} \sin\theta\, d\theta$$
$$= Y_M(-\cos\theta)\,|_0^{\pi}$$
$$= Y_M(-\cos\pi) - Y_M(-\cos 0)$$
$$= 2Y_M$$

and

$$\left.\begin{aligned} y_{\text{avg}} &= \frac{2Y_M}{\pi} \\ &= 0.637Y_M \end{aligned}\right\} \qquad (20.1)$$

Consider the following example.

EXAMPLE 20.1 Determine the average value of the sinusoidal waveform y in Figure 20.3 for $\frac{1}{2}$ cycle, for 1 cycle, and for $1\frac{1}{2}$ cycles. ■

Figure 20.3

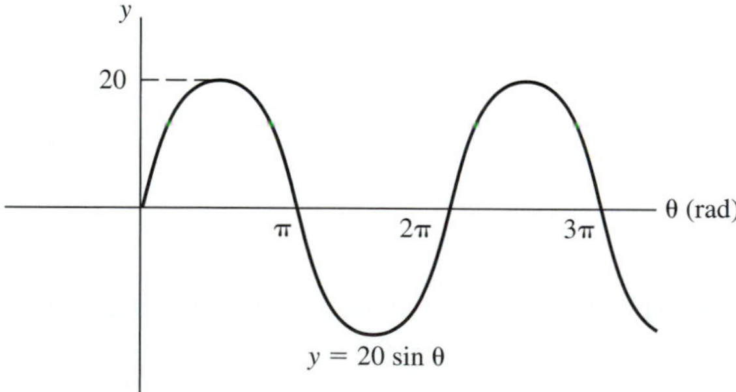

SOLUTION From Equation (20.1) the area under $\frac{1}{2}$ cycle of the sine wave is

$$A = 2Y_M$$
$$= 2(20) = 40$$

For $\frac{1}{2}$ cycle (Figure 20.4)

Figure 20.4

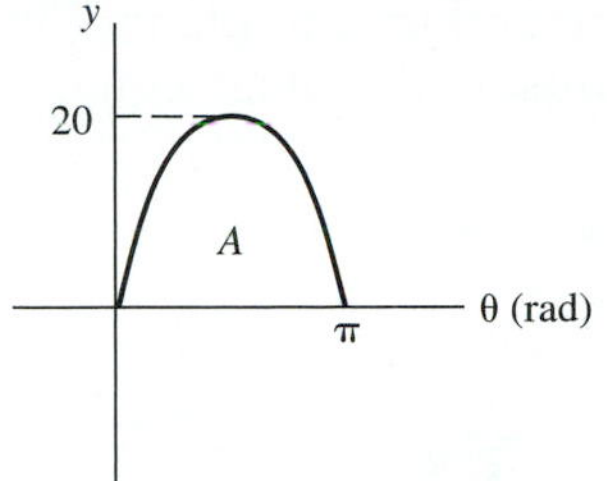

$$y_{\text{avg}} = \frac{40}{\pi}$$
$$= 12.7$$

For 1 cycle (Figure 20.5)

Figure 20.5

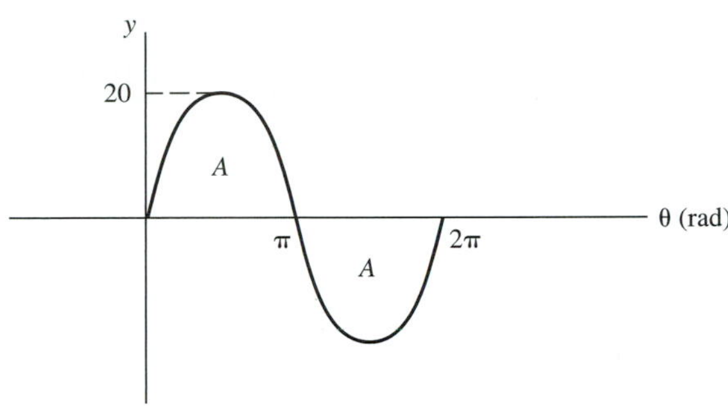

$$y_{\text{avg}} = \frac{40 - 40}{2\pi}$$
$$= 0$$

For $1\frac{1}{2}$ cycles (Figure 20.6)

Figure 20.6

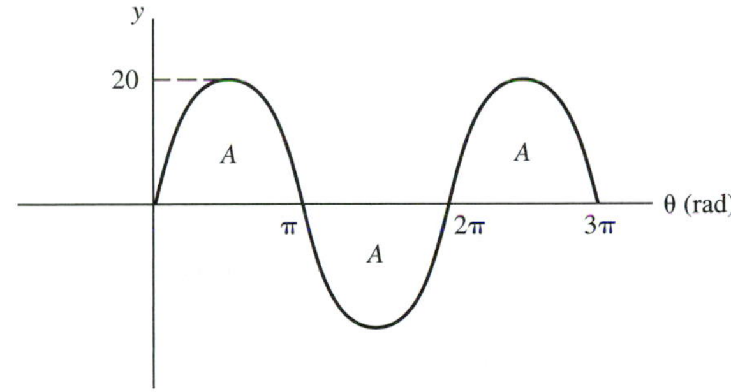

$$y_{\text{avg}} = \frac{40 - 40 + 40}{3\pi}$$
$$= \left\{\frac{80}{3}\right\}\pi$$
$$= 8.49$$

20.2 EFFECTIVE VALUE

Consider the dc and ac circuits in Figure 20.7, where v_{dc} is a constant voltage and v_{ac} is a sinusoidal voltage equal to $V_M \sin(\omega t)$.

Figure 20.7 Effective value.

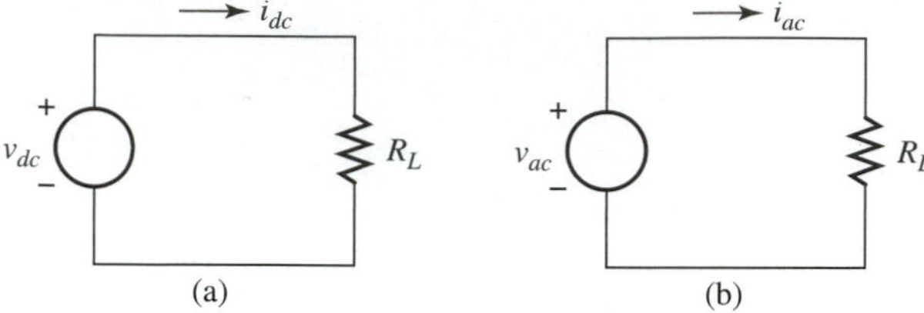

From Equation (2.7) the instantaneous power delivered to R_L by the constant dc source in Figure 20.7(a) is

$$p_{dc} = i_{dc}^2 R_L$$

and the instantaneous power delivered to R_L by the sinusoidal ac source in Figure 20.7(b) is

$$p_{ac} = i_{ac}^2 R_L$$

Because i_{ac} is a sinusoidal current, it can be expressed as

$$i_{ac} = I_M \sin(\omega t)$$

where $I_M = V_M/R_L$ from Equation (17.4).

From the preceding equation for p_{ac}

$$p_{ac} = \left\{ I_M^2 \sin^2(\omega t) \right\} R_L$$

Using the trigonometric identity

$$\sin^2(\omega t) = \frac{1 - \cos(2\omega t)}{2}$$

p_{ac} can be written as

$$p_{ac} = \frac{R_L I_M^2 \left\{ 1 - \cos(2\omega t) \right\}}{2}$$

$$= \frac{\left\{ R_L I_M^2 \right\} - \left\{ R_L I_M^2 \cos(2\omega t) \right\}}{2}$$

In the preceding equation for p_{dc}, i_{dc} is a constant and the average power P_{dc} delivered to R_L in Figure 20.7(a) is

$$P_{dc} = i_{dc}^2 R_L$$

In the preceding equation for p_{ac}, the average value of the first term is

$$\frac{R_L I_M^2}{2}$$

and the average value of the second term is zero, because the average value of $\cos(2\omega t)$ is zero. Therefore, the average power P_{ac} delivered to R_L in Figure 20.7(b) is

$$P_{ac} = \frac{R_L I_M^2}{2}$$

The value of i_{dc} necessary to deliver the same power to R_L that is delivered by the ac source in Figure 20.7 can be determined by equating P_{dc} to P_{ac}, as follows:

$$P_{dc} = P_{ac}$$

$$i_{dc}^2 R_L = \frac{R_L I_M^2}{2}$$

$$i_{dc}^2 = \frac{I_M^2}{2}$$

$$i_{dc} = \sqrt{\frac{I_M^2}{2}}$$

$$= \frac{I_M}{\sqrt{2}}$$

This reasoning can be carried out for voltages with the same result. That is,

$$v_{dc} = \frac{V_M}{\sqrt{2}}$$

The equivalent dc voltage or current necessary to deliver the same average power as delivered by a given ac source to a load is called the *effective value* of the ac voltage or current. That is,

$$\left.\begin{aligned} I_{\text{eff}} &= \frac{I_M}{\sqrt{2}} \\ V_{\text{eff}} &= \frac{V_M}{\sqrt{2}} \end{aligned}\right\} \tag{20.2}$$

for sinusoidal voltages and currents.

Notice that in the preceding mathematical procedure for arriving at Equation (20.2), the currents or voltages are squared and averaged, and then the square root is extracted in that order. Although Equation (20.2) is valid only for sinusoidal waveforms, the mathematical procedure is commonly called the root-mean-square method and can be used to determine the effective value of other periodic waveforms.

Consider the following examples.

EXAMPLE 20.2 Determine V_{eff} for the voltage waveform in Figure 20.8 where $v = 20\sin(\theta)$. ■

Figure 20.8

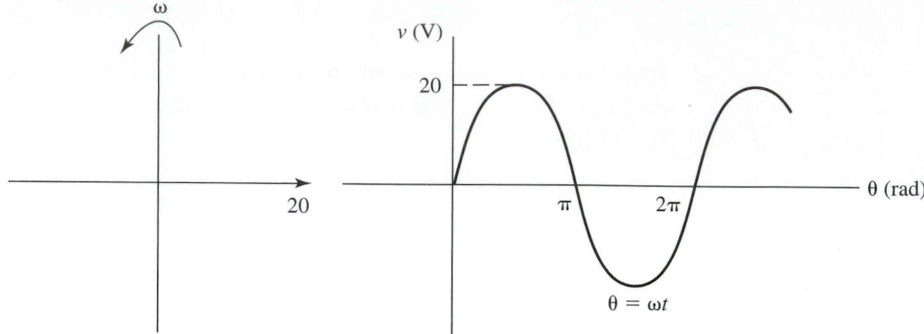

SOLUTION From Equation (20.2)

$$V_{\text{eff}} = \frac{20}{\sqrt{2}}$$
$$= 14.14 \text{ V}$$

Alternatively, following the root-mean-square mathematical procedure to determine effective value, we first square the waveform, which results in the waveform in Figure 20.9, where v^2 can be expressed as

$$v^2 = 400\sin^2(\theta)$$

Figure 20.9

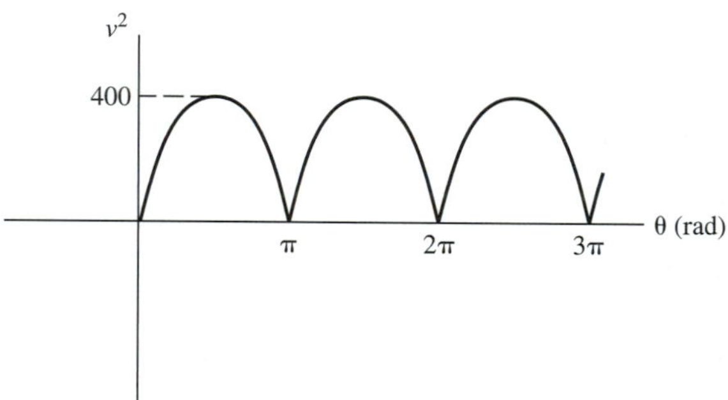

We next determine the area under the v^2 curve for one period using the definite integral, as presented in Chapter 9, which results in

$$\{\text{area under the } v^2 \text{ curve for one period}\} = \int_0^{\pi} v^2\, d\theta = \int_0^{\pi} 400\sin^2(\theta)d\theta$$

Using the trigonometric identity

$$\sin^2(\theta) = \frac{1 - \cos(2\theta)}{2}$$

results in

$$\int_0^{\pi} v^2\, d\theta = \int_0^{\pi} 400\left\{\frac{1 - \cos(2\theta)}{2}\right\}d\theta$$

From integration rules I2 and I7 in Chapter 9,

$$\int_0^{\pi} v^2\, d\theta = 200\int_0^{\pi} d\theta \;-\; 200\int_0^{\pi} \cos(2\theta)\, d\theta$$

From integration rules I2 and I5 in Chapter 9, the antiderivative of unity with respect to θ is θ and the antiderivative of $\cos(2\theta)$ with respect to θ is

$$\frac{\sin(2\theta)}{2}$$

From Equation (9.2)

$$\begin{aligned}\int_0^{\pi} v^2\, d\theta &= 200\theta\Big|_0^{\pi} - \frac{200\sin(2\theta)}{2}\Big|_0^{\pi}\\ &= 200\pi - \{100\sin(2\pi) - 100\sin(0)\}\\ &= 200\pi\end{aligned}$$

which is the area under the v^2 curve for one period of the squared waveform.

The average value of v^2 is obtained by dividing the preceding area by the horizontal base of the area.

$$\text{average value of } v^2 = \frac{200\pi}{\pi} = 200$$

The effective value of v is obtained by taking the square root of the average value of v^2:

$$\begin{aligned}V_{\text{eff}} &= \sqrt{200}\\ &= 14.14 \text{ V}\end{aligned}$$

Notice that the effective value obtained in this manner is the same as the effective value obtained using Equation (20.2).

Equation (20.2) is valid only for sinusoidal waveforms, whereas the alternative mathematical procedure illustrated in Example 20.2 is valid for periodic waveforms in general.

Consider the following example.

EXAMPLE 20.3 Determine I_{eff} for the current waveform in Figure 20.10. ■

Figure 20.10

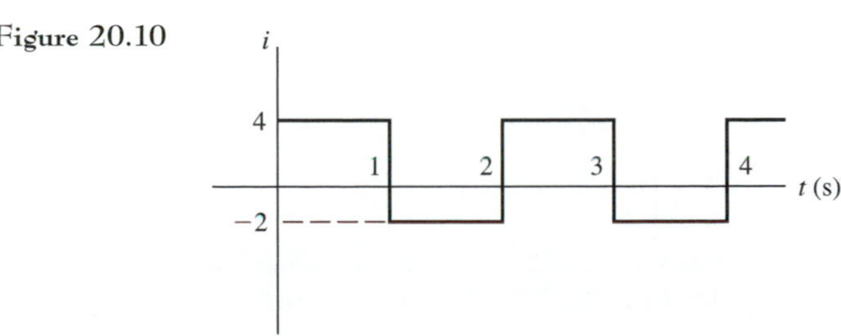

SOLUTION The first step is to determine i^2, as shown in Figure 20.11.

Figure 20.11

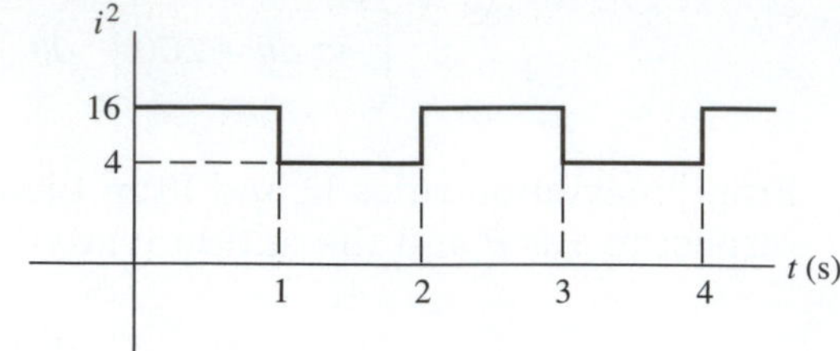

The next step is to determine the average value of i^2 for one period, as shown in Figure 20.12.

Figure 20.12

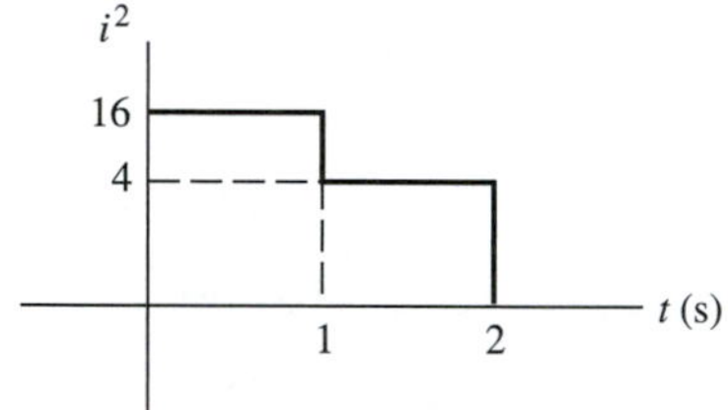

$$\text{average value of } i^2 \text{ for one period} = \frac{A}{2}$$
$$= \frac{16(1) + 4(1)}{2}$$
$$= 10$$

The final step is to take the square root of the average value of i^2:

$$I_{\text{eff}} = \sqrt{10}$$
$$= 3.16$$

20.3 LEAD AND LAG

Consider the ac load **Z** in Figure 20.13, where $\mathbf{Z} = Z\angle\theta$, $i = I_M\sin(\omega t)$, and $v = V_M\sin(\omega t + \theta)$. Recall that for a passive load **Z**, the magnitude (absolute value) of ranges from 0° to 90°, and θ equals 0, 90°, and −90°, respectively for a pure resistor, a pure inductor, and a pure capacitor.

Figure 20.13 Lead and lag circuits.

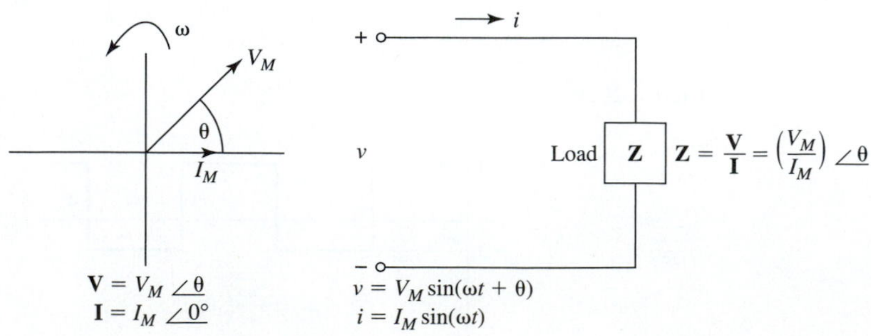

In Figure 20.13 a positive angle θ causes the voltage phasor to lead the current phasor for counter clockwise rotation. θ is always the smaller of the two angles separating the voltage and current phasors. If the current phasor lags the voltage phasor for a pair of terminals as in Figure 20.13 the circuit connected across the terminals is inductive and the circuit is called a *lag circuit*.

If the current phasor leads the voltage phasor for a pair of terminals, the circuit connected across the terminals is capacitive, and the circuit is called a *lead circuit*.

20.4 INSTANTANEOUS POWER

From Equation (2.6) the instantaneous power delivered to **Z** in Figure 20.13 is

$$\begin{aligned} p &= vi \\ &= \{V_M \sin(\omega t + \theta)\}\{I_M \sin(\omega t)\} \end{aligned}$$

From the trigonometric identity

$$\begin{aligned} \sin(\omega t + \theta) &= \sin(\omega t)\cos(\theta) + \cos(\omega t)\sin(\theta) \\ p &= V_M I_M\{\sin(\omega t)\cos(\theta) + \cos(\omega t)\sin(\theta)\}\sin(\omega t) \\ &= V_M I_M\{\sin^2(\omega t)\cos(\theta) + \cos(\omega t)\sin(\theta)\sin(\omega t)\} \end{aligned}$$

From the trigonometric identities

$$\sin^2(\omega t) = \frac{1 - \cos(2\omega t)}{2}$$

and

$$\sin(\omega t)\cos(\omega t) = \frac{\sin(2\omega t)}{2}$$

$$\begin{aligned} p &= V_M I_M\left\{\frac{1}{2}\{1 - \cos(2\omega t)\}\cos(\theta) + \frac{\sin(\theta)\{\sin(2\omega t)\}}{2}\right\} \\ &= \left(\frac{V_M I_M}{2}\right)\{\cos\theta - \cos(2\omega t)\cos(\theta) + \sin(2\omega t)\sin(\theta)\} \\ &= \left(\frac{V_M}{\sqrt{2}}\right)\left(\frac{I_M}{\sqrt{2}}\right)\{\cos\theta - \cos(2\omega t)\cos(\theta) + \sin(2\omega t)\sin(\theta)\} \\ &= V_{\text{eff}}I_{\text{eff}}\cos(\theta) - V_{\text{eff}}I_{\text{eff}}\cos(\theta)\cos(2\omega t) + V_{\text{eff}}I_{\text{eff}}\sin(\theta)\sin(2\omega t) \end{aligned} \quad (20.3)$$

For a pure resistance the angle θ in Figure 20.13 is equal to zero, and the instantaneous power from Equation (20.3) is

$$p_R = V_{\text{eff}}I_{\text{eff}} - V_{\text{eff}}I_{\text{eff}}\cos(2\omega t) \quad (20.4)$$

For a pure inductance the angle θ in Figure 20.13 is equal to 90°, and the instantaneous power from Equation (20.3) is

$$p_L = V_{\text{eff}}I_{\text{eff}}\sin(2\omega t) \quad (20.5)$$

For a pure capacitance the angle θ in Figure 20.13 is equal to $-90°$, and the instantaneous power from Equation (20.3) is

$$p_C = -V_{\text{eff}} I_{\text{eff}} \sin(2\omega t) \tag{20.6}$$

Plotting p_R, p_L, and p_C versus t for Equations (20.4), (20.5), and (20.6) results in the plots in Figure 20.14. p_R has a nonzero average value, which indicates that average power is delivered to a resistor. The average values of p_L

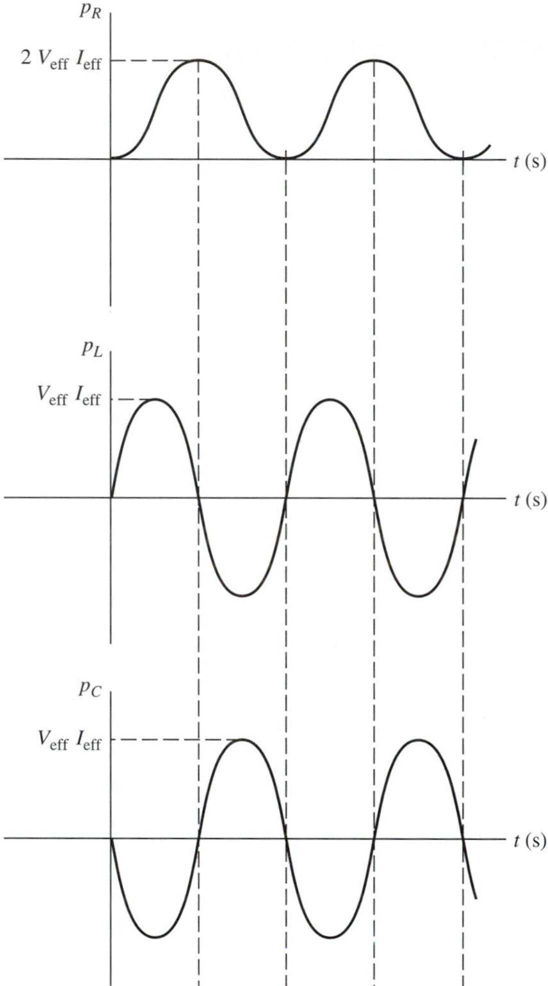

Figure 20.14 Power curves.

and p_C are zero, which indicates that ideal inductors and capacitors cannot dissipate average power; instead, they receive energy from the circuit and return energy to the circuit on a cyclic basis.

20.5 AVERAGE (REAL) POWER *P*

Because V_{eff}, I_{eff}, and θ are constants in Equation (20.3), the first term in the equation is a constant with an average value of $V_{\text{eff}} I_{\text{eff}} \cos(\theta)$, and the second and third terms are cosine and sine waves, respectively, with average values equal to zero. The average value of the instantaneous power expression in Equation (20.3) is, therefore,

$$P = V_{\text{eff}} I_{\text{eff}} \cos\theta \tag{20.7}$$

where P (the symbol for average power) is measured in watts, V_{eff} is in volts, and I_{eff} is in amperes. The average power P given in Equation (20.7) is the actual or real power delivered to the load and lost to the circuit in Figure 20.13.

Because θ equals 0, 90°, and −90° for pure resistors, pure inductors and pure capacitors, respectively, Equation (20.7) results in

$$P_R = V_{\text{eff}} I_{\text{eff}}$$

$$P_L = 0$$

$$P_C = 0$$

for resistors, inductors, and capacitors, respectively.

20.6 POWER FACTOR

The cos θ term in Equation (20.7) is unitless and is called the *power factor* of the load. It is symbolized as

$$F_p = \cos\theta \tag{20.8}$$

F_p equals unity for a pure resistance and zero for a pure inductor or capacitor; in general it ranges from 0 to 1.

Consider the following example.

EXAMPLE 20.4 Determine the average power P delivered to the circuit by the voltage source in Figure 20.15. ■

Figure 20.15

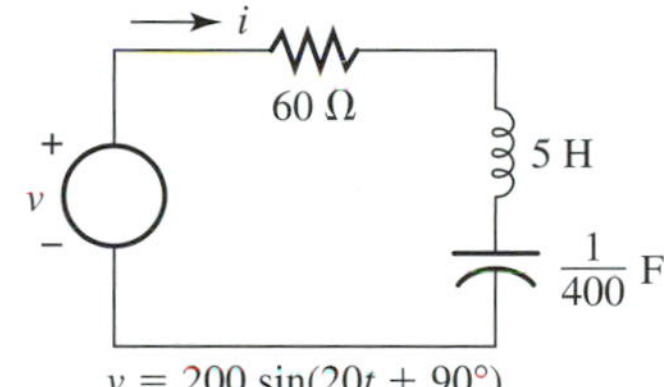

SOLUTION From Equations (17.7) and (17.11),

$$\begin{aligned} X_L &= \omega L \\ &= (20)(5) \\ &= 100\,\Omega \end{aligned}$$

$$\begin{aligned} X_C &= \frac{1}{\omega C} \\ &= \frac{1}{(20)\left(\dfrac{1}{400}\right)} \\ &= 20\,\Omega \end{aligned}$$

From Equations (17.3), (17.8), (17.12), and (17.35), the impedance seen by the voltage source is

$$\begin{aligned} \mathbf{Z} &= 60 + j100 - j20 \\ &= 60 + j80 \\ &= 100\underline{/53.1^\circ} \end{aligned}$$

In phasor form

$$\mathbf{V} = 200\angle 90^\circ$$

And from Equation (17.1)

$$\mathbf{I} = \frac{\mathbf{V}}{\mathbf{Z}} = \frac{200\angle 90^\circ}{100\angle 53.1^\circ} = 2\angle 37.9^\circ$$

From Equation (20.2)

$$V_{eff} = \frac{200}{\sqrt{2}}$$

$$I_{eff} = \frac{2}{\sqrt{2}}$$

From Equation (20.7)

$$P = V_{eff} I_{eff} \cos\theta = \left(\frac{200}{\sqrt{2}}\right)\left(\frac{2}{\sqrt{2}}\right)\cos(53.1^\circ) = 120.1 \text{ W}$$

20.7 APPARENT POWER *S*

The average power delivered to load **Z** in Figure 20.13 is given by Equation (20.7) and is a fraction of the total power ($I_{eff}V_{eff}$) available at the terminals of **Z**. The fractional value of the total available power that is delivered to **Z** is determined by the power factor of **Z**. The total power ($I_{eff}V_{eff}$) available to **Z** is called the *apparent power* and is written as

$$S = V_{eff} I_{eff} \tag{20.9}$$

where S is the apparent power. The units for S are volt-amperes (VA).

20.8 REACTIVE POWER *Q*

A fraction of the apparent power is used to satisfy the power needs of the reactive elements (inductors and capacitors) contained in the load. Although the average power P is zero for pure inductors and capacitors, power is required on a cyclic basis to create the energy fields necessary for the operation of the inductors and capacitors in the load, as illustrated in Figure 20.14. The energy delivered to inductors and capacitors by the circuit is returned to the circuit on a cyclic basis. The power used to furnish the needs of the reactive elements is called *reactive power* and is written as

$$Q = V_{\text{eff}} I_{\text{eff}} \sin\theta \tag{20.10}$$

where Q is the reactive power. The units for Q are volt-amperes reactive (VARS).

20.9 *S, Q,* AND *P* RELATED

From Equations (20.7), (20.9), and (20.10)

$$S = V_{\text{eff}} I_{\text{eff}}$$

$$P = V_{\text{eff}} I_{\text{eff}} \cos\theta$$

$$Q = V_{\text{eff}} I_{\text{eff}} \sin\theta$$

From the trigonometric identity

$$\sin^2\theta + \cos^2\theta = 1$$

it follows that

$$S^2 = Q^2 + P^2 \tag{20.11}$$

From these equations Q and P can also be written as

$$Q = S\sin\theta \tag{20.12}$$

$$P = S\cos\theta \tag{20.13}$$

20.10 POWER TRIANGLE

A triangle called the *power triangle* is drawn in Figure 20.16 to geometrically represent the relationship between S, Q, and P. Recall that θ is the load impedance angle and the angle separating the current and voltage phasors for the load.

Figure 20.16 Power triangles.

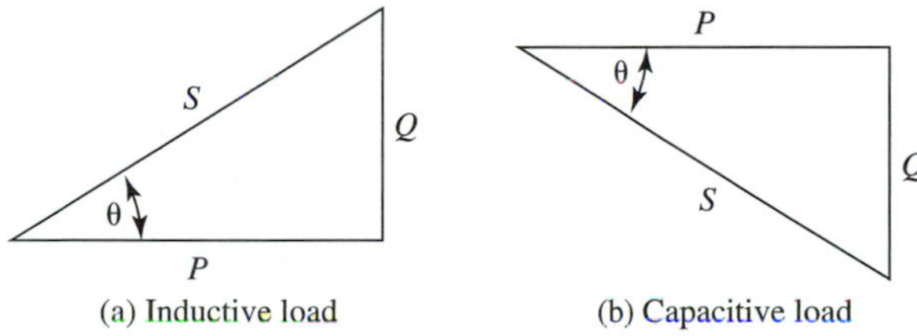

The power triangle is customarily drawn with the base P down if the load is inductive and with base P up if the load $\mathbf{Z}$ is capacitive. Notice from the power triangle (or from Equation (20.13)) that the power factor $\cos\theta$ can be expressed as

$$F_p = \frac{P}{S} \tag{20.14}$$

20.11 EQUATIONS FOR *S*, *Q*, AND *P*

From Equation (17.2) the magnitude of **Z** can be written as

$$Z = \frac{V_M}{I_M}$$

as illustrated in Figure 20.13. Because $V_M = \sqrt{2}V_{\text{eff}}$ and $I_M = \sqrt{2}I_{\text{eff}}$ from Equation (20.2), Z can also be expressed as

$$\begin{aligned} Z &= \frac{\sqrt{2}V_{\text{eff}}}{\sqrt{2}I_{\text{eff}}} \\ &= \frac{V_{\text{eff}}}{I_{\text{eff}}} \end{aligned}$$

From Equation (20.9) and the preceding equation for Z, the apparent power S can be written as

$$S = V_{\text{eff}}I_{\text{eff}}$$

$$S = \frac{V_{\text{eff}}^2}{Z} \tag{20.15}$$

$$S = I_{\text{eff}}^2 Z \tag{20.16}$$

From Equations (17.3), (17.8), and (17.12), Z for pure resistors, inductors, and capacitors can be written as

$$\begin{aligned} Z_R &= R \\ Z_L &= X_L \\ Z_C &= X_C \end{aligned}$$

respectively. From Equations (20.9), (20.15), and (20.16), it follows that the apparent power for pure resistors can be expressed as

$$S_R = V_{\text{eff}}I_{\text{eff}} \tag{20.17}$$

$$= \frac{V_{\text{eff}}^2}{R} \tag{20.18}$$

$$= I_{\text{eff}}^2 R \tag{20.19}$$

The apparent power for pure inductors can be expressed as

$$S_L = V_{\text{eff}}I_{\text{eff}} \tag{20.20}$$

$$= \frac{V_{\text{eff}}^2}{X_L} \tag{20.21}$$

$$= I_{\text{eff}}^2 X_L \tag{20.22}$$

And, the apparent power for pure capacitors can be expressed as

$$S_C = V_{\text{eff}} I_{\text{eff}} \tag{20.23}$$

$$= \frac{V_{\text{eff}}^2}{X_C} \tag{20.24}$$

$$= I_{\text{eff}}^2 X_C \tag{20.25}$$

Because $P = 0$ for pure inductors and capacitors and because $S^2 = P^2 + Q^2$, the reactive power for inductors and capacitors, respectively, can be written as

$$Q_L = S_L = V_{\text{eff}} I_{\text{eff}} \tag{20.26}$$

$$= \frac{V_{\text{eff}}^2}{X_L} \tag{20.27}$$

$$= I_{\text{eff}}^2 X_L \tag{20.28}$$

$$Q_C = S_C = V_{\text{eff}} I_{\text{eff}} \tag{20.29}$$

$$= \frac{V_{\text{eff}}^2}{X_C} \tag{20.30}$$

$$= I_{\text{eff}}^2 X_C \tag{20.31}$$

Because $Q = 0$ for pure resistors and $S^2 = P^2 + Q^2$, the average power for a resistor can be written as

$$P_R = S_R = V_{\text{eff}} I_{\text{eff}} \tag{20.32}$$

$$= \frac{V_{\text{eff}}^2}{R} \tag{20.33}$$

$$= I_{\text{eff}}^2 R \tag{20.34}$$

Consider the following example.

EXAMPLE 20.5 Determine the power factor, the apparent power, average power, and reactive power delivered by the voltage source for the circuit in Figure 20.17. ■

Figure 20.17

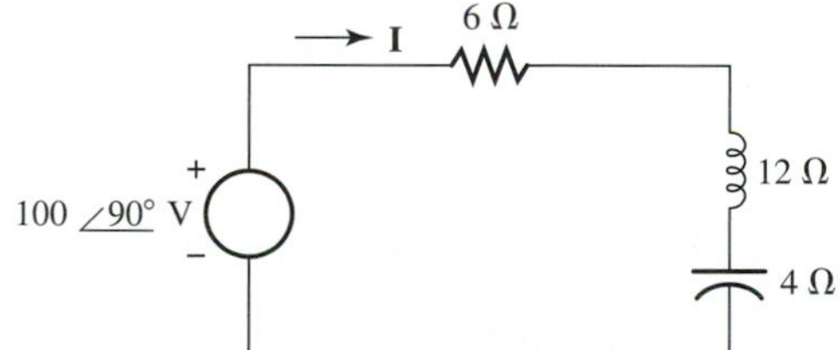

SOLUTION From Equation (17.35) the impedance seen by the voltage source is

$$\mathbf{Z} = 6 + j12 - j4$$

$$= 6 + j8$$

$$= 10\underline{/53.1^\circ}\ \Omega$$

From Equation (17.1)

$$\mathbf{I} = \frac{\mathbf{V}}{\mathbf{Z}}$$

$$= \frac{100\angle 90^\circ}{10\angle 53.1^\circ}$$

$$= 10\angle 36.9^\circ \text{ A}$$

From Equations (20.2) and (20.5)

$$V_{\text{eff}} = \frac{100}{\sqrt{2}}$$

$$I_{\text{eff}} = \frac{10}{\sqrt{2}}$$

$$F_p = \cos(53.1^\circ) = 0.600$$

From Equations (20.9), (20.12) and (20.13) the apparent power, reactive power, and average power delivered by the voltage source are

$$S = V_{\text{eff}} I_{\text{eff}}$$

$$= \left(\frac{100}{\sqrt{2}}\right)\left(\frac{10}{\sqrt{2}}\right)$$

$$= 500 \text{ VA}$$

$$Q = S \sin\theta$$

$$= 500 \sin(53.1^\circ)$$

$$= 400 \text{ VARS}$$

$$P = S \cos\theta$$

$$= 500 \cos(53.1^\circ)$$

$$= 300 \text{ W}$$

The following rules are useful in applying the power equations given previously to the power analysis of ac circuits.

> The total apparent power supplied to a circuit is not generally equal to the sum of the apparent powers delivered to the individual circuit elements.
>
> The total average power supplied to a circuit is equal to the sum of the average powers delivered to the individual circuit elements.
>
> The total reactive power supplied to a circuit is equal to the algebraic sum of the reactive powers of the individual circuit elements.

The reactive powers must be added algebraically because the instantaneous power curves for inductors and capacitors are 180° out of phase, as indicated in Figure 20.14. Inductive reactive power is assigned a positive sign, and capacitive reactive power is assigned a negative sign prior to the algebraic addition. If the result of the algebraic addition is positive, the reactive power is called inductive; if the result is negative, the reactive power is called capacitive. This is demonstrated in the following examples.

EXAMPLE 20.6 Use the previous rules to solve the problem in Example 20.5. ■

SOLUTION As previously determined in Example 20.4,

$$\mathbf{Z} = 10\angle 53.1^\circ$$
$$\mathbf{I} = 10\angle 36.9^\circ$$
$$F_p = 0.600$$

From Equation (20.19)

$$S_R = I_{\text{eff}}^2 R = \left(\frac{10}{\sqrt{2}}\right)^2 6 = 300 \text{ VA}$$

and because $Q_R = 0$,

$$P_R = S_R = 300 \text{ W}$$

From Equation (20.22)

$$S_L = I_{\text{eff}}^2 X_L = \left(\frac{10}{\sqrt{2}}\right)^2 12 = 600 \text{ VA}$$

and because $P_L = 0$,

$$Q_L = S_L = 600 \text{ VARS}$$

From Equation (20.23)

$$S_C = I_{\text{eff}}^2 X_C = \left(\frac{10}{\sqrt{2}}\right)^2 4 = 200 \text{ VA}$$

and because $P_C = 0$,

$$Q_C = S_C$$
$$= 200 \text{ VARS}$$

Because $Q_R = 0$, the total reactive power Q is the absolute value of the difference between Q_C and Q_L and the load is capacitive if $Q_C > Q_L$ and inductive if $Q_L > Q_C$. The total reactive power is the difference between Q_C and Q_L:

$$Q = Q_L - Q_C$$
$$= 600 - 200$$
$$= 400 \text{ VARS} \quad \text{(inductive)}$$

Because P_L and P_C are equal to zero the total average power is

$$P = P_R$$
$$= 300 \text{ W}$$

From Equation (20.11)

$$S^2 = Q^2 + P^2$$
$$= (400)^2 + (300)^2$$
$$= 250{,}000$$
$$S = 500 \text{ VA}$$

Alternatively from Equation (20.16)

$$S = I_{\text{eff}}^2 Z$$
$$= \left(\frac{10}{\sqrt{2}}\right)^2 (10)$$
$$= 500 \text{ VA}$$

EXAMPLE 20.7 Determine the power factor, the apparent power, average power, and reactive power delivered by the voltage source for the circuit in Figure 20.18. ■

Figure 20.18

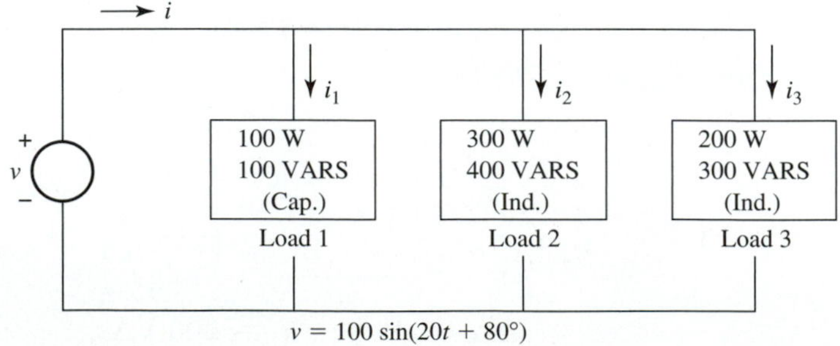

SOLUTION From the given expression for v, the voltage phasor is

$$\mathbf{V} = 100 \underline{/ 80^\circ}$$

The average power delivered by the voltage source is

$$\begin{aligned} P &= 100 + 300 + 200 \\ &= 600 \text{ W} \end{aligned}$$

The total reactive power delivered by the voltage source is

$$\begin{aligned} Q &= -100 + 400 + 300 \\ &= 600 \text{ VARS} \quad \text{(inductive)} \end{aligned}$$

The previous result for Q is arrived at by assigning a negative sign to capacitive reactive power and a positive sign to inductive reactive power. The total Q is not expressed as negative but is called inductive or capacitive based on the sign of the result.

From Equation (20.11)

$$\begin{aligned} S^2 &= P^2 + Q^2 \\ S &= \sqrt{P^2 + Q^2} \\ &= \sqrt{600^2 + 600^2} \\ &= 849 \text{ VA} \end{aligned}$$

From Equation (20.14)

$$\begin{aligned} F_P &= \frac{P}{S} \\ &= \frac{600}{849} \\ &= 0.707 \text{ lag} \end{aligned}$$

EXAMPLE 20.8 Determine i for the circuit in Example 20.7. ■

SOLUTION From Example 20.7 and Equation (20.2),

$$\begin{aligned} V_{\text{eff}} &= \frac{V_M}{\sqrt{2}} \\ &= \frac{100}{\sqrt{2}} \end{aligned}$$

and

$$S = 849$$

From Equation (20.9) the apparent power supplied by the voltage source is

$$S = V_{\text{eff}} I_{\text{eff}}$$

$$849 = \left(\frac{100}{\sqrt{2}}\right) I_{\text{eff}}$$

$$I_{\text{eff}} = \frac{(849)\sqrt{2}}{100}$$

$$= 12.01$$

From Equation (20.2)

$$I_M = I_{\text{eff}}\sqrt{2}$$

$$= 12.01(\sqrt{2})$$

$$= 16.98$$

From Example 20.7 the total average and reactive powers are

$$P = 600 \text{ W}$$

$$Q = 600 \text{ VARS} \quad \text{(inductive)}$$

and the power triangle for the total load is shown in Figure 20.19.

Figure 20.19

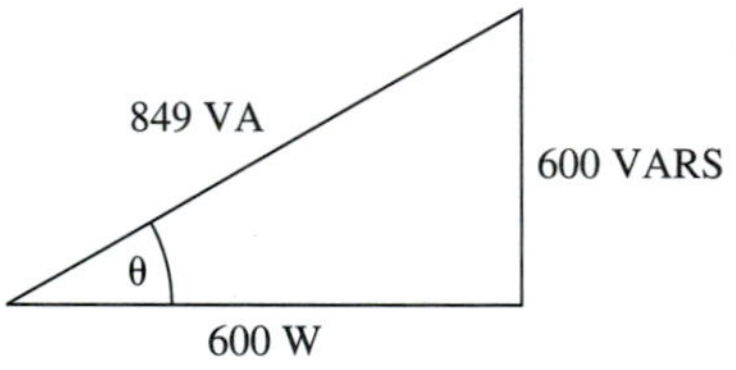

Notice that the base P of the power triangle is drawn down to indicate the circuit is inductive.

Any of the basic trigonometric functions can be used to determine θ from the power triangle. Using the cosine function,

$$\cos\theta = \frac{600}{849}$$

$$\theta = \cos^{-1}\left\{\frac{600}{849}\right\}$$

$$= 45°$$

Recall that θ is the angle separating the voltage and current phasors. Because $I_M = 16.98$, $\mathbf{V} = 100\underline{/80°}$, and the circuit is inductive, it follows that the current phasor lags the voltage phasor by 45°. The current phasor can then be written as

$$\mathbf{I} = 16.98 \underline{/ 80^\circ - 45^\circ}$$
$$= 16.98 \underline{/ 35^\circ}$$

The sinusoidal expression for **I** is

$$i = 16.98 \sin(20t + 35^\circ)$$

EXAMPLE 20.9 Determine i_1, i_2, and i_3 for the circuit in Example 20.7. ■

SOLUTION The power triangles for loads 1, 2, and 3 are shown in Figures 20.20, 20.21, and 20.22. P and Q are given for each load in Example 20.7 and S, F_p, and θ are determined from Equations (20.8) and (20.11).

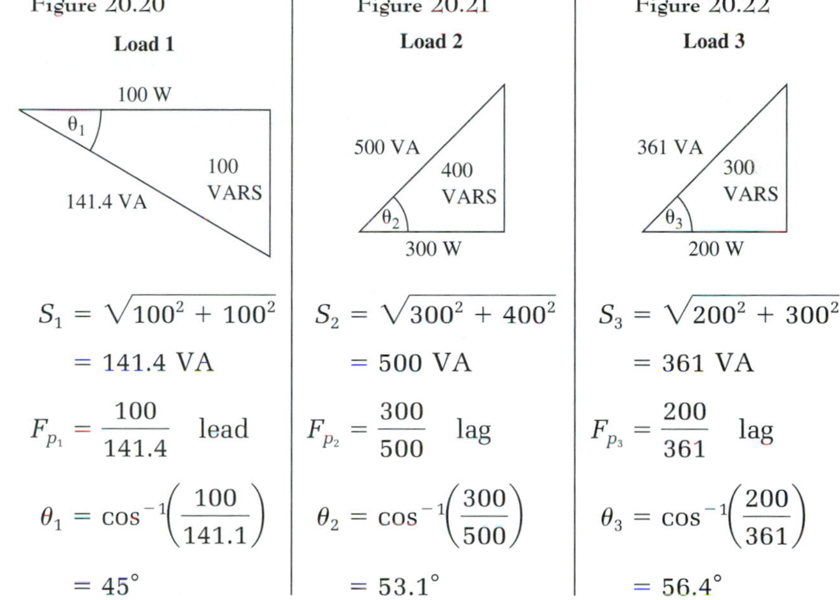

Figure 20.20 | Figure 20.21 | Figure 20.22

$$S_1 = \sqrt{100^2 + 100^2} = 141.4 \text{ VA} \qquad F_{p_1} = \frac{100}{141.4} \text{ lead} \qquad \theta_1 = \cos^{-1}\left(\frac{100}{141.1}\right) = 45^\circ$$

$$S_2 = \sqrt{300^2 + 400^2} = 500 \text{ VA} \qquad F_{p_2} = \frac{300}{500} \text{ lag} \qquad \theta_2 = \cos^{-1}\left(\frac{300}{500}\right) = 53.1^\circ$$

$$S_3 = \sqrt{200^2 + 300^2} = 361 \text{ VA} \qquad F_{p_3} = \frac{200}{361} \text{ lag} \qquad \theta_3 = \cos^{-1}\left(\frac{200}{361}\right) = 56.4^\circ$$

Recognizing that loads 1, 2, and 3 are in parallel with phasor voltage $\mathbf{V} = 100\underline{/ 80^\circ}$ across each load and applying Equation (20.9) results in

$$S_1 = V_{\text{eff}} I_{1_{\text{eff}}} \qquad I_{1_{\text{eff}}} = \frac{S_1}{V_{\text{eff}}} = \frac{141.4}{100/\sqrt{2}} = 2$$

$$S_2 = V_{\text{eff}} I_{2_{\text{eff}}} \qquad I_{2_{\text{eff}}} = \frac{S_2}{V_{\text{eff}}} = \frac{500}{100/\sqrt{2}} = 7.07$$

$$S_3 = V_{\text{eff}} I_{3_{\text{eff}}} \qquad I_{3_{\text{eff}}} = \frac{S_3}{V_{\text{eff}}} = \frac{361}{100/\sqrt{2}} = 5.11$$

From Equation (20.2)

$$I_{1_M} = I_{1_{\text{eff}}}\sqrt{2} = 2\sqrt{2} = 2.83$$

$$I_{2_M} = I_{2_{\text{eff}}}\sqrt{2} = 7.07\sqrt{2} = 10.0$$

$$I_{3_M} = I_{3_{\text{eff}}}\sqrt{2} = 5.11\sqrt{2} = 7.23$$

Because load 1 is capacitive with impedance angle θ_1 equal to 45° and the voltage phase angle is 80°, the current phasor leads the voltage phasor by 45° and has a phase angle equal to 125° The current phasor for $\mathbf{I}_1$ is

$$\mathbf{I}_1 = I_{1_M}\underline{/125^\circ}$$

$$= 2.83\underline{/125^\circ}$$

and the sinusoidal expression for i_1 is

$$i_1 = 2.83\sin(20t + 125^\circ)$$

Because load 2 is inductive with impedance angle θ_2 equal to 53.1° and the voltage phase angle is 80°, the current phasor lags the voltage phasor by 53.1° and has a phase angle equal to 26.9°. The current phasor for $\mathbf{I}_2$ is

$$\mathbf{I}_2 = I_{2_M}\underline{/26.9^\circ}$$

$$= 10.0\underline{/26.9^\circ}$$

and the sinusoidal expression for i_2 is

$$i_2 = 10.0\sin(20t + 26.9^\circ)$$

Because load 3 is inductive with impedance angle θ_3 equal to 56.4° and the voltage phase angle is 80°, the current phasor lags the voltage phasor by 56.4° and has a phase angle equal to 23.6°. The current phasor for $\mathbf{I}_3$ is

$$\mathbf{I}_3 = I_{3_M}\underline{/23.6^\circ}$$

$$= 7.23\underline{/23.6^\circ}$$

and the sinusoidal expression for i_3 is

$$i_3 = 7.23\sin(20t + 23.6^\circ)$$

20.12 REALIZING ELECTRICAL ELEMENTS FOR POWER CONDITIONS

It is often necessary to determine the electrical elements that are included in the boxes, as used in Example 20.7. Although there are generally an infinite number of element combinations that satisfy the given power conditions for a circuit, we shall consider determining the simplest (least number of elements) series and parallel circuits to represent a given box. The methods for arriving at the series and parallel representations are illustrated in the following examples.

EXAMPLE 20.10 Determine a simple series circuit representation for each of the load boxes in Example 20.7. ■

SOLUTION Because loads 1, 2, and 3 are capacitive, inductive, and inductive, respectively, and because each load contains resistance ($P \neq 0$), the three loads can be representated as shown in Figure 20.23, with the respective effective currents determined in Example 20.9.

Figure 20.23

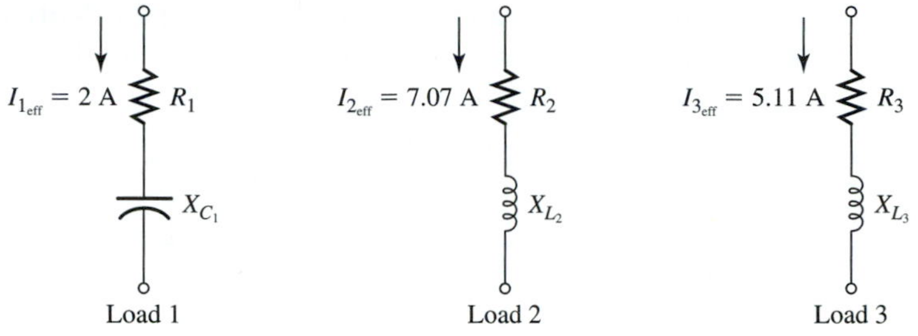

Using Equations (20.28), (20.31), and (20.34) and the values of P and Q for each load given in Example 20.7 results in

$$P = I_{eff}^2 R$$

$$\begin{array}{r|c|c}
P_1 = I_{1_{eff}}^2 R_1 & P_2 = I_{2_{eff}}^2 R_2 & P_3 = I_{3_{eff}}^2 R_3 \\
100 = (2)^2 R_1 & 300 = (7.07)^2 R_2 & 200 = (5.11)^2 R_3 \\
R_1 = 25\,\Omega & R_2 = 6\,\Omega & R_3 = 7.66\,\Omega
\end{array}$$

$$Q = I_{eff}^2 X$$

$$\begin{array}{r|c|c}
Q_1 = I_{1_{eff}}^2 X_{C_1} & Q_2 = I_{2_{eff}}^2 X_{L_2} & Q_3 = I_{3_{eff}}^3 X_{L_3} \\
100 = (2)^2 X_{C_1} & 400 = (7.07)^2 X_{L_2} & 300 = (5.11)^2 X_{L_3} \\
X_{C_1} = 25\,\Omega & X_{L_2} = 8\,\Omega & X_{L_3} = 11.49\,\Omega
\end{array}$$

From Equations (17.7) and (17.11)

$$\begin{array}{r|c|c}
X_{C_1} = \dfrac{1}{\omega C_1} & X_{L_2} = \omega L_2 & X_{L_3} = \omega L_3 \\
25 = \dfrac{1}{2C_1} & 8 = 20L_2 & 11.49 = 20L_3 \\
C_1 = \dfrac{1}{500}F & L_2 = 0.8\,\text{H} & L_3 = 0.574\,\text{H}
\end{array}$$

The load boxes for the circuit in Example 20.7 can be represented as shown in Figure 20.24.

Figure 20.24

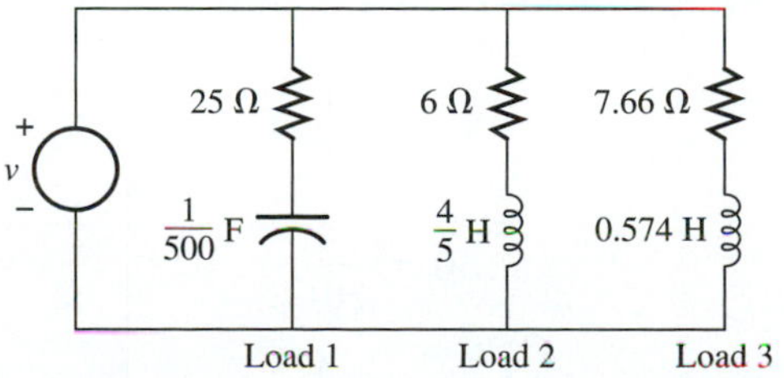

EXAMPLE 20.11 Determine a simple parallel circuit representation for each load box in Example 20.7. ■

SOLUTION Because loads 1, 2, and 3 are capacitive, inductive, and inductive, respectively, and because each load contains resistance ($P \neq 0$), the three loads can be represented as shown in Figure 20.25. Notice that the voltage source v appears across each element of each load, and from Equation (20.2) $V_{eff} = 100/\sqrt{2}$.

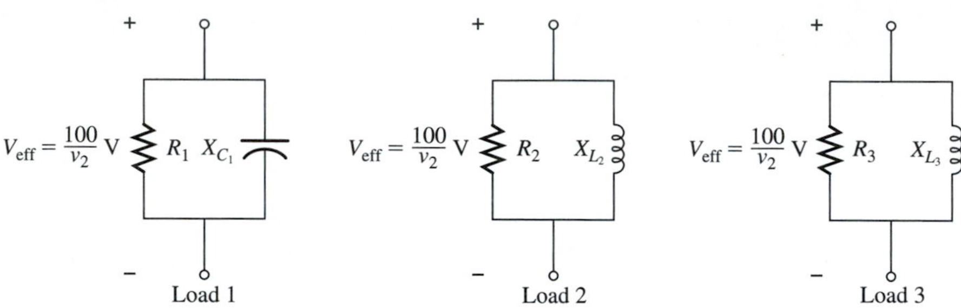

Figure 20.25

Using Equations (20.27), (20.30), and (20.33) and the values of P and Q for each load given in Example 20.7 results in

$$P = \frac{V_{eff}^2}{R}$$

$$\begin{array}{l|l|l} P_1 = \dfrac{V_{eff}^2}{R_1} & P_2 = \dfrac{V_{eff}^2}{R_2} & P_3 = \dfrac{V_{eff}^2}{R_3} \\ 100 = \dfrac{(100/\sqrt{2})^2}{R_1} & 300 = \dfrac{(100/\sqrt{2})^2}{R_2} & 200 = \dfrac{(100/\sqrt{2})^2}{R_3} \\ R_1 = 50\,\Omega & R_2 = \dfrac{50}{3}\,\Omega & R_3 = 25\,\Omega \end{array}$$

$$Q = \frac{V_{eff}^2}{X}$$

$$\begin{array}{l|l|l} Q_1 = \dfrac{V_{eff}^2}{X_{C_1}} & Q_2 = \dfrac{V_{eff}^2}{X_{L_2}} & Q_3 = \dfrac{V_{eff}^2}{X_{L_3}} \\ 100 = \dfrac{(100/\sqrt{2})^2}{X_{C_1}} & 400 = \dfrac{(100/\sqrt{2})^2}{X_{L_2}} & 300 = \dfrac{(100/\sqrt{2})^2}{X_{L_3}} \\ X_{C_1} = 50\,\Omega & X_{L_2} = \dfrac{25}{2}\,\Omega & X_{L_3} = \dfrac{50}{3}\,\Omega \end{array}$$

From Equations (17.7) and (17.11),

$$\begin{array}{l|l|l} X_{C_1} = \dfrac{1}{\omega C_1} & X_{L_2} = \omega L_2 & X_{L_3} = \omega L_3 \\ 50 = \dfrac{1}{20C_1} & \dfrac{25}{2} = 20L_2 & \dfrac{50}{3} = 20L_3 \\ C_1 = \dfrac{1}{1000}\,\text{F} & L_2 = \dfrac{5}{8}\,\text{H} & L_3 = \dfrac{5}{6}\,\text{H} \end{array}$$

The load boxes for the circuit in Example 20.7 can be represented as shown in Figure 20.26.

Figure 20.26

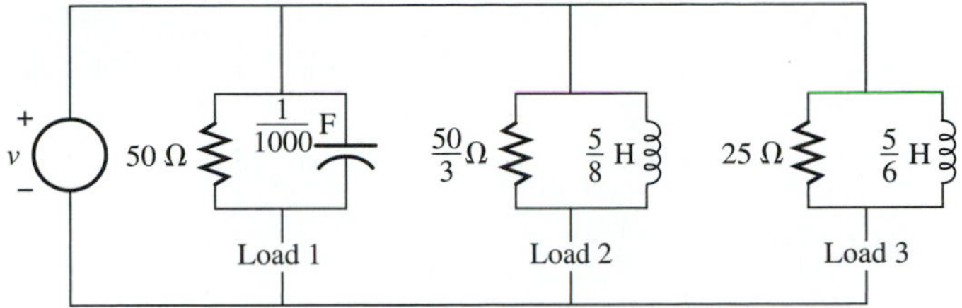

EXAMPLE 20.12 Determine the current i and the voltages v_1 and v_2 for the circuit in Figure 20.27. ■

Figure 20.27

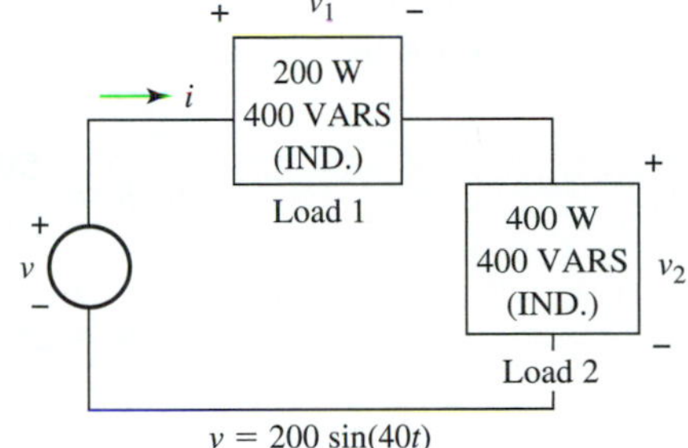

SOLUTION From the given expression for v, the voltage phasor is

$$\mathbf{V} = 200\angle 0^\circ$$

The total power delivered by the voltage source is

$$\begin{aligned} P &= 200 + 400 \\ &= 600 \text{ W} \end{aligned}$$

The total reactive power is

$$\begin{aligned} Q &= 400 + 400 \\ &= 800 \text{ VARS} \quad \text{(inductive)} \end{aligned}$$

From Equation (20.11)

$$\begin{aligned} S^2 &= P^2 + Q^2 \\ S &= \sqrt{P^2 + Q^2} \\ &= \sqrt{600^2 + 800^2} \\ &= 1000 \text{ VA} \end{aligned}$$

From Equation (20.2)

$$\begin{aligned} V_{\text{eff}} &= \frac{V_M}{\sqrt{2}} \\ &= \frac{200}{\sqrt{2}} \end{aligned}$$

From Equation (20.9) the apparent power supplied by the voltage source is

$$S = V_{\text{eff}} I_{\text{eff}}$$
$$1000 = \left(\frac{200}{\sqrt{2}}\right) I_{\text{eff}}$$
$$I_{\text{eff}} = 5\sqrt{2}$$

and

$$I_M = I_{\text{eff}}\sqrt{2}$$
$$= 10 \text{ A}$$

The power triangle for the total load is shown in Figure 20.28, and the impedance angle θ is

$$\cos\theta = \frac{600}{1000}$$
$$\theta = \cos^{-1}\left(\frac{600}{1000}\right)$$
$$= 53.1^\circ$$

Figure 20.28

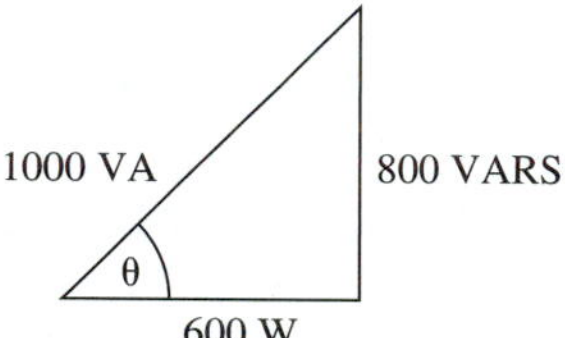

Because the total load is inductive, the phasor current **I** must lag the voltage phasor by 53.1°. Current **I** can be written as

$$\mathbf{I} = I_M \angle 0 - 53.1^\circ$$
$$= 10\angle -53.1^\circ$$

and

$$i = 10\sin(40t - 53.1^\circ)$$

The power triangles for loads 1 and 2 are shown in Figure 20.29; S, F_p, and θ are determined from Equations (20.8) and (20.11).

Figure 20.29

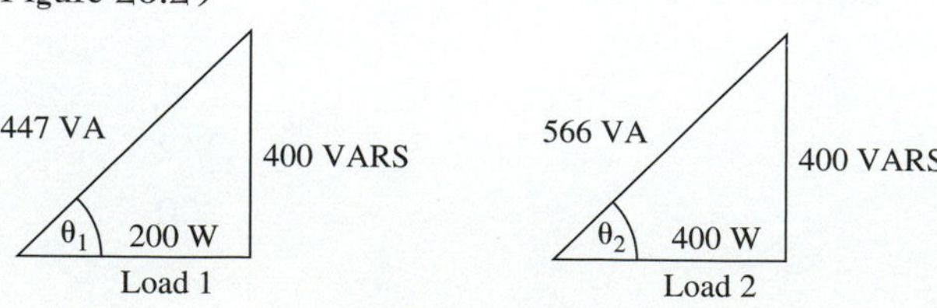

$$S_1 = \sqrt{200^2 + 400^2} = 447 \qquad S_2 = \sqrt{400^2 + 400^2} = 566$$

$$F_{p_1} = \frac{200}{447} \qquad F_{p_2} = \frac{400}{566}$$

$$\theta_1 = \cos^{-1}\left(\frac{200}{447}\right) = 63.4^\circ \qquad \theta_2 = \cos^{-1}\left(\frac{400}{566}\right) = 45^\circ$$

Applying Equation (20.9) to each load and recognizing that current **I** flows through both loads because they are in series,

$$S_1 = V_{1_{\text{eff}}} I_{\text{eff}} \qquad S_2 = V_{2_{\text{eff}}} I_{\text{eff}}$$

$$447 = V_{1_{\text{eff}}}(5\sqrt{2}) \qquad 566 = V_{2_{\text{eff}}}(5\sqrt{2})$$

$$V_{1_{\text{eff}}} = \frac{447}{5\sqrt{2}} \qquad V_{2_{\text{eff}}} = \frac{566}{5\sqrt{2}}$$

From Equation (15.2)

$$V_{1_M} = V_{1_{\text{eff}}}\sqrt{2} = \frac{447}{5} = 89.4 \qquad V_{2_M} = V_{2_{\text{eff}}}\sqrt{2} = \frac{566}{5} = 113.2$$

Because load 1 is inductive with impedance angle θ_1 equal to 63.4° and the current phase angle is −53.1°, the voltage phasor leads the curent phasor by 63.4° and has a phase angle equal to 10.3°. The voltage phasor for $\mathbf{V}_1$ is

$$\mathbf{V}_1 = V_{1_M}\angle 10.3^\circ = 89.4\angle 10.3^\circ$$

and the sinusoidal expression for v_1 is

$$v_1 = 89.4 \sin(40t + 10.3^\circ)$$

Because load 2 is inductive with impedance angle equal to 45° and the current phase angle is at −53.1°, the voltage phasor leads the current phasor by 45° and has a phase angle equal to −8.1°. The voltage phasor for $\mathbf{V}_2$ is

$$\mathbf{V}_2 = V_{2_M}\angle -8.1^\circ = 113.2\angle -8.1^\circ$$

and the sinusoidal expression for v_2 is

$$v_2 = 113.2 \sin(40t - 8.1^\circ)$$

20.13 COMPLEX POWER

The concept of complex power provides an alternative method for determining apparent power, average power, and reactive power. The term *complex power* refers to the expression that results when complex numbers are used to relate apparent power, average power, and reactive power.

Consider the circuit in Figure 20.30, where $\mathbf{Z} = Z\underline{/\theta}$, $\mathbf{V} = V_M\underline{/\alpha}$, and $\mathbf{I} = I_M\underline{/\beta}$.

Figure 20.30 Complex power.

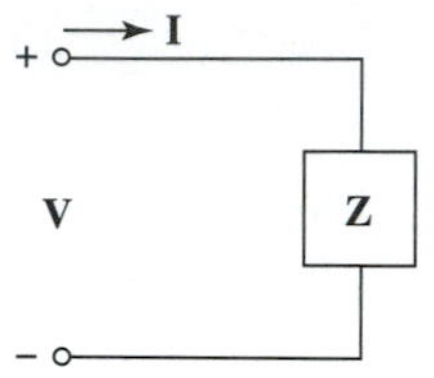

The *complex power* **S** supplied to load **Z** can be written as

$$\mathbf{S} = \frac{\mathbf{V}\tilde{\mathbf{I}}}{2} \tag{20.35}$$

where $\tilde{\mathbf{I}}$ is the conjugate of the complex number representing the phasor **I**. From Chapter 16, if $\mathbf{I} = I_M\underline{/\beta}$, then $\tilde{\mathbf{I}} = I_M\underline{/-\beta}$.

From the complex power expression given in Equation (20.35),

$$\begin{aligned}\mathbf{S} &= \frac{\mathbf{V}\tilde{\mathbf{I}}}{2}\\ &= \frac{(V_M\underline{/\alpha})(I_M\underline{/-\beta})}{2}\\ &= \left\{\left(\frac{V_M}{2}\right)\left(\frac{I_M}{2}\right)\right\}\underline{/\alpha-\beta}\end{aligned}$$

and from Equation (20.2)

$$\mathbf{S} = (V_{\text{eff}}I_{\text{eff}})\underline{/\alpha-\beta}$$

Because the voltage phasor is at an angle α and the current phasor is at angle β at $t = 0$ for the circuit in Figure 20.30, the phase difference angle (the angle separating the voltage and current phasors) is

$$\theta = \alpha - \beta$$

where θ is the angle associated with impedance **Z**. **S** can then be written as

$$\mathbf{S} = (V_{\text{eff}}I_{\text{eff}})\underline{/\theta}$$

Using the polar-to-rectangular conversion rule given in Chapter 16,

$$\mathbf{S} = V_{\text{eff}}I_{\text{eff}}\cos\theta + jV_{\text{eff}}I_{\text{eff}}\sin\theta$$

Comparing Equations (20.7) and (20.10) to the preceding equation, **S** can then be written as

$$\mathbf{S} = P + jQ \tag{20.36}$$

Consider the following example.

EXAMPLE 20.13 Use the complex-power concept to determine the average power P and the reactive power Q delivered by the voltage source in Figure 20.31. ■

Figure 20.31

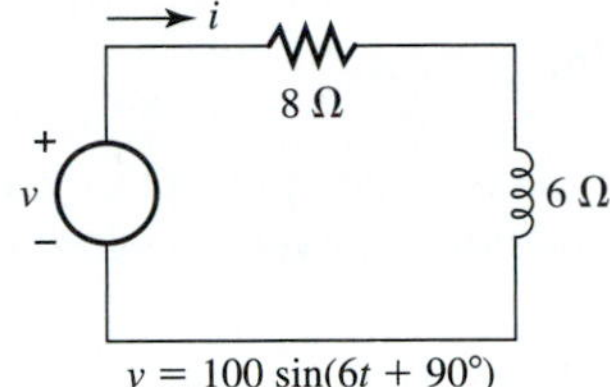

SOLUTION In phasor form the voltage v is

$$\mathbf{V} = 100\angle 90^\circ$$

and from Equations (17.3), (17.8), and (17.35), the impedance seen by the voltage source is

$$\begin{aligned}\mathbf{Z} &= 8 + j6 \\ &= 10\angle 36.9^\circ\end{aligned}$$

From Equation (17.1) the current i in phasor form is

$$\begin{aligned}\mathbf{I} &= \frac{\mathbf{V}}{\mathbf{Z}} \\ &= \frac{100\angle 90^\circ}{10\angle 36.9^\circ} \\ &= 10\angle 53.1^\circ\end{aligned}$$

From Equation (20.35) the complex power delivered by the voltage source is

$$\begin{aligned}\mathbf{S} &= \frac{\mathbf{V}\tilde{\mathbf{I}}}{2} \\ &= \frac{(100\angle 90^\circ)(\widetilde{10\angle 53.1^\circ})}{2} \\ &= \frac{(100\angle 90^\circ)(10\angle -53.1^\circ)}{2} \\ &= 500\angle 36.9^\circ\end{aligned}$$

From the polar-to-rectangular conversion rule given in Chapter 16,

$$\begin{aligned}\mathbf{S} &= 500\cos(36.9^\circ) + j500\sin(36.9^\circ) \\ &= 400 + j300\end{aligned}$$

From Equation (20.36)

$$P = \text{Re}(\mathbf{S}) = 400 \text{ W}$$

$$Q = \text{Im}(\mathbf{S}) = 300 \text{ VARS} \quad \text{(inductive)}$$

The apparent power is the magnitude of the complex power and is

$$S = |\mathbf{S}| = 500 \text{ VA}$$

The rules of the previous section state that the sum of the apparent powers of the individual circuit elements does not generally equal the total apparent power delivered to the circuit. However, the above statement does not hold for complex power.

> The sum of the complex powers for the individual circuit elements is equal to the total complex power delivered to the circuit.

This rule is known as *the conservation of complex power* for electric circuits.

Consider the following example.

EXAMPLE 20.14 Use the complex-power concept to determine the current i and the voltages v_1 and v_2 for the circuit in Figure 20.32. ■

Figure 20.32

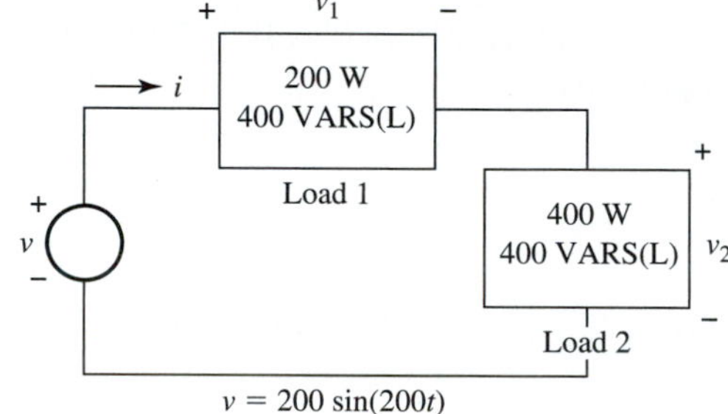

SOLUTION From Equation (20.36) the complex powers for loads 1 and 2 are

$$\mathbf{S}_1 = 200 + j400$$
$$= 447.2\underline{/63.4^\circ}$$

and

$$\mathbf{S}_2 = 400 + j400$$
$$= 565.7\underline{/45^\circ}$$

From the rule for conservation of complex power, the total complex power is

$$\mathbf{S} = \mathbf{S}_1 + \mathbf{S}_2$$
$$= 600 + j800$$
$$= 1000\underline{/53.1^\circ}$$

From Equation (20.35)

$$\mathbf{S} = \frac{\mathbf{V}\tilde{\mathbf{I}}}{2}$$

and from the given voltage expression,

$$\mathbf{V} = 200\underline{/0^\circ}$$

Substituting $\mathbf{S} = 1000\angle 53.1^\circ$ and $\mathbf{V} = 200\angle 0^\circ$ into Equation (20.35) results in

$$(1000\angle 53.1^\circ) = \frac{(200\angle 0^\circ)\tilde{\mathbf{I}}}{2}$$

$$\tilde{\mathbf{I}} = 10\angle 53.1^\circ$$

Because the conjugate of the conjugate of $\mathbf{I}$ equals $\mathbf{I}$,

$$\mathbf{I} = \tilde{\tilde{\mathbf{I}}} = 10\angle -53.1^\circ$$

and the sinusoidal expression for i is

$$i = 10 \sin(40t - 53.1^\circ)$$

Applying Equation (20.35) to load 1 results in

$$\mathbf{S}_1 = 447.2\angle 63.4^\circ$$

$$447.2\angle 63.4^\circ = \frac{\mathbf{V}\tilde{\mathbf{I}}}{2}$$

$$= \frac{\mathbf{V}_1(10\angle 53.1^\circ)}{2}$$

$$\mathbf{V}_1 = \frac{(2)(447.2\angle 63.4^\circ}{10\angle 53.1^\circ}$$

$$= 89.4\angle 10.3^\circ$$

and the sinusoidal expression for v_1 is

$$v_1 = 89.4 \sin(40t + 10.3^\circ)$$

Applying Equation (20.35) to load 2 results in

$$\mathbf{S}_2 = 565.7\angle 45^\circ$$

$$565.7\angle 45^\circ = \frac{\mathbf{V}_2\tilde{\mathbf{I}}}{2}$$

$$= \frac{\mathbf{V}_2(10\angle 53.1^\circ)}{2}$$

$$\mathbf{V}_2 = 113.2\angle -8.1^\circ$$

and the sinusoidal expression for v_2 is

$$v_2 = 113.2 \sin(40t - 8.1^\circ)$$

20.14 POWER FACTOR MODIFICATION

For a given amount of apparent power available to a circuit, the amount of average power delivered to the circuit is determined by the circuit power factor. Because the power factor is equal to the cosine of the angle associated with the circuit impedance, a small impedance angle results in a large power factor and a large impedance angle results in a small power factor. The power factor ranges from zero to unity, because the magnitude of the impedance angle ranges from 0 to 90°.

In many applications it is desirable to reduce the power-factor angle of a load while still furnishing the same average power to the load. A reduced power-factor angle results in an increased power factor. The power-factor-angle reduction reduces the current in the conductors between the source and the load, which is particularly desirable in electric power transmission. If there is less current in the conductors between the source and the load, then less energy is lost prior to reaching the load.

In order to reduce the power-factor angle of a load and maintain the average power delivered to the load, the magnitude of the load reactance or susceptance (the imaginary part of the impedance or admittance) must be altered. This is usually accomplished by adding an inductor or capacitor in parallel or series with the load. Power-factor reduction is most easily accomplished by using the complex-power concept. Consider the following example.

EXAMPLE 20.15 Increase the power factor of the load to 0.9 lag and maintain the average power delivered to the original load for the circuit in Figure 20.33. ■

Figure 20.33

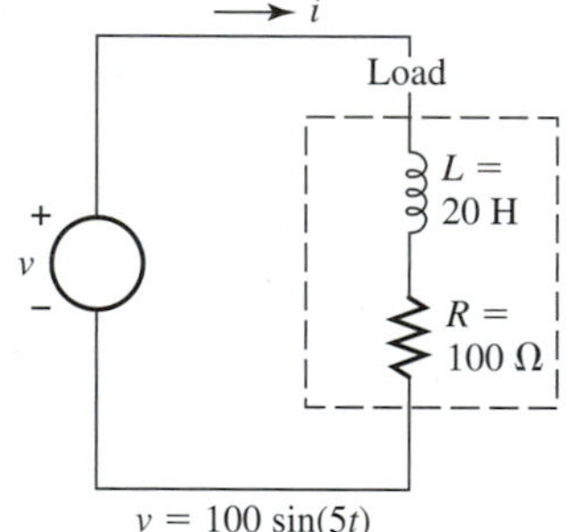

SOLUTION From Equations (17.8) and (17.35), the impedance $\mathbf{Z}$ for the load is

$$\mathbf{Z} = R + j\omega L$$
$$= 100 + j100$$
$$= 141.5\underline{/45^\circ}$$

and the power factor is

$$F_p = \cos 45^\circ$$
$$= 0.707$$

We reduce the power factor angle by adding a capacitor in parallel with the load, as shown in Figure 20.34.

Figure 20.34

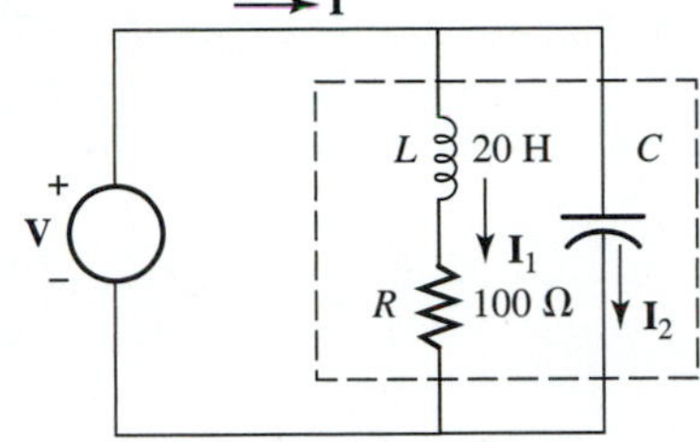

From Equations (17.1) and (17.35)

$$
\begin{aligned}
I_1 &= \frac{100\angle 0^\circ}{100 + j100} \\
&= \frac{100\angle 0^\circ}{141.1\angle 45^\circ} \\
&= 0.707\angle -45^\circ
\end{aligned}
$$

From Equation (20.2)

$$
\begin{aligned}
I_{1_{\text{eff}}} &= \frac{I_1}{\sqrt{2}} \\
&= \frac{0.707}{\sqrt{2}} \\
&= \frac{1}{2}\text{ A}
\end{aligned}
$$

From Equations (20.28) and (20.34), the average and reactive powers for the original load are

$$
\begin{aligned}
P_1 &= I_{1_{\text{eff}}}^2 R \\
&= \left(\frac{1}{2}\right)^2 (100) \\
&= 25\text{ W} \\
Q_1 &= I_{1_{\text{eff}}}^2 X_L \\
&= \left(\frac{1}{2}\right)^2 (100) \\
&= 25\text{ VARS} \quad \text{(inductive)}
\end{aligned}
$$

and from Equation (20.36) the complex power for the original load is

$$
\begin{aligned}
S_1 &= P_L + jQ_1 \\
&= 25 + j25
\end{aligned}
$$

From Equation (20.30) the reactive power for the capacitor is

$$Q_2 = \frac{(V_{\text{eff}})^2}{X_C}$$

$$= \frac{(100/\sqrt{2})^2}{X_C}$$

$$= \frac{5000}{X_C}$$

and from Equation (20.36) the complex power for the capacitor is

$$\mathbf{S}_2 = P_2 + jQ_2$$

$$= 0 - j\left(\frac{5000}{X_C}\right)$$

$$= -j\left(\frac{5000}{X_C}\right)$$

Notice that the negative sign is used for capacitive reactive power.

The total complex power for the corrected load is

$$\mathbf{S} = \mathbf{S}_1 + \mathbf{S}_2$$

$$= 25 + j25 - j\left(\frac{5000}{X_C}\right)$$

$$= 25 + j\left\{25 - \left(\frac{5000}{X_C}\right)\right\}$$

Because the angle associated with complex power is the power factor angle,

$$\tan\theta = \frac{25 - (5000/X_C)}{25}$$

and since the desired power factor is 0.9 lag,

$$\cos\theta = 0.9$$

$$\theta = \cos^{-1}(0.9)$$

$$= 25.8^\circ$$

Substituting $\theta = 25.8^\circ$ into the preceding equation for $\tan\theta$ results in

$$\tan(25.8^\circ) = \frac{25 - (5000/X_C)}{25}$$

$$0.483 = \frac{25 - (5000/X_C)}{25}$$

$$\frac{5000}{X_C} = 25 - 26(0.483)$$

$$X_C = \frac{5000}{25 - (25)(0.483)}$$

$$= 386.8\ \Omega$$

From Equation (17.11)

$$X_C = \frac{1}{\omega C}$$

$$C = \frac{1}{\omega X_C}$$

$$= \frac{1}{(5)(386.8)}$$

$$= 0.000517 \text{ F}$$

$$= 0.517 \text{ mF}$$

20.15 AVERAGE-POWER MEASUREMENT

The device used to measure the average power delivered to an ac load is called a *wattmeter*. The wattmeter senses the voltage across the load and the current flowing through the load and, using the sensed quantities, determines the power delivered to the load according to Equation (20.7) as

$$P = V_{\text{eff}} I_{\text{eff}} \cos\theta$$

where $\cos\theta$ is the load-power factor.

A typical wattmeter is connected to the load by connecting the current-sensing coil in series with the load and the voltage-sensing coil across the load, as illustrated in the symbolic representation of the wattmeter in Figure 20.35.

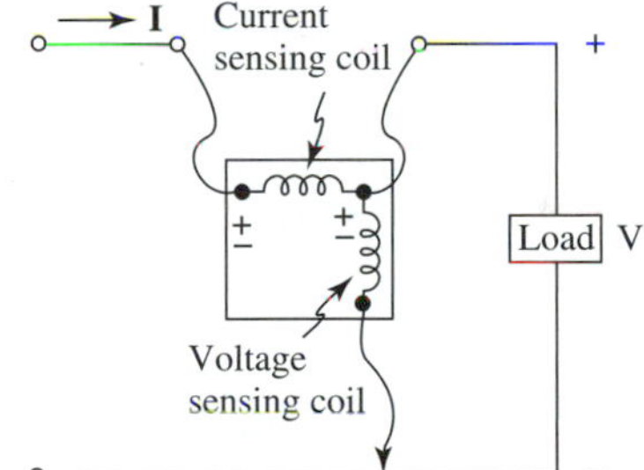

Figure 20.35 Power measurement.

Notice that the current-sensing coil and the voltage-sensing coil each have one terminal marked ±. The ± terminals of the voltage- and current-sensing coils should be connected as shown in Figure 20.35 for the wattmeter to read the average power as

$$P = V_{\text{eff}} I_{\text{eff}} \cos\theta$$

That is, the current **I** measured by the wattmeter should enter the terminal of the current-sensing coil marked ±, and the voltage **V** measured by the wattmeter should be the voltage of the ± terminal with respect to the unmarked terminal of the voltage-sensing coil. If the wattmeter does not read upscale, the terminals of either the voltage-sensing coil or the current-sensing coils can be reversed to obtain an upscale reading.

The voltage-sensing coil is a high-impedance coil, because it parallels the load, and the current-sensing coil is a low-impedance coil, because it is in series with the load. This ensures that the load voltage and current are not changed appreciably by the measuring device and the average power read by the wattmeter is an accurate measurement of the actual power delivered to the load.

20.16 ENERGY ASSOCIATED WITH RESISTORS

For the purely resistive circuit shown in Figure 20.36, the power-factor angle (impedance angle) is zero and the power factor is unity. From Equation (20.4) the instantaneous power delivered to the resistor is

$$p_R = V_{\text{eff}}I_{\text{eff}} - V_{\text{eff}}I_{\text{eff}}\cos(2\omega t)$$

Figure 20.36 Energy for resistor.

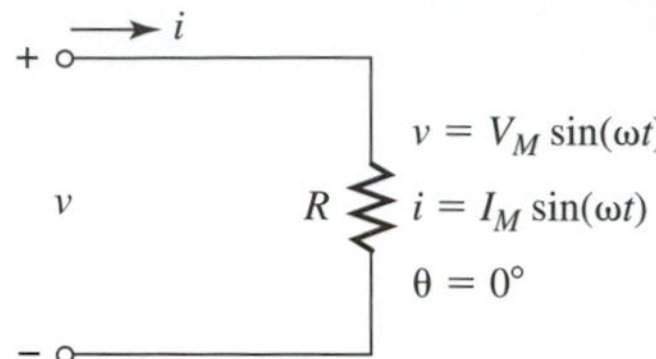

and plotting the waveforms for v, i, and p_R results in the plots shown in Figure 20.37, where T_p is the period of the power waveform and T is the period of the voltage and current waveforms.

Figure 20.37 Power curve for resistor.

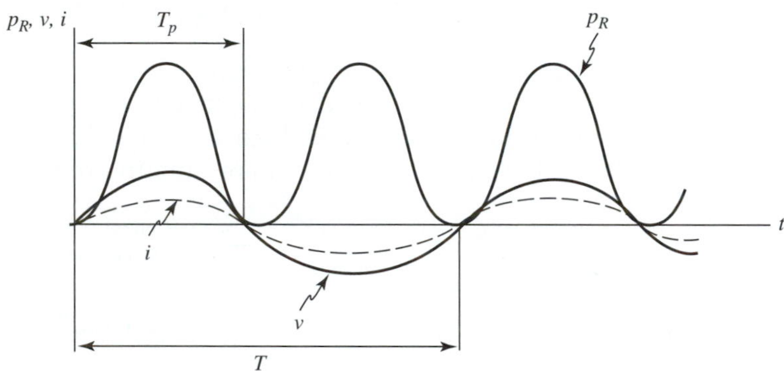

Because the average value of $V_{\text{eff}}I_{\text{eff}}\cos(2\omega t)$ in Equation (20.4) is zero for one period of the power waveform, the average value of p_R for one period (T_p) of the power waveform is $V_{\text{eff}}I_{\text{eff}}$. Because power is the rate of delivering energy, the energy delivered to R for one period of p_R is then

$$w_R = V_{\text{eff}}I_{\text{eff}}T_p \tag{20.37}$$

Because $T_p = T/2$, w_R can be expressed as

$$w_R = V_{\text{eff}}I_{\text{eff}}\left(\frac{T}{2}\right) \tag{20.38}$$

From Equations (16.3), (20.37), and (20.38), w_R can also be expressed as

$$w_R = \frac{V_{\text{eff}}I_{\text{eff}}}{f_p} \tag{20.39}$$

and

$$w_R = \frac{V_{\text{eff}}I_{\text{eff}}}{2f} \tag{20.40}$$

where f and f_p are the frequencies of the voltage (and current) waveforms and the power waveform, respectively, and w_R is the energy delivered to R during one cycle of the power waveform or during one half-cycle of the current or voltage waveform.

20.17 ENERGY ASSOCIATED WITH INDUCTORS

If the load for a circuit is purely inductive, the power-factor angle (impedance angle) is 90° lag, as shown in Figure 20.38.

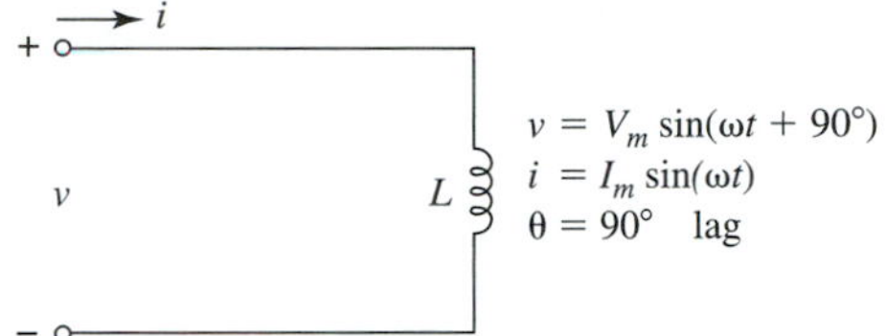

Figure 20.38 Energy for inductor.

From Equation (20.5) the instantaneous power to the inductor is

$$p_L = V_{\text{eff}} I_{\text{eff}} \sin(2\omega t)$$

and plotting the waveforms for v, i, and p_L results in the plots shown in Figure 20.39, where T_p is the period of the power waveform and T is the period of the voltage and current waveforms.

Figure 20.39 Power curve for inductor.

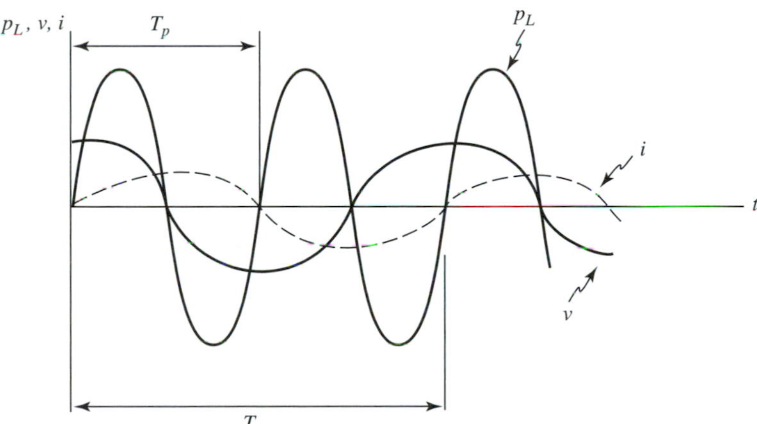

Because the area under the power waveform is positive for one-half cycle and negative for one-half cycle and because the positive area is equal to the negative area, the average power delivered to an inductor for a full cycle of the power waveform is zero. The circuit delivers energy to the inductor for one-half cycle of the power waveform, and the inductor delivers the energy back to the circuit during the next half cycle. The energy exchange continues in a periodic manner, and no energy is lost in the process for the ideal inductor.

It is often desirable to determine the energy stored in an inductor during one-half cycle of the power waveform. From Equation (20.5) p_L is given by

$$p_L = V_{\text{eff}} I_{\text{eff}} \sin(2\omega t)$$

and from Equation (20.1) the average power for one-half cycle of the power waveform is

$$p_{L_{avg}} = \frac{2(V_{eff}I_{eff})}{\pi}$$

Because power is the rate of delivering energy, the energy delivered to L during one-half period of p_L is

$$w_L = p_{L_{avg}}\left(\frac{T_p}{2}\right)$$

$$= \frac{V_{eff}I_{eff}T_p}{\pi} \tag{20.41}$$

Because $T_p = T/2$, w_L can be expressed as

$$w_L = \frac{V_{eff}I_{eff}T}{2\pi} \tag{20.42}$$

From Equations (16.3), (20.41), and (20.42), w_L can also be expressed as

$$w_L = \frac{V_{eff}I_{eff}}{\pi f_p} \tag{20.43}$$

and

$$w_L = \frac{V_{eff}I_{eff}}{2\pi f} \tag{20.44}$$

where f and f_p are the frequencies of the voltage (and current) and power waveforms, respectively.

From Equations (17.5) and (20.2),

$$V_{eff} = I_{eff}(\omega L)$$

and from Equation (16.4)

$$V_{eff} = I_{eff}(2\pi fL)$$

Substituting $I_{eff}(2\pi fL)$ for V_{eff} in Equation (20.44) results in

$$w_L = LI_{eff}^2 \tag{20.45}$$

where w_L is the energy stored in the inductor during one half-cycle of the power waveform.

20.18 ENERGY ASSOCIATED WITH CAPACITORS

If the load for a circuit is purely capacitive, the power-factor angle (impedance angle) is 90° lead, as shown in Figure 20.40. From Equation (20.6) the instantaneous power to the capacitor is

$$p_C = -V_{eff}I_{eff}\sin(2\omega t)$$

Figure 20.40 Energy for capacitor.

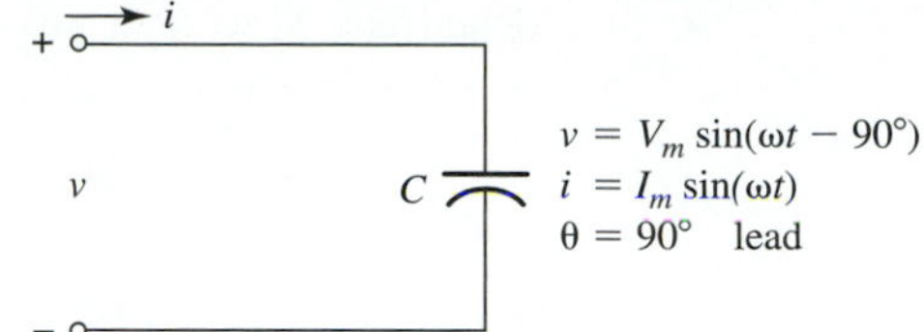

and plotting the waveforms for v, i, and p_C results in the plots shown in Figure 20.41, where T_p is the period of the power waveform and T is the period of the voltage and current waveforms.

Figure 20.41 Power curve for capacitor.

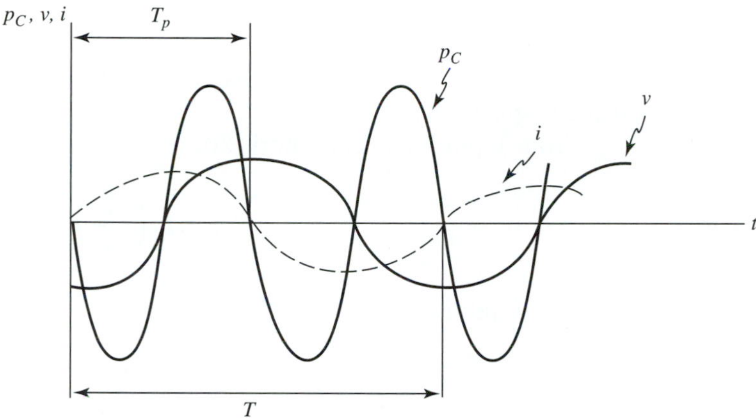

Because the area under the power waveform is negative for one-half cycle and positive for one-half cycle and because the positive area is equal to the negative area, the average power delivered to a capacitor for one full cycle of the power waveform is zero. The circuit delivers energy to the capacitor for one-half cycle of the power waveform and the capacitor delivers the energy back to the circuit during the next half-cycle. The energy exchange continues in a periodic manner, and no energy is lost in the process for the ideal capacitor.

It is often desirable to determine the energy stored in a capacitor during one-half cycle of the power waveform. From Equation (20.6) p_C is given by

$$p_C = -V_{\text{eff}} I_{\text{eff}} \sin(2\omega t)$$

and from Equation (20.1) the average power for one-half cycle of the power waveform is

$$p_{C_{\text{avg}}} = \frac{2(V_{\text{eff}} I_{\text{eff}})}{\pi}$$

Because power is the rate of delivering energy, the energy delivered to C during one-half period of p_C is

$$\begin{aligned} w_C &= p_{C_{\text{avg}}}\left(\frac{T_p}{2}\right) \\ &= \frac{V_{\text{eff}} I_{\text{eff}} T_p}{\pi} \end{aligned} \qquad (20.46)$$

Because $T_p = T/2$, w_C can be expressed as

$$w_C = \frac{V_{\text{eff}} I_{\text{eff}} T}{2\pi} \tag{20.47}$$

From Equation (16.3), (20.46), and (20.47), w_C can also be expressed as

$$w_C = \frac{V_{\text{eff}} I_{\text{eff}}}{\pi f_p} \tag{20.48}$$

and

$$w_C = \frac{V_{\text{eff}} I_{\text{eff}}}{2\pi f} \tag{20.49}$$

where f and f_p are the frequencies of the voltage (and current) and power waveforms, respectively.

From Equations (17.9) and (20.2),

$$V_{\text{eff}} = I_{\text{eff}}\left(\frac{1}{\omega C}\right)$$

and from Equation (16.4)

$$\begin{aligned} I_{\text{eff}} &= \omega C V_{\text{eff}} \\ &= V_{\text{eff}}(2\pi f C) \end{aligned}$$

Substituting $V_{\text{eff}}(2\pi f C)$ for I_{eff} in Equation (20.49) results in

$$w_C = C V_{\text{eff}}^2 \tag{20.50}$$

where w_C is the energy stored in the capacitor during one-half cycle of the power waveform.

EXAMPLE 20.16 Determine the energy stored in the inductor during one-half cycle of the power waveform, the energy stored in the capacitor during one-half cycle of the power waveform, and the energy delivered to the resistor during one cycle of the power waveform for the circuit in Figure 20.42. ■

Figure 20.42

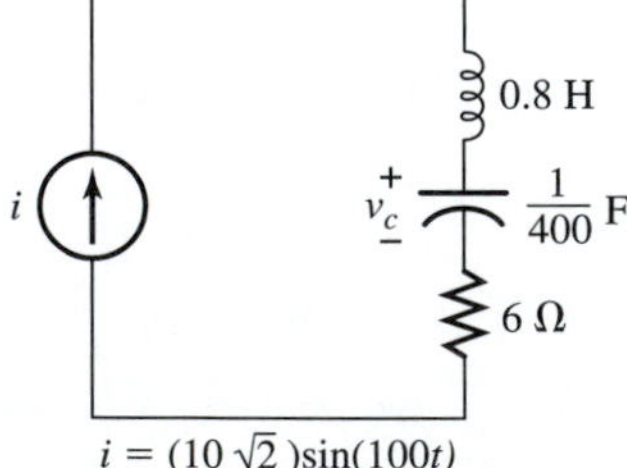

SOLUTION From Equation (20.2)

$$\begin{aligned} I_{\text{eff}} &= \frac{10\sqrt{2}}{\sqrt{2}} \\ &= 10 \text{ A} \end{aligned}$$

From Equation (20.45) the energy stored in the inductor during one-half cycle of the power waveform (or one-fourth cycle of the voltage waveform) is

$$\begin{aligned} w_L &= LI_{\text{eff}}^2 \\ &= 0.8(10)^2 \\ &= 80 \text{ J} \end{aligned}$$

Because $\omega = 100$ rad/s

$$\begin{aligned} f &= \frac{\omega}{2\pi} \\ &= \frac{100}{2\pi} \text{ Hz} \end{aligned}$$

for the current and voltage waveforms. From Equation (20.40) the energy delivered to the resistor during one cycle of the power waveform is

$$w_R = \frac{V_{\text{eff}} I_{\text{eff}}}{2f}$$

and because $V_{\text{eff}} = RI_{\text{eff}}$ from Equations (17.4) and (20.2),

$$\begin{aligned} w_R &= \frac{RI_{\text{eff}}^2}{2f} \\ &= \frac{(6)(10)^2}{2(100/2\pi)} \\ &= 6\,\pi \\ &= 18.84 \text{ J} \end{aligned}$$

From Equation (20.50) the energy delivered to the capacitor during one-half cycle of the power waveform (or one-fourth cycle of the current (or voltage) waveform) is

$$w_C = CV_{\text{eff}}^2$$

and because $V_{\text{eff}} = I_{\text{eff}} X_C$

$$w_C = CI_{\text{eff}}^2 X_C^2$$

From Equation (17.11)

$$\begin{aligned} X_C &= \frac{1}{\omega C} \\ &= \frac{1}{100\left(\frac{1}{400}\right)} \\ &= 4\ \Omega \end{aligned}$$

and

$$w_C = \left(\frac{1}{400}\right)(10)^2(4)$$
$$= 1\ \text{J}$$

20.19 RESONANCE AND POWER CONSIDERATION

A circuit containing resistance, inductance, and capacitance and driven by a sinusoidal source is said to be at resonance when the sinusoidal source is operating at a particular frequency (called the resonant frequency).

When a circuit is at resonance, the voltage and current associated with the driving source are in phase and the power factor is zero. At the resonant frequency the source delivers maximum power to the circuit.

We shall consider the characteristics of the series resonant circuit and the parallel resonant circuit.

20.20 SERIES RESONANCE

Consider the series *RLC* circuit in Figure 20.43, where the frequency ω of the input voltage v is variable. In phasor form the circuit appears as shown in Figure 20.44.

Figure 20.43 Series resonance.

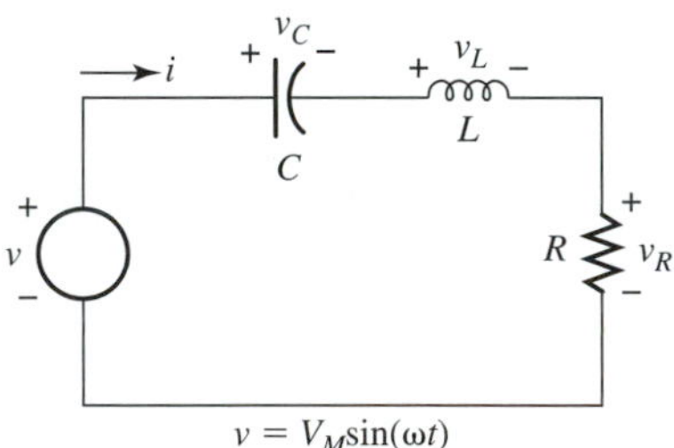

Figure 20.44 Series resonance.

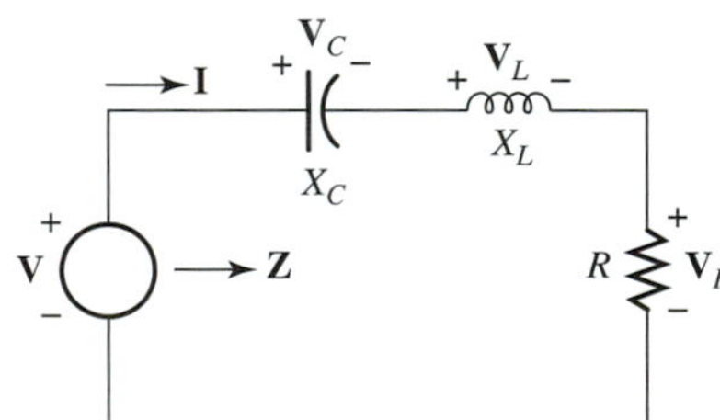

From Equations (17.1) and (17.2), $\mathbf{Z} = \mathbf{V}/\mathbf{I}$ and $Z = V_M/I_M$, where Z is the magnitude of $\mathbf{Z}$. It follows that

$$\mathbf{I} = \frac{\mathbf{V}}{\mathbf{Z}}$$

and

$$I_M = \frac{V_M}{Z}$$

From Equations (17.3), (17.8), (17.12), and (17.35),

$$\mathbf{Z} = R + jX_L - jX_C$$
$$= R + j(X_L - X_C)$$

The magnitude of **Z** is

$$Z = \sqrt{R^2 + (X_L - X_C)^2}$$

and the magnitude of I_M is

$$I_M = \frac{V_M}{\sqrt{R^2 + (X_L - X_C)}} \tag{20.51}$$

From these equations, if $X_L = X_C$, Z is minimum, I_M and therefore I_{eff} are maximum, and **V** and **I** are in phase. Under these conditions maximum average power is delivered to the circuit by the voltage source and the circuit is said to be at resonance.

20.21 RESONANT FREQUENCY OF A SERIES *RLC* CIRCUIT

Because $X_L = X_C$ at resonance

$$\omega L = \frac{1}{\omega C}$$

$$\omega^2 = \frac{1}{LC}$$

and the resonant frequency ω_o is

$$\omega_o = \frac{1}{\sqrt{LC}} \tag{20.52}$$

From Equation (16.4) the resonant frequency f_o for the RLC series circuit is

$$2\pi f_o = \frac{1}{\sqrt{LC}}$$

$$f_o = \frac{1}{(2\pi\sqrt{LC})} \tag{20.53}$$

where f_o is in hertz, L is in henrys, and C is in farads.

20.22 INSTANTANEOUS POWER FOR A SERIES *RLC* CIRCUIT

From the preceding equation for **Z**, the impedance at resonance is

$$\mathbf{Z} = R\angle 0^\circ \tag{20.54}$$

and the angle separating **V** and **I** is, therefore, zero at resonance.

From Equations (17.1), (17.3), (17.8) and (17.12),

$$\frac{\mathbf{V}_R}{\mathbf{I}} = R\angle 0^\circ$$

$$\frac{\mathbf{V}_L}{\mathbf{I}} = X_L\angle 90^\circ$$

$$\frac{\mathbf{V}_C}{\mathbf{I}} = X_C\angle -90^\circ$$

for the circuit in Figure 20.44.

The preceding equations indicate that $\mathbf{V}_R$ is in phase with **I**, $\mathbf{V}_L$ leads **I** by 90°, and $\mathbf{V}_C$ lags **I** by 90° at any frequency including the resonant frequency. Because $\mathbf{X}_L = \mathbf{X}_C$ at resonance, $\mathbf{V}_L$ and $\mathbf{V}_C$ are equal in magnitude and 180° out of phase.

Considering the phase relationships given previously and that the input voltage **V** is given as $\mathbf{V}_M\angle 0°$, the phasor diagram for the circuit in Figure 20.44 at the resonant condition is as shown in Figure 20.45. If the phasors in Figure

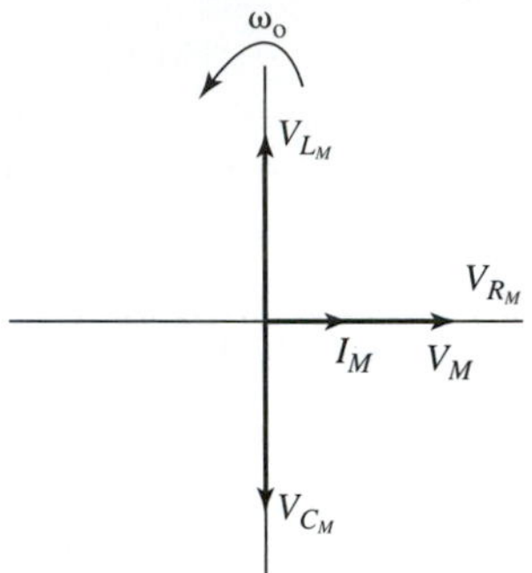

Figure 20.45 Phasor diagram at resonance.

20.45 are rotated at ω_o rad/s counterclockwise, the instantaneous waveforms for v, i, v_R, v_L, and v_C at resonance are generated and the result is shown in Figure 20.46.

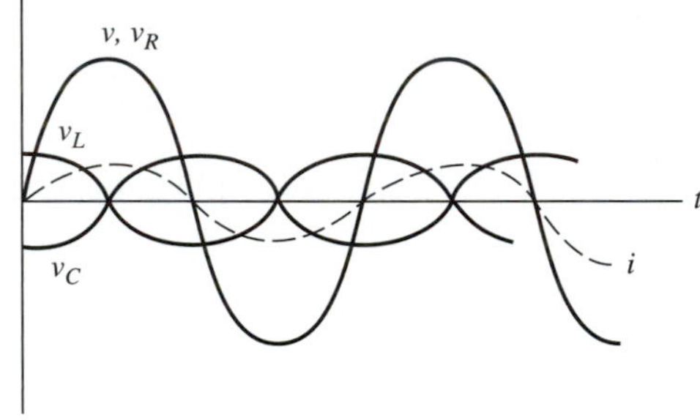

Figure 20.46 Waveforms for voltage.

From Equation (2.6) instantaneous power is

$$p = vi$$

Using the plots in Figure 20.46 and the previous equation, instantaneous powers for the resistor, inductor, and capacitor at the resonant condition are plotted in Figure 20.47.

Figure 20.47 Power waveform.

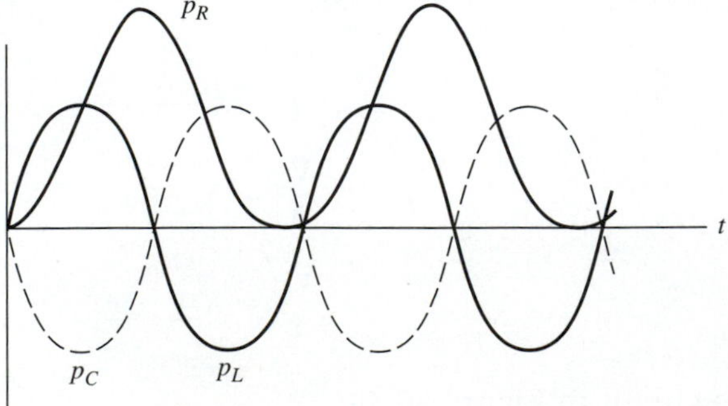

Notice that in Figure 20.47 the total reactive power at any instant is zero because $p_L = -p_C$ at any instant. Equal amounts of energy are alternately absorbed and released by the inductor and capacitor on a cyclic basis and all the average power delivered to the circuit at resonance is delivered to the resistor.

20.23 IMPEDANCE OF A SERIES *RLC* CIRCUIT

As previously determined, the impedance for the circuit in Figure 20.44 is

$$\begin{aligned}\mathbf{Z} &= R + j(X_L - X_C)\\ &= R + j\left(\omega L - \frac{1}{\omega C}\right)\\ &= R + j\left(2\pi f L - \frac{1}{2\,\pi f C}\right)\end{aligned}$$

and the magnitude of **Z** is

$$\begin{aligned}Z &= \sqrt{R^2 + (X_L - X_C)^2}\\ &= \sqrt{R^2 + \left(2\pi f L - \frac{1}{2\pi f C}\right)^2}\end{aligned}$$

Plots of Z and each of the components of Z are shown in Figures 20.48 and 20.49.

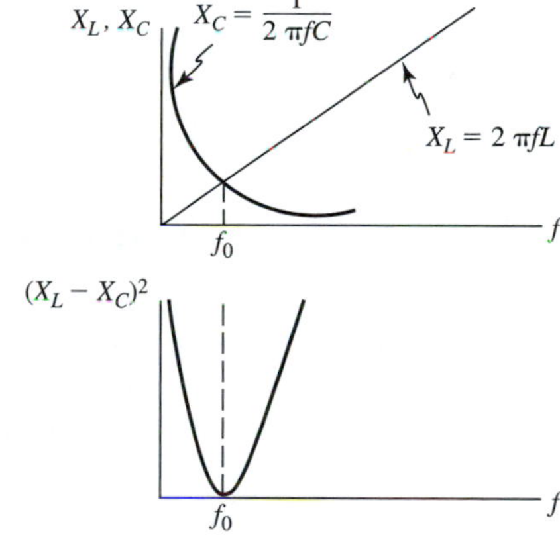

Figure 20.48 Reactance curves.

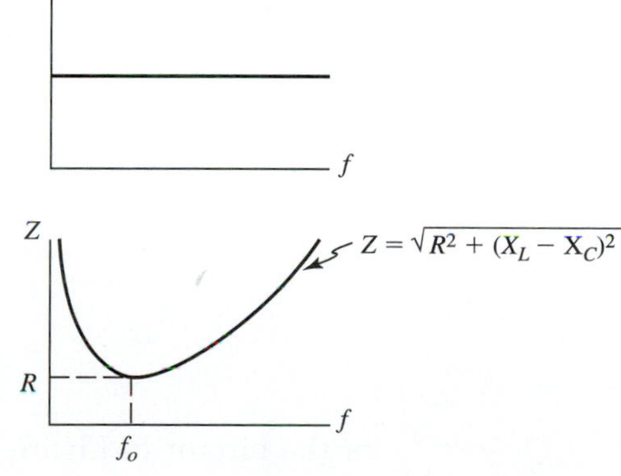

Figure 20.49 Impedance curve.

Notice that the minimum impedance occurs at resonance when $X_C = X_L$ and the impedance at resonance is equal to R, as previously determined.

20.24 PHASE RELATIONSHIP FOR A SERIES *RLC* CIRCUIT

Because the equation for **Z** for the circuit in Figure 20.44 is

$$\mathbf{Z} = R + j(X_L - X_C)$$

the angle associated with **Z** is

$$\left.\begin{aligned}\theta &= \tan^{-1}\left\{\frac{X_L - X_C}{R}\right\}\\ &= \tan^{-1}\left\{\frac{2\pi fL - 1/2\pi fC}{R}\right\}\end{aligned}\right\} \tag{20.55}$$

As f increases, θ approaches 90° and as f decreases, θ approaches -90°. At resonance $X_L = X_C$ and $\theta = 0^\circ$. A plot of θ versus f for the impedance of the circuit in Figure 20.44 is shown in Figure 20.50.

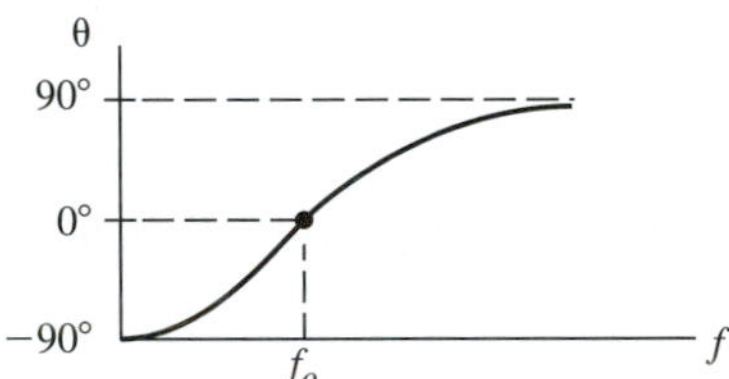

Figure 20.50 Phase angle.

Because $\mathbf{Z} = \mathbf{V}/\mathbf{I}$, the plot in Figure 20.50 indicates that **V** leads **I** for $f > f_o$, **V** lags **I** for $f < f_o$, and **V** is in phase with **I** for $f = f_o$. That is, the circuit is inductive for $f > f_o$, capacitive for $f < f_o$, and purely resistive for $f = f_o$.

20.25 QUALITY FACTOR FOR A SERIES *RLC* CIRCUIT

The *quality factor of a series RLC resonant circuit is designated* Q_S and is defined as the ratio of the reactive power delivered to the inductor (or to the capacitor) to the average power delivered to the resistor at resonance. From Equations (20.28), (20.31), and (20.34),

$$\begin{aligned}Q_S &= \frac{I_{\text{eff}}^2 X_L}{I_{\text{eff}}^2 R}\\ &= \frac{X_L}{R}\end{aligned} \tag{20.56}$$

or

$$\begin{aligned}Q_S &= \frac{I_{\text{eff}}^2 X_C}{I_{\text{eff}}^2 R}\\ &= \frac{X_C}{R}\end{aligned} \tag{20.57}$$

for the circuit in Figure 20.44. Recall that $X_L = X_C$ at resonance.

From Equation (20.53)

$$f_o = \frac{1}{2\pi\sqrt{LC}}$$

and because $X_L = 2\pi f_o L$ and $X_C = 1/(2\pi f_o C)$ at resonance

$$\begin{aligned} Q_S &= \frac{X_L}{R} \\ &= \frac{(2\pi f_o L)}{R} \\ &= \frac{\{(2\pi)\{1/(2\pi\sqrt{LC})\}(L)\}}{R} \\ &= \left\{\frac{2\pi}{R}\right\}\left\{\frac{L}{2\pi\sqrt{L}\sqrt{C}}\right\} \\ &= \frac{1}{R}\sqrt{\frac{L}{C}} \end{aligned} \tag{20.58}$$

at the resonant condition.

Equation (20.58) can also be determined as follows.

$$\begin{aligned} Q_S &= \frac{X_C}{R} \\ &= \frac{\{1/(2\pi f_o C)\}}{R} \\ &= \frac{\left\{\frac{2\pi\sqrt{LC}}{2\pi C}\right\}}{R} \\ &= \frac{1}{R}\sqrt{\frac{L}{C}} \end{aligned}$$

Applying Equations (17.1) and (17.5) to circuit in Figure 20.44

$$\begin{aligned} I_M &= \frac{V_M}{Z} \\ V_{L_M} &= X_L I_M \\ &= \left(\frac{X_L}{Z}\right)V_M \end{aligned}$$

and because $Z = R$ at resonance,

$$\begin{aligned} V_{L_M} &= \left(\frac{X_L}{R}\right)V_M \\ &= Q_S V_M \end{aligned}$$

at the resonant condition.

Dividing both sides of the equation by $\sqrt{2}$ results in

$$V_{L_{\text{eff}}} = Q_S V_{\text{eff}} \tag{20.59}$$

at resonance.

In a similar manner

$$I_M = \frac{V_M}{Z}$$
$$V_{C_M} = X_C I_M$$
$$= \left(\frac{X_C}{Z}\right) V_M$$

and because $Z = R$ at resonance,

$$V_{C_M} = \left(\frac{X_C}{R}\right) V_M$$
$$= Q_S V_M$$

at the resonant condition.

Dividing both sides of the equation by $\sqrt{2}$ results in

$$V_{C_{\text{eff}}} = Q_S V_{\text{eff}} \tag{20.60}$$

at resonance.

The quality factor is a measurement of the energy stored in a circuit on a cyclic basis compared to the energy dissipated.

EXAMPLE 20.17 Determine the resonant values of I_{eff}, $V_{L_{\text{eff}}}$, $V_{R_{\text{eff}}}$, and $V_{C_{\text{eff}}}$ and draw a phasor diagram for the resonant condition of the circuit in Figure 20.51. ■

Figure 20.51

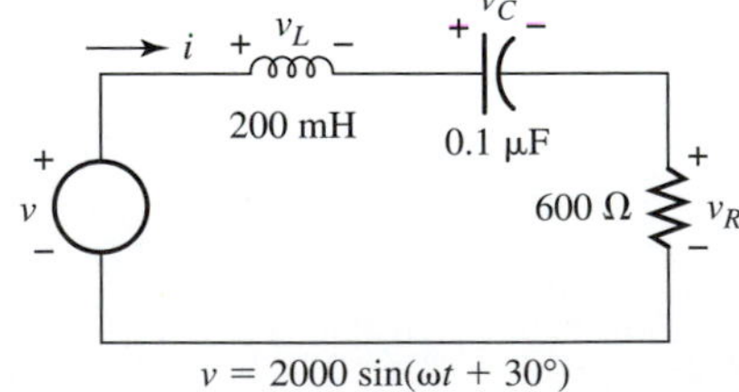

SOLUTION From Equation (20.13) the resonant frequency is

$$f_o = \frac{1}{2\pi\sqrt{LC}}$$
$$= \frac{1}{(2\pi\sqrt{(200)(10^{-3})(0.1)(10^{-6})})}$$

$$= \frac{1}{(2\pi)(\sqrt{200})(10)^{-5}}$$
$$= \frac{10^5}{2\pi\sqrt{200}}$$
$$= 1125.4 \text{ Hz}$$

From Equations (17.7) and (17.11),

$$X_L = 2\pi fL$$
$$= 2\pi(1125.4)(200)(10)^{-3}$$
$$= 1414\ \Omega$$

and

$$X_C = \frac{1}{2\pi fC}$$
$$= \frac{1}{\{2\pi(1125)(0.1)(10)^{-6}\}}$$
$$= 1414\ \Omega$$

at resonance.

The circuit at resonance in phasor form is shown in Figure 20.52.

Figure 20.52

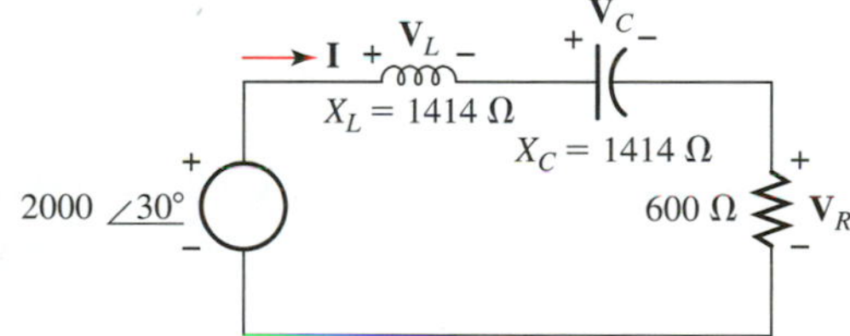

From Equation (20.14), the circuit impedance at resonance is

$$\mathbf{Z} = R\angle 0^\circ$$
$$= 600\angle 0^\circ$$

From Equation (20.17), the quality factor for the circuit is

$$Q_S = \frac{X_L}{R}$$
$$= \frac{1414}{600}$$
$$= 2.36$$

or

$$Q_S = \frac{X_C}{R} = 2.36$$

From Equations (20.59) and (20.60),

$$\begin{aligned} V_{L_{\text{eff}}} &= Q_S V_{\text{eff}} \\ &= (2.36)\left(\frac{2000}{\sqrt{2}}\right) \\ &= 3340 \\ V_{C_{\text{eff}}} &= Q_S V_{\text{eff}} \\ &= (2.36)\left(\frac{2000}{\sqrt{2}}\right) \\ &= 3340 \end{aligned}$$

at resonance.

Because $I_{\text{eff}} = V_{\text{eff}}/R$ at resonance,

$$I_{\text{eff}} = \frac{2000/\sqrt{2}}{600} = 2.357$$

and

$$\begin{aligned} V_{R_{\text{eff}}} &= RI_{\text{eff}} \\ &= (600)(2.357) \\ &= 1414 \end{aligned}$$

at resonance.

The effective values determined here will occur when the value of ω in the expression for v is equal to

$$\begin{aligned} \omega_o &= 2\pi f_o \\ &= 2\pi(1125.4) \\ &= 7071 \text{ rad/s} \end{aligned}$$

From the effective values determined here and Equation (20.2), the maximum values of v_L, v_C, i, and v_R at resonance are

$$\begin{aligned} V_{L_M} &= 3340\sqrt{2} \\ &= 4723 \end{aligned}$$

$$V_{C_M} = 3340\sqrt{2}$$
$$= 4723$$
$$I_M = 2.357\sqrt{2}$$
$$= 3.33$$
$$V_R = 1414\sqrt{2}$$
$$= 2000$$

We now write the currents and voltages in phasor form at resonance. From the given expression for v,

$$\mathbf{V} = 2000\underline{/30^\circ}$$

Because $\mathbf{Z} = R\underline{/0^\circ}$, i is in phase with v at resonance and

$$\mathbf{I} = 3.33\underline{/30^\circ}$$

Because $\mathbf{V}_L$ leads $\mathbf{I}$ by 90°, $\mathbf{V}_C$ lags $\mathbf{I}$ by 90° and $\mathbf{V}_R$ is in phase with $\mathbf{I}$ for all frequencies (including resonance):

$$\mathbf{V}_L = 4723\underline{/120^\circ}$$
$$\mathbf{V}_C = 4723\underline{/-60^\circ}$$

and

$$\mathbf{V}_R = 2000\underline{/30^\circ}$$

at resonance.

The phasor diagram at resonance is shown in Figure 20.53.

Figure 20.53

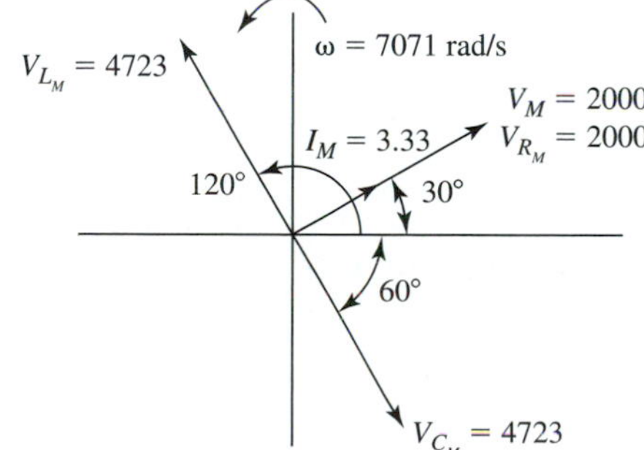

If these phasors are rotated at 7071 rad/s as indicated in the diagram, sinusoidal waveforms for v_i, v_R, i, v_L, and v_C at resonance will be generated.

20.26 PARALLEL RESONANCE

Consider the parallel *RLC* circuit in Figure 20.54, where the frequency ω of

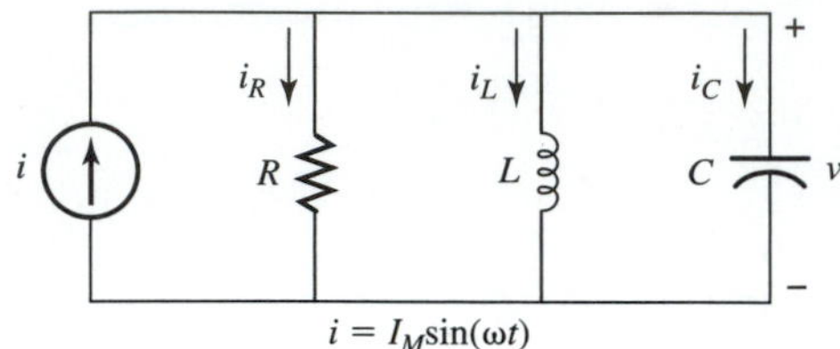

Figure 20.54 *RLC* parallel circuit.

the input current i is variable. In phasor form the circuit appears as shown in Figure 20.55.

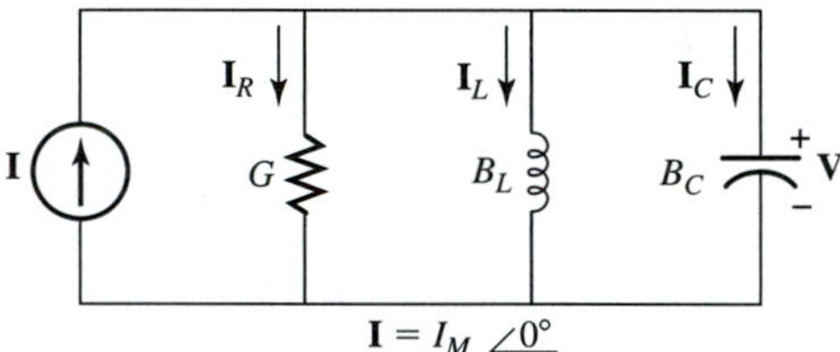

Figure 20.55 *RLC* parallel circuit.

From Equations (17.13) and (17.14), $\mathbf{Y} = \mathbf{I}/\mathbf{V}$ and $Y = I_M/V_M$ where Y is the magnitude of $\mathbf{Y}$. It follows that

$$\mathbf{V} = \mathbf{I}/\mathbf{Y}$$

$$V_M = I_M/Y$$

and dividing both sides of the equation by $\sqrt{2}$ results in

$$V_{\text{eff}} = \frac{I_{\text{eff}}}{Y}$$

From Equations (17.17), (17.20), (17.24), and (17.40),

$$\begin{aligned}\mathbf{Y} &= G + jB_C - jB_L \\ &= G + j(B_C - B_L)\end{aligned}$$

The magnitude of $\mathbf{Y}$ is

$$Y = \sqrt{G^2 + (B_C - B_L)^2}$$

and the magnitude of V_M is

$$V_M = \frac{I_M}{\sqrt{G^2 + (B_C - B_L)^2}} \tag{20.61}$$

From the preceding equations, if $B_C = B_L$, Y is minimum, V_M and, therefore, V_{eff} are maximum, and **V** and **I** are in phase.

Under these conditions maximum average power is delivered to the circuit by the current source and the circuit is at resonance.

20.27 RESONANT FREQUENCY FOR A PARALLEL *RLC* CIRCUIT

From Equations (17.21) and (17.25) and the observation that $B_C = B_L$ at resonance,

$$\omega C = \frac{1}{\omega L}$$

$$\omega^2 = \frac{1}{LC}$$

and the resonant frequency ω_o is

$$\omega_o = \frac{1}{\sqrt{LC}} \tag{20.62}$$

From Equation (16.4) the resonant frequency f_o for the *RLC* parallel circuit is

$$2\pi f_o = \frac{1}{\sqrt{LC}}$$

$$f_o = \frac{1}{2\pi\sqrt{LC}} \tag{20.63}$$

where f_o is in hertz, L is in henrys, and C is in farads.

20.28 INSTANTANEOUS POWER FOR A PARALLEL *RLC* CIRCUIT

From the preceding equation for **Y**, the admittance at resonance is

$$\mathbf{Y} = G\angle 0^\circ$$

and the angle separating **V** and **I** at resonance is zero. From Equations (17.13), (17.17), (17.20), and (17.24),

$$\frac{\mathbf{I}_R}{\mathbf{V}} = G\angle 0^\circ$$

$$\frac{\mathbf{I}_L}{\mathbf{V}} = B_L\angle -90^\circ$$

$$\frac{\mathbf{I}_C}{\mathbf{V}} = B_C\angle 90^\circ$$

These equations indicate that $\mathbf{I}_R$ is in phase with **V**, $\mathbf{I}_L$ lags **V** by 90°, and $\mathbf{I}_C$ leads **V** by 90°. Because $B_L = B_C$ at resonance, $I_{L_M} = I_{C_M}$; and because $\mathbf{I}_L$ and $\mathbf{I}_C$ are 180° out of phase at resonance, $\mathbf{I} = \mathbf{I}_R$.

A phasor diagram for the circuit in Figure 20.55 at the resonant frequency is shown in Figure 20.56.

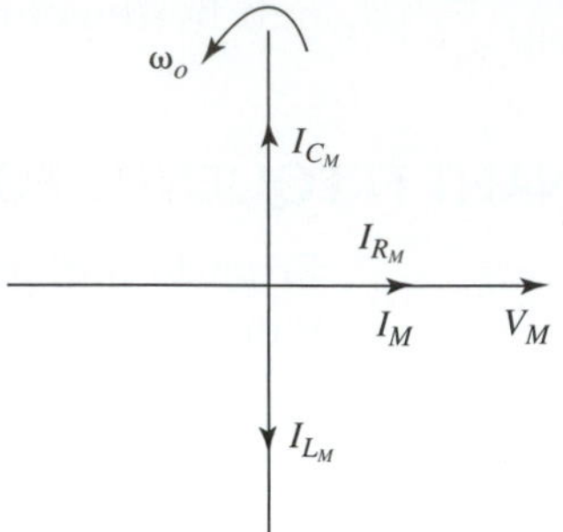

Figure 20.56 Phasor diagram at resonance.

If the phasors in Figure 20.56 are rotated at ω_o rad/s counterclockwise, the instantaneous waveforms for v, i, i_R, i_L, and i_C at resonance are generated; the result is shown in Figure 20.57.

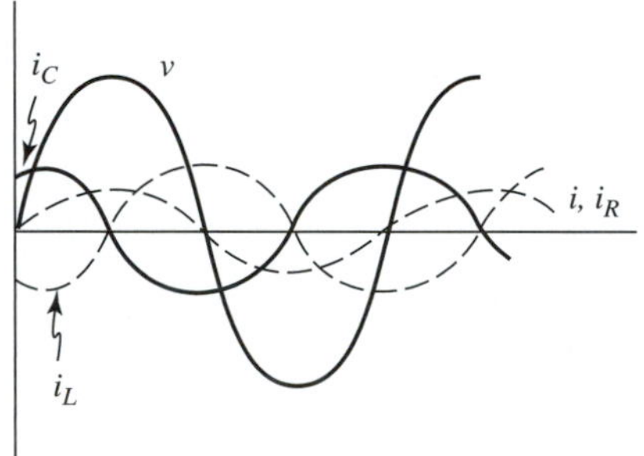

Figure 20.57 Current and voltage at resonance.

From Equation (2.6) instantaneous power is

$$p = vi$$

Using the plots in Figure 20.57 and the preceding equation, instantaneous powers for the resistor, inductor, and capacitor at the resonant condition are shown in Figure 20.58.

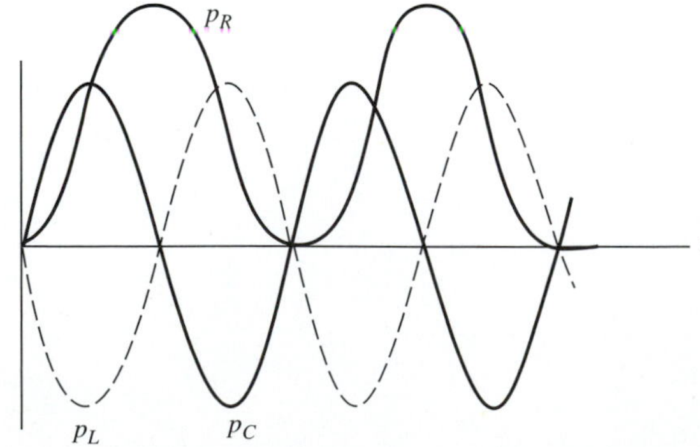

Figure 20.58 Power waveforms.

Notice in Figure 20.58 that the total reactive power at any instant is zero, because $p_L = -p_C$ at any instant. Equal amounts of energy are alternately absorbed and released by the inductor and capacitor on a cyclic basis, and all the average power delivered to the circuit is delivered to the resistor.

20.29 ADMITTANCE OF A PARALLEL *RLC* CIRCUIT

As previously determined, admittance for the circuit in Figure 20.55 is

$$\mathbf{Y} = G + j(B_C - B_L)$$

From Equations (16.4), (17.21), and (17.25),

$$\mathbf{Y} = G + j\left(\omega C - \frac{1}{\omega L}\right)$$
$$= G + j\left(2\pi f C - \frac{1}{2\pi f L}\right)$$

and the magnitude of **Y** is

$$Y = \sqrt{G^2 + (B_C - B_L)^2}$$
$$Y = \sqrt{G^2 + \left(2\pi f C - \frac{1}{2\pi f L}\right)^2}$$

A plot of Y and each of the components of Y appears in Figure 20.59.

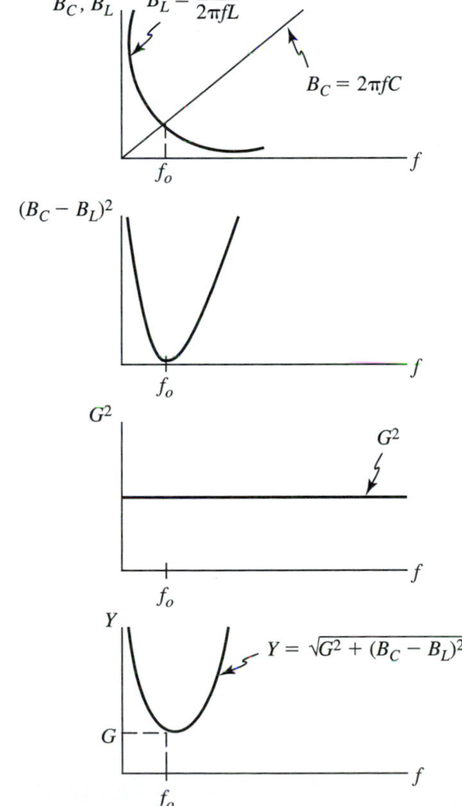

Figure 20.59 Admittance.

Notice that the minimum admittance occurs at resonance when $B_C = B_L$ and the admittance at resonance is equal to G, as previously determined.

20.30 PHASE RELATIONSHIP FOR A PARALLEL *RLC* CIRCUIT

Because the equation for **Y** for the circuit in Figure 20.55 is

$$\mathbf{Y} = G + j(B_C - B_L)$$

the angle associated with **Y** is

$$\theta = \tan^{-1}\left\{\frac{B_C - B_L}{G}\right\} = \tan^{-1}\left\{\frac{2\pi fC - 1/2\pi fL}{G}\right\} \tag{20.64}$$

As f increases, θ approaches 90° and as f decreases, θ approaches −90°. At resonance, $B_C = B_L$ and $\theta = 0°$. A plot of θ versus f for the admittance of the circuit in Figure 20.55 is shown in Figure 20.60.

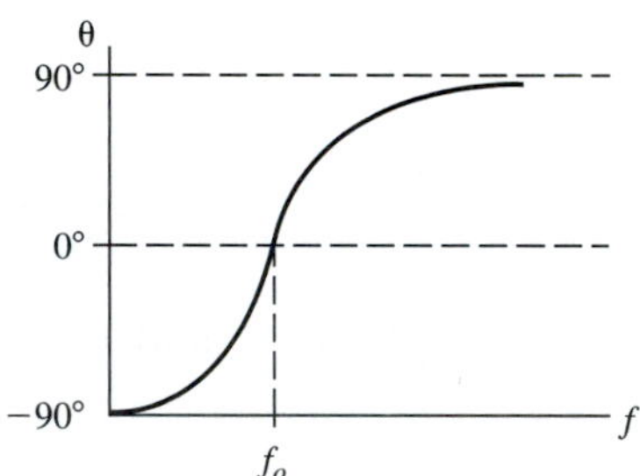

Figure 20.60 Phase angle.

Because $\mathbf{Y} = \mathbf{I}/\mathbf{V}$, **I** leads **V** for $f > f_o$, **I** lags **V** for $f < f_o$, and **I** is in phase with **V** for $f = f_o$. That is, the circuit is capacitive for $f > f_o$, inductive for $f < f_o$, and purely resistive for $f = f_o$.

20.31 QUALITY FACTOR FOR A PARALLEL *RLC* CIRCUIT

The quality factor of a parallel *RLC* circuit is designated Q_p and is defined as the ratio of the reactive power delivered to the inductor (or the capacitor) to the average power delivered to the resistor at resonance. From Equations (20.18), (20.21), and (20.24),

$$Q_p = \frac{V_{\text{eff}}^2/X_L}{V_{\text{eff}}^2/R} = \frac{R}{X_L} \tag{20.65}$$

or

$$Q_p = \frac{V_{\text{eff}}^2/X_C}{V_{\text{eff}}^2/R} = \frac{R}{X_C} \tag{20.66}$$

for the circuit in Figure 20.55.

From Equation (20.63)

$$f_o = \frac{1}{2\pi\sqrt{LC}}$$

and because $X_L = 2\pi f_o L$ and $X_C = 1/(2\pi f_o C)$ at resonance,

$$\begin{aligned} Q_p &= \frac{R}{X_L} \\ &= \frac{R}{2\pi f_o L} \\ &= \frac{R}{2\pi(1/2\pi\sqrt{LC})L} \\ &= \frac{R}{\sqrt{L/C}} \\ &= R\left(\frac{C}{L}\right) \end{aligned} \tag{20.67}$$

at the resonant condition.

Q_p can also be determined as follows:

$$\begin{aligned} Q_p &= \frac{R}{X_C} \\ &= \frac{R}{\dfrac{1}{2\pi f_o C}} \\ &= \frac{R}{\dfrac{2\pi\sqrt{LC}}{2\pi C}} \\ &= R\left(\sqrt{\frac{C}{L}}\right) \end{aligned} \tag{20.68}$$

Using Equations (17.1) and (17.8) for the circuit in Figure 20.24,

$$\begin{aligned} V_M &= \frac{I_M}{Y} \\ I_{L_M} &= \frac{V_M}{X_L} \\ &= \left(\frac{1}{YX_L}\right)I_M \end{aligned}$$

and because $Y = G = 1/R$ at resonance,

$$\begin{aligned} I_{L_M} &= \left\{\frac{1}{(1/R)X_L}\right\}I_M \\ &= \left(\frac{R}{X_L}\right)I_M \\ &= Q_p I_M \end{aligned}$$

at the resonant condition.

Dividing both sides of the equation by $\sqrt{2}$ results in

$$I_{L_{\text{eff}}} = Q_p I_{\text{eff}} \tag{20.69}$$

at resonance.

In a similar manner

$$V_M = \frac{I_M}{Y}$$

$$I_{C_M} = \frac{V_M}{X_C}$$

$$= \left(\frac{1}{YX_C}\right) I_M$$

and because $Y = G = 1/R$ at resonance,

$$I_{C_M} = \left\{\frac{1}{(1/R)X_C}\right\} I_M$$

$$= \left(\frac{R}{X_C}\right) I_M$$

$$= Q_p I_M$$

at the resonant condition. Dividing both sides of the equation by $\sqrt{2}$ results in

$$I_{C_{\text{eff}}} = Q_p I_{\text{eff}} \tag{20.70}$$

at resonance.

The quality factor is a measurement of the energy stored in a circuit on a cyclic basis compared to the energy dissipated.

EXAMPLE 20.18 Determine the resonant values of v_{eff}, $I_{L_{\text{eff}}}$, $I_{C_{\text{eff}}}$, and $I_{R_{\text{eff}}}$ and draw a phasor diagram at resonance for the circuit in Figure 20.61. ■

Figure 20.61

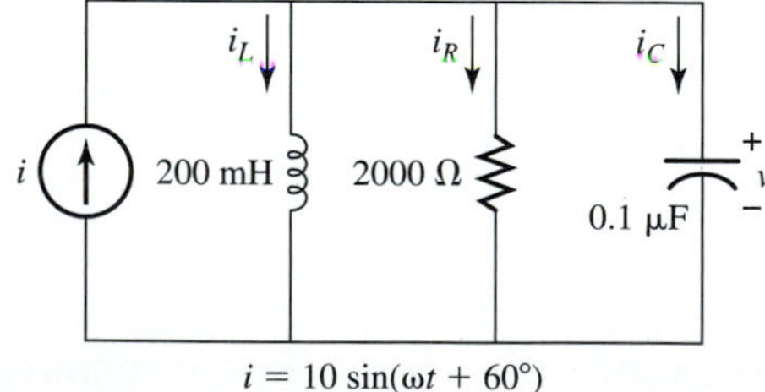

$i = 10 \sin(\omega t + 60°)$

SOLUTION From Equation (20.63) the resonant frequency is

$$f_o = \frac{1}{2\pi\sqrt{LC}}$$

$$= \frac{1}{2\pi\sqrt{(200)(10)^{-3}(0.1)(10)^{-6}}}$$

$$= 1125.4 \text{ Hz}$$

From Equations (17.7) and (17.11),

$$X_L = 2\pi fL$$
$$= 2\pi(1125.4)(200)(10)^{-3}$$
$$= 1414\ \Omega$$

and

$$X_C = \frac{1}{2\pi fC}$$
$$= \frac{1}{2\pi(1125.4)(0.1)(10)^{-6}}$$
$$= 1414\ \Omega$$

at resonance.

The circuit at resonance in phasor form is shown in Figure 20.62. At resonance the admittance is

$$\mathbf{Y} = G\underline{/0^\circ}$$
$$= \left(\frac{1}{2000}\right)\underline{/0^\circ}$$

Figure 20.62

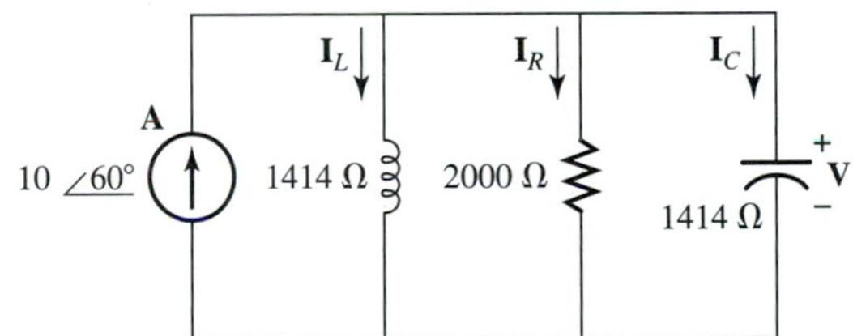

and the impedance at resonance is

$$\mathbf{Z} = \frac{1}{\mathbf{Y}}$$
$$= 2000\underline{/0^\circ}$$

From Equation (20.65) the quality factor for the circuit is

$$Q_p = \frac{R}{X_L}$$
$$= \frac{2000}{1414}$$
$$= 1.414$$

or from Equation (20.66),

$$Q_p = \frac{R}{X_C} = \frac{2000}{1414} = 1.414$$

and from Equations (20.69) and (20.70),

$$I_{L_{\text{eff}}} = Q_p I_{\text{eff}} = (1.414)\left(\frac{10}{\sqrt{2}}\right) = 10$$

$$I_{C_{\text{eff}}} = Q_p I_{\text{eff}} = (1.414)\left(\frac{10}{\sqrt{2}}\right) = 10$$

at resonance.

Because $V_{\text{eff}} = (I_{\text{eff}})(R)$ at resonance,

$$V_{\text{eff}} = \left(\frac{10}{\sqrt{2}}\right)(2000) = 14{,}142$$

and

$$I_{R_{\text{eff}}} = \frac{V_{\text{eff}}}{R} = \frac{14{,}142}{2000} = 7.07$$

at resonance.

The effective values determined here will occur when the value of ω in the expression for i is equal to

$$\omega = 2\pi f_o = 2\pi(1125.4) = 7071 \text{ rad/s}$$

From the effective values determined here and Equation (20.2), the maximum values of i_L, i_C, i_R, and v at resonance are

$$I_{L_M} = 10\sqrt{2} = 14.14$$

$$
\begin{aligned}
I_{C_M} &= 10\sqrt{2} \\
&= 14.14 \\
I_{R_M} &= 7.07\sqrt{2} \\
&= 10 \\
V_M &= 14{,}142\sqrt{2} \\
&= 20000
\end{aligned}
$$

and from the given expression for i,

$$I_M = 10$$

We shall now write the currents and voltages in phasor form at resonance.

From the given expression for i,

$$\mathbf{I} = 10\angle 60^\circ$$

Because $\mathbf{Y} = G\angle 0^\circ$, v is in phase with i at resonance and

$$\mathbf{V} = 20{,}000\angle 60^\circ$$

Because $\mathbf{I}_C$ leads $\mathbf{V}$ by 90°, $\mathbf{I}_L$ lags $\mathbf{V}$ by 90° and $\mathbf{I}_R$ is in phase with $\mathbf{V}$ for all frequencies (including resonance):

$$
\begin{aligned}
\mathbf{I}_C &= 14.14\angle 150^\circ \\
\mathbf{I}_L &= 14.14\angle -30^\circ \\
\mathbf{I}_R &= 10\angle 60^\circ
\end{aligned}
$$

at resonance.

The phasor diagram at resonance is shown in Figure 20.63.

7071

Figure 20.63

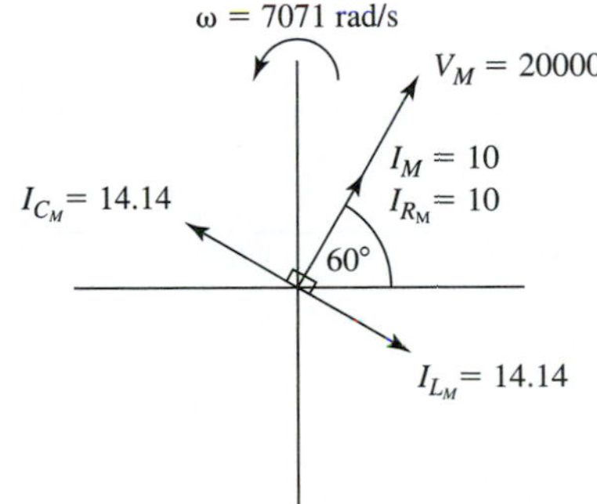

If the preceding phasors are rotated at rad/s as indicated in the diagram, sinusoidal waveforms for i, i_L, i_R, i_C, and v at resonance are generated.

20.32 EXERCISES

1. Determine the average value of v for each periodic waveform in Figure 20.64.

Figure 20.64

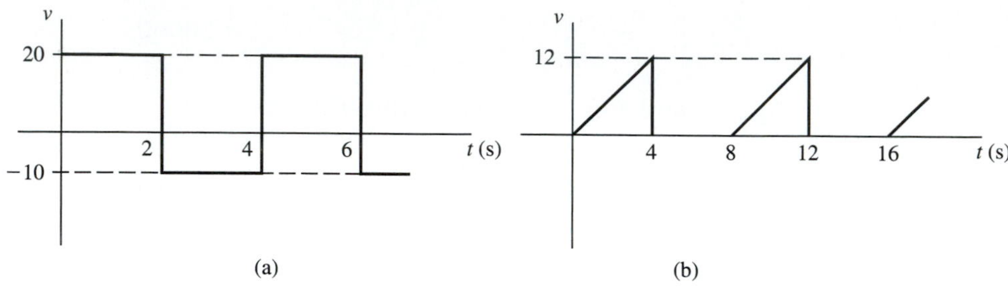

2. Determine the average value of v for each periodic waveform in Figure 20.65.

Figure 20.65

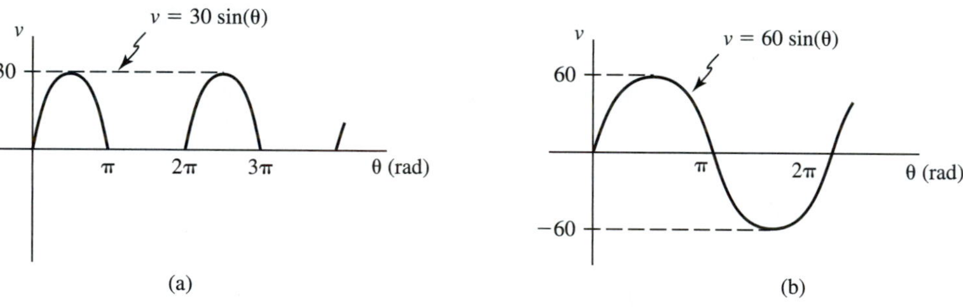

3. Determine the average value of v for each periodic waveform in Figure 20.66.

Figure 20.66

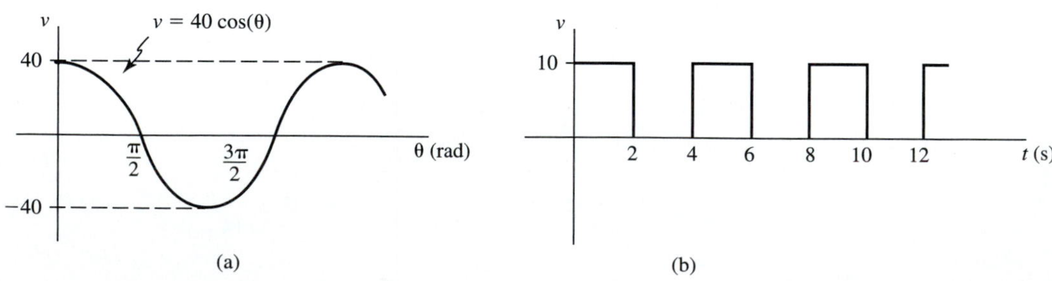

4. Determine the effective value of v for each waveform in Figure 20.64.
5. Determine the effective value of v for each waveform in Figure 20.65.
6. Determine the effective value of v for each waveform in Figure 20.66.

7. Determine the power factor for each circuit in Figure 20.67.

Figure 20.67

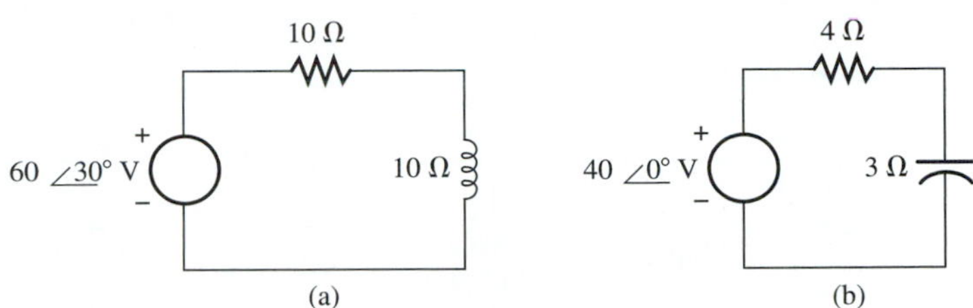

8. Determine the power factor for each circuit in Figure 20.68.

Figure 20.68

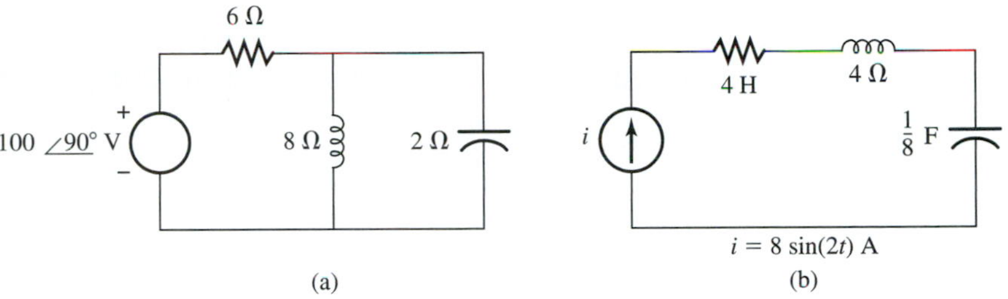

9. Determine the power factor for each circuit in Figure 20.69.

Figure 20.69

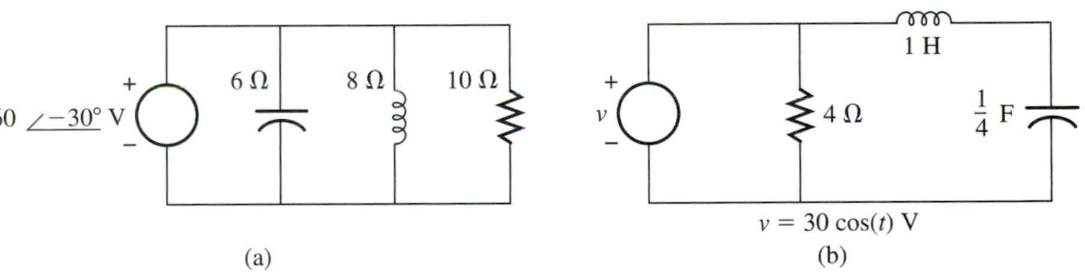

10. Determine the apparent power, the average power, and the reactive power delivered by the source for each circuit in Figure 20.67.

11. Determine the apparent power, the average power, and the reactive power delivered by the source for each circuit in Figure 20.68.

12. Determine the apparent power, the average power, and the reactive power delivered by the source in Figure 20.69.

13. Determine the apparent power, the average power, and the reactive power delivered by the source for each circuit in Figure 20.70.

Figure 20.70

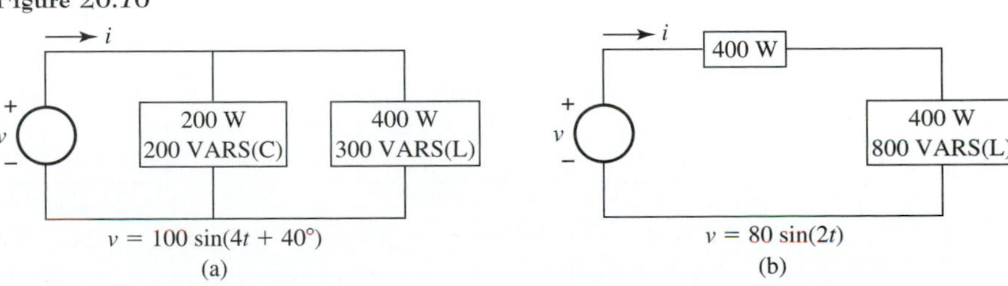

14. Determine the apparent power, the average power, and the reactive power delivered by the source for each circuit in Figure 20.71.

Figure 20.71

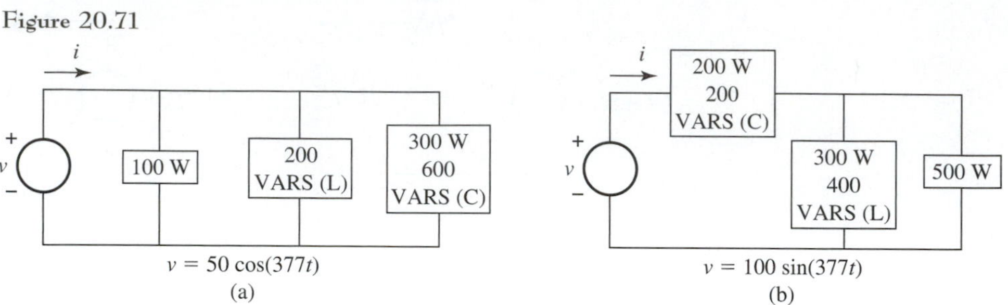

15. Determine the apparent power, the average power, and the reactive power delivered by the source for each circuit in Figure 20.72.

Figure 20.72

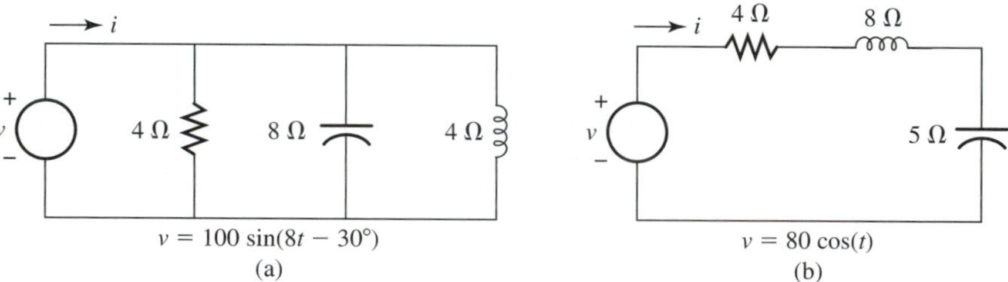

16. Determine the current i for each circuit in Figure 20.70.

17. Determine the current i for each circuit in Figure 20.71.

18. Determine the current i for each circuit in Figure 20.72.

19. Determine a simple series combination of elements for each load box in Figure 20.70. (Recall that simple implies the least number of elements).

20. Determine a simple parallel combination of elements for each load box in Figure 20.71.

21. Determine the complex power, the average power, and the reactive power delivered by the source for each circuit in Figure 20.73.

Figure 20.73

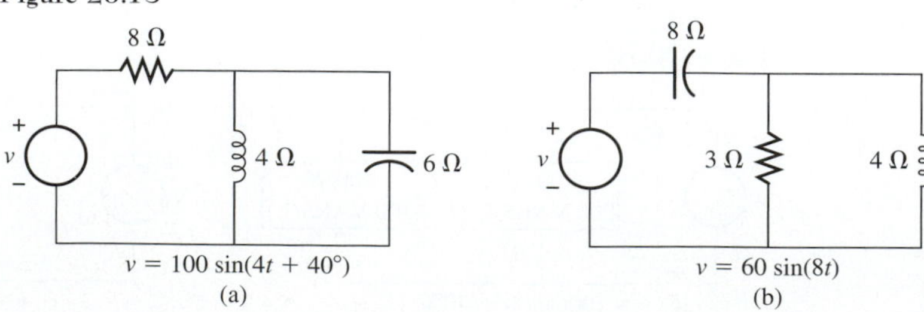

22. Determine the complex power, the average power, and the reactive power delivered by the source for each circuit in Figure 20.74.

Figure 20.74

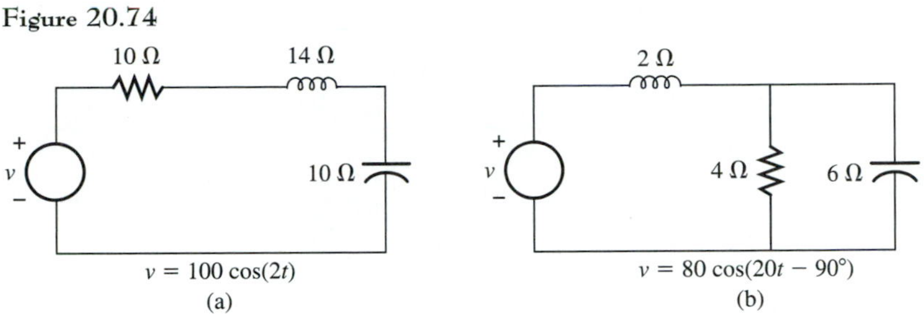

23. Determine the complex power, the average power, and the reactive power delivered by the source for each circuit in Figure 20.75.

Figure 20.75

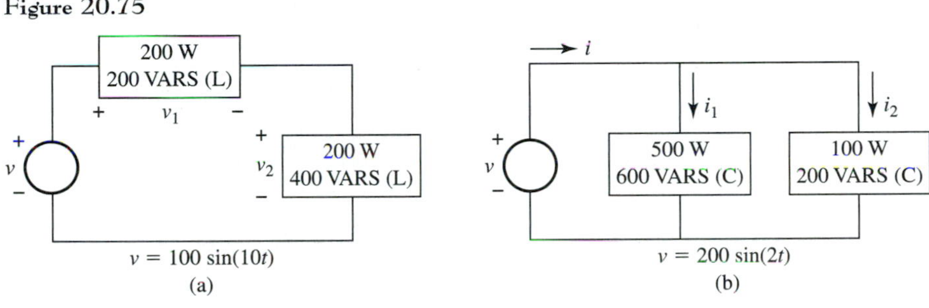

24. Determine the currents i, i_1, and i_2 for the circuit in Figure 20.75(b).

25. Determine the voltage v_1 and v_2 for the circuit in Figure 20.75(a).

26. The voltage source v delivers average power to the circuits in Figure 20.76. It is desired to change the power factor to 0.9 in each circuit while maintaining the same delivery of average power. Determine an element to connect in parallel with each circuit to accomplish this.

Figure 20.76

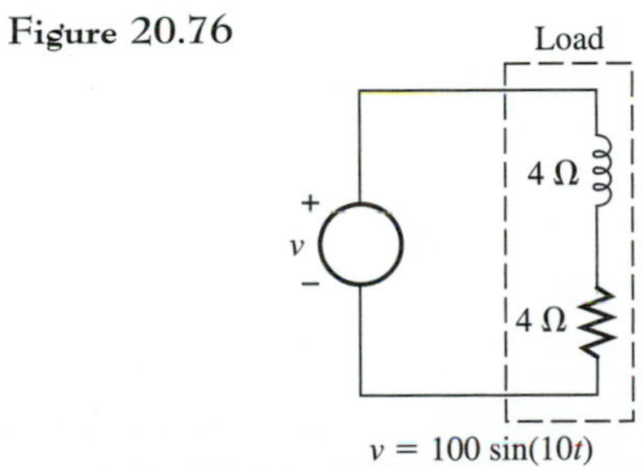

27. The voltage source v delivers average power to the circuits in Figure 20.77. It is desired to change the power factor to 0.9 in each circuit while maintaining the same delivery of average power. Determine an element to connect in parallel with each circuit to accomplish this.

Figure 20.77

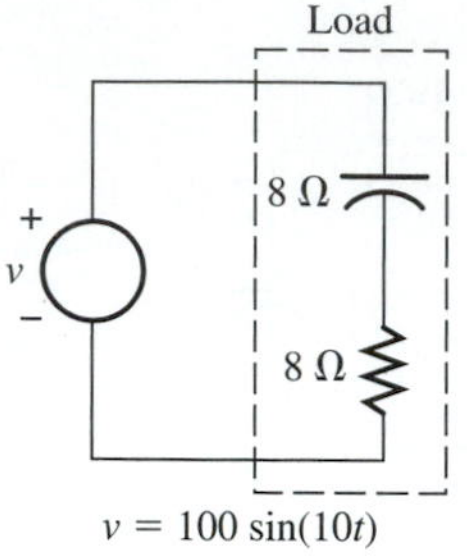

28. Consider the circuit in Figure 20.78.

Figure 20.78

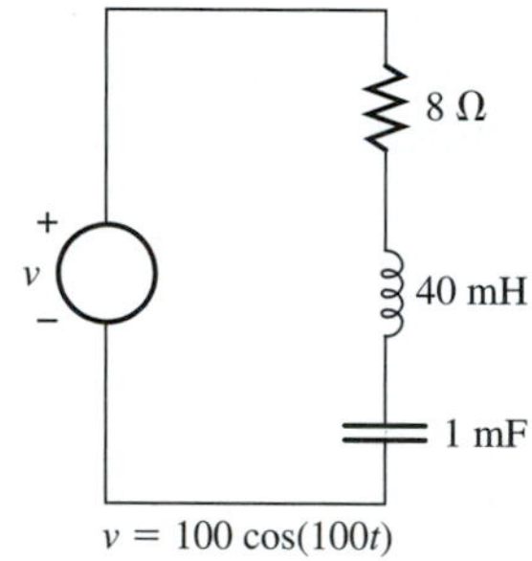

(a) Determine the energy stored in the capacitor during half cycle of the power waveform.
(b) Determine the energy stored in the capacitor during one cycle of the voltage waveform.
(c) Determine the energy stored in the inductor during two cycles of the power waveform.
(d) Determine the energy stored in the inductor during one quarter cycle of the voltage waveform.
(e) Determine the energy stored in the resistor during one half cycle of the voltage waveform.

29. Repeat Problem 28 for the circuit in Figure 20.79.

Figure 20.79

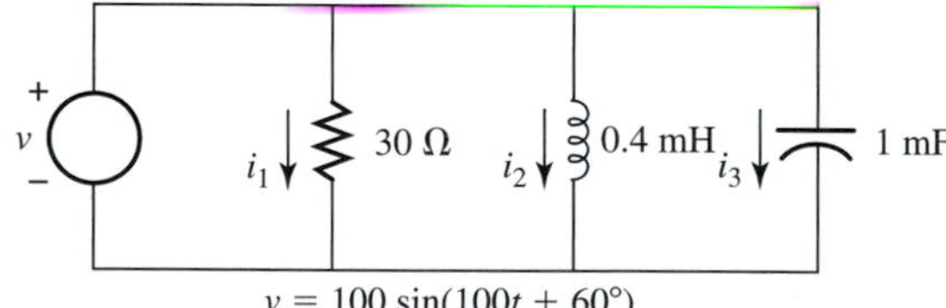

30. Consider the circuit in Figure 20.80.

Figure 20.80

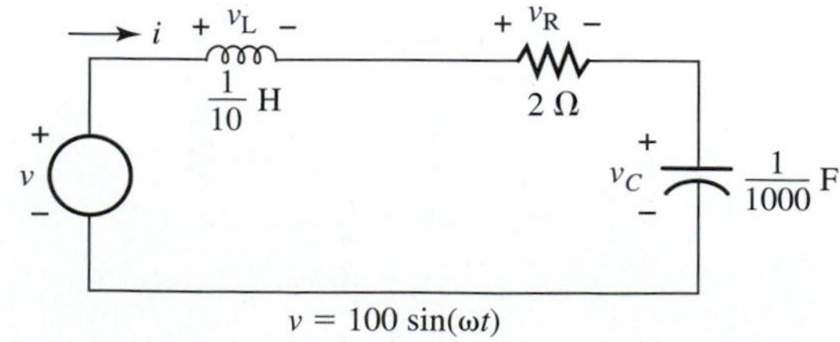

(a) Determine the resonant frequency ω.
(b) Determine I_{eff}, $V_{C_{eff}}$, $V_{L_{eff}}$, and $V_{R_{eff}}$ at resonance.
(c) Determine the circuit power factor at resonance.
(d) Determine the power delivered by the voltage source at resonance.
(e) Determine the quality factor for the circuit.

31. Consider the circuit in Figure 20.81.

Figure 20.81

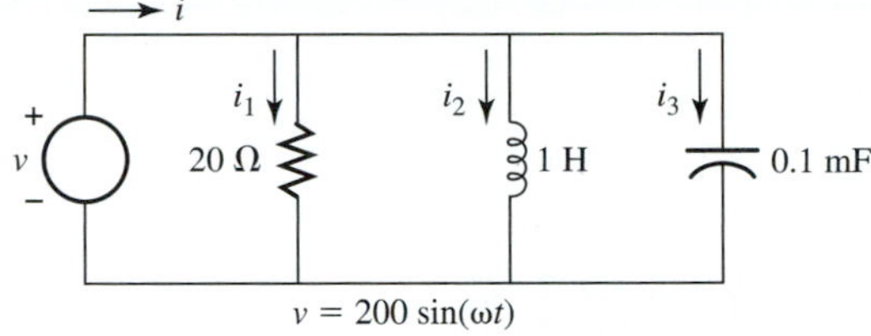

(a) Determine the resonant frequency.
(b) Determine I_{eff}, $I_{1_{eff}}$, $I_{2_{eff}}$, and $I_{3_{eff}}$ at resonance.
(c) Determine the circuit power factor at resonance.
(d) Determine the power delivered by the voltage source at resonance.
(e) Determine the quality factor for the circuit.

CHAPTER

21

Transformers

After completing this chapter the student should be able to

* analyze coupled circuits in the time (t) domain
* analyze coupled circuits in the Laplace (s) domain
* analyze coupled ac circuits
* analyze ideal transformers

When two coils (windings) are placed close enough to each other to allow the changing magnetic flux in one to induce a voltage in the other, the coils are said to be *magnetically coupled*, and the interaction between the coils is called *transformer action*.

A basic *transformer* is a device specifically constructed to allow a varying voltage applied across an input coil to induce a voltage across an output coil through transformer action. A path (usually iron) for the magnetic flux (as shown in Figure 21.2) is provided to improve the transformer efficiency. Transformers are widely used in communications, electronics, power-distribution systems, control systems, and in many other areas of application.

Transformers allow us to transfer energy from one circuit to another without establishing physical contact between the circuits. They can be used to step up (increase) or step down (decrease) ac voltages and currents with little loss of energy. They are very efficient devices that are relatively easy to construct, requiring little maintainance and having no moving parts. The availability of transformers greatly enhances the design possibilities for ac systems.

21.1 SELF-INDUCTANCE

The inductance described in Chapter 11 is called self-inductance and is reviewed in Figure 21.1, where ϕ is the magnetic flux caused by the current i. The quantity ϕ_T is called the total flux (sometimes referred to as the flux linkage) and is defined as the product $N\phi$, where N is the number of turns in the winding.

Figure 21.1 Self-inductance.

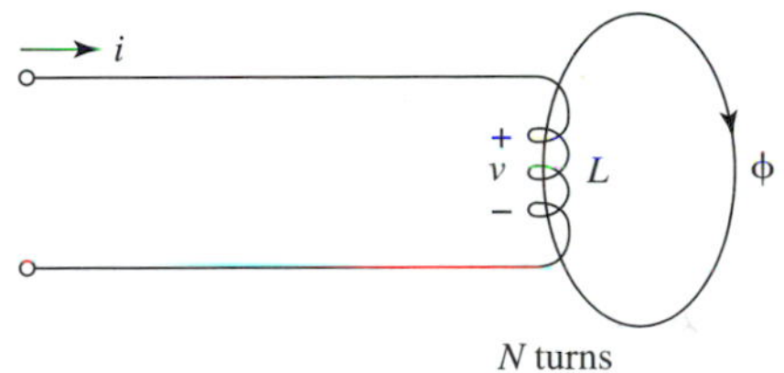

From Equation (11.1) and Faraday's law,

$$\phi_T = Li$$

$$v = \frac{d\{\phi_T\}}{dt}$$

$$= \frac{d\{Li\}}{dt}$$

$$= L\frac{di}{dt}$$

The inductance L is the proportionality constant relating ϕ_T and i, and it depends on the physical characteristics of the inductor. Lenz's law, presented in Chapter 11, is used to determine the polarity of the voltage across the inductor.

21.2 MUTUAL INDUCTANCE

In the case of self-inductance, the voltage in an inductor is caused by a varying flux produced by a current flowing through the inductor. In the case of mutual inductance, the voltage induced in a winding is caused by a varying flux produced by a current in another winding. Whereas L is the proportionality constant relating flux and current for a single coil, M is the proportionality constant relating flux in one coil to current in another coil. M is called the mutual inductance and has the same units as L. The mutual inductance depends on the physical characteristics of the windings—that is, the number of turns in each winding, the closeness of the windings to each other, and the magnetic properties of the core through which the flux flows, as shown in Figure 21.2. The voltage induced in a winding related to the current in another winding is of the form

$$v = M\frac{di}{dt} \tag{21.1}$$

The polarity of the voltage is determined from Lenz's law and the right-hand rule.

Consider the circuit in Figure 21.2. Assume i_1 is increasing in the direction shown. From the right-hand rule presented in Chapter 11, the flux ϕ caused by i_1 will be in the direction shown. From Lenz's law the voltage v_2 generated in L_2 by the current i_1 will have a polarity such that it will cause a current i_2 in a direction that produces a flux opposite to the flux produced in L_2 by i_1. The direction of i_2 is determined by the right-hand rule—that is, by pointing the thumb in a direction that opposes the flux and letting the fingers indicate the direction i_2. In a like manner, i_2 will produce a voltage in L_1, and the polarity can be determined in a similar manner.

Figure 21.2 Mutual inductance.

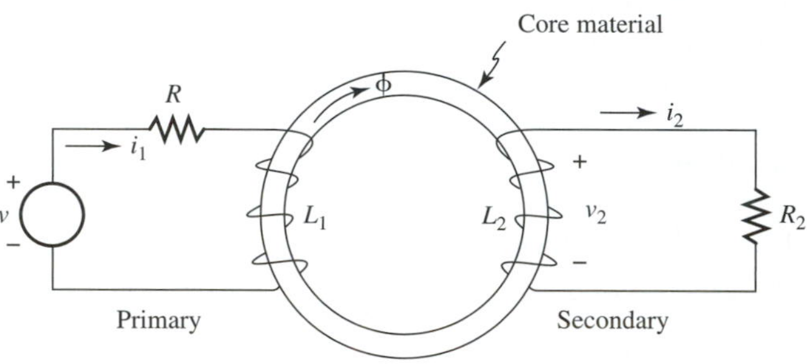

There are four induced voltages to consider for the circuit in Figure 21.2, as shown in Figure 21.3: two caused by self-induction and two caused by mutual induction. The self-induced voltages v_{s1} and v_{s2} are, respectively, the voltages produced in L_1 by i_1 and in L_2 by i_2 and are

Figure 21.3 Mutual inductance.

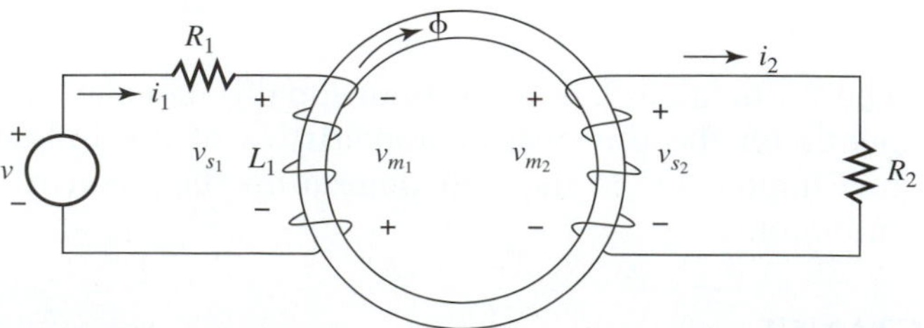

$$v_{s1} = L_1 \frac{di_1}{dt} \tag{21.2}$$

$$v_{s2} = L_2 \frac{di_2}{dt} \tag{21.3}$$

The polarities of v_{s1} and v_{s2} are determined using Lenz's law, as presented in Chapter 11.

The mutually induced voltages v_{m1} and v_{m2} are, respectively, the voltages produced in L_1 by the current i_2 and in L_2 by the current i_1 and are

$$\left.\begin{aligned} v_{m1} &= M\frac{di_2}{dt} \\ v_{m2} &= M\frac{di_1}{dt} \end{aligned}\right\} \tag{21.4}$$

The voltages v_{m1} and v_{s1} are algebraically added, as are the voltages v_{m2} and v_{s2} to determine the voltage across the primary and secondary, respectively.

21.3 DOT NOTATION

Because it is impractical to give a detailed sketch of each winding in a circuit, a procedure known as dot notation allows us to avoid detailed winding sketches and use only the inductor symbol to represent windings.

The dot notation involves placing a dot on each winding to designate the polarities of mutually induced voltages. The windings are usually dotted by the manufacturer. However, if the windings are not dotted and a detailed sketch of the windings is available, the dots can be assigned according to the following procedure, as illustrated in Figure 21.4.

1. Arbitrarily assign a dot to the terminal of one of the windings, as shown in Figure 21.4(a).
2. Assume a current i_x flowing into the dotted terminal and determine the direction of the resulting flux, ϕ_x, using the right-hand rule (Section 11.1), as shown in Figure 21.4(b).
3. Using the right-hand rule, place a dot on one of the terminals of the remaining winding such that a current i_y flowing into the dot causes a flux ϕ_y in the same direction as ϕ_x, as illustrated in Figure 21.4(c).

Figure 21.4 Dot notation.

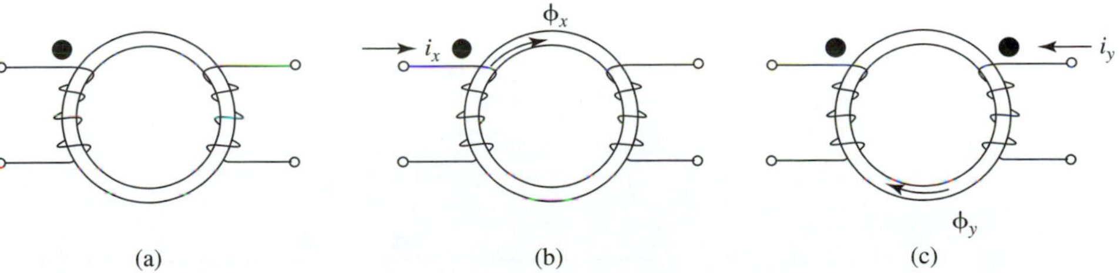

The transformer in Figure 21.4(a) can be represented using dot notation, as shown in Figure 21.5, where M is the mutual inductance relating the windings.

Figure 21.5 Dot notation.

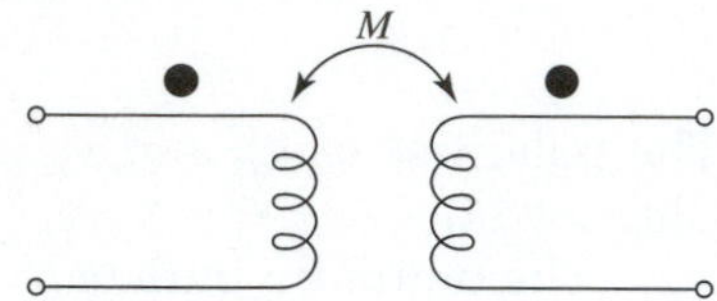

21.4 INTERPRETING DOTS

The dot markings on a transformer are usually permanently placed on the terminals by the manufacturer and are independent of the voltage polarities and current directions assigned for analysis purposes. The rule for determining the polarity of the mutually induced voltages for dotted windings is as follows.

> The assigned current i entering the dotted terminal of one winding induces a voltage $M\,di/dt$ across the other winding, and the positive terminal for the induced voltage is the dotted terminal of the winding across which it appears.
>
> The assigned current i leaving the dotted terminal of one winding induces a voltage $M\,di/dt$ across the other winding and the positive terminal for the induced voltage is the undotted terminal of the winding across which it appears.

For example, in Figure 21.6 the current i_1 causes induced voltages, as indicated, according to these rules. In Figure 21.6(a) the dotted terminal of the

Figure 21.6 Interpreting dots.

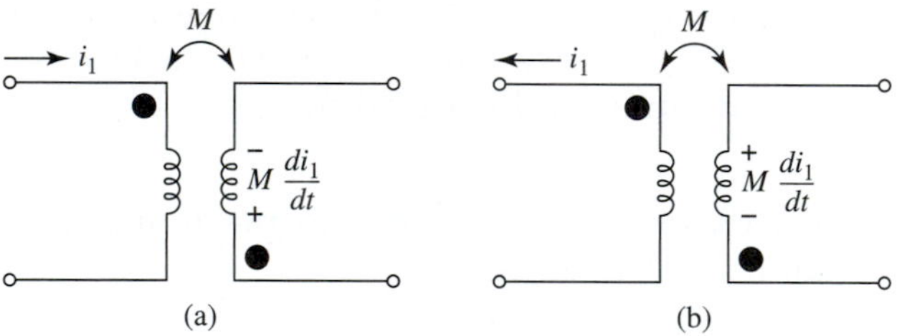

secondary is positive, because i_1 enters the dot on the primary, and in Figure 21.6(b) the dotted terminal of the secondary is negative, because i_1 leaves the dot on the primary.

Consider the following example.

EXAMPLE 21.1 Determine the current i and the voltage v_o after the switch is closed at $t = 0$ for the circuit in Figure 21.7. ■

Figure 21.7

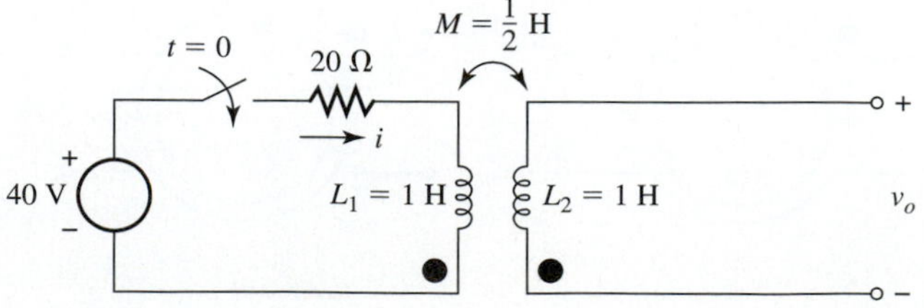

SOLUTION Using the dot notation rule and the assigned current direction, the mutually induced voltage in the secondary and the self-induced voltage in the primary are shown in Figure 21.8.

Figure 21.8

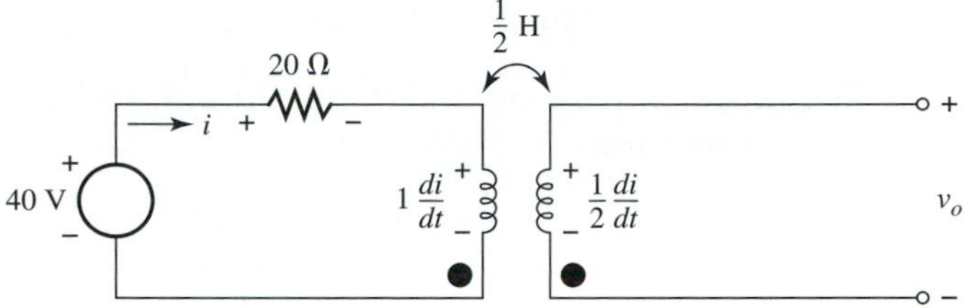

Because the secondary current is equal to zero, there is no mutually induced voltage in the primary and no self-induced voltage in the secondary.

Applying Kirchhoff's voltage law to the primary results in the following first-order, nonhomogeneous differential equation:

$$40 - 20i - \frac{di}{dt} = 0$$

$$\frac{di}{dt} + 20i = 40$$

Using the classical method presented in Chapter 12 for solving differential equations, the complete general solution is

$$i = Ke^{-20t} + 2$$

Because current cannot change instantaneously through an inductor and $i(0^-) = 0$,

$$i(0) = i(0^-) = 0$$

Substituting the condition $i(0) = 0$ for i in the preceding equation results in

$$0 = Ke^{-20(0)} + 2$$

$$K = -2$$

and the particular complete solution for $i(0) = 0$ is

$$i = -2e^{-20t} + 2$$

Applying Kirchhoff's voltage law to the secondary results in

$$v_o = M\frac{di}{dl}$$

$$= \frac{1}{2}\frac{di}{dt}$$

$$= \frac{1}{2}\frac{d\{-2e^{-20t} + 2\}}{dt}$$

$$= \left(\frac{1}{2}\right)\{40e^{-20t}\}$$
$$= 20e^{-20t}$$

21.5 COUPLED CIRCUITS IN THE s (LAPLACE) DOMAIN

As illustrated previously, the mutually induced voltage term for coupled circuits is of the form

$$v = M\frac{di}{dt}$$

If we take the Laplace transform of each side of the preceding equation for $i(0) = 0$, then from rules L1 and L3(a) in Chapter 12,

$$\mathscr{L}\{v\} = \mathscr{L}\left\{M\frac{di}{dt}\right\}$$
$$= M\mathscr{L}\left\{\frac{di}{dt}\right\}$$
$$\mathbf{V} = (sM)\mathbf{I}$$

where **V** is the Laplace transform of the voltage induced by i, **I** is the Laplace transform of i, and M is the mutual inductance between the coupled windings.

We can think of the term (sM) as the quantity that relates the Laplace transform of an induced voltage to the Laplace transform of the current, causing the induced voltage in the same manner that the term (sL) relates the self-induced voltage to the current through an inductor. Recall from Chapter 15 that (sL) is the impedance of an inductor in the s domain.

Let us consider the circuit in Example 21.1 and transform the circuit to the s domain, as shown in Figure 21.9. Notice that the mutual inductance is represented by sM, which is equal to $\frac{1}{2}s$ for the circuit under consideration.

Figure 21.9 Coupled circuit.

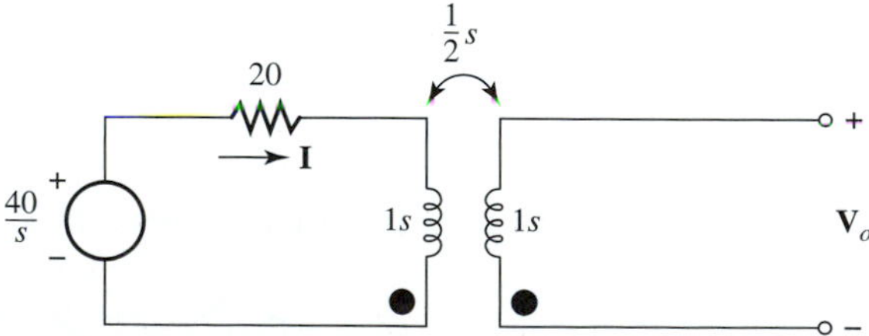

Applying Kirchhoff's voltage law to the primary results in

$$\frac{40}{s} - 20\mathbf{I} - s\mathbf{I} = 0$$

There is no mutually induced voltage term in the primary equation, because the secondary current is equal to zero. Notice that the preceding equation is the Laplace transform of the differential equation $(40 - 20i - di/dt = 0)$ for the primary winding obtained in Example 21.1.

Solving the preceding equation for **I** gives

$$\mathbf{I} = \frac{40}{s(s + 20)}$$

and using the rules in Chapter 15 to expand into partial fractions results in

$$\mathbf{I} = \frac{2}{s} - \frac{2}{s + 20}$$

Using entries 1 and 2 of Table 12.2 and rule L3(b) in Chapter 12 to take the inverse of both sides of the above equation results in

$$i = -2e^{-20t} + 2$$

Applying Kirchhoff's voltage law to the secondary results in

$$\begin{aligned}\mathbf{V}_o &= Ms\mathbf{I} \\ &= \frac{1}{2}(s\mathbf{I})\end{aligned}$$

and because $\mathbf{I} = 40/\{s(s + 20)\}$,

$$\begin{aligned}\mathbf{V}_o &= \left\{\frac{s}{2}\right\}\left\{\frac{40}{s(s + 20)}\right\} \\ &= \frac{20}{s + 20}\end{aligned}$$

From rule L3(b) and entry 2 of Table 12.2,

$$v_o = 20e^{-20t}$$

The preceding expressions for i and v_o are, of course, the same expressions obtained for the analysis method used in Example 21.1.

Consider the following example, where the secondary current is not zero.

EXAMPLE 21.2 Determine the output voltage v_2 for the circuit in Figure 21.10. ■

Figure 21.10

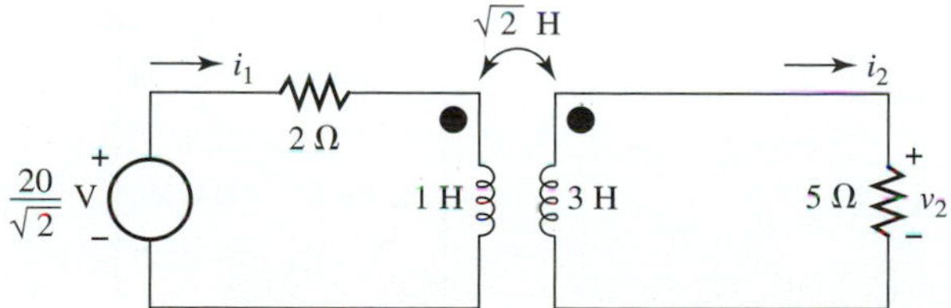

SOLUTION Transforming the circuit to the s domain results in Figure 21.11.

Figure 21.11

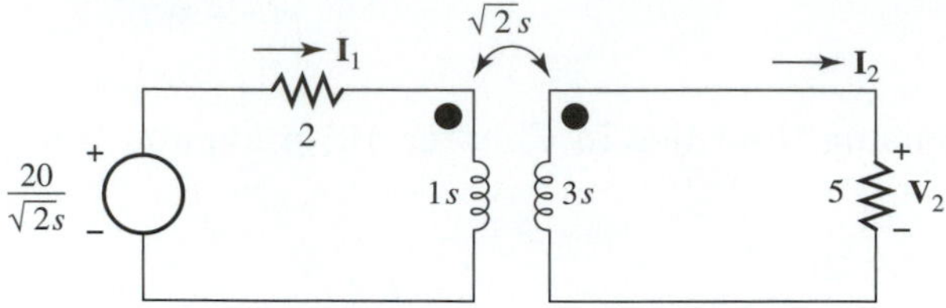

Using the dot notation rules, the mutually induced voltages in the primary and secondary by i_2 and i_1, respectively, are shown in Figure 21.12.

Figure 21.12

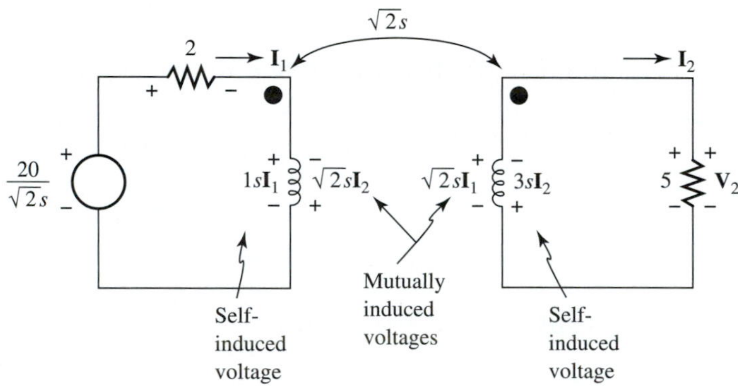

Applying Kirchhoff's voltage law to the primary and secondary windings and recalling that the self-induced and mutually induced voltages are algebraically added in series results in

$$\text{primary:}\quad \frac{20}{\sqrt{2}s} - 2\mathbf{I}_1 - s\mathbf{I}_1 + \sqrt{2}s\mathbf{I}_2 = 0$$

$$\text{secondary:}\quad \sqrt{2}s\mathbf{I}_1 - 3s\mathbf{I}_2 - 5\mathbf{I}_2 = 0$$

Rearranging terms gives

$$(s + 2)\mathbf{I}_1 - (\sqrt{2}s)\mathbf{I}_2 = \frac{20}{\sqrt{2}s}$$

$$-(\sqrt{2}s)\mathbf{I}_1 + (3s + 5)\mathbf{I}_2 = 0$$

Employing Cramer's rule from Chapter 5 to solve for $\mathbf{I}_2$ results in

$$\mathbf{I}_2 = \frac{\begin{vmatrix} s + 2 & \dfrac{20}{\sqrt{2}s} \\ -\sqrt{2}s & 0 \end{vmatrix}}{\begin{vmatrix} s + 2 & -\sqrt{2}s \\ -\sqrt{2}s & 3s + 5 \end{vmatrix}}$$

$$= \frac{\sqrt{2}s(20/\sqrt{2}s)}{s^2 + 11s + 10}$$

$$= \left\{\frac{(\sqrt{2}s)}{(s + 10)(s + 1)}\right\}\left(\frac{20}{\sqrt{2}s}\right)$$

Because $\mathbf{V}_2 = 5\mathbf{I}_2$,

$$\mathbf{V}_2 = \left\{\frac{(\sqrt{2}s)}{(s + 10)(s + 1)}\right\}\left\{\frac{20}{\sqrt{2}s}\right\}\{5\}$$

$$= \frac{100}{(s + 10)(s + 1)}$$

Using the rules in Chapter 12 to expand into partial fractions results in

$$\mathbf{V}_2 = \frac{\left(-\frac{100}{9}\right)}{s + 10} + \frac{\left(\frac{100}{9}\right)}{s + 1}$$

Using Entry 2 of Table 12.2 and rule L3(b) of Chapter 12 to take the inverse Laplace transform of both sides of the above equation results in

$$v_2 = -\left(\frac{100}{9}\right)e^{-10t} + \left(\frac{100}{9}\right)e^{-t}$$

21.6 COUPLED AC CIRCUITS

Following the same reasoning that was used in Chapter 15 to write the impedance of an inductor as $j\omega L$, we can write the impedance of the mutual inductance term as $j\omega M$ for steady-state ac analysis. This allows us to solve coupled circuits driven by sinusoidal sources for steady-state solutions.

The impedance of the mutual inductance term $(j\omega M)$ relates the phasor current causing the induced voltage to the phasor voltage induced.

Consider the following example.

EXAMPLE 21.3 Determine the steady-state value of the current i_2 for the circuit in Figure 21.13. ■

Figure 21.13

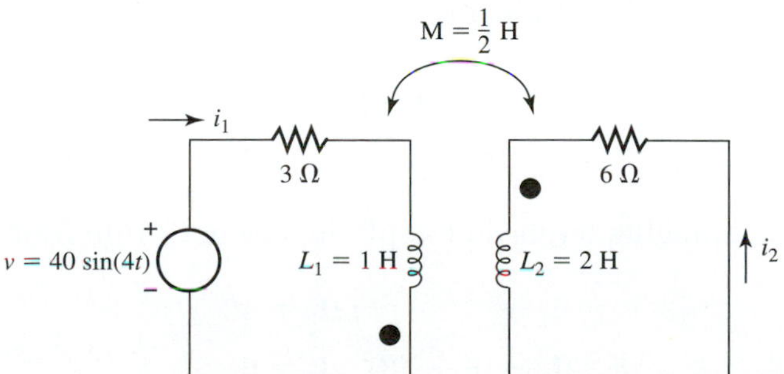

SOLUTION The steady-state ac representation of the circuit in phasor form is as shown in Figure 21.14. Notice that the inductors and the mutual inductance

Figure 21.14

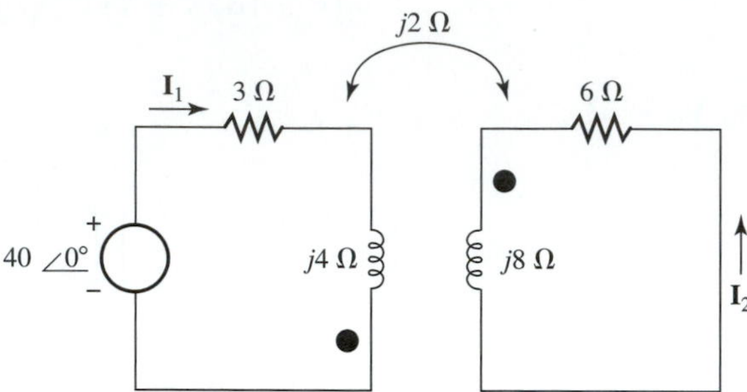

are replaced by $j\omega L_1$, $j\omega L_2$, and $j\omega M$, respectively, the resistors are unchanged, and the voltage source is replaced by its phasor.

Using the dot notation rules for mutual inductance and the voltage polarity rule for self-inductance, the self- and mutually induced voltages for the circuit are shown in Figure 21.15.

Figure 21.15

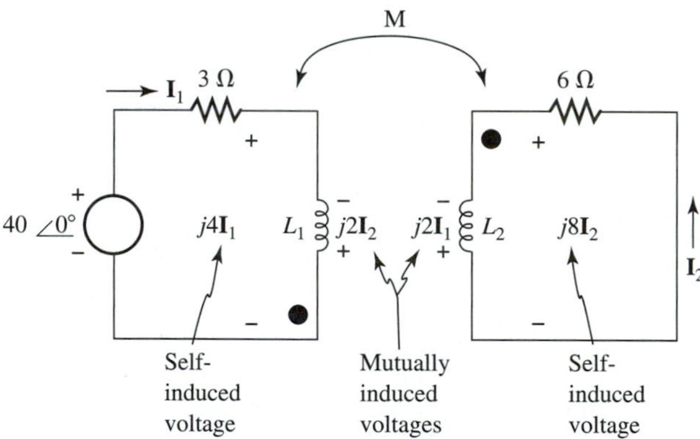

Notice that in accordance with the dot notation rules, the dot on the secondary is negative for the mutually induced secondary voltage, because $\mathbf{I}_1$ leaves the dot of the primary winding; and the dot on the primary winding is positive for the mutually induced voltage, because $\mathbf{I}_2$ enters the dot of the secondary winding. Notice also that the self-induced voltages oppose the current in each winding in accordance with the voltage polarity rule for self-induced voltages presented in Chapter 11.

Applying Kirchhoff's voltage law to the primary and secondary windings, respectively, results in

$$40\angle 0^\circ - 3\mathbf{I}_1 - j4\mathbf{I}_1 + j2\mathbf{I}_2 = 0$$

$$j8\mathbf{I}_2 - j2\mathbf{I}_1 + 6\mathbf{I}_2 = 0$$

Rearranging terms and applying Cramer's rule from Chapter 5 results in

$$(3 + j4)\mathbf{I}_1 - \qquad j2\mathbf{I}_2 = 40\angle 0^\circ$$

$$j2\mathbf{I}_1 - (6 + j8)\quad \mathbf{I}_2 = 0$$

$$\mathbf{I}_1 = \frac{\begin{vmatrix} 40\angle 0^\circ & -j2 \\ 0 & -(6+j8) \end{vmatrix}}{\begin{vmatrix} 3+j4 & -j2 \\ j2 & -(6+j8) \end{vmatrix}} = \frac{(40\angle 0^\circ)(-6-j8)}{(3+j4)(-6-j8)-(-j2)(j2)}$$

$$= 8.16\angle -48.7^\circ$$

$$\mathbf{I}_2 = \frac{\begin{vmatrix} 3+j4 & 40\angle 0^\circ \\ j2 & 0 \end{vmatrix}}{\begin{vmatrix} 3+j4 & -j2 \\ j2 & -(6+j8) \end{vmatrix}} = \frac{-(j2)(40\angle 0^\circ)}{(3+j4)(-6-j8)-(-j2)(j2)}$$

$$= 1.63\angle -11.8^\circ$$

and from the phasor expression

$$i_1 = 8.16\sin(4t - 48.7^\circ)$$
$$i_2 = 1.63\sin(4t - 11.8^\circ)$$

Magnetic coupling occurs between the windings of physically separated circuits (primary and secondary circuits) linked only by a common magnetic flux as illustrated previously or between windings that are part of the same physical circuit as illustrated in the following examples.

EXAMPLE 21.4 Determine the steady-state ac current i_2 for the circuit in Figure 21.16. i_1 and i_2 are mesh currents. ■

Figure 21.16

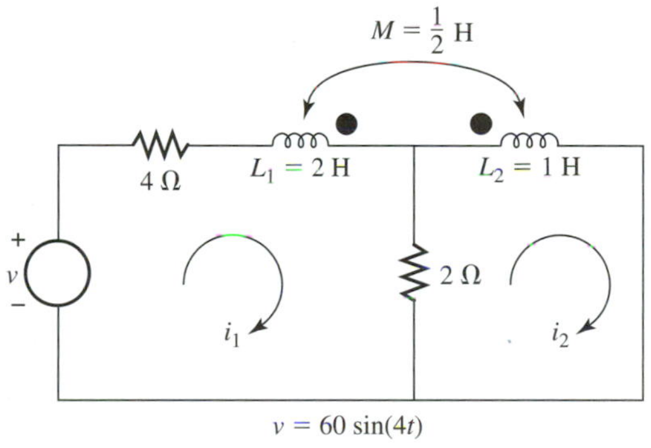

M = 1/2 H
4 Ω
L1 = 2 H
L2 = 1 H
+
v
−
2 Ω
i1
i2
v = 60 sin(4t)

SOLUTION The steady-state ac representation of the circuit is shown in Figure 21.17. Using the dot notation rules and the voltage polarity rule for self-induced

Figure 21.17

j2 Ω
4 Ω
j8 Ω
j4 Ω
+
60 ∠0°
−
2 Ω
I1
I2

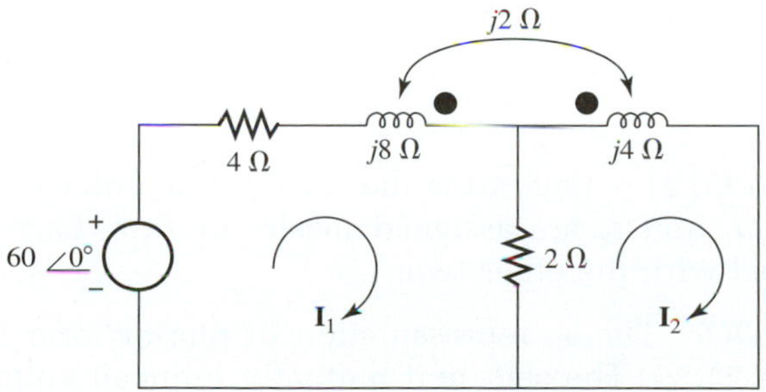

voltages, the self- and mutually induced voltages for the inductors are indicated in Figure 21.18.

Figure 21.18

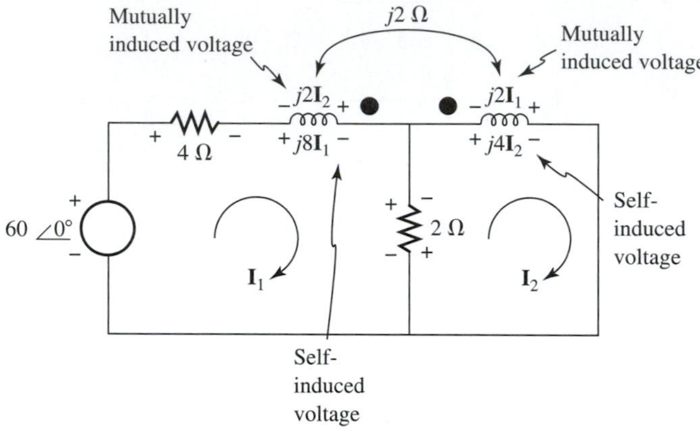

Applying Kirchhoff's voltage law to each mesh results in

$$\text{mesh 1:}\quad 60\angle 0^\circ - 4\mathbf{I}_1 + j2\mathbf{I}_2 - j8\mathbf{I}_1 - 2\mathbf{I}_1 + 2\mathbf{I}_2 = 0$$

$$\text{mesh 2:}\quad -2\mathbf{I}_2 + 2\mathbf{I}_1 + j2\mathbf{I}_1 - j4\mathbf{I}_2 = 0$$

Rearranging terms and applying Cramer's rule from Chapter 5 to solve for $\mathbf{I}_2$ results in

$$(6 + j8)\mathbf{I}_1 - (2 + j2)\mathbf{I}_2 = 60\angle 0^\circ$$

$$-(2 + j2)\mathbf{I}_1 + (2 + j4)\mathbf{I}_2 = 0$$

$$\mathbf{I}_2 = \frac{\begin{vmatrix} 6 + j8 & 60\angle 0^\circ \\ -(2 + j2) & 0 \end{vmatrix}}{\begin{vmatrix} 6 \pm j8 & -(2 + j2) \\ -(2 + j2) & 2 + j4 \end{vmatrix}} = \frac{170\angle 45^\circ}{37.7\angle 122^\circ}$$

$$= 4.50\angle -77^\circ$$

and from the phasor expression for $\mathbf{I}_2$

$$i_2 = 4.50 \sin(4t - 77^\circ)$$

EXAMPLE 21.5 Determine the steady-state voltage v_o for the circuit in Figure 21.19. i_1 and i_2 are assigned mesh currents. Each winding is magentically coupled with the other two. ■

SOLUTION The ac representation in phasor form of the circuit is shown in Figure 21.20. The self- and mutually induced voltages for each winding are

Figure 21.19

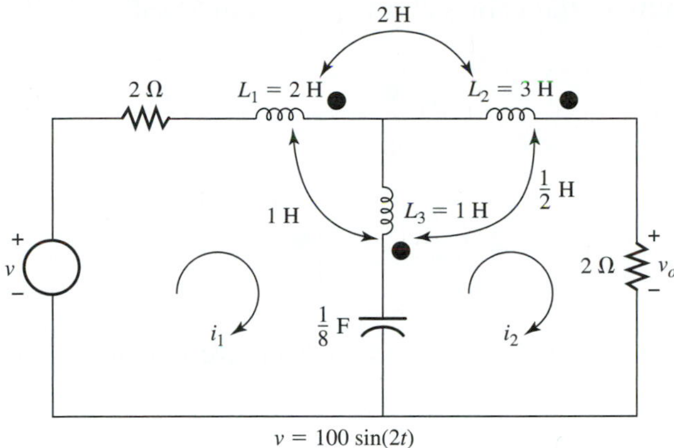

Figure 21.20

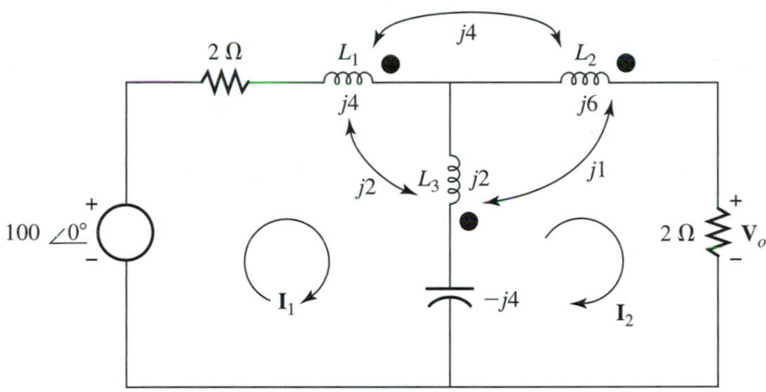

indicated in the partial circuit shown in Figure 21.21. The polarities are determined from the dot notation rules for the mutually induced voltages and from the voltage polarity rule for self-induced voltages.

Figure 21.21

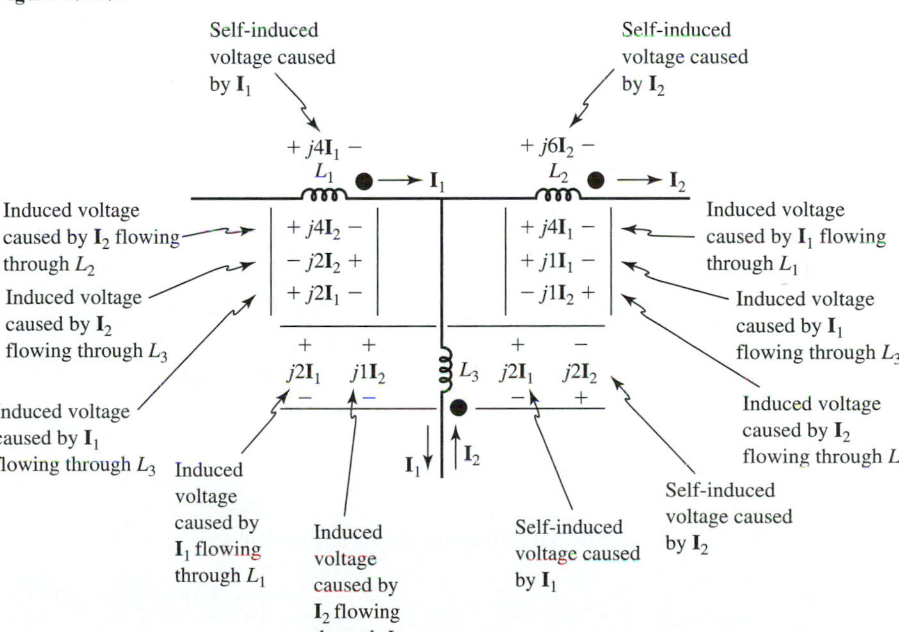

Summing the self- and mutually induced voltages for each inductor results in the values shown in Figure 21.22.

Figure 21.22

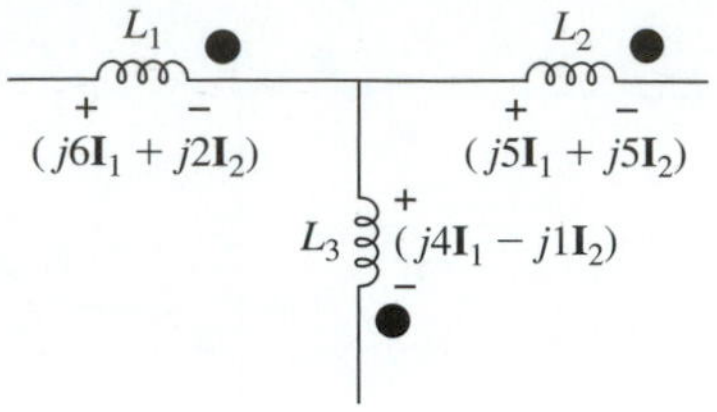

L_1 L_2

$+\quad-$ $(j6\mathbf{I}_1 + j2\mathbf{I}_2)$

$+\quad-$ $(j5\mathbf{I}_1 + j5\mathbf{I}_2)$

L_3 $+$ $(j4\mathbf{I}_1 - j1\mathbf{I}_2)$ $-$

The voltage across each circuit element is shown in Figure 21.23. Employing Kirchhoff's voltage law to each mesh results in

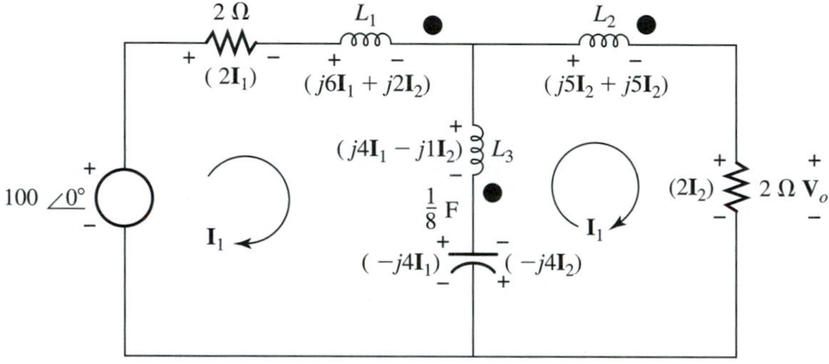

Figure 21.23

$$\text{mesh 1: } 100\angle 0^\circ - 2\mathbf{I}_1 - (j6\mathbf{I}_1 + j2\mathbf{I}_2) - (j4\mathbf{I}_1 - j1\mathbf{I}_2) - (-j4\mathbf{I}_1) + (-j4\mathbf{I}_2) = 0$$

$$\text{mesh 2: } -(-j4\mathbf{I}) + (-j4\mathbf{I}) + (j4\mathbf{I} - j1\mathbf{I}) - (j5\mathbf{I} + j5\mathbf{I}) - (2\mathbf{I}) = 0$$

Rearranging terms and applying Cramer's rule from Chapter 5 results in

$$(2 + 6j)\mathbf{I}_1 + (j5)\mathbf{I}_2 = 100\angle 0^\circ$$

$$(j5)\mathbf{I}_1 + (2 + j2)\mathbf{I}_2 = 0$$

$$\mathbf{I}_2 = \frac{\begin{vmatrix} 2 + j6 & 100\angle 0^\circ \\ j5 & 0 \end{vmatrix}}{\begin{vmatrix} 2 + j6 & j5 \\ j5 & 2 + j2 \end{vmatrix}} = \frac{500\angle -90^\circ}{23.3\angle 43.3^\circ}$$

$$= 21.5\angle -133^\circ$$

and

$$\mathbf{V}_o = 2\mathbf{I}_2$$

$$= 2\{21.5\angle -133^\circ\}$$

$$= 43\angle -133^\circ$$

From the phasor expression for v_o

$$v_o = 43\sin(2t - 133^\circ)$$

21.7 REFLECTED IMPEDANCE

It is often convenient when analyzing coupled circuits to know the impedance of the circuit looking into the primary winding from the driving source. Consider the simple coupled circuit in Figure 21.24 for steady-state ac conditions, where $\mathbf{V}_s$ is the driving source and $\mathbf{Z}_s$ is the impedance of the source.

Figure 21.24 Reflected impedance.

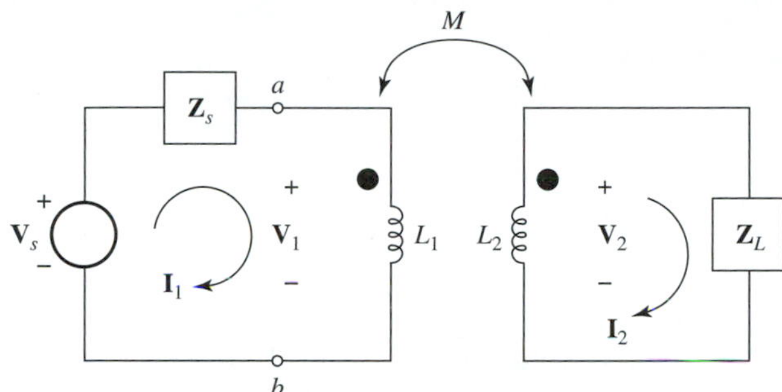

The voltage $\mathbf{V}_1$ is equal to the algebraic sum of the self-induced voltage across L_1 caused by $\mathbf{I}_1$ and the mutually induced voltage across L_1 caused by $\mathbf{I}_2$. That is,

$$\mathbf{V}_1 = j\omega L_1\mathbf{I}_1 - j\omega M\mathbf{I}_2$$

Applying Kirchhoff's voltage law to the secondary circuit results in

$$j\omega M\mathbf{I}_1 - j\omega L_2\mathbf{I}_2 - \mathbf{Z}_L\mathbf{I}_2 = 0$$

where $j\omega M\mathbf{I}_1$ is the voltage induced across the secondary winding by the current $\mathbf{I}_1$ and $j\omega L_2\mathbf{I}_2$ is the voltage induced across the secondary winding by the current $\mathbf{I}_2$. The signs of the induced voltage terms in the preceding equations are determined from the voltage polarity rule and the dot notation rule previously presented.

Combining terms and rearranging the preceding equations results in

$$j\omega L_1\mathbf{I}_1 - j\omega M\mathbf{I}_2 = \mathbf{V}_1$$
$$-j\omega M\mathbf{I}_1 + (j\omega L_2 + \mathbf{Z}_L)\mathbf{I}_2 = 0$$

Using Cramer's rule from chapter 5 to solve for $\mathbf{I}_1$,

$$\mathbf{I}_1 = \frac{\begin{vmatrix} \mathbf{V}_1 & -j\omega M \\ 0 & j\omega L_2 + \mathbf{Z}_L \end{vmatrix}}{\begin{vmatrix} j\omega L_1 & -j\omega M \\ -j\omega M & j\omega L_2 + \mathbf{Z}_L \end{vmatrix}}$$
$$= \frac{(j\omega L_2 + \mathbf{Z}_2)\mathbf{V}_1}{(j\omega L_1)(j\omega L_2 + \mathbf{Z}_L) - (-j\omega M)(-j\omega M)}$$

and solving for the ratio $\mathbf{V}_1/\mathbf{I}_1$ results in

$$\frac{\mathbf{V}_1}{\mathbf{I}_1} = \frac{(j\omega L_1)(j\omega L_2 + \mathbf{Z}_L) - (j\omega M)(j\omega M)}{j\omega L_2 + \mathbf{Z}_L}$$

From Equation (17.1) $\mathbf{V}_1/\mathbf{I}_1$ is the impedance looking into the primary winding from terminals *a-b* in Figure 21.24. Rearranging terms for $(\mathbf{V}_1/\mathbf{I}_1)$ and calling it $\mathbf{Z}_1$ results in

$$\mathbf{Z}_1 = j\omega L_1 + \frac{\omega^2 M^2}{j\omega L_2 + \mathbf{Z}_L} \tag{21.5}$$

as shown in Figure 21.25.

Figure 21.25 Reflected impedance.

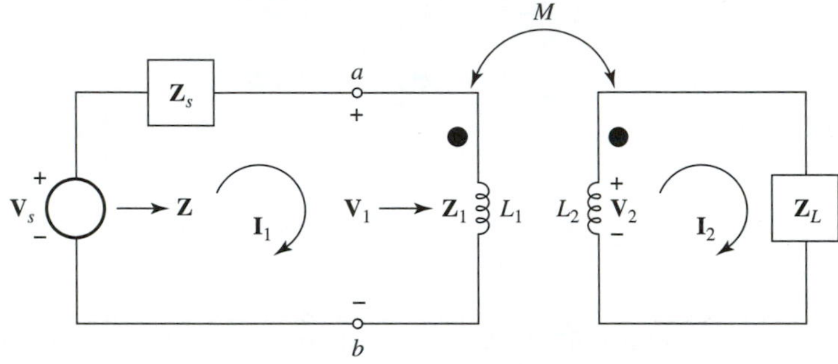

From Figure 21.25 the total impedance $\mathbf{Z}$ seen by the voltage source $\mathbf{V}_s$ is $\mathbf{Z}_s + \mathbf{Z}_1$, and from Equation (17.1),

$$\mathbf{Z} = \frac{\mathbf{V}_s}{\mathbf{I}_1} = \mathbf{Z}_s + \mathbf{Z}_1$$

$$= \mathbf{Z}_s + j\omega L_1 + \frac{\omega^2 M^2}{j\omega L_2 + \mathbf{Z}_L} \tag{21.6}$$

where $\mathbf{Z}_s$ is the source impedance, $j\omega L_1$ is the primary winding impedance, and

$$\left\{\frac{\omega^2 M^2}{j\omega L_2 + \mathbf{Z}_L}\right\}$$

is the impedance reflected into the primary circuit from the secondary caused by the coupling between the primary and secondary windings.

Equation (21.6) is independent of the location of the dots on the windings, because the mutual inductance term M is squared in the equation.

Consider the following example.

EXAMPLE 21.6 Determine an expression for the steady-state current i_1 for the circuit in Example 21.3 using Equation (21.6). The circuit is redrawn in Figure 21.26. ■

Figure 21.26

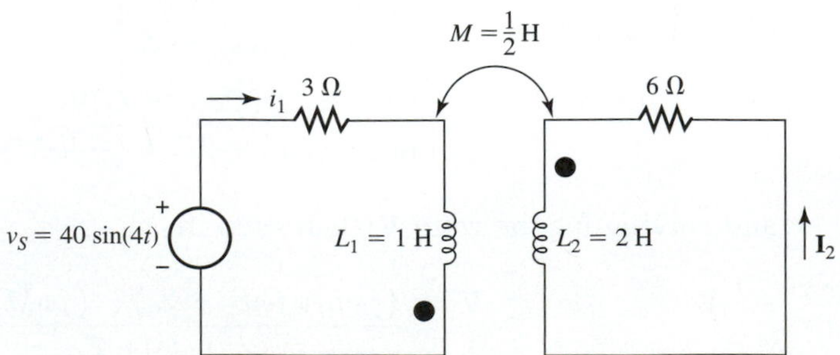

SOLUTION A steady-state ac representation of the circuit is shown in Figure 21.27.

Figure 21.27

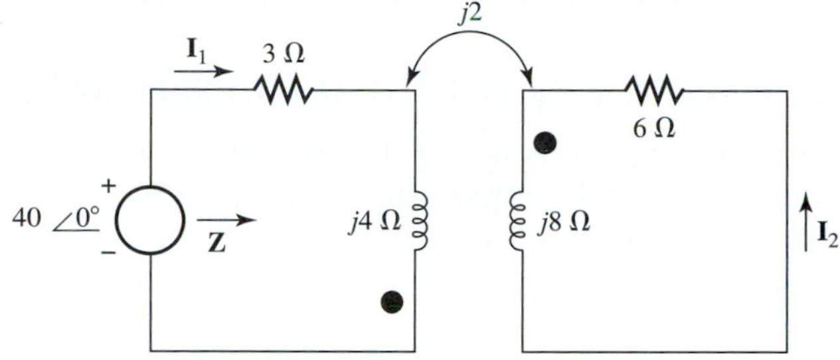

Applying Equation (21.6) results in

$$\mathbf{Z} = \mathbf{Z}_s + j\omega L_1 + \frac{\omega^2 M^2}{j\omega L_2 + \mathbf{Z}_L}$$

$$= 3 + j4 + \frac{(4)^2\left(\frac{1}{2}\right)^2}{j8 + 6}$$

$$= 3 + j4 + \frac{4}{j8 + 6}$$

$$= 4.9\underline{/48.7^\circ}$$

From Equation (17.1)

$$\mathbf{Z} = \frac{\mathbf{V}_s}{\mathbf{I}_1}$$

and

$$\mathbf{I}_1 = \frac{40\underline{/0^\circ}}{4.9\underline{/48.7^\circ}}$$

$$= 8.16\underline{/-48.7^\circ}$$

From the phasor expression for $\mathbf{I}_1$

$$i_1 = 8.16\sin(4t - 48.7^\circ)$$

Notice that the preceding expression for i_1 is the same as obtained in Example 21.3.

21.8 IDEAL TRANSFORMER

The ideal transformer consists of two windings that are coupled in such a manner that there is no leakage flux and no energy loss in the windings or in the magnetic core material. Because there is no leakage flux, both windings are linked by the same flux in the ideal transformer.

Although the ideal transformer cannot be physically realized, the iron-core transformer is a close approximation when the primary and secondary coils are wound on a laminated iron core in such a manner that the energy loss in the core and the leakage flux are minimized. We usually consider the iron-core transformer as an ideal transformer for analysis purposes.

21.9 COEFFICIENT OF COUPLING FOR THE IDEAL TRANSFORMER

The coefficient of coupling (labeled k) between two coils is a measure of the ability of the coils to perform as an ideal transformer. The coefficient of coupling can be written as

$$k = \frac{\phi_{1b}}{\phi_1}$$

where ϕ_1 is the flux caused by the primary current and ϕ_{1b} is the portion of ϕ_1 that links both the primary and secondary windings.

Because there is no leakage flux for the ideal transformer, $\phi_{1b} = \phi_1$ and k is equal to unity; and because ϕ_{1b} cannot exceed ϕ_1, the maximum value of k for any transformer is unity. If there is no coupling between two windings, $\phi_{1b} = 0$ and therefore $k = 0$. It follows that k varies between zero and unity, with $k = 1$ for the ideal transformer and $k = 0$ when the windings are not coupled.

The values of k and M for two windings depends on the physical aspects of the coupled windings—the number of turns in each winding, the dimensions and spacing of the turns, and the properties of the magnetic core material. The coefficient of coupling is very close to unity for most iron-core transformers and is usually less than 0.6 for air-core transformers, where the leakage flux can be significant.

It can be shown that the coefficient of coupling between two windings can be written as

$$k = \frac{M}{\sqrt{L_1 L_2}} \tag{21.7}$$

where M is the mutual inductance between the windings with inductances L_1 and L_2. Because

$$M = k\sqrt{L_1 L_2}$$

and because k varies between 0 and 1,

$$0 \leq M \leq \sqrt{L_1 L_2}$$

Because $k = 1$ for ideal coupling, $M = \sqrt{L_1 L_2}$ for the ideal transformer.

21.10 PRIMARY AND SECONDARY RELATIONSHIPS FOR IDEAL TRANSFORMERS

When there is no energy lost in the core material, no energy lost in the windings, and no leakage flux, the transformer is said to be ideal, and the following relationships exist between the primary and secondary windings.

If N_1 is the number of turns in the primary winding, N_2 is the number of turns in the secondary winding, and

$$n = \frac{N_2}{N_1} \tag{21.8}$$

it can be shown that

$$\frac{i_2}{i_1} = \frac{1}{n} \tag{21.9}$$

$$\frac{v_2}{v_1} = n \tag{21.10}$$

and

$$\frac{L_2}{L_1} = n^2 \tag{21.11}$$

where i_1, v_1, and L_1 are the primary current, voltage, and inductance, respectively, and i_2, v_2, and L_2 are the secondary current, voltage, and inductance, respectively.

21.11 AC CIRCUITS WITH IDEAL TRANSFORMERS

Equations (21.9) and (21.10) involve instantaneous values for current and voltage for ideal transformers. For steady-state ac operation, the equations can be written as

$$\frac{\mathbf{I}_2}{\mathbf{I}_1} = \frac{1}{n} \tag{21.12}$$

and

$$\frac{\mathbf{V}_2}{\mathbf{V}_1} = n \tag{21.13}$$

where i_1, i_2, v_1, and v_2 in Equations (21.9) and (21.10) are generated by phasors $\mathbf{I}_1$, $\mathbf{I}_2$, $\mathbf{V}_1$, and $\mathbf{V}_2$, respectively.

21.12 REFLECTED IMPEDANCE FOR THE IDEAL TRANSFORMER

Consider the ideal transformer in Figure 21.28. The vertical lines between the windings symbolically indicate that the transformer is an iron-core transformer and can be treated as ideal.

Figure 21.28 Ideal transformer.

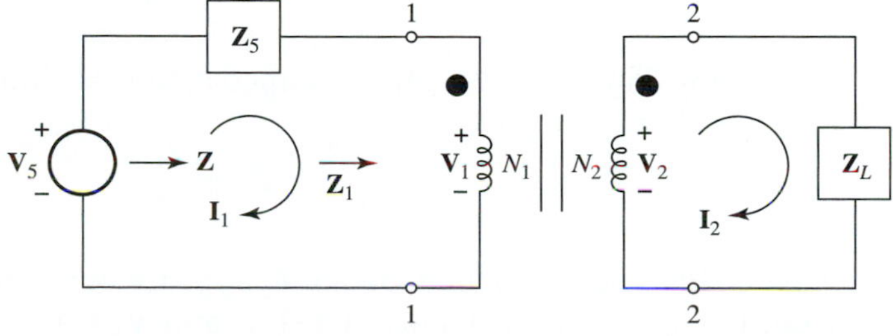

From Equation (17.1)

$$\mathbf{Z}_1 = \frac{\mathbf{V}_1}{\mathbf{I}_1}$$

and from Equations (21.12) and (21.13),

$$\mathbf{Z}_1 = \frac{\{\mathbf{V}_2/n\}}{\{n\mathbf{I}_2\}}$$
$$= \left(\frac{1}{n^2}\right)\left\{\frac{\mathbf{V}_2}{\mathbf{I}_2}\right\}$$

From Equation (17.1)

$$\mathbf{Z}_L = \frac{\mathbf{V}_2}{\mathbf{I}_2}$$

and from the preceding equation for $\mathbf{Z}_1$,

$$\mathbf{Z}_1 = \left(\frac{1}{n^2}\right)\mathbf{Z}_L \tag{21.14}$$

The impedance $\mathbf{Z}$ seen by the voltage source $\mathbf{V}_s$ is

$$\mathbf{Z} = \mathbf{Z}_s + \mathbf{Z}_1$$
$$= \mathbf{Z}_s + \left(\frac{1}{n^2}\right)\mathbf{Z}_L \tag{21.15}$$

where $\mathbf{Z}_s$ is the source impedance.

The impedance $\mathbf{Z}_1$ is called the *reflected impedance*, because it is the secondary load impedance reflected into the primary circuit by the circuit coupling.

21.13 EQUIVALENT CIRCUITS

Consider the circuit in Figure 21.28 as redrawn for convenience in Figure 21.29.

Figure 21.29 Ideal transformer.

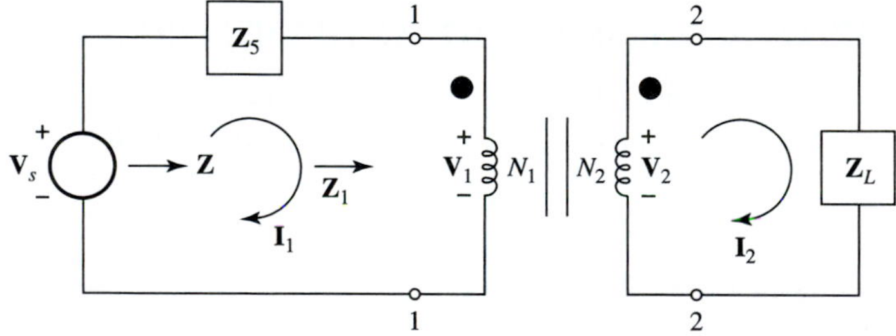

From Equation (21.14) the impedance seen looking into terminals 1-1 is

$$\mathbf{Z}_1 = \frac{\mathbf{Z}_L}{n^2}$$

That is, the secondary impedance $\mathbf{Z}_L$ appears in the primary loop as $\mathbf{Z}_L/n^2$, and from Equations (21.12) and (21.13), $\mathbf{I}_1$ and $\mathbf{V}_1$ can be written as

$$\mathbf{I}_1 = n\mathbf{I}_2$$
$$\mathbf{V}_1 = \frac{\mathbf{V}_2}{n}$$

From the relationships between primary and secondary, an equivalent circuit for the ideal transformer circuit in Figure 21.29 can be drawn as shown in Figure 21.30.

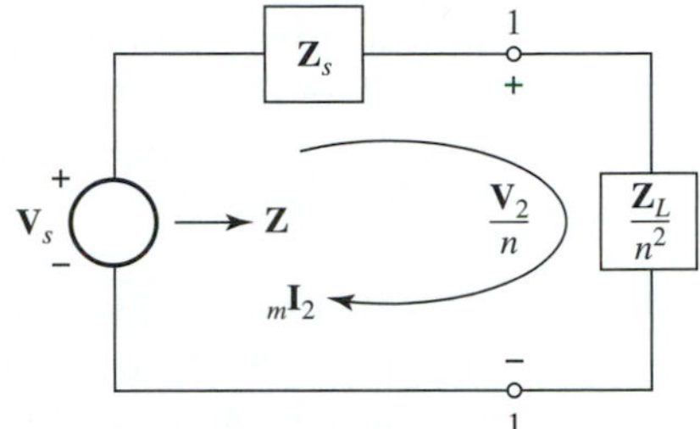

Figure 21.30 Equivalent circuit.

Notice that $\mathbf{Z}$ (the impedance seen by the voltage source $\mathbf{V}_s$) is the same for both circuits. For the circuit in Figure 21.29, $\mathbf{Z} = \mathbf{Z}_s + (1/n^2)\mathbf{Z}_L$ from Equation (21.15); and for the circuit in Figure 21.30, $\mathbf{Z} = \mathbf{Z}_s + (1/n^2)\mathbf{Z}_L$ from an inspection of the circuit. The circuit formed in this manner is called a primary equivalent circuit.

In a manner similar to forming the primary equivalent circuit, a secondary equivalent circuit can be formed, where the primary is reflected into the secondary. Rules for forming the primary and secondary equivalent circuits are given next.

> The *primary equivalent circuit* is formed by removing the ideal transformer, multiplying the secondary impedances by $(1/n^2)$, multiplying the secondary currents by (n), and multiplying the secondary voltages by $(1/n)$, as illustrated in Figure 21.31(b).
>
> The *secondary equivalent circuit* is formed by removing the ideal transformer, multiplying the primary impedances by (n^2), multiplying the primary currents by $(1/n)$, and multiplying the primary voltages by (n), as illustrated in Figure 21.31(c).

Figure 21.31 Primary and secondary equivalent circuits.

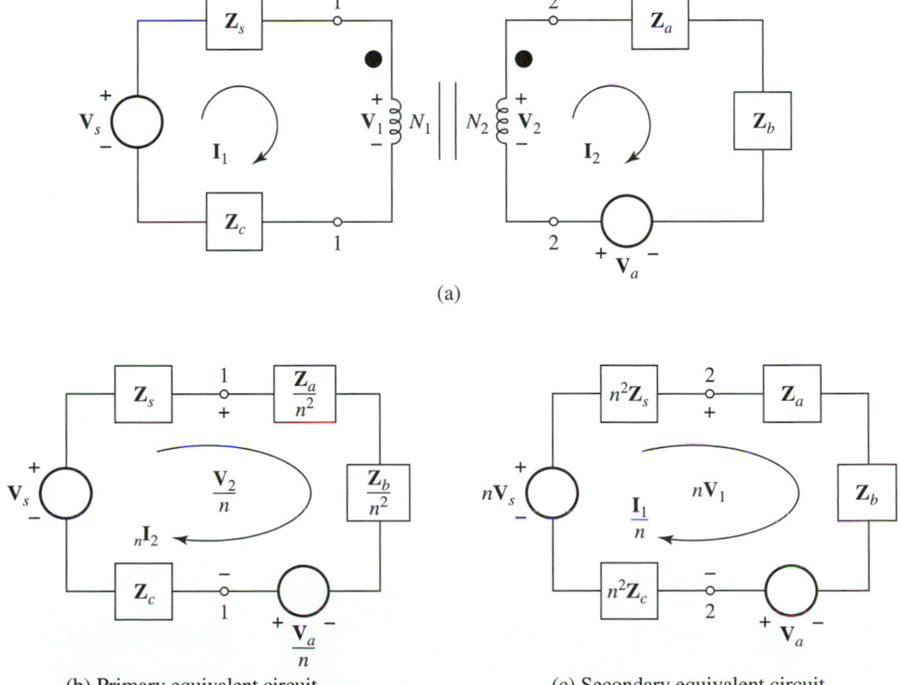

The methods described here for forming primary and secondary equivalent circuits for ideal transformers are based on the ideal transformer with windings dotted and currents and voltages assigned as shown in Figure 21.31. If the dot on one of the windings is reversed, then n must be replaced by $(-n)$ in the rules given for forming the primary and secondary equivalent circuits.

The primary and secondary equivalent circuits provide a useful alternative for analyzing circuits with ideal transformers, as illustrated in the following examples.

EXAMPLE 21.7 Determine steady-state ac expressions for i_1, i_2, v_1, and v_2 for the iron-core transformer shown in Figure 21.32, where $N_2/N_1 = 2$. Consider the transformer as ideal. ■

Figure 21.32

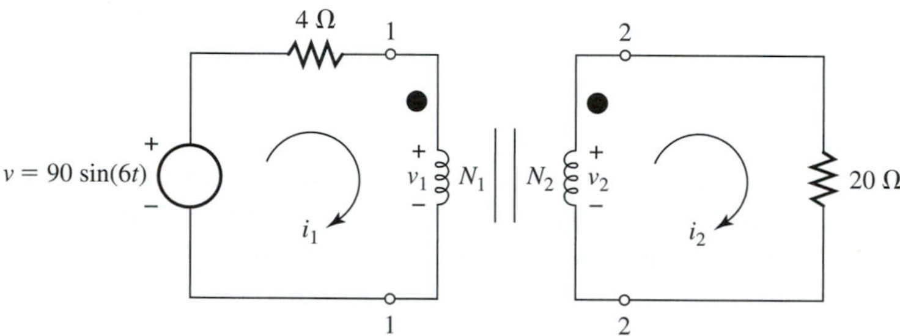

SOLUTION Using the rules for forming a primary equivalent circuit and transforming the circuit to a steady-state ac representation results in the circuit shown in Figure 21.33.

Figure 21.33

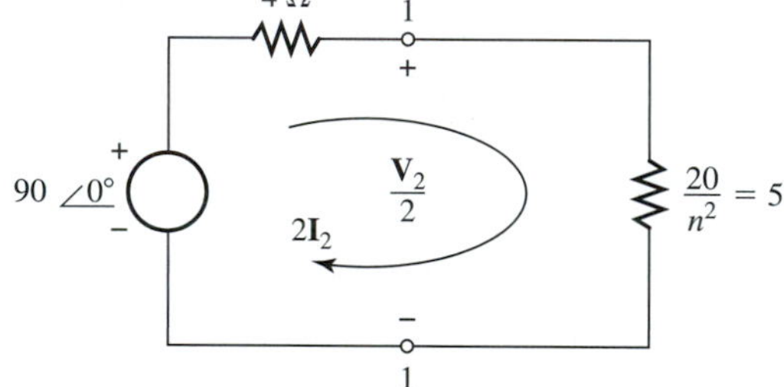

Notice that the secondary impedance, secondary current, and secondary voltage are multiplied by $(1/n^2)$, (n), and $(1/n)$, respectively, when reflected into the primary.

Applying Kirchhoff's voltage law to the primary equivalent circuit results in

$$90\angle 0^\circ - 4(2\mathbf{I}_2) - 5(2\mathbf{I}_2) = 0$$

$$\mathbf{I}_2 = 5\angle 0^\circ$$

and applying Equation (17.1) to the reflected impedance,

$$\frac{\mathbf{V}_2}{2} = 5(2\mathbf{I}_2)$$

$$= 10(5\angle 0^\circ)$$

$$\mathbf{V}_2 = 100\angle 0^\circ$$

From Equations (21.12) and (21.13),

$$\mathbf{I}_1 = n\mathbf{I}_2$$
$$= 2\{5\angle 0°\}$$
$$= 10\angle 0°$$
$$\mathbf{V}_1 = \frac{\mathbf{V}_2}{n}$$
$$= \frac{100\angle 0°}{2}$$
$$= 50\angle 0°$$

From these phasor expressions for $\mathbf{I}_2$, $\mathbf{V}_2$, $\mathbf{I}_1$, and $\mathbf{V}_1$,

$$i_2 = 5\sin(6t)$$
$$v_2 = 100\sin(6t)$$
$$i_1 = 10\sin(6t)$$
$$v_1 = 50\sin(6t)$$

EXAMPLE 21.8 Solve the problem given in Example 21.7 using a secondary equivalent circuit. ■

SOLUTION Using the rules for forming a secondary equivalent circuit and transforming the given circuit to a steady-state ac representation results in the circuit shown in Figure 21.34.

Figure 21.34

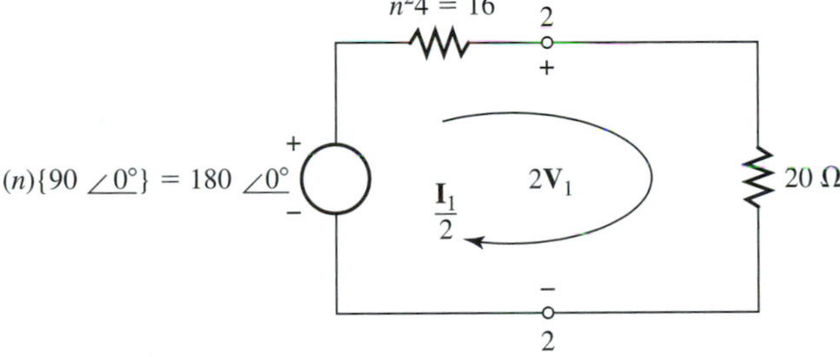

Notice that the primary impedance, primary current, and primary voltage are multiplied by n^2, $(1/n)$, and n, respectively, when reflected into the secondary.

Applying Kirchhoff's voltage law to the secondary equivalent results in

$$180\angle 0° - 16\left\{\frac{\mathbf{I}_1}{2}\right\} - 20\left\{\frac{\mathbf{I}_1}{2}\right\} = 0$$
$$180\angle 0° - 8\mathbf{I}_1 - 10\mathbf{I}_1 = 0$$
$$\mathbf{I}_1 = 10\angle 0°$$

and applying Equation (17.1) to the 20-Ω resistor,

$$2\mathbf{V}_1 = 20\left\{\frac{\mathbf{I}_1}{2}\right\}$$

$$= 20\left\{\frac{10\underline{/0^\circ}}{2}\right\}$$

$$\mathbf{V}_1 = 50\underline{/0^\circ}$$

From Equations (21.12) and (21.13),

$$\mathbf{I}_2 = \left(\frac{1}{n}\right)\mathbf{I}_1$$

$$= \left(\frac{1}{2}\right)10\underline{/0^\circ}$$

$$= 5\underline{/0^\circ}$$

$$\mathbf{V}_2 = n\mathbf{V}_1$$

$$= (2)50\underline{/0^\circ}$$

$$= 100\underline{/0^\circ}$$

From these phasor expressions for $\mathbf{I}_1$, $\mathbf{V}_1$, $\mathbf{I}_2$, and $\mathbf{V}_2$,

$$i_1 = 10\sin(6t)$$

$$v_1 = 50\sin(6t)$$

$$i_2 = 5\sin(6t)$$

$$v_2 = 100\sin(6t)$$

EXAMPLE 21.9 Solve the problem given in Example 21.7 using Equation (21.15) for the impedance seen by the voltage source. ■

SOLUTION A steady-state ac representation for the given circuit is shown in Figure 21.35.

Figure 21.35

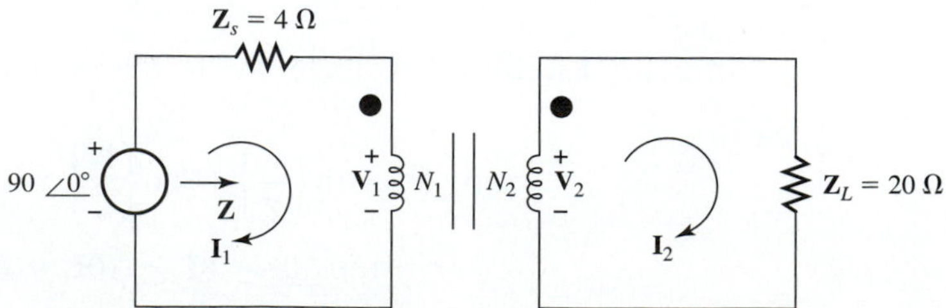

From Equation (21.15)

$$\mathbf{Z} = \mathbf{Z}_s + \frac{\mathbf{Z}_L}{n^2}$$

$$= 4\angle 0° + (20\angle 0°)/(2)^2$$

$$= 9\angle 0°$$

From Equation (17.1) for the terminals of the voltage source,

$$\mathbf{Z} = \frac{90\angle 0°}{\mathbf{I}_1}$$

$$9\angle 0° = (90\angle 0°)\mathbf{I}_1$$

$$\mathbf{I}_1 = 10\angle 0°$$

From Equation (21.12)

$$\mathbf{I}_2 = \frac{\mathbf{I}_1}{n}$$

$$= \frac{10\angle 0°}{2}$$

$$= 5\angle 0°$$

From Equation (17.1) to the terminals of $\mathbf{Z}_L$,

$$\mathbf{V}_2 = 20\mathbf{I}_2$$

$$= 20(5\angle 0°)$$

$$= 100\angle 0°$$

From Equation (21.13)

$$\mathbf{V}_1 = \frac{\mathbf{V}_2}{n}$$

$$= \frac{100\angle 0°}{2}$$

$$= 50\angle 0°$$

From the preceding phasor expressions for $\mathbf{I}_1$, $\mathbf{I}_2$, $\mathbf{V}_2$, and $\mathbf{V}_1$,

$$i_1 = 10\sin(6t)$$

$$i_2 = 5\sin(6t)$$

$$v_2 = 100\sin(6t)$$

$$v_1 = 50\sin(6t)$$

EXAMPLE 21.10 Consider the steady-state ac circuit representation given in Figure 21.36. Draw primary and secondary equivalent circuits. Consider the iron-core transformer as ideal with $n = 2$. ■

Figure 21.36

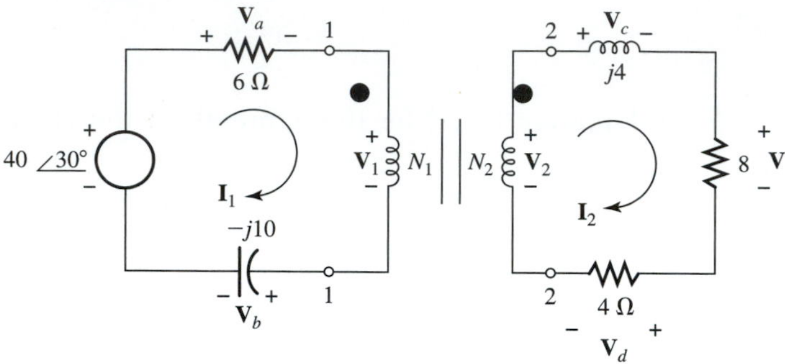

SOLUTION The primary equivalent circuit is formed by removing the ideal transformer, connecting the primary and secondary circuits, and multiplying the secondary voltage, current, and impedances by $(1/n)$, (n) and $(1/n^2)$, respectively, as shown in Figure 21.37.

Figure 21.37

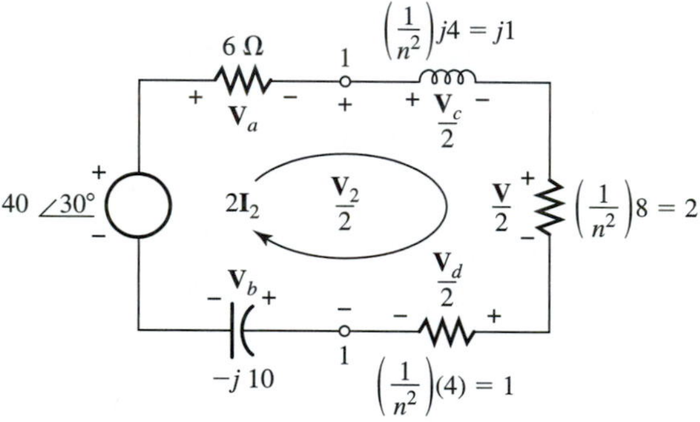

The secondary equivalent circuit is formed by removing the ideal transformer, connecting the primary and secondary circuits, and multiplying the primary voltages, current, and impedances by n, $(1/n)$, and n^2, respectively, as shown in Figure 21.38.

Figure 21.38

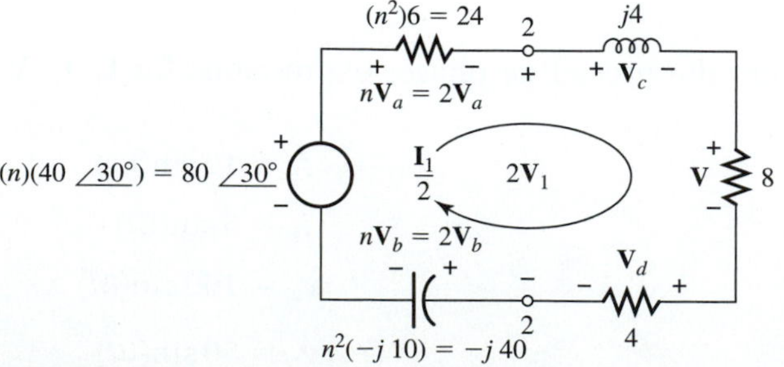

EXAMPLE 21.11 Use the primary and secondary equivalent circuits in the previous example to solve for $\mathbf{I}_1$ and $\mathbf{I}_2$. ■

SOLUTION Applying Kirchhoff's voltage law to the primary equivalent results in

$$40\angle 30^\circ - 6\mathbf{I}_1 - j1\mathbf{I}_1 - 2\mathbf{I}_1 - 1\mathbf{I}_1 - (-j10)\mathbf{I}_1 = 0$$

or

$$40\angle 30^\circ - 6(2\mathbf{I}_2) - j1(2\mathbf{I}_2) - 2(2\mathbf{I}_2) - 1(2\mathbf{I}_2) - (-j10)(2\mathbf{I}_2) = 0$$

From the first equation,

$$\mathbf{I}_1 = \frac{40\angle 30^\circ}{9 - j9} = 3.14\angle 75^\circ$$

and from the second equation,

$$\mathbf{I}_2 = \frac{40\angle 30^\circ}{18 - j18} = 1.57\angle 75^\circ$$

Notice that $\mathbf{I}_2/\mathbf{I}_1 = 1/n = \frac{1}{2}$, as required by Equation (21.12).

Alternatively, applying Kirchhoff's voltage law to the secondary equivalent results in

$$80\angle 30^\circ - 24\mathbf{I}_2 - j4\mathbf{I}_2 - 8\mathbf{I}_2 - 4\mathbf{I}_2 - (-j40)\mathbf{I}_2 = 0$$

or

$$80\angle 30^\circ - 24\left(\frac{\mathbf{I}_1}{2}\right) - j4\left(\frac{\mathbf{I}_1}{2}\right) - 8\left(\frac{\mathbf{I}_1}{2}\right) - 4\left(\frac{\mathbf{I}_1}{2}\right) - (-j40)\left(\frac{\mathbf{I}_1}{2}\right) = 0$$

From the first equation,

$$\mathbf{I}_2 = \frac{80\angle 30^\circ}{36 - j36} = 1.57\angle 75^\circ$$

and from the second equation,

$$\mathbf{I}_1 = \frac{80\angle 30^\circ}{18 - j18} = 3.14\angle 75^\circ$$

21.14 EXERCISES

1. Determine the current i and the voltage v_{ab} after the switch is closed for the circuit in Figure 21.39.

Figure 21.39

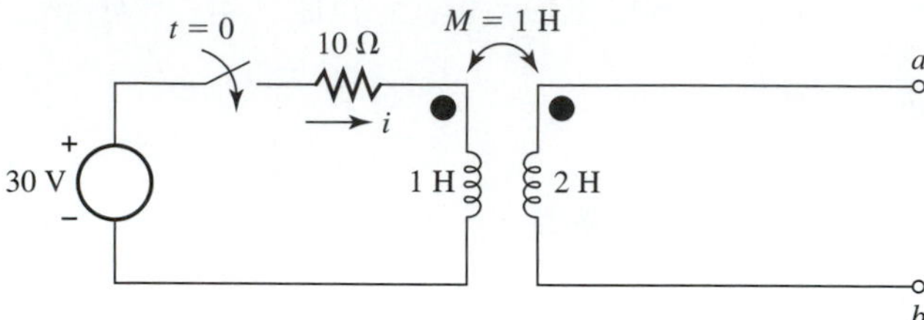

2. Determine the current i and the voltage v_{ab} after the switch is closed for the circuit in Figure 21.40.

Figure 21.40

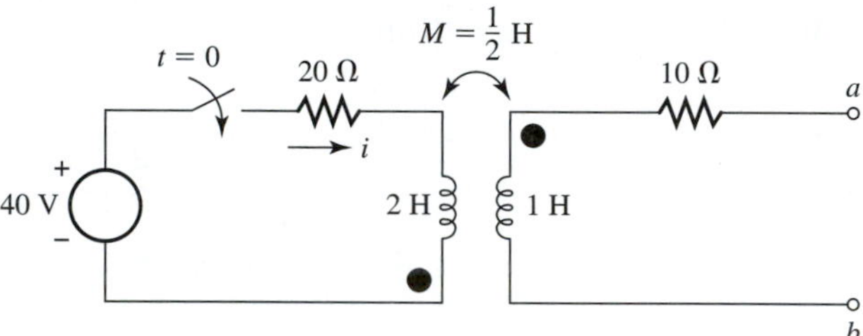

3. Determine the current i and the voltage v_{ab} after the switch is closed for the circuit in Figure 21.41.

Figure 21.41

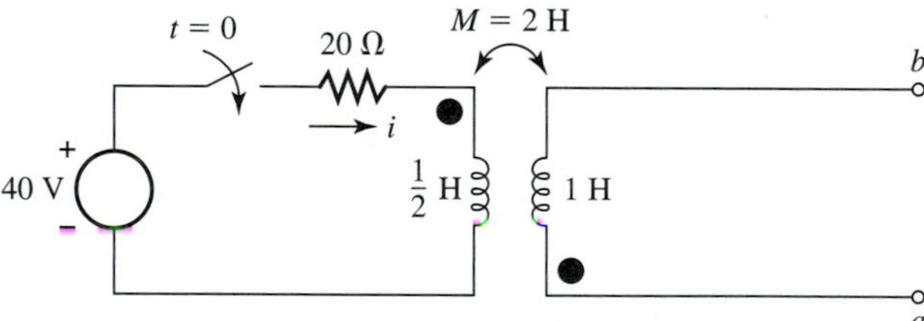

4. Determine the current i and the voltage v_{ab} after the switch is closed for the circuit in Figure 21.42.

Figure 21.42

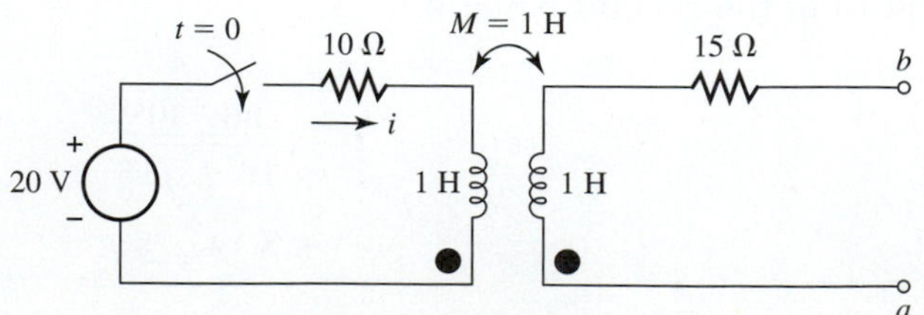

5. Determine the currents i_1 and i_2 and the voltage v_o for the circuit in Figure 21.43.

Figure 21.43

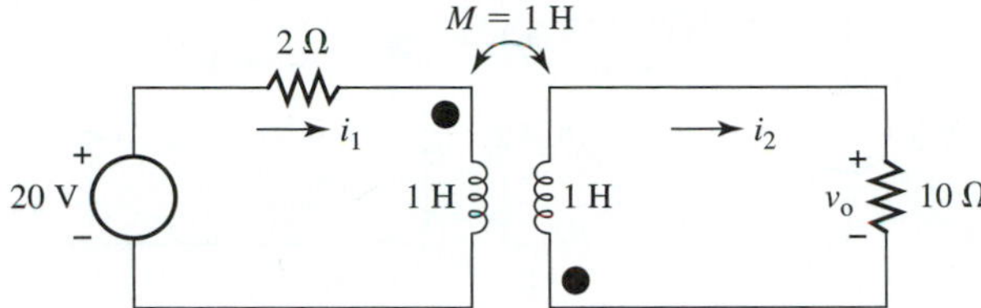

6. Determine the currents i_1 and i_2 and the voltage v_o for the circuit in Figure 21.44.

Figure 21.44

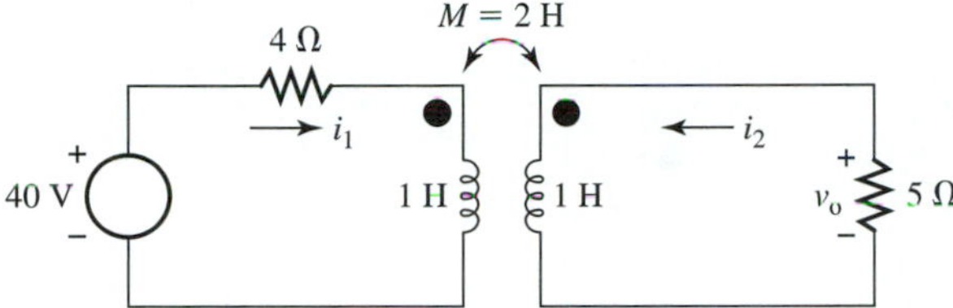

7. Determine the currents i_1 and i_2 and the voltage v_o for the circuit in Figure 21.45.

Figure 21.45

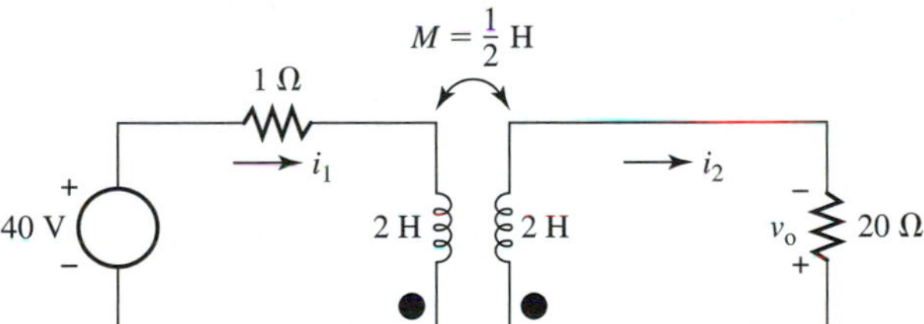

8. Determine the currents i_1 and i_2 and the voltage v_o for the circuit in Figure 21.46.

Figure 21.46

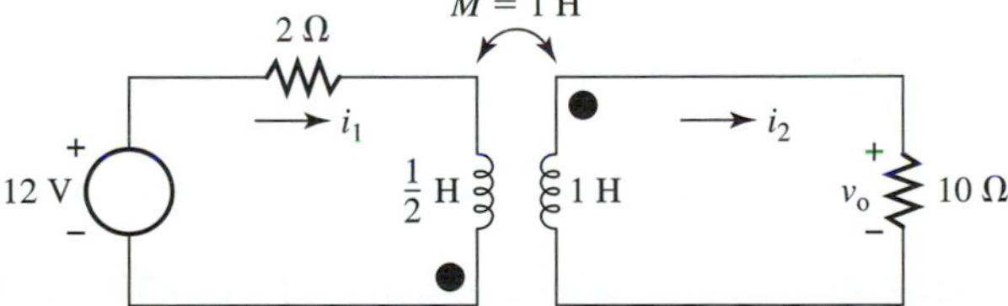

9. Determine the phasor currents $\mathbf{I}_1$ and $\mathbf{I}_2$ for the circuit in Figure 21.47.

Figure 21.47

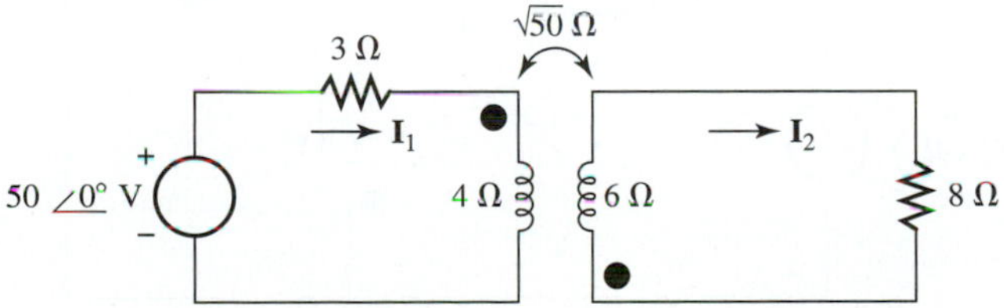

10. Determine the phasor currents $\mathbf{I}_1$ and $\mathbf{I}_2$ for the circuit in Figure 21.48.

Figure 21.48

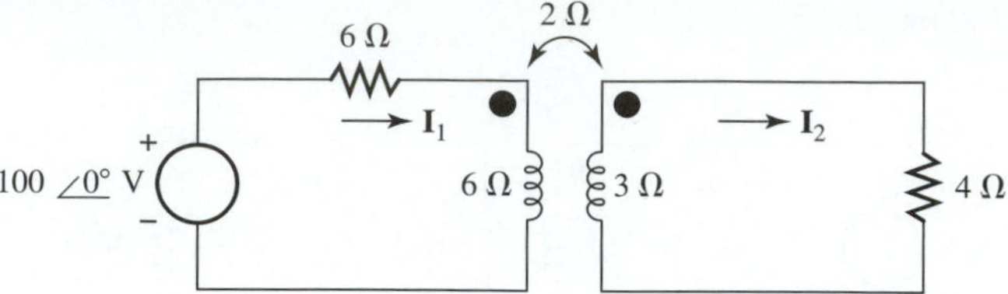

11. Determine the phasor currents $\mathbf{I}_1$ and $\mathbf{I}_2$ for the circuit in Figure 21.49.

Figure 21.49

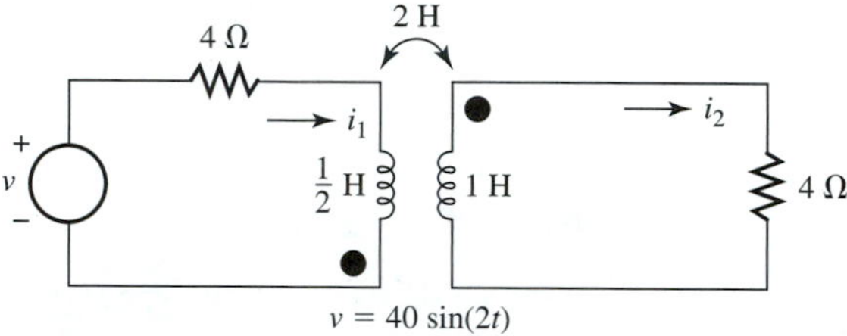

12. Determine the phasor currents $\mathbf{I}_1$ and $\mathbf{I}_2$ for the circuit in Figure 21.50.

Figure 21.50

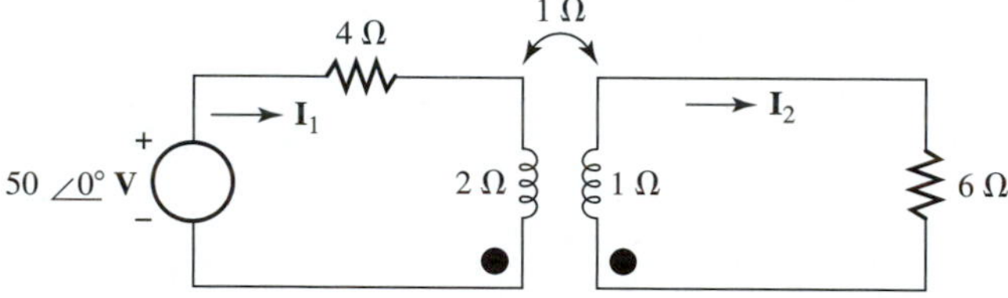

13. Determine the phasor ac currents $\mathbf{I}_1$ and $\mathbf{I}_2$ for the circuit in Figure 21.51.

Figure 21.51

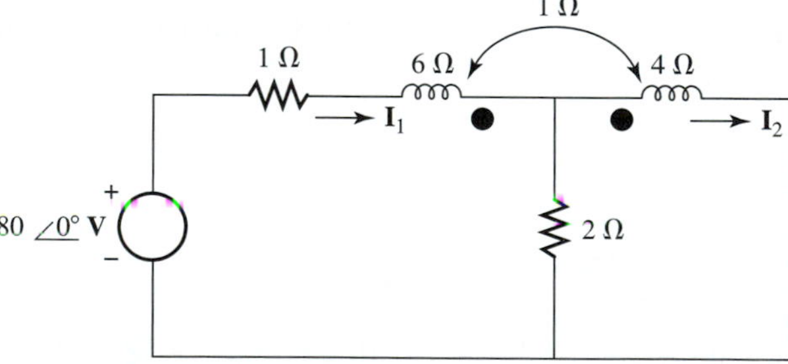

14. Determine the phasor ac currents $\mathbf{I}_1$ and $\mathbf{I}_2$ for the circuit in Figure 21.52.

Figure 21.52

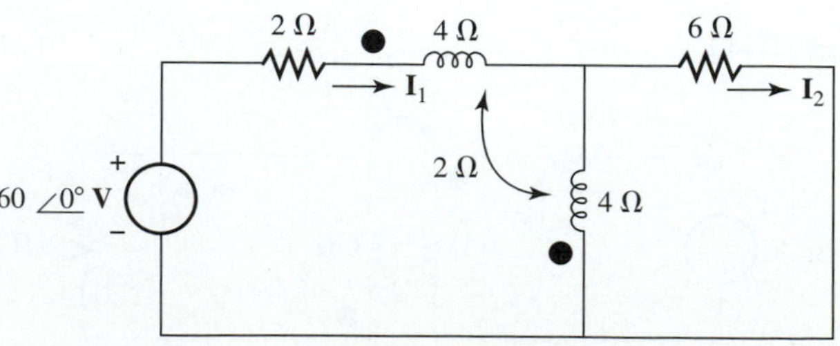

15. Determine the phasor ac currents $\mathbf{I}_1$ and $\mathbf{I}_2$ for the circuit in Figure 21.53.

Figure 21.53

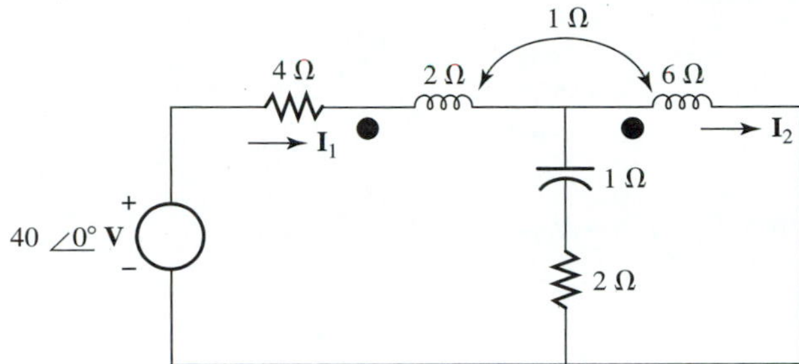

16. Determine the phasor ac currents $\mathbf{I}_1$ and $\mathbf{I}_2$ for the circuit in Figure 21.54.

Figure 21.54

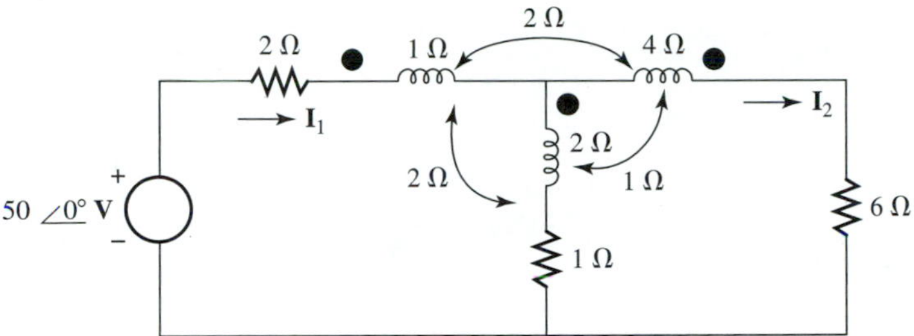

17. Determine the impedance $\mathbf{Z}_{ab}$ and the phasor current $\mathbf{I}$ for the circuit in Figure 21.55.

Figure 21.55

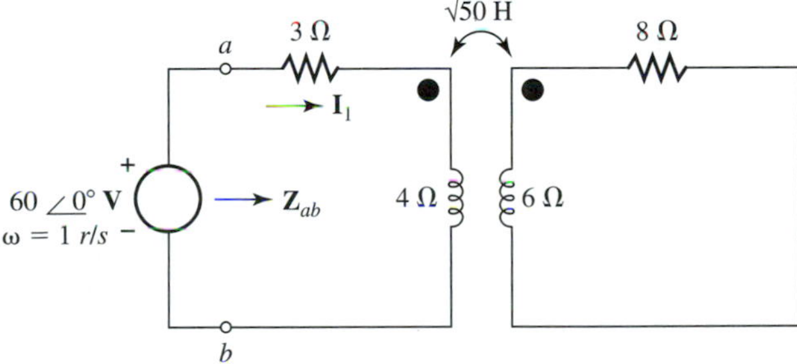

18. Determine the impedance $\mathbf{Z}_{ab}$ and the phasor current $\mathbf{I}$ for the circuit in Figure 21.56.

Figure 21.56

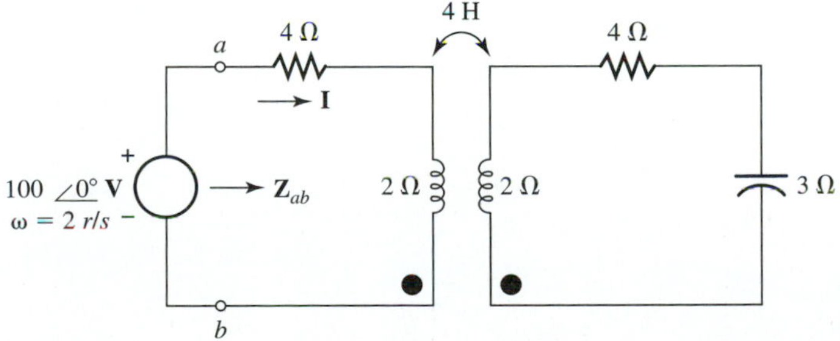

19. Determine the impedance $\mathbf{Z}_{ab}$ and the phasor current $\mathbf{I}$ for the circuit in Figure 21.57.

Figure 21.57

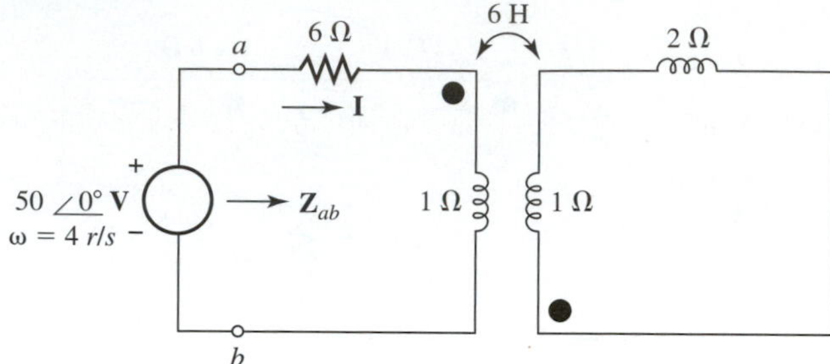

20. Determine the ac phasor currents $\mathbf{I}_1$ and $\mathbf{I}_2$ and the ac phasor voltages $\mathbf{V}_1$ and $\mathbf{V}_2$ for the circuit in Figure 21.58. Consider the transformers ideal and use a primary equivalent circuit.

Figure 21.58

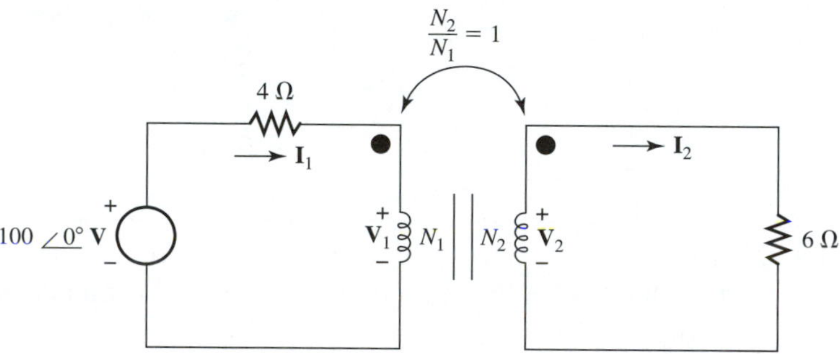

21. Determine the ac phasor currents $\mathbf{I}_1$ and $\mathbf{I}_2$ and the AC phasor voltages $\mathbf{V}_1$ and $\mathbf{V}_2$ for the circuit in Figure 21.59. Consider the transformers ideal and use a secondary equivalent circuit.

Figure 21.59

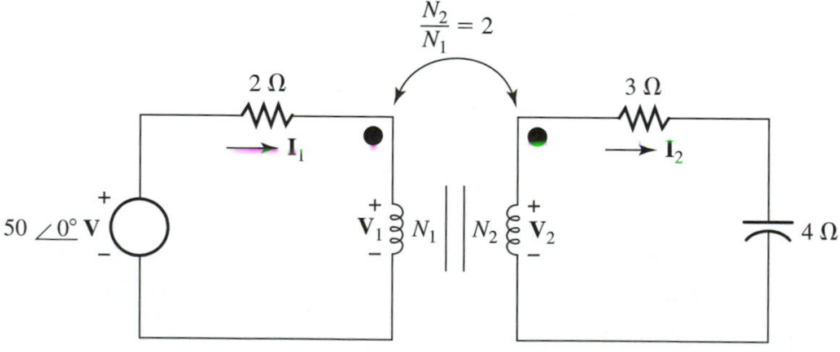

CHAPTER

22

Two-Port Networks

After completing this chapter the student should be able to

* determine h, y, and z parameters for a two-port network
* convert from one set of parameters to another
* determine input and output impedance or admittance for two-port networks
* determine current and voltage gain for two-port networks

We previously discussed equivalent circuits for two terminal networks. Thevenin and Norton equivalents are good examples of equivalent circuits for two terminal networks.

We shall now consider equivalent circuits for networks and devices (such as the transistor) which are characterized by four terminals and called two-port networks. Two-port networks consist of an input port (two terminals) and an output port (two terminals) as illustrated in Figure 22.1. Notice that two of the four terminals for a two-port network can be common (the transistor is an example) as illustrated in Figure 22.1(b)

Figure 22.1 Two-port network.

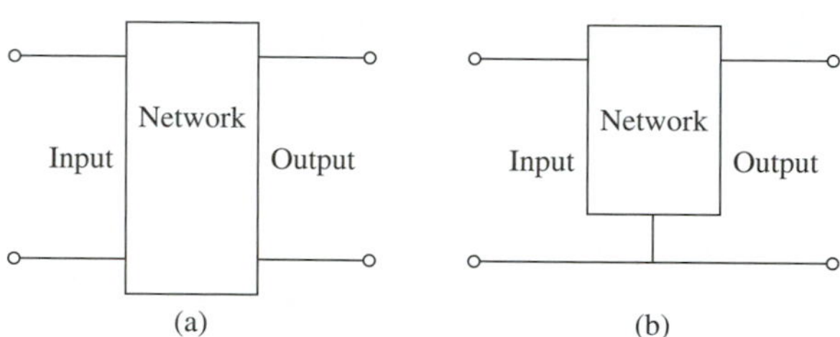

Mathematical models exist that allow two-port networks and devices to be replaced by simple equivalent two-part circuits (called parameter equivalents) consisting of basic circuit elements for analysis purposes. This enables us to analyze a two-port device such as the transistor using basic circuit analysis procedures. Since this is a basic circuits analysis textbook, the two-port networks given as examples in this chapter involve only passive elements. However, it should be pointed out that the parameter equivalent circuits are more significant in the analysis of electronic devices and other two-port active networks. In dealing with electronic devices (such as the transistor) the parameters are usually determined under laboratory conditions.

The mathematical models used to form equivalent two-port circuits relate the currents and voltages at the input and output ports. The elements included in equivalent two-port circuits and in the associated math models are called parameters. We shall consider the following types of equivalent two-port circuits: the impedance parameter equivalent (also called the z parameter or the open-circuit parameter equivalent), the admittance parameter equivalent (also called the y parameter or the short-circuit parameter equivalent) and the hybrid parameter equivalent (also called the h parameter equivalent).

22.1 IMPEDANCE (z) PARAMETERS

Consider the two-port network in Figure 22.2 where $\mathbf{V}_2$ is the output voltage and $\mathbf{V}_1$ is the input voltage. The input and output voltages are affected by $\mathbf{I}_1$

Figure 22.2 Two-port network.

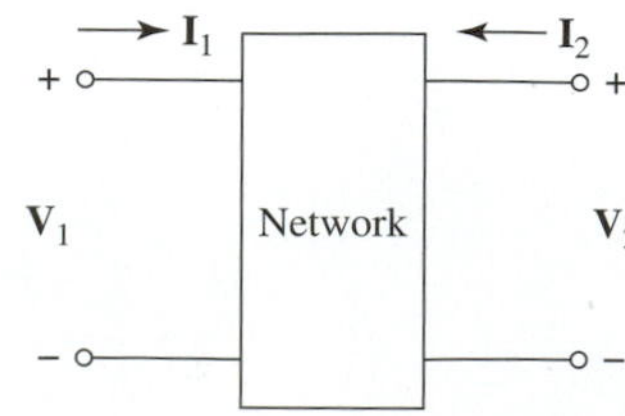

and $\mathbf{I}_2$. The principle of superposition allows us to write the equations for $\mathbf{V}_1$ and $\mathbf{V}_2$ as

$$\mathbf{V}_1 = \mathbf{z}_{11}\mathbf{I}_1 + \mathbf{z}_{12}\mathbf{I}_2 \tag{22.1}$$

$$\mathbf{V}_2 = \mathbf{z}_{21}\mathbf{I}_1 + \mathbf{z}_{22}\mathbf{I}_2 \tag{22.2}$$

where $\mathbf{z}_{11}\mathbf{I}_1$ and $\mathbf{z}_{12}\mathbf{I}_2$ are the contributions to $\mathbf{V}_1$ from $\mathbf{I}_1$ and $\mathbf{I}_2$, respectively, and $\mathbf{z}_{21}\mathbf{I}_1$ and $\mathbf{z}_{22}\mathbf{I}_2$ are the contributions to $\mathbf{V}_2$ from $\mathbf{I}_1$ and $\mathbf{I}_2$, respectively. $\mathbf{z}_{11}$, $\mathbf{z}_{12}$, $\mathbf{z}_{21}$, and $\mathbf{z}_{22}$ are measured in ohms and are the four $\mathbf{z}$ parameters that characterize the network in Figure 22.2. Although several circuits can be formed that are described by Equations (22.1) and (22.2), we shall restrict our discussion to the circuit in Figure 22.3.

Figure 22.3 *z* Equivalent.

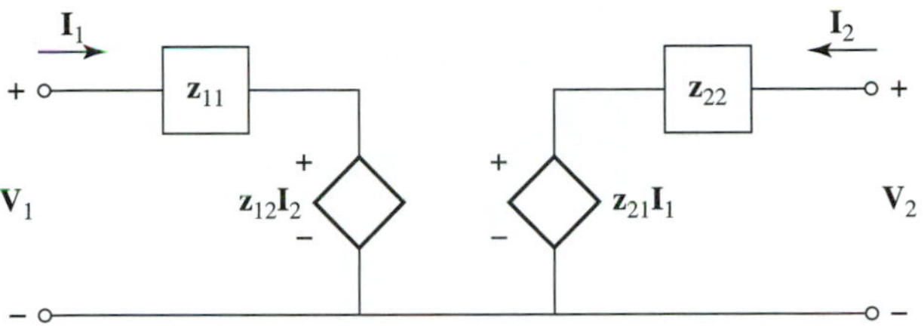

The circuit in Figure 22.3 is described by Equations (22.1) and (22.2), since the application of Kirchhoff's voltage law around the input and output loops results in Equations (22.1) and (22.2), respectively. Notice that the circuit contains two dependent voltage sources that are current controlled. The dependent voltage source in the input circuit is a reflection of the current $\mathbf{I}_2$, and the dependent voltage source in the output circuit is a reflection of the current $\mathbf{I}_1$. We shall now use Equations (22.1) and (22.2) to derive the *z* parameters using particular network conditions as follows.

If the output terminals of the network in Figure 22.3 are open ($\mathbf{I}_2 = 0$) and a voltage source $\mathbf{V}_1$ is connected across the input terminals, Equation (22.1) results in

$$\mathbf{z}_{11} = \left.\frac{\mathbf{V}_1}{\mathbf{I}_1}\right|_{\mathbf{I}_2=0} \tag{22.3}$$

and Equation (22.2) results in

$$\mathbf{z}_{21} = \left.\frac{\mathbf{V}_2}{\mathbf{I}_1}\right|_{\mathbf{I}_2=0} \tag{22.4}$$

If the input terminals of the network in Figure 22.3 are open ($\mathbf{I}_1 = 0$) and a voltage source $\mathbf{V}_2$ is connected across the output terminals, Equation (22.1) results in

$$\mathbf{z}_{12} = \left.\frac{\mathbf{V}_1}{\mathbf{I}_2}\right|_{\mathbf{I}_1=0} \tag{22.5}$$

and Equation (22.2) results in

$$\mathbf{z}_{22} = \left.\frac{\mathbf{V}_2}{\mathbf{I}_2}\right|_{\mathbf{I}_1=0} \tag{22.6}$$

Notice that these equations are arrived at for open-circuit conditions and the z parameters are therefore properly labeled the open-circuit parameters.

Consider the following example.

EXAMPLE 22.1 Replace the circuit in Figure 22.4 with a z parameter equivalent circuit. ■

Figure 22.4

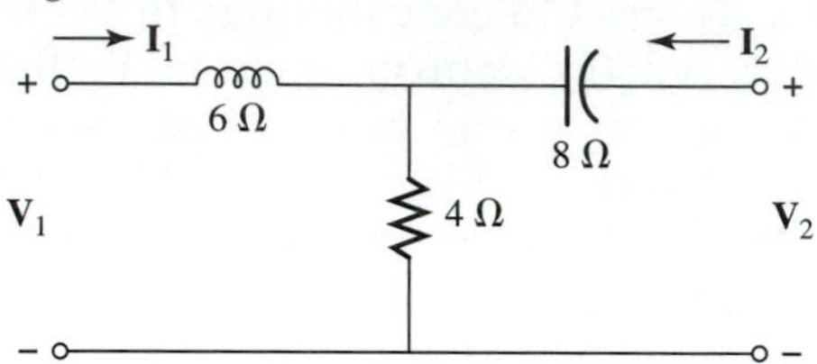

SOLUTION Using Equation (22.3) and the associated open-circuit condition, as shown in Figure 22.5, results in $\mathbf{z}_{11}$. Because $\mathbf{I}_2 = 0$,

$$\mathbf{I}_1 = \frac{\mathbf{V}_1}{4 + j6}$$

$$\frac{\mathbf{V}_1}{\mathbf{I}_1} = 4 + j6$$

Figure 22.5

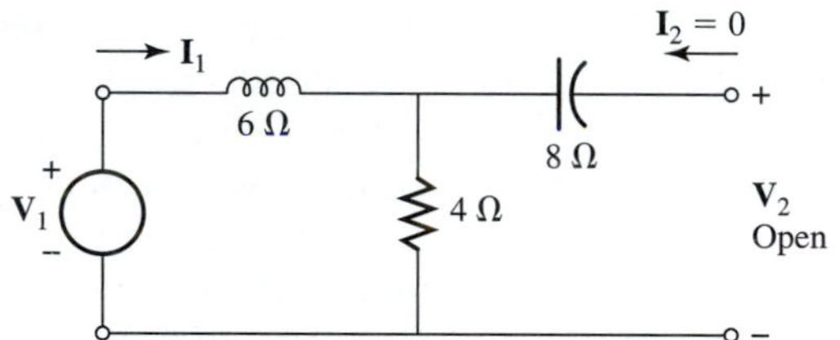

From Equation (22.3)

$$\mathbf{z}_{11} = 4 + j6$$

Using Equation (22.4) and the associated open-circuit condition, as shown in Figure 22.6, results in $\mathbf{z}_{21}$. Because $\mathbf{I}_2 = 0$,

$$\mathbf{V}_2 = 4\mathbf{I}_1$$

$$\frac{\mathbf{V}_2}{\mathbf{I}_1} = 4$$

Figure 22.6

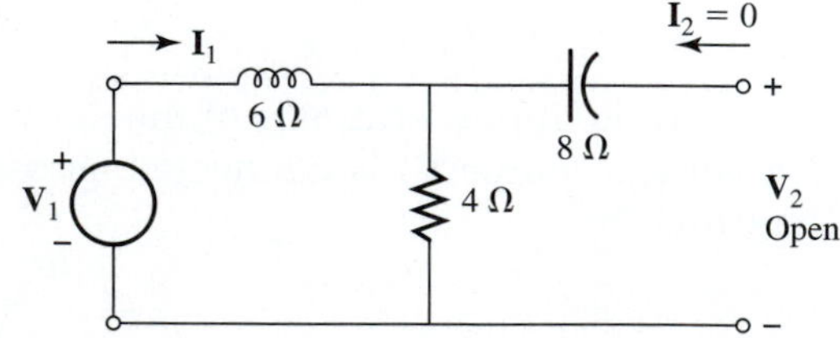

From Equation (22.4)

$$\mathbf{z}_{21} = 4$$

Using Equation (22.5) and the associated open circuit condition, as shown in Figure 22.7, results in $\mathbf{z}_{12}$. Because $\mathbf{I}_1 = 0$,

$$\mathbf{V}_1 = 4\mathbf{I}_2$$

$$\frac{\mathbf{V}_1}{\mathbf{I}_2} = 4$$

Figure 22.7

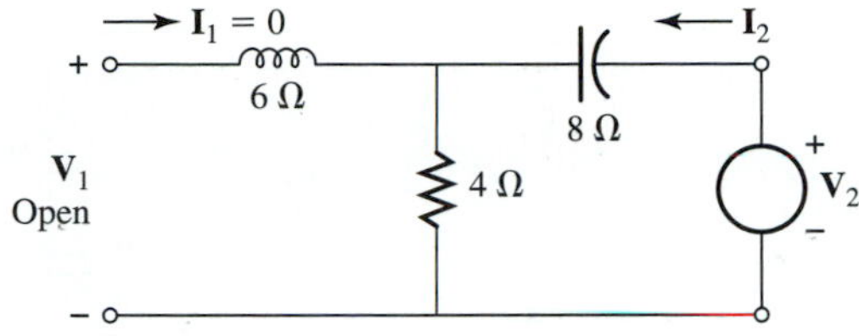

From Equation (22.5)

$$\mathbf{z}_{12} = 4$$

Using Equation (22.6) and the associated open circuit condition, as shown in Figure 22.8, results in $\mathbf{z}_{22}$. Because $\mathbf{I}_1 = 0$,

$$\mathbf{I}_2 = \frac{\mathbf{V}_2}{4 - j8}$$

$$\frac{\mathbf{V}_2}{\mathbf{I}_2} = 4 - j8$$

Figure 22.8

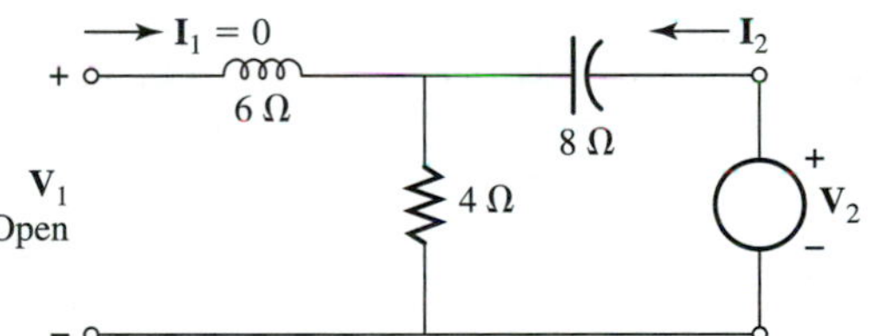

From Equation (22.6)

$$\mathbf{z}_{22} = 4 - j8$$

The z parameter equivalent is shown in Figure 22.9.

Figure 22.9

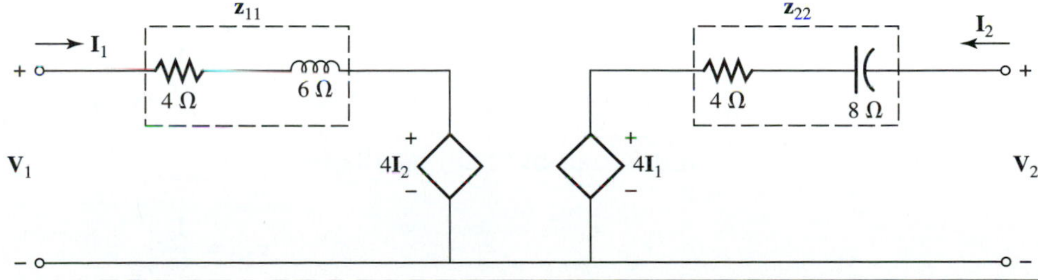

22.2 ADMITTANCE (*y*) PARAMETERS

Consider the two-port network in Figure 22.10, where $\mathbf{I}_2$ is the output current and $\mathbf{I}_1$ is the input current. The input and output currents are affected by $\mathbf{V}_1$ and $\mathbf{V}_2$.

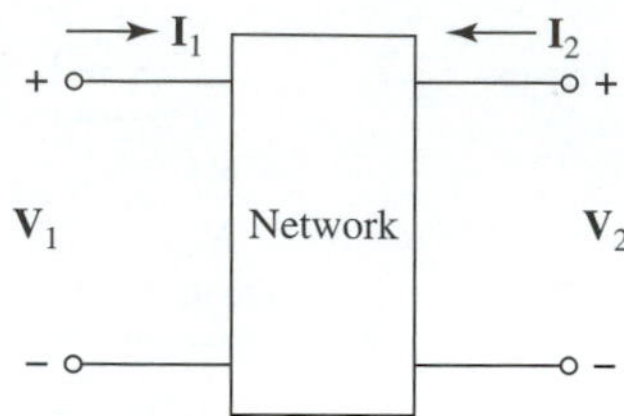

Figure 22.10 Two-port network.

The principle of superposition allows us to write the equations for $\mathbf{I}_1$ and $\mathbf{I}_2$ as

$$\mathbf{I}_1 = \mathbf{y}_{11}\mathbf{V}_1 + \mathbf{y}_{12}\mathbf{V}_2 \tag{22.7}$$

$$\mathbf{I}_2 = \mathbf{y}_{21}\mathbf{V}_1 + \mathbf{y}_{22}\mathbf{V}_2 \tag{22.8}$$

where $\mathbf{y}_{11}\mathbf{V}_1$ and $\mathbf{y}_{12}\mathbf{V}_2$ are the contributions to $\mathbf{I}_1$ from $\mathbf{V}_1$ and $\mathbf{V}_2$, respectively, and $\mathbf{y}_{21}\mathbf{V}_1$ and $\mathbf{y}_{22}\mathbf{V}_2$ are the contributions to $\mathbf{I}_2$ from $\mathbf{V}_1$ and $\mathbf{V}_2$, respectively. $\mathbf{y}_{11}$, $\mathbf{y}_{12}$, $\mathbf{y}_{21}$, and $\mathbf{y}_{22}$ are measured in siemens and are the four *y* parameters that characterize the network in Figure 22.10.

Although several circuits can be formed that are described by Equations (22.7) and (22.8), we shall restrict our discussion to the circuit in Figure 22.11.

Figure 22.11 *y* equivalent.

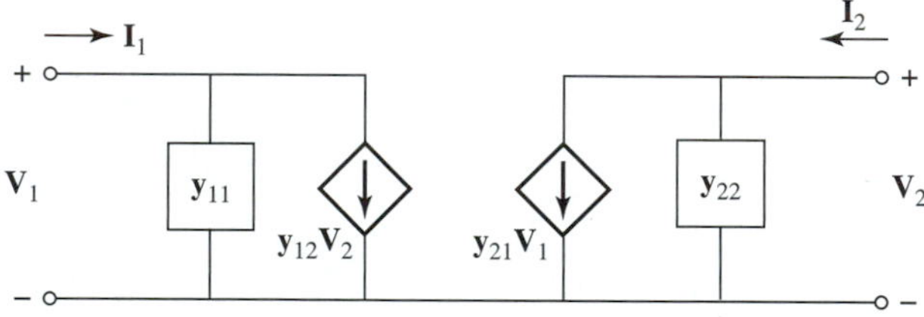

The circuit in Figure 22.11 is described by Equations (22.7) and (22.8), since the application of Kirchhoff's current law at the input and output nodes results in Equations (22.7) and (22.8), respectively. Notice that the circuit in Figure 22.11 contains two dependent current sources that are voltage controlled. The dependent current source in the input circuit is a reflection of the output voltage $\mathbf{V}_2$, and the dependent current source in the output circuit is a reflection of the input voltage $\mathbf{V}_1$.

We use Equations (22.7) and (22.8) to derive the *y* parameters using particular network conditions as follows.

If the output terminals of the network in Figure 22.11 are shorted ($\mathbf{V}_2 = 0$) and a voltage source $\mathbf{V}_1$ is connected across the input terminals, Equation (22.7) results in

$$\mathbf{y}_{11} = \left.\frac{\mathbf{I}_1}{\mathbf{V}_1}\right|_{\mathbf{V}_2=0} \tag{22.9}$$

and Equation (22.8) results in

$$\mathbf{y}_{21} = \left.\frac{\mathbf{I}_2}{\mathbf{V}_1}\right|_{\mathbf{V}_2=0} \tag{22.10}$$

If the input terminals of the network in Figure 22.11 are shorted ($\mathbf{V}_1 = 0$) and a voltage source $\mathbf{V}_2$ is connected across the output terminals, Equation (22.7) results in

$$\mathbf{y}_{12} = \left.\frac{\mathbf{I}_1}{\mathbf{V}_2}\right|_{\mathbf{V}_1=0} \tag{22.11}$$

and Equation (22.8) results in

$$\mathbf{y}_{22} = \left.\frac{\mathbf{I}_2}{\mathbf{V}_2}\right|_{\mathbf{V}_1=0} \tag{22.12}$$

Notice that these equations are arrived at for short-circuit conditions and the y parameters are therefore properly labeled the short-circuit parameters.

Consider the following example.

EXAMPLE 22.2 Replace the circuit in Figure 22.12 with a y parameter equivalent circuit. ■

Figure 22.12

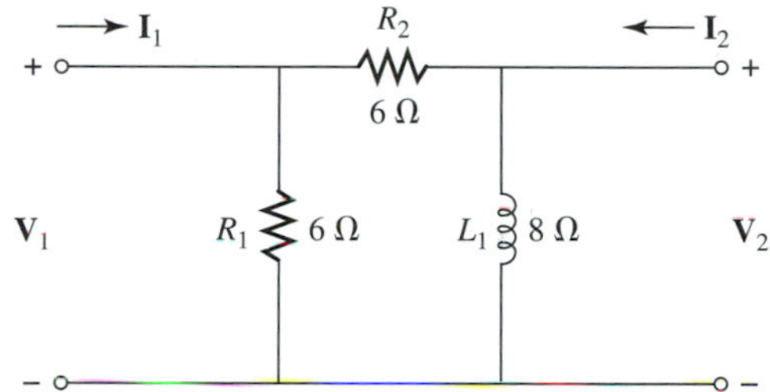

SOLUTION Using Equation (22.9) and the associated short-circuit condition, as shown in Figure 22.13, results in $\mathbf{y}_{11}$. Because L_1 is shorted, the admittance seen by $\mathbf{V}_1$ is

$$\left\{\left(\frac{1}{6}\right)\middle\|\left(\frac{1}{6}\right)\right\} = \frac{1}{3}$$

Figure 22.13

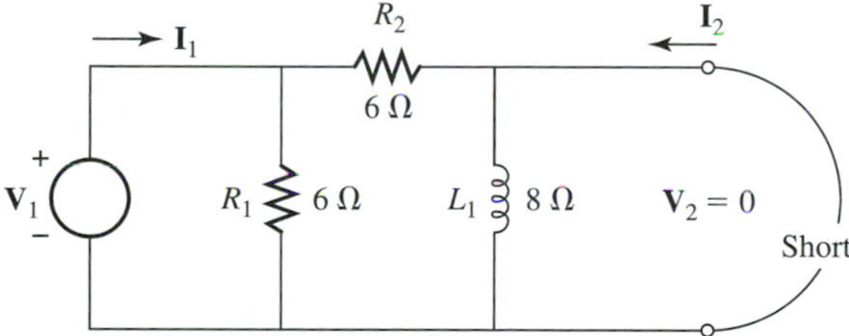

and, therefore,

$$\frac{\mathbf{I}_1}{\mathbf{V}_1} = \frac{1}{3}$$

From Equation (22.9)

$$\mathbf{y}_{11} = \frac{1}{3}\ \mathrm{S}$$

Using Equation (22.10) and the associated short-circuit condition, as shown in Figure 22.14, results in $\mathbf{y}_{21}$. Because L_1 is shorted,

$$\mathbf{I} = \frac{\mathbf{V}_1}{6}$$

Figure 22.14

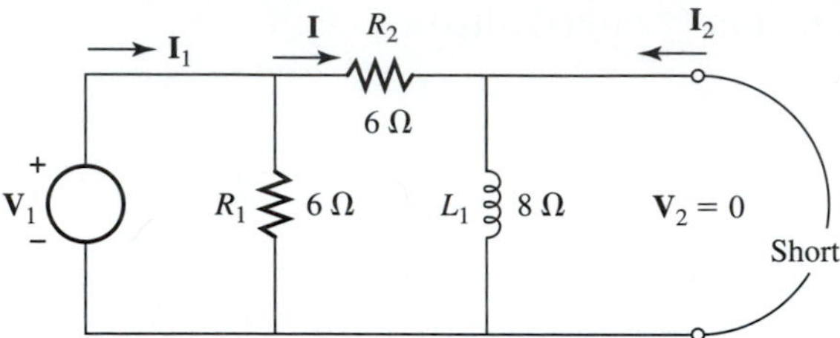

and

$$\mathbf{I}_2 = -\mathbf{I} = \frac{-\mathbf{V}_1}{6}$$

From Equation (22.10)

$$\mathbf{y}_{21} = \frac{\mathbf{I}_2}{\mathbf{V}_1} = -\frac{1}{6}\ \mathrm{S}$$

Using Equation (22.11) and the associated short-circuit condition, as shown in Figure 22.15, results in $\mathbf{y}_{12}$. Because R_1 is shorted,

$$\mathbf{I} = \frac{\mathbf{V}_2}{6}$$

Figure 22.15

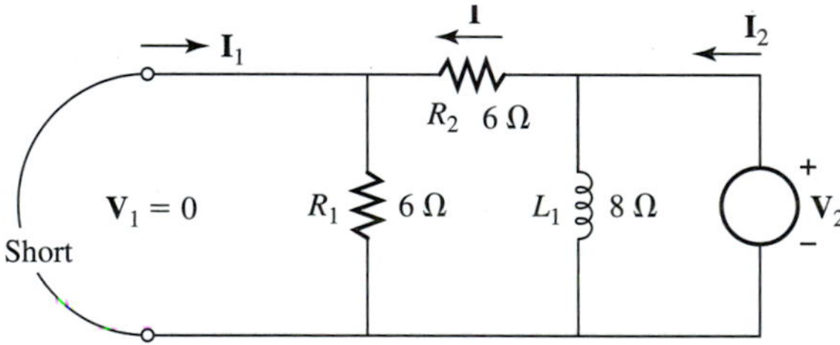

and

$$\mathbf{I}_1 = -\mathbf{I} = \frac{-\mathbf{V}_2}{6}$$

From Equation (22.11)

$$\mathbf{y}_{12} = \frac{\mathbf{I}_1}{\mathbf{V}_2} = -\frac{1}{6}\ \mathrm{S}$$

Using Equation (22.12) and the associated short-circuit condition, as shown in Figure 22.16, results in $\mathbf{y}_{22}$. Because R_1 is shorted, the admittance seen by $\mathbf{V}_2$ is

$$\left\{\left(\frac{1}{6}\right)\middle\|\left(-j\frac{1}{8}\right)\right\}$$

Figure 22.16

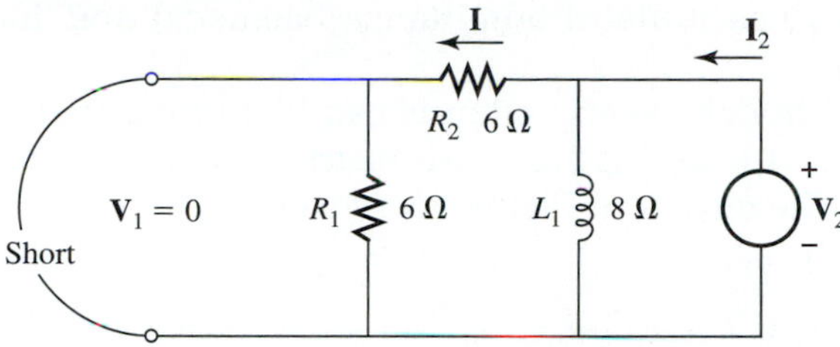

and therefore

$$\frac{\mathbf{I}_2}{\mathbf{V}_2} = \frac{1}{6} - j\left(\frac{1}{8}\right)$$

From Equation (22.12)

$$\mathbf{y}_{22} = \frac{1}{6} - j\left(\frac{1}{8}\right)$$

The y parameter equivalent of the circuit in Figure 22.12 is shown in Figure 22.17.

Figure 22.17

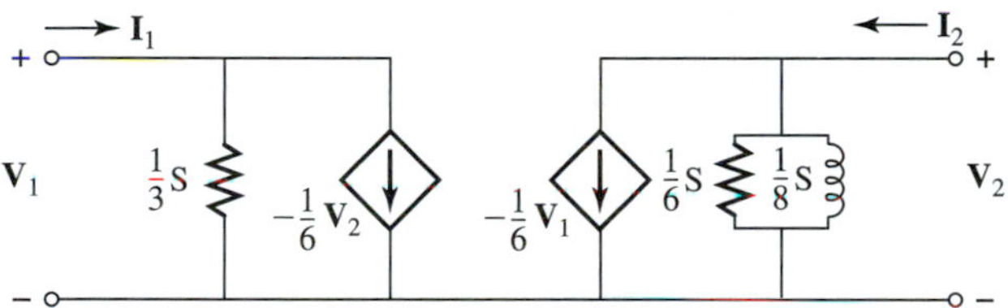

22.3 HYBRID (*h*) PARAMETERS

Consider the two-port network in Figure 22.18, where $\mathbf{V}_1$ is the input voltage and $\mathbf{I}_2$ is the output current. The input voltage and the output current are af-

Figure 22.18 Two-port network.

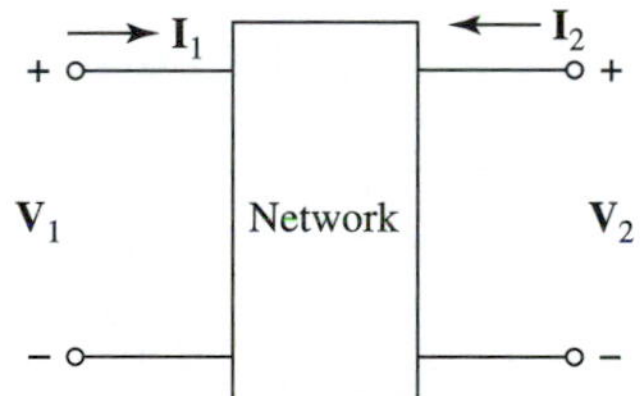

fected by $\mathbf{I}_1$ and $\mathbf{V}_2$. The principle of superposition allows us to write the equations for $\mathbf{V}_1$ and $\mathbf{I}_2$ as

$$\mathbf{V}_1 = \mathbf{h}_{11}\mathbf{I}_1 + \mathbf{h}_{12}\mathbf{V}_2 \tag{22.13}$$

$$\mathbf{I}_2 = \mathbf{h}_{21}\mathbf{I}_1 + \mathbf{h}_{22}\mathbf{V}_2 \tag{22.14}$$

where $\mathbf{h}_{11}\mathbf{I}_1$ and $\mathbf{h}_{12}\mathbf{V}_2$ are the contributions to $\mathbf{V}_1$ from $\mathbf{I}_1$ and $\mathbf{V}_2$, respectively, and $\mathbf{h}_{21}\mathbf{I}_1$ and $\mathbf{h}_{22}\mathbf{V}_2$ are the contributions to $\mathbf{I}_2$ from $\mathbf{I}_1$ and $\mathbf{V}_2$, repectively. $\mathbf{h}_{11}$, $\mathbf{h}_{12}$, $\mathbf{h}_{21}$, and $\mathbf{h}_{22}$ are the four h parameters that characterize the network in

Figure 22.18. Notice from the preceding equations that $\mathbf{h}_{12}$ and $\mathbf{h}_{21}$ are unitless, $\mathbf{h}_{22}$ has units of admittance (siemens) and $\mathbf{h}_{11}$ has units of impedance (ohms).

Although several circuits can be formed that are described by the Equations (22.13) and (22.14), we restrict our discussion to the circuit in Figure 22.19. The circuit in Figure 22.19 is an h parameter equivalent for the network in Figure 22.18.

Figure 22.19 h equivalent.

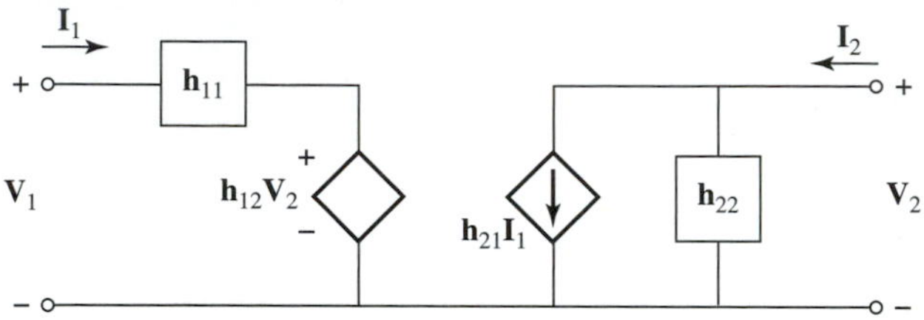

The circuit in Figure 22.19 is described by Equations (22.13) and (22.14), because the application of Kirchhoff's voltage law around the input loop and the application of Kirchhoff's current law at the output node results in Equations (22.13) and (22.14), respectively. Notice that the circuit contains a dependent voltage source that is voltage controlled and a dependent current source that is current controlled. The dependent voltage source in the input circuit is a reflection of the output voltage $\mathbf{V}_2$, and the dependent current source in the output circuit is a reflection of the input current $\mathbf{I}_1$.

We use Equations (22.13) and (22.14) to derive the h parameters using particular network conditions, as shown next.

If the output terminals of the network in Figure 22.19 are shorted ($\mathbf{V}_2 = 0$) and a voltage source $\mathbf{V}_1$ is connected across the input terminals, Equation (22.13) results in

$$\mathbf{h}_{11} = \left.\frac{\mathbf{V}_1}{\mathbf{I}_1}\right|_{\mathbf{V}_2=0} \tag{22.15}$$

and Equation (22.14) results in

$$\mathbf{h}_{21} = \left.\frac{\mathbf{I}_2}{\mathbf{I}_1}\right|_{\mathbf{V}_2=0} \tag{22.16}$$

If the input terminals of the network in Figure 22.19 are open ($\mathbf{I}_1 = 0$) and a voltage source $\mathbf{V}_2$ is connected across the output terminals, Equation (22.13) results in

$$\mathbf{h}_{12} = \left.\frac{\mathbf{V}_1}{\mathbf{V}_2}\right|_{\mathbf{I}_1=0} \tag{22.17}$$

and Equation (22.14) results in

$$\mathbf{h}_{22} = \left.\frac{\mathbf{I}_2}{\mathbf{V}_2}\right|_{\mathbf{I}_1=0} \tag{22.18}$$

Notice that the preceding equations are arrived at for open circuits and short-circuit conditions, so the h parameters are, therefore, properly labeled hybrid parameters.

Consider the following example.

EXAMPLE 22.3 Replace the circuit in Figure 22.20 with an *h* parameter equivalent circuit. ■

Figure 22.20

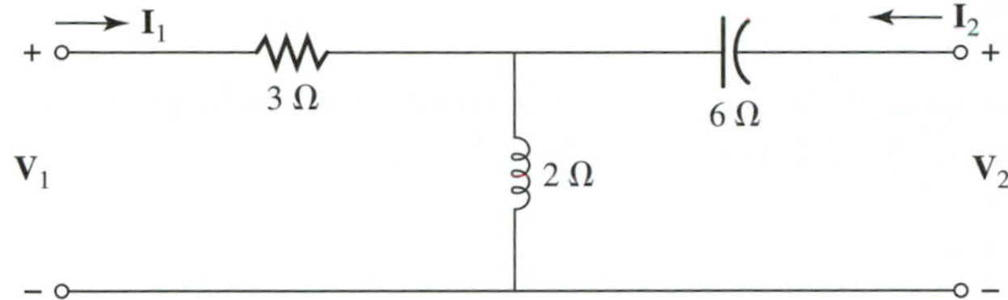

SOLUTION Using Equation (22.15) and the associated short-circuit condition, as shown in Figure 22.21, results in $\mathbf{h}_{11}$. Because the output terminals are shorted,

$$\frac{\mathbf{V}_1}{\mathbf{I}_1} = 3 + \{(j2)\|(-j6)\}$$

$$= 3 + \left\{\frac{j2(-j6)}{j2 - j6}\right\}$$

$$= 3 + j3$$

Figure 22.21

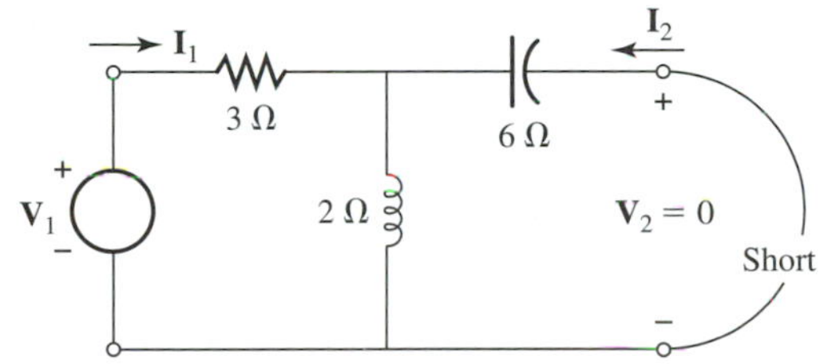

From Equation (22.15)

$$\mathbf{h}_{11} = 3 + j3$$

Using Equation (22.16) and the associated short-circuit condition, as shown in Figure 22.22, results in $\mathbf{h}_{21}$. Using the current-division formula,

$$\mathbf{I}_2 = -\left\{\frac{j2}{j2 - j6}\right\}\mathbf{I}_1$$

$$= \left(\frac{1}{2}\right)\mathbf{I}_1$$

$$\frac{\mathbf{I}_2}{\mathbf{I}_1} = \frac{1}{2}$$

Figure 22.22

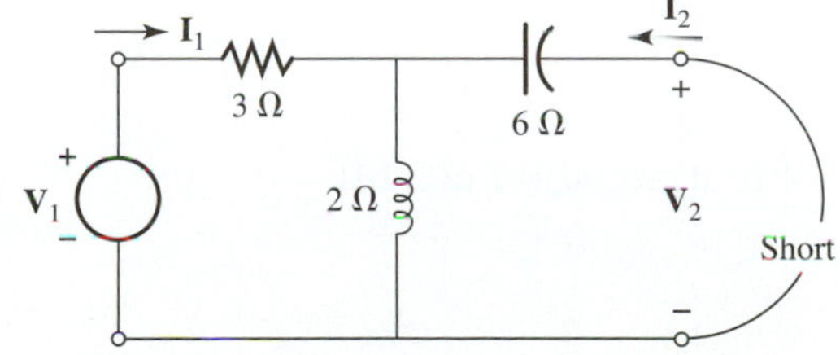

From Equation (22.16)

$$\mathbf{h}_{21} = \frac{1}{2}$$

Using Equation (22.17) and the associated open-circuit condition, as shown in Figure 22.23, results in $\mathbf{h}_{12}$. Because $\mathbf{I}_1 = 0$

$$\mathbf{V}_1 = (j2)\mathbf{I}_2$$

Figure 22.23

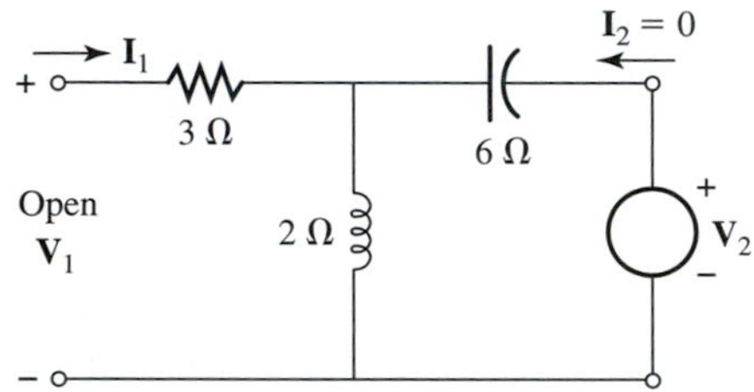

and

$$\mathbf{V}_2 = (j2 - j6)\mathbf{I}_2$$
$$= (-j4)\mathbf{I}_2$$

It follows that

$$\mathbf{V}_1 = (j2)\left\{\frac{\mathbf{V}_2}{-j4}\right\}$$

$$\frac{\mathbf{V}_1}{\mathbf{V}_2} = -\frac{1}{2}$$

From Equation (22.17)

$$\mathbf{h}_{12} = -\frac{1}{2}$$

Using Equation (22.18) and the associated open-circuit condition, as shown in Figure 22.24, results in $\mathbf{h}_{22}$. Because $\mathbf{I}_1 = 0$

$$\mathbf{I}_2 = \frac{\mathbf{V}_2}{j2 - j6}$$

$$\frac{\mathbf{I}_2}{\mathbf{V}_2} = j\left(\frac{1}{4}\right)$$

From Equation (22.18)

$$\mathbf{h}_{22} = j\left(\frac{1}{4}\right)$$

Figure 22.24

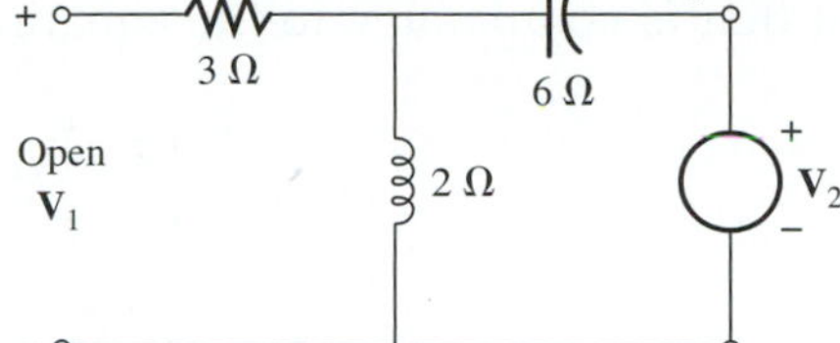

The h parameter equivalent for the circuit in Figure 22.20 is shown in Figure 22.25.

Figure 22.25

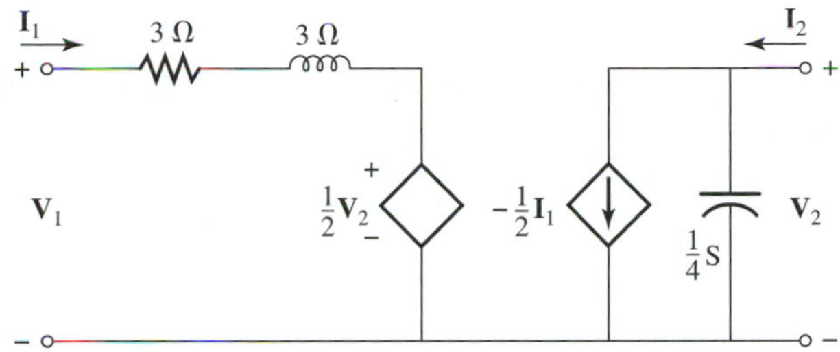

22.4 CONVERSION OF *z* TO *y* AND *z* TO *h* PARAMETERS

Formulas for the conversion of one set of parameters to another can be arrived at by rearranging the equations associated with the parameters to be converted into a form comparable to the equations associated with the desired parameters. A comparison of the rearranged equations and the equations associated with the desired parameters results in a relationship between parameters, as illustrated here.

Consider Equations (22.1) and (22.2) associated with the z parameters:

$$\mathbf{z}_{11}\mathbf{I}_1 + \mathbf{z}_{12}\mathbf{I}_2 = \mathbf{V}_1$$

$$\mathbf{z}_{21}\mathbf{I}_1 + \mathbf{z}_{22}\mathbf{I}_2 = \mathbf{V}_2$$

Using Cramer's rule $\mathbf{I}_1$ and $\mathbf{I}_2$ can be expressed as

$$\mathbf{I}_1 = \frac{\begin{vmatrix} \mathbf{V}_1 & \mathbf{z}_{12} \\ \mathbf{V}_2 & \mathbf{z}_{22} \end{vmatrix}}{\begin{vmatrix} \mathbf{z}_{11} & \mathbf{z}_{12} \\ \mathbf{z}_{21} & \mathbf{z}_{22} \end{vmatrix}} = \frac{\mathbf{z}_{22}\mathbf{V}_1 - \mathbf{z}_{12}\mathbf{V}_2}{\mathbf{z}_{11}\mathbf{z}_{22} - \mathbf{z}_{12}\mathbf{z}_{21}}$$

and

$$\mathbf{I}_2 = \frac{\begin{vmatrix} \mathbf{z}_{11} & \mathbf{V}_1 \\ \mathbf{z}_{21} & \mathbf{V}_2 \end{vmatrix}}{\begin{vmatrix} \mathbf{z}_{11} & \mathbf{z}_{12} \\ \mathbf{z}_{21} & \mathbf{z}_{22} \end{vmatrix}} = \frac{\mathbf{z}_{11}\mathbf{V}_2 - \mathbf{z}_{21}\mathbf{V}_1}{\mathbf{z}_{11}\mathbf{z}_{22} - \mathbf{z}_{12}\mathbf{z}_{21}}$$

Letting $\Delta_z = \mathbf{z}_{11}\mathbf{z}_{22} - \mathbf{z}_{12}\mathbf{z}_{21}$, the preceding equations can be written in a form that is comparable to the equations associated with the y parameters:

$$\mathbf{I}_1 = \left\{\frac{\mathbf{z}_{22}}{\Delta_z}\right\}\mathbf{V}_1 + \left\{\frac{-\mathbf{z}_{12}}{\Delta_z}\right\}\mathbf{V}_2$$

$$\mathbf{I}_2 = \left\{\frac{-\mathbf{z}_{21}}{\Delta_z}\right\}\mathbf{V}_1 + \left\{\frac{\mathbf{z}_{11}}{\Delta_z}\right\}\mathbf{V}_2$$

Comparing the y parameter Equations for $\mathbf{I}_1$ and $\mathbf{I}_2$ (Equations (22.7) and (22.8)), as rewritten here,

$$\mathbf{I}_1 = \mathbf{y}_{11}\mathbf{V}_1 + \mathbf{y}_{12}\mathbf{V}_2$$

$$\mathbf{I}_2 = \mathbf{y}_{21}\mathbf{V}_1 + \mathbf{y}_{22}\mathbf{V}_2$$

with the preceding equations for $\mathbf{I}_1$ and $\mathbf{I}_2$ in terms of the z parameters results in

$$\mathbf{y}_{11} = \frac{\mathbf{z}_{22}}{\Delta_z} \qquad (22.19)$$

$$\mathbf{y}_{12} = \frac{-\mathbf{z}_{12}}{\Delta_z} \qquad (22.20)$$

$$\mathbf{y}_{21} = \frac{-\mathbf{z}_{21}}{\Delta_z} \qquad (22.21)$$

$$\mathbf{y}_{22} = \frac{\mathbf{z}_{11}}{\Delta_z} \qquad (22.22)$$

which are the formulas to convert from the z parameters to the y parameters.

The equation

$$\mathbf{I}_1 = \left\{\frac{\mathbf{z}_{22}}{\Delta_z}\right\}\mathbf{V}_1 + \left\{\frac{-\mathbf{z}_{12}}{\Delta_z}\right\}\mathbf{V}_2$$

can be written as

$$\mathbf{V}_1 = \left\{\frac{\Delta_z}{\mathbf{z}_{22}}\right\}\mathbf{I}_1 + \left\{\frac{\mathbf{z}_{12}}{\mathbf{z}_{22}}\right\}\mathbf{V}_2$$

and Equation (22.2) for the z parameter can be arranged as follows:

$$\mathbf{z}_{22}\mathbf{I}_2 = -\mathbf{z}_{21}\mathbf{I}_1 + \mathbf{V}_2$$

$$\mathbf{I}_2 = \left\{\frac{-\mathbf{z}_{21}}{\mathbf{z}_{22}}\right\}\mathbf{I}_1 + \left\{\frac{1}{\mathbf{z}_{22}}\right\}\mathbf{V}_2$$

Comparing the h parameter equations for $\mathbf{V}_1$ and $\mathbf{I}_2$ (Equations (22.13) and (22.14)) as rewritten here,

$$\mathbf{V}_1 = \mathbf{h}_{11}\mathbf{I}_1 + \mathbf{h}_{12}\mathbf{V}_2$$

$$\mathbf{I}_2 = \mathbf{h}_{21}\mathbf{I}_1 + \mathbf{h}_{22}\mathbf{V}_2$$

with the preceding equations for $\mathbf{V}_1$ and $\mathbf{I}_2$ in terms of the z parameters results in

$$\mathbf{h}_{11} = \frac{\Delta_z}{\mathbf{z}_{22}} \tag{22.23}$$

$$\mathbf{h}_{12} = \frac{\mathbf{z}_{12}}{\mathbf{z}_{22}} \tag{22.24}$$

$$\mathbf{h}_{21} = \frac{-\mathbf{z}_{21}}{\mathbf{z}_{22}} \tag{22.25}$$

$$\mathbf{h}_{22} = \frac{1}{\mathbf{z}_{22}} \tag{22.26}$$

which are the formulas to convert from the z parameters to the h parameters.

22.5 CONVERSION OF *y* TO *z* AND *y* TO *h* PARAMETERS

Consider Equations (22.7) and (22.8) associated with the y parameters:

$$\mathbf{y}_{11}\mathbf{V}_1 + \mathbf{y}_{12}\mathbf{V}_2 = \mathbf{I}_1$$
$$\mathbf{y}_{21}\mathbf{V}_1 + \mathbf{y}_{22}\mathbf{V}_2 = \mathbf{I}_2$$

Using Cramer's rule $\mathbf{V}_1$ and V_2 can be expressed as follows:

$$\mathbf{V}_1 = \frac{\begin{vmatrix} \mathbf{I}_1 & \mathbf{y}_{12} \\ \mathbf{I}_2 & \mathbf{y}_{22} \end{vmatrix}}{\begin{vmatrix} \mathbf{y}_{11} & \mathbf{y}_{12} \\ \mathbf{y}_{21} & \mathbf{y}_{22} \end{vmatrix}} = \frac{\mathbf{y}_{22}\mathbf{I}_1 - \mathbf{y}_{12}\mathbf{I}_2}{\mathbf{y}_{11}\mathbf{y}_{22} - \mathbf{y}_{12}\mathbf{y}_{21}}$$

$$\mathbf{V}_2 = \frac{\begin{vmatrix} \mathbf{y}_{11} & \mathbf{I}_1 \\ \mathbf{y}_{21} & \mathbf{I}_2 \end{vmatrix}}{\begin{vmatrix} \mathbf{y}_{11} & \mathbf{y}_{12} \\ \mathbf{y}_{21} & \mathbf{y}_{22} \end{vmatrix}} = \frac{\mathbf{y}_{11}\mathbf{I}_2 - \mathbf{y}_{21}\mathbf{I}_1}{\mathbf{y}_{11}\mathbf{y}_{22} - \mathbf{y}_{12}\mathbf{y}_{21}}$$

Letting $\Delta_y = \mathbf{y}_{11}y_{22} - y_{12}y_{21}$, these equations can be written in a form that is comparable to the equations associated with the z parameters:

$$\mathbf{V}_1 = \left\{\frac{\mathbf{y}_{22}}{\Delta_y}\right\}\mathbf{I}_1 + \left\{\frac{-\mathbf{y}_{12}}{\Delta_y}\right\}\mathbf{I}_2$$

$$\mathbf{V}_2 = \left\{\frac{-\mathbf{y}_{21}}{\Delta_y}\right\}\mathbf{I}_1 + \left\{\frac{\mathbf{y}_{11}}{\Delta_y}\right\}\mathbf{I}_2$$

Comparing the z parameter Equations for $\mathbf{V}_1$ and $\mathbf{V}_2$ (Equations (22.1) and (22.2)) as rewritten here,

$$\mathbf{V}_1 = \mathbf{z}_{11}\mathbf{I}_1 + \mathbf{z}_{12}\mathbf{I}_2$$
$$\mathbf{V}_2 = \mathbf{z}_{21}\mathbf{I}_1 + \mathbf{z}_{22}\mathbf{I}_2$$

with the preceding equations for $\mathbf{V}_1$ and $\mathbf{V}_2$ in terms of the y parameters results in

$$\mathbf{z}_{11} = \frac{\mathbf{y}_{22}}{\Delta_y} \tag{22.27}$$

$$\mathbf{z}_{12} = \frac{-\mathbf{y}_{12}}{\Delta_y} \tag{22.28}$$

$$\mathbf{z}_{21} = \frac{-\mathbf{y}_{21}}{\Delta_y} \tag{22.29}$$

$$\mathbf{z}_{22} = \frac{\mathbf{y}_{11}}{\Delta_y} \tag{22.30}$$

which are the formulas to convert from the y parameters to the z parameters.

The equation

$$\mathbf{V}_2 = \left\{\frac{-\mathbf{y}_{21}}{\Delta_y}\right\}\mathbf{I}_1 + \left\{\frac{\mathbf{y}_{11}}{\Delta_y}\right\}\mathbf{I}_2$$

can be written as

$$\mathbf{I}_2 = \left\{\frac{\mathbf{y}_{21}}{\mathbf{y}_{11}}\right\}\mathbf{I}_1 + \left\{\frac{\Delta_y}{\mathbf{y}_{11}}\right\}\mathbf{V}_2$$

and Equation (22.7) for the y parameter equivalent can be written as

$$\mathbf{y}_{11}\mathbf{V}_1 + \mathbf{y}_{12}\mathbf{V}_2 = \mathbf{I}_1$$

$$\mathbf{y}_{11}\mathbf{V}_1 = \mathbf{I}_1 - \mathbf{y}_{12}\mathbf{V}_2$$

$$\mathbf{V}_1 = \left\{\frac{1}{\mathbf{y}_{11}}\right\}\mathbf{I}_1 + \left\{\frac{-\mathbf{y}_{12}}{\mathbf{y}_{11}}\right\}\mathbf{V}_2$$

Comparing the h parameter equations for $\mathbf{I}_2$ and $\mathbf{V}_1$ (Equations (22.13) and (22.14)) as rewritten here,

$$\mathbf{V}_1 = \mathbf{h}_{11}\mathbf{I}_1 + \mathbf{h}_{12}\mathbf{V}_2$$

$$\mathbf{I}_2 = \mathbf{h}_{21}\mathbf{I}_1 + \mathbf{h}_{22}\mathbf{V}_2$$

with the preceding equations for $\mathbf{I}_2$ and $\mathbf{V}_1$ in terms of the y parameters results in

$$\mathbf{h}_{11} = \frac{1}{\mathbf{y}_{11}} \tag{22.31}$$

$$\mathbf{h}_{12} = \frac{-\mathbf{y}_{12}}{\mathbf{y}_{11}} \tag{22.32}$$

$$\mathbf{h}_{21} = \frac{\mathbf{y}_{21}}{\mathbf{y}_{11}} \tag{22.33}$$

$$\mathbf{h}_{22} = \frac{\Delta_y}{\mathbf{y}_{11}} \tag{22.34}$$

which are the formulas to convert from the y parameters to the h parameters.

22.6 CONVERSION OF *h* TO *z* AND *h* TO *y* PARAMETERS

Consider Equations (22.13) and (22.14) associated with the *h* parameters:

$$\mathbf{h}_{11}\mathbf{I}_1 + \mathbf{h}_{12}\mathbf{V}_2 = \mathbf{V}_1$$
$$\mathbf{h}_{21}\mathbf{I}_1 + \mathbf{h}_{22}\mathbf{V}_2 = \mathbf{I}_2$$

Using Cramer's rule $\mathbf{I}_1$ can be written as

$$\mathbf{I}_1 = \frac{\begin{vmatrix} \mathbf{V}_1 & \mathbf{h}_{12} \\ \mathbf{I}_2 & \mathbf{h}_{22} \end{vmatrix}}{\begin{vmatrix} \mathbf{h}_{11} & \mathbf{h}_{12} \\ \mathbf{h}_{21} & \mathbf{h}_{22} \end{vmatrix}} = \frac{\mathbf{h}_{22}\mathbf{V}_1 - \mathbf{h}_{12}\mathbf{I}_2}{\mathbf{h}_{11}\mathbf{h}_{22} - \mathbf{h}_{12}\mathbf{h}_{21}}$$

Letting $\Delta_h = \mathbf{h}_{11}\mathbf{h}_{22} - \mathbf{h}_{12}\mathbf{h}_{21}$ results in

$$\mathbf{I}_1 = \frac{\mathbf{h}_{22}}{\Delta_h}\mathbf{V}_1 - \frac{\mathbf{h}_{12}}{\Delta_h}\mathbf{I}_2$$
$$\mathbf{V}_1 = \frac{\Delta_h}{\mathbf{h}_{22}}\mathbf{I}_1 + \frac{\mathbf{h}_{12}}{\mathbf{h}_{22}}\mathbf{I}_2$$

From Equation (22.14)

$$\mathbf{I}_2 = \mathbf{h}_{21}\mathbf{I}_1 + \mathbf{h}_{22}\mathbf{V}_2$$

and $\mathbf{V}_2$ can be written as

$$\mathbf{V}_2 = \left\{\frac{-\mathbf{h}_{21}}{\mathbf{h}_{22}}\right\}\mathbf{I}_1 + \left\{\frac{1}{\mathbf{h}_{22}}\right\}\mathbf{I}_2$$

Comparing the *z* parameter equations for $\mathbf{V}_1$ and $\mathbf{V}_2$ (Equations (22.1) and (22.2)) as rewritten here,

$$\mathbf{V}_1 = \mathbf{z}_{11}\mathbf{V}_1 + \mathbf{z}_{12}\mathbf{V}_2$$
$$\mathbf{V}_2 = \mathbf{z}_{21}\mathbf{V}_1 + \mathbf{z}_{22}\mathbf{V}_2$$

with the preceding equations for $\mathbf{V}_1$ and $\mathbf{V}_2$ in terms of the *h* parameters results in

$$\mathbf{z}_{11} = \frac{\Delta_h}{\mathbf{h}_{22}} \tag{22.35}$$

$$\mathbf{z}_{12} = \frac{\mathbf{h}_{12}}{\mathbf{h}_{22}} \tag{22.36}$$

$$\mathbf{z}_{21} = \frac{-\mathbf{h}_{21}}{\mathbf{h}_{22}} \tag{22.37}$$

$$\mathbf{z}_{22} = \frac{1}{\mathbf{h}_{22}} \tag{22.38}$$

which are the formulas to convert from the *h* parameters to the *z* parameters.

Using Cramer's rule to solve Equations (22.13) and (22.14) associated with the h parameters for $\mathbf{V}_2$ results in

$$\mathbf{V}_2 = \frac{\begin{vmatrix} \mathbf{h}_{11} & \mathbf{V}_1 \\ \mathbf{h}_{21} & \mathbf{I}_2 \end{vmatrix}}{\begin{vmatrix} \mathbf{h}_{11} & \mathbf{h}_{12} \\ \mathbf{h}_{21} & \mathbf{h}_{22} \end{vmatrix}} = \frac{\mathbf{h}_{11}\mathbf{I}_2 - \mathbf{h}_{21}\mathbf{V}_1}{\mathbf{h}_{11}\mathbf{h}_{22} - \mathbf{h}_{12}\mathbf{h}_{21}}$$

$$\mathbf{V}_2 = \left\{\frac{\mathbf{h}_{11}}{\Delta_h}\right\}\mathbf{I}_2 - \left\{\frac{\mathbf{h}_{21}}{\Delta_h}\right\}\mathbf{V}_1$$

$$\mathbf{I}_2 = \left\{\frac{\mathbf{h}_{21}}{\mathbf{h}_{11}}\right\}\mathbf{V}_1 + \left\{\frac{\Delta_h}{\mathbf{h}_{11}}\right\}\mathbf{V}_2$$

and from Equation (22.13)

$$\mathbf{h}_{11}\mathbf{I}_1 + \mathbf{h}_{12}\mathbf{V}_2 = \mathbf{V}_1$$

$$\mathbf{I}_1 = \left\{\frac{1}{\mathbf{h}_{11}}\right\}\mathbf{V}_1 - \left\{\frac{\mathbf{h}_{12}}{\mathbf{h}_{11}}\right\}\mathbf{V}_2$$

Comparing the y parameter equations for $\mathbf{I}_1$ and $\mathbf{I}_2$ (Equations (22.7) and (22.8)) as rewritten here,

$$\mathbf{I}_1 = \mathbf{y}_{11}\mathbf{V}_1 + \mathbf{y}_{12}\mathbf{V}_2$$

$$\mathbf{I}_2 = \mathbf{y}_{21}\mathbf{V}_1 + \mathbf{y}_{22}\mathbf{V}_2$$

with the preceding equations for $\mathbf{I}_1$ and $\mathbf{I}_2$ in terms of the h parameters results in

$$\mathbf{y}_{11} = \frac{1}{\mathbf{h}_{11}} \tag{22.39}$$

$$\mathbf{y}_{12} = \frac{-\mathbf{h}_{12}}{\mathbf{h}_{11}} \tag{22.40}$$

$$\mathbf{y}_{21} = \frac{\mathbf{h}_{21}}{\mathbf{h}_{11}} \tag{22.41}$$

$$\mathbf{y}_{22} = \frac{\Delta h}{\mathbf{h}_{11}} \tag{22.42}$$

which are the formulas to convert from the h parameters to the y parameters.

Consider the following example.

EXAMPLE 22.4 The h parameters for a particular network are $\mathbf{h}_{11} = 1000\ \Omega$, $\mathbf{h}_{12} = 4 \times 10^{-4}$, $\mathbf{h}_{21} = 50$, and $\mathbf{h}_{22} = 25 \times 10^{-6}$ S. Determine h, y, and z parameter equivalents for the network. ■

SOLUTION The quantity Δ_h is determined as follows:

$$\begin{aligned}\Delta_h &= \mathbf{h}_{11}\mathbf{h}_{22} - \mathbf{h}_{12}\mathbf{h}_{21} \\ &= (10^3)(25)(10^{-6}) - (4)(10^{-4})(50) \\ &= 5(10^{-3})\end{aligned}$$

Using Equations (22.35) through (22.38), the z parameters are

$$\mathbf{z}_{11} = \frac{\Delta_h}{\mathbf{h}_{22}} = \frac{5(10^{-3})}{25(10^{-6})}$$
$$= 200$$
$$\mathbf{z}_{12} = \frac{\mathbf{h}_{12}}{\mathbf{h}_{22}} = \frac{4(10^{-4})}{25}$$
$$= 16$$
$$\mathbf{z}_{21} = \frac{-\mathbf{h}_{21}}{\mathbf{h}_{22}} = \frac{-50}{25(10^{-6})}$$
$$= -(2)(10^{6})$$
$$\mathbf{z}_{22} = \frac{1}{\mathbf{h}_{22}} = \frac{1}{25(10^{-6})}$$
$$= (4)(10^{4})$$

Using Equations (22.39) through (22.42), the y parameters are

$$\mathbf{y}_{11} = \frac{1}{\mathbf{h}_{11}} = \frac{1}{(10)^{3}}$$
$$= 10^{-3}$$
$$\mathbf{y}_{12} = \frac{-\mathbf{h}_{12}}{\mathbf{h}_{11}} = \frac{-(4)(10^{-4})}{(10^{3})}$$
$$= -(4)(10^{-7})$$
$$\mathbf{y}_{21} = \frac{\mathbf{h}_{21}}{\mathbf{h}_{11}} = \frac{50}{(10^{3})}$$
$$= (5)(10^{-2})$$
$$\mathbf{y}_{22} = \frac{\Delta_h}{\mathbf{h}_{11}} = \frac{(5)(10^{-3})}{10^{3}}$$
$$= (5)(10^{-6})$$

Referring to Figures 22.3, 22.11, and 22.19, the z, y, and h equivalent circuits for the given network are shown in Figures 22.26, 22.27, and 22.28, respectively.

Figure 22.26

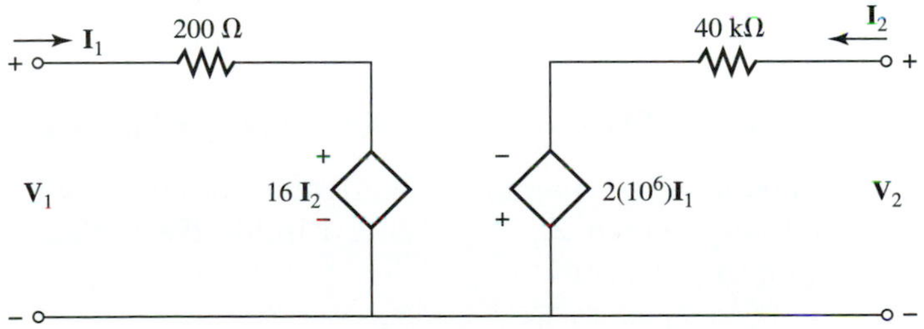

z parameter equivalent

Figure 22.27

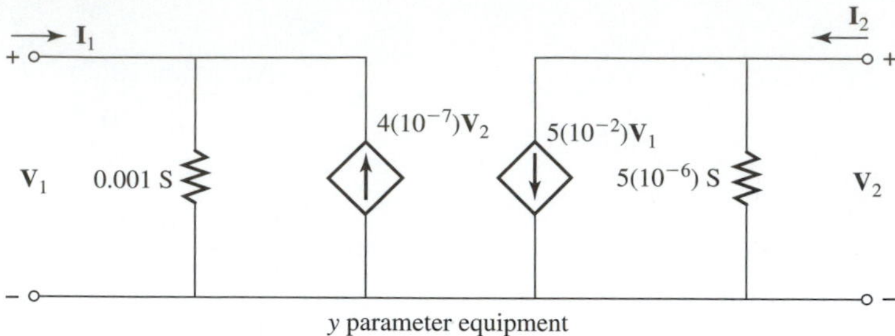

Figure 22.28

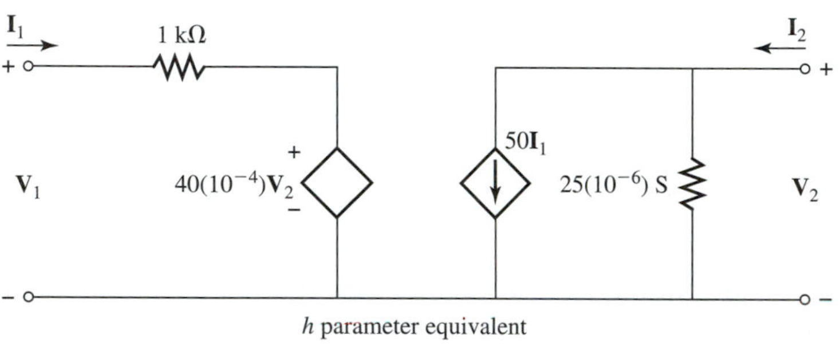

22.7 INPUT AND OUTPUT IMPEDANCE

When using the z, y, and h parameter equivalent circuits to analyze networks it is often convenient to know the input and output impedance for the equivalent circuits.

We consider the input impedance $\mathbf{Z}_{\text{in}}$ as the impedance seen from the input terminals with a load impedance $\mathbf{Z}_L$ connected across the output terminals, as shown in Figure 22.29(a). The output impedance $\mathbf{Z}_o$ is the impedance seen from the output terminals with a source impedance $\mathbf{Z}_s$ connected across the input terminals, as illustrated in Figure 22.29(b). From Equation (17.1) $\mathbf{Z}_{\text{in}} = \mathbf{V}_1/\mathbf{I}_1$ and $\mathbf{Z}_o = \mathbf{V}_2/\mathbf{I}_2$, and from Equation (17.13) $\mathbf{Y}_{\text{in}} = \mathbf{I}_1/\mathbf{V}_1$ and $\mathbf{Y}_o = \mathbf{I}_2/\mathbf{V}_2$.

Figure 22.29 Input and output impedance.

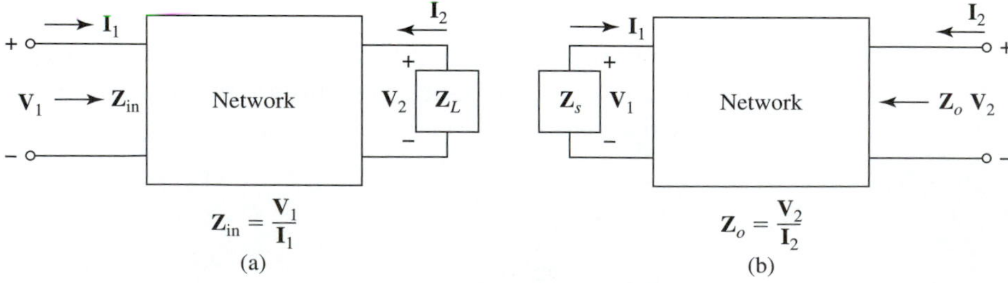

22.8 INPUT AND OUTPUT IMPEDANCE IN TERMS OF z PARAMETERS

Consider the z parameter equivalent circuit in Figure 22.30 to obtain the input impedance $\mathbf{Z}_{\text{in}}$. Applying Kirchhoff's voltage law to the input loop results in

$$\mathbf{V}_1 = \mathbf{z}_{11}\mathbf{I}_1 + \mathbf{z}_{12}\mathbf{I}_2$$

Figure 22.30 Input impedance.

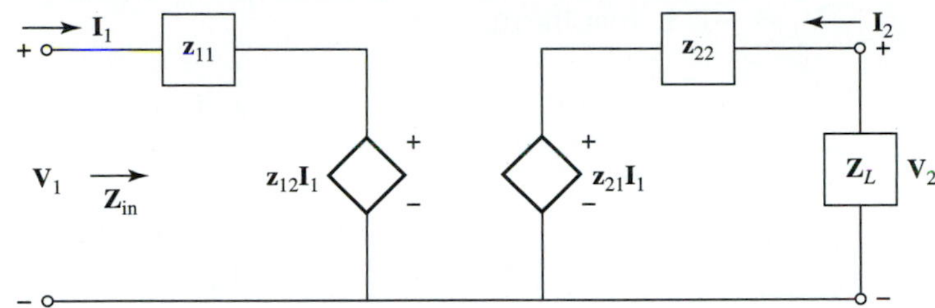

Applying Kirchhoff's Voltage law to the output loop and recognizing that $\mathbf{V}_2 = -\mathbf{Z}_L\mathbf{I}_2$ results in

$$\mathbf{V}_2 = \mathbf{z}_{21}\mathbf{I}_1 + \mathbf{z}_{22}\mathbf{I}_2$$

$$-\mathbf{Z}_L\mathbf{I}_2 = \mathbf{z}_{21}\mathbf{I}_1 + \mathbf{z}_{22}\mathbf{I}_2$$

$$\mathbf{I}_2 = \left\{\frac{-\mathbf{Z}_{21}}{\mathbf{z}_{22} + \mathbf{Z}_L}\right\}\mathbf{I}_1$$

Substituting the expression for $\mathbf{I}_2$ into the preceding expression for $\mathbf{V}_1$ results in

$$\mathbf{V}_1 = \mathbf{z}_{11}\mathbf{I}_1 + \mathbf{z}_{12}\left\{\frac{-\mathbf{z}_{21}}{\mathbf{z}_{22} + \mathbf{Z}_L}\right\}\mathbf{I}_1$$

$$\mathbf{V}_1 = \left\{\mathbf{z}_{11} - \frac{\mathbf{z}_{12}\mathbf{z}_{21}}{\mathbf{z}_{22} + \mathbf{Z}_L}\right\}\mathbf{I}_1$$

$$\frac{\mathbf{V}_1}{\mathbf{I}_1} = \mathbf{z}_{11} - \frac{\mathbf{z}_{12}\mathbf{z}_{21}}{\mathbf{z}_{22} + \mathbf{Z}_L}$$

and the input impedance is

$$\mathbf{Z}_{in} = \mathbf{z}_{11} - \frac{\mathbf{z}_{12}\mathbf{z}_{21}}{\mathbf{z}_{22} + \mathbf{Z}_L} \quad (22.43)$$

Consider the z parameter equivalent circuit in Figure 22.31 to obtain the output impedance $\mathbf{Z}_o$. Applying Kirchhoff's voltage law to the output loop results in

$$\mathbf{V}_2 = \mathbf{z}_{21}\mathbf{I}_1 + \mathbf{z}_{22}\mathbf{I}_2$$

Figure 22.31 Output impedance.

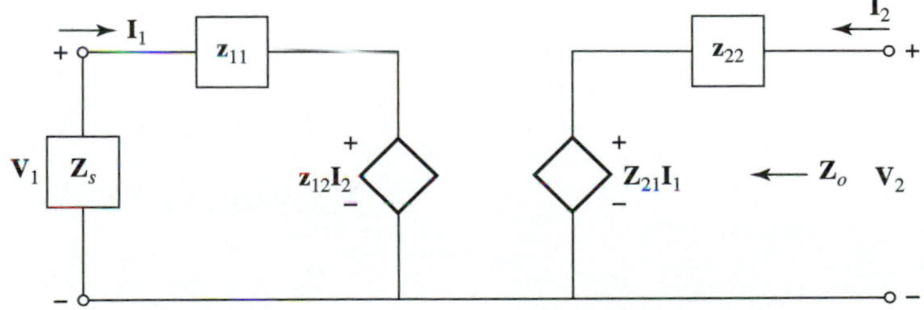

Applying Kirchhoff's voltage law to the input loop and recognizing that $\mathbf{V}_1 = -\mathbf{Z}_s\mathbf{I}_1$ results in

$$\mathbf{V}_1 = \mathbf{z}_{11}\mathbf{I}_1 + \mathbf{z}_{12}\mathbf{I}_2$$

$$-\mathbf{Z}_s\mathbf{I}_1 = \mathbf{z}_{11}\mathbf{I}_1 + \mathbf{z}_{12}\mathbf{I}_2$$

$$\mathbf{I}_1 = \left\{\frac{-\mathbf{z}_{12}}{\mathbf{z}_{11} + \mathbf{Z}_s}\right\}\mathbf{I}_2$$

Substituting the expression for $\mathbf{I}_1$ into the preceding expression for $\mathbf{V}_2$ results in

$$\mathbf{V}_2 = \mathbf{z}_{21}\left\{\frac{-\mathbf{z}_{12}}{\mathbf{z}_{11} + \mathbf{Z}_s}\right\}\mathbf{I}_2 + \mathbf{z}_{22}\mathbf{I}_2$$

$$\mathbf{V}_2 = \left\{\mathbf{z}_{22} - \frac{\mathbf{z}_{12}\mathbf{z}_{21}}{\mathbf{z}_{11} + \mathbf{Z}_s}\right\}\mathbf{I}_2$$

$$\frac{\mathbf{V}_2}{\mathbf{I}_2} = \mathbf{z}_{22} - \frac{\mathbf{z}_{12}\mathbf{z}_{21}}{\mathbf{z}_{11} + \mathbf{Z}_s}$$

and the output impedance is

$$\mathbf{Z}_o = \mathbf{z}_{22} - \frac{\mathbf{z}_{12}\mathbf{z}_{21}}{\mathbf{z}_{11} + \mathbf{Z}_s} \qquad (22.44)$$

22.9 INPUT AND OUTPUT ADMITTANCE IN TERMS OF *y* PARAMETERS

Consider the *y* parameter equivalent circuit in Figure 22.32 to obtain the input admittance $\mathbf{Y}_{\text{in}}$. Applying Kirchhoff's current law to the input node results in

$$\mathbf{I}_1 = \mathbf{y}_{11}\mathbf{V}_1 + \mathbf{y}_{12}\mathbf{V}_2$$

Figure 22.32 Input admittance.

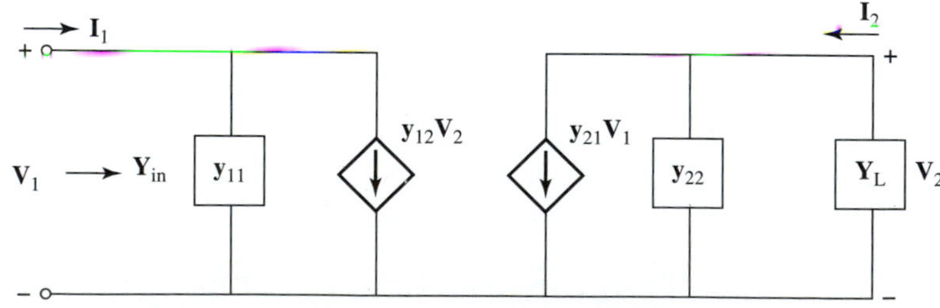

Applying Kirchhoff's current law to the output node and recognizing that $\mathbf{I}_2 = -\mathbf{Y}_L\mathbf{V}_2$ results in

$$\mathbf{I}_2 = \mathbf{y}_{22}\mathbf{V}_2 + \mathbf{y}_{21}\mathbf{V}_1$$

$$-\mathbf{Y}_L\mathbf{V}_2 = \mathbf{y}_{22}\mathbf{V}_2 + \mathbf{y}_{21}\mathbf{V}_1$$

$$\mathbf{V}_2 = \left\{\frac{-\mathbf{y}_{21}}{\mathbf{y}_{22} + \mathbf{Y}_L}\right\}\mathbf{V}_1$$

Substituting the expression for $\mathbf{V}_2$ into the preceding expression for $\mathbf{I}_1$ results in

$$\mathbf{I}_1 = \mathbf{y}_{11}\mathbf{V}_1 + \mathbf{y}_{12}\left\{\frac{-\mathbf{y}_{21}}{\mathbf{y}_{22} + \mathbf{Y}_L}\right\}\mathbf{V}_1$$

$$\mathbf{I}_1 = \left\{\mathbf{y}_{11} - \frac{\mathbf{y}_{12}\mathbf{y}_{21}}{\mathbf{y}_{22} + \mathbf{Y}_L}\right\}\mathbf{V}_1$$

$$\frac{\mathbf{I}_1}{\mathbf{V}_1} = \mathbf{y}_{11} - \frac{\mathbf{y}_{12}\mathbf{y}_{21}}{\mathbf{y}_{22} + \mathbf{Y}_L}$$

and the input admittance is

$$\mathbf{Y}_{\text{in}} = \mathbf{y}_{11} - \frac{\mathbf{y}_{12}\mathbf{y}_{21}}{\mathbf{y}_{22} + \mathbf{Y}_L} \tag{22.45}$$

Consider the y parameter equivalent circuit in Figure 22.33 to obtain the output admittance $\mathbf{Y}_o$. Applying Kirchhoff's current law to the output node results in

$$\mathbf{I}_2 = \mathbf{Y}_{22}\mathbf{V}_2 + \mathbf{y}_{21}\mathbf{Y}_1$$

Figure 22.33 Output admittance.

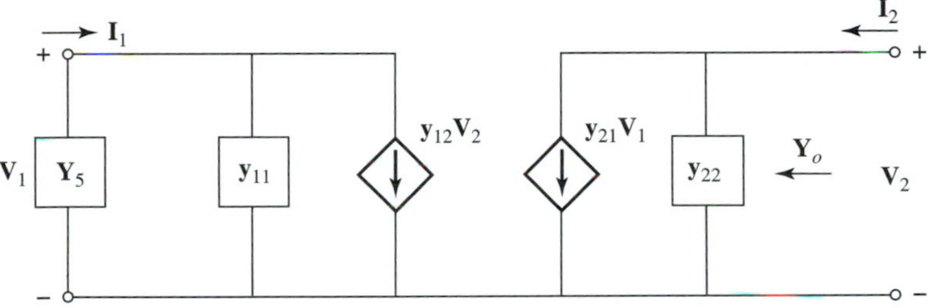

Applying Kirchhoff's current law to the input node and recognizing that $\mathbf{I}_1 = -\mathbf{Y}_s\mathbf{V}_1$ results in

$$\mathbf{I}_1 = \mathbf{y}_{11}\mathbf{V}_1 + \mathbf{y}_{12}\mathbf{V}_2$$

$$-\mathbf{Y}_s\mathbf{V}_1 = \mathbf{y}_{11}\mathbf{V}_1 + \mathbf{y}_{12}\mathbf{V}_2$$

$$-\mathbf{Y}_s\mathbf{V}_1 - \mathbf{y}_{11}\mathbf{V}_1 = \mathbf{y}_{12}\mathbf{V}_2$$

$$\mathbf{V}_1 = \left\{\frac{-\mathbf{y}_{12}}{\mathbf{y}_{11} + \mathbf{Y}_s}\right\}\mathbf{V}_2$$

Substituting the expression for $\mathbf{V}_1$ into the preceding expression for $\mathbf{I}_2$ results in

$$\mathbf{I}_2 = \mathbf{y}_{22}\mathbf{V}_2 + \mathbf{y}_{21}\left\{\frac{-\mathbf{y}_{12}}{\mathbf{y}_{11} + \mathbf{Y}_s}\right\}\mathbf{V}_2$$

$$= \left\{\mathbf{y}_{22} - \frac{\mathbf{y}_{12}\mathbf{y}_{21}}{\mathbf{y}_{11} + \mathbf{Y}_s}\right\}\mathbf{V}_2$$

$$\frac{\mathbf{I}_2}{\mathbf{V}_2} = \mathbf{y}_{22} - \frac{\mathbf{y}_{12}\mathbf{y}_{21}}{\mathbf{y}_{11} + \mathbf{Y}_s}$$

and the output admittance is

$$\mathbf{Y}_o = \mathbf{y}_{22} - \frac{\mathbf{y}_{12}\mathbf{y}_{21}}{\mathbf{y}_{11} + \mathbf{Y}_s} \tag{22.46}$$

22.10 INPUT AND OUTPUT IMPEDANCE IN TERMS OF *h* PARAMETERS

Consider the *h* parameter equivalent circuit in Figure 22.34 to obtain the input impedance $\mathbf{Z}_{\text{in}}$. Applying Kirchhoff's voltage law to the input loop results in

$$\mathbf{V}_1 = \mathbf{h}_{11}\mathbf{I}_1 + \mathbf{h}_{12}\mathbf{V}_2$$

Figure 22.34 Input impedance.

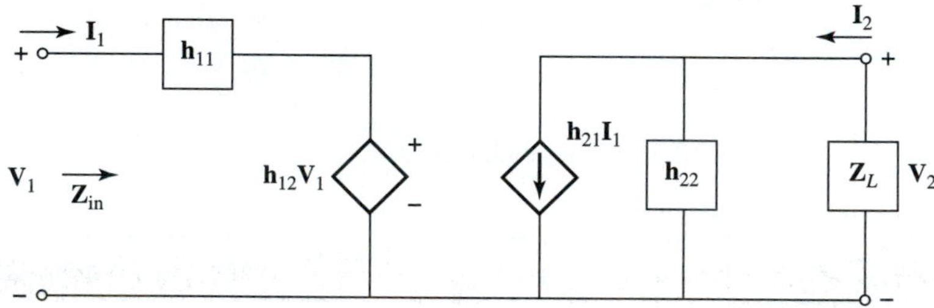

Applying Kirchhoff's current law to the output node and recognizing that $\mathbf{I}_2 = -\mathbf{Y}_L\mathbf{V}_2$ results in

$$-\mathbf{Y}_L\mathbf{V}_2 = \mathbf{h}_{22}\mathbf{V}_2 + \mathbf{h}_{21}\mathbf{I}_1$$

$$\mathbf{V}_2 = \left\{\frac{-\mathbf{h}_{21}}{\mathbf{h}_{22} + \mathbf{Y}_L}\right\}\mathbf{I}_1$$

Substituting $\left\{\frac{-\mathbf{h}_{21}}{\mathbf{h}_{22} + \mathbf{Y}_L}\right\}\mathbf{I}_1$ for $\mathbf{V}_2$ in the equation for $\mathbf{V}_1$ results in

$$\mathbf{V}_1 = \mathbf{h}_{11}\mathbf{I}_1 + \left\{\frac{-\mathbf{h}_{12}\mathbf{h}_{21}}{\mathbf{h}_{22} + \mathbf{Y}_L}\right\}\mathbf{I}_1$$

$$\frac{\mathbf{V}_1}{\mathbf{I}_1} = \mathbf{h}_{11} - \frac{\mathbf{h}_{12}\mathbf{h}_{21}}{\mathbf{h}_{22} + \mathbf{Y}_L}$$

and the input impedance is

$$\mathbf{Z}_{\text{in}} = \mathbf{h}_{11} - \frac{\mathbf{h}_{12}\mathbf{h}_{21}}{\mathbf{h}_{22} + \mathbf{Y}_L} \tag{22.47}$$

Consider the h parameter equivalent circuit in Figure 22.35 to obtain the

Figure 22.35 Output impedance.

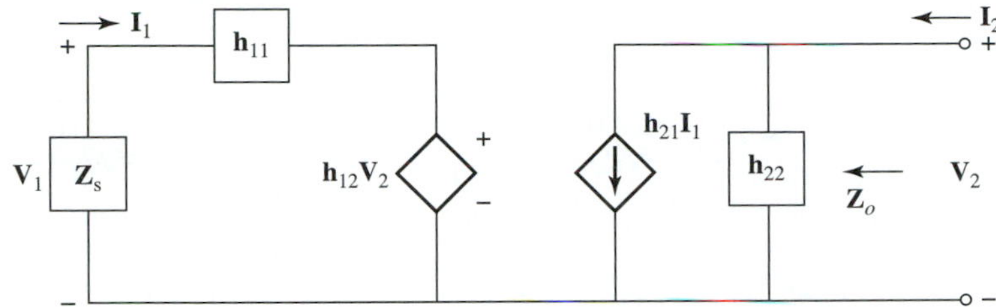

output impedance $\mathbf{Z}_o$. Applying Kirchhoff's voltage law to the input loop and recognizing that $\mathbf{V}_1 = -\mathbf{Z}_s\mathbf{I}_1$ results in

$$\mathbf{V}_1 = \mathbf{h}_{11}\mathbf{I}_1 + \mathbf{h}_{12}\mathbf{V}_2$$

$$-\mathbf{Z}_s\mathbf{I}_1 = \mathbf{h}_{11}\mathbf{I}_1 + \mathbf{h}_{12}\mathbf{V}_2$$

$$\mathbf{I}_1 = \left\{\frac{-\mathbf{h}_{12}}{\mathbf{h}_{11} + \mathbf{Z}_s}\right\}\mathbf{V}_2$$

Applying Kirchhoff's current law to the output node and substituting

$$\left\{\frac{-\mathbf{h}_{12}\mathbf{V}_2}{\mathbf{h}_{11} + \mathbf{Z}_s}\right\}\mathbf{V}_2$$

for $\mathbf{I}_1$ results in

$$\mathbf{I}_2 = \mathbf{h}_{21}\mathbf{I}_1 + \mathbf{h}_{22}\mathbf{V}_2$$

$$\mathbf{I}_2 = \left\{\frac{-\mathbf{h}_{12}\mathbf{h}_{21}}{\mathbf{h}_{11} + \mathbf{Z}_s}\right\}\mathbf{V}_2 + \mathbf{h}_{22}\mathbf{V}_2$$

$$\frac{\mathbf{I}_2}{\mathbf{V}_2} = \mathbf{h}_{22} - \frac{\mathbf{h}_{12}\mathbf{h}_{21}}{\mathbf{h}_{11} + \mathbf{Z}_s}$$

$$\frac{\mathbf{V}_2}{\mathbf{I}_2} = \frac{\mathbf{h}_{11} + \mathbf{Z}_s}{\mathbf{h}_{11}\mathbf{h}_{22} + \mathbf{h}_{22}\mathbf{Z}_s - \mathbf{h}_{12}\mathbf{h}_{21}}$$

and the output impedance is

$$\mathbf{Z}_0 = \frac{\mathbf{h}_{11} + \mathbf{Z}_s}{\mathbf{h}_{11}\mathbf{h}_{22} + \mathbf{h}_{22}\mathbf{Z}_s - \mathbf{h}_{12}\mathbf{h}_{21}} \quad (22.48)$$

EXAMPLE 22.5 Determine $\mathbf{Z}_{in}$ and $\mathbf{Z}_o$ for the network given in Example 22.4 if $\mathbf{Z}_L = 24\text{ K}\,\Omega$ and the network is driven by a voltage source with internal impedance $\mathbf{Z}_s = 100\ \Omega$. ■

SOLUTION Because the z, y, and h parameters are all available from Example 22.4, Equation (22.43), (22.45), or (22.47) can be used to determine $\mathbf{Z}_{in}$, and Equation (22.44), (22.46), or (22.48) can be used to determine $\mathbf{Z}_o$.

We use the z parameter equivalent and employ Equations (22.43) and (22.44) to obtain $\mathbf{Z}_{\text{in}}$ and $\mathbf{Z}_o$, respectively. From Example 22.4, the z parameter equivalent with $\mathbf{Z}_s$ connected across the input as required to determine $\mathbf{Z}_o$ is as shown in Figure 22.36, where $\mathbf{Z}_o = \mathbf{V}_2/\mathbf{I}_2$.

Figure 22.36

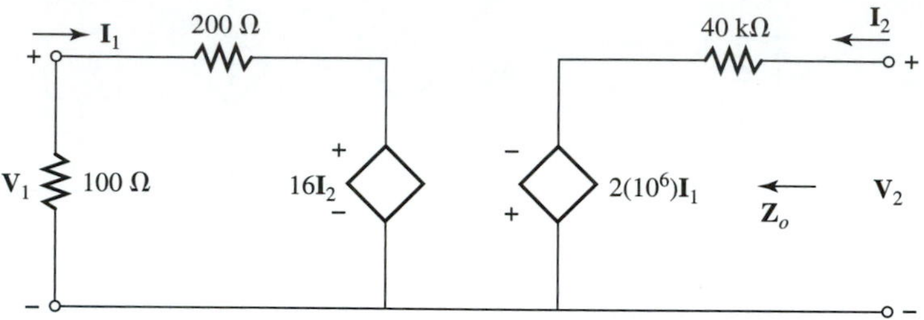

The z parameter equivalent with $\mathbf{Z}_L$ connected across the output as required to determine $\mathbf{Z}_{\text{in}}$ is shown in Figure 22.37, where $\mathbf{Z}_{\text{in}} = \mathbf{V}_1/\mathbf{I}_1$.

Figure 22.37

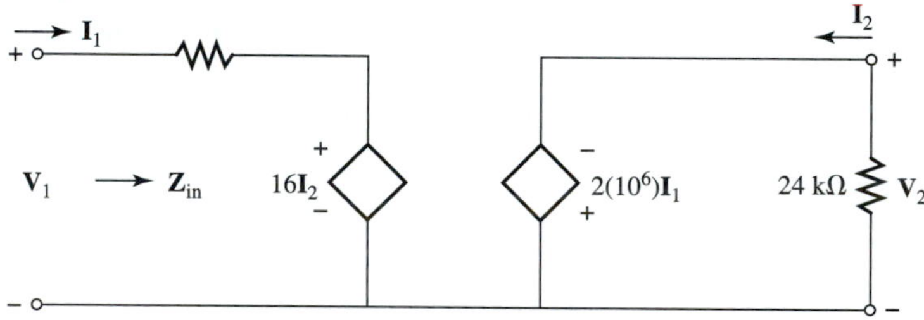

From Equation (22.43)

$$\begin{aligned}\mathbf{Z}_{\text{in}} &= \mathbf{z}_{11} - \frac{\mathbf{z}_{12}\mathbf{z}_{21}}{\mathbf{z}_{22} + \mathbf{Z}_L} \\ &= 200 - \frac{(16)(-2)(10^6)}{(4)(10^4) + (24)(10^3)} \\ &= 700\ \Omega\end{aligned}$$

From Equation (22.44)

$$\begin{aligned}\mathbf{Z}_o &= \mathbf{z}_{22} - \frac{\mathbf{z}_{12}\mathbf{z}_{21}}{\mathbf{z}_{11} + \mathbf{Z}_s} \\ &= 4(10^4) - \frac{(16)(-2)(10^6)}{200 + 100} \\ &= 146.7\ \text{k}\Omega\end{aligned}$$

22.11 VOLTAGE AND CURRENT GAINS

When using the z, y, and h parameter equivalent circuits to analyze networks, it is often convenient to know the voltage and current gains for the equivalent circuits. We consider the voltage gain as the ratio of the output voltage to the input voltage and the current gain as the ratio of the output current to the input current, as illustrated in Figure 22.38, where the voltage gain $\mathbf{A}_V$ equals $\mathbf{V}_2/\mathbf{V}_1$ and the current gain $\mathbf{A}_i$ equals $\mathbf{I}_2/\mathbf{I}_1$.

Figure 22.38 Voltage gain.

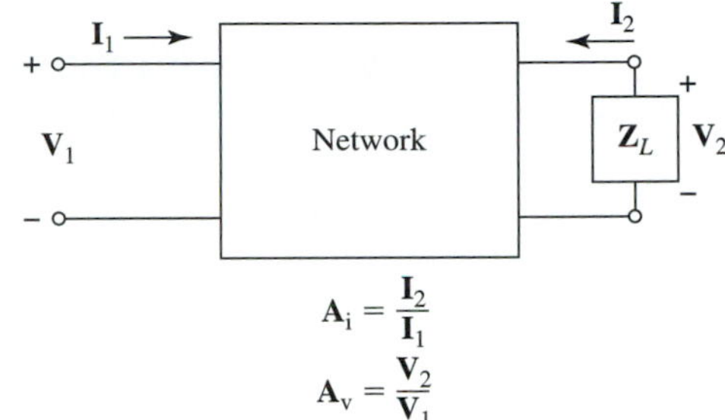

22.12 CURRENT AND VOLTAGE GAIN IN TERMS OF z PARAMETERS

Consider the z parameter equivalent in Figure 22.39 to obtain the voltage and

Figure 22.39 Current gain.

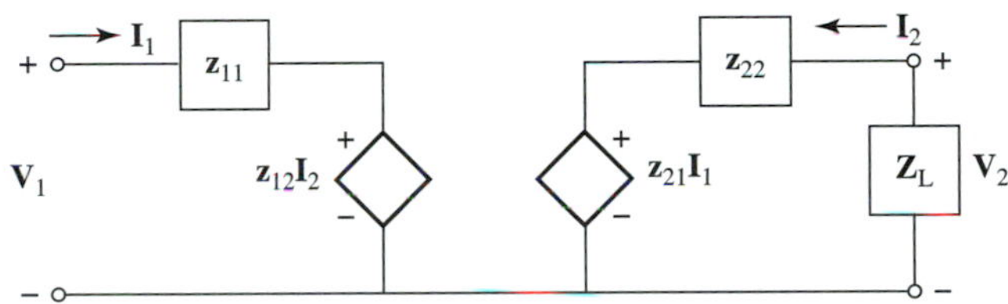

current gains $\mathbf{A}_v$ and $\mathbf{A}_i$. Applying Kirchhoff's voltage law to the output loop and recognizing that $\mathbf{V}_2 = -\mathbf{Z}_L\mathbf{I}_2$ results in

$$\mathbf{V}_2 = \mathbf{z}_{21}\mathbf{I}_1 + \mathbf{z}_{22}\mathbf{I}_2$$

$$-\mathbf{Z}_L\mathbf{I}_2 = \mathbf{z}_{21}\mathbf{I}_1 + \mathbf{z}_{22}\mathbf{I}_2$$

$$(-\mathbf{z}_{22} - \mathbf{Z}_L)\mathbf{I}_2 = \mathbf{z}_{21}\mathbf{I}_1$$

$$\frac{\mathbf{I}_2}{\mathbf{I}_1} = \frac{-\mathbf{z}_{21}}{\mathbf{z}_{22} + \mathbf{Z}_L}$$

and the current gain is

$$\mathbf{A}_i = \frac{-\mathbf{z}_{21}}{\mathbf{z}_{22} + \mathbf{Z}_L} \tag{22.49}$$

Applying Kirchhoff's voltage law to the output loop and substituting $(-\mathbf{V}_2/\mathbf{Z}_L)$ for $\mathbf{I}_2$ results in

$$\mathbf{V}_2 = \mathbf{z}_{21}\mathbf{I}_1 + \mathbf{z}_{22}\mathbf{I}_2$$

$$= \mathbf{z}_{21}\mathbf{I}_1 - \left\{\frac{\mathbf{z}_{22}}{\mathbf{Z}_L}\right\}\mathbf{V}_2$$

$$\mathbf{I}_1 = \left\{\frac{1}{\mathbf{z}_{21}}\right\}\mathbf{V}_2 + \left\{\frac{\mathbf{z}_{22}}{\mathbf{z}_{21}\mathbf{Z}_L}\right\}\mathbf{V}_2$$

$$= \left\{\frac{1}{\mathbf{z}_{12}} + \frac{\mathbf{z}_{22}}{\mathbf{z}_{21}\mathbf{Z}_L}\right\}\mathbf{V}_2$$

Applying Kirchhoff's voltage law to the input loop and substituting

$$= \left\{\frac{1}{\mathbf{z}_{21}} + \frac{\mathbf{z}_{22}}{\mathbf{z}_{21}\mathbf{Z}_L}\right\}\mathbf{V}_2$$

for $\mathbf{I}_1$ and $\{-\mathbf{V}_2/\mathbf{Z}_L\}$ for $\mathbf{I}_2$ results in

$$\mathbf{V}_1 = \mathbf{z}_{11}\mathbf{I}_1 + \mathbf{z}_{12}\mathbf{I}_2$$

$$\mathbf{V}_1 = \mathbf{z}_{11}\left\{\frac{1}{\mathbf{z}_{21}} + \frac{\mathbf{z}_{22}}{\mathbf{z}_{21}\mathbf{Z}_L}\right\}\mathbf{V}_2 + \mathbf{z}_{12}\left\{-\frac{\mathbf{V}_2}{\mathbf{Z}_L}\right\}$$

$$\mathbf{V}_1 = \left\{\frac{\mathbf{z}_{11}}{\mathbf{z}_{21}} + \frac{\mathbf{z}_{11}\mathbf{z}_{22}}{\mathbf{z}_{21}\mathbf{Z}_L} - \frac{\mathbf{z}_{12}}{\mathbf{Z}_L}\right\}\mathbf{V}_2$$

$$= \left\{\frac{\mathbf{z}_{11}\mathbf{Z}_L + \mathbf{z}_{11}\mathbf{z}_{22} - \mathbf{z}_{12}\mathbf{z}_{21}}{\mathbf{z}_{21}\mathbf{Z}_L}\right\}\mathbf{V}_2$$

$$\frac{\mathbf{V}_2}{\mathbf{V}_1} = \frac{\mathbf{z}_{21}\mathbf{Z}_L}{\mathbf{z}_{11}\mathbf{Z}_L + \mathbf{z}_{11}\mathbf{z}_{22} - \mathbf{z}_{12}\mathbf{z}_{21}}$$

and the voltage gain is

$$\mathbf{A}_v = \frac{\mathbf{z}_{21}\mathbf{Z}_L}{\mathbf{z}_{11}\mathbf{Z}_L + \mathbf{z}_{11}\mathbf{z}_{22} - \mathbf{z}_{12}\mathbf{z}_{21}} \tag{22.50}$$

If we divide the numerator and denominator of $\mathbf{A}_v$ by $\mathbf{Z}_L$, the result is

$$\mathbf{A}_v = \frac{\mathbf{z}_{21}}{\mathbf{z}_{11} + \left\{\frac{\mathbf{z}_{11}\mathbf{z}_{22}}{\mathbf{Z}_L}\right\} - \left\{\frac{\mathbf{z}_{12}\mathbf{z}_{21}}{\mathbf{Z}_L}\right\}}$$

Because the terms

$$\left\{\frac{\mathbf{z}_{11}\mathbf{z}_{22}}{\mathbf{Z}_L}\right\} \quad \text{and} \quad \left\{\frac{\mathbf{z}_{12}\mathbf{z}_{21}}{\mathbf{Z}_L}\right\}$$

are equal to zero if $\mathbf{Z}_L$ is infinitely large, the voltage gain is

$$\mathbf{A}_v = \frac{\mathbf{z}_{21}}{\mathbf{z}_{11}} \tag{22.51}$$

when the output terminals are open.

22.13 CURRENT AND VOLTAGE GAIN IN TERMS OF y PARAMETERS

Consider the y parameter equivalent in Figure 22.40 to obtain the voltage and

Figure 22.40 Voltage and current gain.

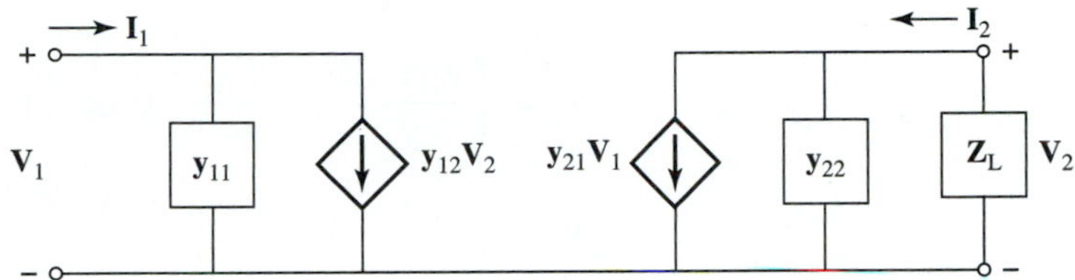

current gains $\mathbf{A}_v$ and $\mathbf{A}_i$. Applying Kirchhoff's current law to the output node and recognizing that $\mathbf{I}_2 = -\mathbf{V}_2/\mathbf{Z}_L$ results in

$$\mathbf{I}_2 = \mathbf{y}_{21}\mathbf{V}_1 + \mathbf{y}_{22}\mathbf{V}_2$$

$$\frac{-\mathbf{V}_2}{\mathbf{Z}_L} = \mathbf{y}_{21}\mathbf{V}_1 + \mathbf{y}_{22}\mathbf{V}_2$$

$$\mathbf{V}_2\left\{-\mathbf{y}_{22} - \frac{1}{\mathbf{Z}_L}\right\} = \mathbf{y}_{21}\mathbf{V}_1$$

$$\frac{\mathbf{V}_2}{\mathbf{V}_1} = \frac{-\mathbf{y}_{21}}{\mathbf{y}_{22} + 1/\mathbf{Z}_L}$$

and the voltage gain is

$$\mathbf{A}_v = \frac{-\mathbf{y}_{21}}{\mathbf{y}_{22} + 1/\mathbf{Z}_L} \tag{22.52}$$

Because the term $(1/\mathbf{Z}_L)$ is zero if $\mathbf{Z}_L$ is infinitely large, the voltage gain is

$$\mathbf{A}_v = \frac{-\mathbf{y}_{21}}{\mathbf{y}_{22}} \tag{22.53}$$

when the output terminals are open.

Applying Kirchhoff's current law to the output node and recognizing that $\mathbf{V}_2 = -\mathbf{Z}_L\mathbf{I}_2$ results in

$$\mathbf{I}_2 = \mathbf{y}_{21}\mathbf{V}_1 + \mathbf{y}_{22}\mathbf{V}_2$$

$$\mathbf{I}_2 = \mathbf{y}_{21}\mathbf{V}_1 + \mathbf{y}_{22}(-\mathbf{Z}_L\mathbf{I}_2)$$

$$\mathbf{V}_1 = \left\{\frac{\mathbf{y}_{22}\mathbf{Z}_L}{\mathbf{y}_{21}} + \frac{1}{\mathbf{y}_{21}}\right\}\mathbf{I}_2$$

Applying Kirchhoff's current law to the input node, again recognizing that $\mathbf{V}_2 = -\mathbf{Z}_L\mathbf{I}_2$, and substituting

$$\left\{\frac{\mathbf{y}_{22}\mathbf{Z}_L}{\mathbf{y}_{21}} + \frac{1}{\mathbf{y}_{21}}\right\}\mathbf{I}_2$$

for $\mathbf{V}_1$ results in

$$\mathbf{I}_1 = \mathbf{y}_{11}\mathbf{V}_1 + \mathbf{y}_{12}\mathbf{V}_2$$

$$\mathbf{I}_1 = \mathbf{y}_{11}\left\{\frac{\mathbf{y}_{22}\mathbf{Z}_L}{\mathbf{y}_{21}} + \frac{1}{\mathbf{y}_{21}}\right\}\mathbf{I}_2 + \mathbf{y}_{12}\{-\mathbf{Z}_L\mathbf{I}_2\}$$

$$\mathbf{I}_1 = \left\{\frac{\mathbf{y}_{11}\mathbf{y}_{22}\mathbf{Z}_L}{\mathbf{y}_{21}} + \frac{\mathbf{y}_{11}}{\mathbf{y}_{21}} - \mathbf{y}_{12}\mathbf{Z}_L\right\}\mathbf{I}_2$$

$$\mathbf{I}_1 = \left\{\frac{\mathbf{y}_{11}\mathbf{y}_{22}\mathbf{Z}_L + \mathbf{y}_{11} - \mathbf{y}_{12}\mathbf{y}_{21}\mathbf{Z}_L}{\mathbf{y}_{21}}\right\}\mathbf{I}_2$$

$$\frac{\mathbf{I}_2}{\mathbf{I}_1} = \frac{\mathbf{y}_{21}}{\mathbf{y}_{11} + \mathbf{y}_{11}\mathbf{y}_{22}\mathbf{Z}_L - \mathbf{y}_{12}\mathbf{y}_{21}\mathbf{Z}_L}$$

and the current gain is

$$\mathbf{A}_i = \frac{\mathbf{y}_{21}}{\mathbf{y}_{11} + \mathbf{y}_{11}\mathbf{y}_{22}\mathbf{Z}_L - \mathbf{y}_{12}\mathbf{y}_{21}\mathbf{Z}_L} \tag{22.54}$$

22.14 CURRENT AND VOLTAGE GAIN IN TERMS OF *h* PARAMETERS

Consider the *h* parameter equivalent in Figure 22.41 to obtain the voltage and

Figure 22.41 Voltage and current gain.

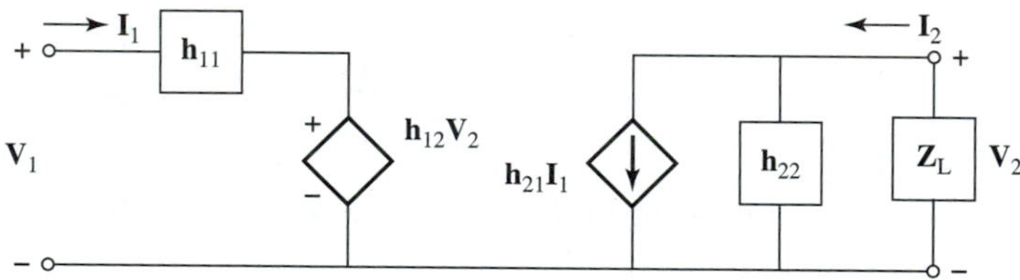

current gains $\mathbf{A}_v$ and $\mathbf{A}_i$. Applying Equation (17.41) for current division at the output node results in

$$\mathbf{I}_2 = \left\{\frac{(1/\mathbf{h}_{22})}{(1/\mathbf{h}_{22}) + \mathbf{Z}_L}\right\}\mathbf{h}_{21}\mathbf{I}_1$$

$$= \left\{\frac{\mathbf{h}_{21}}{1 + \mathbf{h}_{22}\mathbf{Z}_L}\right\}\mathbf{I}_1$$

$$\frac{\mathbf{I}_2}{\mathbf{I}_1} = \frac{\mathbf{h}_{21}}{1 + \mathbf{h}_{22}\mathbf{Z}_L}$$

and the current gain is

$$\mathbf{A}_i = \frac{\mathbf{h}_{21}}{1 + \mathbf{h}_{22}\mathbf{Z}_L} \tag{22.55}$$

Applying Kirchhoff's current law to the output node and recognizing that $\mathbf{I}_2 = -\mathbf{V}_2/\mathbf{Z}_L$ results in

$$\mathbf{I}_2 = \mathbf{h}_{21}\mathbf{I}_1 + \mathbf{h}_{22}\mathbf{V}_2$$

$$\frac{-\mathbf{V}_2}{\mathbf{Z}_L} = \mathbf{h}_{21}\mathbf{I}_1 + \mathbf{h}_{22}\mathbf{V}_2$$

$$\mathbf{I}_1 = \left\{\frac{-1}{\mathbf{h}_{21}\mathbf{Z}_L} + \frac{-\mathbf{h}_{22}}{\mathbf{h}_{21}}\right\}\mathbf{V}_2$$

Applying Kirchhoff's current law to the input node and substituting

$$\left\{\frac{-1}{\mathbf{h}_{21}\mathbf{Z}_L} + \frac{-\mathbf{h}_{22}}{\mathbf{h}_{21}}\right\}\mathbf{V}_2$$

for $\mathbf{I}_1$ results in

$$\mathbf{V}_1 = \mathbf{h}_{11}\mathbf{I}_1 + \mathbf{h}_{12}\mathbf{V}_2$$

$$\mathbf{V}_1 = \mathbf{h}_{11}\left\{\frac{-1}{\mathbf{h}_{21}\mathbf{Z}_L} + \frac{-\mathbf{h}_{22}}{\mathbf{h}_{21}}\right\}\mathbf{V}_2 + \mathbf{h}_{12}\mathbf{V}_2$$

$$= \left\{\mathbf{h}_{12} - \frac{\mathbf{h}_{11}}{\mathbf{h}_{21}\mathbf{Z}_L} - \frac{\mathbf{h}_{11}\mathbf{h}_{22}}{\mathbf{h}_{21}}\right\}\mathbf{V}_2$$

$$= \left\{\frac{\mathbf{h}_{12}\mathbf{h}_{21}\mathbf{Z}_L - \mathbf{h}_{11} - \mathbf{h}_{11}\mathbf{h}_{22}\mathbf{Z}_L}{\mathbf{h}_{21}\mathbf{Z}_L}\right\}\mathbf{V}_2$$

$$\frac{\mathbf{V}_2}{\mathbf{V}_1} = \frac{\mathbf{h}_{21}\mathbf{Z}_L}{\mathbf{h}_{12}\mathbf{h}_{21}\mathbf{Z}_L - \mathbf{h}_{11}\mathbf{h}_{22}\mathbf{Z}_L - \mathbf{h}_{11}}$$

and the voltage gain is

$$\mathbf{A}_v = \frac{\mathbf{h}_{21}\mathbf{Z}_L}{\mathbf{h}_{12}\mathbf{h}_{21}\mathbf{Z}_L - \mathbf{h}_{11}\mathbf{h}_{22}\mathbf{Z}_L - \mathbf{h}_{11}} \tag{22.56}$$

If we divide the numerator and denominator by $\mathbf{Z}_L$, the result is

$$\mathbf{A}_v = \frac{\mathbf{h}_{21}}{\mathbf{h}_{12}\mathbf{h}_{21} - \mathbf{h}_{11}\mathbf{h}_{22} - \mathbf{h}_{11}/\mathbf{Z}_L}$$

Because the term $\mathbf{h}_{11}/\mathbf{Z}_L$ is equal to zero if $\mathbf{Z}_L$ is infinitely large, the voltage gain is

$$\mathbf{A}_v = \frac{\mathbf{h}_{21}}{\mathbf{h}_{12}\mathbf{h}_{21} - \mathbf{h}_{11}\mathbf{h}_{22}} \tag{22.57}$$

when the output terminals are open.

EXAMPLE 22.6 Determine $\mathbf{A}_v$ and $\mathbf{A}_i$ for the network given in Example 22.4 if $\mathbf{Z}_L = 24\ \text{k}\Omega$. ■

SOLUTION Because the z, y, and h parameters are all available from Example 22.4, Equation (22.50) (22.52), or (22.56) can be used to determine $\mathbf{A}_v$ and Equation (22.49), (22.54), or (22.55) can be used to determine $\mathbf{A}_i$.

We use the z parameter equivalent and employ Equations (22.50) and (22.49) to determine $\mathbf{A}_v$ and $\mathbf{A}_i$, respectively. From Example 22.4, the z parameter equivalent with $\mathbf{Z}_L$ connected across the output as required to determine $\mathbf{A}_v$ and $\mathbf{A}_i$ is shown in Figure 22.42.

Figure 22.42

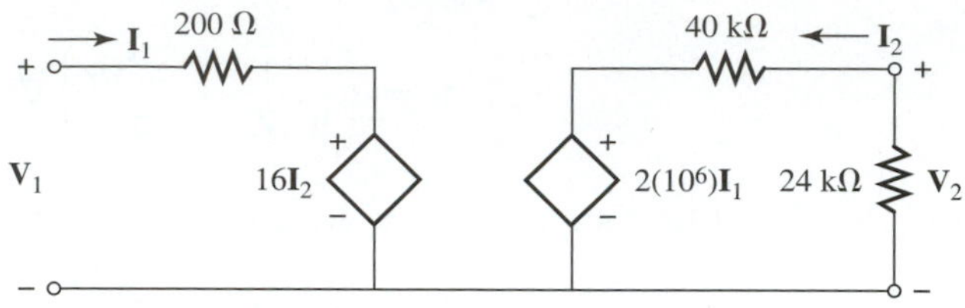

From Equation (22.50)

$$\mathbf{A}_v = \frac{\mathbf{z}_{21}\mathbf{Z}_L}{\mathbf{z}_{11}\mathbf{Z}_L + \mathbf{z}_{11}\mathbf{z}_{22} - \mathbf{z}_{12}\mathbf{z}_{21}}$$

$$= \frac{(-2)(10^6)(24)(10^3)}{(200)(24)(10^3) + (200)(4)(10^4) - (16)(-2)(10^6)}$$

$$= -1071$$

$$= 1071\angle 180^\circ$$

Because $\mathbf{A}_v = \mathbf{V}_2/\mathbf{V}_1$, the magnitude of the voltage gain is 1071, and $\mathbf{V}_2$ and $\mathbf{V}_1$ are 180° out of phase.

From Equation (22.49)

$$\mathbf{A}_i = \frac{-\mathbf{z}_{21}}{\mathbf{z}_{22} + \mathbf{Z}_L}$$

$$= \frac{(-2)(10^6)}{4(10^4) + (24)(10^3)}$$

$$= -31.25$$

22.15 EXERCISES

1. The z parameters for one two-port network are $\mathbf{z}_{11} = 6\text{ k}\Omega$, $\mathbf{z}_{12} = 1\text{ k}\Omega$, $\mathbf{z}_{21} = 1\text{ k}\Omega$, and $\mathbf{z}_{22} = 2\text{ k}\Omega$.

(a) Convert the z parameters to y parameters.
(b) Draw the equivalent y parameter circuit.
(c) Convert the z parameters to h parameters.
(d) Draw the equivalent h parameter circuit.

2. The y parameters for one two-port network are $\mathbf{y}_{11} = 8\text{ S}$, $\mathbf{y}_{12} = -8\text{ S}$, $\mathbf{y}_{21} = -8\text{ S}$, and $\mathbf{y}_{22} = 10\text{ S}$.

(a) Convert the y parameters to z parameters.
(b) Draw the equivalent z parameter circuit.
(c) Convert the y parameters to h parameters.
(d) Draw the equivalent h parameter circuit.

3. The h parameters for a two-port network are $\mathbf{h}_{11} = 5\text{ k}\Omega$, $\mathbf{h}_{12} = 4 \times 10^{-4}$, $\mathbf{h}_{21} = 100$, $\mathbf{h}_{22} = 10 \times 10^{-6}\text{ S}$.

(a) Convert the h parameters to z parameters.
(b) Draw the equivalent z parameter circuit.

(c) Convert the h parameters to y parameters.
(d) Draw the equivalent y parameter circuit.

4. Determine a z parameter equivalent for the circuit in Figure 22.43 and sketch the z equivalent circuit.

Figure 22.43

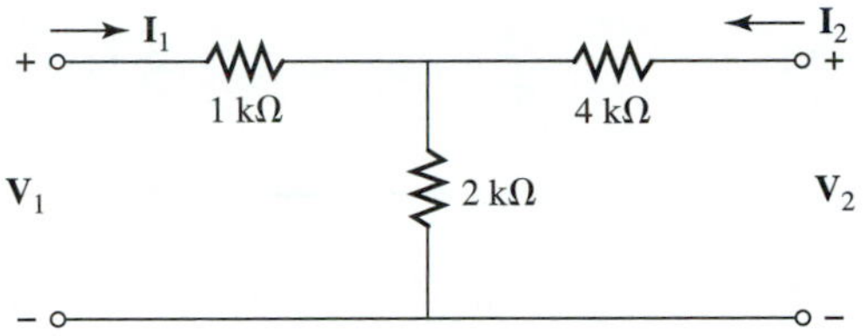

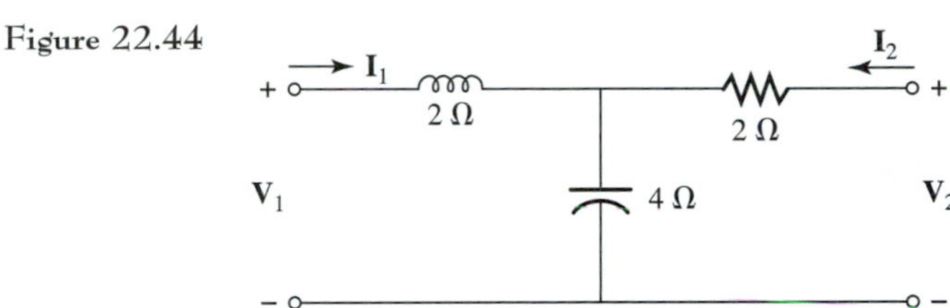

5. Determine a z parameter equivalent for the circuit in Figure 22.44 and sketch the z equivalent circuit.

Figure 22.44

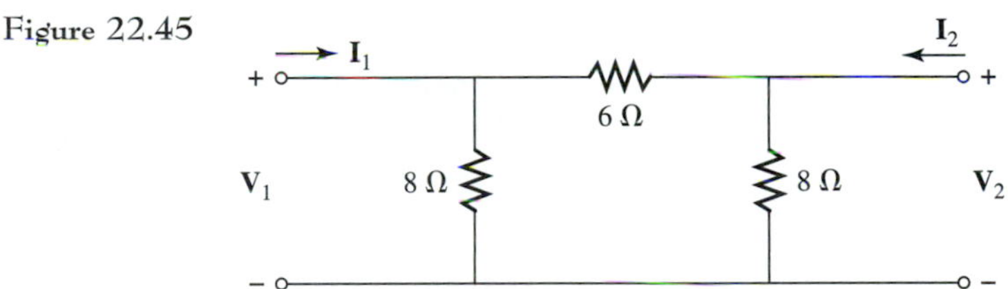

6. Determine a y parameter equivalent for the circuit in Figure 22.45 and sketch the y equivalent circuit.

Figure 22.45

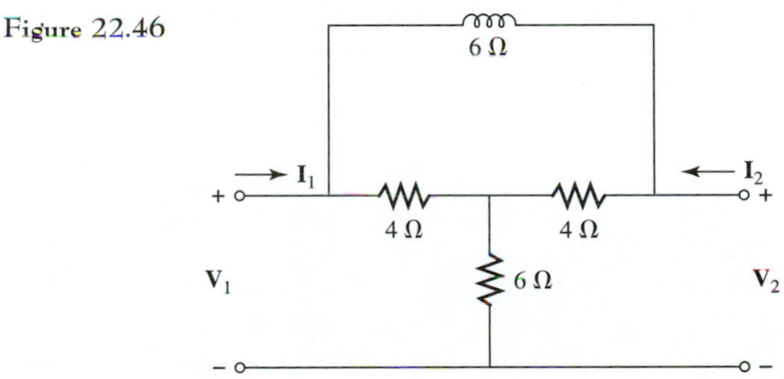

7. Determine a y parameter equivalent for the circuit in Figure 22.46 and sketch the y equivalent circuit.

Figure 22.46

8. Determine an h parameter equivalent for the circuit in Figure 22.47 and sketch the h equivalent circuit.

Figure 22.47

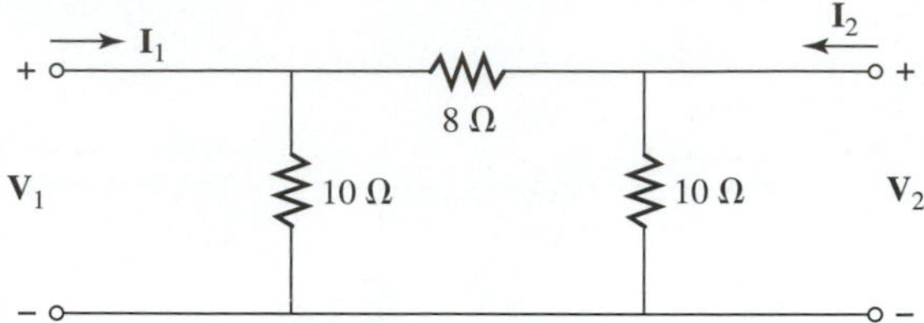

9. Determine an h parameter equivalent for the circuit in Figure 22.48 and sketch the h equivalent circuit.

Figure 22.48

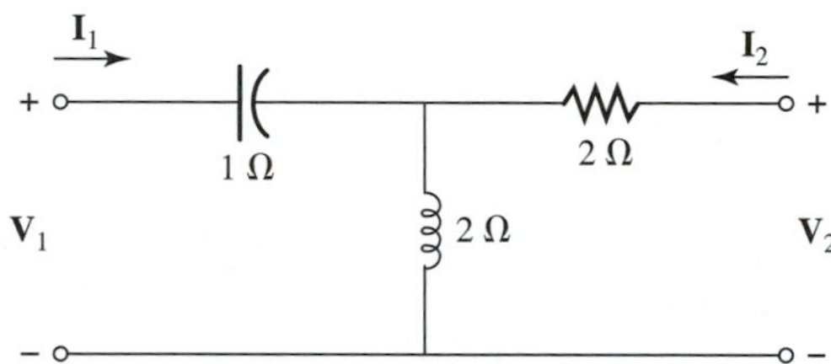

10. Determine a z parameter equivalent for the circuit in Figure 22.49 and sketch the z equivalent circuit.

Figure 22.49

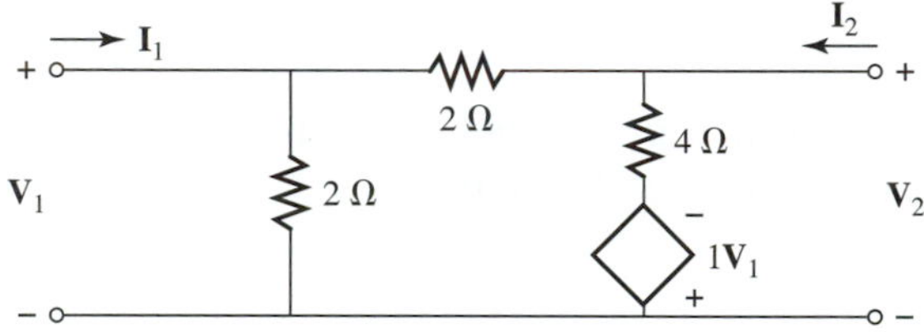

11. Determine a y parameter equivalent for the circuit in Figure 22.50 and sketch the y equivalent circuit.

Figure 22.50

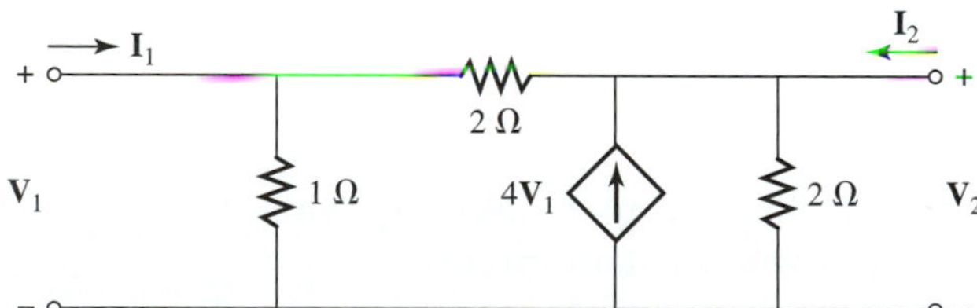

12. Determine an h parameter equivalent for the circuit in Figure 22.51 and sketch the h equivalent circuit.

Figure 22.51

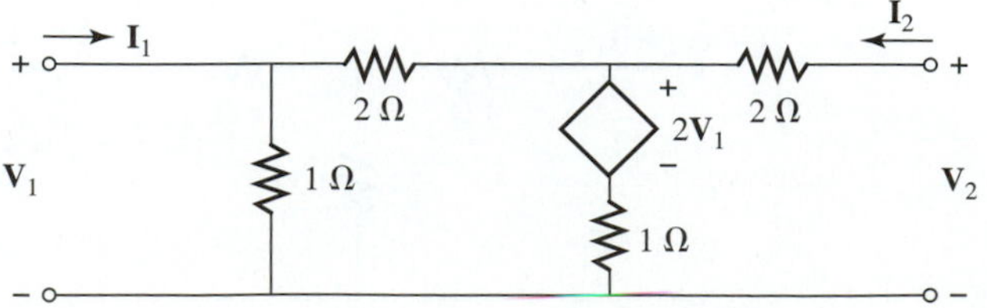

13. Consider the circuit in Figure 22.52 with the following z parameters: $\mathbf{z}_{11} = 4\ \Omega$, $\mathbf{z}_{12} = 4\ \Omega$, $\mathbf{z}_{21} = 4\ \Omega$, $\mathbf{z}_{22} = 6\ \Omega$.

Figure 22.52

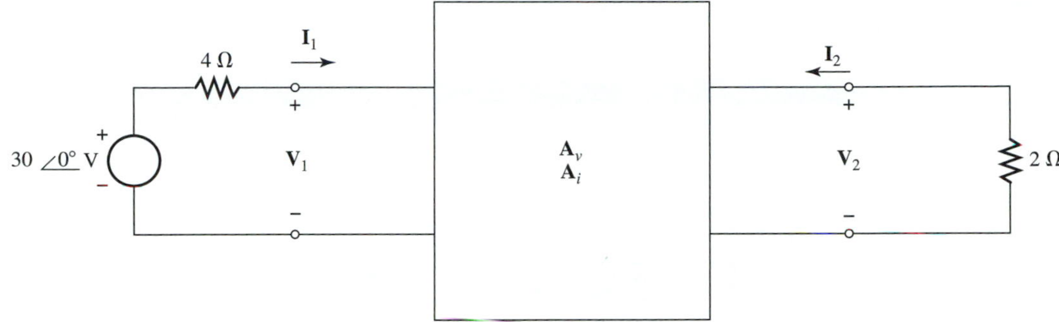

(a) Determine $\mathbf{A}_v$ for the 2-Ω resistor load as shown.
(b) Determine $\mathbf{A}_v$ for no load.
(c) Determine $\mathbf{A}_i$ for the 2-Ω resistor load as shown.
(d) Determine $\mathbf{Z}_{\text{in}}$ for the 2-Ω resistor load as shown.
(e) Determine $\mathbf{Z}_o$ for the 4-Ω source resistor as shown.
(f) Determine $\mathbf{V}_1$. (g) Determine $\mathbf{V}_2$.
(h) Determine $\mathbf{I}_1$. (i) Determine $\mathbf{I}_2$.

14. Consider the circuit in Figure 22.53 with the following y parameters: $\mathbf{y}_{11} = 2$ S, $\mathbf{y}_{12} = 4$ S, $\mathbf{y}_{21} = 4$ S, and $\mathbf{y}_{22} = 6$ S.

Figure 22.53

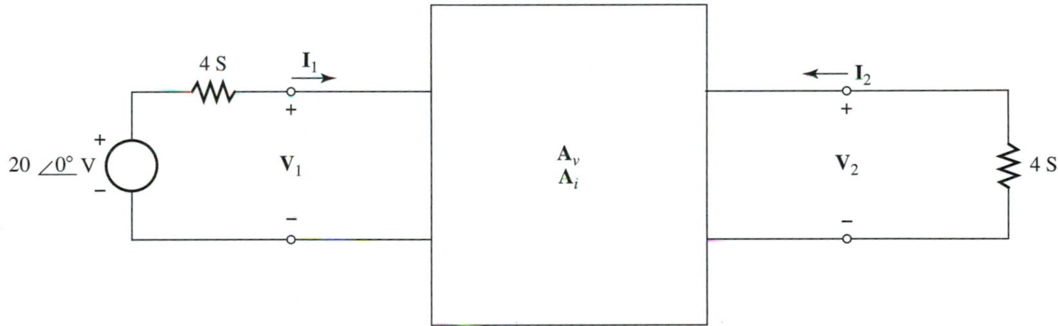

(a) Determine $\mathbf{A}_v$ for the 4-S resistor load as shown.
(b) Determine $\mathbf{A}_v$ for no load. (c) Determine $\mathbf{A}_i$.
(d) Determine $\mathbf{Z}_{\text{in}}$. (e) Determine $\mathbf{Z}_o$.
(f) Determine $\mathbf{V}_1$. (g) Determine $\mathbf{V}_2$.
(h) Determine $\mathbf{I}_1$. (i) Determine $\mathbf{I}_2$.

15. Determine $\mathbf{A}_v$ and $\mathbf{A}_i$ for the circuit in Figure 22.44 if a 4-Ω resistor is connected across the output terminals.

16. Determine $\mathbf{A}_v$ and $\mathbf{A}_i$ for the circuit in Figure 22.46 if a 4-Ω resistor is connected across the output terminals.

17. Determine $\mathbf{A}_v$ and $\mathbf{A}_i$ for the circuit in Figure 22.50 if a 4-Ω resistor is connected across the output terminals.

18. Determine $\mathbf{Z}_o$ for the circuit in Figure 22.49 if a 4-Ω resistor is connected across the input terminals.

19. Determine $\mathbf{Z}_{\text{in}}$ for the circuit in Figure 22.49 if a 2-Ω resistor is connected across the output terminals.

CHAPTER

23

Three-Phase Circuits

After completing this chapter the student should be able to

* determine line and phase currents for delta connections
* determine line and phase currents for wye connections
* determine power measurements for three-phase loads

23.1 THREE-PHASE GENERATORS

A conceptual view of a simple ac voltage generator is illustrated in Figure 23.1. It basically consists of a conducting coil (winding), an armature, a magnetic field, a device for connecting the generated voltage to an external load, and an external driver.

Figure 23.1 AC generator.

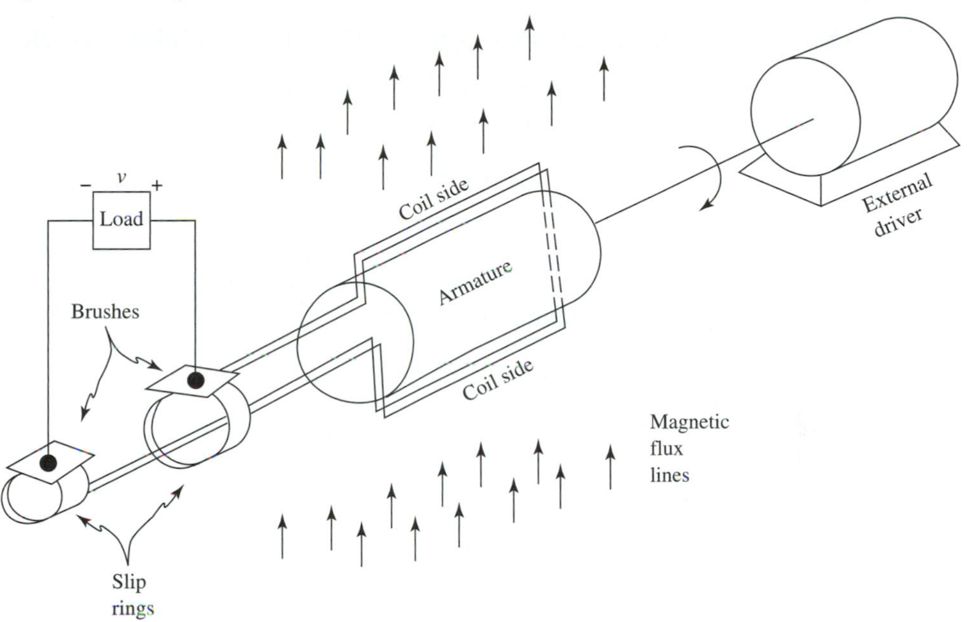

The external driver (steam turbine, electric motor, or internal combustion engine) turns the armature, which is a cylindrical drumlike device that physically supports the conducting coil. The coil sides cut lines of magnetic flux as the armature rotates, and a voltage is developed in the coil in accordance with Faraday's law. The ends of the coil are hard-connected to slip rings, which rotate with the coil and the armature. A stationary brush rides on each slip ring, and the voltage across the coil appears across the brushes and finally across the load. The magnetic flux lines are provided by a permanent iron-core magnet or, for larger generators, an iron core magnetized by a dc voltage source. The polarities of the induced voltages on each side of the coil are additive and together form the output voltage, as shown. Because the coil rotates in a circular path at a constant angular velocity and the flux is along straight lines, the coil cuts across flux lines at a rate that is a function of its position. The voltage induced in the coil and the resulting output voltage is sinusoidal with a frequency that depends on the angular velocity of the armature.

The ac voltage generator in Figure 23.1 is considered a single-phase voltage generator, because it has a single sinusoidal output voltage waveform. The winding in Figure 23.1 can be expanded to include parallel and series paths to increase the voltage and power output capabilities of the generator, but only a single voltage waveform appears across the output of a single-phase voltage generator.

If additional independent windings are placed on the armature of the voltage generator in Figure 23.1 and result in more than one output voltage, the generator becomes a polyphase voltage generator, with each independent winding behaving as a single-phase generator. Although two-phase (two windings) and other polyphase configurations are found in control systems and other applications, the single-phase (one winding) and the three-phase (three windings) are by far the most widely used configurations. Single-phase voltages and currents have been encountered in all the ac circuits previously considered, and three-phase voltages and currents are considered in this chapter.

Almost all the significant electrical power generation in the civilized world involves three-phase and single-phase ac voltages and currents. The major advantages of three-phase ac generators compared to single-phase are smoother operation, more efficient performance, and lighter weight.

23.2 WINDING CONNECTIONS FOR THREE-PHASE GENERATORS

Consider three independent windings physically placed on the armature of an ac voltage generator in such a manner that the three voltages generated are 120° out of phase with each other and equal in magnitude. The three windings (coils) can be thought of as three separate single-phase ac generators and are represented symbolically in either of the two ways shown in Figure 23.2(a) and (b). Phasors with the associated voltage waveforms are shown in Figure 23.2(c) for each winding. Notice that the phasors are equal in magnitude and are separated by 120°.

Figure 23.2 Three-phase voltages.

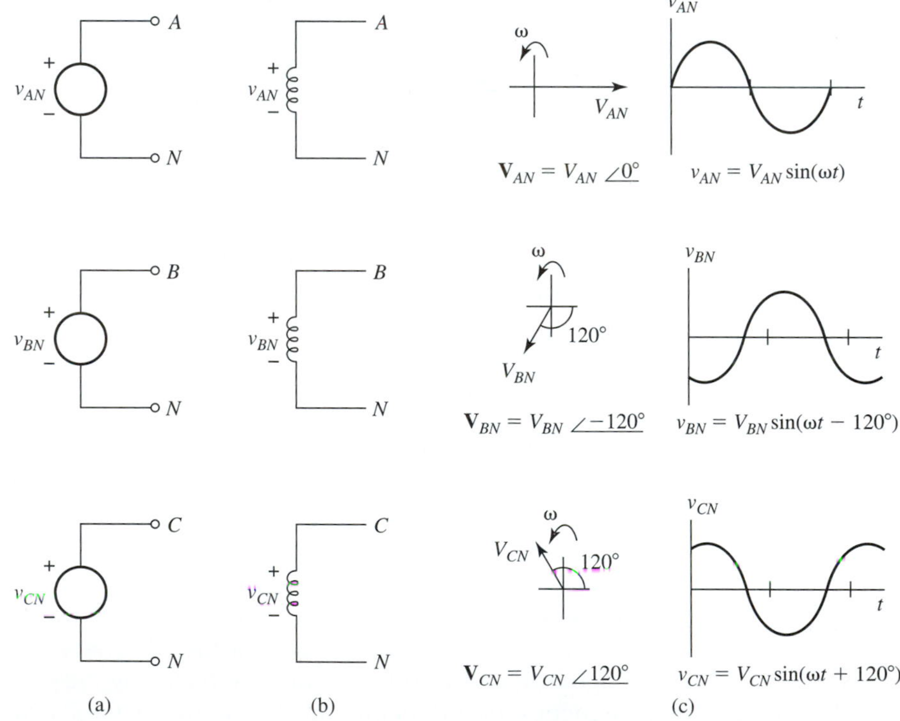

The three windings in Figure 23.2 act as three independent single-phase ac generators. However, if they are connected internally to the generator in a delta (Δ) or wye (Y) configuration, the generator is called a three-phase voltage generator. We shall use the inductor (coil) symbol in Figure 23.2(b) to symbolically represent the generated voltages for the Δ and Y connections.

23.3 THE FOUR-WIRE Y CONNECTION

If the three windings in Figure 23.2 are connected as shown in Figure 23.3, the generator is called a four-wire, Y-connected, three-phase generator. The three terminals labeled N in Figure 23.2(b) are connected to form the single terminal labeled N in Figure 23.3. Notice that the windings are symbolically placed in a Y configuration. Notice also that the output includes point N (called the neutral point) as an output terminal.

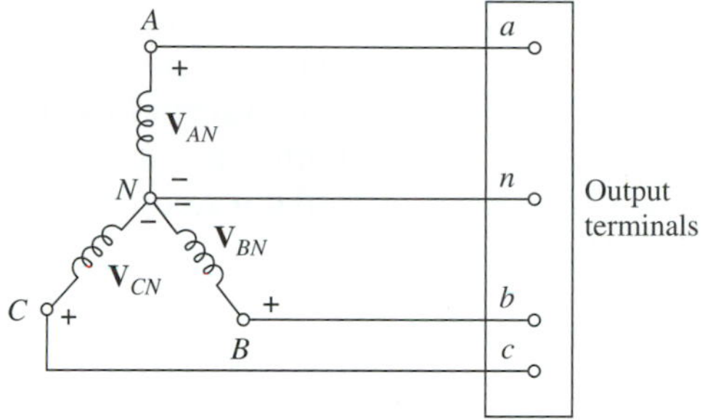

Figure 23.3 Four-wire Y connection.

23.4 THE THREE-WIRE Y CONNECTION

If the three windings in Figure 23.2 are connected as shown in Figure 23.4, the generator is called a three-wire, Y-connected, three-phase voltage generator, and as before, the windings are symbolically placed in a Y configuration. Notice that there are three output terminals and the neutral point N is not available at the output, as it is for the four-wire connection.

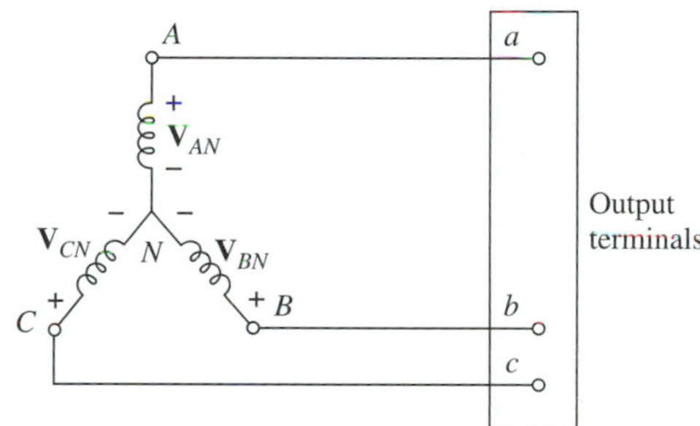

Figure 23.4 Three-wire Y connection.

23.5 THE Δ CONNECTION

If the three windings in Figure 23.2 are connected as shown in Figure 23.5, the generator is called a Δ-connected, three-phase voltage generator; and as before, the windings are symbolically placed in a Δ configuration. The Δ connection can have only three output terminals. Notice that the delta is formed by connecting point N of winding A-N to point B on winding B-N, connecting point N of winding B-N to point C on winding C-N, and connecting point N of winding C-N to point A on winding A-N.

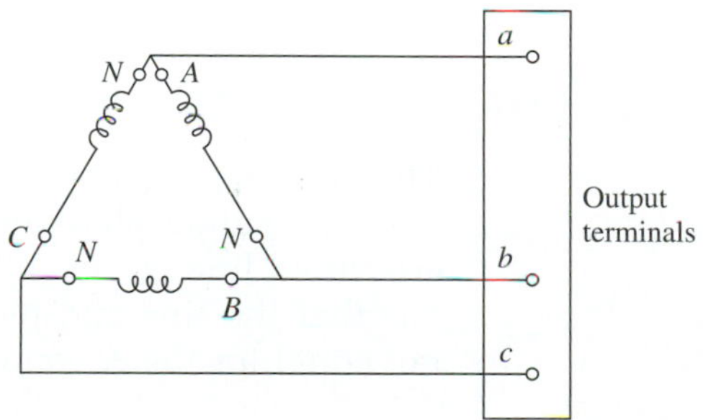

Figure 23.5 Delta connection.

23.6 BALANCED LOADS

Three-phase loads (passive elements, electric motors, or other devices) are connected in delta and wye configurations in the same manner as three-phase voltage generators. Three-phase loads are considered balanced if the impedances of the three phases of the load are equal. That is, if $\mathbf{Z}_1 = \mathbf{Z}_2 = \mathbf{Z}_3$, the Y-connected load in Figure 23.6(a) is balanced; in a like manner, if $\mathbf{Z}_a = \mathbf{Z}_b = \mathbf{Z}_c$, the Δ-connected load in Figure 23.6(b) is balanced. The Y-connected load may or may not contain a neutral line, as shown, and the Δ-connected load does not have a neutral line.

Figure 23.6 Three-phase loads

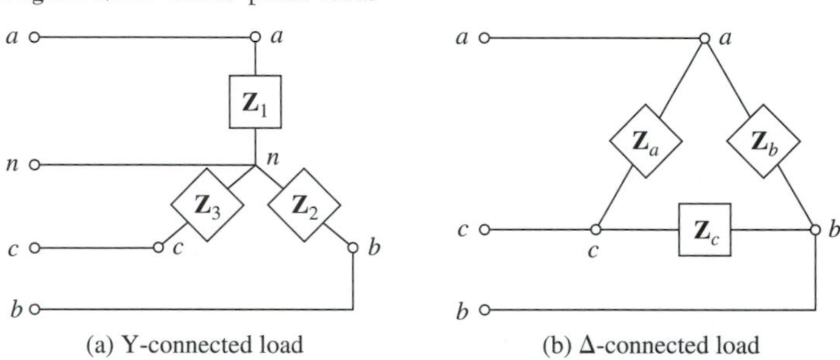

(a) Y-connected load (b) Δ-connected load

23.7 UNBALANCED LOADS

If the three impedances ($\mathbf{Z}_1$, $\mathbf{Z}_2$, and $\mathbf{Z}_3$) in Figure 23.6(a) are not equal, the Y-connected load is said to be unbalanced. In a like manner, if the three impedances ($\mathbf{Z}_a$, $\mathbf{Z}_b$, and $\mathbf{Z}_c$) in Figure 23.6(b) are not equal, the Δ-connected load is said to be unbalanced. The multiloop circuit-analysis procedures previously presented can be employed to analyze most three-phase circuits with balanced or unbalanced loads. This chapter is concerned with presenting systematic procedures for analyzing the widely encountered balanced three-phase systems.

23.8 LINE AND PHASE VOLTAGES

The voltages across the phase windings of the Δ- and Y-connected generators are called generator phase voltages, and the voltages across the output lines are called line voltages. Referring to Figures 23.4 and 23.5, it should be obvious that the line and phase voltages are equal for the Δ connection and are not equal for the Y connection.

23.9 LINE AND PHASE CURRENTS

The currents flowing in the phase windings of the Δ- and Y-connected generators are called phase currents, and the currents flowing in the output lines are called line currents. Referring to Figures 23.4 and 23.5, it should be obvious that the line and phase currents are equal for the Y connection and are not equal for the Δ connection.

23.10 PHASE SEQUENCE

The rotating sequence for the phasors associated with the generator phase voltages in Figure 23.2(c) are shown in Figure 23.7.

For counterclockwise rotation the phasors in Figure 23.7 pass a given point *X* in the sequential order *ABC*, which is called a positive phase sequence. The positive phase sequence can also be written *BCA* or *CAB*. In this text the generator phase windings that are Δ or Y connected are always connected using the positive phase sequence.

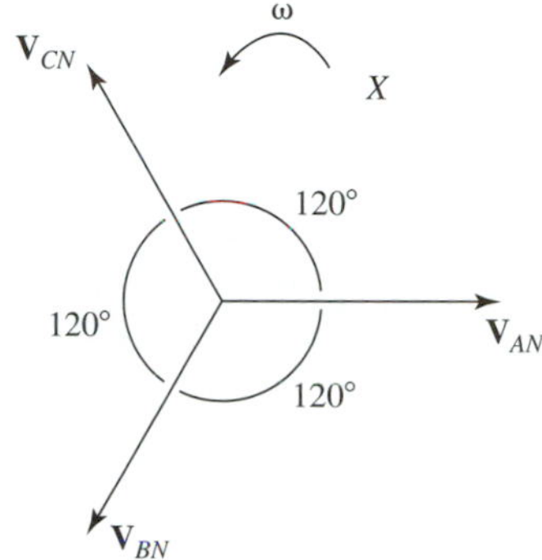

Figure 23.7 Phase sequence.

23.11 THE Y-CONNECTED GENERATOR

Consider the Y-connected three-phase voltage generator in Figure 23.8. The voltages $\mathbf{V}_{AN}$, $\mathbf{V}_{BN}$, and $\mathbf{V}_{CN}$ are the generator phase voltages, and $\mathbf{V}_{AB}$, $\mathbf{V}_{BC}$, and $\mathbf{V}_{CA}$ are the line voltages. $\mathbf{I}_{Aa}$, $\mathbf{I}_{Bb}$, and $\mathbf{I}_{Cc}$ are the line currents and $\mathbf{I}_{Nn}$ is the current in the neutral line. Notice that the generator phase currents are the

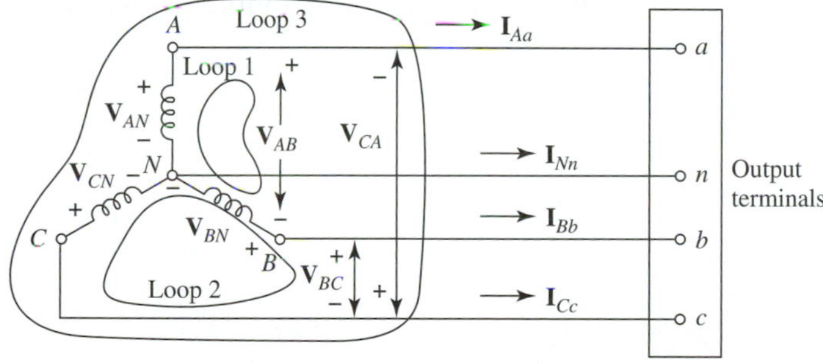

Figure 23.8 Four-wire, Y-connected generator.

same as the line currents, as previously stated. Applying Kirchhoff's voltage law to loops 1, 2, and 3, respectively, in Figure 23.8 results in

$$\mathbf{V}_{AB} = \mathbf{V}_{AN} - \mathbf{V}_{BN}$$

$$\mathbf{V}_{BC} = \mathbf{V}_{BN} - \mathbf{V}_{CN}$$

$$\mathbf{V}_{CA} = \mathbf{V}_{CN} - \mathbf{V}_{AN}$$

A phasor diagram including the generator phase voltages and the line voltages is shown in Figure 23.9. The phase voltages are added using the parallelogram law (Chapter 16) to form the line voltages in accordance with the preceding equations. Refer to Figure 23.2 for the locations of $\mathbf{V}_{AN}$, $\mathbf{V}_{BN}$, and $\mathbf{V}_{CN}$ on the phasor diagram.

Figure 23.9 Phasor diagram.

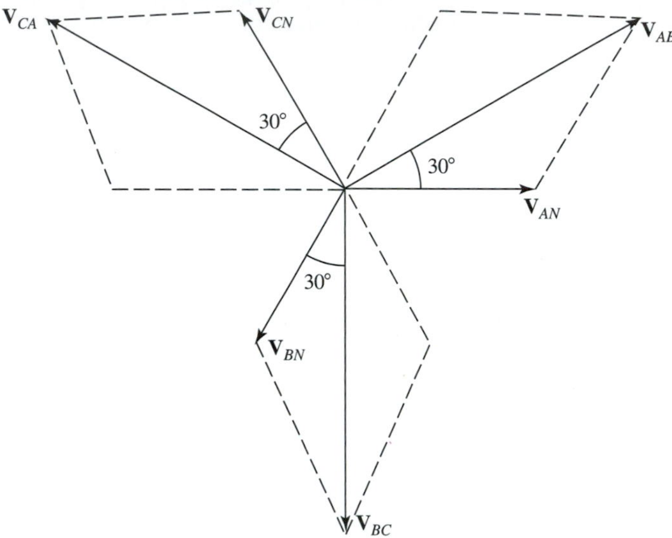

Let us determine $\mathbf{V}_{AB}$ in Figure 23.9 from the phase voltages as an illustration and note that $\mathbf{V}_{BC}$ and $\mathbf{V}_{CA}$ can be determined in a similar manner. From the preceding equations,

$$\mathbf{V}_{AB} = \mathbf{V}_{AN} - \mathbf{V}_{BN}$$
$$= \mathbf{V}_{AN} + (-\mathbf{V}_{BN})$$

The phasors $\mathbf{V}_{AN}$ and $(-\mathbf{V}_{BN})$ are added as phasors using the parallelogram law, as shown in Figure 23.10.

Figure 23.10 Phasor addition.

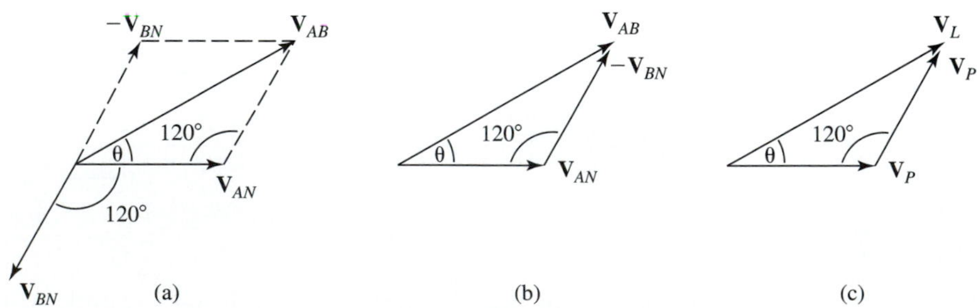

The addition of phasors $\mathbf{V}_{AN}$ and $(-\mathbf{V}_{BN})$ to form $\mathbf{V}_{AB}$ by the parallelogram law (Chapter 16) is illustrated in Figure 23.10(a), and the triangular relationship between the phasors is shown in Figure 23.10(b). Because the magnitudes of all of the generator phase voltages are equal,

$$V_P = V_{AN} = V_{BN} = V_{CN}$$

where V_P is the magnitude of the generator phase voltages. Also, because the magnitudes of the line voltages are equal,

$$V_L = V_{AB} = V_{BC} = V_{CA}$$

where V_L is the magnitude of the line voltages. Substituting V_P for V_{AN} and V_{BN} and V_L for V_{AB} results in the triangle in Figure 23.10(c). Applying the law of cosines (Chapter 16) to the triangle in Figure 23.10(c) results in

$$\begin{aligned} V_L^2 &= V_P^2 + V_P^2 - 2V_PV_P\cos 120^\circ \\ &= V_P^2 + V_P^2 - 2V_P^2\left(-\frac{1}{2}\right) \\ &= 3V_P^2 \\ V_L &= \sqrt{3}V_P \end{aligned} \qquad (23.1)$$

where V_L is the magnitude of the line voltages and V_P is the magnitude of the generator phase voltages.

Applying the law of sines (Chapter 16) to the triangle in Figure 23.10(c) results in

$$\begin{aligned} \frac{V_L}{\sin 120^\circ} &= \frac{V_P}{\sin\theta} \\ \frac{\sqrt{3}V_P}{\sin 120^\circ} &= \frac{V_P}{\sin\theta} \\ \sin\theta &= \frac{\sin 120^\circ}{\sqrt{3}} \\ &= \frac{(\sqrt{3}/2)}{\sqrt{3}} \\ &= \frac{1}{2} \\ \theta &= \sin^{-1}\left(\frac{1}{2}\right) \\ &= 30^\circ \end{aligned}$$

It follows that the angle separating the generator phase voltage and line voltage phasors is 30°, as shown in Figure 23.9.

An inspection of the circuit in Figure 23.8 reveals that the generator phase current is the same as the line current for each phase, and it follows that

$$I_L = I_P \qquad (23.2)$$

for the Y-connected generator.

23.12 THE Δ-CONNECTED GENERATOR

Consider the delta-connected, three-phase voltage generator in Figure 23.11. The voltages $\mathbf{V}_{AN}$, $\mathbf{V}_{BN}$, and $\mathbf{V}_{CN}$ are the generator phase voltages and are equivalent to the line voltages, which are labeled $\mathbf{V}_{AB}$, $\mathbf{V}_{BC}$, and $\mathbf{V}_{CA}$, respectively.

The point N is indicated simply to relate the connections in Figure 23.11 to

Figure 23.11 Delta-connected generator.

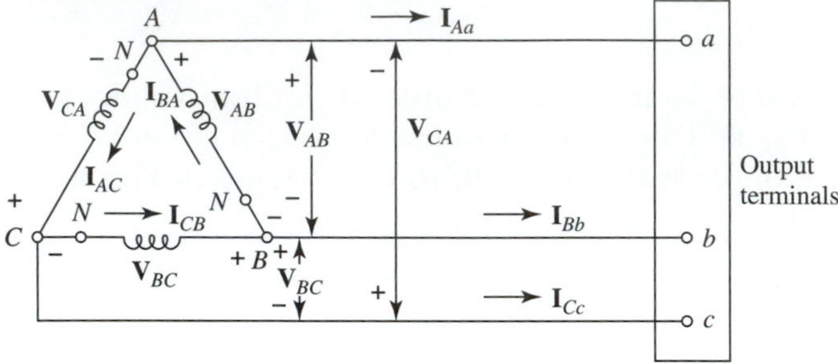

the single-phase windings in Figure 23.2. The line currents are $\mathbf{I}_{Aa}$, $\mathbf{I}_{Bb}$, and $\mathbf{I}_{Cc}$ and the generator phase currents are $\mathbf{I}_{BA}$, $\mathbf{I}_{CB}$, and $\mathbf{I}_{AC}$.

Applying Kirchhoff's current law to nodes A, B, and C in Figure 23.11 results in

$$\mathbf{I}_{BA} - \mathbf{I}_{Aa} - \mathbf{I}_{AC} = 0$$

$$\mathbf{I}_{AC} - \mathbf{I}_{Cc} - \mathbf{I}_{CB} = 0$$

$$\mathbf{I}_{CB} - \mathbf{I}_{Bb} - \mathbf{I}_{BA} = 0$$

A phasor diagram indicating the relationship between the generator phase currents and the line-currents is shown in Figure 23.12. For balanced loads the phase angle differences between the generator phase currents and generator phase voltages are equal for each phase.

It follows that the phase currents form a balanced set of phasors separated by 120° and equal in magnitude, as shown in Figure 23.12. The significance of the phasor diagram in Figure 23.12 is the phase relationship between the line and phase currents. The relationship between the current phasors in Figure 23.12 and the voltage phasors in Figure 23.2 depends on the generator load. The generator phase currents are added using the parallelogram law (Chapter 16) to form the line currents in accordance with the preceding equations.

Figure 23.12 Phasor diagram.

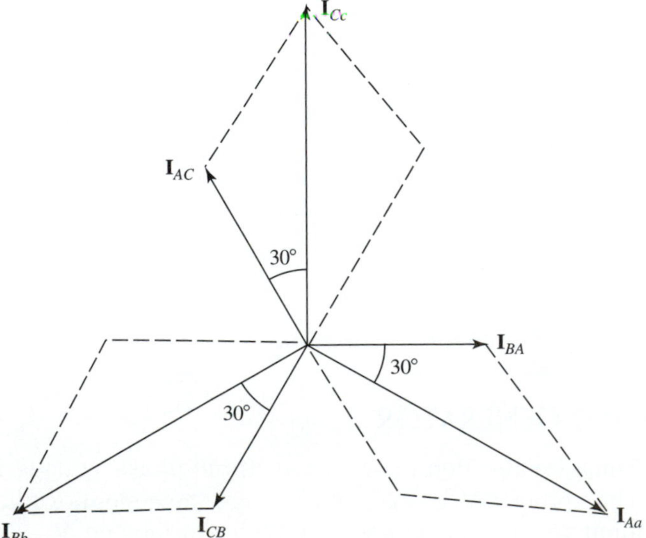

Let us determine $\mathbf{I}_{Aa}$ in Figure 23.12 from the phase currents as an illustration and note that $\mathbf{I}_{Bb}$ and $\mathbf{I}_{Cc}$ can be determined in a similar manner. From the preceding equations.

$$\mathbf{I}_{Aa} = \mathbf{I}_{BA} - \mathbf{I}_{AC}$$
$$= \mathbf{I}_{BA} + (-\mathbf{I}_{AC})$$

The phasors $\mathbf{I}_{AB}$ and $(-\mathbf{I}_{AC})$ are added as phasors using the parallelogram law, as shown in Figure 23.13.

Figure 23.13 Phasor addition.

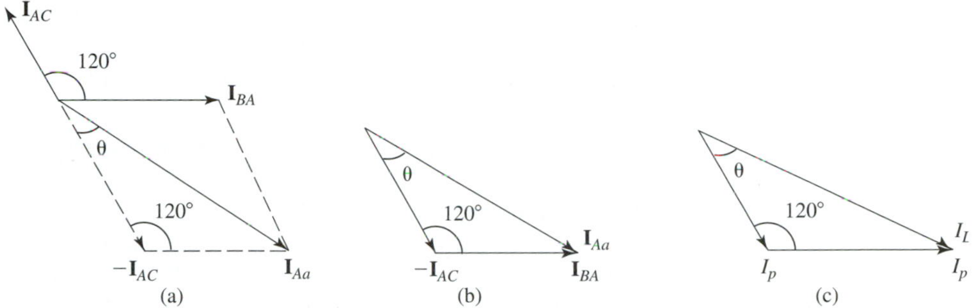

The addition of phasor currents $\mathbf{I}_{BA}$ and $(-\mathbf{I}_{AC})$ to form $\mathbf{I}_{Aa}$ by the parallelogram law (Chapter 16) is illustrated in Figure 23.13(a), and the triangular relationship between the phasors is shown in Figure 23.13(b). Because the magnitudes of all the generator phase currents are equal,

$$I_P = I_{BA} = I_{AC} = I_{CB}$$

where I_P is the magnitude of the generator phase currents. Also, because the magnitudes of the line currents are equal,

$$I_L = I_{Aa} = I_{Bb} = I_{Cc}$$

where I_L is the magnitude of the line currents.

Substituting I_P for I_{BA} and I_{AC} and I_L for I_{Aa} results in the triangle in Figure 23.13(c). Applying the law of cosines (Chapter 16) to the triangle in Figure 23.13(c) results in

$$I_L^2 = I_P^2 + I_P^2 - 2I_P I_P \cos 120°$$
$$= I_P^2 + I_P^2 - 2I_P^2\left(-\frac{1}{2}\right)$$
$$= 3I_P^2$$
$$I_L = \sqrt{3}I_P \qquad (23.3)$$

where I_L is the magnitude of the line current and I_P is the magnitude of the generator phase current.

Applying the law of sines (Chapter 16) to the triangle in Figure 23.13(c) results in

$$\frac{I_L}{\sin 120°} = \frac{I_P}{\sin \theta}$$
$$\frac{\sqrt{3}\,I_P}{\sin 120°} = \frac{I_P}{\sin \theta}$$

$$\sin\theta = \frac{(\sqrt{3}/2)}{\sqrt{3}}$$

$$= \frac{1}{2}$$

$$\theta = \sin^{-1}\left(\frac{1}{2}\right)$$

$$= 30°$$

It follows that the angle separating the generator phase current and the line current phasors is 30°, as shown in Figure 23.12.

An inspection of the circuit in Figure 23.11 reveals that the generator phase voltage is the same as the line voltage for each phase, and it follows that

$$V_L = V_P \tag{23.4}$$

for a Δ-connected generator.

The Δ-connected generator is not as widely used in commercial applications as the Y-connected generator. The principal reasons are the circulating current in the Δ-connected generator windings that results if the phase voltages do not form a perfectly balanced set and the unavailability of a neutral line.

EXAMPLE 23.1 Three independent windings (A-N, B-N, and C-N) are placed on the armature of an ac generator in such a manner that the generated voltages v_{AN}, v_{BN}, and v_{CN} are equal in magnitude and 120° out of phase with each other. The single-phase generated voltages are

$$v_{AN} = 100\sin(\omega t)$$

$$v_{BN} = 100\sin(\omega t - 120°)$$

$$v_{CN} = 100\sin(\omega t + 120°)$$

Determine the line voltages if the three independent voltages are (a) Y-connected and (b) Δ-connected to form three-phase voltage generators. ■

SOLUTION We shall first consider the Y connection, which is symbolically drawn in Figure 23.14. From the given expressions of v_{AN}, v_{BN}, and v_{CN}, phasor expressions for the generator phase voltage are

$$\mathbf{V}_{AN} = 100\underline{/0°}$$

$$\mathbf{V}_{BN} = 100\underline{/-120°}$$

$$\mathbf{V}_{CN} = 100\underline{/120°}$$

Figure 23.14

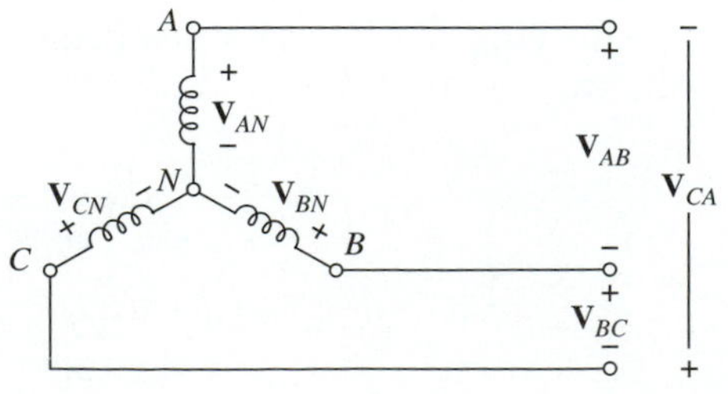

From Equation (23.1)

$$V_L = \sqrt{3}\,V_P$$
$$= \sqrt{3}(100)$$

where V_L is the magnitude of the line voltages $\mathbf{V}_{AB}$, $\mathbf{V}_{BC}$, and $\mathbf{V}_{CA}$, and V_P is the magnitude of the generator phase voltages $\mathbf{V}_{AN}$, $\mathbf{V}_{BN}$, and $\mathbf{V}_{CN}$.

From Figure 23.9, the line voltages lead the phase voltages by 30°, as indicated. It follows that

$$\mathbf{V}_{AB} = V_{AB}\underline{/0° + 30°}$$
$$= \sqrt{3}(100)\underline{/30°}$$
$$\mathbf{V}_{BC} = V_{BC}\underline{/-120° + 30°}$$
$$= \sqrt{3}(100)\underline{/-90°}$$
$$\mathbf{V}_{CA} = V_{CA}\underline{/120° + 30°}$$
$$= \sqrt{3}(100)\underline{/150°}$$

From these phasor expressions, the generator line voltages in the t domain are

$$v_{AB} = \sqrt{3}(100)\sin(\omega t + 30°)$$
$$v_{BC} = \sqrt{3}(100)\sin(\omega t - 90°)$$
$$v_{CA} = \sqrt{3}(100)\sin(\omega t + 150°)$$

We shall now consider the Δ connection, which is symbolically drawn in Figure 23.15. The point N is included only to indicate the method of connecting the three windings.

Figure 23.15

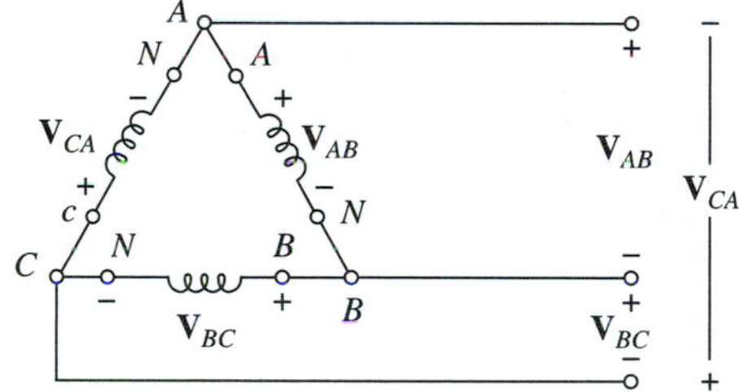

From the given expressions for v_{AN}, v_{BN}, and v_{CN}, the generator phase voltages are

$$\mathbf{V}_{AN} = 100\underline{/0°}$$
$$\mathbf{V}_{BN} = 100\underline{/-120°}$$
$$\mathbf{V}_{CN} = 100\underline{/120°}$$

For the Δ-connected generator, the line voltages and the generator phase voltages are equal. It follows that

$$\mathbf{V}_{AB} = \mathbf{V}_{AN} = 100\underline{/0°}$$
$$\mathbf{V}_{BC} = \mathbf{V}_{BN} = 100\underline{/-120°}$$
$$\mathbf{V}_{CA} = \mathbf{V}_{CN} = 100\underline{/120°}$$

From the phasor expressions, the generator line voltages in the t domain are

$$v_{AB} = 100\sin(\omega t)$$
$$v_{BC} = 100\sin(\omega t - 120°)$$
$$v_{CA} = 100\sin(\omega t + 120°)$$

23.13 Y-CONNECTED LOAD

Consider the balanced three-phase, Y-connected load in Figure 23.16, where $\mathbf{Z} = Z\angle\theta$, Z is the magnitude of the impedance, and θ is the impedance angle separating the load phase voltage and load phase current for each phase of the load.

Figure 23.16 Y-connected load.

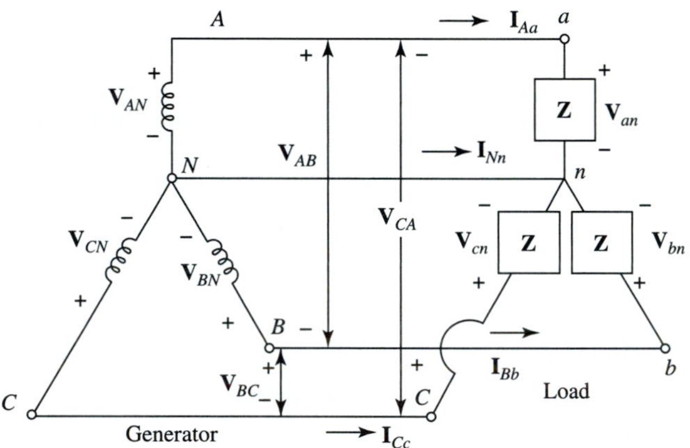

The line voltages are a balanced three-phase set produced by the Y-connected generator, as previously presented. The phasor diagram for the generator phase voltages and the line voltages is shown in Figure 23.9 and reproduced for convenience in Figure 23.17.

Figure 23.17 Phasor diagram.

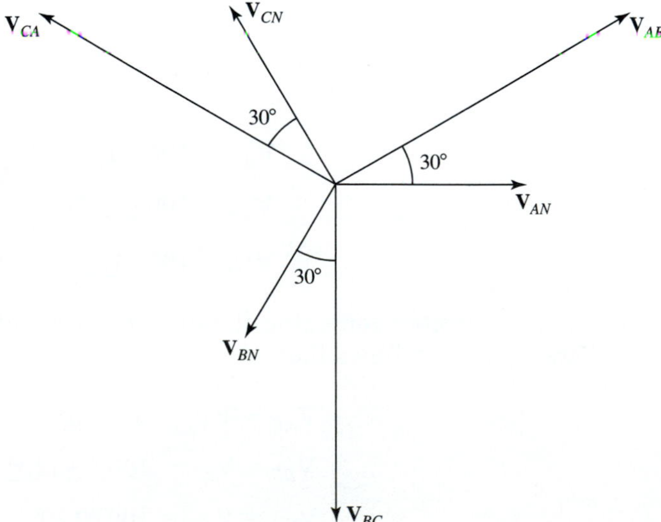

An inspection of the circuit in Figure 23.17 reveals that the load phase voltages are equal to the generator phase voltages and that the load phase currents and line currents are also equal. That is,

$$\mathbf{V}_{an} = \mathbf{V}_{AN} = V_P\underline{/0^\circ}$$

$$\mathbf{V}_{bn} = \mathbf{V}_{BN} = V_P\underline{/-120^\circ}$$

$$\mathbf{V}_{cn} = \mathbf{V}_{CN} = V_P\underline{/120^\circ}$$

where V_P is the magnitude of both the generator phase and the load phase voltage.

From these equations and Equation (17.1),

$$\mathbf{I}_{an} = \mathbf{I}_{Aa} = \frac{\mathbf{V}_{an}}{\mathbf{Z}} = \frac{V_P\underline{/0^\circ}}{Z\underline{/\theta}} = \left(\frac{V_P}{Z}\right)\underline{/0^\circ - \theta}$$

$$\mathbf{I}_{bn} = \mathbf{I}_{Bb} = \frac{\mathbf{V}_{bn}}{\mathbf{Z}} = \frac{V_P\underline{/-120^\circ}}{Z\underline{/\theta}} = \left(\frac{V_P}{Z}\right)\underline{/-120^\circ - \theta}$$

$$\mathbf{I}_{cn} = \mathbf{I}_{Cc} = \frac{V_{cn}}{\mathbf{Z}} = \frac{V_P\underline{/120^\circ}}{Z\underline{/\theta}} = \left(\frac{V_P}{Z}\right)\underline{/120^\circ - \theta}$$

where **Z** is the impedance of each phase of the load. The angle θ is positive for a load with a lag power factor and negative for a load with a lead power factor. A lag power factor is assumed for the phasor diagram in Figure 23.18, where the line voltages and currents and the phase voltages and currents are indicated. (If θ were negative (lead power factor), the phase and line currents would lead the phase voltages by the angle θ in Figure 23.18.)

Kirchhoff's current law applied at point n in Figure 23.16 results in

$$\mathbf{I}_{an} + \mathbf{I}_{bn} + \mathbf{I}_{cn} + \mathbf{I}_{Nn} = 0$$

Figure 23.18 Phasor diagram.

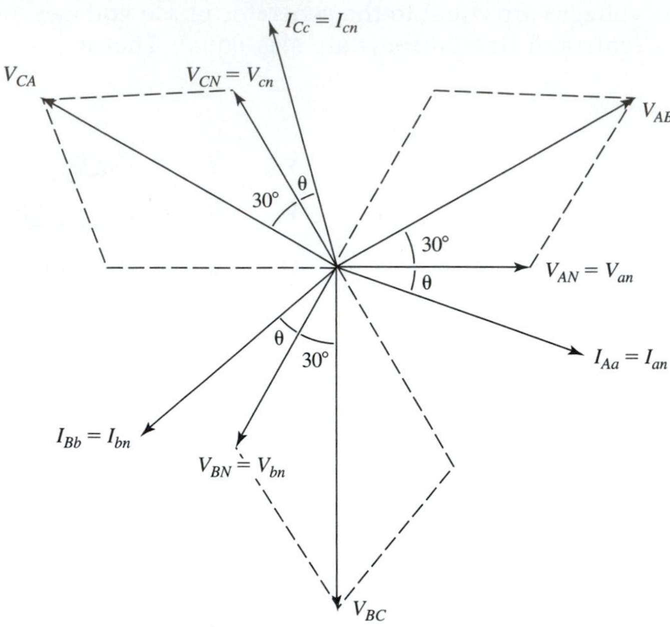

Because the three phasors ($\mathbf{I}_{an}$, $\mathbf{I}_{bn}$, and $\mathbf{I}_{cn}$) are equal in magnitude and separated by 120°, their sum is zero and it follows from the preceding equation that

$$\mathbf{I}_{Nn} = 0$$

as long as the load is balanced.

Because the line voltages and load phase voltages in Figure 23.18 bear the same relationship as the line voltages and generator phase voltages in Figure 23.17, the relationship between the magnitudes of the load phase voltages and the line voltages is the same as given by Equation (23.1). That is,

$$V_L = \sqrt{3}\, V_P \tag{23.5}$$

where V_L is the magnitude of the line voltage and V_P is the magnitude of the load phase voltage. Because the line currents and load phase currents are equal,

$$I_L = I_P \tag{23.6}$$

where I_L is the magnitude of the line current and I_P is the magnitude of the load phase current.

If the load for the circuit in Figure 23.16 is balanced, the neutral line current $\mathbf{I}_{Nn}$ is zero as previously stated and the removal of the neutral line has no effect on the current-voltage relationships developed in this section.

EXAMPLE 23.2 A balanced *Y*-connected load is driven by a three-phase generator with line voltages as shown in Figure 23.19. Determine expressions for the phase voltages, phase currents, and line currents. ■

Figure 23.19

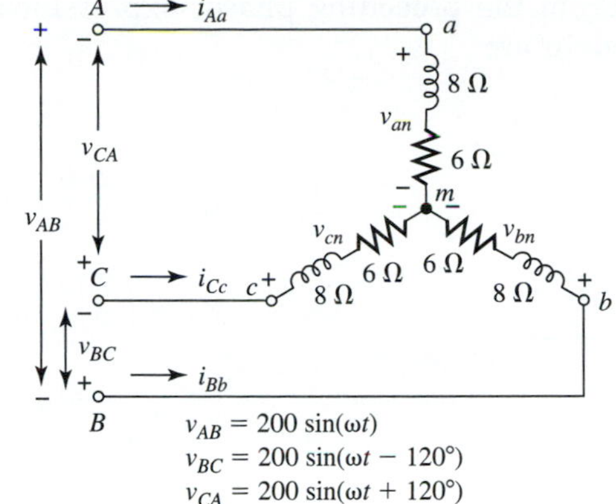

SOLUTION From the given expressions for the line voltages, the phasor line voltages are

$$\mathbf{V}_{AB} = 200\angle 0°$$

$$\mathbf{V}_{BC} = 200\angle -120°$$

$$\mathbf{V}_{CA} = 200\angle 120°$$

From Equation (23.5)

$$V_P = \frac{V_L}{\sqrt{3}} = \frac{200}{\sqrt{3}}$$

where V_L is the magnitude of the line voltages $\mathbf{V}_{AB}$, $\mathbf{V}_{BC}$, and $\mathbf{V}_{CA}$ and V_P is the magnitude of the load phase voltages $\mathbf{V}_{an}$, $\mathbf{V}_{bn}$, and $\mathbf{V}_{cn}$.

From Figure 23.18, the line voltages lead the phase voltages by 30°, as indicated. That is, $\mathbf{V}_{AB}$ leads $\mathbf{V}_{an}$ by 30°, $\mathbf{V}_{BC}$ leads $\mathbf{V}_{bn}$ by 30°, and $\mathbf{V}_{CA}$ leads $\mathbf{V}_{cn}$ by 30°. It follows that

$$\mathbf{V}_{an} = V_{an}\angle 0° - 30° = \left(\frac{200}{\sqrt{3}}\right)\angle -30°$$

$$\mathbf{V}_{bn} = V_{bn}\angle -120° - 30° = \left(\frac{200}{\sqrt{3}}\right)\angle -150°$$

$$\mathbf{V}_{cn} = V_{cn}\angle 120° - 30° = \left(\frac{200}{\sqrt{3}}\right)\angle 90°$$

From the preceding phasor expressions, the load phase voltages in the t domain are

$$v_{an} = \left(\frac{200}{\sqrt{3}}\right)\sin(\omega t - 30°)$$

$$v_{bn} = \left(\frac{200}{\sqrt{3}}\right)\sin(\omega t - 150°)$$

$$v_{cn} = \left(\frac{200}{\sqrt{3}}\right)\sin(\omega t + 90°)$$

From Equation (17.1), the phase currents $\mathbf{I}_{an}$, $\mathbf{I}_{bn}$, and $\mathbf{I}_{cn}$ are

$$\mathbf{I}_{an} = \frac{\mathbf{V}_{an}}{6 + j8} = \frac{(200/\sqrt{3})\angle -30°}{10\angle 53.1°} = \left(\frac{20}{\sqrt{3}}\right)\angle -83.1°$$

$$\mathbf{I}_{bn} = \frac{\mathbf{V}_{bn}}{6 + j8} = \frac{(200/\sqrt{3})\angle -150°}{10\angle 53.1°} = \left(\frac{20}{\sqrt{3}}\right)\angle -203.1°$$

$$\mathbf{I}_{cn} = \frac{\mathbf{V}_{cn}}{6 + j8} = \frac{(200/\sqrt{3})\angle 90°}{10\angle 53.1°} = \left(\frac{20}{\sqrt{3}}\right)\angle 36.9°$$

Because the line currents and load phase currents are equal for the Y-connected load, the line and phase currents in the t domain are as follows.

$$i_{an} = i_{Aa} = \left(\frac{20}{\sqrt{3}}\right)\sin(\omega t - 83.1°)$$

$$i_{bn} = i_{Bb} = \left(\frac{20}{\sqrt{3}}\right)\sin(\omega t - 203.1°)$$

$$i_{cn} = i_{Cc} = \left(\frac{20}{\sqrt{3}}\right)\sin(\omega t + 36.9°)$$

23.14 Δ-CONNECTED LOAD

Consider the balanced three-phase, Δ-connected load in Figure 23.20, where $\mathbf{Z} = Z\angle\theta$, Z is the magnitude of the impedance, and θ is the impedance angle separating the load phase voltage and the load phase current for each phase of the load.

Figure 23.20 Δ-connected load.

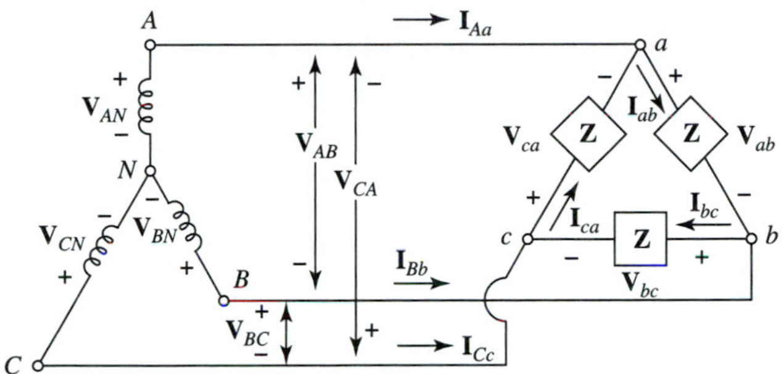

The line voltages are a balanced three-phase set produced by the Y-connected generator, as previously presented. The phasor diagram for the generator phase voltages and the line voltages is shown in Figure 23.9 and is reproduced for convenience in Figure 23.21.

Figure 23.21 Phasor diagram.

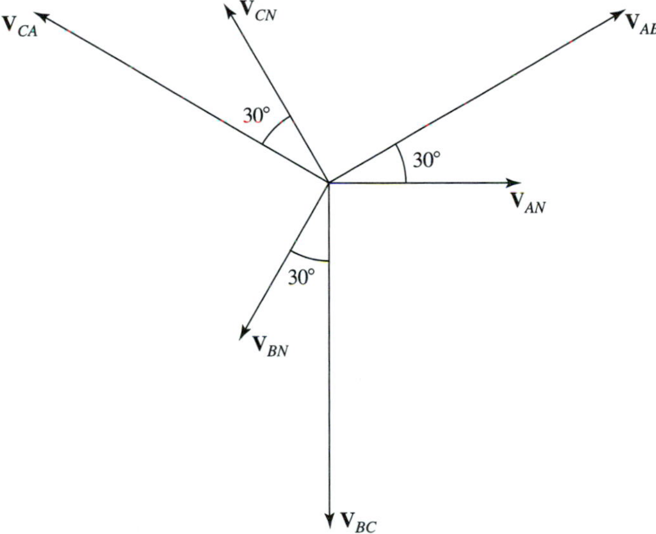

An inspection of the circuit in Figure 23.17 reveals that the load phase voltages are equal to the line voltages. That is,

$$\mathbf{V}_{ab} = \mathbf{V}_{AB} = V_L\angle 30^\circ$$

$$\mathbf{V}_{bc} = \mathbf{V}_{BC} = V_L\angle -90^\circ$$

$$\mathbf{V}_{ca} = \mathbf{V}_{CA} = V_L\angle 150^\circ$$

where V_L is the magnitude of both the line voltages and the load phase voltages. It follows that

$$V_L = V_P$$

where V_P is the magnitude of the load phase voltage.

From Equation (17.1)

$$\begin{aligned}
\mathbf{I}_{ab} &= \frac{\mathbf{V}_{ab}}{\mathbf{Z}} \\
&= \frac{V_L\angle 30^\circ}{Z\angle\theta} \\
&= \left(\frac{V_L}{Z}\right)\angle 30^\circ - \theta \\
\mathbf{I}_{bc} &= \frac{\mathbf{V}_{bc}}{\mathbf{Z}} \\
&= \frac{V_L\angle -90^\circ}{Z\angle\theta} \\
&= \left(\frac{V_L}{Z}\right)\angle -90^\circ - \theta \\
\mathbf{I}_{ca} &= \frac{\mathbf{V}_{ca}}{\mathbf{Z}} \\
&= \frac{V_L\angle 150^\circ}{Z\angle\theta} \\
&= \left(\frac{V_L}{Z}\right)\angle 150^\circ - \theta
\end{aligned}$$

where $\mathbf{Z}$ is the impedance of each phase of the load. The angle θ is positive for a load with a lag power factor and negative for a load with a lead power factor. A lag power factor is assumed for the phasor diagram in Figure 23.22, where the line voltages and currents and the phase voltages and currents for the load are indicated. If θ were negative (lead power factor), the phase currents would lead the line voltages by the angle θ in Figure 23.22.

The line currents $\mathbf{I}_{Aa}$, $\mathbf{I}_{Bb}$, and $\mathbf{I}_{Cc}$ in Figure 23.22 are obtained by applying Kirchhoff's current law at points a, b, and c of the load. Kirchhoff's current law applied at nodes a, b, and c results in

$$\begin{aligned}
\mathbf{I}_{Aa} &= \mathbf{I}_{ab} - \mathbf{I}_{ca} \\
&= \mathbf{I}_{ab} + (-\mathbf{I}_{ca}) \\
\mathbf{I}_{Bb} &= \mathbf{I}_{bc} - \mathbf{I}_{ab} \\
&= \mathbf{I}_{bc} + (-\mathbf{I}_{ab}) \\
\mathbf{I}_{Cc} &= \mathbf{I}_{ca} - \mathbf{I}_{bc} \\
&= \mathbf{I}_{ca} + (-\mathbf{I}_{bc})
\end{aligned}$$

Figure 23.22 Phasor diagram.

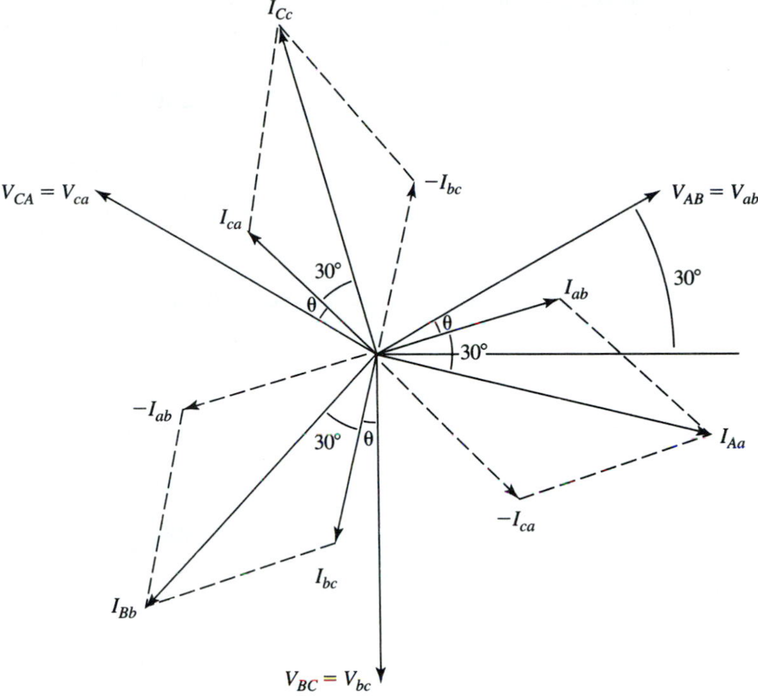

The phase currents are added using the parallelogram law, as indicated in Figure 23.22, to form the line currents. This type of addition was demonstrated in the last section and results in the 30° angle between the line and load phase currents, as shown in Figure 23.22, and

$$I_L = \sqrt{3}\, I_P \tag{23.7}$$

where I_L is the magnitude of the line current and I_P is the magnitude of the load phase current. As stated previously, the line voltages and load phase voltages are equal for a Δ-connected load; that is,

$$V_L = V_P \tag{23.8}$$

EXAMPLE 23.3 A balanced Δ-connected load is driven by a three-phase generator with line voltages as shown in Figure 23.23. Determine the expressions for the phase voltages, phase currents, and line currents. ■

Figure 23.23

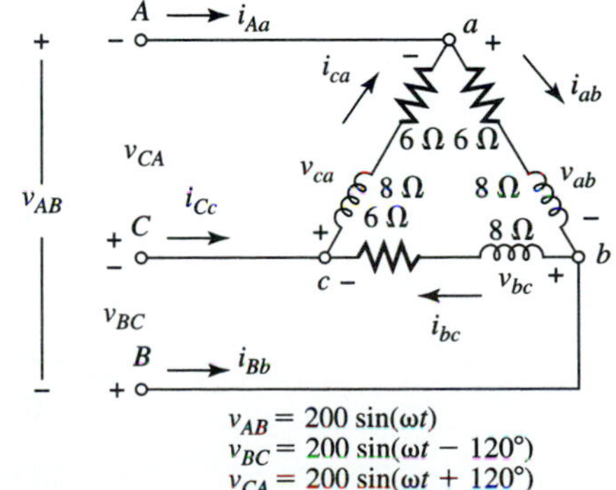

SOLUTION From the given expressions for the line voltages, the phasor line voltages are

$$\mathbf{V}_{AB} = 200\underline{/0^\circ}$$
$$\mathbf{V}_{BC} = 200\underline{/-120^\circ}$$
$$\mathbf{V}_{CA} = 200\underline{/120^\circ}$$

Because the line and phase voltages are equal for the Δ-connection, the load phase voltages are

$$\mathbf{V}_{ab} = 200\underline{/0^\circ}$$
$$\mathbf{V}_{bc} = 200\underline{/-120^\circ}$$
$$\mathbf{V}_{ca} = 200\underline{/120^\circ}$$

and in the t domain

$$v_{ab} = 200 \sin(\omega t)$$
$$v_{bc} = 200 \sin(\omega t - 120^\circ)$$
$$v_{ca} = 200 \sin(\omega t + 120^\circ)$$

From Equation (17.1), the phase currents $\mathbf{I}_{ab}$, $\mathbf{I}_{bc}$, and $\mathbf{I}_{ca}$ are

$$\mathbf{I}_{ab} = \frac{\mathbf{V}_{ab}}{6 + j8}$$
$$= \frac{200\underline{/0^\circ}}{10\underline{/53.1^\circ}}$$
$$= 20\underline{/-53.1^\circ}$$
$$\mathbf{I}_{bc} = \frac{\mathbf{V}_{bc}}{6 + j8}$$
$$= \frac{200\underline{/-120^\circ}}{10\underline{/53.1^\circ}}$$
$$= 20\underline{/-173.1^\circ}$$
$$\mathbf{I}_{ca} = \frac{\mathbf{V}_{ca}}{6 + j8}$$
$$= \frac{200\underline{/120^\circ}}{10\underline{/53.1^\circ}}$$
$$= 20\underline{/66.9^\circ}$$

and in the t domain

$$i_{ab} = 20\sin(\omega t - 53.1°)$$
$$i_{bc} = 20\sin(\omega t - 173.1°)$$
$$i_{ca} = 20\sin(\omega t + 66.9°)$$

From Equation (23.7)

$$\begin{aligned} I_L &= \sqrt{3}\,I_P \\ &= \sqrt{3}(20) \end{aligned}$$

where I_L is the magnitude of the line currents $\mathbf{I}_{Aa}$, $\mathbf{I}_{Bb}$, and $\mathbf{I}_{Cc}$ and I_P is the magnitude of the load phase currents $\mathbf{I}_{ab}$, $\mathbf{I}_{bc}$, and $\mathbf{I}_{ca}$.

From Figure 23.22, the line currents lag the load phase currents by 30°, as indicated. That is, $\mathbf{I}_{Aa}$ lags $\mathbf{I}_{ab}$ by 30°, $\mathbf{I}_{Cc}$ lags $\mathbf{I}_{ca}$ by 30°, and $\mathbf{I}_{Bb}$ lags $\mathbf{I}_{bc}$ by 30°. It follows that

$$\begin{aligned} \mathbf{I}_{Aa} &= I_{Aa}\underline{/-53.1° - 30°} \\ &= \sqrt{3}(20)\underline{/-83.1°} \\ \mathbf{I}_{Bb} &= I_{Bb}\underline{/-173.1° - 30°} \\ &= \sqrt{3}(20)\underline{/-203.1°} \\ \mathbf{I}_{Cc} &= I_{Cc}\underline{/66.9° - 30°} \\ &= \sqrt{3}(20)\underline{/36.9°} \end{aligned}$$

From the preceding phasor expressions, the line currents in the t domain are

$$i_{Aa} = \sqrt{3}(20)\sin(\omega t - 83.1°)$$
$$i_{Bb} = \sqrt{3}(20)\sin(\omega t - 203.1°)$$
$$i_{Cc} = \sqrt{3}(20)\sin(\omega t + 36.9°)$$

23.15 POWER IN THREE-PHASE LOADS

The power delivered to a three-phase load is measurable using wattmeters under laboratory or field conditions and can also be calculated analytically using the power formulas presented here.

The total power delivered to a three-phase load (balanced or unbalanced) is the sum of the powers delivered to each of the three load phases; that is,

$$P = P_1 + P_2 + P_3 \tag{23.9}$$

where P is the total power and P_1, P_2, and P_3 are the individual load phase powers.

If the load is balanced, $P_1 = P_2 = P_3$ and Equation (23.9) can be written as

$$P = 3P_\phi$$

where P_ϕ is the power delivered to each phase of the load. From Equation (20.7)

$$P_\phi = V_{P_{\text{eff}}} I_{P_{\text{eff}}} \cos\theta$$

where $V_{P_{\text{eff}}}$, $I_{P_{\text{eff}}}$, and $\cos\theta$ are the effective phase voltage, the effective phase current, and phase power factor, respectively, for each phase of the load.

If a balanced load is connected in a Δ configuration, $V_{P_{\text{eff}}} = V_{L_{\text{eff}}}$, $I_{P_{\text{eff}}} = I_{L_{\text{eff}}}/\sqrt{3}$, and the equation for total power, $P = 3P_\phi$, can be written as

$$\begin{aligned} P &= 3V_{L_{\text{eff}}}\left(\frac{I_{L_{\text{eff}}}}{\sqrt{3}}\right)\cos\theta \\ &= \sqrt{3}\,V_{L_{\text{eff}}} I_{L_{\text{eff}}} \cos\theta \end{aligned}$$

The same result is obtained for Y-connected loads. If a balanced load is connected in a Y configuration, $V_{P_{\text{eff}}} = V_{L_{\text{eff}}}/\sqrt{3}$, $I_{P_{\text{eff}}} = I_{L_{\text{eff}}}$, and the equation $P = 3P_\phi$ can be written as

$$\begin{aligned} P &= 3\left(\frac{V_{L_{\text{eff}}}}{\sqrt{3}}\right)(I_{L_{\text{eff}}})\cos\theta \\ &= \sqrt{3}\,V_{L_{\text{eff}}} I_{L_{\text{eff}}} \cos\theta \end{aligned}$$

It follows that the total average power delivered to a balanced three-phase Δ- or Y-connected load is

$$P = \sqrt{3}\,V_{L_{\text{eff}}} I_{L_{\text{eff}}} \cos\theta \tag{23.10}$$

From Equation (23.10) the power factor for each phase of a three-phase balanced load can be written as

$$\cos\theta = \frac{P}{\sqrt{3}\,V_{L_{\text{eff}}} I_{L_{\text{eff}}}} \tag{23.11}$$

23.16 POWER MEASUREMENT (THREE-WATTMETER METHOD)

Three wattmeters can be employed to measure the power delivered to a three-phase load (balanced or unbalanced), as shown in Figure 23.24. The proper connection of wattmeters is presented in Chapter 20. The sum of the wattmeter readings P_1, P_2, and P_3 is the total power delivered to the load for the Δ and Y connections in Figure 23.24. Notice that each wattmeter senses the phase voltage and the phase current and determines the phase power. Wattmeter P_1 senses i_{an} and v_{an}, wattmeter P_2 senses i_{bn} and v_{bn}, and wattmeter P_3 senses i_{cn}

and v_{cn} for the Y-connected load, and wattmeter P_1 senses v_{ab} and i_{ab}, wattmeter P_2 senses v_{bc} and i_{bc}, and wattmeter P_3 senses v_{ca} and i_{ca} for the Δ-connected load.

Figure 23.24 Three-wattmeter method.

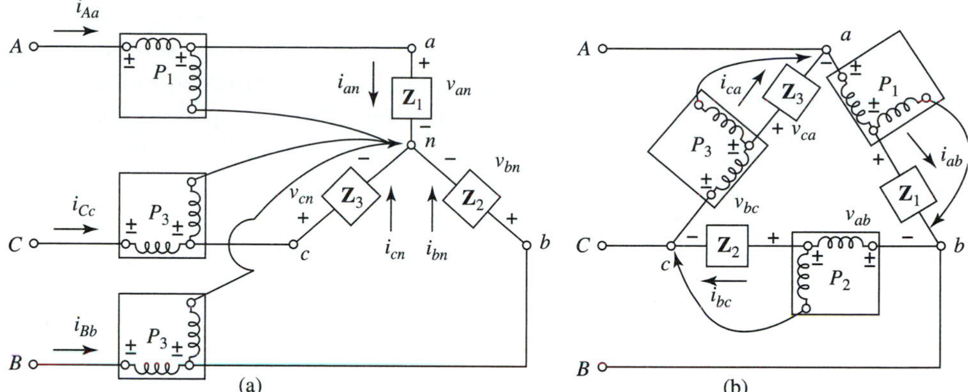

Because the neutral point n for three-wire Y-connected loads and the individual phase terminals for Δ-connected loads are not always physically available for the wattmeter connections as shown in Figure 23.24, the three-wattmeter method is not widely used. It is far more desirable to measure the power delivered to a three-phase load using only the lines delivering power to the load for the wattmeter connections and avoiding the necessity of accessing the internal parts of the load. This can be accomplished by using the two-wattmeter method presented next.

23.17 POWER MEASUREMENT (TWO-WATTMETER METHOD)

Consider the three-wire, Y-connected load in Figure 23.25(a). The instantaneous voltages sensed by wattmeters P_1, P_2, and P_3 are v_{an}, v_{bn}, and v_{cn}, respectively. The instantaneous currents sensed by wattmeters P_1, P_2, and P_3 are

Figure 23.25 Two-wattmeter method.

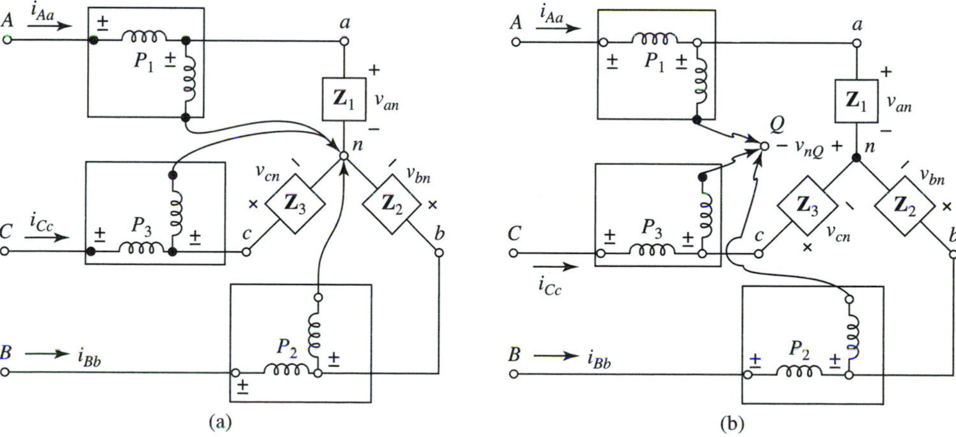

i_{Aa}, i_{Bb}, and i_{Cc}, respectively, where $i_{an} = i_{Aa}$, $i_{bn} = i_{Bb}$, and $i_{cn} = i_{Cc}$. It follows from Equation (2.6) that the instantaneous phase powers seen by wattmeters P_1, P_2, and P_3 are $v_{an} i_{Aa}$, $v_{bn} i_{Bb}$, and $v_{cn} i_{Cn}$, respectively, and the total instantaneous power p for the three phase load is

$$p = p_1 + p_2 + p_3$$
$$= v_{an} i_{Aa} + v_{bn} i_{Bb} + v_{cn} i_{Cc}$$

Consider the Y-connected load in Figure 23.25(b), where the neutral point n is moved to an arbitrary point Q. The instantaneous voltages sensed by wattmeters P_1, P_2, and P_3 are $(v_{an} + v_{nQ})$, $(v_{bn} + v_{nQ})$, and $(v_{cn} + v_{nQ})$, respectively, and the instantaneous currents sensed by the wattmeters are unchanged from Figure 23.25(a). It follows from Equation (2.6) that the instantaneous powers seen by the wattmeters P_1, P_2, and P_3 are $(v_{an} + v_{nQ})i_{Aa}$, $(v_{bn} + v_{nQ})i_{Bb}$, and $(v_{cn} + v_{nQ})i_{Cc}$, respectively, and the total instantaneous power for the three phase load is

$$p = (v_{an} + v_{nQ})i_{Aa} + (v_{bn} + v_{nQ})i_{Bb} + (v_{cn} + v_{nQ})i_{Cc}$$
$$= v_{an} i_{Aa} + v_{bn} i_{Bb} + v_{cn} i_{Cc} + v_{nQ}(i_{Aa} + i_{Bb} + i_{Cc})$$

Applying Kirchhoff's current law at point n results in

$$i_{Aa} + i_{Bb} + i_{Cc} = 0$$

and the preceding equation for p can be written as

$$p = v_{an} i_{Aa} + v_{bn} i_{Bb} + v_{cn} i_{Cc}$$

Because the total instantaneous power seen by the three wattmeters in Figure 23.25(a) with common connection n is equal to the total instantaneous power seen by the three wattmeters in Figure 23.25(b) with common connection Q, the same total average power will be indicated for both wattmeter configurations. Because point Q is an arbitrary point in Figure 23.25(b), it can be placed on any one of the three input lines, resulting in a zero reading for one of the wattmeters, as shown in Figure 23.26. Wattmeter 3 will read zero in Figure 23.26, because the voltage coil of wattmeter 3 senses 0 V.

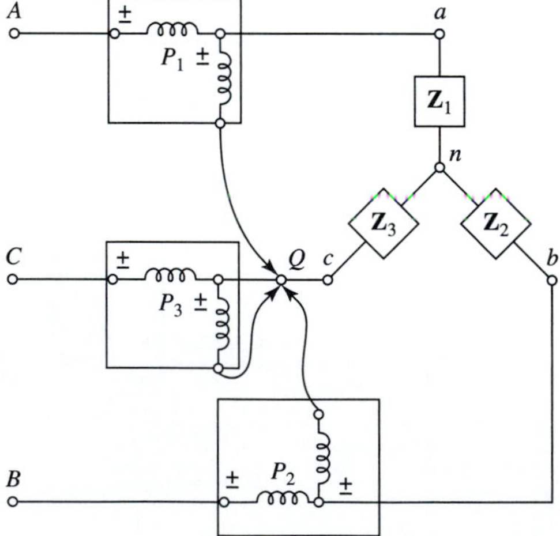

Figure 23.26 Two-wattmeter method.

Because wattmeter 3 reads zero, it can be removed, and the sum of wattmeters 1 and 2 will be the total average power delivered to the three-phase load. Because the load power factor can cause one of the wattmeters to read negative, the two wattmeter readings must be algebraically summed to determine the total power delivered to the three-phase load.

The three possible connections for measuring three-phase power with two wattmeters are shown in Figure 23.27. The two-wattmeter method presented here is applicable to delta and wye loads.

Figure 23.27 Two-wattmeter method.

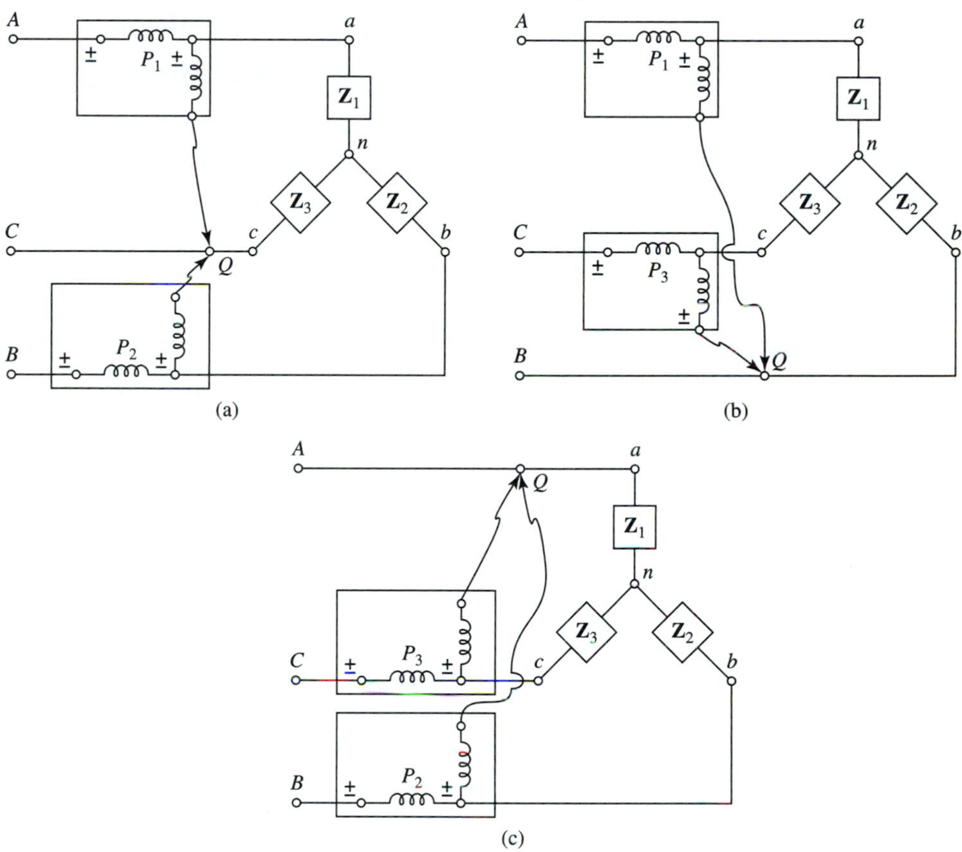

Consider the following examples.

EXAMPLE 23.4 If the power delivered to the three-phase load in Figure 23.28 is measured using the two-wattmeter method, as shown in the figure, determine the wattmeter readings and the total power delivered to the load. ■

Figure 23.28

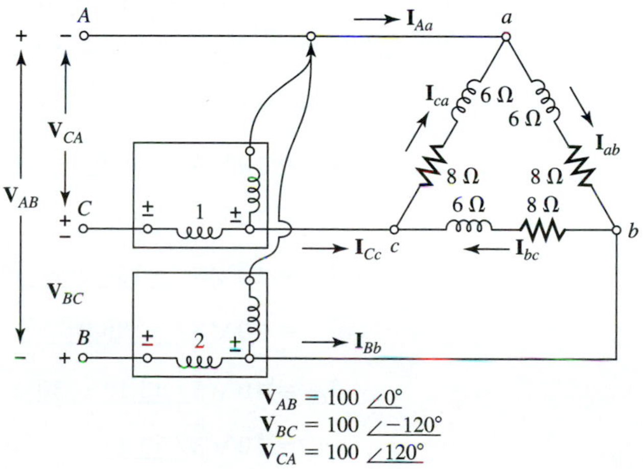

I_{Aa}, I_{ca}, I_{ab}, I_{Cc}, I_{bc}, I_{Bb}, V_{AB}, V_{CA}, V_{BC}, 6 Ω, 6 Ω, 8 Ω, 8 Ω, 6 Ω, 8 Ω

$\mathbf{V}_{AB} = 100 \angle 0°$
$\mathbf{V}_{BC} = 100 \angle -120°$
$\mathbf{V}_{CA} = 100 \angle 120°$

SOLUTION Let us first locate the voltages and currents sensed by the wattmeters by drawing a phasor diagram. Because the phase voltages are equal to the line voltages for the Δ-connected load,

$$\mathbf{V}_{ab} = 100\angle 0^\circ$$
$$\mathbf{V}_{bc} = 100\angle -120^\circ$$
$$\mathbf{V}_{ca} = 100\angle 120^\circ$$

From Equation (17.1)

$$\mathbf{I}_{ab} = \frac{\mathbf{V}_{ab}}{8 + j6} = \frac{100\angle 0^\circ}{10\angle 36.9^\circ} = 10\angle -36.9^\circ$$

$$\mathbf{I}_{bc} = \frac{\mathbf{V}_{bc}}{8 + j6} = \frac{100\angle -120^\circ}{10\angle 36.9^\circ} = 10\angle -156.9^\circ$$

$$\mathbf{I}_{ca} = \frac{\mathbf{V}_{ca}}{8 + j6} = \frac{100\angle 120^\circ}{10\angle 36.9^\circ} = 10\angle 83.1^\circ$$

As shown previously, the line currents lag the phase currents by 30° for a balanced Δ-connected load. That is, $\mathbf{I}_{Aa}$ lags $\mathbf{I}_{ab}$ by 30°, $\mathbf{I}_{Bb}$ lags $\mathbf{I}_{bc}$ by 30°, and $\mathbf{I}_{Cc}$ lags $\mathbf{I}_{Ca}$ by 30∘; and from Equation (23.7) $I_L = \sqrt{3}I_P$. It follows that

$$\mathbf{I}_{Aa} = 10\sqrt{3}\angle -36.9^\circ - 30^\circ = 10\sqrt{3}\angle -66.9^\circ$$

$$\mathbf{I}_{Bb} = 10\sqrt{3}\angle -156.9^\circ - 30^\circ = 10\sqrt{3}\angle -186.9^\circ$$

$$\mathbf{I}_{Cc} = 10\sqrt{3}\angle 83.1^\circ - 30^\circ = 10\sqrt{3}\angle 53.1^\circ$$

A phasor diagram consisting of the line currents and line voltages is shown in Figure 23.29. This phasor diagram contains the voltages and currents sensed by the wattmeters in the two-wattmeter configuration given here.

Figure 23.29

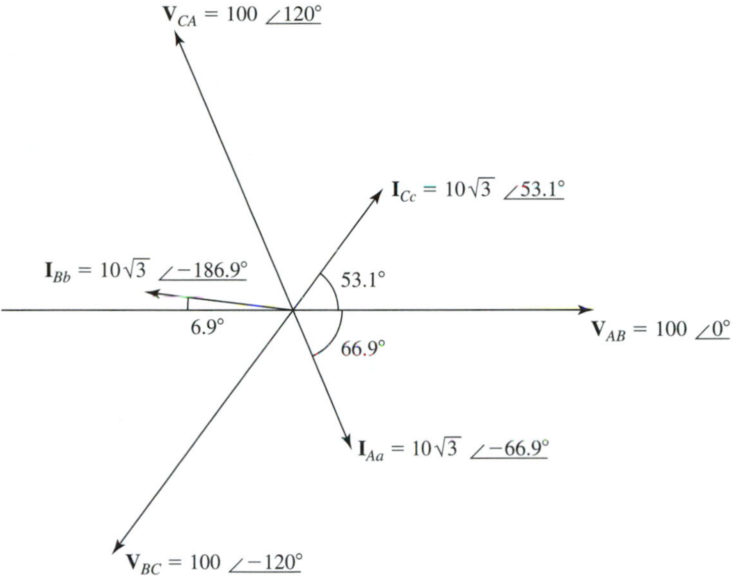

Wattmeter 1 senses line current $\mathbf{I}_{Cc}$ and line voltage $\mathbf{V}_{CA}$. Therefore, the power read by wattmeter 1 is

$$P_1 = V_{CA_{\text{eff}}} I_{Cc_{\text{eff}}} \cos\theta$$

where θ is the angle separating $\mathbf{V}_{CA}$ and $\mathbf{I}_{Cc}$. From the phasor diagram, the preceding equations, and Equation (20.2),

$$\theta = 120^\circ - 53.1^\circ = 66.9^\circ$$

$$V_{CA_{\text{eff}}} = \frac{100}{\sqrt{2}}$$

$$I_{Cc_{\text{eff}}} = \frac{10\sqrt{3}}{\sqrt{2}}$$

and substitution of these values into the equation for P_1 results in

$$\begin{aligned} P_1 &= \left(\frac{100}{\sqrt{2}}\right)\left(\frac{10\sqrt{3}}{\sqrt{2}}\right)\cos(66.9^\circ) \\ &= 340 \end{aligned}$$

Wattmeter 2 senses line current $\mathbf{I}_{Bb}$ and line voltage $\mathbf{V}_{BA}$. Therefore, the power read by wattmeter 2 is

$$P_2 = V_{BA_{\text{eff}}} I_{Bb_{\text{eff}}} \cos\theta$$

where θ is the angle separating $\mathbf{V}_{BA}$ and $\mathbf{I}_{Bb}$. From the phasor diagram, the preceding equations, and Equation (20.2),

$$\theta = 6.9°$$

$$V_{BA_{\text{eff}}} = \frac{100}{\sqrt{2}}$$

$$I_{Bb_{\text{eff}}} = \frac{10\sqrt{3}}{\sqrt{2}}$$

and substitution of these values into the equation for P_2 results in

$$\begin{aligned} P_2 &= \left(\frac{100}{\sqrt{2}}\right)\left(\frac{10\sqrt{3}}{\sqrt{2}}\right)\cos(6.9°) \\ &= 860 \end{aligned}$$

The total power P delivered to the load is

$$\begin{aligned} P &= P_1 + P_2 \\ &= 340 + 860 \\ &= 1200 \text{ W} \end{aligned}$$

EXAMPLE 23.5 If the power delivered to the three-phase load in Example 23.4 is measured using the three-wattmeter method shown in Figure 23.30, determine the wattmeter readings and the total power delivered to the load. ■

Figure 23.30

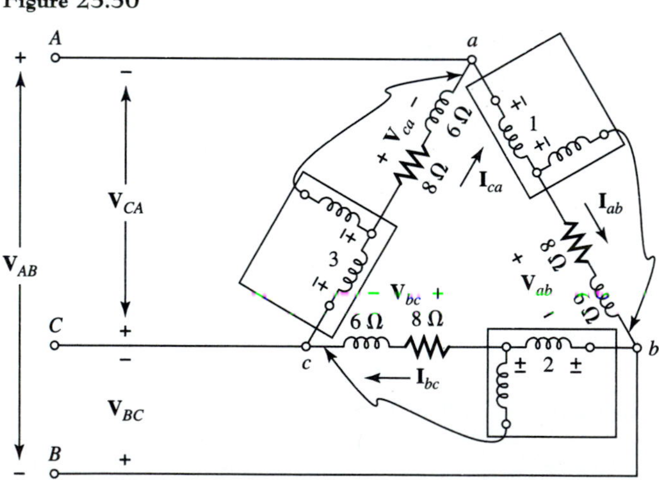

SOLUTION Wattmeter 1 senses phase current $\mathbf{I}_{ab}$ and phase voltage $\mathbf{V}_{ab}$. Wattmeter 2 senses phase current $\mathbf{I}_{bc}$ and phase voltage $\mathbf{V}_{bc}$. Wattmeter 3 senses phase current $\mathbf{I}_{ca}$ and phase voltage $\mathbf{V}_{ca}$. From Equation (20.7) the wattmeter readings for P_1, P_2, and P_3 are

$$P_1 = V_{ab_{\text{eff}}} I_{ab_{\text{eff}}} \cos\theta$$

$$P_2 = V_{bc_{\text{eff}}} I_{bc_{\text{eff}}} \cos\theta$$

$$P_3 = V_{ca_{\text{eff}}} I_{ca_{\text{eff}}} \cos\theta$$

Here, $\cos\theta$ is the load power factor and $\theta = 36.9°$ from the given load conditions. The phase voltages each have a maximum value equal to 100 V, and the phase currents each have a maximum value of 10 A from Example 23.4. From Equation (20.2)

$$V_{ab_{\text{eff}}} = V_{bc_{\text{eff}}} = V_{ca_{\text{eff}}} = \frac{100}{\sqrt{2}}$$

$$I_{ab_{\text{eff}}} = I_{bc_{\text{eff}}} = I_{ca_{\text{eff}}} = \frac{10}{\sqrt{2}}$$

and substituting these values in the preceding equations results in

$$P_1 = \left(\frac{100}{\sqrt{2}}\right)\left(\frac{10}{\sqrt{2}}\right)\cos(36.9°) = 400$$

$$P_2 = \left(\frac{100}{\sqrt{2}}\right)\left(\frac{10}{\sqrt{2}}\right)\cos(36.9°) = 400$$

$$P_3 = \left(\frac{100}{\sqrt{2}}\right)\left(\frac{10}{\sqrt{2}}\right)\cos(36.9°) = 400$$

The total power P delivered to the load is

$$\begin{aligned} P &= P_1 + P_2 + P_3 \\ &= 400 + 400 + 400 \\ &= 1200 \text{ W} \end{aligned}$$

EXAMPLE 23.6 Determine the total power delivered to the three-phase load in Example 23.4 using Equation (23.10). ■

SOLUTION Because the load is balanced, Equation (23.10) can be applied, and the result is

$$P = \sqrt{3}\, V_{L_{\text{eff}}} I_{L_{\text{eff}}} \cos\theta$$

where $V_L = 100$ and $\theta = 36.9$ are given and $I_L = 10\sqrt{3}$, as determined in Example 23.4. From Equations (20.2) and (23.10),

$$V_{L_{\text{eff}}} = \frac{100}{\sqrt{2}}$$

$$I_{L_{\text{eff}}} = \frac{10\sqrt{3}}{\sqrt{2}}$$

and

$$\begin{aligned} P &= \sqrt{3}\left(\frac{100}{\sqrt{2}}\right)\left(\frac{10\sqrt{3}}{\sqrt{2}}\right)\cos(36.9°) \\ &= 1200 \text{ W} \end{aligned}$$

23.18 EXERCISES

1. A Y-connected, three-phase voltage generator operating at $100/2\pi$ Hz has the following line voltages:

$$\mathbf{V}_{AB} = 200\angle 30^\circ \text{ V}$$
$$\mathbf{V}_{BC} = 200\angle -90^\circ \text{ V}$$
$$\mathbf{V}_{CA} = 200\angle 150^\circ \text{ V}$$

The generator is connected to a balanced y load. Each phase of the y load consists of a 100-Ω resistor and a 1-H inductor in series.

(a) Determine the load phase voltages in phasor form.
(b) Determine the load phase currents in phasor form.
(c) Draw a phasor diagram.
(d) Determine the power delivered to the load.
(e) Express the load phase voltages and load phase currents in the t domain.

2. Repeat Problem 1 for the same load-connected delta.

3. A balanced Y-connected generator has a line current $\mathbf{I}_{Aa} = 10\angle -30^\circ$ A and a phase voltage $\mathbf{V}_{AN} = 100\angle 0^\circ$ V. Find the phase impedance and the power delivered to a balanced Y-connected load.

4. Repeat Problem 3 for a balanced Δ-connected load.

5. A balanced Y-connected load is energized as shown in Figure 23.31.

Figure 23.31

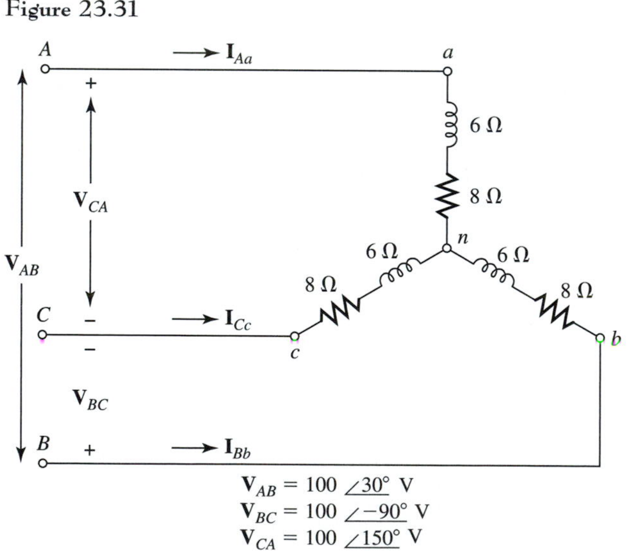

(a) Determine $\mathbf{I}_{Aa}$, $\mathbf{I}_{Bb}$, and $\mathbf{I}_{Cc}$.
(b) Determine $\mathbf{V}_{an}$, $\mathbf{V}_{bn}$, and $\mathbf{V}_{cn}$.
(c) Determine the power delivered to the load.

6. A balanced Δ-connected load is energized as shown in Figure 23.32.

Figure 23.32

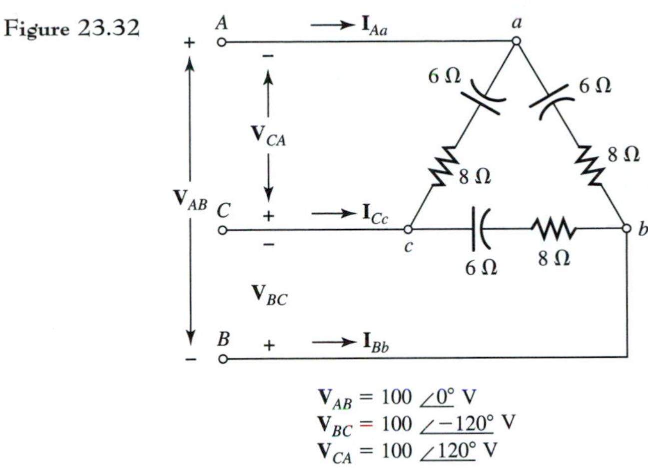

(a) Determine $\mathbf{I}_{Aa}$, $\mathbf{I}_{Bb}$, and $\mathbf{I}_{Cc}$.
(b) Determine $\mathbf{I}_{ab}$, $\mathbf{I}_{bc}$, and $\mathbf{I}_{ca}$.
(c) Determine the power delivered to the load.

7. The power delivered to a balanced three-phase load equals 1000 W at a power factor equal to 0.5 lag. The maximum value of line voltage is 100 V. Determine the effective line current.

8. Two wattmeters are used to measure the power delivered to the load in Figure 23.33.

Figure 23.33

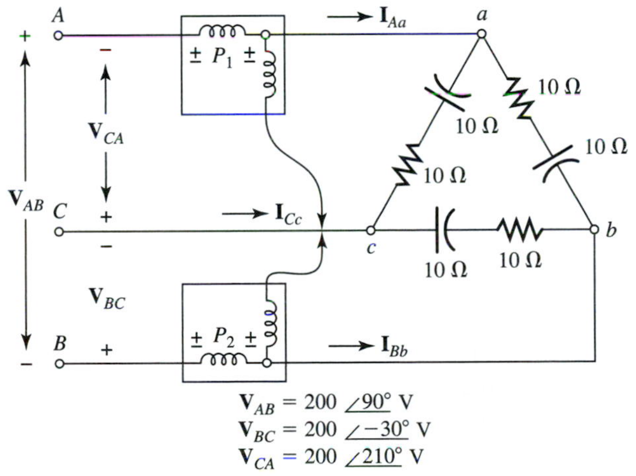

$\mathbf{V}_{AB} = 200 \angle 90°$ V
$\mathbf{V}_{BC} = 200 \angle -30°$ V
$\mathbf{V}_{CA} = 200 \angle 210°$ V

(a) Determine the reading of meter 1.
(b) Determine the reading of meter 2.
(c) Determine the total power delivered to the load.

9. Repeat Problem 8 for the wattmeters connected as shown in Figure 23.34.

Figure 23.34

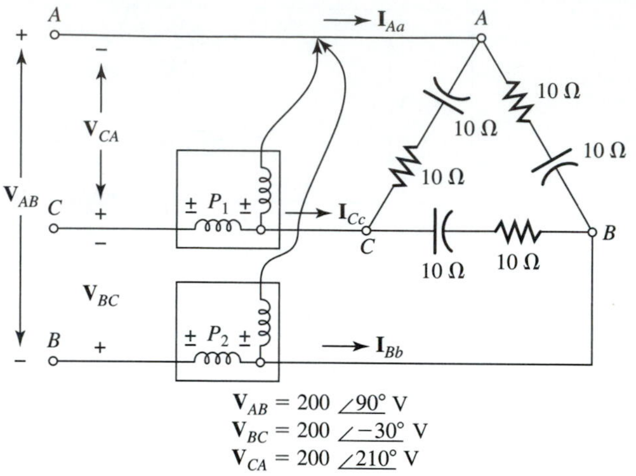

$\mathbf{V}_{AB} = 200\ \angle 90°\ \text{V}$
$\mathbf{V}_{BC} = 200\ \angle -30°\ \text{V}$
$\mathbf{V}_{CA} = 200\ \angle 210°\ \text{V}$

10. Consider the circuit connected as shown in Figure 23.35.

Figure 23.35

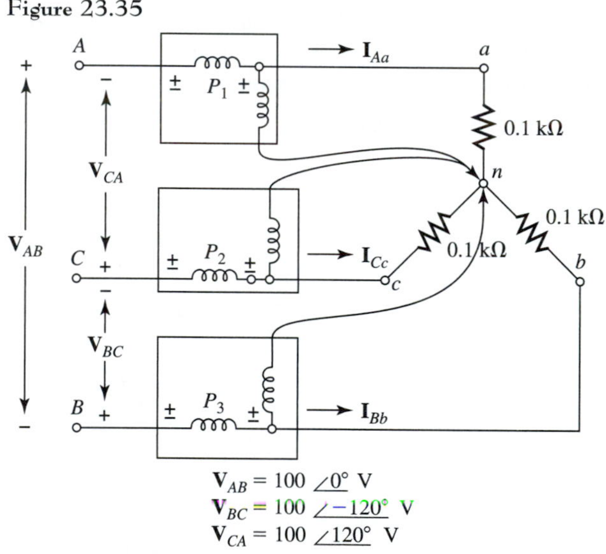

$\mathbf{V}_{AB} = 100\ \angle 0°\ \text{V}$
$\mathbf{V}_{BC} = 100\ \angle -120°\ \text{V}$
$\mathbf{V}_{CA} = 100\ \angle 120°\ \text{V}$

(a) Determine $\mathbf{V}_{an}$.
(b) Determine $\mathbf{V}_{bn}$.
(c) Determine $\mathbf{V}_{cn}$.
(d) Determine $\mathbf{I}_{Aa}$.
(e) Determine $\mathbf{I}_{Bb}$.
(f) Determine $\mathbf{I}_{Cc}$.
(g) Determine the reading of meter 1.
(h) Determine the reading of meter 2.
(i) Determine the reading of meter 3.
(j) Determine the total power delivered to the load.

11. Three wattmeters are used to measure the power delivered to the load, as shown in Figure 23.36.

Figure 23.36

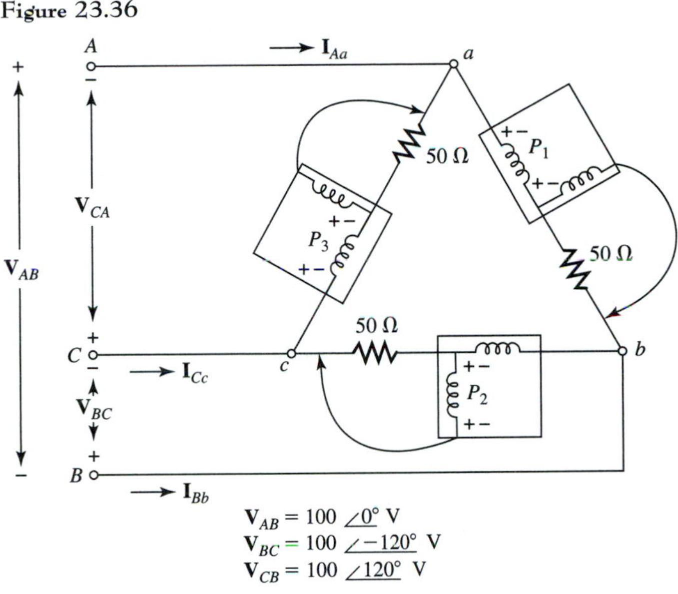

(a) Determine $\mathbf{I}_{ab}$.
(b) Determine $\mathbf{I}_{bc}$.
(c) Determine $\mathbf{I}_{cb}$.
(d) Determine $\mathbf{I}_{Aa}$.
(e) Determine $\mathbf{I}_{Bb}$.
(f) Determine $\mathbf{I}_{Cc}$.
(g) Determine the reading of meter 1.
(h) Determine the reading of meter 2.
(i) Determine the reading of meter 3.
(j) Determine the total power delivered to the load.

12. What is the advantage of the two-wattmeter method for measuring power to a three-phase load?

13. A balanced three-phase delta generator with a positive phase sequence furnishes a line current $\mathbf{I}_{Aa} = 10\angle 0°$ A to a Y-connected load. Each phase of the load has an impedance $\mathbf{Z} = 10 + j10\ \Omega$.
(a) Determine the line voltages in phasor form.
(b) Determine the total power delivered to the load.

14. A balanced three-phase Y generator with a positive phase sequence furnishes a line voltage $\mathbf{V}_{AB} = 100\angle 30°$ V to a Δ-connected load. Each phase of the load has an impedance $\mathbf{Z} = 4 - j3\ \Omega$.
(a) Determine the line currents in phasor form.
(b) Determine the total power delivered to the load.

15. Consider the three-phase load shown in Figure 23.37.

Figure 23.37

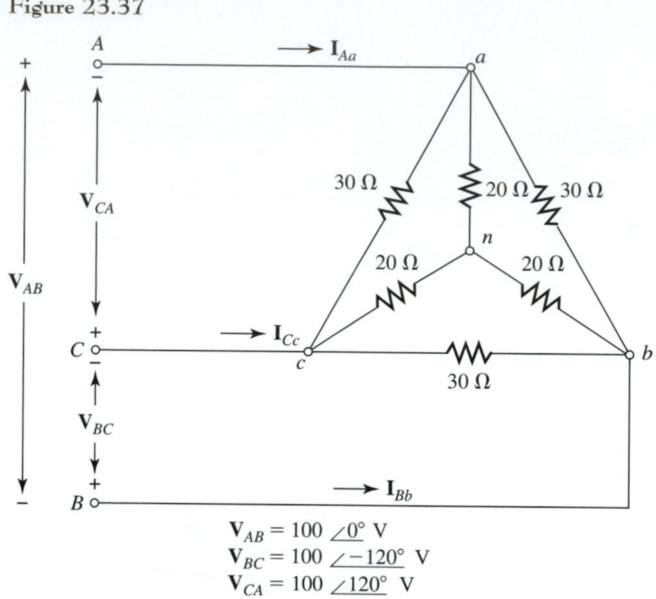

(a) Determine $\mathbf{V}_{an}$, $\mathbf{V}_{bn}$, and $\mathbf{V}_{cn}$.
(b) Determine $\mathbf{I}_{an}$, $\mathbf{I}_{bn}$, and $\mathbf{I}_{cn}$.
(c) Determine $\mathbf{I}_{ab}$, $\mathbf{I}_{bc}$, $\mathbf{I}_{ca}$.
(d) Determine $\mathbf{I}_{Aa}$, $\mathbf{I}_{Bb}$, and $\mathbf{I}_{Cc}$.
(e) Determine the power delivered to the 20-Ω Y load.
(f) Determine the power delivered to the 30-Ω Δ load.

APPENDIX A

PSPICE Demonstration

High computational speed and extreme accuracy make the digital computer very desirable for analyzing electric circuits. This is especially true for very large, complex circuits that include electronic devices. Manufacturers of integrated circuits—which can contain millions of active devices—depend on the computer to determine the workability of their circuits.

Unfortunately, the use of the digital computer does not relieve the electrical engineering technologist from the necessity of understanding the principles of electric circuit analysis. This understanding is necessary to effectively utilize the digital computer to analyze electric circuits. The electrical engineering technologist must be well versed in the analytical methods and principles presented in this text and familiar with the computer methods for analyzing electric circuits.

The software most widely used today for analyzing electric circuits is called SPICE (Simulation Program with Integrated Circuit Emphasis). The development of SPICE was funded by the United States government and, therefore, is public-domain software; that is, it is free to any U.S. citizen. However, SPICE is not designed for small computers and it is not very user-friendly. Software designers have modified and improved SPICE through the years to make it more user-friendly and adaptable to smaller computers. The modified forms of SPICE are generally called PSPICE.

Two forms of PSPICE are presently available—the DOS version and the standard Windows version. The main difference is in the way the circuit is presented to the computer. The Windows version allows us to actually draw the circuit to be analyzed on the computer screen; that is, it allows for the graphical input of the circuit to be analyzed. The DOS version of PSPICE requires that the circuit be introduced by text statements rather than drawing the circuit, as provided for in the Windows version. The Windows version is more widely used today, because it is easier to use and it is being advanced by the software manufacturers.

In both the DOS and Windows version of PSPICE the assignment of a zero reference point (reference node), as we discussed in Chapter 6 (in the section on nodal analysis), is required. All voltages that PSPICE determines in the process of analyzing a circuit are with respect to the assigned reference point.

A component of the Windows version of PSPICE called Schematics (Capture in some versions) provides for the graphical input of the circuit to be analyzed. Schematics provides various menus and dialogue boxes that allow us to draw the circuit, assign values to the circuit components, and select the type of analysis we wish the computer to perform.

After the circuit is completely inputted (using either the DOS or the Windows version), it is stored on the computer disk using the file menu. We can then initiate the analysis of the circuit using the analysis menu. When the analysis is complete, the results are saved in an output file, which can then be called up for view.

The following examples are offered to demonstrate the analysis of electric circuits using PSPICE and not to provide the details of the computer operation associated with PSPICE. PSPICE operates under a standard Windows environment, and the details of operation are easily learned from one of the many PSPICE manuals available.

DEMONSTRATION A. 1

The following is a PSPICE analysis of the dc circuit in Figure A.1. In the DOS

Figure A. 1

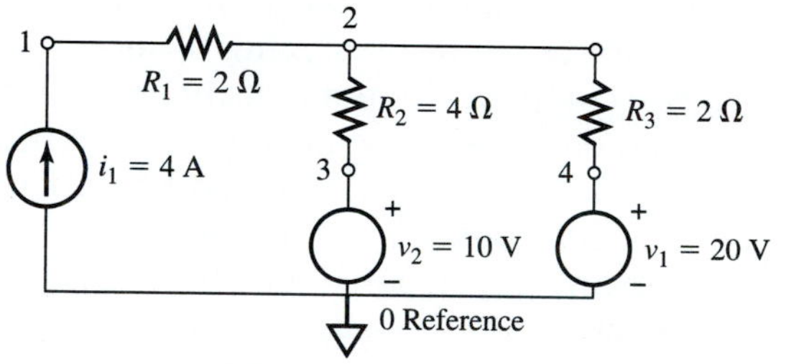

version of PSPICE, the circuit is inputted to the computer using the following statements.

PSPICE	dc		Example		(Program name)
I1	0	1	dc	4	(Locates I1 between nodes 0 and 1)
V1	4	0	dc	20	(Locates V1 between nodes 4 and 0)
V2	3	0	dc	10	(Locates V2 between nodes 3 and 0)
R1	1	2	2		(Locates R1 between nodes 1 and 2)
R2	2	3	4		(Locates R2 between nodes 2 and 3)
R3	2	4	2		(Locates R3 between nodes 2 and 4)
.End					(Marks the end of program)

The computer output will include the following:

Node Voltages

Node	Voltage	Node	Voltage
1	30.000	2	22.000

Voltage Source Currents

Name	Current
V1	1.000
V2	3.000

In the Windows version of PSPICE the circuit is inputted graphically by opening the Schematics window, which makes available a toolbar that includes all the menus and parts necessary for drawing the circuit to be analyzed on the computer screen.

From the parts menu, arrows are placed above nodes 1 and 2 and ammeters are placed in series with the voltage sources, as shown in Figure A.2. This indicates to the computer the voltages and currents that we wish to view in the output, which is as shown in Figure A.2.

Another option for viewing the output is to enable the schematic display, which results in the circuit shown in Figure A.3, where the voltages and currents are displayed.

Figure A.2

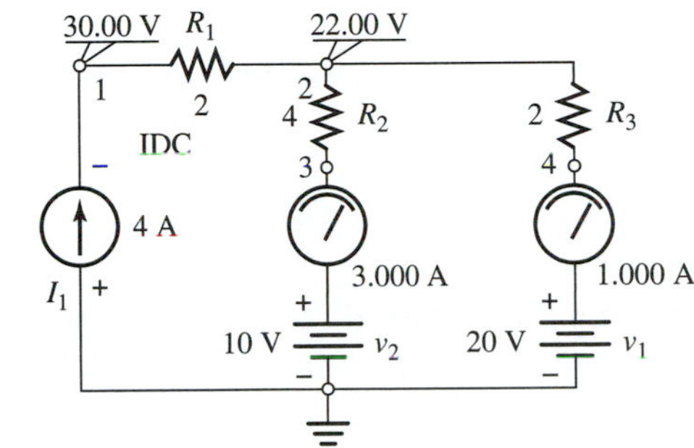

Figure A.3

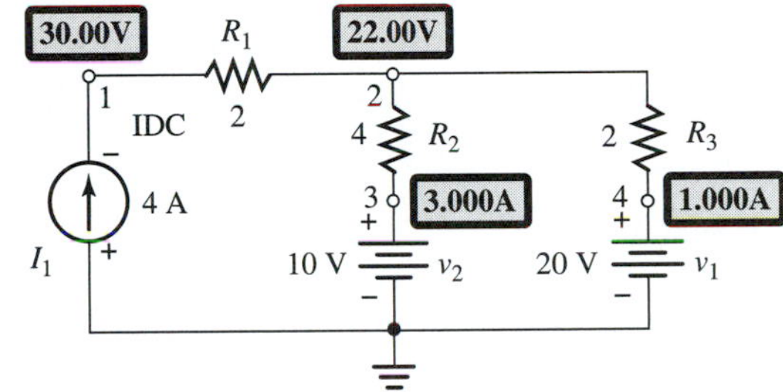

DEMONSTRATION A.2

The following is a PSPICE analysis of the ac circuit in Figure A.4. In the DOS

Figure A.4

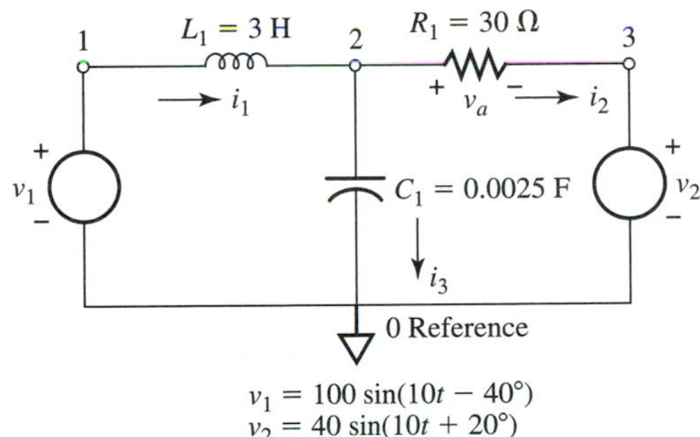

version of PSPICE, the circuit is inputted to the computer using the following statements.

```
PSPICE  ac  Example              (Program name)

V1   1   0   AC   100   −40      (Locates V1 between nodes 1 and 0)
L1   1   2   3                   (Locates L1 between nodes 1 and 2)
```

```
C1    2   0   .0025                  (Locates C1 between nodes 2 and 0)
R1    2   3   30                     (Locates R1 between nodes 2 and 3)
V2    3   0   AC   40   20           (Locates V2 between nodes 3 and 0)
.AC   LIN   1   1.591   1.591        (Control statement indicating a steady-
                                     state ac analysis at a single frequency
                                     of 1.592 Hz)
.Print AC VM(2, 3)   VP(2, 3)        (Output Statement calling for the
                                     magnitude and phase of the voltage
                                     between nodes 2 and 3)
.End                                 (Marks the end of program)
```

The computer output will include the following:

Freq	VM(2,3)	VP(2,3)
1.591	9.08E01	−1.212E02

Comparing the circuit in Figure A.4 to the computer output, VM(2, 3) and VP (2, 3) are the magnitude and phase of v_a. Therefore,

$$v_a = 90.8 \sin(10t - 121.2^\circ)$$

In the Windows version of PSPICE the circuit is inputted graphically using the Schematics component of PSPICE, as shown in Figure A.5.

Figure A.5

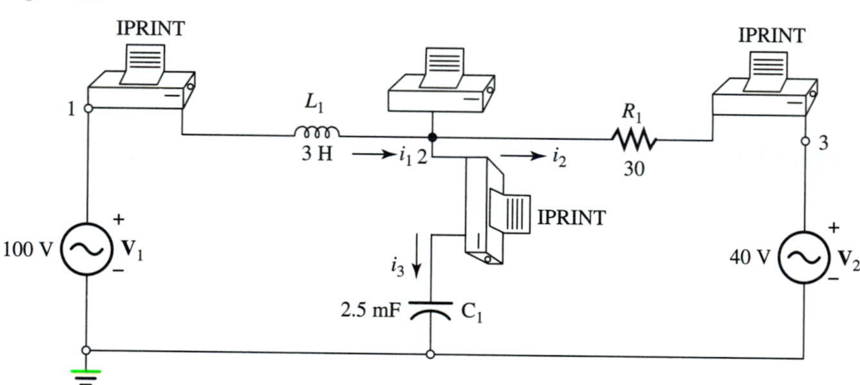

The print symbols in Figure A.5 indicate the currents and voltage to be viewed (i_1, i_2, i_3, and nodal voltage v_2). The output for a single frequency (159.2 Hz) is shown next, where i_1, i_2, i_3, and nodal voltage v_2 are shown in phasor form.

```
 ****  AC ANALYSIS
***********************************************
    Phasor Current I1

  FREQ            IM (V_PRINT1)     IP(V_PRINT1)

  1.592E+00       2.898E+00         -8.919E+01
  1.592E+00       2.898E+00         -8.920E+01
```

```
**** AC ANALYSIS
***************************************************
    Phasor Voltage V2

  FREQ                VM($N_0002)         VP($N_0002)

   1.592E+00          6.631E+01           -9.895E+01
   1.592E+00          6.631E+01           -9.895E+01
```

□

```
**** AC ANALYSIS
***************************************************
    Phasor Current I2

  FREQ                IM(V_PRINT3)        IP(V_PRINT3)

   1.592E+00          3.085E+00           -1.212E+02
   1.592E+00          3.085E+00           -1.212E+02
```

□

```
**** AC ANALYSIS
***************************************************
    Phasor Current I3

  FREQ                IM(V_PRINT2)        IP(V_PRINT2)

   1.592E+00          1.658E+00           -8.947E+00
   1.592E+00          1.658E+00           -8.952E+00
```

□

From this printout the currents i_1, i_2, and i_3 are

$$i_1 = 2.898 \sin(10t - 89.1^\circ)$$

$$i_2 = 3.085 \sin(10t - 121.2^\circ)$$

$$i_3 = 1.658 \sin(10t - 8.94^\circ)$$

and the nodal voltage is

$$v_2 = 66.3 \sin(10t - 98.9^\circ)$$

for the single frequency 159.2 Hz (10 rad/s).

If it is desired to view the currents and voltages as the frequency of the driving voltage sources vary, we can utilize the *probe* component of PSPICE. For example, we can view the currents and voltages (i_1, i_2, and i_3, and nodal voltage v_2) for the circuit in Figure A.6 as the frequency varies from 0.1 Hz to 10 Hz using probe. The output will appear as shown in Figures A.7, A.8, A.9, and A.10.

Figure A.6

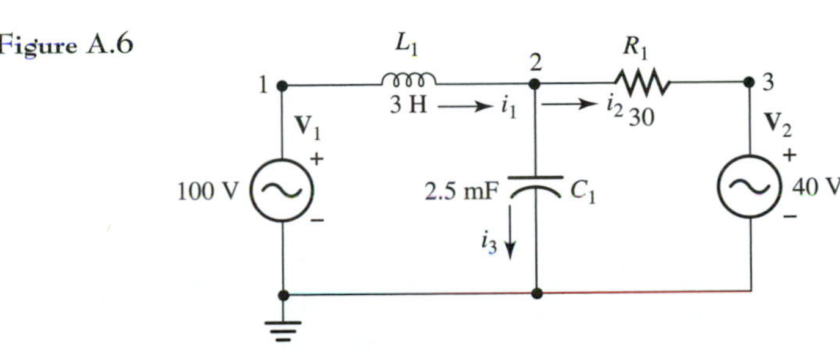

Figure A.7

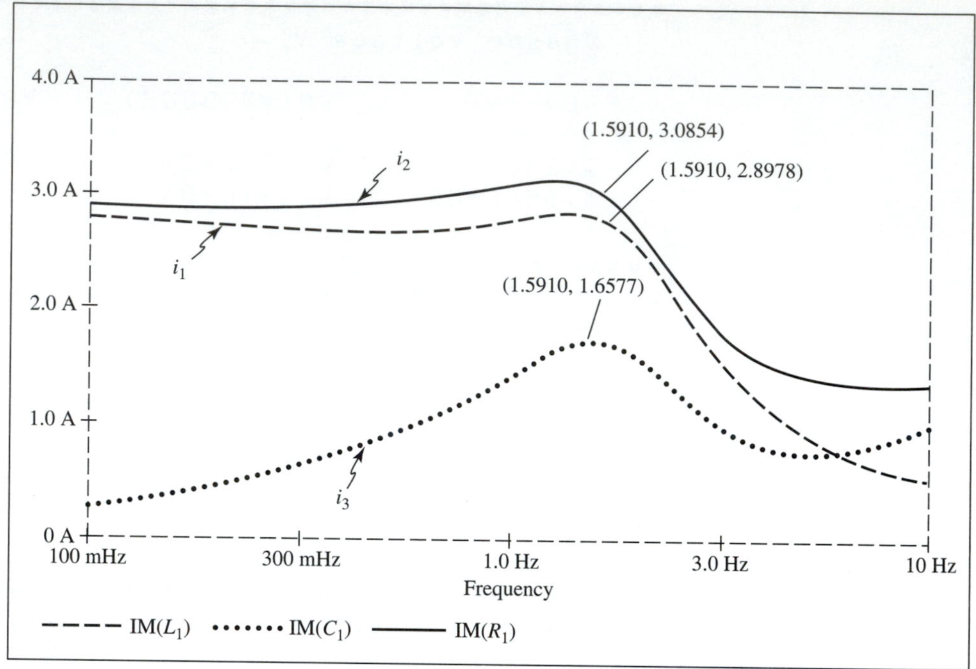

Figure A.8

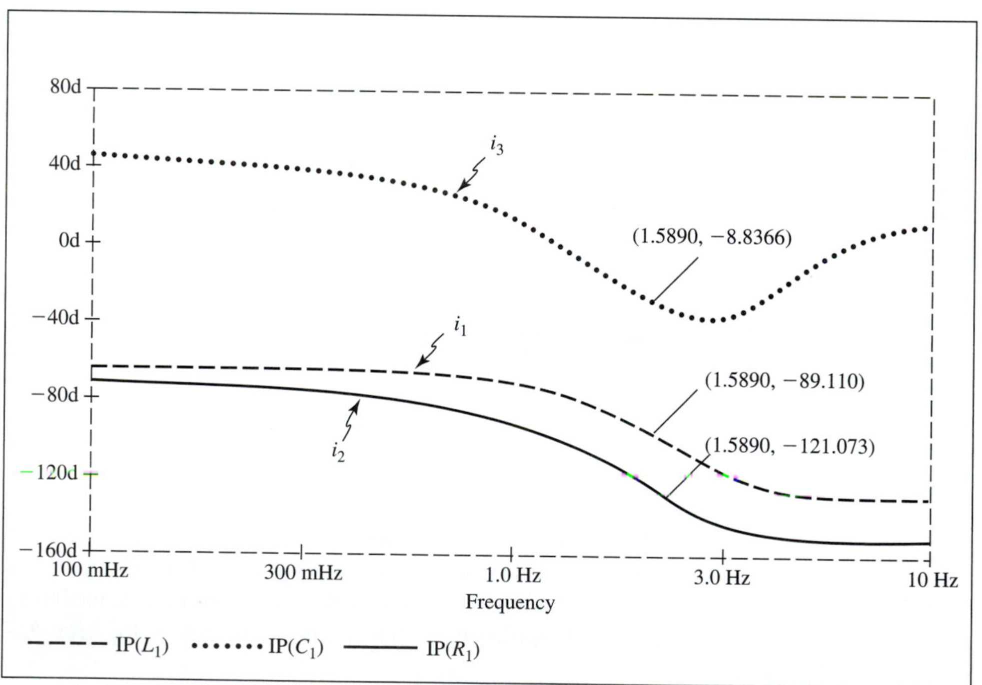

Figure A.9

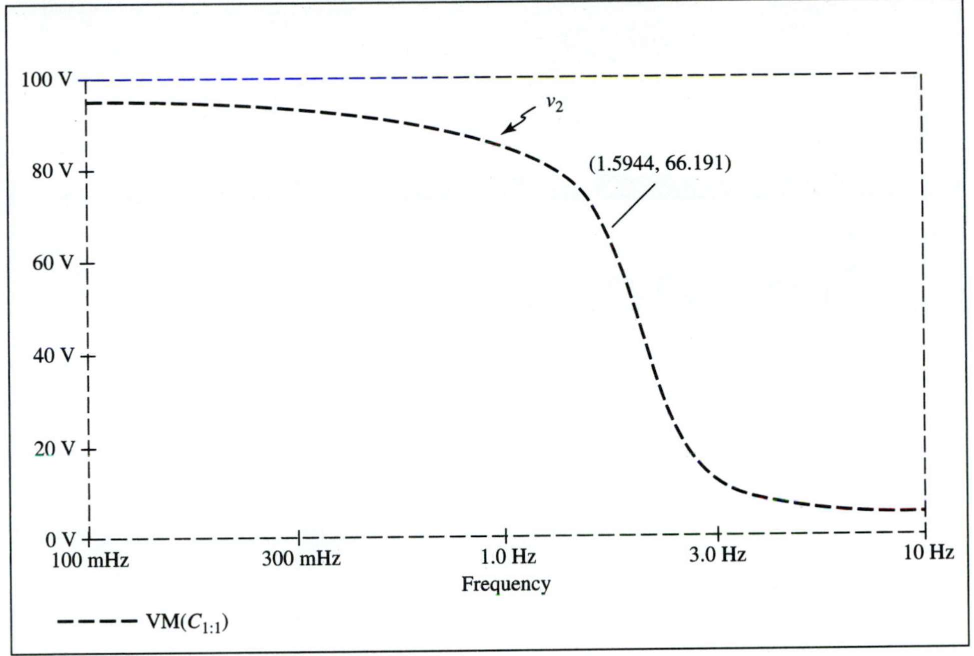

Figure A.10

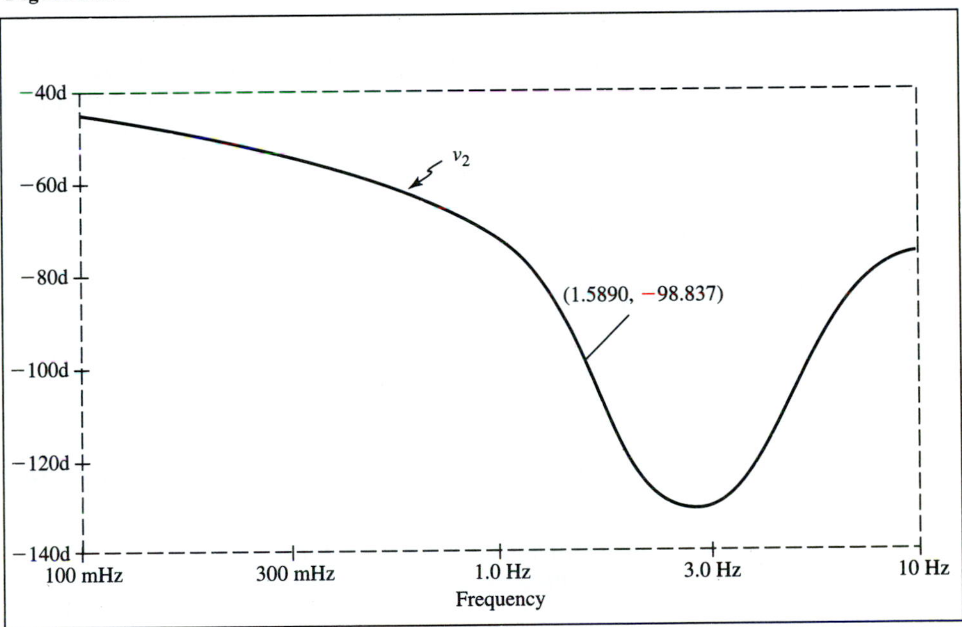

References

Boctor, S. A. *Electric Circuit Analysis*. Englewood Cliffs, N.J.: Prentice Hall, Inc., 1987.

Bogart, Theodore F., Jr. *Electric Circuits*. New York: MacMillan Publishing Co., 1988.

Boylestad, Robert L. *Introductory Circuit Analysis*. 9th ed. Upper Saddle River, N.J.: Prentice Hall, Inc. 2000

Goody, Roy W. *MicroSim PSpice for Windows, Vol. I: DC, AC, and Devices and Circuits*, 2nd ed. Upper Saddle River, N.J.: Prentice Hall, Inc., 1998.

Irwin, J. David *Basic Engineering Circuit Analysis*, 5th ed. Upper Saddle River, N.J.: Prentice Hall, Inc., 1996.

Jackson, Herbert W., and Preston A. White III. *Introduction to Electric Circuits*, 7th ed. Englewood Cliffs, N.J.: Prentice Hall, Inc., 1989.

Johnson, David E., John L. Hilburn, and Johnny R. Johnson. *Basic Electric Circuit Analysis*, 2nd ed. Englewood Cliffs, N.J.: Prentice Hall, Inc., 1984.

Monssen, Franz. *MicroSim PSpice with Circuit Analysis*, 2nd ed. Upper Saddle River, N.J.: Prentice Hall, Inc., 1998.

Paynter, Robert T. *Introductory Electric Circuits*, Upper Saddle River, N.J.: Prentice Hall, Inc., 1999.

Tse, Chi Kong. *Linear Circuit Analysis*. Harlow, England: Addison-Wesley, 1998.

Answers to Selected Problems

CHAPTER 2

6. (a) 3 A ↻ (b) 2 A ↺ (c) $\frac{1}{2}$ A ↺
(d) 5 A ↻

7. (a) 8 mA (b) −2 mA
(c) −0.1 mA (d) 200 mA

8. (a) ± 12 V (b) ∓ 12 V
(c) ∓ 12 V (d) ± 9 V

9. (a) $\frac{1}{20}$ V ± (c) 0.4 mA ∓

10. (a) 10^{-9} S (c) 13.9 mS

14. $i_1 = 10$ A
$i_2 = 15$ A
$i_3 = -3$ A

18. 300 W
450 W
90 W

19. 288 W
144 W
96 W

20. 3 W

21. 7.74 V

22. 0.387 A

CHAPTER 3

3. −3 A

4. −2.94 A

5. $v_1 = 8$ V
$v_2 = 20$ V

6. 3 A

7. $v_1 = 18$ V
$v_2 = -12$ V
$v_{ab} = 0$

9. −5 mA

10. $i_1 = 10$ mA
$i_2 = 4$ mA

11. 1.67 mA

12. 3 V

13. $v_{ab} = 40$ V
$v_{bc} = 40$ V
$v_{ac} = 80$ V

14. $i = \frac{1}{7}$ A
$v = 3$ V

15. $v_{ab} = 21.3$ V
$i = \frac{4}{3}$ A

16. 18 A

17. 13 kΩ

18. 0.077 mS

19. 0.09 S

20. 11.1 S

21. 20 Ω

22. 0.05 S
23. 12 Ω
24. $i = 2$ A
$i_1 = 1$ A
$i_2 = 1$ A
$i_3 = 0.5$ A
$i_4 = 0.5$ A
26. 8 V
27. 38 V
28. 8 Ω
29. 3 A

CHAPTER 4

1. (a) $i_1 = 1.6$ A
$i_2 = 2.4$ A
2. (a) $i_1 = 2$ A
$i_2 = 6$ A
$i_3 = 4$ A
3. (a) $i_1 = 1$ A
$i_2 = 2$ A
$i_3 = 1$ A
$i_4 = 2$ A
4. (a) 4 A
5. (a) 3 A
6. (b) $v_1 = 5$ V
$v_2 = 5$ V
7. (b) $v_1 = 32$ V
$v_2 = 5.33$ V
8. (a) 42 V
9. (c) 12.7 V
10. (a) 10 V
11. (a) 13.3 V
12. (a) $i_1 = 4$ A
$i_2 = 10$ A
$i_3 = 7$ A
13. (a) $v = 12$ V
$i = 4$ A
15. (a) 4.09 A
20. −3.33 A

CHAPTER 5

1. (a) 14 (b) 32
2. (a) −12 (b) 500
5. (a) 72 (b) 32
7. (a) 128
8. (a) 270
9. (a) 111 (b) −108
10. (a) −1300
11. (b) 5408
12. (a) $x_1 = 18$
$x_2 = 3$
13. (c) $x_1 = \frac{5}{12}$
$x_2 = \frac{1}{3}$
$x_3 = \frac{7}{12}$
14. (c) $x_1 = 0$
$x_2 = 2$
$x_3 = 1$
15. (b) $x_1 = \frac{10}{9}$
$x_2 = \frac{2}{3}$
$x_3 = \frac{11}{9}$
$x_4 = -\frac{8}{9}$
(c) $x_1 = \frac{13}{18}$
$x_2 = -\frac{29}{3}$
$x_3 = -\frac{49}{18}$
$x_4 = \frac{41}{18}$

CHAPTER 6

1. 20 mA

3. −3.56 A

5. 0.78 A

8. $$i_1 = \frac{\begin{vmatrix} 20 & -2 & 0 & -6 & 0 \\ 10 & 9 & -2 & -5 & 0 \\ -15 & -2 & 10 & 0 & 0 \\ 40 & -5 & 0 & 11 & -6 \\ -40 & 0 & 0 & -6 & 10 \end{vmatrix}}{\begin{vmatrix} 12 & -2 & 0 & -6 & 0 \\ -2 & 9 & -2 & -5 & 0 \\ 0 & -2 & 10 & 0 & 0 \\ -6 & -5 & 0 & 11 & -6 \\ 0 & 0 & 0 & -6 & 10 \end{vmatrix}}$$

12. $v_1 = -0.57$ V
$v_2 = 0.14$ V
$v = 0.71$ V

19. $$v_1 = \frac{\begin{vmatrix} 2 & -0.17 & 0 & 0 \\ 4 & 1 & -0.33 & -0.5 \\ 2 & -0.33 & 1.33 & -1 \\ -1 & -0.5 & -1 & 2 \end{vmatrix}}{\begin{vmatrix} 0.42 & -0.17 & 0 & 0 \\ -0.17 & 1 & -0.33 & -0.5 \\ 0 & -0.33 & 1.33 & -1 \\ 0 & -0.5 & -1 & 2 \end{vmatrix}}$$

CHAPTER 7

1. $v_1 = 1.6\ V$
$v_2 = 4$ V

3. 0.25 A

6.

a
3.5 Ω
+
4 V
−
b

7. 0.8 A

9.

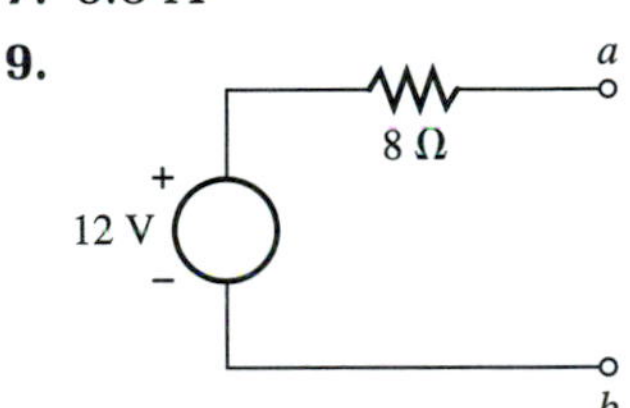

10.

a
1.33 A
4.5 Ω
b

13.

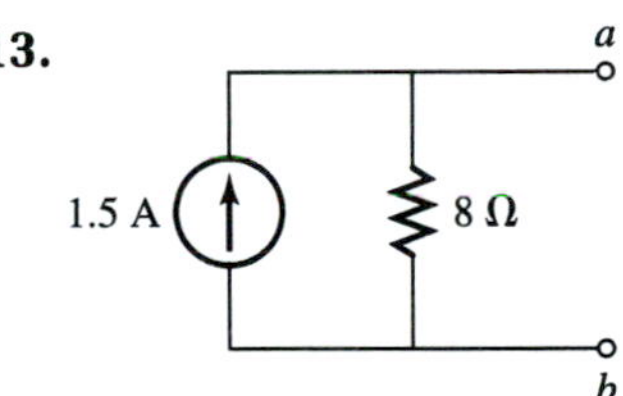

24. 6 A

CHAPTER 8

1. −16 V

4. 60.7 V

5. 4 kΩ

6. 13.3 V

7. −84 V

9. −1.67 A

CHAPTER 9

1. (a) $-4t^{-5}$

2. (a) 0

3. (a) $-420e^{-30t}$

4. (a) $400\cos(4t + 20°)$

5. (a) $18t^2 - e^{-t}$

6. (a) $-8e^{2t} + 16e^{t}$

7. (a) $6e^{-t}\cos(t) - 6e^{-t}\sin(t)$

8. (b) $10t^{-3}$

9. 10 ft/s
10. 25 ft/s
13. (a) $10t + K$
14. (c) $\left(\frac{7}{4}\right)e^{4t} + K$
(f) $-5\cos(t + 20) + K$
17. (a) 48

CHAPTER 10

1. (b) $0.6e^{-t}$ mA (e) -2.4 mA
2. (a) 0.26 J
3. (a) 2 μF (b) 0.5 μF
4. (a) 10 μF (b) 3 μF
5. (a) 4 μF
7. (b) 48 μC on C_1
9. (a) $i_1 = 1$ A
$i_2 = 1$ A
11. (a) $v_1 = 9.33$ V
$v_2 = 14$ V
15. 6 μF
16. 1.86 μF

CHAPTER 11

1. (a) $20e^{-t}$ V (b) $6e^{-2t}$ V
2. (c) 40 J
3. (a) 2.7 H (b) 0.855 H
4. (a) 1 H (b) 0.1 H
5. (a) 0.8 H
10. (a) $i_1 = 8/3$ A
$i_2 = 4/3$ A
11. $v_1 = 3.2$ V
$v_2 = 1.6$ V
15. 0.3 H

CHAPTER 12

2. (a) $-2e^{-6t}$
3. (b) $2e^{-3t} - 6e^{-t}$
5. (a) $1.5e^{-4t} + 2.5$
8. (a) $\dfrac{\frac{7}{6}}{S+8} - \dfrac{\frac{1}{6}}{S+2}$
12. (a) $\left(\frac{7}{6}\right)e^{-8t} - \left(\frac{1}{6}\right)e^{-2t}$

CHAPTER 13

1. (a) $10e^{-(1/4)t}$
4. (a) 4
5. $i_1 = 0$
$i_2 = 0.83$ A
8. (a) $v = 50e^{-(8/5)t}$
$i_1 = 10e^{-(8/5)t}$
$i_2 = 7.5e^{-(8/5)t}$
10. (a) $i = 2e^{-(4)t}$
$v = 8e^{-(4)t}$
13. (a) 0.25
14. (a) $i_1 = 2.5$ A
$i_2 = 0.83$ A
15. (a) $i_1 = -5e^{-(\frac{3}{4})t}$
$i_2 = -\left(\frac{5}{4}\right)e^{-(\frac{3}{4})t}$
$v = -\left(\frac{15}{4}\right)e^{-(\frac{3}{4})t}$

CHAPTER 14

1. $v = -10e^{-(\frac{1}{2})t} + 20$
$i = \left(\frac{5}{4}\right)e^{-(\frac{1}{2})t}$
5. $v = -30e^{-(\frac{4}{3})t} + 30$
$i = 5e^{-(\frac{4}{3})t}$
6. $v = 24e^{-4t}$
$i = -3e^{-4t} + 5$
9. $v = 34e^{-2t}$
$i = -17e^{-2t} + 15$

17. (a) $i = -\left(\frac{10}{7}\right)e^{-8t} + \left(\frac{10}{7}\right)e^{-t}$

19. (a) $-\left(\frac{2}{3}\right)t^3$

(b) $100e^{-t}$

CHAPTER 15

1. (a) $\frac{I}{V} = \frac{s}{6s + 4}$

(b) $\frac{I}{V} = \frac{s}{4s + 2}$

3. (a) $\frac{8s^3 + 6s^2 + 6s + 6}{2s^2 + 1}$

5. (a)

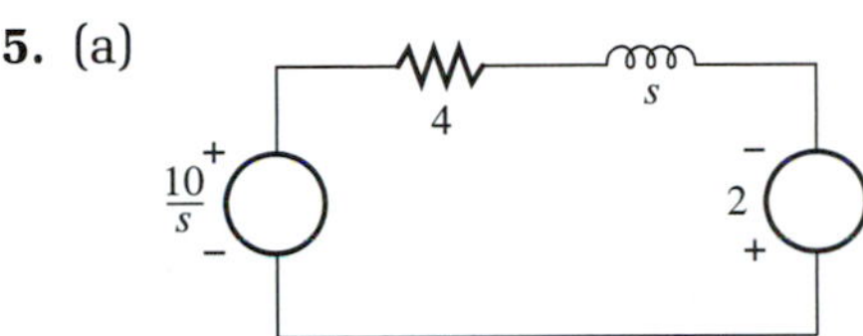

8. (a) $I = \frac{10 - s}{s(s + 3)}$

$V = \frac{20 - 2s}{s + 3}$

10. (b) $v_o = \frac{40}{s(s + 4)}$

$V_o = 10 - 10e^{-4t}$

22. (b)

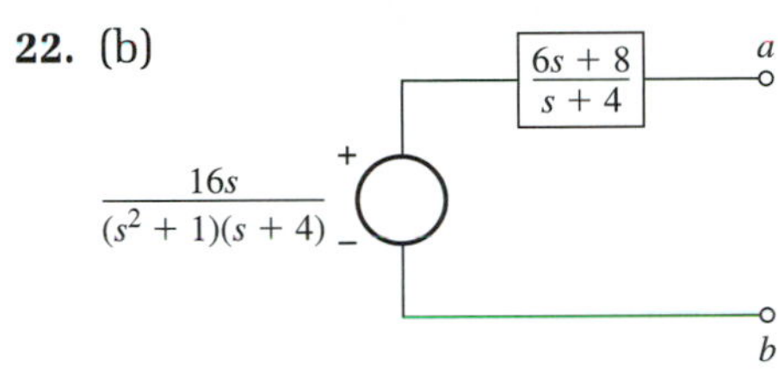

24. (b)

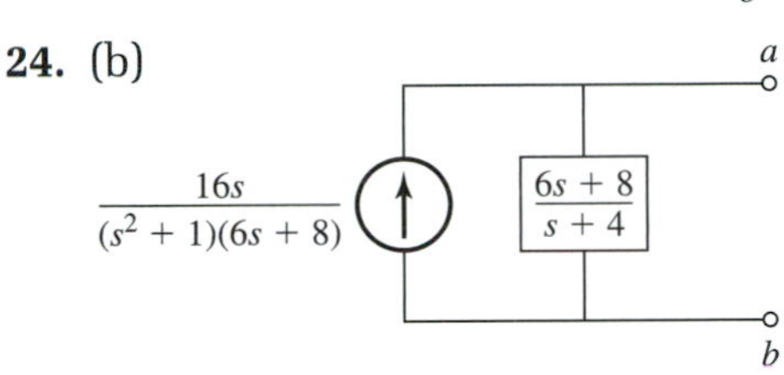

CHAPTER 16

1. (a) $\frac{1}{2}$ Hz (b) 14 Hz
2. (a) 377 rad/s (e) 3.14 rad/s
4. (a) 4 s, $\frac{1}{4}$ Hz (c) 2 ms, 500 Hz
5. 0.2 s, 5 Hz
6. (a) 10 Hz (b) 10 Hz
7. (a) 0.1 s (b) 1 s
8. (a) 180° (e) 90°
9. (a) 0.523 rad (d) π rad
10. (a) 30°
11. (a) $100 \sin(20t + 40°)$
 (b) $50 \sin(100t - 200°)$
13. (a) $40\angle 30°$ (b) $80\angle 140°$
15. (a) 80° (c) 65°
16. (a) $50 \sin(\omega t - 30°)$
 (b) $20 \sin(\omega t + 150°)$
18. (a) $100 \cos(20t - 50°)$
20. (a) $50\angle 270°$ (b) $20\angle 0°$
24. (a) 5.7 (c) 10
26. (a) $5\angle -53.1°$ (c) $14.1\angle 45°$
 (d) $14.1\angle 135°$
27. (a) $14.1 + j14.1$
 (e) $-13 + j7.5$
28. (a) $2j$ (e) 16
29. (a) $6 + j4$ (b) $40\angle 50°$
30. (a) $9 + j4$ (e) $40 + j30$
31. (a) $18\angle 153°$ (c) $11.7\angle -149°$
32. (a) $3.4\angle 53°$ (e) $571\angle 30°$
34. (a) $70\angle -20°$ (b) $10\angle 210°$
35. $\mathbf{I}_2 = 14.14\angle 135°$
 $i_2 = 14.14 \sin(\omega t + 135)$
36. $\mathbf{V}_3 = 30\angle 180°$
 $v_3 = 30 \sin(\omega t + 180°)$

CHAPTER 17

2. $1.5\angle -30°$
3. (a) $0.5\angle -60°$ (c) $8\angle -30°$
4. (a) $80 \sin(\omega t + 30°)$
 (b) $120 \cos(\omega t)$
5. (a) $2.5 \sin(\omega t + 60°)$
 (d) $1 \sin(\omega t + 150°)$
6. (a) 188.5 Ω (b) 3 Ω
7. (a) 0.95 H (c) 1 H

8. (c) $12.5 \sin(8t - 30°)$
(e) $20 \sin(t - 90°)$

9. (c) $5 \sin(t + 130°)$
(e) $150 \sin(60t + 90°)$

10. (a) $4\underline{/90°}$ (c) $16\underline{/90°}$

11. (a) $2\underline{/90°}$ (b) $1\underline{/90°}$

12. (a) $0.80\ \Omega$ (b) $39.79\ \Omega$

13. (d) 0.05 F

15. (a) $25 \sin(4t - 90°)$
(c) $400 \sin(2t - 180°)$

16. (a) $0.80\underline{/-90°}$ (f) $10\underline{/-90°}$

19. $0.4\underline{/-30°}$

20. (a) $0.02\underline{/-30°}$

21. (a) $0.0025\underline{/0°}$ (b) $2\underline{/0°}$

22. (c) $160 \sin(\omega t)$

23. (a) $0.005\underline{/-90°}$

24. (a) 0.8 H

25. (b) $1000\underline{/-90°}$ (c) $500\underline{/-90°}$

26. (a) 1.25 S

27. (a) 0.4 F

28. (a) $0.025\underline{/90°}$ (c) $0.001\underline{/90°}$

29. (a) $10\underline{/53°}$
(d) $10 \sin(\omega t + 53°)$

30. (a) $2\underline{/-113°}$ (b) $20\underline{/-23°}$
(c) $44\underline{/-203°}$
(d) $2 \sin(\omega t - 113°)$
(e) $20 \sin(\omega t - 23°)$
(f) $44 \sin(\omega t - 203°)$

31. (a) $11.3\underline{/135°}$
(b) $11.3 \sin(2t + 135°)$

32. (a) $10\ \Omega$ (b) $17.32\ \Omega$
(c) $20\underline{/60°}$

33. $6 + j8$

34. $0.02 - j0.02$

35. (a) $5\underline{/-32°}$ (d) $5\underline{/-53°}$

36. (a) $\mathbf{I} = 40\underline{/37°}$
$i = 40 \sin(\omega t + 37°)$

37. (a) $2.86\underline{/18.4°}$

38. (a) $\mathbf{I} = 34.97\underline{/11.6°}$
$i = 34.97 \sin(\omega t + 11.6°)$

43. (a) $\mathbf{Z}_{ab} = 44.7\underline{/26.6°}$

44. (a) $R = 21.7\ \Omega$
$X_L = 12.5\ \Omega$
(c) $25\underline{/30°}$ (d) $0.04\underline{/-30°}$

45. (a) $4\underline{/90°}$

46. (a) $\mathbf{I}_1 = 5.16\underline{/11.3°}$
$i_1 = 5.16 \sin(\omega t + 11.3°)$

47. (a) $\mathbf{V}_1 = 108\underline{/27°}$
$v_1 = 108 \sin(\omega t + 27°)$

CHAPTER 18

1. $j50\mathbf{I}_1 - j40\mathbf{I}_2 = 100 - j50$
$-j40\mathbf{I}_1 + j20\mathbf{I}_2 = -j50$
$\mathbf{I}_1 = 2.5\underline{/90°}$

2. $\mathbf{I}_1 = 2\underline{/30°}$

3. $-j2\mathbf{I}_1 - j4\mathbf{I}_2 = -150$
$-j4\mathbf{I}_1 + j2\mathbf{I}_2 = 50$
$\mathbf{I}_1 = 5\underline{/-90°}$

7. $$I_1 = \frac{\begin{vmatrix} 40 & 0 & -j2 & -j8 & 0 \\ -50 & 90 & 0 & 0 & j6 \\ 100 & 90 & 0 & 12 + j2 & j6 \\ 0 & & j6 & j6 & -j2 \end{vmatrix}}{\begin{vmatrix} 8 + j10 & -j2 & -j8 & 0 \\ -j2 & 0 & 0 & j6 \\ -j8 & 0 & 12 + j2 & j6 \\ 0 & j6 & j6 & -j2 \end{vmatrix}}$$

CHAPTER 19

1. $\mathbf{I} = 8.47\underline{/45°}$

3. $\mathbf{V}_1 = 12\underline{/180°}$
$\mathbf{V}_2 = 42\underline{/180°}$

5. $\mathbf{V}_T = 26\underline{/-60°}$
$\mathbf{Z}_T = 3 - j3$

9. $\mathbf{I}_N = 3.2\underline{/53°}$
$\mathbf{Z}_N = 3.33\underline{/37°}$

10. $\mathbf{I}_N = 7.50\underline{/0°}$
$\mathbf{Z}_N = 8\underline{/90°}$

11. $\mathbf{I}_{ab} = 3.33\underline{/0°}$

15. $\mathbf{I} = 7.5\underline{/90°}$

CHAPTER 20

1. (a) 5
2. (b) 0
3. (b) 5
4. (a) 15.8
5. (b) 42.4
6. (b) 15.8
7. (a) 0.71 lead (b) 0.8 lag
10. (b) $S = 160$
$Q = 96.3$
$P = 127.8$
13. (b) $S = 1131$
$Q = 800$
$P = 800$
18. (b) $16 \sin(t + 53)$
24. (b) $i = 10 \sin(2t - 53°)$

CHAPTER 21

1. $i = 3 - 3e^{-10t}$
$v_{ab} = 30e^{-10t}$
5. $i_2 = (\frac{5}{3})e^{-(\frac{5}{3})t}$
$v_o = -41.6e^{-(\frac{5}{3})t}$
9. $\mathbf{I}_2 = 5\angle 45°$
17. $\mathbf{Z}_{ab} = 6\angle 0°$
20. $\mathbf{I}_1 = 10\angle 0°$
$\mathbf{I}_2 = 10\angle 0°$
$\mathbf{V}_1 = 60\angle 0°$
$\mathbf{V}_2 = 60\angle 0°$

CHAPTER 22

1. (a) $\mathbf{y}_{11} = (2/11)\text{S}$
$\mathbf{y}_{12} = -(1/11)\text{S}$
$\mathbf{y}_{21} = -(1/11)\text{S}$
$\mathbf{y}_{22} = (6/11)\text{S}$

(b)

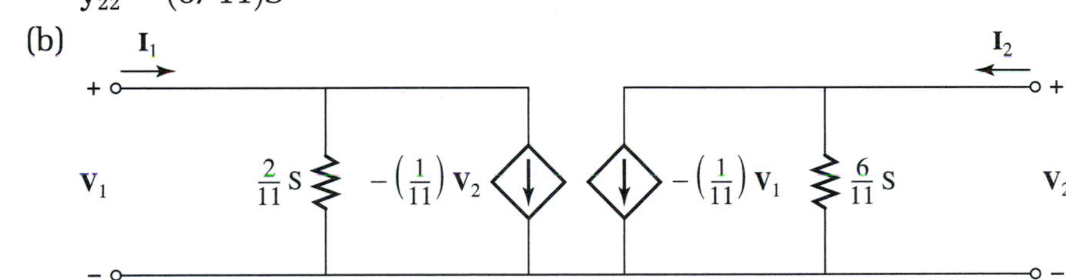

2. (a) $\mathbf{z}_{11} = \frac{5}{8}\ \Omega$
$\mathbf{z}_{12} = 2\ \Omega$
$\mathbf{z}_{21} = 2\ \Omega$
$\mathbf{z}_{22} = \frac{5}{16}\ \Omega$

(b)

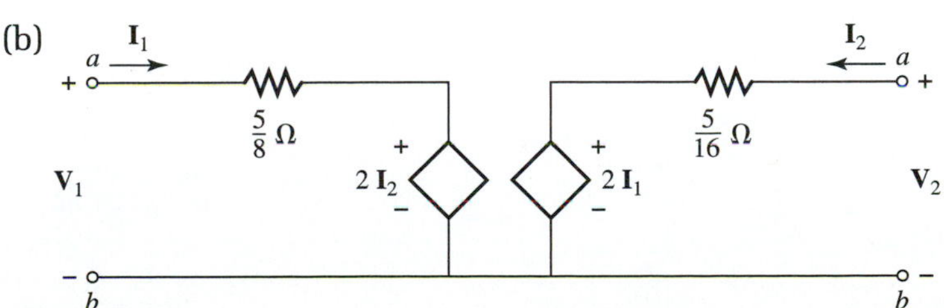

3. (a) $\mathbf{z}_{11} = 1\ \text{k}\Omega$

$\mathbf{z}_{12} = 40\ \Omega$

$\mathbf{z}_{21} = -(10^7)\ \Omega$

$\mathbf{z}_{22} = (10^5)\ \Omega$

8. $\mathbf{h}_{11} = 4.44\ \Omega$

$\mathbf{h}_{21} = -\dfrac{5}{9}$

$\mathbf{h}_{12} = \dfrac{5}{9}$

$\mathbf{h}_{22} = 6.43\text{S}$

13. (a) $\mathbf{A}_i = -\left(\dfrac{1}{2}\right)$

(b) $\mathbf{A}_V = \dfrac{1}{2}$

(c) $\mathbf{A}_V = 1$ for no load

(d) $\mathbf{Z}_{in} = 2\ \Omega$

(e) $\mathbf{Z}_o = 4\ \Omega$

CHAPTER 23

1. (a) $\mathbf{V}_{an} = 115\angle 0^\circ$

$\mathbf{V}_{bn} = 115\angle -120^\circ$

$\mathbf{V}_{cn} = 115\angle 120^\circ$

(b) $\mathbf{I}_{an} = 0.81\angle -45^\circ$

$\mathbf{I}_{bn} = 0.81\angle -165^\circ$

$\mathbf{I}_{cn} = 0.81\angle 75^\circ$

(d) 98.8

3. $\mathbf{Z} = 10\angle 30^\circ$

$P = 1259$

5. (a) $\mathbf{I}_{an} = 5.77\angle -53^\circ$

(b) $\mathbf{V}_{an} = 100\angle 0^\circ$

(c) $P = 10.4$

7. 16.3

10. (a) $100\angle -30^\circ$ (b) $100\angle -150^\circ$

(c) $100\angle 150^\circ$ (d) $1\angle -30^\circ$

(e) $1\angle -150^\circ$ (f) $1\angle 150^\circ$

(j) 150

Index

Q

R

S

T

U

V

W

Y